Reconciling Fisheries with Conservation:

Proceedings of the Fourth World Fisheries Congress

Volume I

Cover art: "Respect" created for the Fourth World Fisheries Congress by Calvin Hunt.

The theme of the Congress is reflected by the title—the reconciliation of fisheries is the expression of "respect." Embracing the colors of the nations of the world, the woman in the center embodies the spirit of Mother Earth. The eagle on her chest is a crest recognized for its strength and courage. Unity, coming together with one voice, is represented in the circles she holds. The number four, sacred in Northwest Coast native culture, is reflected in the four salmon surrounding the woman, symbolizing four directions, four seasons, and four stages of life.

Reconciling Fisheries with Conservation:

Proceedings of the Fourth World Fisheries Congress

Volume I

Edited by

Dr. Jennifer L. Nielsen
USGS Alaska Science Center
1011 East Tudor Road
Anchorage, Alaska 99503, USA

Dr. Julian J. Dodson
Department of Biology, University Laval
Ste-Foy, Quebec G1K 7P4, Canada

Dr. Kevin Friedland
National Marine Fisheries Service
Blaisdell House, University of Massachusetts
Amherst, Massachusetts 01003, USA

Dr. Troy R. Hamon
National Park Service
Post Office Box 7, King Salmon Mall Suite 101
King Salmon, Alaska 99613, USA

Dr. Jack Musick
Virginia Institute of Marine Science
7036 Sassafras Landing
Gloucester, Virginia 23061, USA

Dr. Eric Verspoor
Freshwater Fisheries Programme
FRS Freshwater Laboratory
Pitlochry PH16 5LB, Scotland, UK

American Fisheries Society Symposium 49

Proceedings of the Fourth World Fisheries Congress
Held in Vancouver, British Columbia, Canada
2 May–6 May 2004

American Fisheries Society
Bethesda, Maryland

2008

The American Fisheries Society Symposium Series is a registered serial.

Suggested citation formats follow.

Entire Book

Nielsen, J. L., J. J. Dodson, K. Friedland, T. R. Hamon, J. Musick, and E. Verspoor, editors. 2008. Reconciling fisheries with conservation: proceedings of the Fourth World Fisheries Congress. American Fisheries Society, Symposium 49, Bethesda, Maryland.

Chapter within the Book

Adlerstein, S., F. Alvarez, and R. Goñi. 2008. Quality data versus analysis? Abundance indices from a Mediterranean trawl fishery as case study. Pages 1107–1120 *in* J. L. Nielsen, J. J. Dodson, K. Friedland, T. R. Hamon, J. Musick, and E. Verspoor, editors. Reconciling fisheries with conservation: proceedings of the Fourth World Fisheries Congress. American Fisheries Society, Symposium 49, Bethesda, Maryland.

Printed in the United States of America on acid-free paper

Library of Congress Control Number 2008920748
ISBN 978-1-888569-80-3
ISSN 0892-2284

American Fisheries Society Web site:
http://www.fisheries.org

American Fisheries Society
5410 Grosvenor Lane, Suite 110
Bethesda, Maryland 20814-2199
USA

Table of Contents—Volume I

Preface

The Fourth World Fisheries Congress held in Vancouver, B.C. in May 2004 provided a forum for a broad range of discussions on "Reconciling Fisheries with Conservation," the theme of this congress. Enormous demand for fish and fisheries products coupled with growing concerns over sustainability of the world's fish populations have created a contentious dialogue among fishers, scientists, management, and different factions of the general public. While fisheries resources suffer depletion across the globe on many scales, extractive practices continue to expand and enter new areas. With increasing human population growth, especially in coastal areas, many remaining fish habitats are altered or destroyed; yet the demand for protein sources derived from fish grows rapidly. Controversial dialogue on sustainability of aquaculture hounds development and implementation of alternatives to fisheries on wild stocks. Across the globe, in developed and developing countries, the conflict is clear. Economic and social issues involving fish and fisheries are at odds with conservation and sustainability.

The Fourth World Fisheries Congress dedicated four days to scientific presentations and posters addressing key questions covering the range of global fisheries issues in a diversity of aquatic ecosystems:

- What should we care about when attempting to reconcile fisheries with conservation?
- Who owns the fish and what are they worth to society?
- Can we get more fish or benefits from fish and still reconcile fisheries with conservation?
- How can we manage aquatic ecosystems to reconcile fisheries with conservation?

Plenary speakers provided succinct coverage of each question as it addressed the theme. These keynote talks were followed by multiple concurrent sessions that debated each issue in specific terms. Comprehensive coverage was provided for cutting edge issues in management, human dimensions, economics, and ecosystem dynamics attempting to reach reconciliation on these questions. Two full-day forums dedicated to sustainable seafood and reconciliation within the fisheries community were offered. Hundreds of poster presentations provided further opportunities for participants to see and engage scientists and participants in ongoing research and management in fisheries and conservation. This eventful and inspired program contributed significantly to the global community involved in reconciling fisheries with conservation. This dialogue is by no means over and the questions voiced in Vancouver remain viable today. Reconciliation of these two paths, fisheries and conservation, remains elusive years after the Fourth World Fisheries Congress addressed this issue.

Many products and publications have evolved from scientific presentations and discussions that took place in Vancouver during this congress. This book represents a compendium of peer-reviewed papers submitted by participants with a diversity of perspectives on reconciling fisheries and conservation from around the world. In my experience, there are two ways to address and publish proceedings from a world congress. One is to publish a limited number of papers from a selection of esteemed contributors that reflect important discussions on the theme; the other is to give voice to as many congressional participants as possible in a comprehensive volume. I chose the latter path for the proceedings of the Fourth World Fisheries Congress. This policy was not without costs in terms of both time and resources in our efforts to reach a final document. It takes a great deal of time and effort to review and edit submissions from many parts of the world where English is not the author's first language and where fisheries science and management remain highly localized or regionally specific. These perspectives are very important, however, and reading them together with professional scientific findings presented by sophisticated, well-educated scientists casts a relevant light on the goals of this congress.

The results are clearly as diverse as the contributing authorship. The perspectives provided are as broad as the divergent population attending the congress. The contents herein present the diversity of important local or regional perspectives on the congress theme. Reading these submissions will provide unconditional evidence that the objective of the Fourth World Fisheries Congress—to reconcile fisheries with conservation—is still achievable and remains the desired outcome for many people of the world regardless of background, education, or management structure.

In closing, I want to thank the congress co-chairs, Dana Schmidt and Bruce Ward, for organizing such a comprehensive meeting. Special thanks need to go to Tony Pitcher, chair and Paul Hart, vice-chair of the program committee for providing the expansive platform of materials leading to submissions for these proceedings. I include my sincere thanks to the associate editors and the book production staff at the American Fisheries Society who so patiently showed great concern for the diversity of authors placed in their care. I also must express my sincere apologies to authors who have waited so long to get their submissions into press. My only excuse is that comprehensive inclusion takes significant time and effort in any endeavor, including these proceedings.

Jennifer L. Nielsen
Chair, Publications Committee
Anchorage, Alaska
November 2, 2007

Symbols and Abbreviations

The following symbols and abbreviations may be found in this book without definition. Also undefined are standard mathematical and statistical symbols given in most dictionaries.

Symbol	Meaning
A	ampere
AC	alternating current
Bq	becquerel
C	coulomb
°C	degrees Celsius
cal	calorie
cd	candela
cm	centimeter
Co.	Company
Corp.	Corporation
cov	covariance
DC	direct current; District of Columbia
D	dextro (as a prefix)
d	day
d	dextrorotatory
df	degrees of freedom
dL	deciliter
E	east
E	expected value
e	base of natural logarithm (2.71828...)
e.g.	(exempli gratia) for example
eq	equivalent
et al.	(et alii) and others
etc.	et cetera
eV	electron volt
F	filial generation; Farad
°F	degrees Fahrenheit
fc	footcandle (0.0929 lx)
ft	foot (30.5 cm)
ft^3/s	cubic feet per second (0.0283 m^3/s)
g	gram
G	giga (10^9, as a prefix)
gal	gallon (3.79 L)
Gy	gray
h	hour
ha	hectare (2.47 acres)
hp	horsepower (746 W)
Hz	hertz
in	inch (2.54 cm)
Inc.	Incorporated
i.e.	(id est) that is
IU	international unit
J	joule
K	Kelvin (degrees above absolute zero)
k	kilo (10^3, as a prefix)
kg	kilogram
km	kilometer
l	levorotatory
L	levo (as a prefix)
L	liter (0.264 gal, 1.06 qt)
lb	pound (0.454 kg, 454g)
lm	lumen
log	logarithm
Ltd.	Limited
M	mega (10^6, as a prefix); molar (as a suffix or by itself)
m	meter (as a suffix or by itself); milli (10^{-3}, as a prefix)
mi	mile (1.61 km)
min	minute
mol	mole
N	normal (for chemistry); north (for geography); newton
N	sample size
NS	not significant
n	ploidy; nanno (10^{29}, as a prefix)
o	ortho (as a chemical prefix)
oz	ounce (28.4 g)
P	probability
p	para (as a chemical prefix)
p	pico (10^{-12}, as a prefix)
Pa	pascal
pH	negative log of hydrogen ion activity
ppm	parts per million

qt	quart (0.946 L)
R	multiple correlation or regression coefficient
r	simple correlation or regression coefficient
rad	radian
S	siemens (for electrical conductance); south (for geography)
SD	standard deviation
SE	standard error
s	second
T	tesla
tris	tris(hydroxymethyl)-aminomethane (a buffer)
UK	United Kingdom
U.S.	United States (adjective)
USA	United States of America (noun)
V	volt
V, Var	variance (population)
var	variance (sample)
W	watt (for power); west (for geography)
Wb	weber
yd	yard (0.914 m, 91.4 cm)
α	probability of type I error (false rejection of null hypothesis)
β	probability of type II error (false acceptance of null hypothesis)
Ω	ohm
μ	micro (10^{26}, as a prefix)
′	minute (angular)
″	second (angular)
°	degree (temperature as a prefix, angular as a suffix)
%	per cent (per hundred)
‰	per mille (per thousand)

American Fisheries Society Symposium 49:1–4

Reconciling Fisheries with Conservation: The Challenge of Managing Aquatic Ecosystems

DANIEL PAULY*

University of British Columbia, Fisheries Centre, 2329 West Mall, Vancouver, British Columbia V6T 1Z4, Canada

Grand Conference Keynote
3 May 2004

The concept of "conservation" is deeply embedded in the history of fisheries science, as documented, for example, in the name of fisheries management organization such as the International Commission for the Conservation of Atlantic Tuna. In this, conservation is similar to "sustainability," incorporated deep inside the quantitative models – recall maximum sustainable yield—which, since the 1950s, have inspired fisheries management.

Thus, all the three basic models proposed in that period—the recruit versus parent stock relationship of Ricker (1954), the surplus-production model of Schaefer (1954, 1957) and the yield-per-recruit model of Beverton and Holt (1957)—involve a convex curve, whose optimum identifies a population (or stock) size ensuring its conservation, while providing us with high, but sustainable, catches.

We now know that fixed target such as maximum sustainable yield are not appropriate for managing variable fish populations, but we will still briefly follow on one implication of our earlier "aim-for-the-maximum" approach. That is, given these models, fishing effort beyond that needed to generate the maximum implied overfishing, and thus lack of sustainability and conservation. Thus, given an observation of excess effort as defined here, fisheries scientists should have been able to agree with other biologists, including those involved with conservation-orientated groups, that the excess effort in question must be reduced. Also, the implication of the aim-for-the-maximum approach should have been that fisheries scientists working for governments, and those working for conservation-orientated groups should have been able to jointly develop tactics and strategies for maintaining fishing effort at optimum levels, and develop, in the process, a unified science of what may be called "fisheries conservation."

What we have seen instead is 50 or more years of isolation of fisheries science from conservation biology, as reflected, for example, in distinct journals, whose contributions usually fail to acknowledge the existence of the sister discipline. The Fourth World Fisheries Congress is probably the first major event in fisheries science that acknowledges the problem that the chasm between these two disciplines represents, if mainly through its motto of "Reconciling Fisheries with Conservation."

*Corresponding author: d.pauly@fisheries.ubc.ca

Building on the solid achievements of the classic phase of fish population dynamics—the key papers cited above—I propose that the basic models developed during that period, while not any more useful for tactical management of fisheries, could still be used to identify the areas of reconciliation (i.e., potential collaboration) between fisheries (management) and conservation. This would amount, in practice, in identifying those stocks that have fallen below 30–40% of their original biomass (and which thus are overfished in terms of the classical models), and agreeing on a range of approaches to reduce the fishing mortality impacting on them.
The task will be huge, as the overwhelming majority of commercial stocks, throughout the world, have been reduced by a factor of 10 or more in the last 50 years, accelerating trends initiated much earlier (Jackson et al. 2001). This was documented, in the keynote, through a series of maps illustrating biomass declines in the North Atlantic (Christensen et al. 2003a), Southeast Asia (Christensen et al. 2003b), Northwest Africa (Christensen et al. 2004), and globally (Myers and Worm 2002), and all included in the PowerPoint presentation documenting my keynote, available at www.seaaroundus.org/WFC4F.HTM.

These maps, and the strong biomass decline they illustrate, document the emergence of a new way of looking at fisheries, which have so far tended to be perceived as local affairs, involving predictions for one, or a small suite of, resource species, and shorter time spans, the equivalent of predicting the "weather" of fisheries. The maps, rather, describe long-term, planet-wide trends, equivalent to the "climate" of fisheries. Fishing down marine food webs (Pauly et al. 1998), now a well-established phenomenon (Pauly and Palomares 2004), is the most conspicuous of the trends paralleling the biomass declines mentioned above. This goes along with 1) size reduction among the fish landed, and those remaining in marine ecosystems, 2) reductions of the length and complexity of the food webs structuring those ecosystems, and 3) in shelf systems impacted by trawling, a gradual transition from the benthos to the open water column as the major site for consumption of primary and detrital production. Jointly, these features imply increased variability of, and uncertainty about the ecological processes leading to fisheries catches, which, as well, are likely to continue their global decline (Watson and Pauly 2001).

Global mapping of fishery catches (Watson et al. 2004), besides being a requirement for the mapping of biomass presented above, also allows direct identification of another set of major trends characterizing the global 'climate' of fisheries, and its changes over the last fifty years.

Thus, mapping catches as function of water depth and latitude shows that fisheries, since 1950, have been relying increasingly on fish caught over, then at, great depths, and particularly so in the Southern Hemisphere. This reflects the geographic expansion of fisheries, and also the fact that increasingly, fish consumed in the developed countries of the Northern Hemisphere (notably the USA, Japan, and some countries of the European Union) originate from developing countries, and southern hemisphere waters. These trading patterns have long masked, for consumers in the developed countries of the north, the effects of the large-scale depletion of marine resources in the traditional fishing grounds (e.g., of the North Atlantic). Nevertheless, these effects are gradually becoming visible to the public at large—often via the media campaigns of environmental nongovernmental organizations (NGOs), and the government agencies in charge of fisheries, throughout the world, are forced to respond.

One of the issues that regulatory agencies will have to address is the limited use so far of ocean zoning as a tool for regulating fisheries, especially the creation of large marine protected areas (MPAs, including no take reserves at their core), the obvious analog to the national parks used to conserve the biodiversity of terrestrial systems. The geographic expansion of fisheries alluded to above should be seen, in this context, as the invasion, by increasingly sophisticated and powerful fishing vessels, of the natural marine reserves which, through their depth or distance from various ports, or rocky grounds, had protected populations previously exploited at the shallow, nearshore, soft bottom ends of their distributions (Pauly et al. 2002; 2003).

Indeed, one can show the geographic expansion, over time, of the fished areas of the world (i.e., of what may be called marine unprotected areas [MUA]). Thus, demands for MPAs do not introduce an arbitrary, novel concept into the debate about fishing management, but rather seek to re-establish the balance between MUA and MPA that prevailed a few decades ago, and which now appear to be a prerequisite to sustainable fisheries and their reconciliation with conservation.

Acknowledgments

I thank my colleagues of the *Sea Around Us* project, notably Jackie Alder, Villy Christensen, and Reg Watson, for the figures illustrating my keynote and for sharing with me the first 5 years of our voyage of discovery.

References

Beverton, R. J. H., and S. J. Holt. 1957. On the dynamics of exploited fish populations. Chapman and Hall, London, facsimile reprint 1993.

Christensen, V., S. Guénette, J. J. Heymans, C. J. Walters, R. Watson, D. Zeller, and D. Pauly. 2003a. Hundred-year decline of North Atlantic predatory fishes. Fish and Fisheries 4:1–24.

Christensen, V., L. Garces, G. T. Silvestre and D. Pauly 2003b. Fisheries impact on the South China Sea large marine ecosystem: a preliminary analysis using spatially-explicit methodology. Pages 51–62 *in* G. T. Silvestre, L. R. Garces, I. Stobutzki, M. Ahmed, R. A. Valmonte-Santos, C. Z. Luna, L. Lachica-Aliño, P. Munro, V. Christensen and D. Pauly, editors. Assessment, management and future directions for coastal fisheries in Asian countries. WorldFish Center Conference Proceedings 67.

Christensen, V., P. Amorim, I. Diallo, T. Diouf, S. Guénette, J. J. Heymans, A. Mendy, M. Ould Taleb Ould Sidi, M. L. D. Palomares, B. Samb, K. A. Stobberup, J. M. Vakily, M. Vasconcellos, R. Watson and D. Pauly. 2004. Trends in fish biomass off northwest Africa, 1960–2000. In M. Ba, P. Chavance, D. Gascuel, M. Vakily and D. Pauly, editors. Pêcheries maritimes, écosystèmes et sociétés: un demi-siècle de changement. Institut de recherche pour le developpement, Paris.

Jackson, J. B. C., M. X. Kirby, W. H. Berger, K. A. Bjorndal, L. W. Botsford, B. J. Bourque, R. Cooke, J. A. Estes, T. P. Hughes, S. Kidwell, C. B. Lange, H. S. Lenihan, J. M. Pandolfi, C. H. Peterson, R. S. Steneck, M. J. Tegner, R. R. Warner. 2001. Historical overfishing and the recent collapse of coastal ecosystems. Science 293:629–638.

Myers, R. A., and B. Worm. 2003. Rapid worldwide depletion of predatory fish communities. Nature (London) 423:280–283.

Pauly, D., J. Alder, E. Bennett, V. Christensen, P. Tyedmers, and R. Watson. 2003. The future for fisheries. Science 302:1359–1361.

Pauly, D., V. Christensen, J. Dalsgaard, R. Froese and F. C. Torres Jr. 1998a. Fishing down marine food webs. Science 279:860–863.

Pauly, D., V. Christensen, S. Guénette, T. Pitcher, U. R. Sumaila, C. Walters, R. Watson, and D. Zeller. 2002. Toward sustainability in world fisheries. Nature (London) 418:689–695.

Pauly, D. and M. L. Palomares. In press. Fishing down marine food web: it is far more pervasive than we thought. Marine Science Bulletin 74/3(S).

Ricker, W. E. 1954. Stock and recruitment. Journal of the Fisheries Research Board of Canada 11:559–623.

Schaefer, M. B. 1954. Some aspects of the dynamics of populations important to the management of the commercial marine fisheries. Bulletin of the Inter-American Tropical Tuna Commission 1:27–56.

Schaefer, M. B. 1957. A study of the dynamics of the fisheries for yellowfin tuna in the eastern tropical Pacific Ocean. Bulletin of the Inter-American Tropical Tuna Commission 2: 247–268.

Watson, R., A. Kitchingman, A. Gelchu, and D. Pauly. 2004. Mapping global fisheries: sharpening our focus. Fish and Fisheries 5:168–177.

Watson, R. and D. Pauly. 2001. Systematic distortions in world fisheries catch trends. Nature (London) 414:534–536.

American Fisheries Society Symposium 49:5–24

What Should We Care About When Attempting to Reconcile Fisheries with Conservation?

KEVERN COCHRANE*

Food and Agriculture Organization of the United Nations, via delle Terme di Caracalla, Rome 00100, Italy

Abstract.—The key considerations in trying to reconcile fisheries and conservation are related to the range of motives that give rise to both primary goals. This paper will examine the motives that can give rise to the need or desire to fish, ranging from a basic need for food to a less urgent, but still important, desire for recreation. Which particular incentives are dominant in any case will be closely related to the social and economic context in which the fishery exists. Similarly, the need and desire to conserve may range from a local or global recognition of the link between conservation and longer-term survival to more extreme protectionist views that could effectively lead to exclusion of human use. Knowledge and understanding of the underlying motives and goals for both conservation and fisheries are a prerequisite for reconciliation. This paper will consider some of the biological, social, economic, and psychological factors that have led to overfishing and how these are addressed in some of the recent international fisheries instruments, such as the Code of Conduct for Responsible Fisheries, the Convention on Biological Diversity, and the Plan of Implementation of the 2002 World Summit on Sustainable Development. It will explore the obstacles to effective implementation of the instruments. These obstacles can, in many instances, be summarized as the temporal and socioeconomic mismatch between short-term costs and realizing the full, long-term benefits of conservation. Approaches to bridging this mismatch will be considered.

Introduction

The global problems confronting fisheries and aquatic ecosystems are well known and have been the subject of much scientific debate and discussion (e.g., Ludwig et al. 1993; FAO 1994; Cochrane 2000) and considerable media attention, albeit not always well informed or balanced. The essence of the problem is that the human demand for fish and fishery products exceeds the sustainable production of aquatic systems, both natural and culture systems, and the methods used in the capture and culture of fish when used in excess have negative impacts on the ecosystem as a whole. As a result, there is a global attempt to control both the direct harvesting of target species to maintain extraction at sustainable levels and the impact of fisheries on affected and dependent species, ecological relationships, and the environment. To date, these efforts have led to very little success. In general, many fish stocks are currently overexploited or depleted (e.g., NMFS 2002; Garcia and De Leiva 2003) and many aquatic habitats continue to deteriorate (e.g., Hall 1999).

*Corresponding author: kevern.cochrane@fao.org

Blame for the crisis flows as freely as fish once flowed into markets. The fishers, fishery managers, scientists, politicians, and natural predators have all been blamed, usually with some justification, for causing or contributing to overfishing and its ecological consequences (McGoodwin 1990; Symes 1996; Walters and Maguire 1996; Hutchings et al. 1997; Cochrane et al. 1998; Cochrane 2000; Tamura 2003; Trites 2003). However, the underlying cause of overfishing and damage to aquatic ecosystems and of the conflicts between different users, consumptive and noncon-sumptive, lies with conflicting uses of and demands from aquatic ecosystems and their biota. All who impact directly or indirectly on these systems must accept a share of the responsibility. This paper will consider the different desires and demands directed at aquatic ecosystems and their implications for reconciling the goals and objectives of fisheries with those of conservation. It will focus particularly on capture fisheries.

The Issues

The organizers of the Fourth World Fisheries Congress provided a preliminary list of priority societal goals in relation to fisheries and conservation when they decided on the sessions that would fall within this topic in the Congress program. The sessions they identified under the broad heading "What should we care about when attempting to reconcile fisheries with conservation?" can be summarized as follows:

- Ethics and justice in fisheries,
- The human dimension,
- The ecological dimension,
- Economics and trade,
- Maintaining biodiversity,
- Jurisdictional equity,
- Sport fisheries, and
- Lessons from history, which is more a tool than a consideration.

This list provides a useful overview of some of the current key issues in fisheries but reflects a strong bias towards bureaucratic and academic perspectives and does not provide much guidance on the underlying human motives or drives that have led to the current problems.

A widespread problem among fisheries scientists from all disciplines is that fisheries are all too often treated as a stand alone, essentially isolated, issue with unique problems and unique solutions. An explicit or implicit assumption is made that both cause and solution must be sought within the fisheries sector. This limited view explains the emphasis within fisheries for treating the secondary causes or even the symptoms of the problem through measures such as improved scientific knowledge and methods, systems of regulating access ranging from, for example, zoning to individual transferable quotas (ITQs), marine protected areas (MPAs), comanagement and community-based management, adoption of an ecosystem approach to fisheries, and other well-intentioned, necessary, but, on their own, insufficient measures.

A more realistic approach requires recognition that fish resources are, in effect, simply one more resource in limited supply being competed for by different stakeholders. In the words of McGoodwin (1990), fisheries management is no more nor less than "an arena in which diverse societal, political, and market interests participate in an age-old struggle for the allocation and control of scarce resources" and the motives that drive humans in the struggle for fish will have the same basic driving forces as they have in all such struggles. Identifying those driving forces will assist in identifying what we should care about as we struggle to reconcile fisheries and conservation.

National Perspectives

Humans have been engaged in fisheries for millennia and most, if not all, governments now have legislation in place to help to regulate their national fisheries in order to contribute to wider national policies. This legislation provides insight into the priorities, at least at the national level, that states have assigned to their fisheries and the utilization of their living marine resources. It should therefore indicate what countries have chosen to care about in relation to conservation and fisheries. For this overview, the primary legislation of three countries, differing considerably in their developmental state and cultural context, was examined. These three were India (medium development, human development index [HDI] = 0.56), South Africa (medium development,

HDI = 0.684), and the United States of America (high development, HDI = 0.937). Only the primary legislation and those acts that had fisheries or marine living resources as a clear and dominant priority were examined. The results are summarized in Table 1.

As could be expected, the structure of the legislation varied considerably, from a single instrument, the Living Marine Resources Act No. 18 from South Africa, to the four and five separate but closely related instruments from India and the United States respectively. The information shown in Table 1 should be seen as being indicative only as it is probable that additional information that may fill some of the apparent gaps for some countries would be found in other acts or in more focused, secondary legislation within each country. The Guidelines for Fishing Operations in Indian Exclusive Economic Zone (as circulated with order No. 21005/1/2001-FY, 1 November 2002) refer to the obligation of deep sea vessels to comply with the Code of Conduct for

Table 1. The priorities for fisheries of India, South Africa and the United States of America as indicated by their primary fisheries legislation. The sources used are India—The Indian Fisheries Act 1897; Maritime Zones of India (Regulation of Fishing by Foreign Vessels) Act, 1981; Coast Guard Act, 1978 and Guidelines for Fishing Operations in Indian Exclusive Economic Zone 01 November 2002; South Africa—Marine Living Resources Act No. 18, 1998; and the USA—Coastal Zone Management Act Of 1972, Endangered Species Act of 1973, Magnuson–Stevens Fishery Conservation Management Act, National Marine Sanctuaries Act, and the Marine Mammal Protection Act of 1972.

South Africa	United States of America	India
i) Sustainable development	Take into account the importance of fishery resources to fishing communities	Maintain fisheries as important source of food supply. Development of deep sea fisheries.
ii) Optimum utilisation	Optimum yield	
iii) Conservation of marine living resources	Prevent overfishing. Preserve, protect and enhance the resources of the coastal zone.	Compliance with the FAO Code of Conduct. Conservation of fish resources.
iv) Sound ecological balance and protection of ecosystem as a whole	Protection of natural resources and their habitat within the coastal zone; Conservation of ecosystems upon which endangered and threatened species depend; restore and enhance natural habitats, populations and ecological processes (through marine sanctuaries).	Prohibition on use of dynamite and poisons.
v) Preserve marine diversity	Minimize bycatch and mortality through bycatch, protection marine mammals—not permitted to diminish below optimum sustainable population. Conservation endangered species.	

Table 1. Continued.

South Africa	United States of America	India
Human resource development		Maintain fisheries as important source of food supply (food security)and as source of employment (within state legislation, S. Mathew, personal communication)
Economic growth	Efficiency of utilisation but economic allocation shall not be the sole purpose	
Capacity building		Capacity building and modernisation of deep sea fishing industry
Employment creation		Protection of fishers from hardships caused by poaching by foreign vessels
		Avoid conflicts between deep sea fishers and other stakeholders.
Minimize pollution		
Participatory approach	Measures shall provide for sustained participation of fishing communities	
Equity in access and redressing historical imbalance	Fair and equitable distribution of fishing privileges	
	Minimize costs of conservation and management measures	
	Promote safety of life at sea.	Offer protection to fisherman, including assistance by Coastguard when in distress.

Responsible Fisheries, which demonstrates that the very broad requirements of the Code are implicit in Indian fisheries law, substantially increasing the scope otherwise indicated in the four Indian instruments examined. Notwithstanding these difficulties in interpreting and comparing the legislation, this analysis showed that all of the countries gave priority to

• Developing or sustaining fisheries and other utilization of living marine resources and ecosystems for human needs,

• Conservation of target resources,

• Protection of the ecosystem as a whole, and

• The welfare of the fishers and others dependent on fisheries for their food security and livelihoods.

Other priorities identified by at least two of the three countries included preservation of biodiversity, equity in access to the resources, capacity building (in the two developing countries), and the participation of stakeholders in the management of their activities. This list is consistent with the breakdown provided by the Congress organizers but, unsurprisingly given that two of the three countries considered were developing countries, puts an emphasis on development that was not explicit in the Congress list.

The International Perspective

Countries do not exist in isolation any more than fish stocks do, and in fisheries, all countries are influenced by a range of global and regional laws and agreements. The overarching global agreement dealing with fisheries is the United Nations Convention on the Law of the Sea (UN-LOS), which was agreed to in 1982. The most significant aspect of the UN-LOS in relation to fisheries was the allowance for establishment of the exclusive economic zones (EEZ) up to 200 mi from the shore for all coastal countries and allocation to them of the resources within that zone and responsibility for their sustainable use of those resources (Articles 56 and 57). The articles and paragraphs covering fisheries in UN-LOS give some useful indications of the global concerns related to fisheries. For example, Article 56 gives sovereign rights to the coastal state for "exploring and exploiting, conserving and managing natural resources." Article 61 calls for the maintenance of or restoration to levels that can produce the maximum sustainable yield, consideration for the economic needs of coastal communities, and special requirements of developing states. Article 61 also requires that consideration is given to the effects of management measures on "species associated with or dependent upon harvested species." Article 62 points out that coastal states shall promote the objective of optimum utilization of the living resources of their EEZ.

Many other international instruments, most of them established after UN-LOS, have reinforced and developed the UN-LOS issues further. Included in these instruments are the 1973 Convention on International Trade in Endangered Species of Wild Fauna and Flora (CITES) that predates the UN-LOS; the 1992 UN Conference on Environment and Development and, arising from that, the Convention on Biological Diversity and Agenda 21: Program of Action for Sustainable Development; the 1995 UN Agreement for the Implementation of the Provisions of the United Nations Convention on the Law of the Sea of 10 December 1982 Relating to the Conservation and Management of Straddling Fish Stocks and Highly Migratory Fish Stocks; and most recently, the 2002 World Summit on Sustainable Development Plan of Implementation that reinforced the existing global instrumental framework and provided a good, if very optimistic, plan for making progress towards achieving the intentions of that framework.

Arguably, the most comprehensive, in terms of fisheries, of these global instruments is the FAO Code of Conduct for Responsible Fisheries (FAO 1995). The Code is a nonbinding instrument, and probably because of the greater openness and flexibility allowed within a nonbinding agreement, it goes beyond the binding instruments. It addresses not only the immediately attainable, but also longer-term goals and aspirations of countries and the global community. It therefore provides very good insights into the international desires and priorities in relation to fisheries and conservation.

There are seven substantive articles of the Code that cover general principles, fisheries management, fishing operations, aquaculture development, integration of fisheries into coastal area management, postharvest practices and trade, and fisheries research (FAO 1995). The priorities contained within the Code are summarized in Table 2. They are taken from the first substantive article, Article 6 General Principles. The Code places heavy emphasis on conserving, maintaining, and restoring, where necessary, living aquatic organisms and the ecosystems of which they form a part. These are requirements in their own right but are also valued because of the benefits for humankind that can be obtained from those resources, now and in the future, and for their contribution to sustainable development. The rights and conditions of work and living of those involved in fisheries are expected to be of an acceptable standard, with particular attention given to those groups, subsistence, small-scale and artisanal fishers and fishworkers, considered to be most vulnerable to the whims of nature and of the more powerful human groupings. For reasons of both ethics and effectiveness, participatory and transparent decision making is called for as is the recognition of the interactions between fisheries and other users of coastal zones. Finally, efforts should be made to ensure optimum use of the products from fisheries.

These priorities again cover the same general issues found

Table 2. Global priorities for fisheries and conservation as indicated by Article 6 General Principles of the FAO Code of Conduct for Responsible Fisheries (FAO 1995).

Code of Conduct for Responsible Fisheries
1. Sustainable development (6.2)
2. Maintenance of the quality, diversity, and availability of fishery resources (6.2)
3. Maintain biodiversity (6.6)
4. Conserve aquatic ecosystems, including species other than the target species and the aquatic environment (6.2 and 6.5). Conserve the population structure (6.6)
5. Protect and rehabilitate critical fish habitats (6.8)
6. Maintenance of fishery resources for present and future generations in the context of food security, poverty alleviation, and sustainable development (6.2).
7. Fishing facilities and equipment should allow for safe, healthy, and fair working and living conditions (6.17).
8. The rights of fishers and fishworkers, particularly those involved in subsistence, small-scale, and artisanal fisheries should be protected (6.18).
9. Promotion of awareness of responsible fisheries through education and training.
10. Industry, fishworkers, environmental, and other interested organisations should be included in decision-making in transparent processes (6.13).
11. Ensure compliance with and enforcement of conservation and management measures (6.10).
12. Ensure fisheries interests, including conservation, taken into account in multiple uses of the coastal zone (6.9).
13. Post-harvest practices should ensure nutritional value, quality and safety of products, also minimizing negative impacts on environment (6.7).

in the three national policies examined in the previous section but go into greater detail on most issues.

Reconciling Fisheries and Conservation: The Global Cares

The perspectives of the science-dominated program committee, the sampled national governments, and the global community as represented by the Code of Conduct on priorities in reconciling fisheries and conservation can be summarized as

- The role of fisheries and other uses of living marine resources and ecosystems for sustainable development to meet human needs, including food, security, and poverty alleviation;

• Conservation of target resources;

• The maintenance and restoration where required of biodiversity and aquatic ecosystems, including the environment and critical habitats;

• The rights and welfare of the fishers and others dependent on fisheries for their food security and livelihoods and the provision of safe, healthy, and fair working and living conditions for them; and

• Availability to the consumer of food products from fisheries that provide nutritional value and quality and are safe.

Successes and Failures

The preceding analyses indicate that there is widespread, explicit agreement at the policy level on the priority concerns and interests relating to fisheries and conservation of marine living resources and ecosystems. While there have been some encouraging indications of success in recent years (e.g. Mace 2004), study after study has concluded that, at the national, regional, and global levels, there has been insufficient progress in achieving these objectives and that many fish stocks and ecosystems are under varying degrees of stress caused by a range of human activities, including fisheries (Ludwig et al. 1993; Rosenberg et al. 1993; FAO 1994; Cochrane 2000; Balmford et al. 2002; Christensen et al. 2003; Garcia and De Leiva 2003; and Cochrane and Doulman 2005). In 2002, based on available information, FAO (2002) estimated that 25% of fish stocks or species groups are underexploited or moderately exploited, approximately 47% are fully exploited, 18% overexploited, and 10% of stocks are substantially depleted or are recovering from depletion.

Garcia and De Leiva (2003) reported that there are no comprehensive and fully reliable data available on global fleet size but that the data available indicated an increase in fleet size between the 1950s and 1990s, with signs that growth has stopped in recent years with the fleet size stable and possibly decreasing. When technological improvements are also taken into account, there was an estimated increase in fishing power of more than 400% between 1960 and 1990. The current potential fishing power is substantially more than required for capture of the global sustainable yields and over-capacity is reported by the same authors as being between 30% and 50% too high, depending on the reference points used.

More directly relevant to this paper are the trends in number of people employed in fisheries worldwide. Again, Garcia and De Leiva (2003) reported that data on this are scarce and incomplete, but the best available estimate is that in 1998 about 36 million people were employed in fisheries and aquaculture, of which approximately 15 million were employed full-time, 13 million part-time, and 8 million as occasional workers (FAO 2000). Based on the average number of persons per household, this translates to more than 100 million people being directly dependent on fisheries for their livelihoods. The number of people employed in fisheries appears still to be growing (Figure 1). If this trend cannot be halted and ultimately reversed, the task of reconciling fisheries and conservation will become even more difficult.

In an analysis of the progress made by countries in meeting international obligations and goals, Cochrane and Doulman (2005) concluded that progress was slow and that much remained to be done. For example, the 25th Session of the Committee on Fisheries (COFI) recognized that although there had been some progress in implementing the Code of Conduct for Responsible Fisheries, much still needed to be done (FAO 2003a). Similar problems are being experienced in other international bodies and as a further example, the relevant range states of CITES are experiencing substantial difficulties in responsibly and sustainably managing fishing and export of the Acipenseriformes sturgeon and paddlefish species and of queen conch *Strombus gigas* found in the Caribbean Sea (CITES 2002, 2003; Cochrane and Doulman 2005). Both these taxonomic groups are listed on CITES Appendix II.

The factors leading to this widespread failure of fisheries management have been discussed widely. Cochrane (2000) proposed that there were four basic underlying causes:

• High levels of biological and ecological uncertainty that frequently hinder good management decisions and performance;

• The conflict between short-term economic and social objectives and the longer-term objectives of

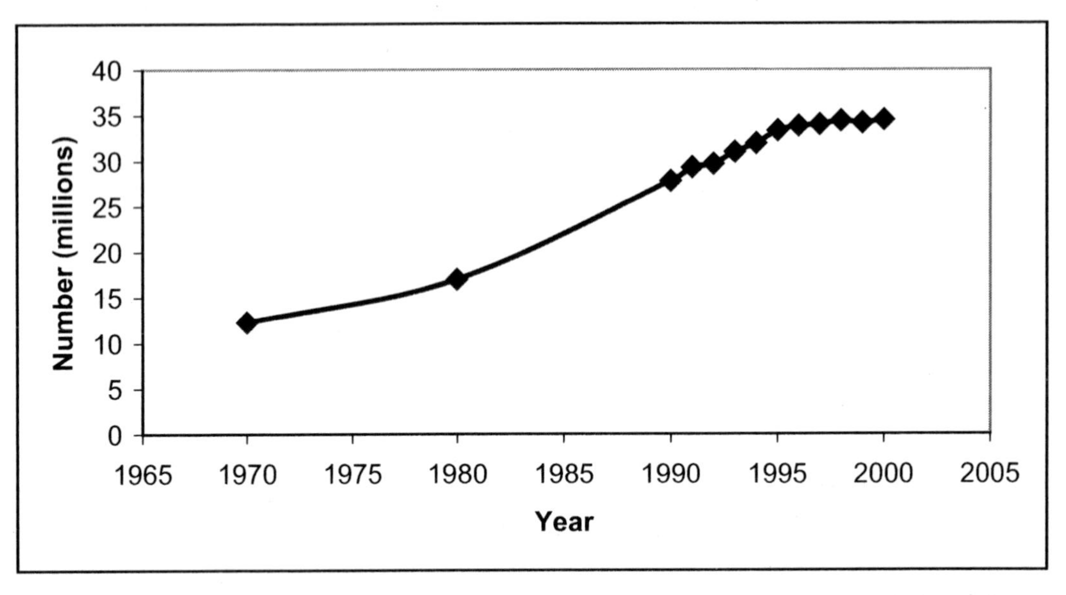

Figure 1. The estimated number of people employed in fisheries and aquaculture, including full-time, part-time, and occasional workers (from FAO 2000 and Garcia and De Leiva 2003).

sustainability, with the short-term goals usually being favored;

• The absence of or poorly defined objectives leading to ad hoc decisions focused on immediate problems and short-term solutions; and

• Institutional weaknesses especially weak and inappropriate systems of user rights, including open access fisheries, and inadequate participation by stakeholders in management.

Cochrane and Doulman (2005) elaborated on this list and included the prevalence of illegal, unreported, and unregulated (IUU) fishing; the importance of social attitudes in cultivating political decisions favoring short-term goals; the still widespread prevalence of subsidies for fishing; and the low levels of capacity for fisheries management, especially but not exclusively in developing countries. They particularly emphasized the important role of societal attitudes that favor short-term social and economic goals over long-term sustainability in driving political will and the need to address this as a high priority.

Taking a broader look at the factors hindering sustainable development, WSSD (2002) identified three essential requirements for sustainable development, poverty eradication, changing unsustainable patterns of production and consumption, and managing the natural resource base upon which economic and social development depends. We live in a world with very high technological ability, sophisticated codes of ethics and governance (albeit practices that frequently depart from those codes), and high and rapidly increasing global linkages with a resulting breakdown of national barriers and constraints. We therefore appear to have the ability and the theory to address instantly all of these requirements, but we have consistently failed to do so and the three WSSD overarching objectives apparently remain as inaccessible as they have ever been. Political will, largely driven by societal will, must underlie this failure, and the factors driving societal will therefore warrant further exploration.

The critical need to attain sustainable utilization cannot be challenged. Unless humankind can achieve the conservation goals of maintaining productive populations of target resources and the maintenance and restoration, where required, of biodiversity and ecosystems, there is no hope of achieving the goals related to human welfare. In whatever way the problems being confronted in reconciling fisheries and conservation are grouped, the underlying constraints reside in inadequate political will or opportunity and in the social pressures that influence political attitudes and decisions. The rest of this paper considers the threats to achieving these conservation goals and, in particular, the reasons for the failure to date to achieve them at a global level. The next section therefore examines some of the factors determining social attitudes, in order to consider solutions in the search for reconciliation.

Back to the People: Social Attitudes and Fisheries

Cochrane (2000) discussed the central role of societal dependency on fisheries in influencing fishery management decisions and actions and suggested that strong resistance to change by fishers is an important determinant of political will. Resistance by fishers to pressure to change their occupations is a feature of both developing and developed countries, as demonstrated by the large excess in fishing capacity at a global level. The rest of this section, under three subheadings, focuses on factors driving this resistance, mainly in developing countries. The subsequent section explores some additional psychological factors, uncovered by prospect theory, which help to explain strong resistance to change in developed countries as well, where one would have expected the presence of more options and alternatives to have reduced that resistance.

Basic Needs of Mankind

As stated previously, the cares of the world, in relation to fisheries and conservation, are not isolated and unique but are simply examples or manifestations of more generic needs and desires. Understanding those generic needs provides a better route to identifying solutions than will be obtained by treating fisheries as a closed system. An American psychologist, Dr. Abraham Maslow (1908–1970), developed such a generic guide to human needs and desires. Maslow challenged the then prevailing view of human behavior as being a perpetual struggle to address tensions or to compensate for something lacking. He proposed that humans also experience and respond to positive needs, often pursuing goals for their own sake such as seeking beauty in art and nature or taking on a task for the pleasure of accomplishment. Maslow therefore developed an integrated hierarchy of needs, in which it was proposed that humans strive to climb upwards from the most basic needs of requiring food and water to an ultimate goal of self-actualization. Gleitman (1981) describes seven levels in Maslow's hierarchy (Table 3). According to Maslow, individuals will, in general, only attempt to meet needs higher in the hierarchy once their needs at lower levels have been met. Therefore the position of one's dominating needs in the hierarchy will be a major determinant of one's priorities and perspectives at a given time.

The needs and goals described by Maslow are the underlying drivers of what individuals actually care about in relation to fisheries and conservation. Clearly, when an individual has inadequate food or water, he or she will be able to think of little else but satisfying that need. In the case of the chronically poor and malnourished, there will be little if any opportunity to escape from a preoccupation with level 1. When physiological needs have been met, the individual can turn their attention to safety needs and to seeking stability, security, and routines. The next step is for belongingness and love. Fulfilling those needs will normally require, in addition to the prerequisites for levels 1 and 2, a functioning and supportive family within a stable social structure and community. Level 4 involves the need to feel valued and respected by oneself and by those around one at home and at work. It flows into the next two levels; cognitive and aesthetic needs and must be related to them. In fact, Davidoff (1987) lists only five levels, omitting Gleitman's levels 5 and 6, presumably including them as a part of esteem needs (Table 3). These three levels, esteem, cognitive, and aesthetic, must include a longer-term view, a process of growth and enrichment, as opposed to the immediate and urgent needs included in levels 1 to 3. When preoccupied with needs on levels 1 to 3, an individual, society, or government will be forced to focus on the short-term and meeting those immediate needs. Food security, safety, health, and housing will dominate the agenda and longer-term goals related to, for example, recreation, aesthetics, personal development, and the environment will tend to be dreams that must wait until better days. While such a short-term view is ultimately counterproductive and severely constraining, physiological and emotional survival impose it.

Development and Need

The threats and opportunities available for people are heavily influenced by the development status of the country in which they live. The United Nations Development Program (UNDP) defines the human development index (HDI) as a composite index incorpo-

Table 3. Maslow's hierarchy of needs: starting from 1, the most basic needs up to 7, the highest order needs, as described by Gleitman (1995) and Davidoff (1987).

Needs and desires	
Gleitman	**Davidoff**
1. Physiological: sufficient oxygen, food, and water.	1. Physiological
2. Safety: obtaining comfort, security, and freedom from fear	2. Safety
3. Belongingness and love: affiliation, acceptance, belongingness	3. Belongingness and love
4. Esteem: competence, approval, recognition.	4. Esteem: achievement, approval, competence, and recognition
5. Cognitive needs: knowledge, understanding, novelty.	5. Self-actualization
6. Aesthetic needs: experiencing symmetry, order, and beauty	
7. Self-actualization	

rating three primary human conditions: health as measured by mean life expectancy at birth; knowledge as reflected as a weighted average of adult literacy and the combined primary, secondary, and tertiary gross enrolment ratio; and the mean standard of living as indicated by the mean gross domestic product per capita in the country (Figure 2).

The development status of countries calculated in this way varies enormously, as indicated by the distribution of the HDI for 175 countries of the world (Figure 3). Such large differences will result in enormously different perspectives and priorities. The UNDP (2003) classified the 175 countries for which it provided statistics into three basic development states: high, medium, and low. An additional layer of resolution can be added to this by separating out the countries that are considered to be low income and food deficit countries (LIFDC) from those that are not. The classification of a country as (LIFDC) [1] is based on three criteria:

[1] http://www.fao.org/countryprofiles/lifdc.asp?lang=en.

1) The country should have a per capita income below the "historical" ceiling used by the World Bank to determine eligibility for International Bank for Reconstruction and Development (IBRD) or the International Development Association (IDA) assistance (US$1,435 in 2001);

2. The country should be a net food importer, where food imports and exports are calculated from the aggregated calorific content of a selection of basic foodstuffs (cereals, roots and tubers, pulses, oilseeds, and oils other than tree crop oils, meat, and dairy products), averaged over the most recent 3 years; and

3. Countries may opt not to be classified as LIFDC.

Cuba and Belarus are classified as high development but still as LIFDC. A number of middle development countries are also LIFDC. Examples include Macedonia, Western Samoa, Philippines, Azerbaijan, Ecuador, and Egypt. All low development countries are also LIFDC. With the added consideration of LIFDC sta-

DIMENSION	A LONG AND HEALTHY LIFE	KNOWLEDGE		A DECENT STANDARD OF LIVING
INDICATOR	Life expectancy at birth	Adult literacy rate	Gross enrollment ratio (GER)	GDP per capita US$
DIMENSION INDEX	Life expectancy index	Adult literacy index [0.67]	GER Index [0.33]	
		Education index		GDP index
Human development index				

Figure 2. Composition of the human development index (HDI). The numbers in square brackets after the knowledge indicators show the relative weighting given to the two indicators in calculating the composite education index. The final HDI is a simple average of the three dimension indices

tus, a finer scale of resolution can be achieved for examination of broad national priorities.

Table 4 shows the sum of the population sizes of all countries in each HDI group (total population by nation). Those numbers could, however, be misleading as there are substantial differences between the richest and poorest within each country. The ratio in income between the poorest 20% and the richest 20% in a country ranges from 3.4 in Japan to 57.6 in Sierra Leone. A simple geometric mean of the ratio between the poorest and richest 20% in each of the 115 countries for which this statistic was available was 8.6 (UNDP 2003, Table 13). Using an example of a middle devel-

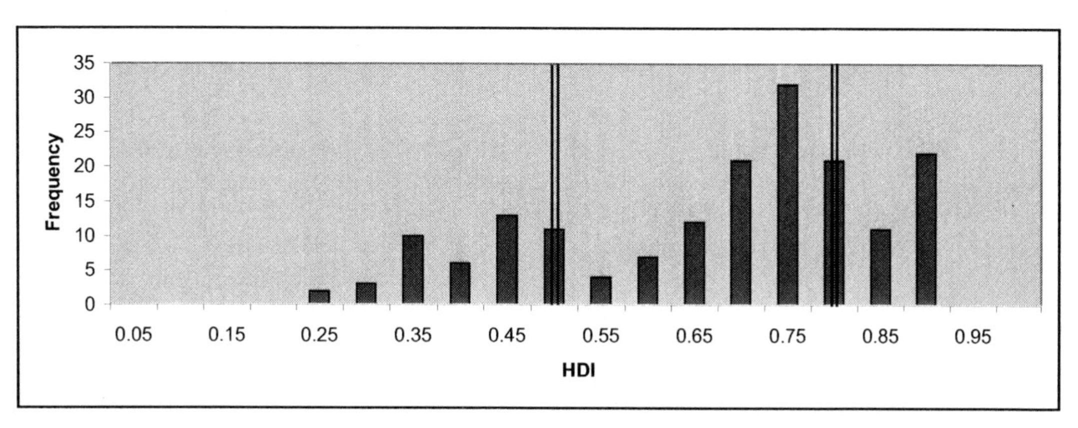

Figure 3. Distribution of the human development index for 175 countries of the world. (data from UNDP 2003). The vertical double lines indicate the boundaries between low and medium development (left) and medium and high development (right).

Table 4. Mean development indicators for high, medium and low development countries. Total population by nation shows combined population numbers for all countries at the specified human development stage and total population: by individual shows the number of individuals estimated to actually live under the specified human development conditions through consideration of the distribution within countries (see text for explanation). Information from UNDP (2003), pp 237–240.

Human develop-ment	Total population by nation (millions)	Total population: distribution by individual (millions)	Life ex-pectancy at birth (years)	Adult literacy (% age 15 and above)	Combined 1°, 2°, 3° gross enrollment ratio (%)	GDP per capita (PPP US$)
High	1,173	932	77	96	90	23, 482
High + LIFD	21	759	73	98	81	6,373
Medium	1,044	1,227	68	89	75	6,196
Medium + LIFD	3,072	1,587	67	73	61	3,318
Low	737	1,542	49.4	54.0	41	1,195

opment country, Thailand, for which the ratio was 8.3 and the mean gross domestic product (GDP) per capita was US$6,400, this translates to a range in GDP per capita from $1,952 for the poorest 20% to $16,000 for the richest 20%. From Table 4, this indicates a wide distribution of individuals about the national mean and, for example, that the lower 20% of the population in Thailand would be experiencing a lifestyle comparable with the mean in a low development country, while the richest 20% would be comparable to the lower end of a highly developed country.

In order to account for the distributions within each country, the global figures were recalculated for individuals on the assumption that 40% of the population of any country would be at the average standard of living of the country, as indicated by developmental stage, 20% at a standard of living one category above or below that, and 10% two categories above and below the average. The recalculations (total population: distribution by individual), not surprisingly, result in a more even distribution of individuals across the development range than is indicated by using just the sum of national populations at a given level. Nevertheless, they still show a strong weighting towards the lower end of the development scale.

Table 4 also shows the arithmetic averages of each of the HDI subcomponents calculated across all countries, weighted by population size, in each category. These figures reveal substantial differences in the various components of the HDI. Of particular note are the big falls in all components between the medium LIFDC and low development countries, the big fall in enrollment ratio and GDP per capita between medium and medium LIFD countries, and the large drop in GDP per capita when moving from high development to high LIFD and medium development countries.

These aggregated indices reflect substantive differences in the situation and cares of people living within different development stages (UNDP 2003). A key component of the indices is, obviously, poverty. Some 1.17×10^9 people live on less than US$1.00 per day and about the same number do not have access to improved water sources. Nearly 830 million of these people are malnourished. Approximately 900 million of them live in extreme poverty in rural areas where they are largely dependent on natural resources for their survival. They live, overwhelmingly, in developing countries and there can be little doubt that they will be essentially fully occupied with addressing Maslow level 1 needs.

In addition, an estimated 2.4×10^9 people do not have access to adequate sanitation. This contributes to infant mortality rates in 2001 of 81 deaths per 1,000 births, nearly 10%, in low development countries and 31 per 1,000 in medium development countries, markedly higher than the 5 per 1,000 experienced in high development countries. 2×10^9 people do not enjoy access to electricity, depriving them of the many fundamental benefits that electricity can provide, including light, refrigeration, and mechanical power, which are essential components of effective health and education systems.

Within this group of more than 2×10^9 people, are the 1.2×10^9 people described above who are trapped into addressing level 1 needs, while the remainder are almost certainly going to be focused on level 2 or, at best, level 3 needs in a struggle for safety, security, and a sense of belonging. These individuals will be found mainly among the 3.1×10^9 people estimated to experience low and medium LIFD development conditions (Table 4). The number of people actually preoccupied with addressing the immediate needs of levels 1–3 is likely to be higher than this, though. The GPD per capita is markedly lower for medium LIFD countries ($3,318) than it is for other medium development countries ($6,196) as are adult literacy and gross enrollment ratios (Table 4). In addition to the higher infant mortality rates referred to in the previous paragraph, medium development countries experience tuberculosis infection rates of 137 per 10,000 compared to 12 per 10,000 in high development countries. These and comparable statistics, suggest that 2×10^9 can be seen as a minimum estimate of the percentage of people constrained to focus on short-term needs, with the true figure probably being closer to 3×10^9, 50% of the global population, or even higher. Referring specifically to fisheries, FAO (2002) estimated that 20% of the world's fishers, nearly 6 million people, are probably small-scale fishers earning less than US$1.00 per day, but that if related activities are also included, the total number of people earning less that US$1.00 per day through small-scale fisheries is approximately 23 million.

The consequences of this widespread poverty for sustainable development are well known and explain the top priority given by WSSD (2002) to its eradication. Nevertheless, its profound implications are all too often neglected in practice, and the more powerful countries and lobbies able to give greater attention to high, longer-term Maslow needs frequently act and apply international pressure on the assumption that their priorities and perspectives are inherently appropriate for all of mankind. These attempts will inevitably fail to achieve sustained compliance and will lead to a continuation of the conflicts and confrontations being experienced over access to and conservation of limited living resources throughout the world. As agreed by WSSD (2002), in order to achieve sustainable development, poverty must be eradicated and all people of the world must be given the opportunity to give strong attention to the needs higher up the Maslow hierarchy, so that they too can share the right to devote their energies to fulfilling higher cognitive, esthetic, and environmental goals.

Insecurity and the Discount Rate

The implications for sustainable use of being constrained to the lower level needs in Maslow's hierarchy are clear and profound. They can also be demonstrated by reference to classical fisheries economics based on the standard surplus production model (Clarke 1990; Seijo et al. 1998). The aim of a single stakeholder with a long-term right to use that resource would normally be to maximize his or her long-term return from the resource. According to economic theory, the value of any return from the resource to the user will be a function of how far in the future that return will be made and different users will have different time preferences. The time preference can be expressed as a discount rate, ∂, such that the value of a return from the resource at time t in the future, V_t, is

$$V_t = V_0 \times \frac{1}{(1+\partial)^t}$$

where V_0 is the value of the return today.

Assuming a Schaefer production model and that the individual fisher will strive to maximize the net present value of his or her catch, the optimum equilibrium biomass level is given by (Clarke 1990)

$$B_{OPT} = \frac{1}{4}\{B_{BE} + k(1-\frac{\partial}{r}) + \sqrt{[B_{BE} + K(1-\frac{\partial}{r})]^2 + \frac{8K \cdot B_{BE} \cdot \partial}{r}}\}$$

where B_{OPT} = the optimum biomass level

B_E = bioeconomic equilibrium = $c/(pq)$

c = cost of a unit of effort

p = price per unit of the resource

q = catchability coefficient

∂ = discount rate.

The standard view of the discount rate is that it indicates the rate of return required by an investor in order to achieve an acceptable profit. Under this explanation, as the discount rate ∂ increases, the optimum biomass level B_{OPT} decreases, approaching the open-access bioeconomic equilibrium biomass. Therefore, a high discount rate increases the risk of overexploitation, threatening any conservation objectives (Figure 4). It can also be argued that a high discount rate reflects the opportunity costs of capital and therefore that a high discount value implies high operational costs that will tend to reduce effort and therefore fishing mortality (Seijo et al. 1998). Both of these views, however, are still based on the assumption that the discount rate is purely a function of economic opportunity costs and therefore that there are alternatives available for the capital that is being invested in the fishery. In contrast, where a fisher is trapped in a fishery with no alternatives in which to invest his or her capital or time, that fisher may be totally dependent on catches to satisfy immediate and basic needs. In such cases, if faced with a choice between taking an unsustainable catch today or leaving the fish in the water, the discount rate will be independent of longer-term financial implications and will be related to the magnitude of the threat to the immediate survival of the fisher and his or her dependants. Within this context, the discount rate will be highest among those struggling to meet the lowest needs in the Maslow hierarchy, including the 800 million estimated to be chronically undernourished for whom the discount rate will be enormous.

Poverty also has the potential to increase the pressure to overexploit through its impact on the bioeconomic equilibrium, B_E. The bioeconomic equilibrium is affected by both the cost of a unit of effort and the price per unit of resource: the lower the cost of effort and the higher the price realized for the catch, the lower will be

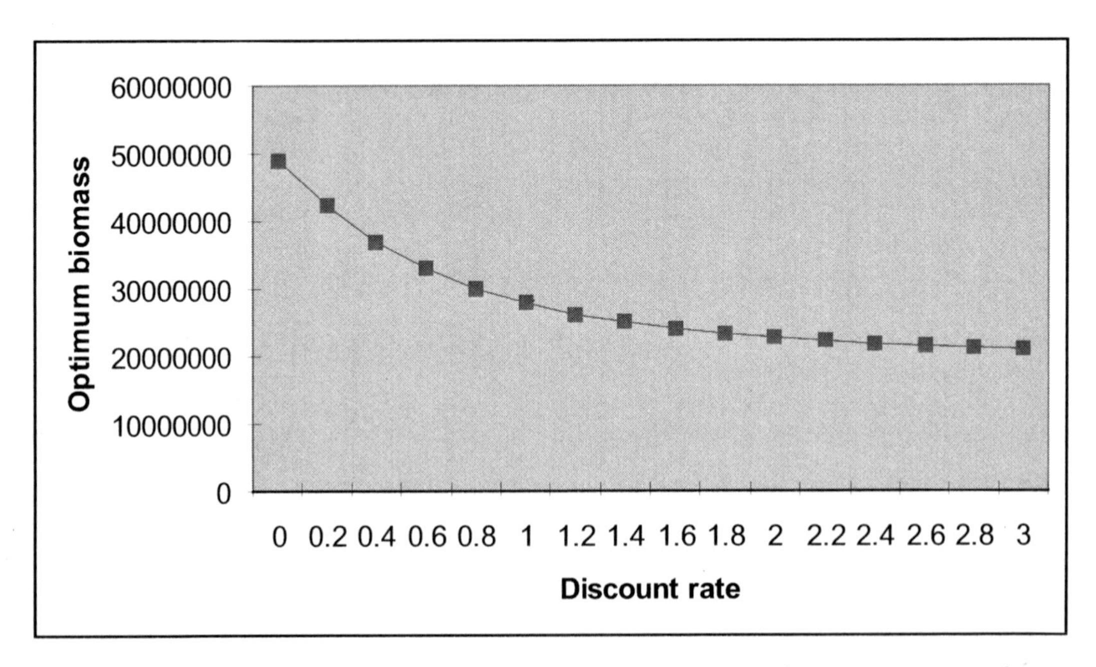

Figure 4. The relationship between optimal biomass (B_{opt}), expressed as a percentage of mean unexploited biomass (K) and discount rate for a hypothetical fishery. The example used is of the Pacific halibut fishery (Clarke 2000), where B_E is 17,500 metric tons, K = 80,500 metric tons, and r = 0.71.

the bioeconomic equilibrium. In a subsistence fishery, the cost of effort, often effectively just the fishers' time, will often be seen by them as being very low, as the fishers will have no more profitable use of their time than fishing. At the same time, the value of the fish will be very high as there will be very limited alternatives. The net result will be that the bioeconomic equilibrium will be very low and in extreme cases it can be zero. The prevailing, top-down approach to solving this problem is to exclude fishers from the fishery or to increase costs through management measures that effectively increase the costs, pushing up the bioeconomic equilibrium. In contrast, the ethical approach for those facing low Maslow needs, and the approach more likely to result in sustained compliance, is to decrease the value of the fish products to the fishers and increase the value of their time by providing them with higher value and quality alternatives.

Prospect Theory and Status Quo Bias

The preceding section focused on the extent of poverty at a global level and the implications of poverty for the prevailing attitudes towards fisheries and conservation. If Maslow's theory of a hierarchy of needs is correct, and common experience suggests that it is very plausible, then one could expect a broad commonality of views on fisheries and conservation among people at similar development stages.

Developing countries and poverty are not the only causes of overfishing, however, and may not even be the major causes. Overexploitation of resources is as much a feature of developed as it is of developing countries. Developing countries were responsible for 60% of global fish production in 1998 (Mathew 2003), with the remaining 40% coming from developed countries. Developed countries, however, are responsible for the bulk of fishery imports and in 2000 accounted for 80% of global imports (FAO 2002). The status of resources in, for example, the United States and the European Union (Cochrane 2000) is as much cause for concern as it is in most developing countries. In searching for solutions, it is important to recognize that the fundamental factors causing overexploitation in the two developmental categories are likely to differ. While fishing is contributing to meeting some of the lower level needs of Maslow's hierarchy in developed countries as well as developing ones, the former do have greater potential for providing alternative sources of livelihood, and the hierarchy alone does not provide much insight into the fundamental human motives driving overexploitation in wealthy, developed countries.

Some relevant insights to this problem can be obtained from the study of decision making, especially when considered from the perspective of risk management in a branch of economics known as "prospect theory" (Kahneman and Tversky 2000a). Prospect theory arose out of observations that the assumptions of expected utility theory, in which the outcomes of decisions are evaluated in terms of total wealth (e.g., welfare), frequently fail in practice. Under conventional utility theory, all reasonable people would be expected to follow the axioms and, able to predict future outcomes with complete accuracy, make their choices accordingly. Among the assumptions of utility theory are those of invariance and dominance. Invariance requires that the preference given to a particular option should not depend on the manner in which that option is described, while dominance requires that if an option A is at least as good as an option B in all respects and better than B in at least one respect, then A will be preferred to B.

There have been a number of results obtained in recent decades that demonstrate that these two axioms do not always hold. Some of these results are pertinent to this paper and as a supplement to the inferences made from Maslow's hierarchy and are summarized here (Kahneman and Tversky 2000a, 2000b; Kahneman et al. 2000).

1) In 1738, Bernoulli wrote an essay that attempted to explain why people are commonly risk averse and why risk aversion increases with decreasing wealth.

2) Bernoulli suggested that people do not evaluate prospects (or options) by considering their monetary value but instead evaluate them on the basis of the subjective value they expect from these options.

3) Arising from point 2, Kahneman and Tversky (2000b) proposed in prospect theory that choices are made on the basis of gains, losses and neutral outcomes, not on the basis of final total assets.

4) Conventional utility theory assumes that people tend to be risk averse but prospect theory, working from observation and experimentation, proposes that people tend to be risk averse in relation to gains but risk seek-

ing in relation to losses. In other words, individuals will tend to take a higher risk to avoid a loss than they would to make the equivalent gain.

As an example of the last point, if faced with a choice between an 85% chance of loosing $1,000 (with a 15% chance of losing nothing) or a sure loss of $800, a "large majority" of people would select the 85% option. This tendency, which the authors describe as being a robust effect that has been observed in many cases, including nonmonetary choices, leads to an asymmetrical value function that is concave for gains and convex for losses and is substantially steeper for losses than for gains, with the ratio of the slopes for small to moderate gains and losses equal to about 2:1 (Figure 5). In a fisheries context, this attitude would lead, for example, to a fisher preferring to continue struggling to eke out a livelihood from an overexploited resource, resting their hopes on the low probability that conditions in the fishery will improve, rather than opting to accept the risk, albeit with a higher probability of success, of leaving the fishery to take up another source of livelihood.

Adding to the results obtained from psychophysical analysis of choices and captured in prospect theory, Kahneman et al. (2000) described two important "anomalies" from economic theory. The first of these is the "endowment effect," the common trait that people will often demand more compensation for losing or giving up an object than they would be prepared to pay for it. They also suggest, based on experimental results, that the endowment effect is not a consequence of possession increasing the value of the endowment that is held, but of increasing the pain involved in giving it up. The second, complementary anomaly they referred to as the "status quo bias," which is the tendency for people to prefer the status quo because they weight the disadvantages of any change as outweighing the advantages. The preference for the status quo increases with the number of alternative choices.

These two anomalies are clearly consistent with the aversion to loss described in prospect theory; people are far more sensitive to losses than they are to gains.

The implications for this work in the psychophysics of decision making for fisheries and conservation is clear and striking. Wherever stakeholders are heavily dependent on fish and fisheries resources for their livelihoods, lifestyles or both, any threats to their access to those resources are likely to be met by

1) A bias towards the status quo, which is likely to mean that they will perceive the potential change as a painful loss and will oppose any changes that ap-

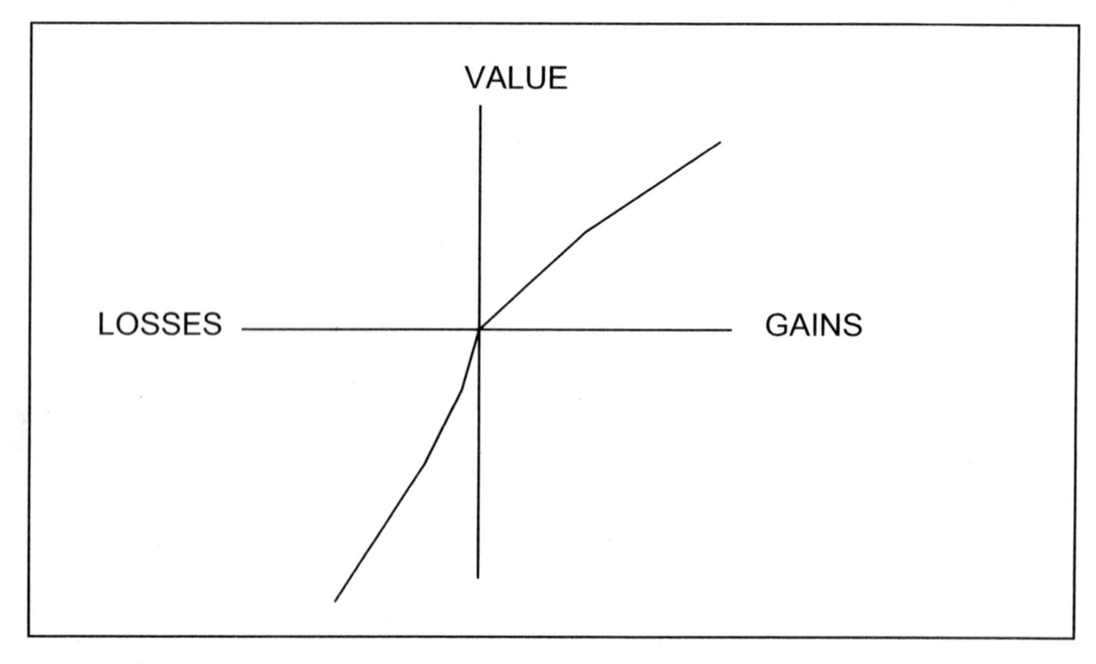

Figure 5. A hypothetical value function within prospect theory (Kahneman et al. 2000).

pear to threaten their lifestyle, even when that lifestyle is relatively poor;

2) A manifestation of the endowment effect in that the pain of the loss is likely to exceed the actual replacement costs for the loss; and

3) A risk-seeking attitude to the potential loss in that they will tend to select options with a relatively small likelihood of avoiding the loss, rather than settling for the certainty of partial compensation. This will exacerbate the resistance to change and the willingness to take risks to avoid change.

While these responses may seem irrational, it is important to appreciate, as indicated by prospect theory, that they are real psychological responses, not stubbornness, awkwardness, or a bargaining stance.

The above psychological factors would tend to hinder movement of fishers out of an overexploited fishery in cases even where there was reasonable compensation or viable alternatives to the loss of livelihoods, a situation most likely to be found in developed countries. In the cases of many developing countries where there are no social-service safety nets and no alternative sources of livelihoods, the reactions will be an even greater, and justified, sense of insecurity and an even greater desire to retain the status quo, however inadequate that might be.

Satisfaction Guaranteed

Despite the depressing statistics and the inherent psychological needs and attitudes that have contributed to them, there have been success stories where concerted action by the responsible authorities and stakeholders has led to improvements in the status of the resources and of the performance of the fisheries. Examples include the improvements in species important to fisheries on the George's Bank, the Norwegian spring herring and the Icelandic summer herring, the South African sardine, and others (Mace 2004). In these cases the socioeconomic environment has been such that a strict control in effort has been possible and ethically defensible. The successes were facilitated by the socioeconomic environment, the existence of the political will to implement positive change, and good underlying science.

Without challenge to the current emphasis on an ecosystem approach to fisheries (WSSD 2003), these examples suggest that considerable progress towards successful fisheries management is possible within the conventional, predominantly single-species paradigms, given the necessary prerequisites. This suggests that the basic problems in fisheries do not lie in inadequate science or technology; they lie in human needs and human behavior. The solutions must therefore lie in addressing legitimate needs and influencing reasonable but still obstructive behavior in order to achieve the desired goal of sustainable and equitable use of our living resources. The solutions, therefore, need to include the following goals:

- Eliminating poverty and thereby releasing people from preoccupation with lower Maslow needs will allow them to focus on achieving higher level goals, including increased opportunity to address the sustainable use of ecosystem goods and services.

- The removal of market distortions that discourage sustainable utilization of living resources and ecosystems, as well as the development of systems of incentives that will actively encourage stakeholders to contribute to sustainable use.

- Education and training to increase awareness and a sense of responsibility for the life support systems of the planet and the need to work actively towards sustaining these systems and restoring them where necessary.

- Creating appropriate institutional arrangements that provide for equitable, long-term rights and appropriate levels of participation of stakeholders in policy formulation and management.

- Effective methods of monitoring and control to deal with any residual transgressors.

Of the above list, the two most important but also difficult tasks will be the elimination of poverty and the creation of incentives for sustainable use. The problem of elimination of poverty has been thoroughly discussed in many contexts, but unfortunately, effective action has rarely flown out of wise words. Elimination of poverty requires substantial improvements in governance in many developing countries, which are ultimately responsible for improving their own conditions. However, real progress will only be possible if there are also significant policy changes in developed countries that must include removing barriers to trade, the provision of debt relief, and support to developing countries to assist them in tackling fundamental problems (UNDP 2003). Without real progress in these directions leading to major reductions in poverty, there can be no hope for sustainable utilization of natural resources. This reality has been demonstrated in many examples ranging from poaching of highly valuable, endangered species to on-going habitat destruction, even in cases where governments were seriously attempting to manage the living resources.

Recognition of the need for positive incentives for sustainable use is relatively recent and only beginning to receive serious attention. Principle 4 of the Convention on Biological Diversity recognizes the need to take the economic context into account when managing ecosystems and proposes three economic requirements for effective management (CBD 2000). These are 1) to remove all market distortions that negatively impact on biodiversity; 2) to provide incentives to promote "biodiversity, conservation, and sustainable use"; and 3) as far as possible, to internalize the costs and benefits from utilizing the ecosystem so that the beneficiaries also pay the costs. Positive incentives could include market-based approaches such as ecolabeling, as well as nonmarket-based incentives such as taxes and appropriate subsidies (FAO 2003b). Point 3 could, for example, entail making sure that all activities, including fishing, recreational use, maritime transport, the impacts of land-based activities, and others that impact on marine ecosystems, cover the true costs of those activities. Further, the wider beneficiaries such as consumers of fish products and those benefiting from other ecosystem services would also need to pay the full costs of the benefits that they receive (Cochrane and Doulman 2005).

Conclusions

The question posed in the title of this paper was what should we should care about when attempting to reconcile fisheries and conservation. This analysis suggests that the cares are generally well understood and, at the conceptual and policy levels, well defined. They can be summarized, as reflected in the FAO Code of Conduct, as

- The role of fisheries and other uses of living marine resources and ecosystems for sustainable development to meet human needs, including food security and poverty alleviation;
- Conservation of target resources;
- The maintenance and restoration where required of biodiversity and aquatic ecosystems, including the environment and critical habitats;
- The rights and welfare of the fishers and others dependent on fisheries for their food security and livelihoods and the provision of safe, healthy and fair working and living conditions for them; and
- Availability to the consumer of food products from fisheries that provide nutritional value and quality and are safe.

This list implicitly strives to reconcile fisheries and conservation at the policy level, but, as observed in practice, that reconciliation is rarely achieved. Piecemeal attempts to achieve reconciliation seem inevitably doomed to failure, and lasting, global reconciliation of natural resource use and conservation will only be possible if the basic prerequisites are in place. These are

- The elimination of poverty and more equitable distribution of benefits to all the people of the world;
- The creation of a global economic system that is effectively free from distortions favoring short-term gain over sustainable use and includes appropriate and sufficient incentives for sustainable use;
- Provision of thorough education and training to all on the need for global stewardship and the role of individuals and institutions in achieving this;
- Equitable systems of access to and responsibility for sustainable use of living resources;
- Recognition of the rights of stakeholders to participate in policy formulation and management of those resources; and
- Effective monitoring and control systems.

Of these, the first two are the most fundamental but have been, to date, hindered by totally inadequate political will. Unless that political will can be mobilized, reconciliation will be as far out of reach as it always has been.

Acknowledgments

The author gratefully acknowledges the valuable assistance given by Sebastian Mathew and Ramya Rajagopalan on the laws in India related to fisheries and Ray Hilborn and Andre Punt for the information and guidance they provided on fisheries laws in the United States of America. Hilary Cochrane, Serge Garcia, Jean-Jacques Maguire, Derek Staples and Rolf Willmann are thanked for helpful comments on earlier versions of this manuscript.

References

Balmford, A., A. Bruner, P. Cooper, R. Costanza, S. Farber, R. Green, M. Jenkins, P. Jefferiss, V. Jessamy, J. Madden, K. Munro, N. Myers, S. Naeem, J. Paavola, M. Rayment, S. Rosendo, J. Roughgarden, K. Trumper, and R. Turner. 2002. Economic reasons for conserving wild nature. Science 297:950–953.

CBD (Convention on Biological Diversity). 2000. Decision 6 of the Fifth Conference of the Parties to the Convention on Biological Diversity. May 2000, Nairobi, Kenya.

Christensen, V., S. Guénette, J. J. Heymans, C. J. Walters, R. Watson, D. Zeller, and D. Pauly. 2003. Hundred-year decline of North Atlantic predatory fishes. Fish and Fisheries 4:1–24.

CITES. 2002. Interpretation and implementation of the Convention. Significant trade in Appendix-II species: Caspian Sea sturgeons. Forth-seventh Meeting of the Standing Committee, Santiago, Chile, 1–2 November 2002. SC47 Document 11. CITES, Geneva.

CITES. 2003. Report of the Queen Conch Significant Trade Review, International Queen Conch Initiative, 11–12 June 2003, Montergo Bay, Jamaica. Caribbean Fishery Management Council, Puerto Rico.

Clarke, C. W. 1990. Mathematical bioeconomics. The optimal management of renewable resources. 2nd edition. Wiley and Sons, New York.

Cochrane, K. L. 2000. Reconciling sustainability, economic efficiency and equity in fisheries: the one that got away? Fish and Fisheries 1:3–21.

Cochrane, K. L., D. S. Butterworth, J. A. A._de Oliveira, and B. A. Roel. 1998. Management procedures in a fishery based on highly variable stocks and with conflicting objectives: experiences in the South African pelagic fishery. Reviews in Fish Biology and Fisheries 8:177–214.

Cochrane, K. L. and D. J. Doulman. 2005. The rising tide of fisheries instruments and the struggle to keep afloat. Philosophical Transactions of the Royal Society Series B 360: 77–94.

Davidoff, L. L. 1987. Introduction to psychology. 3rd edition. McGraw-Hill, New York.

FAO (Food and Agriculture Organization of the United Nations). 1994. Review of the state of world marine fishery resources. FAO Fisheries Technical Paper 335.

FAO (Food and Agriculture Organization of the United Nations). 1995. Code of Conduct for Responsible Fisheries. FAO, Rome.

FAO (Food and Agriculture Organization of the United Nations). 2000. The state of world fisheries and aquaculture. FAO, Rome.

FAO (Food and Agriculture Organization of the United Nations). 2002. The state of world fisheries and aquaculture. FAO, Rome.

FAO (Food and Agriculture Organization of the United Nations). 2003a. Report of the Twenty-fifth session of the Committee on Fisheries. Rome, 24–28 February 2003. FAO Fisheries Reports 702.

FAO (Food and Agriculture Organization of the United Nations) 2003b. The ecosystem approach to fisheries. FAO technical guidelines for responsible fisheries. 4, Supplement 2. FAO, Rome.

Garcia, S., and I. De Leiva. 2003. Global overview of marine fisheries. Pages 1–24 *in* M. Sinclair, and G. Valdimarsson, editors. Responsible fisheries in the marine ecosystem. Food and Agriculture Organization of the United Nations, Rome and CABI Publishing, Wallingford, UK.

Gleitman, H. 1981. Psychology. 4th edition. W.W. Norton, New York.

Hall, S. J. 1999. The effects of fishing on marine ecosystems and communities. Blackwell Scientific Publications, Oxford, UK.

Hutchings, J. A., C. Walters, and R. L. Haedrich. 1997. Is scientific enquiry incompatible with government information control? Canadian Journal of Fisheries and Aquatic Sciences 54:1198–1210.

Kahneman, D., and R. Tversky. 2000a. Choices, values and frames. Pages 1–16 *in* D. Kahneman and A. Tversky, editors. Choices, values and frames. Cambridge University Press, Cambridge, UK.

Kahneman, D., and R. Tversky. 2000b. Prospect theory: an analysis of decision under risk. Pages 17–43 *in* D. Kahneman and A. Tversky, editors. Choices, values and frames. Cambridge University Press, Cambridge, UK.

Kahneman, D., J. L. Knetsch, and R. H. Thaler. 2000. Anomalies: the endowment effect, loss aversion and status quo bias. Pages 159–170 *in* D. Kahneman and A. Tversky, editors. Choices, values and frames. Cambridge University Press, Cambridge, UK.

Ludwig, D. E., R. Hilborn, and C. Walters. 1993. Uncertainty, resource exploitation and conservation: lessons from history. Science 260:17–36.

Mace, P. 2004. In defense of fisheries scientists, single species models and other scapegoats: confronting the real problems. Marine Ecological Progress Series 274:269-303.

Mathew, S. 2003. Small-scale fishery perspectives on an ecosystem-based approach to fisheries management. Pages 47–63 *in* M. Sinclair and G. Valdimarsson, editors. Responsible fisheries in the marine ecosystem. Food and Agriculture Organization of the United Nations, Rome and CABI Publishing, Wallingford, UK.

McGoodwin, J. R. 1990. Crisis in the world's fisheries. People, problems and policies. Stanford University Press, Stanford, California.

NMFS (National Marine Fisheries Service). 2002. Towards rebuilding America's marine fisheries. Annual report to Congress on the status of U.S. fisheries 2001. NMFS, Silver Spring, Maryland.

Rosenberg, A., M. J. Fogarty, M. P. Sissenwine, J. R. Beddington, and J. G. Shepherd. 1993. Achieving sustainable use of renewable resources. Science 262:828–829.

Seijo, J. C., O. Defeo, and S. Salas. 1998. Fisheries bioeconomics. Theory, modeling and management. FAO Fisheries Technical Papers 368.

Symes, D. 1996. Fishing in troubled waters. Pages 3–16 *in* K. Crean and D. Symes, editors. Fisheries management in crisis. Blackwell Scientific Publications, London.

Tamura, T. 2003. Regional assessments of prey consumption and competition by marine cetaceans in the world. Pages 143–170 *in* M. Sinclair and G. Valdimarsson, editors. Responsible fisheries in the marine ecosystem. Food and Agriculture Organization of the United Nations, Rome.

Trites, A. W. 2003. Food webs in the ocean: who eats whom and how much? Pages 125–141 *in* M. Sinclair and G. Valdimarsson, editors. Responsible fisheries in the marine ecosystem. Food and Agriculture Organization of the United Nations, Rome.

UNDP (United Nations Development Program). 2003. Human development report 2003. United Nations Development Program, Oxford University Press, New York.

Walters, C., and J. J. Maguire. 1996. Lessons for stock assessment from the northern cod collapse. Reviews in Fish Biology and Fisheries 6:125–137.

WSSD (World Summit on Sustainable Development). 2002. Plan of implementation. World Summit on Sustainable Development, Johannesburg, South Africa.

American Fisheries Society Symposium 49:25–28

Session Summary

Achieving Reconciliation of Fisheries with Conservation Through an Ethical Approach

HAROLD COWARD AND MELANIE D. POWER-ANTWEILER
University of Victoria, Center for Studies in Religion and Society
SED B102, Box 1700 STN CSC, Victoria, British Columbia V8W 2Y2, Canada

If the schedule of concurrent sessions for the *Fourth World Fisheries Congress* is any indication, ethical considerations in fisheries are a growing concern. The first concurrent session on the Congress schedule focused on the question of how to achieve reconciliation of fisheries with conservation using ethical approaches. Convened on the first day of the Congress, this session served to introduce the notion of ethics in fisheries. A rather new application of ethical analysis (see, for instance, Coward et al. 2000; Garcia and Bartley 2001; Collet 2002), the concept of ethics in fisheries offers a novel approach to endemic issues in fisheries management and new avenues of research.

Four oral papers were presented in the session and subsequently submitted for inclusion in the conference proceedings. These collectively identified a selection of ethical issues inherent in fisheries. Through case studies, representing experiences in Australia, Canada, and the European Union, examples of ethical concerns were enumerated, and possible solutions were indicated.

The papers given, in order of presentation, were

1. *Ecosystem Justice as an Ethical Basis for Fisheries Management* by Harold Coward (session chair and presenting author), Conrad Brunk, and Melanie Power-Antweiler;

2. *Why Examine Court Decisions When Talking About Allocation?* by Kelly Crosthwaite;

3. *Social Resistance to the Obvious Good* by Kenneth Patterson; and

4. *Reconciling Fisheries and Allocation Using a Justice-Based Approach: Troll Fishers Score Best* by Melanie Power-Antweiler (presenting author) and Tony J. Pitcher

Of these four, the first and last presented Canadian examples and found their roots in an earlier volume (Coward et al. 2000). These began with theoretical principles, which were subsequently illustrated using case studies. The second and third papers, by contrast, presented Australian and European case studies, respectively, and drew ethical principles from specific examples.

Drawing on examples from Canada's Pacific coast, the paper presented by Power-Antweiler described several macro-level ethical issues inherent in fisheries. This was accomplished in two directions: by identifying nine ethical criteria against which fisheries ought to be evaluated, and by subsequently reconsidering these ethical attributes and ascribing them to each of the five types of justice first described in *Just Fish: Ethics and Canadian Marine Fisheries* (Coward et al. 2000), namely creative, distributive, ecosystem, productive, and restorative justice. Following on their contribution to the previously published volume (Pitcher and Power 2000), Power-Antweiler and Pitcher evaluated on the basis of justice a selection of commercial fisheries for Pacific salmon, as prosecuted in British Columbia, Canada. Using Rapfish, a rapid appraisal method based on multidimensional scaling, they assessed 24 fisheries subdivided by species and gear type. The nine ethical criteria represented an overall assessment; these were further subdivided as representative of each of the five types of justice, and additional evaluations were conducted using these subsets to assess the fisheries on the basis of each justice typology. Based on these assessments, Power-Antweiler and Pitcher determined that the troll sector was more ethically sound than either gill-net or seine sectors and hence should be given preferential consideration in policy decisions. For example, in recent fleet reduction efforts, the troll fleet was reduced disproportionately (23%) relative to the other sectors (18% in the gill-net sector, and 9% of seines); this occurred despite a stated goal of a 20% reduction across all sectors. According to Power-Antweiler and Pitcher's findings, allocation and fleet size decisions should favor the troll fleet over the net fleet.

Coward, Brunk, and Power-Antweiler also drew on *Just Fish*, but focused exclusively on ecosystem justice. Moreover, where Power-Antweiler and Pitcher applied ethical principles to commercial fisheries for wild salmon, Coward, Brunk, and Power-Antweiler examined aquaculture, particularly salmon farming in British Columbia. Ecosystem justice embodies a holistic approach in which the interconnectedness and interdependencies of a marine ecosystem and those who rely on it, human and nonhuman alike. As such, a fishery must be understood as being embedded within an ecosystem, rather than detached from it, and the well-being of the whole must be considered. A fishery, then, cannot be regarded only as meeting the needs of humans, but as one way in which humans interact with the rest of the ecosystem. Coward, Brunk, and Power-Antweiler find the ecosystem justice approach evident in the worldviews of both Christians and the Haida First Nation of British Columbia. The authors write that, according to the Haida worldview, "...each creature is thought to play an important part in keeping the rest of nature alive," and interactions between the Ocean People (sea creatures) and Haida people occur for sustenance or spiritual purposes. From this understanding, a duty of care arises in the form of stewardship; clan chiefs controlled access to a fishery and bore the responsibility of ensuring that the resource was not overexploited. In Christian thinking, the principles of Sabbath and Jubilee, which grant a period of rest to humans and the rest of creation, also embody stewardship. Evident in both worldviews is the duty to ensure bounty for future generations. Considering the example of salmon farming in British Columbia, the authors questioned whether the practice of farming fish in the seas may harm the rest of the marine ecosystem and impact on the interdependent members of the ecosystem, and especially whether present aquaculture activities may harm wild salmon runs and reduce the bounty for future generations.

Crosthwaite identified transparency and accountability as essential components of ethical decision making with regard to fisheries. Through an examination of three Australian case studies relating to court decisions in matters of allocation, she articulated that a clear decision-making process reduces the likelihood of legal action following allocation announcements. She described two situations in which court action is likely to occur: when one or more fishery user groups find a newly imposed allocation agreement to be unacceptable, such as reallocation from one sector to another; or to force an allocation decision, for instance to ensure consideration of a user group in allocation decisions. Crosthwaite provided two case studies to highlight the first type of situation: the banning of the main commercial gear type used in South Australia's River Murray fisheries, and the closure of commercial fishing in Botany Bay and Lake Macquarie in New South Wales. Regarding the second situation, Crosthwaite described Australian court challenges regarding Aboriginal fishing rights. She stated that allocation decisions, fraught with sacrifices and emotions, can be political considerations that politicians may prefer to avoid unless forced to consider. Matters of allocation are increasingly before the courts, and Crosthwaite maintained that there are no positive outcomes arising from court involvement in these decisions. She concluded that a decision-making process that identifies and addresses potential allocation issues in a timely manner will best serve transparency and accountability, thereby improving conditions that lead to better and more acceptable decisions.

A European perspective was provided by Patterson, who spoke of stakeholder resistance to proposed reforms to the European Community's Common Fisheries Policy (CFP). Patterson described the CFP and its framework regulation as forming the backbone for the management of commercial fisheries and aquaculture in European waters and by European vessels. The framework regulation is reviewed on a decadal basis; the most recent review, published in a 2001 Green Paper, judged the CFP unsuccessful, with biological and economic sustainability at risk. Reforms recommended responsively by the European Commission in 2002 were intended to promote long-term fisheries planning (and thereby improve stability and reduce the appearance of arbitrariness of decision making) and improve transparency in enforcement of regulations. As part of the review process, the European Commission conducted a consultation with member states and stakeholders; it is on these consultations that Patterson reported. The author focuses on the concept of "obvious good," in the form of long-term benefits to society. He determined that "...stakeholders have lost faith in their own industry, and in the equitability of fisheries enforcement...," and as a result no longer believe that short-term sacrifices will bring about long-term benefits—in other words, employment in the industry must be a primary concern in decision making. This, Patterson concluded, arose from a mismatch of values between fisheries managers and stakeholders. That is, the primary concern of stakeholders is the short-term maintenance of employment rather than long-term conservation and sustainability of fisheries.

Each of these authors presented novel considerations of an ethical approach in fisheries management and conservation. The use of ethical approaches, and the implementation of evaluation and decision-making techniques arising from these methods, presents a fresh perspective to fisheries problems. With seemingly never-ending crises in fisheries around the globe, innovative outlooks are particularly invaluable. The holistic nature of ethical assessments provides one more means of improving our comprehension of fisheries issues, and of applying that understanding for the benefit of fisheries ecosystems and the communities that rely on marine resources.

References:

Collet, S. 2002. Appropriation of marine resources: from management to an ethical approach to fisheries governance. Social Science Information 41(4):531–553.

Coward, H., R. Ommer, and T. Pitcher, editors. 2000. Just fish: ethics and Canadian marine fisheries, social and economics papers, volume 23. ISER Books, St. John's, Newfoundland.

Garcia, S. M., and D. Bartley. 2001. Ethics in fisheries. World fisheries and aquaculture atlas CD-ROM. Food and Agriculture Organization of the United Nations, Rome.

Pitcher, T. J., and M. D. Power. 2000. Fish figures: quantifying the ethical status of Canadian fisheries, east and west. Pages 225–253 *in* H. Coward, R. Ommer, and T. Pitcher, editors. Just fish: ethics and Canadian marine fisheries, social and economics papers, volume 23. ISER Books, St. John's, Newfoundland.

American Fisheries Society Symposium 49:29–41

Ecosystem Justice as an Ethical Basis for Fisheries Management

HAROLD COWARD, CONRAD BRUNK, AND MELANIE D. POWER-ANTWEILER*
University of Victoria, Centre for Studies in Religion and Society and the Coasts Under Stress Project SED B102, Box 1700TN CSC, Victoria, British Columbia V8W 2Y2, Canada

Abstract.—While salmon were once a seasonal and expensive luxury item, the development of salmon farming has made them a staple food for many. The argument is often made that aquaculture will produce more fish and feed more people; although this claim is arguably dubious, the fact remains that more salmon are available in the marketplace. However, the advent and expansion of salmon aquaculture has given rise to ethical concerns, including the impact of farming activities on the broader ecosystem. Here, we describe and explore the concept of ecosystem justice, particularly as represented in Christian and Haida worldviews. We describe current practices in salmon farming and subsequently evaluate the ecosystem justice factors evident in the industry. Ecosystem justice requires that one must consider not only the benefits (and harms) to humans, but also to the fish and the ecosystem as a whole. We ask, does aquaculture violate the principles of ecosystem justice? If so, can fish farming take place in a manner that is in keeping with ecosystem justice?

Introduction

Modern industrial society developed with the assumption that natural resources were inexhaustible. Under this assumption, there are few questions of how to distribute these natural resources and fewer questions of productive justice—that is, questions of how to manage the ecosystem so that it continues to produce the resources we need. Instead, it is assumed that the ecosystem can continue producing resources such as fish regardless of the impacts of human exploitation. In Canada, however, the collapse of the cod fishery on the East Coast and the deterioration of the salmon fishery on the West Coast have gone a long way toward challenging our assumption that the supply of fish is inexhaustible. This has devastated local communities that have been economically dependent on the fishery and has led to much debate over how the fisheries are being managed. This debate is raising questions of justice and fairness in the production and distribution of fisheries resources. The language used in these public debates—over rights, entitlements, and equal opportunities—makes clear the ethical nature of the issues raised. The aim of this paper is to identify some of the ethical choices that are made, often unconsciously, in establishing fishery policy and to suggest that an ecosystem justice approach could play a valuable role when fishery policy decisions are made.

*Corresponding author: melanie.antweiler@shaw.ca

The Concept of Ecosystem Justice in Fisheries Management

An ecosystem justice approach enables us to think of the various species and physical components of the marine environment as an interdependent system with its own properties (Costanza et al. 1997). This holistic feature of the ecosystem approach has practical significance for management issues. It requires that one look beyond specific species of interest such as cod or salmon to a wider focus of the fish as part of a marine system that is a holistic unity. The interdependencies in the system require those who manage the fishery to keep its wholeness in view. An ecosystem approach forces us to look at all of the interdependent elements of the fishery from the point of view of their value to the functioning of the whole ecosystem. This is different from the usual way we have dealt with the fishery in Canada. When conflicts have arisen around a diminishing or a depleted resource, such as that seen on Canada's east and west coasts, the usual management approach has been to arbitrate between the human claims to a certain share of the fishery in question by examining (1) the relative merits of the claims of the various groups to a share of the fishery, and (2) the management techniques necessary to sustain the continued harvesting of the fishery. The management problem is defined around a specific fishery (e.g., the cod or the salmon) and its division among the competing groups (e.g., aboriginal fishers, commercial fishers, sports fishers, etc.) and does not take into account impacts of the division and harvesting upon the marine ecosystem as a whole. This traditional way of making policy choices and managing the Canadian marine fishery is seen to be fundamentally flawed when viewed from the perspective of an ecosystem justice approach.

Rather than focusing on a single species such as cod or salmon, an ecosystem justice approach to management looks at the fishery from the broader perspective of the ecosystem in which it is embedded. Such an approach takes us away from a preoccupation with the production and distribution of a single fishery in isolation from the larger biological and social system of which it is a part (Pitcher and Power 2000). Our traditional focus on a single fishery and its distribution among competing groups has been at the root of mistaken assessments of the real state of the fisheries. How does an ecosystem justice approach help? Plato regarded the concept of justice as a kind of harmonious ordering of all the parts of a community. The just community was one in which all members of the community performed their various roles in ways that contributed to the well-being of the whole. Plato's holistic approach to justice in terms of the good of the whole community has much in common with modern ecosystem thinking. A just marine ecosystem is one in which the various interdependent members (e.g., the fish, the humans, and the physical habitat) interact with each other in a way that contributes to the good of the others and to the good of the whole system. The concept of a just marine ecosystem is a normative ethical concept that allows us to judge some states as better than others. Applied to fisheries management, this approach maintains that a fishery is not simply a matter of what the human managers of that fishery want it to be to serve human needs, but rather a management approach which considers what is good for the cod or salmon, the herring, and the seabed (the whole marine ecosystem), along with what is good for the humans involved. Applying an ecosystem justice approach involves analyzing the fishery in terms of the ethical issues raised by human involvements with the ecosystems in which they are embedded. Just management requires that we make explicit as many of these ethical values as possible and understand the implications for the whole system of handling them in one way rather than another.

Support for Ecosystem Justice in Haida and Christian Thinking

The Haida Approach

The Haida writers Russ Jones and Terri-Lynn Williams-Davidson in their essay "Applying Haida Ethics in Today's Fishery" (Jones and Williams-Davidson 2000) argue that the Haida fishing communities on Canada's west coast followed a traditional approach that fostered ecosystem justice in the fishery. According to Jones and Williams-Davidson, "First Nations around the world demonstrate interwoven cultural, spiritual, and ecological values derived from spiritual principles embedded in nature" (Jones and Williams-Davidson 2000:102), and the Haida approach to the fishery is based on a spiritual relationship with their natural environment and all its elements, including fish, birds, land animals, creeks, and places. The Haida place animals at an equal and sometimes higher level than humans. This contrasts strongly with western European policy and law, which is human-centered and values the fish in terms of their use by humans. The Haida believe that all animate and inanimate beings have a spirit, which translates into a holistic ethical approach to the utilization of the resources of the land and sea. The creatures of the oceans are regarded as Ocean People, and each creature is thought to play an important part in keeping the rest of nature alive. Interactions between the Haida and the Ocean People are for sustenance or spiritual reasons. Thus, fish are seen as making an important sacrifice to keep humans alive. Thus, "a fisher would talk respectfully to the halibut, referring to it as a *k'aagaay*, or elder, while asking it to bite his hook" (Jones and Williams-Davidson 2000:103).

Two important principles of Haida ethics are respect and reciprocity. Respect is taught for oneself, others, and the environment that keeps us alive. This is manifested in a variety of ways. Offerings are made to honor the spirits of the fish that we intend to kill. All fish that are killed must be treated respectfully, which means it must not be wasted (all parts must be used). Sports fishing, where a fish is played (caught and then released), is offensive to the Haida (Jones and Williams-Davidson 2000). Food must be handled with care. Ceremonies are held, such as a gathering at the mouth of the Yakoun River to welcome back the first sockeye salmon (Jones and Williams-Davidson 2000).

Haida individuals or clans also follow a stewardship or management responsibility with regard to fish that seeks to actualize a holistic or ecosystem relationship with nature. Clans traditionally controlled fishery access within their territory; on behalf of the clan, the chief was the steward of the resource and had the responsibility to see that it was not overfished. Another guiding principle for Haida conduct is a sacred quest for balance that is embodied in the Haida proverb "The world is as sharp as the edge of a knife," a story in which a man, proud of walking along a narrow board just above the ground, nevertheless slipped and fatally fell (Jones and Williams-Davidson 2000). This speaks to the narrow relationship not only between life and death, but also between humans and nature, analogous to the two sides of a knife, and it teaches the importance of finding balance in all of our activities, including stewardship of the fishery.

Current Haida management of fisheries incorporates traditional spiritual values with modern scientific techniques in management and decision making. For example, the Haida fishery on the Copper River on Haida Gwaii takes place over a short period of time in April and May. The river is the traditional territory of the Gitksan clan, who owned the single trap on the river—others had to request permission to fish there. Today, all Haida living in Skidegate can fish the river. The fishery is managed as follows.

"The target escapement is 10,000 spawners to the river, and the Haida Fisheries Program operates a counting fence and makes recommendations on fisheries openings. Public meetings are called each year to appoint a management committee to make decisions on openings ... The public meeting involves all interested people and includes a mix of elders, men and women. Decisions are made by consensus but may be delegated in season to the management committee. Fence counts and catches are followed closely by everyone and reported in the local village newsletter. If an opening is called, fishing effort might consist of up to 50 nets that take annual catches averaging 3,300 sockeye ... The Copper River fishery is an example of a carefully managed fishery that uses a public community consensus process and traditional and scientific knowledge to make decisions" (Jones and Williams-Davidson 2000:109).

By following Haida values, which treat the fish and people as spiritual beings that are parts of a holistic interdependent environment, the Haida follow a model of ecosystem justice in their approach to the fishery. Their community-based decision making is one example of how a comanagement model based on the ethical assumptions of ecosystem justice can work effectively.

Christian Approach

To begin with, it must be admitted that Christianity, in its modern western forms, bears much of the responsibility for having fostered destructive patterns of development throughout the planet. It is not that Christianity has the worst ideas for the environment and the consumption of its resources, but that the modern western Christian cultures developed the ideological framework for unprecedented domination in the political, economic, and cultural spheres (Keller 1995). It is this aggressive domination of peoples and nature by the ideological framework of the modern West that seems to be one of the root causes of overconsumption of the earth's resources in general and of the destruction of fisheries in particular. That being said, let us see if there is a basis for an ecosystem justice approach in Christian thought.

Rosemary Ruether finds support within Christianity for seeing the entire planet as a living organism or unified ecosystem. Ruether points out that 19th century European Protestantism made the Bible seem to be more anti-nature than it is by interpreting the scriptures with a dualism between history (the true realm of human activity) and nature. Any spirituality associated with nature was seen to be a false god. More recent interpretation, however, has shown that this dualism distorts the biblical perspective. Nature in the Bible is not subhuman, nor is history set against nature. "The God made present in historical acts of deliverance," says Ruether, "is at the same time the God who 'made heaven and earth'" (Ruether 1992:208). God's love is present when God makes the earth and the waters, fish, animals, and humans, all of which God sees as good and gives to it blessing. Ruether finds the Hebrew understanding of God's covenantal relationship with Israel to include the land (nature) as well as human history and to relate humans to the land in terms of Sabbath and jubilee. While scholars doubt that the jubilee laws were ever applied—laws that serve to put right unjust relationships that have developed—such laws do offer us today a model of restorative ecosystem justice (Ruether 1992). The messianic vision in Isaiah 65:17–22 and 24–25 also offers a model of the restoration of right relationships between people and nature, including ourselves and the fish. In the New Testament, says Ruether, this intimate unity between justice and right-relationship is often lost in the singular focus on social justice. It seems to have been present in Jesus' own teachings in the gospels, where strong elements of the jubilee vision of renewal are present (Ruether 1992). The vision of a renewed right-relationship between humans, nature, and God is also found in Paul's childbirth image in Romans 8:18–39 in which the whole of creation is seen to be groaning in hope towards a rebirth of ecosystem justice. The groan of nature recognizes that creation is a dynamic equilibrium of competition in which the blood and guts of killed creatures (and eaten plants) provide food for the sustaining of life. The Christian ethic in this regard recognizes our necessary predation of other forms of life (including fish) but demands that we be ecologically respectful and just in our killing. This requires redemption from our selfish desires that drive our unjust decimation of nature. This will happen only when we opt for an ecosystem concept in which humans and fish are interdependent parts of nature—a nature infused by God's spirit. Ruether concludes that the challenge for Christians is to use their God-given reflective consciousness, which distinguishes humans from other life forms, not to separate ourselves from nature, but rather to recognize our kinship with other beings and join them in the cosmic dance. Part of that dance may involve the catching and eating of fish, along with other animals and plants. But this must be done in a way that embodies a sacramental spirit of kinship, thanksgiving, and respect in consideration of what will nourish the whole community of life. This sacramental spirit, together with the covenantal ethic of the Torah, gives Christians a firm foundation for an ecosystem justice response to the challenges of Canada's marine fisheries (Ruether 1992).

How does this Christian basis for ecosystem justice translate into principles for fisheries management? Dunham and Coward (2000) have summarized these principles. (1) *The world's people and resources are interdependent and human activity needs to be regulated accordingly.* The principle of Sabbath—the need for regular periods of rest—must be applied to the fishery as well as to people. Decision-making needs to give effective Sabbath rest to marine ecosystems. An example can be seen in Northern Ireland where Father Kennedy, a Catholic priest, organized a cooperative of more than 500 eel fishers in 1969. The cooperative virtually eliminated illegal fishing and trawling, a practice that had caused much damage. The cooperative's work brought self-restraint to fishery practices, which in turn brought rest from overfishing to the fish and their ecosystem—without sacrificing the ability of humans to live healthy economic lives. (2) *A just policy will encourage an equitable approach to ensuring the economic success of a fishery.* Since all the people of the earth have been given the planet as their home, and must make their living here, economic guidelines that account for a shared planet and, thus, a justly shared fishery must be part of public policy. Like the Haida approach, this principle requires Christians to look to the needs of the nonhuman community—namely the fish and the seabed—in the ecosystem. (3) *The historical values and wisdom of families and communities need to be taken seriously and allowed to flourish.* The Bible places a limit on the accumulation of resources and economic power (the jubilee principle) so as to safeguard the social values and family structures upon which Israel was built. In the book of Leviticus, when families were split apart by debt and loss of land, the judgment was that dehumanizing factors had usurped the situation and redress was required. As the Coasts Under Stress (www.coastsunderstress.ca) research project shows, for many who have been involved in

Canada's fishery in Newfoundland and British Columbia, there is an unambiguous association of life itself with the land and the sea. Such values are life-giving and cannot be replaced. (4) *Restorative justice is a necessary goal for policymakers and a healthy attitude for conceiving an ecosystem.* Restorative justice is a healthy principle not only for people, but for all parts of the fishery ecosystem. The decimation of fish stocks for human employment has been pursued throughout Canadian society. A concentrated effort to make sure that clean oceans, rivers, and lakes are given back, as well as a Sabbath rest for nature, is part of restoration. It is done expressly for the fish at society's expense—as in marine protected areas. What we have taken must be given back for future flourishing. (5) *The protection of the fishery for future generations is a priority for just policy-making.* Intergenerational justice comes with the realization that the values of the past are not only worth maintaining, but are rightfully inherited by those who will come after us—our children and grandchildren. The same applies to the physical and economic benefits of the fishery. Christian values here agree with the Aboriginal rule of thumb that management decisions should safeguard the flourishing of the ecosystem for seven generations into the future.

In the above, we have seen that the ethical teachings of Plato, the Haida, and Christianity can all be used to generate policy decisions that result in ecosystem justice in the fishery. So far, we have considered mainly the fishery for wild stocks. Let us now see if these same ethical approaches can be applied to policy-making in aquaculture.

Industrial Salmon Aquaculture and Ecosystem Justice

In both published literature (see, for instance, Roberts 2000; Tacon 2003) and in recent research interviews[1], the argument has been put forth that aquaculture is necessary to increase the amount of fish available for human consumption and to meet growing demand for seafood. Indeed, a recent document from the UN Food and Agriculture Organization states that aquaculture "...has achieved a reputation as a significant contributor to poverty alleviation, food security and income generation" (Subasinghe 2003). Aquaculture is the fastest-growing animal-based food production sector, accounting for 32.2% of total fisheries landings (by weight) in 2000, up from 5.3% in 1970 (Tacon 2003). Excluding China[2], in 2000, aquaculture production totaled 11 million metric tons and capture fisheries totaled 77.9 million metric tons (FAO 2002). Indeed, with the global expansion of salmon aquaculture, prices for both farmed and wild salmon have fallen (Babcock and Weninger 2004)[3]. Concurrent with a drop in production costs and increased production (Guttormsen 2002; Asche et al. 2003), the rate of salmon consumption has risen. Anderson (2002) notes that between 1987 and 2000, consumption in the United States (British Columbia's primary market for exported seafood products (British Columbia 2004)) increased by 265%. Since 1990, the aquaculture industry has been growing at about 10% a year. In the United States, roughly half of all fresh and frozen sea-

[1] Research interviews were conducted in British Columbia (Greater Vancouver, Victoria, and Campbell River and by telephone to remote coastal areas), Newfoundland (St. John's), and Ottawa, beginning in June 2004, through March 2005. Interviewees included government officials, representatives of both industry and environmental organizations, and academics.

[2] Note that the figure for capture fisheries—52 million metric tones—excludes Chinese fisheries catches as there are significant doubts regarding the credibility of these data (Pauly and Watson 2003; Weber 2003).

[3] Asche et al. (1999) discerned that the price of farmed salmon sets the price for wild Pacific salmon and concluded that "...one might expect further productivity improvements in farmed salmon production, such that prices for farmed Atlantic and wild Pacific salmon are likely to continue to decline unless there are significant shifts in demand."

food is farmed. Expectations are that by 2030, aquaculture will become the dominant source of fish products (Promise of a blue revolution 2003).

Ecosystem justice requires that one must consider not only the benefits (and harms) to humans, but also to the fish and the ecosystem as a whole. One must therefore ask, Does aquaculture violate the principles of ecosystem justice? If so, can fish farming take place in a manner that respects ecosystem justice? Salmon aquaculture, a controversial issue on Canada's west coast and elsewhere, will serve as an example.

When one evaluates aquaculture from the perspective of ecosystem justice, a number of issues become readily apparent. The interdependence of the ecosystem, central to the concept of ecosystem justice, can be challenged through present aquaculture practices. The potential ecosystem impacts of aquaculture are felt at local and distant scales, both of which will be discussed below.

The most obvious potential for ecosystem impacts presents locally. In the case of salmon aquaculture, impacts may be had on the seabed in the direct vicinity of the farm or on water quality. These can result from chemical and pharmaceutical residues, excrement, and unconsumed feed (Gardner and Peterson 2003). The stability of the marine ecosystem may be compromised by increased algal blooms (and subsequent oxygen depletion) or the use of harmful copper-based anti-foulant solutions to clean salmon farm net-pens.

During migrations, wild salmon may be at risk of encountering and subsequently contracting any infections present in nearby salmon farms. Recent research (Morton and Symonds 2002; Krkosek et al. 2005) examined the prevalence, abundance, and intensity of sea lice *Lepeophtheirus salmonis* in British Columbia. Both studies, which looked at wild pink salmon *Oncorhynchus gorbuscha* and chum salmon *O. keta*, found a relationship between sea lice infections and the proximity of salmon farms and raised concern over the impact on wild salmon populations. Whereas the science surrounding sea lice has previously been contested (see Gardner and Peterson 2003), these new studies have established a connection that necessitates further scientific assessment to determine with scientific certainty the cause for the increased presence of sea lice. If the cause is found to be linked to human activities (whether salmon aquaculture or otherwise), then it would be necessary to evaluate whether the activity resulting in the increased sea lice infection provides sufficient benefits to justify the risk to other species and the broader ecosystem.

For animals and birds that feed on salmon, fish farm pens present a tempting buffet. Predation by wildlife not only results in loss of salmon by direct consumption and fatal wounding, but the damage caused to nets during predation can allow salmon to escape (British Columbia Environmental Assessment Office 1997). In response, "...salmon farmers go to considerable lengths to prevent or deter predation losses," including trapping or even shooting the predators such as river otters, minks, and seals (British Columbia Environmental Assessment Office 1997). Farmers may also try to scare off or otherwise deter the animals through the use of lights or guard dogs, or even acoustic harassment devices (noisemakers), which have been found to have displaced marine mammals, including whales (Morton and Symonds 2002) and porpoises (Johnston 2002) from the areas in which they are used.

When salmon escape from farms, they pose a range of threats to wild salmon. In British

Columbia, Atlantic salmon are predominant. In 2003, the latest year for which figures are available, 76% of farmed salmon were Atlantics, down from 85% in both 2001 and 2002 (British Columbia Ministry of Agriculture and Lands 2003). Since the species are so different, the chance of interbreeding is extremely low, and the risk to the genetic stability of British Columbia's wild stocks is mitigated accordingly (Volpe 2001); yet, the risk persists where farmed and wild species are the same and interbreeding between these conspecifics is possible. Some British Columbia fish farms raise Chinook salmon *O. tshawytscha* or coho salmon *O. kisutch*; escapes from these populations could reduce the genetic fitness of the wild stocks through outbreeding depression. Sterilization of farmstock through triploidisation reduces the genetic risks caused by interbreeding of wild and escaped farmed conspecifics (see Power 2003).

Aside from the potential for interbreeding, the risk of establishment of self-sustaining colonies of escaped farmstock persists. In September 1998, Volpe confirmed that escaped Atlantic salmon had successfully reproduced over several age-classes in a stream on Vancouver Island (see Volpe et al. 2000; preface in Volpe 2001). A large, self-sustaining colony may compete with wild salmon for food and for spawning space, thus directly challenging the viability of the wild runs.

The impacts of salmon aquaculture extend beyond the waters in which the farms operate. Although "[m]any believe that fish farming will relieve the pressure on [wild marine] fish stocks" (Pauly and Watson 2003:47), salmon are carnivores and, as such, feed on other fish. In aquaculture operations, salmon are fed fish meal, which requires

"...turning small pelagics—including fish that are otherwise perfectly fit for human consumption, such as herring, sardines, anchovies and mackerels—into animal fodder. In fact, salmon farms consume more fish than they produce: it can take three pounds of fish meal to yield one pound of salmon." (Pauly and Watson 2003:47)

Is it appropriate to transfer potential food sources from one region or ecosystem to another? International trade results in transfers between regions of food, resources, and goods; fish have long been caught in one region for sale in another. Is it really all that different to take fish from one ecosystem to artificially produce fish in another? This does not appear to meet the principle of equitable sharing of the resource between users and within the ecosystem itself.

Many of the fish used to produce feed for aquaculture are caught in developing waters, to produce premium farmed fish in the developed world. In a keynote address to the Fourth World Fisheries Congress, Daniel Pauly noted that not only are these forage fish edible, but they are even enjoyed by some diners. The governments of Chile and Peru, leading producers of fish meal and oil (Tyedmers 2000), are reported (Tuominen and Esmark 2003) to advocate direct human consumption rather than reduction of the forage fish caught in their waters.

Issues of both ecosystem and distributive justice are raised by the practice of harvesting forage fish for reduction into meals and oils to feed farmed fish. Fish such as cod and salmon are at the upper (trophic) levels of food webs; as these and other high trophic level fish have become more scarce, fishing efforts have shifted to other fish (forage fish) at lower levels of the food webs (Pauly et al. 1998, 2000; Power and Newlands 1999). From the perspective of ecosystem justice, the capture of these lower trophic level fish, if sufficiently aggressive, may destablise an ecosystem by reducing the food available for predatory fish. In effect, this activity facilitates a conversion of the fish from low-value wild fish to marketable, high-value farmed fish.

From the perspective of distributive justice, the catch and reduction of forage fish into feed reduces the energy available for direct human consumption, and since these fish are often caught in waters distant from where the aquaculture occurs, the food is also taken away from the local people. Economic benefits can be realized for local people that may be involved in the fishery, but fish caught in such reduction fisheries is transferred out of the area, despite the fact they could be useful source of nutrients.

This raises a more abstract concern: these edible fish are removed from waters in one region to produce a luxury product in another. People enjoy eating salmon, but consumption of salmon is generally a matter of want rather than need. In fact, "It takes 2–5 kg of wild fish just to produce 1 kg of a farmed [carnivorous] fish such as salmon" (Powell 2003, citing Weber 2003); the actual amount of fish available for direct consumption is greatly reduced. This dependence on wild fisheries to produce feed, as well as transportation, processing, and other direct and indirect energy inputs, expands the ecological footprint of intensive aquaculture (Troell et al. 2004). Ecosystem justice is violated when wants trump needs; the harms to the ecosystem do not correlate with the benefits to the few who enjoy the premium fish product. Alternative, plant-based fish feeds are being evaluated but are not yet feasible due to the physiological needs of the fish and the potential impacts on product taste and texture resulting from feeds not derived from fish (see Power 2003).

Both Haida and Christian approaches to ecosystem justice compel consideration of future generations. Should aquaculture operations compromise the viability—or even the flourishing—of wild salmon runs, descendents of the present generation will be denied the intrinsic joys and instrumental values of those salmon and perhaps even the ecosystem of which they are integral component. This also holds true in the distant waters where forage fish are caught to produce fish meal. Christian and Haida teachings remind one that these resources (and the potential benefits of these) are, in effect, borrowed from future generations; should our generation undermine an ecosystem through either greed or need, we deny potential benefits to our descendents. It is hubristic to presume to know what future generations may require or desire. To keep within the principles of ecosystem justice, and also distributive justice, it is unacceptable to preclude options that may be desired or needed for the future flourishing of the ecosystem and our descendents.

Humans, like plants and nonhuman animals, are components of the interdependent ecosystems in which they reside and on which they depend. Individuals therefore have a tangible interest in maintaining the viability of their home ecosystem. Both Haida and Christian teachings profess respectful stewardship of creation (Jones and Williams-Davidson 2000; Roach 2000). While many would agree that stewardship is a suitable, even noble, goal, the challenge lies in how to operationalize this goal in the policy realm. The Haida approach to comanagement demonstrates a means by which stewardship can be achieved in fisheries management; this can be translated into aquaculture management, through the incorporation of traditional or local ecological knowledge (T/LEK) in various ways. Those who are most closely connected with an ecosystem will also have the most detailed local knowledge of it. Such information can be passed between and accumulated through generations (Brunk and Dunham 2000; Haggan 2000; Neis and Morris 2000). T/LEK can inform aquaculture policy for such decisions as appropriate siting of salmon farms: those most familiar with wild salmon runs could provide advice on areas not suitable for farming because of the proximity of wild runs. Anecdotal reports by commercial and sports fishers of catches of escaped farmed salmon (see, for instance, The Associated Press 2003) provide notice that an escape has occurred, and can help determine the number of escaped fish and the geographic range affected. From such reports, close monitoring can be planned to evaluate whether self-sustaining colonies of escaped fish are established and what if any impact is had on the wild fish and the environment. Monitoring of rivers by streamkeeping organizations can also fill a similar role. Such measures help to protect the ecosystem and provide one mechanism through which affected people can help decide how to protect and share the ecosystem. The value of T/LEK holds true, too, for those people and communities that rely on the ecosystems from which the forage fish are taken for reduction into fishfeed.

Conclusion

It is necessary to revisit the rationale for fish farming: to produce more fish for human consumption. Ecosystem justice recognizes the value of each element of an ecosystem. The simple anthropocentric instrumental value of an individual component of an ecosystem does not trump the intrinsic value of that element or of the ecosystem as an interdependent whole. A salmon can feed a human, but it is also of value to the bears and eagles that eat it and the streams and trees that are fertilized by the nutrients released by the decaying carcass; together, these contribute to the healthy, continued functioning of the ecosystem. While a tasty source of human nutrition, to view salmon (or arguably any fish) from this perspective alone is to reduce the fish's value to the instrumental. This contravenes not only the principles of ecosystem justice, but also those Haida and Christian principles described above. Moreover, salmon are a cultural and spiritual touchstone for First Nations and other cultures, with not only economies developing from salmon and salmon-bearing waters, but also identities and communities (Meggs 1991; Jones and Williams-Davidson 2000; Power 2003). Haida teachings of respect—for the broader ecosystem and for the noble fish sacrificing itself to meet human needs—remind us that humans are one interdependent component of a holistic ecosystem, and our reliance on the ecosystem brings with it responsibilities to it. This too is echoed in Christian principles.

While it is true, as often asserted, that more salmon are produced and hence available for human consumption[4], there is no overall increase in the amount of fish available. Aquaculture does help to meet the demand for salmon, but for the sector to operate in a manner respectful of ecosystem justice, aquaculture policies must reflect the need for ecosystem stability. These encompass a number of considerations:

1. All activities—including aquaculture and supporting industries—must be conducted in a manner that does not compromise the continued functioning of the ecosystem. Respect for the ecosystem must be a primary concern.

2. The flourishing of the ecosystem must be encouraged not only in the present, but also

[4] Pauly, in his address to the 2004 World Fisheries Congress, cited the example of American actor Ted Danson, who likes to eat anchovies and who expressed surprise at learning that anchovies are used to produce fishmeal.

for generations to come. Current decisions must not preclude the choices available to future generations. Long-term considerations must be part of any policy calculus; the ecosystem must retain the capacity to function at levels able to support itself and future human generations.

3. Those individuals who are part of an ecosystem are best placed to recognize its unique subtleties and are therefore a valuable source of information as to how to care for that ecosystem (for examples relating to wild fisheries, see Haggan 2000; Neis and Morris 2000; Ommer 2000). Moreover, often they also have the most immediate stake in the ongoing health of the ecosystem. Aquaculture policies (like fisheries policies) must be informed by information provided by lay experts along with traditional formal experts, and there needs to be a mechanism to include ecosystem-wide considerations in decision making. Avenues for inclusion range from fact-finding consultation exercises and inclusion of T/LEK in scientific assessments to inclusion of fishery and other workers in research (see, for instance, Morton and Volpe 2002, for an example of including commercial fishermen in a study tracking Atlantic salmon escapes) to devolution of decision-making authority to local communities.

4. Aquaculture of carnivorous species such as salmon does not result in a net gain of fish. Instead, it represents a transfer of nutrients from one ecosystem to another. These activities may provide employment opportunities for those individuals harvesting the forage fish needed to produce fish meal; however, the needs of the local ecosystem must not be compromised. This means that acceptable food must not be denied to those people who could best use it, nor should the level of fishing for forage fish be allowed to destabilize the food web in the local marine ecosystem. Again, the choices made to produce the salmon must not be allowed to preclude choices available to other affected ecosystems. Research into alternative (nonfish) feed sources is therefore to be encouraged, and continued reliance on forage fish reduced accordingly. Should reduced forage fish remain a necessary component of fishfeed, the impact on the ecosystems from which they are taken must be evaluated and harvest levels set at a level that cannot harm the source ecosystem. Tangible benefits must accrue to communities connected to that ecosystem, and ideally, they too should be involved in the management of their local fisheries. If the communities wish to catch fish to sell to feed producers, rather for their direct consumption, that wish should be respected. However, the needs of the local people should be elevated over the wishes of people elsewhere.

It must be emphasized that ecosystem justice does not preclude human involvement in an ecosystem; quite the contrary, it recognizes that humans are members of interdependent ecosystems. As with other animals, human survival necessitates the consumption of other creatures. Both Haida and Christian teachings remind us, though, that respect for human and nonhuman components of the ecosystem, and the ecosystem as a whole, is fundamental. For the good of the whole ecosystem, it must be treated carefully rather than wantonly, with respect, and with a goal of encouraging the ongoing flourishing of the ecosystem for future generations.

References

Anderson, J. L. 2002. Aquaculture and the future: why fisheries economists should care. Marine Resource Economics 17(2):133–151.

Asche, F., T. Bjørndal, and E. H. Sissener. 2003. Relative productivity development in salmon aquaculture. Marine Resource Economics 18(2):205–210.

Asche, F., H. Bremnes, and C. R. Wessels. 1999. Product aggregation, market integration, and relationships between prices: an application to world salmon markets. American Journal of Agricultural Economics 81(3):568–581.

Babcock, B. A., and Q. Weninger. 2004. Can quality revitalize the Alaskan salmon industry? Iowa State University, Center for Agricultural and Rural Development Working Papers, Ames.

British Columbia. 2004. The 2004 British Columbia seafood industry year in review. Ministry of Environment and Ministry of Agriculture and Lands, Victoria.

British Columbia Environmental Assessment Office. 1997. Salmon Aquaculture Review final report—volume 1: report of the Environmental Assessment Office. British Columbia Environmental Assessment Office, Victoria.

British Columbia Ministry of Agriculture and Lands. 2003. Salmon aquaculture in British Columbia: 2003 quick facts. Province of British Columbia, Victoria.

Brunk, C., and S. Dunham. 2000. Ecosystem Justice in the Canadian Fisheries. Pages 9–33 *in* H. Coward, R. Ommer, and T. Pitcher, editors. Just fish: ethics and Canadian marine fisheries, volume 23. ISER Books, St. John's, Newfoundland, Canada.

Costanza, R., R. d'Arge, R. de Groot, S. Farber, M. Grasso, B. Hannon, K. Limburg, S. Nacem, R.V. O'Neill, J. Paruelo, R.G. Raskin, P. Sutton, and M. van den Belt. 1997. The value of the world's ecosystem services and natural capital. Nature(London) 387:253–260.

Dunham, S., and H. Coward. 2000. World religions and ecosystem justice in the marine fishery. Pages 47–66 *in* H. Coward, R. Ommer, and T. Pitcher, editors. Just fish: ethics and Canadian marine fisheries, volume 23. ISER Books, St. John's, Newfoundland, Canada.

FAO (Food and Agriculture Organization of the United Nations). 2002. The state of world fisheries and aquaculture 2002. FAO, Rome.

Gardner, J., and D. L. Peterson. 2003. Making sense of the salmon aquaculture debate: analysis of issues related to netcage salmon farming and wild salmon in British Columbia. Pacific Fisheries Resource Conservation Council, Vancouver.

Guttormsen, A. G. 2002. Input factor substitutability in salmon aquaculture. Marine Resource Economics 17(2):91–102.

Haggan, N. 2000. "Back to the future" and creative justice: recalling and restoring forgotten abundance in Canada's marine ecosystems. Pages 83–99 *in* H. Coward, R. Ommer, and T. Pitcher, editors. Just fish: ethics and Canadian marine fisheries, volume 23. ISER Books, St. John's, Newfoundland, Canada.

Johnston, D. W. 2002. The effect of acoustic harassment devices on harbour porpoises (*Phocoena phocoena*) in the Bay of Fundy, Canada. Biological Conservation 108(1):113–118.

Jones, R., and T.-L. Williams-Davidson. 2000. Applying Haida ethics in today's fishery. Pages 100–115 *in* H. Coward, R. Ommer, and T. Pitcher, editors. Just fish: ethics and Canadian marine fisheries, volume 23. ISER Books, St. John's, Newfoundland, Canada.

Keller, C. 1995. A Christian response to the population apocalypse. Pages 73–108 *in* H. Coward, editor. Population, consumption, and the environment. State University of New York Press, Albany.

Krkosek, M., M. A. Lewis, and J. P. Volpe. 2005. Transmission dynamics of parasitic sea lice from farm to wild salmon. Proceeding of the Royal Society of London Series B 272(1564):689–696.

Meggs, G. 1991. Salmon: the decline of the British Columbia fishery. Douglas and McIntyre, Vancouver.

Morton, A., and J. Volpe. 2002. A description of escaped farmed Atlantic salmon *Salmo salar* captures and their characteristics in one Pacific salmon fishery area in British Columbia, Canada, in 2000. Alaska Fishery Research Bulletin 9(2):102–110.

Morton, A. B., and H. K. Symonds. 2002. Displacement of *Orcinus orca* (L.) by high amplitude sound in British Columbia, Canada. ICES Journal of Marine Science 59(1):71–80.

Neis, B., and M. Morris. 2000. Creative, ecosystem, and distributive justice and Newfoundland's capelin fisheries. Pages 174–200 *in* H. Coward, R. Ommer, and T. Pitcher, editors. Just fish: ethics and Canadian marine fisheries, volume 23. ISER Books, St. John's, Newfoundland, Canada.

Ommer, R. 2000. The ethical implications of property concepts in a fishery. Pages 117–39 *in* H. Coward, R. Ommer, and T. Pitcher, editors. Just fish: ethics and Canadian marine fisheries, volume 23. ISER Books, St. John's, Newfoundland, Canada.

Pauly, D., A. Beattie, A. Bundy, N. Newlands, M. Power and S. Wallace (2000). Not just fish: value of marine ecosystems on the Atlantic and Pacific coasts. Pages 34–46 *in* H. Coward, R. Ommer, and T. Pitcher, editors. Just fish: ethics and Canadian marine fisheries, volume 23. ISER Books, St. John's, Newfoundland, Canada.

Pauly, D., and R. Watson. 2003. Counting the last fish. Scientific American 289:42–47.

Pauly, D., V. Christensen, J. Dalsgaard, R. Froese, and F. Torres Jr. 1998. Fishing down marine food webs. Science 279:860–863.

Pitcher, T. J., and M. D. Power. 2000. Fish figures: quantifying the ethical status of Canadian fisheries, east and west. Pages 225–53 *in* H. Coward, R. Ommer, and T. Pitcher, editors. Just fish: ethics and Canadian marine fisheries, volume 23. ISER Books, St. John's, Newfoundland, Canada.

Powell, K. 2003. Eat your veg. Nature(London) 426:378–379.

Power, M. D. 2003. Lots of fish in the sea: salmon aquaculture, genomics and ethics. University of British Columbia, W. Maurice Young Centre for Applied Ethics, Paper No. DEG 004, Vancouver.

Power, M. D., and N. Newlands. 1999. A report on historical, human-induced changes in Newfoundland's fisheries ecosystem. Pages 391–404 *in* B. Baxter, editor. Ecosystem approaches for fisheries management. Alaska Sea Grant College Program, AK-SG-99–01, Anchorage.

Roach, C. M. 2000. Stewards of the sea: a model for justice? Pages 67–82 *in* H. Coward, R. Ommer, and T. Pitcher, editors. Just fish: ethics and Canadian marine fisheries, volume 23. ISER Books, St. John's, Newfoundland, Canada.

Roberts, R. J. 2000. Salmonids—positive outlook for salmon industry. Aquaculture Magazine 26(5).

Ruether, R. R. 1992. Gaia and God: an ecofeminist theology of earth healing. Harper Collins, New York.

Subasinghe, R. P. 2003. An outlook for aquaculture development: major issues, opportunities and challenges. Pages 31–36 *in* Review of the state of world aquaculture. Food and Agriculture Organization of the United Nations, FAO Fisheries Circular No. 886, Rome.

Tacon, A. J. 2003. Aquaculture production trends analysis. Pages 5–29 *in* Review of the state of world aquaculture. Food and Agriculture Organization of the United Nations, FAO Fisheries Circular No. 886, Rome.

The Associated Press. 2003. Adult Atlantic salmon caught south of Petersburg. Anchorage Daily News, Anchorage.

Promise of a blue revolution. 2003. The Economist 363(8336):20–23.

Troell, M., P. Tyedmers, N. Kautsky, and P. Rönnbäck. 2004. Aquaculture and energy use. Pages 97–108 *in* Encyclopedia of energy, volume 1(Issue). Elsevier Academic Press, Amsterdam.

Tuominen, T-R., and M. Esmark. 2003. Food for thought: the use of marine resources in fish feed. WWF-Norway, 02/03, Oslo.

Tyedmers, P. H. 2000. Salmon and sustainability: the biophysical cost of producing salmon through the commercial salmon fishery and the intensive salmon culture industry. Doctoral dissertation. University of British Columbia, Vancouver.

Volpe, J. 2001. Super un-natural: Atlantic salmon in BC waters. David Suzuki Foundation, Vancouver.

Volpe, J. P., E. B. Taylor, D. W. Rimmer, and B. W. Glickman. 2000. Evidence of natural reproduction of aquaculture-escaped Atlantic salmon in a coastal British Columbia river. Conservation Biology 14(3):899–903.

Weber, M. L. 2003. What price farmed fish: a review of the environmental & social costs of farming carnivores. SeaWeb, Providence, Rhode Island.

American Fisheries Society Symposium 49:43–52

Social Resistance to the Obvious Good: A Review of Responses to a Proposal for Regulation of European Fisheries

KENNETH PATTERSON*
European Commission, Directorate-General for Fisheries and Maritime Affairs
B-1049 Brussels, Belgium

Abstract.—Classical science-based fisheries management offers a number of "obvious goods" for society that can accrue from avoiding growth overfishing and recruitment overfishing. These include an increase in yield, income, and profitability of fishing operations, a decrease in fishing effort, costs, time spent at sea, and capital expenditure. It offers a better stability of yield, better resistance to environment-driven instability and shorter working hours. Lower fishing mortality rates can mean larger fish in the catches with better market values and a decrease in discarding, as well as lower impacts on the marine ecosystems and on nontarget species and a lower risk of stock depletion events. These constitute the "obvious good" referred to in the title. In preparing a legislative package offering to deliver benefits such as these in European fisheries in 2002, the European Commission sought consultations with member states, with various representative elements of the fishing sector, and a number of other representational bodies (globally, "stakeholders").This paper summarizes a sample of these responses that are in the public domain and compares how the demands of the stakeholders compare with the "obvious good" that science-based management can offer.Stakeholders' responses each represented a unique viewpoint, but some common themes could be identified. The most important request from the stakeholders was for a policy that maintains employment, even in the short term, and with state aid if necessary. Meeting conservation goals was generally allocated a secondary importance, though the importance of such goals varied among stakeholders. A high value was also placed on stability in catching opportunities, but this was usually not linked to conservation objectives.Stakeholders placed generally little value on other "obvious goods" such as improving profitability of fishing (so long as it could be maintained at a "viable" level), improving working conditions or reducing ecosystem impacts.The conclusion from this analysis is that more policy attention should be given to management strategies that reconcile high employment with mitigation of biological risks to the sustainable exploitation of the resources

Introduction

Fisheries Management in the European Community has largely not worked over the last 10 years. Many important stocks are depleted and some are close to collapse. This was recognized by the European Commission in a green paper reviewing the state of the marine fishing industry and the resources on which it depends (Anonymous 2001a). The Commission proposed a reform of the management policy in 2002 by proposing a new basic regulation concerning marine commercial fishing in European waters. This was adopted in December 2002 (Anonymous 2002a), and implementation measures have followed thereafter.

This paper describes briefly the main conservation problems that were identified prior to 2000, the solutions that were proposed, and the response

*Corresponding author: kenneth.patterson@ec.europa.eu

by stakeholders to the proposed strengthening of conservation measures. The issues covered in the reform were wide-ranging and complex. This paper focuses on the issue of reducing fishing mortality in the main European demersal fisheries and does not address many important issues that were also addressed in the reform process (such as subsidies, access to resources, third-country agreements, etc.), nor does it address the contributions by nongovernmental organizations outside the fisheries sector. Only documents in the public domain are reviewed here.

The purpose of the paper is to contrast the objectives of the reform, in generating an "obvious good" for society, and the views of the sectors most affected, and to explore some of the reasons for resistance to conservation-objective based management.

The Common Fisheries Policy and the Reform Process

The Common Fisheries Policy (CFP) is a body of European Community legislation that has its legal basis in the "Treaty Establishing the European Community." The most important regulation is the "Framework Regulation," which defines, inter alia, the legal instruments that can be used to manage fisheries, their scope, and how they shall be used. This regulation is reviewed and rewritten once every ten years.

The Framework Regulation and the Common Fisheries Policy form the basis for the management of aquaculture and commercial fisheries by European vessels and in European seas. It affects, directly or indirectly, nearly 7 million metric tons of marine catch (1999), 530,000 (1997) jobs, and 100,000 (1998) fishing vessels. A further 1.4 million metric tons (1999) from aquaculture production is covered. The total value of the output from the processing sector is about 11 billion (10^9) euros (1997) (Anonymous 2001b). The total value of subsidies is about 1.1 billion euros annually.

Effects of the Common Fisheries Policy

The outcome of the CFP over the previous 10 years was evaluated in a green paper (Anonymous 2001a), the relevant parts of which are summarized briefly in this section.

The CFP was judged in the Green Paper to have been largely, though not entirely, unsuccessful. Sustainable exploitation of resources had not been achieved in many cases, with many demersal stocks being outside safe biological limits, stocks too heavily exploited, and with too low quantities of mature fish. The fishing sector was economically fragile due to overinvestment, rapidly rising costs, and a shrinking resources base. Employment and profitability were both declining.

Catches and spawning stock sizes in the late 1990s were just over half the levels they had been before the inception of the CFP. The main cause of this situation was identified as the annual setting of total allowable catches (TACs) at too high values, coupled with poor enforcement of catches and fishing fleets that are too large to exploit the marine resources. Furthermore, employment in the catching sector declined by 22% (by 66 000 jobs) and by 14% in the processing sector.

An overall reduction in fishing effort of the order of 40% from the levels of the mid-1990s had been identified as needed for prudent management, but instead, fishing mortality had continued to increase. Despite recent increases in fishing mortality, low stock size, and poor recruitment, the biological situation was categorized as not immediately catastrophic, but that the absence of remedial measures would lead to stock collapses.

Nevertheless, the CFP had at least been partially successful in that it had avoided conflicts at sea, had largely avoided total stock collapses (arguably with the exception of West of Scotland cod), and had provided some degree of stability for the industry.

Consultation and Review procedure

Based on the green paper, the Commission organized a round of consultation with member states and with interested parties. This paper presents a brief appraisal of a selection of the conservation and stock management aspects of those submissions. Necessarily, large parts of the discussions concerning fishing rights, access, fleet structure, fishing in third countries' waters, safety, and other issues are not discussed here.

Commission Proposals

The Commission prepared an extensive package of legislative reform proposals covering many aspects of fisheries management (Anonymous 2002b). The key conservation elements of these proposals had as their aims

• to refocus management on a more long-term approach to securing sustainable fisheries with high yields;

• to manage fishing effort in line with sustainable catching opportunities, which would require an immediate and significant reduction of fishing effort;

• to incorporate environmental concerns into fisheries management, in particular by contributing to biodiversity protections;

• to move towards an ecosystem-based approach to fisheries management;

• to make the best use of harvested resources and avoid waste; and

• to support the provision of high-quality scientific advice.

The Commission proposed a legal instrument to facilitate the attainment of these aims (Anonymous 2002c). The key part of the proposal was the possibility of establishing multiannual management plans for stocks that would be consistent with the precautionary principle, take account of impacts on nontarget species, and provide high and stable yields. The plans would also include harvest rules, reference points, and targets.

The Obvious Good

The Commission proposals were intended to lay the basis for the "obvious good" that could be achieved from classical fisheries management based on sound biological management. These are, synoptically and with respect to fishing at a high level of fishing mortality and a depleted biomass, as follows:

• an increase in yield and income;

• a decrease in the fishing effort, fishing costs, time at sea, and capital expenditure;

• improved stability of yield, with lower fluctuations in stock size due to recruitment variations;

• improved profitability of operations, higher incomes and shorter working hours for those in the fishery;

• larger sizes of fish in the catches, with reduced discarding and improved value of the catches;

• lower impact on the marine ecosystem, on nontarget species, and on species of no commercial value; and

• reduced biological risk of stock depletion events.

Stakeholders' Responses to the Reform Proposals

The Commission arranged a comprehensive consultation exercise with member states, stakeholders, and various interested parties. A synoptic selection of the contributions made in this consultation is presented below.

The "Friends of Fishing"

The "Friends of Fishing" was an informal group of fishing nations (Spain, France, Greece, Ireland, Italy, and Portugal) that presented a common response to the Commission's reform proposals through coordinated political and public actions. Their position can be summarized as follows (Anonymous 2002a):

The reform of the CFP should result in a balance between sustainable management of the resources and the equally important social, economic, and geographic aspects of fishing (i.e., biological sustainability of fishing is not a higher priority than other aspects). In addition, the CFP should take account of additional needs of the Mediterranean and outlying regions.

The "Friends of Fishing" objected to

• general application of long-term management plans, because stock management methods should be differentiated according to the state of the stocks concerned;

• treatment of the Mediterranean and of outlying regions on the same basis as other areas. Here, the application of scientific advice should be moderated and social and economic considerations should be taken into account;

• only a single management instrument should be used: either TACs and quotas or else effort management—TACs covering several years and several species were preferred; and

• any further reductions in fleet capacity.

Some additional points were made as follows.

• Aid for fleet modernization and renewal should continue, within a framework of capacity limitations. The aid should not be limited to measures improving safety, working conditions or product quality. The capacity limitations are subject to adaptations and may increase. Aid for the export of vessels to third countries should continue.

• Recovery plans should apply only where stocks are outside safe biological limits.

• Multiannual, multispecies plans based on the "precautionary approach," should keep stocks within "reasonable" biological limits and within social and economic constraints.

• State aid should be maintained both for decommissioning and for building new vessels, though overall fleet capacity should not decrease.

Netherlands

The national response from the Netherlands (Anonymous 2001c) was broadly supportive of the Commission's views. In specific issues, Netherlands supported a reduction in state aids for fishing capacity, management according to multiannual strategies respecting target levels for fishing mortality and spawning stock size, and a general application of the precautionary approach. Netherlands also favored closures for spawning areas and juvenile areas, and developments to technical measures regulations favoring increased selectivity. Netherlands favored the introduction of effort management as a primary instrument in managing fisheries.

France

The submission by France (Anonymous 2000) appears to place a higher priority on employment than conservation.

The conservation elements of this text were that TACs should be the primary management instrument (possibly as TACs over several years and covering several species). Discards and catches of juveniles should be reduced and vulnerable species and habitats should be protected. However, the fisherman's "proper" place should be kept in the ecosystem and environmental protection should be made compatible with continuing economic activity.

Spain

Spain agreed with the principle of rebuilding fish stocks but also stressed the social and economic importance of fishing in European coastal regions, many of which are economically disadvantaged.

The main objective would seem to be to ensure sustainable exploitation, specifically the continuation of fishing activity. Spain would agree with conservation measures to reduce fleet capacity in order to achieve sustainable activity, and mentioned a maximum sustainable yield criterion in fisheries management. The precautionary approach was supported, and fishing activity should make "reasonable and sustainable" use of fisheries resources and should take account of environmental criteria, ecosystem integrity, and intergenerational equity.

However, conservation measures based on the precautionary approach should be accompanied by measures to eliminate or diminish the social and economic costs in fishing-dependent areas. Overall, the precautionary approach should be applied taking account of economic, social and territorial repercussions of such measures.

Ireland

The presentation by Anonymous (2001 d) does not address the choice of conservation objectives or reference points, but recommends that reference points need to be re-evaluated and that "management decisions should be based on both biological and socioeconomic evaluations." More support for research in order to obtain more precise estimates is required, in order to reduce the margin of safety needed in applying the precautionary approach. This implies a higher fishing mortality.

The report places a high value of technical conservation measures. A recent conference was convened to address to promote "environmentally friendly fishing" that would allow more intensive fishing activity.

Economic and Social Committee of the European Community: Section for Agriculture, Rural Development, and the Environment

The Chagas Report (Anonymous 2001e) recognizes that fisheries are in a fragile state with a bleak outlook in some areas and are being weakened by the depletion of stocks. Remedial measures should be balanced and decided in dialogue with the sector. The measures must provide support to cushion any socioeconomic consequences. There should be more support to the outermost regions.

Dissatisfaction felt by sector interests does not call into question the principle that resource conservation has top priority.

However, the Committee places a very high value on fisheries policy as an instrument to maintain distant and isolated fishing communities.

The Committee supports a strong fleet policy and better compliance.

Contribution on the Green Paper by the Advisory Committee on Fisheries and Aquaculture (ACFA)

The ACFA is a community body of representatives from the fishing sector, the processing sector, and also some NGOs. It provided a formal consultation to the Commission on the CFP reform (Anonymous 2001f).

The ACFA held the view that the "precautionary approach" should apply not only to the stocks but also to the fishermen, the processing industry, and the consumers. It endorsed a TAC-based management system with limits on interannual changes.

The ACFA requested the continuation of the state aid to build ships, even though most of the fleet is profitable. Aid should be targeted at improving safety, working conditions and product quality. It considered that the Commission should have an active social policy to accompany the restructuring measures under way, particularly in view of the imbalance between fishing capacity and resources.

The Committee also requested special treatment should be available for given to small-scale coastal fishing, and there should be support for employment, training, safety, and working conditions at sea.

Shetland Oceans Alliance response to the Green Paper

The Shetlands Ocean Alliance (SHOAL) is a body established to represent the fishing industry at the Shetland Islands. A report (Anonymous 2001 g) provided a response to the Commission's proposals. The main emphasis of this report was the need to continue fishing in order to maintain community structure in a remote area where there are few alternatives.

As management instruments, SHOAL preferred multiannual and multispecies quotas because "it is not possible for North Sea trawlers to avoid catching a specific species. The fishery is mixed by its very nature. Reducing a quota for a species in a mixed fishery does not protect that species, it merely increases discards of that species."

Where emergency conservation measures become required, they should be accompanied with financial compensation.

Scottish Fishermen's Federation

The Scottish Fishermen's Federation (SFF) represents one of the larger demersal fishing sectors in Northern European waters. Their views on the CFP reform are summarized briefly below, from Anonymous (2001 h).

The SFF recognized the need for a change in fishing régimes in the North Sea, stressing that no effort should be spared to restore the North Sea to its historical productivity. The SFF supported a rebuilding plan to achieve a reduction in the fishing mortality for cod and species caught together with cod, according to a defined rebuilding trajectory.

However, the SFF thought it unrealistic to rebuild stocks to the sizes seen in the 1970s, which were influenced by several unusually large recruitments ("gadoid outburst"). Therefore, the SFF considered that stock rebuilding should only be intended to restore stocks to recent levels and did not necessarily want risk-based or objective-based management.

The management plans should include tight stability criteria for TACs.

Overall, management should avoid excessive caution and application of the precautionary approach should not be a substitute for good science and analysis. Fishermen are a part of the marine ecosystem with co-equal value alongside other marine creatures, and an endangered species.

For the longer term, the SFF view was that after fisheries have been brought within precautionary levels they should then be "worked at their maximum sustainable yield indefinitely." This would ensure long term stability for the industry and its dependent communities.

The SFF approach to securing cod recovery was a 20% permanent decommissioning of the whitefish fleet together with temporary layups with compensation.

A number of views concerning fleet capacity are expressed in the SFF contribution. An imbalance between fishing opportunities and capacity is recognized, but capacity controls are unnecessary if the quota system is well conceived and properly enforced. The SFF did not support the expansion of control and enforcement capabilities. Fleet capacity should be retained in the short term (with support) but it is necessary for the capacity of all national fleets to be adjusted to the level appropriate to working the recovered stock. In principle, reductions in activity should attract compensation: there should be compensatory or palliative mechanisms for cases where biological events cause a need for restrictive mechanisms, which cause short-term losses.

However, the SFF also acknowledge the need to reduce total capacity permanently and to put in place an effort management system.

As many other stakeholders, the SFF preferred that fishermen should participate in TAC decisions along with scientists and government officials. The primacy of interest of coastal communities was stressed.

Europêche/Cogeca

The Europêche submission (Anonymous 2001i) recommended, inter alia,

• better scientific research, in order to reduce the area of uncertainty around the precautionary approach;

• TACs as the main conservation instrument, but not altered by more than 15% annually;

• Europêche recognizes the need to ensure sustainability and that a good number of stocks need recovery measures. Overcapacity in some fleet segments was recognized.

In 2003, Europêche's position statement (Anonymous 2003a) concerning fishing possibilities for 2004 included

• rejection of scientific advice for zero catches in some stocks, because this "would cause chaos and bankruptcies in harbours;"

• rejection of the principle of TAC reductions for species that are caught together with species at high biological risk;

• rejection of management measures directed exclusively at fisheries when other factors also affect stocks (seals, wind and tidal generation of electricity, consumption by birds, pollution, etc.);

• a demand for a limit of TAC variation of ± 15% from 1 year to the next; and

• a demand for evaluation of the effects of measures already in place, and an a priori social and economic evaluation to be made before introducing new conservation measures.

European Parliament

The European Parliament prepared an extensive opinion of 122 points concerning the green paper and the CFP reform (Anonymous 2002d). The key point is perhaps the recital stating that "the fundamental aim of the common fisheries policy is to balance ensuring the viability of an economic sector ... with maintaining sustainable marine ecosystems .... Although Parliament stresses its concern about the depleted state of fish stocks, the extent to which it supports restrictive conservation measures to restore stocks is unclear. Parliament does stress a need to promote employment in fisheries where fisheries-dependent communities face serious problems, with diversification away from fishing to be supported when necessary.

Overall, it appears that Parliament's opinion calls for both conservation-based management and the maintenance of employment in coastal regions and deplores the present poor state of fish stocks, without explicitly recognizing that achieving biological sustainability may not be compatible with maintaining stability of fishing activity. However, Parliament also supported reductions in fishing capacity through reducing the size of the fleet and by limiting time spent fishing.

Synopsis of Stakeholders' Contributions

Although each stakeholder submission represented a unique and special viewpoint on the reform process, some general themes could be identified.

A primary importance was placed on social and economic factors, even in short time-frames. Such factors were often not spelled out in detail, but the most important elements were usually the maintenance of employment and fishing capacity in the short term (with aid if necessary), and most especially in remote coastal regions.

The balance between maintaining employment and meeting conservation goals varied. Some stakeholders clearly place a higher value on the former, stating that compliance with precautionary criteria should be done only insofar as it does not prejudice fishing activity. Others placed a more equal importance on the two criteria, stressing only that some balance must be found between attaining conservation goals and maintaining fishing activity. In contrast, there did not seem to be any support for the scientific view that fisheries management decisions should only be taken within the area of "safe" population dynamics defined by the implementation of the precautionary approach made by the relevant providers of scientific advice (Anonymous 2003b) (i.e., the primacy of the application of scientific advice to maintain a low risk of stock collapse does not find support among the stakeholders).

A high value was placed on stability of catching opportunities, with several references to a ±15% limit on changes in TACs from one year to the next. It was not explicitly recognized that stability in TACs depends substantially on stock stability, and hence on reduced fishing mortalities.
Support for reducing fleet capacity was mixed according to area and experience. Some contributors recognized a crisis in fish stocks and a need to reduce capacity. However, even where scientific advice implied zero catches or very large reductions, only much smaller reductions were found acceptable by stakeholders. In other areas, no reduction in capacity was acceptable other than temporary cessation of activity (with financial compensation) during periods of stock depletion.

Impacts on nontarget species and ecosystem considerations were generally not addressed to the level of operational substance.

When issues of yield and profitability were discussed, the principal concern was usually on maintaining a viable fishing activity. In other words, profitability of the fleets was not explicitly very interesting unless it were to fall so low as to prejudice employment. However, two submissions referred to a maximum sustainable yield as a plausible long-term objective to be pursued after recovery of the stocks.

In oral consultations, less moderate opinions have also been heard, such as forthright denials of the validity of scientific advice, a belief that stocks that have been depleted will be replaced by other resources, and a belief in the greater importance of environmental and ecosystem effects than the directs effects of fishing.

Conclusions: The Obvious Good Reconsidered

With this context, it is now useful to reconsider the list of benefits that traditional fish stock management could bring, as were cited above:

1. An increase in yield and income.—So long as fishing activity is viable and fishermen's incomes are sufficient to sustain employment, any subsequent increase in yield or in income does not count highly as a priority in the stakeholders' presentations.

2. A decrease in the fishing effort, fishing costs, time at sea, and capital expenditure.—Stakeholders report a strong reluctance to reduce fishing effort (and employment). Even where effort reductions are imperative in conservation terms, temporary cessation of activity with compensation rather than permanent withdrawal of capacity is often preferred.

3. Improved stability of yield, with lower fluctuations in stock size due to recruitment variations.—A high value was placed on limiting TAC fluctuations, but it was not recognized that it is necessary to have fish stocks in a stable state in order to achieve such stability.

4. Improved profitability of operations, higher incomes, and shorter working hours for those in the fishery. —Only one of the stakeholders' submissions placed an improvement in working conditions among its highest priorities. Achieving high income was usually not a stated objective.

5. Larger sizes of fish in the catches, with reduced discarding and improved value of the catches.—Reduction in discarding through changes to fishing gear was indicated in some sectors, but by technical measures and not through stock rebuilding. Achieving better prices for catches through landing larger fish was not a priority area.

6. Lower impact on the marine ecosystem, on non-target species, and on species of no commercial value.—This achieved only low priority among the fishing sectors and national administrations.

7. Reduced biological risk of stock depletion events.—The approach to mitigating risk was often not to reduce exploitation rates, but instead to improve and update the scientific advice and the stock assessments.

The conclusion of this simple review will not be a surprise to many fisheries managers. It is that the primary concern of the stakeholders appears to be to maintain employment and traditional community structure. Other objectives are generally subsidiary to this. This conclusion is in line with Hilborn and Walters' (1992) conclusion about societal objectives for fisheries management, and also with the conclusions of Cochrane (2004, this volume) about societal resistance to change in the face of a shortfall in available natural resources.

Biologically based fisheries management offers a number of benefits upon which stakeholders place much less value, and it does so by incurring large decreases in employment. Formal recognition of this mismatch may be required, possibly accompanied by a refocusing of management effort on strategies to maintain high employment while mitigating biological risks to the resources.

However, acceptance by stakeholders of strategies to deliver long-term benefits—in whatever currency—remain dependent on the acceptance of the basic scientific tenets and on the credibility of administrations to deliver a credible system of compliance. The problem identified by Clark (1990), as follows, remains still to be solved:

"*Thoughtful social analysts have proposed that a fundamental duty of the state is a concern for the welfare of future generations. How this duty is to be fulfilled in a democratic society whose individuals are concerned largely with the demands of the present is a dilemma that has yet to be resolved. But resolved it must be if our vital renewable resources are to be preserved for the future.*"

References

Anonymous 2000. French memorandum on the reform of the common fisheries policy in 2002. Council Document 13490/00, Brussels.

Anonymous 2001a. Green Paper: the future of the Common Fisheries Policy. Office of the Official Publications of the European Communities, Luxembourg.

Anonymous 2001b. Facts and figures on the CFP. Basic data on the Common Fisheries Policy. Office for Official Publications of the European Communities, Luxembourg.

Anonymous 2001c. Position of the Netherlands on the green paper on the Common Fisheries Policy. Paper submitted by the government of the Netherlands to the European

Parliament and the European Commission. Government of the Netherlands, Antilles.

Anonymous 2001d. Proposals and recommendations in response to the Commission's Green Paper on the Common Fisheries Policy 2002. Irish national strategy review group on the Common Fisheries Policy, Dublin.

Anonymous 2001e. Draft opinion of the Section for Agriculture, Rural Development and the Environement on the Green Paper on the future of the Common Fisheries Policy (COM(2001) 135 final). Economic and Social Committee, NAT/110, Brussels.

Anonymous 2001f. Opinion. Green Paper on the future of the Common Fisheries Policy. Advisory Committee on Fisheries and Aquaculture, Brussels.

Anonymous 2001g. Response to the Green Paper on the future of the Common Fisheries Policy. Shetland Oceans Alliance, Development Departtment, Greenhead, Lerwick, Shetland.

Anonymous 2001h. European Commission Green Paper: the future of the Common Fisheries Policy. Response by the Scottish Fishermen's Federation, Aberdeen.

Anonymous 2001i. Prise de position d'Europêche et du COGECA concernant le livre vert presenté par la Commission sur l'avenir de la politique commune de la pêche (COM(2001) 135 final). Association des organisations nationales d'entreprises de pêche de l'Union Européenne. Comité Général de la Coopération Agricole de l'Union Européenne, Bruxelles.

Anonymous 2002a. Réforme de la politique commune des pêches. Conclusions ministérielles communes sur la proposition de la Commission. Press release issued by fisheries ministries of Spain, France, Greece, Ireland, Italy, and Portugal, Paris.

Anonymous 2002b. Communication from the Commission on the reform of the Common Fisheries Policy. Office for Official Publications of the European Communities, Luxembourg.

Anonymous 2002c. Proposal for a council regulation on the conservation and sustainable exploitation of fisheries resources under the Common Fisheries Policy. COM(2002) 185 final, Brussels.

Anonymous 2002d. Future of the Common Fisheries Policy. European Parliament resolution on the Commission green paper on the future of the common fisheries policy (COM [2001] 135 - C5–0261/2001–2001/2115[COS]), P5-TA(2002)0016, Brussels.

Anonymous 2003a. Position statement on the ICES opinion concerning management of fish stocks within the framework of Commission proposals for TACs and quotas for 2004. Association of National Organisations of the Fishing Enterprises of the EU/ General Committee for Agricultural Cooperation in the EU, Brussels.

Anonymous 2003b. Report of the ICES Advisory Committee on Fishery Management. ICES Cooperative Research Report No. 261, Copenhagen.

Clark, C. W. 1990. Mathematical bioeconomics, 2nd edition. Wiley, New York.

Cochrane, K. 2008. What should we care about when attempting to reconcile fisheries with conservation? Pages 7 to26 *in* J. L. Nielsen, J. J. Dodson, K. D. Friedland, T. R. Hamon, J. A. Musick, and E. Verspoor, editors. Reconciling fisheries with conservation: proceedings of the Fourth World Fisheries Congress. American Fisheries Society, Symposium 49, Bethesda, Maryland.

Hilborn, R. and C. J. Walters 1992. Quantitative fisheries stock assessment: choice, dynamics and uncertainty. Chapman and Hall, New York and London.

Availability of documents: Proposals and Communications from the European Commission and documents from the European Parliament are available at www.europa.eu.int.

The views expressed in this paper are the views of the author. This paper does not express a position of the European Commission.

American Fisheries Society Symposium 49:53–61

Making the Judge the Fall Guy: Reliance on the Court System When Processes for Allocation of Fisheries Resources Between User Groups Are Mismanaged or Nonmanaged

KELLY CROSTHWAITE*

Agriculture, Food and Fisheries
Department of Primary Industries and Resources South Australia (PIRSA)
Post Office Box 1625, Adelaide, South Australia 5000, Australia

Abstract.—Fisheries management is highly litigious, particularly in relation to allocation decisions. But decisions made by courts invariably create winners and losers, and in the context of fisheries management, an area that is characterized by the pursuit of multiple competing objectives, polarization of winners and losers is not a desirable result. Nevertheless, persons on the end of allocation decisions are often motivated to seek relief from the courts.

As overall pressures on natural resources become more acute, conservation imperatives are becoming more stringent and allocation between user groups is rapidly forcing its way onto the political agenda. However, the ascendancy of the issue has outpaced development of tools to value fisheries resources and the establishment of decision-making frameworks for making allocations. As a result, politicians and managers are making decisions in the absence of a coherent objective framework and therefore without the information and structured debate needed to help constrain the politicization of what are inherently political decisions.

An examination of court decisions, and the questions they have been asked to answer, suggests that the motivation to have courts resolve issues stems from a number of factors including an unwillingness to make hard decisions at the right time.

This paper identifies the types of circumstances where courts become involved in allocation issues, and attempts to distil some lessons for designing frameworks for allocating between uses and users of a fishery.

Why Examine Court Decisions When Talking about Allocation?

Allocation of access to fisheries resources between user groups is a political decision that brings into focus the central tension in fisheries management—long-term benefits require short-term sacrifices. It involves situations in which the relative value of resources to different user groups is often unknown, hard to quantify, and therefore difficult to compare (Hundloe 2002). It necessitates value judgments and imposes costs that can be disproportionate for some participants in the fishery. Allocation issues are emotive and easily misrepresented using simplistic arguments.

It is not surprising therefore that decisions about access are unattractive to politicians and

*Corresponding author: crosthwaite.kelly@saugov.sa.gov.au

are often avoided until forced, either when allocation becomes critical to sustainability or when other factors intervene to set the political agenda. However, as increases in human populations are driving demand for fish, creating compounding environmental pressures and expanding recreational catches, allocation is becoming more important as a fisheries management tool because impacts of competing uses must be limited to the capacity of the existing resource base (Regier and Grima 1985; Edwards 1991; Charles 1992).

As pressures on fisheries escalate, courts are being called on to consider allocation-related matters either directly or indirectly. In such cases, examining the factual circumstances and the issues the courts have been asked to resolve helps identify the factors leading to a court's involvement and may provide insight into deficiencies in decision-making processes.

Case Studies

Courts become involved in allocation of access to a fishery between uses in two broad categories of case. The first is when a decision has been imposed by government about who has access to a fishery and the outcome is unacceptable to a user of the fishery. Such decisions are most likely to end up in court when access is reallocated from one use to another. For example, the closure of commercial fisheries for the benefit of enhanced recreational access has resulted in two recent Australian court actions, as discussed below.

The second category of case in which courts become involved in allocation is when a legal mechanism is available for a user of a fishery to indirectly force an allocation issue to be considered. Such cases usually involve a smaller or nonconsumptive user group that has been recognized as having value but has not been specifically incorporated into the management framework of a fishery. For example, in some contexts, conservation laws may be able to be used to override the direct management of a fishery to ensure that conservation interests are taken into account in allocating access to users of a fishery. The litigation in relation to the Hawaiian Longline Fishery is a case in point[1]. In other situations, access rights have been recognized but poorly defined in legislation, so the meaning of that right is not tested until a fisheries offense is prosecuted. The regulation of fishing by indigenous fishers for customary purposes in Australia is an example of this type of situation and is discussed below.

South Australian River Fishery

This case[2] involved the closure of the commercial fishery for native species in the River Murray in South Australia (SA). The fishery was historically a limited entry commercial fishery and an open-access recreational fishery. The entire river system (which extends through three other state jurisdictions before it reaches SA) has been dramatically modified since European settlement, which has had a profoundly negative impact on the overall environmental health of the entire riverine environment (Cadwallader 1978, 1986; Pierce and Doonan 1999; Norris et al. 2001). The most valued species in the fishery are the native Murray cod *Maccullochella peeli* and callop *Macquaria ambigua*. In particular, Murray cod is big, tasty, and long-lived and has been described

[1] *Center for Marine Conservation v. NMFS ("CMC")*, Civ. No. 99–00152 (D. Haw. 2000). In that case, an action was brought under environmental laws, which resulted in a court ordering the closure of a commercial fishery due to bycatch considerations. Subsequent court-ordered negotiations about access to the fishery by commercial longliners have included green groups as a party.

[2] *South Australian River Fishery Association & Warrick v The State of South Australia* [2003] SASC 38 (14 February 2003). *South Australian River Fishery Association & Warrick v The State of South Australia* [2003] SASC 174 (6 June 2003).

as Australia's icon freshwater fish (Kearney and Kildea 2001).

The SA government's decision to exercise powers under the Fisheries Act 1982 to effectively close the commercial fishery as of mid-2003 was highly political. The decision was made to partially fulfill an extensive compact between an independent Member of Parliament and the Labor Party, which resulted in the independent member aligning himself with the Labor Party to allow them to form government following an election (consequently with a majority of one). The independent politician represents an electorate located on the River Murray. One of the terms of this compact was the immediate prohibition on the use of gill nets (the primary commercial method) and the phasing out within a year of commercial fishing for native species.

The signing of this compact, and the subsequent closure of the fishery for native species, occurred in the wake of growing discontent among users of the river and the wider riverland community. In particular, the restructure of the fishery in 1997 galvanized opposition to commercial fishing. It ignited debate about its impact on recreational fishing opportunities and fishing-based tourism (ERD Committee 1999). This discontent continued to grow as the conditions in the River Murray continued to deteriorate (Parliamentary Select Committee of the Parliament of South Australia on the Murray River 2001). Notably, there was no formal public policy response to this identified problem.

Commercial license holders initiated litigation to challenge the regulations that gave effect to the closure. Two decisions have been handed down[3], a trial judge decision in favor of the license holders[4] and a full court appeal decision in favor of the state[5]. The court had the task of determining whether parliament had overstepped its powers in removing gill nets, Murray cod, and callop from the list of permitted devices and species. On almost all counts[6], both courts were in agreement in rejecting the arguments presented by the commercial fishermen. It was held that a reallocation is certainly within the scope of the Fisheries Act 1982, and in the absence of "clandestine abuse," a government is permitted, and expected, to take into account political considerations in making regulations. It was recognized that fisheries management can give effect to social decisions, however controversial, as long as there is a basis for doing so connected to the purpose of the Fisheries Act 1982.

New South Wales Fisheries

The second case study relates to the prohibition of commercial fishing in Botany Bay and Lake Macquarie from May 2002. The process that led to the closures was initiated by the release in January 2000 of a government discussion paper, "Sustaining Our Fisheries," in which, as highlighted in the ensuing litigation, "the government recognized the potential environmental impacts of

[3] Application for leave to appeal to the High Court of Australia was heard on 30 April 2004 and rejected.

[4] *South Australian River Fishery Association & Warrick v The State of South Australia* [2003] SASC 38 (14 February 2003).

[5] *South Australian River Fishery Association & Warrick v The State of South Australia* [2003] SASC 174 (6 June 2003).

[6] In coming to a different conclusion, the trial judge instead largely relied on a peculiar provision in the SA *Acts Interpretation Act* 1915 as a basis for examining the content (as opposed to the general legality) of the regulatory scheme. In doing so, he concluded that the regime could be taken to have an element of permanency that implied a right to reasonable notice prior to closure and probably the payment of fair compensation (although he stopped short of adopting this as a ground for deciding in the plaintiff's favour). However, the Full Court rejected the basis for examining these issues to the detailed extent that the trial judge did and hence rejected the findings.

change in fisheries management and the uncertainty about the consequences of these impacts, and, secondly, that one of its key initiatives had been to amend the relevant legislation to provide for environmental assessment of fishing activities." The paper also canvassed the introduction of recreational fishing licenses and buy-out of commercial licenses to create a series of recreational fishing areas.

Between January 2000 and the announcement in August 2001 of the decision to prohibit commercial fishing in Botany Bay and Lake Macquarie, an environmental assessment was not conducted, although a consultation process was undertaken by the NSW Department of Fisheries. This process has been criticized by the commercial industry (Howard 2002) and the conservation sector (NCCNSW 2001) for being conducted in an unrealistic timeframe and failing to facilitate the proper scrutiny of the costs and benefits of the proposal, particularly the impact of recreational fishing.

In this context, the Professional Fishers Association initiated litigation and argued that existing environmental legislation should have been utilized to conduct an impact assessment prior to making the allocation decision. They argued that there was no evidence to support the conclusion that the reallocation from the commercial to the recreational sector would not negatively impact on the environment and that an impact assessment regime existed to answer these types of questions. Therefore, the parliament must have intended that the regime extend to these types of decisions, and the legislation should be interpreted accordingly. It was also argued that the mechanism in the fisheries legislation used to implement the closure was not intended to be used to reallocate access.

Lengthy litigation ensued, with three courts involved[7]. Each court found that the government action was legally valid. The cases largely turned on technical matters of statutory interpretation. An impact assessment was found to not be mandatory for this type of action. In relation to the fisheries legislation, it was also held that the mechanism was not restricted in its application so as to exclude reallocations. This was in part based on the fact that allocation of access to a fishery is clearly within the ambit of the NSW Fisheries Act.

Aboriginal Fishing

In recent decades, Australian law has followed other Western nations such as New Zealand and Canada in recognizing the rights of indigenous peoples. In relation to the ownership and use of land and natural resources, these rights were first described by the term "native title" by the High Court in 1992[8] and further articulated in Commonwealth legislation in 1993[9]. In 2001, the High Court provided its first enunciation of principle in relation to native title rights over the sea[10]. Essentially, these principles provide that, upon examination of the evidence presented in a formal claim, native title to fishing could be found to exist over the waters of the sea but not to the exclusion of public fishing rights.

Other native title rights have been recognized that are not dependent on the determination of a statutory claim, including the right to

[7] Two court decisions were handed down—a single judge (*Professional Fishers Association Inc v Minister for Fisheries* [2002] NSWLEC 15) and an appeal decision (*Professional Fishers Association v Minister for Fisheries* [2002] NSWCA 145). The commercial fishers also unsuccessfully sought leave to appeal to the High Court (transcript of proceedings available at http://www.austlii.edu.au/au/other/hca/transcripts/2002/S208/1.html).

[8] *Mabo v The State of Queensland (No. 2)* (1992) 175 CLR 1.

[9] *Native Title Act 1993 (Cth)*

[10] *Commonwealth of Australia v Yarmirr* (2001) 208 CLR 1.

hunt and fish for personal and communal purposes in accordance with native title tradition[11] and other similar rights that have been recognized in the fisheries legislation of some states and the Northern Territory (LRC 1985). Generally, the relevant legislative provisions provide an exemption to fisheries laws. Mostly, they are only able to be asserted as a defense, rather than a positive right. Defenses based on a native title right to fish or hunt have been considered in a number of cases around Australia[12]. In most cases relating to fishing, the evidentiary burden for establishing native title has not been met[13].

The significance of these cases is the context in which charges are laid. If a criminal court is left to determine a person's access to a fishery, this is because the law has defined entry criteria to be met on an individual basis, but does not provide a positive mechanism for recognizing that those criteria have been met (such as a license). Instead, the law provides for a retrospective assessment of whether the entry criteria are met. This leaves access criteria to be addressed post facto, but in full recognition that the activity does, and can legally, occur.

The Impact of Court Decisions

In relation to the first category of case, such as the River Murray and the NSW case studies, the legal effect of the court decisions is essentially neutral. No new law came out of those cases. The courts only had jurisdiction to determine whether the government had complied with the minimum legal standards applicable to exercising authority. They did not attempt to endorse the decisions as the best outcome and distanced themselves from involvement in political choices. This will always be the legal situation where access rules are imposed on a user group as a whole, which is generally done on the basis of parliamentary authority[14]. In contrast, the practical effect of going to court in both cases was, on balance, negative. In both examples, the initiating parties had costs orders made against them. Although the governments had decisions in their favor, they also had significant litigation and political costs in terms of the ongoing fallout related to the conflict.

In relation to the second category of case, the regulation of Aboriginal fishing creates a system whereby difficult determinations (which include a range of historical, social, and evidentiary issues) are made by courts on a case-by-case basis. Irrespective of the wider political and justice aspects of this approach (which are widely debated), from a resource management perspective, this is an inadequate way to allocate access to a fishery. It does not provide any certainty to any of the users of the fishery and provides no basis for managing the impact of fishing by Aboriginal people (Fletcher and Curnow 2002; Franklyn 2003).

Why Do Courts Become Involved?

On this analysis, there are no positive impacts per se from having courts involved in alloca-

[11] Codified in section 211 of the *Native Title Act 1993 (Cth)*.

[12] The High Court decision in *Yanner v Eaton* in 1998[12] (in relation to hunting crocodiles) confirmed that, generally, where a regime was in place for licensing the take of a species an aboriginal person who was hunting pursuant to their native title did not require a permit or license to do so. Notably, the defendant, Yanner, was able to establish that he was acting according to native title in undertaking the activity for which he was charged.

[13] For example, Dershaw, Clifton and Murphy v Sutton (1997) 2 *Australian Indigenous Law Reporter* 53; *Mason v Triton* (1994) 34 NSWLR 572; *Dillon v Davies* (1998) 156 ALR 142. Compare *Wilkes v Johnsen* [1999] WASCA 74.

[14] For example *Dighton v State of South Australia & Anor* [2000] SASC 194; *Professional Fishers Association Incorporated v Minister for Fisheries* [2002] NSWCA 145; *Tucker v Canada (Minister of Fisheries and Oceans)* [2001] F.C.J. No. 1862 (F.C.A.)

tion issues. Governments cannot take a win in court to be an endorsement of a decision or action. Rather, the factors that can be identified as leading to court involvement reflect the absence of a systematic approach to forward planning for allocation. It may not be possible to avoid court involvement in future allocations between user groups, but the absence of litigation usually reflects well on the long-term management framework of a fishery.

For example, three factors contributed to the River Murray and NSW cases ending up in court. First, the allocation had severe and disproportionate consequences for the commercial fishermen (commercial loss) compared to other users of the fishery. Second, they were relatively sudden decisions, and the decision-making process was over very quickly without any further recourse but to go to court. Third, the decisions (particularly the River Murray decision) were played out in the political arena without a clearly defined long-term public policy framework. There was therefore an absence of a structure and supporting information for making the allocation decision. This fuelled speculation and allowed public debate to be oversimplified and distorted, even more so than usual.

These three factors, in combination, provided the commercial fishermen with the motivation to litigate and injected enough uncertainty and controversy into the situation to at least make the legal position arguable. In terms of legal principle, a structured and inclusive decision-making process, and an outcome supported by the information generated by such a process, will always be easier to justify and defend (ALRC 2002). Conversely, a political decision and a sudden outcome will, at a minimum, provide more scope for legal argument and, depending on the legal context, can provide grounds for invalidity.

The regulation of Aboriginal fishing also demonstrates the need to plan an allocation framework to ensure that issues are dealt with in a considered way rather than leaving the difficult determinations to courts. The ad hoc method of allocating access to fisheries resources for indigenous fishers has evolved in response to developments in the law that have accelerated the recognition of some rights to fish and interposed those rights over fishery management regimes. However, in most jurisdictions, processes have not yet been implemented for characterizing the issues in fisheries management terms (as opposed to legal terms). The potential reallocation of access from nonindigenous to indigenous fishers that is perceived to be an inevitable outcome of exploring Aboriginal fishing rights is a politically charged issue. As with the allocation issues involved in the other case studies, they involve complex interactions, values, and impacts that can be easily oversimplified and misrepresented. As such, in the absence of factual information and structured debate, public perception about the correct outcome does not necessarily have any relationship to genuine consideration of fishery management objectives and how to best achieve them. The issues are politically unpalatable and hence avoided if possible.

How to Minimize Court Involvement

A systematic approach to tackling access issues has three key elements: (1) a process for assessing a fishery to identify potential allocation issues and characterize them in fisheries management terms, (2) a process for addressing the identified issues, and (3) a timetable for addressing those issues. These elements help establish transparency and accountability. This will not only strengthen the legal validity of subsequent decisions, but will also increase the potential for a more accepted and better result. If conducted in accordance with a justifiable timetable for addressing the allocation issue, they will further reduce the conditions that lead to political interference. In this way, processes serve to make the allocation less vulnerable to the considerable and legitimate pressures that politicians are under to take a short-term view.

A process, in itself, will not provide a solution; however, "process is both a guide and a way to achieve more disciplined and consistent results that are more likely to be right" (Blunn 2003). It can serve two important interrelated functions. First, a process facilitates the injection of information. If information about the complex relationships between the users of a fishery can be incorporated into decision making, entrenched ideological positions will be less sustainable because information militates against reliance on opinion or assumption. This assists in reducing uncertainty and conflict and achieving a more balanced resolution (Edwards 1991; Charles 1992). For example, in the River Murray fishery, factual information in relation to the relative value and impact of commercial and recreational fishing was noticeably absent from public debate, which was emotive and speculative (see ERD Committee 1999; Parliamentary Select Committee 2001). In relation to indigenous access to fisheries resources, examination of the issues at a practical level can reveal an absence of conflict where it is assumed to be inherent. For example, some traditional fishing activities may actually already be taking place within existing recreational fishing limits and simply recognizing that as traditional fishing may not impact on the stock or on other fishery users.

The second benefit of a decision-making process is that it provides a series of opportunities for information to be tested by various people, including the users of a fishery. This can reduce a party's motivation to go to court if they have had an opportunity to be heard (ALRC 2002). Conversely, if a party feels that only token opportunities for input are provided, then this can increase motivation to litigate. Furthermore, a process that has a number of steps is more likely to produce an outcome based on accurate and accepted information. In relation to the River Murray, if a process had been put in place for characterizing and addressing access issues, the independent member of parliament may have been less motivated to respond to discord in his electorate by imposing a closure (especially if that process had a realistic timetable). Certainly, even if a political decision were imposed, it would have been clear that it was a political decision, rather than being able to be perceived as an obvious common sense approach to managing the fishery's resources.

The challenge for fisheries managers and user groups is to harness these benefits of a process. Depending on the immediacy of the need to allocate access, a process might simply involve identifying information likely to be required in the future and starting to collect it[15]. In other circumstances, a more complex decision-making framework may be required (see for example, McLeod and Nicholls 2002). Regardless, experience suggests that to be of most use, a process must include broad participation (Jentoft and McCay 1995; Solomon 2000). Consultation brings accountability if it is genuine and requires a person or group to articulate and justify their position in the context of facts or an acknowledged absence of necessary facts. This reduces but does not eliminate the scope for participants, lobbyists, or political representatives to attempt to manage perception rather than managing a fisheries resource (Fletcher and Curnow 2002).

[15] For example, population growth forecasts can be routinely incorporated into management processes.

Discussion—An Ethical Responsibility

The challenge of dealing with allocation of access is being addressed to varying degrees in Australia[16]. Different circumstances exist in each jurisdiction, and bureaucracies are constrained by the resources available to manage fisheries. However, there are clear benefits from government taking a systematic approach to make itself aware of allocation issues and attempting to avoid uninformed decisions. Being on the front foot helps ensure that allocation is used as a deliberate tool for achieving fisheries management objectives, rather than a reactive measure used to attenuate conflict in the short term. A systematic approach may be implemented with varying degrees of complexity and should not be discarded as being inherently too difficult or too expensive.

In this context, fisheries managers (be they government employees, management advisory committees, or other structures, depending on the responsibility established by legislation) must take on some responsibility for ensuring that decisions about allocating access to a fishery are as depoliticized as possible by putting in place processes that clearly link allocation to fisheries management objectives and providing the best available information for assessing how to achieve those objectives. To assign this responsibility to fisheries managers is not to usurp the proper role of an elected decision maker. Rather, fisheries managers should be seeking to generate processes that provide the conditions for the best advice to political decision makers to inform outcomes. There is an opportunity now to learn from the past and take the initiative to systematically tackle allocation before allocation of access inevitably comes to the fore.

[16] The Australian Commonwealth Fisheries Research and Development Corporation has recognized that access to fisheries must be based on objective criteria, and has directed research funding into establishing allocation frameworks in accordance with this priority. Various projects are supported in this context, including a comprehensive project in Western Australia (see McLeod and Nicholls 2002).

Acknowledgments

The author would like to thank Will Zacharin, Jon Presser, Sean Sloan, and Guy Wright for their valuable comments on drafts of this paper.

References

ALRC (Australian Law Reform Commission). 2002. Report number 95. Principled regulation: civil and administrative penalties in Australian federal regulation. Commonwealth of Australia, Sydney. Available: http://www.alrc.gov.au (January 2004)

Blunn, A. S. 2003. Administrative decision-making—an insider tells. Australian Institute of Administrative Law (AIAL) Forum 37:35, Canberra.

Cadwallader, P. 1986. The Macquarie perch of Lake Darmouth. Australian Fisheries 45:14–16.

Cadwallader, P. L. 1978. Some causes of the decline in range and abundance of native fish in the Murray-Darling River system. Proceedings of the Royal Society of Victoria 90:211–224.

Charles, A.T. 1992. Fishery conflicts: a unified framework. Marine Policy 16:379–393.

Edwards, S. F. 1991. A critique of three "economics" arguments commonly used to influence fishery allocations. North American Journal of Fisheries Management 11:121–130.

ERD Committee (Environment, Resources and Development Committee of the Parliament of South Australia). 1999. Thirty-first report of the committee: fish stocks in inland waters. Parliament of South Australia, Adelaide.

Fisheries Research and Development Corporation. 2000. Investing for tomorrow's fish: the FRDC's research and development plan, 2000 to 2005. Fisheries Research and Development Corporation, Canberra.

Fletcher, W., and Curnow, I., editors. 2002. Scoping paper: processes for the allocation, reallocation and governance of resource access in connection with a framework for the future management of fisheries in Western Australia. Government of Western Australia, Fisheries Management Report No. 7, Perth.

Franklyn, E. M. 2003. Fisheries Management Paper No.

168: Aboriginal fishing strategy. Draft report to the Minister for Agriculture, Forestry and Fisheries. Government of Western Australia, Perth.

Howard, M. 2002. Towards good governance in fisheries management in NSW: making environmental and social impact assessment count. Proceedings of Coast to Coast 2002. Coastal CRC, Brisbane. Available: http://www.coastal.crc.org.au/coast2coast2002 (December 2003).

Hundloe, T., editor. 2002. Valuing fisheries: an economic framework. University of Queensland Press, Brisbane, Queensland, Australia.

Jentoft, S., and B. McCay. 1995. User participation in fisheries management: lessons drawn international experiences. Marine Policy 19(3):227.

Kearney, R. E., and M. A. Kildea. 2001. The status of Murray cod in the Murray-Darling basin. Commonwealth Department of Environment & Heritage, Canberra, Australia.

LRC (Law Reform Commission). 1985. Report number 31: The recognition of aboriginal customary law. Australian Government Publishing Service, Canberra. Available: http://www.alrc.gov.au

McLeod, P., and J. Nicholls. 2002. A socio-economic valuation of resource allocation options between recreational and commercial fishing uses. Fisheries Research and Development Corporation, Perth, Western Australia.

Norris, R. H., P. Liston, N. Davies, F. Dyer, S. Linke, I. Prosser, and B. Young. 2001. Snapshot of the Murray-Darling basin river condition. Murray-Darling Basin Commission, Canberra, Australia.

NCCNSW (Nature Conservation Council New South Wales). 2001. Environment loses out in recreational fishing area decisions. Media release 31 August 2001. Available: http://www.nccnsw.org.au/water/news/media/20010831_lmbbrfa.html (December 2003).

Parliamentary Select Committee of the Parliament of South Australia on the Murray River. 2001. Parliament of South Australia, Final Report, Adelaide.

Pierce, B. E., and A. M. Doonan. 1999. A summary report on the status of selected species in the River Murray and Lakes and Coorong Fisheries. South Australian Research and Development Institute (SARDI), South Australian Fisheries Assessment Series 99/1, Adelaide.

Regier, H. A., and A. P. Grima. 1985. Fishery resource allocation: an exploratory essay. Canadian Journal of Fisheries and Aquatic Sciences 42:845.

Solomon, F. 2000. Zen and the art of stakeholder involvement. Australian Minerals & Energy Environment Foundation, Melbourne.

American Fisheries Society Symposium 49:63–78

Reconciling Fisheries and Allocation Using a Justice-Based Approach: Troll Fisheries Score Best

MELANIE D. POWER-ANTWEILER*

University of Victoria, Centre for Studies in Religion and Society and the Coasts Under Stress Project SED B102, Box 1700 STN CSC, Victoria, British Columbia V8W 2Y2, Canada

TONY J. PITCHER

University of British Columbia, Fisheries Centre, Aquatic Ecosystems Research Laboratory 2202 Main Mall, Vancouver, British Columbia V6T 1Z4, Canada

Abstract.—The British Columbia salmon fishery is exceptionally complex, with three mutually exclusive gear-based commercial fleets, Aboriginal, and sports sectors competing for access to five species of Pacific salmon. Additionally, British Columbia's salmon migrate through Canadian, American, and international waters. Conflict between sectors is evident, and illegal fisheries occur with increasing frequency. These conflicts raise ethical issues that are becoming widespread in many fisheries, such as the question as to how to share the resource—who is permitted to fish and who is not—as well as how, and by whom, such decisions are made. Extending our work published in Coward et al. (*Just Fish: Ethics and Canadian Marine Resources*, ISER 2000), we have assessed the ethical status of 24 commercial salmon fisheries in British Columbia, measured against eight ethical criteria representative of five types of justice: creative, distributive, ecosystem, productive, and restorative. The fisheries were first evaluated for overall ethical status and then reevaluated against each of the five types of justice. Justice-based policy recommendations for the commercial salmon fishery are offered. Our assessments show that the troll fishery is the most ethically sustainable fleet in the commercial sector and the purse seine the least. In recent allocation and fleet restructuring policies, the purse-seine fleet appears to have been favored, whereas our justice-based analysis suggests that the favoring the troll fleet is more ethical and sustainable.

Introduction

More than two decades after his report on Canada's Pacific coast fisheries, the "bleak and problematical picture" described by Pearse (1982) persists, and the "improved policy framework" he called for has yet to materialize (see also McRae and Pearse 2004). In fact, Canadian fisheries seem to have fallen into a state of perpetual crisis, and policies have tended to be reactionary rather than proactive, perhaps an example of "governance by coping" (Tansey 2003). Recent policy documents, such as the Canada Oceans Act (Parliament of Canada 1996) and Canada's Ocean Strategy (DFO 2002a; DFO 2002b) demonstrate first steps towards improved flexibility, proactiveness, precaution, and integrated management.

*Corresponding author: melanie.antweiler@shaw.ca

Following Coward et al. (2000) and Pitcher and Power (2000), we apply principles of justice to fisheries evaluation and policy-making. To this end, we define justice as the fair distribution of benefits and harms, following the five justice typology developed by Coward et al. We have conducted a semiquantitative justice-based evaluation of British Columbia's commercial salmon fishery, using a rapid appraisal method (Rapfish method: Pitcher and Preikshot 2001) modified for use with a set of justice-based attributes (Pitcher and Power 2000). Using eight justice-based ethical attributes, the ethical status of 24 salmon fisheries was evaluated. A comprehensive justice-based assessment is complemented by subevaluations based on the five justice typology; each uses a relevant subset of the eight attributes. From these evaluations, we offer policy recommendations for British Columbia's commercial salmon fisheries and compare and contrast these with actual policies governing these fisheries.[1]

Methods

Evaluating Modalities of Justice

Coward et al. (2000) sought to identify, make explicit, and assess ethical considerations relating to the fisheries ecosystem (Ommer 2000). The interdisciplinary group of scholars who contributed to Coward et al. identified five forms of justice as being central to fisheries issues:

1. Creative justice involves bringing together disparate voices, especially those that have typically been kept apart, and to share knowledge and understanding of ecosystems (Haggan 2000; Ommer 2000);

2. Distributive justice is concerned with sharing, of both the resource and access to it (Ommer 2000);

3. Ecosystem justice recognizes the intrinsic value of all members of an ecosystem (Ommer 2000);

4. Productive justice addresses issues of husbandry of the resource, such that it continues to produce at the desired levels of abundance (Brunk and Dunham 2000; Ommer 2000); and

5. Restorative justice recognizes the depletion of the ecosystems over time and is thus concerned with rebuilding the ecosystem to restore lost richness (Brunk and Dunham 2000; Ommer 2000).

Each of these forms of justice must be considered in evaluating the ethical status of a fishery and in developing ethical fisheries policies; they are strongly interlinked and best served when considered in conjunction with one another. Creative justice and ecosystem justice can be helpful in deciding issues of distributive justice and, without restorative justice and productive justice, there will be less of the resource for all, including perhaps what is necessary to ensure the continued functioning of the ecosystem itself.

The Rapfish Method Applied to Ethics

Rapfish permits rapid and robust evaluations of fisheries based on a coherent and interlinked set ("discipline" or "field") of discrete criteria ("attributes"). Rapfish serves a heuristic role, such that it provides a structure that prompts a consideration of how one thinks about fisheries (Power 2003). By systematizing assessments of discrete but inter-related facets of fisheries issues, Rapfish requires researchers to contemplate why those elements are of significance and encourages the examination of detail that otherwise might be overlooked. Rapfish is a framework for a holistic and cumulative appraisal of

[1] This paper is based on research conducted by M. D. Power (now Power-Antweiler) and presented in her doctoral thesis, *Fishing for Justice: An Ethical Framework for Fisheries Policies in Canada* (University of British Columbia 2003).

fisheries that subsequently provides a basis for policy.

Rapfish employs a multidimensional scaling (MDS) statistical ordination that is locked into a numerical framework that ranges from the worst to best possible scores. Multidimensional scaling is a nonparametric statistical technique that represents spatially (or "ordinates") the evaluated data (Stalans 1995). Multidimensional scaling assumes only an ordinal level of data (Stalans 1995). The Rapfish MDS ordinates fisheries along an axis ranging between defined extremes that constrain the assessment. Extremes are constructed by creating one dummy fishery that receives the worst possible score on each attribute (the "bad" fishery), and another that receives the best possible scores on each attribute (the "good" fishery). Raw results are translated into a percentage scale, so that the "bad" fishery scores 0% and the "good" fishery scores 100%, with the actual fisheries ordinating between these two extremes. Assessments are done in Microsoft Excel, using a purpose built automation routine (Kavanagh and Pitcher 2004) that expedites the process and reduces the likelihood of user error (Power 2003).

Four evaluation fields each with 8–12 scored attributes were initially developed for Rapfish: ecological, economic, social, and technological (Pitcher et al. 1998a; Preikshot et al. 1998). Attributes must be able to be clearly assigned "good" and "bad" extremes, be readily and objectively scored, and, with improved information, allow for scores that can be easily updated without disrupting the whole analysis (see Pitcher et al. 1998b, Alder et al. 2000; Pitcher and Power 2000). The original Rapfish fields and attributes covered sustainability, but several new evaluation fields have been developed (for example: Gregory 2000; Pitcher 1999, 2003; Power 2003).

In this work, we have employed a set of ethical attributes (Table 1), originally published in Pitcher and Power (2000) and further refined as eight attributes by Power (2003). We developed the attributes through consultation with scholars contributing to Coward et al. (2000), including iterative workshop discussions by the Just Fish team, with members of fishing communities in Newfoundland and British Columbia. The final set of attributes reflect concerns raised by fishers and community members, ethicists, and other researchers. As with the original Rapfish attribute fields, the ethical attributes explore a range of fishery considerations. The set of ethical attributes may be further subdivided, as we have done here.

British Columbia's Salmon Fisheries

The British Columbia salmon fishery is exceptionally complex. More than 9,600 stocks of five species of Pacific salmon (sockeye salm on *Oncorhynchus nerka*, pink salmon *O. gorbuscha*, chum salmon *O. keta*, Chinook salmon *O. tshawytscha*, and coho salmon *O. kisutch*) have been identified in the waters of Canada's westernmost province (Canada 2001). These five species are targeted by commercial, Aboriginal, and sport sectors. The commercial sector is further subdivided into hook-and-line (troll) and purse-seine and gill-net fleets. Chinook and coho salmon are customary targets of both the troll fleet and sports fishers, while the net fleets have traditionally preferred sockeye, pink, and chum (Gislason et al. 1996b; Pearse 1982). As perceived and actual shortages of salmon have increased and annual spawning runs have failed, these traditional preferences have shifted (Gislason et al. 1996b; Pearse 1982). Moreover, the recent salmon allocation policy of Fisheries and Oceans Canada (DFO) grants primacy of access to coho and Chinook to the recreational sector and sockeye, pink, and chum to the commercial sector, with the possibility of access to the other stocks "when abundance permits" (DFO 1999). Competition between and within sectors is further complicated by the lengthy

Table 1. Ethical attributes used in the Rapfish analysis, slightly modified from those published in Coward et al. 2000. Note that non-integer scores may be given. Current fields and attributes are available at http://www.fisheries.ubc.ca/projects/rapfish.php

Attribute	Definition and scoring scheme	Good	Bad
Adjacency and reliance	Geographical proximity and historical connection to the fishery: not adjacent/no reliance [0]; not adjacent/some reliance [1]; adjacent/some reliance [2]; adjacent/strong reliance [3]	3	0
Alternatives	Alternative (non-fishery) sources of employment within community? None [0]; some [1]; lots [2]	2	0
Equity in entry to fishery	Is entry based on traditional fishery activity? traditional access is not considered [0]; traditional access is considered [1]; the fishery is a traditional indigenous fishery [2]	2	0
Just management	Level of inclusion of fishers in management of fishery: none [0]; consultations [1]; co-management/government leading [2]; co-management/community leading [3]; genuine co-management where all parties are equal or N/A [4]	4	0
Mitigation of habitat destruction	Damage to fish habitat: much ongoing damage [0]; some ongoing damage [1]; no ongoing damage and no mitigation [2]; some mitigation [3]; much mitigation [4]	4	0
Mitigation of ecosystem depletion	Fisheries-induced changes in ecosystem: much ongoing damage [0]; some ongoing damage [1]; no ongoing damage and no mitigation [2]; some mitigation [3]; much mitigation [4]	4	0
Illegal fishing	Is there evidence of illegal activity (ex. catching, poaching, trans-shipments)? none [0]; some [1]; lots [2]	0	2
Discards and wastes	Is there discarding or wasting of fish? none [0]; some [1]; lots [2]	0	2

migrations undertaken by the anadromous Pacific salmon (Groot and Margolis 1991). Migratory routes cross borders between Canadian, American (both Alaska and the "lower 48"), and international waters, and salmon returning to spawn must run a gauntlet of fishing gears to reach their natal waters. Distinct fish stocks intermingle in the open ocean and river runs may coincide; weak or endangered stocks or species may be compromised when coincident with strong, targeted stocks or species (see DFO 2001). The 1990 Regina v. Sparrow decision (Supreme Court of Canada 1990) compels DFO to ensure primacy of fishing access to First Nations (DFO 2001). As Aboriginal fisheries are typically prosecuted in rivers upstream of commercial and recreational fisheries, this requirement compels conservative allocation to the other sectors and has resulted in legal challenges to DFO's allocation decisions and conflict between Aboriginal and non-Aboriginal fishers (see Thibault 2003).

A Rapfish Assessment of BC's Commercial Salmon Fisheries

We conducted six Rapfish assessments of British Columbia's commercial salmon fisheries: an overall evaluation using all eight ethical attributes and five subevaluations, each representing one of the five types of justice and including only those attributes indicative of the specific form of justice. These are described in Table 2.

Included in the analyses were a total of 24 commercial fisheries for Pacific salmon.[2] These fisheries are listed in Table 3, together with sources of data and the abbreviations used in the results (Table 4a–b). Eighteen of these fisheries consisted of each of the three gear types in combination with each of the five species of Pacific salmon, plus one aggregate fishery in recognition of the multispecies nature of the commercial salmon fishery; these data were supplemented with six additional gill-net fisheries from northern areas of coastal British Columbia[3]. In cases where multiple sources provided information for a fishery, the data rarely varied; where it did, the average value was used in this evaluation. Error values on the attribute scores, used in a Monte Carlo analysis of uncertainty (Kavanagh and Pitcher 2004), were applied to each attribute, to reflect the quality of data available for each; these are presented in Table 5. Leverage assessments (Pitcher and Preikshot 2001) were run to check the multivariate status and relative statistical strength of each attribute in each evaluation.

Six assessments were conducted: an overall evaluation of the 24 fisheries using all eight attributes and five "subevaluations," one for each type of justice[4].

---

[2] These 24 fisheries represent a subset of a 62-fishery data set; the evaluations here represent the relevant segments of the ethical assessments conducted on all 62 fisheries.

[3] This research was conducted under the aegis of the UBC Fisheries Centre's Back to the Future project in the Hecate Strait (Pitcher 2004; Pitcher et al. 2002), with support from the Coasts Under Stress project. The supplementary data set did not include trollers or seiners due to the strong participation of gillnetters (see: Power 2002) in the project's activities.

[4] Problematically, only three attributes (see Table 2) are indicative of creative justice, and therefore form the basis of this evaluation. As a result, the Rapfish ordination for creative justice was unreliable, as demonstrated by extremely high attribute leverages 2000, citing Pitcher, 1999). Thus, the original scores by each individual attribute for each fishery were scaled (0–100) so that in all cases 0 is Bad and 100 is Good. These scores were then averaged to produce a rough percentage-based score for each fishery, with a subset of attributes used to approximate the creative justice score. These scaled and averaged results have proven comparable to the Rapfish scores.and a stress score of 0.28. An important measure of the credibility of any MDS ordination is the stress score. While Stalans (1995) notes that a stress score of below 0.15 is particularly good, as a general guideline a stress score up to 0.25 is acceptable (Gregory 2000, citing Pitcher 1999). Thus, the original scores by each individual attribute for each fishery were scaled (0–100) so that in all cases 0 is bad and 100 is good. These scores were then averaged to produce a rough percentage-based score for each fishery, with a subset of attributes used to approximate the creative justice score. These scaled and averaged results have proven comparable to the Rapfish scores.

Table 2. Five types of justice, as represented in the Rapfish ethical attributes[a]

Ethical attributes	Creative justice	Distributive justice	Ecosystem justice	Productive justice	Restorative justice
Adjacency and reliance	✓[b]	✓	✓		
Alternatives		✓			✓
Equity in entry	✓	✓			
Just management	✓	✓	✓		
Mitigation of habitat destruction			✓	✓	✓
Mitigation of ecosystem depletion			✓	✓	✓
Illegal fishing		✓		✓	✓
Discards and waste		✓	✓	✓	✓

[a] Adapted from Pitcher and Power, 2000:Table 1.
[b] Indicates that the attribute represents this form of justice.

Table 3. Assessed fisheries and sources of fishery data

Fishery	Abbreviation	Time period	Sources of fishery data	
			Personal communications	Published sources
1 Multi-species Gill net	SalGil	Mid-1990s	J. Sutcliffe (United Fishermen and Allied Workers' Union [UFAWU]); C. Young (fisher)	Gislason et al. (1996a); BC Salmon Marketing Council Web site
2 Chinook gill net	ChinGil	Mid-1990s		
3 Chum gill net	CmGil	Mid-1990s		
4 Coho gill net	CohGil	Mid-1990s		
5 Pink gill net	PkGil	Mid-1990s		
6 Sockeye gill net	SokGil	Mid-1990s		
7 Inshore gill net (Northern BC)	InshGil	Early 2000s	Anonymous Respondent #16 (fisher)	
8 Gill net (Skeena River, #1)	SknaGil1	Early 2000s	W. Thompson Sr. (fisher)	
9 Gill net (Skeena River, #2)	SknaGil2	Early 2000s	F. Hawkshaw (fisher)	
10 Gill net (Skeena River, #3)	SknaGil3	Early 2000s	R. Warren (fisher)	
11 Gill net (Skeena and Nass rivers)	SknaNsGil	Early 2000s	Anonymous respondent #40 (fisher)	
12 Gill net (Hecate Strait north to Alaska)	HecAkaGil	Early 2000s	Anonymous respondent #12 (fisher)	

Table 3. continued

Fishery	Abbreviation	Time period	Sources of fishery data	
			Personal communications	Published sources
13 Multi-species Seine	SalSen	Mid-1990s	J. Sutcliffe (UFAWU); C. Young (fisher)	Gislason et al. (1996a); BC Salmon Marketing Council Website
14 Chinook Seine	ChiSen	Mid-1990s		
15 Chum Seine	CmSen	Mid-1990s		
16 Coho Seine	CohSen	Mid-1990s		
17 Pink Seine	PkSen	Mid-1990s		
18 Sockeye Seine	SokSen	Mid-1990s		
19 Multi-species Troll	SalTrol	Mid-1990s	C. McKee (fisher); J. Sutcliffe (UFAWU); C. Young (fisher)	Gislason et al. (1996a); BC Salmon Marketing Council Website
20 Chinook Troll	ChiTrol	Mid-1990s		
21 Chum Troll	CmTrol	Mid-1990s		
22 Coho Troll	CohTrol	Mid-1990s		
23 Pink Troll	PkTrol	Mid-1990s		
24 Sockeye Troll	SokTrol	Mid-1990s		

Table 4a. Results for six Rapfish evaluations of commercial salmon fisheries in British Columbia; 100% represents the best possible score. This table presents the results for overall ethical evaluation field, including ranges for the 95% confidence limits (abbreviations are described in Table 3).

Overall ordination		95% confidence limits[c]	
Relative rank[a]	**% score[b]**	**Lower**	**Upper**
SokTrol	58.4	48.9	65.5
SalTrol	51.7	46.6	59.2
ChiTrol	51.6	46.5	60.0
PkTrol	51.1	46.1	56.4
CohTrol	49.9	41.9	56.6
CmTrol	49.6	45.0	59.8
SknaGil2	49.4	42.9	59.0
CmGil	47.3	42.8	52.3
SokGil	45.1	39.7	51.8
SalGil	45.0	41.0	50.2
PkGil	44.3	39.8	52.8
ChinGil	43.3	38.9	48.1
CohGil	43.3	40.7	50.1
SknaGil1	41.9	31.8	51.7
HecAkaGil	39.2	32.9	47.3
CmSen	39.1	34.5	44.1
PkSen	39.1	33.5	44.7
SalSen	36.2	31.7	40.1
SokSen	35.3	31.1	40.8
ChiSen	32.5	25.0	37.6
CohSen	32.5	27.3	36.4
InshGil	27.7	22.6	31.4
SknaNsGil	25.7	21.2	29.5
SknaGil3	16.6	12.4	25.3

[a] The relative ranking of fisheries within this Rapfish evaluation.
[b] The percentage Rapfish score achieved by fisheries in this evaluation.
[c] Error distributions were assumed asymmetrically triangular between upper and lower bounds calculated for each score from the error structure in Table 5. For each overall Rapfish evaluation, upper and lower 95% tiles were calculated from an array of 50 Monte Carlo runs of the MDS.

Rapfish Results

The table of results is shown in Table 4a–b. Leverage analysis provided two statistical checks on the analysis. First, there were no cases where one or two individual attributes dominated the analyses. Second, individual attributes varied from 2% to 4.8% influence on the overall ordination scores; this range of values is much less than the 12–15%, which might be considered to bias a Rapfish analysis. Moreover MDS stress scores were always less than 0.25; higher values may invalidate the ordination technique (Gregory 2002, citing Pitcher 1999). Using the error structure reported in Table 5, 95% confidence limits on each Rapfish evaluation were assessed; these are given for the overall ethical evaluation are given in Table 4a. The combined results for all troll fisheries are significantly different from the combined scores for both gill-net and seine fisheries. Across all evaluations, the troll fisheries consistently produce the best sustainability results, although in the subevaluations

Table 4b. Results for six Rapfish evaluations of commercial salmon fisheries in British Columbia; 100% represents the best possible score. This table gives the results for the five ethical modalities described in the text. (Abbreviations are described in Table 3)

Creative justice[a]		Distributive justice		Ecosystem justice		Productive justice		Restorative justice	
Relative Rank	%	Relative Rank	%	Relative Rank	%	Relative Rank	%	Relative Rank	%
SknaNsGil	58.3	SokTrol	59.1	SokTrol	62.1	SokTrol	71.7	SknaGil2	65.7
SokTrol	54.2	ChiTrol	53.4	SknaGil2	53.2	SknaGil2	69.5	SokTrol	65.1
ChiTrol	48.6	CohTrol	53.4	ChiTrol	50.6	PkTrol	62.5	PkTrol	58.3
CohTrol	48.6	SalTrol	52.4	SalTrol	50.5	SalTrol	61.3	SalTrol	57.0
SalTrol	43.1	CmTrol	49.8	PkTrol	49.6	CmTrol	59.0	CmTrol	55.3
CmGil	38.9	PkTrol	49.8	CohTrol	48.4	ChiTrol	59.0	ChiTrol	55.3
SokGil	38.9	SokGil	47.6	CmTrol	47.8	CohTrol	56.7	CohTrol	52.7
CmTrol	31.9	CmGil	46.0	CmGil	46.6	PkGil	51.6	PkGil	50.1
PkTrol	31.9	SalGil	45.0	HecAkaGil	44.8	CmGil	51.6	CmGil	50.1
SalGil	30.0	ChinGil	44.2	SokGil	44.2	SalGil	49.6	SknaGil1	49.1
ChinGil	27.8	CohGil	44.2	SknaGil1	43.5	SokGil	49.6	SalGil	47.8
CohGil	27.8	SknaNsGil	42.1	SalGil	43.4	SknaGil1	48.9	PkSen	47.8
SknaGil3	27.8	SknaGil2	41.3	PkGil	42.0	HecAkaGil	48.7	CmSen	47.7
SknaGil1	22.2	SokSen	40.7	ChinGil	41.0	PkSen	48.3	SokGil	47.1
SknaGil2	22.2	PkGil	40.0	CohGil	40.9	CmSen	48.3	CohGil	45.8
InshGil	19.4	CmSen	38.6	CmSen	36.3	CohGil	47.7	ChinGil	45.8
HecAkaGil	19.4	PkSen	38.5	PkSen	36.3	ChinGil	47.7	HecAkaGil	45.4
PkGil	16.7	SalSen	38.0	SalSen	32.2	SokSen	45.4	SalSen	44.7
SalSen	9.7	SknaGil1	37.3	SokSen	31.6	SalSen	45.1	SokSen	44.2
ChiSen	9.7	ChiSen	34.9	InshGil	27.6	CohSen	40.7	CohSen	40.9
CmSen	9.7	CohSen	34.9	ChiSen	26.6	ChiSen	40.7	ChiSen	40.9
CohSen	9.7	HecAkaGil	31.4	CohSen	26.5	InshGil	37.5	InshGil	31.3
PkSen	9.7	InshGil	26.1	SknaNsGil	20.7	SknaNsGil	24.5	SknaNsGil	20.0
SokSen	9.7	SknaGil3	14.0	SknaGil3	18.0	SknaGil3	15.1	SknaGil3	11.8

[a] Creative justice values were obtained through scaling and averaging of Rapfish scores, rather than full Rapfish assessment.

Table 5. Error values applied to attributes in the Canadian fisheries ethical analyses.

Attribute	E_{min}	E_{max}	Rationale
Adjacency and reliance	15%	15%	Easily quantifiable; well documented; reliable sources
Alternatives	25%	20%	Well documented; reliable sources; but may overestimate capacity of alternative sectors to survive without fishery
Equity in entry to fishery	20%	20%	
Just management	30%	30%	Higher error limits to account for different perceptions of degree of comanagement
Mitigation of habitat destruction	20%	20%	
Mitigation of ecosystem depletion	20%	30%	May underestimate extent of damage to ecosystem due to uncertainty
Illegal fishing	10%	40%	Difficult to quantify with high degree of accuracy
Discards and wastes	25%	30%	Difficult to quantify with high degree of accuracy

Notes:
1. Assumed baseline error of 20%; adjusted upwards or downwards based on confidence in aggregate data.
2. Assumed symmetry of error; adjusted for asymmetry when data were thought to be either over- or under-estimated.
3. Rationale given when error differed from above assumptions.

for creative and restorative justice, traditional, or local ecological knowledge (T/LEK) salmon gill-net fisheries (SknaNsGil and SknaGil2, respectively) scored highest. In all instances but the creative justice evaluation, six of the top six or seven positions are occupied by troll fisheries; in the case of the creative justice assessment, the chum and sockeye gill nets receive better scores than the chum and pink trolls, while in the overall ethical ordination and the assessment for distributive justice, the top six fisheries are all troll fisheries. Through the evaluations, it becomes clear that those fisheries that produce extreme results tend to be either uniformly good or bad. In comparison, those with variable results are characterized by one or two extreme attribute scores and the relative ranking of such fisheries is therefore affected to a significant degree by the attributes applied within an evaluation.

Throughout the evaluations, the six northern gill-net fisheries (numbers 7 through 12 in Table 3) that supplemented the original data set demonstrate greater variability than the other commercial salmon fisheries. The original data set included coast-wide fisheries, and the data were broader and more vague than those in the supplementary data set. By comparison, the northern gillnetters who shared knowledge of their fisheries were able to provide more specific information, which better discriminated between the fisheries. Such subtle distinctions are nicely exploited by the MDS routines. For example, within the creative justice evaluation, the coefficient of variation ("CoV"; CoV = SD/mean as a percentage) for the supplemental (northern) gill nets was 54%, compared with 27% for the other six gill-net fisheries and 40% for the trolls.[5] The highest scoring fishery in this appraisal, with a sustainability score of 58.3%, was a northern gill-net fishery, conducted in the Skeena and Nass rivers ("SknaNsGil"). Yet, the remaining supplemental gill-net fisheries scored lower than all troll fisheries and all original gill-net fisheries save for the pink gill net, which, other than the seine fisheries, returned the worst score (16.7%). In all six evaluations, the six coast-wide gillnetters produce better sustainability results than the six northern gillnetters. For example, in the overall ethical ordination, the original gillnetters score an average of 44.5%, while the supplementary gillnetters averaged 33.5%. The six northern gill-net fisheries were evaluated solely by fishers active within the specific fishery; the nuanced data and the varied results reflect their detailed awareness of that fishery.

The troll fisheries are consistently good through all assessments. Any variability is between the seiners and the gillnetters, as well as between the coast-wide gillnetters and the northern gillnetters. Furthermore, in all evaluations, the original gillnetters produce better results, on average, than the seines. We consider that the patterns evident in these six evaluations can inform fisheries policy.

Discussion: Justice-Based Policies for British Columbia's Commercial Salmon Fisheries

How, then, could the commercial salmon fisheries be made more ethical? Based on our results, what would a just commercial salmon fishery look like?

First, consider the sector's fleet composition. According to the assessments presented herein, the troll sector has the highest justice score of the three. Yet fleet reduction efforts, notably the so-called Davis Plan (announced in 1968) and the Mifflin Plan (1996), had the effect of removing from the commercial salmon fishery a disproportionate number of trollers. Gillnetters, which ranked between trollers and seiners in the overall ethical evaluation, were also disproportionately removed from the fishery. Through the two licence retirement rounds of the Mifflin Plan, the troll fleet was reduced by 23%, the gill-net fleet by 18%, and the seine fleet by just 9%, despite a stated goal of removing 20% of licences from each sector (Gregory 2000). Furthermore, as Pearse (1982) notes with regard to the 1968 Davis Plan, while the size of the commercial salmon fleet was reduced, capacity actually increased through capital investment (Pearse 1982, citing Pearse and Wilen 1979).

A just policy for the commercial salmon fishery should address this imbalance between gear sectors. A proactive policy would have produced a more proportionate fleet reduction and addressed the concerns associated with the increasing relative dominance of

[5] All seine fisheries presented the same result, because of the limited number of attributes in the creative justice assessment.

seiners in the gear mix. For a more just fishery, the policy goal would involve shifting the mix, such that trollers would comprise a greater percentage of the fleet and seiners a smaller percentage. Under a program such as Mifflin, the number of licenses could still have been reduced and capacity increases discouraged, while explicitly encouraging the continuation of the more just components of the fishery.

Allocation is another ethical issue. Fisheries and Oceans Canada's Allocation Policy for Pacific Salmon provides a framework for determining distribution among the three commercial gear types, with an original catch target allocated 42% to the seine fleet, 34% to the gill nets, and 24% to the trollers (DFO 1999). This allocation subsequently shifted to give the gill-net fleet 4% more of the allocation, at the expense of both the seine and troll fleets; in 2001, the mix realized in the actual coast-wide catch was 41% seine, 50% gill net, and 9% troll (DFO 2002, 2002d). The catch allocations effectively marginalizes the more just troll gear within the commercial salmon fishery and are in contravention of DFO's assertion that "Over time, allocations by gear... may be adjusted to favour those that can demonstrate their ability to fish selectively" (DFO 1999). The policy framework for rectifying the imbalance exists; we suggest that it should be applied.

A just policy for the commercial Pacific salmon fisheries would cultivate creative justice. The establishment of a Consultations Secretariat (DFO 2002c, 2002d) and the Framework for Improved Decision-Making in the Pacific Salmon Fishery (DFO 2000) are encouraging. For creative justice to flourish, it is essential that different voices be heard in decision making. This may in part be accomplished at a community level. Communities that are dependant upon local small-scale fisheries have strong incentives to steward the resource and "economic, cultural, and social" conditions that may encourage community involvement (Glavin 1996). Shared interests are more apparent at a local scale, for instance: wishing to maintain a vibrant community; achieving conditions that will reduce out-migration; and attracting new prospects (for instance, tourism) while cultivating the existing opportunities (such as the fishery). Unfortunately, at the local level, conflicts may also be magnified and more deeply entrenched; however, the community may also be inclined to draw together against what may be perceived as a common enemy.

Also at the community level, it may be most appropriate to call upon T/LEK. This knowledge, typically built up over generations and detailed on a fine scale, complements scientific knowledge, which may be geographically broader and quantifiable; indeed, the differing types of knowledge are evident in the variation between the gill-net fisheries in our data set. A just policy then would seek to cultivate collaborative efforts between DFO (and other) scientists and local fishers. It is not enough for DFO scientists to simply collect T/LEK; such would represent a very low rung on Arnstein's (1969) ladder of participation. A dialogue is necessary. Fishers should be encouraged to share their knowledge, but the value of such an exchange, as well as the respect and trust it requires, must be made plain.

However, for such collaboration to occur—whether the parties involved include the community, DFO scientists or officials, academic researchers, or others—terms of reference for respectful interaction ought to be developed. For instance, academic researchers must be mindful of community sensitivity regarding what C. Menzies (personal communication) refers to as "hit and run research," when researchers descend upon the community, seize whatever data they require, and disappear forever without any follow-up. Instead, collaborative research requires a long-term commitment to cultivation of dialogue, as well as aiming for mutual sharing, rather than a one-way data "grab."

The ethical issues associated with fisheries are expansive, and fisheries complex. This study does not claim to be exhaustive of either. It does, however, represent a basis for a justice-based approach to fisheries policy making. Future work will permit additional development and testing of the approach and more detailed, concrete policy proposals for fisheries.

Acknowledgments

We thank all who participated in the Just Fish project, including all researchers and community members, as well as Canada's Social Science and Humanities Research Council for the doctoral fellowship that funded this research. We also thank two anonymous referees for their helpful comments.

References

Alder, J., T. J. Pitcher, D. Preikshot, K. Kaschner, and B. Ferriss. 2000. How good is good? A rapid appraisal technique for evaluation of the sustainability status of fisheries of the North Atlantic. Pages 136–182 *in* D. Pauly and T. J. Pitcher, editors. Methods for assessing the impact of fisheries on marine ecosystems of the North Atlantic. Fisheries Centre Research Report 8(2). Fisheries Centre, University of British Columbia, Vancouver. Available: http://www.seaaroundus.org/report/method/alder11.pdf. (October 2005).

Arnstein, S. R. 1969. A ladder of citizen participation. Journal of the American Institute of Planners 35(July):216–224.

BC Salmon Marketing Council. 2004. Welcome to BC Salmon.ca. BC Salmon Marketing Council, Vancouver. Available: http://www.bcsalmon.ca. (October 2005).

Brunk, C., and S. Dunham. 2000. Ecosystem justice in the Canadian fisheries. Pages 9–33 *in* H. Coward, R. Ommer, and T. Pitcher, editors. Just fish: ethics and Canadian marine fisheries, social and economics papers, volume 23. ISER Books, St. John's, Newfoundland and Labrador, Canada.

Coward, H., R. Ommer, and T. Pitcher, editors. 2000. Just fish: ethics and Canadian marine fisheries, Social and Economics Papers, volume 23. ISER Books, St. John's, Newfoundland and Labrador, Canada.

DFO (Fisheries and Oceans Canada). 1999. An allocation policy for Pacific salmon—a new direction: the fourth in a series of papers from Fisheries and Oceans Canada. Fisheries and Oceans Canada. Available: http://www-comm.pac.dfo-mpo.gc.ca/publications/allocation/AllocationPolicyoct201.htm. (October 2005)

DFO (Fisheries and Oceans Canada). 2000. A framework for improved decision-making in the Pacific salmon fishery—discussion paper. Fisheries and Oceans Canada. Available: http://www-comm.pac.dfo-mpo.gc.ca/publications/idm_e.pdf. (October 2005)

DFO (Fisheries and Oceans Canada). 2001. Fish stocks of the Pacific coast. Fisheries and Oceans Cananda, Pacific Region, Vancouver.

DFO (Fisheries and Oceans Canada).2002a. Canada's cceans strategy. Fisheries and Oceans Canada, Oceans Directorate, Fs23–116/2002E, Ottawa. Available: http://www.dfo-mpo.gc.ca/Library/264675.pdf. (October 2005).

DFO (Fisheries and Oceans Canada). 2002b. Canada's oceans strategy: policy and operational framework for integrated management of estuarine, coastal and marine environments in Canada. Fisheries and Oceans Canada, Oceans Directorate, Fs77–2-2002E, Ottawa. Available: http://www.dfo-mpo.gc.ca/Library/264678.pdf. (October 2005).

DFO (Fisheries and Oceans Canada). 2002c. Pacific region integrated fisheries management plan—salmon: northern BC, April 1, 2002 - March 31, 2003. Fisheries and Oceans Canada, Pacific Region, Vancouver.

DFO (Fisheries and Oceans Canada). 2002d. Pacific region integrated fisheries management plan—salmon: southern BC, April 1, 2002 - March 31, 2003. Fisheries and Oceans Canada. Available: http://www.pac.dfompo.gc.ca/ops/fm/mplans/plans02/SSalmon02pl.PDF.

Gislason, G., E. Lam, and M. Mohan. 1996a. Fishing for answers: coastal communities and the BC salmon fishery—final report. The ARA Consulting Group

Inc. for the BC Job Protection Commission, Vancouver.

Gislason, G., E. Lam, J. Paul, and E. Battle. 1996b. Economic value of salmon: Chinook and coho in British Columbia. The ARA Consulting Group Inc., Vancouver.

Glavin, T. 1996. Dead reckoning: confronting the crisis in Pacific fisheries. Greystone Books, Vancouver.

Gregory, M. C. 2000. Fleet reduction through licence removal: an application of auction theory. Doctoral Thesis. University of British Columbia, Vancouver. Available: http://www.nlc-bnc.ca/obj/s4/f2/dsk1/tape3/PQDD_0027/NQ50305.pdf. (October 2005).

Groot, C., and L. Margolis. 1991. Preface. Pages ix–xi *in* C. Groot and L. Margolis, editors. Pacific salmon life histories. UBC Press, Vancouver.

Haggan, N. 2000. "Back to the Future" and Creative Justice: Recalling and Restoring Forgotten Abundance in Canada's Marine Ecosystems. Pages 83–99 *in* H. Coward, R. Ommer, and T. Pitcher, editors. Just Fish: Ethics and Canadian Marine Fisheries, Social and Economics Papers, volume 23. ISER Books, St. John's.

Kavanagh, P., and T. J. Pitcher. 2004. Implementing Microsoft Excel software for Rapfish: s technique for the rapid appraisal of fisheries status. UBC Fisheries Centre, Vancouver. Available: http://www.fisheries.ubc.ca/publications/reports/12–2.pdf. (Novemeber 2005).

McRae, D. M., and P. H. Pearse. 2004. Treaties and transitions: towards a sustainable fisheries on Canada's Pacific coast. Fisheries and Oceans Canada, Vancouver. Available: http://www-comm.pac.dfo-mpo.gc.ca/publications/jtf/Prs-McRae-Report.pdf. (November 2005).

Ommer, R. 2000. Just fish: ethics and Canadian marine fisheries. ISER Books, St. John's and Newfoundland, Canada.

Parliament of Canada. Oceans Act. S.C. 1996, C-31.URL: http://www-comm.pac.dfo-mpo.gc.ca/publications/oceans/oact_e.pdf.

Pearse, P. H. 1982. Turning the tide: a new policy for Canada's Pacific fisheries. The Commission on Pacific Fisheries Policy, Fs 23–18/1982E, Vancouver.

Pitcher, T., S. Mackinson, M. Vasconcellos, L. Nøttestad, and D. Preikshot. 1998a. Rapid appraisal of the status of fisheries for small pelagics using multivariate, multidis-ciplinary ordination. Pages 759–782 *in* F. Funk, T. J. Quinn II, J. Heifetz, J. N. Ianelli, J. E. Powers, J. F. Schweigert, P. J. Sullivan, and C. I. Zhang, editors. Fishery Stock Assessment Models, Lowell Wakefield Symposium. Alaska Sea Grant College Program, Fairbanks.

Pitcher, T. J. 1999. Rapfish, a rapid appraisal technique for fisheries, and its application to the Code of Conduct for Responsible Fisheries. Food and Agriculture Organization of the United Nations, FAO Fisheries Circular No. 947, Rome. Available: http://ftp.fao.org/docrep/fao/005/x4175e/X4175E00.pdf. (November 2005).

Pitcher, T. J. 2003. The Compleat angler and the management of aquatic ecosystems. Pages 3–7 *in* A. P. M. Coleman, editor. Regional experiences for global solutions: Proceedings of the 3rd World Recreational Fisheries Conference, Northern Territories Fisheries Report, volume 67. Northern Territories, Darwin, Australia.

Pitcher, T. J., editor. 2004. Back to the future: advances in methodology for modelling and evaluating past ecosystems as future policy goals. Fisheries Centre Research Report, volume 12(1). University of British Columbia, Fisheries Centre, Vancouver. Available: http://www.fisheries.ubc.ca/publications/reports/report12_1.php. (October 2005).

Pitcher, T. J., A. Bundy, D. Preikshot, T. Hutton, and D. Pauly. 1998b. Measuring the unmeasurable: a multivariate and interdisciplinary method for rapid appraisal of the health of fisheries. Pages 31–54 *in* T. J. Pitcher, P. J. B. Hart, and D. Pauly, editors. Reinventing fisheries management. Kluwer Academic Publishers, Dordrecht, The Netherlands.

Pitcher, T. J., and M. D. Power. 2000. Fish figures: quantifying the ethical status of Canadian fisheries, east and west. Pages 225–53 *in* H. Coward, R. Ommer, and T. Pitcher, editors. Just fish: ethics and Canadian marine fisheries, social and economics papers, volume 23. ISER Books, St. John's, Newfoundland and Labrador, Canada.

Pitcher, T. J., M. D. Power, and L. Wood, editors. 2002. Restoring the past to salvage the future: report on a community participation workshop in Prince Rupert, BC. Fisheries Centre Research Reports 10(7). University of British Columbia, Fisheries Centre, Vancouver. Available: http://www.fisheries.ubc.ca/publications/reports/10–7.pdf .(November 2005).

Pitcher, T. J., and D. Preikshot. 2001. Rapfish: a rapid appraisal technique to evaluate the sustainability status of fisheries. Fisheries Research (49):255–270.

Power, M. D. 2002. The thoughtful use of words. Page 35 *in* T. J. Pitcher, M. Power, and L. Wood, editors. Restoring the past to salvage the future: report on a community participation workshop in Prince

Rupert, BC. Fisheries Centre Research Reports 10(7). University of British Columbia, Fisheries Centre, Vancouver. Available: http://www.fisheries.ubc.ca/publications/reports/10–7.pdf. (November 2005).

Power, M. D. 2003. Fishing for justice: an ethical framework for fisheries policies in Canada. Doctoral dissertation. University of British Columbia, Vancouver. Available: http://www.fisheries.ubc.ca/grad/abstracts/mpower_thesis.pdf. (November 2005).

Preikshot, D., E. Nsiku, T. Pitcher, and D. Pauly. 1998. An interdisciplinary evaluation of the status and health of African lake fisheries using a rapid appraisal technique. Journal of Fish Biology 53(Supplement A):381–393.

Regina v. Sparrow (1990) 1. S. C. R. 1075 (1990), 104 S. C. C. Available: http://www.canlii.org/ca/cas/scc/1990/1990scc49.html. (October 2005).

Stalans, L. J. 1995. Multidimensional scaling. Pages 137–168 *in* L. G. Grimm, and P. R. Yarnold, editors. Reading and understanding multivariate statistics. American Psychological Association, Washington, D.C.

Tansey, J. 2003. Did God speak in codons and bases? The prospects for governing biotechnology in Canada. W. Maurice Young Centre for Applied Ethics (Democracy, Ethics, and Genomics), Vancouver. Available: http://www.ethics.ubc.ca/workingpapers/deg/deg001.pdf. (October 2005).

Thibault, R. G. 2003. Fisheries and Oceans Canada: statement by the Honourable Robert G. Thibault Minister of Fisheries and Oceans Canada. Available: http://www2.ccnmatthews.com/scripts/ccn-release.pl?/2003/08/19/0819126n.html. (October 2005).

American Fisheries Society Symposium 49:79

Session Summary

The Human Dimension in Achieving the Reconciliation of Fisheries with Conservation

Evelyn Pinkerton
Simon Fraser University, School of Resource and Environmental Management British Columbia, Canada

The session began with an overview of the state of social science knowledge regarding the conditions under which conservation is achievable through community–government partnerships. Such partnerships are able to go farther than simply tightening regulations or setting aside large protected areas; they set up appropriate institutional mechanisms for successful comanagement. This discussion focused particularly on complex multiparty partnerships, considering issues of temporal and geographic scale (especially related to the fit of the scale of certain ecosystem functions with human communities). Forty-eight hypotheses about the conditions under which co-management operates best were grouped under five general categories: (1) the nature/characteristics of the resource, (2) the nature of the community/groups that depend on/have an interest in the resource, (3) the nature of the government agency mandated to regulate resource use, (4) the nature of the institution (i.e. the agreement among the partners, the rights, duties, and roles created by the agreement, as well as the norms, values, and patterns of behavior), and (5) the nature of the relationship of groups and governments to external forces such as markets, technology, the university, and political power-holders, such as corporations with resource rights.

The oral presentations that followed this overview were from Argentina, Denmark, Chile, the United States, and Canada, and focused on various aspects of the nature of the institution (territorial control as a management tool for artisanal fishermen in Argentina; the response of artisanal fishermen's organizations in Chile to territorial seabed rights), the nature of the co-managing group (adaptive strategies of fishermen as a component of management in Denmark, community assessment methods under the National Marine Fisheries Service [NMFS] in the United States), and the nature of the relationship to external forces (joint social science research with universities in Canada).

This session had by far the greatest number of applicants at the conference and thus offered 63 posters, covering the full gamut of categories.

American Fisheries Society Symposium 49:81–94

Institutional Sustainability of Aquaculture and Fishing—A Case of the Archipelago Sea Region, Finland

TIMO MÄKINEN
Finnish Game and Fisheries Research Institute
Post Office Box 6, FIN-00721 Helsinki, Finland

PEKKA SALMI*
Finnish Game and Fisheries Research Institute, Saimaa Fisheries Research and Aquaculture
Laasalantie 9, FIN-58175 Enonkoski, Finland

JUHANI SALMI
Finnish Game and Fisheries Research Institute, Reposaari Unit
Konttorikatu 1, FIN-28900 Pori, Finland

JUHANI KETTUNEN
Finnish Game and Fisheries Research Institute
Post Office Box 6, FIN-00721 Helsinki, Finland

Abstract.—This paper addresses the problems in achieving sustainable institutional solutions in the management of fish farming and archipelago fisheries. Institutional sustainability refers to the sets of management rules and the organizations that implement those rules. We study sustainability from environmental, socioeconomic and community perspectives. A special attention was paid to conflict between fish farming and environmental protection. Semistructured interviews with the stakeholders were used as method for data collection. The fish farmers and fishermen possess a strong local identity and feel that the local people support their livelihood. The governance system was considered distant and bureaucratic. The interviewed fish farmers and fishermen were suspicious over the support coming from the state administration and from the EU. They would not necessarily take a turn to self-management but would like to increase communication and authorities' awareness of the fishermen's situation and problems. Both the causes of problems and the paths for sustainable development in the area are often seen in contradictory ways among local fish farmers, leisure dwellers, and environmental officers. There are no clear-cut means to reach a balance between the socioeconomic, ecological, and community dimensions, but a way forward is to study the options for existing or new institutions to enhance collaboration for better regulation of the conflicts. The aim of spatial planning, when relocating the fish farming sites, should be to balance between different interests and mitigate conflicts in advance and thus create improved legitimacy of the regulation system and acceptability of the decisions. The present system has created a vicious circle, which not only hinders positive development in other dimensions of sustainability than the environmental, but also causes poorer results than it probably would be without the present regulation system. Three main recommendations given in the Finnish

*Corresponding author: pekka.salmi@rktl.fi

AQCESS synthesis for developing the institutional sustainability of fisheries in the study areas are 1) improved definition of goals for state governance; 2) development of cooperation between governance sectors; and 3) development of participatory planning.

Introduction

Sustainable governance of fisheries depends on the ability of institutions to balance between different dimensions of sustainability. The governance approach "... takes into account that the fishing industry forms an interactive socio-economic and ecological system embedded in institutions, social networks and cultures" (Kooiman et al. 1999). The complex, dynamic, and diverse character of fisheries is a challenge for fisheries governance, which encompasses the interactions between the state, the market, and civil society. New partnerships and other institutional designs are needed in order to mitigate conflicts and to find sustainable solutions in fisheries.

According to Charles (2001), ecological sustainability refers to the fundamental task of maintaining or enhancing the resilience and overall health of the ecosystem while socioeconomic sustainability focuses on maintaining the overall long-term socioeconomic welfare. In fishery context this means, for example, a focus on the generation of sustainable net benefits, a reasonable distribution of those benefits among the fishery participants and maintenance of the system's overall viability within local and global economies. The community dimension of sustainability emphasizes the "micro" level: maintaining or enhancing the group welfare of human communities in the fishery system by maintaining or enhancing its sociocultural and economic well-being, its overall cohesiveness, and the long-term health of the relevant human systems (Charles 2001).

Jentoft (2003) emphasizes that a broad concept of an institution is needed. It includes the social and cultural underpinnings of management systems and one that captures the social processes and governance mechanisms that are essential to fisheries management in the broadest sense. Institutional sustainability is a prerequisite for the simultaneous achievement of ecological, socioeconomic, and community sustainability. Institutions interact with and underlie the pursuit of the other three sustainability dimensions. Institutional sustainability involves maintaining suitable financial, administrative, and organizational capability over the long term and refers to the sets of management rules by which fishery is governed, and the organizations that implement those rules at the different levels of the management regime (Charles 2001).

This study focuses on the institutional sustainability of small-scale commercial fisheries and fish farming and especially the question of achieving a balance between differing interests and values in defining the management rules. The material results from an international project, which focused on aquaculture and coastal economic and social sustainability (AQCESS) and was funded by the EU. Three areas in the Archipelago Sea region, Southwest Finland, were chosen as the Finnish case study sites of the project (Figure 1). Each site covers three municipalities. The studied region is a prominent area for fish farming and commercial fishing in Finland.

The utilization of natural resources has remained important in the Archipelago Sea region, although the recreational and service sectors have increased. The density of summer houses is very high in the case study areas. There are slightly over 7,000 inhabitants

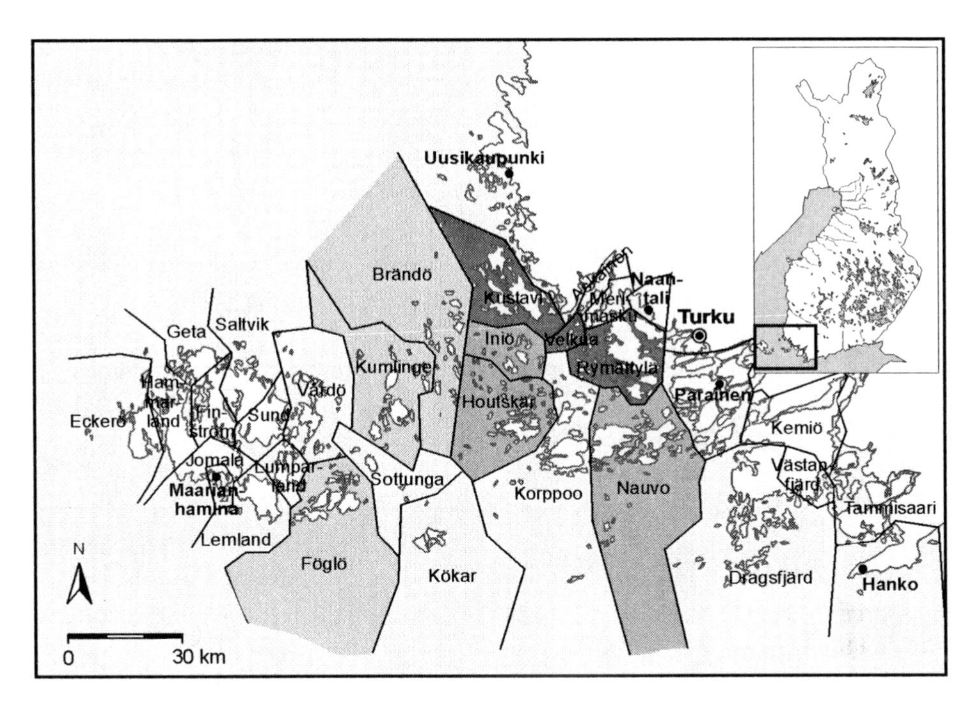

Figure 1. The three case study areas in the Archipelago Sea region in soutwestern Finland.

and a total of 9,000 summerhouses in the study area (Statistics Finland 2004). At the same time with the increasing recreational pressure, the increased emphasis in nature protection has produced resistance from the side of local people who stress their opportunities for utilization of local natural resources. In the study area private water ownership forms an institution, which affects the management of both fish farming and commercial fishing.

We study household economy, fish farmers, and fishermen's attitudes towards their work and the support from other stakeholder groups for them and to understand their situation. The major aim is to declare the preconditions for enhancing management practices and institutional sustainability. A special attention will be paid to a conflict between fish farming and environmental protection in the Finnish Archipelago Sea. The fish farming debate reflects the various aspects and problems in the attempts to reach a balance between the fish production and other interests, such as environmental production and recreation.

Fisheries, Fish Farming, and Institutions in the Archipelago Sea Region

Finnish commercial fishery has historically been characterized by maritime peasant economy, which combines agriculture with coastal fishing. Particularly in the Archipelago Sea region shipping and transportation have also formed an important economic sphere (Andersson and Eklund 1999). A clear majority of Finnish Baltic fishermen use only coastal fishing methods. Today only 10% of Finnish marine fishermen operate in the open sea by using trawls. However, the value of catch is an equal level both in the open sea and coastal Finnish fisheries. The open sea fisheries have high importance especially in the supply of Baltic herring *Clupea harengus membras* (a subspecies of Atlantic herring). The coastal fishermen harvest generally high priced zander (also known as pikeperch) *Stizostedion lucioperca*, Eurasian perch *Perca fluviatilis*, and powan *Coregonus lavaretus* (Finnish Fishery Time Series 2001).

In the Archipelago Sea region a decline of employment in the primary sector has first and foremost taken place in agriculture, but fishing employment has also decreased drastically. For example, the number of full-time fishermen in the Swedish speaking parts of the Archipelago Sea dropped from 2000 in 1934 to the current 100 in the 1970s. Correspondingly, the number of part time fishermen decreased from 1450 to 80 (Åbolands Fiskarförbund 1977). Although the importance of fish industries has decreased, it has remained an important employer in many municipalities of the archipelago. This is partly due to the expansion of fish farming in the 1980s.

The small-scale coastal fishing in Finland is traditionally operated by fishermen's households and using relatively small boats. The Finnish fisherman typically owns his fishing equipment and especially in the archipelago areas harvests often the near waters of their home island on a seasonal basis. This part-time nature of fishing reflects the seasonally varying availability of target fish species in the fishing grounds. In addition to the flexibility in fishing strategies, the sustainability of fisheries employment is often connected to the availability of additional income sources (Salmi et al. 1998; Salmi 2005).

Fisheries often compete with other uses of coastal areas and especially fish farming seems to be such an activity. Already at the end of the 1970s the growth of aquaculture brought about conflicts with other forms of water utilization in the Archipelago Sea region. Especially owners of summer cottages began to complain about harms caused for their beaches and for their fishing equipment. Since the mid 1980s, pollution caused by fish farms grew from a local issue to a national, even an international question of environmental politics. The main interests and values in collision include water conservation claims and recreational needs on the one side and fish farming livelihood and vitality of local community on the other. The disagreements have been fertilized by different knowledge especially concerning the fish farms' contribution on water eutrophication. The fish farming conflict reflects wider tensions concerning local control of natural resources versus interventions by the state authorities in the name of public interest.

The Finnish aquaculture is rearing of the rainbow trout *Oncorhynchus mykiss*. In the Archipelago Sea region only four out of 37 enterprises studied reared also whitefish on a small scale. The end product of rainbow trout rearing is a big fish weighing about one and half kilogram, often commercially called salmon trout (Saarni et al. 2003). The production amount as well as the demand on the domestic market was steadily growing till the beginning of 1990s. After that time the decline in domestic production has taken place while the amount of imported Atlantic salmon *Salmo salar* and rainbow trout has grown very rapidly (Setälä et al. 2003).

During one generation the Archipelago Sea region has transformed from a production landscape of fishing peasants towards one of leisure, recreation, and consumption. The recreational use of the area has various dimensions from dwelling in own summerhouses to boating and organized tourism. Along with this development the stakeholder groups have multiplied and especially the interests of summerhouse owners and other recreational users of the archipelago area have become more decisive.

Institutions in Fisheries Governance

The fisheries governance system is a combination of local decision-making by the water owners and a top-down management system by the state. Archipelago water areas, like most of the Finnish coastal and inland waters, are under private ownership in conjunction with possession of land. The decision maker is commonly a collective, a shareholders association, which jointly controls the interests of individual owners in fishery matters (Salmi and Muje 2001). In addition to these fishery associations, there are also a large number of waters managed solely by individual owners. The associations and individual owners are often unwilling to rent waters to fishermen outside their community. Consequently, commercial fishermen have had problems with acquiring fishing opportunities for the small and scattered privately owned water areas in the Archipelago Sea.

Fisheries Regions form geographically larger management units, often covering water areas of at least one municipality. These organizations offer a wider forum for decision making among water owners and fishermen. Although commercial fishermen are represented in the Fisheries Regions, these organizations seldom have a role in providing access to fishing waters for commercial fishermen. Waters outside the privately owned areas in the open sea are state-owned and state-governed.

In the governance of commercial fishing the role of the state has increased during the last 20 years. The authorities of state government are divided into two sublevels. The regional level of the fisheries administration is a part of the Employment and Economic Development Centres. These authorities operate under the auspices of the Ministry of Agriculture and Forestry, which handles official tasks within the fisheries in Finland. The coastal and archipelago fisheries are chiefly regulated according to national principles, the Fisheries Act. The property rights of the state are mostly used in the governance of open sea fishery. The funding of the commercial coastal fisheries is based on the guidelines by the EU. The Common Fisheries Policy by the EU has emphasized renewal and improving of efficient fishing units (e.g. trawl fisheries, which compete in the market with the small-scale fishermen). The financial support is coupled with the proportion of fishing incomes, which must cover at least 30% of the total personal incomes.

In the province of Åland the structure of private water ownership is similar to the other coastal areas in Finland. The Fisheries Office of the Provincial Government is the main manager of the fisheries and aquaculture activities. The Provincial Government has autonomy in most issues, but it co-operates also with the fisheries authorities in the Ministry of Agriculture and Forestry. However, the Fisheries Office (Fiskeribyrån) has the right to issue local fishing regulations through a provincial decree; (e.g. it grants licenses for commercial fishing and maintains the registers of the fishing monitoring system of CFP).

Governance Institutions of Fish Farming

The regulation of fish farming is in the hands of environmental authorities, which decide over the environmental permits for each fish farming site. The Ministry of the Environment is the highest environmental authority. The implementation of environmental regulations and policies is delegated to Regional Environment Centres. The permissions for aquaculture are granted by the Regional Environmental Permit Authorities (Varjopuro et al. 2000). The permit issued is a pollution and construction permit, not a licence to farm fish as an explicit production license known in some other industries. Fish farming activity requires also a permit granted by the owner of the water area. The fish farmer may be a water owner himself, but a majority of the farms operate on hired sites.

The structural funds of the EU are granted to the fish farming industry by the regional administration of the Ministry of Agriculture and Forestry. Spatial regional planning of the sea areas regarding fish farming sites is a task of environmental as well as fisheries regional authorities. In the first trials the environmental viewpoint has dominated meanwhile the co-operative spatial planning, which utilizes the engagement of the stakeholders, is still in a pilot study phase.

The effects of fish farming on the environment have been monitored since the 1970s. Nutrients and chlorophyll-content in water, benthos and periphyton sampling are most commonly used as indicators in the mandatory monitoring paid by the farmers. Monitoring is obligatory, ordered in fish farm permits and financed by fish farmers (Varjopuro et al. 2000). In addition to environmental monitoring a farmer is obliged to keep a diary of daily operations at the farm. The protocols are needed when authorities inspect the farms and when farmers prepare their annual self-monitoring report (Varjopuro et al. 2000).

In Åland the Provincial Government acts as environmental and fishery authority as well as permission authority (Ålands landskapsstyrelse 2001). The Environmental Licensing Board of Åland grants the fish farming licenses. The appeals concerning the licenses, however, are issued at the Provincial Government. Monitoring is basically similar as in rest of Finland. Fish farmers run monitoring programs, which are supervised by the Provincial Government. Permits in the Åland Islands have allowed greater average production than in other areas of Finland. For example, in 1998 the average annual production level permitted to a single fish farm was 108 metric tons (mt) in the Åland Islands and 46 mt in the Archipelago Sea area.

Methods

The study area, Archipelago Sea region, is composed of two main parts: Archipelago Sea (total surface area 8,300 km^2) closer to the mainland and the province of Åland (6,785 km^2) near the Swedish border. There are nearly 30,000 islands in the area, which makes the connections between the islands slow and complicated. The islands closest to the shoreline are connected by bridges to the mainland, but traveling to other islands requires using of ferry or boat connections. During winter the sea is ice-covered for 50–70 d in the outer archipelago and for over 100 d in the inner parts. The number of commercial fishermen in the Archipelago Sea region was 263 in 2002. Slightly over 100 of the fishermen earned at least 30% of their incomes from fishing. The 37 fish farm enterprises in study area employed over 200 persons.

The material of this study comprises of 1) personal interviews conducted with fish farmers and commercial fishermen and 2) key person interviews. Personal interviews were conducted with 113 fishermen and 56 fish farmers in 2002. The sampling frames were the registers of commercial fishermen and fish farming enterprises. One half of the fishermen and at least one owner of every fish farm were interviewed. This resulted in 40 interviews with fish farm owners (i.e. persons who had a stake in the enterprise they worked at) and 16 interviews with salary workers (without ownership in the firm). Three of the interviewed fishermen were salary workers and 110 self-employed. Interviews with fishermen were typically made at the fishermen's home. One half of the fish farmer interviews were made in their working place and one half at their home.

Most of the questions were harmonized with the other four partner countries of the AQCESS project to allow comparisons. The questionnaires collected background information about the interviewed person, basic quantitative data about fishing activities and economy of fishing in addition to its relative importance in the household. Attitudes towards work and the support from other stakeholder groups were inquired mostly using a structured questionnaire form or as a combination of structured and qualitative method.

In this study we categorize the interviewed fish farmers in three groups. The first group is called expanders (n = 6) and defined as the owner-workers of enterprises with a turnover over 2 million euros. The second group, family enterprises (n = 34), may posses more than one farming site, but they are mostly orientated in securing the livelihood of their families. Salary workers (n = 16) form the third category. In the AQCESS project also a case study focusing on Finnish summerhouse dwellers' perspectives on fish farming was done. For this purpose a total of 17 summerhouse dwellers were interviewed (Salmi et al. 2004).

The key person interviews (n = 16) were made during September 2002 and January 2003. The informants represented 10 different stakeholder groups in three main categories:

1) resource users and owners;

2) residents and political actors; and

3) authorities and regulators (Table 1).

The interviews were made using an interview guidance specified with open-ended questions. The answers were manually written down, but also a tape recorder was used for storing the answers. In addition to the preformed main questions open conversation and responses to the raised issues were encouraged. Transcriptions were made within 2 months after the interview, on the average.

Results

Perspectives of Work and Support from Stakeholders

The results show that a majority of fishermen (73%) have a life long commitment to their occupation and nearly all (94%) are reluctant to leave their home islands. The commitment to the occupation was often grounded by traditions and the way of life near to nature provided by fishing. Also lack of alternatives was put forward as a reason for continuing fishing for lifetime. Those who were not sure about their future as fishermen often referred to the uncertainty of the profession, which, for instance, created the need for additional income sources. Switching to fish farming was not any more seen as a feasible alternative. The rare willingness to move from the area even if fishing opportunities would not exist was motivated by general reluctance to move from birthplace, additional income sources in the household, family ties and possessed real property. The interviewees stressed the independence and variety of the life mode and had a strong conception of a membership in the archipelago community, where the nature, sea and fish are important elements.

Also fish farmers running their family enterprises had low willingness for occupational and geographical mobility, which reflected strong local identity, appreciation of the way of life and nature of the household economy. Reasons to stay in the area—even if there would be no more fish farming opportunities—were motivated, for example, by continuing agriculture activities and fish trade or by wife's job in the area or pension. They stressed the way of life; in spite of the low incomes from fish farming they would continue the industry. The expanders (who have made large investments in the business and possess large

Table 1. Stakeholder groups represented by conducted key person interviews.

Stakeholder groups
Resource users and owners
Commercial fishermen's association
Recreational fishermen's association
Fish farmers' association
Fishery advisory organization
Residents and political actors
Municipalities
Voluntary organizations
Authorities and regulators
Local Environmental Centre/Provincial environment administration in Åland
Environmental Permit Office
Regional/provincial fishery administration
Fisheries regions

number of fish farming sites) were most motivated to move to other regions. Salary workers in fish farms are more willing for occupational and geographical mobility, for example they have not made investments in fish farming. Yet many salary workers hold the opinion that their work is too hard and exhausting to continue for the whole life. Although one-half of the interviewed fish farmers had sometimes considered quitting their fish farming activities due to economic uncertainty and decreased profitability, they generally believed in the future of aquaculture.

Both fishing and fish farming are components of the changing pluriactivity, with a growing emphasis of wage work in the public service sector (Figure 2). Fish farming is, compared with the fishermen, of greater economic importance for the families involved, on the average more than 60% of the income. The less is the proportion of the fish farming income in the household the more is needed other "pluriactive" income sources, wage work being the most important. In fishing the traditional peasant strategy combines fishing incomes with those from agriculture, forestry and horticulture. Employment in a combination boat or a ferry or in some other occupation in the public sector is typical for the households using the wage work strategy. The service-oriented fishermen's households acquire part of their incomes straight from the tourist industry or from an own firm providing services for the leisure sector, for example, building summer cottages for the urban dwellers (Salmi 2005).

People in the local community are considered generally to support both fishing and fish farming activities (Figure 3). Using the local natural resources is an essential part in the identity and traditional fisherman-peasant way of life in the archipelago region. Although fish farming is a relatively new innovation it has been welcomed as a continuation of the local fish-based activities. Fishermen's and fish farmers' attitudes towards the recreational sector reveal the problems related to the environmental conflict in fish farming and conflict over access of fishermen to privately owned waters. The recreational use of the area is substantial and thus differing interests and values concerning the rules of resource use, and the groups who are allowed to define the rules, are crucial issues.

Both fish farmers and fishermen are suspicious over the aims and practices of the official branch of governance (Figure 3). They find

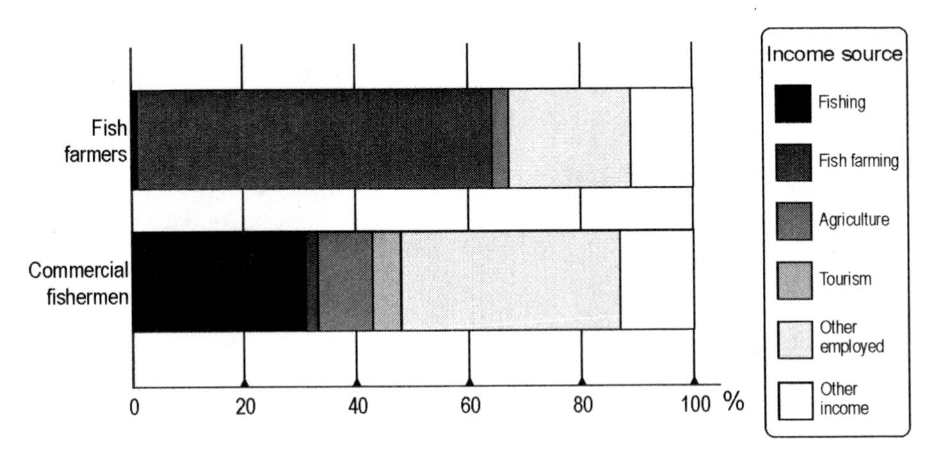

Figure 2. The distribution of fish farmers' and commercial fishermen's households' income sources.

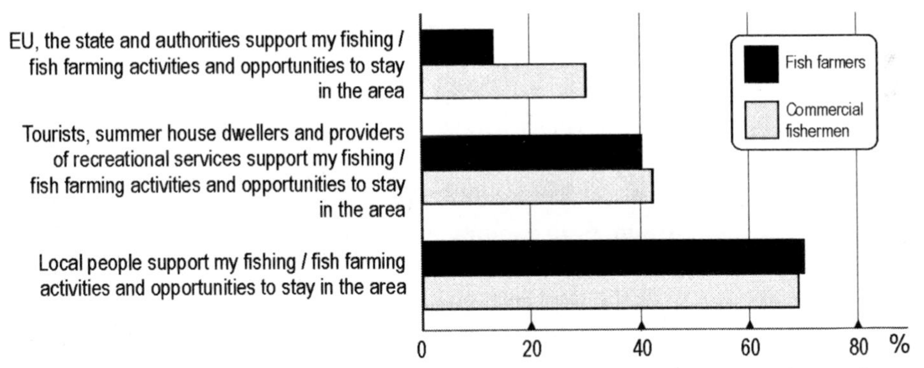

Figure 3. Fish farmers'and commercial fishermen's opininons concerning the support from other stakeholder groups. The proportions of agreement to an argument. The informants were asked to choose one alternative of the following: Agree, Neutral/ don't know, Don't agree.

the system bureaucratic and distant. The fish farmers claim that the officials care only about the environmental protection and do not listen to the producers. Small-scale archipelago fishermen feel that the authorities support only the big professional units and the exclusion of subsidies from part time fishermen is not reasonable. For instance, an interviewed fishery advisor would like to see a change in the rigid 30% limit due to the inherited multiplicity of the livelihoods. The fishermen claimed that authorities support only big industries and thus, due to exhaustive harvesting, fish will become extinct. They stated also that bureaucracy and paper work demanded from the fishermen has increased and authorities concentrate on making new laws, quotas and restrictions, which diminish the fishermen's freedom.

Fish Farming Conflict

The fish farming conflict in the Archipelago Sea region is here used as an example of institutional problems in coping with sustainability. Three larger forces lie behind the conflict, namely the eutrophication development of the Baltic Sea together with growing environmental ethos in the society, globalization of the trade and the expansion of the recreational sector (Figure 4). From the fish farmers' perspective the core problems are twofold. Firstly,

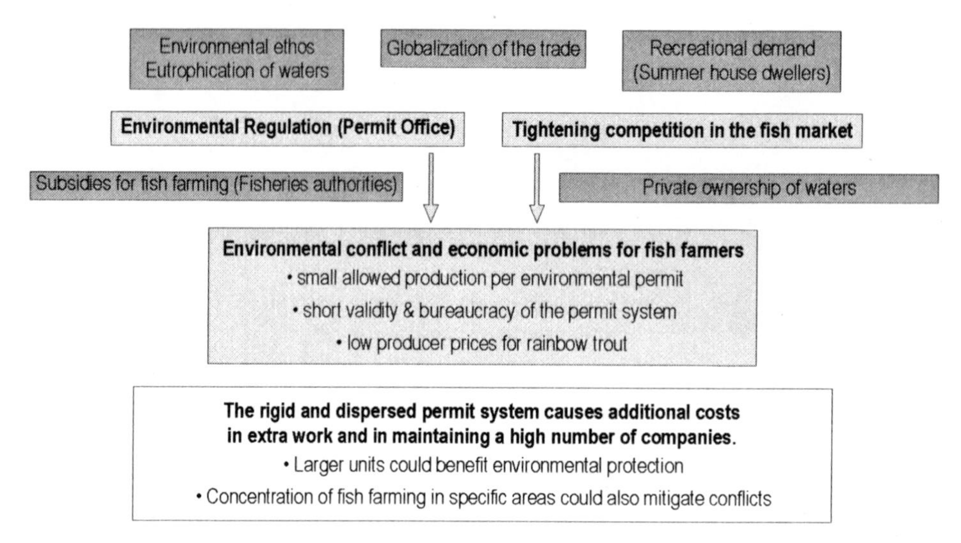

Figure 4. The main factors affecting and the consequences of the fish farmers' situation in Finland.

the environmental regulation concerning fish farming by the environmental authorities have become increasingly tighter. It is difficult for a fish farmer to renew his production permit not to talk about heightening the limit for allowed production. The site-specific permissions are valid for relatively short periods, which hampers the planning of the livelihood. Secondly, fish farming has had to cope with declining prices. Price drop is mainly a consequence of tightened competition in the Finnish fish market caused by increased import.

The development of fish farming was initially seen as a means to harness the natural strengths of the region. Economic subsidies granted by regional state authorities, in the late phase channeling also the EU subsidies, enhanced this expansion. The private ownership of the water areas affects the location of the fish farms. Majority of the farms operate on hired sites, but the availability of contracts depends on the relations to the local land (and thus water) owners. The sites regarded by permission authorities as most suitable for fish farming are often unavailable for this purpose.

In cases when fish farm effluents are supposed to have an impact on quality of a water body owned by a summer house owner he has a legal standing in permitting procedure. Summer house dwellers have used this right actively, although most of them have a positive attitude towards fish farming. Those who have claimed against fish farming have often acquired high education and social positions, which can be used against the livelihood. The summer house dwelling provides nature experiences and opportunities for recreation. Many of the part-time dwellers interpret the decreases in the water quality as consequences of fish farming activities. On the other hand, the summer cottage owners support the archipelago culture and local livelihoods. They typically hold positive attitudes towards family-based fish farming run by the local people. An interviewed representative of the fish farmers' association held a position that summer cottage owners have an idyllic perspective on the archipelago area in which fish farming does not fit. A fishery authority in the Åland islands stressed the economic importance of recreational use of archipelago areas and means to adapt fish farming activities with tourism and summer cottage dwelling.

In the interviews both the key persons and the fish farmers named the small allowed production, the long-lasting bureaucracy of the permit system and short validity of the permits as serious problems for fish farming livelihood (Figure 4). In the studied area the validity time of the permit has been shortened: in the beginning of 1990s a ten year permit was quite possible but after that time only shorter periods for the permits has been granted. This has happened not only against the will of the farmers but also against opinions and plans of some authorities. According to an interviewed person from the environmental administration, the Environmental Permit Office constricts the nutrient emission by about 30% at each decision round in order to maintain the state of the environment and the water quality. As a consequence, the average production decreases and the cost-per-unit increases, according to a key person from the fishery administration. The farmer may close down or adopt the expander strategy: buy other fish farming sites and thus increase the company size. As the third strategy the entrepreneur tries to strengthen the role of other income sources to compensate the loss.

In the province of Åland the specific loads from aquaculture are markedly lower and average farm sizes higher than in the other parts of the study area. The reason for the differences between the province of Åland and the Archipelago Sea area was interpreted in the key person interviews in a way that the employing of the permit system begun later in the Åland province than in the other parts of Finland. If this is the case, the environment is suffering more than it would in a situation with governance leading to fewer but bigger farms. However, the expanders usually keep to the existing farming sites and companies, because new permissions for new companies are hard to get. According to one interviewee the applicant wants to take no risk on the acceptance of the permit. The stiff permit system may enhance maintaining a needlessly high number of companies, which otherwise could be smelted together.

Discussion

Sustainable development is a broad and contradictory concept. The image of what is sustainable varies between groups of people. In this study we have highlighted the perspectives of various stakeholder groups concerning fish farming and commercial fishing in a Finnish archipelago area. Especially the importance of institutional arrangements was emphasized. For this purpose we studied fisheries in the Archipelago Sea region in general and especially a conflict between fish farming activities and environmental protection. Both the causes of problems and the paths for sustainable development in the area were often seen in contradictory ways between local fish farmers, leisure dwellers, and environmental officers. Analysis of the fish farming conflict highlighted various ecological, socioeconomic, and community aspects and outlined a scheme of the main components affecting the overall sustainability concerning the chosen case.

Separate evaluation of one sustainability dimension alone does not reveal the complexity and interactions connected to problems in the use and protection of natural resources (Charles 2001). There are no clear-cut means to reach a balance between the sustainability dimensions, but a way forward is to study the options for existing or new institutions to enhance collaboration for better regulation of the conflicts. The institutional development needs knowledge and understanding of the preconditions for specific types of fisheries and fish farms and options for their development, which calls for supporting research and development actions.

Community Support and Identity

One of the central tenets of sustainable development that emerged from the famous Brundtland Report was that communities should have greater access to and control over decisions affecting their resources, in cooperation with government, economic and administrative functions (Noble 2000). The fish farmers and fishermen mostly possess a strong local identity and feel that the local people support their livelihood. The governance system is considered distant and bureaucratic. The strong identity connected to the archipelago area seems to be a reason for the observed low willingness for geographical and occupational mobility of the islanders. Firmness of the social fabric, strong identity, and the archipelago way of life seem to support keeping people and livelihoods in the Archipelago Sea region.

The strength of fisheries in the study area is connected more with the local support, skills, and services than with the volume of production. The flexibility of fish farming and fishing is connected to small-scale family enterprises and a new variety of strategies for household income combinations. The image of the livelihood is important for marketing and commercial archipelago fishing has inherited a commonly positive one. In order to reach sustainability, environmental, and fishery administration, together with the producers and their organizations, should contribute to change the prevailing image of fish farming as an environmentally detrimental activity.

Planning and Collaboration for Institutional Development

The private water ownership is a special feature of the decision-making regime in the Finnish archipelago. Changes in the ownership structure and society have limited the access of commercial fishermen and fish farmers to suitable fishing waters. On the other hand, people in the study area are accustomed in managing the water areas locally and thus it is not surprising that the increased nonlocal involvement in governance has aroused strong resistance. This development is parallel with rapid rise of environmental ethos and recreational interests in the late modern society. The growing scale and expertise in planning and decision making are grounded by the new interests held especially by urban people. The interviewed fish farmers and fishermen were suspicious over the support coming from the state administration and from the EU. They would not necessarily take a turn to self-management but increase communication and authorities' awareness of the fishermen's situation and problems.

The spatial planning of fish farming sites should take into account all relevant features, especially the suitability of the area for economic activity, environmental aspects and other (recreational and commercial) use of the area. The aim of this spatial planning is to balance between different interests and mitigate conflicts in advance and thus create improved legitimacy of the regulation system and acceptability of the decisions. On the other hand, for the expander type of fish farming companies, the creation of big but remote fish farming parks could form basis for a solution to increase the profitability of the bulk fish production and to diminish the conflicts around their activities. However, the consequence of this arrangement should not be the exclusion of small family based enterprises from other areas, which is important for sustaining the socioeconomic sustainability of the area and guaranteeing the continuation of local support for the livelihood.

Although the Finnish Archipelago Sea region is not capable of even competition in production volumes with for example the Norwegian fish farming industry, environmental regulations should provide means to take the economic and social sustainability into account, in parallel with aiming at limiting the environmental impacts. The present system has created a vicious circle which not only hinders a positive way of development in other dimensions of sustainability than environmental but also causes a poorer result than it probably would be without the present regulation system. This paradox situation is illustrated in Figure 4. The solution for the livelihood might be sought in a way to benefit of global market but in a way, which is based on the local identity and way of life. The institutions should change and coordinate their activities to this end.

Recommendations

Three main recommendations given in the Finnish AQCESS synthesis for developing the institutional sustainability of fisheries in the study areas are:

- Improved definition of goals for state governance;

- Development of cooperation between governance sectors; and

- Participatory planning.

Transparent definitions of goals are needed to give national and regional administration the legitimacy and tools to act. The emphasis in governance should be changed from managing and supporting the professionals in one sector towards cooperation in enhancing the sustainability of local multi-dimensional livelihoods.

The use of social networks and knowledge of the stakeholder groups and their organizations should be enhanced by increased participation of these groups. Participatory spatial planning is needed to strike a balance between different interests and mitigate conflicts in advance and thus create an improved legitimacy of the regulation system in both fishing and fish farming. While the preconditions of the local livelihoods are increasingly determined from outside the local level, sustainable employment in fisheries needs means for understanding the contextual situation and requirements in the regional, national and international level of the decision making regime. In a favorable situation the livelihoods are regulated and new opportunities created through understanding the diversity and dynamics of the system.

In fish farming the most serious indications of the legitimacy problem of the permit system are the observed faults in consent compliance (Mäkinen 1998). The aim of the participatory planning system would be to recreate the legitimacy of the control system among the fish farmers. In fishing there is a need for improved planning and discussion with water owners to ensure commercial fishermen's access to the scattered privately owned water areas.

Acknowledgment

This study is part of the EU funded AQCESS, (Aquaculture and Coastal, Economic and Social Sustainability http://www.abdn.ac.uk/aqcess/index.html), QLRT-1999–31151.

References

Åbolands Fiskarförbund. 1977. Femtio år I Åboland. (Fifty years in Åboland). Åbolands Fiskarförbund 1926–1976. Ab Sydvästkusten, Åbo. Fishermen Association of Åboland, Åboland. (In Swedish).

Ålands landskapsstyrelse. 2001. Available: http://www.aland.fi/virtual/eng/frame.html. (September 2006).

Andersson, K., and E. Eklund. 1999. Tradition and innovation in coastal Finland: the transformation of the Archipelago Sea region. Sociologia Ruralis 39:377–393.

Charles, A. 2001. Sustainable fishery systems. Fish and Aquatic Resources Series 5, Blackwell Scientific Publications, Oxford, UK.

Finnish Fishery Time Series. 2001. SVT Agriculture, forestry and fishery 2001:60. Finnish Game and Fisheries Research Institute, Helsinki.

Jentoft, S. 2003. Institutions in fisheries: what they are and what they do, and how they change. Marine Policy 28(2):137–149.

Kooiman, J., M. van Vliet, and S. Jentoft. 1999. Creating opportunities for action. Pages 259–272 *in* J. Kooiman, M. van Vliet, S. Jentoft, editors. Creative governance. Opportunities for fisheries in Europe. Ashgate, Aldershot, England.

Mäkinen, T., editor. 1998. Kalankasvatuksen ympäristökuormitustavoitteet ja oikeudellinen ohjaus Saaristomerellä ja Ahvenanmaalla (Environmental goals and juridic control of fish farming in the Archipelago Sea and Åland). Suomen ympäristökeskuksen moniste 133. Finnish Environment Institute, Helsinki.

Noble, B. 2000. Institutional criteria for co-management. Marine Policy 24:69–77.

Saarni, K., J. Setälä, A. Honkanen, and J. Virtanen. 2003. An overview of salmon trout aquaculture in Finland. Aquaculture Economics and Management 7(5/6):335–343.

Salmi, J., J. Nordquist, and T. Mäkinen. 2004. Kalankasvatus saaristoelinkeinona. Kriittisesti kalankasvatukseen suhtautuneiden kesäasukkaiaden näkemyksiä elinkeinosta (Fish farminga as a coastal livelihood – opinions of the summer house dwellers critical on the livelihood, in Finnish with an English summary). Kala ja riistaraportteja 317.

Salmi P. 2005. Rural pluriactivity as a coping strategy in small-scale fisheries. Sociologia Ruralis 45(1/2): 22–36.

Salmi, P., J. Salmi, and P. Moilanen. 1998. Strategies and flexibility in Finnish commercial fisheries. Boreal Environment Research 3(4):347–359.

Salmi, P., and K. Muje. 2001. Local owner-based management of Finnish lake fisheries: social dimensions and power relations. Fisheries Management and Ecology 8:435–442.

Setälä, J., P. Mickwitz, J. Virtanen, A. Honkanen, and K. Saarni. 2003. The effect of trade liberation on the salmon market in Finland. Proceedings of the eleventh biennial conference of the International Institute of Fisheries Economics and Trade (IIFET). Bruce Shallard and Associates, Wellington, New Zealand.

Statistics Finland 2004. StatFin –Online service. Available: http://statfin.stat.fi/statweb/. (September 2006).

Varjopuro, R., E. Sahivirta, T. Mäkinen, and H. Helminen. 2000. Regulation and monitoring of marine aquaculture in Finland. Journal of Applied Ichtyology 16:148–156.

American Fisheries Society Symposium 49:95–104

A Pilot Program for Local Development of State Species Recovery Plans in California: The Shasta-Scott River Coho Salmon Recovery Team

Lisa C. Thompson*

University of California, Davis, Wildlife, Fish, and Conservation Biology Department 1 Shields Avenue, Davis, California 95616, USA

Abstract.—The Shasta-Scott River Coho Salmon Recovery Team (SSRT) is an innovative pilot program to develop a local coho salmon *Oncorhynchus kisutch* recovery plan for the Shasta and Scott River watersheds, part of the Klamath River basin in northern California. The SSRT plan is part of the state-level plan for recovery of coho salmon. A state-wide Coho Salmon Recovery Team developed plans for other watersheds and for the nonagricultural areas of the Shasta and Scott River watersheds. Residents of the Shasta and Scott River watersheds requested permission to develop a local recovery plan, on the basis of the large agricultural land use sector in the watersheds (in contrast to forestry in most other coho salmon-bearing watersheds), and because of previous landowner efforts to improve fish habitat. The multi-stakeholder SSRT includes representatives of local landowners, the environmental community, sport fishers, and agencies, plus a science advisor from the University of California Cooperative Extension. Facilitated meetings were held from January to December 2003, during which the SSRT developed recovery action recommendations for agricultural water use, habitat management and restoration, fish passage, monitoring and assessment, and education. A unique feature of the process, and critical to continued landowner participation, is the concurrent development of new state processes for programmatic permitting of normal ranching and agricultural activities, including streambed alteration agreements and state incidental take permits. The SSRT approach relies mainly on voluntary, incentive-based actions, publicly funded through grants. The SSRT recommendations were part of the state coho salmon recovery strategy that was approved by the California Fish and Game Commission (Commission) in February 2004.

Introduction

Coho salmon *Oncorhynchus kisutch* in California are being considered for listing under the California Endangered Species Act. On 30 August 2002, the Commission found that coho salmon warranted listing, but delayed regulatory action and gave the California Department of Fish and Game (DFG) 12 months to prepare a coho salmon recovery strategy. There are numerous factors that have contributed to the decline of coho salmon in the Shasta and Scott River watersheds (CDFG 2003a, 2003b, 2004). Key factors influencing both watersheds include declines in water flow and quality, loss of riparian habitat, passage barriers, and unscreened water diversions (Table 1).

Pilot Program

The anticipated state listing of coho salmon and the development of a state coho salmon recovery strategy were motivating factors in the formation of the Shasta-Scott-River Coho Salmon Recovery Team (SSRT). Stakeholders in the Shasta and Scott River watersheds, located within the Klamath River Basin, requested permission from DFG to develop a local recovery plan for the agricultural areas of the watersheds, due to the large agricultural land use sector in the watersheds (in contrast to forestry in most other coho salmon-

*Corresponding author: lcthompson@ucdavis.edu

Table 1. Problems facing coho salmon in the Shasta and Scott River watersheds (adapted from CDFG 2003a, Table 7; CDFG 2003b, section 8; CDFG 2004, sections 6.1.5 and 6.1.6).

Limiting factor	River-specific issues	
	Shasta River	Scott River
Water flow	-Reduced summer flows due to climate, diversion, development -Loss of channel maintenance flows -Groundwater use depleting surface flows -Fish access limitations	-Reduced stream flows resulting from drought and diversions -Limitation of rearing areas -Limitation of access to spawning areas -Disconnect between tributaries and mainstream inhibits movement of rearing fish -Stranding of juveniles
Water quality	-High water temperatures -Low levels of dissolved oxygen -Elevated nutrient levels -Turbidity	-High summer water temperatures
Habitat	-Limitation on spawning gravel quantity -Loss of spawning gravel quality -Loss of riparian habitat (trees) -Microhabitat limitations—lack of depth, suitable substrate, cover, holding habitat, etc. -Minor barriers to passage -Major barriers to passage (Dwinnell and Greenhorn dams)	-Poor spawning gravel quality significantly reduces egg to fry survival -Sedimentation of rearing pools -Reduction of riparian zones, lack of riparian cover in some tributary reaches -Lack of sufficient summering habitat in tributaries -Lack of instream structure for rearing
Protection	-Unscreened diversions -Legal and illegal harvest and predation	-Unscreened water diversions
Other	-Lack of funding for planning and studies necessary to precede restoration or fill data gaps -Lack of on-the-ground access for studies -Urban impacts -Dangerously low population numbers for recovery of sustained population	

bearing watersheds), and because of previous landowner efforts to improve fish habitat. A 21 member range-wide Coho Salmon Recovery Team (CRT) developed plans for all other watersheds that have or historically contained coho salmon, and for nonagricultural areas of the Shasta and Scott River watersheds. The anticipated outcome of having a separate recovery team for the agricultural areas of the Shasta and Scott River watersheds was that SSRT recommendations would receive greater local "buy in" and more rapid implementation, due to the greater local involvement in their development. However, there was also the risk that there would be a disconnect between SSRT recommendations and those developed for implementation range-wide and for nonagricultural areas of the Shasta and Scott River watersheds.

Pilot Program Process

Program Criteria

The SSRT focused on developing a strategy that would meet the following criteria

- Conserve, protect, restore and enhance coho salmon populations;

- Be scientifically, technologically, and economically feasible;

- Be supported by the best available data;

- Balance public and private funding;

- Balance regulatory and incentive-based approaches;

- Maintain a healthy agricultural industry; and

- Address water management in the two watersheds.

The first five criteria come directly from the California Endangered Species Act (California Fish and Game Code, Division 3, Chapter 1.5, Article 7, Section 2111). The criteria regarding the maintenance of a healthy agricultural industry and addressing water management were unique goals of the SSRT.

Team Members

The 13 member SSRT included representatives of local landowners, resource conservation districts (RCDs), environmental groups, sport fishers, and agencies, plus a science advisor (the author) from the University of California Cooperative Extension (UCCE) (Figure 1). A DFG representative headed the team, and was the liaison between the SSRT and DFG, and between the SSRT and CRT. The science advisor attended and actively participated in all SSRT meetings, but did not vote on recommendations. The Yurok Tribe requested to have membership on the SSRT, but this request was declined on the grounds that tribal interests were already represented on the CRT, and because the Yurok Tribe was not a stakeholder group located within the agricultural areas of the Shasta or Scott River watersheds.

Facilitated Team Meetings

More than 20 d of facilitated meetings were held from January to December, 2003, during which the SSRT developed recovery action recommendations. Two professional facilitators worked together to manage the SSRT process, leaving SSRT members free to focus on the content of recommendations. Facilitators assisted group members in seeking compromise, despite frequent fundamental differences of opinion, and sometimes worked behind the scenes to draw out concerns of different stakeholders so that they could be addressed by the SSRT. Professional note-takers recorded the meetings, and distributed minutes for review, providing an unbiased record of the SSRT's deliberations. Facilitation and note-taking was funded by DFG.

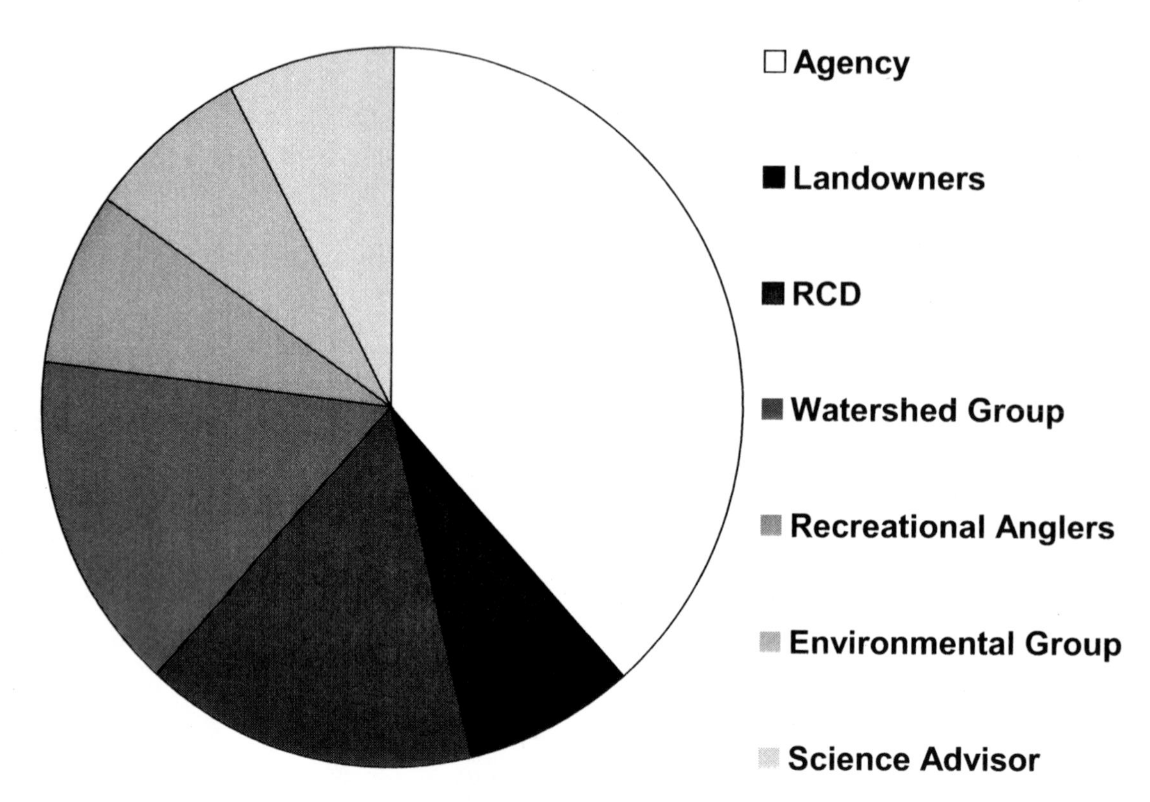

Figure 1. Affiliation of members of the Shasta-Scott River Coho Salmon Recovery Team.

Role of University of California Cooperative Extension in SSRT Process

The University of California Cooperative Extension is viewed favorably by the ranching community due to local presence of UCCE farm advisors who have worked in the community for over 20 years. The science advisor (the author) attended the February 2003 meeting of the SSRT on the advice of the UCCE farm advisor and was asked to join the team in March 2003. The science advisor is a fisheries extension specialist from the University of California at Davis, and was starting a coho salmon and steelhead *O. mykiss* field research program on the Shasta River at the time of her appointment to the SSRT.

As science advisor, I was asked to review the available literature and data reports available for the two watersheds and present a summary of data and data gaps to the SSRT. While a wide range of data had been collected over the past century, and numerous recovery efforts had been made (e.g., screening of water diversions), very little of this could be directly related to coho salmon populations, since fish were rarely surveyed, and even more rarely were other studies or recovery actions done in conjunction with fish studies. Given the level of uncertainty regarding the potential response of coho salmon to proposed recovery actions, I also presented the rudiments of adaptive management experiments (Walters 1986; Lee 1993) to the SSRT, as a prelude to embarking on the recovery strategy.

Public Involvement

During the development of the recovery plan the SSRT's meetings were open to the public and the media, and public comments were taken two times each day. The SSRT also held two "town hall" style public meetings at which the SSRT presented its proposals, and received and answered questions from the public. Each public meeting also included informal time for community members to interact with SSRT members. The highly interactive public meetings were viewed by the SSRT as being essential to the development of local acceptance of the recovery strategy recommendations.

Relationship Between SSRT and CRT

During the development of the SSRT and CRT's recommendations there was often tension between the two groups. Some SSRT members were concerned that the CRT would have the authority to change SSRT recommendations after the SSRT recovery strategy was submitted to DFG. Numerous communications between the SSRT and DFG assured the SSRT that its recommendations would be independent of those of the CRT. While this process was time-consuming for some SSRT members, notably the DFG members, it was essential to ensure the continued participation of SSRT members, who did not want their time and efforts to be wasted.

Components and Features of Recovery Strategy

Incentive-Based Approach

The SSRT approach relied mainly on voluntary, incentive-based actions, publicly funded through grants, as opposed to a more regulatory approach. The CRT also used a voluntary, incentive-based approach. However, this does not preclude the responsibility of DFG to take regulatory action where warranted by the Fish and Game Code. In the overall recovery strategy (CDFG 2004) it is noted that the strategy "recognizes the need for...maintaining a balance between regulatory and voluntary efforts." The intent of the SSRT was to develop recommendations that were unanimously supported by all stakeholders in the group, in the hope that regulatory approaches would be unnecessary. To this end, recommendations were painstakingly "word-smithed" by the group.

Implementation of the recovery strategy is broadly estimated to cost US$5 billion over the next 20 years and is subject to availability of adequate funding and staffing (CDFG 2004). This value includes the estimated $466 million for actions specific to the Shasta and Scott River watersheds, including water acquisition. Some of these funds are expected to come from existing programs such as the DFG Fisheries Restoration Grants Program, Farm Bill Grants, and the NOAA Community-based Restoration Program. However, the recovery strategy did not specify how much additional funding will be needed to accomplish all the strategy actions.

SSRT Coho Salmon Recovery Recommendations

The SSRT produced 148 recommendations, in seven categories, designed to recover coho salmon populations in the two watersheds (Table 2). The categories corresponded to the limiting factors and issues identified in Table 1, such as water flow and quality, habitat degradation, and protection, and also addressed needs for monitoring and assessment and education and outreach. The recommendations were presented to DFG as a "stand alone" strategy report (SSCSRT 2003) and were also incorporated into the Recovery Strategy for California Coho Salmon (CDFG 2004). The SSRT's recommendations had all received a status of preliminary favorable regard (PFR), short of final approval. Preliminary favorable regard is a status adopted by the SSRT to indicate that a recommendation was acceptable to all SSRT members in principal, but members reserved the right to reconsider approval in the context of the final report and once the programmatic permitting issues were resolved. The SSRT hoped to revisit all its recommendations prior to submission in order to give them final approval, but time was insufficient. Recommendations involving water management tended to be most controversial. In the case of issues like groundwater use and instream flow requirements of coho salmon, the SSRT opted to endorse studies, rather than immediate restoration actions. Some stakeholders favored a morato-

Table 2. Shasta-Scott River Coho Salmon Recovery Team recommendations.

Recommendation category	Number of recommendations	Examples	Relative controversy (1 = high, 3 = low)
Water management	26	Verification of water use, irrigation rotation, instream flow study, groundwater study	1
Water augmentation	17	Small offstream water storage, conjunctive groundwater use program	2
Water use efficiency	17	Irrigation system efficiency, tailwater reclamation	3
Habitat management	40	Remove barriers to fish passage, decrease summer water temperature	2
Protection	7	Remove passage barriers, screen water diversions	2
Monitoring and assessment	19	Habitat assessment, coho population assessment	2
Education and outreach	22	Demonstration projects on public lands, brochures, newsletter	3

rium on groundwater well drilling, and the establishment of minimum instream flows for fish. However, the consensus of the SSRT was that such recommendations would meet with enough disfavor with landowners that they would refuse to participate in the recovery strategy.

Not surprisingly, the SSRT produced many more recommendations for its two watersheds than the average produced by the CRT for a particular watershed. SSRT recommendations also tended to be more detailed than recommendations of the CRT, reflecting the depth of local experience and knowledge on the SSRT. Potentially there may be more "buy in" by local stakeholders on SSRT recommendations, since recommendations were shaped with the input and feedback of SSRT members representing landowners and riparian water rights holders. However, in other watersheds dealt with by the CRT for which there was considerable input and feedback, there may also be considerable local buy-in.

Implementation Phase

Programmatic Incidental Take Permits

A critical feature for continued landowner participation is the concurrent development of new state processes for programmatic permitting of normal ranching and agricultural activities, including streambed alteration agreements and state incidental take permits (ITPs). In order for an ITP to be approved, permit holders must agree to minimize and fully mitigate the impacts of their activities that may cause harm to coho salmon. In other words, the permit holder must take actions to avoid "take" or must take actions that will compensate for take that does occur. The permit holder must be able to show financial capacity to do the mitigation, and must show how many fish the actions will conserve, to be counted against the "take" due to agricultural activities. The take estimate being used in the Scott and Shasta River watersheds is the number of young-of-year coho salmon leaving the rivers, estimated from catches in rotary screw traps operated by DFG. It is assumed that the normal life cycle of a coho salmon involves rearing in the stream until the fish is 1 year old. For example, in Waddell Creek, California, Shapovalov and Taft (1954) observed that of 18,362 coho salmon migrating to the sea, only 106 were young-of-year coho salmon, while the rest were age-1. Thus, any fish that leaves the Scott and Shasta River watersheds as a young-of-year is assumed to be leaving early due to poor habitat conditions or insufficient habitat and is expected to perish prematurely. Other more indirect measures could have been chosen to estimate take, such as the number of fish below an expected minimum number of outmigrating age-1 fish based on adequate spawning and rearing habitat. However, the number of young-of-year outmigrants was chosen because it provided a directly observable estimate of fish that were expected to be killed. According to Section 86 of the California Fish and Game Code, "take" means hunt, pursue, catch, capture, or kill, or attempt to hunt, pursue, catch, capture, or kill." This definition does not include "harm" as is the case in federal regulations, and which would more readily allow indirect measures of take. The use of young-of-the-year outmigrants to estimate take requires further validation for the Shasta and Scott Rivers, since outmigrating young-of-year coho salmon may rear in other tributaries, or in the Klamath River estuary prior to entering the ocean at age-1. The fate of these fish is the subject of pending investigations.

There are several motivating factors for development of a programmatic ITP. Department of Fish and Game, the state agency responsible for the programmatic ITP, is short-staffed and cannot adequately deal with the multitude of individual ITP applications that would arise in an area such as the Scott or Shasta River watershed. Landowners have the option to apply for an individual permit, but this route is likely to be more costly and time-consuming than joining a group that is applying for a programmatic permit. In the Shasta River watershed, it appears that the local RCD may be the holder of the ITP.

Current Status of Coho Salmon Recovery Process

The Recovery Strategy for California Coho Salmon (CDFG 2004), which incorporated the SSRT's local recommendations, was presented to and approved by the Commission on 4 February 2004. On the same day, the Commission decided to publish notice of the Commission's intent to amend section 670.5, Title 14, California Code of Regulations (CCR), to add coho salmon to the state list of threatened and endangered species. The notice of proposed regulatory action relative to Section 670.5, Title 14, CCR appeared in the California Regulatory Notice Register on 12 March 2004. In light of the anticipated coho salmon listing, some members of the SSRT are continuing to develop an application for an ITP to cover agricultural activities in the Shasta River watershed. However, this process is meeting with considerable resistance in the Scott River watershed. The SSRT reconvened for two meetings in 2005 and, on 26 October 2005, voted to give its final approval to the recommendations in the Recovery Strategy for California Coho Salmon for the agricultural areas of the Scott and Shasta River watersheds, pending revisions to be submitted to DFG for the annual review of recommendations, and approval of these modifications by the Commission, as stipulated in the recovery strategy (see CDFG 2004, section 12.2).

Risk and Common-Pool Resources

A fundamental dilemma in the Scott and Shasta River watersheds is that the restoration of coho salmon, a common-pool resource (Dolšak and Ostrom 2003), is in potential conflict with riparian water rights, a private resource. If studies indicate that coho salmon require more instream flows, then flow requirements may exceed the water available after riparian water rights holders have taken their allotments. Legal water rights may conflict with the needs of a species listed under the California Endangered Species Act, the federal Endangered Species Act, or both. Thus, private property holders may stand to lose water in order to restore a common-pool resource. The SSRT recommendations include provisions for the purchase of water from willing sellers for instream use in coho salmon restoration. Arrangements such as this may avert the potential conflict between "water for fish" versus "water for farms."

Opportunities to Improve Recovery Strategy Implementation

Use of Science in Recovery Efforts

The SSRT's recovery recommendations are, by definition, focused at the subwatershed scale. Assessment of the effectiveness of local coho salmon restoration efforts will require scientists to separate the impacts of local habitat conditions from impacts that affect coho salmon when they are in the mainstem Klamath River, estuary, and ocean. It is possible that poor ocean conditions could negate the positive impacts of watershed-scale habitat improvements (including instream water quality and quantity) on coho salmon populations, while more productive ocean conditions could mask a lack of improvement in freshwater habitat. Counts of outmigrants and returning spawners for both the Shasta and Scott River watersheds are currently conducted by DFG, allowing the calculation of spawner–smolt relationships for the two watersheds, and these efforts need to be supported in future. Without this verification of the relative contribution of watersheds (versus mainstem, estuary, and ocean habitat) to coho salmon survival and production it will be difficult to justify the costs of local restoration actions, and to maintain stakeholder support for the recovery effort. Monitoring of the numerous local recovery actions will clarify the impact of these actions on coho salmon populations and guide scientists, managers, and stakeholders in assessing the efficacy of the actions and adapting them as experience is gained. In addition, restoration actions are likely to gain increased support if monitoring data are subject to independent peer interpretation, including sharing of data sets. While this happens informally at present, it may be beneficial to hold an annual Shasta-Scott coho salmon recovery/watershed restoration workshop at which coho salmon status data and restoration projects are presented.

Public and Stakeholder Involvement

There has been ongoing public and stakeholder input to the development of the recovery strategy, within the SSRT and CRT processes, and through public review of the recovery strategy. The SSRT recommendations will be more likely to succeed if there continues to be meaningful participation of the public and stakeholders in recovery efforts. Improvement of science education in the Scott and Shasta River watershed communities will increase public capacity to understand monitoring results, and to contribute to the development of future watershed management policies. Efforts by agency staff and independent scientists to communicate results to the public in lay terms will also aid this process. Stakeholders and the public could be involved in activities such as adaptive management workshops, spawner counts, and watershed monitoring. Resources will be needed to train volunteers, and to pay coordinators who would liaise with volunteers, and encourage continued participation. The public and stakeholders will probably be interested in continuing to participate in setting research and restoration priorities, potentially through continued meetings of the SSRT.

Potential Benefits of Public and Stakeholder Involvement

Involving the public and stakeholders in the recovery process may complicate the job of agency staff, but may also result in benefits such as increased stakeholder knowledge about coho salmon. Participants will also broaden their understanding of science, uncertainty, and management policies. This understanding may in turn lead to increased support of restoration efforts. While the various governmental agencies maintain their regulatory authority over coho salmon issues, the implementation and success of the recovery strategy would be enhanced by broad stakeholder support, particularly in areas such as the Shasta and Scott River watersheds where a high proportion of riparian lands are in private ownership.

Conclusions

The SSRT process is unique, and may represent the future in recovery planning. However, it is too soon to determine whether the local approach used in the SSRT process will result in improved implementation of the coho salmon recovery in the Shasta and Scott River watersheds, and whether a voluntary, incentive-based approach will provide beneficial results for coho salmon, in terms of habitat restoration, appropriate water flows, and increased coho salmon population. The Shasta and Scott River watersheds constitute a fairly small proportion of coho salmon habitat in the state, comprising two out of some 39 hydrologic units and subareas subject to recovery planning in the California portion of the Southern Oregon/Northern California Coast Evolutionarily Significant Unit. Nevertheless, the outcome of coho salmon recovery efforts in these two watersheds may have broad ramifications for how future recovery planning and implementation is undertaken throughout the coho salmon range in California and elsewhere. Should recovery efforts be seen as successful, the SSRT may be the first of many local recovery teams. However, if outcomes are seen as inferior to or no better than those of other watersheds, the chances of support and funding for the establishment of other local recovery teams would be diminished. In summary, the key elements in the SSRT process thus far have been the presence of a potential listed species, local stakeholder membership on the team, professionally facilitated meetings, the addition to the team of an independent science representative, regular public meetings to keep the local community informed of the process, and the use of an incentive based approach for recovery actions.

Acknowledgments

I thank the members of the SSRT for inviting me to participate in their coho salmon recovery planning process. Due to severe state budget cuts affecting University of California Cooperative Extension, the California Department of Fish and Game provided funding for my travel expenses related to SSRT meetings in 2003 and 2004. Carolyn Penny, one of the two professional facilitators for the SSRT, improved the manuscript with many thoughtful comments. I thank two anonymous reviewers for their knowledgeable and thorough critiques.

References

California Department of Fish and Game. 2003a. Klamath River hydrologic unit: Scott River hydrologic area. Public Review Draft. California Department of Fish and Game, Sacramento.

California Department of Fish and Game. 2003b. Klamath River hydrologic unit: Shasta Valley hydrologic area–subarea. Public Review Draft. California Department of Fish and Game, Sacramento.

California Department of Fish and Game. 2004. Recovery strategy for California coho salmon. Report to the California Fish and Game Commission, Species Recovery Strategy 2004–1, Sacramento.

Dolšak, N., and E. Ostrom. 2003. The challenges of the commons. Pages 3–34 *in* N. Dolšak and E. Ostrom, editors. The commons in the new millennium: challenges and adaptations. The MIT Press, Cambridge, Massachusetts.

Lee, K. N. 1993. Compass and gyroscope: integrating science and politics for the environment. Island Press, Washington, D.C.

Shapovalov, L., and A. C. Taft. 1954. The life histories of the steelhead rainbow trout (*Salmo gairdneri gairdneri*) and silver salmon (*Oncorhynchus kisutch*): with special reference to Waddell Creek, California, and recommendations regarding their management. State of California Department of Fish and Game, Fish Bulletin No. 98, Sacramento.

Shasta-Scott Coho Salmon Recovery Team. 2003. Shasta and Scott River Pilot Program for Coho Salmon Recovery: with recommendations relating to agriculture and agricultural water use. Prepared for the California Department of Fish & Game, July 28, 2003, Redding.

Walters, C. J. 1986. Adaptive management of renewable resources. Macmillan Publishing Company, New York.

American Fisheries Society Symposium 49:105–122

Social Barriers to Sustainable Recreational Fisheries Management Under Quasicommon Property Fishing Rights Regime

Robert Arlinghaus*,[1]

Leibniz-Institute of Freshwater Ecology and Inland Fisheries, Department of Biology and Ecology of Fishes Müggleseedamm 310, Post Office Box 850119, Berlin 12561, Germany

Abstract.—When debated publicly, the challenge to reconcile fishery resource use and resource conservation is often only discussed with regard to marine commercial fisheries. Recreational fisheries are often not considered despite their high social and economic importance across the industrialized world. This paper addresses key obstacles to the reconciliation of fishery resource use and resource conservation with respect to freshwater recreational fisheries management under quasicommon property regimes of Central Europe. It is argued that it is crucial to understand the human dimension of fisheries. Nine human dimensions related obstacles operating either at the societal level or at the level of individuals (or groups of stakeholders) were identified and are briefly summarized. These include: (1) lack of social priority; (2) lack of integrated approaches; (3) lack of cooperative institutional linkages; (4) lack of systems thinking; (5) lack of research and monitoring; (6) lack of shared values and dominance of stereotyped perceptions; (7) lack of consideration of the demand side of fish-angler interactions; (8) lack of self-criticism among angler lobbyists; and (9) lack of self-criticism among individual anglers. Potential solutions to overcome the identified obstacles are presented. It is concluded that although an understanding of the biological and ecological dimensions of recreational fisheries remains at the heart of any effort towards sustainability, the success of recreational fisheries management crucially depends on the human dimension. If some of the elements presented in this review are considered, the prospect for reconciliation of resource use and resource conservation in traditional recreational fisheries management will be enhanced.

Introduction

Society must seek ways to govern fisheries systems without causing irreversible losses of biodiversity, biomass and aquatic ecosystem resilience (Folke et al. 2002). To achieve this goal, fisheries researchers and managers need to develop strategies to manage seemingly irrational fisher behavior (Roslin 2002; Sullivan 2003) and maintain the economic, social and ecological benefits healthy fisheries provide (Arlinghaus et al. 2002). All this must be achieved in the face of uncertainty caused by global climate change, human population trends, competing habitat and water demands, and the expressed desire for a future world of diverse aquatic ecosystems (Ludwig et al. 1993; Folke 2003). The golden rule of natural resource management in general should be to retain critical types and ranges of natural variation in ecosystems (i.e., man-

*Corresponding author: arlinghaus@igb-berlin.de

[1] Second affiliation: Humboldt University, Institute of Animal Sciences, Faculty of Agriculture and Horticulture, Invalidenstrasse 42, Berlin 10115, Germany.

agement should facilitate existing processes and variation rather than altering or trying to control them) (Holling and Meffe 1996; Carpenter and Gunderson 2001) or rebuild entire aquatic ecosystems and food webs if they have been degraded by human activities (Pitcher 2001). This task should be pursued by integrating most, if not all, stakeholders and using all types of available information (ecological, economic, political, and sociocultural) in decision making to achieve goals and objectives established for fish resources (Costanza et al. 1998; Krueger and Decker 1999; Arlinghaus et al. 2002).

When debated publicly, the challenge to reconcile fishery resource use and resource conservation is often only discussed with regard to marine commercial fisheries. Recreational fisheries, which in most industrialized societies and particularly in freshwater ecosystems of the temperate regions have long represented the major use of aquatic wildlife (Welcomme 2001; Arlinghaus et al. 2002) are often not considered. For example, recent review papers dealing with the sustainability of world's fisheries (Pauly et al. 2002), the state of the world's fisheries resources (Hilborn et al. 2003) or the future for fisheries (Pauly et al. 2003) largely omitted any commentary on recreational fisheries and almost did not cite any relevant work on noncommercial fishing. This is partly the result of the limited research efforts on recreational fisheries in freshwater as compared to marine commercial fisheries (Arlinghaus et al. 2002). On the other hand, the great socioeconomic and ecological dimension of recreational fisheries (see Arlinghaus et al. 2002; Arlinghaus 2004a, 2004b, for references and discussion) may often remain unnoticed because marine commercial fisheries are highly visible in the media, while recreational fishing is a regionally dispersed activity with millions of people exploiting hundreds or thousands of different fish stocks (Cox 2000; Post et al. 2002). However, fishing activity of any kind, whether commercial or recreational, has the potential to negatively affect fisheries resources and entire aquatic ecosystems (McPhee et al. 2002; Cooke and Cowx 2006). It is expected that freshwater recreational fisheries can collapse much in the manner that commercially exploited species such as Atlantic cod *Gadus morhua* already has worldwide, at least in areas with high angling effort and mortality (e.g., near densely populated centers) (Schindler 2001; McPhee et al. 2002; Post et al. 2002). As a consequence, the normative concept of sustainable development equally applies to both commercial and recreational fisheries (Arlinghaus et al. 2002). Recreational fisheries should increasingly be discussed and studied in plane with commercial marine fisheries in an effort to be put on a path to reconcile fishery resource use with conservation. Particularly in densely-populated countries of Central Europe, there is an urgent need for concerted efforts to prevent and reduce ecosystem degradation—as well as conservation of freshwater fish and fisheries as renewable common pool resources or entities in their own right—to move towards sustainability (Arlinghaus et al. 2002). Against this background, recreational fisheries and its traditional management must recognize that conservation of the resource is one of the essential domains of natural resource management (Leopold 1933). There is no need for a "new" player in the arena of applied ecosystem management called conservationists. Instead, what are needed are practitioners of traditional resource management to become conservation biologists (Aplet et al. 1992).

This paper will argue that the key to reconciliation of recreational fisheries with conservation values is to understand the human dimension of fisheries because wildlife management today is "90% people, and 10% natural resource management" (e.g., Decker et al.

2001). Thus, this article from the human dimension perspective discusses key obstacles to reconcile recreational fisheries management (RFM) with conservation. This discussion is intended to increase awareness of the topic among (mostly biologically trained) fisheries researchers and managers, and should help to develop strategies to overcome the identified constraints. Hence, this paper focuses not only the main obstacles to reconcile recreational fisheries with fishery resource conservation, but also on potential solutions (see Table 1 for an overview). Emphasis is placed on the private property system of fishing rights in Central Europe where the duty to manage fishery resources lies in hand of private fishing rights holders and not within public agencies as for example in North America (Nielsen 1999; Arlinghaus et al. 2002). This distinction is needed as different property-rights regimes have a substantial influence on the probability of overexploitation of common-pool-resources such as fisheries (Hardin 1968; Feeny et al. 1990).

Property of Recreational Fishing Rights

In Central Europe, ownership of freshwater fishery resources is usually dependent on ownership of land adjacent to the water body. There are also traditional fishing rights that have survived since the Middle Ages, where feudal kings and land barons declared themselves owners of the land (and hence the fishery resources) (Wolter et al. 2003). Today, fishing rights can be purchased or leased from fishing rights (i.e., land) owners. In the latter case, the person or the group (e.g., an angling club or an angling organization) that leases the fishing rights for a certain time period also have the duty to manage the fishery resources without causing harm to the entire ecosystem. This duty is formulated in fisheries legislation and nature conservation laws and bylaws. Legislation is enforced by public agencies either at the national or federal-state level. Thus, the Central European recreational fisheries systems can be characterized as joint community-federal state cooperative management regimes (cf. Pinkerton 1994). That is, government and federal states at the public RFM level set the larger institutional framework (e.g., fisheries laws). Fishing rights holders (e.g., an angling club) or more generally angling communities at the private RFM level then implement and plan local management. They can also enforce, supplement and complement state-wide regulations (e.g., by local angling club rules). Therefore, fishing rights in Central Europe are private entities. However, in recreational fisheries the fishery resources are typically jointly used by all anglers belonging to a club or purchasing a license by paying a fee to the fishing rights holder (e.g., a commercial fisher). Thus, recreational fishery resources usually exhibit quasi-common properties within the realm of private fishing rights (cf. Feeny et al. 1990). For example, within the community of an angling club, even if outsiders are sometimes excluded (e.g., other anglers willing to enter a private angling club), access rights to the fishery resources for the purpose of angling are equal to all club members. Mostly, however, clubs are not restricted by a certain amount of memberships. Hence, access rights are equal to all anglers willing to purchase a license.

Within the quasicommon property rights system of inland fishing in Central Europe, in recreational fisheries there are three types of common-pool-resources. Such resources are characterized by the difficult or very costly possibility to exclude users and a rivalry in consumption (Berkes et al. 1989; Ostrom 1990). The latter means that joint use involves sustractability of the resource (i.e. one unit consumed by one user is not available, at least temporarily to another). Common-pool-resources in recreational fisheries include fish

Table 1. Overview of human dimensions obstacles limiting the reconciliation of traditional recreational fisheries management and resource conservation under quasi-common-property regimes in Central Europe. Obstacles are grouped according to the level (either societal or individual scales) they are acting on. Some consequences and potential solutions mentioned in the text are also presented.

Obstacle	Consequences	Potential solutions
	Societal scale	
Lack of social priority	interests of anglers rarely considered by water and ecosystem managers	thorough evaluation of the socioeconomic benefits of angling on regional and national scales
	fishery stakeholders not involved in consultation processes	offensive lobbying based on "hard scientific data"
	diffucult or impossible to attract developments for fishery development	proactively seeking input in water and ecosystem management
Lack of integrated approaches	socially suboptimal management solutions	rehabilitate ecosystem structure and function on larger scales
	often disruption of ecosystem structure and function	create win-win situation for all stakeholders
Lack of cooperative institutional linkages	severe intrasectoral and intersectoral conflicts	facilitation of structured communication
	nonsustainable management measures	top down expert advice
		management plans
Lack of systems thinking	understanding of system's behavior rudimentary	shift in thinking
	short-sighted management	incorporate anglers in system-analytical studies and communication results
	shifted perceptions	
Lack of research and monitoring	invisible declines of fish stocks	educate about overfishing
	reduced ability to reacte appropriately	cooperative research, funding for long-term monitoring
	black-box management	precaution

Table 1. Continued.

Obstacle	Consequences	Potential solutions
	Individual or group scale	
Lack of shared values and dominance of stereotyped perceptions	inter- and intrasectoral conflicts	foster common values such as fairness and justice
	low level of cooperation and mutal acceptance	facilitate face-to-face interactions, sometimes facilitator may be needed
	consumer attitude, free-riding behavior	
Lack of consideration of the demand side of fish-angler interactions	homogenising angling quality on regional scales	active adaptive management
	overexploitation	partially limiting angler effort, first indirectly by "soft paths"
	low angler satisfaction levels on regional scales	
	management impossible	protected areas
Lack of self-criticism among angler lobbyists	unawareness about potential negative impacts	increase communication of research results about positive/negative impacts
	low awareness to be partly responsible for declining fish stocks	convince anglers to meet their own targets by more restrictive regulations
	low level of environmental concern	facilitate personal experiences
	support of nonsustainable management measures	reduce fear against new approaches

stocks, fishing sites and catchability of fish (Policansky 2001, 2002), the latter due to the increasingly more difficult recapture of previously caught-and-released fish (Raat 1985). Although overexploitation of such common-pool-resources is much more likely under open-access (Hardin 1968) than under common-property regimes (Feeny et al. 1990), Hardin (1998) noted that degradation of the resource is inevitable as long as effective management is not in place. Thus, tragedy of the commons phenomena are also likely to occur in recreational fisheries in both open-access and common-property situations. However, anglers as other fishing communities (Feeny et al. 1990; Ostrom 1990) should have the ability to develop resource preserving institutions (i.e., ways of organizing angling systems), for example by monitoring resource use or devising resource conserving informal (i.e., voluntary) institutions. But to achieve this self-regulating system, several obstacles have to be overcome to reconcile RFM with resource conservation. Some of the most important barriers that limit the development of sustainable RFM strategies will therefore be discussed below. It is important to note, however, that the following is the subjective opinion of the author and does not necessarily reflect the true situation all over Europe. The list of social barriers is therefore subjected to further change in future as more information becomes available.

Obstacles to Reconcile Recreational Fisheries with Conservation

Barriers to reconcile recreational fisheries goals with conservation operate at two different scales; either at the societal level or at the level of individual stakeholders (or groups of stakeholders). Therefore, in the following these two divergent scales are treated separately (cf. Table 1). However, it is important to realize that some obstacles are interrelated, mutually reinforcing each other and are therefore difficult to categorize.

Societal Scale

Lack of Social Priority

The multipurpose nature and use pattern of freshwater ecosystems has created a climate in developed societies in which recreational fisheries suffer in the face of economically and socially higher priorities such as agriculture, hydro-electric power production, navigation and flood prevention – activities that often contradict conservation of biodiversity and fishery resources worldwide (Dynesius and Nilsson 1994; Collares-Pereira et al. 2002; Wolter and Arlinghaus 2003). This lack of public acceptance, inter alia, results from the rarely assessed socioeconomic importance of recrea-tional fisheries (Arlinghaus et al. 2002; Arlinghaus and Mehner 2003a). Therefore, recreational fisheries are often given low priority in any consultation process and it is difficult to attract investments for development of a fishery (Cowx 2002). The result is that nonfishery stakeholders such as water management authorities rarely, if at all, solicit the input and consider the interests and demands of fishery stakeholders leading to management actions that are often detriment to ecosystem structure and function (Cowx 2000). This calls for thorough socioeconomic evaluation of recreational fisheries and fish resource conservation at local, regional and national scales to ensure that these values are well represented in all development activities concerning freshwater ecosystems (Arlinghaus et al. 2002; Arlinghaus and Mehner 2003a). Moreover, fishery stakeholders should constantly try to increase their input by proactive instead of reactive lobbying based on "hard scientific data."

Lack of integrated approaches.—Increasing pressures on and multiple uses of freshwater ecosystems (Jackson et al. 2001; Schindler 2001) dictate that recreational fisheries and resource conservation can no longer be treated in isolation (Cowx 1998; Arlinghaus et al. 2002). Therefore, an integrated approach to aquatic resource management is required spanning all stakeholders potentially affected by that management action (Decker et al. 1996). Integrated management is characterized as management across scientific disciplines, stakeholder groups and considering, at least in concept, future generations (Costanza et al. 1998). Such approaches are currently rarely pursued in traditional RFM (Arlinghaus et al. 2002). The most important reason is that many fishery agencies and private fishing rights holders responsible for management of an extensive and diverse array of recreational fisheries often do not have adequate human and financial resources (and political power) for monitoring either the fishery or the fish stocks supporting the fishery (Post et al. 2002; Pereira and Hansen 2003). Furthermore, freshwater fishery resources are strongly dependent on nonfishery activities such as land-use changes and water management activities (Arlinghaus et al. 2002). The inability of RFM to alter nonfishing activities affecting fishery resources, inter alia, has led in the past to focus on single-species management of recreationally valuable fish stocks (Arlinghaus et al. 2002). This has been mainly achieved by stocking practices of hatchery-reared fish, translocation of fish species across catchments and introductions of exotics, which today is perhaps the greatest concern regarding conservation of genetic, species and community biodiversity of freshwater fish (Meffe 1992; Arlinghaus et al. 2002; Myers et al. 2004). Worldwide, however, degradation of the environment and loss of fishery habitat are the preeminent barriers for the sustainability of inland fisheries (FAO 1999). This is particularly relevant in densely-populated countries of Central Europe (Arlinghaus et al. 2002). Rehabilitation of ecosystem structure and function and the associated fishery resources is therefore a goal that fishery and conservation stakeholders should pursue together in an integrated approach to create true win-win situations for all stakeholders (Pinkerton 1994). This is necessary as large-scale environmental engineering or habitat rehabilitation techniques need a full consultation with water resource managers and environmental experts (Arlinghaus et al. 2002). However, if anglers and fishery managers cannot carry out complex aquatic ecosystem rehabilitation and restoration projects independently, conservation and traditional RFM can hardly be reconciled. The reasoning behind this statement is that the most sustainable management strategy to conserve fishery resources in the long-term (i.e. habitat rehabilitation on larger scales) usually lies outside the domain of traditional RFM. Thus, there is no scale-matching of problem causing and problem solving institutions.

Lack of Cooperative Institutional Linkages

As was already mentioned above, public fisheries authorities and private fishing rights holders are often politically too weak to take effective actions to deal with policy problems in RFM. Many problems affecting fishery resources and conservation in general, can only be addressed by nonfishery bodies such as water management agencies. However, there is seldom a structured communication demanded formally (e.g., a by-law) between water or nature conservation agencies and fisheries agencies before actions are taken (Raat 1990; Welcomme 2001). Moreover, in many recreational fisheries systems traditionally management takes place locally without approval of or advice on management plans by experts trained in ecology or sociology. For example, in Germany many angling clubs as fishing rights holders expand the legal state-wide regulations such as minimum-size limits without scientific evaluation of the biological and social effects of such measures, and enforcement of regulations is inadequate, at best (McPhee et al. 2002). Other examples include that most stocking events take place without a priori appraisal of the ecological and evolutionary risks or social and economic cost and benefits (Arlinghaus et al. 2002). Evaluation of such stocking practices is rarely conducted such that RFM becomes an oxymoron. Lack of institutional linkages and communication, however, is not only prevalent between fishery and nonfishery stakeholders intersectorally, but can also be found intrasectorally, (e.g. in an angling club where the private fishery managing board seldom incorporates inputs from the anglers). There are also often not well-developed institutional linkages between angling clubs and public fishery authorities. Thus, in some European countries such as in Germany and The Netherlands, the local RFM remains more or less in the hands of angling clubs and organizations, which are often not well or not at all trained in fisheries science or conservation biology (Walder and van der Spiegel 1990). However, legislation is changing, at least in Germany, where fishing rights holders have to set up management plans that then have to be approved by experts in public authorities. This is a promising way forward. In an attempt to reconcile conservation with resource use, local community-driven recreational fishery management may need some kind of top-down expert advice, productive communication, education and control to better manage "unsustainable knowledge" of (often voluntary) private fisheries managers. However, it is doubtful whether public fishery authorities may be able to accurately validate hundreds of management plans in their jurisdictions given limited human resources available.

Lack of Systems Thinking

Although RFM is highly complex involving uncertainty, information gaps and the correlations between slow (e.g., evolution, habitat change) and fast (e.g., fish stock development, angler responses) variables, voluntary fishery managers of angling organizations in Central Europe often do not accept that some of their traditional management practices are partly unsustainable. This may result from their limited scientific background and the typical lack of exposure to scientific findings. In fact, RFM has traditionally been sectoral in orientation (Arlinghaus et al. 2002). However, recreational fisheries as socio-ecological systems are itself only nested elements of aquatic ecosystems, society, and ultimately the biosphere. Therefore, the sectoral approach has run its course and should be substituted by a system approach, at least from an academic standpoint (Arlinghaus et al. 2002). Nonetheless, given the complexity of all socialecological systems, including recreational fisheries, it is often difficult for fishery (and other) stakeholders and managers to see and understand the larger picture of the system as a whole and be aware of processes occurring at scales larger and longer than their own experience (Post et al. 2002). In practice, most people are overtaxed to make reasonable decisions given complex, uncertain and slowly changing (time lag between cause and effect) processes of dynamic socioecological systems (Dörner 1996; Folke et al. 2002; Folke 2003). Thus, they tend to focus on local conditions and short-term solutions (Fehr 2002). This problem is not unique to recreational fisheries, but is a common and central feature of the whole sustainability debate (Folke 2003). This is difficult even for scientists, and is much more difficult for fishery stakeholders and other "laypeople" not trained in systems thinking. The solution to overcome this problem is to incorporate fishery stakeholders in scientific, system-analytical studies and communicate results in an understandable manner. In this respect, traditional ecological knowledge can be extremely useful for fishery researchers. This necessitates that researchers and managers foster the intellectual input of anglers (e.g., Olsson and Folke 2001). However, over the generations, expectations about what a healthy fish community or an "natural" aquatic ecosystem constitute may shift towards lower optima, i.e. anglers, researchers and managers may perceive a degraded status of the aquatic ecosystem and the impoverished fish stocks as being the natural state and may use this perception to define conservation or management targets (Pauly 1995; Arlinghaus and Mehner 2003b).

Lack of Research and Monitoring

Post et al. (2002) stated that there were few documented instances of declines in fish stocks that could clearly be attributed to recreational fisheries. At the same time, however, the authors identified four high profile fisheries that showed evidence of angling-induced declines in Canada. These declines were largely unnoticed by the angling public, a characteristic that may be widespread in recreational fisheries (Schindler 2001; Sullivan 2003) as long-term monitoring of thousands of regionally highly dispersed fish stocks is lacking (Cox 2000; McPhee et al. 2002). It has also repeatedly been shown that angling exploitation can force the length- and age-frequencies of fish populations towards smaller and younger fishes, in particular under minimum-size-limit regulations and high harvest levels (e.g., Goedde and Coble 1981; Anderson and Nehring 1984; Olson and Cunningham 1989). To increase our understanding of angler behavior and the effects of angling activities on the ecology and evolutionary biology of fish stocks, long-term monitoring programs coupled with rigorous experimentation of regulations are needed (active adaptive management, Pereira and Hansen 2003). At the moment, funding for such applied research is limited and the private property systems of Europe necessitate the willingness of fisheries owners to allow such research to take place. Research initiated by fishing rights holders is practically nonexistent. Consequently, there is limited basic and applied research to guide management, which severely reduces the system's ability to correspond appropriately to the demands set by the sustainability concept (Aas 2002; Arlinghaus et al. 2002). Furthermore, many of the modern management concepts (e.g., human dimensions research on the diversity of angler subgroups) are often not accepted as being critical for management (Ditton 1996; Arlinghaus 2004a). Nonetheless, the scope for cooperative initiatives between academic professionals and private fishing rights owners is huge, if the awareness of the necessity to apply scientific rigor to day-to-day management is increased and funding is made available. In the face of incomplete scientific facts to aid RFM, decisions and actions should be taken with precaution.

Individual and Group Scale

Lack of Shared Values and Dominance of Stereotyped Perceptions

Recreationally valued species are often at the heart of management efforts by fishery stakeholders. That utilitarian bias in wildlife management in general has profoundly affected the public perception of the wildlife management profession and has created strong barriers to public support for fishery and wildlife agencies (Decker et al. 2001) because the extractive use of wild living resources is often opposed by those who object to the killing or collecting of fish on ethical and moral grounds (Hutton and Leader-Williams 2003). Confusion and strong conflicts arise, in part at least, because of the different visions or worldviews of use, conservation and management under the umbrella of sustainable development. For example, conservation biologists often only consider ecological sustainability, whereas consumptive/extractive stakeholders often emphasize the socio-economic domain of the sustainability concept. In the past decades, there has been a gradual shift away from traditional wildlife values that emphasize the use of wildlife for human benefit towards wildlife conservation and protection (Manfredo et al. 2003). Thus, one of the greatest challenge for recreational fisheries taken place in freshwater ecosystems characterized by multiple uses and high degree of anthropogenic degradation (Vitousek et al. 1997) is to make sound management decisions to ensure viable recreational fisheries are compatible with esthetic and nature conservation values in the 21st century (Arlinghaus et al. 2002). Only if all stakeholders respect and at least try to understand each other based on values such as justice and fairness, can RFM and conservation be reconciled at larger scales. However, this might be very difficult as stereotyping may for example result in a view among anglers that conservation per se is authoritarian and threatening. Alternatively, stereotypic attitudes among conservationists may lead to the perception that anglers are *always* a threat to the protection of aquatic ecosystems (Stoll-Kleemann 2001). This stereotyping can reduce the potential for cooperation to extremely low levels or inhibit it at all and encourage group processes of social identity (Stoll-Kleemann 2001). Stereotypical thinking within angler groups can also lead to increased free-riding and tragedy of the commons phenomena if certain anglers consider other anglers in a stereotypic sense as "not belonging to their group," aggravating rivalry in consumption. Stereotypic thinking can also occur in angler-fishery manager interactions if the manager does not achieve or is not willing to integrate the anglers in the process of decision taking and perceives anglers as laypeople not able to contribute to resource management. The angler may then develop a feeling of frustration, which will often result in rule breaking behavior and the development of a "consumer attitude" (e.g., the attitude that a certain amount of fish has to be taken out of the water to balance the fishing license cost). Again, the result will be that overfishing is more likely on larger scales with potential detrimental impacts on fishery resource conservation. The overcoming of this lack of shared value systems acting at different levels in RFM is particularly difficult, as once developed values and beliefs are notoriously resistant to change (Manfredo et al. 2003). Managers are envisaged to provide a productive environment for face-to-face interaction and facilitate conflict resolution processes to achieve mutual acceptance. Sometimes a facilitator may be needed to achieve this aim.

Lack of Consideration of the Demand Side of Fish-Angler Interactions

Modern anglers are highly mobile, potentially shifting effort, angling mortality and other impacts at regional scales if angling quality declines (Carpenter et al. 1994; Johnson and Carpenter 1994; Walters and Cox 1999; Cox and Walters 2002; Cox et al. 2002; Lester et al. 2003). Until now, most RFM decisions are made on a case-by-case basis wherein quality on individual fisheries is assumed to be independent of management actions and fishery dynamics elsewhere (Cox et al. 2003). However, dynamic relationships between anglers and angling quality seem to be the rule rather than the exception. Two general types of fish-angler dynamics are conceivable. First, fish populations can respond to fishing impacts and fisheries management actions (e.g., stocking, regulations), and second, anglers may respond by functional and numerical responses to changes in the abundance of fish, in the quality of the angling experience per se or in regulatory schemes (e.g., Beard et al. 2003). Recreational fisheries management has traditionally concentrated on the supply side of this dynamic relationship (fishing quality), with a tacit assumption that the demand side (angling effort) will somehow be self-regulating (Walters and Cox 1999; Cox and Walters 2002; Post et al. 2002). However, under conditions of low to moderate access costs (e.g., time, energy, money) and particular under open or quasiopen-access, anglers will regionally shift angling effort, hence mortality from "bad" to "good" fishing waters (Walters and Cox 1999; Cox et al. 2002, 2003; Post et al. 2002; Arlinghaus 2004a, 2004b). This can result in declining fish stocks on regional scale and in a competitive situation from which recreational species might not recover when angling effort and mortality declines (Walters and Kitchell 2001; Post et al. 2002, 2003). Moreover, angler satisfaction may be severely reduced in the long term, with potentially detrimental effects on the environmental concern of anglers in the case of dissatisfaction (see Arlinghaus and Mehner 2005). To overlook dynamics between prey (fish) and predator (angler) on regional scales may ultimately lead to serious losses in total socioeconomic benefits (Cox et al. 2003). One possible solution is active management of angling effort/mortality, which is rarely pursued by contemporary RFM (Pereira and Hansen 2003). Partial control of angling effort (and indirectly angler harvest and mortality) in waters needing protection may take place by direct access restrictions, increases of access cost (time, money), lottery systems of access, annual rotating access schemes (e.g., between angling club members), license price increases, implementations of total allowable angling effort (e.g., days) schemes, or a combination of the options. However, it is by far more advisable to first try to change angler behavior indirectly, e.g. by soft paths (education, zoning etc.) instead of implementing more restrictive regulations aiming at controlling angler directly (Arlinghaus 2004a). Other strategies may include the implementation of large-scale protected areas that receive none or only limited angler effort to help species to recover. However, as Walters and Cox (1999) noted, fishery managers are typically hit from all sides when they suggest effort limitation. However, opposition by anglers may quickly die away if quality increases become evident (Walters and Cox 1999).

Lack of Self-Criticism Among Angler Lobbyists

Ecological impacts of angling are cumulative and can be substantial, whereas there is a tendency for consideration of impacts in isolation, if at all (McPhee et al. 2002). Usually, there is the public or political perception that recreational fishing is more a benign activity than commercial fishing (Kearney 1999; Cooke and Cowx 2006). Recreational fishing lobbyists have been successful in focusing public and political attention on other impacts such as commercial fishing in the past or cormorant predation in the present. In some countries such as in Australia, recreational fishing has not come under close scrutiny from conservation groups (McPhee et al. 2002). However, in other countries such as in Germany there is a fierce debate between fishery and nature conservation and animal welfare groups surrounding potential negative impacts of RFM actions such as stocking or angling practices such as catch-and-release fishing (Aas et al. 2002; Arlinghaus 2004b). Both trends, however, stimulate one outcome at the level of fisheries lobbying: angling impacts are attenuated or not accepted, and research to analyze potential ecological or evolutionary effects of selective angling mortality is not being funded by angling organizations. Instead, research is funded that analyses the economic benefits of recreational fisheries, and the results are used to lobby about the (indeed overwhelming) socioeconomic importance of angling (cf. Arlinghaus 2004b). From the political perspective, this procedure is understandable and by no means immoral. However, if not even the angling lobbyists and the angling media try to inform their constituencies about the potential negative effects of selective angling expoitation, awareness, and environmental concern among anglers will very likely not develop. The implication is for fisheries researchers to increase communication with angler organizations and angler media to inform about all aspects of the activity, including the negative effects of certain angling practices such as biological and evolutionary effects of selective angling mortality, lethal and sublethal effects of catch and release or the effects of excessive nutrient inputs by groundbaiting (cf. Arlinghaus et al. 2002; Cooke and Cowx 2006).

Lack of Self-Criticism Among Individual Anglers

An awareness among anglers of the potential of angling exploitation or RFM practices to negatively affect fish populations and a feeling of responsibility is paramount to serve as an antecedent to change angler behavior directly or indirectly comply with more restrictive regulations (Fulton et al. 1996; Arlinghaus and Mehner 2005). For many years since the seminal paper of Dunlap and Heffernan (1975), it was assumed that simply the involvement in angling leads to increased environmental concern, because people are exposed to instances of ecosystem deterioration, thus creating a commitment to the protection of habitats, cultivating an esthetic taste for a natural environment, and fostering a general opposition to environmental degradation. Although this assumption seems reasonable and was often uncritically cited as an "ecological benefit" of recreational fishing (e.g., Kearney 1999), empirical results were weak, at best (see Theodori et al. 1998; Tarrant and Green 1999; Bright and Porter 2001 and references therein for details). Available evidence to date suggests that the environmental concern of anglers appears high if their environmental attitudes are solicited about very general ecological aspects (e.g., the limited nature of fishery resources per se, the equal rights of animals and plants and humanity or the moral justification that humankind is allowed to rule over nature; Gill et al. 1999; Arlinghaus 2004b). Some angler populations are aware of the possibility that angling can overharvest fish stocks (Schramm et al. 1999). However, a recent study from Germany (Arlinghaus 2004b) indicated that the majority of the anglers surveyed agreed or strongly agreed that the balance of aquatic ecosystems is strong enough to cope with angling impacts. Less than half of the anglers surveyed indicated that they would be willing to change current behavior for the protection of aquatic ecosystems. Most anglers thought that their learning and observational capabilities will result in fish stocks not being overfished. Therefore, at least some angler populations show a low level of self-criticism and critical self-reflection about the potential effects of their own behavior. The consequence of the low awareness among angler to be part of the problem of declining fish stocks (Reed and Parsons 1999), inter alia, can result in lack of support of more restrictive regulations (cf. Arlinghaus 2004a for discussion and references). If regulations are in direct conflict with fishing practices that are familiar and enjoyed, optimistic biases about the risks of overfishing (cf. Weinstein 1982) may ultimately result in low support of conservation goals. For managers the challenge ahead is to convince anglers about ways to meet their own targets by conserving the resource. Opposition to conservation goals may occur because of little experience with the ecosystem management concept, which may be perceived by anglers as an untested theory or threat to continued enjoyment of the activity (Jacobson and Marynowski 1997). To overcome this situation, anglers need to be included in the process of fishery and ecosystem management decision making. However, the angling public is often in a "show me" mood and does not necessarily trust people in authorities, whether scientists or government and agency officials (Smith et al. 1997). Accepting angling impacts as crucial for sustainability will very likely only occur if personal experiences are gained as many anglers tend to rely on personal experiences or knowledge of peers that are known and respected.

Prospect with Emphasis on the Human Dimension

Fish and aquatic ecosystems in general exist with or without human intervention; but the fishery resource and the goal of conservation of these resources is a human construct. Consequently, many obstacles to reconcile conservation with recreational fisheries discussed above are human dimensions issues. Although an understanding of the biology, ecology, and evolutionary dimension of fish stocks exploited by fisheries and the rebuilding of ecosystem structure and function remains at the heart of any effort to achieve sustainability, we should be reminded that solutions to this crucially depend, inter alia, on appropriate angler behavior, effective institutions (see Ostrom 1990 for details), stakeholder (in particular angler) involvement in decision making, productive communication, environmental education, facilitation of bottom-up processes, consideration of angler diversity in their human dimensions, precaution, active adaptive management systems designed to "learning by doing," and harmonization of divergent world views (Arlinghaus 2004a). All of these are human dimensions issues that are rarely fully addressed in RFM (at least in Central Europe) in which managers have traditionally relied on information from the biological sciences, if at all. However, it makes little sense to develop and test human dimension understandings in recreational fisheries, if managers are either not adequately trained and prepared to make use of fundamental insights, or even worse, reject their applicability (Ditton 1996; Wilde et al. 1996; Sharp and Lach 2003). The consequence of the lack of knowledge about human aspects of RFM is the possibility that myths, personal views, and the opinions of strong interest groups guide management decisions in this arena, replacing or circumventing scientific knowledge.

For the near-term, it is doubtful that anglers/angling clubs or more generally fishing rights holders or fishery agencies will be able to properly address all these challenging issues surrounding sustainable RFM alone. It is a matter of societal values whether is it judged necessary to increase management efforts and investments of funds and human resources related to RFM on a broader scale to allow that RFM is able to better address the multiple levels demanded by the sustainability concept (cf. Forsgren and Loftus 1993; Fisher et al. 1998; Epifanio 2000). Unfortunately, in many parts of Central Europe the idea of sustainable RFM is still in its infancy somewhere between the innovators and early adopters stage according to the adoption-diffusion theory (cf. Decker and Krueger 1999). In the case the early majority or late majority stages will be achieved, i.e. it is agreed societal value to manage for sustainability in recreational fisheries, it might be advisable to increasingly involve experts in practical management of recreational fisheries (e.g., by extension services of public agencies supporting anglers or fishing rights holders in general) that are trained in interdisciplinary scientific disciplines (and not only in biology, limnology or ecology). This is necessary, as complete devolution of management to angling communities may not be appropriate in Central European RFM (cf. Feeny et al. 1990). It therefore makes sense for the public hand to continue to play a role in resource conservation. Thus, shared governance of state regulations coupled with user self-management may be a viable option (Feeny et al. 1990). Such comanagement could capitalize on the local knowledge and long-term self-interest of anglers, while providing coordination with relevant uses and users over a wide geographic area at potentially lower transaction (e.g., policy compliance and enforcement) cost. What is needed as agents of change in management approach and understanding are "barefoot managers" (taken from the barefoot ecologist advocated by Prince 2003) as holistic sociobiological, ecological-evolutionary thinkers. Their role will be to motivate and empower anglers to research, monitor, and manage their own localized fishery resources. For each new angler community and fishery stock, the starting point for a "barefoot manager" will be the application of data less management, gleaning local knowledge (Olsson and Folke 1998), reading the comparative literature, offering basic information to anglers, and recommending sensible rule-of-thumb management (Prince 2003) including the human dimension. This type of advisor role is well developed and accepted in the agricultural sector, but almost nonexistent and even frowned upon, in (recreational) fisheries (Prince 2003).

Thinking longer term, many traits of RFM systems in Central Europe provide excellent conditions for effective natural resource management. For example, well-defined access rights are conditions for environmentally sound management because the tragedy of the commons is less likely to occur (Hardin 1968). Moreover, the small-scale structures in European recreational fisheries embedded in quasi-common property together with the fact that the participation in angling is less dependent on high physical yield as compared to commercial fisheries constitute beneficial prerequisites for reconciling conservation with resource use. This increases the hope that the "race for fish phenomena" may be less pronounced in recreational fisheries as compared to open-access commercial fisheries. Lastly, the small-scale structures in recreational fisheries systems also increase the probability of regular interaction between anglers, which is one condition for resource conserving (informal) institutions to evolve (see Feeny et al. 1990; Ostrom et al. 1999; Dietz et al. 2003; Fehr and Fischbacher 2003 for details). To achieve this, anglers' and other stakeholders' abilities to perceive, understand, and act must be developed if we are to approach sustainability (Ludwig 2001). However, it is a utopian view to assume that reconciliation of recreational fisheries and conservation will progress quickly and immediately. There are severe constraints to recreational fisheries development originating from outside the fishery that will not be circumvented and resolved in the near future. A shift in approach and thinking might sometimes be needed to reconcile conservation and use values in freshwater RFM. Before this becomes reality, it is important to note that anglers or RFM in general are not guilty that sustainable management approaches are sometimes lacking. This is a societal problem and results from the traditional way by which fish resources are managed. In fact, voluntary fisheries managers at the private level in Central Europe are often highly active, willing to improve fisheries management and have contributed to effective management in many cases. However, lack of financial and operational power, limited research in particular as regards the human dimension, limited communication across scientific disciplines and between stakeholders and a high degree of ideologically driven conflicts between nature conservation and animal welfare on the one hand and fishery stakeholders on the other hand can produce deleterious outcomes and suboptimal conditions that do not reward the tremendous efforts of anglers and private fisheries managers.

Acknowledgments

Comments by Steven J. Cooke and Thomas Mehner on an earlier manuscript version helped to describe the content in a more concise way. I also thank two referees for their encouraging comments.

References

Aas, Ø. 2002. The next chapter: multicultural and cross-disciplinary progress in evaluating recreational fisheries. Pages 252–263 *in* T. J. Pitcher and C. E. Hollingworth, editors. Recreational fisheries: ecological, economic and social evaluation. Blackwell Scientific Publications, Oxford, UK.

Aas, Ø., C. E. Thailing, R. B. Ditton. 2002. Controversy over catch-and-release recreational fishing in Europe. Pages 95–106 *in* T. J. Pitcher and C. E. Hollingworth, editors. Recreational fisheries: ecological, economic and social evaluation. Blackwell Scientific Publications, Oxford, UK.

Anderson, R. M., and R. B. Nehring. 1984. Effects of a catch-and-release regulation on a wild trout population in Colorado and its acceptance by anglers. North American Journal of Fisheries Management 4:257–265.

Aplet, G. H., R. D. Laven, and P. L. Fiedler. 1992. The relevance of conservation biology to natural resource management. Conservation Biology 6:298–300.

Arlinghaus, R. 2004a. A human dimensions approach towards sustainable recreational fisheries management. Turnshare Ltd., London.

Arlinghaus, R. 2004b. Recreational fisheries in Germany – a social and economic analysis. Berichte des IGB 18:1–160.

Arlinghaus, R., and T. Mehner. 2003a. Testing the reliability and construct validity of a simple and inexpensive procedure to measure the use value of recreational fishing. Fisheries Management and Ecology 11:61–64.

Arlinghaus, R., and T. Mehner. 2003b. Management preferences of urban anglers: habitat rehabilitation versus other options. Fisheries 28(6):10–17.

Arlinghaus, R., and T. Mehner. 2005. Determinants of management preferences of recreational anglers in Germany: habitat management versus fish stocking. Limnologica 35:2–17.

Arlinghaus, R., T. Mehner, and I. G. Cowx. 2002. Reconciling traditional inland fisheries management and sustainability in industrialized countries, with emphasis on Europe. Fish and Fisheries 3:261–316.

Beard, T. D: Jr., S. P. Cox, and S. R. Carpenter. 2003. Impacts of daily bag limit reductions on angler effort in Wisconsin walleye lakes. North American Journal of Fisheries Management 23:1283–1293.

Berkes, F., D. Feeny, B. McCay, and J. M. Anderson. 1989. The benefits of the commons. Nature (London) 340:91–93.

Bright, A. D., and R. Porter. 2001. Wildlife-related recreation, meaning, and environmental concern. Human Dimensions of Wildlife 6:259–276.

Carpenter, S. R., A. Muñoz-del-Rio, S. Newman, P. W. Rasmussen, and B. M. Johnson. 1994. Interactions of anglers and walleyes in Escabana Lake, Wisconsin. Ecological Applications 4:822–832.

Carpenter, S. R., and L. H. Gunderson. 2001. Coping with collapse: ecological and social dynamics in ecosystem management. BioScience 51:451–457.

Collares-Pereira, M. J., I. G. Cowx, and M. M. Coelho, editors. 2002. Conservation of freshwater fishes: options for the future. Fishing News Books, Blackwell Scientific Publications, Oxford, UK.

Cooke, S. J., and I. G. Cowx. 2006. Contrasting recreational and commercial fishing: searching for common issues to promote unified conservation of fisheries resources and aquatic environments. Biological Conservation 128:93–108.

Costanza, R., F. Andrade, P. Antunes, M. van den Belt, D. Boersma, D. F. Boesch, F. Catarino, S. Hanna, K. Limburg, B. Low, M. Molitor, J. G. Pereira, S. Rayner, R. Santos, J. Wilson, and M. Young. 1998. Principles for sustainable governance of the oceans. Science 281:198–199.

Cowx, I. G. 1998. Aquatic resource planning for resolution of fisheries management issues. Pages 97–105 *in* P. Hickley and H. Tompkins, editors. Recreational fisheries: social, economic and management aspects. Fishing News Books, Blackwell Scientific Publications, Oxford, UK.

Cowx, I. G., editor. 2000. Management and ecology of river fisheries. Fishing News Books, Blackwell Scientific Publications, Oxford, UK.

Cowx, I. G. 2002. Recreational fishing. Pages 367–390 *in* P. J. B. Hart and J. D. Reynolds, editors. Handbook of fish biology and fisheries, volume 2, Fisheries. Blackwell Scientific Publications, Oxford, UK.

Cox, S. 2000. Angling quality, effort response, and exploitation in recreational fisheries: field and modeling studies on British Columbia rainbow trout lakes. Doctoral dissertation. University of British Columbia, Vancouver.

Cox, S., and C. Walters. 2002. Maintaining quality in recreational fisheries: how success breeds failure in management of open-access sport fisheries. Pages 107–119 *in* T. J. Pitcher and C. E. Hollingworth, editors. Recreational fisheries: ecological, economic and social evaluation. Blackwell Scientific Publications, Oxford, UK.

Cox, S., C. L. Walters, and J. R. Post. 2003. A model-based evaluation of active management of recreational fishing effort. North American Journal of Fisheries Management 23:1294–1302.

Cox, S., T. D. Beard, and C. Walters. 2002. Harvest control in open-access sport fisheries: hot rod or asleep at the reel? Bulletin of Marine Science 70:749–761.

Decker, D. J., T. L. Brown, and W. F. Siemer, editors. 2001. Human dimensions of wildlife management in North America. The Wildlife Society, Bethesda, Maryland.

Decker, D. J. and C. C. Krueger. 1999. Communication for effective fisheries management. Pages 61–81 *in* C. C. Kohler and W. A. Hubert, editors. Inland fisheries management in North American, 2nd edition). American Fisheries Society, Bethesda, Maryland.

Decker, D. J., C. C. Krueger, R. A. Baer Jr., B. A. Knuth, and M. E. Richmond. 1996. From clients to stakeholders: a philosophical shift for fish and wildlife management. Human Dimensions of Wildlife 1:70–82.

Dietz, T., E. Ostrom, and P. C. Stern. 2003. The struggle to govern the commons. Science 302:1907–1912.

Ditton, R. B. 1996. Human dimensions in fisheries. Pages 74–90 *in* A. W. Ewert, editor. Natural resource management: the human dimension. Westview Press, Oxford, UK.

Dörner, D. 1996. Der Umgang mit Unbestimmtheit und Komplexität und der Gebrauch von Computersimulationen. Pages 489–515 *in* A. Diekmann and C. C. Jaeger, editors. Umweltsoziologie. Sonderheft 36 der Kölner Zeitschrift für Soziologie und Sozialpsychologie. Westdeutscher Verlag, Opladen, Germany.

Dunlap, R. E., and R. B. Heffernan. 1975. Outdoor recreation and environmental concern: an empirical examination. Rural Sociology 40:18–30.

Dynesius, M., and C. Nilsson. 1994. Fragmentation and flow regulation of river systems in the northern third of the world. Science 266:753–762.

Epifanio, J. 2000. The status of coldwater fishery management in the United States. Fisheries 25(7):13–27.

FAO (Food and Agriculture Organization of the United Nations). 1999. Review of the state of world fishery resources: inland fisheries. FAO Fisheries Circular 942.

Feeny, D., F. Berkes, B. J. McCay, and J. M. Ancheson. 1990. The tragedy of the commons: twenty-two years later. Human Ecology 18:1–18.

Fehr, E. 2002. The economics of impatience. Nature (London) 415:269–272.

Fehr, E., and U. Fischbacher. 2003. The nature of human altruism. Nature (London) 425:785–791.

Fisher, W. L., A. F. Surmont, and C. D. Martin. 1998. Warmwater stream and river fisheries in the southeastern United States: are we managing them in proportion to their values? Fisheries 23(12):16–24.

Folke, C. 2003. Freshwater for resilience: a shift in thinking. Philosophical Transactions Royal Society London B 358:2027–2036.

Folke, C., S. Carpenter, T. Elmqvist, L. Gunderson, C. S. Holling, B. Walker. 2002. Resilience and sustainable development: building adaptive capacity in a world of transformations. Ambio 31:437–440.

Forsgren, H., and A. J. Loftus. 1993. Rising to a greater future: forest service fisheries program accountability. Fisheries 18(5):15–21.

Fulton, D. C., M. J. Manfredo, and J. Lipscomb. 1996. Wildlife value orientations: a conceptual and measurement approach. Human Dimensions of Wildlife 1:24–47.

Gill, D. A., H. L. Schramm, Jr., J. T. Forbes, and G. S. Bray. 1999. Environmental attitudes of Mississippi catfish anglers. Pages 407–415 *in* E. R. Irwin, W. A. Hubert, C. F. Rabeni, H. L. Schramm, Jr., and T. Coon, editors. Catfish 2000: Proceedings of the International Ictalurid Symposium. American Fisheries Society, Symposium 24, Bethesda, Maryland.

Goedde, L. E., and D. W. Coble. 1981. Effects of angling on a previously fished and an unfished warmwater fish community in two Wisconsin lakes. Transactions of the American Fisheries Society 110:594–603.

Hardin, G. 1968. The tragedy of the commons. Science 162:1243–1248.

Hardin, G. 1998. Extensions of "the tragedy of the commons." Science 280:682–683.

Hilborn, R., T. A. Branch, B. Ernst, A. Magnussson, C. V. Minte-Vera, M. D. Scheuerell, and J. L. Valero. 2003. State of the world's fisheries. Annual Review of Environment and Resources 28:359–399.

Holling, C. S., and G. K. Meffe. 1996. Command and control and the pathology of natural resource management. Conservation Biology 10:328–337.

Hutton, J. M., and N. Leader-Williams. 2003. Sustainable use and incentive-driven conservation: realigning human and conservation interests. Oryx 37:215–226.

Jackson, R. B., S. R. Carpenter, C. N. Dahm, D. M. McKnight, R. J. Naiman, S. L. Postel, and S. W. Running. 2001. Water in a changing world. Ecological Applications 11:1027–1045.

Jacobson, S. K., and S. B. Marynowski. 1997. Public attitudes and knowledge about ecosystem management on Department of Defense land in Florida. Conservation Biology 11:770–781.

Johnson, B. M., and S. R. Carpenter. 1994. Functional and numerical response: a framework for fish-angler interactions. Ecological Applications 4:808–821.

Kearney, B. 1999. Evaluating recreational fishing: managing perceptions and/or reality. Pages 9–14 *in* T. J. Pitcher, editor. Evaluating the benefits of recreational fisheries. University of British Columbia, Fisheries Centre, Fisheries Centre Research Reports 7(2), Vancouver.

Krueger, C. C., and D. J. Decker. 1999. The process of fisheries management. Pages 31–59 *in* C. C. Kohler and W. A. Hubert, editors. Inland fisheries management in North America, 2nd edition. American Fisheries Society, Bethesda, Maryland.

Leopold, A. 1933. The conservation ethic. Journal of Forestry 31:634–643.

Lester, N. P., T. R. Marshall, K. Armstrong, W. I. Dunlop, and B. Ritchie. 2003. A broad-scale approach to management of Ontario's recreational fisheries. North

American Journal of Fisheries Management 23:1312–1328.

Ludwig, D. 2001. The era of management is over. Ecosystems 4:758–764.

Ludwig, D., R. Hilborn, and C. Walters. 1993. Uncertainty, resource exploitation, and conservation: lessons from history. Science 260:17/36.

Manfredo, M. J., T. L. Teel, and A. D. Bright. 2003. Why are public values toward wildlife changing? Human Dimensions of Wildlife 8:287–306.

McPhee, D. P., D. Leadbitter, and G. A. Skilleter. 2002. Swallowing the bait: is recreational fishing in Australia ecologically sustainable? Pacific Conservation Biology 8:40–51.

Meffe, G. K. 1992. Techno-arrogance and halfway technologies: salmon hatcheries on the Pacific Coast of North America. Conservation Biology 6:350–354.

Myers, R. A., S. A. Levin, R. Lande, F. C. James, W. W. Murdoch, and R. T. Paine. 2004. Hatcheries and endangered salmon. Science 303:1980.

Nielsen, L. A. 1999. History of inland fisheries management in North America. Pages 3–30 *in* C. C. Kohler and W. A. Hubert, editors. Inland fisheries management in North America, 2nd edition. American Fisheries Society, Bethesda, Maryland.

Olson, D. E., and P. K. Cunningham. 1989. Sport-fisheries trends shown by an annual Minnesota fishing contest over a 58-year period. North American Journal of Fisheries Management 9:287–297.

Olsson, P., and C. Folke. 2001. Local ecological knowledge and institutional dynamics for ecosystem management: a study of Lake Racken watershed, Schweden. Ecosystems 4:85–104.

Ostrom, E. 1990. Governing the commons: the evolution of institutions for collective action. Cambridge University Press, Cambridge, England.

Ostrom, E., J. Burger, C. B. Field, R. B. Norgaard, and D. Policansky. 1999. Revisiting the commons: local lessons, global challenges. Science 284:278–282.

Pauly, D. 1995. Anecdotes and the shifting baseline syndrome of fisheries. Trends in Ecology and Evolution 10:430.

Pauly, D., J. Alder, E. Bennett, V. Christensen, P. Tyedmers, and R. Watson. 2003. The future for fisheries. Science 302:1359–1361.

Pauly, D., V. Christensen, S. Guénette, T. J. Pitcher, U. Rashid Sumaila, C. J. Walters, R. Watson, and D. Zeller. 2002. Towards sustainability in world fisheries. Nature (London) 418:689–695.

Pereira, D. L., and M. J. Hansen. 2003. A perspective on challenges to recreational fisheries management: summary of the symposium on active management of recreational fisheries. North American Journal of Fisheries Management 23:1276–1282.

Pinkerton, E. W. 1994. Local fisheries co-management: a review of international experiences and their implications for salmon management in British Columbia. Canadian Journal of Fisheries and Aquatic Sciences 51:2363–2378.

Pitcher, T. J. 2001. Fisheries managed to rebuild ecosystems? Reconstructing the past to salvage the future. Ecological Applications 11:601–617.

Policansky, D. 2001. Recreational and commercial fisheries. Pages 161–173 *in* J. Burger, E. Ostrom, R. B. Noorgard, D. Policansky and B. D. Goldstein, editors. Protecting the commons: a framework for resource management in the Americas. Island Press, Washington, D.C.

Policansky, D. 2002. Catch-and-Release recreational fishing: a historical perspective. Pages 74–94 *in* T. J. Pitcher and C. E. Hollingworth, editors. Recreational fisheries: ecological, economic and social evaluation. Blackwell Scientific Publications, Oxford, UK.

Post, J. R., C. Mushens, A. Paul, and M. Sullivan. 2003. Assessment of alternative harvest regulations for sustaining recreational fisheries: model development and application to bull trout. North American Journal of Fisheries Management 23:22–34.

Post, J. R., M. Sullivan, M., S. Cox, N. P. Lester, C. J. Walters, E. A. Parkinson, A. J. Paul, L. Jackson, and B. J. Shuter. 2002. Canada's recreational fisheries: the invisible collapse? Fisheries 27(1):6–15.

Prince, J. D. 2003. The barefoot ecologist goes fishing. Fish and Fisheries 4:369–371.

Raat, A. J. P. 1985. Analysis of angling vulnerability of common carp, *Cyprinus carpio* L., in catch-and-release angling in ponds. Aquaculture and Fisheries Management 16:171–187.

Raat, A. J. P. 1990. Fisheries management: a global framework. Pages 344–356 *in* W. L. T. van Densen, B. Steinmetz and R. H. Hughes, editors. Management of freshwater fisheries. Pudoc, Wageningen, Netherlands.

Reed, J. R., and B. G. Parsons. 1999. Angler opinions on bluegill management and related hypothetical effects on bluegill fisheries in four Minnesota lakes. North American Journal of Fisheries Management 19:515–519.

Roslin, T. 2002. Fishy behaviour. Trends in Ecology and Evolution 17:547.

Schindler, D. W. 2001. The cumulative effects of climate warming and other human stresses on Canadian freshwaters in the new millennium. Canadian Journal of Fisheries and Aquatic Sciences 58:18–29.

Schramm, H. L. Jr., J. T. Forbes, D. A. Gill, and W. D. Hubbard. 1999. Fishing environment preferences and attitudes toward overharvest: are catfish anglers unique? Pages 417–425 *in* E. R. Irwin, W. A. Hubert, C. F. Rabeni, H. L. Schramm, Jr., and T. Coon, editors. Catfish 2000: Proceedings of the International Ictalurid Symposium. American Fisheries Society, Symposium 24, Bethesda, Maryland.

Sharp, S. B., and D. Lach. 2003. Integrating social values into fisheries management: a Pacific Northwest study. Fisheries 28(4):10–15.

Smith, C. L., J. D. Gilden, J. S. Cone, and B. S. Steel. 1997. Contrasting views of coastal residents and coastal coho restoration planners. Fisheries 22(12):8–15.

Stoll-Kleemann, S. 2001. Barriers to nature conservation in Germany: a model explaining opposition to protected areas. Journal of Environmental Psychology 21:369–385.

Sullivan, M. 2003. Active management of walleye fisheries in Alberta: dilemmas of managing recovering fisheries. North American Journal of Fisheries Management 23:1343–1358.

Tarrant, M. A., and G. T. Green. 1999. Outdoor recreation and the predictive validity of environmental attitudes. Leisure Sciences 21:17–30.

Theodori, G. L., A. E. Luloff, and F. K. Willis. 1998. The association of outdoor recreation and environmental concern: re-examining the Dunlap-Heffernan thesis. Rural Sociology 63:94–108.

Vitousek, P. M., H. A. Mooney, J. Lubchenco, and J. M. Melillo. 1997. Human domination of earth's ecosystems. Science 277:494–499.

Walder, J., and A. van der Spiegel. 1990. Education for fisheries management in The Netherlands. Pages 372–381 *in* W. L. T. van Densen, B. Steinmetz, and R. H. Hughes, editors. Management of freshwater fisheries. Pudoc, Wageningen, Netherlands.

Walters, C. J., and J. F. Kitchell. 2001. Cultivation-depensation effects on juvenile survival and recruitment: a serious flaw in the theory of fishing? Canadian Journal of Fisheries and Aquatic Sciences 58:39–50.

Walters, C. J., and S. Cox. 1999. Maintaining quality in recreational fisheries: how success breeds failure in management of open-access sport fisheries. Pages 22–28 *in* T. J. Pitcher, editor. Evaluating the benefits of recreational fisheries. University of British Columbia, Fisheries Centre, Fisheries Centre Research Reports 7(2), Vancouver.

Weinstein, N. D. 1982. Optimistic biases about personal risks. Science 246:1232–1233.

Welcomme, R. L. 2001. Inland fisheries: ecology and management. Blackwell Scientific Publications, Oxford, UK.

Wilde, G. R., R. B. Ditton, S. R. Grimes, and R. K. Riechers. 1996. Status of human dimensions surveys sponsored by state and provincial fisheries management agencies in North America. Fisheries 21(11):12–17.

Wolter, C., R. Arlinghaus, U. A. Grosch, and A. Vilcinskas. 2003. Fische & Fischerei in Berlin. VNW Verlag Natur & Wissenschaft, Solingen.

Wolter, C., and R. Arlinghaus. 2003. Navigation impacts on freshwater fish assemblages: the ecological relevance of swimming performance. Reviews in Fish Biology and Fisheries 13:63–89.

American Fisheries Society Symposium 49:123–138

Socioeconomic Factors in Fisheries Management: The Ocean Prawn Trawl Fishery in New South Wales, Australia

Ramana Rallapudi* and Donald F. Gartside
Southern Cross University, School of Environmental Science and Management
Lismore, New South Wales 2480, Australia

Abstract.—We have used the Ocean Prawn Trawl Fishery in New South Wales, Australia as a case study to examine how fishers' expectations for management of their fishery compare with the management approach and measures applied by the state management agency. This paper gives a brief overview of the fishery and discusses the possibilities and constraints for sustainable management. The Ocean Prawn Trawl Fishery is the most commercially valuable and regionally important fishery in New South Wales. Despite complex input regulations, increasing fishing effort and ineffective management strategies are threatening the sustainability of the fishery. Although there is a long history of biological research on prawns, socioeconomic factors have thus far been neglected. Most stakeholders in the fishery agree that the fishery needs to be restructured, but they have differing, and often contradictory, views on how to achieve this goal. The most pressing problems facing management of the fishery appear to be of an institutional, economic, and social nature rather than primarily biological. The findings of the survey are discussed in the context of the importance of economic and social factors in underpinning the sustainability of the fishery and providing improved outcomes for fishers. If there is no improvement in the policies to address these issues, the fishery may reach an "irreversible" decline.

Introduction

Within less than a century, marine fisheries, which constitute the last major world industry exploiting wild animal resources, have reached many ecological and economic limits and they face many uncertainties (Cury and Cayre 2001). The search for sustainable, efficient, and equitable ways of managing fisheries has been a long and difficult one and most existing management concepts have been largely ineffective (Abdullah et al. 1996; Young 1999).

Traditional fisheries management policies were based on biological and ecological research findings and mostly aimed at maintaining fish stocks. However, there is increasing awareness that the human welfare outcomes of the application of various policy instruments also must be taken into account (Davis and Gartside 2001). As Jentoft (2000) noted, viable fishing communities require viable fish stocks and viable fish stocks require viable fishing communities; one cannot succeed without the other.

*Corresponding author: rallapudi@hotmail.com

Fisheries management is frequently argued to be highly complex and this complexity is commonly invoked as a reason for widespread failure of fisheries management (Cochrane 2000). Achieving multiple, and often conflicting, fishery management objectives is a challenge for modern policymakers. If fisheries management is to be effective and if all role players in a fishery are to be accountable to the public for their decisions and actions, then the political considerations and approaches for dealing with them in fisheries need to be explicitly identified and clearly and transparently stated (Cochrane 2000). Without understanding the social constraints and the differences among fishery stakeholder groups, even the best available scientific information will not achieve the goal of developing a sustainable fishing industry. Fisheries and fisheries research need to be modernized in a way amenable to integrating new objectives, paradigms, and ethical concerns (Cury and Cayre 2001). Fisheries scientists could begin by studying fisheries as complete systems rather than as disaggregated, discipline specific bits (Healey 1997). In order to examine these issues, the ocean prawn trawl fishery of New South Wales (NSW), Australia, has been chosen as a case study.

In the NSW ocean prawn trawl fishery in 1999 and 2000, there were 329 endorsed vessels and a total catch of 1048 metric tons, worth an estimated AUD (Australian dollar) $17.5 million (Montgomery and Craig 2001). The average annual catches over the past 5 years of the three main target species were 686 metric tons of eastern king prawns *Penaeus plebejus*, 117 metric tons of school prawns *Metapenaeus macleayi* and 156 metric tons of royal red prawns *Haliporoides sibogae* (Montgomery and Craig 2001). This makes the ocean prawn trawl fishery the most commercially valuable fishery in NSW. The distribution of the fishing grounds away from the major metropolitan areas also means the fishery is important for regional economies.

A number of factors are contributing towards the NSW ocean prawn trawl fishery becoming unsustainable. Fishing effort has increased rapidly and there is no limit on the number of nights which can be fished. Catches are static or declining. There is a poor understanding of socioeconomic issues in the fishery by both fishers and the management agency, ineffective management strategies based largely around input controls and lack of an integrated approach to policy development. As a result of these factors, the fishery is moving towards becoming an "artisanal" fishery: small scale, low cost, and labor intensive fisheries in which most of the catch is consumed locally and there is little if any surplus of revenue from catches over costs, rather than a commercial fishery which generates economic rent (King 1995).

Sustainable prawn fisheries in NSW can be realized only with policy formulation that integrates biological and socioeconomic approaches and involves all interest groups.

Our study sought to analyze the views and perceptions of those involved in the fishery about their industry and its future, examine major socioeconomic and institutional issues and constraints, and identify policy instruments that may have more positive outcomes for the future of the fishers and sustainability of the fishery.

Development of the Ocean Prawn Trawl Fishery

Overview of the Fishery

Prawn fishing in Australia began in the early 1800s in NSW estuaries (Ruello 1975a). From its commencement until 1926, prawn trawling was limited to estuarine and inshore waters. The prawn fishery entered a new phase in 1926 after the controversial introduction of otter trawling in Port Jackson (Ruello 1974). Otter trawling was opposed by other fishers because it allegedly destroyed weed beds and spawning areas. The fishery rapidly expanded to offshore waters and to other States over the following decades as many new prawning grounds and new species were discovered and more efficient fishing technology was developed. An incidental discovery of schools of prawns in offshore waters in the late 1940s led to the expansion of trawling activities in ocean waters off NSW (Racek 1957).

The sole method of fishing presently used is otter prawn net trawling. Vessels range from 11 to 19 m in length with nets ranging from 33 to 57 m headline length (NSW Fisheries 1999).

The catches of the prawn species appear to be influenced by several different ecological factors apart from the level of fishing effort. Many scientists have investigated the biological and ecological aspects of the fishery and identified major biological reference points for management. These include discovering new species, new prawning grounds, stock assessment, resource abundance, species distribution, habitat requirements and bycatch issues. There were, however, no comparable socioeconomic studies on the fishery.

Bycatch is an important issue which presently is being addressed cooperatively by NSW fisheries and the fishers. Approximately 80 species (1782 metric tons in 1999 and 2000) of fish, crustaceans, and mollusks are recorded as bycatch in this fishery, most of which is discarded (Kennely et al. 1998; Montgomery and Craig 2001). The main bycatch species retained for sale are trawl whiting, red mullet, flathead, flounder, blue swimmer and sand crabs, octopus, squid, and shovel-nosed lobsters.

In 1979, commercial fishers sought to limit access to the ocean prawn fishery off northern NSW on the basis that catches were declining and the fishery was becoming less viable (NSW Fisheries 1999). Recent fishery reviews indicate that the fishery is fully fished: more effort will not result in increased catches. The distribution of catch among vessels is extremely uneven. McIlgorm and Campbell (1998) found that 30% of endorsed vessels take 80% of the gross value of production of the fishery.

Management Arrangements

The first plan for management of the ocean prawn trawl fishery off NSW was prepared in 1985 by the Commonwealth government and introduced a unitization system based on vessel size and engine power, to control increasing fishing effort (NSW Agriculture and Fisheries 1990). Responsibility for managing the ocean prawn trawl fishery off NSW was handed over to the NSW government in 1991 by the Commonwealth as part of the Offshore Constitutional Settlement. The fishery is now being managed under the Fisheries Management Act 1994 of NSW.

The ocean prawn trawl fishery is administered by NSW Fisheries. NSW Fisheries has established an Ocean Prawn Trawl Advisory Committee (OPTMAC) provides a forum for stakeholders to consult with NSW Fisheries on the development of a management plan (NSW Parliament 1997). Under the Fisheries Management Act 1994, the functions of each management advisory committee are to:

• Advise the Minister on the preparation of any management plan or regulations for the fishery;

• Monitor whether the objectives of the management plan or those regulations are being attained;

• Assist in a fishery review in connection with any new management plan or regulations; and

• Advise on any other matter relating to the fishery.

The OPTMAC comprise five prawn fishers, one person from a conservation group, a Director's nominee and observers from NSW Fisheries. Until recently, the OPTMAC reported to the Director of NSW Fisheries, but now it reports to the NSW Minister for Fisheries.

The major criterion for granting a license endorsement to operate a vessel in the fishery was based on actual and substantial participation (catch history) in the fishery, between mid-1983 and 31 December 1989 (NSW Fisheries 1997).

Currently, the ocean prawn trawl fishery comprises three sectors: inshore (from the coast to 3 nautical miles [5.56 km] to sea); offshore (3–80 nautical miles) and deepwater (offshore waters for taking deepwater prawn species). In 1991, four different classes of license were introduced in the offshore sector of the prawn fishery with differing fishery and transfer rights (NSW Fisheries 1999):

P1—Boats can be upgraded and the entitlements are transferable;

P2—Boats cannot be upgraded, but the entitlements are transferable;

P3—Boats cannot be upgraded and the entitlement cannot be transferable; and

P4—Boats are restricted to operate in a particular area and entitlements are transferable subject to conditions.

The current management system is based solely on input controls (Montgomery and Craig 2001). The input controls include license endorsements, hull length, engine power, time and area closures, net sizes, and sizes of mesh. The use of bycatch reduction devices is now mandatory in this fishery. There are currently seven closures located in ocean waters off northern and central NSW. Compliance in this fishery is carried out primarily onshore with net measurement and in near shore waters by boating patrols enforcing closures (NSW Fisheries 1999).

Methods

Our study involved surveys of stakeholders in the fishery to ascertain their perceptions and attitudes towards the management of the fishery and its future needs. The surveys were conducted during 2000 and 2001. Four main stakeholder groups were surveyed:
1) The catching sector—ocean prawn trawl fishery endorsement-holders;
2) The postharvest and marketing sector;
3) The conservation sector; and
4) The principal government regulatory agency—NSW Fisheries.

However, the main focus of the study was on the catching sector—the fishers.

A structured printed questionnaire survey form was developed and refined using preliminary personal and telephone discussions with fishers. The survey was then mailed to each of the endorsement-holders. Because of privacy concerns, printing of address labels and the mailing of the surveys were undertaken by NSW Fisheries on behalf of the researchers. The survey forms were clearly identified as being from Southern Cross University. The surveys were accompanied by reply-paid return envelopes addressed to the independent researcher's address.

The questionnaire contains 46 questions in four clusters: career history in fishing, how the fishers view their industry, the impact of present policies, and fishers' hopes for the future of their industry. Fishers were also requested to express any other concerns that were not covered in the questionnaire. Surveys were mailed to 329 endorsement-holders in the NSW ocean prawn fishery. Of these 329 questionnaires, 12 were returned undelivered because of incorrect addresses.

The other major stakeholders, whose views on the fishery and its future were sought, were: Sydney Fish Market, which is the major wholesale and retail seafood outlet in NSW; the Nature Conservation Council of NSW, a peak conservation group; and NSW Fisheries. For the Sydney Fish Market and the Nature Conservation Council of NSW, personal interviews were conducted using questions from the survey questionnaire as a basis for discussion. NSW Fisheries preferred that its position be determined from the policy documents and papers prepared on management of the ocean prawn trawl fishery.

Results

A total response rate for the survey of around 14% was achieved (45 out of 317). Approximately 20% (45 of 220) of total active fishers were responded to the survey. This rate is consistent with those of other blind mail surveys and in part may reflect uncertainty over the level of independence of the survey from the government because it was mailed by the government department (NSW Fisheries). Further references to the "fishers" in the results section refers to those responding to the survey.

Stakeholder Views and Perceptions

Fishers

Fishers' objectives.—The most common objectives of present ocean prawn trawl fishers were

- Making a good day-to-day living (64%);
- Building up enough assets to support them in retirement (56%);
- Security for their families over the next 5 to 10 years (49%); and
- Building up a business for their sons and daughters to carry on (40%).

Only a few fishers were interested in building a major fishing business (11%) and making enough assets to set them up in their next business (9%).

Around 60% of fishers expressed a willingness to leave the fishery if there were opportunities available for them in other sectors, but most (71%) of the fishers also thought that there were no alternative employment opportunities in their local areas. Even if some areas had alternative employment opportunities, fishers did not have skills, expertise or desire in those areas to leave the fishery.

Views on Current Status of the Fishery

Most fishers (78%) thought the present zones were appropriate for the species of prawns they caught and that present closures were effective (71%) in protecting juveniles and breeding grounds and providing a long-term future for the fishery.

About equal numbers thought that since they had entered the fishery, the catch per unit effort (CPUE) had stayed about the same (47%) or had decreased (40%). The majority (57%)

thought fishing effort should be reduced by application of more controls. Between 53% and 60% of fishers identified global positioning systems, net design, navigation equipment, engine power, and sonars and sounders as having increased fishing effort.

Clear majorities of fishers thought there had been no change in the species or size of prawns caught since they entered the fishery (67%) and there had been no change in bycatch species caught (62%). Also, the majority thought that present fishing operations had no harmful effects on prawn and other fish habitats.

Views on Management

About 60% of fishers did not feel that they had been, directly or indirectly, involved in decision making and 73% of fishers thought their professional knowledge had not been incorporated into management strategies. Fishers were also concerned that the studies done by scientists were very seldom passed on to them. The processes of policy formulation and implementation were too lengthy, policies were constantly changing, and fishers were consequently unable to plan their businesses.

A majority (53%) of fishers did not think that existing policies would maintain an industry that would survive in the long term. Clear majorities thought that existing policies would not allow them to make enough profits to set up their next business (62%), help build up enough assets to support them in retirement (56%), and help build up a business for their sons and daughters to carry on (58%).

Large majorities considered that the following regulations were needed for effective management of the fishery

• Limitations on vessel numbers, sizes and characteristics (89%);

• Minimum legal mesh sizes (78%);

• Closed areas (78%);

• Use of bycatch reduction devices (73%); and

• Net length regulations (69%).

However, closed seasons, catch limitations, and prohibitions on catching and selling certain types of species were not supported.

The most favored personal preferences for managing the prawn fishery were

• Controlling the number of boats (87%);

• Closed areas (76%); and

• Controlling the sizes of boats (69%).

While 57% of fishers considered that effort should be reduced in both the ocean and estuarine fisheries, 40% of fishers thought the fishery should maintain the present level of effort. A large majority of fishers (89%) were in favor of buy-backs to reduce the number of boats.

Very few fishers preferred using catch quotas (7%) or personal shares (13%) as means of managing the fishery. Fishers believe that these methods restrict their capacity to maximize the prawn catch from their fishing operations.

Only 16% of fishers thought that the fishery should be managed solely by the government. The majority (56%) of fishers thought that it should be managed by fishers and 47% referred a fisheries comanagement system in which all stakeholders were involved. Most (76%) thought the government (general revenue) should pay for or share the cost of managing the fishery.

Fishers obtained information on regulations or policies mainly from NSW Fisheries meetings and letters (84%), but also from fishers' meetings and associations (53%), fishers' co-operatives (40%), and other fishers (40%).

Issues for the Future

Fishers appear to be concerned and uncertain about the future of their industry. The majority (60%) indicated they would leave the fishery now if they had the opportunity. The majority (60%) also considered that their license was not secure under present fisheries policies and (80%) thought their license could not be used as security to obtain loans for capital improvements to their fishing operations. The major difficulty in obtaining funds to upgrade the fishing operation was that lending institutions perceived that fishing was too risky.

Although 69% of fishers thought that future prawn fishers in the next 5–10 years would come from fishing families, 58% did not expect their sons or daughters to be in the fishing industry in the next 5–10 years.

The Postharvest Sector

The post harvest and marketing sector's view was that it had not been involved in all stages of decision-making. For example, influential fisheries advisory bodies such as the Fisheries Resource Conservation and Assessment Council which is the principal group advising the minister on fisheries resource issues, do not have members from the postharvest sector.

The postharvest sector considered that there is no evidence that commercial fishing is the only activity leading to overfishing; hence, reducing effort in commercial fishing alone is not the answer to the overfishing problem. They considered that land-based activities are destroying the habitats and should be controlled. Rapidly increasing recreational fishing also should be regulated effectively. Moreover, the present impact of recreational fishing is largely unknown.

The postharvest sector considers current fisheries policies are discriminatory towards a large majority of consumers and believes food security will never be achieved if policies keep encouraging recreational fishing by diverting commercial fishery resources into the recreational fishery. The postharvest sector expressed these concerns as a result of recent initiatives by NSW Fisheries to create recreational fishing havens to enhance recreational fishing opportunities in coastal areas, mostly estuaries. Commercial fishing is generally prohibited in these areas although limited commercial fishing of some species is permitted in some recreational fishing havens. Although government compensates any loss of commercial fishing opportunities due to establishment of recreational fishing havens, a number of commercial fishers were not willing to accept these packages, as commercial fishing has been their lifestyle, not just a business opportunity.

The postharvest sector believes that financial institutions are not interested in investing in commercial fisheries for three reasons: fisheries may be closed at any time, government policies are changing constantly, and the licenses are not secure. The seafood industry requires a legislative framework that allows for long-term business decisions. In contrast to the fishers, the postharvest sector supports property rights-based fishery management strategies. Resources should be shared fairly among all user groups. The management costs should also be shared by both the commercial and recreational sectors.

Conservation Sector

The Nature Conservation Council of NSW considered that it was viewed in the fisheries context as a minor stakeholder and so felt unable to influence policymakers. The peak conservation group considered that NSW Fisheries was acting on public perceptions rather than real conservation needs. Conservation plans for fisheries that addressed these real issues were needed.

The conservation group considered that fishers were able to circumvent management restrictions aimed at limiting effort. Fishing effort continues to increase due to constant refinements in fishing practices and to improvements in technology, such as the use of global positioning systems. They consider that net productivity of the fishery is decreasing. Trawling is considered to be a highly destructive fishing practice, removing a huge amount of seabed life and resulting in huge amounts of bycatch.

They consider that NSW Fisheries statistics are unreliable and the department should collect fishery-independent data. Fishers' skills such as awareness of seasonal changes in species distributions should also be taken into account in estimating effort. A recent meeting on protecting small and juvenile prawns does not seem to have been effective in improving their protection. Existing closures are not in the proper places to conserve stocks and breeding areas, prawn fishing areas are expanding and changing.

The conservation group considered that the fishery needs to be restructured by reducing fishing effort, restricting the number of fishing nights, and identifying genuine fishers. Local communities should be given access to the fishery. A full financial plan needs to be developed for the fishery to assist it in obtaining financial aid for restructuring. This should involve development of a 10-year plan, addressing all issues of profitability and conservation for the fishery. A comanagement strategy should be introduced and all parties that benefit from the fishery should share the costs of management and research. A buyback scheme should be established to reduce the existing level of effort, marine protected areas should be established to protect breeding and nursery areas, and designated fishing grounds proclaimed to reduce the spread of fishing over wider areas of habitat. Responsible fishing should be encouraged, including the adoption of lower impact fishing methods and an improved vessel monitoring system introduced to ensure that vessels stay out of closed areas. More studies into selective fishing were needed and input measures devised to address the above issues.

NSW Fisheries

NSW Fisheries seeks to develop plans and strategies for the long term sustainable use of fisheries resources by promoting the principles of ecologically sustainable development to enhance biological diversity and to take a precautionary approach to decision making (NSW Fisheries 1999). Details of institutional arrangements are contained in the introduction (above). The current management system for this fishery is based solely on input controls.

NSW Fisheries has prepared a number of documents on the management of the Ocean Prawn Trawl Fishery over recent years, either providing information or seeking comment, or both. Since 1994, these include:

• New South Wales Offshore Prawn Trawling Management Rules Version 1.2 (NSW Fisheries 1994) (information document);

• Ocean Prawn Trawl Fishery Specific Options Paper (NSW Fisheries 1997) (seeking comment);

• Draft Ocean Prawn Trawl Strategies Paper: Ocean Prawn Trawl Restricted Fishery, NSW (Fisheries 1998a) (information document);

• Ocean Prawn Trawl Fishery Issues Paper (NSW Fisheries 1998b) (seeking comment);

• NSW Fisheries Profile - Ocean Prawn Trawl (NSW Fisheries 1999) (information document); and

• 2001 Ocean Prawn Trawl Fishery Report (Montgomery and Craig 2001) (information document).

Most of the formal interactive aspects of management in this fishery between NSW Fisheries and the fishers have been through various versions of management advisory committees (MACs) and port meetings.

According to recent reviews by NSW Fisheries:

•The exploitation status of the fishery is fully fished and current stock levels of both oceanic and estuarine prawns (eastern king and school) show evidence of long-term declines in stock levels. The total annual landings of prawns have fallen since about the mid-1980s and the ocean component of annual landings has also fallen since 1973–1974;

•The Ocean Prawn Trawl Management Advisory Committee members believe that quota management or output controls on prawn species are unrealistic options for the ocean prawn trawl fishery because there are high levels of seasonal variation in available prawn stocks due to factors such as prevailing ocean currents and local rainfall. High grading of prawns may result from the use of output controls. Compliance in quota management of this fishery would be difficult and expensive, due to the many ports of landing;

• Input controls have proven to be effective in capping fishing effort, with the result that catches have remained generally constant over time. Input controls already used in the fishery reduce unwanted bycatch, but the level of understanding of bycatch species is very small;

• The reliability of assessment procedures is low. However, the quality of information available for management continues to increase;

• Fishers probably report a greater proportion of their catches during discussions on restructuring to stake higher claims;

• There is no information about the economics of vessels operating in the ocean prawn trawl fishery; hence, the optimal conditions for economic efficiency within the fishery are not known; and

• There is no information to assess whether trawling has impacted upon habitat and the biota not caught but affected by the trawling process.

Discussion

The Ocean Prawn Trawl Fishery is the most commercially valuable and regionally important fishery in New South Wales, Australia. The fishery has a strong base in regional communities and is facing the problems common to many other capital-intensive coastal marine fisheries around Australia.

Despite a long history of fisheries biological research, complex input-based regulations and intense efforts in community consultation, management has been largely ineffective in limiting effort increases, maintaining sustainable levels of average annual landings, reducing costs of the industry, protecting deteriorating prawn nursery grounds and essential habitats, resolving user conflicts, and protecting the livelihood of the people who entirely depend on the prawn fishery. However, some progress has been made in converting a "free and open-access" fishery to a "restricted" fishery, improving consultation through a MAC and reducing bycatch.

Change and Uncertainty

The NSW Ocean Prawn Trawl Fishery is facing a period of fundamental change, reflected by the number of papers on its management and future prepared by NSW Fisheries over the last decade. Most commercial prawn fishers have a long fishing tradition and come from fishing families, and some of them depend entirely on fishing for their day-to-day living. The objective of most prawn fishers was to make a good day-to-day living, with only a few seeing their operations as purely commercial in a business sense. There is, however, increasing concern in the community over impacts of fishing operations, both for sustainability of catches and the wider effects of fishing on the marine environment and there is uncertainty by fishers about their futures.

Some factors feeding uncertainty and insecurity of the prawn fishers are:

• A recent court case in the NSW Land and Environment Court (Sustainable Fishing and Tourism Inc v the NSW Minister for Fisheries and others, 2000) which resulted in NSW Fisheries having to conduct environmental impact statements for all fisheries;

• The Commonwealth Government's *Environment Protection and Biodiversity Conservation Act 1999*, which requires that fisheries satisfy ecologically sustainable development principles before the Commonwealth will issue export permits for fishery products;

• Introduction by the NSW Government of exclusive recreational fishing zones in 29 bays and estuaries and buy-out of commercial fishers; and

• Buy-out of commercial fishers by the NSW Government from inshore marine waters to establish noncommercial fishing zones in marine parks.

Perceptions of Management

In this climate of uncertainty about their future, our survey suggests that there is little trust and confidence by fishers in the management of their fishery. This lack of confidence is reflected in the views of fishers that they were not fully involved in decision making; that scientific knowledge was not passed on to them; that poor policy development and dissemination meant that they were unable to make long-term plans for their businesses; that their licenses could not be used as security for building their fishing operations, and that constant changes to policies and management personnel inhibited their ability to plan for the future. In this framework, a high proportion of survey respondents indicated they were willing to leave the fishery, but recognized that there were no alternative employment opportunities in their communities.

There were similar themes in the postharvest and conservation sectors: that they had little involvement in fisheries management processes, or were marginalized; that fishing licenses lacked security; and that managers were unable to address important issues.

NSW Fisheries, for its part, seems intent on continuing to manage through input controls, although it recognizes that data used for fisheries assessment may be unreliable and there are little, if any, fishery independent data.

Overcapacity and Effort Reduction

A pressing problem for the Ocean Prawn Trawl Fishery is the large overcapacity of effort compared with that needed to take the available catch. Despite a management strategy moving towards more complex, input-based controls to limit fishing effort, the fishing effort is still increasing. This is partly due to the ability of most fishers to increase their days fished (McIlgorm and Campbell 1998). In addition, there is some latent effort in the fishery as some license holders fishing only occasionally or not at all. These fisheries either do not depend on prawn fishery or opted to be out of the fishery as their fishing operations are not commercially viable. If the latent effort is reactivated, it will further increase the level of active effort in the fishery. Fishers recognized that technological factors such as global positioning systems and improved fishing methods were continuing to result in increases in effort. The most common response of fishers was that this required more input controls.

It is clear that latent effort, potential effort, and the total effort applied to the fishery must be reduced in the next 5–10 years (McIlgorm and Campbell 1998). The excess fishing effort in the fishery could be reduced by adopting output-based management strategies.

An ideal approach would be to assist the fishers in voluntarily working out appropriate fishing effort levels and adopting responsibility for maintaining sustainable fishing practices by providing incentives rather than ineffective regulatory controls. Around 60% of fishers expressed a willingness to leave the fishery if there were opportunities available for them in other sectors. An important element in effort reduction, then, could be introducing a buy-back scheme to assist some fishers in leaving the fishery. Campbell (1989) suggested that any reduction in effort generates an economic benefit which must be weighed against increased costs. This buy-back would be complicated by fishers remaining in the fishery because of lack of alternative employment, a lack of necessary skills and expertise to start another business, or both. To address this problem, the outgoing fishers could be given training in their alternative professional interests to prepare them for entering new businesses.

Effort reduction also could involve placing a cap on total fishing nights at a level lower than the present one and then allocating fishing nights among license-holders.

Stakeholder Involvement

NSW Fisheries is attempting to improve stakeholder participation in management through publishing discussion papers and establishment of MACs. Our survey suggests that these initiatives are not presently effective. The majority of prawn fishers did not feel they had been directly or indirectly involved in decision making. While everyone agrees that all stakeholder groups should be involved in decision making, the question is, what is the appropriate level of involvement? This is where the interest groups differ. The postharvest sector, represented by the Sydney Fish Market, is not a member of some influential fishery management committees. The peak conservation group, Nature Conservation Council of NSW, considers that it is a minority group among all stakeholders and so is unable to influence policymakers for any major policy change.

The challenge for NSW Fisheries lies in recognition that fisheries management increasingly is being viewed more as people management than fish management (Ditton and Hunt 2001). Gaining industry support for management decisions is one of the major issues, because fishers feel that the regulations are thrust on them. The inclusion of fishers' knowledge in fisheries management plans is important to achieve full support for successful implementation of regulations. As Lauber and Knuth (2000) noted, citizen participation can :

1) Improve relationships between stakeholders which can increase their ability to work together toward management objectives,

2) Increase the capacity of citizens or agencies to participate constructively in management by providing them with skills, experience, or knowledge, or

(3) Change beliefs, attitudes, or behavior to help management processes occur more smoothly.

Stakeholders' Understanding of the Problems

All stakeholder groups agree that the fishery needs to be restructured, but they have differing, and often contradictory, views on who is responsible for unsustainable fishing practices and how to overcome these difficulties and make the prawn fishery sustainable. Lack of reliable data for fishery assessment and communication between the interest groups seems to be encouraging a debate on stakeholder c
laims rather than achieving a consensus. Stakeholders are divided on several issues, such as:

• Which factors are contributing to overfishing and unsustainable fishing practices?

Conservation groups claim that trawling is a highly destructive fishing practice and removes huge amounts of seabed life and bycatch. Most prawn fishers thought that present fishing operations had no harmful effects either on prawn stocks or their habitats. The postharvest sector considers that commercial fishing is one of many activities leading to overfishing and habitat destruction. The postharvest sector believes land management practices and recreational fishing should also be considered as major issues when addressing overfishing and habitat destruction.

• Which type of management system should be in place to address key issues in the fishery?

While there is apparent agreement over there being excess fishing effort, stakeholders differed on effort reduction strategies. Fishers preferred input-based regulations and did not support management measures like closed seasons, catch quotas, and prohibitions on catching and selling certain types of bycatch species. This shows that the fishers do not perceive benefits of different policy instruments; particularly there is a lack of understanding of the incentives of output-based management systems such as individual transferable quotas (ITQs). Both the postharvest and conservation sectors favored improved property-rights based systems.

• Who should manage the fishery?

The government's absolute authority to manage the fishery is also controversial. Stakeholders want to share both the responsibility and the authority for management through comanagement arrangements. During NSW Fisheries consultation processes, the option of a comanagement system was not discussed. Our study shows that both fishers and the conservation sector are in favor of a comanagement system.

Contradictory Views among Fishers

There are not only differences among the stakeholder groups: fishers also hold contradictory views on several issues in the fishery. These include:

• Fishers agree that new fishing technology and better fishing practices have increased fishing effort without increasing total annual catches and that the fishing fleet now works 50% more nights than 10 years ago. But they still opt for input-based controls to reduce effort, while many fishers think that they must be allowed to increase the horsepower of their engines;

• Most fishers did not want catch limits, but most thought that existing policies would not maintain the fishery;

• Fishers agree that NSW Fisheries hold full and frank port meetings, but they think they have not been adequately involved in decision making and policymakers should talk more to experienced fishers and use their knowledge;

• Similar proportions of fishers considered that effort should be reduced and that effort should be maintained at present levels; and

• Although fishers thought that the future fishers will come from fishing families and existing crew and that they wanted to build a business for their sons or daughters, a majority also thought that their family is unlikely to be involved in the fishery.

Reducing Differences and Conflicts among Stakeholder Groups

All stakeholders with an interest in the prawn fishery—fishers, politicians, scientists, managers, processors, marketers, distributors, conservationists, boat builders, net makers, financiers, fuel suppliers, and transporters—must be able to make more informed choices and be considered in policy decision making. As long as serious differences remain between major stakeholders, it is unlikely that management strategies will achieve their declared objectives.

In order to reduce conflict within the fishery and between fishers and other user groups, it is important to understand the motives and concerns of the respective parties (Somers 1990). It is important for fisheries managers to reconcile fishers' opinions and attitudes with management goals, both to gain public support for regulations and to avoid legal challenges. Fishers should not feel that the new regulations are thrust on them without consideration of the socioeconomic implications of changing policies. Means for achieving consensus could include:

• Clearly stating management goals and objectives and addressing any mismatches between policy objectives and the needs of stakeholders in the fishery;

• Passing scientific findings to all resource users and other interested groups in the form of nontechnical reports and paying close attention to fishers' knowledge and its use in improving the management database;

• Seeking the views and involving all stakeholder groups in sharing the responsibility and the authority for decision making and implementing the decisions; and

• Considering the individual social and economic constraints and differences between stakeholder groups.

Strengthening the Socioeconomic Database for the Fishery

Traditionally, the management policies of NSW Fisheries have been mainly based on biological research findings and are primarily aimed at maximizing catch—annual landings of prawns. Fisheries research is dedicated to biological aspects of known species and to identify new commercial prawn species and fishing grounds in NSW coastal waters. Core socioeconomic issues have long been neglected in fisheries research and, hence, the management advisory committees have not been effective in providing adequate management advice to policymakers. There has been virtually no consideration of the cost of taking catch, which is fundamental information for assessing the economic state of the fishery. Similarly, there has been no management consideration of taking catch when it will provide its highest returns for the fishers. Other neglected areas are the social impact of fisheries policies on communities and the impact of great differences in distribution of catch among fishers as occurs in the Ocean Prawn Trawl Fishery.

There is a need to strengthen the socioeconomic database of the fishery by collecting data on economic and social issues in the fishery. In addition, fishers have detailed knowledge of their resources, their environment, and their fishing practices, that is rarely systematically collected (Neis et al. 1999). Most fishers thought their professional knowledge had not been incorporated into management policies. Fishers are among the first group to notice, and be affected by, any change that occurs in the fishery. Collection of socioeconomic data could be supplemented by cost-effective methods, where necessary, such as the use of postgraduate students and studies (Ruello 1975a).

Security of Access Rights and Funding

The fishery is presently characterized by poorly developed property-rights, and constant changes to management policies and managers, which make it difficult for fishers to plan for the future. Fishers need improved security of access rights for planning and developing their businesses in the long-term. These rights could be the basis of a new management plan addressing the needs of the fishery over the next decade. Development of a simple formula for trading in these access rights would assist this process.

Another issue that should be addressed is how management and restructuring of the fishery will be funded. The recovery of management costs is under a moratorium for the duration of the current management plan and no community contribution is payable. This constitutes a subsidy for the fishery from public funds. Fisheries must be able to generate resource rent and cover their management costs as part of the proper utilization of valuable fish stocks. The sustainability of the fish stocks and the well being of fishers is impeded by stocks being harvested so inefficiently that the fishers require a public subsidy in order to survive.

The vision of a sustainable prawn fishery will be realized only when the above issues attract urgent attention by fisheries managers; otherwise the fishery may reach an "irreversible" decline.

Acknowledgments

We are grateful for the support of New South Wales Ocean Prawn Trawl Fishers, NSW Fisheries, the Nature Conservation Council of New South Wales, and Sydney Fish Market Pty

Limited in conducting this study. This study was supported by funds from the Southern Cross University research vote to Professor Donald F. Gartside.

References

Abdullah, N. M. R., K. Kuperan, and R. S. Pomeroy. 1996. Transaction costs and fisheries comanagement. Fisheries Co-management Working Paper 15. International Center for Living Aquatic Resources Management, Manila, Philippines.

Campbell, H. F. 1989. Fishery buy-back programmes and economic welfare. Australian Journal of Agricultural Economics 33(1):20–31.

Cochrane, K. L. 2000. Reconciling sustainability, economic efficiency and equity in fisheries: the one that got away? Fish and Fisheries 1:3–21.

Cury, P., and P. 2001. Hunting became a secondary activity 2000 years ago: marine fishing did the same in 2021. Fish and Fisheries 2:162–169.

Davis, D., and D. F. Gartside. 2001. Challenges for economic policy in sustainable management of marine natural resources. Ecological Economics 36:223–236.

Ditton, R. B., and K. M. Hunt. 2001. Combining creel intercept and mail survey methods to understand the human dimensions of local freshwater fisheries. Fisheries Management and Ecology 8:295–301.

Fisheries Management Act 1994. Available: http://www.austlii.edu.au/au/legis/nsw/consol_act/fma1994193/. (July 2002).

Healey, M. C. 1997. Comment: the interplay of policy, politics, and science. Canadian Journal of Fisheries and Aquatic Sciences 45:1427–1429.

Hynd, J. S. 1975. The management of Australian prawn fisheries from the viewpoint of a fisheries biologist. Pages 246–251 *in* P.C. Young, editor. First Australian National Prawn Seminar, Australian Government Publishing Service, Canberra.

Jentoft, S. 2000. The community: a missing link of fisheries management. Marine Policy 24:53–59.

Kennely, S. J., Liggins, G. W., and M. K. Broadhurst. 1998. Retained and discarded by-catch from oceanic prawn trawling in New South Wales, Australia. Fisheries Research 36:217–236.

King, M. 1995. Fisheries biology, assessment and management. Fishing News Books, Blackwell Publishing, London.

Lauber, T. B., and B. A. Knuth. 2000. Citizen participation in natural resource management: a synthesis of HDRU research. HDRU Series No. 00–7. Cornell University, Department of Natural Resources, Human Dimensions Research Unit, HDRU Series No. 00–7, Ithaca, New York.

McIlgorm, A., and H. Campbell. 1998. Ocean prawn trawl fishery adjustment study: modelling of effort and latent effort in the ocean prawn trawl fishery. A report to NSW Fisheries by Dominion Consulting Pty. Ltd., Hurstville, New South Wales, Australia.

Montgomery, S. S., and J. Craig. 2001. Ocean prawn trawl fishery report. Pages 105–119 in S. Kennelly and T. McVea, editors. Status of fisheries resources 2000/2001. NSW Fisheries Institute, Cronulla, New South Wales, Australia.

Neis, B., D. C. Schneider, L. Felt, R. L. Haedrich, J. Fischer, and J. A. Hutchings. 1999. Fisheries assessments: what can be learned from interviewing resource users? Canadian Journal of Fisheries and Aquatic Sciences 56:1949–1963.

NSW Agriculture and Fisheries. 1990. NSW east coast trawl fishery (Part: A): offshore prawn fishery. NSW Agriculture and Fisheries, Fisheries management plan No. 1.

NSW Fisheries. 1994. NSW offshore prawn trawling management rules - version 1.2. 9/94. NSW Fisheries Research Institute, Cronulla, Australia.

NSW Fisheries. 1997. Ocean prawn trawl: a fishery specific options paper. December. NSW Fisheries Research Institute, Cronulla, Australia.

NSW Fisheries. 1998a. Draft Ocean Prawn Trawl Strategies Paper: Ocean Prawn Trawl Restricted Fishery. NSW Fisheries Research Institute, Cronulla, Australia.

NSW Fisheries. 1998b. Ocean prawn trawl fishery: an issues paper. August. NSW Fisheries Research Institute, Cronulla, Australia.

NSW Fisheries. 1999. Fisheries profile: ocean prawn trawl. NSW Fisheries Research Institute, Cronulla, Australia.

NSW Parliament. 1997. Standing committee on state development report on fisheries management and resource allocation in New South Wales. Report No. 17, NSW Government, Sydney, Australia.

Racek, A. A. 1957. Penaeid prawn fisheries of Australia with special reference to New South Wales. Research Bull. St. Fish, 3, Chief Secretary's Department, New South Wales, Australia.

Ruello, N. V. 1974. History of prawn fishing in Australia. Australian Fisheries. February, 2:(33).

Ruello, N. V. 1975a. A historical review and annotated bibliography of prawns and prawning industry in Australia. In P. C. Young, editor. First Australian National Prawn Seminar. Australian Government Publishing Service, Canberra, Australia.

Somers, I. F. 1990. Manipulation of fishing effort in Australia's penaeid prawn fisheries. Australian Journal of Marine Freshwater Research 41(1):1–12.

Young, M.D. 1999. The design of fishing-right systems - the NSW experience. Ecological Economics, 31:305–316.

Additional Works of Importance

Bowen, B. K., and D. A. Hancock. 1985. Review of penaied prawn fishery management regimes in Australia. Pages 247–265 *in* P. C. Rothlisberg, B. J. Hill, D. J. Staples, editors. Second Australian National Seminar, Cleveland.

Dall, W. 1985. A review of penaeid prawn biological research in Australia. Pages 11–21 *in* P. C. Rothlisberg, B. J. Hill, D. J. Staples, editors, Second Australian National Seminar, Cleveland.

Galaister, J. P., S. S. Montgomery, and V. C. McDonall. 1990. Yield-per-recruit analysis of eastern king prawns *Penaeus plebejus* Hess, Eastern Australia. Australian Journal of Marine and Freshwater Research 41(1):175–197.

Gorman, T. B. S., and K. J. Graham. 1975. Deep water prawn survey off New South Wales. Pages 162—173 *in* P.C. Young, editor. First Australian National Prawn Seminar. Australian Government Publishing Service, Canberra.

Graham, K. J., and T. B. Gorman. 1985. New South Wales deepwater prawn fishery research and development. Pages 231–243 *in* P. C. Rothlisberg, B. J. Hill, D. J. Staples, editors.Second Australian National Seminar, Cleveland.

Haysom, N. M. 1985. Review of the penaeid prawn fisheries of Australia. Pages 195–203 *in* P. C. Rothlisberg, B. J. Hill, D. J. Staples, editors. Second Australian National Seminar, Cleveland.

Kennely, S. J., and M. K Broadhurst. 1998. Development of by-catch reducing prawn-trawl and fishing practices in NSW's prawn–trawl fisheries (and incorporating an assessment of the effect of increasing mesh size in fish trawl gear). NSW Fisheries Final Report Series No. 5, NSW Fisheries and FRDC, Sydney.

Loneragan, N. R., and S. E. Bunn. 1999. River flows and estuarine ecosystems: implications for coastal fisheries from a review and a case study of the Logan River, Southern Queensland. Australian Journal of Ecology. 24:431–440.

Montgomery, S. S. 1988. Trends in New South Wales prawn catch. Australian Fisheries 47(8):24–30.

Montgomery, S. S. 1990. Possible impacts of the greenhouse effect on commercial prawn populations and fisheries in New South Wales. Wetlands (Australia)10:1–2.

Montgomery, S. S., and R. H. Winstanley. 1982. Prawns (East of Cape Otway). CSIRO Marine Laboratories, South Eastern Fisheries Committee, Fishery Situation Report 6. Hobart, Australia.

Pollard, D. A. 1976. Estuaries must be protected. Australian Fisheries 36(6):6–10.

Racek, A. A. 1959. Prawn investigations in eastern Australia. Research Bull. St Fish N.S.W 6:1–57.

Ruello, N. A. 1967. $2 million a year from N.S.W. prawning industry. Australian Fisheries Newsletter 26(9): 23- 25:27.

Ruello, N. V. 1975b. Biological research and the management of prawn fisheries in New South Wales. Pages 222–233 *in* P. C. Young, editors. First Australian National Prawn Seminar. Australian Government Publishing Service, Canberra, Australia.

Ruello, N. V. 1975c. Geographical distribution, growth and breeding migration of the eastern Australian king prawn *Penaeus plebejus* Hess. Australian Journal of Marine Freshwater Research (26):343–354.

Ruello, N. V. 1977. Migration and stock studies on the Australian school prawn *Metapenaeus macleayi.* Marine Biology (41):185–190.

Staples, D. J., D. J. Vance, and D. S. Heales. 1985. Habitat requirements of juvenile penaeid prawns and their relationship to offshore fisheries. Pages 47–54 *in* P. C. Rothlisberg, B. J. Hill, D. J. Staples, editors. Second Australian National Seminar, Cleveland.

Walker, R. H. 1975. Australian prawn fisheries. Pages 284–304 *in* Young, P.C. editors, First Australian National Prawn Seminar. Australian Government Publishing Service, Canberra, Australia.

American Fisheries Society Symposium 49:139–148

Building Capacity for 21st Century Fisheries Management: A Global Initiative

Laura W. Jodice*

Department of Parks, Recreation, and Tourism Management
263 Lehotsky Hall, Box 340735, Clemson University, Clemson, South Carolina 29634, USA

Gilbert Sylvia

Coastal Oregon Marine Experiment Station, Oregon State University, Hatfield Marine Science Center
2030 Marine Science Drive, Newport, Oregon 97365, USA

Michael Harte

Marine Resource Management, College of Oceanic and Atmospheric Sciences & Oregon Sea Grant
Oregon State University, 104 COAS Admin Building, Corvallis, Oregon 97331, USA

Susan Hanna

Coastal Oregon Marine Experiment Station, Oregon State University, Hatfield Marine Science Center
2030 Marine Science Drive, Newport, Oregon 97365, USA

Kevin Stokes

New Zealand Seafood Industry Council Ltd.
Private Bag 24-901, Wellington, 6001, New Zealand

Abstract.—The world's nations confront significant and complex challenges in managing fisheries resources in the 21st century. Given evolution toward ecosystem management, rights-based management, and international ocean governance, is the current supply of fishery managers capable of meeting the demand for effective and sustainable 21st century fishery management? The international initiative *Training Managers for 21st Century Fisheries* addresses these challenges. The initiative was first conceived in Queenstown, New Zealand, in December 2001, when 63 government, industry, academic, and nongovernmental organization leaders from Oceania, North America and Europe discussed the challenges for fishery management. Workshop participants defined the training, education, and professional working environments necessary to produce fishery managers who are problem solvers, leaders, and innovators. They also produced a collective vision and a comprehensive list of skill and knowledge requirements. Participant consensus produced the following eight priority strategies for developing training capacity at international, national, and regional levels 1) develop creative partnerships within and among institutions, sectors, and nations; 2) include the management process as a learning experience; 3) broaden and lengthen career paths; 4) identify the gaps between training supply and demand; 5) create a Web site that shares information about training opportunities and resources; 6) encourage industry scholarships; 7) develop a case study library; and 8) establish a network of training providers. The workshop

*Corresponding author: jodicel@clemson.edu

concluded with establishing an international steering committee charged with facilitating implementation of these strategies and guiding the initiative. To progress the global initiative, steering committee members have begun fostering partnerships and communications with other training efforts and are initiating training in case study teaching methods. The intended outcome of this global initiative is a coordinated international fisheries management training framework that supports integration, partnership, cooperation, and exchange. The workshop report and additional information is available at http://oregonstate.edu/dept/trainfishmngr.

Introduction

The developed and developing nations of the world confront complex challenges in managing fisheries resources in the 21st century. Achieving sustainable fisheries will depend on the strength and resilience of the fisheries management process as well as ecosystem structure and function. While attention focuses on the need for new institutional ideas, designing and implementing effective governance may be imperiled by inadequate investment in the human capital needed to lead, innovate, and manage fisheries and aquatic ecosystems. To address this challenge, an international initiative, *Training Managers for 21st Century Fisheries,* was established with an inaugural workshop in Queenstown, New Zealand, in December 2001 (Jodice et al. 2003). This paper summarizes areas of consensus and collective recommendations of the 63 government, industry, academic, and nongovernmental organization (NGO) leaders from Oceania, North America, and Europe who participated in the 2001 workshop and initiative formation. The *Training Managers for 21st Century Fisheries* initiative defines strategies for progressing capacity building through planned, integrated, cooperative, and coordinated international fisheries management training programs that will support fisheries that can fully utilize as well as sustain marine ecosystems at regional, national, and international levels.

Collective Vision for the Future

The initiative is guided by the collective vision that 21st century fishery managers should be capable of balancing utilization and sustainability mandates, structuring and allocating property rights, advancing self-governance, defining and implementing ecosystem management, designing cooperative research and management, contending with risk and uncertainty, addressing international management, integrating fisheries within ocean governance and conservation regimes, and reducing oppressive bureaucracy and litigation. These capabilities are critical as fisheries management evolves from a single species to a broader and more complex marine ecosystem management focus. The immediacy of this challenge is also highlighted by rising demand for marine reserves and protected areas. Ultimately, success will rely on the capacity of fisheries managers and specialists, including industry, government, and stakeholder representatives, to develop, participate in, and support, effective fisheries governance at regional, national, and international levels. Cooperation among stakeholders and management jurisdictions is significant to success. But more important will be the ability of managers and stakeholders to design institutions that internalize adverse ecosystem impacts and create incentives that drive the learning processes and innovation necessary for successful adaptive comanagement.

Current Investment

Participants in the *Training Managers for 21st Century Fisheries* initiative unanimously agree that governments and institutions have not adequately invested in the human capital capable of codesigning and implementing the institutional structures which will lead to rational management of 21st century fisheries. The initiative illustrates this lack of investment with the following observations

1. Fishery managers worldwide have received little formal training in fisheries management.

2. An insufficient number of education programs exist that provide professional development, fishery management curricula that integrate leadership, critical decision-making, and systems level thinking while providing significant hands on experiential learning opportunities (Figure 1).

3. Recruitment and retention of quality managers is difficult and will intensify with retirement of upper level managers.

4. There is no single vision of fisheries manager.

5. The definition of fisheries manager has broadened in response to evolving institutions and increased participation in management to include many types of players who significantly participate in the management process.

a. A participant's role in management depends upon the institutional setting (sci entific centralized, pluralistic and comanagement, rights-based, self-governing), sector (industry, government, NGO), management scale (international, national, local), type of fishery, and level of economic development.
b. Depending on the governance system, there are many classes of managers, with varying levels of accountability and authority: i) stakeholders, ii) stakeholder representatives, iii) directors of private sector and NGO groups, iv) mid-level government managers, v) lead managers of government agencies, vi) elected policymakers, and vii) policy analysts and institutional designers.

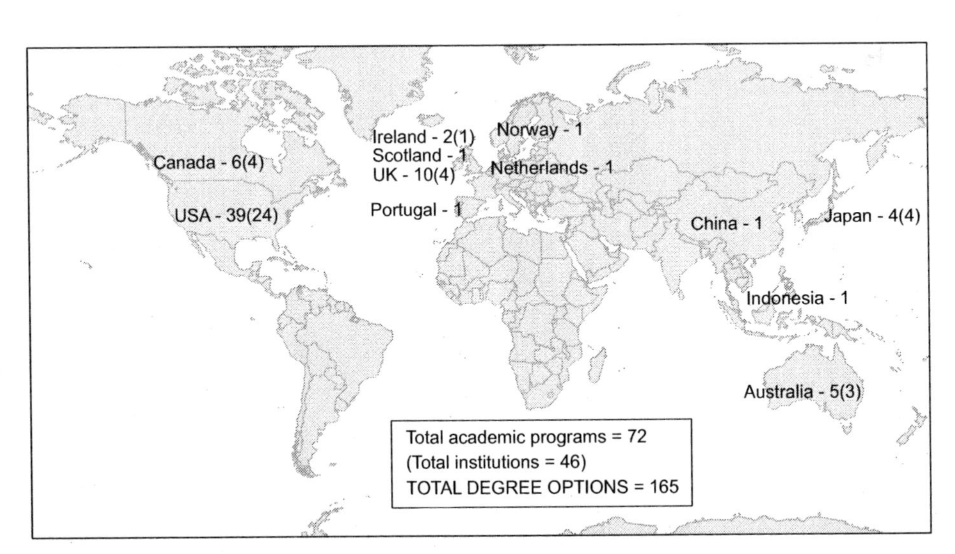

Figure 1. Distribution of fisheries management academic programs from 2001 review of programs with available online literature in English (Jodice et al. 2003).

For example, the management performance of many U.S. fisheries suggests managers may not have the required knowledge base to cope with the burden of new regulations layered on old, "leaving the way to redirect fishery management toward ecosystem-based management unclear" (Hanna 2002). The past practice of promoting biological scientists to key management positions also appears insufficient. This is illustrated by Rassam and Eisler's (2001) survey of North American (United States and Canada) fisheries administrators, where fishery and aquatic science professionals were found to be in need of continuing education in fisheries management (64%), public administration (62%), management theory (52%), natural resource economic theory (40%), fisheries law (43%), technical writing (79%), and communications (41%). Furthermore, 30–45% of NOAA Fisheries employees who are upper level scientists, managers, and administrators are estimated to be eligible for retirement in the next few years (Ocean Studies Board 2000; Holliday 2001; NAPA 2002). The U.S. situation is mirrored in other nations. The current investment in human capacity and excellence is failing traditional single species fisheries management and provides little basis for meeting the complex management challenges implied when reconciling fisheries with conservation and ecosystem management.

What Are the Needs? (Further Defining the Gap)

Based on the 2001 workshop consensus, the initiative created a comprehensive list of skills and knowledge requirements for fisheries managers, summarized in Table 1 (see Jodice et al. 2003 for full list). Underlying this effort is the collective understanding that fisheries management is a dynamic process requiring team-based analysis and implementation and involving individuals with strengths in different disciplines and skill sets that must be melded together to address complex problems. Therefore, training that meets the broadened definition of a 21st century fisheries manager must recognize all significant participants in the fisheries management process and should be guided by the following observations:

• All necessary skills and knowledge (Table 1) cannot exist in one manager.

Table 1. Summary of skill and knowledge needs for 21st century fisheries management (*Training Managers for 21st Century Fisheries* Workshop, Queenstown, New Zealand, December 5–7, 2001).

Basics	Leadership and management	Fisheries specific
• Sciences –Biology –Ecology –Conservation biology • Economics • Social science • Policy and law • Business	• Communications –Conflict resolution –Consensus building –Facilitation –People skills • Intercultural skills • Systems thinking • Critical thinking • Decision making • Problem solving • Risk analysis	• Fisheries science –Fisheries management tools –Risk analysis –Stock assessment –Ecosystem management • Knowledge of all stakeholder groups • Managing specialist and decision maker interface • Incorporating Indigineous knowledge • Incorporating industry knowledge

• All participants in fisheries management require a basic minimum level of common skills and knowledge.

• Each class of manager needs different levels of competency.

• Minimum competency levels should be defined specific to each managerial class (Figure 2).

• Training needs assessment should rely on benchmarking of current knowledge, attitudes, skills, and abilities against an idealized profile for a manager's role in the process (Table 2).

Capacity building efforts must include improved retention of well-trained managers. Identified barriers to retention include limited professional development opportunities; low job satisfaction; perceived conflict between science and management disciplines; the political and, in some jurisdictions, litigious nature of fisheries management; and low morale. Retention efforts must also address decreasing involvement of nonagency stakeholders who may disengage from the process due to negative outcomes or more success through other political or legal strategies.

Building Human Capital

Building human (intellectual) capital is the dynamic process of discovering, collecting, and synthesizing knowledge that directs human action in extending existing systems or the creation of new systems (Resilience Alliance 2003). Society's well-being depends upon developing institutions that compel learning, build infrastructure to create, store, and disseminate knowledge and stimulate flexibility in problem solving. Institutional commitment to building human capital is a long-term and challenging goal given the contemporary crisis inherent in fisheries management. Strengthening the management process requires building human capacity to deal with change, surprise, and uncertainty through learning and adaptation (Folke et al. 2003). Ultimately, achieving sustainable fisheries through more effective institutions that internalize ecosystem principles will depend upon leadership and sustained commitment to building the adaptive capacity and resilience of the fisheries management process. True commitment to adaptive management requires understanding that unlearning and new learning are imperative prerequisites for effective management

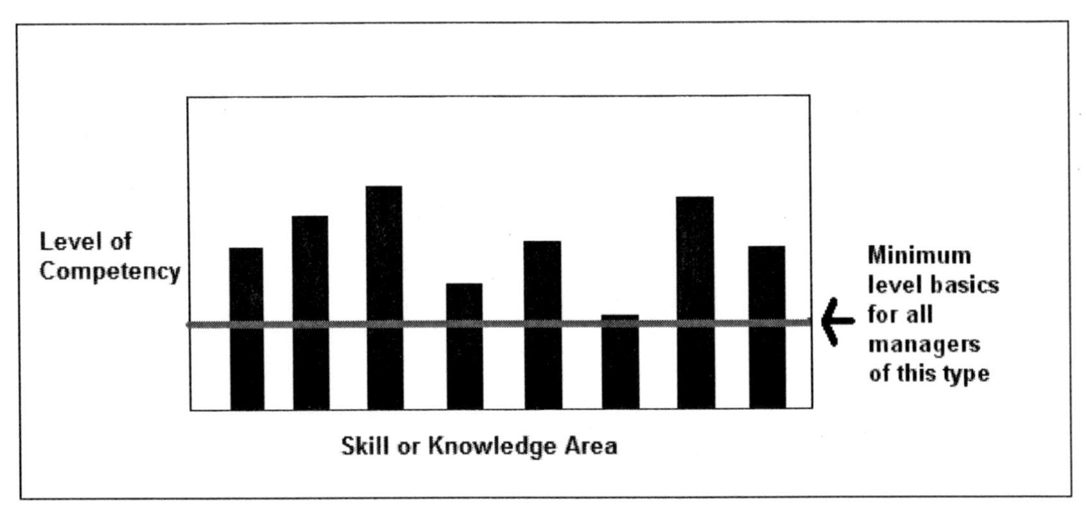

Figure 2. Conceptual model of fisheries manager competencies profile.

Table 2. Sample individual benchmark assessment based on an ideal profile (Harte 2003).

Skill/knowledge	Current score	Ideal score	Training needs/notes
RATING KEY: 1 = Training, 2 = Developing, 3 = Competent, 4 = Advanced, 5 = Expert			
Stock assessment	1	2–3	Limited formal training in stock assessment or related discipline. Key training need.
Cost benefit analysis	3	3–4	Sound innate skills, overall score restricted by limited knowledge/ experience of formal analytical methodologies (e.g., multicriteria analysis).
Leadership ability	2	4	Needs training and opportunities to be able to lead diverse groups through to successful outcomes. Particular emphasis required on transformational leadership skills.
Ecosystem management	1–	3	Limited understanding and experience principles of legal mandate for ecosystem management and resource management tools to incorporate ecosystem concepts into fisheries management.
Communication skills	3	4	Good skills; requires taking to next level to be persuasive and influential communicator.
And more…			

(Michael 1995). Building adaptive learning capacity should be intrinsic to 21st century fisheries management training.

The *Training Managers for 21st Century Fisheries* initiative outlines an initial framework for curricula focused on meeting 21st century needs. In addition to understanding basics (Table 1), meeting fisheries management challenges requires knowledge and skills in 1) innovative problem solving and institutional design; 2) designing and participating in cooperative research, management, and organizational learning networks; and 3) leadership, trust building, and communications skills. To assure managers engage effectively in making long-term decisions and taking action to reduce uncertainty, training strategies should be coupled with an incentive-based environment that is compatible with the effective application of knowledge, provides transparency, and promotes consistent expectations and learning (Hanna 2002). Training programs should be flexible and supportive of individual and organizational learning needs (Johnsen 2001; Cloughesy 2003). Managers and administrators must also acknowledge their role as learners and that learning and training must be continuous (these are prerequisites to learning how to engage in adaptive learning).

Priority Strategies for Progressing Capacity Building

The *Training Managers for 21st Century Fisheries* initiative has eight priority strategies for creating excellence in fisheries management.

These are fully described in the workshop report *An International Workshop, Training Managers for 21 st Century Fisheries, Queenstown, New Zealand, December 5–7, 2001*, available at http://oregonstate.edu/dept/trainfishmngr. These strategies are:

1. *Develop creative partnerships within and among academia, government, NGOs, and industry at regional, national, and international levels*—Examples include peer exchanges for managers and agency personnel, teaching fellowships to support government or industry involvement in academia, dual degree programs, and internet-based specialized training opportunities that link management and academic organizations nationally and internationally.

2. *Use the management process as a participatory learning experience*—Conduct ongoing evaluation of management outcomes and feed the results back into the management and learning process by providing access to the management process and the primary data. Allow stakeholders (including agencies) to participate in developing the learning experience. For example, Degnbol and Nielsen (2002) describe how institutional structures, such as comanagement arrangements, can contribute to education through research projects coupled to fisheries management in an adaptive context, where results are disseminated to participants as an ongoing part of the management decision-making process.

3. *Broaden and lengthen the fishery management career path*—Recruit from a broader range of disciplines and provide long-run management career paths including supportive working environments and educational opportunities.

4. *Conduct needs assessment and gap analyses*—Evaluate existing and potential programs involved in training all classes of fishery managers relative to needs identified by managers and employers. Results will aid recognition and definition of a minimum level of skills and knowledge and ultimately competencies and benchmarking processes for each managerial class.

5. *Develop an international fishery management education and training website* —This should be a comprehensive, multilingual, and searchable database or clearinghouse with links to all training programs and opportunities, including notices of internships, mentorships, and fellowships suitable for all levels, sectors, and types of fishery managers.

6. *Encourage industry scholarships for fishery managers*—This will allow industry to cooperatively ensure that students have a basic familiarity with industry.

7. *Develop a library of fishery management case studies*—Catalog existing teaching and research cases used by educators that may have already been published; create new case studies; establish standards and templates; use written, video, and simulation software as supportive materials; include a teaching methodology and evaluation component; select best cases relative to defined managerial competencies; and facilitate special case study development conference sessions (e.g., International Institute of Fisheries Economics and Trade [IIFET]) and workshops or special sections or issues of fisheries management journals.

8. *Establish a training providers network*—Identify an organization to coordinate the network and provide accountability, consider tasking the network with coordinating an international needs assessment by sector, use internet based coordination tools (Web site, online database, and listserv), and integrate network development with needs assessment and benchmark-

ing to more efficiently align academic and government training efforts.

Stakeholders, managers, and administrators from all levels and sectors should participate in implementation of these strategies to assure ownership of and commitment to the learning process, particularly if the goal is evolution toward adaptive comanagement or self-governance. Furthermore, intra- and cross-sectoral collaborative partnership will be essential to creating integrated, efficient, well-funded, and effective training programs.

Building the Global Initiative

Many of these strategies are being implemented to a limited extent and scale. The New Zealand Ministry of Fisheries has engaged in needs assessment and associated training development of agency managers. In the United States, NOAA Fisheries has established fellowship and academic partnership programs and has evaluated need for expertise in economics and social science (Ocean Studies Board 2001) and National Environmental Policy Act (NEPA) implementation (NAPA 2002), partly in response to political and legal pressure. Sea Grant programs in New England have begun delivering a regional fisheries science and policy workshop series for stakeholders and decision makers. The April 20, 2004 draft of the U.S. Commission on Ocean Policy Preliminary Report recommends establishing a national ocean workforce database for tracking trends in ocean related human resource development and needs and providing multidisciplinary training for Regional Fisheries Management Council members. At the international level, the United Nations Food and Agriculture Organization (FAO) Fisheries Department has made human capacity building in fisheries an integral component of all its assistance programs; for example, their ACP Fish II initiative focuses on creating a more powerful fisheries management knowledge dissemination and mentoring system for African, Caribbean, and Pacific Island nations. Additional examples exist at regional and national levels. These efforts represent important, but isolated building blocks to excellence in fisheries management.

While on the ground action can likely occur more quickly within classes and levels of managers at the regional level, existing fisheries training programmes are fragmented and far less coordinated relative to the training strategy envisioned by the *Training Managers for 21st Century Fisheries* initiative. Therefore, current activities under the umbrella of the initiative include

• Providing expert input to FAO fisheries training initiative development.

• Facilitating communication among training providers and users through an international e-mail listserv (trainfishmngr@lists.oregonstate.edu) and Web site (http://oregonstate.edu/dept/trainfishmngr/).

• Developing a case study writing workshop for IIFET 2004 in Tokyo, Japan.

• Dissemination and presentation of 2001 workshop results and report.

Ultimately, perpetuating broader action requires leadership, cooperation, and commitment from the highest levels of national management authorities and international bodies. Thus, a stronger coordinating mechanism is necessary.

The *Training Managers for 21st Century Fisheries* initiative envisions an international level organization or consortium charged with facilitating fishery management capacity building within, and among, classes of managers at a variety of scales. The organization could perhaps be formed using existing institutions, a network of academic institution-based centers co-located with the top fisheries management training programs, a network of industry- or government-based training centers, or both. Responsibilities could include 1) facilitating more accurate and collaboratively developed definition of competencies and training gaps at all levels, with special attention to conservation and adaptive ecosystem management, 2) providing qualitative and quantitative data and recommendations useful in planning and development of future training proposals and partnerships, 3) facilitating strengthening of training and partnership building efforts, 4) communicating about successful adaptive learning programs (including outcome evaluation results), 5) helping define training frameworks, 6) building the provider network, and 7) providing other resources and ideas. An international coordinating organization could also play a larger role in bringing academics, agencies, industries, NGO's, and so forth, focused on building capacity for innovation in fishery management policy, governance, and adaptive ecosystem management, together into an annual or biennial international forum. These activities would ultimately improve alignment and efficiency of training efforts and progress the global initiative.

Acknowledgments

For making this project a reality, we thank the Oregon State University Coastal Oregon Marine Experiment Station, NOAA Fisheries, and the New Zealand Seafood Industry Council. We also thank the following additional contributors for their support of the 2001 workshop—New Zealand Ministry of Fisheries; Te Ohu Kai Moana; New Zealand Foundation for Research, Science and Technology; and the American Fisheries Society Marine Division.

References

Cloughesy, M. J., and A. S. Reed. 2003. Appendix G: flexible learning pathways: a forestry extension perspective on curriculum development for continuing professional education of natural resource managers. In L. W. Jodice, G. Sylvia, M. Harte, S. Hanna, and K. Stokes, editors. An international workshop: training managers for 21st century fisheries, Queenstown, New Zealand, December 5–7, 2001. Oregon State University, Agricultural Experiment Station, Special Report 1045. Available: http://oregonstateedu/dept/trainfishmngr/report/APPENDIXG.pdf. (May 2004).

Degnbol, P., and J. R. Nielsen. 2002. Educating fisheries managers as an integral part of management. In Proceedings of the IIFET 2002 Conference: Fisheries in the Global Economy, August 19–22, Wellington, New Zealand.

Folke, C., J. Colding, and F. Berkes. 2003. Synthesis: building resilience and adaptive capacity in social—ecological systems. In F. Berkes, J. Colding, and C. Folke, editors. 2003. Navigating social-ecological systems: building resilience for complexity and change. Cambridge University Press, Cambridge, UK.

Hanna, S. 2002. Educating US fishery managers and policy makers: the problem of incentives. In Proceedings of the IIFET 2002 Conference: Fisheries in the Global Economy, August 19–22, Wellington, New Zealand.

Harte, M. 2003. Appendix H: revolutionizing fisheries management training—putting workshop ideas into action. In L. W. Jodice, G. Sylvia, M. Harte, S. Hanna, and K. Stokes. 2003. An international workshop: Training Managers For 21st Century Fisheries, Queenstown, New Zealand, December 5–7, 2001. Oregon State University, Agricultural Experiment Station, Special Report 1045. Available: http://oregonstateedu/dept/trainfishmngr/report/APPENDIXG.pdf. (May 2004).

Holliday, M. 2001. Linking capacity to opportunity. Presentation for the Training Managers for 21st Century Fisheries Workshop, Queenstown, New Zealand, December 5–7, 2001. Available: oregonstate.edu/dept/trainfishmngr/presentations/Holliday.ppt. (May 2004).

Jodice, L.W., G. Sylvia, M. Harte, S. Hanna, and K. Stokes. 2003. An international workshop, Training Managers For 21st Century Fisheries, Queenstown, New Zealand, December 5–7, 2001. Oregon State University, Agricultural Experiment Station, Special Report 1045. Available: http://oregonstateedu/dept/trainfishmngr/report/APPENDIXG.pdf. (May 2004).

Johnsen, B. 2001. Flexible Learning Pathways. Presentation for the Training Managers for 21st Century Fisheries Workshop, Queenstown, New Zealand, December 5–7, 2001. Available: oregonstate.edu/dept/trainfishmngr/presentations/Johnsen.ppt (May 2004).

Michael, D. N. 1995. Barriers and bridges to learning in a turbulent human ecology. Pages 461–485 *in* L. H. Gunderson, C. S. Holling, and S. S. Light, editors. Barriers and bridges to the renewal of ecosystems and institutions. Columbia University Press, New York.

NAPA (National Academy of Public Administration). 2002. Courts, congress, and constituencies: managing fisheries by default. Available: www.napawash.org/publications.html. (May 2004).

Rassam, G. N., and R. Eisler. 2001. Continuing education needs for fishery professionals: a survey of North American fisheries administrators. Fisheries 26(7):24–28.

Resilience Alliance. 2003. Available: www.resalliance.org. (May 2004).

American Fisheries Society Symposium 49:149–156

A Participatory Approach to Identifying Research Needs for Community-Based Fishery Management

MELANIE WIBER*
University of New Brunswick, Anthropology Department
Fredricton, New Brunswick E2B 5A2, Canada

FIKRET BERKES
University of Manitoba, Natural Resources Institute
Winnipeg, Manitoba R3T 2N2, Canada

ANTHONY CHARLES
Saint Mary's University, Management Science/Environmental Studies
Halifax, Nova Scotia B3H 2C3, Canada

JOHN KEARNEY
5064 Doctor's Brook, Rural Route #3
Antigonish, Nova Scotia B2G 2L1, Canada

Abstract.—This paper reports on a project to engage researchers and fishers together in adapting social science approaches to the purposes and the constraints of community-based fisher organizations. The work was carried out in several locations across Canada's Maritime Provinces, with an underlying rationale based on three major arguments. First, effective community-based management requires that managers are able to pose and address social science questions. Second, participatory research involving true cooperation at all stages can support this process. Third, there is a need to overcome practical and methodological barriers faced in developing participatory research protocols to serve the needs of community-based management while not demanding excessive transaction costs. This paper reports on work with fisher organizations, both aboriginal and nonaboriginal, in identifying social science priorities and undertaking small-scale research projects to meet these needs. Several research themes proved crucial, notably, power sharing, defining boundaries of a community-based group, access and equity, designing effective management plans, enforcement, and scaling up for effective regional and ecosystem-wide management. The research results demonstrate the effectiveness of extending participatory methods to challenge traditional scientific notions of the research process.

* Corresponding author: wiber@unb.ca

Introduction

More and more governments are devolving fisheries management responsibilities and costs to fisher organizations and other nongovernmental bodies—whether to international industrial players, local coast-specific organizations, comanagement boards, or community-based management. This paper focuses on the latter approach, community-based management of coastal fisheries, which is seen to have the potential to produce benefits such as better information flow from fishers to managers, better compliance with rules, and—if carried out properly—lower transaction costs (see, e.g., Kearney 1999; Charles 2001).

In shifting to comanagement and community-based management, small-scale fisher organizations are being expected to take on additional management responsibilities. But as Jentoft (1989) points out, any pattern of cooperative management places heavy demands on the organizations to which management responsibilities are devolved. Problem areas that have been identified in such arrangements include conflicts over rule generation, monitoring difficulties, undermining of local regulations by higher authorities, and conflicting management goals. Dealing with these requires an understanding of the social, economic and policy context within which the organizations must operate—and a corresponding need for capability in social science research. Thus, there has been a growing interest in social science information as a way of managing the fishery rather than managing the fish stock (Mahon 1997; Berkes et al. 2001) and of addressing more focused social objectives such as livelihood needs (Allison and Ellis 2001).

The Challenge of Improving Participatory Research in Fisheries

Successful community-based fishery management requires managers and other fishery participants to pose and address social science questions, so there is a need to overcome practical and methodological barriers to developing research protocols that are participatory (Barnsley and Ellis 1992). However, research protocols must also serve broadly defined needs of community-based management and must not demand excessive transaction or information costs. In our view, there is considerable scope for improving participatory research initiatives in fisheries.

Consider, for example, three common levels of engagement that are meant to improve on top-down, completely researcher-driven approaches: (a) including fishers as subjects of research planned and undertaken by academics; (b) training fishers to become research assistants and to collect data, while academics plan what data to collect, as well as undertaking the analysis and dissemination of results; and (c) having fishers identify research questions, then having the collection of data and subsequent analysis carried out by others, such as graduate students. While each of these has its merits relative to past forms of research, none fully embraces the ideals of participatory research, and in particular, they often (1) fail to allow full development of the relationship between the academic team (with its varied expertise) and the practice-engaged target group, and (2) fail to produce research viewed as valid simultaneously by government bureaucrats and scientists and by fishers.

Participatory research must grow to allow for true cooperation in each stage of the research. One requirement is to adapt science to deal with process in a way that broadens the definition of validity both for recipient groups and for the academic community. Any such expanded definition of validity has to include some privileging of knowledge in action and must produce coherence between the research conclusions and fishers' experiences, as well as with the scientific body of knowledge.

An Experiment in Participatory Research

This paper reports on a 3-year experiment to address the need to adapt social science approaches to the purposes and the constraints of local managers, through the development of participatory research activities driven by the identification of research needs at a local level, in the context of community-based fishery management. The goal of the experiment is to examine approaches in which social scientists and fisher-oriented organizations work together to address social science related problems. A unique aspect of this was the opportunity, created through a research development initiative grant from the Social Science and Humanities Research Council of Canada to provide relatively untied research funding to the participating fishery and community organizations. (A more extensive report on the project may be found in Wiber et al. 2004.)

The participants in this experiment included a set of academic researchers and a set of partners in the Scotia-Fundy region of the Canadian Maritimes (i.e., the Bay of Fundy and the Scotian Shelf off Nova Scotia). The partner organizations included (1) five inshore, multispecies fishermen's associations, with clear ties to their home communities and of comparable scale and attributes that are embedded in communities, and (2) four First Nation native communities from Nova Scotia and Prince Edward Island, each of which has a rather wider mandate than that of the fisher organizations. Two of the five nonnative fisher groups are located on either side of the Bay of Fundy (one in Nova Scotia, one in New Brunswick), while the other three are located on the eastern (Scotian Shelf) side of Nova Scotia. The four Mi'kmaq First Nation partners are located in southwest Nova Scotia (two) and Prince Edward Island (two).

In this experiment, a modest level of funding was provided to each of the partner groups, which were given full control over developing all stages of the research process, including problem identification, research design, data control and analysis, and dissemination of results. The limitations imposed on them were few: the project had to involve social science questions, the focus had to involve fisheries management issues, and the results should be shared with all the other partners at the end of the process. The academic team restricted itself to providing support, expertise, and advice as desired by the partner in carrying out the work. This paper describes the process involved and the results obtained.

Identifying Research Needs

This paper focuses on a key aspect of the experiment—identifying social science research needs. With each partner organization, the process began by exploring what the organization needed to better understand in order to do its job of internal management and indeed to play its part in managing the relevant fisheries. It was rapidly discovered that most groups tended to focus initially on natural science, rather than social science, research possibilities. This reality relates to the fact that state agencies have tended to set the research agenda, that this has primarily focused on natural science data (stock-related information such as abundance and year-class data), and that fishers have either been incorporated into this natural science focus or have had to do so to correct faulty natural science analyses. In any case, there is a clear need to engage in extensive discussions over what might be included in the scope of social science research and what problems might be appropriate for a social science project.

After many meetings with project partners, a variety of research problems of interest to fishermen's associations, and a variety of potential approaches to research collaboration, became apparent. Several key themes arose (see Table 1) relating to power sharing, effectively defining boundaries of a community-based group, optimum organizational structures for community based management, access and equity issues, designing effective management plans, fishery enforcement, managing fishing careers over the life cycle, new commercial species development, and scaling up for effective regional and ecosystem-wide management.

This first step of identifying problems rapidly underwent further refinement, subject to pressures of budget and research time frame, as fishers met and designed projects. As a result, a constellation of related interests began to emerge.

In the case of the First Nation partners, the focus of attention lay on the aftermath of the 1999 Marshall Decision of the Supreme Court of Canada, which recognized a community-based right of access to fish stocks, including involvement in commercial fishing. The Marshall Decision has led to the need to free space in the commercial fishery for native entrants (Davis and Jentoft 2001; Wiber and Kennedy 2001). Three of the four participating Mi'kmaq communities were engaged in a process of negotiating with the federal government so-called McKenzie agreements that provided post-Marshall funding to facilitate access to the commercial fisheries. The other First Nation refused to sign a McKenzie agreement but was engaged in a separate process with local nonnative fishers to access the fishery. Thus, all the native partners shared an interest in assessing the impact of past McKenzie agreements, which led to a joint project on the subject. The variable nature of the agreements across First Nation communities, and the variation in implementation of the agreements within First Nation communities, led to many questions to explore. These relate, for example, to (a) the ownership arrangements of the fishing boats within a First Nation, (b) the interaction between First Nation communities and their individual fishery participants, (c) the manner by which First Nations deal with their food fishery vis-à-vis commercial fishing involvement, and (d) the government practice of buying licenses for fishing in one area and providing those licenses to First Nations in other areas (which is seen as contrary to the adjacency principle and to community-based management approaches).

Three of the partner fisher associations focused on questions of sustainable livelihoods, specifically research topics involving access to the resource, flexibility in a multispecies fishery, and the resulting viability of the inshore sector. Two of these groups, combining together for the research project, identified an interest in examining the process of allocation decision making when fisheries currently under a moratorium reopen, particularly the impact of various allocation choices. The argument here is that the socioeconomic impact of such allocation decisions needs to be better understood and taken into account in decisions made by government on this highly political issue. The third association, located on the Bay of Fundy, recognized that the groundfish quota allocated by government to fishers within that association was insufficient to support those relying on it and, accordingly, used this opportunity to study the socioeconomic impact of purchasing more quotas to support their fishing industry.

Table 1. Fisher research priorities.

Research problems	Options explored by fishers
Broad questions of power sharing	• Licensing their own member's boats • Relating stock assessment to policy development • Developing management plans that are multi-species friendly • Reducing DFO inconsistency in applying policy
Effective boundaries for community-based groups	• Those with history in the resource base • Members (present and future) of the geographical community • Gear, boat-size or 'resource dependence'
Organizational needs	• Accrediting organizations (enabling legislation) • Managing liability for organization directors
Allocation and equity issues	• Tracking the socio-economic consequences of allocation decisions both within and across gear type and regional groups
Designing effective management plans	• Balancing local knowledge with DFO demands for scientific levels of assessment • Managing people as well as fish stocks
Enforcement	• By-law committees and limiting DFO interference • Addressing the lack of federal monitoring and enforcement • Finding effective ways to punish infractions
Managing fishing careers over the lifecycle	• Facilitating intergenerational transfer in the fisheries • Measuring the impact of women moving into boat crew roles • Managing for variable levels of fishing effort over a fishing career
New commercial species development	• Identifying and testing for commercial development • Managing temporary quota and the transition to permanent rights

Table 1. Continued.

Research problems	Options explored by fishers
Scaling up for effective regional and ecosystem management	• Integrating CBM to allow for collaboration with other managers in a region (including First Nations). • Developing regional management boards
The role of mediating organizations	• Tapping into nonfisher resources in the community or region, including NGOs and universities
Adjacency, fishing history, property rights and core status	• Monitoring consequences of property rights-based fishing • Monitoring the effects of complex leasing arrangements including those between First Nations and non-natives • Balancing individual rights against community needs • Options for ring fencing quota (restricting transferability) • Effect of the above on prices of licenses and quota
Local empowerment	• Finding and furthering management skills within the community

They particularly hope to show that the cost of obtaining more quota—for example, by taking out a loan and repaying it over time—would be offset by the economic benefits accruing to them and their community.

The remaining two fisher associations chose projects focusing specifically on capacity-building and internal education. In particular, a fisher organization from New Brunswick chose to conduct a survey of its membership to determine members' priorities for the association's activities and policy positions. The second organization, on Nova Scotia's Scotian Shelf, opted to tackle a number of policy issues at a broad level of consciousness-raising among its membership. A survey of the membership identified five particular themes on which members wanted more information. The association then developed information sessions for which it brought in experts to interact with the membership, in order to answer questions about how policy was developed, what the fit was between scientific data and new policy directions, and how fisher groups could better integrate their views in policy discussions.

Cross-Cutting Research Themes

Several insights relating to participatory research have emerged from the work, based on the above research topics. Three of these are noted here.

Intermediary Organizations

Organizations such as the Bay of Fundy Marine Resource Centre and the Fishermen and

Scientists Research Society play an important role in community-based fisheries, in connecting a range of other organizations and communities. Two of the partner projects involved a role for mediating, intermediary organizations, which were involved at the direct request of partner fisher groups, to support the social science research activities. Other organizations chose to work with local people trained in basic research skills, while yet others employed a graduate student. This reflects the need for variable arrangements in participatory research, both to lower transaction costs for fisher organizations and to facilitate the quite different research agendas of local fisher groups. Participatory research that empowers local managers must also recognize and accommodate these variable needs.

Sharing Experiences

First Nation and nonnative fishers share a keen interest in how other groups are solving common problems. When all our partners gathered together in one room to discuss their research ideas and objectives, ethnic and regional boundaries were not obstacles to communication. In particular, there is clear interest in evaluating where and when community-based management is innovative, where it works, and under what conditions it can flourish.

Scaling-Up

Both First Nation and nonnative fishers agree that community-based management must reach beyond particular localities and utilize linking (scaling-up) mechanisms to allow fishers to participate in region-wide management arrangements. In most cases, the groups involved in this project explicitly asked that important linkages with other groups be recognized within the project design, so that other organizations could be involved in the research. Fishers want to explore the most effective means of unifying their management arrangements on a region-wide basis. Indeed, many of the groups we worked with have been expanding their definition of the relevant community members who can and should be involved in management decisions by incorporating larger geographical areas and/or incorporating different kinds of actors such as academics, scientists, business-people, and community workers.

Conclusion

The process of identifying research problems when the entire research process is controlled by fisher groups proved time-consuming but also demonstrated the effectiveness of extending the participatory method. Two points became apparent in the process. First, our partners arrived at a rich diversity of approaches for their respective research projects, ranging from targeted social/economic research through to a series of policy workshops or an internal membership survey. Second, few of the topics chosen by the fisher partners fit neatly within the predefined interests of the academic team. There was a need for flexibility on all sides.

We found that there are intense discussions throughout the inshore sector concerning (1) the balance that fishers hope to see between enabling wise individual economic decisions and protecting wider community interests, and (2) the appropriate organizational structure for managing fishing regimes and whether sector-based, geographical, or ecosystem-based organizations offer the best platform for effective fisheries management.

Devolution of management roles and responsibilities will place heavy demands on local organizations. Many authors have already pointed to the potential for failure for the devolution process if governments that hope to benefit from devolution do not support local organizations that are willing to take on management duties. These organizations not only need some space to take on fishery management functions, but also need to develop capabilities—including the development of research capabilities—that many currently lack. Governments need to think seriously about providing the necessary resources during this capacity-building stage.

Acknowledgments

The authors are deeply grateful to the community partners who participated in the project on which this paper is based, for their ongoing support, insights, and rewarding interaction. We are also grateful to the Social Science and Humanities Research Council of Canada and the Pew Fellows Program in Marine Conservation for financial support.

References

Allison, E. H., and F. Ellis. 2001. The livelihoods approach and management of small-scale fisheries. Marine Policy 25:377–388.

Barnsley, J., and D. Ellis. 1992. Research for change. Participatory action research for community groups. The Women's Research Center, Vancouver.

Berkes F., R. Mahon, P. McConney, R. Pollnac, and R. Pomeroy. 2001. Managing small-scale fisheries: alternative directions and methods. International Development Research Centre, Ottawa.

Charles, A. T. 2001. Sustainable fishery systems. Blackwell Scientific Publications, Oxford, UK.

Davis, A., and S. Jentoft. 2001. The challenge and promise of indigenous peoples' fishing rights—from dependency to agency. Marine Policy 25:223–237.

Jentoft, S. 1989. Fisheries co-management. Delegating government responsibility to fishermen's organizations. Marine Policy 13:137–154.

Kearney, J. 1999. Community-based management as an alternative to privatization; from maritime fishing communities to a new globalization. The 1999 W.C. Desmond Pacey Memorial Lecture. University of New Brunswick, Fredericton, New Brunswick, Canada.

Mahon, R. 1997. Does fisheries science serve the needs of managers of small stocks in developing countries? Canadian Journal of Fisheries and Aquatic Sciences 54:2207–2213.

Wiber, M., F. Berkes, A. T. Charles, and J. Kearney. 2004. Participatory research supporting community-based fishery management. Marine Policy 28:459–468.

Wiber, M., and J. Kennedy. 2001. Impossible dreams: reforming fisheries management in the Canadian Maritimes in the wake of the Marshall decision. Law & Anthropology 11:282–297.

American Fisheries Society Symposium 49:157–166

Compensable Property Interests and Takings: A Case Study of the Louisiana Oyster Industry

James G. Wilkins*
Louisiana State University, Louisiana Sea Grant Program
Baton Rouge, Louisiana 70803, USA

Walter R. Keithly
Louisiana State University, School of the Coast and Environment
2143 Energy, Coast, and Environment Building, Baton Rouge, Louisiana 70803, USA

Abstract.—In its efforts to combat serious coastal erosion, the state of Louisiana has initiated some large freshwater diversion projects to reintroduce water from the Mississippi River into wetlands and shallow bays that received periodic floods prior to levees being erected in the early 20th century. The diversion projects, carried out by the state and federal governments, have had the unfortunate side effect of damaging some cultivated oyster beds on water bottoms leased from the state. Louisiana's oyster leasing laws have been interpreted by state and federal courts to grant a compensable property interest on which suits may be filed against the government for the taking of private property. In two such suits, the issue of assessing lease values has been hotly contested. The fair market value standard in federal takings law was ignored by state courts in favor of a much more generous assessment allowed by the state constitution. This paper provides a chronological history of the major issues involved in the litigation, with emphasis being given to determination of fair market value.

Introduction

Eastern oyster *Crassostrea virginica* production in Louisiana has averaged 5.3 million kg since 1990. This production, valued annually at approximately $30 million (dockside), is derived from a combination of private leases and public grounds. The public grounds, while primarily utilized for the provision of oyster seed for movement to leases for growth to market size, also provide a source of market oysters when there is an overabundance of seed. Private leases, encompassing more than 162,000 ha (400,000 acres), are rented from the state for an annual fee of $4.94/ha ($2.00/acre). Production from these private leases, while historically representing about 80% of the state's total oyster output, has approximated 50% since the early 1990s.

The large production attributable to the state is the result of the large estuarine environment along Louisiana's coast. The area, because of both its size and variability, assures large production in some areas even when environmental factors are not conducive to the production of oysters in other areas (Dugas 1988). This landscape, while conducive to both a high

*Corresponding author: jwilkins@lsu.edu

and stable production of oysters, is fragile and an estimated 15.5–21.7 km^2 (25–35 mi^2) of estuarine habitat (wetlands) are lost annually due to a combination of erosion and subsidence.

In its efforts to combat this coastal erosion, the state has initiated some large freshwater diversion projects to reintroduce water from the Mississippi River into wetlands that received periodic floods before levees were erected in the early 20th century. The diversion projects, carried out by the state and federal governments, have had the unfortunate side effect of damaging some cultivated oyster beds. Louisiana oyster leasing laws have been interpreted by state and federal courts to grant compensable property interest on which suits may be filed against the government for taking of private property. In two such suits, associated with the operation of the Caernarvon Freshwater Diversion Structure, the issue of assessing lease values has been hotly contested. The fair market value standard in federal takings law was ignored by the state courts in favor of a much more generous assessment allowed by the state constitution. This paper provides a chronological history of the major issues involved in the litigation with emphasis given to determination of fair market value.

Leasing: History and Legal Implications

Characterized by extensive estuarine areas created over the past 5,000 years by deltaic processes of the Mississippi River, coastal Louisiana is blessed with some of the most productive oyster grounds throughout the United States. While undoubtedly exploited since prehistoric times, the earliest commercial operations, occurring in the estuaries near the present day Mississippi River Delta, are traced back to the early 1800s (Dugas 1990). Though harvesting techniques were rudimentary by today's standards, a decline in productivity of some oyster reefs near New Orleans had been reported as early as 1870 (Wicker 1979). Additionally, immigrant fishermen from Dalmatian were increasingly realizing that the transfer of seed from the natural reefs near the delta to bedding grounds closer to the Gulf of Mexico produced high quality oysters. These higher salinity areas, while not able to support substantial natural populations, were ideal for fattening and growth of transplanted seed (Dugas 1990). Increased demand for these transplanted oysters in the later 1800s prompted fishermen to recommend legislation that would support their practices.

In response to the industry's request, and to address the issue of declining productivity, the Louisiana Legislature made early attempts to establish leasing of state water bottoms in 1886 and 1892. These lease rights were limited both in leased acreage and the nature of the right granted. With continued decline of the natural reefs, however, the state commissioned a study by the U. S. Fish Service in 1897 to determine the best course of action to promote sustainable development of the Louisiana oyster fishery. Two of the recommendations of the report were to grant oyster growers permanent tenure of their beds and to remove the 10-acre size limit on leases (Moore 1898). In 1902, by Act 153, the Louisiana Legislature established a leasing regime that allowed each person or corporation to lease up to 20 acres for up to 15 years and to transfer leases to other parties. Act 52 of 1904 further strengthened the proprietary nature of oyster leases by increasing the size limit for each person or corporation to 1000 acres, making the leases heritable, and granting the right of first refusal to the current lease holder at the expiration of lease terms. Finally in 1981, by Act 985, the Legislature directed that the Secretary of the Department of Wildlife and Fisheries (LDWF) "shall renew" oyster leases at the expiration of the lease term, provided that the lease is still capable of producing oysters. Act 985 also codified the right of an oyster lessee to recover damages from injuries to the beds or grounds under a lease caused by the wrongful or negligent actions of any party. In short, legislation refined over the years now gives the lessee exclusive use of water bottoms for oyster cultivation and provides that leases are both heritable and transferable. While these provisions give lease holders an incentive to develop and manage their leases in an "optimal" manner, the provisions created a system such that, over time, the oyster lease system evolved into a "quasi-ownership" of state water bottoms.

Caernarvon: History, Conflict Resolution, and Compensation

Caernarvon Project History

Louisiana, in an effort to retard coastal erosion, has recently embarked on some ambitious diversion projects. The primary purpose of these projects is to reverse the trend in increased salt water intrusion into the fragile wetlands by siphoning or diverting freshwater from the leveed Mississippi River into the marshes. While some minor diversion projects have been operational for more than 50 years, the Caernarvon Freshwater Diversion Structure (CFDS) represented the first large-scale diversion project undertaken by the state (primarily the Louisiana Department of Natural Resources), with assistance from the federal government (primarily the U.S. Army Corps of Engineers). The CFDS, located in the Breton Sound Basin in the most southeastern parish of the state (Plaquemines Parish), was constructed to achieve optimal target salinity conditions (from 5 parts per thousand [ppt] at the upper end to 15 ppt for the lower seaward end of the basin) based primarily on conducive oyster growing conditions. As stated in the U.S. Army Corps of Engineers Report (1982) which examined the feasibility of such a diversion "[e]ssentially all of the monetary benefits of the plan are attributable to enhancement of commercial fishing (page 60)" and the preponderance of these benefits were related to increased oyster production. Other, intangible, benefits recognized in the preparation of the report included improved habitat for nongame and noncommercial species, improved recreation potential, and the buffer zone for hurricane tides "derived from the marsh acting as an interface between urban areas and open water areas (page 58)."

While monetary benefits recognized during plan preparation were primarily related to enhancement of commercial fishing, monetary costs were primarily related to construction, routine operation, and monitoring of the structure. Some intangible costs, however, were also recognized. For example, several hundred acres of wetlands would need to be excavated for construction of the structure. In addition, and of considerable relevancy to the current study, it was recognized that some inland oyster leases would be lost due to decreased salinities and that more frequent closures of leases in other areas may be required due to increased coliform concentrations. As such, the report states that "[a] potential issue which will require resolution prior to implementation of the project includes possible compensation for inland oyster leases (page EIS-3)." In 1986, after more than a decade involving public hearings and the convening of several advisory panels to determine site selection and discuss feasibility and the effects of the project, Congress authorized and appropriated the funds to build the Caernarvon structure.

Louisiana Efforts to Avoid Conflicts

By the late 1970s, as a result of the public hearing and the advisory panel process, the state began to realize that Caernarvon had potentially serious liability implications. This realization was only reinforced with the publication and release of the U.S. Army Corps of Engineers Draft Environmental Impact Statement in 1982 which included discussion of possible compensation for leases damaged as a result of operation of the Caernarvon structure. In response, the LDWF began refusing renewal of leases in coastal restoration project impact areas causing alarm among lessees and prompting discussions on alternative solutions. As a compromise, the LDWF, in 1989, began inserting "hold harmless" clauses in new and renewal oyster leases. The clauses, state experts believed, would relieve the state of liability for damages caused by coastal restoration activities.

At a construction cost of $26 million, the Caernarvon diversion project was dedicated on 12 April 1991 and became fully operational in September of that year. The Caernarvon Interagency Advisory Committee (CIAC), comprised of agency representatives and stakeholders, was formed to determine the flow rate of the Caernarvon Diversion Project. In 1993, the CIAC decided to greatly increase the flow of the diversion allowing more freshwater to reach Breton Sound. This greatly improved production on the public oyster seed grounds but caused damage to many of the leases at issue in this case.

In 1994, the plaintiff oystermen brought a class action lawsuit against the State of Louisiana and the Louisiana Department of Natural Resources (LDNR), and subsequently against the United States government, for damage caused to their oyster leases by coastal restoration projects. The plaintiffs claimed that the Caernarvon Freshwater Diversion Project had lowered the salinity of their leases, making them unsuitable for oyster cultivation and was, in effect, a taking of their leases without just compensation. The Louisiana Legislature responded to the Caernarvon lawsuits in 1995 by enacting Act 936, holding the state harmless from liability for damages arising from coastal restoration activities and directing all state agencies to include hold harmless clauses in all leases, permits, or licenses on state lands or water bottoms granted after 1 July 1995. In Act 107 of 2000, the Legislature reiterated that the hold harmless clauses applied to oyster leases from 1 July 1995 and that the liability waiver was intended to be retroactive to all oyster leases. Additionally, in 1997, the Legislature enacted a bobtail oyster leasing system that allowed the LDWF flexibility in issuing leases from 1 to 15 years in areas permanently or temporarily affected by coastal restoration projects or in areas of projected coastal restoration effects. The lessee assumes all risks involved from leasing in impact or projected impact areas.

The validity of the hold harmless clauses came into question in the 1999 Louisiana Supreme Court case of *Jurisich v. Jenkins* (1999) which held that the administrative agency, LDWF, exceeded its authority by inserting a clause in oyster leases making them subservient to navigation dredging necessary for mineral leases executed prior to oyster leases. The court ruled that statutes governing oyster leasing clearly required the LDWF to renew oyster leases for areas suitable for oyster cultivation and can only alter leases in ways that benefit the industry. While the court did not address the coastal restoration hold harmless clause, and made clear that its holding applied only to the navigation clause, the implication was that LDWF initiated clauses, inserted prior to 1 July 1995, would be invalid because they were not legislatively authorized and did not promote the oyster industry.

Lawsuits Seeking Compensation

The Federal Suit

After the plaintiffs filed a class action lawsuit in state court they filed suit in federal court (*Avenal v. United States* 1995) against the United States, based on the Corps' involvement in the Caernarvon project, claiming their property rights had been taken by the federal agency's action and they were owed just compensation under the Fifth Amendment of the U.S. Constitution. The Court of Federal Claims found that the plaintiffs "had no compensable expectancy in the continued artificially elevated saline levels" caused by the flood control mechanisms built in and around the Mississippi River beginning in 1927. In other words there was no property interest in the water quality above their oyster beds. The Federal court of Appeals for the Federal Circuit affirmed the verdict of the lower court, but on different grounds. The federal appeals court found that the plaintiffs did have a compensable property interest in the water quality over their oyster beds, but they were barred from recovering because they did not have a reasonable investment-backed expectation that their leases would not be interfered with by the Caernarvon Diversion Project, which was planned in 1959. The oyster farmers did not invest in their leases until the 1970s and knew or should have known the project would significantly change the area.

The State Suit

In March 1994, a class action was initiated on behalf of all persons holding oyster leases on state-owned water bottoms in Breton Sound. The plaintiffs claimed that the state's action of lowering salinity levels in Breton Sound resulted "in a permanent and substantial interference with plaintiffs' use and enjoyment of their land amounting to a taking of an interest" in their property rights without just compensation, as required by the Article I Section 4 of the Louisiana Constitution. The state moved for summary judgment on the grounds that the takings issue had already been litigated in the federal suit and the plaintiffs were barred from relitigating it by collateral estoppel. The trial court denied the motion and the Fourth Circuit Court of Appeals exercised its supervisory jurisdiction; first holding for the state and then reversing itself and affirming the trial court denial of summary judgment. The main reason for the court ruling, and the one most important to our discussion, was that the legal standard for takings under the Louisiana Constitution is different from federal law. The Court said the only elements in Louisiana takings law are (a) that a recognized property right (b) has been taken or damaged (c) for a public purpose. These three conditions, according to the Court, were met. The Court went further to say that bad faith was the only bar to compensation and that reasonable investment backed expectations are irrelevant under Louisiana takings law. The Court also noted that the Louisiana Constitution requires a property owner be compensated to the full extent of his loss rather than merely just compensation under federal law.

The trial court ruled that the hold harmless clauses could not be entered into evidence and that ruling was upheld by the Fourth Circuit Court of Appeals. The trial court also granted the plaintiffs motion to exclude the LADNR's use of biological assessments and side-scan sonar to show the productivity of the plaintiff's leases and refused to compel the plaintiffs to show actual oyster production or actual income from their leases. Some evidence of income produced from the plaintiffs' leases was produced at trial, but according to the appeals court, this evidence fell short of proving the actual damages incurred by the lessees.

In December 2000, the court rendered judgment in favor of the plaintiffs. The awarded damages, equal to $52,722/ha ($21,345/acre), were based on replacement costs for damaged leases. In essence, this was determined by estimating the costs of building new oyster reef in another location by placement of necessary cultch (installed). For the five class representatives, this awarded amounted to $48 million which, when expanded to all potential litigants in the class suit, could total $1.3 billion.In a second suit, plaintiffs with leases in Lake Borgne alleged damages to leases and filed suit in district court in St. Bernard Parish (Alonzo et al. versus State of Louisiana). In a bench trial held in 2002 and 2003, the court found that a taking had occurred, citing the federal appeals court characterization of leases as property rights. Again the court did not address the element of reasonable investment backed expectations or the fact that the plaintiffs had ample notice of the impending changes that would occur in the affected areas as a result of Caernarvon when they sought leases in those areas. Instead, the court seemed to attach particular importance to the fact that the state had secured funds to compensate oyster lessees in the Davis Pond impact area (another large-scale diversion project that became operational in 2003) but had not bothered to do so for Caernarvon. The judge in the second trial, stating that the Louisiana Constitution affords more compensation than the Federal Constitution (i.e., more than fair market value), awarded $67,162/ha ($27,191/acre) to 53 leaseholders in Lake Borgne. Awards in this second suit totaled $662 million.

The state of Louisiana appealed the Avenal decision and, in October 2003, a five judge panel of the Fourth Circuit Court of Appeal, in a 4–1 decision, affirmed the verdict of the trial court even though some leaseholders admitted that their leases had never been productive. The court of appeals held that "so long as plaintiffs proved generally that their leases were productive before Caernarvon came online, ...that they were not productive after Caernarvon came online, and ... Caernarvon caused the loss of productivity" the plaintiffs were entitled to recover.

The Louisiana Supreme Court Decision

The Louisiana Supreme Court reversed the findings of the lower courts, determining that the terms of most of the leases prevented the oystermen from suing the state and that in the remainder of the cases, the oystermen had waited to long to bring suit. The first group of plaintiffs held leases that were granted in 1989 or later and contained one version or another of a hold harmless clause which absolved the state from liability for damages to oyster leases resulting from coastal restoration efforts. The second group held leases that were granted before 1989 that did not contain a hold harmless clause. The Louisiana Department of Wildlife and Fisheries (LDWF) began inserting a hold harmless clause into its leases in 1989 that stated, "The lessee hereby agrees to hold and save that the State of Louisiana, its agents or employees, free and harmless from any claims for loss or damages to rights under this lease from...the creation or restoration of coastal wetlands." In 1995 the LADNR began inserting a more detailed indemnity clause that was mandated by the legislature.

The Louisiana Supreme Court found that where either the 1989 or 1995 hold harmless clauses were contained in the lease, the State of Louisiana was not liable for damages. This part of the decision is based on the interpretation of the legislatively created oyster leasing system and the authority of the administrative agency, in this case the LDWF, to alter the way the state contracts with oyster lease holders. The Louisiana Supreme Court found that the court of appeals had erroneously relied on the case of *Jurisich v. Jenkins* in determining that the hold harmless clauses at issue in this case were invalid.

In *Jurisich*, the Louisiana Supreme Court concluded that the LADNR could not unilaterally insert a "navigation and oil field activity clause into oyster leases" but explicitly refrained from extending that ruling to other clauses such as those dealing with coastal restoration. The Court found that the Fourth Circuit read *Jurisich* too expansively and distinguished that opinion from the facts in *Avenal.* The Court noted that LDWF was only obligated to renew oyster leases that were capable of producing oysters and when it was recognized that the long anticipated effects of Cearnarvon would make the subject leases unproductive, the department was within its authority to decline renewing the leases. The hold harmless clauses were the result of a compromise to allow the oyster leases to continue in the Caernarvon impact area and were actually a concession to the oyster lessees. The Court also found that the Caernarvon project and the hold harmless clauses that allowed it to operate without exposing the state to liability were in furtherance of developing the oyster industry as a whole and in line with the state's responsibilities to protect her natural resources under the public trust doctrine. Therefore the court found that the 1989 hold harmless clauses were a valid exercise of LDWF's administrative authority without the need for legislative action. Those lessees whose leases did not contain a hold harmless clause until 1995 were also barred from recovery because the court said they did not allege and prove specific damages to their leases that occurred between the dates Caernarvon began operating and the renewal of their leases in 1995. Instead they chose to base their claims on their post 1995 leases containing the more detailed, legislatively mandated hold harmless clause. Of the approximately 204 leases that were damaged by the Caernarvon project, 12 did not contain hold-harmless clauses, having been granted prior to 1989 and expiring between 2000 and 2005. After a detailed analysis of Louisiana's constitutional categories of property "taken" versus that "damaged" for a public purpose the court found that these plaintiffs had filed their suit too late for relief. The plaintiffs argued that their claim was not for damages, but for a taking. The Court found that since the oyster farmers' exclusive right to use the bottom for oyster culture was not terminated and the leases were apparently still producing some oysters, and the state, being owner of the water, water bottoms and oysters, could not take its own property, the Caernarvon project had damaged the leases but not taken them. The distinction is important because Louisiana law provides a 3-year prescriptive period for taking of property but only a 2-year prescription for damaging of property for a public purpose. The Court found that since the property was damaged and not taken, the action had prescribed by early November 1993, 2 years after freshwater began flowing through the Caernarvon structure. The suit was not filed until March of 1994 and thus invalid.

The Louisiana Supreme Court's holding in Avenal was based on the terms of the contract between the state and the oyster lessees and on failure to bring suit in a timely manner. However, the court's dicta indicates a much stronger support for the state's ability to avoid liability for damaging property in its coastal restoration efforts. The Court in *Avenal*, while basing its holding on the narrow grounds discussed above, seemed to go out of its way to expound on general principles of takings law and police power. The court discussed the plaintiffs' takings claim under the Fifth Amendment of the U.S. Constitution and determined that a taking had not occurred, agreeing with the federal appeals court that the "[oystermen] could not have had reasonable investment backed expectations that their oyster leases would give them rights protected from the planned freshwater diversion projects of the state and federal governments." Then the Court went further and postulated that even ".... if Caernarvon did entirely deprive them of all economically beneficial and productive use of their property rights, the plaintiffs are still not entitled to compensation as Caernarvon was a valid exercise of the state's police power under federal law." Power to act in the "actual necessity" to prevent "grave threats to the lives and property of others" is generally termed a state's "police power." The Court stated that even if Caernarvon eliminated all economically beneficial uses of the plaintiffs' oyster leases, no compensation would be owed because the "right of the state to disperse freshwater from the Mississippi River over saltwater marshes to prevent coastal erosion is derived from a background principle of Louisiana law" "as early as the 1950s and 1960s and "were certainly a part of the environment in which the raising of oysters were conducted" and "the freshening of these waters in order to prevent further coastal erosion and save Louisiana's coast is a matter of 'actual necessity' as it will forestall [a] grave threat to the lives and property of others." While the court was discussing federal takings law the same principles can be applied to Louisiana takings law. Article I § 4 of the Louisiana Constitution makes property rights subject to "reasonable statutory restrictions and the reasonable exercise of police power." The Louisiana Constitution has been interpreted to provide greater protection of property rights than the Fifth Amendment of the U.S. Constitution, but the principle that individual property rights must sometimes bow to the cause of protecting the public form grave and imminent threats is not totally foreign to Louisiana law, although courts have been somewhat schizophrenic in balancing police power against compensable losses. The boundaries between government actions requiring compensation and those that do not, whether or not those actions are termed "police power," are not clear and may depend on the court's perception of the level of risk to the public good or safety. The court in *Avenal* seems to be convinced of the seriousness of coastal land loss and the absolute necessity of government action to combat it. Were the facts different, if the public risk is considered less serious, and if the property owners' prior knowledge of their assumed risks is lacking, the court might be less inclined to allow application of police power without compensation. In 2003, the electorate ratified a constitutional amendment to allow the state to limit compensation to owners of property that is taken or damaged for coastal restoration purposes to that allowed under the Fifth Amendment of the U.S. Constitution, which in most cases will amount to the fair market value rather than the "full extent of the loss" as normally required under the Louisiana constitution. In light of *Avenal*, future compensation to oyster farmers damaged by coastal restoration activities will probably be based on the fair market valuation now allowed by the Louisiana constitution.

The Issue of Compensation

The issue of appropriate compensation is often a contentious issue in legal proceedings. While the Louisiana's Supreme Court's ruling negated the need to compensate in the *Avenal* suit, the issue, for a number of reasons, remains relevant in light of ongoing and future coastal restoration projects. For example, the state of Louisiana responded to the damage awards in *Avenal and Alonzo* in 2003 by proposing a constitutional amendment to the state takings protections allowing limitations on recovery for damages from coastal restoration activities. The amendment was approved by the voters in the fall of 2003. The same year, before the plebiscite on the amendment, the Legislature enacted a statute that limits recovery for damages from coastal restoration activities to that allowed under the Fifth Amendment of the U.S. Constitution. Recovery for takings under the U.S. Constitution is fair market value. The legislation purports to be retroactive and apply to existing and future claims. While it is unlikely that the courts will allow retroactive application in light of the Louisiana constitution prohibition of ex post facto laws, the legislation should be binding for damages related to future restoration project. Similarly, it is likely that Federal appropriations for compensating oyster lessees will be incorporated into all future major diversion projects if the valuation method is acceptable under federal appropriation rules. Federal appropriations tend to be limited to fair market value. Thus, determination of fair market value remains highly relevant.

The economist hired by the state in the *Avenal* trial (one of the authors of this paper) argued that compensation should be based on the loss of income rather than replacement cost. Compensation, based on lost income, he argued, would make plaintiffs "whole"; the primary principle for compensation. He concluded, based on lease sales, tax records, and other information, that the value of leases in the affected area of Breton Sound was, on average, approximately $494/ha ($200/acre). He further suggested that replacement cost would be appropriate compensation only if it were less than lost income. His argument follows the "lesser of" rule generally favored by economists and grounded on the basis of economic efficiency (see Ward and Duffield 1992; for details).

In the promulgation of regulations to guide natural resource damage under CERCLA, justification for the use of the "lesser of" rule by the Department of Interior (DOI) included the following (quoted by Ward and Duffy):

"[i]f use value (in this case lost income from damage to oyster leases) is higher than the cost of restoration or replacement, then it would be more rational for society to be compensated for the cost to restore or replace the lost resource than to be compensated for the lost use. Conversely, if restoration or replacement costs are higher than the value of uses foregone (i.e., lost income from damage to oyster leases), it is rational for society to compensate individuals for their lost uses rather than the cost to restore or replace the injured natural resource."

While justification for the "lesser of" rule proposed by the DOI relates to damages to natural resources for which the U.S. government is the trustee, the same argument can be advanced for damages to oyster leases. Certainly, awards that exceed lost income by a factor of 100 or more (e.g., $52,722/ha compared to about $500/ha) would serve no social purpose since an award of a significantly lesser amount would make lessees "whole."

As suggested by the expert economist by the state in his report prepared for the *Avenal* trial, if one were to use restoration costs of $52,722/ha ($21,345/acre) in assessing damages, it would take more than 350 years to recoup the restoration costs (in terms of income that could be generated from the leases). The report goes on to state that "[n]eedless to say, without the Caernarvon Project, there would be no production in the impacted area in sixty years, much less 350 years, due to erosion and subsidence."

Adding to the complexity of cases previously discussed, however, is the fact that the water bottoms are owned by the state and leased for the purpose of cultivating oysters. Hence, plaintiffs (lessees) brought suit against the state for damages to a resource (i.e., water bottoms) owned by the state and rented to the lessees for the nominal fee of $4.94/ha ($2.00/acre). In essence, use of replacement cost rather than lost income would therefore require the state to compensate lease holders for replacement of an asset owned by the state. This circular reasoning, as suggested by Judge Love in the Appellate ruling on *Avenal*, does "shock the conscience."

As the state develops and promulgates rules for determining fair market value, the basis for compensation appears to closely follow that suggested by the expert economist in the *Avenal* trial. Simply stated, procedures for assessing fair market value are being tied to potential losses in income rather than replacement costs. Remuneration based on lost income will allow lessees to make the needed investments to maintain and, potentially, enhance this important sector without unduly constraining ongoing and future restoration projects.

Discussion

The *Avenal* decision seems to have settled the issue regarding oyster leases with hold harmless clauses, the state will not be liable to lessees for coastal restoration related damages to their leases if they contain the clauses. Leases contracted before 1989 should have all been renewed by now but there may remain some claims stemming from other coastal restoration projects' effects on preclause leases. Those claims face an uphill challenge in the face of *Avenal* especially if the lessees knew or should have known of the coming projects. The Louisiana Supreme Court seems inclined to view coastal restoration as a more urgent and imperative species of police power and, coupled with the Federal *Avenal* decision, will make it easier for state and federal agencies to complete coastal erosion projects.

References

Avenal v. State. 2003. WL 22501685, 2001–0843 (La. App. 4 Cir. 2003).

Avenal v. United States. 1995. 33 Fed. Cl. 778 (1995).

Avenal v. United States. 1996. 95–5149100 F. 3d 933 (Fed. Cir. 1996).

Dugas, R. J. 1988. Administering the Louisiana oyster industry. Journal of Shellfish Research 7(3):493–499.

Dugas, R. J. 1990. Enhancement of oyster seed production on Louisiana public reefs. Report prepared for the National Marine Fisheries Service under Interjurisdictional Grant NA88WC D-IJ150, Baton Rouge, Louisiana.

Inabnet v. Exxon Corp. 1993. 614 So. 2d 336 (La. App. 4 Cir., 1993).

Jurisich v. Hopson Marine. 1993. 619 So. 2d 1111(La. App. 4 Cir. 1993).

Jurisich v. Jenkins. 1999. 99–0076 749 So. 2d 597 (La. App. 1 Cir. 1998).

Moore, H. F. 1899. Report on the Oyster-Beds of Louisiana. Government Printing Office (extracted from U.S. Fish Commission Report for 1898; pp. 45–100), Washington D.C.

U.S. Army Corps of Engineers (New Orleans District). 1982. Louisiana coastal area, Louisiana: Freshwater diversion to Barataria and Breton Sound Basins. Volume 1, Draft Main Report, Draft Environmental Impact Statement, New Orleans, Louisiana.

Ward, K. M., and J. W. Duffield. 1992. Natural resource damages: law and economics. Wiley Law Publications, Wiley New York, New York.

Wicker, K. M. 1979. The development of the Louisiana oyster industry in the 19th century. Doctoral dissertation. Louisiana State University, Baton Rouge.

American Fisheries Society Symposium 49:167–176

Incorporating Stakeholders into the Development of Fisheries Ecosystem Plans: A North Sea Case Study

JENNY HATCHARD* AND TIM GRAY

School of Geography, Politics & Sociology, The University of Newcastle 40-42 Great North Road, Newcastle upon Tyne, NE1 7RU, United Kingdom

KNUT MIKALSEN

Department of Political Sciences, University of Tromsø, 9037 Tromsø, Norway

Abstract.—The development of fisheries ecosystem plans (FEPs) represent an opportunity to reconcile fisheries with conservation. Key to achieving this reconciliation is the inclusion of stakeholders in the development of FEPs, as well as in the latter stages of decision making and implementation. The fishing industry, in particular, must be consulted. A process of stakeholder consultation should thus take place alongside analysis of the chosen ecosystem through which the economic, societal, ecological, and functional importance of different ecosystem components is determined. This paper examines the role of stakeholders in establishing a FEP for the North Sea. Our European Union funded case study, which has developed a North Sea FEP, is the result of an interdisciplinary project involving both marine and social scientists. This research process has revealed that the incorporation of stakeholders can facilitate reconciliation between fisheries and conservation objectives. However, the case study also highlights the dependency of a FEP, and indeed any kind of fisheries management, on the political context within which it operates. Fishers and conservationists will only experience the mutual benefits of ecosystem-based management if the political context is one in which stakeholders are consulted, in which their consent is sought and in which they are actively involved in decisions related to methods of implementation. Therefore, reconciliation between stakeholders, and between them and the political system of fisheries management, as well as between fisheries and conservation, is required.

Introduction

The European Fisheries Ecosystem Plan (EFEP), a European Union funded research project, has incorporated a broad range of stakeholders directly into the development of ecosystem-based fisheries management for the North Sea. This level of stakeholder involvement, unprecedented in the European fisheries context, has facilitated a reconciliation of fisheries and the conservation objectives of European fisheries management within the final output of the project—the North Sea Fisheries Ecosystem Plan (FEP). This paper charts that process of reconciliation and considers the "real-world" implications of introducing a FEP.

Thus, the paper is divided into two sections. First, the iterative process by which EFEP has worked to reconcile fisheries and conservation objectives, via its incorporation of stakehold-

*Corresponding author: j.l.hatchard@ncl.ac.uk

ers and their views into project, is examined. Second, the paper discusses the inherent problems of the political context, which will be faced by any attempt to implement an ecosystem approach within the North Sea. This is accomplished by considering the views of stakeholders on political conflicts within North Sea fisheries management and their solutions for managing and addressing such conflicts. The paper concludes by summarizing the strengths of a FEP for reconciling fisheries and conservation objectives and its potential for facilitating a different kind of reconciliation within and between the institutions of European fisheries management. Communication can promote solution building.

Stakeholders and the Development of a North Sea FEP

The North Sea FEP is the product of a process of collaboration between fisheries scientists, social and political researchers, ecologists and modelers. Playing a central role in the project, the social science team conducted a two-stage process of stakeholder consultation. Defining stakeholders as any "persons who have an interest in, and/or an influence upon, the North Sea and/or its fisheries" (Hatchard et al. 2003), more than 90 North Sea stakeholders, of varying nationalities (Danish, English, Dutch, Norwegian, and Scottish) and interests (fishing industry, environmental groups, regulators, and scientists), were interviewed twice during the course of the three-year project (Table 1). The central role of this process within the project reflects the emphasis given to participatory mechanisms in theories of ecosystem-based management (Sissenwine and Mace 2001; Degnbol 2002; WWF 2002).

A qualitative, semistructured approach was employed to conduct the two phases of stakeholder consultation (Reinharz 1992; Wengraf 2001) and to analyze the data. The choice to adopt a qualitative, rather than quantitative, methodology was grounded in the overall objectives of the consultation which were to develop an understanding of participants' preferences, and their reasons for holding those preferences.

The first phase of the consultation process was conducted at the beginning of the project. A standardized interviewing pro forma was employed to invite stakeholders to talk at length about two broad themes: 1) the health of the North Sea and its commercial fish stocks; and 2) North Sea fisheries management. First, with regard to the North Sea, stakeholders were asked about their optimism or pessimism regarding the health of fish stocks and the ecosystem; what they regarded as the main threats to both fish stocks and the ecosystem; and

Table 1. Breakdown of research participants by stakeholder category and location.

Stakeholder category	Denmark	England	Netherlands	Norway	Scotland	Total
Fishers/fishing industry	10	16	6	12	11	55
Regulators and administrators	2	8	4	4	1	19
Natural and social scientists	2	3	3	3	2	13
Environmentalists	2	4	2	2	0	10
Total	16	31	15	21	14	97

how the health of stocks and the ecosystem could best be secured. Second, with regard to fisheries management, stakeholders were asked whether they were happy or unhappy with the current management structure and how that structure could be improved; with management styles and how they could be improved; with management instruments and how they could be improved; and with management controls and how they could improved, strengthened, or both.

The aim of this first consultation was to develop a detailed picture of how the research participants viewed fisheries management, how much legitimacy they accredited it with, how compliant they were willing to be, and how much they identified with the (twin) goals of sustaining fisheries and sustaining the ecosystem. This information provided a detailed context within which to set and understand stakeholders' particular preferences regarding management instruments, which were essential to help determine the direction of the project's ecological modeling process, which was to take place subsequent to the consultation. These management instruments broadly included input and output restrictions and technical measures, along with spatial restrictions.

The data gathered during this first consultation process was analyzed at both aggregated and disaggregated levels using a qualitative interpretative process whereby the data were coded according to themes and subthemes. From this analysis, the management instrument preferences of stakeholders were assessed according to two criteria. First, the level of consensus or dissensus regarding each type of management instrument (Table 2). This approach was adopted as stakeholders had made it clear that they regarded consensus as an important requirement for successful management. Second, the degree to which stakeholders considered particular instruments to be appropriate means to achieving particular goals was examined (Table 3). This second approach was employed due to the emphasis stakeholders had placed on objective-based management, rather than "one-size-fits-all" sweeping measures.

In this way, it was the stakeholders themselves who were helping to determine the direction of the research. When combined, these two qualitative analyses provided an indication of the direction the project's modeling process would need to take. In particular, stakeholders' emphasis on objective-oriented measures—both fisheries and conservation-oriented, thereby enabling some internal reconciliation of what can be construed as conflicting goals—became central to the project. And, effort, technical, and spatial measures received particular attention within the modeling process as these were identified as the most appropriate means by which key objectives—also identified by stakeholders—could be met.

Thus, the management instrument preferences of stakeholders were used, in conjunction with work to identify significant components of the North Sea ecosystem, to drive a process of ecological modeling of various management scenarios (Paramor et al. 2002; Ragnarsson et al. 2003; Piet el al 2003). These management scenarios focused on spatial effort management; fleet effort reduction; technical measures to reduce discards of undersized target stocks and bycatch of nontarget species; and protected areas for fish and for habitats.

The results of the ecological modeling of management scenarios were fed back to research participants during the second stakeholder consultation, using a series of computerized slides illustrating the basis and implications of the modeling work (Hatchard et al. 2004;

Table 2. First EFEP consultation: consensus/dissensus analysis results (Hatchard et al 2003).

Management instrument	Strong consensus	Consensus	Dissensus	Strong dissensus
Technical management	On all technical measures, particularly selective gears		Dissensus on when, how, and where they should be applied	Minority of six stakeholders rejected technical measured outright
Effort management			Dissensus on how decommissioning should be implemented	Considerable dissensus on days at sea, which received roughly equal support and rejection
Quota management		Weak consensus on the failure of quotas to protect fish stocks	Dissensus on possible solutions: multi-species/multi-annual quota or replacement with effort restrictions? Dissensus on individual transferable quotas (ITQs)	
Area closures	Real-time closures received wide-spread support from those who advocated area closures	Little support for closures with non-fisheries objectives or for ad hoc seasonal closures		
Discard reduction		The EU compulsory discard policy is detrimental to fish stocks and the industry	Dissensus on possible solutions: both support and opposition for a discard ban	

Paramor et al. 2004). Stakeholders were asked how acceptable, implemental, and enforceable each scenario was; what they believed the impact of the measures would be on fisheries and fisheries dependent communities; and if they had any suggestions of additional scenarios that could be modeled (Table 4).

This second phase brought a further degree of reconciliation between fisheries and conservation objectives than had the first. This was partly due to the reduced abstraction of the subject matter of the consultation. The modeling results provided tangible scenarios and associated implications, clearly illustrated to stakeholders, which enabled them to address key conflicts between objectives and between objectives and measures to determine their relative acceptability. This approach works from the principle that there is often more than one way to achieve a goal, and that the implications of each may be differ-

Table 3. First EFEP consultation: objective-appropriate measures results (Hatchard et al 2003).

Management instrument	Objectives	Particular instrument	Stakeholder support
Technical management	Protect juvenile stocks	Minimum mesh sizes, minimum landing size regulations	Pan-stakeholder support as these measures are seen as improving stock sustainability and catch quality.
	Protect nonbenthic fish species	Gear selectivity	Environmentalists and some scientists supportive of measures to protect marine biodiversity.
	Protect charismatic species	Escape hatches and alerting devices	Environmentalist support, with public backing; some scientific support.
Effort management	Match fleet capacity to fishing opportunities	Decommissioning	Regulators and scientists focused on achieving a sustainable balance between stocks and catches.
	Limit amount of fish killed	Days at sea and/or restrictions on vessel specification	Widespread support, but concerns from fishers about introducing even more restrictions.
	Protect the North Sea fleet and sustain fisheries-dependent communities	Days at sea, compensation for short-term restrictions on fishing, decommissioning	Mentioned by all sectors, but indication from scientists and environmentalists that heavy restrictions may still have to be imposed.
Quota management	Secure a regular supply of particular commercial species and stabilise prices	Multi-annual quota	Support from the fishing industry who would like to be able to plan their businesses in the longer term.
	Limit amount of fish that is landed	Quota	Favored by Norwegian and Danish stakeholders, but EU compulsory discards policy poses a problem.
	Match fleet capacity to fishing opportunities	Quota	Regulators and scientists focused on achieving a sustainable balance between stocks and catches.
Area closures	Protect nonbenthic fish species	Closure of areas to trawling	Environmentalists and some scientists supportive of measures to protect marine biodiversity.
	Protect spawning stocks	Real-time area closures, seasonal closures in accordance with spawning periods	Stakeholders from all sectors endorsed the measures to ensure a continued supply of fish.

ent, thereby making a measure either more or less acceptable. These views were communicated, in turn, back to the modeling team, thereby completing an

Table 4. Second EFEP consultation: stakeholder responses regarding acceptability of modelled scenarios (Hatchard et al 2004).

Scenario type	Yes	Maybe	No
Effort	High-frequency effort restrictions to reduce discards of undersized target stocks	General effort reduction to reduce direct effects of fishing on target stocks	Fleet effort restrictions increase prey availability for other fish predators
	Fleet effort restrictions to reduce mortality of lower trophic levels to increase feed availability for high-value stocks	Low frequency effort restrictions to reduce bycatch of non-target stocks	
Technical measures	Minimum mesh size regulations to reduce discards of undersized fish	Discard ban to reduce discards	
	Technical bycatch regulations to reduce bycatch of charismatic species and reduce by-catch of unwanted fish		
Spatial measures	Fisheries protected areas to reduce direct fishing impacts juvenile stocks	Fisheries protected areas to reduce direct fishing impacts on spawning stocks	Conservation-protected areas to protect 20–30% habitat types
	Conservation-protected areas to protect unique habitats	Conservation-protected areas to protect essential fish habitats Conservation-protected areas to protect significant sessile species	

iterative loop between science, social science and stakeholders. They were then employed to shape the final output of the project—a fisheries ecosystem plan for the North Sea, which focused on the key recommendations of stakeholders—effort, technical, and spatial management instruments as the most appropriate means for achieving particular management objectives, both conservation and fisheries-oriented.[1]

Overall, the decision to incorporate a wide range of stakeholders into EFEP, and the subsequent two-phase consultation process, facilitated reconciliation of fisheries and conservation through the opening up of a channel of iterative communication between researchers and those on whom research has an impact. Thus, with social science researchers acting as a conduit, effective two-directional communication was possible between scientists and stakeholders. This took the form of both informing and learning on both sides and caused the gap that existed between the EFEP team and North Sea fisheries stakeholders at the outset of the project to be narrowed.

On the one hand, throughout the consultation process, EFEP has been disseminating, and inviting discussion of, the concept of eco-

[1] To be found at www.efep.org.

system-based fisheries management. Stakeholders have also consistently been kept up to date with project developments. And on the other, stakeholders have provided the EFEP team with valuable, practical advice concerning the viability of different management scenarios and the likely impact of them on fisheries dependent communities.

The benefits of this process have manifested themselves in both expected, and unexpected, ways. Stakeholders have had the opportunity to shape how EFEP's research agenda has developed. First, the initial consultation results informed the process of ecological modeling. Second, the subsequent consultation influenced additional modeling and the content of the draft North Sea FEP itself. The outcome of this, as was hoped, is that the North Sea FEP is as practicable and appropriate as possible.

Less predictable was the sea-change that we have witnessed during the course of the two consultations regarding stakeholders' opinions of ecosystem-based fisheries management. There has been a considerable attitudinal shift on the part of the majority of stakeholders consulted from skepticism of an ecosystem approach which was perceived as prioritizing environmental objectives over economic ones, to interest in the potential applications offered by this new way of managing fisheries, particularly for meeting particular objectives.

This last point illustrates effectively that the reconciliation of fisheries and conservation under an ecosystem banner which has taken place within the EFEP-stakeholder community has done so somewhat in isolation from the rest of the stakeholder community. A genuine reconciliation of fisheries and conservation requires that this communication between stakeholders and scientists must be translated outside of the project context. The next section of this paper explores the potential of FEPs for promoting reconciliation in a policy context.

Stakeholders, Their Political Context, and a North Sea FEP

The premise described at the beginning of this paper that the full range of interested parties involved in fisheries will only experience the mutual benefits of ecosystem-based fisheries management if and when they are actively involved in fisheries management and its decision-making processes is one that has been borne out, to a limited degree, by the recent implementation of a stakeholder-led regional approach to fisheries advice in the North Sea context.

The first phase of the EFEP consultation—with its focus on management structures and styles, as well as the issue of compliance—highlighted the inherent importance of the political and management context within which fisheries operate. An argument that was made repeatedly to the social science team during the course of the first consultation phase was that the North Sea fisheries crisis is as much a crisis of management as of stocks, with criticism being directed mainly towards the European Union's Common Fisheries Policy. Stakeholders perceived this to have been politicized to a dangerous degree, as far as the sustainability of both fish stocks and fisheries are concerned.

During the course of the first phase of consultation, stakeholders were critical of two key political factors in operation in the European context. First, stakeholders were unhappy about the damaging effects of the European Common Fisheries Policy (CFP) on the North Sea fishing industry and on fisheries dependent communities. For example, one fishers' representative stated, "...a council of fifteen member states, together with the European Commission, make very important decisions for the livelihood of hundreds of thousands of fishermen with their families and a lot of those decisions are made in a political way; we don't think that for real and honest management that Brussels is going the right way." (Hatchard et al. 2003)

Second, stakeholders rejected the way in which fisheries management structures have been politicized and used to further national, rather than fisheries or conservation, interests. In particular, it was widely felt that politicos in Brussels have consistently ignored scientific advice on the health of key commercial fish species, at the expense of both fish stocks and the fishing industry. In the past, this meant that scientific recommendations on total allowable catch (TAC) were overshot significantly at the decision-making table, to allow for nation-stated economic interests in fisheries. More recently, the pressure of the environmental lobby has meant that recommended catch levels have been treated with extreme caution. As one environmentalist put it, "European management is dominated by political, not environmental or scientific, concerns."

Overall, the general consensus among North Sea stakeholders was that the Common Fisheries Policy (CFP) had, up to that time, failed to deliver either sustainable fisheries or an effective conservation policy, and had certainly not achieved a reconciliation of these two goals. The politicization of fisheries management had merely served to reinforce the uncertainty and skepticism with which both resource users and other stakeholders viewed North Sea fisheries management. This negative impression is reflected by the fact that, at this stage of the project, stakeholders' most constructive solutions to this problem related to the, then imminent, decision to adopt a policy of regionalization of management advice. More concrete, specific proposals for making fisheries management more effective were not forthcoming.

The decision to introduce a Regional Advisory Council for the North Sea was taken in between the two EFEP consultations (EC 2002). The establishment of the North Sea Regional Advisory Council (NSRAC), with its broad membership taking in the full range of stakeholders from the fishing industry to environmentalists and from community groups to industrial enterprises, has provided the first opportunity in the European context to consider the benefits to be gained in terms of reconciliation from broad participation. The NSRAC has not, as yet, moved as far as proposing an ecosystem approach to fisheries management, but it has taken on the challenge of addressing both fisheries and conservation goals.

This institutional change in the policy arena, alongside the more specific, management scenario approach taken by the second consultation, created a climate in which stakeholders were able to offer more specific suggestions for mitigating political constraints on fisheries management, than had been the case in the first consultation. Thus, during the second consultation, there was a clear consensus among stakeholders of the need for an FEP to institutionalize transparency and stakeholder participation. Without these two factors, it was felt that management would continue to be characterized by the same political uncertainty, which had undermined potentially successful policies in the past.

Independently of each other, stakeholders described remarkably similar management frameworks that they argued should be incorporated into a North Sea FEP. First, the objectives of management measures should be agreed collectively. Second, ecological and socio-economic assessments should be conducted prior to measures being introduced to ascertain the probable impact of regulations and pilot studies should be conducted if necessary. Third, once implemented, the impact of measures must be regularly monitored. Fourth, measures should have a pre-agreed timetable, including a scheduled review process. Fifth, there should be a mechanism for adapting measures in response to monitoring results, if agreed, at any time. Finally, measures that have demonstrably not worked should be abandoned. The flexibility and inherent responsiveness of such a system should make effective management of the dynamic processes of the North Sea ecosystem more possible. There was general agreement that broad and open stakeholder participation at all stages in policy development, implementation and review is vital to this process.

Thus, in political terms, this North Sea case study has provided three key insights into the essential features of a North Sea FEP for a persistent reconciliation of fisheries and conservation to be achieved. First, stakeholders, from the fishing industry to the environmental lobby, must be involved at all stages of the policy process and there should be effective and consistent channels of communication in place between the different parties. Second, for the management regime to be successful, the FEP must be transparent to prevent political interests over-taking those of fisheries and conservation. Finally, the FEP must incorporate a system of agreement, monitoring and review of management instruments to enable a flexible system of ecosystem-based fisheries management.

Overall, there was consensus among stakeholders regarding both the political obstacles facing any future North Sea FEP and, latterly, the solutions for managing those obstacles. Further, these structural issues were generally regarded as being of greater importance to the success of that FEP than the choice of individual measures. This consensus across nationalities and interests indicates the potential for the North Sea FEP, with its inherent support for pan-stakeholder involvement, to achieve a working reconciliation of stakeholders and fisheries management, which will hopefully aid the reconciliation of fisheries and conservation objectives.

Conclusion

The development of a North Sea FEP has facilitated a reconciliation of fisheries and conservation objectives in two ways. First, we have learned that fisheries management instruments can be used to achieve wider conservation goals, without that success being at the expense of fisheries or operating in conflict with the fishing industry. Second, the North Sea FEP represents a management framework that encourages bringing fisheries and environmental interests together, as well as other stakeholder groups. We conclude that, for a FEP to be implemented within the North Sea context it will need to be stakeholder-led and objective driven and that its introduction could offer a means by which reconciliation of two kinds can be furthered within fisheries management—between fisheries conservation on the one hand and between stakeholders and management institutions on the other.

The North Sea FEP has twin objectives: sustainable fish stocks (conservation) and a viable fishing industry (fisheries). We have learned, from the consultation process; that the success of the North Sea FEP in consistently achieving these twin objectives will be dependent upon its capacity to effectively manage political forces. However, FEPs, with their inherent participative processes, do offer an opportunity to create an enduring reconciliation between the twin-objectives of fisheries management—sustainable fisheries and marine conservation.

Acknowledgments

The EFEP project team would like to thank all the stakeholders who have kindly consented to be interviewed for the purposes of this project. EFEP has been funded by the EU Quality of Life and Management of Living Resources program. Contract number: QLRT-CT-2001–01685.

Reference

Degnbol, P. 2002. The ecosystem approach and fisheries management institutions: the noble art of addressing complexity and uncertainty with all on board and on a budget. IIFET paper no. 171. IFM, Hirtshals, Norway.

EC (European Commission). 2002. Council Regulation (EC) No 2371/2002, December 20 2002, on the conservation and sustainable exploitation of fisheries resources under the Common Fisheries Policy (CFP). European Community, Brussels.

Hatchard, J. L., T. S. Gray, and K. M. Mikalsen. 2003. European fisheries ecosystem plan: stakeholder consultation. EU Project number: Q5RS-2001–01685. Newcastle University, Newcastle upon Tyne, UK.

Hatchard, J. L., T. S. Gray, K. M. Mikalsen, and K. Brookfield, 2004, 'European Fisheries Ecosystem Plan: Second stakeholder consultation' in Paramor, O.A.L., Scott, C.L. and Frid, C.L.J. editors, European Fisheries Ecosystem Plan: Producing a Fisheries Ecosystem Plan. EU Project number QLRT-CT-2001–01685. Newcastle University, Newcastle upon Tyne, UK.

Paramor, O. A. L., C. L. J. Frid, and C. L. Scott, editors. 2002. European fisheries ecosystem management plan: The North Sea ecosystem. EU Project number: Q5RS-2001–01685. Newcastle University, Newcastle upon Tyne, UK.

Paramor, O. A. L., C. L. J. Frid, and C. L. Scott, editors. 2004. European fisheries ecosystem plan: producing a fisheries ecosystem plan. EU Project number: Q5RS-2001–01685. Newcastle University, Newcastle upon Tyne, UK.

Piet, G. J., L. Hill, A. Jaworski, O. A. L. Paramor, S. A. Ragnarsson, and C. L. Scott. 2003. European fisheries ecosystem management plan: Fisheries and the removal of ecosystem components. EU Project number: Q5RS-2001–01685. RIVO, Ijmuiden, The Netherlands.

Ragnarsson, S. A., L. Hill, A. Jaworski, G. J. Piet, O. A. L. Paramor, and C. L. Scott. 2003. European fisheries ecosystem management plan: the North Sea significant web. EU Project number: Q5RS-2001–01685. MRI, Reykjavik, Iceland.

Reinharz, S. 1992. Feminist methods in social research. Oxford University Press, New York.

Sissenwine, M. P., and P. M. Mace. 2001. Governance for responsible fisheries: an ecosystem approach. Reykjavik Conference on Responsible Fisheries in the Marine Ecosystem, Iceland 1–4 October 2001. Available: http://www.refisheries2001.org. (February 2004).

Wengraf, T. 2001. Qualitative research interviewing: biographic narrative and semi-structured methods. Sage, London.

WWF (World Wildlife Federation). 2002. Policy proposals and operational guides for ecosystem-based management of marine capture fisheries. Available: http://www.panda.org. (February 2004).

American Fisheries Society Symposium 49:177–185

Bridging the Gap Between Social and Natural Fisheries Science: Why Is This Necessary and How Can It Be Done?

Rosemary E. Ommer*
CEOR, University of Victoria
Post Office Box 1700 STN CSC, Victoria, British Columbia V8W 2Y2, Canada

R. Ian Perry
Fisheries and Oceans Canada, Pacific Biological Station
Nanaimo, British Columbia V9T 6N7, Canada

Barbara Neis
SafetyNet and Department of Sociology, Memorial University
St. John's, Newfoundland A1C 5S7, Canada

Abstract.—This paper examines some of the reasons why social and natural fisheries scientists need to work together with fish harvesters to improve our understanding of communities of fish and fishery communities. It identifies some tools and methods with the potential to promote such collaborations and hence our capacity for better management of our marine resources as well as to support the legitimate skills and requirements of fishing communities. In this paper, we seek to do three things. First, we identify the need for a more holistic system of fisheries management; then, we identify the challenges this presents; and last, we propose a solution to this issue. The story of the collapse of the cod stocks in the northwest Atlantic provides a useful case study of inadequate management of marine fisheries and its consequences for fish and fishing communities. On the east coast of Canada, international high-technology fleets were in operation by the 1960s, resulting in a massive dragger effort and serious depletion of fish stocks. As the international and Canadian offshore fleets expanded in the 1970s, the small inshore fisheries of the nation were increasingly seen as out of date, inefficient, and of low productivity, labels that gained in credence as dragger fleets intercepted stocks causing inshore landings to declined. Also, by the 1970s, fisheries management had come to rely heavily on a combination of biological and economic modeling, which was used to estimate the optimal catch that could be taken in any given fishery, based on measures of the biomass of commercial species in conjunction with the economic returns to the fishery sector.

After the imposition by most nations of 200-nautical-mile exclusive economic zones (1977), national fishing effort in most industrialized nations came to be increasingly dominated by offshore wetfish trawlers. In Canada, stock assessment scientists provided advice to the federal Minister of Fisheries and to international fisheries organizations. Managers drew upon this advice when setting quotas for different species, sectors, and zones. However, missing from stock assessment science and fisheries

*Corresponding author: ommer@uvic.ca

management initiatives instituted during this period were two things: an awareness of the importance of variability in the dynamics of marine ecosystems and of the dynamics of fisheries, and fishery communities that mediate the data used in fisheries science and the outcomes of management initiatives.

Until the 1920s, in Canada and elsewhere, the knowledge of local people, reported to fisheries officers, provided the detail that was used by governments in their admittedly limited observation and management of fisheries. This is clear from, for example, the Canadian fisheries reports on the fisheries of the Gulf of St Lawrence, which contain local reports and observations made to the officer as he patrolled the Gulf each year. His annual reports to government were a compilation of local events, explanations, and observations that are an invaluable historical source for the state of both fishing communities and the fish themselves (Government of Canada/United Province of Canada 1854–1857; Hutchings et al. 2002). After the First World War, as western economies sought to stabilize, marine biologists started to be regularly employed by governments, and by mid-century, government management of fisheries had become much more science-based–dependent, that is, on expert knowledge (Smith 1994), but, with the exception of economists after the 1960s, social scientists were largely excluded from these government agencies. An unintended result of the shift to research vessels and laboratory-based fisheries science was that the latter became increasingly remote from the local fishing communities and local ecosystems it sought to serve. This was occurring just as fisheries management was recognizing variability in the dynamics of marine ecosystems and the need to take both social and ecological factors into account in, for example, setting catch limits for fisheries. Some fisheries social science developed in most industrialized nations starting in the 1960s, but such social scientists were rare and almost exclusively based in universities, with only economists represented in government science branches.

As commercial fisheries around the world began to face the prospect of frequent, multiple, and serious stock collapses, in the late 1980s and early 1990s, fishing communities in many countries that were dependent on local fisheries came under increasing economic and social stress. In 1992, with the Canadian declaration of moratoria on catching certain species of groundfish in Canadian waters in the northwest Atlantic, most fishing communities in Atlantic Canada, and particularly in Newfoundland and Labrador, went into crisis. Fish processing workers and young people were particularly hard hit by the moratoria and efforts to downsize the industry. In the short term, many fish harvesters were able to shift effort to other species benefiting from expanding quotas of high value such as crab, shrimp, and seals, which were predominantly controlled by the under-65-ft fleet. Fish harvesting and processing companies were seriously affected, but many managed to survive by selling off their groundfish trawlers, diversifying investments, and finding new supplies of fish from other parts of the globe.

It was this combination of pressure on fish, fishers, and fishing communities that really started to bring social and marine scientists together in Atlantic Canada and elsewhere. Some university researchers realized that it was time to bridge the intellectual gap between marine science and social scientific studies and between expert and local knowledge, since there was a serious disconnect between fisheries policy (often developed by natural scientists and economists who live at some distance from the fishery), the focus and results of fisheries social science, and the lived experience of local people, based on lay, local, or traditional ecological knowledge. Lay knowledge is nonacademic science-based and is a generic term that covers local, traditional, gender-based, technician-derived, and other forms of nonexpert data. Traditional ecological knowledge (TEK) is that of First Nations whose information is centuries deep. Local ecological knowledge (LEK) has less generational depth but is based on the experience of people who live adjacent to, and depend on, a resource.

Scholars began to work in teams of natural, social, and health scientists, seeking to interpret local and traditional knowledge, to identify and fill gaps in scientific knowledge, and to find ways of bringing new forms of data and new concerns into the more formal scientific, management, and policy development processes (Neis et al. 1999; Ommer 2002). Government scientists began to incorporate local fishers into their consultations and to involve them in watchdog or sentinel fisheries and government-established intersectoral committees. These latter involve different departments and industry and labor stakeholders to address issues related to professionalization and fishing safety (Hilborn et al. 2004; www.fsrs.ns.ca). The problem is how to effectively bridge these various solitudes of the natural, social, and health sciences between experts and local people in order to promote the potential for healthy people and recovered communities and marine ecosystems.

The Challenges

The challenges inherent in bridging the disciplinary gaps among social, natural, and health science research and between lay and expert knowledge are considerable. This is perhaps why the initial efforts, in both government and the universities, at devising models for fisheries management were built by economists whose models and methodologies, being highly quantitative, were compatible with those used in stock assessments. The results of bioeconomic modeling, however, concerned many social scientists because such models did not, and could not, take into account the complexities of human behavior in different social, cultural, and natural contexts. Moreover, as ecological knowledge began to build, it became increasingly clear how hugely complex marine ecosystems were and the human–marine ecosystem interactions that had to be understood, since their functioning is central to the evolving state of global fisheries, but had been completely ignored in most of the bioeconomic modeling of the day. As many world fisheries fell more and more into disarray, it became increasingly clear that working with the human problems in the social sciences, and with the fish population problems in the separate solitude of the marine biological sciences, was proving seriously dysfunctional. Within the university research community, therefore, interdisciplinary research partnerships between social and natural scientists began to be developed and fostered, and within government, attention shifted from stock assessment science towards integrated coastal zone management.

The first Canadian university-based experiments were in natural and social interdisciplinary work on fisheries with studies that sought to combine the expert knowledge of marine biologists with the local knowledge found in fishing communities. This work started in the mid-1990s and has continued to this day (Coward et al. 2000; Ommer and Newell 2000; Ommer and the Coasts Under Stress Team 2007). Recognizing and reconciling different taxonomic systems was an obvious place to start work of this kind. When dealing with lay and expert knowledge, it is important to think about the purpose for which each type of knowledge has been created. Fishers are harvesters. Their knowledge is about where, when, and how to catch fish.

That requires knowing a great deal about the fish themselves and their behavior on the fishing grounds and in relation to fishing gear and practices. Small-scale fishers know about assemblages of different species and some things about predator–prey relations—but only about local ones. They know (for example) that when in their area and during the fishing season, codfish tend to arrive at particular times and to behave *this* way in spring and *that* way in summer and are therefore to be found on *this* part of the coast in *this* depth of water in the spring and in different parts and depths and perhaps in association with different species in the fall. Unless they have a winter fishery, they are unlikely to know much about cod behavior during that season. In the language of Newfoundland fishers, a mother fish is the large old fish with lots of eggs, the harvesting of which, many believe, threatened fish stocks. A herring fish is not a herring but a codfish at that season when it is chasing herring for food, and so on.

Local taxonomies speak to local ecosystem dynamics. They also speak to local complexities. Because fishers classify fish according to catching information built up over generations, for those species they have a long history of targeting, they have some capacity to distinguish between normal variability and variability that appears to reflect deeper, longer term change and may indicate that something is going wrong. "There's no herring fish this year" in Newfoundland means that there are no codfish in the place where they usually follow the herring and get caught by local fishers. Herring fish may not appear in a particular area every year, but if this happens several years in a row or an abrupt change occurs that appears to be associated with a change in fishing activity or in the local ecosystem (development of a purse seine herring fishery), it may indicate more substantial change.

Academic scientists are also classifiers, but of a different kind. Their task is to understand the biology of the species as a whole, in all its physical and related behavioral complexity and then describe that to science. They deal, therefore, with the species generically, so to speak—they are global generalists where small-scale fishers are local specialists. Government scientists come halfway between these two because they work with the specialist scientific language and methodology of classification, but they then apply the knowledge so derived to issues of species (and, increasingly, ecosystem) health in particular areas and contexts in order to create the knowledge base for wise harvest management. They bridge two worlds, one regional (where management regulations will be applied) and the other much wider—perhaps covering several ecosystems or a whole coastline and engaging national and international communities of scientists. They are not social scientists, and they do not always have the information or the conceptual tools they need to take into account the relationship between science and society, including the way their science might be mediated by the actions of people and the consequences of their science for those people, their options, and behavior. This is a problem because fishery management is not merely, or even primarily, the management of fish—it is the management of fishers and fishing communities. It would make sense, then, for social scientists (not just economists) to be directly involved in fisheries management and to participate with marine scientists in studying fisheries and listening to harvesters and their communities—what they know and what they need—because this information is essential to any harvesting strategy that is to be effective and also because their knowledge will be of use to scientists interested in species behaviors and evolutionary distinctions among subspecies.

With the involvement of fish harvesters, contemporary LEK can be documented in a systematic and ethical fashion. Data from systematic career history interviews with expert harvesters selected by their peers can be aggregated to construct a larger-scale, finely textured picture of regional fisheries extending back several decades. Fish harvesters' knowledge is perhaps most likely to be sought after where scientific research is weak and where more precise knowledge is required to protect the remaining fish. A feature of recent research involving LEK has been a tendency to combine methodologies and, by triangulating, to increase the potential reliability of the findings. Scientific research including tagging studies and genetic research can help verify fishers' observations and interpretations of fish behavior. In some cases, harvesters and scientists have begun to work together to devise and test hypotheses (Neis et al. 1999; Neis and Felt 2000; Stanley and Rice 2003).

Working with local people starts to reveal the complex dynamics that exist *among* communities of fish, fishing communities, science, and policy; it illuminates ways these interact, and it makes us aware of the unfortunately still controversial point that the divide between people and nature is artificial (Murdoch 2001). It is precisely the failure to recognize that humans are part of nature (a construct that challenges current conventional wisdom) and that human behavior and human and environmental health are products of individual, social, and ecological dynamics that has kept the social and natural sciences apart for a long time. Humans depend in an absolute sense on the social and natural environments in which we are embedded, although urban society in particular forgets this when protected by the web of trade and technology that surrounds urban life.

The Solution

In keeping the natural and social sciences apart, we have fostered this kind of removed thinking. This must change because it is contributing to the declining sustainability of the planet and risking the health of people and environments. We therefore define fish, fishers, and fishery communities as existing in nested social-ecological webs composed of human institutions and natural physicochemical environments. This idea of social-ecological webs derives from work done by scholars on traditional and local knowledge; the terminology is derived from the work of Fikret Berkes and others (Berkes and Folke 1998; B. Neis, I. Perry, and R. E. Ommer, presentation to the International Human Dimensions Program Conference, 2003). We also recognize the different scales that are involved in both the human and the natural world parts of this complex and do so without involving the artificial divide between humans and nature. The concept of the marine resource system (MRS) has been proposed (Bakun and Broad 2002) to include the fishery resource itself, the regional marine ecosystem, the physical-chemical habitat, the fisheries on the resource, the associated economic activities and social values, the management framework and institutions, and the political contexts in which the management and economic contexts operate. The more narrowly focused factors of the fishery resource, regional marine ecosystem, and physical-chemical habitat—what are usually referred to as the resource system—have been called the fish–habitat system (Bakun and Broad 2002). With this as our point of departure, we can then ask what characteristics of the webs in the MRS contribute to high (or low) resilience (or vulnerability) to disturbances for people, societies, *and* marine environments and begin to explore what these webs might teach us about resilience and vulnerability and, thus, how we might identify wiser ways to handle the interactions between marine resources and human communities. There are significant sets of interacting characteristics in marine ecosystems and human communities, some of which can be thought of as parallel and all of which can be sources of change.

In marine systems, these sets of characteristics are

• Climate variability and trends (biophysical characteristics); internal ecosystem dynamics (predator/prey behaviors, diseases, population structures)

• Fishing

• Habitat degradation

• Pollution (introduction of exotic species, new diseases, etc.)

In human systems, the significant sets of characteristics are

• Environmental change

• Demographic change

• Technical innovation

• Law and property relations

• Policy

• Relations of production and reproduction

• Gender/ethnicity

• Shifting values

Note that changes in these social and ecological webs all

• Have spatial, temporal, and institutional dimensions,

• Operate at a variety of scales,

• Are bidirectional (fish assemblage changes alter fishing, fishery communities and fishery policies, which can further alter the composition of fish assemblages), and

• Can have multiple causes.

We can take, for example, the collapse of several of the groundfish fisheries in the Northwest Atlantic, and the shift in marine ecosystems and in fishing effort to shrimp, lobster, and crab in that area, and explore how these interactions have worked together to affect the health of people, communities, and the ecosystems (Dolan et al. 2005). Indeed, we can identify pathways between the human and natural ends of this social-ecological web that demonstrate the bi-directionality and interactive effects that we have been positing. Thus, the interacting impact of overfishing groundfish and adverse environmental conditions (e.g., Drinkwater 2002) appears to have been associated with increased catches of shrimp, snow crab *Chionoecetes opilio*, and lobster in Atlantic Canadian fisheries. Shifting effort to these expanded fisheries in new places has led to (i.e., is a pathway to) increased fishing costs (enhanced by fisheries policies that affect fishing vessel design) and changing occupational health risks for fish harvesters (crossing shipping lanes) and processing workers (shellfish occupational asthma) (Pauly et al. 1998; Howse et al. 2006). As effort shifts across species, it tends to move towards species and areas where scientific uncertainty is greater. As science shifts to these expanded target species, the capacity to monitor recovery of depleted stocks and to protect remaining fish can be jeopardized.

Industrial changes, triggered by fisheries closures and the shifting of effort to new target species and places, will produce spatial and temporal shifts in work and possibly changing power dynamics within social networks with potentially serious consequences for the health of communities and regions. Depending on the availability of alternative employment opportunities and appropriate social programs, overfishing can jeopardize access to employment, income, social support, and food security and increase social and gender inequality, all known health determinants (Neis and Kean 2003; M. Macdonald, B. Grzetic, and B. Neis, presentation to the annual meetings of the Canadian Sociology and Anthropology Association, 2003). In turn, reduced employment, income, and social support and increased costs may encourage harvesters to intensify their fishing effort, reduce social constraints on cheating, and reduce the willingness of harvesters to comply with management initiatives, thus potentially further jeopardizing the health of marine ecosystems, driving up the costs of fisheries management, and potentially further undermining the knowledge base required for stock assessment science (Copes 1996a, 1996b).

This is not a pretty picture. It is, however, an accurate portrayal of what is revealed when scientific analysis adopts an interdisciplinary approach, draws on local knowledge, starts at the small scale, and works up from there. This approach is in contradistinction to that of working within disciplinary boundaries, using top-down analyses that blur the details of lived reality, and the ways it varies in different times, places, and for different groups. It is also what emerges when one ties human experience to its impact on the natural world—what people can end up doing to the marine ecosystems upon which they depend with often severe repercussions for themselves or others. For example, there are several stakeholder groups involved in the Peruvian fishing sector, each with their own goals. The decisions that these stakeholders take when presented with information on impending El Niño events can differ markedly depending on their different goals (Broad et al. 2002). Industrial purse-seine fishers and processors aim for large industrial catches and profits; their responses to pending El Niño-Southern Oscillation (ENSO) events include changing ratios of their product mixes (e.g., between fish meal and canning) and diversifying into other industries. Artisanal fishers have the goal of increasing catches and maintaining fishing traditions; their responses to information on pending ENSOs includes changing fishing gears to target new species and changing household production options. Regulatory administrators and scientists have the goal of sustainable fisheries and job security (human community sustainability); their responses to information on pending ENSOs include changing management regulations, permitting experimental fishing, and increased sampling and observations. The goal of the media is sales; their responses to ENSO predictions tend to be to exaggerate the impacts of ENSO events and often to inject a false certainty about the information presented or to raise uncertainties about otherwise confident information.

There are many other examples of these kinds of interactions that set nested social-ecological webs in a global context and demonstrate the resilience and vulnerability of human and natural communities operating together as complex adaptive systems (Holling 1973, 1986, 2003; Gunderson and Holling 2002). The issue of scale, in particular of different and often unrecognized choices of analytical scales between natural and social scientists, is crucial since the scales chosen for studies of marine systems and human interactions can constrain recognition of the drivers and responses of these systems to global changes (Perry and

Ommer 2003). We advocate an approach which focuses on communities of fish and fishing communities, which makes explicit the need to manage marine resources so that global to local scales are considered; identification and use of appropriate natural science scales in the development of management policies; and the need to be aware of shifting temporal baselines and the representative nature of the data over time. More work also needs to be done on issues of value (Coward et al. 2000) (how to assign values to different ecosystems—because policy makers do, and we need to understand how they do it), and on different kinds of knowledge and how they can be married (science and TEK/LEK as an example) (Murray et al. 2006 and in press). Most importantly, all of this needs to be done in the context of thinking about the matches and mismatches among life cycles of fish people, fishing communities, fisheries institutions, and governments and with awareness of the wider context, and spatiotemporal features, of environmental change, diversity, species composition, and the like.

As scientists, policymakers, and people in fishing industries and communities, unless we understand not only what we and others do, and how we and others do it, but also why we and others do it, there is little hope that things will change. In other words, when researchers, policymakers, and local people understand the complex ways in which initiatives stemming from many different sources (natural and social) can interact to affect different groups of people, communities, and marine ecosystems across time and space, then they will finally have a better framework for understanding the factors contributing to overfishing and marine ecosystem degradation as well as its costs and consequences. We suggest, then, that in light of the incapacity of nonholistic management processes to deal with the growing crisis in world fisheries, natural and social science research should jointly provide a basis for comanagement structures in the future, with fishing people and their communities being a welcome and respected part of the overall management system. The likelihood that initiatives designed to protect these ecosystems as well as the people who depend upon them will have positive consequences should also increase. There is a great deal of work to do and not much time in which to do it because global marine ecosystems and small fishery-based rural communities are seriously stressed at local, regional, and global levels. What is encouraging is that many researchers are now devoting time and energy to developing new frameworks and methods for studying social-ecological systems and for drawing on local and expert knowledge. We can only hope that those whose task it is to develop the institutions and governance systems that promote sustainable small and large-scale fisheries will, in the light of their increasing awareness of the nature of the problems they face, be able to put in place the kind of policies that support, guide, and direct an ecologically and ethically sensitive exploitation of the world's oceans and the welfare of its coastal peoples.

References

Bakun, A., and K. Broad. 2002. Climate and fisheries; interacting paradigms, scales, and policy approaches. Columbia Earth Institute, International Research Center for Climate Prediction, IRI-CW/02/1. Palisades, New York.

Berkes, F., and C. Folke., editors 1998. Linking social and ecological systems: management practices and social mechanisms for building resilience. Cambridge University Press, Cambridge, UK.

Broad, K., A. Pfaff, and M. H. Glantz. 2002 Effective and equitable dissemination of seasonal-to-interannual climate forecasts: policy implications from the Peruvian fishery during El Niño 1997-98. *Climatic Change* 54(4):415–438.

Copes, P. 1996. 1996a. Adverse impacts of individual quota systems on conservation and fish harvest productivity. Simon Fraser University, Institute of Fisheries Analysis, Discussion Paper 96–1, Vancouver, B.C.

Copes, P. 1996b. Social impacts of fisheries management regimes based on individual quotas. Simon Fraser University, Institute of Fisheries Analysis, Discussion Paper 96–2, Vancouver, B.C.

Coward, H., R. Ommer, and T. Pitcher, editors. 2000. Just fish: ethics and Canadian fisheries management. ISER Books, St John's, Newfoundland.

Dolan, A. Holly, M. Taylor, B. Neis, J. Eyles, R. Ommer, D. C. Schneider, and W. Montevecchi. 2005. Restructuring and health in Canadian coastal communities: a social-ecological framework of restructuring and health. Eco-Health 2:1–14.

Drinkwater, K. F. 2002. A review of the role of climate variability in the decline of northern cod. Pages 113–130 *in* N. A. McGinn, editor. Fisheries in a changing climate. American Fisheries Society, Symposium 32, Bethesda, Maryland.

Government of Canada/United Province of Canada. 1854–1887. Sessional papers: "Report on the fisheries in the Gulf of St Lawrence," Appendices, 1854–87. National Archives of Canada, Ottawa.

Gunderson, L. H., and C. S. Holling. 2002. Panarchy: understanding transformations in human and natural systems. Island Press, Washington, D.C.

Hilborn, R., J. DeAlteris, R. Deriso, G. Graham, S. Iudicello, M. Lundsten, E. Pikitch, G. Sylvia, P. Weeks, J. Williamson, and K. Zwanenburg. 2004. Cooperative research with the National Marine Fisheries Service. The National Academies Press, Washington, D.C.

Holling, C. S. 1973. Resilience and stability of ecological systems. Annual Review of Ecology and Systematics 4:1–23.

Holling, C. S. 1986. The resilience of terrestrial ecosystems: local surprise and global change. Pages 292–317 *in* W. C. Clarkland and R. E. Munn, editors. Sustainable development of the biosphere. Cambridge University Press, Cambridge, UK.

Holling, C. S. 2003. From complex regions to complex worlds. Ecology and Society 9(1):11.

Howse, D., D. Gautrin, B. Neis, A. Cartier, L. Horth-Susin, M. Jong, and M. Swanson. 2006. Gender and snow crab occupational asthma in Newfoundland and Labrador, Canada. Environmental Research 101:163–274.

Hutchings, J. A., B. Neis, and P. Ripley. 2002. The nature of cod, *Gadus Morhua*. Pages 140–185 *in* R. Ommer, editor. The resilient outport. ISER Books, St. John's, Newfoundland, Canada.

Murdoch, J. 2001. Ecologising sociology: actor-network theory, co-construction and the problem of human exemptionalism. Sociology 35(1):111–133.

Murray, G., B. Neis, and J. P. Johnsen. 2006. Lessons learned from reconstructing interactions between local ecological knowledge, fisheries science and fisheries management in the commercial fisheries of Newfoundland and Labrador, Canada. Human Ecology 34:549–571.

Murray, G., B. Neis, D. C. Schneider, D. Ings, K. Gosse, J. Whalen, and C. Palmer. In press. Opening the black box: methods, procedures and challenges in the historical reconstruction of marine socio-ecological systems. In J. Lutz and B. Neis, editors. Making and moving knowledge. McGill-Queen's University Press, Montreal.

Ommer, R. E., editors. 2002. The resilient outport. ISER Books, St John's, Newfoundland.

Ommer, R. E., and D. Newell, editors. 2000. Fishing places, fishing people. University of Toronto Press, Toronto.

Ommer, R. E., and the Coasts Under Stress Team. 2007. Coasts under stress. Restructuring and social-ecological health. McGill-Queen's University Press, Montreal.

Neis, B., and L. Felt, editors. 2000. Finding our sea legs: linking fishery people and their knowledge with science and management. ISER Books, St. John's, Newfoundland, Canada.

Neis, B., D. C. Schneider, L. F. Felt, R. L. Haedrich, J. A. Hutchings, and J. Fischer. 1999. Northern cod stock assessment: what can be learned from interviewing resource users? Canadian Journal of Fisheries and Aquatic Sciences 56:1949–1963.

Neis, B, and R. Kean 2003. Why fish stocks collapse: an interdisciplinary approach to the problem of 'fishing up'. In R. Byron, editor. Retrenchment and regeneration in rural Newfoundland. University of Toronto Press, Toronto.

Pauly, D., Christensen, V., Dalsgaard, J., Forese, R., and Torres, F. 1998. Fishing down marine food webs. Science 279:860–863.

Perry, R. I., and R. Ommer. 2003. Scale issues in marine ecosystems and human interactions. Fisheries Oceanography 12:513–522.

Smith, T. D. 1994. Scaling fisheries: the science of measuring the effects of fishing, 1855–1955. Cambridge University Press, Cambridge, England.

Stanley, R. D., and J. C. Rice. 2003. Participatory research in the British Columbia groundfish fishery. In N. Haggan, C. Brignall, and L. Wood, editors. Putting fishers' knowledge to work. Fisheries Center Research Reports 11(1):44–56.

American Fisheries Society Symposium 49:187–196

Territorial Control as a Fisheries Management Instrument: The Case of Artisanal Fisheries in the Estuary of the Patos Lagoon, Southern Brazil

TIAGO ALMUDI* AND DANIELA C. KALIKOSKI
Federal University of Rio Grande, Department of Oceanography
Caixa Postal 474, RS 96201-900, Brazil

JORGE P. CASTELLO
Federal University of Rio Grande, Department of Geosciences
Caixa Postal 474, RS 96201-900, Brazil

Abstract.—The objective of this work is to characterize the territoriality of artisanal fishers in the estuary of the Patos Lagoon with the goal to propose an action plan for fisheries comanagement in the region that takes into account local forms of fisheries control through territories. Semistructured interviews and document analysis were used in the evaluation of the conditions and events that led to the development, maintenance, and transformation of territorial control. Results show that at the local level fishers do share the notion of boundaries and do respect each other. They carry out a territorial control on the use of resources. The sense of care and respect is not shared by outsiders that do not apply the body of informal rules and territorial control developed by local fishers in the use of common pool resources. Territorial control has faced transformation over the years and, in some cases, has been disrupted due to lack of recognition of its importance by industrial fishers and the government. Territorial control at the local level has not prevented overexploitation of resources. The neglect of such informal institution coupled with government–industry-driven interests lead to the transformation and breakdown of sustainable informal rules that in the past were key in controlling overexploitation of fisheries in the estuary of the Patos Lagoon.

Introduction

Human territoriality is a fundamental method to control space and resources (Malberg 1985) and has been analyzed in terms of controlling and defending an area with resources (Begossi 1998). Territoriality can happen in different scales (individual, family, race, community) and under different forms of control of the resource (schools, points, bays) (Begossi 1998). Territoriality is not necessarily an aggressive behavior; in many cases, the resources are obtained through disputes, but in other cases, there are agreed local norms that regulate resources access (Begossi 1998). Territoriality is an important mechanism of monitoring because it deals with rules concerning location and rights to use specific areas. In this sense, knowledge of territoriality is considered a requirement to accomplish a sound community-based management where informal rules such as territories could be legitimized and accepted as an institution to protect the resources (Achenson et al. 1998).

*Corresponding author: tiagoalmudi@yahoo.br

This study analyses the territoriality of artisanal shrimp fisheries in the estuary of the Patos Lagoon. Three major questions are discussed: Do fishers in the estuary of the Patos Lagoon develop a territorial control appropriate to deal with the challenges in managing fisheries resources? What are the conditions and events that led to the development of territorial control and that have influenced its maintenance, its breakdown, or its transformation? And why, for this particular case, has the presence of a territorial control not avoided the overexploitation of fisheries resources?

Methods

Field work in the estuary of the Patos Lagoon was carried out in 2003 and 2004 (the site is described below). Data were obtained from primary and secondary sources. The primary sources of data were (1) in-depth semistructured interviewing with key fishers and officials from the Brazilian Environmental Agency, and (2) a fishing trip to observe the practice of shrimp fisheries, fishing area, and fishers' behavior. Supplemental data included reviews of scientific publications, as well as laws, decrees, and policy statements from the IBAMA and from the former Federal Sub-Secretary for Fisheries Development (SUDEPE). The survey addressed but was not limited to questions concerning issues of mobility of technology, abundance of resources, predictability of the fisheries, density of fishers, and exclusion of outsiders (McCay and Acheson 1987; Feeny et al. 1996; Begossi 2001).

The Patos Lagoon

The estuarine region of the Patos Lagoon is located in the southern Brazilian coastal zone (Rio Grande do Sul State). With an area of approximately 10,000 km^2, the Patos Lagoon is recognized as the world's largest choked lagoon, stretching from 30°30'S to 32°12'S near the city of Rio Grande where the lagoon connects to the Atlantic Ocean (Figure 1). The estuarine region encompasses approximately 10% of the lagoon and is characterized by a shallow body of water (mean depth of 7 m) with variable temperature and salinity depending on local climatic and hydrological conditions (Castello 1985). The Patos Lagoon system connects with the ocean via the channel between a pair of jetties, about 4 km long and 740 m apart at the mouth. All the estuarine dependent marine organisms enter and leave the estuary through this channel for nursery, reproductive, and feeding purposes. Four important species of fish and crustacean have sustained artisanal fisheries activities in the estuary for more than a century. They are Brazilian pink shrimp (also known as São Paulo shrimp) *Farfantepenaeus paulensis*, marine catfish *Netuma barba*, whitemouth croaker *Micropogonias furnieri*, and *Mugil platanus*. The life cycle of these species produces seasonal variability, diversity, and abundance of resources in the estuary to artisanal fishers.

Artisanal fishers reside in small communities along the estuary and are spatially organized into four fishers colonies (colônia de pescadores)[1]. The artisanal fishery operates in estuarine and shallow coastal waters and is characterized by minimal fishing technology. Fishers normally own their vessels and work together in kin groups. Artisanal fishing gear includes gill nets, stow nets, and otter trawls. In 1966, considering all the existing fisheries, artisanal catches accounted for more than 80% of the total catches in southern Brazil. However, landings from artisanal fisheries declined from an historical peak of 43,600 metric tons in 1972 to about 6,000

[1]The Fisher Colony is a professional organization of fishers of a given municipality, which is legitimized by the Federal Constitution as one form of working union. Although affiliation to the Colonies is compulsory there are fishers (illegal or underage) operating in the estuary that are not registered in any Colony.

Figure 1. Map of study area.

in the late 1990s. Today, the main artisanal resources are either fully exploited, overexploited, or depleted and catches are close to subsistence levels, with the exception of mullet and Brazilian pink shrimp, which provide sporadic economic returns during ideal environmental conditions (IBAMA 1995).

Fisheries institutional arrangements

Governmental institutional arrangements for regulating fisheries activities in Brazil have been changing over the years. The role of the federal government in marine fisheries management became particularly influential in the mid-1960s with the creation of the Federal Fisheries Agency (SUDEPE) of the Ministry of Agriculture. Up to this period, fisheries in the estuary were mostly governed informally by the small-scale fishers communities, but never recognized as important local informal institution and were largely affected by the implementation of this top-down management system. This agency governed fisheries until 1989 when SUDEPE became extinct due to a recognition of its failure to manage fisheries in a sustainable fashion (Dias Neto 1999). In 1989, fisheries management became one of the agendas of the IBAMA, a subsidiary of the Ministry of Environment. From 1998 to 1902, authority for fisheries management in Brazil has been split between two agencies: the Ministry of Agriculture (Department of Fisheries and Aquaculture [DPA]) and the Ministry of Environment (IBAMA). Since 2003, the DPA was substituted by the new Special Secretariat of Aquaculture and Fisheries (Secretaria Especial de Aqüicultura e Pesca).

In the estuary of the Patos Lagoon, a certain type of devolution and/or delegation of power from the IBAMA to small-scale fisheries communities and/or to local-based institutions is taking place via a comanagement arrangement that regulates fisheries in the estuary of the Patos Lagoon. This comanagement was initiated in 1996 with the creation of the Forum of Patos Lagoon as an institutional response to the crisis of estuarine fisheries (Reis and D'Incao 2000; Kalikoski 2002). This new regime establishes rules regarding how much, when, and how different resources can be harvested via management regulations such as licensing, timing, location, and vessel or gear restriction to prevent overexploitation inside the estuary and within the 3-mi zone (Decrees 171/98 and 144/2001modified by the norm IN MMA/SEAP 03/2004) (Kalikoski 2002; Kalikoski and Satterfield 2004).

Territoriality in the Estuary of the Patos Lagoon

General characteristics

Fisheries communities had their rights exercised in the management of the resources through different mechanisms. One such mechanism was the definition of fishing territories. In the estuary of the Patos Lagoon, there were fishing practices used by artisanal fishers that had important significance to the definition of fishing territories and rights. Fishing practices involved interrelated mechanisms such as the definition of fishing areas, periods, and technologies.

Technology

Territoriality is highly related to the type of technology in place to capture fish resources. This study has identified the existence of an informal level of protection of the fishing spots related to the capture of Brazilian pink shrimp. Stow nets, introduced in the 1970s, are the dominant type of gear used in the estuarine Brazilian pink shrimp fishery. Stow nets are fixed in shallow areas of the lagoon and operate by attracting shrimp to the net with light produced by gas lamps. The nets are placed mostly in the shallows where the young shrimp are caught before migrating from the nursery areas to the canal. Since stow nets were introduced, they became the only fishing technology legally allowed in the shrimp fishery. The rules that define the number of allowed nets, the space between them, and the exact location of stow nets have remained mostly determined informally among fishers based on agreed fishing territories. Government controls only the number of nets allowed to a fisher in a fishing season (10 per fisher). The "andainas" (placement of wood stakes that fasten stow nets) facilitates the demarcation of territories. We identified a kinship among fishers related to the definition and respect to fishing grounds with a certain level of help from the IBAMA that acts as mediator of conflicts over fishing territories when conflict happens. Government's norm (IN MMA/SEAP 03/2004) formally regulates the exact position of the andainas in the estuary and the allowed number of andainas per fishers. Fishers generally respect regulation related to the position of the andainas because this position represents their historical informal fishing spots. However, when there is a misplacement of the andainas and consequently invasion of fishing territoriality, the IBAMA is notified and usually mediates the situation favoring the appropriate predetermined respect position and re-establishing respect for regulation. According to officials and fishers there is a lack of legal sanctions for fishers that misplace their andainas. This factor, if not changed, may weaken efforts to control territoriality in the future.

Fishing for Brazilian pink shrimp is done during the night. Fishers start preparing for the fishing during the afternoon and fish the whole night. During the morning, they pick up the stow nets. Although the shrimp fishing nets are fixed nets, on some occasions, fishers do move from their fishing spots according to changes in shrimp abundance. Therefore, during the season, from one day to another, there may be a change in the fishing spot if there is space available for the fishers to move. This mobility is part of the informal territorial accords made informally by fishers. During the day, when fishers are selling the shrimp, they need to leave the nets hung in the place of fishing. There is an exchange of favors, and generally, when one fisher is selling his shrimp, the others help to take care of the fishing equipment and spots to make sure that no invasion and/or robbery occurs. This study has identified that informal territorial control was only identified for the Brazilian pink shrimp fishery due to its fixed type technology. In turn, our results demonstrate that there is no territoriality for the mobile fisheries. The protection of the fishing territories and fishing equipments was mainly identified at level of individuals and little groups (i.e., it happens through the protection and monitoring of fishing territories among fishers that know each other, the ones that are relatives or friends).

Abundance of the Resource and Density of Fishers

Fishing for Brazilian pink shrimp is the most disputed fishery in the study area. In addition, during the shrimp season, there are many fishers (some illegal) that are called occasional, that participate in this fishing. Occasional fishers fish only for shrimp during the summer, while the rest of the year, they have other economic occupation that is not related to fishing at all (e.g., taxi driver).

There is a lack of accurate information on the number of fishers that fish in the estuary of the Patos Lagoon, and according to Haimovici et al. (2006), around 15,000 people are involved in the activity. It is considered that if the number of exclusive fishers operating in the estuary is about 5,000, this number triplicates during the Brazilian pink shrimp season. The density of fishers increases depending on the fishing spots as well. Since the introduction of stow nets during the 1970s, there has been an observed decrease in abundance and an increase in fishing effort. With the introduction of this technology, a rearrangement of territories took place in the artisanal fisheries. There were a considerable number of new entrants in this fishery. This technology opened access to occasional fishers, people who work in the cities and farmers, that have never lived from the fishery activity but that began to catch shrimp as an alternative source of income (Asmus 1989). According to Asmus (1989), three factors led to this: the high prices of shrimp on the market, the easy access to fishing grounds (the city of Rio

Grande, for instance, is surrounded by two important shallow water bays used as shrimp fishing grounds by city workers), and the considered simple fishing technique that requires only one fisher and a small boat to operate the nets. During this period, there were no rules to prevent new entrants to this fishery. The open access combined to noncompliance and nonenforcement of rules regarding the maximum number of nets per fisher resulted in the overcrowding of the lagoon shallows during the shrimp season. The opening of access to new entrants, and the spreading of stow nets around the lagoon, redefined fishing territories and created unintended access limitation problems to some artisanal fishers, mainly to those that continued using other technologies (e.g., trawling and bag nets). This was due to the lack of space for new shrimp nets, the difficult navigation in the fishing ground overcrowded with nets, and the fact that most shrimp are caught before migrating out of nursery grounds in shallow waters. Stow nets are highly efficient in capturing shrimp as a result of the effect of light attraction, and few shrimp escape to be caught by other fishing techniques.

Fishing effort control started to be established by Decree 171/98 that limited access to fish inside the estuary to local fishers that depend on this activity for a living in order to control the high density of fishers and to guarantee the territoriality control of this fishery.

Predictability of the Fisheries

It is rather difficult to predict Brazilian pink shrimp's abundance for the upcoming fishing season. A good season is highly dependent on the rainfall and wind regime that influences the entrance of salted water in the lagoon when it arrives during the winter. According to fishers' knowledge, if estuarine waters are salted (and consequently clear) from the middle of October until middle of January, the season fishing will be good. Fishers informally apply their knowledge and predict when and where a good fishing spot could be found and develop their territoriality accordingly. This knowledge is informally applied to define when the fishing season in fact starts, despite official season openings.

Capacity to exclude outsiders and kinships

Outsiders fishing in the estuary of the Patos Lagoon during the shrimp season multiplies because many fishers that are occasional try to fish in this area. In some cases, there are problems related to robbery of fishing equipments and even of the shrimp that is captured in the nets. In some fishing spots inside of the lagoon, conflicts over fishing spots commonly happen.

We identified the creation of fishing groups that devised informal fishing accords to protect their fishing spots. If somebody disrespects these accords, the role of the IBAMA is limited to asking the invader to leave the place where he is not allowed to fish. There are no sanctions to enforce territoriality, and this may cause a problem of controlling efficiently the existence of informal fishing accords and territories among fishers. The registration of the andainas (in number to support at maximum 10 stow nets) with their specific fishing location began when the federal government formally institutionalized this new type of gear. The location of the andainas follows to some extent the traditional access rights and territories determined informally by fishers. The function of the IBAMA is to grant the license and not to monitor the fishing spots that are controlled among fishers. In fact, we could observe that there are no formal means for fishers to protect their fishing spots besides the territory's agreement made informally by them to respect each others' fishing spots. This lack of formal control sometimes causes some conflicts related to invasion of fishing territories.

Since the introduction of stow nets as the allowed technology to fish shrimp during the 1970s, many fishers from Santa Catarina state would come to fish in the estuary of the Patos Lagoon (now forbidden by IIN MMA/SEAP 03/2004). Local fishers did not agree with this situation; however, they could not expel them because at that time, there was no law preventing fishers from other places to fish in the Patos Lagoon (in fact, the Brazilian constitution considers fishing open to any Brazilian with a valid fishing license). In addition, fishers from Santa Catarina state used to come in great number, making it extremely difficult for locals to fight for protecting fishing spots, even those spots that were not used by anyone but were considered protection areas. Identified conflicts with outsiders happen when resource abundance in certain places of the estuary does not support more fishers than the allowed licensed ones. These shrimp fishing conflicts happen when several occasional or opportunist fisher fish as an alternative source of income corroborating the problem that Pauly (1997) defined as the neomalthusianism, where many people from other activities in the coastal zone migrate to fisheries as a source of income or food.

Discussion and Conclusions

Table 1 summarizes the factors that have been influencing the maintenance and evolution of territoriality in the estuary of the Patos Lagoon.

This work has identified that fishers in the study site develop a behavior of protecting fishing territories. This territoriality is informally enforced and regulated by fishers. The IBAMA, the governmental agency responsible for regulating and enforcing fishing activity in Brazil, has control over who is fishing, the technology being used accordingly the species being captured, and the fishing period, but the IBAMA has no control over enforcing fishing territories. The IBAMA does informally recognize and legitimize the placement of andainas, according to the informal fishing accords, on fishing territories made informally by fishers. The IBAMA, therefore, recognizes the informal territoriality of fishers.

The beginning of the territorial behavior started when a great part of the resources explored in the fishing area started to become scarce by overfishing. The increase of the density of local fishers and the absence of formal mechanisms of exclusion of new entrants influenced the emergence of territoriality among fishers. This in turn created mechanisms of protecting fishing spots, as a way to avoid the problem of invasion and of robbery of their fishing equipment. In spite of the decrease in the fishing production and even the collapse of some species (e.g., marine catfish), fishers agree that the number of fishers is still growing.

We identified that territorial behavior was developed for shrimp fishery mainly because it is done with fixed nets. Some factors have determined the absence of behavior of territorial control in other fisheries. The mobility of the technology (gill net) is one of those factors. Territoriality was not observed during the past fishing seasons for species such as whitemouth croaker, marine catfish, and mullet fished inside the estuary and/or on the adjacent coast. However Kalikoski and Vasconcellos (2007) identified territoriality for mullets during the 1960s. Another factor would be the small density of fishers involved in the fishing activity of these species compared to the shrimp fishery. An increase in density is shown to be an important factor triggering fishing territoriality.

Although shrimp fishing effort has been increasing, few minor conflicts over fishing spots were

Table 1. Factors affecting the territoriality in the fisheries of Patos Lagoon estuary.

CRITERIA	CHARACTERISTICS OF FISHERIES IN THE ESTUARY
Mobility of technology	Stow nets, a fixed gear introduced in the 1970s, were key to the creation of territoriality for pink shrimp fishery. The development of territoriality was not observed for the current mobile fisheries. Transformation in fishing practices during the late 1960s broke forms of territoriality in these mobile fisheries.
Abundance of resources	Increasing scarcity of shrimp intensified territorial control to sustain fishers livelihoods. Technology was key to the development of territoriality in a situation of resources depletion. Despite the decrease in CPUE for finfish (croackers and catfish), territoriality was not observed for the current fishing practice.
Predictability of fisheries	Fish behavior (schooling) increases predictability of good fishing spots and increase conflict in the use of fishing territories (e.g., mullet fishery)
Density of fishers	Increasing number of fishers competing for the same resources is triggering the development of territoriality. This effect is more prominent in shrimp fisheries because of difficulty to exclude outsiders. The extent that this factor influence territoriality in other fisheries is yet to be better analyzed.
Exclusion of outsiders	Enforcement of legislation to exclude outsiders is fundamental to guarantee the observed forms of territoriality informally developed by fishers and legitimized by IBAMA.

observed so far. Nevertheless, there is a consensus that this fact will change, and even the IBAMA expects that in the near future, there will be no place for new entrants in the shrimp fishery due to the lack of space to fish in the estuary of the Patos Lagoon.

An action plan for territorial control in the study area should be developed and involve the sharing of responsibilities between fishers and officials to monitor and protect fishing territories through the Forum of Patos Lagoon co-management arrangement. It is interesting to note that sometimes there are some difficulties in protecting the territories, at least based on the current governance system. An example of this is the relationship between artisanal and industrial fisheries. Although they occupy most of the time different territories, they do fish for mullets and white-mouth croaker in the same areas around the mouth of the estuary. In these cases, we can observe that the industrial fisheries compete for the same resources in an uneven way and sometimes destroy the nets of the artisanal fishers when they are fishing. Based on this observed problem, artisans were able via the Forum of Patos Lagoon to create a decree in December 2003 (curently regulated under the norm IN MMA/SEAP/03/2004) that forbids industrial purse seiners to fish in these areas. Another conflict includes the large number of occasional fishers that fish in the estuary of the Patos Lagoon during the shrimp season. On the one hand, local fishers never accepted these outsiders, but on the other hand, locals did not have enough force to avoid their invasion of the estuary.

This work indicates that the main modification in the fishing territories in the estuary of the Patos Lagoon relates to the changes in technology that affected how, where, and by whom the resources are captured. Two factors contribute to this. First, territoriality is shown to be associated more with guaranteeing profitability than securing fish conservation; hence, territoriality has not avoided fisheries over-exploitation. Second, in spite of the fact that shrimp technology, (i.e., stow nets) is key to the development of territoriality, this gear facilitates new entrants to the Brazilian pink shrimp fishery because this gear does not demand specialized fishing skills, in turn increasing conflicts and the race for fish. In spite of the existence of respect for fishing spots among the fishers that know each other (family and partners), we could observe the difficulty in devising a body of formal rules related to territories, given the high concentration of fishers and the lack of infrastructure from the IBAMA to enforce such rules. However, territory control must be better enforced by cooperation between fishers and the governmental enforcement agency in order to avoid further territorial disruption in the future.

Acknowledgments

The author thanks CNPq for the scholarship received and financial assistance to conduct the work. Daniela Kalikoski thanks FAPERGS (Fundação de Amparo a Pesquisa do Estado do Rio Grande do Sul) for the support received to develop this work (Process No. 02/1122.6—ARD).

References

Achenson, J., J. A. Wilson, and R. S. Steneck. 1998. Managing chaotic fisheries in linking social and cultural systems for resilience. Pages 390–413 *in* C. Folke and F. Berkes, editors. Cambridge University Press, Cambridge, England.

Asmus, M. L. 1989. Pradarias de gramíneas marinhas (*Ruppia maritima*) como áreas vitais na região estuarial da lagoa dos Patos. Pages 291–299 *in* III Encontro Brasileiro de Gerenciamento Costeiro, 1989, Fortaleza. Anais do III Encontro Brasileiro de Gerenciamento Costeiro. Fortaleza, Ceará, Brasil.

Begossi, A. 2001. Protecting the commons. A framework for resource management in the Americas. Island Press, Washington, D.C.

Begossi, A. 1998. Cultural and ecological resilience among caiçaras of the Atlantic Forest and caboclo of the Amazon, Brazil. Pages 129–57 *in* Carl Folke and Fikret Berkes, editors. Linking social and cultural systems for resilience. Cambridge University Press, Cambridge, England.

Castello, J. P. 1985. The ecology of consumers from dos Patos Lagoon estuary, Brasil. Pages 383–406 *in* A. Yañez-Arancibia, editor. Fish community ecology in estuaries and coastal lagoons: towards an ecosystem integration. UNAMN Press, Mexico City.

Dias Neto, J. 1999. Pesca nacional: Anarquia oficilizada. Boletim da Sociedade Brasileira de Ictiologia 55:9–10.

Feeny, D., S. Hanna, and A. F. McEvoy. 1996. Questioning the assumptions of the 'tragedy of the commons' model of fisheries. Land Economics 72:187–205.

Haimovici, M., M. C. Vasconcellos, D. C. Kalikoski, P. Abdalah, J. P. Castello, and D. Hellebrandt. 2006. Diagnóstico da pesca no litoral do estado do Rio Grande do sul. Pages 157–180 *in* V. J. Isaac, A. S. Martins, M. Haimovici, and J. M. Andrigueto, organizers. A pesca marinha e estuarina do Brasil no início do século XXI: recursos, tecnologias, aspectos sócio-econômicos e institucionais. Universidade Federal do Pará—UFPA, Belém, Brazil.

IBAMA (Instituto Brasileiro de Meio Ambiente). 1995. Peixes demersais. Ministério do Meio Ambiente, dos Recursos Hídricos e da Amazônia Legal. Coleção Meio Ambiente. Séries Estudos de Pesca

Kalikoski, D. 2002. The Forum of the Patos Lagoon: an analysis of co-management arrangement for conservation of coastal resources in southern Brazil. Ph.D. thesis. University of British Columbia, Vancouver.

Kalikoski, D. C., and M. Vasconcellos. 2007. Fishers knowledge role in the management of artisanal fisheries in the estuary of Patos lagoon, southern Brazil. In N. Haggan, B. Neis, and I. G. Baird, editors. Fishers' knowledge in fisheries science and management. Coastal Management Sourcebooks 4. UNESCO, Paris.

Kalikoski, D. C., and T. Satterfield. 2004. On crafting a fisheries co-management arrangement in the estuary of Patos Lagoon (Brazil): opportunities and challenges faced through implementation. Marine Policy 28:503–522.

Malberg T. 1985. Territoriality at sea: preliminary reflections on marine behavioral territories in view of recent planning. Man-Environment Systems 15:15–18.

McCay, B., and J. M. Acheson. 1987. The question of the commons. University of Arizona Press, Tucson.

Pauly, D. 1997. Small-scale fisheries in the tropics: marginality, marginalization, and some implications for fisheries management. Pages 40–49 *in* E. K. Pikitch, D. D. Huppert, and M. P. Sissenwine, editors. Global trends: fisheries management. American Fisheries Society, Symposium 20, Bethesda, Maryland.

Reis, E. G., and F. D'Incao. 2000. The present status of artisanal fisheries of extreme southern Brazil: an effort towards community based management. Ocean and Coastal Management 43(7).

American Fisheries Society Symposium 49:197–208

Salmon Farming, Genomics, and Ethics

MELANIE D. POWER-ANTWEILER
University of Victoria, Centre for Studies in Religion and Society
SED B102, Box 1700 STN CSC, Victoria, British Columbia V8W 2Y2, Canada

Abstract.—Aquaculture in its most basic form utilizes simple methodology and technology. By comparison, contemporary Atlantic salmon *Salmo salar* aquaculture incorporates advanced research that has driven the development and application of research into salmon genomics. Such applications are evident in all aspects of the farming process, including chromosomal manipulation to produce sterile farmed salmon, the use of genetically-modified products in salmon feed, the development of vaccines for disease prevention, and even the creation of a strain of transgenic salmon to produce desired characteristics. In British Columbia and elsewhere, aquaculture is increasingly driving research and development in salmon genomics. As the field advances, hopes regarding potential benefits and concerns regarding possible negative impacts are raised. Policy must be developed to address the realities of genomic research and applications within Atlantic salmon aquaculture. For this policy to be developed in an ethical manner, policy makers must be respectful of public and expert interests, in keeping with the democratic values of transparency, accountability, and representation. The Genome BC-funded Democracy, Ethics, and Genomics (DEG) project at the W. Maurice Young Centre for Applied Ethics seeks to evaluate alternative methodologies of ethical analyses regarding genomics research. Three streams of research explore the methodologies of public consultation, deliberative democracy, and computer modeling as means of ethical analysis. Through these activities, values relating to genomics and salmon aquaculture are identified. This paper, which represents a portion of the background research conducted in support of the DEG project, describes aquaculture-related elements of salmon genomics research and applications, and the ethical issues associated with these.

Introduction

Broadly, "genomics" means genes and their function. With regard to salmon, applications of genomics may benefit wild stocks through improved understanding of the basic biology and life histories of different salmonids and the unique genetic variations characterizing natal stocks[1]. However, genomics data are often intended for use in salmon aquaculture, where research is driven by promises of increased yields, decreased production costs, and solutions to harms commonly ascribed to conventional salmon aquaculture methodologies.

In this paper, I provide an overview of salmon aquaculture, emphasizing the industry in British Columbia, Canada. From this context, I describe a range of potential or actual applications of genomics within the salmon aquaculture industry. Finally, I discuss ethical is-

*Corresponding author: melanie.antweiler@shaw.ca

[1] Aubin-Horth and her colleagues have found, for instance, that expression of some genes in male Atlantic salmon varies with the rearing environment (Aubin-Horth et al. 2005a).

sues arising from genomics applications in salmon farming.

Genomics, Salmon, Aquaculture, and Ethics?

Governance of genomics research represents what Tansey (2003, citing P. 6, Nottingham Trent University, unpublished data) describes as governance through coping, meaning that research often outpaces the development of relevant policies. As a result, policies are developed as a response to new findings, rather than as a result of proactive foresight. Obviously, this impinges on the possibility of considered reflection and painstaking determination of the appropriate directions of and uses of the products resulting from this research[2], and little if any opportunity can be given to receive and thoughtfully incorporate views of those with a less-direct interest than the researchers and companies directly involved in aquaculture genomics work. Yet the nature of genomics research is such that virtually every person is likely to be affected in some way and to some extent, and the democratic values of transparency, accountability, and representation ought to be represented in policy decision making.

The Democracy, Ethics and Genomics project (DEG) seeks to evaluate alternative consultation methods for incorporating diverse opinions on genomics, particularly with regard to governance (Burgess 2003a, 2003b)[3]. Three streams of research explore the methodologies of public consultation, deliberative democracy, and computer modeling as means of ethical analysis. To facilitate this research, two case studies have been selected, one relating to human genomics (DNA banking) and a second regarding salmon aquaculture.

This paper, which represents a portion of the background research conducted in support of the DEG project and is intended to stimulate discussion within the project and provide research direction, describes aquaculture-related elements of salmon genomics research and applications, and the ethical issues associated with these.[4]

Salmon Farming in British Columbia

Why do we farm fish? To bridge the gap between the wild catch and the demand for fish (http://www2.marineharvest.com/faq/why-do-we-farm-fish-.html).

Global trade in cultured salmon totaled approximately 1 million metric tons (mt) in 2001 (FAO 2002, Box 7, "Trade in aquaculture products"). Eighty to 90% of farmed salmon was produced in four countries: Norway and Scotland (the two countries in which the industry was pioneered); together with Chile and Canada (Asche et al. 2001; Bjørndal and Aarland 1999; FAO 2002). Salmon farming is the single largest segment of Canada's aquaculture industry (Canada 2001, 2003e), with activities occurring on both the Atlantic and Pacific coasts.

The salmon farm industry on Canada's Pacific coast initially focused on two species of indigenous Pacific salmon, coho *Onchorhynchus kisutch* and Chinook *O. tshawytscha*. In 1985 the still-nascent industry began to import Atlantic salmon *Salmo salar* eggs (Volpe

[2] A notable exception is the Royal Society of Canada expert panel report on food biotechnology (The Royal Society of Canada 2001).

[3] See also the Democracy, Ethics, and Genomics project website at http://gels.ethics.ubc.ca/.

[4] A comprehensive background document has been posted as a pre-publication electronic working paper on the website of the W. Maurice Young Centre for Applied Ethics (see: Power 2003).

2001), and 3 years later the first harvest of Atlantic salmon was realized (Tyedmers 2000, citing British Columbia Environmental Assessment Office 1997c). While coho and Chinook continue to be farmed in British Columbia, more than 70% of salmon farmed in Canada's Pacific Ocean province are Atlantics rather than Pacifics (Gardner and Peterson 2003:75)

In 1995, British Columbia's provincial government placed a moratorium (of indefinite duration) on the issuing of fish farm licenses, effectively capping the number of tenures along Canada's Pacific coast at 121 (Gardner and Peterson 2003). Note, however, that the ceiling on farm site licenses did not limit growth of the sector, and in fact employment and production in the salmon farming industry continued to expand throughout the moratorium (Gardner and Peterson 2003). Simultaneously, the province's Ministry of Environment, Lands and Parks commissioned a review of salmon aquaculture; the findings of the *Salmon Aquaculture Review* were published as a five-volume set 2 years into the moratorium (British Columbia Environmental Assessment Office 1997a, 1997b, 1997c,1997d, 1997e). On September 12, 2002, and following a change in provincial government, the moratorium was lifted (British Columbia Ministry of Agriculture Food and Fisheries 2002; Gardner and Peterson 2003).

Salmon Aquaculture and Genomics

Genomics research with applicability to salmon aquaculture takes two approaches: to ameliorate those problems inherent in conventional salmon farming; and to improve efficiency (and hence profitability) within the industry. Note that often these two approaches are not mutually exclusive, particularly since criticism on conventional methods may negatively impact the market for the farmed fish and give added incentive for improvement. A summary of genomics approaches is provided in Table 1.

Disease and Vaccines

A frequent criticism of conventional salmon aquaculture involves the use of pharmaceuticals to prevent and/or control disease outbreaks, particularly where these could be transferred to wild salmon. (See, for example, Gardner and Peterson 2003; CBC News British Columbia 2004; Hume 2004) According to a report by the USDA's Agricultural Research Service, "[b]iotechnology holds great promise in controlling diseases in aquaculture" (USDA Agricultural Research Service 2001) through DNA vaccines and novel pathogen-testing methods. Researchers with Fisheries and Oceans Canada are likewise investigating the genomics of salmonid pathogens to better understand how a given pathogen affects specific stocks (Canada 2003a, 2003b).

The development of DNA vaccines would work in a manner similar to traditional attenuated vaccine agents, but instead of containing the virus of interest, this type of vaccine would instead contain protein-coding genes (Canada 2003a). Gene-based vaccines are expected to be more stable than attenuated virus vaccines, require lower doses, and present a reduced risk of accidental infection. To date, only genes from mammalian viruses have been used in this research; Fisheries and Oceans Canada researchers are working to develop "all-salmon" DNA vaccine constructs (Canada 2003a), which would be expected to be more acceptable (or at least less threatening) to consumers.

Feed

Feed represents one of the largest expenses associated with salmon farm operations, ac-

Table 1. Genomic research to address challenges of conventional salmon aquaculture.

Conventional salmon aquaculture	Genomic response
Disease transfer between wild and farmed salmon; use of antibiotics	DNA vaccines
Use of fish meals/oils in feed; expense of fish meals/oils for use in feed; ecological impacts of using forage fish to feed carnivorous farmed fish	Development of plant-based feeds; genomic research to ensure plant-based feeds are digestible for salmon
Need for synthetic carotenoids to ensure flesh of farmed salmon is the expected shade of pink	Genomic research to: a)modify plants to contain the necessary pigments; b) modify the fish to produce the pigments themselves; and/or c) modify the fish to more efficiently use the pigments contained within the feed.
Time to marketable size	Transgenic fish, enhanced for growth, to reduce time to marketable size
Genetic impacts of interbreeding with wild conspecific stocks	Sterility through triploidy; sex-reversal

counting for between 35% and 60% of the total costs of production (Naylor et al. 2000; Tyedmers 2000). Of the feed costs, about 50% covers the protein component of the feed (mainly fish oil and/or fish meal) and 15–25% pays for the synthetic carotenoids used to colour the flesh the pinkish shade expected by consumers (Canada 2003c). Hence, "...finding cheaper, but equally palatable and digestible alternatives to fish meal would significantly reduce one of the major costs of salmon aquaculture production" (Canada 2003c). The aquaculture industry, which accounts for an estimated 30–50% of global fish oil consumption (Jystad 2001), competes with health food, pharmaceutical sectors, and other sectors for fish oils, leading to concerns of potential shortages. Furthermore, since the fish farming industry continues to expand, the cost of the protein component of fish feed can reasonably be expected to increase.

Given the high cost of fish meals and oils, as well as perceived and potential shortages of these products, research is being conducted into alternate feed components, notably soy- and canola-based products. Canola oil, already used in feed for poultry, swine, and cattle is a leading contender since it is comparable to fish oil in fatty acid composition (Jystad 2001; Canada 2003c), and is "...less than half the cost of fish meal on a per kilogram of protein basis" (Naylor et al. 2000). However, digestive ailments have been suffered by salmonids fed feed containing either of these vegetable-based feeds (Canada 2003c; Jystad 2001). The role for genomics research is to overcome the indigestibility of the plant-based feeds. The approach is two-fold: firstly, genomic manipulation of the fish (to enable digestion of plant-based feed) and/or of the plants themselves (to improve digestibility), such that alternative plant-based feeds become feasible (Canada 2003c); and secondly, genetic modification to improve feed conversion efficiency, to reduce the amount of feed required to bring the fish to marketable size (Roberts 2000).

Growth Enhancement

The commercial advantage of higher growth rates serves as the primary motivation for this type of research (The Royal Society of Canada 2001). For salmon farmers, faster growth means less time to market (Lewis 2001), and reducing the risk of escape or disease (MacKenzie 1996, quoting Garth Fletcher of A/F Protein) and reducing feed requirements (Niiler 2000; CEQ/OSTP 2001; Lewis 2001; Institute for Social Economic and Ecological Sustainability 2003). By extension this means increased overall production due to expedited (and therefore more) production cycles (CEQ/OSTP 2001; Pew Initiative on Food and Biotechnology 2003), lower costs through fewer fish losses, and reduced feed inputs (Pew Initiative on Food and Biotechnology 2003).

Increased growth rates have been realized in Atlantic salmon through the transgenic insertion of genes from other fish species. The *AquAdvantage* transgenic Atlantic salmon, developed by A/F Protein, received two genes: a Chinook salmon growth hormone gene and an ocean pout anti-freeze gene that acts as a promoter sequence. This combination results in year-round, rather than seasonal, growth (Union of Concerned Scientists 2001), thereby reducing from 3 years to 2 the length of time for individuals to reach marketable size (MacKenzie 1996). The *AquAdvantage* salmon received a United States patent in 1996 (Hew and Fletcher 1996), and is presently under review by the U.S. Food and Drug Administration (Union of Concerned Scientists 2001). Having completed all major studies for the review, the company expects that the salmon will be ready to market in 2009 (Anonymous 2006). The FDA could have chosen to assess the *AquAdvantage* salmon as a food product, at which point substantial equivalence would have been the default for evaluation. That is, the American *Federal Food, Drug, and Cosmetic Act* presumes the safety of new varieties of food already available in the market, and considers them to be "substantially equivalent" to those varieties already available, and consequently requiring no supplementary evaluation; only if exceptional risk is present does substantial equivalence not apply. New animal drugs and food additives are not granted this presumption of safety and must be evaluated (Pew Initiative on Food and Biotechnology 2003). Instead, the FDA has elected to consider the salmon as a "new animal drug" (Union of Concerned Scientists 2001), on the basis that "...substances are being added [to the animal] for the purpose of improving animal health or productivity" (Matheson 1999), and thereby triggering a strict level of evaluation.

Genetic Impacts on Wild Stocks

In British Columbia, it is agreed that fish escape, although the extent of the escapes, as well as what happens to the escapees—whether they survive, reproduce, interfere with spawning of wild salmon, or colonize provincial waterways—remains contentious (Gardner and Peterson 2003). The most prevalent concerns arise from the release or escape of farmed stocks and the potential for interbreeding between cultured and wild fish. Consequently, concern is greatest where the farmed fish are of the same species as the indigenous wild salmon. In some Norwegian streams, farm escapees outnumber wild salmon (FAO 2000), and one estimate indicates that up to 40% of Atlantic salmon caught in the North Atlantic originated in salmon farms (Naylor et al. 2000, citing Hansen et al. 1993).

Since the majority of British Columbia's salmon farms culture Atlantic salmon, the genetic risk is mitigated. As distinct species, the probability that escaped Atlantic salmon will breed with wild Pacific salmon is believed to be extremely low (Volpe 2001), but to date experimental crossings of Atlantic and Pacific salmon species have not been extensively conducted (Gardner and Peterson 2003). Yet the risk persists where farmed and wild species are the same and interbreeding between conspecifics is possible. Some British Columbia fish farms raise Chinook or coho, and escapes from these populations could reduce the genetic fitness of the wild stocks through outbreeding depression. Similar would be true in aquaculture zones where Atlantic salmon are indigenous. Two genomics-based responses address this risk: firstly, triploidy to render infertile the farm stock; and secondly, the use of all-female farm stocks.

Triploidy can be achieved through either heat treatment of or hydrostatic pressure on fertilised salmon eggs, which results in retention of a third set of chromosomes. While an adult triploid fish will produce sperm and eggs, those are not viable due to the extra chromosomal material (Canada 2003d); additionally, although the triploid salmon may breed, their offspring will not be viable and thus will not survive (Benfey 2001; CEQ/OSTP 2001). Note as well that the success rate of this process is short of 100% (The Royal Society of Canada 2001), with an estimated success rate of anywhere from 10 to 95% (Maclean and Laight 2000). Nothing short of complete sterility of potential farm escapees can guarantee elimination of all genetic risks to wild salmon (Wilkins et al. 2001).

All-female populations are created by first exposing juvenile salmon to hormones (androgens) through feed (Benfey 2001; USDA Agricultural Research Service 2001; Wilkins et al. 2001). All fish receiving the hormone-containing feed are chromosomally altered, such that all of the fish are genetically XX (female) although phenotypically some fish are female and some are male; the physical sexual characteristics of the fish are not altered. Though outwardly male, these fish are genetically female and carry no Y-chromosome. These genetically-female but phenotypically-male fish (Benfey 2001) can produce only female offspring. The second part of the process is to then mate XX males to normal females to produce broodstock that is chromosomally and physiologically female. Since female triploids typically remain sexually immature, if these all-female fish are also triploid, the genetic risk arising from intermingling with wild conspecifics is reduced.

Discussion: Ethics and Salmon Genomics

The promises of genomics are substantial, and if realized can produce tremendous benefits. However, genomics research is not without risks, and the unknown quality of these risks provokes worries. The tangible risks have been described above; here, I direct your attention to the potential harms experienced by individuals, societies, and ecosystems, and the ethical implications of these.

Respect of Cultures and Individuals, of Beliefs and Choices

Salmon fill a range of diverse roles: economic, recreational, food, and even as a "symbol of environmental quality" (Aarset 2002). Even "[t]oday, salmon has a symbolic position, representing heritage and continuity, even for nontribal communities" (Aarset 2002, citing Scarce 2000). They are cultural touchstones for many other cultures, including Celtic Ireland (Charron 2000), Sami, Kven, Norwegian, Swedish and Finnish societies, and Pacific salmon remain central to the worldviews of many of British Columbia's First Nations (Newell 1993). The cultural aspect is especially noteworthy: Scarce (2000) argues that nature, including salmon, is socially constructed, while Collet (2002) maintains that biotechnology has the capacity to create a "techno-nature".

Culturally, farmed and wild salmon are simply not interchangeable. The cultural and spiritual ceremonies of British Columbia's First Nations people center around wild salmon, for instance the range of 'first salmon' ceremonies evident in many nations (see, for example, Duff 1952; Barnett 1955; Stewart 1977; Newell 1993). Although some First Nations individuals may exercise the choice to eat cultured salmon, and/or to work in the aquaculture industry, farmed salmon are perceived by some Nations to be a serious threat to the viability of wild stocks. This perceived menace increases with genetically modified farm fish. The potential risks of these fish to wild stocks can only be hypothesized at present, and this uncertainty can only increase trepidation. Given the potential apprehension arising from genomic applications and First Nations historical, cultural, spiritual, and legal relationship to wild salmon, consultation and shared decision-making authority will become increasingly important. First Nations perspectives on the application of genomics to salmon aquaculture will need to be considered and respectful approaches to such consultation must be developed.

Including exotic genetic materials in foodstuffs will present challenges to those who abstain from consuming certain plants or animals due to religious or cultural beliefs. Most obvious is the concern that genetic material from a proscribed plant or animal could be included in an otherwise-acceptable food. In the early 1990s, for instance, DNA Plant Technology inserted a flounder antifreeze gene into a tomato, giving the plant increased cold resistance (Schmidt 2005) and to potentially allow freezing of the fruit while preserving satisfactory texture (Hightower et al. 1991). While the transgenic tomato was never marketed (Schmidt 2005), its development raises the question, when a plant contains an animal's genetic material, does it remain acceptable to vegetarians?

The challenge then is whether this type of harm to a subset of individuals can be addressed without limiting the choices of others, and to what level of risk should these subsets be exposed. It seems infeasible that all potential groups could be given a guarantee that none of the items forbidden by their beliefs would be present in any way. This is particularly true given that, while any list of proscribed items will be consistent for a subset (although not necessarily between individuals within that group), there could be substantial variation among subsets.

Labeling food is the most commonly suggested solution to this problem, and would satisfactorily respect the principle of autonomy and the right to informed choice (FAO 2001). Proponents argue that labeling will enfranchise choice at an individual level, allowing the development, production, and marketing of genetically modified foods to be balanced by the individual's choice of whether or not to purchase and/or consume those foods. Reticence towards labeling is based on fears that it may bias consumers against these products and worries that the information will not be meaningful to consumers. *Prima facie*, labeling would enable those with food restrictions to avoid proscribed foods. However, while this would likely work for novel material added directly into a foodstuff, it would probably prove unworkable with regards to animal feed. Likely, it would also be impractical in foodservice facilities.

While the harms associated with belief-based food prohibitions would be relatively limited, the benefits associated with providing all consumers with the choice through labeling of whether or not to consume genetically modified material would probably exceed the potential harms to producers due to consumer bias against their modified products. In fact, labeling would allow those who specifically wish to purchase and/or consume genetically modified products the option of doing so.

Harms to the Environment, Wild Fish Stocks, and Fisheries

One criticism of conventional aquaculture centers on actual or perceived harms to the ecosystem, wild stocks, and fisheries, proximate to farm operations. While certain genomic interventions (such as triploid mono-sex farm stocks and DNA vaccines) seek to address these harms, new challenges become apparent. For instance, genomic applications to improve disease resistance among farm stocks may potentially allow those fish to carry disease without becoming symptomatic. Lacking enhanced resistance, wild stocks coming into contact with farmed fish (whether through contact with farm escapees or during migrations past farm sites) may continue to be at increased risk of disease transmitted from infected farms.

The scope of the potential harm is disputed, as is the actual likelihood of disease transmission. However, a run failure could have ecosystem-wide effects—salmon are prey for bears and eagles, and the rotting spent carcasses release nutrients into freshwaters which sustain other aquatic species. In terms of fisheries harvest allotments, beyond conservation needs, First Nations ceremonial/social fisheries receive priority allocation, followed by commercial and, lastly, recreational fisheries.

While the value of the economic contribution of the commercial fishery is relatively limited province-wide, in local communities it is substantial. Small communities would be devastated by a run failure, particularly those that are not economically diversified or those whose other resource-based sectors have already diminished. These effects would be localized, but locally substantial. Recreational fishing is economically valuable, but the benefits are of a particularly narrow geographic scope and employment is comparatively limited; while the effects would be felt, these would not be as widespread as for coastal communities, due to the relative employment rates in both sectors and the geographic dispersion of the fishing activities.

Indeed, it is worth noting that even those researchers most active in the field are cautious with regard to genetically modified species. A leading triploidy researcher cautions that triploid fish should be treated as a novel species (Benfey 2001). Moreover, the creators of the *AquAdvantage* transgenic salmon state plainly that it "...is essential that transgenic broodstock be maintained in secure, contained, land-based facilities" (Fletcher et al. 2000). Conventional salmon aquaculture is contentious, as is genetic modification. Neither is universally accepted by consumers or the public; in fact, both are often vociferously rejected. When those two issues are combined, the differences of opinion are sure to be magnified.

Food Security, Sustainability, and Justice

Aquaculture makes possible an abundant year-round supply of fresh salmon, regardless of traditional fishing seasons. Furthermore this abundance has had the effect of depressing prices such that the fish has become less a luxury item than a staple (what Hannesson [2003] refers to as "chicken of the sea") and, hence, accessible to a greater number of consumers. Food security (that is, the sufficient availability of food which is nutritionally adequate and culturally acceptable) is arguably enhanced through aquaculture, and the application of genomics within aquaculture is intended, at least in part, to expand farm production. However, the use of fish products in farm feed affects the marine food web, often those distant from the actual fish farm operations. Although farming does produce more salmon, the reliance on fish feed derived from forage fish to feed carnivorous farmed salmon, aquaculture is not wholly detached from the natural marine food web (Naylor et al. 2000; Pauly and Watson 2003).

It is not an appropriate use of limited marine resources to convert less-preferred but edible forage fish to more "palatable" fish, such as salmon. The reduction fisheries needed to feed the production of fish meal represent a form of "fishing down marine food webs" (Pauly et al. 1998). Since many of the forage fish are caught in developing waters and used to produce premium farmed fish for sale in developed waters, Garcia was prompted to describe this practice as representing "'a transfer of fish food resources from the poor strata of developing countries to the population of the rich ones'" (1994, quoted in Collet 2002). Genomic research into fish and alternative (plant-based) diets may address these significant challenges arising from conventional salmon aquaculture.

Opportunity Costs

Opportunity costs may be manifested as either economic (the costs associated with electing to pursue a given path of research or development at the exclusion of another) or temporal. It is these temporal costs that are particularly salient to the present discussion, and which will be the focus here.

Temporal opportunity costs describe the price resulting from decisions in the present that introduce or preclude future options. Temporal opportunity costs are particularly salient in the area of transgenic salmon because of the high levels of uncertainty surrounding these novel fish.

While nobody knows for certain what will happen to wild salmon if transgenic salmon are released into wild salmon habitat, it is certain that should these fish enter the environment, it will be essentially impossible to "recall" them. If they colonize waterways, or if successive waves of salmon are released into the ecosystem, wild salmon may be permanently affected and runs diminished.

The right to democratic participation requires that no options for future generations be foreclosed for the sake of the present generation (FAO 2001). Consequently, discussions on transgenic salmon need to address not only how their introduction might affect the present generation, but also future generations. While contemporary individuals may be willing to sacrifice wild salmon for the potential benefits associated with genetically modified farmed salmon, are we willing and, most especially, entitled to make this decision for all future generations?

Acknowledgments

This research was conducted in support of the Democracy, Ethics and Genomics project at the W. Maurice Young Centre for Applied Ethics, at the University of British Columbia. Project funding was provided by Genome Canada through the office of Genome

British Columbia. I thank Mike Burgess for his guidance in discerning the ethical the issues described herein and especially for the opportunity to explore this research, and Judy Maxwell and Anji Samarasekera for research support.

Reference

Aarset, B. 2002. Pitfalls to policy implementation: controversies in the management of a marine salmon-farming industry. Ocean & Coastal Management 45(1):19–40.

Anonymous. 2006. Aqua Bounty successfully completes key study for AquAdvantage Salmon. Special Publications, The Oban Times Ltd. Available: http://www.fishupdate.com/news/fullstory.php/aid/6010/Aqua_Bounty_successfully_completes_key_study_for_AquAdvantage_Salmon.html. (January 2007).

Asche, F., T. Bjørndal, and J. A. Young. 2001. Market interactions for aquaculture products. Aquaculture Economics and Management 5(5/6):303–318.

Aubin-Horth, N., C. R. Landry, B. H. Letcher, and H. A. Hofmann. 2005a. Alternative life histories shape brain gene expression profiles in males of the same population. Proceeding of the Royal Society of London Series B 272:1655–1662.

Aubin-Horth, N., B. H. Letcher, and N. A. Hofmann. 2005b. Interaction of Rearing Environment and Reproductive Tactic on Gene Expression Profiles in Atlantic Salmon. Journal of Heredity 96(3):261–278.

Barnett, H. G. 1955. The Coast Salish Of British Columbia. University of Oregon Press, Eugene.

Benfey, T. J. 2001. Use of sterile triploid Atlantic salmon (*Salmo salar* L.) for aquaculture in New Brunswick, Canada. ICES Journal of Marine Science 58(2):525–529.

Bjørndal, T., and K. Aarland. 1999. Salmon aquaculture in Chile. Aquaculture Economics and Management 3(3):238–253.

British Columbia Environmental Assessment Office. 1997a. Salmon Aquaculture Review Final Report - Bibliographic Listings Volume 5. British Columbia Environmental Assessment Office, Victoria.

British Columbia Environmental Assessment Office. 1997b. Salmon Aquaculture Review Final Report - Socio-Economic Impacts and Related Technical Papers Volume 4. British Columbia Environmental Assessment Office, Victoria.

British Columbia Environmental Assessment Office. 1997c. Salmon Aquaculture Review Final Report - Technical Advisory Team Discussion Papers Volume 3. British Columbia Environmental Assessment Office, Victoria.

British Columbia Environmental Assessment Office. 1997d. Salmon Aquaculture Review Final Report - Volume 1: Report of the Environmental Assessment Office. British Columbia Environmental Assessment Office, Victoria.

British Columbia Environmental Assessment Office. 1997e. Salmon Aquaculture Review Final Report - Volume 2: First Nations Perspectives. British Columbia Environmental Assessment Office, Victoria.

British Columbia Ministry of Agriculture Food and Fisheries. 2002. New standards allow sustainable growth in aquaculture. British Columbia Ministry of Agriculture, Food and Fisheries, Campbell River.

Burgess, M. 2003a. Starting on the right foot: public consultation to inform issue definition in genome policy. W. Maurice Young Centre for Applied Ethics, University of British Columbia, Vancouver. Available: http://www.ethics.ubc.ca/workingpapers/deg/deg002.pdf. (May 2004).

Burgess, M. 2003b. What difference does public consultation make to ethics? W. Maurice Young Centre for Applied Ethics, University of British Columbia, Vancouver. Available: http://www.ethics.ubc.ca/workingpapers/deg/deg003.pdf. (May 2004).

Canada. 2001. Legislative and Regulatory Review of Aquaculture in Canada. Office of the Commissioner for Aquaculture Development, Fisheries and Oceans Canada, DFO/6144, Fs23–402/2001E, Ottawa. http://www.ocad-bcda.gc.ca/elegalreview.pdf. (May 2004).

Canada. 2003a. Biotechnology improves vaccination options for disease control in aquaculture. Fisheries and Oceans Canada Science. Available: http://www.dfo-mpo.gc.ca/science/aquaculture/biotech/fact18_e.htm. (June 2003).

Canada. 2003b. Biotechnology sheds new light on fish pathogens. Fisheries and Oceans Canada - Science. Available: http://www.dfo-mpo.gc.ca/science/aquaculture/biotech/fact16_e.htm. (June 2003).

Canada. 2003c. Biotechnology to help develop better salmon feeds. Fisheries and Oceans Canada - Science. Available: http://www.dfo-mpo.gc.ca/science/aquaculture/biotech/fact3_e.htm. (June 20 2003).

Canada. 2003d. Biotechnology to help protect British Columbia's wild salmon stocks - the all-female approach. Fisheries and Oceans Canada - Science. Availble: http://www.dfo-mpo.gc.ca/science/aquaculture/biotech/fact12_e.htm. (June 2003).

Canada. 2003e. Biotechnology to help protect wild salmon

stocks - the triploid approach. Fisheries and Oceans Canada - Science. Available: http://www.dfo-mpo.gc.ca/science/aquaculture/biotech/fact7_e.htm. (June 2003).

CBC News British Columbia. 2004. Sea lice threatens salmon run, says researcher. CBC News, Vancouver. Available: http://vancouver.cbc.ca/regional/servlet/View?filename = bc_lice20040428. (April 2004).

CEQ/OSTP. 2001. Case Study No. 1: growth enhanced salmon. Office of Science and Technology Policy, Washington D.C.

Charron, B. 2000. An IntraFish.com Industry Report on Salmon farming and the environment: an overview of some of the attitudes encountered. IntraFish, Bodø, Norway.

Collet, S. 2002. Appropriation of marine resources: from management to an ethical approach to fisheries governance. Social Science Information 41(4):531–553.

Duff, W. 1952. The upper Stalo Indians of the Fraser Valley, British Columbia. British Columbia Provincial Museum, Victoria.

FAO (Food and Agriculture Organization of the United Nations). 2000. The State of World Fisheries and Aquaculture 2000. FAO, Rome.

FAO (Food and Agriculture Organization of the United Nations). 2001. Genetically modified organisms, consumers, food safety and the environment. FAO Ethics Series 2, Rome.

FAO (Food and Agriculture Organization of the United Nations). 2002. The state of world fisheries and aquaculture 2002. FAO, Rome.

Fletcher, G. L., S. V. Goddard, and C. L. Hew. 2000. Current status of transgenic Atlantic salmon for aquaculture. Pages 179–184 *in* C. Fairbairn, G. Scoles, and A. McHughen, editors. 6th International Symposium on The Biosafety of Genetically Modified Organisms. University of Saskatchewan, Saskatoon, Canada.

Gardner, J., and D. L. Peterson. 2003. Making sense of the salmon aquaculture debate: analysis of issues related to netcage salmon farming and wild salmon in British Columbia. Pacific Fisheries Resource Conservation Council, Vancouver.

Genome Canada. 2004. Glossary. Genome Canada, Ottawa. Available: http://www.genomecanada.ca/GCglossaire/glossaire/index.asp?l = e. (April 2004).

Hannesson, R. 2003. Aquaculture and fisheries. Marine Policy 27(2):169–178.

Hansen, P., J. A. Jacobson, and R. A. Und. 1993. High numbers of farmed Atlantic salmon, *Salmo salar*, observed in oceanic waters north of the Faroe Islands. Aquaculture Fisheries Management 24:777–781.

Hew, C. L., and G. L. Fletcher. 1996. Transgenic salmonid fish expressing exgenous salmonid growth hormone. U.S. Patent. HSC Research and Development Limited Partnership, Toronto, and Seabright Corporation, St. John's.

Hightower, R., C. Baden, E. Penzes, P. Lunda, and P. Dunsmuir. 1991. Expression of antifreeze proteins in transgenic plants. Plant Molecular Biology 17(5):1013–1021.

Hume, M. 2004. Fish farming threatens wild salmon, scientist says. Available: http://www.theglobeandmail.com/servlet/ArticleNews/TPStory/LAC/20040428/FISHFARMS28/TPNational/Canada. (June 2007).

Institute for Social Economic and Ecological Sustainability. 2003. Marine GEOs: products in the pipeline. Marine Biotechnology Briefs: Science, Policy and Marine Conservation Implications 1. Available: http://www.fw.umn.edu/isees/MarineBrief/1/brief1.htm. (June 2007).

Jystad, P. T. 2001. An IntraFish.com industry report on fish meal and oil or vegetable alternatives: Will high volume production spoil premium fish products? IntraFish, Bodø, Norway.

Lewis, C. 2001. A new kind of fish story: the coming of biotech animals. FDA Consumer magazine 35(1). Available: http://www.fda.gov/fdac/features/2001/101_fish.html. (June 2007).

MacKenzie, D. 1996. Can we make supersalmon safe?: The transgenic salmon in task by a Scottish Loch are becoming a problem. Should the British government allow the fish to grow up? New Scientist 149(2014):14.

Maclean, N., and R. J. Laight. 2000. Transgenic fish: an evaluation of benefits and risks. Fish and Fisheries 1(2):146–172.

Matheson, J. 1999. Will transgenic fish be the first ag-biotech food-producing animals? FDA Veterinarian Newsletter 9(3).

Naylor, R. L., and coauthors. 2000. Effect of aquaculture on world fish supplies. Nature 405:1017–1024. URL.

Newell, D. 1993. Tangled webs of history: indians and the law in Canada's Pacific Coast fisheries. University of Toronto Press, Toronto.

Niiler, E. 2000. FDA, researchers consider first transgenic fish. Nature Biotechnology 18(2):143.

Pauly, D., V. Christensen, J. Dalsgaard, R. Froese, and F. Torres, Jr. 1998. Fishing down marine food webs. Science (279):860–863.

Pauly, D., and R. Watson. 2003. Counting the last fish. Scientific American 289(1):42–47.

Pew Initiative on Food and Biotechnology. 2003. Future fish: issues in science and regulation of transgenic fish.

Pew Initiative on Food and Biotechnology, Washington, D.C.

Power, M. D. 2003. Lots of fish in the sea: salmon aquaculture, genomics and ethics. W. Maurice Young Centre for Applied Ethics, University of British Columbia, Electronic Working Paper No. DEG 004, Vancouver. Available: http://ethics.ubc.ca/workingpapers/deg/deg004.pdf (August 2007).

Roberts, R. J. 2000. Salmonids - positive outlook for salmon industry. Aquaculture Magazine 26(5).

Scarce, R. 2000. Fishy business: salmon, biology, and the social construction of nature. Temple University Press, Philadelphia.

Schmidt, C. W. 2005. Genetically modified foods: breeding uncertainty. Environmental Health Perspectives 113(8):A526-A533.

Stewart, H. 1977. Indian fishing: early methods on the northwest coast. Douglas and McIntyre, Vancouver.

Tansey, J. 2003. The prospects for governing biotechnology in Canada. W. Maurice Young Centre for Applied Ethics (Democracy, Ethics, and Genomics), University of British Columbia, Electronoic Working Paper No. DEG 001, Vancouver. Available: http://ethics.ubc.ca/workingpapers/deg/deg001.pdf (August 2007).

The Royal Society of Canada. 2001. Elements of precaution: recommendations for the regulation of food biotechnology in Canada. The Royal Society of Canada for Health Canada, Canadian Food Inspection Agency, and Environment Canada, Ottawa.

Tyedmers, P. H. 2000. Salmon and sustainability: the biophysical cost of producing salmon through the commercial salmon fishery and the intensive salmon culture industry. University of British Columbia, Vancouver.

Union of Concerned Scientists. 2001. Genetically engineered salmon. Union of Concerned Scientists. Available: www.ucsusa.org/food_and_environment/biotechnology/page.cfm?pageID = 327. (May 2004).

USDA Agricultural Research Service. 2001. Report from the Biotechnology-Aquaculture Interface: The Site of Maximum Impact Workshop. U.S. Department of Agriculture, Shepherdstown, West Virginia.

Volpe, J. 2001. Super un-natural: Atlantic salmon in BC waters. David Suzuki Foundation, Vancouver.

Wilkins, N. P., D. Cotter, and N. O'Maoiléidigh. 2001. Ocean migration and recaptures of tagged, triploid, mixed-sex and all-female Atlantic salmon (*Salmo salar* L.) released from rivers in Ireland. Genetica 111:197–212.

American Fisheries Society Symposium 49:209–210

Session Summary

The Ecological Dimensions in Achieving the Reconciliation of Fisheries with Conservation

Doug S. Butterworth
University of Cape Town, Department of Mathematics and Applied Mathematics
Marine Resources Assessment and Management Group, Rondebosch 7701, South Africa

Given its broad title, this session unsurprisingly attracted a widely ranging set of contributions, with 26 scheduled posters as well as six oral presentations. Most of these may, however, be conveniently categorized under three broad headings:

1) Reference points, with a range from single-species to whole ecosystem models as a framework for computations;

2) Marine mammal and sea bird top predators; and

3) Salmon-related contributions (no surprise given the conference venue!).

It is interesting to search for common themes, within these sets of contributions, for indications of the areas of focus of current research initiatives. First though, there are some themes that spanned these headings, of which two were particularly evident.

1) Increasing attempts, and importantly at last it appears reasonably *successful* attempts, to incorporate climate-related and other environmental factors into models (and especially predictive models) for fisheries.

2) Trade-offs as a fundamental component of fisheries management, whether in choosing between risk and reward for managing single-species fisheries, or balancing the interests of user groups with conflicting objectives.

Reference point and related issues received the widest attention. An encouragingly common feature was the stress laid in moving away from the practice of basing management decisions on a single "best" assessment alone, to instead also include consideration of uncertainties about the structures of the underlying assessment models used (in the spirit of the precautionary approach). Also encouraging were moves to simulation test reference point based management algorithms to check for robustness given these uncertainties, rather than to rely on deterministic concepts. Such simulations also provide the performance statistics that are the core information required for the trade-off decisions that managers need to make. A difficulty is the data-intensive basis for such approaches, so that welcome complements to these contributions were analyses related to the use of more simply obtainable life-history or habitat related parameters to aid management in data-poor situations.

Contributions extending from single- to multispecies considerations included focus on both technological (often linked to bycatches) and biological interactions. Because of the former, the achievement of some single-species based reference targets may be compromised by restrictions applying to catches of other species. While management typically seeks to minimize bycatches (and hence also likely associated discarding), some valuable lateral thinking on this last problem was provided by accounts of attempts to improve the potential for sales of products from bycatch species. Perhaps the most challenging task addressed was the use of multispecies and ecosystem models to investigate the impact of fishing on the catastrophic and sudden ecosystem reorganizations that can result naturally from climatic variations. Ideally, strategic fisheries management should seek to maintain resilience to such shifts, but are species interactions sufficiently well understood as yet to allow the provision of reliable scientific advice at this level?

Contributions on interactions with marine mammals (on this occasion chiefly seals and sea lions) and sea birds also spanned both the technological and biological, and highlighted trade-off issues. Mitigation measures to reduce the former may decrease fishing efficiency. The latter related to conflicts between those who wish to harvest forage fish and those who seek increased predator populations, but at current levels of scientific understanding it is difficult to quantify the extent to which the one affects the other.

The issues considered in the salmon-related contributions, although naturally specific to this species grouping, also included elements of a more generic nature. Common themes were resource monitoring methods, seeking biological explanations of changes in migratory behavior and natural mortality that have had important economic consequences, and improving forecasting models by taking account of environmental covariates.

An overview impression of the subject area conveyed by the contributions is of steady, though perhaps slow, progress across a broad front.

American Fisheries Society Symposium 49:211–213

Temporal Trends in Age-Specific Recruitment of Sockeye Salmon: Implications for Forecasting

CARRIE A. HOLT* AND RANDALL M. PETERMAN

Simon Fraser University, School of Resource and Environmental Management
Fisheries Science and Management Group, 8888 University Drive
Burnaby, British Columbia V5A 1S6, Canada

For management of Pacific salmon, accurate forecasts of adult recruits, or fish returning to spawn, can be crucial. Sibling models, which are one type of forecasting model, predict age-specific recruitment based on recruitment of the same cohort in the preceding year. The abundance of 5-year-old sockeye salmon *Oncorhynchus nerka* recruits from a given brood year, for instance, can be forecasted by a sibling model from the abundance of its 4-year-old siblings that were estimated in the previous year. Normally, this stock-specific model assumes a constant linear relationship between age-classes over time:

$$\log_e\left(R_{x.i,y}\right) - a + b\log_e\left(R_{x.i-1,y-1}\right) + v_t$$

where R is abundance of adult recruits that spent x winters in freshwater and i winters in the ocean, y is the year of recruitment, a and b are parameters, and v_t is a normally distributed random error term. However, increases in mean age at maturity of sockeye salmon over the last 40 years (Pyper et al. 1999) suggest that it may be more appropriate to assume time-varying parameters in sibling models.

Our research objectives were threefold. First, we identified whether there are long-term trends or persistent changes in parameters of sibling models for British Columbia (BC) and Alaska sockeye salmon stocks, as opposed to just year-to-year variation. Second, we examined potential physical environmental and biological drivers of changes in parameters of sibling models. Third, we used historical data in a retrospective analysis to compare the performance of forecasts of the model that allows for temporal changes in its parameters with forecasts of the standard sibling model.

We analyzed 41 time series of age-specific recruitment data (catch and escapement) from 24 sockeye salmon stocks from eight different management regions in BC and Alaska. We analyzed recruitment data for both the age-1.2-to age-1.3 relation, or stanza, and the age-2.2-to-age-2.3 stanza.

For objective one, we estimated temporal trends in the *y*-intercept (*a* parameter) of the sibling model for each stock using a Kalman filter:

$$\log_e\left(R_{x.3,y}\right) - a_t + b\log_e\left(R_{x.2,y-1}\right) + v_t$$

which is the same as equation (1) except that a_t values can vary by brood year, t; the b parameter remains constant. We assumed a random walk for a_t as in Peterman et al. (2003). The Kalman filter estimates temporal trends in observed age-specific recruitment due to systematic process variation separately from

*Corresponding author: cholt@sfu.ca

random sources of variation that are independent of that trend (Chatfield 1996). The Kalman filter approach thus permits annual estimation of effects of underlying processes influencing age-specific recruitment with reduced confounding from random sources of variability that are independent of the trend.

The majority of sockeye salmon stocks that we examined showed evidence for long-term trends, reflecting increasing numbers of fish maturing at later ages for a given abundance maturing at the preceding age (e.g., Figure 1).

Covariation in sibling-model a_t parameters among stocks can help identify the spatial scale of processes driving these temporal trends (e.g., local, regional or ocean-basin scale). Thus for objective two, we calculated pairwise correlation coefficients between all stocks. We also calculated correlations between model parameters and physical environmental and biological factors that vary at spatial scales similar to those of sibling-model parameters.

The time series of sibling-model *y*-intercepts showed strong positive correlations both within and among regions (Figure 2). Positive covariation was apparent across both regional and ocean-basin scales, the latter indicated by correlations among BC and Bristol Bay, Alaska stocks, for example. We found significant positive correlations between sibling-model *y*-intercepts and physical environmental and biological factors that also vary at these large spatial scales such as Pacific Decadal Oscillation, sea surface temperature in the Gulf of Alaska, total sockeye salmon abundance, and Bristol Bay sockeye salmon abundance. These positive correlations occurred for more cases than expected by chance alone. In contrast, there were numerous significant negative correlations between sibling-model *y*-intercepts for

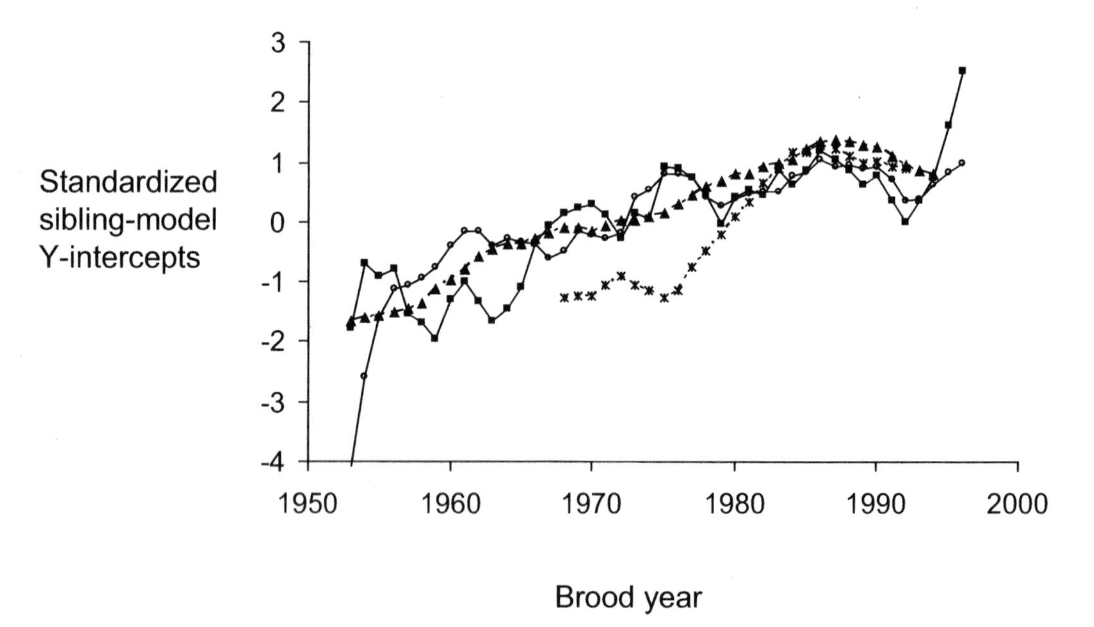

Figure 1. Annual *y*-intercepts (a_t) for the age-1.*x* sibling model for two Bristol Bay, Alaska, sockeye salmon stocks, Igushik (solid line with open circles) and Togiak (solid line with closed squares), and two northern BC stocks, Skeena River (dashed line with closed triangles) and Nass River (dashed line with asterisks), each standardized to a mean of zero and standard deviation of one.

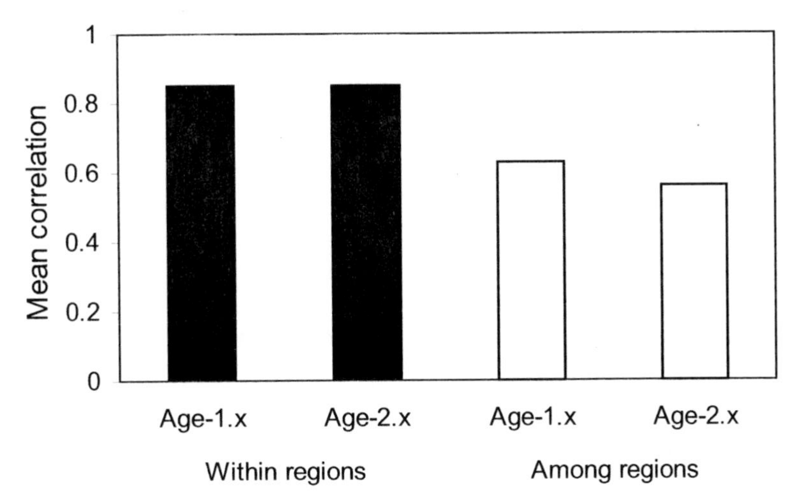

Figure 2. Mean pairwise correlations between sockeye salmon stocks in their time series of Kalman-filter a_t values, within (black bars) and among (white bars) the eight regions. Both age-1.x and age-2.*x* stanzas are shown.

the age-1.*x* stanza and body size of age-1.2 recruits. In multiple regressions, total sockeye salmon abundance was the variable most frequently associated with trends in sibling-model *y*-intercepts, suggesting that density-dependent processes may be driving these changes. In years of increased competition for food and slow growth, salmon may delay maturation, resulting in a higher *y*-intercept. However, it is difficult to attribute causality because of multicollinearity among environmental factors and autocorrelation in their time series.

For the third objective, we used a retrospective analysis and compared the mean squared error and mean percent bias of forecasts of age-specific recruits from the standard sibling models to those from our new Kalman filter sibling models. The Kalman filter models reduced mean squared forecasting errors compared to the standard sibling models in 29–39% of the stocks (age-1.*x* and age-2.*x* stanzas respectively), and reduced mean percent biases for 54–94% of the stocks. However, for the Bristol Bay region, the Kalman filter models reduced mean-squared forecasting errors and mean percent biases for the majority of stocks, suggesting that our new version of the sibling model may be appropriate for this region.

References

Chatfield, C. 1996. The analysis of time series: An introduction. Chapman and Hall, New York.

Peterman, R. M., B. J. Pyper, and B. MacGregor. 2003. Use of the Kalman filter to reconstruct historical trends in productivity of Bristol Bay sockeye salmon (*Oncorhynchus nerka*). Canadian Journal of Fisheries and Aquatic Sciences 60:809–824.

Pyper, B. J., R. M. Peterman, M. F. Lapointe, and C. J. Walters. 1999. Patterns of covariation in length and age at maturity of British Columbia and Alaska sockeye salmon (*Oncorhynchus nerka*) stocks. Canadian Journal of Fisheries and Aquatic Sciences 56:1046–1057.

American Fisheries Society Symposium 49:215–222

Why Fisheries Reference Points Miss the Point

Doug S. Butterworth*

University of Cape Town, Department of Mathematics and Applied Mathematics Marine Resources Assessment and Management Group, Rondebosch, South Africa, 7701

Abstract.—Prompted by the fundamental tenet underlying Heisenberg's reformulation of the foundations of quantum mechanics in 1925, the suggestion is made that fisheries management should be based on quantities that are directly measurable in practice. Such an approach would accommodate quantities such as catches and measures proportional to abundance, but likely exclude fishing mortality (*F*) and maximum sustainable yield (MSY). This runs counter to customary management practice, which is based on reference points. These are frequently related to *F*, about which a number of problems are raised. Instead, it is proposed that management be based upon a single limit reference point related to abundance (B_{lim}), augmented by feedback control rules. Such rules would be simulation tested for robustness to uncertainties, with the selection between alternative rules based upon preferred trade-offs among catches, abundance trends, and interannual catch variability, together with their socioeconomic implications. Among other advantages, such an approach, based upon directly measurable quantities, is more readily understood by managers and industry. The reference point approach misses the point by failing to address the core issue of structural uncertainty in modeling fishery dynamics. The approach advocated does take account of this issue, in sympathy with the precautionary principle. However, more attention needs to be given to guiding the process of according relative plausibility weights to structurally different models, which is required to effect the associated calculations.

Introduction

In 1925, Werner Heisenberg, best known for his association with the uncertainty principle, published a paper (Heisenberg 1925) that laid the foundations for a reformulation of quantum mechanics. A key tenet of Heisenberg's approach was to focus on "observables."

"The present paper seeks to establish a basis for theoretical quantum mechanics founded exclusively upon relationships between quantities which in principle are observable" (Heisenberg 1925).

*Corresponding author: doug.butterworth@uct.ac.za

This paper seeks to explore the implications of adopting an analogous approach to fisheries management. These implications will be shown to augment reservations concerning the appropriateness of the traditional "reference point" basis for fisheries management, and in particular of the advocacy of reference points based on fishing mortality (*F*).

Arguments will be offered that there is only one reference point of any fundamental importance, and that in other respects the fisheries management paradigm should be refocused upon anticipated performance under feedback control rules—the operational management procedure (OMP) or, equivalently, the management strategy evaluation (MSE) approach (Butterworth and Punt 1999; Smith et al. 1999). This is, in particular, to be able to take proper account of structural uncertainty in modeling population dynamics, is in sympathy with the precautionary principle and approach. The general failure of the reference point approach to address this core aspect of fisheries management is the basis on which it will be argued to have missed the point.

Observable (Directly Measurable) Quantities

The following is offered as an analog of Heisenberg's "observables" tenet in the context of fisheries management (alterations italicized):

"The present paper seeks to establish a basis for *fisheries management* founded exclusively upon *consideration of* quantities which *in practice* are *directly measurable.*"

Examples of pertinent fisheries quantities that are directly measurable are *catch* and *indices of abundance*[1] (such as from surveys). *Effort* might also be pertinent in circumstances where its relationship to catches or to abundance (*via catch per unit effort* [*CPUE*]) is clear. Clearly *catch-at-size* (or *-at-age*) data fall into this directly measurable category, but they are not referenced further below as the other quantities considered provide a sufficient basis to formulate the arguments that follow.

These quantities are not only directly measurable, but also of key importance to stakeholders in the management process. Short- and medium-term expectations for catch and possibly effort are fundamental determinants of the probable success or otherwise of an industry. Similarly, a fisheries manager perhaps concerned more with conservation in the medium to long term would be interested primarily in likely abundance (index) trends.

In contrast, examples of quantities that are *not* directly measurable are fishing mortality (F), maximum sustainable yield (MSY), and the corresponding fishing mortality (F_{MSY})[2]. These can be estimated only given the combination of directly measurable quantities in the form of data and some population dynamics model. The "Heisenberg's observables" tenet offered would imply that such quantities should not be featured when formulating a basis for fisheries management policy or its implementation in the field.

The "Reference Point" Basis for Fisheries Management

A key role for reference points in fisheries management has become deeply ingrained in the fisheries literature and even legal instruments.

The FAO Code of Conduct for Responsible Fisheries (http://www.fao.org.DOCREP/005/v9878e/v9878e00.htm) advocates the determination of stock specific target reference points (to which management is intended to drive stocks) and limits reference points (which management should seek to ensure that stocks avoid approaching).

The 1995 UN Fish Stocks Agreement[3] (http://www.un.org/depts/los/

[1] Reference to "indices of abundance", rather than "abundance" *per se*, is deliberate; bias factors (usually poorly known) mean that even scientific surveys do not provide direct measurements of resource abundance in absolute terms, but they nevertheless can provide consistent relative measures.

[2] See footnote 4.

[3] A colloquial title for the "Agreement for the implementation of the provisions of the United Nations Convention on the Law of the Sea of 10 December 1982 relating to the conservation and management of straddling fish stocks and highly migratory fish stocks".

convention_agreements/texts/fish_stocks_agreement/CONF164_37.htm) goes further in specifically advocating F_{MSY} as a minimum standard for limit reference points in its guidelines.

The International Council for the Exploration of the Sea (ICES) equates F_{lim} to the lowest fishing mortality at which, in the long term, the stock would collapse through recruitment failure, and then establishes an associated precautionary reference point:

$$F_{pa} = \hat{F}_{lim} e^{-2\sigma} \quad (1)$$

where σ is the standard error of the estimate of F_{lim} (i.e., F_{pa} is effectively the lower 95% confidence bound on the estimate of F_{lim}). A precautionary reference point for biomass B_{pa} is established likewise (Garcia 2000). The North Atlantic Fisheries Organization has established similar reference points (Garcia 2000).

The heavy emphasis on fishing mortality (F) in establishing such reference points is notable. While this does have some merit in the sense that constant proportion harvesting strategies (related to constant effort if the CPUE proportional to abundance assumption holds) have better deterministic stability properties than, say, constant catch strategies, it also leads to numerous problems.

•F is not directly measurable; it requires the specification of a model for estimation; different models may be advocated and will yield different results (with concomitant potential implementation and possible legal problems)[4].

•Given that F estimates vary with age (or length), it is not obvious how a single value of F pertinent to a particular year should be defined; over which range of ages should an average be taken; and what happens if the pattern of F with age (i.e., selectivity) changes over time.

• Lay interpretability: numerical values for F are seldom meaningful to managers or industry—this is in contrast to projections of catch, effort, and abundance (indices), whose implications (both as regards themselves and in relation to socioeconomic considerations) are readily grasped.

• Estimates of F are often sensitive to the value of natural mortality M; frequently this is a fixed input for which there is only relatively weak supporting evidence; yet estimates of an F_{MSY} reference point can be very sensitive to the value assumed for M.

Furthermore, at a more fundamental level, it is not fishing mortality per se that determines reproductive output and hence surplus production capability (except in circumstances where disturbance of spawning aggregations by fishing may reduce reproductive success). Rather, it is abundance that matters and so merits primary attention from a conservation perspective.

An Alternative Approach

First, I assert that fundamentally only one reference point is needed for fisheries management: a minimum abundance (abundance index value) reflecting (in the words of the

[4] Tag–recapture exercises might be argued to provide a direct measurement of F. One must bear in mind, however, that often such estimation is less than straightforward, with models remaining necessary to take account of factors such as tag loss, tagging-induced mortality, the extent of dispersal of tagged fish throughout the entire population under management, tag-recovery reporting rates and age- and length-specific selectivity effects on tag recovery rates. Thus while tagging data may usefully be incorporated in fitting the assessment models that provide the basis for simulation testing of feedback control rules, it seems doubtful that tag return results alone could provide a basis for the operational implementation of management policy. (See also footnote 5 and the text to which it refers.)

1995 UN Fish Stocks Agreement) a "level at [below] which reproduction may become seriously threatened." Notwithstanding the fact that this level will not be straightforward to specify quantitatively in many cases, such a "B_{lim}" is clearly a level which management should seek to avoid that a stock approach.

Other reference points are replaced by selected feedback control rules, designed (naturally) to respect the B_{lim} reference point, but also in particular to achieve desired trade-offs between predicted catch and abundance (index) trends in the short and medium term, and between abundance trend and year-to-year catch variability.

The first of these trade-offs is the classic instance of more catch now versus higher catch rates later, with its concomitant socioeconomic implications. The second involves signal–noise considerations: one does not want to vary catch levels unnecessarily to react to what might be noise rather than a genuine trend in the abundance index; but, equally, one must not overly delay catch adjustments given a series of index values that could reflect a genuine trend. Similar trade-offs involving effort may also be relevant.

Note that these trade-offs involve only Heisenberg "observables" (no *F*, MSY). This is not to dispense entirely with such concepts. To quote Dirac (1930) on Heisenberg's tenet,

"One may introduce auxiliary quantities not directly observable for the purpose of mathematical calculations; but variables not observable should not be introduced merely because they are required for the description of phenomena according to ordinary classical notions, e.g. orbital frequencies in Bohr's theory."

One may formulate the correspondences of ordinary classical notions ↔ deterministic population dynamics, and atomic electron orbital frequencies ↔ *F*, MSY, and F_{MSY}, which then suggests that concepts such as *F*, MSY, and F_{MSY} may be admitted in (certain limits of) models of fishery population dynamics, and may certainly therefore play some role in the scientific computations used to develop a recommendation for management (e.g., a total allowable catch [TAC]), but they should nevertheless be excluded at the implementation level of management policy formulation and selection (and any associated legal instruments), which should be restricted to reference "observables."[5]

An example may help clarify the concept. Figure 1 shows predictions for future catches and abundances for southern bluefin tuna *Thunnus thynnus* under a range of feedback control rules ranging from less to more conservative. The assertion then is that management should be based on a selection between these options (those that are consistent also with the B_{lim} reference point), with the information contained in such plots providing a sufficient basis for stakeholders to make an informed trade-off decision (which takes account also of the associated social and economic implications). Note that

- Such plots are more readily understood by managers and industry than, say, *F*-based reference points[6];

- There is no need to impose constraints such as $F < F_{MSY}$, which can compromise objectives

[5] Thus, for example, management policy for a resource should be formulated in terms of an index of abundance (or an average over a number of recent measurements thereof) reaching a specified value, rather than (say) recovery to an abundance estimated to correspond to MSY.

[6] The appeal here is against the like of pronouncements by a fisheries modeling priesthood of the form: "*F* is 0.28 and needs to be reduced to a target reference point level of 0.21", which are meaningless to the layity.

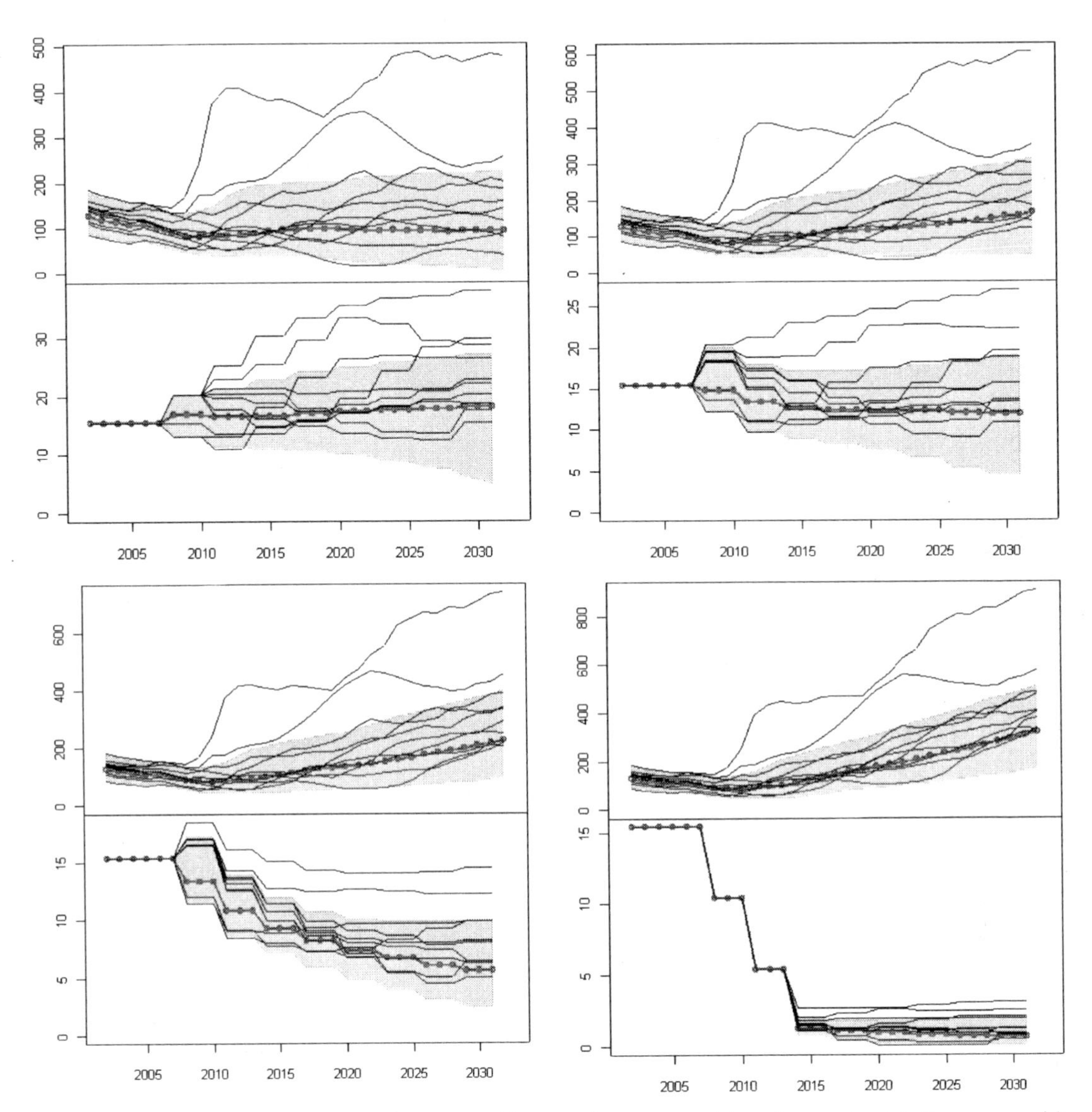

Figure 1. Stochastic projections of spawner biomass (upper panels) and annual catch (lower panels), both in thousand tons, for southern bluefin tuna under different TAC control rules. The shaded areas indicate 95% probability interval envelopes, the heavy line shows the medians of projected distributions, and the light lines show 10 individual realizations. The control rule options range from least conservative (least spawner biomass recovery) on the top left to most conservative on the bottom right. The control rules include limitations on the frequency (every three years) and extent of TAC changes, and the results are computed from models reflecting relative-plausibility-weighted integration over six alternative scenarios for resource dynamics spanning the primary sources of uncertainty. (Figure from Butterworth, D.S. and M. Mori. 2004. Application of variants of a Fox-model based MP to the "Christchurch" SBT trials. Commission for the Conservation of Southern Bluefin Tuna unpublished document CCSBT-MP/0404/07.) In a strict "observables" context, this plot could be criticized for reporting spawning biomass, which is not directly observable in this fishery; however the plot could readily have been developed instead to show CPUE predictions, which for this resource closely reflect anticipated spawning biomass trends, though somewhat advanced in time.

to reduce interannual variability in circumstances where the attainment of conservation-related objectives (related to B_{lim}) remains guaranteed; and

• Classifying MSY as irrelevant may seen questionable at first sight, but on reflection it is evident that medium-term (±1–3 decades) catch projections under a range of control rules incorporate the same information, and cast this in a form more relevant and meaningful to stakeholders; it might be countered that this assertion is questionable for long-lived resources with slow dynamics, but the medium term is as long as the planning horizon of industry and managers will extend, and also as long as any claim of reliability can realistically be accorded to resource projections.

Some Aspects of Control Rules

It is important to appreciate that the specification of a feedback control rule alone is not sufficient. As emphasized in the OMP/MSE approach, it is also critical that the resource monitoring data to be collected be prespecified, together with the manner in which these are to be used in implementing the control rule to provide, say, a TAC recommendation (i.e., to use statistical terminology, the "estimator" that converts the data into a TAC recommendation must be specified).

In 1976, the International Whaling Commission (IWC) introduced its New Management Procedure (NMP), which was, in essence, a catch control rule based on the MSY concept (IWC 1977). The perceived subsequent failure of the NMP was not a result of any flaw in the control rule approach, but rather that the NMP was incomplete, having omitted to also specify data requirements and estimators—a deficiency later corrected in the IWC's Revised Management Procedure (RMP; Kirkwood 1992).

A further important consideration is whether control rules should be generic or case-specific. I assert that experience with OMP/MSE implementations over the past decade overwhelmingly indicates the latter, if only because case-specific simulation testing of generic control rule candidates is always eventually demanded anyway, so that case-specific options can readily be tested as well. Why insist on introducing generic policy considerations (e.g., $F < F_{MSY}$) into control rules if case-specific variants without such constraints offer improved tradeoffs among "observables" (see also Punt et al. 2006)?

Structural Model Uncertainty and the Precautionary Principle

A key feature of the OMP/MSE approach is that control rule performance be tested for robustness to uncertainties on how best to model the underlying population dynamics. Such uncertainties extend not only to imprecision in the estimated values of model parameters, but particularly to alternative structural assumptions (pertinent when fundamentally different models are all compatible with the available data).

In my experience, it is such structural considerations rather than ones of estimation precision that dominate uncertainties in fisheries assessments and projections, particularly in the more controversial cases. Some examples are

• South African west coast rock lobster *Jasus lalandii*: likely future trends in somatic growth rate and recruitment (Johnston and Butterworth 2005);

• Southern bluefin tuna: forms of the stock–recruitment and longline CPUE-abundance relationships (CCSBT 2003);

• Northwest Pacific minke whales: number and spatio–temporal distribution of reproductively (relatively) isolated population components (IWC 2004).

Reference points are generally computed based upon a "best assessment" of a resource. Occasionally they take considerations of estimation imprecision into account (e.g., ICES—see equation 1). But it is in their general failure to address the core fisheries management issue of model structure uncertainty that they miss the point.

Checking for robustness to uncertainty dovetails with the requirements of the precautionary principle. Chris Hedley (Department for Environment, Food and Rural Affairs, UK, personal communication) comments that the principle "has become such a common component of international legal instruments related to fisheries management as to have achieved the status of customary international law in this area." This is rather daunting, given that there is hardly clarity or consensus in fisheries circles as to what is implied by operational implementation of the precautionary principle. In the context of potential international litigation, this may not matter. The traditional concept of a legal principle is that it provides an argument for a particular direction, rather than a rule for determining a specific outcome (Dworkin 1976). Thus it would likely prove a sufficient defense in legal proceedings to present evidence that minds had been applied to take due account of alternative possible outcomes of the action intended, given the scientific uncertainties.

This, however, would not seem to provide sufficient guidance for a responsible operational basis for fisheries management that is consistent with the intent underlying the precautionary principle. Restricting evaluations to the "best" model or assessment does not go far enough; equally, a worst-case scenario approach of basing decisions on the most pessimistic appraisal of resource status and productivity that is arguably compatible with available information is unrealistic and impractical (Butterworth 2000).

It is impossible to design feedback control rules that will self-correct perfectly for all plausible model scenarios, so that there is a need to assign "relative plausibility weights" to different scenarios to be able to integrate over them when computing probability distributions for future catch and abundance projections (as in Figure 1). Note that Bayes' theorem hardly helps in this exercise, as generally the available data and hence likelihoods are unable to discriminate between such alternative scenarios. Hence, one is left with effectively specifying "priors" based upon scientific judgment, which in turn involves an element of subjectivity that is impossible to circumvent.

How might such a process (upon the results of which any ultimate scientific recommendation for management will likely be critically dependent) be best enacted? The exercises of developing operational management procedures for both the South African west coast rock lobster and the southern bluefin tuna resources referenced above both employed a Delphi-like process in scientific working group discussions to reach concen-sus[7] on relative weights to be accorded to each member of a set of alternative hypotheses spanning the few axes of major structural uncertainty identified (CCSBT 2003; Johnston and Butterworth 2005). However, a similar approach effectively failed in the IWC's scientific committee, where key groups each considered the other's favored stock structure hypothesis to have minimal plausibility (IWC 2004).

[7] Naturally, attainment of consensus is no guarantee that such consensus best reflects reality; the verdict thereon can rest only with posterity.

The primary problem facing fisheries stock assessment scientists today is how to give guidance to assist future processes of this nature to reflect a broad consistency in their outcomes. A promising initiative towards this end, which offers a process to be used in evaluating the results of simulation trials of different feedback control rules across a range of alternative scenarios, is reported in Section 4.1 of IWC (2005).

Conclusion

The key message of this paper is that future initiatives in fisheries assessment and management related research and its applications need to focus much less on reference points (particularly ones related to fishing mortality *F*). Instead they should concentrate further on the development of robust feedback control rules and of processes to accord relative plausibility weights to structurally different models of resource dynamics in a broadly consistent manner.

Acknowledgments

The comments of Jon Schnute and an anonymous reviewer of an earlier version of this paper are gratefully acknowledged.

References

Butterworth, D. S., and A. E. Punt. 1999. Experiences in the evaluation and implementation of management procedures. ICES Journal of Marine Science 56:985–998.

Butterworth, D. S., and M. Mori. 2004. Application of variants of a Fox-model based MP to the "Christ-church" SBT trials. Commission for the Conservation of Southern Bluefin Tuna document CCSBT-MP/0404/06.

Butterworth, D. S. 2000. Science and fisheries management entering the new millennium. Pages 37–54 *in* M. N. Nordquist and J. N. Moore, editors. Current fisheries issues and the Food and Agriculture Organisation of the United Nations, Centre for Oceans, Law and Policy. University of Virginia School of Law, Charlottesville.

Commission for the Conservation of Southern Bluefin Tuna (CCSBT). 2003. Report of the Eighth Meeting of the Scientific Committee, Christchurch, New Zealand, September 2003. CCSBT, Canberra, Australia.

Dirac, P. A. M. 1930. The principles of quantum mechanics, 1st edition. Clarendon Press, Oxford, UK.

Dworkin, R. 1976. Taking rights seriously. Harvard University Press, Cambridge, Massachusetts.

Garcia, S. M. 2000. The precautionary approach to fisheries: progress review and main issues (1995–2000). Pages 479–560 *in* M. N. Nordquist and J. N. Moore, editors. Current fisheries issues and the Food and Agriculture Organisation of the United Nations, Centre for Oceans, Law and Policy. University of Virginia School of Law, Charlottesville.

Heisenberg, W. 1925. Über quanten theoretische umdeutung kinematischer und mechanischer beziehung. Zeitschrift fur Physik 33:879–883, Berlin.

International Whaling Commission. 1977. Chairman's Report of the Twenty-Eighth Meeting. Reports of the International Whaling Commission 27:22–35, Cambridge, UK.

International Whaling Commission. 2004. Report of the Scientific Committee, Annex D. Report of the Sub-Committee on the Revised Management Procedure. Journal of Cetacean Research and Management 6 (Supplement):75–184.

International Whaling Commission. 2005. Report of the Scientific Committee, Annex D. Report of the Sub-Committee on the Revised Management Procedure. Journal of Cetacean Research and Management 7 (Supplement):84–92.

Johnston, S. J., and D. S. Butterworth. 2005. Evolution of operational management procedures for the South African West Coast rock lobster (Jasus lalandii) fishery. New Zealand Journal of Marine and Freshwater Research 39:687–702.

Kirkwood, G. P. 1992. Background to the development of revised management procedures. Reports of the International Whaling Commission 42:236–243, Cambridge, UK.

Punt, A. E., J. M. Cope, and M. A. Haltuch. 2008. Reference points and decision rules in U.S. Federal fisheries: west coast groundfish experiences. Pages 1343–1356 *in* . L. Nielsen, J. J. Dodson, K. Friedland, T. R. Hamon, J. Musick, and E. Verspoor, editors. Reconciling fisheries with conservation: proceedings of the Fourth World Fisheries Congress. American Fisheries Society, Symposium 49, Bethesda, Maryland.

Smith, A. D. M., K. J. Sainsbury, and R. A. Stevens. 1999. Implementing effective fisheries management systems—management strategy evaluation and the Australian partnership approach. ICES Journal of Marine Science 56:967–979.

American Fisheries Society Symposium 49:223–239

Evaluating Conservation Reference Points and Harvest Strategies in Pacific Herring and the Impact of Dispersal

Jake F. Schweigert*, Caihong Fu, and Chris C. Wood
*Fisheries and Oceans Canada Pacific Biological Station
Nanaimo, British Columbia V9T 6N7, Canada*

Abstract.—The risk assessment framework of Fu et al. (2004) for evaluating reference points for resource management was applied to the five Pacific herring *Clupea pallasi* stocks within the metapopulation managed in British Columbia, Canada. A suite of performance indicators reflecting both fisheries and conservation considerations were investigated as a basis for evaluating appropriate harvesting policy for this resource. The existing policy of a 20% annual removal rate as long as the forecast abundance exceeds the fishing threshold (25% of estimated unfished equilibrium spawning biomass) was determined to be precautionary, given the performance indicators, for all but the Queen Charlotte Islands stock. However, it was noted that dispersal rates of herring among the five stocks has a dramatic impact on the estimate of stock resilience and consequently on the appropriate stock reference points. It is recommended that the effect on harvest policy of the magnitude and consistency of dispersal be further investigated as part of any policy review.

Introduction

The nature and magnitude of the uncertainties in fisheries management provides the principal justification for a precautionary approach (Caddy 1998; Richards and Maguire 1998) and stock-specific reference points (RPs) provide the primary mechanism for its application. Reference points may be expressed in terms of fishing mortality rate (F) or stock biomass (B) (Caddy 1998). Two types of RPs have been identified, conservation or limit RPs intended to constrain harvesting within safe biological limits and management or target RPs intended to meet management objectives (Caddy and Mahon 1995). In general, scientists are responsible for identifying both RPs and determining the current stock biomass with associated uncertainties.

The international agreement relating to the conservation and management of straddling and highly migratory fish stocks clearly specifies that F_{MSY}, the F that produces the maximum sustainable yield (MSY), is a conservation reference point that should not be exceeded. However this implicitly assumes equilibrium environmental conditions. If environmental conditions change systematically creating trends in fish productivity, then F_{MSY} derived from recent conditions may no longer be sustainable. The question of which RPs are appropriate for a variable environment has received only limited scientific attention. On the other hand, the application of F_{MSY} to complicated situations involving natural variabil-

*Corresponding author: schweigertj@dfo-mpo.gc.ca

ity or parameter uncertainty has not been successful (Thompson 1999).

In a changing environment, the precautionary approach is challenged by uncertainties associated with establishing and measuring RPs, and these uncertainties should be expressed in probabilistic terms when providing scientific advice for fishery management (McAllister et al. 1999). The precautionary approach is also challenged by practical difficulties in the development and application of management objectives. Practical fishery management decision making has lacked a coherent set of quantified objectives. Yet in the process of conflict resolution, parties interested in fishery management are forced to state their objectives explicitly (Richards and Maguire 1998). Fu et al (2004) developed a generic Monte Carlo simulation framework for evaluating conservation RPs and harvest strategies in fisheries management to address some of these uncertainties and this report presents its application to five stocks of Pacific herring *Clupea pallasi* in British Columbia, Canada.

Methods

Life History and Vital Rates

Pacific herring in British Columbia spawn in March and April and then spend several months inshore as age-0 juveniles. During the fall and winter some juveniles migrate offshore and join schools of immature age-1 and age-2 herring, others remain inshore until the following summer. The majority of herring mature for the first time at age 3 and join the schools of adult fish (ages 4–10) on their fall inshore migration to the spawning grounds (Hourston and Haegele 1980; Ware 1996).

Tagging and genetic studies indicate that Pacific herring exist as a metapopulation (McQuinn 1997; Ware and Schweigert 2001, 2002; Beacham et al. 2001, 2002). In British Columbia, the metapopulation consists of five major "stocks" denoted by geographical area (Figure 1): the Queen Charlotte Islands (QCI), the Prince Rupert District (PRD), the Central Coast (CC), the Georgia Strait (GS), and the west coast of Vancouver Island (WCVI). Nevertheless, each of the five major migratory stocks has been assessed and managed independently assuming they are demographically isolated (Schweigert 2001). Since 1983, the current herring roe fisheries in each stock have been managed to achieve a constant harvest rate with the quota set at 20% of forecasted stock size. In 1986, a threshold biomass or "cutoff" level for each stock was introduced to further restrict harvest at depressed stock abundance (Schweigert 2001). In addition, the current age-structured stock assessment model assumes a natural mortality value that is constant over time and age but allows for the estimation of year- and age-specific availability to the fishing gears. The constant natural mortality assumption has prevailed in most stock assessments due to confounding with other parameters, such as fishing mortality and recruitment (Schnute and Richards 1995). However, assuming constant natural mortality in the face of salient interannual, and particularly directional variations in natural mortality, can greatly bias the estimation of other parameters (Fu and Quinn 2000) and F_{MSY} as well.

In developing the current framework, it was necessary to modify the existing age-structured stock assessment model. To simplify the descriptions, we have used the subscripts i, t, and a to represent stock, year, and age, respectively. First, we estimated interannual variation in natural mortality. In addition, to reduce the number of estimated parameters,

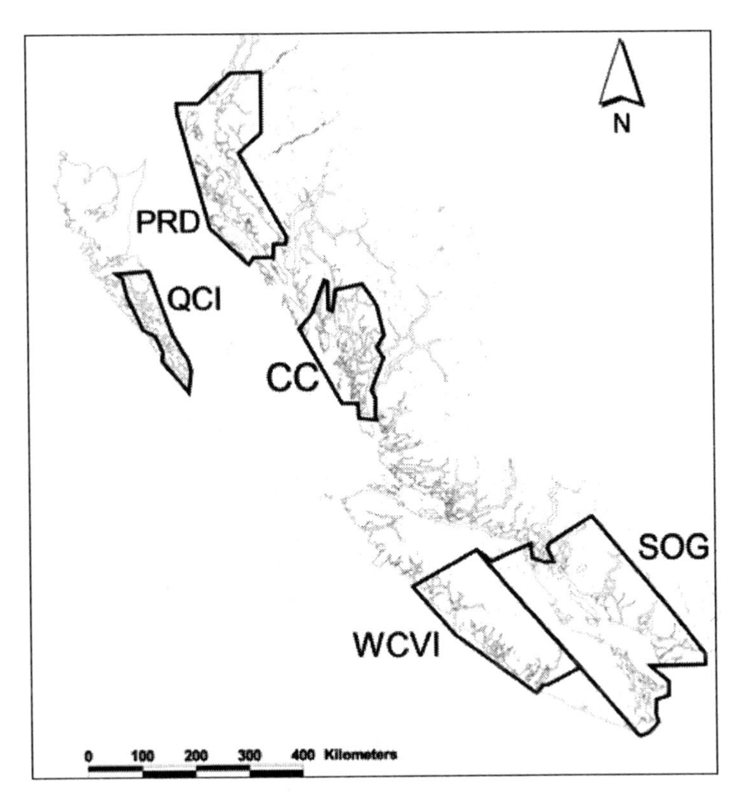

Figure 1. The five major British Columbian herring stock assessment regions: Prince Rupert District (PRD), Queen Charlotte Islands (QCI), Central Coast (CC), west coast Vancouver Island (WCVI), and the Strait of Georgia (SOG).

the availability parameters were approximated by year- and age-specific maturity ogives $O_{i,t,a}$ that were calculated based on the weight-at-age in each year $w_{i,t,a}$:

$$O_{i,t,a} = \frac{1}{1+e^{-\phi\left(w_{i,t,a}-w_{50}\right)}}, \quad (1)$$

where ϕ is the shape parameter for the logistic curve and w_{50} is the weight at which 50% of individuals become mature. The two parameters were estimated independently of the stock assessment model with the values of 0.088 and 85.696, respectively, and assumed to be identical for each stock. Because of the change in the fishery from an interception reduction fishery to a spawning ground roe fishery in the 1970s and questions about data consistency, we only used data collected after 1971 in this analysis.

For each stock, the assessment model estimates a time series of natural mortality and thus survival rate $S_{i,t,a}$, recruits at age-2 over time $R_{i,t}$, spawning stock biomass $SB_{i,t}$, and thus recruit-per-spawner,

$$S_{i,t,0} = \frac{R_{i,t}}{SB_{i,t-2}}.$$

Survival rate is assumed not to change with age after age 2 ($S_{i,t,a} = S_{i,t,}$ *for* $a \geq 2$) . The average recruitment rate $r_{i,a}$, defined as recruits at age-2 per individual from each age was calculated based on the average maturity ogive at age, weight at age, and recruit-per-spawner (i.e., $r_{i,a} = \bar{O}_{i,a}\bar{w}_{i,a}\bar{S}_{i,0}$, where $2 \leq a \leq 10$. The estimated $\bar{O}_{i,a}$, $\bar{w}_{i,a}$, and $r_{i,a}$ for the period from 1972 to 2000 $\left(r_{i,a}^{\text{avg}}\right)$ and that averaged for the last 5 years from 1996 to 2000 $\left(r_{i,a}^{\text{avgL5}}\right)$ are listed in Table 1. The estimated average recruit-per-spawner for the entire period 1972–2000 $\left(\bar{S}_{i,0}^{\text{avg}}\right)$ and for just 1996–2000 $\left(\bar{S}_{i,0}^{\text{avgL5}}\right)$, and the estimated average survival rate over the entire period 1972–2000 $\left(\bar{S}_{i,a}^{\text{avg}}\right)$ and 1996 to 2000 $\left(\bar{S}_{i,a}^{\text{avgL5}}\right)$ are listed in Table 2. The values of $r_{i,a}^{\text{avg}}$ $\left(\text{or } r_{i,a}^{\text{AvgL5}}\right)$, $\bar{S}_{i,a}^{\text{avg}}$ $\left(\text{or } \bar{S}_{i,a}^{\text{avgL5}}\right)$, and their relative changes were then used as input to the risk assessment software package RAMAS Metapop (Akçakaya 1998).

RAMAS Model Structure

RAMAS Metapop allowed us to model the age (or stage) structure of each population in a metapopulation with a transition matrix (e.g., a Leslie matrix when the model is age-structured) (Akçakaya 1998). Abundance at each age in the next time step (the vector N_{t+1})

Table 1. Average weight at age *a* (W_a), maturity ogive at age *a* (O_a), long-term average recruitment rate (from 1972 to 2000, $r_{i,a}^{\text{avg}}$), and recent average recruitment rate (from 1996 to 2000, $r_{i,a}^{\text{avgL5}}$) for the QCI, PRD, CC, GS, and WCVI herring stocks.

	Age	2	3	4	5	6	7	8	9	10
QCI	W_a	0.065	0.095	0.120	0.141	0.160	0.174	0.185	0.194	0.197
	O_a	0.134	0.694	0.953	0.992	1.000	1.000	1.000	1.000	1.000
	$r_{i,a}^{\text{avg}}$	0.052	0.399	0.693	0.847	0.964	1.050	1.120	1.180	1.200
	$r_{i,a}^{\text{avgL5}}$	0.138	1.052	1.817	2.231	2.534	2.773	2.949	3.092	3.140
PRD	W_a	0.052	0.080	0.106	0.127	0.143	0.156	0.168	0.176	0.186
	O_a	0.048	0.373	0.853	0.973	0.994	1.000	1.000	1.000	1.000
	$r_{i,a}^{\text{avg}}$	0.03	0.357	1.080	1.480	1.710	1.870	2.01	2.110	2.230
	$r_{i,a}^{\text{avgL5}}$	0.036	0.427	1.291	1.763	2.035	2.236	2.408	2.523	2.666
CC	W_a	0.053	0.085	0.108	0.128	0.146	0.161	0.173	0.185	0.194
	O_a	0.052	0.474	0.878	0.977	1.000	1.000	1.000	1.000	1.000
	$r_{i,a}^{\text{avg}}$	0.027	0.394	0.936	1.240	1.430	1.580	1.710	1.820	1.910
	$r_{i,a}^{\text{avgL5}}$	0.029	0.419	0.995	1.309	1.518	1.686	1.812	1.937	2.032
GS	W_a	0.057	0.083	0.108	0.128	0.147	0.160	0.171	0.179	0.185
	O_a	0.074	0.444	0.872	0.976	1.000	1.000	1.000	1.000	1.000
	$r_{i,a}^{\text{avg}}$	0.160	1.400	3.560	4.720	5.530	6.050	6.480	6.780	7.010
	$r_{i,a}^{\text{avgL5}}$	0.185	1.615	4.107	5.473	6.392	7.005	7.486	7.837	8.099
WCVI	W_a	0.064	0.093	0.120	0.143	0.162	0.176	0.185	0.193	0.198
	O_a	0.128	0.652	0.954	0.994	1.000	1.000	1.000	1.000	1.000
	$r_{i,a}^{\text{avg}}$	0.073	0.544	1.030	1.280	1.460	1.580	1.670	1.740	1.780
	$r_{i,a}^{\text{avgL5}}$	0.077	0.570	1.084	1.348	1.527	1.659	1.743	1.819	1.866

is given by the matrix multiplication $N_{t+1} = L(t)N_t$, where $L(t)$ is the transition matrix, r is the recruitment per individual from each age (or stage), and S is the survival rate at each age (or stage):

$$L(t) = \begin{pmatrix} r_1 & r_2 & \cdots & \cdots & r_A \\ S_1 & 0 & 0 & 0 & 0 \\ 0 & S_2 & 0 & 0 & 0 \\ 0 & 0 & \ddots & 0 & 0 \\ 0 & 0 & 0 & S_{A-1} & 0 \end{pmatrix}.$$

RAMAS Metapop extends the original transition matrix to allow elements of the matrix to fluctuate in time, thereby representing the impact of environmental fluctuations on vital rates. Stochasticity in vital rates including recruitment and survival is generated from a matrix of standard deviations assuming log-normal distributions. Each population in the metapopulation can have its own age (or stage) matrix, and its own standard deviation matrix to allow for population specific vital rates and stochasticity. Once the transition matrix and initial population size are input for each component of a metapopulation, forward projections are possible for any specified time period.

Density-Dependence

A population modeled with just the transition matrix would either increase or decrease exponentially. However, density-dependence (d-d) resulting from limitation of food or space, or depensation at low abundance (Allee effects) may operate during part or all of the life span to moderate this effect. In Metapop, specific d-d effects for a population can be modeled by defining three components: (1) the form of the d-d function (exponential, ceiling, scramble, or contest); (2) the stages that are affected; and (3) the parameter values of the d-d function (carrying capacity K, maximum population growth rate, and Allee effect parameter A).

Except for the PRD stock, the relationship between age-2 abundance (i.e., recruits) and the spawning biomass for British Columbia herring resembled the "ceiling" model (often called the "hockey-stick" model [e.g., Barrowman and Myers 2000]), with recruits increasing linearly with spawning biomass until a threshold biomass is achieved (Ware and Schweigert 2001). Therefore, the ceiling d-d model was used for all stocks and all life stages.

The unfished equilibrium abundance $\left(N_i^{\mathrm{EQ}}\right)$ used to set the cutoff levels (currently 25% of N_i^{EQ}) was estimated to be nearly half of the historical maximum abundance (Hist Max, Figure 2). We arbitrarily set the carrying capacity for each stock (K_i) at twice N_i^{EQ} (Table 2) so that it was biologically consistent with Hist Max and the basis of current fisheries management policy. Extensive simulations with different K_i indicated that our results would be relatively insensitive to the choice of K_i within limits of ±30%.

In RAMAS Metapop, Allee effects are modeled by multiplying each vital rate with $\frac{N_t}{A+N_t}$, where A is the number of individuals at which the vital rate is half of what it would have been without the Allee effect (Akçakaya 1998). The Allee parameter (A_i) for each herring stock was set arbitrarily at 1% of K_i. Other levels of A_i were also tried with lower A_i resulting in more optimistic stock projections and vice versa.

Dispersal

Dispersal is a key component of population regulation within the context of a metapopulation model. Although dispersal or migration among the five herring stocks has not been incorporated explicitly in the age-structured stock assessment model, it is at least

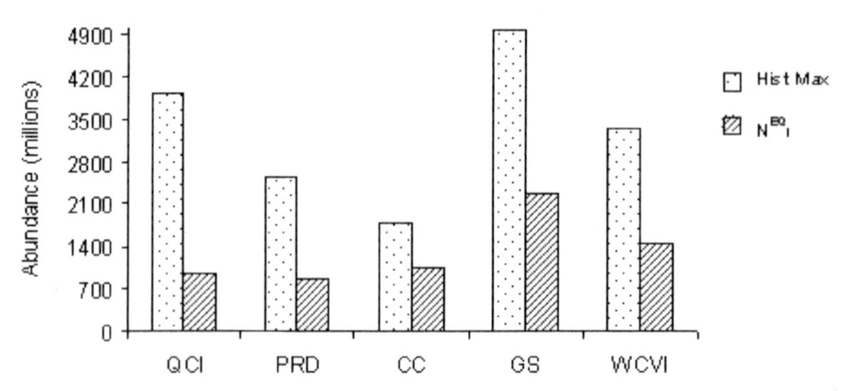

Figure 2. Historical maximum abundance (Hist Max) and equilibrium abundance $\left(N_i^{\mathrm{EQ}}\right)$ for the QCI, PRD, CC, GS, and WCVI herring stocks.

logically consistent to ignore the possibility of dispersal among the five stocks in the population projection simulations. However, because tagging studies have suggested that movement of fish among herring stocks can be quite extensive it was desirable to investigate the possible impacts of population migration and dispersal on stock resilience and RP estimates.

Dispersal rate ($m_{i,j}$) in RAMAS Metapop is defined as the proportion of organisms that move from population i to population j (Akçakaya 1998). Ware and Schweigert (2001) concluded that the dispersal pattern for the British Columbia herring metapopulation was consistent with an isolation by distance model with most herring straying to nearby stocks and fewer to the more distant ones, but it appears to be confounded by variations in spawner biomass. Nevertheless, we assumed that dispersal rates from the five stocks of British Columbia herring were dependent only on the distance $D_{i,j}$ between stocks so that

Table 2. Carrying capacity (K_i millions), estimated abundance in 1985 ($N_{i,1985}$, millions), estimated long-term average rate of recruitment at age 2 (from 1974 to 2000, $\bar{S}_{i,0}^{\mathrm{avg}}$), estimated long-term average survival rate (from 1974 to 2000, $\bar{S}_{i,a}^{\mathrm{avg}}$), estimated abundance in 2000 ($N_{i,2000}$), estimated recent average rate of recruitment (from 1996 to 2000, $\bar{S}_{i,0}^{\mathrm{avgL5}}$), and estimated recent average survival rates (from 1996 to 2000, $\bar{S}_{i,a}^{\mathrm{avgL5}}$), for the QCI, PRD, CC, GS, and WCVI herring stocks.

	K_i	$N_{i,1985}$/ % of N_i^{EQ}	$\bar{S}_{i,0}^{\mathrm{avg}}$	$\bar{S}_{i,a}^{\mathrm{avg}}$	$N_{i,2000}$/ % of N_i^{EQ}	$\bar{S}_{i,0}^{\mathrm{avgL5}}$	$\bar{S}_{i,a}^{\mathrm{avgL5}}$
QCI	1,896	268/28%	0.06	0.88	139/15%	0.16	0.80
PRD	1,768	540/61%	0.12	0.74	863/98%	0.14	0.76
CC	2,071	317/31%	0.10	0.81	545/53%	0.10	0.82
GS	4,536	2,600/115%	0.38	0.54	4,450/196%	0.44	0.55
WCVI	2,890	543/38%	0.09	0.77	479/33%	0.09	0.69

$m_{i,j} = \exp\left(\frac{-D_{i,j}}{b}\right)$ (Table 3). We adopted the lower bound of dispersal estimates of Ware and Schweigert (2001) as the maximum rate of dispersal among the stocks to be conservative about the prospects for additions from adjacent stocks. The maximum dispersal rate was assumed to be 14% and resulted in an estimated b value of 100.

Population Management Strategies

RAMAS Metapop version 4.0 offers the capability to investigate dual population management strategies in that harvest rate (μ_i) can vary with stock-specific abundance. In this study, we used the following harvest control rule:

$$\mu = \begin{cases} 0, \text{ if } N_i \leq N_i^{\text{cutoff}} \\ \text{linearly increases from 0 to } \mu_i, \\ \quad \text{if } N_i^{\text{cutoff}} < N_i \leq 1.05 \cdot N_i^{\text{cutoff}} \\ \mu_i, \text{ if } N_i > 1.05 \cdot N_i^{\text{cutoff}} \end{cases}$$

The bound N_i^{cutoff} is the predefined cutoff abundance for stock i, below which no fishing is allowed. When abundance N_i exceeds N_i^{cutoff}, harvest rate is allowed to increase linearly with N_i to a predefined maximum value when abundance exceeds N_i^{cutoff} by 5%. We used this restricted transition zone $\left(N_i^{\text{cutoff}} \text{ to } 1.05 \cdot N_i^{\text{cutoff}}\right)$ to reflect current policy and to ensure clear contrast between scenarios with different levels of $N_{i,a}^{\text{cutoff}}$ and μ_i used to evaluate the performances of different harvest strategies.

To account for partial recruitment to the fishery, age-specific harvest rate was calculated as: $\mu_{i,a} = \mu_i \delta_{i,a}$, where $\delta_{i,a}$ is the gear selectivity parameter for age a in stock i as estimated from the assessment model and μ_i is the fully selected harvest rate for stock i. Each age-specific cutoff abundance $\left(N_{i,a}^{\text{cutoff}}\right)$ was calculated as

$$N_{i,a}^{\text{cutoff}} = \delta_{i,a}\sigma_{i,a}N_i^{\text{cutoff}} = \delta_{i,a}\sigma_{i,a}N_i^{\text{EQ}}\pi_i,$$

where $\sigma_{i,a}$ is the relative abundance of the five age groups from 2 to 6+, and π_i is the cutoff coefficient (percentage). For example, under current policy, π_i is set at 25%, so that N_i^{cutoff} is 25% of N_i^{EQ}.

Stochasticity

Demographic and environmental stochasti-cities were included in the simulations, and the effect was measured over 200 replications for each simulation condition. Although 1,000 replications would have been desirable, the smaller number of replications was necessary to reduce the computation time given the large number of conditions evaluated. We argue that our general conclusions about conservation RPs and

Table 3. Distance (km) and dispersal rates ($m_{i,j}$) among five stocks.

	QCI	PRD	CC	GS	WCVI
QCI	0	197/0.140	202/0.133	622/0.002	596/0.003
PRD	197/0.140	0	303/0.048	776/0.0004	713/0.0008
CC	202/0.133	303/0.048	0	434/0.013	447/0.011
GS	622/0.002	776/0.0004	434/0.013	0	274/0.065
WCVI	596/0.003	713/0.0008	447/0.011	274/0.065	0

harvest strategies will not be seriously affected. The levels of coefficient of variation (CV) for recruitment rate and survival rate were set at 0.5 and 0.2, respectively, consistent with their estimated variability in the assessment model. The CV levels for dispersal, sampling error for abundance, and K_i were set at 0.2, 0.3, and 0.1, respectively, based on our best guess because empirical estimates were not available.

Population Projections

For Pacific herring and other marine fish species with very high reproductive potential, the average age of maturity is about 3 years. To evaluate risk of extinction, the IUCN uses a time scale of 10 years or three generations, whichever is longer (Baillie and Groombridge 1996). Following Restrepo et al. (1998), generation time was calculated accordingly as

$$\sum_{a=1}^{a_A} a r_a N_a \Big/ \sum_{a=1}^{a_A} r_a N_a ,$$

where a_A, the oldest age in an unfished population, is about 10, resulting in an average value of 4.98 (approximated at 5) for the five herring stocks. In our simulations, we used a timescale of 15 years (i.e., three generations) to be consistent with the IUCN criteria and used vital rates estimated for the years from 1985 to 2000 to make population projections.

To evaluate the effects of initial population size $N_{i,1}$, harvest rate μ_i, and cutoff coefficient π_i on population dynamics, our program looped over values of these three variables with a step value of 5%, ranging from $5\% \cdot N_i^{EQ}$ to $100\% \cdot N_i^{EQ}$ for $N_{i,1}$, 0% to 95% for μ_i, and 0% to 65% for π_i. The scenarios examined are listed in Table 4.

Performance Indicators

An extensive literature exists on performance indicators as measures of management strategy efficacy relating primarily to fisheries (Francis and Shotton 1997; Butterworth and Punt 1999; Punt and Smith 1999). However, we were also interested in exploring the probability of triggering the IUCN critically endangered listing because of its possible application to other species under the pending Canadian Species at Risk Act. Following IUCN Criterion A, a reduction of 80% or more in population size within 10 years or three generations, whichever is the longer,

Table 4. Summary of Monte Carlo simulations presented in Figures 3 and 4. The scenarios in Figure 3 were designed to investigate the effect of initial population size for various combinations of harvest rate and cutoff level using the estimated vital rates series. Scenarios in Figure 4 were designed to investigate the effect of both harvest and cutoff from a retrospective point of view using initial population size in 1985. In combination, results in Figures 3 and 4 can be used to examine probable parameter combinations by comparing values at the crosshairs with historical experience.

	Assumptions				
Purpose	Variables	Dispersal	π_i	$N_{i,1}$	Vital rates
Collapse—recovery probabilities		Without	0% 25%	Variable	Estimated time series
Retrospective from 1985	π_i, μ_i	Without	Variable	$N_{i,1985}$	Estimated time series

the first performance indicator we used is $P_{80\%terdec}$, the probability of 80% terminal decline in simulations of 15 years duration (three generations).

The current harvest rule would trigger fishery closure if the population size goes below 25% of N_i^{EQ}. Therefore, our second performance indicator, the probability of fishery closure was computed as the probability of quasi-extinction in three generations, where the "quasi-extinction" threshold was set at 25% of N_i^{EQ}. This indicator is denoted as P_{lowN} to make it clear that the threshold is much higher than normally used for quasi-extinction.

Our third performance indicator is the probability of recovery from initial population size to 90% of N_i^{EQ}, the quasi-explosion threshold within three generations. This is denoted $P_{recovery}$. Other performance indicators related to catch include annual catch from the metapopulation (all stocks) averaged over the entire temporal duration (C_{avgann}) and the probability of catch falling below 1% of N_i^{EQ} for the metapopulation ($P_{lowcatch}$).

These performance indicators were used to measure the effectiveness of alternative reference points and harvest strategies for achieving the conservation objective of maintaining populations and species within bounds of natural variability. Fishery benefits and losses are quantified as catch and probability of low catch.

Results

Effects of Cutoff with Allee Effect and No Dispersal

With no cutoff rule (i.e., $cutoff_i = 0$), the probability of triggering an IUCN critically endangered (CE) listing for any specified μ_i varied greatly among the five stocks (Figure 3A, column 1). If the initial population size was kept above 25% of N_i^{EQ}, a CE listing was unlikely for the most resilient stock (GS) unless μ_i exceeded 0.6. On the other hand, the chance of triggering a CE listing exceeded 90% for the least productive stock (QCI) for $\mu_i = 0.3$.

The $P_{80\%terdec}$ contours changed significantly when $cutoff_i$ was set at 25% of N_i^{EQ} as specified by current policy (Figure 3A, column 2). Nevertheless, $P_{80\%terdec}$ still exceeded 0 for low μ_i and $N_{i,1}$. To avoid triggering a CE listing at high $N_{i,1}$, μ_i had to remain below 0.2 in QCI, 0.4 in PRD and CC, 0.75 in GS, and 0.6 in WCVI.

The same pattern of vulnerability among stocks was evident for P_{lowN} but isopleths were more tilted than for $P_{80\%terdec}$ so that P_{lowN} got larger as $N_{i,1}$ decreased (Figure 3B, column 1). The greatest difference between results for $P_{80\%terdec}$ and P_{lowN} was that increasing the $cutoff_i$ to 25% of N_i^{EQ} greatly reduced $P_{80\%terdec}$ but had a much smaller effect on P_{lowN}, particularly when $N_{i,1}$ was lower than 5% of N_i^{EQ}. However, imposing a $cutoff_i$ = 25% of N_i^{EQ} reduced P_{lowN} at higher $N_{i,1}$ levels (Figure 3B, column 2). Again, QCI was clearly the most vulnerable stock, and the probability of closing the QCI fishery under the current prevailing management policy of $\mu_i = 0.2$ was high (>80%) unless the stock size remained above 90% of N_i^{EQ}.

$P_{recovery}$ contours also differed significantly among the five stocks. With no cutoff, the GS stock had a 90% chance of recovery if the stock falls to 25% of N_i^{EQ} provided μ_i remained below 0.4 (Figure 3C, column 1).

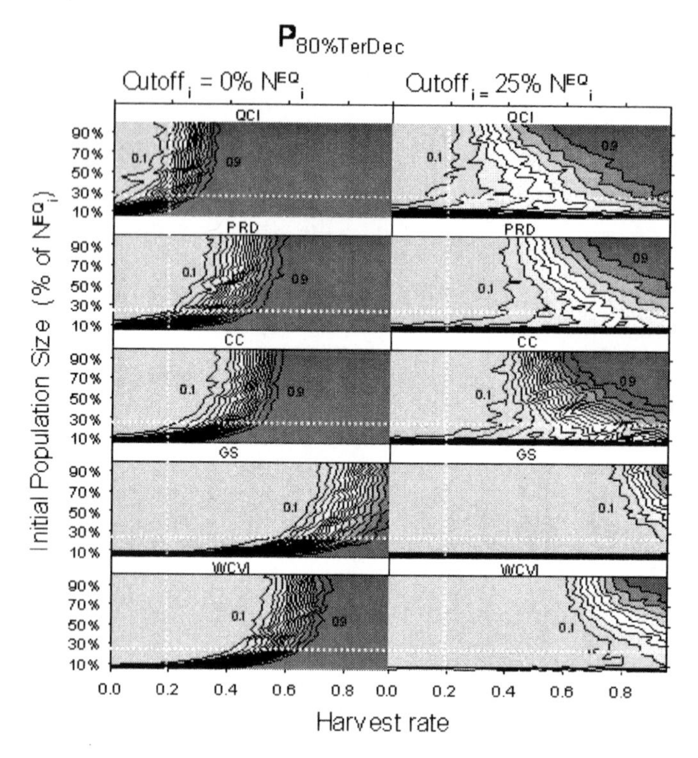

Figure 3A. Contour plots for $P_{80\%terdec}$, the probability of exceeding an 80% decline rate within 15 years (three generations) for the five British Columbia herring stocks (QCI, PRD, CC, GS, and WCVI) under two possible cutoff levels (0% and 25% of N_i^{EQ} in columns 1 and 2) and various combinations of harvest rate and initial population size (% of N_i^{EQ}). White lines represent the harvest rate and cutoff under current harvest rule. The Allee parameter A_i was set at 1% of carrying capacity K_i with no dispersal among stocks.

In contrast, even with no fishing, the QCI stock had to remain above 40% of N_i^{EQ} to be able to recover to 90% of N_i^{EQ} within three generations. Overall, none of the five stocks had a 90% chance of recovery from $N_{i,1}$ less than 5% of N_i^{EQ}. Adding the cutoff had less effect on $P_{recovery}$ (Figure 4C, column 2) than P_{lowN}.

Retrospective Simulations Without and With Dispersal

For these simulations, we set $N_{i,1}$ at the abundance estimated for year 1985, projected the stocks forward for three generations to the year 2000 using the time series of vital rates estimated in the stock assessment model, and then computed the performance indicators looping over a grid of cutoff$_i$ and μ_i values under two scenarios: without and with dispersal.

Without dispersal, the vulnerable QCI stock had over 10% chance of triggering an IUCN CE listing under current prevailing management policy of $\mu_i = 0.2$ and a cutoff$_i$ (Figure 4A, column 1). For other stocks, a decline rate of 80% over the last 15 years would have been improbable for any of the stocks under the prevailing management policy.

Allowing dispersal among the five stocks again reduced $P_{80\%terdec}$ and tended to reduce dif-

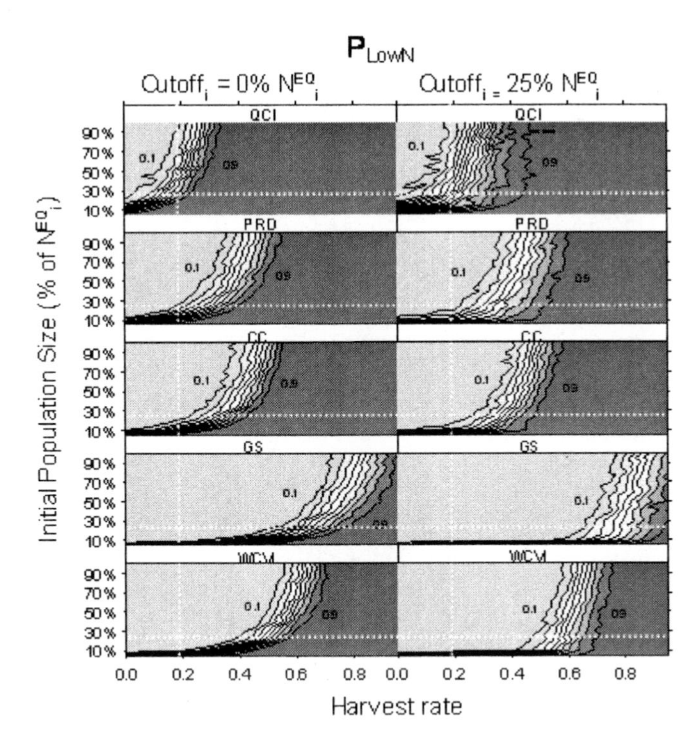

Figure 3B. Contour plots for P_{lowN}, the probability that abundance will fall below the current cutoff level of 25% of N_i^{EQ} within 15 years (three generations) for the five British Columbia herring stocks (QCI, PRD, CC, GS, and WCVI) under two possible cutoff levels (0% and 25% of N_i^{EQ} in columns 1 and 2) and various combinations of harvest rate and initial population size (% of N_i^{EQ}). White lines represent the harvest rate and cutoff under current harvest rule. The Allee parameter A_i was set at 1% of carrying capacity N_i and no dispersal among stocks was allowed.

ferences among stocks. Contours for $P_{80\%terdec}$ were similar for the QCI and PRD stocks, and slightly higher than those for the GS and WCVI stocks. Dispersal also reduced the differences between the GS and WCVI stocks (Figure 4A, column 2).

The P_{lowN} and $P_{recovery}$ indicators revealed the same patterns among stocks as $P_{80\%terdec}$ and were similarly affected by changing assumptions about dispersal (Figure 4B, 4C). Only for the QCI stock was there any appreciable probability of a fishery closure (i.e., low N) or failure to rebuild to 90% of N_i^{EQ} under the prevailing management policy, assuming no dispersal; the likelihood of these outcomes was reduced significantly by allowing dispersal.

Discussion

Canada's objectives for the evaluation of conservation reference points and of a harvesting strategy developed in Fu et al. (2004) and applied here is part of the Department of Fisheries and Oceans Canada's Objectives based fishery management initiative. The intent of the framework is to identify a flexible approach for the development of conservation reference points that would be applicable to a wide range of both short and long-lived species. The RAMAS Metapop software provided a flex-

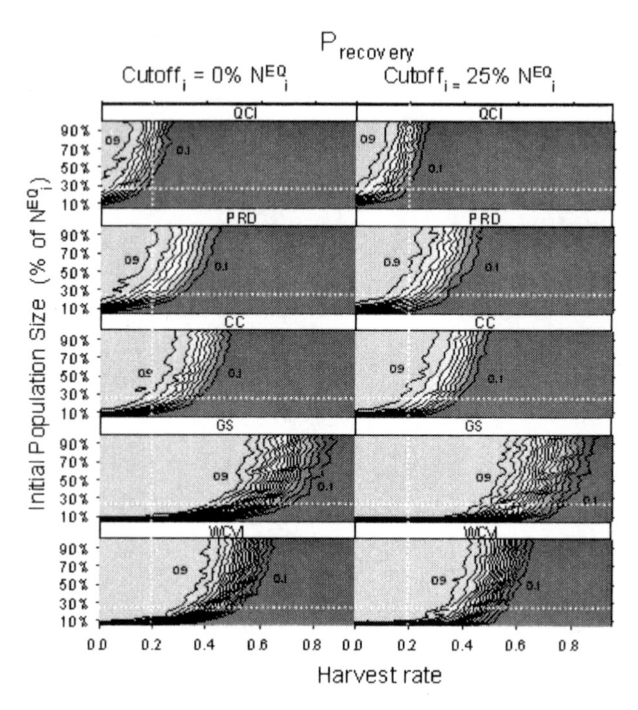

Figure 3C. Contour plots for $P_{recovery}$, the probability of recovering to 90% of N_i^{EQ} within 15 years (three generations) for the five British Columbia herring stocks (QCI, PRD, CC, GS, and WCVI) under two possible cutoff levels (0% and 25% of N_i^{EQ} in columns 1 and 2) and various combinations of harvest rate and initial population size (% of N_i^{EQ}). White lines represent the harvest rate and cutoff under current harvest rule. The Allee parameter A_i was set at 1% of carrying capacity K_i and no dispersal among stocks was allowed.

ible and convenient tool for population risk assessment. Its flexibility and convenience lies mainly in three areas. First, RAMAS Metapop provides options for modeling age (or stage) and spatial structure in a metapopulation, demographic and environmental stochasticit-ies and measurement errors, and temporal variations in factors such as vital rates and carrying capacity. Second, RAMAS Metapop provides various options to diagnose risk. Indicators such as the average annual catch (C_{avgann}), the probability of a very small catch ($P_{lowcatch}$), the probability of the population declining by 80% over three generations ($P_{80\%terdec}$), the probability of the population declining below the fishery threshold (P_{lowN}), and the probability of the population rebuilding to 90% of the unfished equilibrium level ($P_{recovery}$) are performance measures that are useful in fisheries management. Third, RAMAS Metapop can be used in tandem with other computer programs to examine the performance of conservation reference points and harvest strategies over a wide range of scenarios involving combinations of state variables (e.g., $N_{i,1}$, cutoff$_i$, and μ_i), thus facilitating a higher level of synthesis.

Nevertheless, one must be careful interpreting results because they are affected by assumptions about vital rates estimated from a stock assessment model, carrying capacity, Allee effects, dispersal rates among the stocks, measurement errors, and fishery implemen-

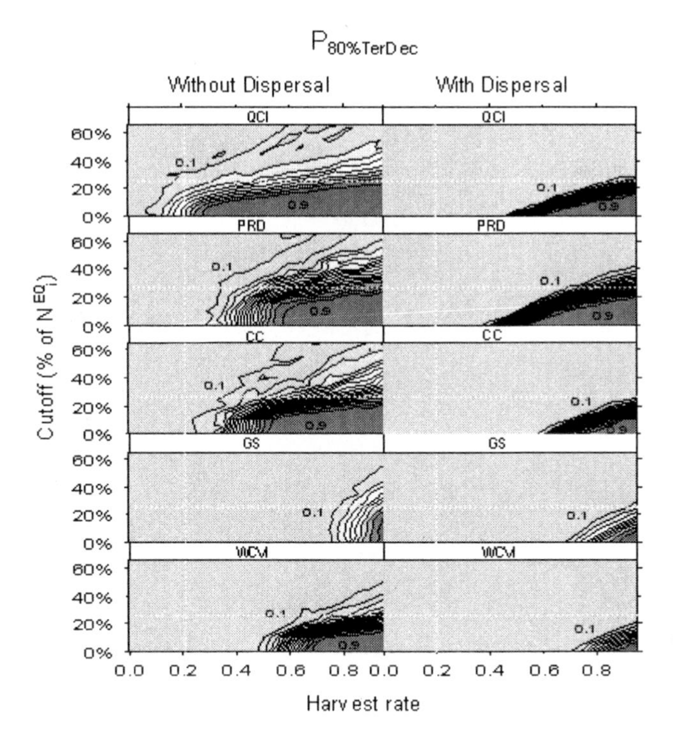

Figure 4A. Contour plots for $P_{80\%terdec}$, the probability of exceeding an 80% decline rate within 15 years (three generations) for the five British Columbia herring stocks (QCI, PRD, CC, GS, and WCVI) under various combinations of harvest rate and cutoff levels (% of N_i^{EQ}) and two scenarios (column 1: without dispersal; column 2: with dispersal). White lines represent the harvest rate and cutoff under current harvest rule. The initial population size was set at the estimate in 1985.

tation errors. All of these uncertainties should be considered when making recommendations on conservation RPs and optimal harvest strategies. For example, Pacific herring are impacted by a wide array of predators such as Pacific hake and mackerel thought to be the major source of natural mortality (Ware 1996). Inter-annual variation in natural mortality is therefore a critical factor in accurately modeling population dynamics of herring and other species (Fu and Quinn 2000). In fact, relaxing the assumption of constant survival rate over time dramatically improved the retrospective analysis for these stocks. The decreasing trend in estimated survival and abundance for the WCVI herring stock is consistent with the observed dramatic increase in abundance of Pacific hake in association with changes in temperature in the 1990s (Ware 1996). Thus, knowledge of recent temporal variations in estimated vital rates should improve the basis for forecasting vital rates and abundance for these stocks. Similarly, without dispersal among stocks, population projections, and performance indicators ($P_{80\%terdec}$, P_{lowN}, and $P_{recovery}$) were all significantly affected by the choice of vital rates (Figure 4A–C). Stock projections based on the historical average estimates of vital rates indicated that the QCI stock was least resilient and the GS stock was most resilient of the metapopulation. Allowing dispersal reduced differences in performance ($P_{80\%terdec}$, P_{lowN}, and $P_{recovery}$) among stocks under the various scenarios. For example, the plots for QCI, PRD, and CC became more similar, and those for GS and WCVI became more similar

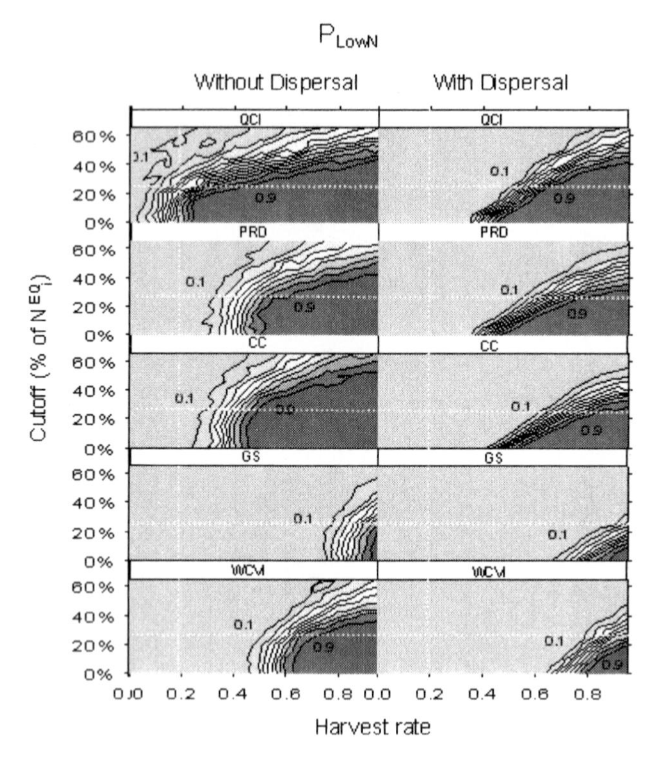

Figure 4B. Contour plots for P_{lowN}, the probability that abundance will fall below the current cutoff level of 25% of within 15 years (three generations) for the five British Columbia herring stocks (QCI, PRD, CC, GS, and WCVI) under various combinations of harvest rate and cutoff levels (% of N_i^{EQ}) and two scenarios (column1: without dispersal, column 2: with dispersal). White lines represent the harvest rate and cutoff under current harvest rule. The initial population size was set at the estimate in 1985.

(Figure 4A–C). However, the observed low population abundance in the 1990s for QCI and WCVI (Schweigert 2001) seemed to contradict the simulation results based on a constant dispersal rate. Although Ware and Schweigert (2001, 2002) did not explicitly consider the possibility of increased natural mortality for QCI and WCVI stocks, they concluded that the much reduced abundance in the 1990s may have been caused by a reduction of dispersal from GS, CC, and PRD, shortly after the 1992–1993 El Niño–Southern Oscillation, which was followed by an anomalously warm period. If this is true, temporal patterns in dispersal should be incorporated into the risk assessment model to make more realistic projections. However, this would require a better understanding of the biological or environmental basis of migration and dispersal. In addition, the inclusion of dispersal in the risk assessment is only legitimate when dispersal is explicitly incorporated into the stock assessment model where the vital rates are determined. Nevertheless, the magnitude and consistency of dispersal among stocks within a metapopulation clearly has an overwhelming impact on stock resilience and the choice of RPs and harvest policy. We recommend efforts to explicitly incorporate dispersal into the stock assessment model.

The importance of the Allee parameter A_i is difficult to assess because it is almost impossible to measure. Exclusion of Allee effects in

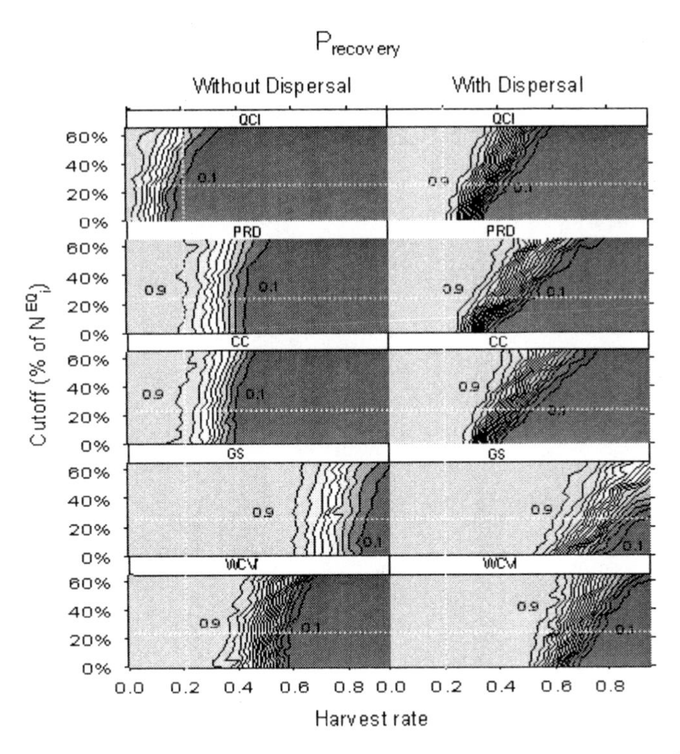

Figure 4C. Contour plots for $P_{recovery}$, the probability of recovering to 90% of N_i^{EQ} within 15 years (three generations) for the five British Columbia herring stocks (QCI, PRD, CC, GS, and WCVI) under various combinations of harvest rate and cutoff levels (% of N_i^{EQ}) and two scenarios (column1: without dispersal; column 2: with dispersal). White lines represent the harvest rate and cutoff under current harvest rule. The initial population size was set at the estimate in 1985.

population dynamics generally gives a more optimistic view of stock resilience when a stock is at low abundance. This can be quite misleading in fisheries management, because the dynamics of fish stocks are least understood and least predictable at low abundance (Quinn and Deriso 1999). To avoid the possibility of irreversible harm at low stock abundance, it is precautionary and conceptually sound to assume that Allee effects exist. However, because A_i is not a very intuitive parameter, another option is to use the management threshold or cutoff (R. Akcakaya, Applied Biomathematics, personal communication) to guard against the possibility of the stock being reduced to very low levels where Allee effects might become important. In the case of Pacific herring, a cutoff of 25% of N_i^{EQ} has been adopted by various agencies since the mid-1980s (Hall et al. 1988; Stocker 1993; Zheng et al. 1993). Some recent authors suggest that this cutoff level (25% of N_i^{EQ} or 12.5% of K_i) is rather high (Francis and Shotton 1997; Butterworth and Punt 1999). On the other hand, assuming no cutoff and even a very low level of Allee effect suggests a rather pessimistic picture of stock resilience (Figure 3A–C). Assuming a more intermediate level for the Allee effect (A_i = 1% of K_i) seems to be more consistent with the results of previous attempts to assess harvest strategy for Pacific herring (Hall et al. 1988; Zheng et al. 1993).

On the basis of this preliminary assessment of a risk assessment framework for Pacific herring the existing harvest rule (20% harvest rate and cutoff of 25% of N_i^{EQ}) appears to be conservative and ensures that there is negligible risk of triggering an IUCN CE listing for any stock except perhaps for QCI.

The risk assessment framework of Fu et al. (2004) presented here is promising in terms of providing insight into the impact of various harvest strategies on stock resilience and vulnerability for a suite of performance indicators. Unfortunately, some limitations were noted within RAMAS for investigating more complex harvest policies and their investigation may require customized software. Additionally, a major challenge remains to develop a process for synthesizing the results for all the performance indicators into a simple vulnerability index that all stakeholders can use as a basis for discussion of the acceptable risk to the resource and the decision as to what constitutes an appropriate harvest policy.

Acknowledgments

The evaluation framework developed here is in support of and sponsored in part by the Objective Based Fisheries Management initiative of Fisheries and Oceans Canada.

References

Akçakaya, H. R. 1998. RAMAS GIS: linking landscape data with population viability analysis (version 3.0). Applied Biomathematics, Setauket, New York.

Baillie, J., and B. Groombridge. 1996. IUCN Red List of threatened animals. IUCN Species Survival Commission, IUCN, Gland, Switzerland and Cambridge, UK.

Barrowman, N. J., and R. A. Myers. 2000. Still more spawner-recruit curves: the hockey stick and its generalizations. Canadian Journal of Fisheries and Aquatic Sciences 57:665–676.

Beacham, T. D., J. F. Schweigert, C. MacConnachie, K. D. Le, K. Labaree, and K. M. Miller. 2001. Population structure of herring (*Clupea pallasi*) in British Columbia: an analysis using microsatellite loci. Canadian Science Advisory Secretariat Research Document 2001/128, Ottawa.

Beacham, T. D., J. F. Schweigert, C. Mac-Connachie, K. D. Le, K. Labaree, and K. M. Miller. 2002. Population structure of herring (*Clupea pallasi*) in British Columbia determined by microsatellites, with comparisons to southeast Alaska and California. Canadian Science Advisory Secretariat Research Document 2002/109, Ottawa.

Butterworth, D. S., and A. E. Punt. 1999. Experiences in the evaluation and implementation of management procedures. ICES Journal of Marine Science 56:985–998.

Caddy, J. 1998. A short review of precautionary reference points and some proposals for their use in data-poor situations. FAO Fisheries Technical Paper No. 379.

Caddy, J. F., and R. Mahon. 1995. Reference points for fisheries management. FAO Fisheries Technical Paper No. 347.

Caddy, J. F., and R. McGarvey. 1996. Targets or limits for management of fisheries? North American Journal of fisheries management 16:479–487.

Francis, R. I. C. C., and R. Shotton. 1997. "Risk" in fisheries management: a review. Canadian Journal of Fisheries and Aquatic Sciences 54:1699–1715.

Fu, C., and T. J. Quinn, II. 2000. Estimability of natural mortality and other population parameters in a length-based model: *Pandalus borealis* Krøyer in Kachemak Bay, Alaska. Canadian Journal of Fisheries and Aquatic Sciences 57:2420–2432.

Fu, C., C. C. Wood, and J. F. Schweigert. 2004. Pacific herring (*Clupea pallasi*) in Canada: generic framework for evaluating conservation limits and harvest strategies in fisheries management. Pages 256–269 *in* H. R. Akçakaya, M. A. Burgman, O. Kindvall, C. C. Wood, P. Sjögren-Gulve, J. S. Hatfield, and M. A. McCarthy, editors. Species conservation and management: case studies. Oxford University Press, New York.

Hall, D. L., R. Hilborn, M. Stocker, and C. J. Walters. 1988. Alternative harvest strategies for Pacific herring (*Clupea harengus pallasi*). Canadian Journal of Fisheries and Aquatic Sciences 45:888–897.

Hourston, A. S., and C. Haegele. 1980. Herring on Canada's Pacific coast. Canadian Special Publication of Fisheries and Aquatic Sciences 48.

McAllister, M. K., P. J. Starr, V. R. Restrepo, and G. P. Kirkwood. 1999. Formulating quantitative methods to evaluate fishery-management systems: what fishery processes should be modelled and what trade-offs should be made? ICES Journal of Marine Science 56:900–916.

McQuinn, I. H. 1997. Metapopulations and the Atlantic herring. Reviews in Fish Biology and Fisheries 7:297–329.

Punt, A. E., and A. D. M. Smith. 1999. Harvest strategy evaluation for the eastern stock of gemfish (*Rexea solandri*). ICES Journal of Marine Science 56:860–875.

Quinn, T.J. II, and R. B. Deriso. 1999. Quantitative fish dynamics. Oxford University Press, New York.

Restrepo, V. R., G. G. Thompson, P. M. Mace, W. L. Gabriel, A. D. MacCall, R. D. Methot, J. E. Powers, B. L. Taylor, P. R. Wade, J. F. Witzig. 1998. Technical guidance on the use of precautionary approaches to implementing national standard 1 of the Magnuson-Stevens fishery conservation and management act. National Marine Fisheries Service, NOAA Technical Memorandum NMFS-F/SPO-31, Washington, D.C.

Richards, L. J., and J. J. Maguire. 1998. Recent international agreements and the precautionary approach: new directions for fisheries management science. Canadian Journal of Fisheries and Aquatic Sciences 55:1545–1552.

Schnute, J. T., and L. J. Richards. 1995. The influence of error on population estimates from catch-age models. Canadian Journal of Fisheries and Aquatic Sciences 52:2063–2077.

Schweigert, J. 2001. Stock assessment for British Columbia herring in 2001 and forecasts of the potential catch in 2002. Canadian Science Advisory Secretariat Research Document 2001/140.

Stocker, M. 1993. Recent management of the British Columbia herring fishery. Pages 267–293 *in* L. S. Parsons, and W. H. Lear, editors. Perspectives on Canadian marine fisheries management. Canadian Bulletin of Fisheries and Aquatic Sciences 226:267–293.

Thompson, G. G. 1999. Optimizing harvest control rules in the presence of natural variability and parameter uncertainty. National Marine Fisheries Service, NOAA Technical Memorandum NMFS-F/SPO-40.

Thompson, G. G., and P. M. Mace. 1997. The evolution of precautionary approaches to fisheries management, with focus on the United States. National Marine Fisheries Service, Northwest Atlantic Fisheries Organization SCR Document 97/26, Washington, D.C.

Ware, D. M. 1996. Herring carrying capacity and sustainable harvest rates in different climate regimes. Pacific Scientific Advice Review Committee Working Paper H96–3, Ottawa.

Ware, D. M., and J. Schweigert. 2001. Metapopulation structure and dynamics of British Columbia herring. Canadian Science Advisory Secretariat Research Document 2001/127, Ottawa.

Ware, D. M., and J. Schweigert. 2002. Metapopulation dynamics of British Columbia herring during cool and warm climate regimes. Canadian Science Advisory Secretariat Research Document 2002/107, Ottawa.

Zheng, J., F. C. Funk, G. H. Kruse, and R. Fagen. 1993. Threshold management strategies for Pacific herring in Alaska. Pages 141–165 *in* G. Kruse, D. M. Eggers, R. J. Marasco, C. Pautzke, and T. J. Quinn II, editors. Proceedings of the International Symposium on Management Strategies for Exploited Fish Populations. University of Alaska, Alaska Sea Grant Report 93–02, Fairbanks.

American Fisheries Society Symposium 49:241–258

Bycatch Reduction of Multispecies Bottom Trawl Fisheries: Some Approaches to Solve the Problem in the Northwest Pacific

ALEXEI M. ORLOV*
Russian Federal Institute of Fisheries and Oceanography (VNIRO)
17, V. Krasnoselskaya, Moscow 107140, Russia

Introduction

The term bycatch commonly refers to the part of a catch that is not the target species. Bycatch includes those fish captured that are undersized, prohibited, inedible, or unsalable. Different parts of the bycatch are sometimes called trash, discard, and incidental catch. Global estimates of discarded fish ranged from 17.9 to 39.5 million metric tons per year that constitutes roughly one-third of the total marine catch (Alverson 1996, 1997). Bycatch, the capture of unmarketable or restricted species during commercial fishing, is a world economic, environmental, and political problem (Alverson and Hughes 1996; Alverson 1999; Hall et al. 2000).

Bycatch problem attracted the attention of society, which may be demonstrated by a number of specific meetings that were held during the past decade: the California Sea Grant Workshop on Effects of Different Fishery Management Schemes on Bycatch, Joint Catch, and Discards (January 1990; San Francisco, USA); the National Industry Bycatch Workshop (February 1992; Newport, USA); the American Fisheries Society Symposium on Global Trends: Fisheries Management (June 1994; Seattle, USA); the Alaska Sea Grant Solving Bycatch Workshop (September 1995; Seattle, USA); the East Coast Bycatch Conference (April 1995; Rhode Island, USA); the American Fisheries Society Symposium on Consequences and Management of Fisheries Bycatch (August 1996; Deaborn, USA); the New York Sea Grant Bycatch Workshop (March 1999; Port Jefferson, USA), and so forth (Dewees and Ueber 1990; Schoning et al. 1992; Baxter and Keller 1996, 1997; Alverson 1996, 1997; Malchoff 1999). Many countries consider the bycatch problem to be a vital national issue. For instance, the United States developed a national bycatch strategy and the European Commission implemented action plans to reduce discards of fish (COM(2002)656) and to integrate environmental protection requirements into the common fisheries policy (COM(2002)186). Bycatch studies are of special interest to some international and regional fisheries management organizations. Thus, the Food and Agriculture Organization recently engaged in activity concerned with bycatch issues (Alverson et al. 1994; Grainger 1997). Several years ago, the International Council for the Exploration of the Sea established the Study Group on Discards and By-Catch Information. Recently, a bycatch subgroup within the frame of the Working Group on

*Corresponding author: orlov@vniro.ru

Fish Stock Assessment of the Commission for the Conservation of Antarctic Marine Living Resources was organized. Many fisheries research and commercial organizations conduct bycatch studies. The Commonwealth Scientific and Industrial Research Organization (Australia) completed a bycatch reduction project, the Mississippi State University (USA) carries out fisheries bycatch monitoring programs in the Gulf of Mexico, the Southwest Fisheries Science Center (USA) implements international collaboration–observer programs on bycatch reduction in South American fisheries (Chile, Peru, Ecuador) and in the U.S. Pacific longline and driftnet fisheries, the Southern and Northern Fisheries Centers (Australia) conduct research project on bycatch weight, composition, and preliminary estimates of the impact of bycatch reduction devices on Queensland's trawl fishery, and the Southern Fisheries Association, Inc. (USA) accomplished a bycatch reduction research program in the Gulf of Mexico and South Atlantic shrimp fisheries.

The problem of reduction of bycatches began to be addressed relatively recently, though most of this work has to do with the improvement of fishing gear and development of bycatch reduction devices (Perra 1992; Hickey et al. 1993; Keegan et al. 1998; Kennelly and Broadhurst 1998; Broadhurst et al. 2002; Davis 2002; Fonteyne and Polet 2002). Only a few recently published papers dealt with the more ample utilization of the species caught incidentally by developing processing techniques (Feidi 1989; Clucas 1999; Isaac and Braga 1999). Other biological and behavioral aspects of the problem have been examined to a lesser extent (Adlerstein and Trumble 1993; Anderson and Clark 2002; Walker et al. 2002).

Although our knowledge of discards in world fisheries has increased rapidly in the past two decades, the quality and detail of information are still lacking for many regions of the world (Alverson 1996). The highest quantities of discards are from the northwest Pacific, and catches of the bottom trawl fishery according to bycatch values are second in rank after that of the shrimp trawl fishery. Until now, bycatch studies of bottom trawl fisheries of the northwestern Pacific were not conducted. In Russian waters of the Far East seas, such investigations are at the initial stage only. Just recently, a regional research program entitled "Biostatistical bycatch studies of Far East specialized fisheries and development of management mechanisms for multispecies fishery" was established (Vinnikov and Terentiev 1999). However, results of this research are uncertain yet.

This paper, based on an independent author's study, provides some possible approaches to a solution of the bycatch reduction problem by example of a multispecies bottom trawl fishery in the Pacific waters off the northern Kuril Islands and southeastern Kamchatka, the northwest Pacific Ocean, Russia. The objectives of this study are to identify the prospective fishing species, which are not being utilized at present, to specify procedures for their processing and to optimize the areas, depths, seasons, and time of fishing in terms of minimizing the noncommercial bycatch.

Methods

This paper is based on catch data obtained from 19 bottom trawl surveys (a total of 1,481 bottom trawl stations between 1993 and 2000) conducted in the Pacific waters off southeastern Kamchatka and the northern Kuril Islands. The total area investigated was between 47°30' N and 52°20'N and between 154°30' E and 158°50' E (Figure 1). Catch data were obtained from three chartered commercial Japanese trawlers (*Tomi-Maru 53*,

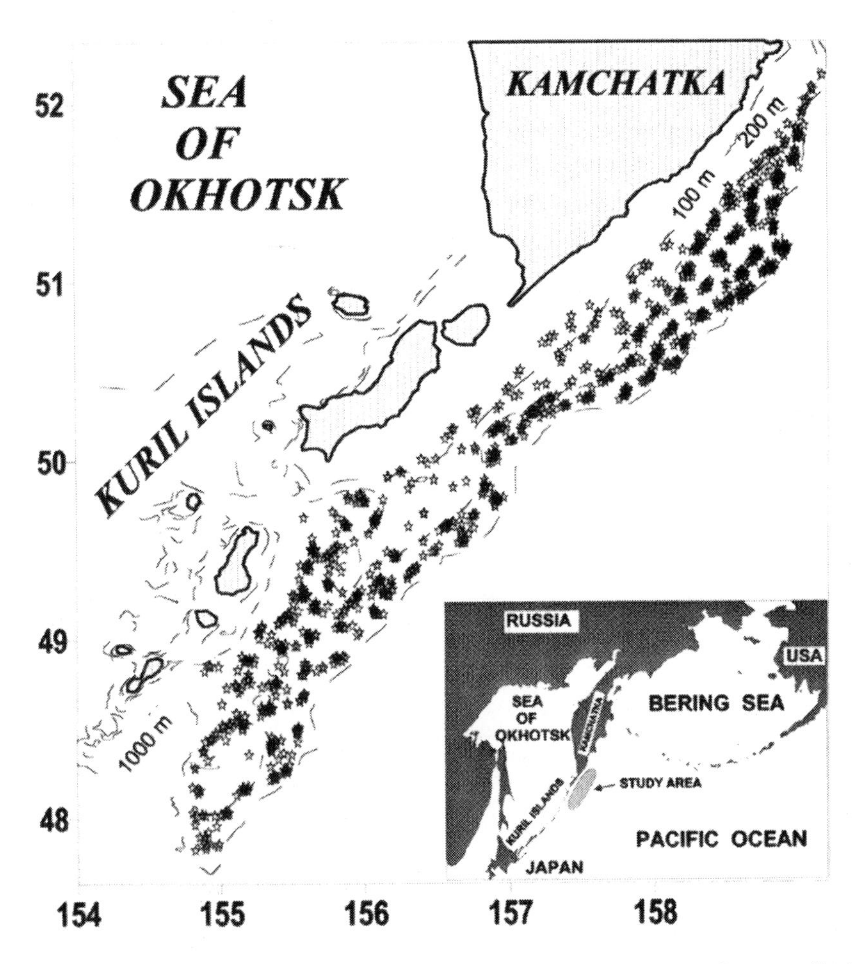

Figure 1. Locations of bottom trawl stations in the Pacific waters off the northern Kuril Islands and southeastern Kamchatka, 1993–2000 (dashed lines show isobaths).

Tomi-Maru 82, and *Tora-Maru 58*). Bottom trawl surveys were conducted during the daytime (0300–1900 hours, Tokyo time in spring; 0200–2000 hours in summer; and 0400–1900 hours in autumn) at depths between 78 and 836 m. Commercial types of polyethylene bottom trawl nets with 100–120 mm mesh size in cod end were used. The former mesh size was used during surveys conducted in the southern part of the study area, which is characterized by deeper depths and rough bottom. The latter mesh size was used for a survey of the northern part of study area, with rather shallower depths and a relatively flat bottom. According to current fishery regulation, the mesh size of cod end of commercial bottom trawls in the area surveyed should not be smaller than 100 mm. The trawl mouth, some 25–30 m wide and 5–7 m high, was rigged with a steel and rubber ball roller. Only samples from successful trawls were used for analysis in this study. This meant that the horizontal and vertical dimensions of the trawl mouth were within the normal range, the roller was constantly in contact with the bottom, the net suffered little or no damage during the tow, and no derelict fishing gear was encountered. All the hauls were made along isobaths. The duration of the hauls varied from 20 min to 1 h. All catches were recalculated by 1-h hauls. Percentage of the fishing species was determined for each catch, which included commercially important species (targeted species and commercial bycatch), as well as prospective unmarketable species and noncommercial bycatch. Those data were used for further analysis. To reveal geographic variations of catch composition, the study area was divided into three subareas (Figure 2): northern (52°20'–51°N), central (51°–49°30'N), and southern (49°30'–47°30'N). To detect fine local differences in catch composition, the study area was divided into 31 squares, 0.5° in latitudinal and longitudinal sides. Data on appropriateness of currently unmarketable species for direct consumption and processing in human, pet, and animal food are obtained from published papers, unpublished sources of Pacific Fisheries Research Center (Vladivostok, Russia), observations aboard Russian and Japanese trawlers, conversations with some fisheries managers, and attendance of Japanese fish restaurants.

Results and Discussion

The waters of the Pacific off the North Kurils and South Kamchatka feature a great diversity of species. More than 150 fish species have been recorded in bottom trawl catches (Orlov 1998), of which 20 are of commercial importance (Table 1). There are six species targeted by bottom trawls. The species complex of

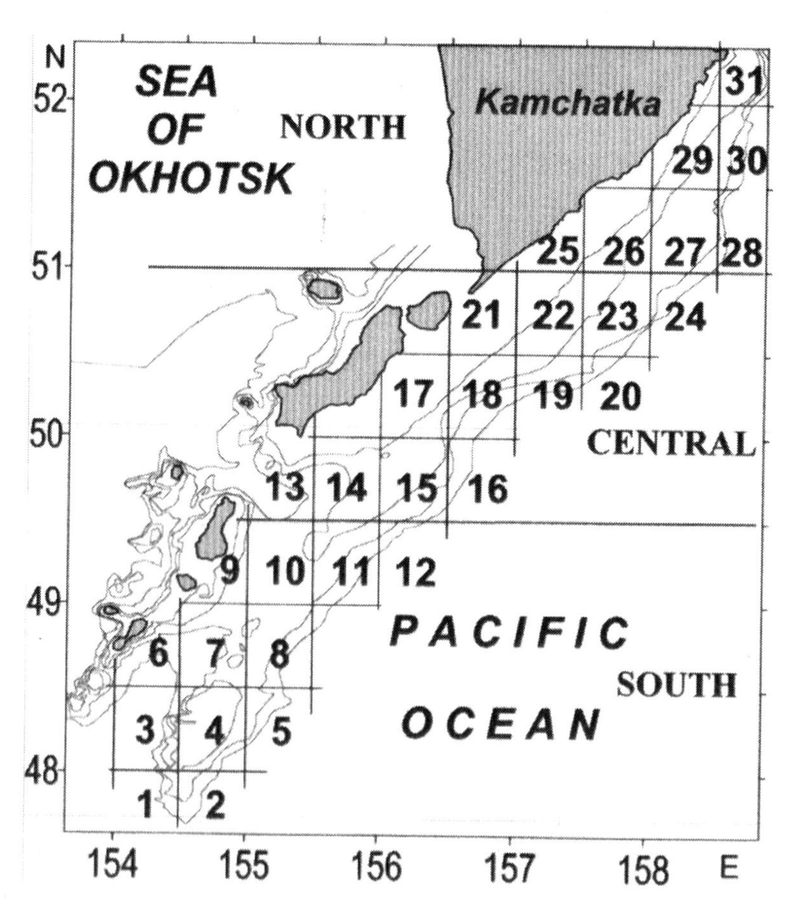

Figure 2. Map of study area showed borders of the northern, central, and southern subareas and distinguished squares (bold numbers) with 0.5° on side.

Table 1. Composition of catches during bottom trawl surveys in the Pacific waters off the northern Kuril Islands and southeastern Kamchatka, 1993–2000.

Common name	Scientific name	Abbreviation	Proportion of catch (%)
Walleye pollock	*Theragra chalcogramma*	WP	23.43
Pacific cod	*Gadus macrocephalus*	PC	2.04
Northern rock sole	*Lepidopsetta polyxystra*	NRS	3.53
Atka mackerel	*Pleurogrammus monopterygius*	AM	1.41
Pacific ocean perch	*Sebastes alutus*	POP	1.15
Red squid	*Berryteuthis magister*	RS	13.15
Shortraker rockfish	*Sebastes borealis*		
Rougheye rockfish	*S. aleutianus*		
Broadbanded thornyhead	*Sebastolobus macrochir*		
Shortspine thornyhead	*S. alascanus*	DWS	5.40
Kamchatka flounder	*Atheresthes evermanni*		
Greenland halibut	*Reinhardtius matsuurae*		
Sablefish	*Anoplopoma fimbria*		
Roughscale sole	*Clidoderma asperrimum*		
Pacific halibut	*Hippoglossus stenolepis*		
Flathead sole	*Hippoglossoides elassodon*		
Rock greenling	*Hexagrammos lagocephalus*	CBC	1.13
Grey rockfish	*Sebastes glaucus*		
Yellowfin sole	*Limanda aspera*		
Alaska plaice	*Pleuronectes quadrituberculatus*		
Giant grenadier	*Albatrossia pectoralis*		
Popeye grenadier	*Coryphaenoides cinereus*		
Pacific grenadier	*C. acrolepis*	PS	28.94
Longfin codling	*Laemonema longipes*		
Twoline eelpout	*Bothrocara brunneum*		
Whitebar eelpout	*Lycodes albolineatus*		
Tawnystripe eelpout	*L. brunneofasciatus*		
Soldatov's eelpout	*L. soldatovi*		
Giant sculpin	*Myoxocephalus polyacanthocephalus*		
Irish lords	*Hemilepidotus* spp.		
Spectacled sculpin	*Triglops scepticus*		
Armorhead sculpins	*Gymnocanthus* spp.		
Prowfish	*Zaprora silenus*		
Hawk poacher	*Podothecus sturioides*		
Dragon poacher	*Percis japonica*		
Aleutian skate	*Bathyraja aleutica*		
Whiteblotched skate	*B. maculata*		
Alaska skate	*B. parmifera*		
Matsubara skate	*B. matsubara*		
Whitebrow skate	*B. minispinosa*		
Okhotsk skate	*B. violacea*		
Sandpaper skate	*Rhinoraja interrupta*		
Mud skate	*R. taranetzi*		
Snailfishes	Liparidae		
Lumpsuckers	Cyclopteridae		
Soft sculpins	Psychrolutidae	NBC	19.82
Small sculpins	Cottidae		
Small poachers	Agonidae		

specialized deepwater fishes consists of at least eight species. Some species caught in small quantities constitute commercial salable bycatch that includes at least six species. Virtually half of the catch weight consists of presently unutilized though prospective species. These species are grenadiers, longfin codling *Laemonema longipes*, eelpouts, large and medium-sized sculpins, prowfish *Zaprora silenus*, skates, and large poachers. More than 20% of catches from the study area consisted of noncommercial species. These are snailfishes (*Liparis* spp., *Polypera simushirae*, *Careproctus* spp, *Elassodiscus* spp., *Crystallichthys* spp., *Paraliparis* spp., etc.), small sculpins (*Icelus* spp., *Artediellus* spp., etc.), soft sculpins (darkfin sculpin *Malacocottus zonurus*, spinyhead sculpin *Dasycottus setiger*), small poachers (*Sarritor* spp., blackfin poacher *Bathyagonus nigripinnis*), lumpsuckers (smooth lumpsucker *Aptocyclus ventricosus*, *Eumicrotremus* spp.), and some other species.

Involving Currently Unmarketable Species into Fisheries

Bycatch can be reduced significantly by engaging in fish of the stocks of species, which are actually not being utilized now. This would require the development of respective processing techniques.

Most promising in this connection are resources of grenadiers. Their biomass estimation along the Kuril Islands is several thousand metric tons, and daily average catches of large trawlers may comprise 20–30 metric tons (mt). All grenadiers may serve as source of delicious canned liver (Figure 3). Giant grenadiers *Albatrossia pectoralis* and Pacific grenadiers *Coryphaenoides acrolepis* have large orange eggs that can be used for processing to canned goods. Popeye grenadiers *C. cinereus* and Pacific grenadiers are characterized by high percentage of fillet (Table 2) and can be used for direct human consumption due to high content of protein in their meat (Table 3). Giant grenadier's flesh contains too much water and therefore is unsuitable for direct consumption but can be used for processing to animal and pet food. According to opinion of some Japanese specialists, the development of respective processing technology may allow to use grenadiers' flesh for treatment to surimi, which is currently manufactured from walleye pollock *Theragra chalcogramma* and Pacific cod *Gadus macrocephalus*. Current market prices of grenadiers are rather low (Table 4) but are comparable with those of arrowtooth flounder *Atheresthes stomias* and Dover sole *Microstomus pacificus*, which recently became harvested species in Alaskan waters and off the west coast of the United States (Orlov 2005).

Longfin codling is important target of Russian trawl fishery. This species is fished mostly in the Pacific off Japan and southern Kuril Islands, where recent total annual catch exceeds 30,000 mt. In study areas, mostly juveniles of longfin codling feed during the summer time (Savin 1993), and fish caught there are discarded. This species is in certain demand at Russian fish markets; it is used for direct human consumption. Since mass and chemical composition of longfin codling is similar to that of gadids (Tables 2 and 3), this species may be probably processed to surimi.

Other important idle fishery resources are skates, which biomass in the study area consisting of more than 10% of total fish biomass, recently amounting to about 40, 000 mt. Skates are traditional fishery targets in Southeast Asia, where mostly skate wings are used for consumption (Ishihara 1990). Skates are realized in fish markets in dried, frozen, and cooled conditions. In Russian fish markets, these off-center products are unpopular. However, Russian fishermen who have caught skates in study areas may supply skates to Japan,

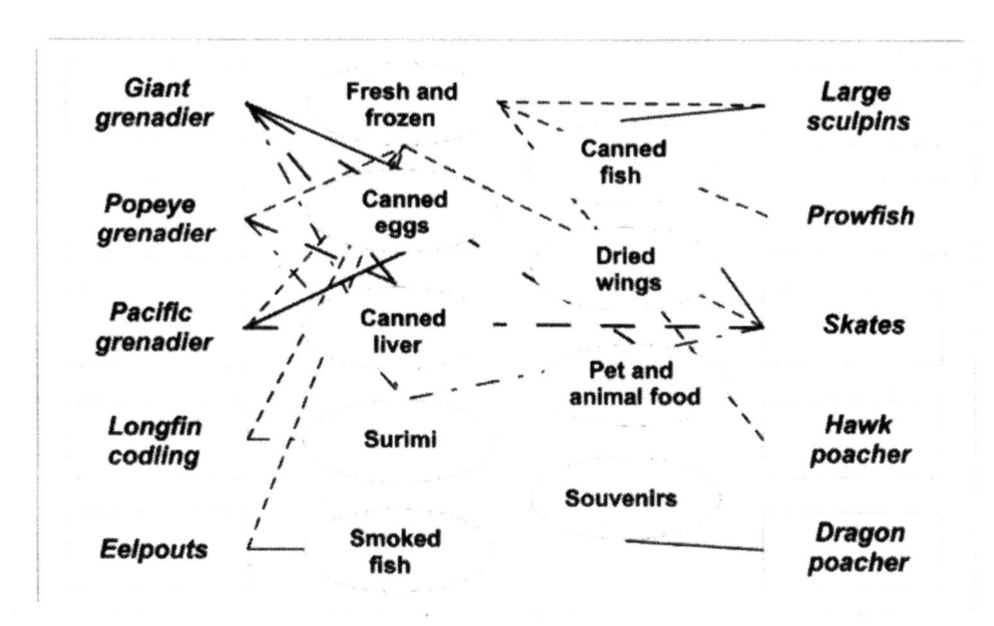

Figure 3. Scheme of possible processing directions of currently unmarketable prospective fish species.

Table 2. Mass composition (% by weight) of some presently unutilized species in the northwestern Pacific (N/A = not available).

Fish species	Head	Trunk	Filet	Bones	Fins	Skin with scale	Viscera
Pacific grenadier	30.6–37.2	48.4–53.2	31.5–40.1	7.2–10.5	1.6–1.4	5.8–6.4	13.0–14.6
Popeye grenadier	27.2–39.2	52.5–54.8	35.2–36.2	12.5–12.6	1.1–1.7	3.8–5.3	7.2–18.0
Giant grenadier	23.0–38.1	58.5–62.5	43.6	12.5	0.6–2.7	5.8	8.4–15.0
Longfin codling	21.0–25.0	61.0–67.0	40.0–52.6	12.0		N/A	9.0
Eelpouts	27.5–30.0	59.5–65.0	39.0–43.0	N/A	N/A	N/A	N/A

China, or Korea. Taking into account the rather high price of skates (Table 4), fishing of these species may be profitable. Skates' liver is rich in nutrients (though to a lesser extent in comparison with sharks) and therefore may be used for the extraction of oil. Skates have a strong and thick skin that, according to historical experience, may be used for the manufacture of various articles. According to results of Japanese studies (Ishihara 1990), the flesh of skates is suitable for making fish meat jelly and is used for the preparation of various Japanese national dishes. Recent Russian studies (Ershov et al. 2002) showed the appropriateness of skates for processing into human foodstuffs.

Table 3. Chemical composition (% by weight) of meat of some presently unutilized species in the northwestern Pacific

Fish species	Moisture	Fat	Protein	Ash
Pacific grenadier	80.2–86.0	0.2–0.4	13.1–16.5	0.5–1.5
Popeye grenadier	81.0–89.0	0.1–1.4	8.5–17.6	1.0–1.5
Giant grenadier	90.4–92.9	0.3–0.4	7.3–8.1	1.0–1.5
Longfin codling	83.0–86.5	0.5–4.6	11.0–14.0	1.3–1.6
Eelpouts	80.6–81.8	1.1–2.9	15.0–17.0	1.1–1.9

Table 4. Recent average market prices of of some presently unutilized species in the northwestern Pacific (according to World Wide Web information).

Species	Processing type	Japan, JPY per kg	USA, USD per lb
Grenadiers	Round		0.10–0.14
Skates	Round	650	0.10–0.30
	Fresh	800–1,800	

Large sculpins (family Cottidae) are important reserve of inshore fisheries. The most promising are abundant in the study area: yellow *Hemilepidotus jordani*, banded *H. gilberti* and longfin *H. zapus* Irish lords, purplegrey *Gymnocanthus detrisus* and armor-head *G. galeatus* sculpins, giant sculpin, and spectacled sculpin (Tokranov 1985, 2002). They comprise considerable bycatch of cod and flatfish fisheries (biomass of about 4,000 mt in the study area), but their fishery is undeveloped probably due to lack of customers' popularity. Though technology of processing of some northwestern Pacific sculpins were almost developed two decades ago (Didenko et al. 1983).

Various eelpouts (family Zoarcidae) are of certain interest for fishery. They are rather abundant in some areas and comprised a considerable part of the bycatch. Their biomass in the study area recently comprised about 2,500 mt. Results of technologic studies conducted by Pacific Fisheries Research Center (Vladivostok) showed the acceptability of eelpouts for processing into human food. Their flesh is suitable for smoking and processing to various foodstuffs.

Flesh of prowfish, having dense consistence and white coloration, looks like that of Pacific cod or Pacific halibut *Hippoglossus stenolepis*. Some Japanese fishermen used prowfish for preparation of sashimi, the national Japanese dish. Caught in the study area, prowfish that is associated with the low abundance of the species and their unappetizing appearance is mostly discarded.

Large poachers of genus *Podothecus* are of certain interest for fishery. Despite their relatively low abundance (biomass estimation in study area is about 600 mt), these poachers are very popular in Japanese fish markets. They are delivered to Japanese fish restaurants, where they are used for cooking of delicious dishes. Some poachers (for instance dragon poacher *Percis japonica*) having an exotic appearance may be used for manufacturing of souvenirs.

Bycatch in Various Parts of Study Area

Some geographic differences in bycatch quantities were observed (Figure 4). Minimum bycatch values are observed in 9–17 (off the fourth Kuril Strait), 20–22 (off the first Kuril Strait), and

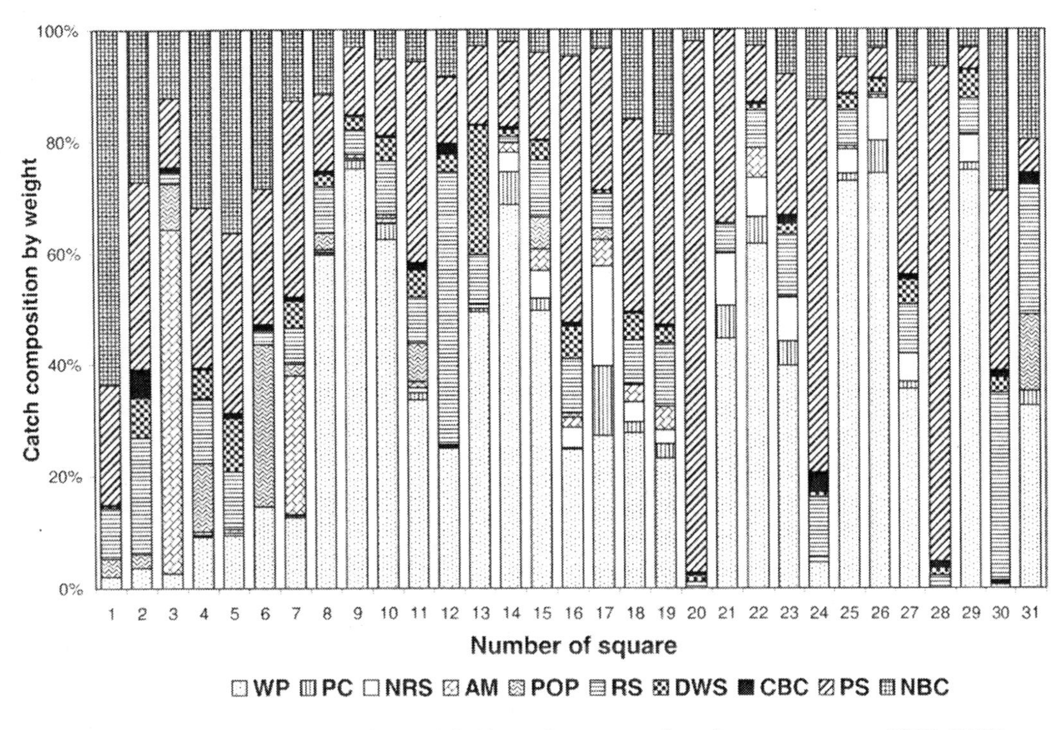

Figure 4. Catch composition (% by weight) in various part of study area, average 1993–2000.

25–29 (off the southeastern Kamchatka) squares. Bycatch values in the rest squares exceed 10% (mean 24.7%), and the share of commercially important species there comprised 46.5% of total catch weight. Maximum bycatch quantities were registered in 1–5 and 30–31 squares. These differences related mostly to distinctions in bottom relief and respective composition of ichthyofauna. Maximum values of noncommercial bycatch are observed within the areas with narrow shelf and sharp continental slope, where snailfishes and small sculpins are most abundant. In the southern part of the area (squares 1–8), several seamounts exist, where richness of ichthyofauna is essentially higher in comparison with that of the rest study area (Orlov 2003). This results in high proportion of bycatch in catches came from this area because the more versatile the species composition in the fishing area is, the greater the number of species is harvested, and as a rule, the smaller number of species is utilized accordingly.

Catch Composition Versus Area and Depth

Catch composition is changed with increasing of trawling depth that is due to vertical zonality of ichthyofauna (Shuntov 1965). Since noncommercial bycatch is composite group of various fish species, its proportion in catches varies with depth changes. Maximum (42.9%) and minimum (0.1%) bycatch values are noted in the northern subarea at depths of 450–500 and 100–150 m, respectively (Figure 5). In all three subareas, maximum bycatch values was registered at a 350–600 m depth range. In the northern subarea, a significant amount of noncommercial species were caught at depths of 250–350 m (12.6–15.9%) and 650–700 m (16.0%). In the southern subarea, a considerable share of bycatch was at 150–250 m depth (22.8–15.4%) as well. Presence of high quantities of noncommercial species at depths of 350–600 m is connected to the fact that overwhelming majority of noncommercial

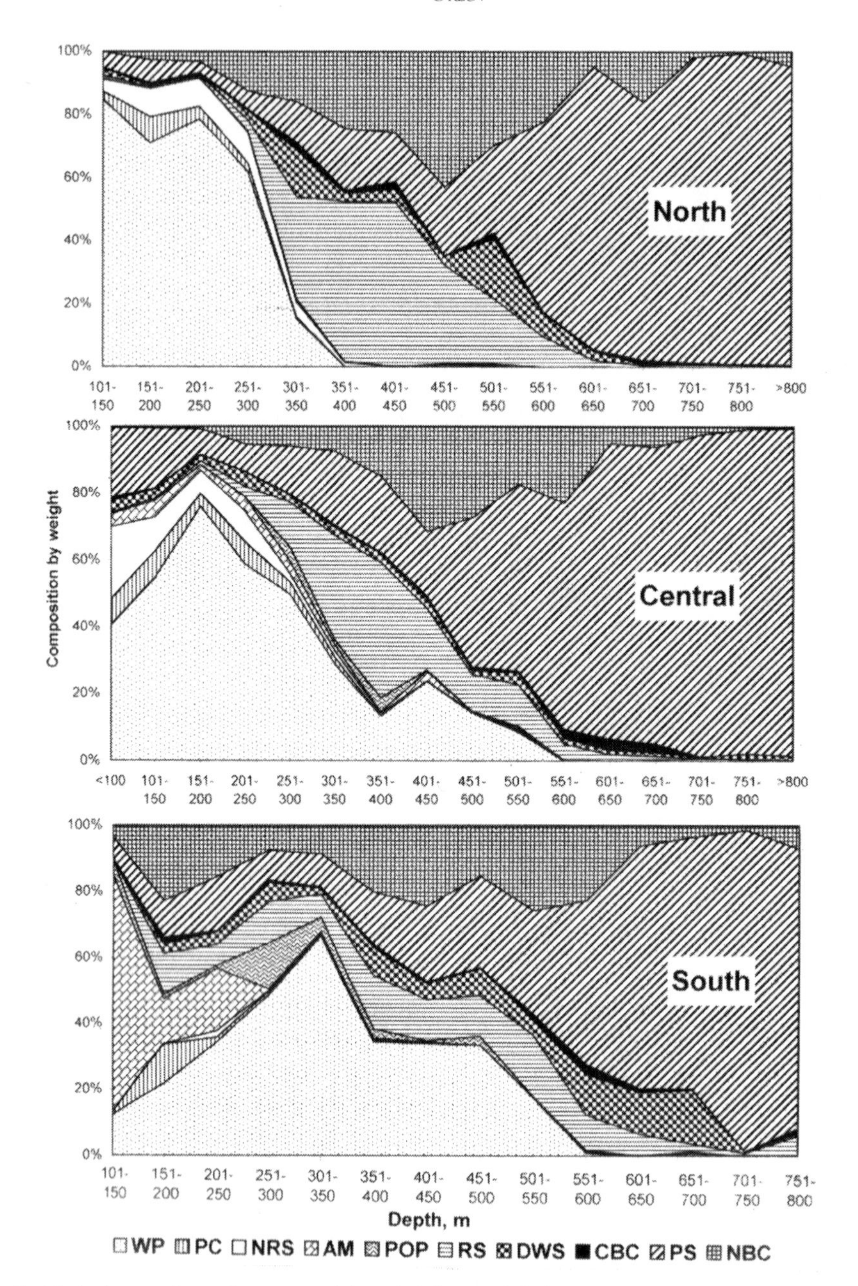

Figure 5. Variations of catch composition (% by weight) in study area by depths and subareas, average 1993–2000 (WP—walleye pollock, PC—Pacific cod, NRS—northern rock sole, AM—Atka mackerel, POP—Pacific Ocean perch, RS—red squid, DWS—deepwater species, CBC—commercial bycatch, PS—prospective species, NBC—noncommercial bycatch).

bycatch in study area is represented by snailfishes, which are mostly mesobenthal species (Orlov 1998) inhabited the above mentioned depth range.

Catch Composition Versus Season and Depth

Some differences in bycatch proportions at various depths were observed also in seasonal aspect (Figure 6). Maximum bycatch value

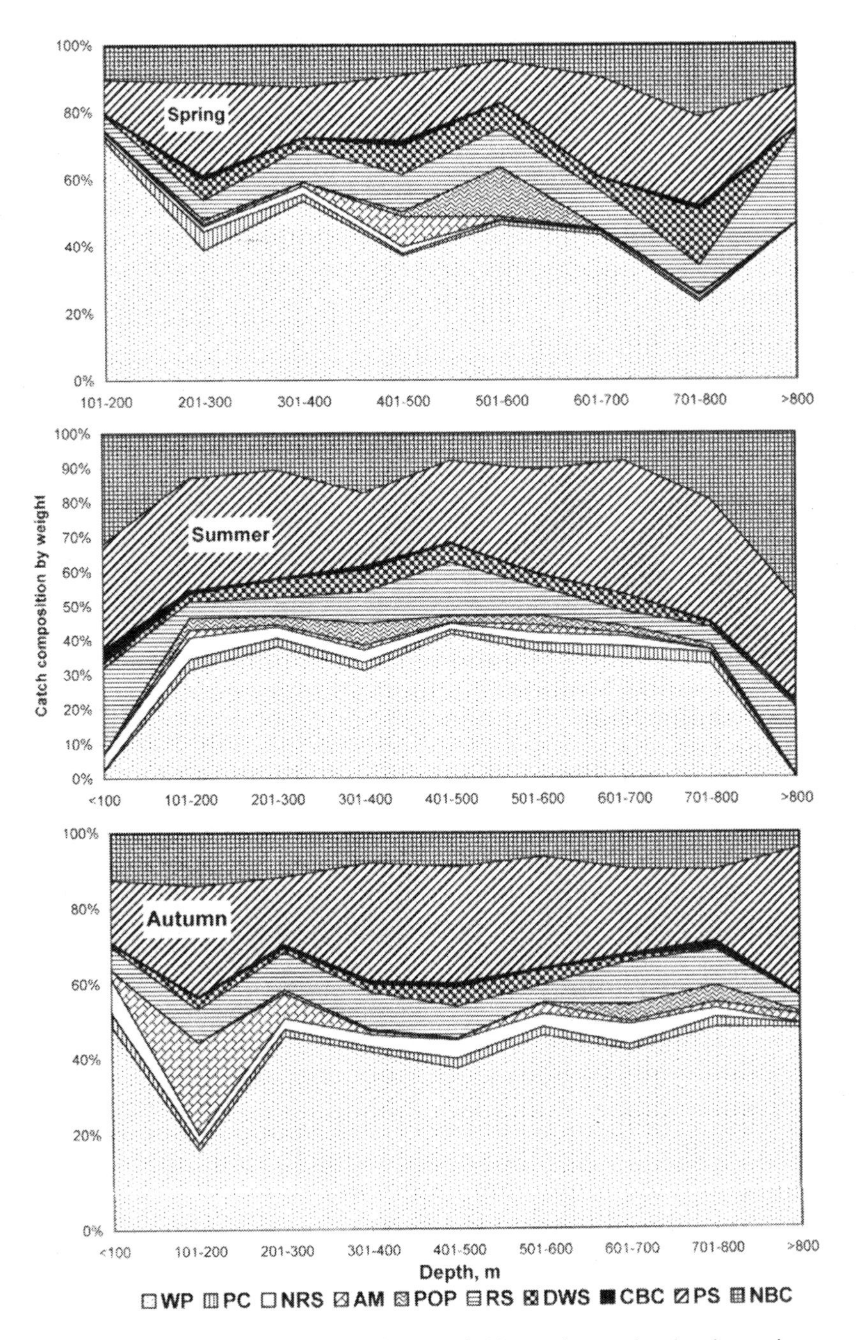

Figure 6. Variations of catch composition (% by weight) in study area by depths and seasons, average 1993–2000 (fish legends are the same as in figure 5).

(48.2%) was noted in summer at depth of greater than 800 m; minimum bycatch value (4.2%) was at the same depth but in autumn. In spring, minimum bycatch values (<10%) were noted at depths of 400–700 m, in summer—400–500 m, and in autumn—deeper than 300 m. These differences are probably related mostly to changes of seasonal vertical distribution affected by vertical migrations of fishes. In the study area, due to geographic location, biological and calendar periods do not coincide. Therefore, autumn may be considered to be a warm period and the proportion of noncommercial fishes at shallower depths is high.

Catch Composition Versus Area and Season

Analysis showed that seasonal changes of bycatch quantities are observed in various parts of study area as well (Figure 7). In the northern subarea, summer and autumn bycatch values were almost equal. In the central subarea, the maximum proportion of bycatch was in spring (20.9%) while in the southern part, the maximum proportion was in summer and autumn (16.8–15.0%). This suggests probable horizontal migrations of noncommercial species from the central to southern subarea during the spring–summer period. Seasonal migrations of fishes in the area under consideration are poorly studied; though migrations of adult walleye pollock from the northern subarea to the southern one are known (Buslov 2001), that is probably due to better feeding conditions in the latter area (Orlov 2003).

Catch Composition Versus Season and Time of Day

Some groundfishes perform diurnal vertical migrations. Such behavioral pattern is typical mostly for planktophage species such as the Atka mackerel *Pleurogrammus monop-terygius* and the Pacific ocean perch *Sebastes alutus* (Figures 8 and 9). Behavioral patterns of noncommercial species are little understood. The analysis showed that there are no well-pronounced changes in bycatch quantities during the daytime (Figure 8). Minimum bycatch values (10%) were noted in spring from 0500 to 0900 hours and in autumn from 0700 to 1000 hours and from 1600 to 1900 hours. There were no patterns found in the summer time, though maximum bycatch values in this period were observed from 0200 to 0500 hours.

Catch Composition Versus Area and Time of Day

Overall changes of bycatch values depending on time of the day in various parts of the study area were not pronounced (Figure 9). However, in the central subarea, there were two periods with minimum bycatch proportions (<10%): 0400–0700 hours. and 1300–1600 hours.

Annual Variations of Catch Composition

The analysis of multiannual data of catch composition showed that proportions of different species and species groups have changed during the study period (Figure 10). The share of noncommercial species in catches increased almost twice (from 7.1% in 1993 to 13.4% in 2000). The proportion of prospective species also increased substantially (from 16.5% to 27.4%). There are two possible reasons of such changes. The first cause is related to a decline in abundance of walleye pollock, Pacific cod, and northern rock sole *Lepidopsetta polyxystra*, which comprised more than one-fourth of the total catch by weight. Since the annual harvest of these species did not exceed recommended total allowable catch values, their declining abundance is most likely due to natural fluctuations. On the other hand, the rise of proportion of prospective and noncommercial species in catches may relate to increasing of abundance of some species within these two groups. Thus, during the study period, a dramatic increase in skate abundance occurred (Orlov 2004) that probably substantially contributed to the rise in overall proportion of prospective species in catches. Also, a significant increase in abundance of snailfishes and sculpins was observed (Orlov 2004), which is a possible reason for the recent increase in proportion of noncommercial bycatch.

Conclusions

Bycatch can be reduced significantly by engaging into the fishery of the stocks of grena-

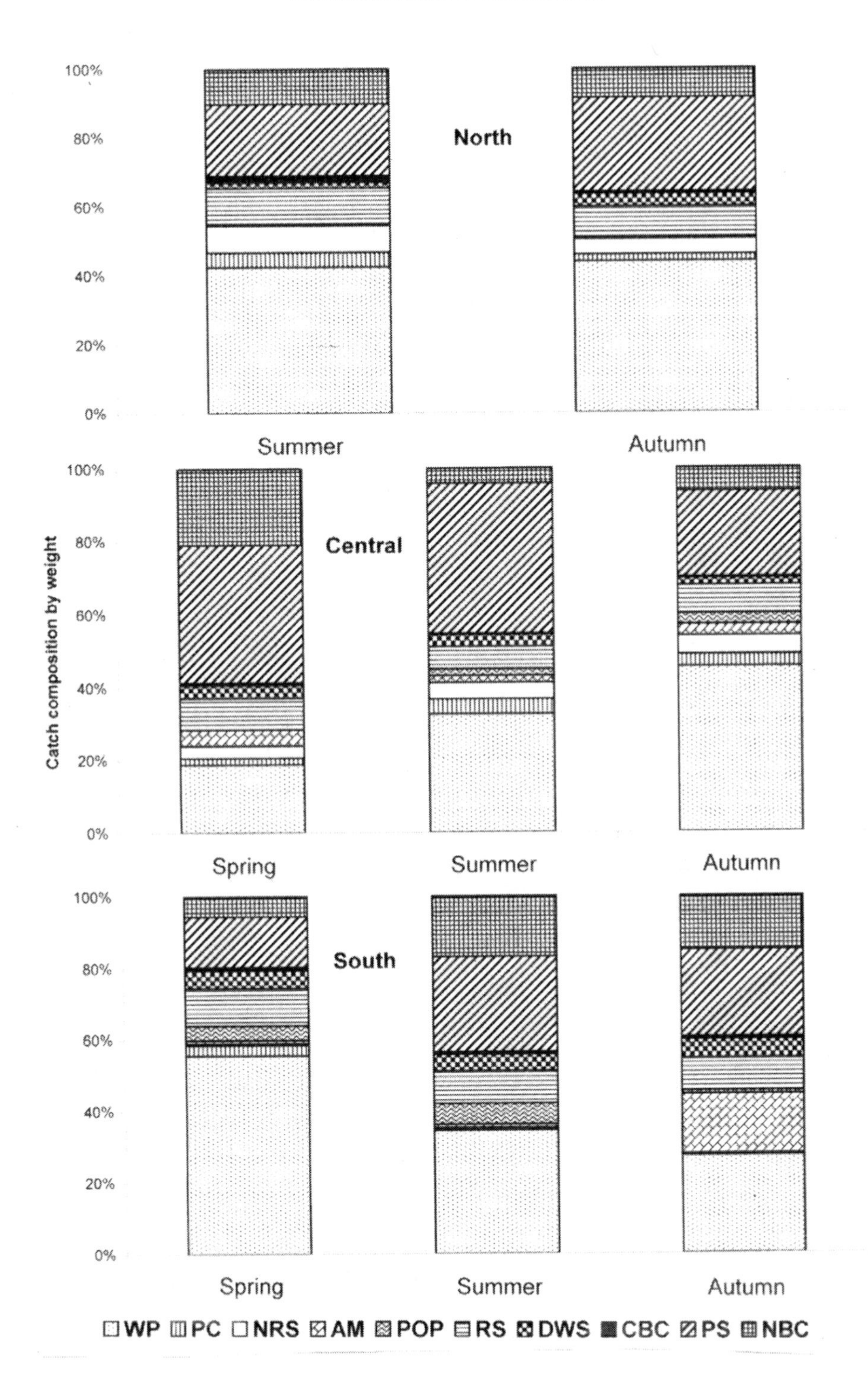

Figure 7. Variations of catch composition (% by weight) in study area by seasons and subareas, average 1993–2000 (fish legends are the same as in Figure 5).

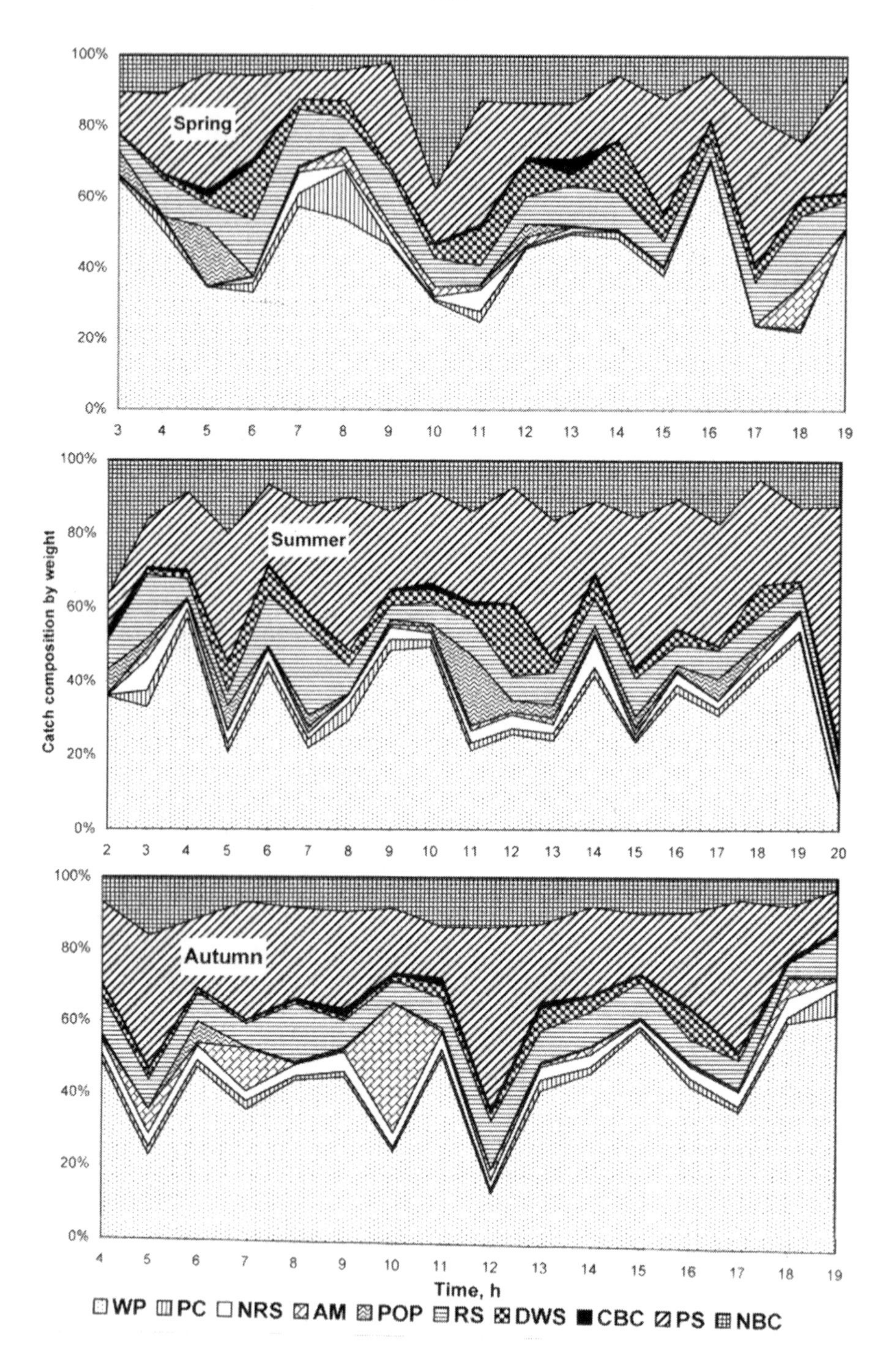

Figure 8. Variations of catch composition (% by weight) in study area by time of day and seasons, average 1993–2000 (fish legends are the same as in Figure 5).

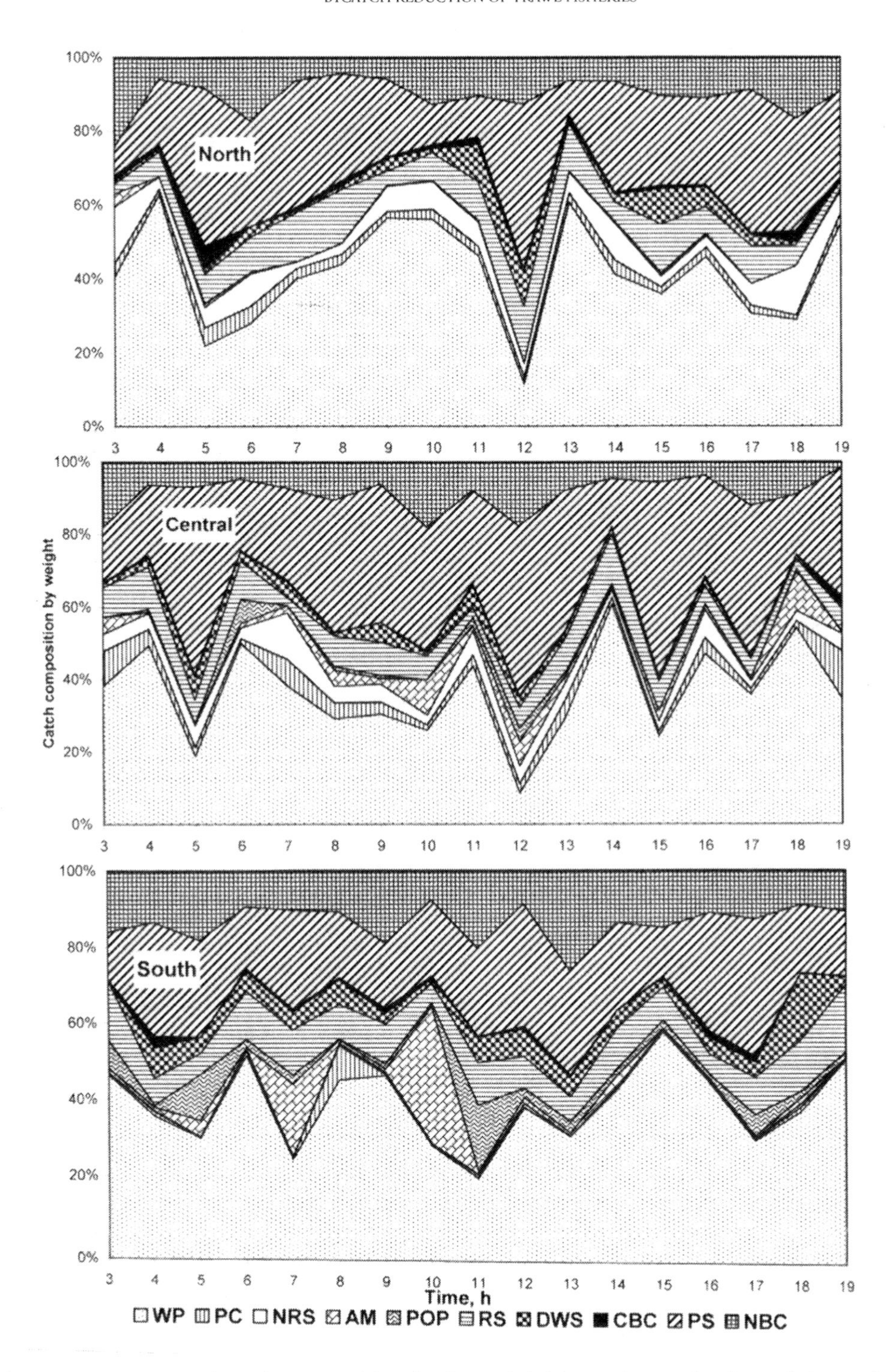

Figure 9. Variations of catch composition (% by weight) in study area by time of day and subareas, average 1993–2000 (fish legends are the same as in Figure 5).

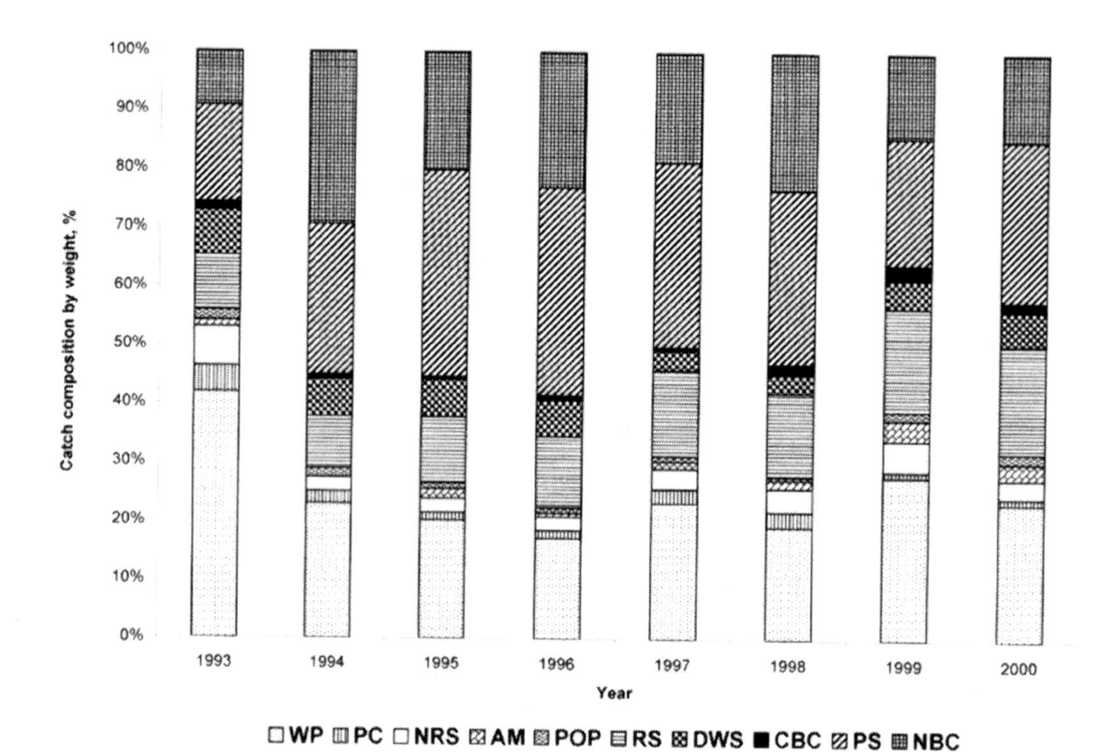

Figure 10. Annual variations of catch composition (% by weight) in study area, average 1993–2000 (fish legends are the same as in Figure 5).

diers, eelpouts, skates, large sculpins, poachers, and some other species that are actually not being utilized now. These species cannot compete with traditional harvested species, but most of them may be used for direct human consumption; this would be required to develop respective processing techniques. Taking into account the abundance of many decreases in commercially important species, the involvement of currently unutilized species in fisheries becomes very real.

Possibilities for reducing bycatch are opened up by the right choice of optimum fishing areas, depths, seasons, and time of the day, which would minimize the share of noncommercial species in catches.

Since this paper only defines some approaches to the problem of diminishing bycatch, future studies should continue to search for promising fishing objects to be involved in fisheries, conduct bottom trawl surveys in winter and spring (December through April), and during periods of darkness, and carry out more detailed analyses of existing and newly obtained data.

References

Adlerstein, S. A., and R. J. Trumble. 196. Management implications of changes in by- catch rates of Pacific halibut and crab species caused by diel behaviour of groundfish in the Bering Sea. Pages 211–215 *in* C. S. Wardle and C. E. Hollingworth, editors. Fish behaviour in relation to fishing operations. ICES Marine Science Symposia 196.

Alverson, D. L. 1996. Discarding: a part of the management equation. Pages 11–13 *in* C. Castro, T. Corey, J. DeAlteris, and C. Gagnon, editors. Proceedings of the East Coast Bycatch Conference. University of Rhode Island, Rhode Island Sea Grant, RIU-W-95-001, Narragansett.

Alverson, D. L. 1997. Global assessment of fisheries bycatch and discards: a summary overview. Pages 115–125 *in* E. K. Pikitch et al., editors. Global trends: fishery management. American Fisheries Society, Symposium 20, Bethesda, Maryland.

Alverson, D. L. 1999. Some observations on the science of bycatch. Marine Technology Society Journal 33(2):6–12.

Alverson, D. L., and S. E. Hughes. 1996. Bycatch: from emotion to effective natural resources management. Reviews in Fish Biology and Fisheries 6(4):443–462.

Alverson, D. L., M. H. Freeberg, S. A. Murawski, and J. G. Pope. 1994. A global assessment of fisheries bycatch and discards. FAO Fisheries Technical Paper 339:1–233.

Anderson, O. F., and M. R. Clark. 2002. Analysis of bycatch in the fishery for orange roughy, *Hoplostethus atlanticus*, on the south Tasman Rise. Marine and Freshwater Research 54(5):643–652.

Baxter, B., and S. Keller editors. 1996. Solving bycatch: considerations for today and tomorrow. University of Alaska, Alaska Sea Grant College Program Report No. 96–03, Fairbanks.

Baxter, B., and S. Keller, editors. 1997. Fisheries bycatch: consequences and management. University of Alaska, Alaska Sea Grant College Program Report No. 97–02, Fairbanks.

Broadhurst, M. K., S. J. Kennelly, and C. A. Gray. 2002. Optimal positioning and design of behavioural type by-catch reduction devices involving square-mesh panels in penaeid prawn-trawl codends. Marine and Freshwater Research 53(4):813–823.

Buslov, A. V. 2001. New data on distribution and migrations of walleye pollock in the Pacific waters off the northern Kuril Islands and southeastern Kamchatka. Izvestiya TINRO 128:177–187, Vladistok (In Russian).

Bykov, V. P., G. P. Ionas, G. N. Golovkova, et al. 1998. Guide to chemical composition and technological properties of marine fishes. VNIRO Publishing, Moscow.

Clucas, I. 1999. By-catch: is it a bonus from the sea? INFOPESCA Internacional 1:33–37, Montevideo, Uruguay (In Spanish with English summary).

Davis, M. W. 2002. Key principles for understanding fish bycatch discard mortality. Canadian Journal of Fisheries and Aquatic Sciences 59:1834–1843.

Dewees, C. M., and E. Ueber, editors. 1990. Effects of different fishery management schemes on bycatch, joint catch, and discards. Summary of a national workshop, San Francisco, California, January 1990. University of California, California Sea Grant College Program Report Series No. T-CSGCP-019, La Jolla.

Didenko, A .P., G. A. Borovskaya, L. I. Drozdova, and L. A. Lavrova. 1983. Techno-chemical characteristic and recommendations on rational use of sculpins. Izvestiya TINRO 108:13–19 (In Russian).

Ershov, A. M., B. F. Petrov, V. V. Korchunov, and O. A. Nikolaenko. 2002. Problems of using of thorny skate for human food purposes. Pages 101–102 *in* Prospects of development of Russian fisheries complex—XXI century. Abstracts of Presentations at the Scientific-Practical Conference, Moscow, 27–28 June 2002. VNIRO Publishing, Moscow.

Feidi, I. 1989. Economic utilization of fish by-catch and by-products in the Arab gulf region. Paper prepared for Seminar on economic utilization of waste. INFOSAMAK/FAO, Jeddah, Saudi Arabia.

Fonteyne, R., and H. Polet. 2002. Reducing the benthos by-catch in flatfish beam trawling by means of technical modifications. Fisheries Research 55(1–3):219–230.

Grainger, R. 1997. Recent FAO activities related to by-catch and discard issues. Pages 188–191 *in* V. T. Sulit, editor. Fishery and aquaculture statistics in Asia. Proceedings of the FAO/SEAFDEC Regional Workshop on Fishery Statistics. Volume 2. SEAFDEC, Bangkok.

Hall, M. A., D. L. Alverson, and K. I. Metuzals. 2000. By-catch: problems and solutions. Marine Pollution Bulletin 41(1–6):204–219.

Hickey, W. M., G. Brothers, and D. L. Boulos. 1993. By-catch reduction in the northern shrimp fishery. Canadian Technical Report of Fisheries and Aquatic Sciences 1964.

Isaac, V. J, and T. M. P. Braga. 1999. By-catch in the marine fisheries off northern Brazil. Arquivos de Ciencias do Mar 32:39–54 (In Portuguese with English summary).

Ishihara, H. 1990. The skates and rays of the western North Pacific: an overview of their fisheries, utilization, and classification. Pages 485–497 *in* H. L. Pratt, Jr, S. H. Gruber, and T. Taniuchi, editors. Elasmobranchs as living resources: advances in the biology, ecology, systematics, and the status of the fisheries. NOAA Technical Report NMFS 90:485–497.

Keegan, B. F, B. van Marlen, M. J. N. Bergman, W. Zevenboom, R. Foteyne, K. Lange, and J. Browne. 1998. Reduction of adverse environmental impact of demersal trawls. In Third European Marine Science and Technology Conference (MAST Conference), Lisbon 23–27 May 1998. Luxembourg, European Commission. Project Synopses 6:313–314.

Kennelly, S., and M. Broadhurst. 1998. Reducing by-catch: we show the world. Fisheries NSW 4:4–5.

Malchoff, M. C., editor. 1999. Proceedings of the Sea Grant bycatch workshop. New York Sea Grant Extension Program, Riverhead, New York.

Orlov, A. M. 1998. Demersal ichthyofauna of Pacific waters around the Kuril Islands and southeastern Kamchatka. Biologiya Morya 24(3):144–160 (In Russian).

Orlov, A. M. 2003. Impact of eddies on spatial distributions of groundfishes along waters off the northern Kuril Islands, and southeastern Kamchatka (north Pacific Ocean). Indian Journal of Marine Sciences 32(2):95–113.

Orlov, A. M. 2004. Present state, periodical changes of composition, fishery potential, and prospects of commercial exploitation of fish communities of the upper bathyal of Pacific Ocean off Kuril Islands and Kamchatka. Pages 2–34 *in* Aquatic biological resources, their state and use: analytical and abstract information. Issue 1. VHIERKh, Moscow (In Russian).

Orlov, A. M. 2005. Groundfish resources of the northern North Pacific continental slope: from science to sustainable fishery. Pages 139–150 *in* Proceedings of the Seventh North Pacific Rim Fisheries Conference. May 18–20, 2004. Busan, Republic of Korea. Alaska Pacific University, Anchorage.

Perra, P. 1992. By-catch reduction devices as a conservation measure. Fisheries 17(1):28–30.

Savin, A.B. 1993. Distribution and migrations of longfin codling *Laemonema longipes* (Moridae) in the northwestern Pacific Ocean. Voprosy Ikhtiologii 33(2):190–197 (In Russian).

Schoning, R. W., R. W. Jacobson, D. L. Alverson, T. G. Gentle, and J. Auyong, editors. 1992. Proceedings of the National Industry Bycatch Workshop February 4–6, 1992, Newport Oregon. Natural Resources Consultants, Seattle.

Shuntov, V. P. 1965. Vertical zonality in distribution of fishes in the upper bathyal of the Sea of Okhotsk. Zoologicheskii Zhurnal 44(11):1678–1689 (In Russian).

Tokranov, A. M. 1985. Sculpins—prospective targets of inshore fishery. Rybnoye Khozyaistvo 5:28–31 (In Russian).

Tokranov, A.M. 2002. Untraditional fishery resources: is the use of their stocks realizable today? Rybnoye Khozyaistvo 6:41–43 (In Russian).

Vinnikov, A. V., and D. A. Terentiev. 1999. Main directions of bycatch studies during conduction of various fisheries in coastal Kamchatkan waters. Pages 117–119 *in* Fisheries studies of the world oceans. Proceedings of the International Scientific Conference. Far East Fisheries University, Vladivostok, Russia (In Russian).

Walker, T. I., R. J. Hudson, and A. S. Gason. 2002. Catch evaluation of target, by- product, and by-catch species taken by gillnets and longlines in the shark fishery of south-eastern Australia. Northwest Atlantic Fisheries Organization, NAFO Scientific Council Research Document no. 02/114, Dartmouth.

American Fisheries Society Symposium 49:259–260

Session Summary

Fisheries Trade and Reconciling Fisheries with Conservation

MAFUZUDDIN AHMED AND ROEHLANO BRIONES
World Fish Center
Penang, Malaysia

International trade, by integrating fish demand and fish supplies on a global scale, leads to an increase in benefits for competitive producers, while intensifying patterns of exploitation in many fishing areas. Most of the papers in the session contributed to the understanding the relationship of trade and reconciling fisheries with conservation (RFC) by analyzing trends and impacts of global trade as well as the implications of environment and conservation concerns on trade.

During the oral presentations, the paper by Ahmed et al. set the tone by discussing the changing structure of fish supply, demand, and trade, with a focus on developing country needs. A related paper by Dey et al. (presented by Briones) refined the analysis using the AsiaFish model, which examined the impacts of declining fish stocks, technological change, and rising export costs. Meanwhile in the poster session, the papers by Garcia and Rodriguez as well as Kumar presented further applications of the AsiaFish model. The former evaluated the impact of recent tariff reforms, while the latter presented baseline projections for supply, demand, and trade in the case of India. In the case of developed countries, Keithly and Diop showed how rising imports have caused structural changes in the shrimp processing sector of the southwestern United States. Marsden and Sumaila meanwhile inspected data for Canada to trace the evolution of trade trends, linkages with global trading agreements, as well as the role of trade in both hindering and helping the reconciliation of fisheries with conservation in Canadian waters. A more specific issue was highlighted in Hunt and Vincent, who focused on rising global demand for fisheries for pharmacological applications—now a major non-food use of aquatic resources. They highlighted the current gaps of national and international policy instruments in addressing this new form of commercial exploitation. Wabnitz and Collete described their efforts at compiling a global marine aquarian database, which is potentially useful for further studies on the relationship between trade and RFC. Finally, Jensen applied an institutional expository analysis to argue that the community is the sole arbiter of values (e.g., production versus conservation), apparently nullifying the global marketplace as a player in determining those values.

Other papers focused on environmental and conservation concerns and their implications for fisheries trade. During the oral session, Clarke et al. argued that trade-induced intensity of exploitation in the case of shark fin trade is apparently unsustainable. They highlighted the method for extrapolating from fragmentary data to arrive at quantitative information of trade flows that can be used for fisheries management. Foster and Vincent examined the Convention on International Trade in Endangered Species (CITES) regulations on seahorse trade and recommended a management measure in the form of minimum size limits. Wallace and Weeber made a critical evaluation of ecolabelling, in the form of the Marine Stewardship Council (MSC) certification scheme in the New Zealand hoki *Macruronus novaezelandiae* fisheries. Meanwhile, in the poster session, two papers extended the analysis of the role of CITES and ecolabelling. Sack et al. advocated inclusion of Antarctic toothfish *Dissotichus mawsoni* and Patagonian tootfish *D. eleginoides* in the CITES as a tool to prevent overexploitation of this fishery. They emphasized the need for sound information on the gains and circumspect assessment of potential losses arising from CITES inclusions. In the case of ecolabelling, Grieve et al. undertook a review of MSC methods and experience to argue that ecolabelling is, within limits, a viable tool for conservation.

The last set of papers in this session dealt with broader issues relating to the economic analysis of environmental concerns. Modeling and simulation analyses were the methods used in papers by Suzy and Fauzi, Fauzy and Suzy, Granzotto and Laukkanen, as well as Igumnova and Timchenko. The first included pollution in a standard bioeconomic model calibrated to the case of Jakarta Bay. The second looked at alternative specifications of welfare objectives in the optimal control representation of fisheries management problem. The third developed an economic-ecosystem model to value the externalities associated with fishing. The fourth presented a dynamic-stochastic modeling of sea–land ecological–economic systems to show how economic control parameters can be applied to analyze the RFC issue. Finally, a more conceptual approach was taken by Ward and Thunberg, to highlight the distinction between excess capacity and overcapacity and the need for fisheries managers to focus on overcapacity.

American Fisheries Society Symposium 49:261–264

Canada's International Marine Fish Trade Since 1950: Volume, Value, and Implications for Conservation

DALE MARSDEN* AND USSIF RASHID SUMAILA

University of British Columbia, Fisheries Centre, Fisheries Economic Research Unit and the Sea Around us Project 2202 Main Mall, Vancouver, British Columbia V6T 1Z4, Canada

Abstract.—We constructed a database of Canada's imports and exports of fisheries products from 1950 to 2001 using government landings and customs data. We describe how these flows of products have changed over time, and how the flows relate to international trade policy. We found a doubling of export quantity and a fivefold increase in export value over the study period. Import quantity increased 16-fold, while import value increased 59-fold over the same period. In 2001, Canada had a strong net export of fish in value terms (Can$2.1 billion), but imported 7,000 metric tons more than it exported. Much of this increase in trade is at least partly attributable to sharp reductions in tariffs. We discuss how international trade and trade agreements might both help and hinder the reconciliation of Canadian fisheries with conservation.

Introduction

The ability to trade fish internationally will affect the price that fishers and processors can obtain for their produce, and may therefore influence which fisheries are exploited and the intensity of exploitation. This may in turn partly determine the health and sustainability of the fishery and its associated ecosystems. To begin to examine these issues, we compiled detailed data on Canada's international trade in fisheries products from 1950 to 2001, examined how these flows were affected by international trade policy, and how the flows might affect ecosystems in Canada and abroad.

Methods

We compiled a database of Canada's imports and exports of fisheries commodities from official customs data (Statistics Canada 1950–1965; 1950–1967; 1966–1984; 1968–1983; 1984–1987; 1985–1987; 1988–2001). These data specify the commodity traded, the source or destination country, and the weight and value of the commodity traded. Import data also include the amount of customs duty collected on each item. We included all marine finfish, mollusks, crustaceans, and echinoderms in our analysis, but excluded marine mammals, freshwater fish and all plants. Quantities are reported as the traded quantity, as opposed to the live weight equivalent. To account for inflation, dollar values are corrected to real 2002 Canadian dollars using the Consumer Price Index.

Results and Discussion

International Trade Patterns

The quantity of exports approximately

*Corresponding author: d.marsden@fisheries.ubc.ca

doubled between 1950 and 2001, while the real value of exports increased fivefold (Figure 1). Demersal fish was the dominant type of export in terms of quantity and value for most of the study period. However, lobster and crab exports expanded through the 1990s so that they were the largest category in value terms from 1999 onward.

Until recently, Canada's imports were minimal in both quantity and value terms. Through the 1990s, however, imports of demersal fishes, shrimp, and "miscellaneous" items (those that were not specified taxonomically in the data) increased dramatically, leading to a tripling of import quantity and a 2.3-fold increase in import value since 1990. A substantial portion of the increase in miscellaneous imports is attributable to imports of fish meal, which rose from 14,000 metric tons in 1988 to 103,000 metric tons in 2001. Much of this (75% in 2001) is designated for use in manufacturing fish feeds, presumably for use in aquaculture operations, and is imported mostly from Peru and Chile.

We combined import and export data to calculate the trade balance, which is calculated as exports minus imports, or net export (Figure 2). The trade balance in terms of value is strongly positive, and has been for the duration of the study period. There has been a sharp decrease in the trade balance in terms of quantity since the 1990s, however, to the point where in 2000 and 2001 Canada was a net importer of fisheries products (although the magnitude of the net import is quite small).

Links with Trade Policy

International trade policy is related to recent trade patterns in a number of ways. As part of Canada's 1988 Free Trade Agreement with the United States and the 1994 North American Free Trade Agreement, almost all tariffs on fish trade with the United States have been removed. The average duty collected on imports from the United States has declined from 1.1% of import value in 1988 to less than 0.01% in 2001, thereby facilitating the expansion of Canada's imports from the United States. Similarly, as part of the Uruguay Round agreements under the General Agreement on Tariffs and Trade (GATT) and the World Trade Organization (WTO), tariffs collected on imports from other countries have been reduced from 4.7% in 1988 to 0.5% in 2001. National trade policies, such as the export-orientation of Thailand, China, and Chile have also contributed to the increase in Canada's fish imports from these countries in the last 15 years. We recognize, however, that these changes in trade policy are probably not large enough to completely explain the changes in trade patterns observed; other factors affecting supply and demand have surely played a role as well.

Implications for Conservation

The effects that Canada's international fisheries trade might have on marine ecosystems are ambiguous in this study and require more empirical work. The shift away from being a net exporter in quantity terms might be beneficial for Canadian ecosystems if it results in a decreased dependence on domestic resources. If domestic shortfalls of edible protein or fish meal for aquaculture are met with imports from other regions, though, this might be seen as effectively shifting the environmental burden of Canada's fish consumption onto other countries. There is nothing inherently objectionable in such an approach if the fisheries that provide the imported fish are well managed to prevent serious damage to the ecosystems

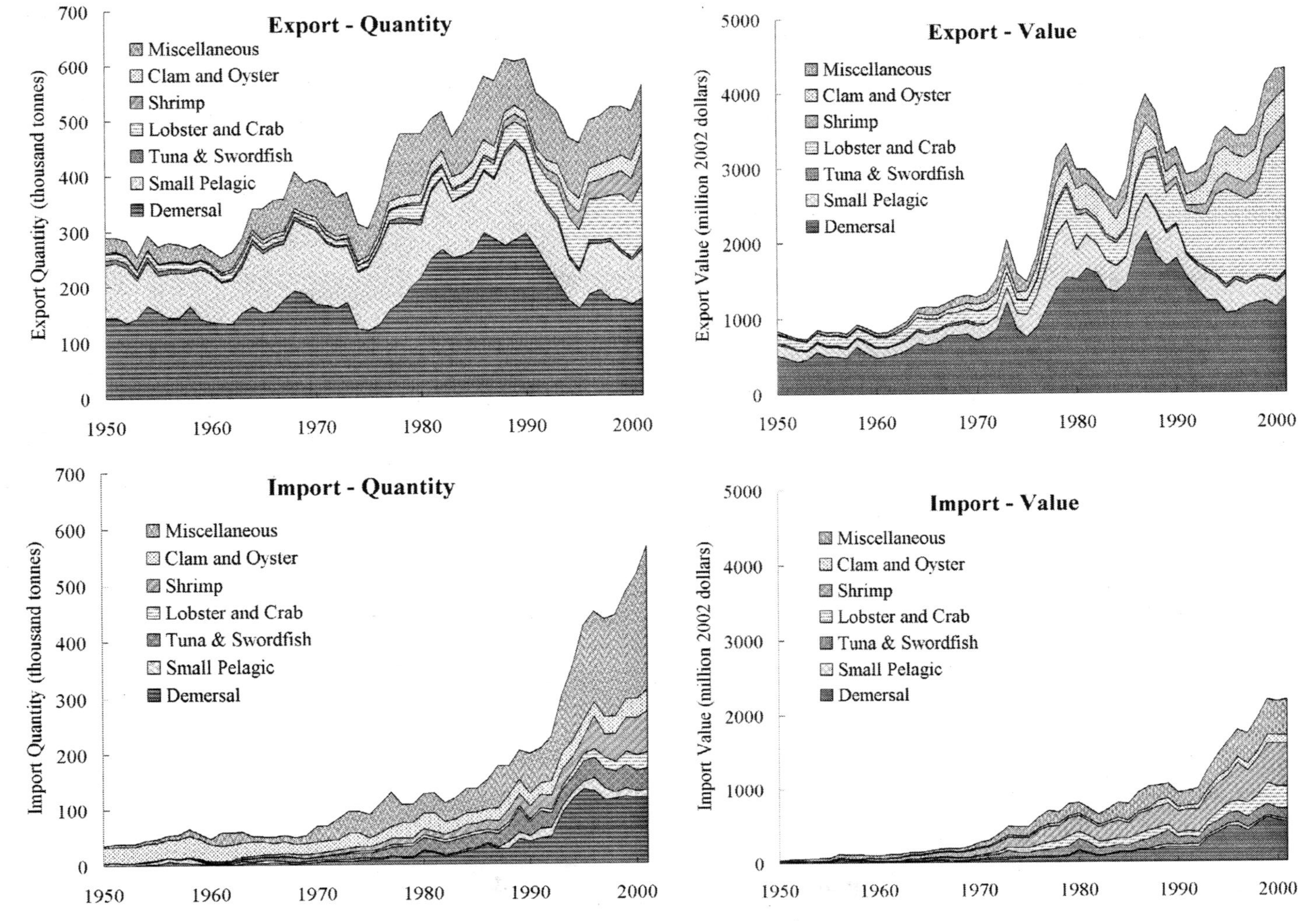

Figure 1. Import and export quantity and value from 1950 to 2001, by major taxonomic group.

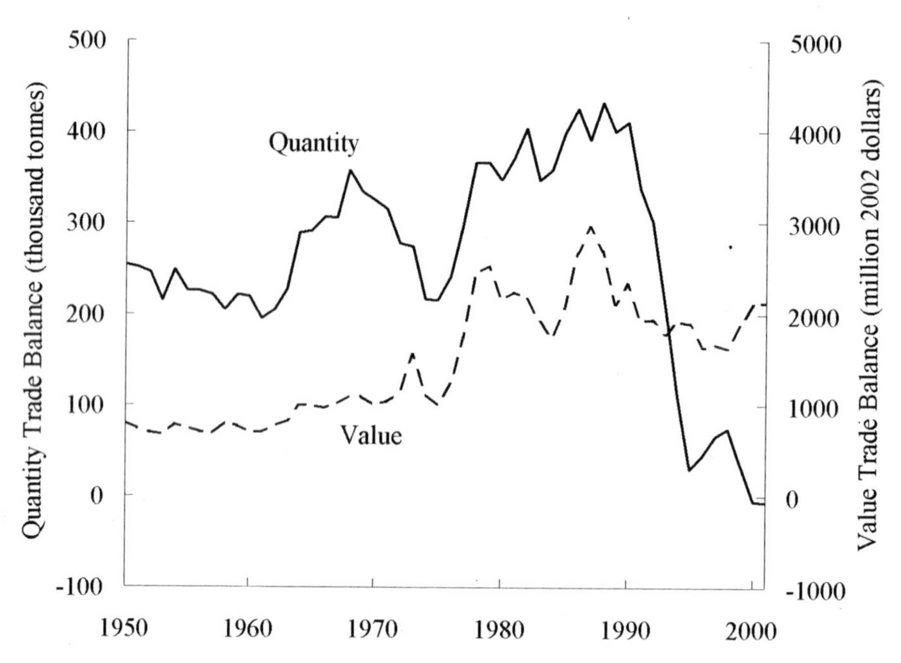

Figure 2. Trade balance from 1950 to 2001, in quantity and value terms. Positive values indicate a net export, while negative values indicate a net import.

in question. However, the sources of many of these growing import flows are developing countries (e.g., Chile, Thailand, and China) which have limited capacity to manage fisheries, and the availability of Canadian and other developed-country markets might provide a strong incentive to overexploit marine resources. In future work we will examine these issues using empirical data on trade and biomass trends.

Acknowledgments

This work was funded by the Pew Charitable Trusts, Philadelphia, through the Sea Around Us Project. We thank an anonymous referee for helpful comments.

References

Statistics Canada 1950–1965. Imports by commodity. Statistics Canada, External Trade Division, Ottawa.

Statistics Canada 1950–1967. Exports by commodity. Statistics Canada, External Trade Division, Ottawa.

Statistics Canada 1966–1984. Imports to Canada. Statistics Canada, External Trade Division, Ottawa. Online database: http://data.library.ubc.ca.

Statistics Canada 1968–1983. Exports from Canada. Statistics Canada, External Trade Division, Ottawa. Online database: http://data.library.ubc.ca.

Statistics Canada 1984–1987. Exports by commodity. Statistics Canada, External Trade Division, Ottawa.

Statistics Canada 1985–1987. Imports by commodity. Statistics Canada, External Trade Division, Ottawa.

Statistics Canada 1988–2001. Canadian International Trade. Statistics Canada, External/International Trade Division, Ottawa. Online database: http://data.library.ubc.ca.

American Fisheries Society Symposium 49:265–271

Economic Loss of Pollution to Fisheries: An Economic Analysis of the Jakarta Bay Fisheries

SUZY ANNA*

Department of Fisheries, Faculty of Fisheries and Marine Science, Padjadjaran University Jl. Raya Jatinangor, Bandung UBR, 40600, Indonesia

AKHMAD FAUZI

Head of Department of Resource and Environmental Economics, Faculty of Economics and Management IPB Jl. Kamper Wing 5 Level 5, Kampus IPB Darmaga Bogor 16680, Indonesia

Introduction

Coastal resources play an important role in the economic development of coastal states. Besides providing goods and services worth more than US$10 trillion a year, coastal resources also provide direct employment for more than 200 million small-scale and commercial fishermen. In addition, more than half a billion of people rely their livelihood indirectly on coastal related activities (Hinrichsen 2002). For example, in Asia alone, of 3.7 billion of total Asia's population, 60% of them live within 400 km of coastal area and engages in coastal dependent activities such as fishing, harvesting, and mining. Consequently, coastal resources have been under tremendous pressure due to rapid economic development. According to a study by World Resource Institute, over one-half of the world's coastlines have suffered from severe development pressure (Fauzi and Anna 2002; Fauzi and Anna 2003).

The pressures on coastal resources are also amplified by economic externalities such as pollution which flows through the river runoff to the coastal areas. It is not doubt that pollution has contributed significantly to the falling catches of the coastal fishing. There are number of cases of externalities which involve environmental degradation for coastal resources. For example, wastes, or even toxic wastes, dumped by households or industries located on the watershed area, may enter into water (or river) causing pollution to the coastal waters once valued for fishing and water recreation. Examples from coastal waters from developing countries reveal how badly pollution problems in coastal areas. Hinrichsen (2002), for example, noted that in Chile, the Bays of Valparaiso and Concepcion receive a combined of total of 244 million metric tons of untreated effluents a year. Worst yet, in India, the cities of Calcutta and Bombay, dump more than 400 million metric tons of raw sewage and 365 million metric tons of other municipal wastes into coastal waters every year causing a huge economic loss to society.

Much analysis of the externalities such as pollution has been focused on the ecosystem in general as well as the environmental effects associated with it. Several analyses, however, have been attempted to study the the effect

*Corresponding author: suzyanna@indo.net.id

of pollution on fishing activities. Collins et al. (1998), for example, have modeled the fishery–pollution interaction using a bioeconomic model by incorporation pollution variable as the driver for stock contamination and its effect on effort displacement. Grigalunas et al. (1988) and Grigalunas et al. (2001) have modeled the fishery–pollution interaction by assuming that pollution will cause economic loss to the fishery through a decline in fishing mortality. Using a slightly different bioeco-nomic approach, this paper develops a fishery–pollution interaction by embedding pollution variable into fishery production function and analyze its effect on the economic loss to the fishing industry.

The Model

To construct the bioeconomic model of fishery-pollution interaction, it is worth considering the specific functional form of biomass growth function as well as the harvest function. Let the biomass growth function be represented by the Gompertz function:

$$F(x) = rx\ln(K / x)$$

where r is the intrinsic growth rate and k is the environmental carryng capacity. Let the harvest function be given by the following Cobb-Douglass type production function:

$$h = qxE$$

where q is catchability coefficient and E is level of effort exerted to fishery. Now suppose, the pollution variable affects directly into the biomass production function, then the dynamic of the biomass can be written as:

$$\chi = (rx\ln(K/x) - \gamma P) - qxE$$

where is P pollution variable and g is a constant coefficient. The estimation of biological parameters as well as pollution parameter can be carried out by transforming equation (1.3),

$$x_{t+1} - x_t = rx_t \ln(K/x_t) - \gamma P_t - qx_t E_t$$

By introducing variable catch per unit effort (CPUE) at time t as $U_t = h_t / E_t$, using algebraic simplification, equation (1).4) can now be modified as:

$$U_{t+1} - U_t = rU_t \ln Kq - rU_t \ln U_t - q\gamma P_t - qU_t E_t$$

Parameters r, q, k, and γ can then be estimated using ordinary least square (OLS) technique from time series data of CPUE, pollution load and effort level. For the purpose of this study, sixteen years of time series data of catch and effort of demersal fishery as well as the data of pollution load were gathered. The pollution load is defined as total load of COD, BOD, and TSS collected from the Jakarta Environmental Impact Agency. The load is an aggregate measure of pollution discharge from 13 rivers runoff.

To calculate the economic loss due to pollution, let assume that the economic rent generated from the fishery without pollution variable is in the form of $\pi(h_t, E_t)$ while that with pollution variable is defined as $\pi(h_t, E_t, P_t)$. The economic loss to the fishery due to pollution is then defined as the difference between these two (i.e., $L = \pi(h_t, E_t, P_t) - \pi(h_t, E_t)$), where L denotes economic loss.

Economic loss was also approximated by calculating the loss in producer's surplus. This welfare effect is an important indicator to determine how the economic loss due to pollution affects the producers (i.e. the fishermen). The measurement of consumer's surplus was derived using backward bending supply curve and perfectly elastic demand curve. The formula for consumer's surplus is described as the following equation:

$$PS = P_0 h_0 - \left(\frac{1}{2} \frac{c\alpha \ln(h)}{\beta} + \frac{c\sqrt{-4\beta h + \alpha^2}}{\beta} \right) \Bigg|_0^{h_0} + \frac{\frac{1}{2} c\alpha \ln(\alpha - 4\beta h + \alpha^2)}{\beta} \Bigg|_0^{h_0} \Bigg|$$

$$- \frac{1}{2} \frac{c\alpha \ln(\alpha + \sqrt{-4\beta h + \alpha^2}}{\beta} \Bigg|_0^{h_0}$$

where parameters α and β are constant coefficients of yield effort curve by solving simultaneously equation (1).1) and equation (1).2) under steady state condition. By employing time series data of catch and effort these parameters can simply be derived using OLS technique. Parameter P and c represent price per kg of fish and cost per unit of effort, respectively. Parameter h represents average landing from the fishery under pollution scenario, while h_0 represents landing under untainted scenario. Data for these landings were obtained from adjacent landing sites with and without polluted waters.

Case Study: The Jakarta Bay Fishery

The Jakarta Bay, located in the north coast of Indonesia capital, Jakarta, has been an important area for both fishing as well as nonfishing activities. The bay is also well known for its high pollution discharge emanating from various river run-off (Anna 1999).

In terms of fishing activities, the Jakarta Bay serves as fishing ground for small-scale fishermen. Since the area is relatively narrow, only some particular fishing gears operate in the area. These include, stationary liftnet, gillnet, traps and hook and line. Some mollusk fishermen also operate along the estuarine area. Besides being fishing area, the Jakarta Bay also serves as the gate for shipping and marine transportation to and from the busiest fishing port in the capital city namely Tanjung Priok Port. Other fishing gears such as liftnet (know as *bagan* to local fishermen) and traps (*bubu*) are often being accused of obstructing the transportation activities around the bay.

The fishery production in the Jakarta Bay was primarily dominated by demersal fish cought by gears described above. The demersal landing accounts for more than 50% of the total landing of the Jakarta area. The demersal fish such as groupers and red snapper are considered as highly economic fish in the market. The prices of these fish are substantially higher than the average pelagic fish. It is not surprising, therefore some fishermen often use destructive fishing practice such as cyanide and bombing, which have caused damage to almost 80% of the coral reefs in the Bay and along the Seribu Island.

Results from analysis can be seen from Table 1 through Table 4. As can be seen from Table 1, pollution has significantly reduced the present value of economic rent that could have been generated from the fishery. This can be seen from the negative number of resource rent generated under pollution scenario. On average, more than Rp 2.5 billion rupiah (approximately US$3 million) of the present value of resource rent has lost due to the pollution. This is a significant loss for small-scale fishery such as the Jakarta Bay fishery.

The economic loss could also be viewed from the loss in economic surplus that could have been secured from the fishery. As can be seen from Table 2, an average of Rp 310.92 million per year would loss because of pollution. This figure is considerably significant for the mostly traditional fishermen of the Jakarta Bay whose income from fishing is relatively low.

Table 3, Table 4, and Figure 1 provide a general overview and summary of total net benefits obtained under the baseline and pollution scenarios. The combined yearly total economic loss to the fishery due to pollution is approximately Rp 700 million rupiah (or US$80,000). It is important to note, how-

Table 1. Yield and present value of rent under baseline and pollution scenario.

Year	SYB (metric tons)	SYP (metric tons)	PVRbase (Rp million)	PVRP (Rp million)
1986	77.8	71.65	1,251.88	1,042.06
1987	91.98	67.88	1,526.22	630.4
1988	38.47	59.29	786.48	1,594.86
1989	39.91	60.49	858.64	1,701.69
1990	82.54	70.77	1,745.24	1,208.62
1991	49.82	66.99	1,285.61	2,149.39
1992	52.72	68.36	1,424.3	2,254.41
1993	98.52	65.08	2,498.11	540.52
1994	110.87	58.31	2,905.38	–495.32
1995	139.07	37.87	3,109.41	–4,063.81
1996	118.51	53.32	3,467.87	–1,488.44
1997	39.67	60.29	2,763.8	5,499.08
1998	53.3	68.61	3,656.12	5,722.46
1999	115.7	55.22	6,760.68	–2,242.79
2000	102.67	63	7,230.3	645.84
2001	124.68	48.95	8,465.28	–5,554.33

Note: SYB = sustainable yield baseline, SYP = sustainable yield with pollution, PVRbase = present value rent baseline, PVRP = present value rent with pollution.

Table 2. Change in producer's surplus due to pollution.

Year	Producer's surplus baseline (Rp million)	Producer's surplus pollution (Rp million)	Δ SP (Rp million)
1986	386.285	364.031	22.254
1987	492.73	376.343	116.387
1988	220.133	342.198	–122.065
1989	241.553	369.121	–127.568
1990	546.477	480.666	65.811
1991	370.074	502.376	–132.302
1992	412.31	540.27	–127.96
1993	823.856	567.509	256.347
1994	1,005.215	561.849	443.366
1995	1,263.005	397.853	865.152
1996	1,240.808	602.698	638.11
1997	776.56	1,190.01	–413.45
1998	1,060.35	1,379.363	–319.013
1999	2,388.563	1,222.899	1,165.664
2000	2,419.975	1,557.172	862.803
2001	3,127.719	1,346.458	1,781.261

Table 3. Comparison of total benefit between the baseline and pollution scenario.

	Baseline (Rp million)			Pollution (Rp million)			ΔTB
	FR	PS	TB	FR	PS	TB	(Rp million)
Year							
1986	187.78	386.29	574.07	156.31	364.031	520.34	–53.73
1987	228.93	492.73	721.66	94.56	376.343	470.9	–250.76
1988	117.97	220.13	338.1	239.23	342.198	581.43	243.32
1989	128.8	241.55	370.35	255.25	369.121	624.38	254.03
1990	261.79	546.48	808.26	181.29	480.666	661.96	–146.3
1991	192.84	370.07	562.92	322.41	502.376	824.78	261.87
1992	213.64	412.31	625.95	338.16	540.27	878.43	252.48
1993	374.72	823.86	1,198.57	81.08	567.509	648.59	–549.99
1994	435.81	1,005.22	1,441.02	–74.3	561.849	487.55	–953.47
1995	466.41	1,263.01	1,729.42	–609.57	397.853	–211.72	–1,941.13
1996	520.18	1,240.81	1,760.99	–223.27	602.698	379.43	–1,381.56
1997	414.57	776.56	1,191.13	824.86	1,190.01	2,014.87	823.74
1998	548.42	1,060.35	1,608.77	858.37	1,379.363	2,237.73	628.96
1999	1,014.1	2,388.56	3,402.67	–336.42	1,222.899	886.48	–2,516.18

Note: FR = fishery rent, PS = producer's surplus, TB = total benefit

Table 4. Summary of some important indicators.

Variable	Baseline			Pollution			% change		
	δ = 6.2%	δ = 15%	Δ = 0	δ = 6.2%	δ = 15%	δ = 0	δ = 6.2%	δ = 15%	δ = 0
Sustainable yield (Hs) (metric tons)	77.38	77.38		60.30	60.30		–22.07	–22.07	
Optimal Effort (E*) (000 days)	2.57	3.09		4.72	6.00		83.66	94.17	
Optimum biomass (X*) (metric tons)	412.25	324.55		217.25	155.38		–47.30	–52.12	
Optimal yield (h*) (metric tons)	176.74	167.63		70.42	64.02		–57.99	–63.78	
Sustanable rent (ð)	466.27	466.27		85.73	85.73		–81.61	–81.61	
Optimal rent (ð*) (Rp million)	2,216.00	2,101.89		882.57	802.18		–60.17	–61.84	
Depreciated rent (Rp million)	529.08	1,280.03		3,234.38	337.50		511.32	4.49	
Producer's surplus (PS)			1,048.48			737.55			–29.66
Total benefit (TB) (Rp million)			1,514.74			823.28			–45.65
Degradation rate (Θ) (%)			18.00			21.00			16.67

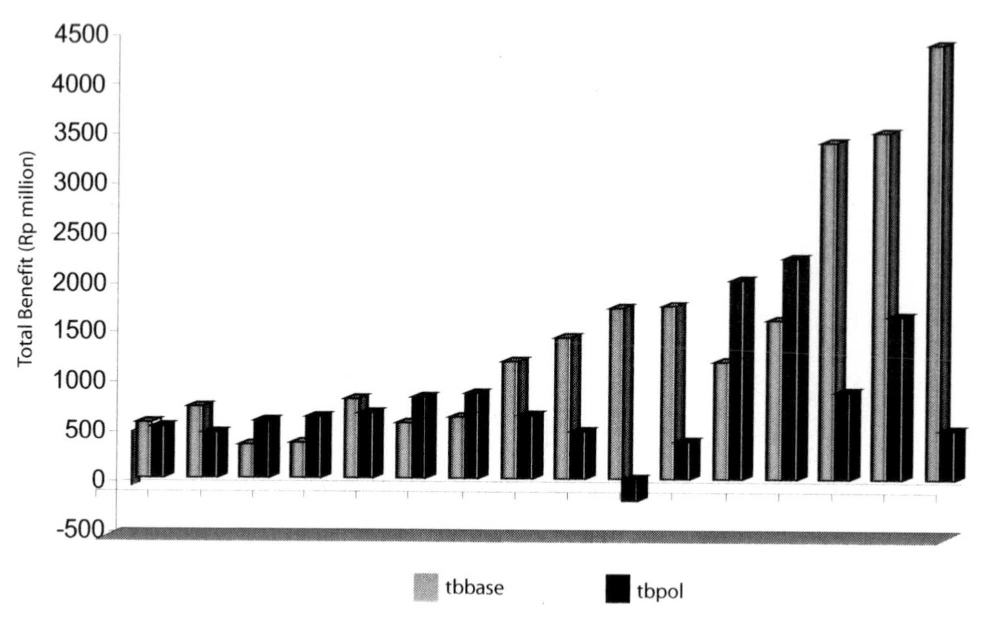

Figure 1. Yearly benefits accrued under baseline scenario and pollution scenario.

ever, that this figure is applicable to the demersal fishery only with two dominant fishing gears. Extrapolating to the fishery of the bay as a whole, the economic loss could have been very much higher than that the number above. As can be seen from Table 4, overall, pollution has caused a significant reduction in the optimum harvest level and some monetary indicators such rent and producer's surplus. The optimum level of harvest or yield is calculated using standard bioeconomic model described in Clark (1990). This reduction in economic surpluses could be attributed to the resource degradation, which is as much as 21% per year due to pollution.

Conclusion

Studies on the economic analysis on the interaction between fishery and pollution are indeed a complex and difficult exercise. Complexity of fishery system and its interaction with environment make the modeling of fishery–pollution interaction a challenging one. Few have attempted such a modeling due to these complexities. This paper, however, offers a simplified modeling of fishery-pollution interaction by imbedding the pollution variable into a standard bioeconomic model of the fishery. By doing this, the model offers an alternative way of analyzing the economic impacts of pollution to the fishery which have long been neglected. In terms of the case study in the Jakarta Bay area, the economic loss due to pollution is quite significant. The total economic loss in terms of net benefit deriving from the loss in resource rent and producer's surplus is approximately Rp 700 million per year. An integrated fishery policy through what so-called Envo-Fishery should be encouraged. That is, pollution control in the Bay should take into account its impact on the fishery in the area, not only in terms of its biological dimension but also in terms of its economic consequences. The study also reveals that due to excessive production and input level in the fishery, there seems no room for further expansion of effort for the demersal fishery in the area. Therefore, management of input and output in the fishery should be directed into efficiency of the industry, through for example, a combination fiscal and environmental policies. Finally, even though this study had attempted to incorporate all possible models with regards to analysis of fishery–pollution interaction, there might be some other things which could have been overlooked, such as incorporating stochastic process in the model. Further study to incorporate such a process, therefore, is strongly encouraged.

References

Anna, S. 1999. Analysis of pollution load and assimilative capacity in the Jakarta Bay. Master's thesis. Bogor Agricultural University, Bogor, Indonesia.

Clark, C.W. 1990. Mathematical bioeconomics: the optimal management of renewable resources, 2nd editon. Wiley, New York.

Collins, A., S. Pascoe, and D. Whitmarsh. 1998. Fishery-pollution interactions, price adjustment and effort transfer in adjacent fisheries: a bioeconomic model. In Proceedings of the First World Congress of environmental and resource economists, Venice, Italy.

Fauzi, A., and S. Anna. 2002. Assessment of fishery resource depreciation for policy considerations. Journal of Coastal and Marine Resources 4(2):36–49.

Fauzi, A., and S. Anna. 2003. Assessment of sustainability of integrated coastal management projects: a CBA-DEA approach. Journal of Coastal and Marine Resources, Special Issue 1:34–48.

Grigalunas, T.A., J. J. Opaluch, D. French, and M. Reed. 1988. Measuring damages to marine natural resources from pollution incidents under CERCLA: application of an integrated ocean system/economic model. Marine Resource Economics 5:1–21.

Grigalunas, T.A., J. J. Opaluch and M. Luo. 2001. The economic cost to fisheries from marine sediment disposal: case study of Providence, RI, USA. Ecological Economics 38:47–58.

Hinrichsen, D., 2002. Ocean planet in decline. People and the Planet. Available: http://www.peopleandplanet.net/doc.php?id=429§ion=6. (January 2007).

American Fisheries Society Symposium 49:273–282

The Effects of Trade Policies on the Philippine Fish Markets

U-Primo E. Rodriguez* and Yolanda T. Garcia

University of the Philippines Los Baños, Department of Economics, College of Economics and Management College, Laguna, Philippines

Abstract.—This paper analyzes the effects of various trade policies that artificially create either an economic protection or penalty to the fisheries sector. It conducts a quantitative analysis of trade regulations using the AsiaFish model, a partial equilibrium model that relates production, consumption, and trade of various fish commodities that are disaggregated by type or species group. The model estimates the magnitude and direction of the effects of current changes in trade policies on the Philippine fish markets, specifically on domestic demand and supply of fish and on importation and exportation of the country's fishery products. The disaggregated treatment of the fisheries sector in the model allows the evaluation of the distribution of benefits/costs of the trade policies on the different stakeholders in fisheries (i.e., producers, consumers, and traders).

Introduction

The Philippines has been pursuing trade policy reforms since the early 1980s. In general, the objectives of these reforms were to reduce the influence of quantitative restrictions, reduce tariff rates, promote the uniformity of tariff rates, and simplify the tariff structure. In some instances, these reforms were undertaken in the context of multilateral arrangements. Examples include the country's participation in the ASEAN Free Trade Area and the World Trade Organization.

Recent reforms have led to substantial changes in tariff rates on fishery products. Costales and Garcia (2002) show that tariff rates have fallen from a range of 10–50% in the 1980s to a range of 3–15% in 2000. Tariff rates on fishery products have since fallen to 5% or less.

The analysis of trade reforms to the fisheries sector is important because these affect the structure of incentives within the country. Such adjustments in turn influence the level and composition of aggregate output, consumption, and trade. Determining the direction and magnitude of these effects will assist in evaluating the suitability of such reforms. Moreover, identifying the losers from the reforms will also aid policymakers in developing and/or instituting instruments that could reduce the pain brought about by the adjustments.

The importance of the analysis is strengthened when one considers the role of the fisheries sector in the Philippines. First, the sector has a significant contribution to the level of economic activity in the country. In 2001,

*Corresponding author: uprime@gmail.com

this sector contributed about 3.9% of gross domestic product (National Economic Development Authority 2002). The sector also directly employs about 1 million people or 4.4% of total employment (National Statistics Office 2004). Second, the Philippines has been and is expected to be a net exporter of fish. Thus, the sector also serves as a source of valuable foreign exchange for the country. Finally, there have been significant structural changes within the fisheries sector. For example, the past four decades witnessed the emergence of aquaculture and the decline of municipal fishing as a source of fish output.

The objective of this paper is to evaluate the impacts of the 2000–2004 tariff changes in the fisheries sector on fish production, consumption, and trade. This is conducted using a multicommodity model of the Philippine fisheries sector. The remainder of the paper is organized as follows. The first section describes a model that will be used as the tool for analysis. The second section presents the results from the tariff experiments. The final section presents the conclusions and limitations of the analysis.

A Supply and Demand model for Philippine Fisheries[1]

The fisheries supply and demand model of the Philippines follows the specification of the AsiaFish model (see Dey et al. 2005). Designed for generating detailed results on supply, demand, trade, and prices for the fisheries sector, it is a partial equilibrium model that assumes that the quantities and prices of nonfish commodities are determined outside of the model.

The general structure of the model specifies that the domestic supply of a particular fish type originates from domestic and foreign (imported) sources. The supply is then allocated among households (consumption), firms (intermediate demand), and foreign agents (exports). The prices of fish types in the domestic market are determined by a series of equilibrium conditions. In instances where the fish types in the supply and demand sides do not match exactly, the model also has a set of linking equations that facilitate the disaggregation of composite sectors for use with the equilibrium conditions.

The demand side of the model separates the Philippines into rural and urban regions. Each of these regions is assumed to have a representative household that earns income and purchases goods and services. It assumes that households consume nine types of fish. These are anchovies, round scad *Decapterus punctatus*, squid, shellfish, milkfish *Chanos chanos*, tilapia, shrimp, processed fish, and others.

Fisheries production data are presented by source (i.e., commercial fisheries, municipal fisheries, and aquaculture[2]). Each source of fish is then assumed to produce/harvest different fish types (Table 1), for example, commercial fisheries is the source of grouper, tuna, anchovies, round scad, squid, shrimp, other shellfish, and other capture species.

Constrained by the availability of detailed consumption data, the base data set was assembled for the year 2000. On the other hand, the parameters used in the model were ob-

[1] For an exhaustive description of the model equations and assumptions, see Dey et.al (2005). Also see Rodriguez et al. (2005) for a detailed discussion on the application of the model to the Philippines.

[2] Commercial fisheries refer to the catch/harvest of operations in areas beyond 15 km from shore and using boats weighing more than three gross tons (Bureau of Agricultural Statistics 2002). It follows naturally that municipal fisheries refers to the catch of boats that weigh three gross tons or less and operating within 15 km from shore. Finally, aquaculture represents all types of fish farming (e.g., fishponds, cages, tanks, and pens, including mariculture) in different environments (freshwater, brackish water, and marine).

Table 1. The fish types in the supply side of the model, classified by source.

Source	Fish types produced
Aquaculture	Carp, catfish, milkfish, mussels and oysters, other aquaculture, other shells, shrimp, tilapia
Commercial fisheries	Anchovies, grouper, other capture, other shells, round scad, shrimp, squid, tuna
Municipal fisheries	Anchovies, carp, catfish, grouper, milkfish, other capture, other shells, round scad, shrimp, squid, tilapia, tuna

tained through a combination of techniques. Elasticities for the demand side were estimated using the quadratic almost ideal demand system (see Garcia et al. 2005). In contrast, elasticities for supply and trade were obtained through consultations with experts and literature search. Following the practice of applied general equilibrium models, the remaining parameters were calibrated to ensure the replication of the base data set.

Using the Generalized Algebraic Modeling Software, a baseline solution of the model is derived from 2000 to 2020. This involves solving the model recursively over which the values of all the relevant variables are continuously updated. Throughout the simulation period, it is assumed that (a) the tariff structure remains at its 2000 level, (b) no technological progress takes place, and (c) all exogenous variables grow at their historical levels. Given these assumptions, the model projects sluggish growth for aggregate fish output and consumption from 2001. It also projects a continued improvement in the net exports of fish as the value of exports is expected to expand faster than the value of imports.

The Effects of Tariff Changes

Salayo (2000) presents the schedule of tariffs for fresh and frozen fish, crustaceans, mollusks, shell products, and processed fish. Because of the seasonal factors in the supply side, fresh and frozen fish products have variable tariff rates for the different months of a year. For the months of March to July, the tariff rate for year 2000 was 7%. This is expected to fall to 5% by 2003. On the other hand, tariff rates will be 3% in the remaining months of 2000–2004. In the case of crustaceans, the pattern is for tariff rates to fall from 7% in 2000–2002 to 5% in 2003. Finally, tariff rates for mollusks, shells, and processed fish are 15% for 2000, 10% for 2001–2002, 7% for 2003, and 5% for 2004.

Since the model is only capable of working with one tariff rate per fish type in each year, an extra set of computations was done for fresh and frozen fish. The variable tariff rates across different months were handled by taking the weighted average of tariff rates for each year. The weights for the two sets of tariffs are based on the number of months in which these are imposed.

Since the Philippines is assumed to be a small, open economy, the direct effect of the tariff changes is to reduce the domestic prices of imported fish. Following the tariff rates discussed earlier, Table 2 shows that the magnitude of the changes in import prices can be classified into three groups of fish types. It also shows that the impacts of the tariffs are likely to be felt more from 2005 onwards. The

Table 2. Implied changes in import prices, per cent per annum.

Fish type	Average deviation from baseline (%) 2001–2004	2005–2020
Other capture fish, carp, milkfish, roundscad, tilapia, tuna	–0.48	–0.80
Shrimp	–1.12	–1.87
Other shells, processed fish, mussels and oysters, squid	–6.61	–8.70

reason for this is the gradual decline in tariffs from 2001 to 2004.

In implementing the experiment, the pattern of changes in import prices was applied to the model. The model was then solved again from 2000 to 2020. In describing the results, the solutions for the different variables from this experiment were compared to their counterparts in the base case.

In order to get a perspective of the potential outcomes from the tariff changes, it is useful to begin by analyzing its potential impacts on a particular fish type. As mentioned earlier, the decline in tariffs rates causes a fall in import prices. This in turn has direct impacts on two variables. First, the decline in import prices raises the demand for imported fish. Second, holding domestic prices constant, the decline in import prices makes it less attractive to buy domestic goods. Hence, it also causes a decline in the demand for the domestically produced/harvested fish type. The decline in the demand for the domestically produced fish type in turn affects its price. Holding supply constant, this causes the price of the domestically produced fish type to decline. These changes eventually affect output, consumption, imports, and exports.

The output of the domestically produced fish is likely to fall because a lower price weakens the incentive to produce/harvest fish. Along with the decline in import prices, the fall in the price of the domestically produced good lowers the consumer price of fish in the domestic market. Holding other things constant, this causes the total consumption (imports and domestic goods) of the fish type to rise.
The effects on trade work through relative prices. The lower price of the domestically produced good makes the relative price of traded goods higher. This reduces the demand for imports and raises the incentive to export. The declines in the prices of imports and domestically produced goods have competing effects on the quantity of imports. However, the net effect is still an increase in imports. This may be inferred from the previous findings that total consumption is higher and that the demand for the domestically produced good is lower. On the other hand, the overall effect on exports is ambiguous. The reason is that the higher relative price and lower domestic output affect this variable in opposite directions. If the impact of the second factor is stronger, then exports will decline. The reverse is true when the impact of the first factor is stronger.

In sum, the previous analysis suggests that the tariff cuts for a particular fish type are likely to cause its (a) domestic output to fall, (b) domestic consumption to rise, and (c) imports to rise.

The present analysis fails to capture other adjustments that might take place in the fisheries sector. One adjustment arises from the possibility that the changes in the market for a particular fish type are likely to affect the markets of other fish types. This is caused mainly by the possibilities for substitution and complementarity in production and consumption. Following the same line of reasoning, the previous analysis also ignores the feedback arising from the responses of the other fish types. Adding further complexity to the analysis is the fact that the tariff rates of other fish types are changing as well. As a whole, these points suggest that the previous analysis is at best interpreted as the initial effects of the tariff changes.

The multicommodity model that was used in this paper attempted to overcome these shortcomings. Not only does it incorporate the relationships among different fish types, it also allows an analysis of simultaneous changes in tariff rates.

The tariff changes are expected to have small but positive effects on the aggregate fisheries output and household demand (Table 3). The quantity of fisheries output is expected to be higher than its baseline value by an average of 0.05%. On the other hand, the quantity of fish consumed by households is expected to be higher by an average of 0.16%. While the effect on the quantity of exports is expected to be negligible, the same cannot be said about imports. On the average, fish imports are projected to be 2.28% higher as a result of the tariff changes.

In some instances, the aggregate results capture the sort of results that one would get from a simple supply and demand analysis. For example, it captures the expansion in consumption and imports. However, the expansion in domestic output appears counterintuitive. In what follows, a disaggregated analysis of the results will be presented. The intent is to shed light on the aggregate results and on the potential changes in the structure of the fisheries sector.

The outputs of the different fish types respond to the tariff changes in different directions (Table 4). Carp (0.20%) and anchovy (0.26%) for example experienced relatively largest increases in their output. In contrast, the outputs of mussels, oysters, and round scad are expected to contract.

An examination of the effects across fish types also reveals the following. First, all the responses in output coincide with the changes in producer prices. That is, fish types that experienced an increase in their producer prices also had higher outputs. This implies that the changes in the (own) producer prices of the respective fish types are either strengthened or cannot be overcome by changes in the prices of other fish types (cross price effects). Second, most of the increase in total output is explained by the expansion of other capture fish. The reasons are that this fish type (a) experienced a relatively large increase in output (0.10%), and (b) has a relatively large initial share of total output. Third, the outputs and producer prices of fresh fish types that experienced the largest tariff cuts (mussels, oysters, shrimp, and squid) are expected to fall. This captures the essence of the simple model that tariff cuts lead to lower output. Finally, the previous discussion also suggests that the fish types that experienced the largest tariff cuts will have a lower share in the total output of fresh fish.

On the demand side, the results are generally consistent with a priori expectations (Table 5). With the exception of round scad, shrimp, and tilapia, the consumption of all fish types expanded. Of these fish types, anchovy (0.26%) and other fish (0.36%) experienced the largest increases.

The expansion in total consumption is explained mostly by the increase in the con-

Table 3. Effects on selected fish aggregates.

	Baseline: 2000 (metric tons)	Effect: 2001–2020 (percent deviation from baseline)
Output	2,633,170	0.05
Imports	154,270	2.28
Exports	131,600	0.03
Consumption	1,355,482	0.16
Intermediate demand	1,300,360	0.06

Table 4. Effects on the output of fish, by source and fish type.

	Baseline: 2000[a]		Effect: 2000–20[b]	
	Quantity	Producer Price	Quntity	Producer Price
Total	2,633,170.0	nc	0.05	nc
Of which				
Fresh fish	2,286,290.0	nc	0.05	nc
Processed fish	346,880.0	25.21	0.06	–0.04
Fresh fish by source				
Aquaculture	393,863.3	nc	0.04	nc
Commercial Fisheries	946,485.0	nc	0.05	nc
Municipal fisheries	945,945.0	nc	0.05	nc
Fresh fish by fish type				
Anchovy	79,630.0	35.75	0.26	0.39
Carp	15,714.0	26.23	0.20	0.23
Catfish	3,135.0	58.59	0.17	0.28
Grouper	12,492.0	201.27	0.10	0.19
Milkfish	210,157.0	56.38	0.12	0.13
Mussels and oysters	31,179.0	6.80	–0.26	–0.15
Other aquaculture Fish	2,043.3	109.90	0.17	0.28
Other capture fish	1,086,869.0	19.08	0.10	0.13
Other shells	127,062.0	56.68	0.02	0.04
Round scad	255,976.0	40.84	–0.07	–0.09
Shrimp	63,287.0	239.85	–0.06	–0.08
Squid	46,778.0	71.48	–0.02	–0.03
Tilapia	121,453.0	46.75	–0.03	–0.03
Tuna	230,518.0	44.71	–0.01	–0.01

[a] Quantities and producer prices in the baseline are expressed in metric tons and pesos per kilo, respectively.
[b] This indicates the average percentage deviation from the baseline solution.
nc = not computed

Table 5. Effects on demand, by region and fish type.

	Baseline: 2000 (metric tons)	Effect: 2001–2020 (percent deviation from baseline)
Total	1,355,480.0	0.16
By region		
Rural	499,880.0	0.09
Urban	855,600.0	0.20
By fish type		
Anchovy	61,168.7	0.26
Milkfish	204,313.2	0.12
Other fish	350,872.7	0.36
Processed fish	294,277.3	0.13
Round scad	242,214.6	-0.07
Shells	20,010.4	0.12
Shrimp	22,752.3	-0.13
Squid	38,420.8	1.60
Tilapia	121,452.2	-0.03

sumption of other fish. The reason is that apart from experiencing the largest increase in consumption, this fish type also accounts for a significant proportion of the total.

The structure of the Philippine model makes analyzing the consumption side a tedious exercise. For each fish type, this requires evaluating the results across and within the urban and rural regions. So as not to diverge too much from the primary objective of this paper, the analysis will only focus on the fish type that expanded the most (other fish) and the three fish types that contracted (round scad, shrimp, and tilapia).

At one level, the expansion in the consumption of other fish is attributable to the increase consumption in the rural (0.20%) and urban (0.44%) regions. Going further, the increase in the demands of two regions is due mostly to the 0.9% decline in its consumer price of other fish. Setting cross price effects and parameter differences aside, the regional disparities in consumption are explained by differences in the response of real fish expenditure. In the case of the urban region, real fish expenditure increased by 0.10%. It therefore strengthens the effect of the lower consumer price. In contrast, the 0.01% decline in the real fish expenditure of the rural region weakens the effect of the lower consumer price. As a result, the increase in the consumption for the rural region is smaller.

The decline in the consumption of round scad, shrimp, and tilapia is explained for the most part by the cross price effects in demand and, to a limited extent, changes in real fish expenditure.[3] The basis of this assertion is that the consumer prices of these fish types actually declined. While the results are not shown here, this may be inferred from the reduction in their respective producer and import prices.[4]

[3] The data in this discussion are available from the author by request.

[4] Recall that the changes in producer prices are shown in Table 4.

The changes in the consumer price and real fish expenditure suggest an increase in the consumption of these fish types in the urban region. However, the fact that their consumption declined suggests that the net impact of the cross price effects was negative and quite strong. On the other hand, the story for the rural region is not as straightforward because the decline in real fish expenditure also causes a decline in consumption. However, it suffices to say at this point that the cross price effects strengthened or were unable to overcome the impact of the change in real expenditure.

Decreasing tariffs cause an across the board increase in imports (Table 6). The largest percentage increases are for mussels and oysters (9.40%), other shellfish (9.98%), processed fish (9.85%), and squid (9.79%) (Table 6). The disaggregated results also suggest the following. First, the changes in imports are consistent changes in relative prices. That is, the lower relative price of imports causes an increase in imports. While changes in consumption also matter in explaining the responses, these changes are in some cases too weak to cause a reversal of the results. For example, the 0.07% decline in the consumption of round scad is not strong enough to overcome the influence of the 0.63% fall in its relative price. In some instances, like milkfish and squid, the response of consumption actually strengthens the impact of the change in relative prices. Second, the fish types that were exposed to the largest tariff cuts also experienced the largest decline in the relative price. This result does not come as a surprise because larger tariff cuts, given exogenously determined foreign prices, imply larger declines in domestic import prices.

Finally, the consequence of the relatively large increase in the imports of mussels and oysters, other shellfish, processed fish, and squid suggests that these fish types will have a larger share of total imports in the future. However, the increased share of mussels and oysters will be hardly noticeable because of its very small initial share in imports.

The effect of the tariff change on the exports of different fish types is mixed (i.e., negative for some fish types and positive on others). This result is consistent with the analysis in the simple model that the impact of tariff change on the exports is ambiguous. The responses of the different fish type also raise three important points.

First, the response of exports is explained by the changes in relative prices and output. With the exception of catfish, mussels, oysters, shrimp, and tuna, the changes in exports coincide with the changes in relative prices. In the case of catfish, mussels, and oysters however, the absence of changes in their exports suggests that the respective changes in their outputs directly offset the effects of changes in their relative prices. On the hand, the effects of changes in output appear to dominate the effects of changes in relative prices in the two other fish types. The higher relative price of shrimp raises the incentive to export more of this fish type. However, the decline in its output overcomes this impact.

Second, the relative large initial share and increase in the exports of processed fish suggests that it is the major source of the increase in total exports.

Third, the relatively high growth rates of tilapia and processed fish suggests that these two fish types will become more important sources of export revenues in the future. Because of its relatively low initial share however, the emergence of tilapia as a source of export revenues is not likely to be as noticeable as processed fish.

Table 6. Effects on trade, by fish type.

	Baseline: 2000[a]		Effect: 2001–20[b]	
	Exports	Imports	Exports	Imports
Quantity				
Total	131,600.0	154,250.0	0.03	2.28
Of which				
Carp	0.9	0.1	-0.15	2.25
Catfish	0.2	–	0.00	na
Grouper	6,642.7	–	–0.07	na
Milkfish	89.9	0.3	–0.08	1.08
Mussels and oysters	0.1	50.8	0.00	9.40
Other shells	5,300.2	23.2	–0.01	9.98
Other fish	5,379.3	100,782.5	–0.07	1.08
Processed fish	53,912.4	1,308.2	0.10	9.85
Round scad	4,546.3	0.1	0.07	0.56
Shrimp	12,058.5	6.2	–0.01	1.66
Squid	1,604.3	17,617.5	0.02	9.79
Tilapia	0.9	0.1	0.10	0.56
Tuna	42,068.0	34,481.8	–0.01	0.77
Relative price				
Carp	5.20	1.00	–0.23	-0.95
Catfish	4.04	na	–0.28	na
Grouper	0.60	na	–0.12	na
Milkfish	1.96	0.78	–0.13	–0.85
Mussels and oysters	12.51	8.58	0.15	–7.30
Other shells	1.98	0.70	–0.02	–0.66
Other fish	6.47	2.99	–0.11	–8.18
Processed fish	2.45	2.32	0.03	–8.05
Round scad	1.87	1.00	0.09	–0.63
Shrimp	1.49	0.25	0.04	–1.60
Squid	2.66	0.39	0.03	–6.68
Tilapia	2.92	1.00	0.03	–0.69
Tuna	1.02	0.89	0.00	–0.15

[a] Quantities in the baseline are expressed in metric tons.
[b] This indicates the average percentage deviation from the baseline solution.
na = not applicable

Concluding Remarks

Using a multicommodity model for fisheries, this paper analyzes the effects of the 2000–2004 tariff changes in different fish types. As a whole, it finds that the tariff changes benefit producers and consumers in the sector. This is captured by the increases in aggregate fish consumption and production. The results also suggest an expansion in trade as imports and exports are expected to rise.

The disaggregated analyses of the results also suggest the following. First, the outputs of fresh fish types that experienced the largest tariff cuts are likely to contract. As a consequence, their share in total output is also expected to decline over time. Second, the consumption of other fish is likely to become more important as result of the tariff changes. This follows from the relatively large increase in the consumption of this fish type. Third, while the increase in imports occurs across the board, fish types that were exposed to the largest tariff cuts experienced the largest changes. Finally, the tariff changes will cause processed fish to become a more important source of export revenues. As export prices are kept constant in the experiment, this assertion is based on the finding that this fish type will experience the largest increase in the quantity of exports.

At this point, it is worth noting some of the limitations of the current analysis. First, the experiment is limited to analyzing the effects of tariff changes in food fish imports only. It ignores tariff changes in imported inputs, like feeds and fertilizer, which could influence the supply side of the market. On the demand side, it ignores tariff changes in other commodities like rice and meat. These effects could also be relevant to the fisheries sector as rice (meat) is seen as a complement to (substitute for) fish. Second, the partial equilibrium nature of the model suggests that it is unable to examine how the tariff changes in the fisheries sector affect the markets for other goods. Consequently, it is unable to trace the feedback of such changes on the fisheries sector.

For all its limitations however, the analysis conducted here is still useful because it provides insights into how the tariff changes affect the different fish types within the sector. To the policymaker, this helps in verifying the consistency of the tariff changes with other fisheries-specific policies and objectives. For example, the decline in the output of tilapia that is caused by the tariff changes is something to think about when taken in the context of the government's expressed desire to promote this fish type.

Acknowledgments

The authors thank the WorldFish Center for funding the construction of the model. They are also grateful to Madan Mohan Dey (WorldFish Center), Roehlano Briones (WorldFish Center), Mahfuzuddin Ahmed (WorldFish Center), and Reynaldo Tan (University of the Philippines Los Baños) for the comments on the model and this paper. Special thanks to Sheryl Navarez for her excellent assistance in the project. The usual disclaimer applies.

References

Bureau of Agricultural Statistics. 2002. Fisheries statistics of the Philippines, 1997–2001. Bureau of Agricultural Statistics, Quezon City, Philippines.

Costales, A., and Y. Garcia. 2002. The evolution of Philippine fishery policies and issues for developing export markets under a global trading environment. Paper presented at the International Institute of Fishery Economics and Trade (IIFET) Meeting, Wellington, New Zealand, August 19–22.

Dey, M., R. Briones, and M. Ahmed. 2005. Disaggregated analysis of fish supply, demand and trade: baseline model and estimation strategy. Aquaculture Economics and Management 9:113–140.

Garcia, Y., M. Dey, and S. Navarez. 2005. Demand for fish in the Philippines: a disaggregate analysis. Aquaculture Economics and Management 9:141–168.

National Economic Development Authority. 2002. Philippine statistical yearbook. National Economic Development Authority, Metro Manila, Philippines.

National Statistics Office. 2004. Labor force survey. Available: census.gov.ph (April 2004).

Rodriguez, U., Y. Garcia, and S. Navarez. 2005. The effects of export prices on the demand and supply for fish in the Philippines. Aquaculture Economics and Management 9:169–194.

Salayo, N. 2000. International trade patterns and policies in the Philippine fisheries. Philippine Institute for Development Studies, Discussion Paper 2000–15, Makati, Philippines.

American Fisheries Society Symposium 49:283–293

An Economic Analysis of the Southeast U.S. Shrimp Processing Industry Responses to an Increasing Import Base

WALTER R. KEITHLY, JR.*
Louisiana State University, Coastal Fisheries Institute
Baton Rouge, Louisiana 70803, USA

HAMADY DIOP
Louisiana State University, Louisiana Sea Grant College Program
Baton Rouge, Louisiana 70803, USA

RICHARD F. KAZMIERCZAK, JR.
Louisiana State University, Center for Natural Resource Economics and Policy
Baton Rouge, Louisiana 70803, USA

MIKE D. TRAVIS
U. S. National Marine Fisheries Service, Southeast Regional Office
263 13th Street, St. Petersburg, Florida 33701, USA

Introduction

U.S. imports of shrimp are known to be large and growing. In 1980, shrimp imports, expressed on a headless shell-on equivalent basis, totaled 117 million kilograms (kg) and accounted for 55% of total U.S. shrimp supply (i.e., U.S. commercial landings plus imports). By 2001, imports had advanced to 510 million kilograms, at which point they represented 85% of total U.S. supply. The vast majority of growth in U.S. shrimp imports represents farmed product. In conjunction with the increasing import base, the composition of imports has also been changing. In 1980, for example, headless shell-on product represented 63% of total imports (by product weight) while peeled product (raw and other) accounted for 36% (the small remaining amount constituted breaded and canned products). By 2001, the share of total imports accounted for by peeled product had advanced to almost 50% while the headless shell-on share had fallen to 50%.

The overall goal of this paper is to examine the role of increasing imports and concomitant change in import composition on the southeastern United States (i.e., North Carolina through Texas) shrimp processing industry. To achieve this goal, the world shrimp supply, including an analysis of farmed and wild production, is first reviewed for the 1980–2001 period. Then, the world export and import situation is briefly examined along with a more detailed analysis of the U.S.

*Corresponding author: walterk@lsu.edu

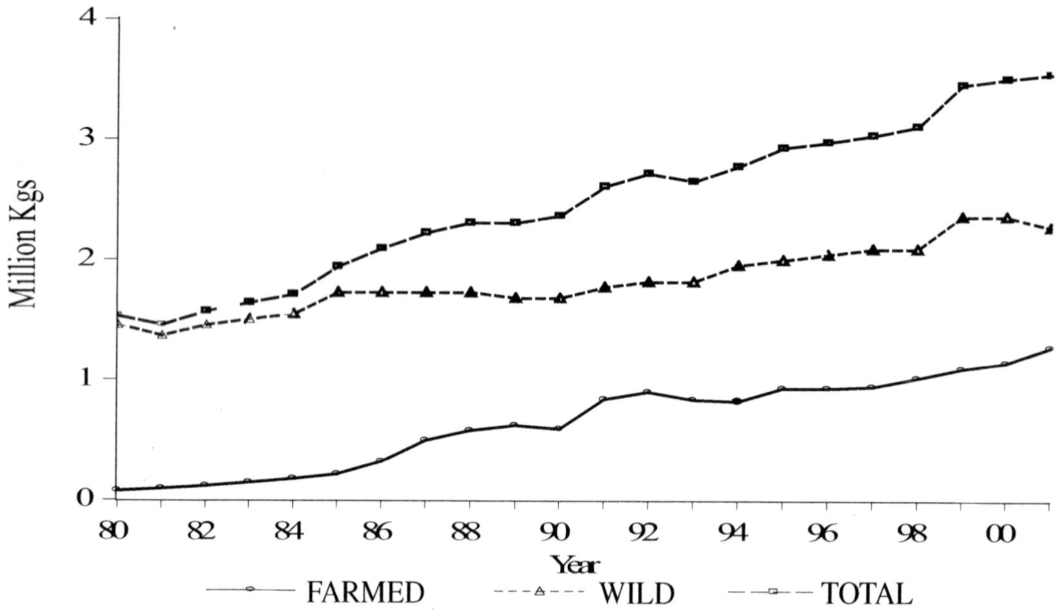

Figure 1. Estimated world production of warm-water wild and farmed shrimp (live weight), 1980–2001.

shrimp import market. The review of world shrimp supply and trade in the product sets the stage for analysis of the southeastern United States shrimp processing industry. Specifically, using secondary data, we explore how the increased imports and changing import composition have impacted processing activities including the number of firms, the marketing margin (and presumably profitability), and the composition of products produced by the southeastern United States shrimp processing sector. Finally, based on this analysis, we provide a brief forecast of changes that can be anticipated over the next several years in the southeastern United States shrimp processing industry.

World Shrimp Supply

Shrimp production, as with many other seafood commodities, is a combination of wild harvest and farming activities. Estimated total annual warmwater shrimp production (i.e., captured and farmed product) throughout the world, as indicated in Figure 1, expanded from 1.5 billion kilograms (live weight) in 1980 to about 3.5 billion kilograms (live weight) in 2001. Overall, the increase in world shrimp production during the 1980 through 2001 period translates into a growth rate of about 100 million kilograms per year. To place this annual growth rate in perspective, southeastern United States shrimp harvests generally fall in the 100 million kilograms to 127 million kilograms (live weight) range. Hence, annual growth in world production of warmwater shrimp can, in some years, approximate total U.S. production of warmwater shrimp.

Much of the growth in world warmwater shrimp production since 1980 has been the result of successful farming activities throughout the world, particularly in Asia and, to a lesser extent, in South and Central America. World production of warmwater farmed shrimp in 1980 equaled about 73 million kilograms (live weight), which accounted for approximately 5% of total world production at the time. By 2001, farmed production had

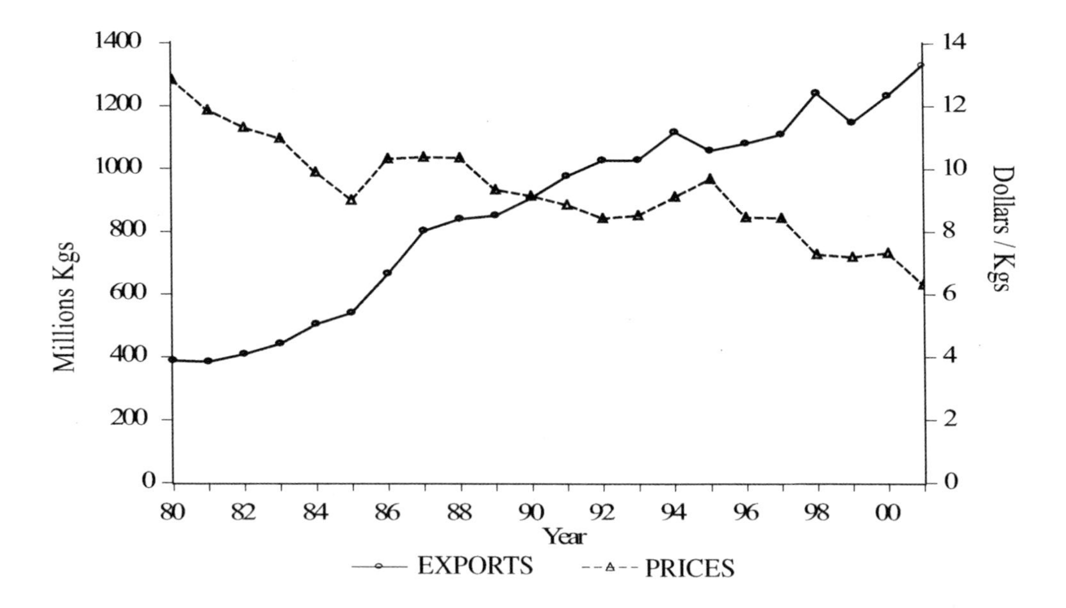

Figure 2. World exports of fresh and frozen shrimp and deflated export price (2001 US CPI), 1980-2001.

advanced to 1.27 billion live-weight kilograms, or more than 35% of total world warmwater shrimp output. Overall, warmwater farmed shrimp production increased by approximately 59 million kilograms per year during the 1980–2001 period. Wild warmwater shrimp output equaled about 1.45 billion pounds in 1980 (Figure 1). Though there has been growth in this sector since 1980, its rate of growth has been substantially less than the rate of growth in farmed production. Specifically, at an annual growth rate of about 39 million kilograms per year, production of wild warmwater shrimp advanced to almost 2.27 billion kilograms in 2001.

World Exports and Imports

World exports of fresh and frozen shrimp (the two categories constituting the overwhelming majority of trade) equaled about 386 million kilograms (product weight) in 1980 (Figure 2). By 2001, exports had almost quadrupled to 1.36 billion kilograms. A minimum of 60% of the total world shrimp production currently enters the trade market. The value of world exports in 1980 equaled $2.3 billion, or about 15% of the total world trade in seafood products. By 2001, the current value of the world shrimp trade in fresh and frozen product had increased to about $8.4 billion and represented close to 20% of the total world trade in seafood products. (In addition to the increase in fresh and frozen trade, there has been a sizeable increase in trade in 'prepared or preserved' shrimp which equaled about $2.0 billion in 2001). Much of the apparent increase in current value is, of course, due to currency inflation. After adjusting for inflation, the value of the world shrimp trade advanced by about 70% (from $4.97 billion to $8.42 billion based on the 2001 U.S. Consumer Price Index). This 70% increase is considerably less than the 240% increase in export quantity, suggesting a sharp decline in the real (i.e., deflated) price of the exported product. Overall, the $6.32 constant-dollar per kilogram price of the exported product in 2001 reflects

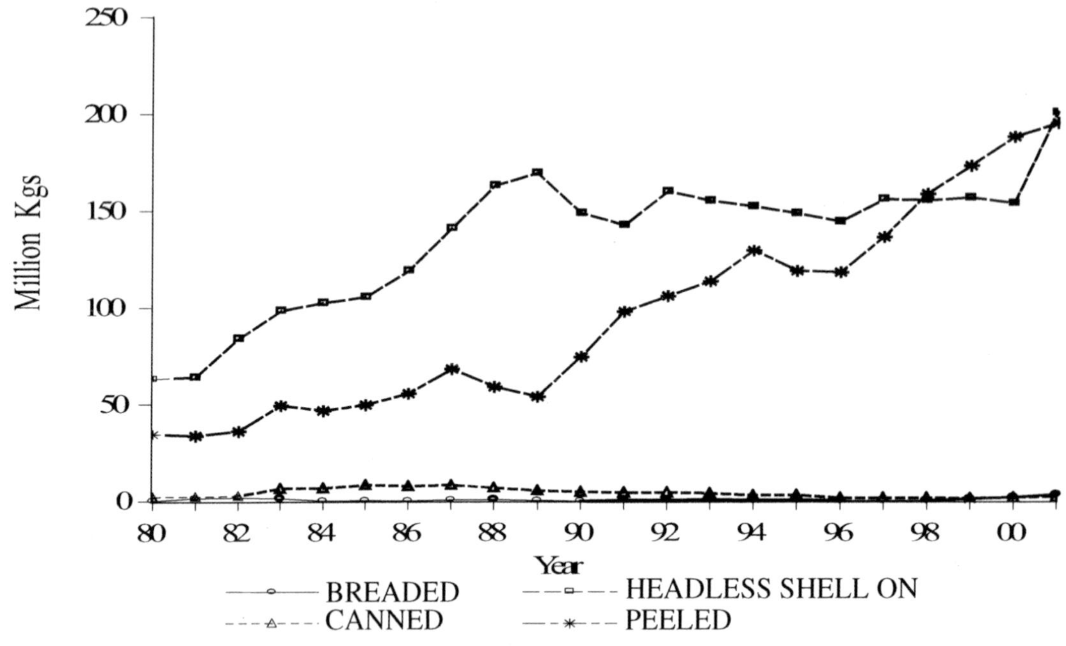

Figure 3. United States imports of shrimp by product form (product weight), 1980–2001.

a 50% decline from the $12.82 per kilogram price in 1980 and about a 40% price decline from the prices observed as recently as the 1986–1988 period (Figure 2). Given the relatively high increase in world income during the 1990s, the decline in price would tend to suggest that growth in shrimp supply has exceeded the growth in demand, resulting in a downward pressure on the real price.

The increased farmed shrimp production, of course, allowed for more product to enter the international trade market. However, it is important to recognize that the rising trade flow reflects not just increased production in total, but also the source (i.e., farmed production versus wild production) of the increased output. As noted by Csavas (1994), the farm-raised shrimp is of greater importance than wild product in world trade. Some of the reasons cited by the author include 1) the farm-raised product has greater "freshness" than the wild product; 2) farmed product is less seasonal in nature, and more reliable, than its wild counterpart; and 3) species and sizes can be controlled better in farm-based system than in a wild-based system.

While the primary exporters of shrimp are many and have changed substantially over time, two countries, the United States and Japan, have long dominated the import market. These two countries, combined, account for upwards of 50% of world shrimp imports by value. The European Union represents a significant portion of the remaining import market, particularly if limited to warmwater shrimp trade.

U.S. Imports: A Closer Look

Annual U.S. shrimp imports, expressed on a headless shell-on equivalent weight basis[1],

[1] The U.S. National Marine Fisheries Service utilizes the following conversion factors: a) Atlantic and Gulf commercial landings are converted to heads-off weight by multiplying them by 0.629; Imports: b) breaded, 0.69; c) shell-on, 1.00; d) canned, 2.52; and e) other, 2.40

more than quadrupled over the 22-year period ending in 2001, from about 118 million kilograms to almost 540 million kilograms. Given the relatively high proportion of the world export market that is destined for the United States, one would anticipate a close relationship between the world export price and the U.S. import price. This relationship is borne out in a comparison of the two price trends (since the U.S. import price is expressed on a headless shell-on basis, one would expect that it would generally be somewhat lower than the world export price, which is given on a product weight basis). Like the world export price, the U.S. import price has been gradually trending down, with the 2001 price exhibiting the lowest level dating back to 1980.

In addition to the increased import base, the composition of the imports has been changing. Specifically, value-added products, particularly peeled product, have been representing an increasing share of total imports. In 1980, for example, headless shell-on shrimp imports, equaling 63 million kilograms, represented 63% of total imports, expressed on a product weight basis (Figure 3). Peeled product (raw and other), equaling 35 million kilograms and representing 35%, accounted for almost all of the remaining imports. Canned and breaded products represented the remaining 2% of total imports. While imports of headless shell-on product increased throughout the period of analysis to 200 million kilograms in 2001, its share fell to 50% of the total import base. By comparison, the share of the total import base represented by peeled product increased to almost 50%.

In general, while there has been a steady growth in peeled product during the 1980–2001 period, the growth since the early 1990s can best be defined as explosive. Specifically, U.S. imports of peeled product advanced by 160% since 1990, from 75 million kilograms to 195 million kilograms in 2001. The increase is being "fueled" by developing countries attempting to garner additional hard currency via value-added activities. As discussed below, the changing import composition has significant ramifications with respect to the domestic processing sector.

U.S. Southeast Shrimp Processing

Shrimp represents the primary component of the southeast seafood processing industry, generally contributing more than 80% of the total edible production activities by value. The southeastern United States processing industry, using a combination of domestic raw material and imported raw material, generated sales of $1.1 billion in 2001. A brief review of the processing sector, with emphasis being given to the impacts of imports, is presented below. Processing information used in the review is based on annual end of the year surveys of processing establishments conducted and maintained by the National Marine Fisheries Service (NMFS). While processing information is available from 1973 forward, the analysis presented in this paper begins in 1980.

Firms and Production Activity

The number of firms engaged in southeast shrimp processing activities declined almost by half, from 173 to 89, during the 1980–2001 period (Figure 4). As indicated, the decline was relatively steady with a reduction of more than 50 firms in the last decade (1991 through 2001). This compares with a reduction of only about 20 firms during the decade of the 1980s. While most agricultural commodities have observed consolidation in recent years, the consolidation in the southeast shrimp processing sector is almost certainly

Figure 4. Reported number of southeast U. S. shrimp processors, 1980–2001.

tied, at least in part, to the increasing import base, including the increasing imports of value-added products (primarily peeled raw and cooked shrimp). Overall, the number of processors in the Gulf fell by about 40%, from 124 to 72, while the reported number of south Atlantic processors fell from 49 to 17, or by almost two-thirds.

Despite the sharp reduction in number of reported southeast shrimp processing establishments, the quantity processed, expressed on a headless shell-on equivalent weight basis, has remained relatively stable since the mid-1980s, fluctuating annually around the 120 million kilograms range (Figure 5). This follows a period of rapid expansion in output during the early to mid-1980s. Overall, Gulf processors have historically accounted for an average of about 85% of this total production.

The deficit in domestic landings relative to southeast shrimp processing needs is well documented (see, for example, Prochaska and Andrew 1974; Roberts et al. 1992; and Keithly and Roberts 1995). Given the significant increase in imports since the early to mid-1980s, one might expect increased import usage by the southeast shrimp processing sector. While NMFS does not ask processors to differentiate output derived from domestic shrimp from that of imported shrimp, some information can be gleaned by comparing total processing activities to domestic landings. Doing so will provide a true estimate of import usage if all domestic production is utilized by the processing sector. Based on analysis by Keithly and Roberts (1995), this assumption appears plausible.

In 1980, imports accounted for an estimated 23 million kilograms, or about one-quarter of total southeastern shrimp processing activities (Figure 5; i.e., the difference between processed poundage and southeastern U.S. landings). Import usage increased rapidly, thereafter, most likely in association with increased Ecuadorian cultured shrimp exports to the U.S. market. By 1986, estimated imports ac-

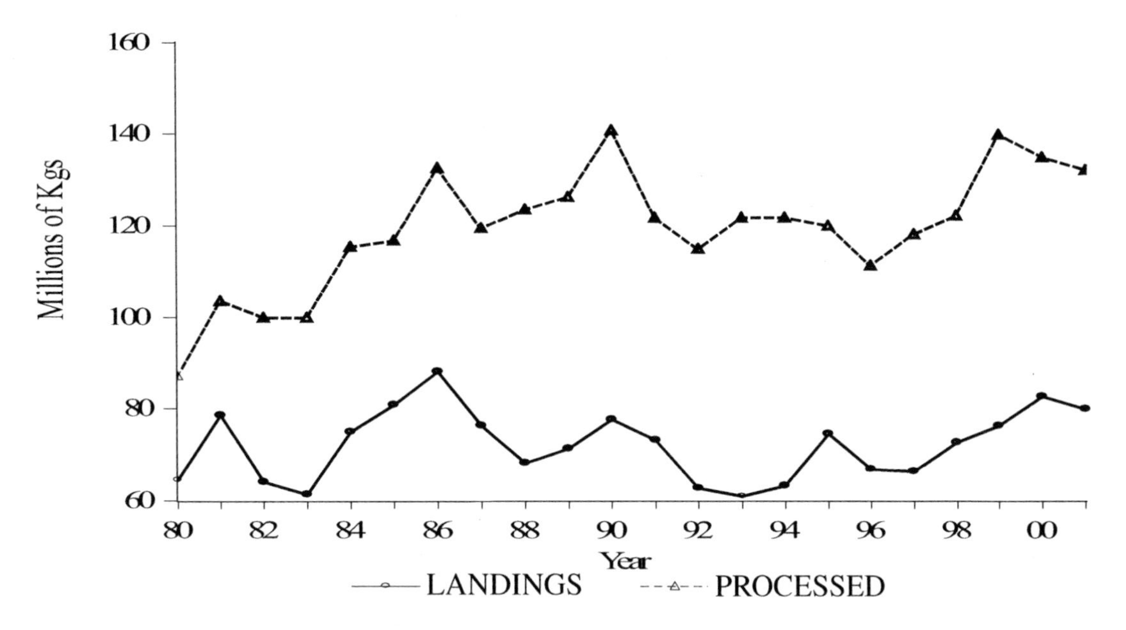

Figure 5. Processed quantities and southeast landings expressed on a headless-shell-on weight basis, 1980–2001.

counted for about 45 million kilograms, or one-third, of the total domestic Southeast shrimp processing activities. During the 1992–1994 period, import usage as a proportion of total processing activities equaled almost 50%, indicating that almost as much imported shrimp was used by the processing sector as domestic shrimp. This period can be characterized as one of very high import usage relative to total processing activities. Since this peak period, however, import usage has fallen, averaging slightly more than 50 million kilograms annually since 1995 (approximately 40% of the total).

To determine why southeast shrimp processing activities have not increased in relation to imports, the value of southeastern shrimp processing activities will be examined first. As the information in Figure 6 suggests, the current value of southeastern U.S. shrimp processing activities, while fluctuating widely on a year-to-year basis, has exhibited no long-term upward trend since the mid-1980s. When adjusted for inflation, the trend has been decidedly downward. Overall, the deflated value of processing activities during the 1999–2001 period averaged only 70% of that observed during the 1983–1985 period. This 30% decline, while significant in and of itself, came during a period of time in which pounds processed increased by more than 20%. This suggests that the deflated price of the processed product has fallen sharply during the 1980–2001 period.

Overall, the deflated price of the processed product fell from well over $15 per kilogram (headless shell-on equivalent weight) during the early 1980s to less than $9 per kilogram during the late 1990s and into the next decade (Figure 7). The decline has been, for the most part, steady with no sign of abatement.

Estimated Marketing Margins

Evaluating only the output price may not provide an accurate depiction of the potential changes in the profitability in the southeastern shrimp processing sector. Specifically, the price of the raw material (i.e., the raw shrimp

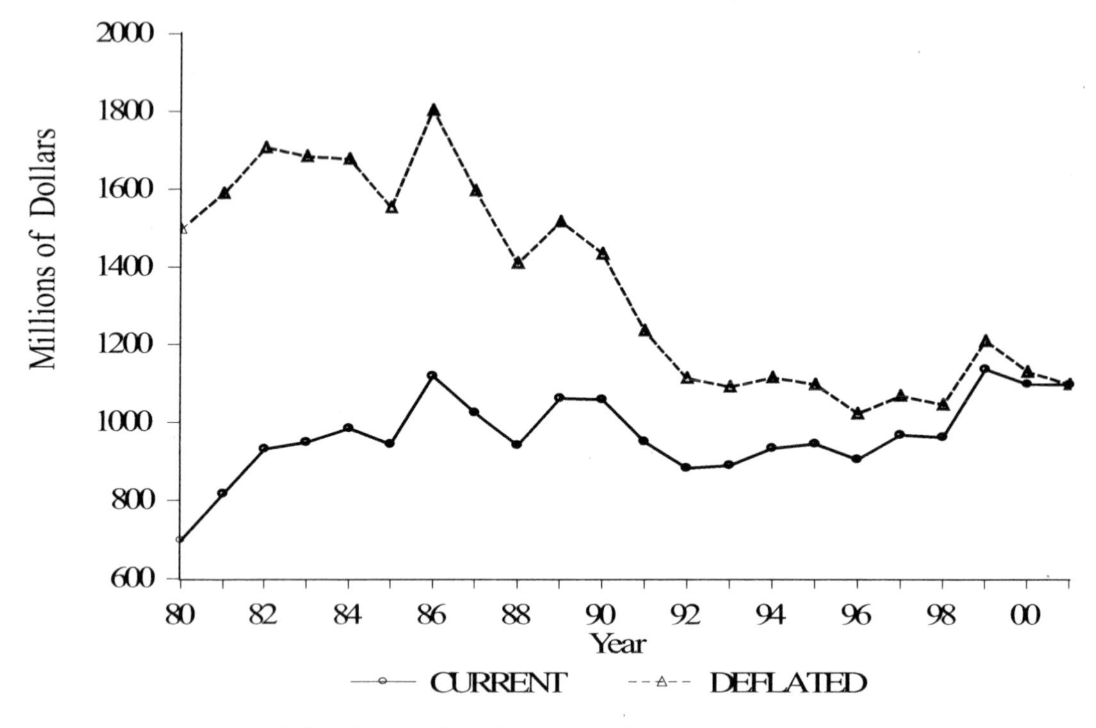

Figure 6. Current and deflated value of southeast U. S. shrimp processing activities, 1980–2001.

product) being used in processing activities may be declining by an equivalent amount. If this is the case, the marketing margin, defined as the difference between the output price and the price of the raw material being used in the production process, would remain constant, suggesting that there may be no change in profitability (The marketing margin is not a direct measure of per unit profitability. Rather, profitability, along with all other production and marketing costs, such as labor and transportation, comprise the marketing margin. Subtracting all production and marketing costs from the marketing margin would result in an estimate of profitability. These costs, however, are not readily available). Unfortunately, the price of the raw product used in processing activities is not collected by NMFS. While the price of the raw product used in processing activities, which is comprised of a combination of both domestic landings and imports, is not collected by NMFS, the dockside price of the domestically harvested product is readily available. To the extent that this price also adequately reflects the price of imported product used by the processing sector, a meaningful estimate of the marketing margin can be derived. Overall, the dockside prices and import prices (converted to a headless shell-on weight) track each other very closely, indicating that the dockside price is likely to be an adequate proxy for the cost of imported product used in domestic processing, subject to an important caveat. This caveat is that one type of imports (say, very small shrimp) does not dominate raw import usage by the southeastern U.S. processing sector.

The difference between the dockside price (expressed on a headless weight basis) and the processed price (expressed on a headless shell-on weight equivalent basis) is illustrated in

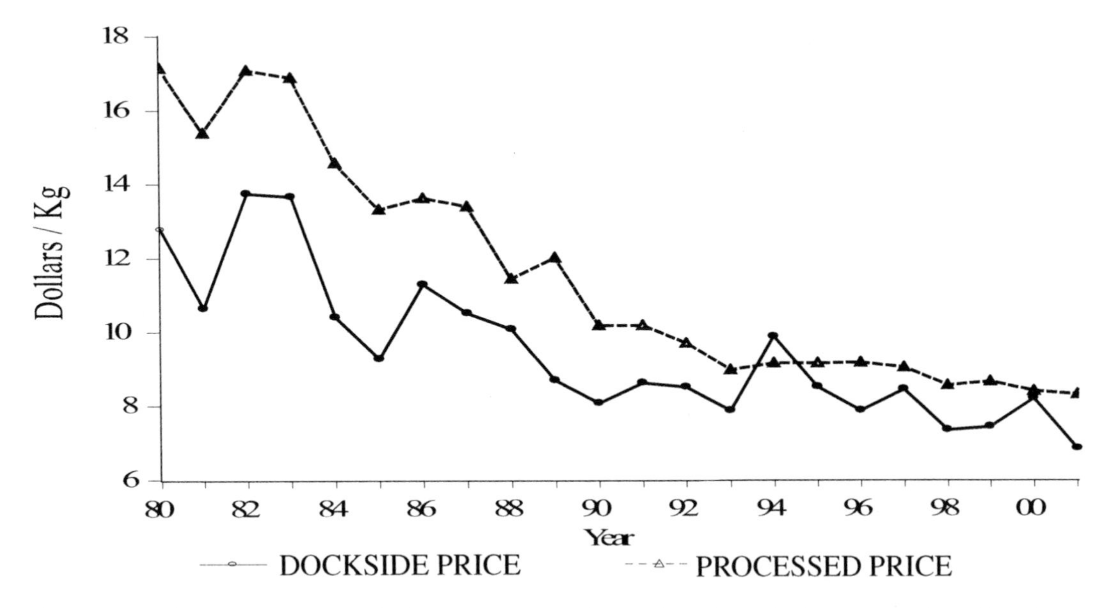

Figure 7. Deflated southeast U.S. shrimp processed price (headless-shell-on weight) and Dockside Price (headless-shell-on weight), 1980–2001.

Figure 8. Overall, the estimated marketing margin has declined substantially with most of the decline occurring since the early 1990s. This is certainly one indication that per unit profitability is falling and provides a rationale for the substantial exit behavior observed since the early 1990s.

Coping with a Declining Marketing Margin

One can hypothesize that the southeastern U.S. shrimp processing establishments have coped with the declining per unit profitability by increasing output. Overall, production averaged about 0.59 million kg per firm during the early 1980s (Figure 9). By 1999–2001, this average had increased to more than 1.45 million kilograms. The deflated value of output per firm averaged about $9.5 million during the 1980–1982 period and advanced to over $12 million during 1999–2001. This most recent 3-year average, however, is substantially above that observed during the previous two decades (by about $2 million) and only additional observations will indicate whether it can be maintained.

Taking the analysis one step further, we can examine changes in the average gross marketing margin per processing firm; defined as the total deflated processed shrimp sales less the cost of raw material. Prior to the early 1990s, the deflated gross margin per firm, with few exceptions, averaged from about $2.0 million to $3.0 million per year. This fell substantially during most of the 1990s, though the 2001 figure does indicate a relatively high gross margin in that year. The relatively high margin in the most recent year is most likely simply an anomaly that can be examined with additional yearly observations not available to the authors at time of publication.

What's in Store?

There is an adage that the only sure things in life are death and taxes. While true, the strong trends discussed in the previous sections of this paper, along with economic theory, al-

Figure 8. Estimated average gross margins per firm.

lows us to make some observations regarding future conditions in the southeastern U.S. shrimp processing sector with a fair amount of confidence. These predictions are predicated, of course, on the assumption that world production, particularly that of farm-raised shrimp, will continue to increase and that a significant proportion of the increase will be directed to the U.S. market. Furthermore, it is assumed that increased pre-export, value-added activities will continue to be the norm. To the extent that these assumptions are not valid, the conclusions derived below become much more tenuous.

The first conclusion is that the processed price for the various products will, in the long run, continue the observed downward trend and the marketing margin will continue to narrow. The continued narrowing of the marketing margin leads directly to the second conclusion: consolidation in the industry will continue. While the degree of consolidation is a matter of extreme speculation, there is considerable less uncertainty around the forecast that consolidation will occur. The final conclusion is that the average production per firm will continue to increase. This conclusion is linked directly to the previous two conclusions. Specifically, a narrowing of the marketing margin implies that increased output per firm will be required to maintain a desired level of profitability. Given the declining number of firms, furthermore, domestic landings will be divided among a fewer number of firms.

The continued expansion in output per firm raises a potentially interesting and unanswerable question. Specifically, beyond some level of output one might anticipate individual firms to reach capacity, diseconomies of scale, or both, in their respective operations. Reaching capacity would place a physical constraint on the ability to increase output and diseconomies of scale, which could occur well before any capacity constraints are realized, and could result in increasing costs per unit of output which could make processing of the domestic product uncompetitive with the value-added imported product. What happens at this point? If capacity constraints are real-

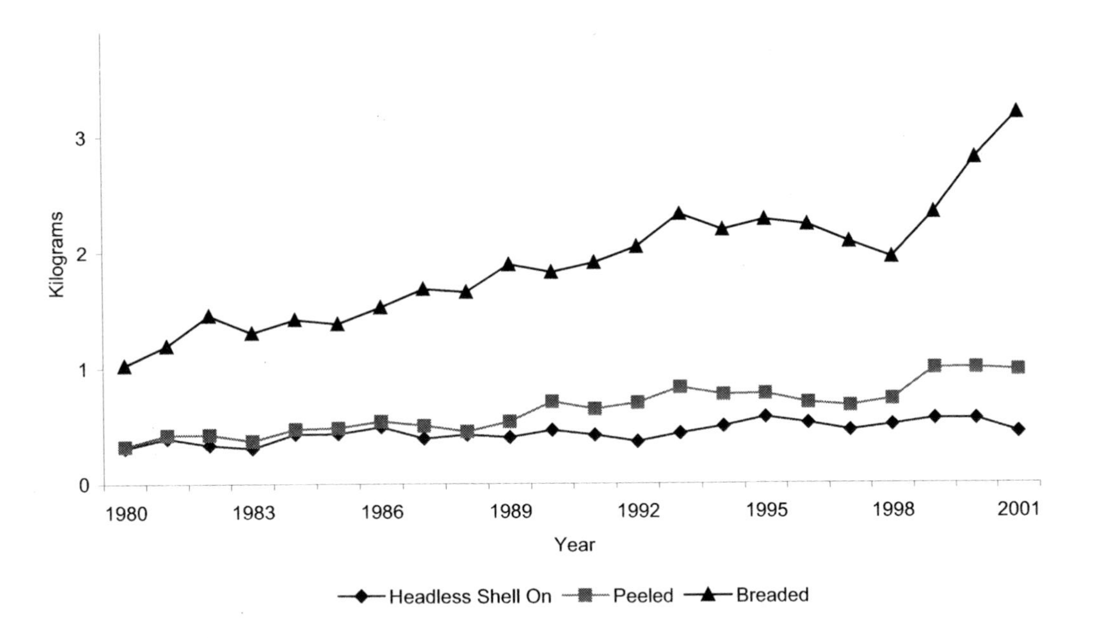

Figure 9. Average per firm production of headless-shell-on, peeled and breaded shrimp—Southeast United States, 1980–2001 (million kilograms, product weight).

ized, one might speculate that this would place an artificial constraint on harvest (i.e, harvest would be artificially constrained to some amount less than that amount which could physically be harvested on a sustainable basis). Diseconomies of scale could yield an analogous artificial constraint on harvest or, alternatively, lead domestic processors to reduce the price of the raw domestic product (i.e., the price received by the domestic harvesters) such that the processed domestic product remains competitive with the processed imported product. Neither of these two scenarios, if they were to come to fruition, bodes well for the harvesting sector.

References

Csavas, L. 1994. Important factors in the success of shrimp farming. World Aquaculture 25(1):34–56.

Keithly, W. R., and K. J. Roberts. 1995. Shrimp closures and their impact on the Gulf region processing and wholesaling sector (expanded to include South Atlantic). Revised final report to National Marine Fisheries Service, Contract No. NA17FF0376–01, Seattle.

Prochaska, F. J., and C. O. Andrew. 1974. Shrimp processing in the southeast: supply problems and structural changes. Southern Journal of Agricultural Economics 6(1):247–252.

Roberts, K. J., W. R. Keithly, and C. M. Adams. 1992. Determinants of imported shrimp and their role in the Southeast shrimp processing sector. NMFS Technical Memorandum (NMFS/SEFC-305), Seattle.

American Fisheries Society Symposium 49:295–311

Disaggregated Projections of Fish Supply, Demand, and Trade in India

PRADUMAN KUMAR*
Division of Agriculture Economics, Indian Agricultural Research Institute New Delhi 110012, India

MADAN MOHAN DEY
World Fish Center, Post Office Box 500, GPO 10670 Penang, Malaysia

Abstract.—The paper provides supply, demand, and trade projections of fish at disaggregated level (species group) for India. AsiaFish model was employed. Producer core system for fish sector is estimated using the dual approach. The multistage budgeting framework with almost ideal demand system (AIDS) model is used for fish demand analysis based on consumer survey data. The Armington approach is used for the trade core. Under various scenarios of technological changes, the projections of fish supply, demand, and export by fish species groups are obtained for the year 2015. Conclusions are drawn for the overall welfare of various stakeholders.

Introduction

Fisheries sector in India is undergoing a transformation and contributes to the livelihood for a large section of economically underprivileged population of the country. The fish production has increased at the rate of 5.1% per year during 1980–2000 and has touched 5.7 million metric tons (mt) in 2000 from a level of 1.76 million mt in 1970–1971 and 2.44 million mt in 1980–1981. The fisheries, a sunrise sector has recorded faster growth as compared to the crop and livestock sector. The share of fisheries in agricultural gross domestic product (GDP) has increased from 1.7% in 1980 to 4.2% in the year 2000. Policy support, production strategies, public investment in infrastructure, research, and extension for fisheries had significantly contributed to increased fish availability.

Available projections of demand and supply for the fish sector are limited by their high degree of aggregation and the lack of empirical basis for the underlying elasticities of supply and demand (Paroda and Kumar 2000; Delgado et al. 2003). A more useful description of fish sector becomes imperative for rational and pragmatic planning for specific fish types. The present paper attempted the analysis of fish demand and supply by species group[1] for India. Attempt has also been made to project the supply, demand, and export of fish by species group by the year 2015 and

*Corresponding author: praduman@cgiar.org

[1] Fish species are grouped into species groups with the help of experts based on biological, commercial value and market destinations.

examine their effects on the welfare of fish producers, consumers, and traders.

The Model

The AsiaFish model (Dey et al. 2003), which consists of producer, consumer, and trade cores, was employed for the disaggregated analysis of the fish sector. The time series data on fish production and farm survey data on fish farming at regional level was used to estimate producer core following the dual approach (Quadratic profit function). The multi-stage budgeting framework with AIDS model was used for fish demand analysis based on consumer survey data (Deaton and Muellbauer 1980; Dey 2000). Armington approach was used for the trade core (Armington 1969). The model is closed with a set of equilibrium conditions between supply, demand, and trade. The model was run under various scenarios of total factor productivity. Projections of supply, domestic demand, and export by species group were obtained.

Fish Species

There are a large number of species of inland and marine fish. These are grouped into Indian major carps (IMC), other freshwater fish (OFWF), shrimp (marine and freshwater), pelagic high value (PHV), pelagic low value (PLV), demersal high value (DHV), demersal low value (DLV), and mollusks and others (mollusks).

Demand Elasticity

The income elasticities vary substantially across fish species by income group. Fish demand is sensitive to price changes. The compensated own price elasticity was estimated for various species. It was –0.97 for mollusks, followed by PLV (–0.95), DHV (–0.92), Shrimp (–0.88), OFWF (–0.97), PHV (–0.86), DLV (–0.86), and least for IMC (–0.52).

Supply Elasticity

Aquaculture fish supply elasticities were estimated to be 1.56 for IMC, 1.72 for OFWF, and 0.73 for shrimp with respect to own-fish price. In case of marine fish, the own-fish price elasticities were found to be highly inelastic for all the fish species ranging from 0.28 to 0.50. The elasticities of fish supply with respect to input prices were negative and inelastic.

Total Factor Productivity

The Divisia–Tornqvist index was used for computing the total factor productivity (TFP) for the inland and marine fisheries sector. The TFP annual growth was estimated to be 4.0% for the aquaculture sector and 2.0% for the marine sector in India.

Projections

Multimarket fish sector model developed at WorldFish Center by Madan et al. 2003 has been used for India. Under the baseline assumption, it is assumed that the growth in the exogenous variables in the projected years will be the same as observed in the past except per capita income and population growth, which has been assumed at 5% and 1.5%, respectively. Given a time horizon (2005–2015), projections for supply, demand, export, and their prices, are generated under the following technological scenarios:

S_1 = baseline assumptions with existing past growth in TFP for marine capture (2%) and aquaculture (4%)

S_2 = baseline assumptions with 25% deceleration in TFP growth by the year 2015

S_3 = baseline assumptions with 50% deceleration in TFP growth by the year 2015

S_4 = baseline assumptions with 75% deceleration in TFP growth by the year 2015

S_5 = baseline assumptions without TFP growth during the projected period.

Supply Projections

The projected growth of fish supply for the period 2000–2015 is given in Table 1. The results revealed that the fish production will grow at the rate of 3.0% corresponding to the baseline scenario (with existing growth in TFP), would decline to 2.2% in scenario 3 (with 50% deceleration in existing TFP) and will stagnate in the absence of technological growth (scenario 5). The supply will steeply decline with the deceleration of TFP growth. Across the species, the growth in supply varies significantly. It was highest for inland fish (Indian major carps and other freshwater fish), which ranges between 2.7% and 3.9% per annum. The supply of shrimp will grow at a faster rate with an annual growth of 2.5–3.4%. The supply of marine fish species is projected to grow in the range of 1.4–1.9% per annum during 2000–2015, except for shrimp. The scenario 5 revealed that the fish production would stagnate if the technological growth does not take place in future. To maintain the supply at desired level, there is a need to put concerted efforts for improving the efficiency of fish production and catches and enhancement of the growth in TFP through appropriate policies of research and development, extension, and so forth.

The rise in supply growth and shift in supply curve towards the right has not declined the price for all the species. A mixed effect was observed on the real prices. For the species

Table 1. Projected growth in fish supply and pricein 2000–2015, India.

	Scenario 1	Scenario 3	Scenario 5
Supply			
Indian major carps	3.88	2.78	–0.09
Other freshwater fish	3.81	2.71	–0.18
Shrimp	3.40	2.51	0.18
Pelagic high value	1.95	1.40	–0.05
Pelagic high value	1.95	1.40	–0.05
Demersal high value	1.92	1.36	–0.09
Demersal low value	1.99	1.43	–0.01
Mollusks	1.98	1.42	–0.03
All	3.04	2.17	–0.06
Producer price			
Indian major carps	–2.85	–1.80	1.03
Other freshwater fish	–2.72	–1.67	1.17
Shrimp	2.07	2.62	4.10
Pelagic high value	–0.06	0.39	1.60
Pelagic high value	–0.76	–0.22	1.21
Demersal high value	1.61	1.99	2.99
Demersal low value	–1.31	–0.75	0.73
Mollusks	1.66	2.09	3.24
All	–0.33	0.38	2.24

that are not entering in the export market, the prices will decline with the increase in supply. These species are IMC and OFWF from inland sources where prices of these species will decline in the projected period at the rate of 1.7–2.8% per annum. Among the marine species which are of low value, fish price will decline by less than 0.76% per annum for PLV and 0.75–1.31% for DLV. The price of export oriented fish species will continue to rise with the increase of their supply. The higher growth in fish supply for the species used in the domestic market will benefit the common man as this fish species will be available at cheaper price in future. In the fish species that are export oriented, the rise in supply will not cut down the price in the domestic market substantially, and the price will keep rising and will benefit the producer. The price of shrimp, which is the most important exportable fish, will rise from 2.1% to 2.6% annually. Other exportable fish species are PHV, DHV and mollusks for which also the price will rise from 1.6% to 2%. But taking all the species together, the impact of positive supply growth on fish prices will move in a narrow band (–0.33–0.38% per annum).

Under the baseline scenario, with the increase of fish supply (as projected in various scenarios), producer prices in the domestic market will decline at an annual rate of 2.9% for IMC, 2.7% for OFWF, 1.3% for DLV, and 0.8% for PLV (Figure 1). These are the species mostly meant for the domestic market. Shrimp, PHV, DHV, and mollusks (high value) are important exportable species. A part of their output is also retained for domestic consumption. Their prices in the domestic market are unlikely to decline even with their increasing supply. Rather their prices are expected to increase. The rate of increase is projected to be 2.1% for shrimp, 1.6% for DHV and 1.7% for mollusks. Prices of PHV group are likely to remain unchanged. Exports will help producers stabilize fish prices in the domestic market (at the aggregate level). Taking all the species together, the price of fish will move in a very narrow band with the annual growth ranging from 0.3% to 0.4% at constant price.

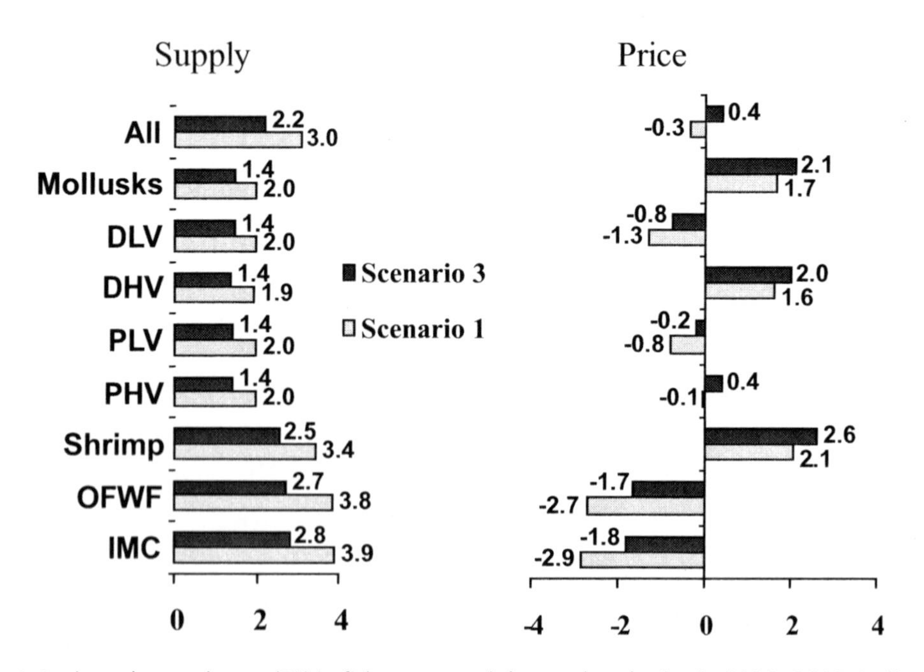

Figure 1. Projected growth rate (%) in fish suppy and demand and price in 2000–2015, India

Table 2. Projected supply, import, and production of fish, India (million kilograms).

Year	Scenario 1	Scenario 2	Scenario 3	Scenario 4	Scenario 5
Supply					
1998 (base)	5481.6	5481.6	5481.6	5481.6	5481.6
2005	6741.8	6669.0	6576.3	6441.2	5460.1
2010	7833.7	7575.0	7275.5	6893.9	5445.0
2015	9119.4	8519.5	7894.1	7199.4	5430.1
Import					
1998 (base)	70.7	70.7	70.7	70.7	70.7
2005	75.6	75.4	75.1	74.7	71.6
2010	79.3	78.7	77.9	76.8	72.3
2015	83.3	81.9	80.3	78.5	73.0
Production					
1998 (base)	5410.9	5410.9	5410.9	5410.9	5410.9
2005	6666.3	6593.6	6501.2	6366.5	5388.5
2010	7754.4	7496.3	7197.6	6817.1	5372.7
2015	9036.1	8437.6	7813.8	7121.0	5357.1

Based on the projected growth rate, the supply of fish is projected under various scenarios using triennium ending (TE) 1998 as the base year supply (Table 2). The scenario 3 will be the most likely scenario to prevail in future, which assumes that the maximum decline in TFP growth of fish production will be 50% in the year 2015. Under the most optimistic scenario (S_1), the production of fish[2] will be 6.67 million mt in the year 2005 and will grow to 9.04 million mt in the year 2015. Considering the other scenarios, the fish production is projected to be 8.4 million mt in scenario 2, 7.8 million mt in scenario 3, and 7.1 million mt in scenario 4. For scenario without TFP growth, the production will be stagnant almost at the current level. The import of fish will be quite marginal and will not vary in the projected years.

As seen in Table 3, annual production of inland fish in the year 2005 will be in the range of 3.6–3.7 million mt and will reach 4.6–5.5 million mt in 2015, with an annual growth rate of 2.9–4.0% under different scenarios. The share of inland fish in total fish production, which was about 50% in the year 2000, will increase to 61% in 2015. The production of marine fish is likely to be in the range of 2.9–3.0 million mt in 2005 and 3.2–3.6 million mt in 2015. The fish production will grow at the annual rate of 2.9–4.0% for inland fish and in the range of 1.2–1.8% for marine fish. The share of marine fish in the total fish production will decline from 50% in 2000 to about 40% in 2015.

The supply of IMC, which contributed 25% of the total supply, will be 1.79–1.85 million mt in the year 2005 and will grow 2.04–2.24 million mt in the year 2010. In the year 2015, the supply will grow to a level of 2.26–2.71 million mt of IMC, 1.6–1.8 million mt of Pelagic fish, 0.7–0.8 million mt of demersal fish, and 0.6–0.7 million mt of mollusks and others. The changes in the share of different spe-

[2] The production projection has been arrived at after subtracting the projected import from the supply projection.

Table 3. Projected growth and supply of fish by source.

	Inland		Marine	
	Scenario	Scenario	Scenario	Scenario
Annual growth (%) During 2000–2015	4.0	2.9	1.8	1.2
Supply of fish (million kg)				
2000	3082	3077	2732	2729
2005	3749	3636	2993	2942
2010	4562	4164	3272	3111
2015	5554	4657	3566	3238
Percent share of inland and marine fish in total production				
2000	53	53	47	47
2005	56	55	43	45
2010	58	57	42	43

cies in total production during the period 2000–2015 can be seen from the Figure 2.

The share of IMC in total fish production will increase to 30% in 2015 from 25% in 2000, and OFWF will increase to 22% from 19%. The share of shrimp will remain almost unchanged. While the share of pelagic, demersal, and mollusks will decline from 24% to 20%, 11–9%, and 9–7%, respectively during this period.

By the year 2015, the incremental production is projected to be 3.3 million mt (Figure 3). Out of this additional production, IMC will contribute maximum (36%) followed by OFWF (26%), pelagic (14%), shrimp (13%), demersal (6%), and mollusks (5%).

A comparison of scenario 1 and 5 provides the effect of TFP growth on fish supply (Table 4). The production of fish will decline substantially with the deceleration in fish tech-

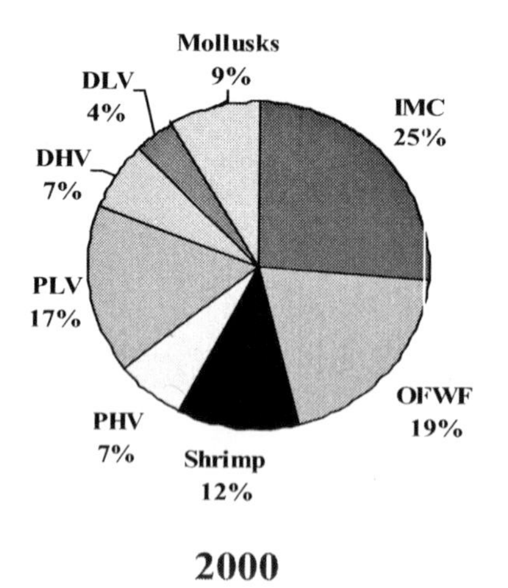

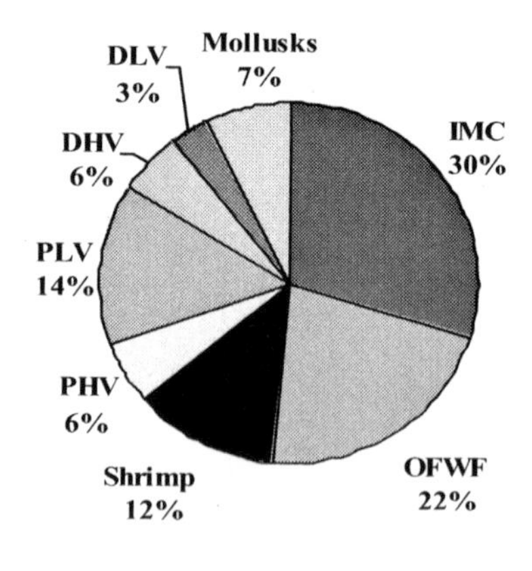

Figure 2. Changes in share of fish species in total population

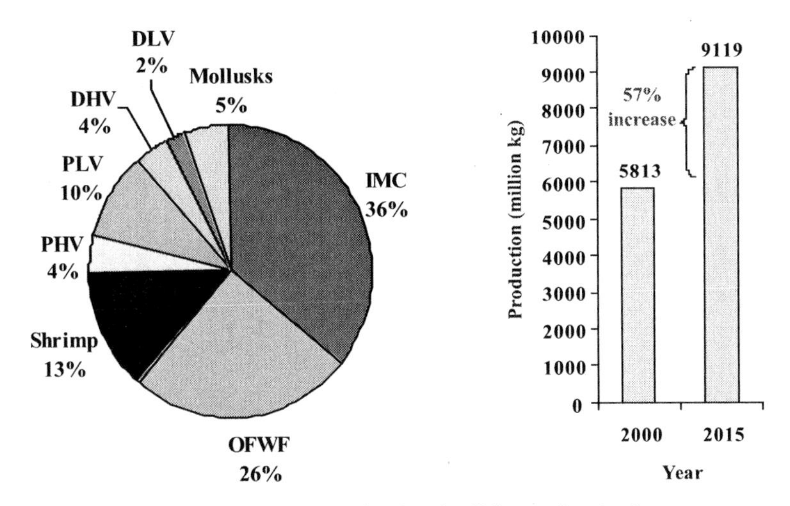

Figure 3. Incremental production and its contribution by fish species, India

nological growth. The contribution of TFP to fish production is projected at 1.3 million mt in 2005, 2.4 million mt in 2010, and 3.7 million mt in 2015. In percentage terms, its contribution in total production will be 19% in 2005 and will grow to 41% in 2015. The contribution of TFP will be the highest in inland fish sector (i.e., 48% in 2015). The technological change (measured in terms of TFP) will contribute about 29% to the total marine fish production except shrimp in 2015. In case of shrimp, it will be about 20% in 2005 and 42% in 2015.

Demand Projections of Fish

The increase in supply will make the fish available to the consumers at a cheaper price, which will increase the fish consumption in their food basket. The demand of fish will grow at the rate of 1.6–2.4% per annum (Table 5 and Figure 4).

Domestic demand for fish under the baseline scenario is likely to grow at an annual rate of 2.4% between 2000 and 2015. Highest growth in demand is projected for IMC (3.9%), followed by OFWF (3.8%), PLV, and DLV (2.0% each). Declining consumer prices are the major drivers of demand growth. However, domestic demand for various species meant for international market is likely to decline due to increase in their prices. Between 2000 and 2015, consumer demand for shrimp will decline at an annual rate of 1.1%, followed by DHV (–0.9 %) and mollusks (–0.7 %).

The domestic demand of fish will be in the range of 5.9–6.0 million mt in the year 2005. It is likely to grow to 6.7–7.7 million mt in 2015 (Tables 6 and 7). Under the baseline scenario, consumption of fish is projected to increase from 5.2 million mt in 1998 to 6.0 million mt in 2005 and 7.7 million mt in 2015. Out of this, in-home consumption will be about 66% and the remaining will be consumed away from home and enter industrial processing. The annual per capita consumption at national level is projected to be 5.6 kg in 2005 and 6.2 kg in 2015. About 35% of the Indian population eats fish. Thus, annual per capita consumption of the fish-eating population will be about 15 kg in 2005 and will increase to 16.7 kg in 2015. In-home annual per capita consumption is projected at 9.8 kg in 2005 and 10.8 kg in 2015. The estimates

Table 4. Projected production of fish by species and TFP contribution, India

Year	Production (million kg)			TFP contribution	
	Scenario 1	Scenario 3	Scenario 5	Quantity (mkg)	%
Indian major carp					
2005	1851.7	1793.6	1409.2	442.5	23.9
2010	2240.2	2039.7	1402.7	837.5	37.4
2015	2710.3	2260.9	1396.2	1314.0	48.5
Other freshwater fish					
2005	1360.6	1317.8	1034.3	326.3	24.0
2010	1640.3	1493.0	1025.1	615.2	37.5
2015	1977.5	1648.4	1016.0	961.5	48.6
Shrimp (marine and freshwater)					
2005	808.2	787.6	647.8	160.4	19.8
2010	955.2	885.1	653.7	301.4	31.6
2015	1128.8	974.3	659.7	469.1	41.6
Pelagic high value					
2005	428.4	421.5	372.9	55.5	13.0
2010	471.9	449.8	372.0	99.9	21.2
2015	519.9	473.8	371.1	148.7	28.6
Pelagic low value					
2005	1066.5	1049.3	928.4	138.0	12.9
2010	1174.8	1119.8	926.3	248.5	21.1
2015	1294.0	1179.7	924.2	369.9	28.6
Demersal high value					
2005	419.9	413.1	365.4	54.5	13.0
2010	461.7	440.1	363.8	97.9	21.2
2015	507.7	462.7	362.3	145.4	28.6
Demersal low value					
2005	248.2	244.2	216.0	32.1	12.9
2010	273.8	261.0	215.9	57.9	21.2
2015	302.1	275.4	215.8	86.4	28.6
Mollusks and others					
2005	558.4	549.4	486.0	72.4	13.0
2010	615.8	587.0	485.4	130.4	21.2
2015	679.1	619.0	484.8	194.3	28.6
All fish categories					
2005	6666.3	6501.2	5388.5	1277.8	19.2
2010	7754.4	7197.6	5372.7	2381.7	30.7
2015	9036.1	7813.8	5357.1	3679.0	40.7

Table 5. Projected growth in fish demand and price, India, 2000–2015.

	Scenario	Scenario	Scenario
Demand			
Indian major carps	3.88	2.78	–0.09
Other freshwater fish	3.81	2.71	–0.18
Shrimp	–1.07	–1.61	–3.05
Pelagic high value	1.07	0.62	–0.56
Pelagic high value	1.95	1.40	–0.05
Demersal high value	–0.86	–1.21	–2.12
Demersal low value	1.99	1.43	–0.01
Mollusks	–0.72	–1.02	–1.82
All	2.40	1.55	–0.60
Consumer price			
Indian major carps	–2.85	–1.80	1.03
Other freshwater fish	–2.72	–1.67	1.17
Shrimp	2.07	2.62	4.10
Pelagic high value	–0.06	0.39	1.60
Pelagic high value	–0.76	–0.22	1.21
Demersal high value	1.61	1.99	2.99
Demersal low value	–1.31	–0.75	0.73
Mollusks	1.66	2.09	3.24
All	–0.33	0.38	2.24

are consistent with the estimates of National Sample Survey (NSS) for the year 1999–2000, which are 3.45 kg/capita at the national level and 9.8 kg/capita for the fish-eating households. As seen in Figure 5, the IMC will continue to consolidate its share in total domestic fish consumption. Its share will increase to 34% in 2015, from 27% in 2000. In 2015, inland fish species will contribute more than 60 % to the total demand. Share of shrimp in total demand will decline from 10 % in 2000 to 6 % in 2015. The marine fish species will contribute about one-third to total demand in 2015.

Under the baseline scenario, the additional fish demand from the year 2000–2015 will be about 2.3 million mt (Figure 6). Out of this, 50 % will be met from IMC, followed by OFWF (36 %), pelagic (14 %), and demersal (3 %). The additional consumption of shrimp, DHV, and mollusks species will be on decline (–9 %).

Export of Fish

Shrimp, PHV, DHV, and mollusks are the major species of fish that are being exported from India. The export of these species will grow at a rate of 4.6–5.5% per annum (Table 8 and Figure 7). The highest growth of export is projected for shrimp (5.8–6.9%), followed by pelagic (4.4–5.4 %), demersal (3.8–4.6 %), and mollusks (3.5–4.3 %). The price of export will also increase at the annual growth rate of 6.1–6.5 % per annum at constant price. The higher export price will benefit the fish producer substantially. The export will stabilize the domestic price of these species at a higher level, safeguarding the interest of pro-

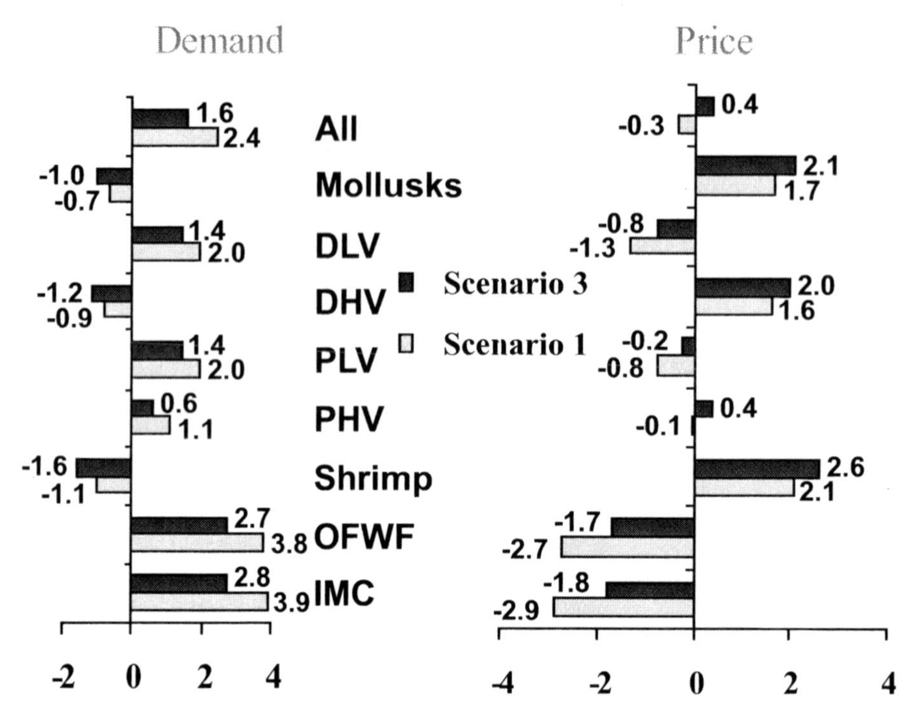

Figure 4. Projected growth in fish demand and price during 2000 to 2015, India.

Table 6. Projected fish domestic demand in India 2005–2015.

Year	Total demand			Household demand	Home away demand
	Scenario 1	Scenario 2	Scenario 3		
Demand					
1998(base)	5174	5174	5174	3350	1824
2005	6040	5899	4945	3911	2129
2010	6813	6342	4801	4411	2402
2015	7741	6719	4671	5012	2729
Annual per capita demand (kg) at the national level					
2005	5.6	5.3	4.6	3.6	2.0
2010	5.8	5.3	4.1	3.8	2.1
2015	6.2	5.4	3.7		
Annual per capita demand (kg) for the fish eating population					
2005	15.1	14.6	12.3	9.8	5.3
2010	15.8	14.7	11.1	10.2	5.5
2015	16.7	14.5	10.1	10.8	5.9

Table 7. Projected demand of fish by species in million killograms, India.

Year	Scenario 1	Scenario 2	Scenario 3	Scenario 4	Scenario 5
1998(base)	1418	1418	1418	1418	1418
2005	1852	1826	1794	1746	1746
2010	2240	2147	2040	1904	1403
2015	2710	2489	2261	2011	1396
Other freshwater fish					
1998(base)	1047	1047	1047	1047	1047
2005	1361	1342	1318	1283	1034
2010	1640	1572	1493	1393	1025
2015	1978	1815	1648	1466	1016
Shrimp (marine and freshwater)					
1998(base)	532	532	532	532	532
2005	494	490	486	479	429
2010	468	458	446	430	367
2015	443	424	404	380	315
Pelagic high value					
1998(base)	349	349	349	349	349
2005	376	374	371	367	336
2010	397	390	381	371	326
2015	418	404	388	369	317
Pelagic low value					
1998(base)	931	931	931	931	931
2005	1066	1059	1049	1035	928
2010	1175	1150	1120	1081	926
2015	1294	1239	1180	1112	924
Demersal high value					
1998(base)	259	259	259	259	259
2005	244	243	241	239	223
2010	234	230	226	221	200
2015	224	217	211	203	180
Demersal low value					
1998(base)	216	216	216	216	216
2005	248	246	244	241	216
2010	274	268	261	252	216
2015	302	289	275	259	216
Mollusks and others					
1998(base)	420	420	420	420	420
2005	399	398	396	393	369
2010	385	381	375	368	337
2015	372	363	353	341	307
Total fish					
1998(base)	5174	5174	5174	5174	5174
2005	6 040	5978	5899	5783	4945

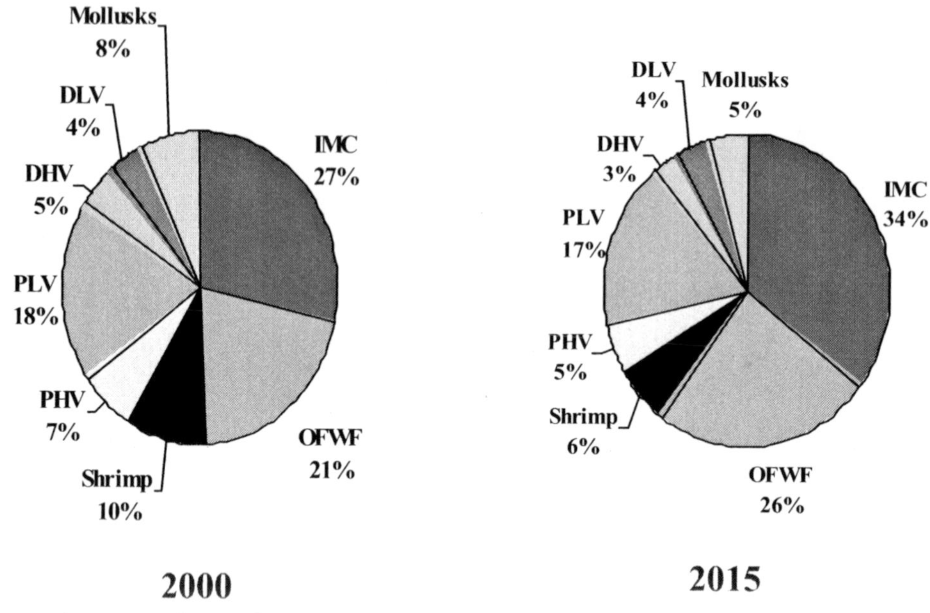

Figure 5. Changes in shape of species in total fish demand, India

ducer. The technological development of fish sector with export orientation will induce higher fish production, demand, and export and will stabilize the domestic market. The benefit of the technological development in the fish sector will be transmitted substantially to both producer and the consumer.

The fish export was 0.31 million mt in the year 1998 (base year). Under the baseline scenario, the export of fish is likely to grow at an annual rate of 5.5 % between 2000 and 2015, and the exports will increase to 0.45 million mt in 2005 and 0.76 million mt in 2015. The export price is likely to in-

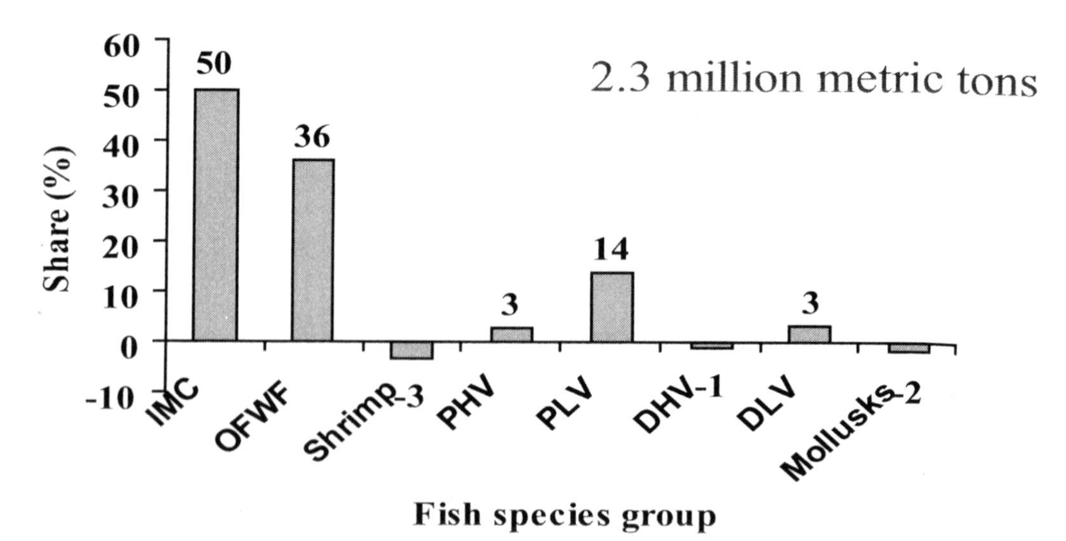

Figure 6. Share of fish species in additional dem and by 2015.

Table 8. Projected export of fish from India, 2000–2015.

	Scenario 1	Scenario 3	Scenario 5
Export growth %			
Shrimp	6.94	5.79	2.77
Pelagic high value	5.36	4.42	1.96
Demersal high value	4.55	3.80	1.85
Mollusks	4.23	3.47	1.49
All	5.49	4.55	2.12
Export price growth (%)			
Shrimp	6.68	6.92	7.56
Pelagic high value	0.81	1.17	2.13
Demersal high value	4.46	4.64	5.12
Mollusks	4.42	4.61	5.13
All	6.08	6.47	7.29
Export			
1998(base)	307.9	307.9	307.9
2005	445.5	433.9	356.3
2010	582.3	538.5	395.7
2015	763.5	655.6	439.9

crease at an annual rate of 6.1% during this period. Under the baseline scenario, shrimp exports will witness the highest annual growth (6.9%), followed by PHV (5.4%), DHV (4.6%), and mollusks (4.2%). The export price of each species is predicted to keep on increasing. A steep rise is predicted in the shrimp price. Exports of shrimp will also increase at a much faster rate, compared to other species. The export projections by

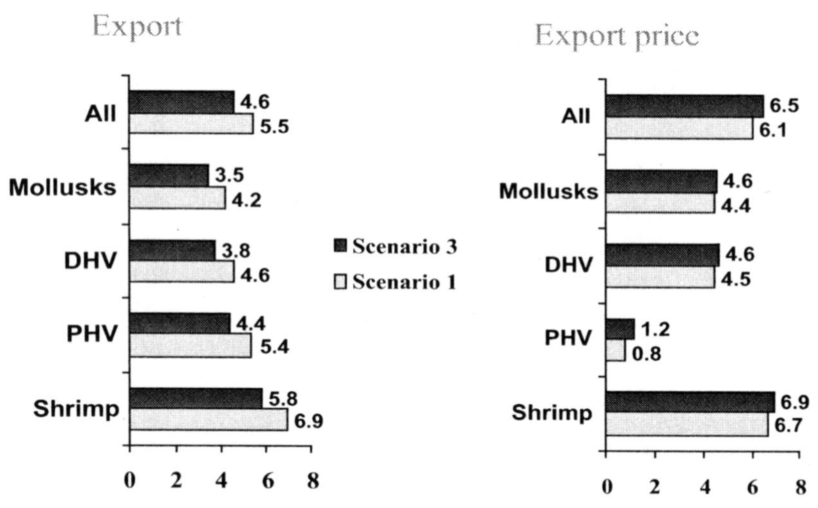

Figure 7. Projected growth in fish export and prices during 2000 –2015, India.

species under various scenarios are given in Table 9.

The shrimp exportable species, generating two-third of the total export value in the past, will continue to hold the major share in future. Its share in total export is likely to increase from 36% in 2000 to 44% in 2015 (Figure 8). The share of other species in the export market will witness a declining trend. The next important exportable species is the demersal fish whose share will be 30% in the year 2015, followed by mollusks (18%), and pelagic (8%). India is likely to increase its fish exports by 0.42 million mt (i.e., nearly 123% by 2015. Out of this, shrimp contribution is projected to be nearly half (Figure 9).

Socioeconomic Impact

The socioeconomic impacts of fish technology on consumers and producers of fish were examined comparing scenario 1 to scenario 4, with scenario 5 (without TFP growth). The salient features are as follows:

Impact on Fish Consumers

- The consumers will be benefited by lower prices. Consumer prices will decline by 13–17% in 2005, and further to 21–35% in 2015, in comparison to the prices the consumer would have paid had there been no growth in TFP.

Table 9. Export projection of fish by species in million killograms, India.

Year	Shrimp*	Pelagic high value	Demersal high value	Mollusks & others	Total
Exports in million kg in base year					
1998	107.4	125.05	108.56	66.84	307.86
Scenario 1: With existing growth in TFP					
2005	171.82	36.11	148.24	89.31	445.48
2010	240.34	46.90	185.18	109.84	582.26
2015	336.18	60.90	231.33	135.10	763.50
Scenario 2: 25 per cent deceleration in existing TFP growth					
2005	169.39	35.69	146.85	88.46	440.39
2010	230.11	45.24	179.94	106.67	561.96
2015	308.13	56.66	218.40	127.40	710.59
Scenario 3: 50 per cent deceleration in existing TFP growth					
2005	166.32	35.15	145.08	87.37	433.93
2010	218.38	43.32	173.83	102.99	538.53
2015	279.33	52.25	204.72	119.28	655.58
Scenario 4: 75 per cent deceleration in existing TFP growth					
2005	161.85	34.37	142.50	85.79	424.51
2010	203.58	40.88	165.98	98.25	508.69
2015	247.87	47.34	189.25	110.12	594.58
Scenario 5: Without growth in TFP					
2005	130.02	28.70	123.43	74.11	356.26
2010	149.04	31.63	135.28	79.79	395.73
2015	170.83	34.86	148.27	85.89	439.85

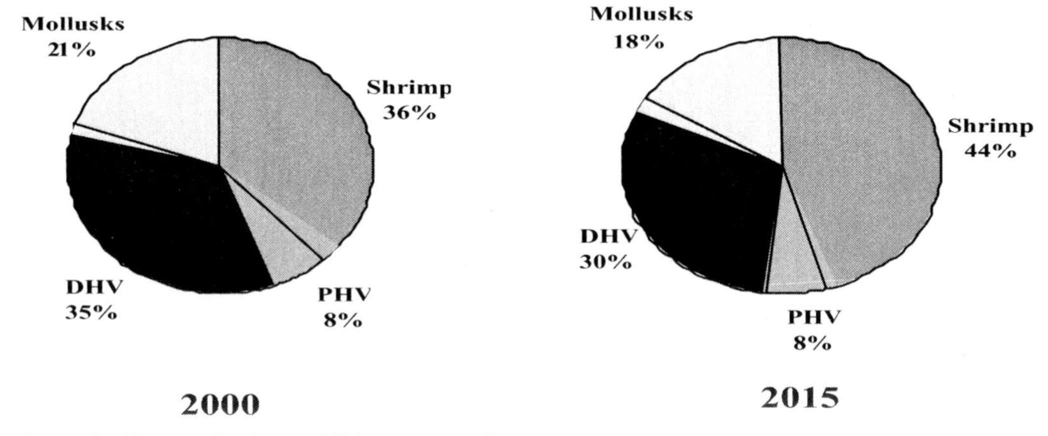

Figure 8. Changes in share of fish export, India.

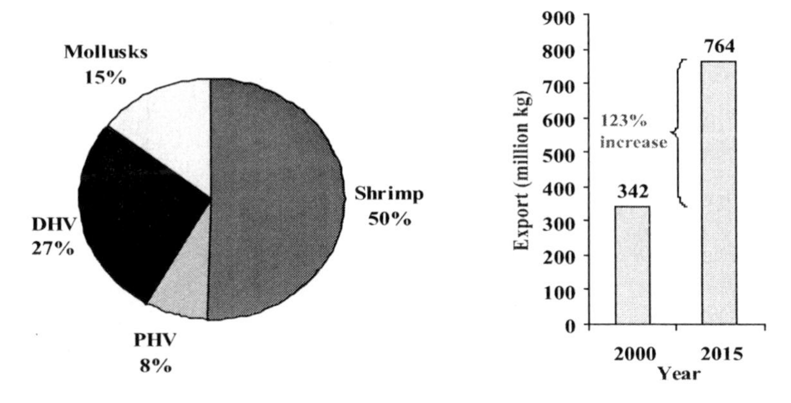

Figure 9. Fish species contribution in incremental export during 2000–2015.

• Technology-driven decline in consumer prices will induce growth in fish consumption. Total fish consumption will increase in the range of 17–22% in 2005 and 31–66 % in 2015, compared to the no TFP growth scenario.

• With technological progress, per capita annual fish consumption of fish-eating households will increase by 2.1–2.7 kg in 2005 and by 3.2–6.6 kg in 2015.

• As a result of technological progress in fish production, annual income gain for fish consumers (as a result of price effect) is expected to be in the range of 76 rupees to 99 rupees (46 rupees = US$1 approximately) per person in 2005, and 132 rupees to 274 rupees per person in 2015.

• At national level, these gains are enormous. Consumer gains are projected under the baseline scenario to the tune of 40 billion (1×10^9) rupees in 2005 and 127 billion rupees in 2015. The consumers will derive maximum benefit from the IMC, followed by OFWF and shrimp. Inland fish will contribute more than 70% to the total gains. Contribution of shrimp is projected to be 13–14%.

Impact of Technology on Fish Producer

• Technological progress will add 15–19% to total fish output in 2005 and 25–40% in

2015 (under different scenarios with different TFP growth rates).

• Addition to output due to technological change will dampen down the producer prices in the domestic market by 15–20% in 2005 and by 27–54% in 2015.

• Nevertheless, producer prices of the exportable fish are likely to remain stable, and fish exports will increase by 16–20% in 2005 and by 26–42% in 2015, under different TFP growth scenarios.

• Technology-driven reduction in producer prices in the domestic market would adversely affect farm profitability. At the aggregate level, the producer will incur losses in the range 154–194 million rupees in 2005 and 295–527 million rupees in 2015. However, the producers will stand to gain from fish exports to the tune of 16–21 billion rupess in 2005 and 74–152 billion rupees in 2015. Thus, taking domestic and export markets together, the fish producer will be benefited substantially with the adoption of fish technology. Shrimp will contribute about 85% in the total income of the producer.

• The annual income to the fisherman (at 1998 prices) will increase in the range of 585–760 rupees per person in 2005. It will increase to 996–1480 rupees per person in the year 2015. The per capita annual income of the fish worker will increase in the range of 727–835 rupees in 2005, and 2615–3889 rupees in the year 2015. The income of fish worker will increase substantially with the adoption of modern fish technology.

• Under the baseline scenario, in 2015 fish producers' gains will be Rs. 152 billion, consumers' gains will be Rs 127 billion and total social gains will be Rs. 279 billion.

• The producers' share in total gain will increase to 54% in 2015 from 25% in 2000. Initially, consumers will benefit more than the producers. However, with the adoption of modern technology and burgeoning export market, the share of producers in total social gains will be higher than that of consumers.

• At national level, technological progress in the fish sector would enhance per capita national income by 56 rupees in 2005 and by 223 rupees in 2015.

Policy Implications

• Development of export market will be crucial to realize the gains from technological progress. Absence of export market will discourage producers to adopt modern technologies, as the fish prices in the domestic market will be descending with increase in fish production, unless domestic fish demand increases at a faster rate. The results also point towards a need for value addition and product diversification for the domestic consumers, thereby enhancing domestic market demand.

Expansion of existing export zones to include other reliable fish exporting zones will go a long way in stimulating exports. Brand promotion in different countries by means of market access initiatives, increased marketing efforts, and high food safety standards will enhance the export potential. Also, there is need to develop and transfer ecofriendly, yet economically viable, technologies that can produce fish conforming to the international standards.

References

Armington, P. S. 1969. A theory of demand for products distinguished by place of production. International Monetary Fund Staff Papers 16:159–178, Washington D. C.

Dey, M. M. 2000. Analysis of demand for fish in Bangladesh. Aquaculture Economics and Management 4:63–79.

Dey, M. M., R. M. Briones, and M. Ahmed. 2003. Modeling the Asian fish sector: issues, framework, and method. The WorldFish Center, Batu Maung, Bayan Lepas, Penang, Malaysia

Deaton, A., and J. Muellbauer. 1980. An almost ideal demand system. American Economic Review 70:312–326, Pittsburgh, Pennsylvania.

Delgado, C. L., N. Wada, M. W. Rosegrant, S.Meijer, N. Wada, and M. Ahmed. 2003. Fish to 2020: supply and demand in changing global markets. International Food Policy Research Institute, Washington, D. C. and WorldFish Center, Penang, Malaysia.

Paroda, R. S., and P. Kumar. 2000. Food production and demand situations in South Asia. Agricultural Economic Research Review 13(1):1–24.

American Fisheries Society Symposium 49:313–315

Session Summary

Reconciling Fisheries with Conservation: Lessons from History

TONY J. PITCHER
University of Bristish Columbia, Fisheries Centre
2202 Main Mall, Vancouver V6T 1Z4, Canada

"He who fails to understand history is condemned to repeat it." The history of fisheries is a dismal record of George Santayana's (1863–1952) dictum. How knowledge of the past may help us deal with the future was the theme of this session at the Fourth World Fisheries Congress. My keynote paper introduced the idea of using past ecosystems, fished with responsible sustainable fisheries, as policy goals for the future; an approach termed "Back to the Future" (BTF). Back to the Future case studies in Newfoundland and northern British Columbia ecosystem were sponsored by the "Coasts Under Stress" project. Detailed results are contained in my own paper, but the general approach is characterized here.

The first step in the BTF process is to describe the relative biomasses of marine organisms in the past in relation to those in the present. This is not a trivial exercise: it requires use of archival data, historical records, archaeology, and local environmental knowledge, an issue set out in a poster by Sheila Heymans (Vancouver, Canada). A shortcut is taken by constructing mass-balance models of the whole ecosystem; whereby the food required to support each trophic level constrains the values that are possible. A series of such ecosystem models are set up for times that span major changes in a region's fisheries. Important issues to be solved in this process include local extinctions and keystone species, the abundance of forage fish, how to model migratory species, and the influence of climate changes.

The next step is to determine what sustainable fisheries might be set up in restored ecosystems that resemble ones from the past, which might be thought of as "Lost Valleys." These imaginary fisheries follow a list of criteria for sustainability and responsibility that reflect the general principles of, and may be evaluated against, the Food and Agriculture Organization (FAO) Code of Conduct for Responsible Fisheries. The amount of fishing allowed by each of these 'Lost Valley' fisheries is optimised at a sustainable level using procedures that employ automated search routines with an objective function using the ecosystem model. Criteria used in the objective function may include profit, employment, or conservation values in mixed or pure forms: in general, we have found that ecological values must predominate if climate change and uncertainty in the model parameters are to be hedged. Solutions found by the optimisation software are further challenged by their impact on biodiversity, resilience, risk of local extinction, depletion of biomass of all elements of the ecosystem, and diversity of employment niches in the fishery. These analyses are completed for each of the historical ecosystems, and the present. Since the "Lost Valley" ecosystems are depleted, somewhat, by their sustainable fisheries, a restoration program starting from the present depleted situation would aim at the end result: a target reference point for fisheries management termed the optimum restorable biomass (ORB), which captures the instruments and costs of restoration, unique to each location.

The final step in the BTF analysis process is to evaluate the relative worth of restoring each of the past virtual ecosystems. This is not just a case of looking for greater biomasses and catches, although almost all past ecosystems would bring those benefits if restored and fished sustainably. Again, utility must be assessed in several ways, in terms of profits, jobs and various ecological indicators. Wai Lung Cheung (Hong Kong) explored the costs of restoration in the South China Sea in a poster. The choice of which past system to adopt as a policy goal, including the costs of restoration, is not a simple one since each alternative future has its own set of trade-offs among economic, social, and conservation indictors. To date, many solutions show win–win options for the restoration of past ecosystems compared to managing the present situation as best we can. This arises because the necessary trade-offs are all less drastic under some restoration scenarios.

As part of the BTF project sponsored by "Coasts Under Stress," Cameron Ainsworth (Vancouver, Canada) explored the concept of Optimal Restorable Biomass as an ecosystem-based policy goal for the restoration of ecosystem and sustainable fisheries, with a worked example from northern British Columbia.

Information about what past ecosystems and fisheries were like is often hard to find; in addition to contemporary letters, books, and historical archives, the local knowledge of fishers and community members can be rich source of material. For example, Barbara Neis and coworkers (Newfoundland, Canada), also partly sponsored by "Coasts Under Stress," presented some impressive examples of painstaking, carefully planned interviews that show how individual fishers change fishing gear, boats, and techniques over the years. Such information can augment logbooks and scientific surveys in reconstructing ecosystem change over time.

Henn Ojaveer (Tallinn, Estonia) described an ambitious project, sponsored by the Census of Marine Life, involving many institutions and a collaboration between historians and natural scientists that aims to gather historical and archival information to enable a reconstruction of the Baltic Sea and its fisheries, quantifying the impact of humans over the past 800 years. Maria Gasalla (São Paulo, Brazil) analysed artisanal fisheries before industrialization in southern Brazil using archives, survey reports, international expeditions, and surveys with veteran fishers. The results revealed huge decreases in sharks, large species of carangids, snappers and mullets, and sardines.

An excellent example of results from a well-documented historical reconstruction was presented for the North Sea as it was in 1880 by Steven Mackinson (Lowestoft, UK) in a poster entitled "Smacks, Smocks and Smokies: the North Sea Wonder Years." In the 1880s, just as industrialization of the fisheries began, the North Sea produced upwards of 400,000 metric tons a year of valuable fish such as herring, cod, tuna, sturgeon, turbot, sole, and halibut, supporting thriving coastal communities. Today's depleted, overfished North Sea does not produce such table fish and marine production has switched down the food chain. In a section headed "What could we have done differently?," Mackinson explored a mix of fisheries that might have saved the North Sea.

Adopting restoration as policy requires people to overcome the "shifting baseline" syndrome and become aware of what was there in former times: research aimed at describing the cognitive maps of fishers in the Bali Strait and Komodo was described in a poster by Eny Buchary (Indonesia, sponsored by International Development Research Centre (IDRC), Canada).

An issue with any restoration agenda is knowing what the ancient past or "pristine" system looked like. Was there some sort of reasonably constant baseline of abundance, or were there large fluctuations driven by climate? Daniel Selbie and colleagues (Kingston, Canada) presented a fascinating account of the 6,000-year history of two sockeye salmon *Oncorhynchus nerka* lakes that show that answer depends on the location. The work takes advantage of paleoindicators in dated lake sediments, including signatures of marine nitrogen. In northern British Columbia, a lake whose climate has been subjected to a large influence from the ebb and flow of glaciers has seen such huge fluctuations that no baseline abundance can be calculated. On the other hand, a lake in the United States exhibited only minor fluctuations around a stable baseline of salmon abundance over thousands of years, until the stock was recently extirpated by dam construction, a topic also covered by Jack Stillwell's (Eugene, Oregon, USA) poster.

American Fisheries Society Symposium 49:317–329

Back to the Future in Northern British Columbia: Evaluating Historic Marine Ecosystems and Optimal Restorable Biomass as Restoration Goals for the Future

CAMERON AINSWORTH* AND TONY PITCHER

University of British Columbia, Fisheries Centre
2202 Main Mall, Vancouver V6T 1Z4, Canada

Abstract.—Where previous work identified candidate whole-ecosystem goals for restoration in Northern British Columbia, this study develops a new technique to achieve those goals. The optimal fishing pattern to bring about recovery is determined through simulation, so that the fisheries modify the current ecosystem over time into one that more closely resembles an ideal ecosystem state based on historic conditions. This study introduces a new conceptual restoration target, a set of optimal restorable biomasses: a whole-ecosystem analogy to B_{MSY}. A number of restoration plans are drafted, drastic (requiring large reductions in harvest from current levels) to moderate (minor reductions). Plans are evaluated using cost–benefit analysis. Socioeconomic and ecological benefits of each plan are described; a convex relationship between fishery profit and ecosystem biodiversity suggests that there may be an optimal rate of restoration.

Introduction

Back to the Future

Back to the Future (BTF) is an integrative, multidisciplinary approach to restorative marine ecology that advocates rebuilding to defined targets as the proper goal for fisheries management (Pitcher 2001; Pitcher and Pauly 1998). Although the need for rebuilding fish populations is acknowledged in world fishery policies (e.g., UNCLOS 2005), BTF extends the rebuilding requirement to the entire ecosystem, including noncommercial and nonfish species. The purpose of BTF is to provide both a strategic and tactical guide for the long-term restoration of depleted marine populations (Pitcher et al. 2004). Fundamentally, the approach can be divided into two parts: setting strategic goals for restoration and developing restoration strategies. The first step has been taken for the ecosystem of northern British Columbia (Ainsworth et al. 2004; Ainsworth and Pitcher 2005), Newfoundland (S. Heymans, unpublished data) and several other areas[1]. This article examines how we may achieve stated restoration goals, and evaluates several candidate restoration plans for northern British Columbia using cost–benefit analysis (CBA).

*Corresponding author: c.ainsworth@fisheries.ubc.ca

[1] BTF world case studies and additional information is available at www.fisheries.ubc.ca/projects/btf/.

Recapturing Historic Productivity

BTF begins with a quantitative assessment of past ecosystems, to determine what has been lost and what restoration may be worth. Based on contributions in Pitcher et al. (2002b), Ainsworth et al. (2002) used Ecopath with Ecosim (EwE) software (Christensen and Pauly 1992; Walters et al. 1997) to model the marine ecosystem of Northern British Columbia as it appeared at various states throughout the history of exploitation prior to European contact (c. 1750 AD), at the onset of industrial fishing (c. 1900), during the heyday of the Pacific salmon fishery (1950), and in the present day (2000).

Ainsworth and Pitcher (2005) used a search routine that maximizes a chosen objective function (EwE policy search routine; Walters et al. 2002) to optimally harvest these modeled past ecosystems using hypothetical fisheries designed according to criteria of sustainability (Pitcher 2004). Long-term benefits were analyzed using a number of indices developed for BTF (e.g., biodiversity: Ainsworth and Pitcher, in press, extinction risk: Cheung and Pitcher 2004; resilience: Heymans 2004; intergenerational equity: Ainsworth and Sumaila 2004a; employment diversity: Ainsworth and Sumaila 2004b). They found, unsurprisingly, that the preEuropean contact (c. 1750) condition offered the greatest potential reward in terms of sustainable harvests, high biodiversity, low extinction risk, and other qualities. However, this ancient system is the least similar to the depleted state we find today (Ainsworth et al. 2002) and so represents the most ambitious restoration target.

This paper demonstrates a new methodology to evaluate the costs and benefits of restoring past ecosystems. We have modified Ecosim's policy search routine to look for the optimal fishing solution that will convert the present-day (2000) ecosystem into one resembling the 1900 period. Future work will generate restoration scenarios for all four historic periods modeled for BTF and determine how far into the past we must reach to maximize the cost-effectiveness of rebuilding.

Optimal Restorable Biomass

The most productive state of a stock is not in its unfished condition, but when older, less productive individuals have been removed from the population. By maneuvering stock abundance to an optimal size, surplus production can be maximized (Graham 1935; Hilborn and Walters 1992). This is pertinent when considering historic ecosystems as rebuilding goals. We would not wish to restore the historic system just to fish it down to a more productive state. Instead, we should restore that optimally productive state directly—the biomass equilibrium that theoretically results after long-term optimal harvests of the historic system (Figure 1). The biomass of each ecosystem component in that new equilibrium is referred to here as the optimal restorable biomass (ORB); an ecosystem containing ORB attuned components is referred to as an ORB ecosystem. Optimal restorable biomass is the species biomass vector that would naturally result after the long-term responsible use of historic ecosystems. Sidestepping the serial depletion of stocks witnessed in reality, it takes into account the activities of fisheries and determines the best compromise between maintaining historic abundance and diversity, while still providing for the needs of humans. However, there is no single solution: the specific ORB configuration we prefer will depend on our harvest objective.

Where stocks interact through predation or competition, it may be impossible to simultaneously achieve B_{MSY} for multiple stocks

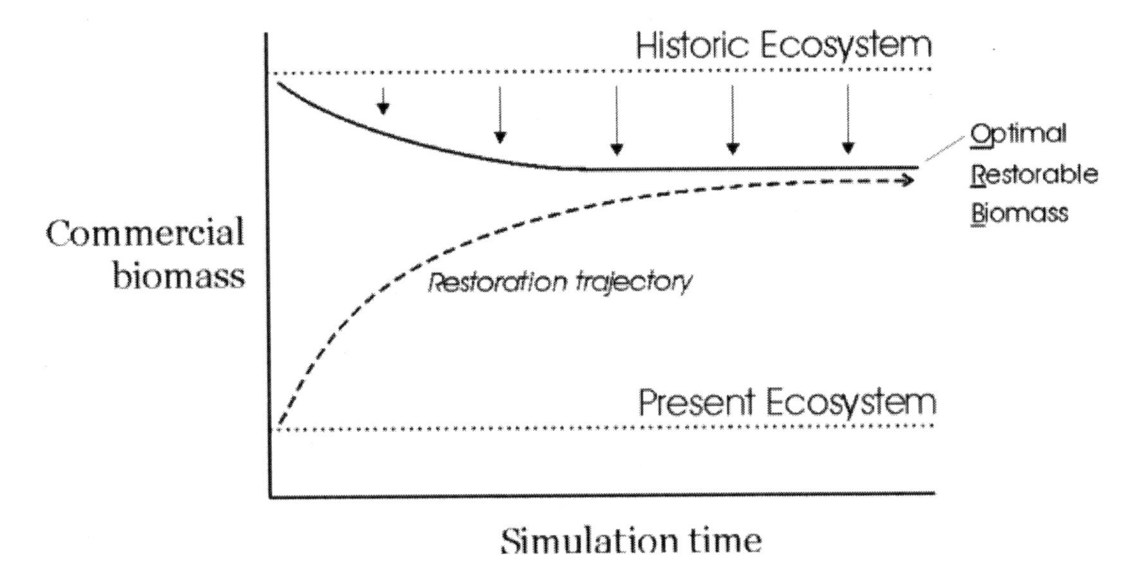

Figure 1. Conceptual diagram showing development of the optimal restorable biomass (ORB) restoration target and a possible restoration trajectory. Optimal restorable biomass is the theoretical biomass equilibrium for each group in the system that would result after long-term optimal harvesting of the historic ecosystem (downward arrows). A restoration plan (dotted line) would see the present ecosystem changed to resemble the ORB state. ORB restoration target for this study is based on the 1900 ecosystem. Note that ORB is an ecosystem-based concept. Simultaneity is not implied between ORB determination and restoration.

(Larkin 1977; Walters et al. 2005). From a whole-ecosystem perspective, it becomes necessary to choose between stocks, holding the biomass of some close to their optimal levels while sacrificing the productivity of others. According to our choice, total catch from the ecosystem can be maximized; the desired modality may be profit, or biodiversity, or any other measure of socioeconomic or ecological utility. If our management goal is simply to maximize catch, for example, B_{MSY} should be sought only for the most productive and massive stocks, while maximum productivity of low-volume fisheries may need to be sacrificed. An alternative goal may maximize system biodiversity. In most cases, a practical management policy will contain a balance between socioeconomic and ecological priorities.

By use of gaming scenarios in our ecosystem models, we can calculate the specific biomass configuration that will yield maximum benefit from our system, though the optimal design may be constrained by additional caveats (e.g., a minimum biomass threshold to inhibit extinctions). If we structure the ecosystem to deliver maximum catch, then ORB becomes an ecosystem analogy to B_{MSY}; if we structure the ecosystem to provide maximum profit, then ORB becomes an analogy to B_{MEY}. Optimal restorable biomass calculation based on historic systems therefore satisfies two requirements: it increases production of commercial groups by reducing their biomass and optimally balances harvests between stocks to provide maximum benefit.

Ainsworth and Pitcher (2005) evaluate various ORB ecosystems based on historical periods. The optimal solutions they present manipulate the past ecosystem to deliver economic and ecological benefits in varying proportion[2] (Figure 2). For the present paper, we chose a moderate restoration target based on the 1900 system; one that provides a mix of economic and ecological benefits (dotted line).

Methods

Modifications to Ecosim

The policy search routine in Ecosim was developed by Walters et al. (2002) to identify harvest patterns that optimally exploit the ecosystem (i.e., provide the highest possible gain to a stated harvest objective). Using a Davidon-Fletcher-Powell search algorithm, the routine iteratively varies fishing mortality (*F*) per gear type, repeating the harvest simulation until an improving step cannot be found. Five objective functions come standard in the software and these can be combined in any proportion using weighting factors. The economic, social and ecological objectives maximize money, jobs, and system biomass and production, respectively. The portfolio utility objective selects risk adverse policies. The fifth standard objective function, mandated rebuilding (MR), allows the user to specify a minimum biomass level for any functional group, which the search routine will try to satisfy through selective fishing.

[2] Ainsworth and Pitcher (2005) optimize the 1900 ORB ecosystem for five harvest objectives. For clarity in the present paper, all policy results have been reduced in Figure 2 to demonstrate only the fundamental tradeoff between profit and biodiversity.

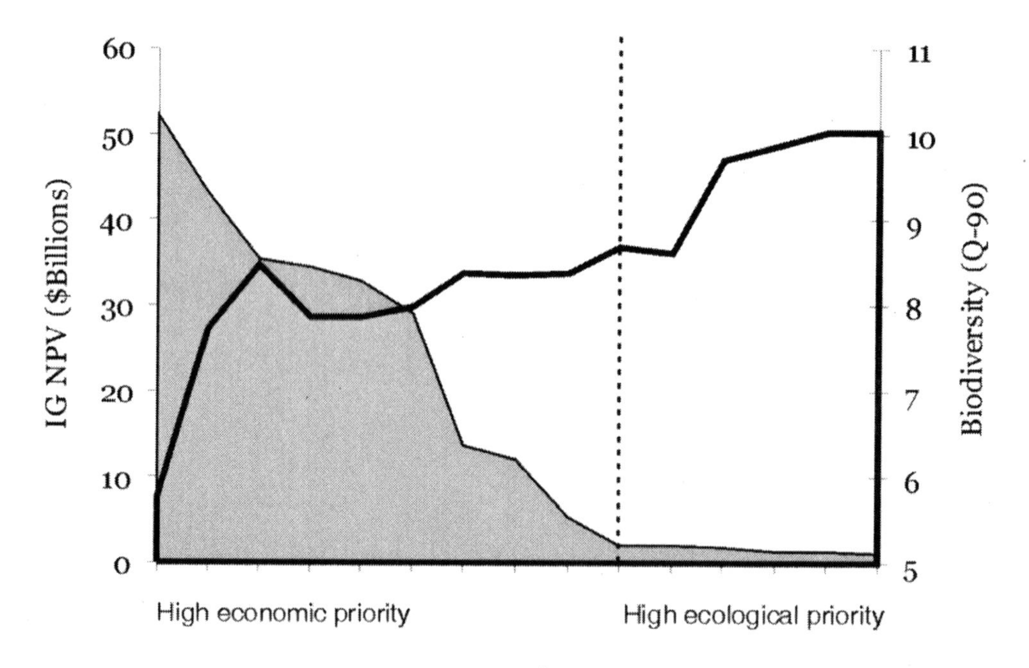

Figure 2. Equilibrium benefits available from the 1900 ecosystem after 50 years of optimal harvest (possible ORB configurations). Along the *x*-axis are various harvest objectives and fleet configurations used to optimally harvest the system. The final ecosystem end-state can deliver revenues[1] (left), maintain biodiversity[2] (right), or a combination of both (center). The economic analysis presented in this paper evaluates an intermediate ORB state (dotted line). [1] Intergenerational Net Present Value (Sumaila and Walters 2005). [2] Q90 biodiversity statistic (Ainsworth and Pitcher in press). From Ainsworth et al. 2004.

However, the MR function was not sufficient for whole-ecosystem restoration. Among other technical problems, it was unidirectional—once the biomass threshold had been reached, no further improvement was recognized. We therefore modified the MR term to seek a specific biomass vector (i.e., biomass goals set for multiple groups). With this modification, the objective function is improved as group biomass approaches its goal, whether from above or below.

The unmodified MR search routine had no way to decide which groups deserve priority in cases where trophic interactions preclude ubiquitous restoration. In its simplest form, the modified MR term also treats every group in the same way, no matter how far it is from its goal biomass (linear model, Figure 3A). Using this method effects a maximum change in system biomass, with malleable groups changing the most. However, a quadratic MR model (Figure 3B) provides the greatest marginal improvement to the objective function when groups first begin to move towards their target biomass so that a maximum number of functional groups will improve. An asymmetric model (Figure 3C) based on the gamma function provides a conservative restoration style that will penalize groups if they fall short of their target biomass by more than if they overshoot. These three rebuilding styles have been incorporated into Ecosim but all plans tested in this report use the conservative gamma function.

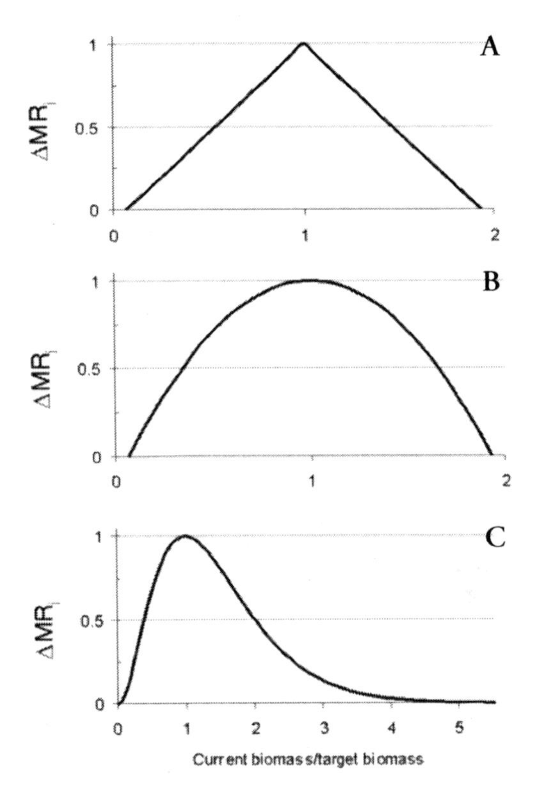

Figure 3. Three models to define restoration objective function. The *y*-axis shows improvement to mandated rebuilding (MR) term for functional group *j*, as current biomass approaches ORB biomass. A) linear model affects a maximum change in system biomass; B) quadratic model affects biomass change in a maximum number of functional groups; C) asymmetric gamma model prefers current biomass to overshoot target biomass, rather than fall short.

Modifications to Base Model

The Ecopath model representing the present day in northern British Columbia is based on Ainsworth et al. (2002). It was modified to provide the policy search routine with maximum leverage in restructuring the ecosystem. The existing fleet, representing today's real-world fisheries with gillnets, trawls, and so forth, was replaced with an idealized one. A single anonymous gear type now pursues each commercial functional group, with all bycatch and discards integrated into total catch. A new fleet, mammal cull, was also included. Under this configuration, the policy search has more dexterity; it can vary catch of each functional group independently to develop an intricate and idealized restoration plan.

Since we have eliminated limitations imposed by the technology of fishing, the potential restoration policies identified here represent best case scenarios limited only by the ecology of the system. This allows us to set the benchmark for restoration and evaluate the effectiveness of candidate fleets to be used for the task of rebuilding. Minor changes were also made to the diet matrix to improve system dynamics. In particular, the diet import of transient salmon (pink salmon *Oncorhynchus gorbuscha*, coho salmon *O. kisutch*, and chum salmon *O. keta*) was reduced to prevent unrealistic population growth (see Martell 2004 for a discussion on migratory species dynamics in Ecosim). Doubtless, our ecosystem model could be further improved[3].

Defining Restoration

The modified policy search routine defines its objective function (OBJ) according to equation (1)

$$\text{OBJ} = W_E \times \Sigma\, \text{NPV}_{ijt} + W_R \times \Sigma\, \text{MR}_i, \quad (1)$$

where W_E and W_R are relative weighting factors for economic and restoration priorities, respectively. By including an economic term we slow down the speed of recovery. NPV is net present value of the harvest forecast during restoration for gear type (i), functional group (j) and simulation year (t). The MR term is based on the gamma distribution and is defined by equation (2).

$$\text{MR}_j = \theta_j \cdot \frac{\left(\frac{x-\mu}{\beta}\right)^{\gamma-1} \cdot e^{\left(\frac{-x-\mu}{\beta}\right)}}{\beta\Gamma(\gamma)} \quad (2)$$

where $\Gamma(\gamma) = \int_0^\infty t^{\gamma-1} e^{-t} dt$ and $x = \frac{B_{j(\text{current})}}{B_{j(\text{ORB})}}$

where γ is the shape parameter, μ is the location parameter, and β is the scale parameter; Γ is the gamma function. B is functional group biomass, and θ is a scaling factor that increases the group's contribution to the cumulative MR term in direct proportion to the absolute difference between current and target biomasses. Upper and lower limits are included on θ so that no one group can override the MR term or be disregarded from it.

Without the scaling factor, the policy search will weigh all groups in the cumulative MR term so that an equal proportional increase in any group yields the same improvement to the objective function. However, since large biomass functional groups are more difficult for the optimization to influence than smaller groups, in terms of a proportional change, optimal policies will tend to abandon large groups from consideration. If it can, the search routine will make only fine scale adjustments in fishing effort seeking to improve a larger number of low-biomass groups. Since we are interested in developing restoration policies to rebuild commercial biomass and since important commercial groups tend to be large in most models, we need the scaling factor to increase their contribution to the MR term.

Measuring Biodiversity

In addition to economic analysis, an important consideration in setting and achieving restoration goals will be the ecological condition of the ecosystem. In this contribution, we measure ecosystem biodiversity using a new statistic developed for use with ecosystem models, where taxonomic species are grouped into aggregate biomass pools of functionally similar organisms. The *Q*-90 biodiversity statistic developed by Ainsworth and Pitcher (in press) is based on Kempton's *Q* index (Kempton and Taylor 1976), and is defined as the slope of the cumulative species abundance curve between the 10th and 90th percentiles (equation 3). In applying Kempton's

[3] Recent advances in Ainsworth (2006) improve model predictions, and expand the optimization algorithm described in this report.

method to the EwE, functional groups in the model are considered equivalent to species, and their biomasses, sorted into bins, are analogous to the number of individuals in field sampling studies.

$$Q90 = 0.8S/\left[\log\left(R_2/R_1\right)\right] \quad (3)$$

S is the total number of functional groups in the model; R1 and R2 are the representative biomass values of the 10th and 90th percentiles in the cumulative abundance distribution. The 10th and 90th percentiles are determined by equations (4) and (5), respectively,

$$\sum_{1}^{R_1-1} n_R < 0.1 \cdot S \leq \sum_{1}^{R_1} n_R \quad (4)$$

$$\sum_{1}^{R_2-1} n_R < 0.9 \cdot S \leq \sum_{1}^{R_2} n_R \quad (5)$$

where n_R is the total number of functional groups with abundance R. Ainsworth and Pitcher (in press) found this biodiversity index to be robust against the effects of subjective model structure.

Rebuilding the Ecosystem

A vector of biomasses representing the 1900 ORB ecosystem is entered into Ecosim's modified mandated rebuilding routine. The policy search identifies the optimal fishing mortality per gear type that will restore the system to that goal. As those mortalities are applied to a harvest simulation of the 2000 system, biomass begins to rebuild. Figure 4 shows the biomass trajectories for a number of restoration policies (gray lines). The solid black line shows the baseline commercial biomass (predicted by the 2000 model). The dotted black

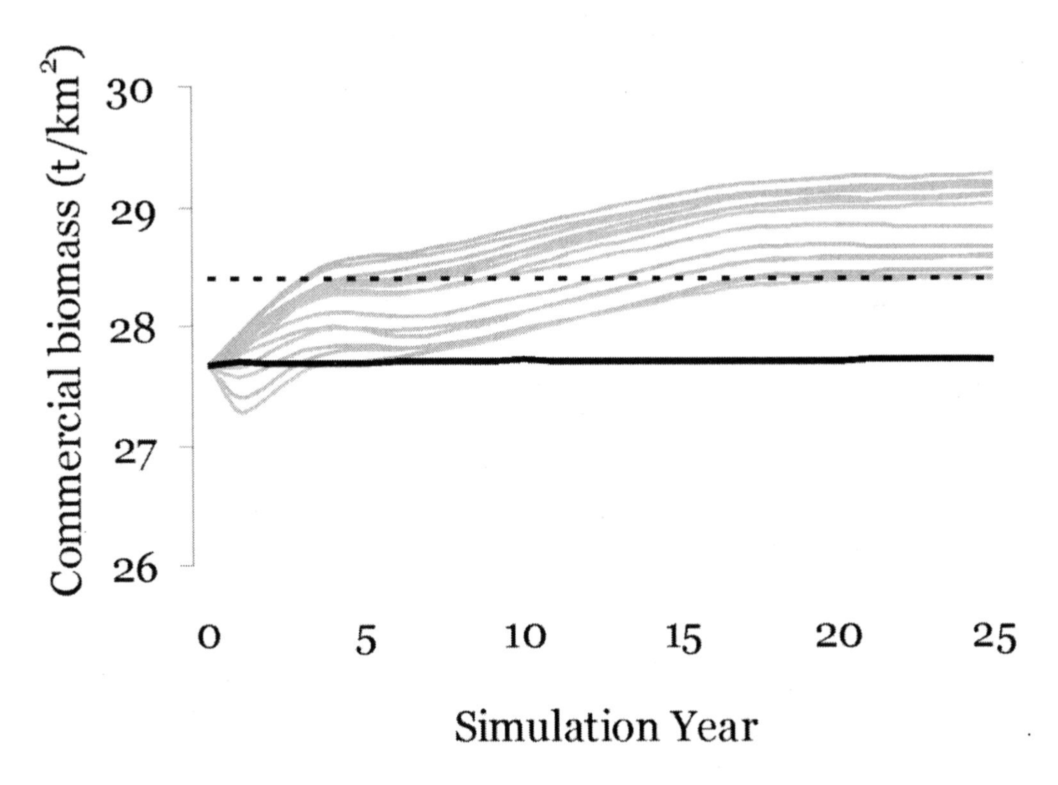

Figure 4. Biomass trajectories of various restoration plans. Solid black line shows model baseline, dotted line shows arbitrary restoration target (28.4 metric tons/km^2 ≈ 103% of current system biomass). The fastest plan achieves restoration quickly but produces negligible income; the slowest plan continues to provide for resource users.

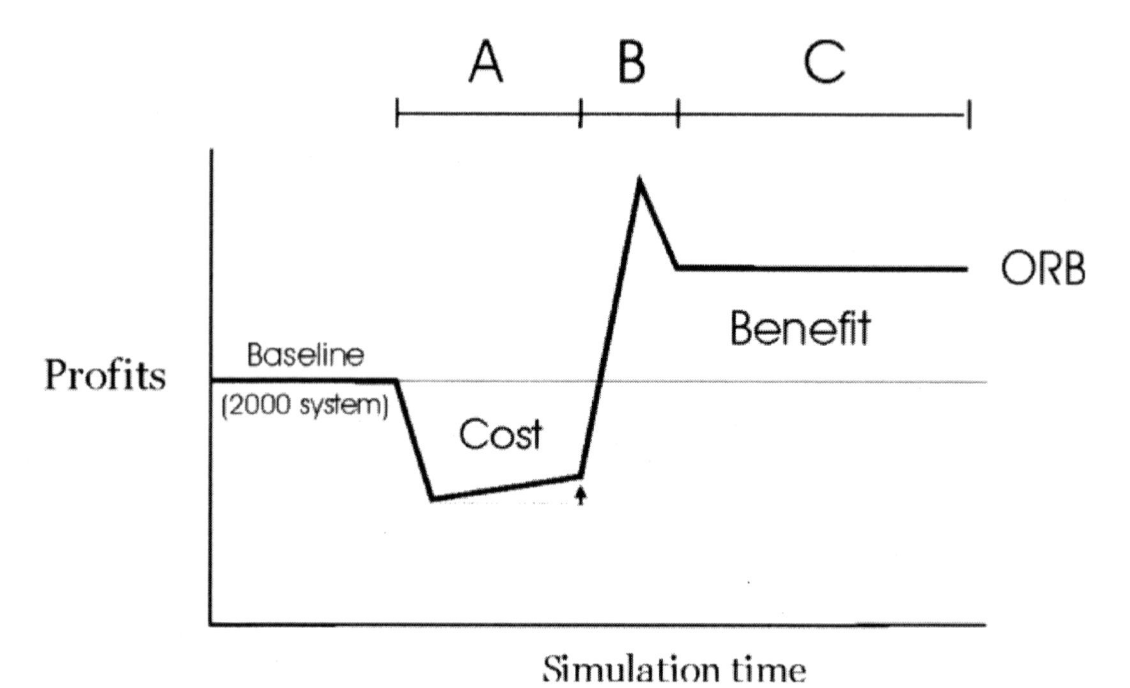

Figure 5. Cost–benefit analysis of restoration. Restoration consists of three phases: A) reduced harvest for biomass rebuilding; B) fishing effort readjustment to cancel biomass accumulation; C) sustained equilibrium harvest. Costs and benefits of restoration are taken in relation to forecasted profits from the 2000 system (baseline). As system biomass rebuilds (phase A), constant harvest rate yields greater catches (up arrow).

line shows the arbitrary threshold of commercial biomass (28.4 tons/km^2) where restoration is considered "achieved." The threshold is equivalent to about 103% of the baseline commercial biomass, a modest improvement. Depending on the relative weighting of economic and rebuilding priorities in the objective function, the goal is attained anywhere in between 6 and 25 years. For restoration to be considered complete, commercial biomass had to be maintained for 36 (monthly) time steps, or 3 years.

Economic Evaluation of Restoration

Figure 5 shows the basis of our cost–benefit analysis of restoration. Costs and benefits are taken in relation to baseline (status quo) exploitation of the 2000 system. The stream of benefits from the restoration plan can be divided into three stages. In the restoration phase, optimal fishing mortalities reduce harvests and allow the system to rebuild. The catch value sacrificed from baseline is considered the cost of restoration. The duration of this period varies from 6 years for the fastest restoration plan, to 25 years for the slowest. As the ecosystem rebuilds, the small medicinal harvest rate is rewarded with a slight increase in catch (upwards arrow). The second phase is transitional; fishing effort is adjusted to cancel biomass accumulation and establish a new equilibrium. This phase lasts 10 years. In the third phase, final equilibrium harvests are maintained until simulation end (year 80). Since the standing level of commercial biomass has increased, we can sustainably draw more from the restored system than from the baseline—the difference is the benefit of restoration. Note that we

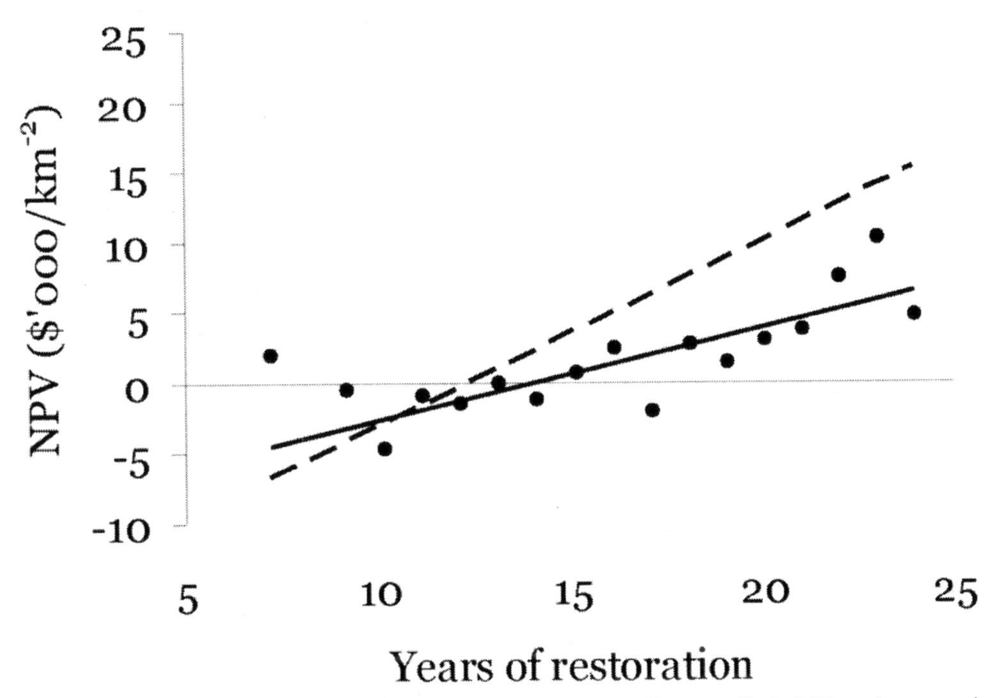

Figure 6. Net present value of various restoration/harvest plans ($\delta = 4.8\%$). Solid line shows evaluation under conventional discounting, dotted line shows intergenerational discounting ($\delta = \delta_{fg}$). Moderate restoration plans outperform bank interest (>16 years under conventional CBA; >15 y earsunder intergenerational CBA).

have assumed completely malleable capital in that there is no penalty or cost associated with fleet restructuring or decommissioning.

Results

Figure 6 shows the net present value of the restoration and harvest plan under conventional discounting (dots, solid regression line) when the standard discount rate (δ) is equal to bank interest or 4.79% (2003 Bank of Canada average 10-year benchmark bond yields). The dotted line shows NPV under intergenerational discounting (Sumaila and Walters 2005), when the intergenerational discount rate (δ_{fg}) is equal to the standard discount rate. Policies that are heavily weighted in favor of restoration (left side) restore biomass quickly but require drastic cuts in fishing. Policies weighted for continued profits (right) tend to spread the costs out for a longer period. Moderate restoration strategies, taking 16 years or longer to achieve restoration, provide a greater return under conventional discounting than the long-term interest rate available from banks. Intergenerational discounting, which leaves more benefits to future generations, advocates more aggressive policies (>15 years).

Figure 7 shows the internal rate of return (IRR) required to make restoration worthwhile under conventional (solid line) and intergenerational discounting (dotted line). Short duration restoration plans, requiring drastic cuts in harvest, are advisable only at very low discount rates. At high discount rates, moderate restoration plans are still advisable (X asymptote ≈ 17.5 years) because the annual profit during restoration is greater than

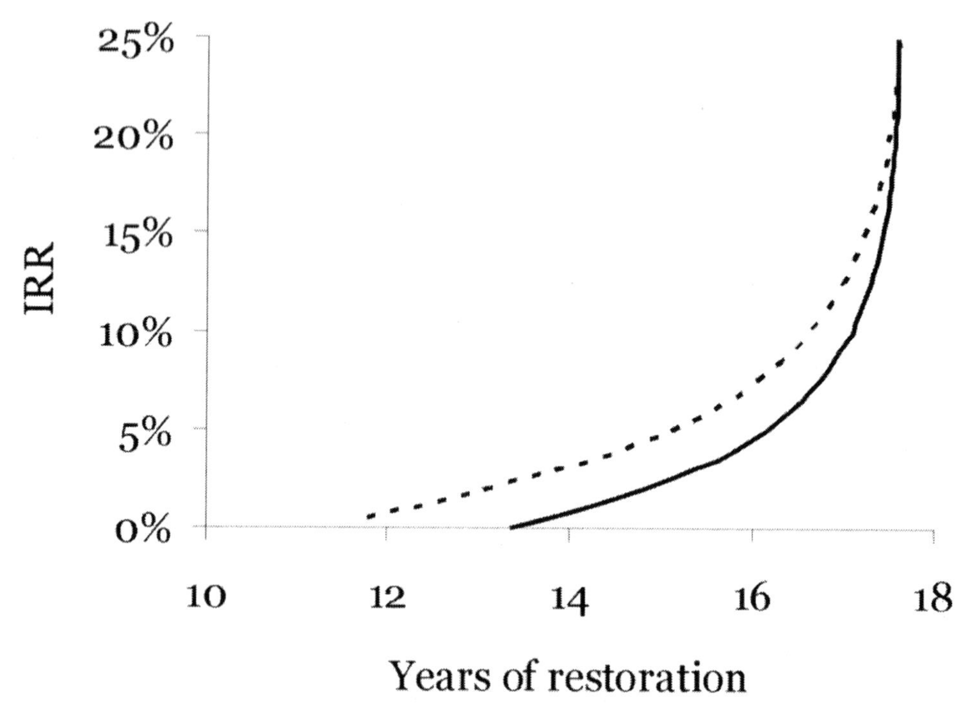

Figure 7. Internal rate of return (IRR) justifying restoration. Solid line evaluates restoration costs and benefits using conventional discounting ($\delta = 4.8\%$), dotted line uses intergenerational discounting ($\delta = \delta_{fg}$). A quick restoration requires drastic cuts in harvest, and is only advisable at low discount rates; slow restoration spreads costs over a longer period.

baseline. The policy search routine has identified optimal harvest levels that will not only restore biomass, but also outperform the current industry. However as mentioned, the fleet configuration used here is ideal. If a real fleet were modeled with technological limitations (as in British Columbia, capturing several species per gear type), the policy search would deliver less optimistic restoration forecasts.

In Figure 8 we have allowed restoration to continue for a full 25 years. Equilibrium profits and biodiversity are shown for a range of restoration plans, from drastic (F greatly reduced from baseline) to moderate (F slightly reduced). Each point represents the end state of one simulation–optimal restorative harvest mortalities applied to the 2000 system: drastic plans appear on the lower right, they achieve great improvement in biodiversity at the expense of profit; moderate plans appear on the upper left, they maintain income but recover less of the ecology. The 2000 system starting point would appear on the far lower left (low profit 0.7, low biodiversity 3.9); the goal 1900 ORB system would appear on the upper right (high profit 1.2, high biodiversity 8.4). The convex relationship between end-state profit and biodiversity suggests that there may be an optimal rate of recovery (which incidentally corresponds to the slowest recovery plans tested in the CBA, those that achieved threshold biomass in 20–25 years).

Discussion

In this first attempt at designing and forecasting restoration scenarios, the end result of the ecosystem manipulation fails to achieve

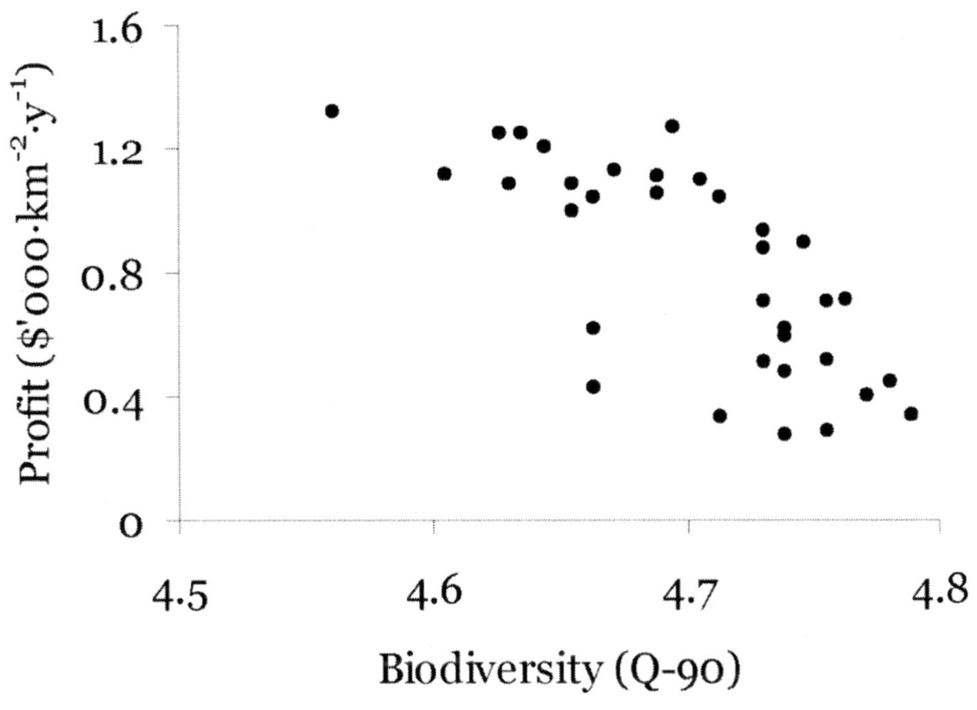

Figure 8. Equilibrium tradeoffs resulting after 25 years of restoration. Each point represents the end state of one simulation—optimal restorative harvest mortalities applied to the 2000 system. Not shown: baseline system provides profit 0.7 and biodiversity 3.9; goal ORB system provides profit 1.2 and biodiversity 8.4. Even aggressive restoration plans (right) can achieve only a fraction of the target biodiversity within 25 years.

the system configuration prescribed by the ORB. However, all the fishing policy optimizations conducted here use a single fishing mortality assigned to each gear type, and these are held constant throughout all simulation years. Revisions to this work could explore the use of complex restorations plans, in which each gear type receives an independent optimal fishing rate for each year in the restoration plan. Staged depletion and succession events may then maneuver the ecosystem more effectively towards the restoration goal. With this improved optimization technique, we could possibly achieve a closer resemblance to the target ecosystem and extend the analysis to compare restoration plans targeting ORBs based on more ancient historical ecosystems. Unfortunately, it seems that equilibrium-level fishing mortalities are able to propel the 2000 system only a short way towards these lofty goals.

Through this work, restoration schemes may be yet identified that could advance the ecosystem into an alternative stable state desirable from a policy perspective. It is theorized that nonlinear and hysteretic change may prevent a complex marine system from reverting to its wilderness state once fishing pressure is removed (Scheffer et al. 2001; Hughes et al. 2005). However, such a recovery may be possible through intentional, directed manipulation of the food web. If trophodynamics, rather than the environment, is responsible for the persistent alternative state, then we could draft ecological recovery plans that use direct and indirect trophic effects as tools for ecosystem restoration. In the case of northern

BC, a preindustrial assemblage may only be achieved through careful manipulation of keystone groups and maintenance of key trophic interactions (e.g., sea otter-kelp-urchin triad; Estes and Duggins 1995; Steneck et al. 2002). However, even a moderate increase in ecosystem health and biodiversity, such as the one demonstrated in this contribution, may be desirable from a policy perspective if a concurrent economic benefit can be shown.

This contribution presents a method to evaluate ambitious restoration policy objectives in a quantitative and predictive way, and the simple economic analysis used here suggests that restoration can be made cost-effective. Moderate plans are shown to provide a harvest value that outperforms the return on bank interest. However, this is a best-case scenario. We have assumed malleable commercial capital, so that no cost or penalty is associated with transferring fishing effort from one fleet to another. Any fleet restructuring or decommissioning costs will reduce the economic feasibility of restoration. Restoration may then only be advisable under very low discount rates, or if some existence value is attached to the standing biomass of stocks.

Acknowledgments

The authors thank Rosemary Ommer and the Coasts Under Stress Major Collaborative Research Initiative and the National Sciences and Engineering Research Council of Canada for providing funding, and to Villy Christensen and Carl Walters for advice on ecosystem modeling. We are also grateful to the many colleagues and partners who made our whole-ecosystem models possible by generously providing information and assistance; their contributions are made explicit in Pitcher et al. 2002a, 2002b and Ainsworth et al. 2002.

References

Ainsworth, C. 2006. Strategic marine ecosystem restoration in northern British Columbia. Doctoral dissertation. Resource Management and Environmental Studies. Faculty of Graduate Studies, University of British Columbia, Vancouver.

Ainsworth, C., and T. J. Pitcher. In press. Modifying Kempton's Species Diversity Index for use with ecosystem simulation models. Ecological Indicators.

Ainsworth, C., and T. J. Pitcher. 2005. Evaluating marine ecosystem restoration goals for Northern British Columbia. Pages 419–438 *in* G. H. Kruse, V. F. Gallucci, D. E. Hay, R. I. Perry, R. M. Peterman, T. C. Shirley, P. D. Spencer, B. Wilson, and D. Woodby, editors. Fisheries assessment and management in data-limited situations. University of Alaska, Alaska Sea Grant College Program, Fairbanks.

Ainsworth, C., and U. R. Sumaila. 2004a. Economic valuation techniques for Back to the future optimal policy searches. Pages 104–107 *in* T .J. Pitcher, editor. Back to the Future: advances in methodology for modeling and evaluating past ecosystems as future policy goals. Fisheries Centre Research Reports 12(1), Vancouver.

Ainsworth, C., and U. R. Sumaila. 2004b. An employment diversity index used to evaluate ecosystem restoration strategies. Page 108 *in* T. J. Pitcher, editor. Back to the Future: advances in methodology for modeling and evaluating past ecosystems as future policy goals. Fisheries Centre Research Reports 12(1), Vancouver.

Ainsworth, C., J. J. Heymans, and T. J. Pitcher. 2002. Ecosystem models of northern British Columbia for the time periods 2000, 1950, 1900 and 1750. Fisheries Centre Research Reports 10(4), Vancouver.

Ainsworth, C., J. J. Heymans, and T. J. Pitcher. 2004. Policy search methods for Back to the Future. Pages 48–63 *in* T. J. Pitcher, editor. Back to the Future: advances in methodology for modelling and evaluating past ecosystems as future policy goals. Fisheries Centre Research Reports 12(1), Vancouver.

Cheung, W. L., and T. J. Pitcher. 2004. An index expressing risk of local extinction for use with dynamic ecosystem simulation ,odels. Pages 94–102 *in* T. J. Pitcher, editor. Back to the Future: advances in methodology for modelling and evaluating past ecosystems. Fisheries Centre Research Reports 12(1), Vancouver.

Christensen, V., and D. Pauly. 1992. ECOPATH II—a software for balancing steady-state models and calculating network characteristics. Ecological Modelling 61:169–185.

Estes, J. A., and D. O. Duggins. 1995. Sea otters and kelp

forests in Alaska: generality and variation in a community ecological paradigm. Ecological Monographs 65(1):75–100.

Graham, M. 1935. Modern theory of exploiting a fishery and application to North Sea trawling. Journal du Conseil International pour l'Exploration de la Mer 10(3):264–274.

Heymans, J. J. 2004. Comparing the Newfoundland marine ecosystem models using information theory. Pages 62–71 *in* J. J. Heymans, editor. Ecosystem models of Newfoundland and southeastern Labrador: additional information and analyses for 'Back to the Future'. Fisheries Centre Research Reports 11(5), Vancouver.

Hilborn, R., and C. J. Walters. 1992. Quantitative fisheries stock assessment: choice, dynamics and uncertainty. Chapman and Hall, New York.

Hughes, T. P., D. R. Bellwood, C. Folke, R. Steneck, and J. Wilson. 2005. New paradigms for supporting the resilience of marine ecosystems. Trends in Ecology and Evolution 20(7):380–386.

Kempton, R. A., and L. R. Taylor. 1976. Models and statistics for species diversity. Nature 262:818–820.

Larkin, P. A. 1977. An epitaph for the concept of maximum sustained yield. Transactions of the American Fisheries Society 106(1):1–11.

Martell, S. 2004. Dealing with migratory species in ecosystem models. Pages 41–44 *in* T. J. Pitcher, editor. Back to the Future: advances in methodology for modelling and evaluating past ecosystems. Fisheries Centre Research Reports 12(1), Vancouver.

Pitcher, T. J. 2001. Fisheries managed to rebuild ecosystems: reconstructing the past to salvage the future. Ecological Applications 11(2):601–617.

Pitcher, T. J. 2004. Why we have to open the lost valley: criteria and simulations for sustainable fisheries. Pages 78–86 *in* T. J. Pitcher, editor. Back to the Future: advances in methodology for modelling and evaluating past ecosystems as future policy goals. Fisheries Centre Research Reports 12(1), Vancouver.

Pitcher, T., and D. Pauly. 1998. Rebuilding ecosystems, not sustainability, as the proper goal of fishery management. Pages 311–329 *in* T. Pitcher, P. Hart, and D. Pauly, editors. Reinventing fisheries. Chapman and Hall, London.

Pitcher, T. J., M. Power, and L. Wood, editors. 2002a. Restoring the past to salvage the future: report on a community participation workshop in Prince Rupert, British Columbia. Fisheries Centre Research Reports 10(7), Vancouver.

Pitcher, T. J., M. Vasconcellos, J. J. Heymans, C. Brignall, and N. Haggan, editors. 2002b. Information supporting past and present ecosystem models of Northern British Columbia and the Newfoundland shelf. Fisheries Centre Research Reports 10(1), Vancouver.

Pitcher, T. J., J. J. Heymans, C. Ainsworth, E. A. Buchary, U. R. Sumaila, and V. Christensen. 2004. Opening the lost valley: implementing a "Back to the Future" restoration policy for marine ecosystems and their fisheries. Pages 165–193 *in* E. E. Knudsen, D. D. MacDonald and J. K. Muirhead, editors. Sustainable management of North American fisheries. American Fisheries Society, Symposium 43, Bethesda, Maryland.

Scheffer, M., S. Carpenter, J. A. Foley, C. Folke, and B. Walker. 2001. Catastrophic shifts in ecosystems. Nature 413:591–596.

Sumaila, U. R., and C. J. Walters. 2005. Intergenerational discounting. Ecological Economics 52:135–142.

Steneck, R. S., M. H. Graham, B. J. Bourque, D. Corbette, J. M. Erlandson, J. A. Estes, and M. J. Tegner. 2002. Kelp forest ecosystems: biodiversity, stability, resilience and future. Environmental Conservation 29(4):436–489.

UNCLOS (United Nations Convention on the Law of the Sea). 2005. 1982 United Nations Convention on the Law of the Sea. Available: www.un.org/Depts/los/convention_agreements/convention_overview_convention.htm. (May 2005).

Walters, C., V. Christensen, and D. Pauly. 1997. Structuring dynamic models of exploited ecosystems from trophic mass-balance assessments. Reviews in Fish Biology and Fisheries 7:139–172.

Walters, C., V. Christensen and D. Pauly. 2002. Searching for optimum fishing strategies for fishery development, recovery and sustainability. Pages 11–15 *in* T. Pitcher and K. Cochrane, editors. The use of ecosystem models to investigate multispecies management strategies for capture fisheries. Fisheries Centre Research Reports 10(2), Vancouver.

Walters, C. J., V. Christensen, S. J. Martell, and J. F. Kitchell. 2005. Possible ecosystem impacts of applying MSY policies from single-species assessment. ICES Journal of Marine Science 62:558–568.

American Fisheries Society Symposium 49:331–348

Historical Policy Goals for Fish Management in Northern Continental Patagonia Argentina: A Structuring Force of Actual Fish Assemblages?

PATRICIO JORGE MACCHI* AND PABLO HORACIO VIGLIANO
Centro Regional Universitario Bariloche, Universidad Nacional del Comahue
Quintral 1250 (8400) Bariloche, Río Negro, Argentina

MIGUEL ÁLBERTO PASCUAL
CONICET, CENPAT, Boulevard Bouchard S/N, Puerto, Madryn, Chubut, Argentina

MARCELO ALONSO, MARÍA AMALIA DENEGRI, DANIELA MILANO,
MARTÍN GARCIA ASOREY, AND GUSTAVO LIPPOLT
Centro Regional Universitario Bariloche, Universidad Nacional del Comahue
Quintral 1250 (8400) Bariloche, Río Negro, Argentina

Abstract.—Early in the 20th century, the fish fauna of Patagonia consisted of 20 native fish species. In 1904, exotic species, mostly salmonids from North America and Europe, were introduced, giving rise to an extensive sport fishery. Scientific literature, stocking and catch logbooks, and registries of organizations in charge of enforcing management practices were searched in relation to effects of probable policy goals upon actual fish assemblage structure of two major river basins of northern Patagonia. Policy goal history in this region can be divided into three periods. Between 1904 and 1910, goals were focused on increasing diversity and sport fishery opportunities through salmonid introductions. From 1910 to 1970, goals shifted towards development of both commercial and sport fisheries, increasing diversity by introducing and restocking salmonids and native fish species and preserving both through fishing regulations. In these two periods, policies were centralized under the federal government. From 1970 on, policy has been characterized by decentralization in relation to provincial and national park jurisdictions. The former fosters the development of sport fisheries, while the latter focuses on the protection of native fish species and opposes new introductions. However, Patagonia's isolated environments, lack of road infrastructure, particular management circumstances, biological capacities of released species, and great dispersion capacity of salmonids were fundamental factors that shaped the artificial and natural dispersion in both the Negro and Manso River drainages.

*Corresponding author: pmacchi@crub.uncoma.edu.ar

Introduction

Many of the studies pertaining to the introduction of exotic species have focused on the impacts upon native species and the ecosystems into which they were introduced. However, understanding the invasion processes is as important (Dunham et al. 2002). Salmonids have both a great capacity of adaptation and of dispersion (Gowan et al. 1994) and they also constitute the freshwater fish family with the greatest world distribution (Welcomme 1988).

Introduction and dispersion of salmonids in Patagonia in general and on its northern region in particular have been a process strongly influenced by shifting and sometimes capricious management policies, both in Argentina and in Chile (Pascual et al. 2002). Not all species of this family were introduced at the same time, nor were all the environments stocked with the same intensity or at the same time. (Marini 1936; González Regalado 1945; Baigún and Quirós 1985). Today, the entire process looks very much like a trial and error route without a template. Decisions were, and still are, made without considering the invasive capacity of different species, the competition among different introduced species that could follow, or their potential impacts upon native biota.

The objectives of the present work are (1) to analyze the introduction of salmonids into northern Patagonia in relation to historical and present data, (2) to evaluate both natural and artificial dispersion of each of the introduced species, (3) to characterize the variation in distribution and abundance of each one of these species, and (4) to determine the degree of success of each species in different water bodies of the Negro and Manso River basins from a historical perspective.

Material and Methods

Study Area

The present work is based on the Negro and Manso River drainage basins, which were chosen because salmonid stocking records and presence–absence data are the most complete of Patagonia. Also, the Negro has been the river basin in Patagonia most strongly and actively affected by plantings with exotic fish. Because of its 70,000-km^2 area, the Negro (Figure 1) drainage basin is one of the most important drainage basins of Argentina and the largest of Patagonia. The effective drainage area is triangular in shape, having been formed by the 36,400 km^2 of the Limay River drainage subbasin in the south and the 32,450 km^2 of the Neuquén River subbasin in the north. They unite at latitude 39°south and longitude 68°west to form the Negro River subbasin with a drainage area of 1,000 km^2 (HIDRONOR, 1978) and a maximum flow of 1,600 m^3/s. The entire basin, fed by rain, snowfall, and ice melting from the Andean range in the west, drains after crossing the Patagonian steppe into the Atlantic Ocean. Climate varies from west to east from cold humid in the Andean range to arid in the steppe, with precipitation decreasing from 4,000 mm to less than 300 mm per year in just 50–100 km and only 200 mm by the seashore (De Aparicio and Difrieri 1958; Mermoz and Martín 1988). The hydrological pattern is characterized by two peak flows, one driven by rainfall rains and the other one by spring snow melting. All lakes and rivers in the basin are well oxygenated and mainly neutral or slightly alkaline with calcium bicarbonate, becoming enriched with sulfates and sodium when they get closer to the ocean (Bonetto and Wais 1995). Conductivity varies among subbasins from 30 to 67 µS/cm in the Limay and from 75 to 250 µS/cm in the

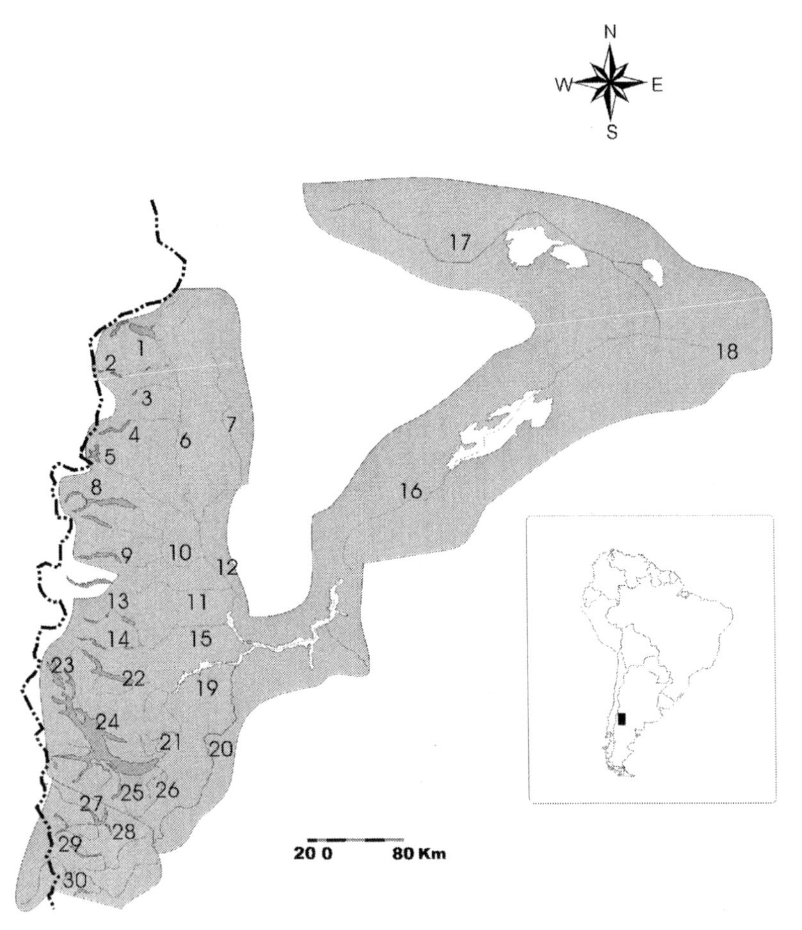

Figure 1. Study area. Negro River drainage basin and Manso River drainage basins. Rivers: 2. Pulmari, 3. Ruca Choroi, 4. Quillén, 5. Malleo, 6. Aluminé, 7. Catan Lil, 9. Quilquihue, 10. Chimehuín, 11. Quemquemtreu, 14, Caléufu, 15. Collon Curá, 16. middle Limay, 17. Neuqén, 18. Negro, 19. and 21. Upper Limay, 20. Pichi Leufú, 22. Traful, 23. Correntoso, 26. Ñiriguau, 29. Middle Manso, 30. Lower Manso. Lakes: 1. Aluminé, 8. Huechulaufquen, 12. Meliquina, 13. Filo Hua Hum, 24. Nahuel Huapi, 25. Gutiérrez, 27. Mascardi, 28. Guillelmo.

Neuquén, with a maximum of 3,600 μS/cm in Lake Pellegrini.

The fish fauna has a low number of species (Pascual et al. 2002), with three siluriforms (otuno *Diplomistes viedmensis*, *Hatcheria macraei*, and *Trichomycterus areolatus*), two Galaxids (inanga *Galaxias maculatus* and *G. platei*), three Percichthids (creole trout *Percichthys colhuapensis*, creole perch *P. trucha*, and *P. vincinguerrai*), two Cyprinodontiformes (onesided livebearer *Jenynsia lineata* and *Cnesterodon decenmaculatus*), a silverside (*Odontesthes hatcheri*) and a lamprey (pouched lamprey *Geotria australis*). Also, mullet (liza *Mugil liza*) can be found in the lower Negro River subbasin area influenced by tides (Arratia et al. 1983; Milano and Vigliano 1997; Cussac et al. 1998; Bello 2002; Pascual et al. 2002). Besides salmonids, other species introduced into the basin by man are Buenosairean pejerrey *Odontesthes bonaerensis*, *Corydoras* sp., *Cheirodon eigenmanniorum*, and common carp *Cyprinus carpio*, which are found only in the

Negro River subbasin. (Pascual et al. 2002; Alvear et al., in press).

The Manso River basin (Figure 1) is formed by three sections: upper, middle, and lower Manso. Fed by the Tronador mountain glaciers (Gallopin 1978), the upper Manso waters are murky, with a high load of sediments. Upon flowing and forming Mascardi lake, sediments deposit and the water becomes clear. The middle Manso connects a series of lakes (Los Moscos, Hess, Fonck Grande, Fonck Chico, Julio A. Roca, Felipe, Martin, and Steffen). The lower Manso flows southeast, from Lake Stephen, joining the Villegas and Foyel rivers and then turning westward across the Andes into Chile. It then flows into the Puelo River, which drains into Pacific Ocean (De Aparicio and Difrieri 1958). The hydrological pattern is characterized by two peak flows, one driven by rainfall and the other one by spring snow melting. The average discharge is 65.2 m^3/s at the end point of the middle Manso (Secretaría de Recursos Naturales y Desarrollo Sustentable. Subsecretaría de Recursos Hídricos 1997). The climate is cold and humid with an annual average rainfall of 1,500 mm (De Aparicio and Difrieri 1958; Drago 1974). In general, the waters, which are near oxygen saturation levels, have a conductivity ranging between 36 and 44.9 $\mu S/cm$ and a neutral pH (Pedrozo et al. 1993). The vegetation cover corresponds mostly to sub-Antarctic forest of *Nothofagus* sp., *Fitzroya cupresoides*, and *Austrocedrus chilensis* (Dimitri 1972). The native fishes are represented only by inanga and *G. platei* (Alonso et al. 1997; Milano et al. 2002), salmonids being the only exotic fish introduced to the basin (Alonso et al. 1997).

For this work, data of both drainage basins were subdivided into strata, which mainly coincide with the natural subbasins (Figure 1).

Data Gathering and Analysis

We compiled a database that covers the 1904–2002 period by using a variety of sources, including historical archives of fish hatcheries and fisher associations, particular log books, and refereed and gray literature, as well as data from our own research projects. Data on stocking and distribution for all species were organized on electronic databases and divided in time periods according to objectives.

Historical Stocking Analysis

We analyzed different official records, of the Aquaculture Division of the Ministerio de Agricultura, Ganadería y Pesca de la República Argentina, MAGPRA (Oficial Archives Ministero de Agricultura y Ganadería de la República Argentina. Pisciculture Nahuel Huapi 1908–1933; Oficial Archives Ministero de Agricultura y Ganadería de la República Argentina 1938–1979; Oficial Archives Ministero de Agricultura y Ganadería de la República Argentina. Piscicultura Nahuel Huapi 1914–1917; Oficial Archives Ministero de Agricultura y Ganadería de la República Argentina 1938–1979; Oficial Archives Ministero de Agricultura y Ganadería de la República Argentina. Piscicultura Río Limay 1950–1979; Oficial Archives Ministero de Agricultura y Ganadería de la República Argentina 1938–1979; Oficial Archives Ministero de Agricultura y Ganadería de la República Argentina. Centro de Salmonicultura Bariloche 1971–1986). We found that records collected at different times had different degrees of reliability, with the oldest information being the most detailed and systematically collected one. Therefore, we used data on stocking site, species, and numbers stocked only up to 1986. Based on this information, we looked at the numbers stocked per species per year, the total number stocked in each basin and the numbers stocked in each subbasin. The 82 years covered by the official records were divided arbitrarily into three periods according to stocking intensity and spread (1) from 1904 to 1930, (2) from 1931 to 1946, and (3) from 1947 to 1986.

Presence–Absence Data

The dispersion of salmonids in both drainage basins was analyzed based on existing data for three periods: 1908–1924, 1937–1943, and 1996–2002 (Oficial Archives Ministero de Agricultura y Ganadería de la República Argentina, Piscicultura Nahuel Huapi 1908–1933; Valette 1924; Bruno Videla 1938, 1941, 1944; GonzÁlez Regalado 1941a, 1941b, 1941c, 1945; Marini 1936, 1942). Pre-1996 presence–absence data were obtained through a variety of sampling methods and interviews conducted by MAGPRA. Data for the 1996–2002 period were obtained from samplings conducted by the "Grupo de Evaluación y Manejo de Recursos Ícticos CRUB-UNC, GEMARI." All analyses were performed by subbasin.

Salmonid Abundance

For this analysis, we compared commercial and exploratory catches in some lakes of both drainage basins conducted between 1937 and 1942, with those conducted in the same environments between 1996 and 2002 (Table 1).

In 1937–1942, catches were made between October and March (Oficial Archives Ministero de Agricultura y Ganadería de la República Argentina. Piscicultura Bariloche 1938–1942), using multifilament gill nets of 45 m in length by 2.5 m in height. Mesh sizes were 52 and 60 mm between knots. Gill nets were set perpendicular to the coastline, operating from dusk to dawn. (Bruno Videla 1938; Oficial Archives Ministero de Agricultura y Ganadería de la República Argentina. Piscicultura Bariloche 1939–1942.). On each occasion species, numbers, and total weight (kg) of fish caught were recorded (Oficial Archives Ministero de Agricultura y Ganadería de la República Argentina. Piscultura Bariloche 1938–1942).

Table 1. National Nahuel Huapi Park Lakes for which salmonid catches of the periods 1937–1942 and 1996–2002 were compared. **a:** 1937–1942 period; **b:** 1996–2002 period. masl = meters above sea level.

Subbasin	Lake	Location	Altitude (masl)	Depth (m)	Sampling date a	Sampling date b
Lake Nahuel Huapi	Moreno	41°05'S–71°32'O	764	112	11/26/40	11/16/97
	Gutiérrez	41°14'S–71°28'O	750	111	11/29/40	11/23/96
	Espejo	40°37'S–71°46'O	772	245	02/20/41	01/20/98
Limay River	Nahuel Huapi	40°55'S–71°30'O	764	438	03/09/41	03/15/02
	Traful	40°25'S–71°27'O	800	339	04/18/40	04/13/02
Upper Manso River	Mascardi	41°19'S–71°35'O	750	218	02/17/41	02/21/97
Middle Manso River	Los Moscos	41°21'S–71°35'O	790	50	01/15/40	02/14/97
	Hess	41°22'S–71°43'O	735	24	02/21/40	02/12/97
	Fonck	41°19'S–71°47'O	780	85	01/28/41	01/29/97
	J. A. Roca	41°22'S–71°46'O	725	38	02/07/41	02/11/97

Fish catches between 1996 and 2002 were obtained using the methodology described by Vigliano et al. (1999), using similar nets and setting procedures as in 1938–1942. The specific composition and the catch per unit of effort in numbers (CPUEN) for each species were estimated for each period. In all cases, the unit effort was defined as 100 m^2 of net and 24 h of operation. Differences in catches between periods and between different basins were compared by one way variance analysis (analysis of variance [ANOVA]), when normality and homoscedasticity of data were proven, or with a Mann-Witney (M-W) test when normality assumptions were not met.

Results

Policy Goals Throughout the Years

It is clear from official documents that the introduction of salmonids in northern Patagonia changed over time. Between 1904 and 1910, the main goals were to increase diversity and sport fishing opportunities. From
1910 to 1970, goals shifted towards the development of both commercial and sport fisheries: fishing regulations were established, new species were introduced, and restocking included both salmonids and native fish species. In these two periods, policies were dictated by the federal government, which also undertook management actions. From 1970 on, there was a decentralization of freshwater management in Patagonia, with the provincial governments and the National Park Administration taking the lead. While provincial governments typically foster the development of sport fisheries, the National Park Administration focuses on the protection of native fish and opposes new introductions (Pascual et al. 2002).

Origin of Salmonids Introduced to Northern Patagonia.

While it is difficult to establish the specific rivers from which original salmonids were brought into Patagonia, it has been possible to establish the country of origin of the initial imports with certain confidence (Table 2), as well as the likely varieties from which actual populations derive.

Brook trout *Salvelinus fontinalis* were introduced in 1904 from the United States (Marini 1936), adapting quickly to the different environments in the region. By 1907, local reproductive stocks were available so as to not require further imports. Therefore, all stocks in both the Negro and Manso River basins are descendants of the fish introduced between 1904 and 1905.

All brown trout *Salmo trutta* inhabiting North Patagonia are descendants from embryos imported from Chile in 1931 (Marini 1936), which in turn descended from eggs brought at the beginning of the 20th century from Hamburg, Germany (de Buen 1959).

The first batches of Atlantic salmon *S. salar* were brought in 1904 from the Great Lakes (MacCrimmon and Gots 1979), and in 1931, more specimens, originally from Germany, were imported from Chile (de Buen 1959).

Rainbow trout *Oncorhynchus mykiss* eggs arrived from the United States for the first time in 1904 (Tulian 1908). These first batches probably came from the McCloud River, California (Pascual et al. 2001). These were sent to hatcheries in the Santa Cruz river, La Cumbre in the Cordoba province and a small number to the Nahuel Huapi hatchery (Marini 1936). The fish in the La Cumbre hatchery were sent to the river Cicerone hatchery in the Tucuman province (Valette 1915),

Table 2. Species introduced in the Negro and Manso river drainage basins since 1904. In the case of established species years of introduction corresponds to the first recorded stocking. On the case of *C. clupeaformis* dates on the three introduction attempts are reported.

Negro River drainage basin				
Species	Introduction year	Origin	Total no. fish introduced	Actual status
Coregonus clupeaformis	1904, 1940, 1965	USA	1,700,000	Nonestablished
Salvelinus fontinalis	1904	USA	4,329,998	Established
S. namaycush	1904	USA	53,000	Nonestablished
Salmo salar	1904	USA, Chile	844,543	Established
Onchorhynchus mykiss	1924	Pisc. Cicerone Argentina, Chile	3,032,092	Established
Salmo trutta	1931	Chile	431,398	Established
Manso River drainage basin				
Species	Introduction Year	Origin	Total no. fish introduced	Actual status
Salvelinus fontinalis	1916	Pisc. Nahuel Huapi, Argentina	480,520	Established
Salmo salar	1931	Chile	47,650	Nonestablished*
Onchorhynchus mykiss	1938	Pisc. Bariloche Argentina, Chile	348,693	Established
Salmo trutta	1931	Chile	341,890	Established

* Presence detected until the 1960s.

from which most of the rainbow introductions at the beginning of the 20th century came (Valette 1924). In 1931, rainbow trout were imported from Chile (Marini 1936). These specimens were descendants from specimens originally imported in 1905 from Hamburg, Germany (de Buen 1959). It is also possible that these rainbow trout were originally from the McCloud River (Pascual et al. 2001). In 1969, rainbow trout from Denmark were brought to the hatchery in Bariloche (Baiz 1973), from where they were extensively distributed to commercial aquaculture facilities throughout Argentina. Even though these were not used to stock natural environments (M. Baiz, Head of Centro de Salmonicultura Bariloche 1971–1993, personal communication), intentional or unintentional releases to water bodies in the region cannot be discarded (Alonso 2003).

Historical Analysis of Salmonid Stocking

Six species of salmonids were introduced in northwestern Patagonia starting in 1904, four of which were able to establish self-sustainable populations (Table 2). Number of species and specimens and dates of introductions and origin were different for the Manso and the Negro basins (Table 2) While introductions in the Negro basin started in 1904 and ended in 1964, those in the Manso basin started 12 years later and ended in 1938. Numbers stocked in both basins differed widely throughout the 82 years (Table 2). Annual stocking rate of all species in the Negro basin was 92,625 fish, while that of the Manso basin was of 17,410 fish. In both basins, the species most intensively stocked were brook trout and rainbow trout. On the Manso basin, as many brown trout were stocked as the other two species. The species that was least intensively stocked was Atlantic salmon (Table 2).

From 1907 to 1930 (Figure 2), brook trout were continuously stocked, and it was the only species disseminated until 1924 when rainbow trout was introduced into the Neuquén drainage basin. Rainbow trout started to be frequently stocked from 1937 on, becoming the preferred species in 1950. Salmon stocking ended in 1906, and in 1931, it was again regularly introduced in small numbers until 1973. Brown trout was sporadically stocked from 1931 to 1958; dissemination was intensified until 1969 and then decreased until 1981 when the last stocking took place.

In the Manso River basin, stocking was more irregular. Brook trout was introduced in 1916 and 1918 and then annually between 1937 and 1971. Rainbow trout was discontinuously propagated for 39 years, since its first introduction in 1938. Stocking of brown trout was more or less regular between 1931 and 1969, usually at numbers similar to those of brook trout. Only seven introductions of Atlantic salmon were identified throughout the 63-year-record for this basin (Figure 3).

Introductions and Presence–Absence Data over Time

During the 1904–1986 period, a total of 603 stocking events of all species were recorded for the Negro and Manso River basins combined. Of these, 371 were in the Negro River subbasins of the Nahuel Huapi, Traful, Pichileufú, and Limay rivers, 134 were in the rest of the Negro river basin, and only 98 were in the Manso River basin.

Brook trout was introduced to all subbasins at different periods, except for the middle and lower Limay subbasins (Figure 4). Presence data throughout the different periods show

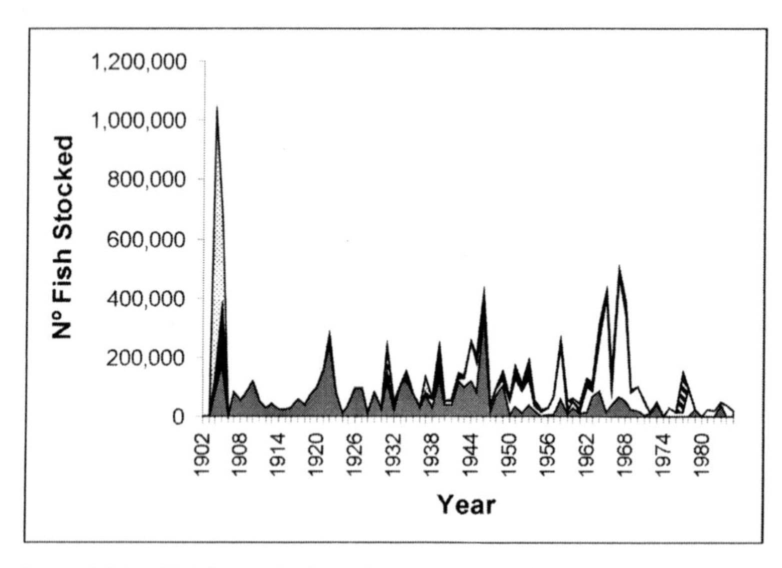

Figure 2. Numbers of fish officially stocked on the Negro River drainage basin between from 1904 to 1986. brook trout, rainbow trout, Atlantic salmon, brown trout, others

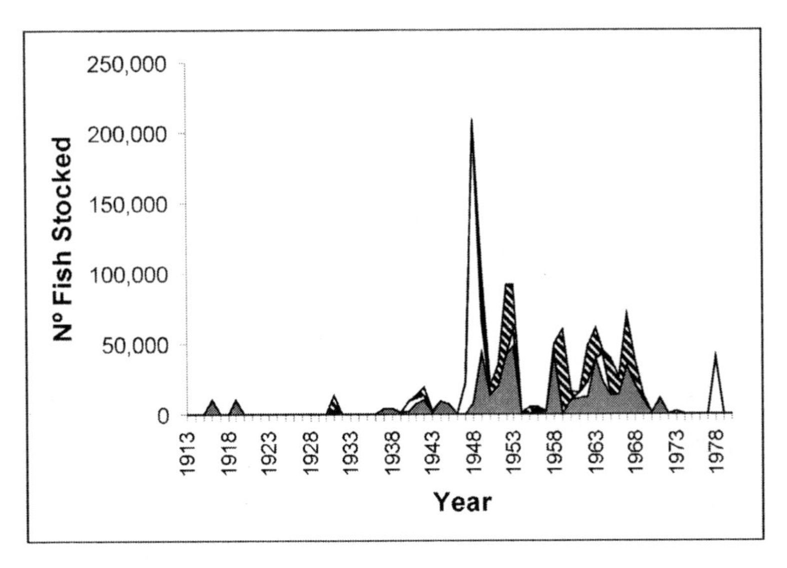

Figure 3. Numbers of fish officially stocked on the Manso River drainage basin from 1916 to 1979. brook trout, rainbow trout, Atlantic salmon, brown trout

that this species retracted towards western environments of the Andean range (Figure 4). Salmon was introduced during all three analyzed periods, but in a geographically more restricted fashion. No stockings have been recorded of this species for the middle and lower Limay or the Neuquén and Negro rivers (Figure 5). After an initial expansion, which lasted approximately until 1945, this species retracted to a small number of water bodies, all of them in the Negro River basin (Figure 5). Before 1930, rainbow trout had been introduced only into Pellegrini Lake in the Negro River basin. Between 1931 and 1945, this species was widely disseminated in locations on all subbasins (Figure 6). A similar scenario is found for brown trout, which after their initial 1931 stocking were introduced to all subbasins (Figure 7). These two last species quickly adapted and dispersed in all considered subbasins (Figures 6–7).

Historic Variation of Salmonid Abundance

An important variation per species catches was detected for both drainage basins between study periods (Table 3). Global CPUEN (all lakes, all species) did not differ significantly between river basins during the 1937–1942 period (ANOVA, $P > 0.05$) or during the 1997–2002 period (ANOVA, $n = 10$, $P > 0.05$).

Even though the numbers of fish caught varied between periods, significant differences were only found for catches at lake Espejo, where catches were higher for the period of 1937–1942 (M-W, $n = 8$, $P < 0.05$).

Important differences were found between periods for particular species. Brook trout catches in the period 1937–1942 did not differ between drainages (ANOVA, $n = 10$, $P > 0.05$), a period where this species was dominant all throughout both river basins. A strong drop in CPUEN values for this species was observed in 1996–2002 (ANOVA, $n = 20$; $P < 0.05$), to the extent that it is absent today in several environments where it occurred and virtually has disappeared from all lakes, while rainbow and brown trout populations increased consider-

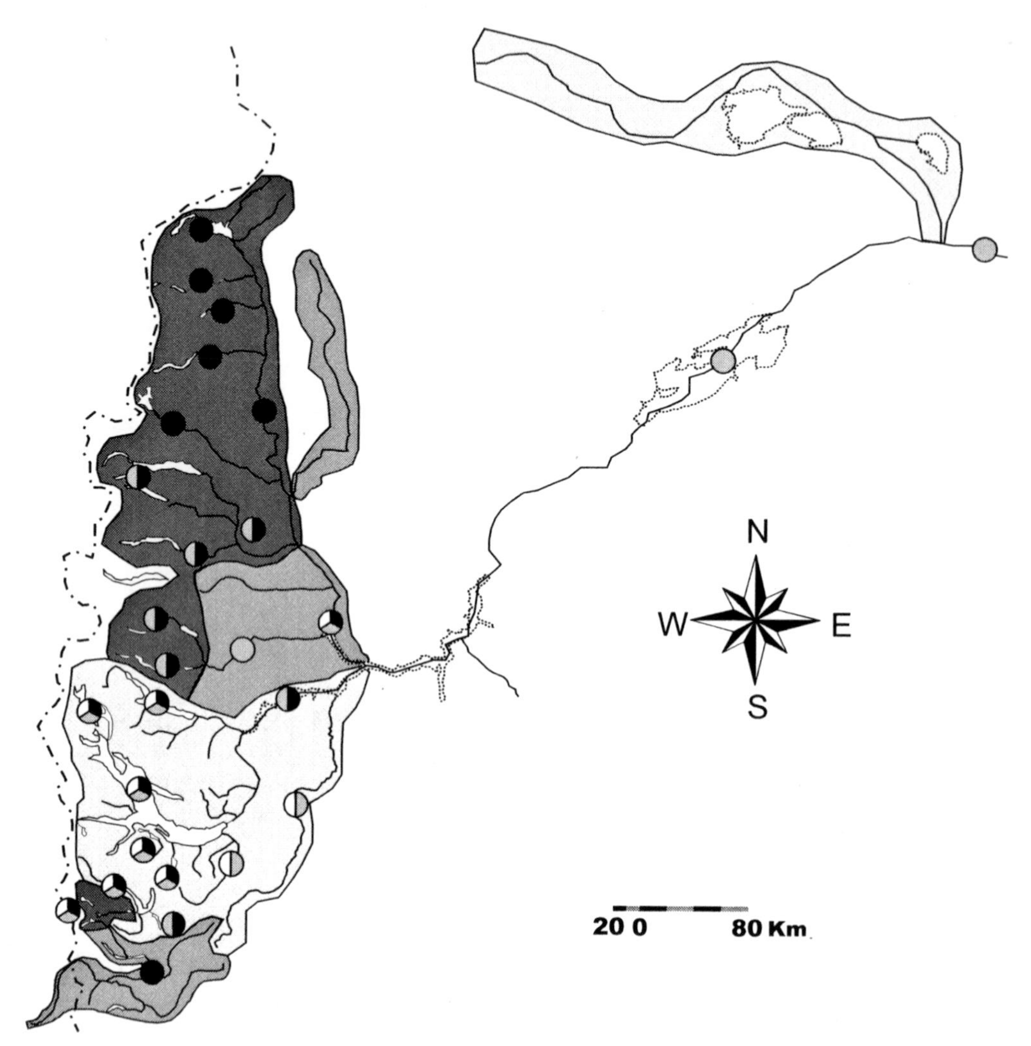

Figure 4. Brook trout stocking distribution and historic presence map. Stocking periods: 1. light gray areas 1904–1930, 2. dark gray areas 1931–1945, 3. medium gray areas 1946–1986. Presence data: white circles 1904–1930, gray circles 1937–1945, black circles 1990–2002.

ably between periods in all studied water bodies. In summary, there has been a replacement of brook trout in both drainage basins by these two species. Thus, in the Manso drainage, brook trout went from representing 100% of the catches on the first period to only 33% of the catches in the second (Table 3).

Discussion

The introduction and acclimatization of salmonids into the Negro and Manso River basins implied a long and complex process, characterized by shifting management goals and species preferences over space and time. Of all species introduced into northern Patagonia, brook, rainbow, and brown trout, as well as Atlantic salmon, managed to adapt successfully to different water bodies in the region. Only lake whitefish *Coregonus clupeaformis*, introduced in 1904, 1937, and 1964, and lake trout *Salvelinus namaycush*, introduced in 1904 and 1905 (Tulian 1908; Marini 1936; Oficial Archives Ministero de

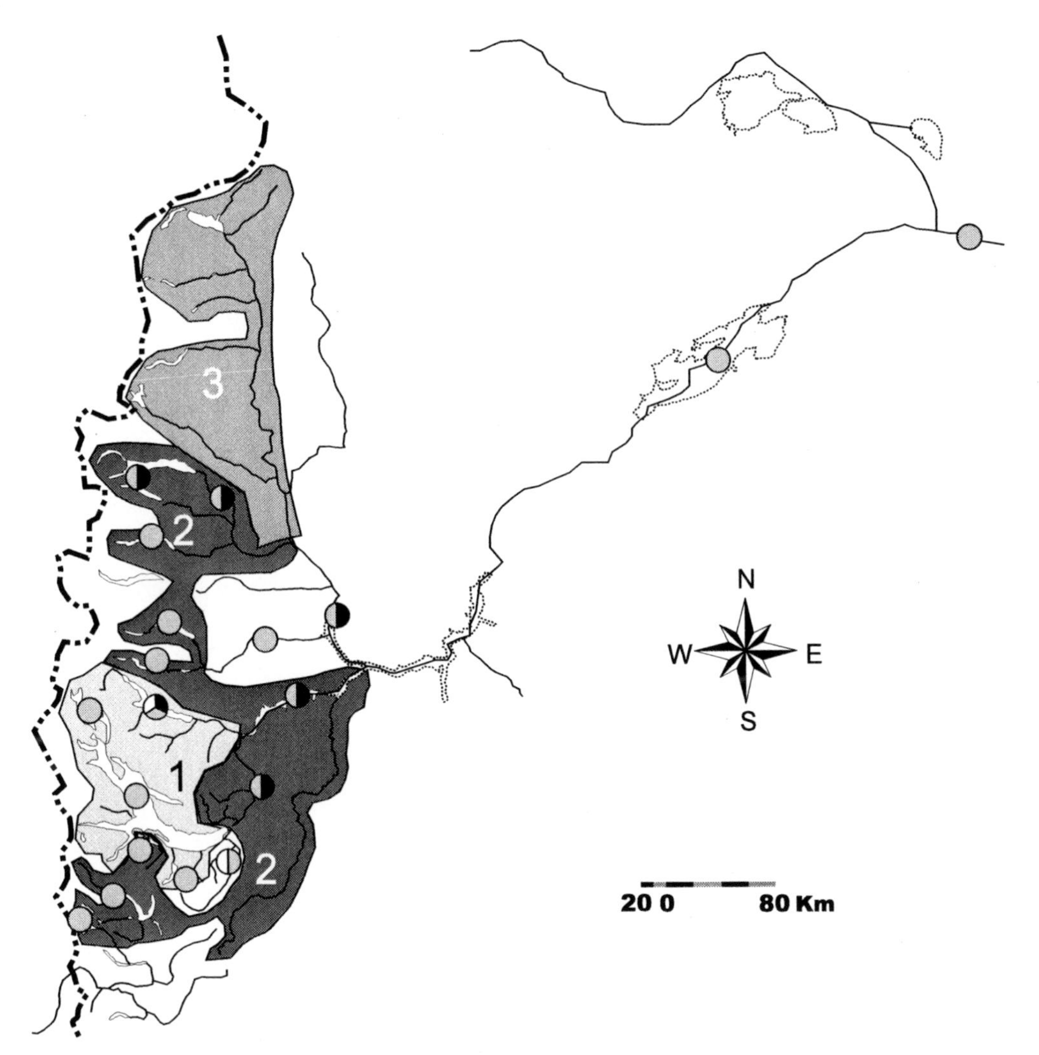

Figure 5. Atlantic salmon stocking periods: 1. light gray areas 1904–1930, dark gray areas 1931–1945, medium gray areas 1946–1986. Presence data: white circles 1904–1930, gray circles 1937–1945, black circles 1990–2002.

Agricultura y Ganadería de la República Argentina. Piscicultura Bariloche 1934–1970), failed to produce self-sustaining populations.

The six species introduced to Patagonia have different success records as introduced species around the world. Lake whitefish was introduced to only seven countries, failing in all of them (Welcomme 1988). Lake trout was introduced into 11 countries in Europe, Oceania, and South America, adapting only to environments in the Alps, New Zealand, and Argentina (Welcomme 1988). In Patagonia, self-sustaining populations exist in the Santa Cruz and Chico basins (Pascual et al. 2002). Atlantic salmon has been the subject of numerous worldwide introductions during the past two centuries, but it has adapted only in New Zealand and Argentina (MacCrimmon and Gots 1979). A new round of dissemination of this species is occurring through escapes from fish farming facilities in Chile (Pascual et al. 2002) that cross the Andes

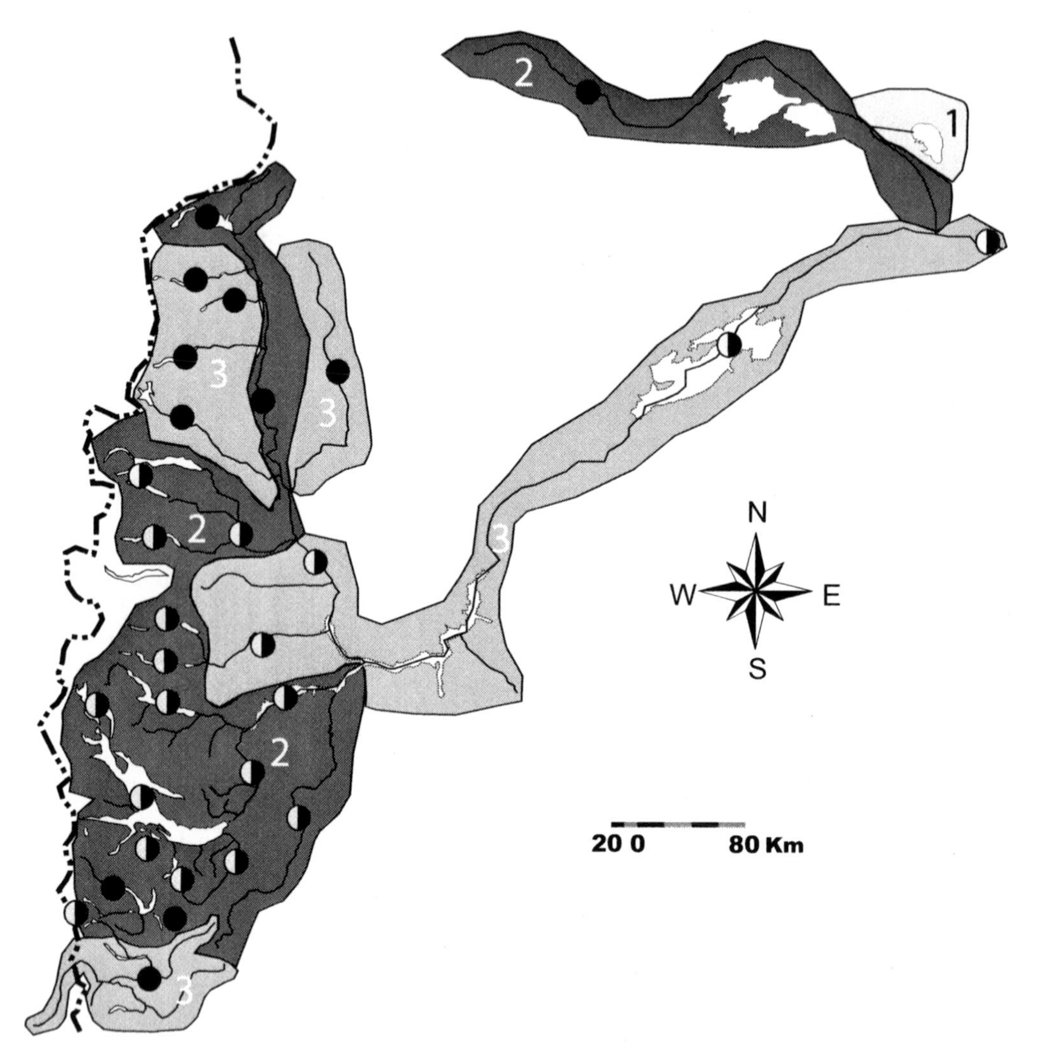

Figure 6. Rainbow trout stocking periods: 1. light gray areas 1904–1930, dark gray areas 2. 1931–1945, 3. medium gray areas 1946–1986. Presence data: gray circles 1937–1945, black circles 1990–2002.

through Pacific river basins of Argentina. Brook, rainbow, and brown trout have widely spread on the five continents with different results (Welcomme 1988). Brook trout established scattered populations in several countries (MacCrimmon and Campbell 1969). Rainbow trout and brown trout are the two salmonid species with the widest world distribution, establishing permanent populations in many countries (MacCrimmon and Marshall 1968; MacCrimmon et al. 1970; MacCrimmon 1971). In Patagonia, these two species have prospered the most among salmonids (Pascual et al. 2002), with anadromous and land-locked populations all throughout their range (Pascual et al. 2001).

Specific policy goals in 1904–1910, 1910–1970, and 1970–2002 undoubtedly influenced the introductions' outcomes. Thus, meanwhile, on the first two periods, four different salmonid species and some native species were introduced. The third period is characterized by provincial administrations focus-

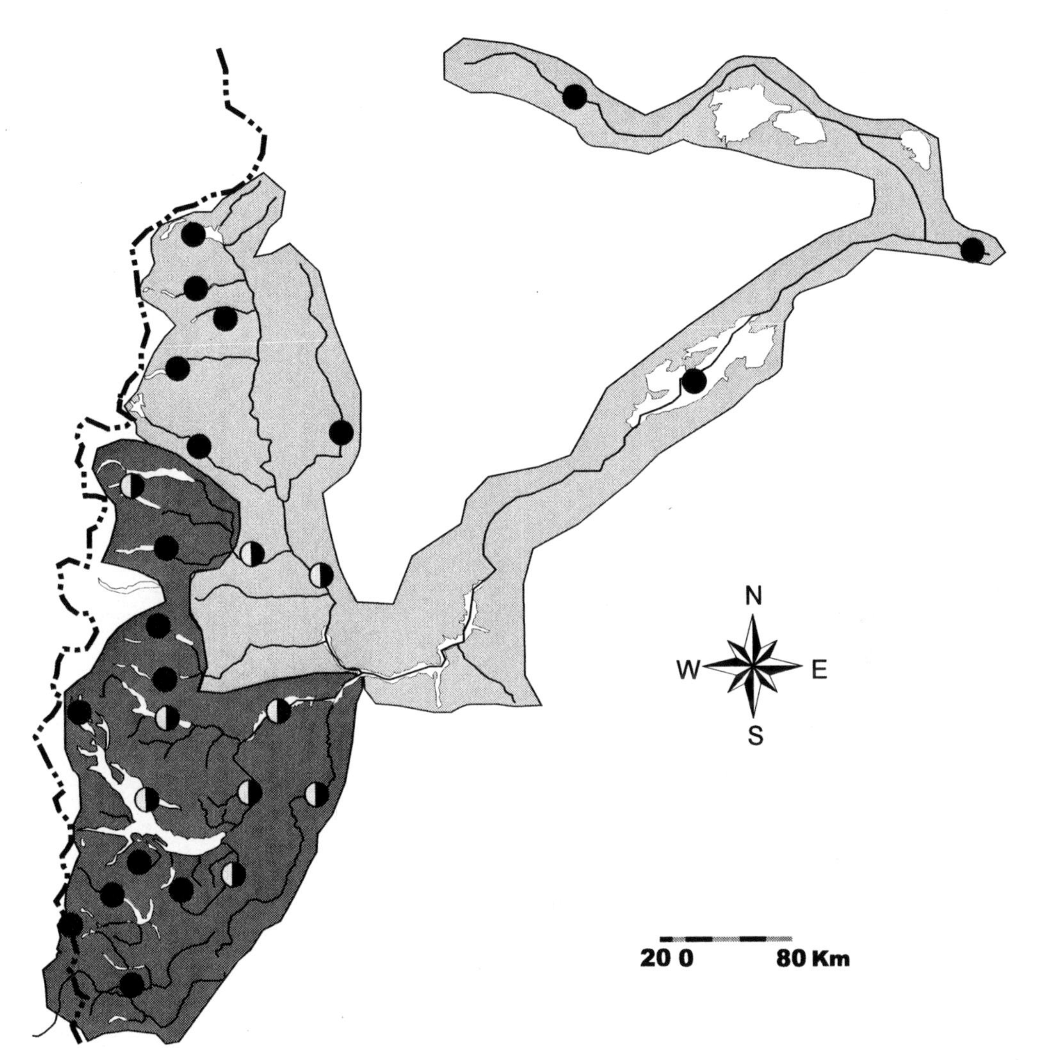

Figure 7. Brown trout stocking periods: 1. dark gray areas, 1904–1930; 2. dark gray areas, 1931–1945; 3. medium gray areas, 1946–1986. Presence data: gray circles 1937–1945, black circles 1990–2002.

ing on restocking of primarily rainbow trout with sport fishing benefits in mind.

Other factors that contributed to shape the artificial and natural dispersion of salmonids in the Manso and Negro basins were the degree of isolation of different locations, with lack of road infrastructure well into the 20th century (Bustillo 1968); a trial and error management route without a template (Marini 1936, 1941); and the biological plasticity of released specimens, including their great dispersive capacity (Gowan et al. 1994; Sakai et al. 2001). Stocking was conducted in all environments, but while western environments were stocked with four species, eastern ones were stocked with only rainbow and brown trout. In 1986, different national and provincial fishing related agencies were closed down and stocking activities continued in the hands of different nongovernmental organizations. The number of fish stocked, as well as the number of stocking events in the different drainage basins, may have influenced the ac-

Table 3. Records of salmonid catch per unit effort in numbers (CPUEN) on lakes of the Andean Range of the Negro river drainage basin (upper Limay subbasin) and Manso river drainage basin (upper and middle Manso subbasins). Catches were compared between same months (see Table 2), of 1937–1942(a) and 1995–2002 period (b).

Negro River drainage basin						
		CPUEN				
Lake		Brook trout	Atlantic salmon	Rainbow trout	Brown trout	Total
Moreno este	a	2,00	0,66	0,66		3,32
	b	2,00		4,00		6,00
Gutiérrez	a	5,33				5,33
	b	1,00		0,50	0,50	2,00
Espejo	a	23,85				23,85
	b	0,52		0,42		0,94
N. Huapi (B. Rincon)	a	5,63	0,15	-		5,78
	b			1,33	4,00	5,33
N. Huapi (Dina Huapi)	a	11,85	0,30			12,15
	b			1,33	0,67	2,00
Traful (Pto. Arrayan)	a	1,33	0,33			1,66
	b	1,00		2,00	2,00	3,00

Manso River drainage basin						
		CPUEN				
Lake		Brook trout	Atlantic salmon	Rainbow trout	Brown trout	Total
Mascardi	a	8,70	0,35			9.05
	b	1,66		1,66	5,33	8,65
Los Moscos	a	10,47	0,15			10,62
	b			0,54	0,54	1,80
Hess	a	1,33				1,33
	b			2,00	2,67	4,67
Fonck	a	16,22				16,22
	b	12,00		8,00	7.00	27,00
J. A. Roca	a	12,17				12,17
	b	2,00		2,25		4,25

tual dispersion of different species, but it is still not clear how these two variables influenced probability of success of both colonization and posterior dispersion of invasive species. Thus, while total stocking of brown trout until 1986 was one order of magnitude less than that of brook trout, its present distribution is much more extended, being also more numerous in several environments.

The different stocking periods have played a key role in the historic distribution of the four adapted species. During the first 30 years, brook trout and salmon were the only salmonids caught in the region. Brook trout was stocked in higher numbers, but both adapted quickly (Ormsby 1908a, 1908b) and initially dispersed naturally in both drainages (Valette 1924; Marini 1942). Brook trout dominated the scene at least until 1945 in most of the Negro River basin and in practically all of the Manso basin (Bruno Videla 1944; GonzÁlez Regalado 1945).

Introduction of rainbow and brown trout produced a dramatic shift in species composition of freshwater bodies in northern Patagonia. Sampling between 1939 and 1942 started to yield rainbow trout in all lakes of the Nahuel Huapi subbasin and upper and middle Manso River. There is also evidence that these changes proceeded fast. Bruno Videla (1944) states that rainbow trout was not dominant in 1941 in the subbasin of the Nahuel Huapi Lake, but by 1944, it had displaced brook trout in all connected water bodies. In short, rainbow trout became dominant in just 8 years from its first report in 1939.

Exotic rainbow trout has been able to displace brook trout in several locations of its endemic range (Larson and Moore 1985). There is, therefore, a great interest to study their interaction to derive management actions for brook trout (reviewed in Fausch 1988). These studies have included field work (Larson and Moore 1985), field removal experiments (Moore et al. 1986), and laboratory experiments (Cunjak and Green 1983, 1984, 1986), which have attempted to reveal those factors that affect coexistence of these two species. In general, rainbow trout appears to be a superior competitor, but the dominance of one or the other species is apparently modulated by several physical factors. Some studies in the Appalachians also indicate that the relative size that the fish have during their first growing season determines the outcome of the interaction (Larson and Moore 1985), but this hypothesis has not been assessed for other environments. The success of rainbow trout and the demise of brook trout in northern Patagonia appear to have followed the same scenario observed in North America. It, therefore, provides a fantastic opportunity to explore the interaction between these species.

Acknowledgments

The present work was done under grants from the "Universidad Nacional del Comahue" CRUB (B940) and PICT 98 SECyT Argentina.

References

Alonso, M. F. 2003. Variaciones temporales en la estructura de los ensambles de peces de los embalses de la cuenca de los ríos Limay y Neuquén: diagnóstico y efecto de los escapes de peces de cultivo. Tesis para optar al grado de maestría de la Escuela para Graduados Alberto Soriano. Universidad de Buenos Aires, Facultad de Agronomía, Buenos Aires.

Alonso, M. F., P. H. Vigliano, P. J. Macchi, D. Milano, A. Denegri, and G. E. Lipppolt. 1997. Extensive fish surveys from lakes of Atlantic and Pacific basins on the Andean region of northern Patagonia. En libro de Resúmenes: 7th International Conference on Lakes Conseration and Management. Lacar 97. Volume II. San Martín d elos Andes, Neuquén, Argentina.

Alvear, P., M. Rechencq, P. Macchi, M. Alonso, A. Denegri, P. Vigliano, G. Navone, G. Lippolt, and E. Zattara. In press. Composición, distribución y relaciones tróficas de la ictiofauna del río Negro, Patagonia, Argentina. Ecología Austral 708.

Arratia, G., M. B. Peñafort, and S. Menú-Marque. 1983. Peces de la región sureste de los Andes y sus probables relaciones biogeográficas actuales. Deserta 7:48–107.

Baigún, C. R. M., and R. Quirós. 1985. Introducción de peces exóticos en la República Argentina. Instituto Nacional de Investigación y Desarrollo Pesquero, Departamento de Aguas Continentales, Informe Técnico N°2, Mar del Plata, Argentina.

Baiz, M. de L. 1973. Crecimiento en cautividad de *Salmo gairdneri* Richardson, 1836 y sus posibilidades comerciales. Ministerio de Agricultura y Ganadería, Dirección Nacional de Recursos Naturales Renovables, Servicio Nacional de Pesca, Buenos Aires.

Bello, M. T. 2002. Los peces autóctonos de la Patagonia Argentina. Distribución natural. Cuadernos Universitarios, Universidad Nacional del Comahue, Centro Regional Bariloche San Carlos de Bariloche, Argentina.

Bonetto, A. A., and I. R. Wais. 1995. Southern South

American streams and rivers. Pages 257–292 *in* C. E. Coshing, K. W. Cummins, and G. W. Minshall, editors. River and streams ecosystems, Elsevier, Amsterdam.

Bruno Videla, P. H. 1938. Informe de las siembras efectuadas en C. Sarmiento, Chubut y estudios de las masas de agua apropiadas para el cultivo de salmónidos. Archivos oficiales del Ministerio de Agricultura y Ganadería de la República Argentina, Buenos Aires.

Bruno Videla, P. H. 1941. La pesca en el parque Nacional Nahuel Huapi. Revista Médica Veterinaria, Buenos Aires.

Bruno Videla, P. H. 1944. Algunos controles efectuados sobre peces existentes en la región de los lagos. Revista de la Facultad de Agronomía y Veterinaria Universidad de Buenos Aires 11:1–33.

Bustillo, E. 1968. El despertar de Bariloche. Editorial Sudamericana, Buenos Aires.

Cunjak, R. A., and J. M. Green. 1983. Habitat utilization by brook char (*Salvelinus fontinalis*) and rainbow trout (*Salmo gairdneri*) in Newfoundland streams. Canadian Journal of Zoology 61:1214–1219.

Cunjak, R. A., and J. M. Green. 1984. Species dominance by brook trout and rainbow trout in a simulated stream environment. Transactions of the American Fisheries Society 113:737–743.

Cunjak, R. A., and J. M. Green. 1986. Influence of water temperature on behavioural interactions between juvenile brook charr, *Salvelinus fontinalis*, and rainbow trout *Salmo gairdneri.* Canadian Journal of Zoology 64:1288–1291.

Cussac, V. E., D. Ruzzante, S. Walde, P. J. Macchi, V. Ojeda, M. F. Alonso, and M. A. Denegri. 1998. Body shape variation of three species of *Percichthys* in relation to their coexistence in the Limay River basin, in northern Patagonia. Environmental Biology of Fishes 53:143–153.

De Aparicio F. O., and H. A. Difrieri. 1958. La Argentina. Suma de geografía. F. De Aparicio and A Difriery, editors. Peuser, Buenos Aires.

de Buen, F. 1959. Los peces exóticos en las aguas dulces de Chile. Investigaciones Zoológicas Chilenas 5:103–137.

Dimitri, M. 1972. La región de los bosques andinos patagónicos. Anales de Parques Nacionales, Buenos Aires.

Drago, E. C. E. 1974. Estructura térmica del lago Mascardi (Provincia de Río Negro, Argentina). Physis 33:207–216.

Dunham, J. B., S. B. Adams, R. E. Schroeter, and D. C. Novinger. 2002. Alien invasions in aquatic ecosystem: toward an understanding of brook trout invasions and potential impacts on inland cutthroat trout in western North America. Reviews in Fish Biology and Fisheries 12:373–391.

Secretaría de Recursos Naturales y Desarrollo Sustentable. Subsecretaría de Recursos Hídricos. 1997. Estadística Hidrológica, tomo II. Secretaría de Recursos Naturales y Desarrollo Sustentable, Subsecretaría de Recursos Hídricos, Buenos Aires.

Fausch, K. D. 1988. Test of competition between native and introduced salmonids in streams: what have we learned? Canadian Journal of Fisheries and Aquatic Sciences 45:2238–2246.

Gallopin, G. 1978. Estudios ecológicos integrales de la cuenca del río Manso superior (Río Negro, Argentina). I Descripción general de la cuenca. Anales de Parques Nacionales 14:161–230.

Gonzáles Regalado, T. 1941a. Informe de la campaña al Parque Nacional Lanín. Archivos oficiales del Ministerio de Agricultura y Ganadería de la República Argentina, Buenos Aires.

Gonzáles Regalado, T. 1941b. Informe de la campaña realizada a Junín de los Andes y alrededores. Archivos oficiales del Ministerio de Agricultura y Ganadería de la República Argentina, Buenos Aires.

Gonzáles Regalado, T. 1941c. Informe de la campaña realizada en el río Limay desde Senillosa a Cipolletti. Archivos oficiales del Ministerio de Agricultura y Ganadería de la República Argentina, Buenos Aires.

Gonzáles Regalado, T. 1945. Peces de los Parques Nacionales Nahuel Huapi, Lanín y Los Alerces. Anales del Museo de la Patagonia 1:121–138, San Carlos de Bariloche, Argentina.

Gowan, C. H., M. K. Young, K. D. Fausch, and S. C. Riley. 1994. Restricted movement in resident stream salmonids: a paradigm lost? Canadian Journal of Fisheries and Aquatic Sciences 51:2626–2637.

HIDRONOR. 1978. Plan de estudios ecológicos de la cuenca del río Negro, tomo III. Convenio HIDRONOR S. A. - M. A. C. N. Informe final de la primera etapa, Buenos Aires.

Larson G. L., and S. E. Moore. 1985. Encroachment of exotic rainbow trout into stream populations of native trout in the southern Appalachian Mountains. Transactions of the American Fisheries Society 114:195–203.

MacCrimmon, H. R. 1971. World distribution of rainbow trout (*Salmo gairdneri*). Journal of the Fisheries Research Board of Canada 28:663–704.

MacCrimmon, H. R., and T. L. Marshall. 1968. World distribution of brown trout, *Salmo trutta.* Journal of the Fisheries Research Board of Canada 25:1527–1548.

MacCrimmon, H. R., and J. S. Campbell. 1969. World distribution of brook trout, *Salvelinus fontinalis*. Journal of the Fisheries Research Board of Canada 26:1699–1725.

MacCrimmon, H. R., J. Scott Campbell, and B. L. Gots. 1970. World distribution of brown trout, *Salmo trutta*. Journal of the Fisheries Research Board of Canada 27:811–818.

MacCrimmon, H. R., and B. L. Gots. 1979. World distribution of Atlantic salmon, *Salmo salar*. Journal of the Fisheries Research Board of Canada 36:422–457.

Marini, T. L. 1936. Los salmónidos en nuestros Parque Nacional de Nahuel Huapi. Anales de la Sociedad Científica Argentina 121:1–24.

Marini, T. L. 1941. La pesca y la piscicultura fuentes inexploradas de riqueza en la República Argentina. Ministerio de Agricultura de la Nación, República Argentina, Buenos Aires.

Marini, T. L. 1942. El landlocked salmón en la Republica Argentina. Ministerio de Agricultura y Ganadería, Dirección Propaganda y Publicaciones, Publicación Miscelánea 117, Buenos Aires.

Mermoz, M., and C. Martín. 1988. Mapa de vegetación del Parque y la Reserva Nacional Nahuel Huapi. Secretaría de Ciencia y Técnica de la Nación, Delegación Regional Patagonia, San Carlos de Bariloche, Argentina.

Milano, D., V. E. Cussac, P. J. Macchi, D. E. Ruzzante, M. F. Alonso, P. H. Vigliano, and M. A. Denegri. 2002. Predator associated morphology in *Galaxias platei* in Patagonian lakes. Journal of Fish Biology 61:138–156.

Moore, S. E., G. L. Larson, and B. Ridley. 1986. Population control of exotic rainbow trout in streams of a natural area park. Environmental Management 10:215–219.

Oficial Archives Ministerio de Agricultura y Ganadería de la RepúblicaArgentina. Centro de Salmonicultura Bariloche. 1971–1986. Reports and correspondence. Ministerio de Agricultura y Ganadería de la República Argentina, Buenos Aires.

Oficial Archives Ministerio de Agricultura y Ganadería de la República Argentina. Piscicultura Bariloche. 1934–1970. Reports and correspondence. Ministerio de Agricultura y Ganadería de la República Argentina, Buenos Aires.

Oficial Archives Ministerio de Agricultura y Ganadería de la República Argentina. Piscicultura Bariloche. 1938–1942. Commercial archives. Ministerio de Agricultura y Ganadería de la República Argentina, Buenos Aires.

Oficial Archives Ministerio de Agricultura y Ganadería de la República Argentina. Piscicultura Bariloche. 1938–1979. Registro de siembras. Ministerio de Agricultura y Ganadería de la República Argentina, Buenos Aires.

Oficial Archives Ministerio de Agricultura y Ganadería de la República Argentina. Piscicultura Bariloche. 1939–1942. Trabajos realizados, resumen mensual. Ministerio de Agricultura y Ganadería de la República, Argentina, Buenos Aires.

Oficial Archives Ministerio de Agricultura y Ganadería de la República Argentina. Piscicultura Nahuel Huapi. 1908–1933. Reports and correspondence. Ministerio de Agricultura y Ganadería de la República Argentina, Buenos Aires.

Oficial Archives Ministerio de Agricultura y Ganadería de la República Argentina. Piscicultura Nahuel Huapi. 1914–1917. Libro diario. Ministerio de Agricultura y Ganadería de la República Argentina, Buenos Aires.

Oficial Archives Ministerio de Agricultura y Ganadería de la República Argentina. Piscicultura Río Limay. 1950–1979. Registro de siembras. Ministerio de Agricultura y Ganadería de la República Argentina, Buenos Aires.

Ormsby, G. 1908a. Informe del mes de enero de la Estación de Piscicultura Nahuel Huapi. Ministerio de Agricultura de la Nación, República Argentina, Buenos Aires.

Ormsby, G. 1908b. Informe del mes de septiembre de la Estación de Piscicultura Nahuel Huapi. Ministerio de Agricultura de la Nación, República Argentina, Buenos Aires.

Pascual, M., P. Bentzen, C. Riva Rossi, G. Mackey, M. T. Kinnison, and R. Walker. 2001. First documented case of anadromy in a population of introduced rainbow trout in Patagonia, Argentina. Transactions of the American Fisheries Society 130:53–67.

Pascual, M., P. Macchi, J. Urbanski, F. Marcos, C. Riva Rossi, M. Novara, and P. Dell' Arciprete. 2002. Evaluating potential effects of exotic freshwaters fish from incomplete species presence–absence data. Biological Invasions 4:101–113.

Pedrozo, F., S. Chillrud, P. Temporetti, and M. Díaz. 1993. Chemical composition and nutrient limitation in rivers and lakes of northern Patagonian Andes (39°5'–42°S; 71Â°W) (Argentina). Verhandlungen Internationale Vereinigung Limnologie 25:207–214.

Sakai, A. K., F. W. Allendorf, J. S. Holt, D. M. Lodge, J. Molofsky, K. A. With, S. Baughman, R. J. Cabin, J. E. Cohen, N. C. Ellstrand, D. E. McCaulei, P. O'Neil, I. M. Parker, J. N. Thompson, and S. G. Weller. 2001. The population biology of invasive species. Annual Review of Ecology and Systematics 32:305–332.

Tulian, E. A. 1908. Acclimatization of American fishes in Argentina. Bulletin of the Bureau of Fisheries 18:957–965.

Valette, L. H. 1915. La trucha cabeza de acero en las

aguas del norte. Boletín del Ministerio de Agricultura, Dirección General de Ganadería, Buenos Aires.

Valette, L. H. 1924. Servicio de piscicultura. Sus resultados hasta 1922 inclusive. Ministerio de Agricultura de la Nación, Circular 338, Buenos Aires.

Vigliano, P., P. Macchi, M. Alonso, A. Denegri, D. Milano, G. Lippolt, and G. Padilla. 1999. Un diseño modificado y procedimiento de calado de redes agalleras para estudios cuali-cuantitativos de peces por estratos de profundidad en lagos araucanos. Natura Neotropicalis 30:1–11, Santa Fe Argentina.

Welcomme, R. L. 1988. International introductions of inland aquatic species. FAO Fisheries Technical Paper 294.

American Fisheries Society, Symposium 49:349–364

Managing Biodiversity of Pacific Salmon: Lessons from the Skeena River Sockeye Salmon Fishery in British Columbia

CHRIS C. WOOD*

Conservation Biology Section, Marine Ecosystem and Aquaculture Division, Science Branch Pacific Biological Station, Nanaimo, British Columbia V9T 6N7, Canada

Abstract.—Mixed-stock harvest of wild and artificially propagated (enhanced) salmon stocks greatly complicates the conservation of salmon diversity and nowhere is this more evident than in the fisheries for sockeye salmon *Oncorhynchus nerka* in the Skeena River, British Columbia, Canada. The total catch and production of sockeye salmon from the Skeena River has set record high levels over the last decade after a century of intensive commercial fishing. However, both species and stock diversity decreased significantly over the course of the fishery. Species diversity has largely been restored through conservation action, but many individual populations remain at very low abundance. Fishery managers have struggled to find an acceptable trade-off between extracting economic benefits from enhanced stocks while protecting less productive wild stocks from extinction. Recent policies promise to provide explicit limits to these trade-offs based on stewardship ethics and conservation principles.

Introduction

Pacific salmon *Oncorhynchus nerka* utilize virtually every freshwater environment that is accessible from the Pacific Ocean within their natural distribution (roughly 40–65° North latitude). Their remarkable ability to home to natal streams where they spawn and die results in partial or complete reproductive isolation of spawning sites. Reproductive isolation facilitates genetic adaptation that improves survival in local environments, sometimes at a surprisingly small spatial scale (Ricker 1972; Taylor 1991). Such local adaptation accounts for the difficulty of transplanting salmon runs from one river to another (Withler 1982; Wood 1995), or of restoring wild salmon populations in modified habitat (Williams 1987). It is now obvious that salmon populations cannot be replaced easily once they have been extirpated (Lichatowich et al. 1999).

Modern conservation policies for Pacific salmon (e.g., the U.S. Endangered Species Act, Waples 1995; Canada's Species at Risk Act [SARA]; and draft Wild Salmon Policy, DFO 2000) strive to protect distinct populations (stocks) to conserve the genetic diversity among populations that is considered essential to evolutionary potential and sustainable production. However, most salmon are still taken commercially in coastal waters before individual stocks have segregated to their natal streams. These mixed-stock fisheries remain entrenched because they are logistically expe-

*Corresponding author: woodc@pac.dfo-mpo.gc.ca

dient and because salmon are commercially most valuable before they lose fat reserves during arduous upstream migrations or change color as they approach maturity.

Fisheries managers are faced with a trade-off that remains unresolved—how to reap the benefits from commercially valuable stocks while maintaining diversity essential for sustainability. Some species and stocks are more productive than others, in part, as a result of natural variation in their freshwater habitat. The harvest rate providing maximum sustainable yield (MSY) from productive stocks will be excessive for less productive, comigrating stocks that are vulnerable to the same fishery. Unless it is possible to selectively harvest productive populations, the overall harvest rate must be reduced to ensure the conservation of less productive stocks.

In some stocks, natural reproduction is supplemented (enhanced) by artificial propagation in hatcheries or spawning channels. Salmon enhancement has important implications for fisheries management. By increasing the abundance and productivity of target stocks, enhancement attracts or provides an opportunity for increased fishing effort while exacerbating natural variations in productivity. Mixed-stock harvest of wild and enhanced salmon stocks greatly complicates the conservation of salmon diversity, and nowhere is this more evident than in the fisheries for sockeye salmon in the Skeena River, British Columbia.

Here I examine what is known about the current status of salmon diversity in the Skeena River, updating information and summarizing conclusions from an earlier unpublished case study sponsored by the United Nations Environmental Program (Wood 2001).

The Skeena River Salmon Resource

The Ecosystem

The Skeena River drainage occupies about 48,000 km^2 in the west-central part of British Columbia between 54° and 57° North latitude. In Canada, it is second only to the Fraser River in its capacity to produce sockeye salmon. At least 70 distinct spawning sites and 27 lakes are utilized by sockeye salmon within the watershed (Smith and Lucop 1966). These nursery lakes are distributed from the coast to the high interior regions and vary widely in size and productivity (Figure 1). The Babine-Nilkitkwa lake system is the largest natural lake entirely within British Columbia (500 km^2) and supports the largest single sockeye salmon population in Canada.

Six other species of *Oncorhynchus* inhabit the Skeena River including four Pacific salmon (pink *O. gorbuscha*, chum *O. keta*, coho *O. kisutch*, and Chinook *O. tshawytscha*) and two anadromous trout (rainbow *O. mykiss* and coastal cutthroat *O. clarkii*). Management of Pacific salmon remains a federal responsibility (Fisheries and Oceans Canada [DFO]), whereas management of the trout species has been delegated to the Province of British Columbia. Pink and sockeye salmon are the most abundant salmon species, followed in order by coho, Chinook and chum salmon (DFO 1985; Riddell 2004).

Pacific salmon migrate to sea as smolts in April through July after spending several months or years in freshwater, or in the case of pink and chum salmon, within a few weeks of emergence as free-swimming fry. Smolts typically move northward along the coast and offshore into the North Pacific Ocean. The extent of seaward movement and duration of ocean

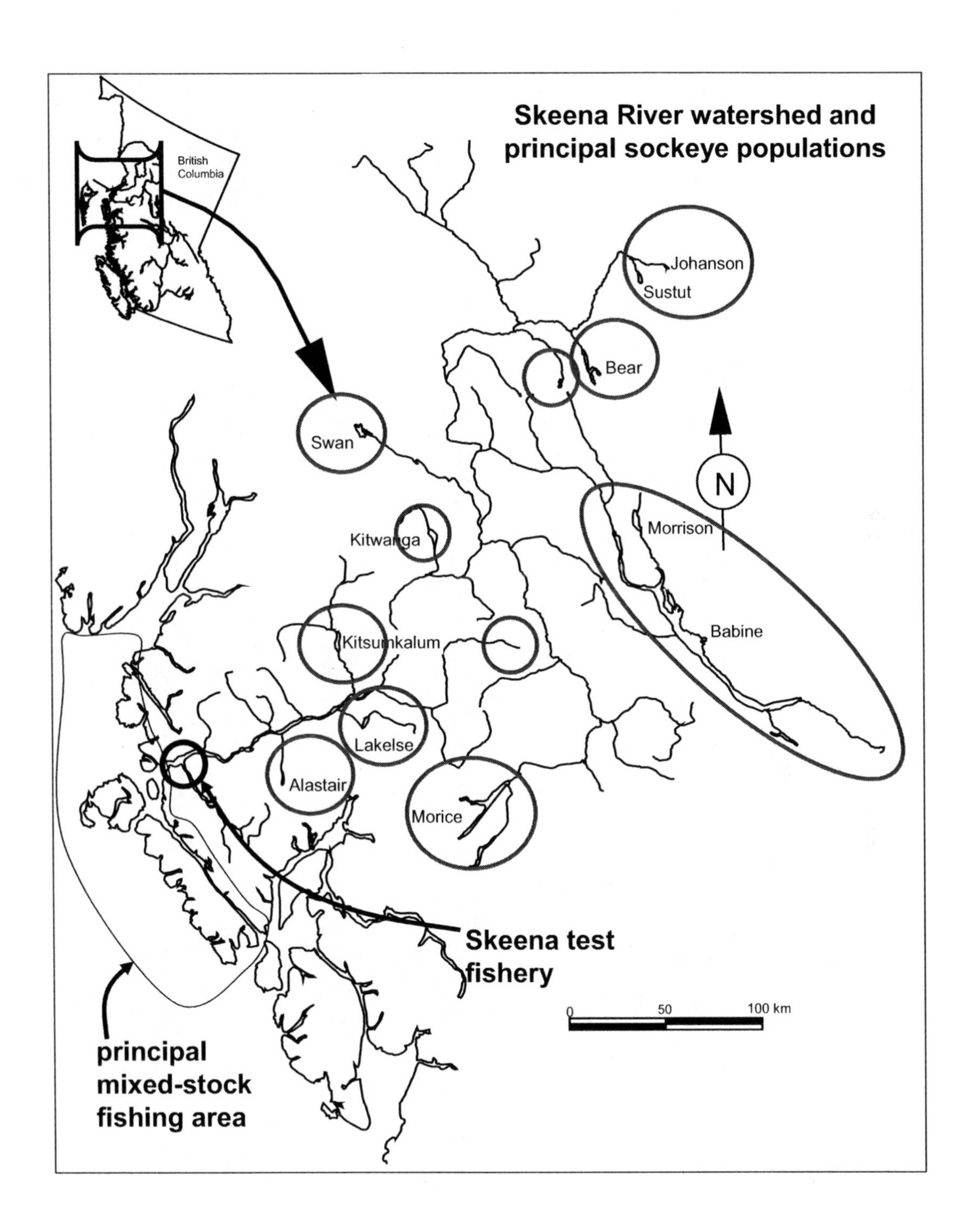

Figure 1. Map of the Skeena River showing principal nursery lakes for sockeye salmon.

residence varies among species (Groot and Margolis 1991), but all species return to the Skeena River, predominantly in summer between May and September. The timing of river entry varies among species and stocks within species but there is broad overlap (Figure 2).

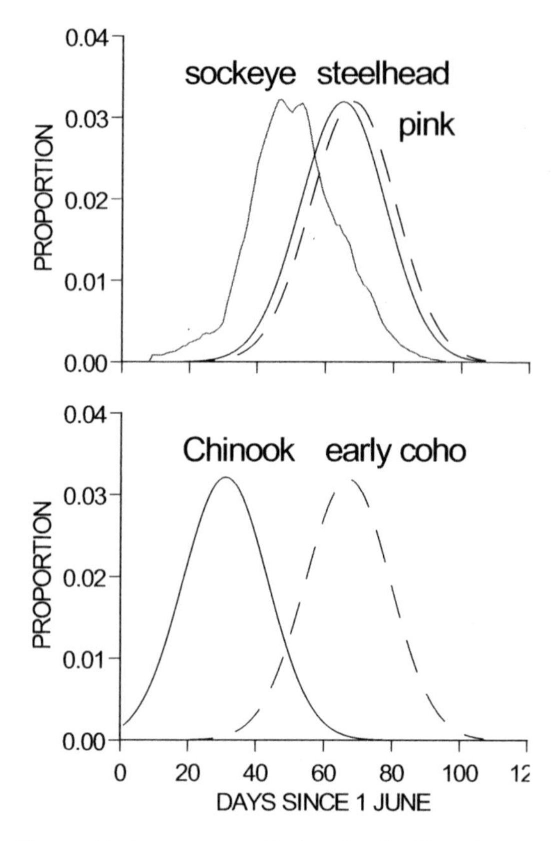

Figure 2. Average run-timing for Pacific salmon and steelhead trout entering Area 4 at the mouth of the Skeena River during 1985–1991 (from Cox-Rogers 1994).

The Salmon Fisheries

Aboriginal fisheries have operated in the Skeena River for at least 5,000 years. Three aboriginal societies (recognized as First Nations in Canada), comprising 17 aboriginal communities, harvest Skeena sockeye salmon: the Carrier-Sekani (Babine Lake area), Gitksan Wet'suwet'en (middle and upper Skeena), and Tsimshian (lower Skeena and adjacent ocean areas). Catches for food, social, or ceremonial purposes have averaged about 150,000 fish in recent years. Since 1993, new opportunities have also developed for First Nations to selectively harvest sockeye salmon that are surplus to Skeena spawning requirements.

The commercial salmon fishery on Skeena sockeye salmon began with the first cannery operations in 1877. Sockeye salmon were harvested predominantly by gill nets in the Skeena River until the 1930s when powered vessels moved out to ocean fishing areas. In recent times, 200–1000 gill-net vessels have fished from the Skeena River mouth to outside fishing areas 70 km distant, accounting for about 75% of the harvest of Skeena sockeye salmon. A seine fishery was introduced in the 1950s and grew rapidly through the next two decades. Since 1996, the number of eligible licenses in the gill-net and seine fleets has been reduced from over 1000 to 502 through fleet restructuring initiatives (Don Radford, DFO, personal communication). The Canadian commercial catch of Skeena sockeye salmon has generally increased since 1970 to a record high of 3.7 million fish in 1996.

A significant proportion of the total run is harvested in Alaskan gillnet and seine fisheries. Since 1985, the Canada–U.S. Pacific Salmon Treaty has limited catch in Alaskan fisheries directed at Skeena sockeye salmon, but other interceptions occur as incidental harvests in Alaskan fisheries directed on pink and chum salmon. The recreational fishery for sockeye salmon remains very small with catches estimated to be only a few thousand fish.

Resource Status

Data Sources and Methods

Historical data on abundance have been documented recently by Wood (2001). Trends in spawning escapement, catch, and total abundance of Skeena sockeye salmon were reconstructed from records of the canned salmon pack (Milne 1955). Trends since 1950 are based on escapement data and estimates of catch from run reconstructions maintained for DFO; complete data since 1970, including fry and smolt abundance estimates were documented by Wood et al. (1998); approximate data (excluding Alaskan catch) prior to 1970 were documented by Macdonald et al. (1987). Reliable escapement data were also available from a weir operating on the Sustut River since 1992 (Dana Atagi, provincial Ministry of Water, Lands and Air Protection). Salmon abundance and survival rates are shown on a logarithmic scale in most figures to better reveal trends; this is appropriate because random year to year variations in salmon survival tend to follow a lognormal distribution (Peterman 1981).

Trends in Sockeye Salmon Production

Skeena sockeye abundance declined steadily from the beginning of the last century to the 1960s, then increased to historic levels by the late 1970s (Figure 3). Total abundance and catches continued to increase to unprecedented levels in the mid-1990s, then collapsed in 1998 and 1999 as predicted because of parasite (*Ichthyophthirius* and *Loma*) epizootics that caused extensive prespawning mortality within Babine Lake in 1994 and 1995 (Traxler et al. 1998). Abundance also remained far below average in 2002 and 2003 primarily because of poor escapements in

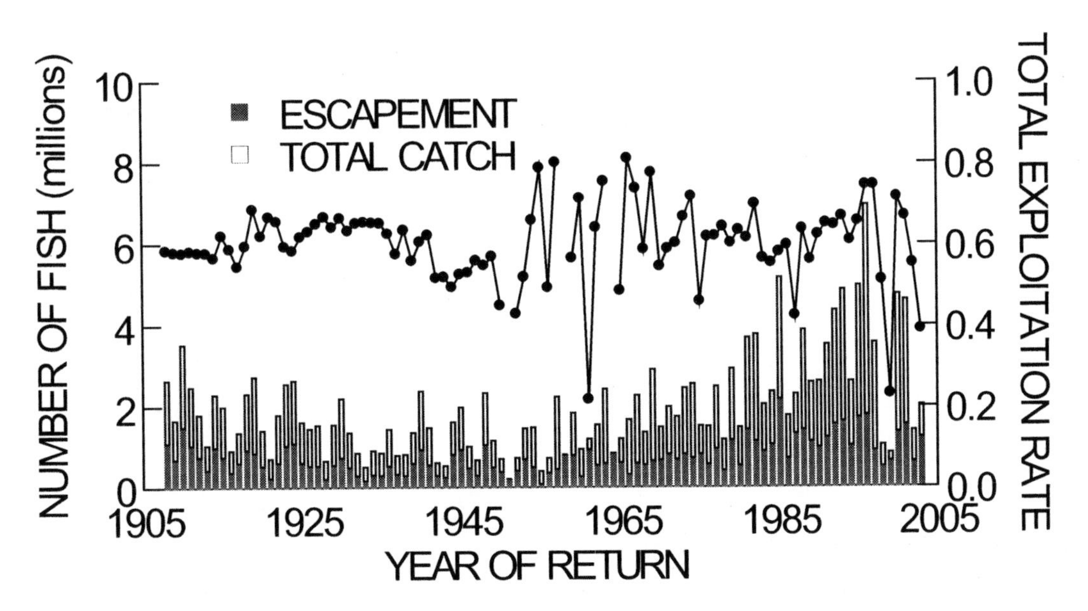

Figure 3. Trends in total stock (bars), shown as both catch (open) and escapement (solid), and total exploitation rate (line) for Skeena River sockeye salmon. Records prior to 1943 involve assumptions about exploitation rate and conversions from canned salmon pack to numbers caught (see Wood 2001).

1998 and 1999. Sockeye returning in 2000 and 2001 were less affected by the parasite epizootic and their abundance remained high.

The increase in abundance during the 1970s can be attributed largely to the construction of spawning channels that increased fry recruitment and smolt production in the main basin of Babine Lake (Figure 4). The relationship between smolt production and adult returns is not linear however; smolt survival appears to decline with increasing smolt abundance, presumably because of density-dependent ecological interactions (Peterman 1982; McDonald and Hume 1984; Wood et al. 1998).

Marine survival of Babine smolts fluctuates randomly from year to year but there appears to have been a long-term declining trend from the beginning of the smolt enumeration program in 1959 to the early 1980s, followed by an increasing trend to the present (Figure 5). This trend is evident even after taking density-dependent effects of smolt abundance into account (Wood et al. 1998). To some extent, these long-term changes in marine survival must also have contributed to the long-term trends in abundance. Sea-entry years 1996 and 1999 (most adults return 2 or 3 years later at age 4 or 5) stand out as ones of anomalously poor marine survival within the recent period of high survival.

Trends in Sockeye Salmon Diversity

From a production perspective, Skeena sockeye appear to be in good shape. However, the diversity of the Skeena sockeye escapement has changed dramatically. The nonBabine escapement (i.e., the number of Skeena sockeye spawning in areas not associated with Babine Lake) declined by an order of magnitude from 1950 to 1976 (Figure 5). Following management action to restrict the fishing season,

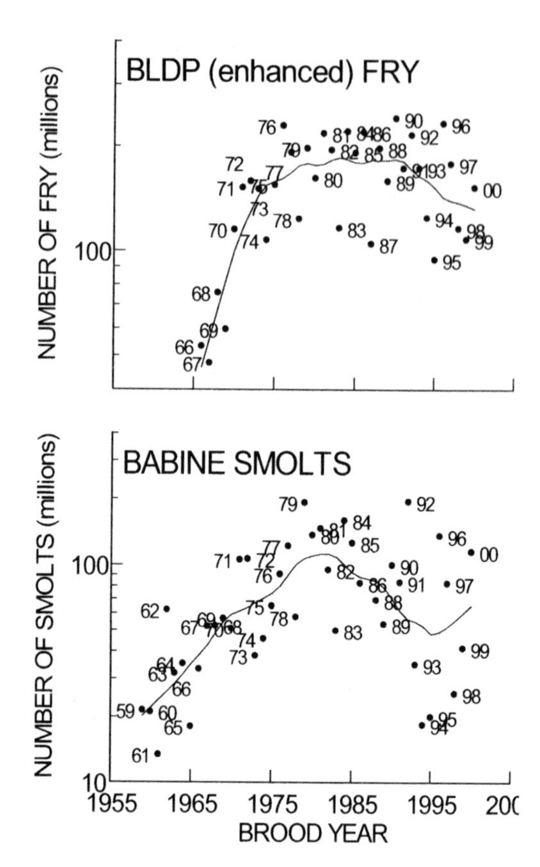

Figure 4. Trends in the abundance of sockeye salmon fry produced by the Babine Lake Development Project facilities and total sockeye smolt abundance from the Babine-Nilkitkwa lake system (enhanced + wild). Numbers refer to brood (parental) years.

nonBabine sockeye (most of which return earlier than the channel-enhanced Babine sockeye) increased steadily over the next 2 decades, almost regaining historic (1950s) levels by 1995. After 1995 however, nonBabine escapements declined alarmingly, even though marine survival (measured for Babine smolts) remained high in most years. A good rebuilding opportunity occurred in 2003 when the overall exploitation rate was restricted to less than 40%, coincident with very favorable marine survival (7.8%, the second highest year on record); exploitation rate was restricted both to increase sockeye salmon escapements

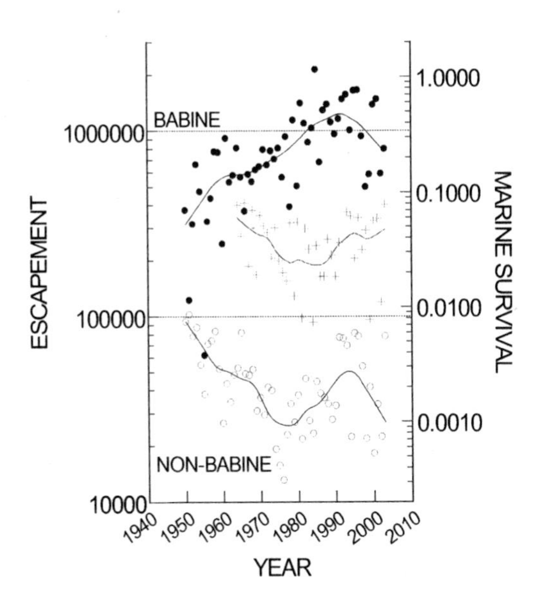

Figure 5. Trends in Babine (closed symbols) and nonBabine (open symbols) sockeye salmon escapements and marine survival of Babine smolts (crosses) lagged to correspond to adults returning at age 5.

to Morice and Kitwanga lakes, and to achieve spawning targets for the enhanced Babine component that had been reduced by the parasite epizootic (S. Cox-Rogers, DFO, Prince Rupert, personal communication) Consequently, the nonBabine escapement in 2003 almost matched that achieved in 1995.

The changes in stock diversity are more dramatic expressed as proportional composition. Between 1950 and 1980, the nonBabine proportion of the Skeena sockeye salmon escapement declined from 30% to 3% using nominal visual estimates, or from about 40% to 5% after doubling visual estimates relative to the Babine fence count to allow for underestimation, a calibration recommended by Milne (1955). Samples collected from the test fishery in the lower Skeena River indicate a similar decline in the proportion of age 2.* (or sub-3) sockeye, which should provide a good index of escapements to Morice Lake, but not to other lakes (McKinnell and Rutherford 1994). From 1987 to 1998, the nonBabine proportion averaged 4% (range 1–7%) based on nominal visual estimates and 7% (2–12%) based on adjusted visual estimates. Over the same period, stock composition analysis of test fishery samples using various biological markers including DNA indicated a much greater average nonBabine proportion (mean 24%, range 14–37%) suggesting that visual estimates have underestimated spawning escapements by more than 50%, or that Babine sockeye are not properly represented in the test fishery samples, perhaps because of gear saturation at peak abundance (Rutherford et al. 1999; Beacham et al. 2000).

The decline (before 2003) in the aggregate nonBabine escapement index, shown in Figure 5, may be exaggerated because of declining survey coverage since 1994. Even so, the same pattern of decline since the early to mid-1990s, followed by a strong escapement in 2003, is evident in all three subareas with continuous records of escapement data and in reliable counts at the Sustut fence, part of the Bear subarea (Figure 6). Furthermore, escapement indices have fallen below provisional limit reference points[1] in all subareas in at least some recent years. Shortreed et al. (2001) concluded that average escapements during the 1990s were far below (only 0.5–20% of) levels required to fully utilize rearing capacity in all nine of the nonBabine lakes surveyed. Some of these estimates are undoubtedly biased low because most visual surveys underestimate spawning abundance, but the Sustut estimate (20%) was based on reliable fence counts.

The decline in nonBabine escapements prior to 1980 has been attributed to overfishing of

[1] Wood, C. 1999. Provisional limit reference points for Skeena River sockeye in 1999. Memorandum to R. Kadowaki, Stock Assessment Division, Science Branch, Fisheries and Oceans Canada, Nanaimo, 14 April 1999.

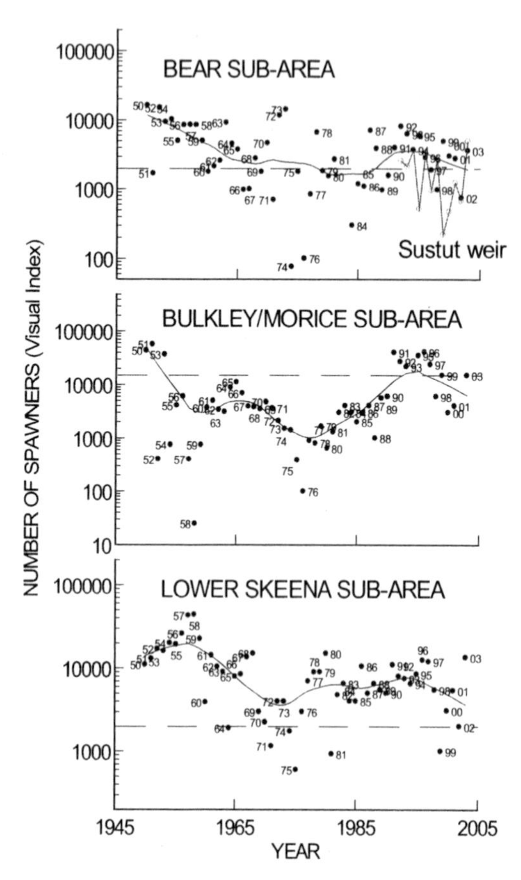

Figure 6. Trends in nonBabine sockeye salmon escapements by subarea. The open circles in the top frame indicate total counts at the Sustut weir; horizontal dashed lines indicate provisional limit reference points[1].

these naturally less productive stocks. Conversely, their recovery after 1980 has been attributed to reduced exploitation through better inseason management and more selective harvest of enhanced Babine sockeye based on differences in run timing (Sprout and Kadowaki 1987) and more terminal fishing at the Babine River fence (Wood et al. 1998). However, this cannot be confirmed because no independent measure of harvest rate is available for any of the nonBabine stocks. An alternative (or complementary) explanation is that nonBabine sockeye have been chronically overexploited since the beginning of intensive commercial fishing, and that the decline and recovery between 1950 and 1995 reflect changes in marine survival (Figure 5) rather than (or in addition to) success in managing the mixed-stock fisheries. Recent declines in nonBabine escapements might be attributed to the fact that overall harvest rates continued to increase through the 1990s, exceeding 70% in several recent years (1996, 1997 and 2000) whereas marine survival had reached a peak in sea entry years 1990–1991 (return years 1992 and 1994) and did not increase again until sea entry year 2000.

Trends in Species Diversity

Escapements of other salmon species have also declined dramatically during the history of the Skeena River sockeye salmon fishery, which is considered a contributing, rather than primary, source of fishing mortality. Chinook salmon escapements to all subareas declined steadily through the late 1970s and then recovered in the 1980s, following restrictions to Canadian troll fisheries, synchronously with nonBabine sockeye (Figure 7). Although Chinook escapements to the Babine subarea continued to recover through 2002, those to the Bear and Bulkley subareas have been declining, again in synchrony with nonBabine sockeye. Steelhead trout escapements declined during the late 1980s and early 1990s, arousing much concern among recreational fishermen, which prompted conservation action. Skeena steelhead now appear to have recovered to record high abundance based on both the aggregate index from the Skeena test fishery and the total fence count for the vulnerable, low productivity Sustut steelhead population (Figure 8). Coho salmon escapements had been declining steadily, especially in the upper Skeena, until all fisheries were greatly restricted in 1998

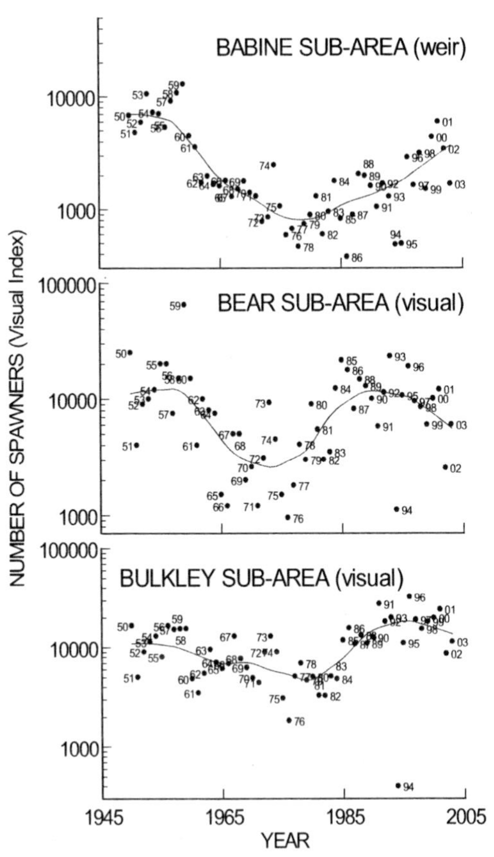

Figure 7. Trends in Chinook salmon escapement to the Skeena River by subarea.

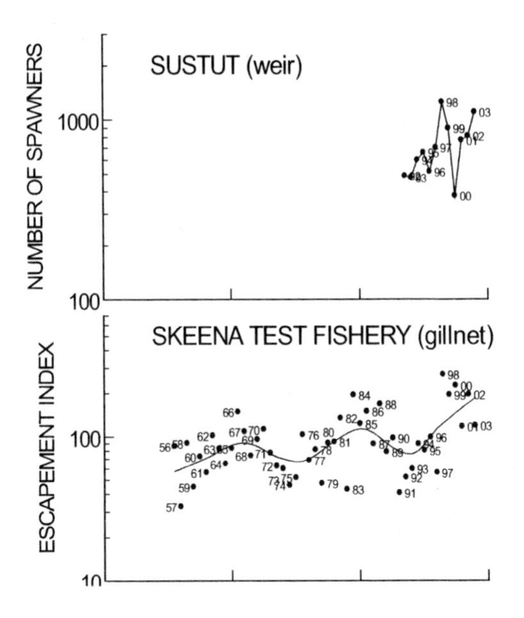

Figure 8. Trends in steelhead trout escapement to the Skeena River inferred from total counts at the Sustut weir (Bear subarea) and the overall test fishery index in the lower Skeena River.

(Figure 9). Coho escapements have now recovered to levels generally observed when records began in the 1950s. An exception is the poor coho escapement to the Babine subarea in 2003, but this anomaly followed record high escapements in the preceding 2 years.

Management History

Sprout and Kadowaki (1987) defined three periods in the evolution of salmon management in the Skeena River, each characterized by different influences of politics and science: the preresearch period (1876–1942), the research period (1943–1971), and the current period (1972–1987). Wood (2001) extended their categories by relabeling the current period as the "mixed-stock management period" (1972–1997) and adding a fourth—the "New Direction period" (1998 to present).

Key events in each period are summarized in Table 1 (see Wood 2001) for a full description). In short, considerable effort has been made to conserve the diversity of Pacific salmon in the Skeena River (and elsewhere), but no official policy exists at the time of writing to resolve the trade-off between total sockeye harvest and sustainability of less productive comigrating populations.

Wood (2001) remarked that "it remains to be seen whether the new approach and major conservation initiatives of recent years will endure sufficiently to warrant recognition as a period in future chronicles." This skepticism still seems warranted at the time of writing three years later, given that a final Wild Salmon Policy has not yet been released, and

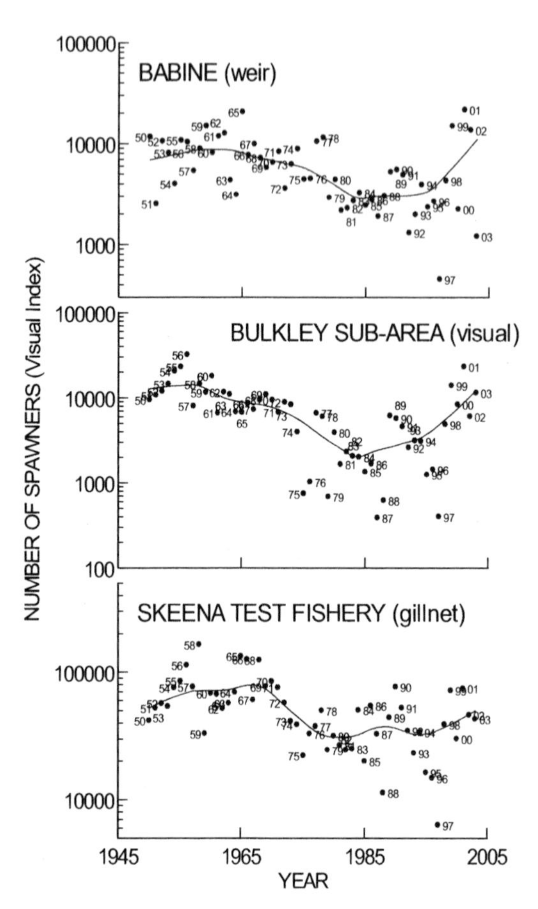

Figure 9. Trends in coho salmon escapement inferred from partial counts at the Babine Babine River fence, visual estimates for the Bulkley/Morice Sub-area and the overall test fishery index in the lower Skeena River.

that none of the three Pacific salmon populations listed as endangered by the Committee on the Status of Endangered Wildlife in Canada (COSEWIC) in 2002 (two as emergency designations) has yet received legal protection under SARA[2].

[2] "A Policy Framework for the Conservation of Pacific Salmon" was released for public consultation in December 2004 (available at http://www-comm.pac.dfo-mpo.gc.ca/publications/wspframework/WSP_e.pdf). It was announced in January 2005 that two endangered sockeye populations (Sakinaw Sockeye and Cultus Sockeye) would not be legally listed under SARA (http://www.ec.gc.ca/press/2005/050121_n_e.htm)

Lessons Learned

Salmon enhancement activities have been widely criticized for their adverse effects on wild populations (e.g., Meffe 1992). The greatest impact has always been attributed to excessive harvest rate on mixtures of enhanced and wild fish. The strident complaints about restricted fishing opportunity given a surplus of enhanced fish at Babine Lake in 2000 are testimony to this problem, and illustrate why enhancement within the context of mixed-stock fisheries is incompatible with the conservation of wild salmon diversity. Fisheries and Oceans Canada has responded to this problem by not authorizing any new production enhancement facilities since 1983. New policies have shifted the focus towards more appropriately scaled, shorter-term supplementation to rebuild wild stocks. Nevertheless, many stakeholders including aboriginal people, recreational fishermen and community groups still press for enhancement and ocean ranching as a way to increase benefits to local communities. This is not to say that enhanced production from the Babine Lake Development Project cannot be turned to advantage. From a fish culture perspective, the project was highly successful. Indeed, there are still opportunities to harvest more of the enhanced surplus terminally, and economic benefits would still accrue from mixed stock fisheries managed at a lower harvest rate.

The most important lesson from the last 30 years is that nonselective mixed-stock fishing in tidal waters must be reduced to conserve salmon diversity in the Skeena and elsewhere. The resulting increase in escapements should allow unenhanced Skeena populations to rebuild, leading to increased total returns in the future. Thus, the reduced catch of enhanced fish in aggressive mixed-stock fisheries would be offset to some extent by the increased catch from a greater abundance of unenhanced sock-

Table 1. Summary of key historical events in the management of the Skeena River sockeye fishery.

Pre-research period (1876–1942)	
1900–1910	Number of fishing boat licenses restricted to 850 per season
1916	Fisheries inspectors expressed first concerns about overexploitation
1925	Recognized as the beginning of chronic overexploitation (Shepard and Withler 1958)
1933	Maximum number of boat licenses (1,218)
Research period (1943–1971)	
1946	Babine counting fence began operation (Aro 1961)
1955	Lower Skeena gill-net test fishery was developed (and is still used) for in-season management (Cox-Rogers 1994)
1958	Aggregate maximum sustainable yield (MSY) for wild Skeena sockeye estimated to be 1.3 million fish at an equilibrium exploitation rate of 57% (Shepard and Withler 1958). Continuing decline in non-Babine sockeye documented (Shepard and Withler 1958) and later attributed to overfishing given natural differences in productivity among populations (Larkin and McDonald 1968)
1959	Babine River smolt trap and mark-recapture enumeration program began operation (MacDonald and Smith1980); continued until 2001
1965	First of three spawning channels constructed in Babine Lake after extensive limnological surveys (Johnson 1956, 1958) revealed that Babine Lake was "underutilized" by sockeye
1970	First "enhanced" sockeye return to Babine Lake spawning channels (West and Mason 1987)

Table 1. Continued.

Mixed-stock fisheries management period (1972–1997)	
1973	Fishing period and net restrictions imposed to protect early runs of Chinook and non-Babine sockeye
1985	Canada-US Salmon Treaty signed with annexes to limit US catch of Skeena salmon
1991	Fishing period and net restrictions modified to protect late runs of steelhead trout
1993–1997	The Skeena-Kitimat Sustainable Fisheries Program ("Skeena Green Plan") sought to develop management strategies for sustainable fisheries through a consensus-based process involving the Skeena Watershed Committee and significant new funding for research, monitoring and assessment (DFO 1998a).
New Directions Period (1998–present)	
1998	DFO released *A New Direction for Canada's Pacific Salmon Fisheries* (DFO 1998b), which identified conservation as its primary objective
1998–2000	Unprecedented fishing restrictions prompted by conservation concern for upper Skeena (and interior Fraser) coho populations (Holtby *et al*. 1999) and reduced sockeye abundance following parasite epizootic in Babine Lake
2000	Draft *Wild Salmon Policy* was released for public consultation (DFO 2000); this policy is intended to provide an operational framework for delivering on commitments made in the "New Direction" policy
2003	Species At Risk Act (SARA) came into force (http://www.sararegistry.gc.ca). Three (southern BC) populations of Pacific salmon designated as *Endangered* by COSEWIC (http://www.cosewic.gc.ca)

eye, sustainable because of lower overall harvest rates. Of course, the lower harvest rates would result in a greater surplus of enhanced fish available to terminal fisheries in the Babine River. It may seem unlikely that this harvest strategy could maximize economic value from the Skeena aggregate, but to my knowledge, a rigorous economic analysis including the potential production from nonBabine lakes at full capacity has not yet been undertaken. Certainly this strategy would increase ecosystem benefits by maximizing the return of marine-derived food and nutrients to each lake system, and to the Skeena ecosystem as a whole. In addition, it would "diversify the portfolio" of sockeye-production in the Skeena and help buffer the fishery against catastrophes like the Babine River slide in 1951 and recent epizootic events in Babine Lake. At present, the Skeena sockeye fishery is very much dependent on a monoculture of enhanced fish that has become increasingly vulnerable. Two years of poor smolt production from the spawning channels, following high prespawning mortality in 1994 and 1995, has had cascading effects on adult returns through two subsequent generations, reducing the total catch of Skeena sockeye in 4 of the last 6 years to low levels not seen since 1978.

Scientific research has played a crucial role in the evolution of Skeena salmon fisheries and in the protection of salmon diversity in the face of commercial exploitation. Before 1943, the motivation and ability to harvest salmon far exceeded biological understanding (Sprout and Kadowaki 1987). After only 1 decade of research, much had been discovered about the distribution and diversity of salmon in the Skeena: reliable monitoring programs had been initiated, optimal rates of harvest had been estimated to a reasonable approximation (57% compared with the median harvest rate of 65% imposed in the 1990s), and it was determined that harvest rates had been excessive prior to 1950. Rebuilding opportunities and potential enhancement techniques were identified over the subsequent decade. A similar flurry of research and discovery was initiated by the Skeena-Kitimat Sustainable Fisheries Plan in the 1990s (DFO 1998a).

Research has also been essential to the conservation of salmon diversity in the Skeena River. The primary role of research has been to identify diversity and take inventory of the populations that require conservation. A second operational role has been to monitor status against reference points to determine when special conservation actions are required. A third technological role has been to provide or clarify options for reducing the social and economic cost of conservation actions so that action will be taken.

While necessary, research alone is clearly not sufficient to promote conservation. The decline of the nonBabine sockeye populations in the Skeena was identified as early as 1958, and the cause was understood by 1968. Yet the nonBabine populations continued to decline and remained at low abundance until the late 1980 and 1990s. Similarly, coho populations in the upper Skeena were allowed to decline until the late 1990s. These declines in diversity were apparently acceptable tradeoffs to managers at the time.

Conservation actions in the Skeena have been most successful in response to crises where DFO has had obvious support from stakeholders. Restoration efforts following the Babine River slide in 1951 are a good example (Godfrey et al. 1954). Similarly, conservation actions have typically been more vigorous and successful in response to crises of declining abundance in highly-valued species like Chinook and steelhead, as compared with chronic declines in less-valued weak stocks within species. In part this is because fewer technological options are available for selective harvest to avoid weak stocks in mixed-stock fisheries.

DFO has clearly had more difficulty reacting to conservation issues involving trade-offs between short-term and long-term economic interests, or conflicts between extraction and stewardship ethics. These decisions are complicated by considerations of catch allocation or "distributive justice" (Ommer 2000), both internationally under the Pacific Salmon Treaty, and domestically in treaty negotiations with aboriginal people, and disputes among commercial and recreational fishery sectors. Conservation in these cases ultimately depends on a strong conservation ethic being defined in policy. Policy is required for regulatory agencies to defend decisions to forego short-term opportunities for groups with vested interests in favor of longer-term benefits for society.

Finally, the history of salmon populations in the Skeena River demonstrates that given the opportunity, Pacific salmon can recover from chronic overfishing provided their habitat remains intact. There can be little doubt that chronic overfishing remains the primary threat to salmon diversity in northern British Columbia where watersheds (like the Skeena) are relatively pristine and marine climate has remained favorable.

Acknowledgments

I am very grateful to Steve Cox-Rogers for providing the recent data required to update analyses by Wood (2001). Helpful comments by two anonymous reviewers were also appreciated.

References

Aro, K. V. 1961. Summary of salmon enumeration and sample data, Babine River counting weir, 1946–1960. Fisheries Research Board of Canada Manuscript Report 708.

Beacham, T. D., C. C. Wood, R. E. Withler, and K. M. Miller. 2000. Application of microsatellite DNA variation to estimation of stock composition and escapement of Skeena River sockeye salmon (*Oncorhynchus nerka*). North Pacific Anadromous Fish Commission Bulletin 2:263–276.

Cox-Rogers, S. 1994. Description of a daily simulation model for the Area 4 (Skeena) commercial gillnet fishery. Canadian Manuscript Report of Fisheries and Aquatic Sciences No. 2256.

DFO (Department of Fisheries and Oceans Canada). 1985. Pacific region salmon resource management plan. Discussion draft: Volume II. Appendices. Department of Fisheries and Oceans, Vancouver.

DFO (Department of Fisheries and Oceans Canada). 1998a. Skeena-Kitimat Sustainable Fisheries Program: final report 1993–1997. Fisheries and Oceans Canada, Vancouver.

DFO (Department of Fisheries and Oceans Canada). 1998b. A new direction for Canada's Pacific salmon fisheries. Fisheries and Oceans Canada, Vancouver.

DFO (Department of Fisheries and Oceans Canada). 2000. Wild Salmon Policy discussion paper. Fisheries and Oceans Canada, Vancouver.

Godfrey, H., W. R. Hourston, J. W. Stokes, and F. C. Withler. 1954. Effects of a rock slide on Babine River salmon. Fisheries Research Board of Canada Bulletin 101.

Groot, C., and L. Margolis. 1991. Pacific salmon life histories. University of British Columbia Press, Vancouver.

Holtby, L. B., B. Finnegan, D. Chen, and D. Peacock. 1999. Biological assessment of Skeena River coho salmon. DFO Canadian Stock Assessment Secretariat Research Document 99/140, Vancouver.

Johnson, W. E. 1956. On the distribution of young sockeye salmon (*Oncorhynchus nerka*) in Babine and Nilkitkwa lakes, B. C. Journal of the Fisheries Research Board of Canada 13:695–708.

Johnson, W. E. 1958. Density and distribution of young sockeye salmon (*Oncorhynchus nerka*) throughout a multibasin lake system. Journal of the Fisheries Research Board of Canada 15:961–982.

Larkin, P. A., and J. G. McDonald. 1968. Factors in the population biology of the sockeye salmon of the Skeena River. Journal of Animal Ecology 37:229–258.

Lichatowich, J., L. Mobrand, and L. Lestelle. 1999. Depletion and extinction of Pacific salmon (*Oncorhynchus* spp.): a different perspective. ICES Journal of Marine Science 56:467–472.

Macdonald, P. D. M., and H. D. Smith. 1980. Mark-recapture estimation of salmon smolt runs. Biometrics 36:401–417.

Macdonald, P. D. M., H. D. Smith, and L. Jantz. 1987. The utility of Babine smolt enumerations in management of Babine and other Skeena River sockeye salmon (*Oncorhynchus nerka*) stocks. Pages 280–295 *in* H. D. Smith, L. Margolis, and C. C. Wood, editors. Sockeye salmon (*Oncorhynchus nerka*) population biology and future management. Canadian Special Publication of Fisheries and Aquatic Sciences 96.

McDonald, J., and J. M. Hume. 1984. Babine lake sockeye salmon (*Oncorhynchus nerka*) enhancement program: Testing some major assumptions. Canadian Journal of Fisheries and Aquatic Sciences 41:70–92.

McKinnell, S., and D. Rutherford. 1994. Some sockeye salmon are reported to spawn outside the Babine Lake watershed in the Skeena drainage. Fisheries and Oceans Canada, Pacific Stock Assessment Review Committee Working Paper S94–11, Vancouver.

Meffe, G. 1992. Techno-arrogance and halfway technologies: salmon hatcheries on the Pacific coast of North America. Conservation Biology 6:350–354.

Milne, D. J. 1955. The Skeena River salmon fishery, with special reference to sockeye salmon. Journal of the Fisheries Research Board of Canada 12:451–485.

Ommer, R. E. 2000. Just fish: ethics and Canadian marine fisheries. Institute of Social and Economic Research, Memorial University of Newfoundland, St. John's, Cananda.

Peterman, R. M. 1981. Form of random variation in salmon smolt-to-adult relations and its influence on production estimates. Canadian Journal of Fisheries and Aquatic Sciences 38:1113–1119.

Peterman, R. M. 1982. Nonlinear relation between smolt and adults in Babine Lake sockeye salmon (*Oncorhynchus nerka*) and implications for other salmon populations. Canadian Journal of Fisheries and Aquatic Sciences 39:904–913.

Ricker, W. E. 1972. Hereditary and environmental factors affecting certain salmonid populations. Pages 27–160 *in* R. C. Simon and P. A. Larkin, editors. The stock concept in Pacific salmon. H .R. MacMillan Lectures in Fisheries, University of British Columbia, Vancouver.

Riddell, B. 2004. Pacific salmon resources in central and north coast British Columbia. Pacific Fisheries Resource Conservation Council, Vancouver.

Rutherford, D. T., C. C. Wood, M. Cranny, and B. Spilsted. 1999. Biological characteristics of Skeena River sockeye salmon (*Oncorhynchus nerka*) and their utility for stock composition analysis of test fishery samples. Canadian Technical Report of Fisheries and Aquatic Sciences No. 2295.

Shepard, M. P., and F. C. Withler. 1958. Spawning stock size and resultant production for Skeena sockeye. Journal of the Fisheries Research Board of Canada 15:1007–1025.

Shortreed, K. S., K. F. Morton, K. Malange and J. M. B. Hume. 2001. Factors limiting juvenile sockeye production and enhancement potential for selected B.C. nursery Lakes. Canadian Science Advisory Secretariat Research Document 2001/098.

Smith, H. D., and J. Lucop. 1966. Catalogue of salmon spawning grounds and tabulation of escapements in the Skeena River and Department of Fisheries Statistical Area 4. Fisheries Research Board of Canada Manuscript Report Series (Biol.) Nanaimo 882:1–7.

Sprout, P. E., and R. K. Kadowaki. 1987. Managing the Skeena River sockeye salmon (*Oncorhynchus nerka*) fishery - the process and the problems. Pages 385–395, *in* H. D. Smith, L. Margolis, and C. C. Wood, editors. Sockeye salmon (*Oncorhynchus nerka*) population biology and future management. Canadian Special Publication of Fisheries and Aquatic Sciences 96.

Taylor, E. B. 1991. A review of local adaptation in Salmonidae, with particular reference to Pacific and Atlantic salmon. Aquaculture 98:185–207.

Traxler, G. S., J. Richard, and T. E. MacDonald. Ichthyophthirius multifiliis (ich) epizootics in spawning sockeye salmon in British Columbia, Canada. Journal of Aquatic Animal Health 10:143–151, 1998.

Waples, R. 1995. Evolutionarily significant units and the conservation of biological diversity under the Endangered Species Act. Pages 8–27 *in* J. L. Nielsen, editor. Evolution and the aquatic ecosystem: defining unique units in population conservation. American Fisheries Society, Symposium 17, Bethesda, Maryland.

Williams, I. V. 1987. Attempts to re-establish sockeye salmon (*Oncorhynchus nerka*) populations in the upper Adams River, British Columbia, 1949–84. Pages 385–395 *in* H. D. Smith, L. Margolis, and C. C. Wood, editors. Sockeye salmon (*Oncorhynchus nerka*) population biology and future management. Canadian Special Publication of Fisheries and Aquatic Sciences 96.

Withler, F. C. 1982. Transplanting Pacific salmon. Canadian Technical Report of Fisheries and Aquatic Sciences 1079.

West, C. J., and J. C. Mason. 1987. Evaluation of sockeye salmon (*Oncorhynchus nerka*) production from the Babine Lake Development Project. Pages 176–190 *in* H. D. Smith, L. Margolis, and C. C. Wood, editors. Sockeye salmon (*Oncorhynchus nerka*) population biology and future management. Canadian Special Publication of Fisheries and Aquatic Sciences 96.

Wood, C. C. 1995. Life history variation and population structure in sockeye salmon. Pages 195–216 *in* J. L. Nielsen, editor. Evolution and the aquatic ecosystem: defining unique units in population conservation. American Fisheries Society, Symposium 17, Bethesda, Maryland.

Wood, C. 2001. Managing biodiversity in Pacific salmon: the evolution of the Skeena River sockeye salmon fishery in British Columbia. United Nations Environmental Program. Available: http://www.unep.org/bpsp/HTML%20files/TS-Fisheries2.html. (January 2006).

Wood, C. C., D. T. Rutherford, D. Bailey, and M. Jakubowski. 1998. Assessment of sockeye salmon production in Babine Lake, British Columbia with forecast for 1998. Canadian Technical Report of Fisheries and Aquatic Sciences 2241:50

American Fisheries Society Symposium 49:365–383

Back to the Future: A Candidate Ecosystem-Based Solution to the Fisheries Problem

TONY J. PITCHER* AND CAMERON H. AINSWORTH

University of British Columbia, Fisheries Centre
2202 Main Mall, Vancouver V6T 1Z4, Canada

Abstract.—Back to the Future (BTF) seeks to solve the fisheries crisis by employing past ecosystems with sustainable, clean fisheries as policy goals for a restored future. The first step is to construct whole-ecosystem models of present and selected past times. This process uses scientific, archeological, traditional, and local environmental knowledge in a way that encourages the support from coastal communities. Next, fisheries are chosen to meet sustainability and other criteria—termed "Lost Valley" fisheries—which are assessed for their impacts on biomass, biodiversity, and risk of local extinctions, and their benefits in producing jobs and profit. Optimum restorable biomass (ORB) offers a new ecosystem-based policy goal for the restoration of fisheries. Viability analysis, including risks from climate variability, shows the results of different trade-offs between benefits and costs. The third step is to evaluate the total benefits and costs that might derive from each alternative ecosystem, along with the costs of achieving each restoration and the emplacement of sustainable fisheries. Finally, the ecosystem that maximizes benefits over costs is an optimal policy goal that embodies ecological, economic, and social criteria, and, within the limits of uncertainty and knowledge of critical ecosystem processes, provides for sustainable fisheries in a restored future. Some restoration trajectories forecast by the model may offer direct economic benefits. Experience with this system so far suggests that, while all fisheries affect biodiversity, acceptable options include only small amounts of weighting on economic or social goals. And although some quantitative details remain to be solved, Back to the Future represents a coherent, viable set of ecosystem-based tools and policy goals that may allow for sustainable fisheries and restored ecosystems while avoiding the depletions that have dogged the past. Preliminary results with a few case studies suggest that trade-offs may offer win–win solutions, offering the exciting possibility of a genuine reconciliation of fisheries with conservation.

Introduction: The Rationale for Back to the Future

Back to the Future (BTF) employs past marine ecosystems as policy goals for restoration and sustainable management of fisheries (Figure 1, Pitcher et al. 1999; Pitcher 2005). Back to the Future is a restoration ecology for the oceans that is based on whole ecosystems: this paper reviews the theory, methodology, and current status of Back to the Future research and shows how it relates to the Fourth World Fisheries Congress theme of reconciling fisheries with conservation.

The last century has seen a huge reduction in the biomass of large table fish in the Atlantic Ocean and elsewhere (e.g., fish biomass:

*Corresponding author: t.pitcher@fisheries.ubc.ca

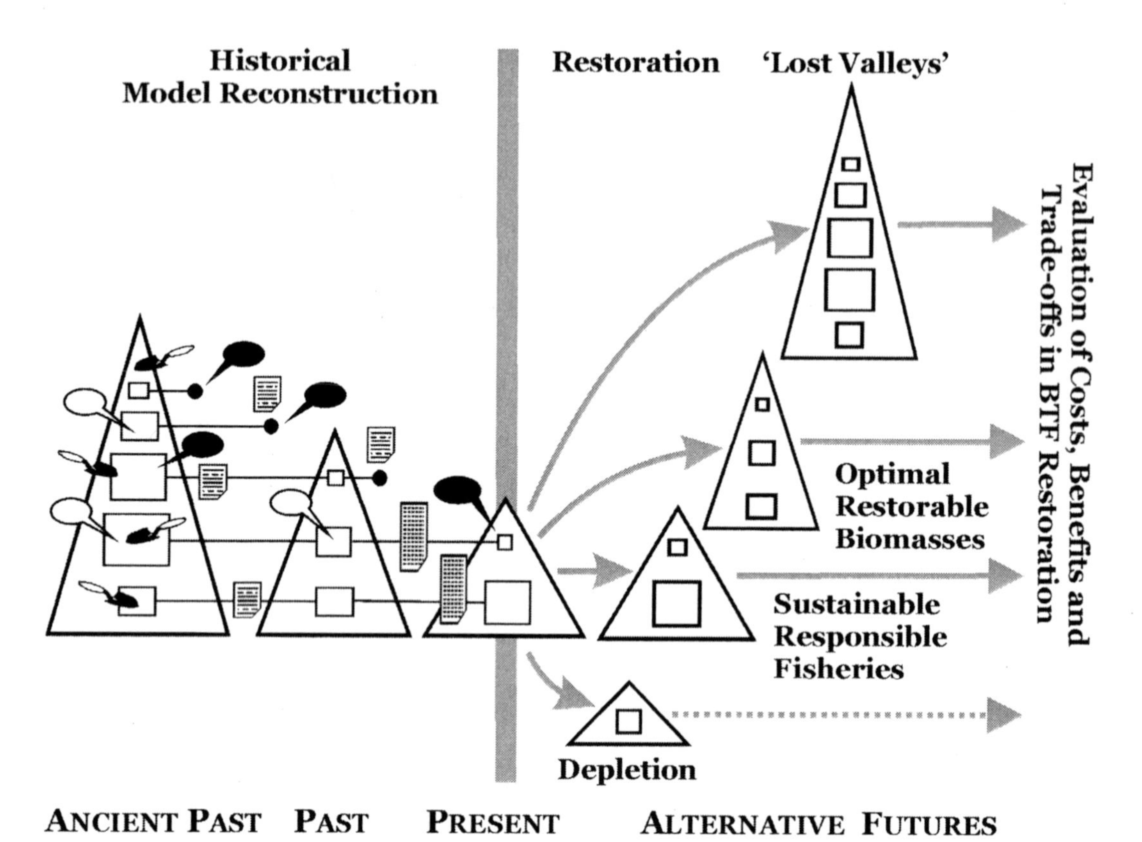

Figure 1. Diagram illustrating the 'Back-to-the-Future' concept for the adoption of past ecosystems as future policy goals. Triangles at the left represent a time series of historical ecosystem models, constructed at appropriate past times before the present (gray vertical line), where the vertex angle is inversely related and the height directly related to biodiversity and internal connectance. Time lines of some representative species in the ecosystems are indicated; size of the boxes represents relative abundance and solid circles represent local extinctions (=extirpations). Sources of information for constructing and tuning the ecosystem models are illustrated by symbols for historical documents (*paper sheet symbol*), data archives (*tall data table symbol*), archaeological data (*trowel*), the traditional environmental knowledge of indigenous Peoples (*open balloons*) and local environmental knowledge (*solid balloons*). At right are alternative future ecosystems, representing further depletion, or restored 'Lost Valleys' that may be used as alternative policy goals. Restored 'Lost Valleys' may be fished with sustainable, responsible fisheries set up according criteria described herein, and aiming at Optimal Restorable Biomasses found using objective policy searches. Final choice of BTF policy goal is made by comparing trade-offs, cost and benefits among possible futures. Diagram does not show necessary challenges by climate change and uncertainty. (Modified from Pitcher et al. 1999; Pitcher 2001; Pitcher et al. 2005a).

Christensen et al. 2003; whales: Roman and Palumbi 2003; large fish: Myers and Worm 2003; deep sea fish: Morato et al. 2006; sharks: Schindler et al. 2002; Baum et al. 2002; turtles: Hays et al. 2003). The result has been loss of livelihoods, lifestyles, and traditional knowledge (Coward et al. 2000), requiring a major shift in public perception to emplace policies that have any hope of turning the tide (Pauly et al. 2002; Pauly et al. 1998a). Therein lies a dilemma: sustainability in a depleted system is an inappropriate policy goal (Pitcher and Pauly 1998); restoration and rebuilding of fish stocks are essential to ecosystem health, and determine the wealth and sustainable benefits that may be taken from them (Pitcher 2001). The present bleak situation has produced a series of calls for more effective management, such as "more MPAs," "more property rights in fishing," and the like, but clear, easy to understand policy goals that these instruments might be employed to achieve have been lacking. On the contrary, BTF offers an integrated, quantitative, and coherent policy agenda to fill this vacuum.

The basic theory of BTF is simple: a number of possible futures are based quantitatively and qualitatively on ecosystems restored to resemble those described by a series of snapshots of the past (termed "Lost Valleys," Pitcher et al. 2004; Pitcher 2004d), each fished in a fashion judged sustainable and responsible according to set criteria (these are termed "fished Lost Valleys"). The costs and benefits of each alternative future are evaluated, for several policy goals, in terms of ecological economics, employment opportunities, and impacts on biodiversity (concept summarized in Figure 1). The different policy goals range from purely economic to purely conservationist through a series of intermediate weightings, and comparison among them quantifies the trade-offs that would have to be made to achieve each goal. Choice of the best policy goal is the future scenario that, allowing for the costs and time span of restoration, will produce the highest rewards.

Walters et al. (2004) have recently demonstrated what has been suspected for a long time; that management guided by a set of single-species multiple sustainable yield (MSY) goals is not viable for most ecosystems, since unacceptable depletions of trophic structure may result. Hence, the methods used in BTF are ecosystem-based. Quantitative work is made possible by whole-ecosystem modeling, at present implemented using the Ecopath-with-Ecosim suite of modeling tools (Christensen and Walters 2004), although other whole-ecosystem modeling systems are being explored (e.g., Fulton et al. 2003). The results may be challenged numerically by uncertainty of model parameter values and by possible climate shifts and fluctuations, in order to evaluate the robustness of each policy goal. Restoration may adopt a short cut between the Lost Valley itself and the fished Lost Valley: the biomass goals in this policy we have termed optimum restorable biomass' (ORB, Ainsworth and Pitcher 2004b). The most recent summary of these procedures may be found in Pitcher (2005); further detail of methods applied to a range of case studies may be found in Pitcher et al. (2005).

It is interesting that the BTF procedure allows us to compare quantitatively possible futures based on restoration, with a form of status quo represented by fishing the present day ecosystem in a more sustainable and responsible fashion; presumably the policy adopted implicitly or explicitly by most of the world's fisheries management agencies today. Hence BTF can provide an evaluation of the benefits (or otherwise) of restoration, which may entail a costly recovery and adjustment period, compared to incremental good management of today's fisheries and ecosystems. Moreover, we will show how trade-offs between exploitation and conservation made explicit for the best ORB policy may often be less extreme than the trade-offs faced in the present day; a nice example of a sustainable development policy that addresses a genuine reconciliation of exploitation with conservation.

Pitcher (2004a) outlines the considerable methodological challenges that have been overcome in order to implement a practical version of this innovative Back to the Future policy agenda. A recent project sponsored by

Coasts Under Stress (CUS) focused on the marine ecosystems of northern British Columbia and the Newfoundland shelf: the issues are fully discussed in 25 papers in Pitcher (2004b). The present account provides a brief review and summary of what has been achieved and what remains to be done.

Building Models of the Past

The "back" element of BTF consists of constructing models of past systems that capture their trophic structure, abundance and diversity. This is not an easy job (Christensen 2002); it entails integrating knowledge from multiple sources as diverse as scientific data, historical, archaeological and archival information, traditional and local knowledge (Heymans and Pitcher 2004). Following two small pilot projects (Pauly et al. 1998b; Haggan and Beattie 1999), in a recent major test of the BTF method sponsored by the Coasts under Stress project, the enormous scope of building models of the ecology and fisheries in Newfoundland shelf and in northern British Columbia made it essential to enlist the help of many scientific partners on both Canadian coasts. Early in the project, workshops were held with DFO and University partners on both coasts in order to review data and ideas on all the major ecosystem groups, from birds and whales to plankton and jellyfish. Published papers from these workshops were authored by over 35 individual research scientists (e.g., Bundy 2002; Lilly 2002, Burke et al. 2002; and other contributions in Pitcher et al. 2002a).

Following these science workshops, four preliminary ecosystem models were constructed for both coasts. For Newfoundland the snapshot BTF dates were 2000, representing the present day; 1985, prior to the major cod collapse; 1900, prior to the expansion of steam trawlers; and 1450, before contact of aboriginal peoples with European fishing (Pitcher et al. 2002b). For northern British Columbia, the dates were 2000; 1950, before expansion of the trawl fisheries; 1900 before mechanization of the salmon fisheries; and 1750, prior to contact of indigenous fishers with Europeans (Ainsworth et al. 2002). The choice of snapshot dates reflects major changes in the fisheries, but it would have been helpful to have had twice as many snapshots on both coasts so that subsequent work with policy choice algorithms would have had a more continuous time series of ecosystems to work with.

The ecosystem models have inevitably been improved and enhanced in an ongoing fashion; ideally, further science workshops would be held to perfect them. In practice, for the Newfoundland models, considerable improvement of the models of the recent past was obtained through a partnership with a government ecosystem modeling project (Heymans 2003c), and, in the case of local environmental knowledge, from east coast CUS partners (Alcock et al. 2003; Rogers 2003). For the oldest Newfoundland model (1500), ancient aboriginal catches were estimated (Heymans 2003b). The two most recent ecosystem models were fitted to time series of biomasses from VPA and other survey information (Heymans 2003c, 2003d).

Likewise, in British Columbia, additional local and environmental knowledge for improvement of the ecosystem models was obtained in community workshops and from interviews held in Prince Rupert (Ainsworth and Pitcher 2005; and contributions in Pitcher et al. 2002c) and Haida Gwaii (Ainsworth and Pitcher 2005a), with support from the local communities, and the Tshimtsiam and Haida First Nations respectively. Archaeological input was received from Orchard and Mackie (2004). Contributions from indigenous fishers are summarized by Simeone (2004) and by Lucas (2004).

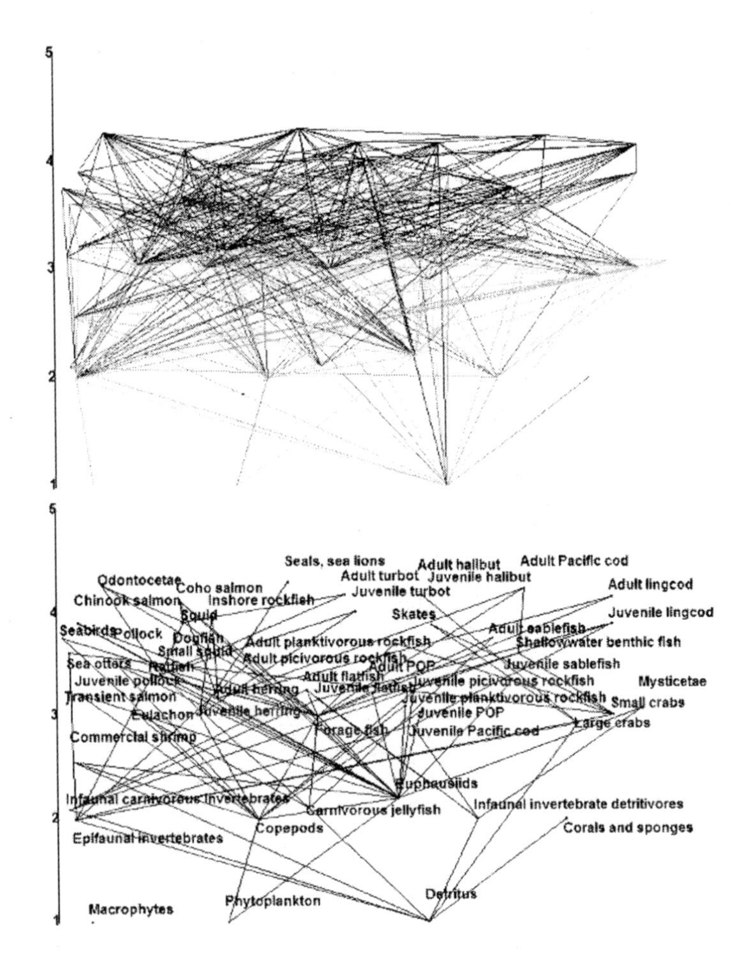

Figure 2. Example of the trophic web in a whole-ecosystem model used for Back-to-the-Future analysis, taken from the northern British Columbia precontact 1750 model. Trophic level is shown on vertical axis (primary producers are defined as 1). *Upper diagram:* all trophic links in the model, as determined by the diet matrix entered as data and modified during model balancing. *Lower diagram:* shows food web linkages forming more than 15% of the diet; names of the 57 functional groups in the model are given: note important trophic nodes for euphausids, copepods, benthic invertebrates, forage fish, and juvenile and adult herring.

Diagrams showing the complex food web in our northern British Columbia ecosystem model are presented in Figure 2; a number of important trophic nodes are visible, especially for copepods, euphausiids, herring, forage fish (mainly sandlance), and small benthic invertebrates. Flow pyramids summarizing the 4 historical models on each coast are illustrated in Figure 3. The effects of depletion are apparent on both coasts, but, not surprisingly given the collapse of Atlantic cod stocks, are more apparent in Newfoundland, where an entire trophic level is absent in the present-day ecosystem. An example of changes in biomass of model groups through time for the northern British Columbia suite of ecosystem models is shown in Figure 4.

Participatory Elements in BTF

The scientific methodology in BTF relies on the ability of advances in ecosystem modeling

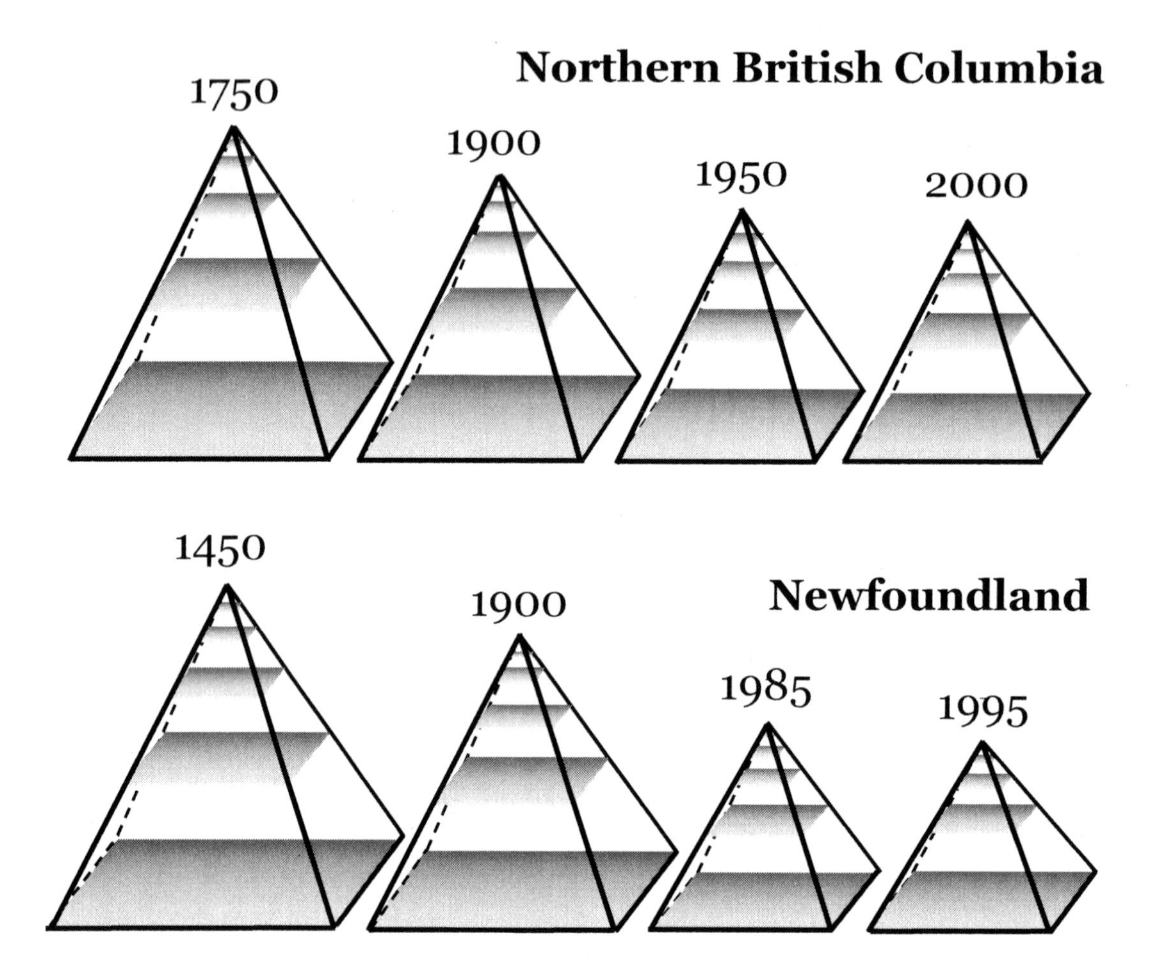

Figure 3. Flow pyramids for the series of Back-to-the-Future models in Northern British Columbia and Newfoundland shelf marine ecosystems. Horizontal shelves are trophic levels (pyramid floor is 1). Pyramids drawn approximately to scale; position and size of shelves are in proportion to flows between levels. Note higher flows and more levels in more ancient ecosystems; in particular the 1995 Newfoundland ecosystem is missing an entire trophic level (greatly reduced— so not obvious on plot—but still present in the 1985 system). Newfoundland models comprised 52 functional groups, of which 3 were ontogenetically linked as juvenile/adults; northern British Columbia model comprised 53 functional groups of which 10 were linked as juveniles/adults. Full details of ecosystem models are given in Ainsworth et al. (2002), Pitcher et al. (2002b), and Heymans (2003a).

to integrate quantitative and qualitative information in the creation of ecosystem models. But in addition, BTF has a strong participatory element based on respect for different traditions of knowledge, commitment to shared knowledge, and aspiration to rebuild (Haggan et al. 1998), that can be thought of as forming a cognitive map of BTF (Pitcher and Haggan 2003). Such factors may encourage successful implementation and compliance, and lead to a restored system and greater benefits for all. Haggan (2000) points out that BTF can facilitate knowledge-sharing by acting as a neutral forum and an honest broker of information, building and iterating local ecosystem models with locally

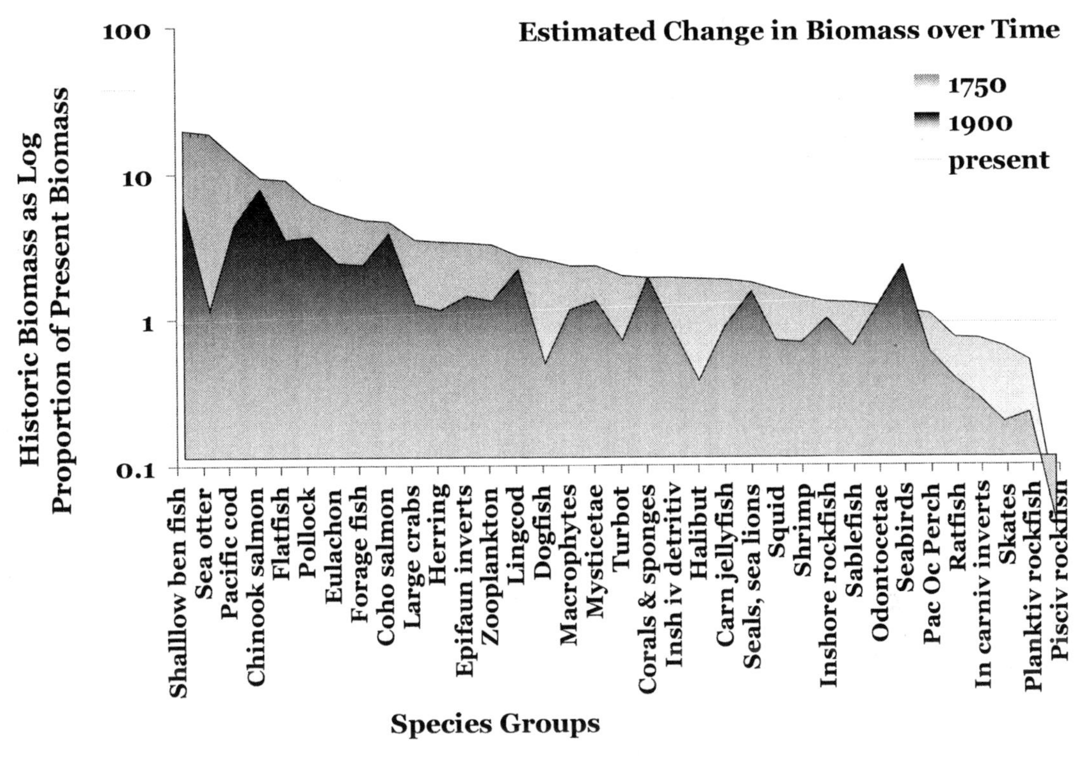

Figure 4. Biomasses of selected northern British Columbia ecosystem model groups in 1750 and 1900 compared to present biomass (horizontal line).

derived data builds intellectual capital in the models themselves, and social capital among the collaborators.

Opening Sustainable, Responsible Fisheries: The Lost Valley

The "future" element of BTF consists of a series of future possible ecosystems (Lost Valleys), each of which resembles one of the snapshots from the past. Back to the Future assumes these past systems may be restored and then fished (Pitcher 2004 d). Benefits from these Lost Valleys are not taken by fleets that resemble those of today, or those of the past, but by fisheries that are designed to be both sustainable and responsible according to defined criteria (see Figure 1). In the portfolio of fisheries opened in the Lost Valleys, each fishery must meet nine stated criteria of sustainability and responsibility viz. minimal bycatch discards, minimal damage to habitat by gear, include Aboriginal fisheries, include traditional target species, minimize risk to charismatic species, exclude fisheries on juveniles, participatory vetting of fisheries, simulations show fishery is sustainable, adaptive management plan is in place (a full discussion of these items is provided in Pitcher et al. 2004). In general, sustainability is judged against impacts on biodiversity, risk of local extinction, and the integrity of ecosystem structure and biomass pools using 50 or 100 year simulations, while responsibility may be judged against compliance with the Food and Agriculture Organization of the United Nations Code of Conduct for Responsible Fisheries (FAO 1995) using a rapid appraisal technique (Pitcher 1999). Lost Valley fisheries used in the BTF analysis of the two Canadian coastal ecosystems are listed in Table 1.

Table 1. Responsible fishing gears used in the fished Lost Valley BTF analyses. Trawls, dredges, and so forth. are assumed designed to minimize habitat damage and by-catch as far as technologically possible. Initial catches for the policy search routine were set to 2.5 % of biomass. X indicates gears eliminated from some scenarios.

Northern BC	Newfoundland
Shrimp trawl	Shrimp trawl
Shrimp trap	Cod traps
Herring seine	Capelin
Halibut longline	Longline
Salmon freezer troll	Snowcrab traps
Salmon wheel	Inshore crab traps
Rockfish live	Lobster traps
Crab trap	Salmon wheel
Clam dredge	Clam dredges
Aboriginal	
Recreational	
Groundfish trawl (x)	

After an ideal set of fisheries have been selected according to these criteria, model simulations are used to forecast fishing and its effect over a long time period. Relative fishing mortalities for each gear type in the fisheries are adjusted until catches are sustainable and impacts on the ecosystem meet specified criteria. The adjustment is carried out automatically using an automated search routine in Ecosim (Walters et al. 2002; Christensen and Walters 2004). Alternative fishery objectives may be selected, including economic value, numbers of jobs, or the biomass of long lived species. In practice, many model searches have to be performed to make sure a true optimum is found, and in some cases a wide scatter of similar-valued peaks can sometimes result (Ainsworth and Pitcher 2005b). The results of the search provide forecasts of fishery catches, biomass, economic values, numbers of jobs, and biomass changes in all other groups in the fished Lost Valley ecosystem. Any scenarios that cause specified risks of extirpation, severe depletion of species, or reductions in biodiversity are eliminated at this stage. The search procedure is repeated for the range of policy objectives and for each candidate restored ecosystem, producing a number of forecast scenarios that may be compared. Changes to the biomass of model groups after being subjected to sustainable, responsible fishing designed by the automated search procedure with a range of policy objectives are illustrated in Figure 5. Further details and examples of the policy search process are detailed in Pitcher et al. (2004). An example of the fishing mortalities from the fished Lost Valley procedure are compared in Figure 6 with fishing mortalities, including discard mortalities that obtain in the present day; it is evident that our BTF procedure has chosen dramatically lower fishing mortalities for most groups.

In addition, we may seek to challenge these results with climate changes that might realistically be expected for the locality in question, and taking account of the principal uncertainties in model parameters used in the simulations. This can be achieved by driving the simulations with various types of climate forcing functions, and with semiBayesian Monte Carlo simulations (Pitcher and Forrest 2004; Pitcher et al. 2005b).

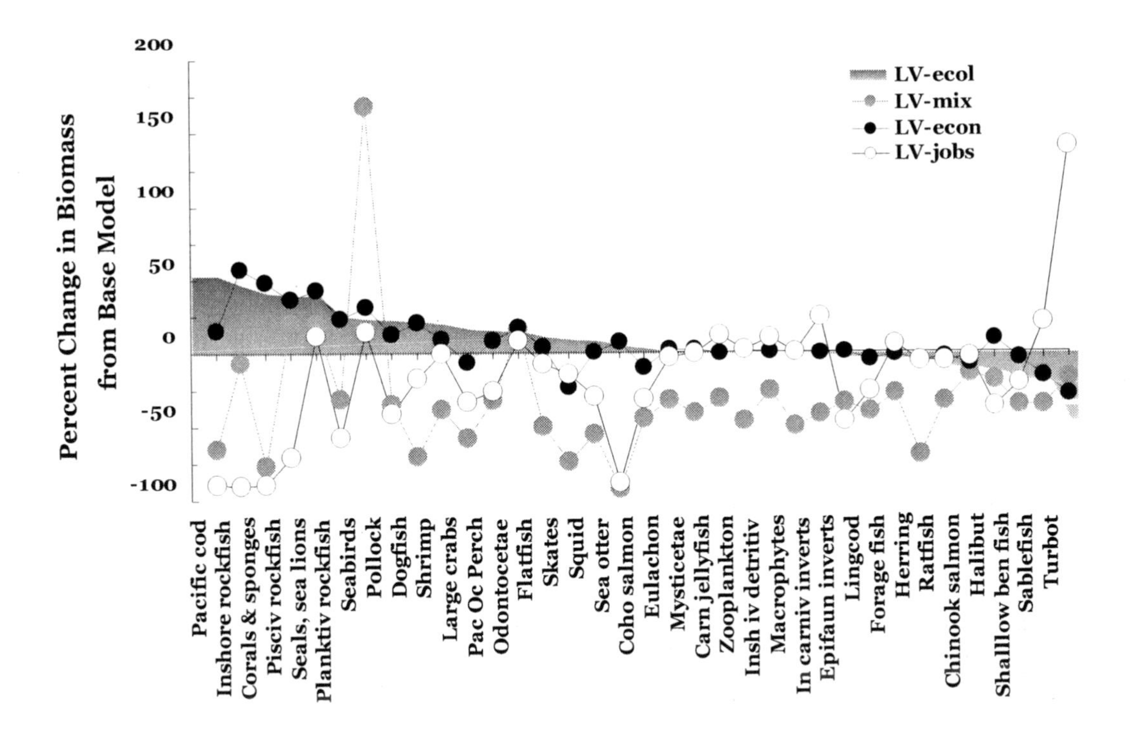

Figure 5. Percentage changes of selected biomasses in fished Lost Valley ecosystems compared to the northern British Columbia 1750 historical ecosystem (baseline). Results represent the Optimal Restorable Biomass (ORB) targets for management of BTF restoration obtained after 50-year simulations. Graph shows four different policy objectives used in the searches for maxima: LV-ecol: the ecological objective alone (shading); L V-econ: the economic (NPV) objective alone (solid circles); LV-jobs: the social (employment) objective alone (open circles); LV-mix: a mixed ecological, social and economic objective (gray circles). *x*-axis shows abbreviated names of functional groups in the model: pisc = piscivorous; planktiv = planktivorous; carn = carnivorous; detritiv = detritivorous; Oc = Ocean; ben = benthic; insh = inshore; epifaun = epifaunal.

Choosing Restoration Goals: Targeting Optimum Restorable Biomass

The result of the Lost Valley process provides us with set of alternative restored ecosystems, complete with sustainable fisheries, and the remaining problem is to find an objective way to choose a rational policy goal from among them. This may be done by comparing the benefits that will accrue to society from each alternative future represented by a fished Lost Valley ecosystem. In order to show the full range of options that may be considered, the present day ecosystem is included in this process (see Figure 1), albeit in an analogous way as a pseudo lost valley with a new portfolio of fisheries designed to be sustainable (Pitcher 2004b; Pitcher et al. 2004).

One fundamental way to evaluate the benefits of alternative restored ecosystems is to compare the net present economic values of their fisheries, information that is readily estimated from the Ecosim simulations mentioned above. In general, experience with this technique shows that evaluating economic objectives using conventional discounting will cause unacceptable depletion and loss of biodiversity, a problem referred to as the conservationist dilemma that has been well-

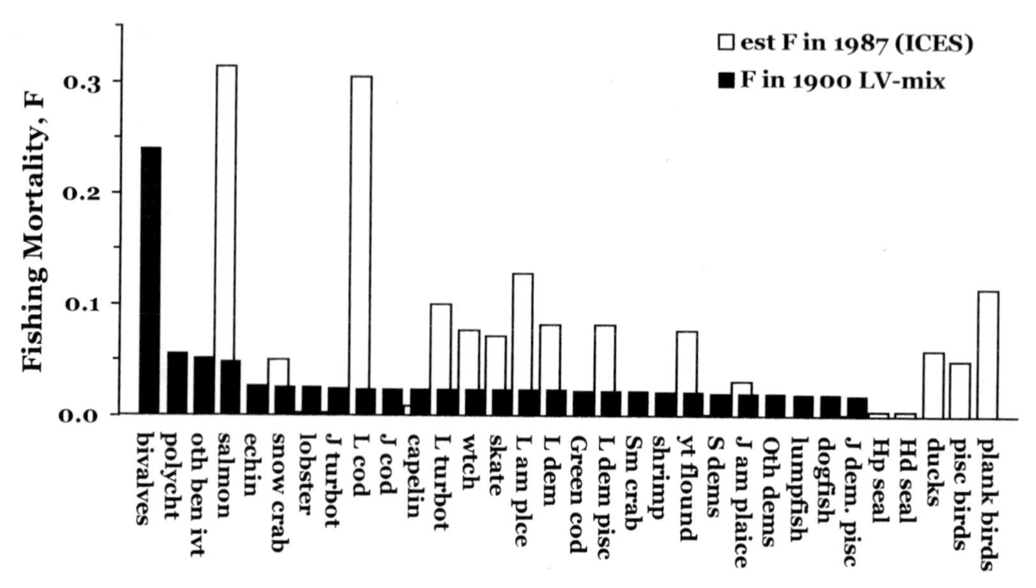

Figure 6. Comparison of fishing mortalities in a sustainably and responsibly fished restored 1900 ecosystem of the Newfoundland shelf (responsible fisheries listed in Table 1), compared to estimated fishing mortalities today (A. Bundy, DFO, personal communication, 1997). LV values are from an optimization that attempted to balance ecological values, profits, and jobs (redrawn from Pitcher et al. 2004).

known since the time of Clark (1973). To take account of this intergenerational externality (Padilla 2002), various forms of low or zero discount rates algorithms have been argued to be more appropriate for natural resources (e.g., Fearnside 2002), although these attempts have generated some controversy (e.g., Goulder and Stavins 2002). In the BTF work, Ainsworth and Sumaila (2004a) show how an intergenerational discounting equation is employed that takes into account both conventional discounting and a societal view of the need to recognize the rights of the next generation (in a Pigouvian fashion, Sumaila et al. 2001). The discount clock is partially reset each generation, and both market and societal intergenerational discount rates are used (Sumaila and Walters 2004, 2003). For example, using this algorithm, Ainsworth and Sumaila (2003) analyze how much of the collapse of Newfoundland cod may be attributed to a failure to take account of value to future generations. In BTF analysis, either intergenerational or conventional values may be calculated as options in the ecosystem simulations.

The simulation technique also provides an estimate of the number of jobs in the fishery, at least those directly involved: the processing and marketing sectors are not included but might, for a particular fishery, also be estimated from the total catch. In addition, however, a job diversity index may also be used for evaluation (Ainsworth and Sumaila 2004b)). In BTF the aim is to compare the ecological and social as well as economic value of past and present ecosystems (Angelsen et al. 1994; Angelsen and Sumaila 1996).

Since the simulations cover all biological components of the ecosystem, it is also possible to examine the effect of the candidate fisheries on a number of measures of integrity and diversity. For example, system resilience may be estimated by a whole-ecosystem index (Heymans 2004; Ulanowicz 1999). Comparisons may also be made using a biodiversity index modified for use with the functional groups in this type of modeling, rather than the samples of species for which they were designed (a modified Kempton index, Q-90, Ainsworth and Pitcher 2005), or a fuzzy logic index based on life history and fishery parameters that expresses the risk of local extinction (Cheung and Pitcher 2004; Cheung et al. in press).

Comparisons among the candidate restoration scenarios may now be made. One useful technique is to sort them into order using each of the criteria in turn (see example in Ainsworth and Pitcher, this volume). In this way, trade offs among conflicting objectives, such as economic value and biodiversity, are made explicit. Policy choices for fisheries inevitably require such trade–offs, but they are rarely made transparent such that industrial, union, and conservation stakeholders may attempt to achieve a consensus.

In classic fisheries population dynamics, stock abundance is manipulated to an optimal size so that surplus production can be maximized (Pitcher and Hart 1982). In an analogous ecosystem-based process, fishing the lost valley ecosystem with sustainable and responsible fisheries depletes these historic ecosystems into a more productive state, according to the chosen policy goal. However, in practice, we would be unlikely to go to the trouble, time, and expense of restoration only to deplete the system. Hence, our goal should not be to recreate the historic lost valley ecosystem, but to aim directly for the equilibrium condition after long-term responsible harvests. We have described this harvest–equilibrium abundance of ecosystem components the ORB. (Ainsworth and Pitcher 2006, this volume; Pitcher et al. 2005a). Manipulating the ecosystem so that each functional group approaches its ORB is the BTF policy goal to which restoration efforts are directed.

There is no unique ORB solution: the specific ORB configuration of biomasses that is preferred will depend on what benefit modalities were used in the lost valley simulations of the historic system, monetary or otherwise. Optimal restorable biomass provides an ecosystem-based solution to the problem that, where stocks interact through predation or competition, it may be impossible to achieve B_{MSY} simultaneously for multiple stocks (Walters et al. 2004). From a whole-ecosystem perspective, it becomes necessary to choose between stocks, holding the biomass of some close to their optimal levels while sacrificing the productivity of others. According to our choice, total catch from the ecosystem can be maximized—or profit, or biodiversity, or any other measure of socioeconomic or ecological utility. If our management goal is simply to maximize catch, for example, B_{MSY} should be sought only for the most productive and massive stocks, while maximum productivity of low-volume fisheries may need to be sacrificed. An alternative goal may maximize system biodiversity. In most cases, a practical management policy will contain a balance between socioeconomic and ecological priorities. An example of ORB calculation in northern British Columbia is provided by Ainsworth and Pitcher (2006, this volume); these authors modified an existing routine in Ecosim to accept a vector of ORB biomasses as target rebuilding goals.

Implementing the BTF Policy

The intention of the BTF process is to provide a policy goal to which all may agree and aspire: a cognitive map of a future ecosystem that, as a far as possible, resembles one from the past. But there is a significant problem here in that estimating the costs of restoration may depend on precisely what techniques are adopted and the actual instruments of restoration may themselves generate conflict. For example, MPAs set up adjacent to a traditional fishing community may trigger protests from inshore and local fishers if they cannot be convinced of the long-term advantages. Moreover, reducing quotas for some sectors as fisheries are modified to become more sustainable is bound to generate conflict. A second issue is that there are considerable psychological advantages in emphasizing the possibility of restored biomass, which may be considerable in long-lived target species like halibut, wreckfish, or cod. Hence, it may sometimes be easier to achieve agreement by putting forward an attractive long-term goal rather than by diverting attention to the means by which one might get there. Agreement on one element of a policy generally eases subsequent steps, although final agreement may always be difficult. This collective vision of "a better tomorrow" underpins the "Future" element of BTF.

Implementation of BTF might cause some difficulties for managers and conservationists as the ecosystem moves towards the ORB in fished Lost Valley objective. Aiming for higher biodiversity may entail reductions in some species, while others increase. For example, seabird populations in the North Sea, mainly fulmar petrels, have increased at least twofold as a result of discards by fishing vessels over the last 100 years (Furness 2002). On the other hand, resident humpback whales in the Strait of Georgia, British Columbia, were reduced to zero by whaling almost a century ago, but are gradually recolonizing the area (Winship 1998). Sea otters, extirpated by the fur trade 150 years ago, have been reintroduced to northern British Columbia (Watson et al. 1997). So it is likely that attempts to restore older systems might involve active or passive management to reduce, encourage, or to reintroduce some species.

In this manner, it is interesting to contemplate some of the trade-offs that may have to be faced in a BTF restoration process. For example, restoration of habitat and wild populations in terrestrial systems usually has to be strictly managed (e.g., Sinclair et al. 1995). The common taboo on killing mammals and birds in marine ecosystems does not extend to terrestrial systems, and we see many examples of elephants, kangaroos, elk, wolves, and crocodiles whose populations are controlled by active culling at the same time as they are protected from hunting. So it is important to realize that once marine mammal, bird, fish, or carnivorous turtle populations have recovered to sustainable levels within the meaning of the target ORB fished Lost Valley ecosystem, they may have to be controlled within the management boundaries. Similarly, Walters and Martell (2004) show that active culling may lead to higher fishery values when predator and prey are inversely linked. In BTF restoration, it is expected that monitoring programs should be set up to ensure that all changes, including charismatic fauna, are within the expected bounds of the transitional path towards new management objectives.

Trade-offs between economic and ecological values for a range of policy goals from the four fished Lost Valley ecosystem models of northern British Columbia and from Newfoundland are shown in Figure 7. In most cases, it is clear that restoration provides gains in both the economic and ecological modalities, offering a win–win solution for restoration. Gains or losses in the trade-offs are made explicit in Figure 8, which plots relative economic and biodiversity values for 15 equivalent optimal policy scenarios. The upper right quadrant signifies profit and biodiversity ra-

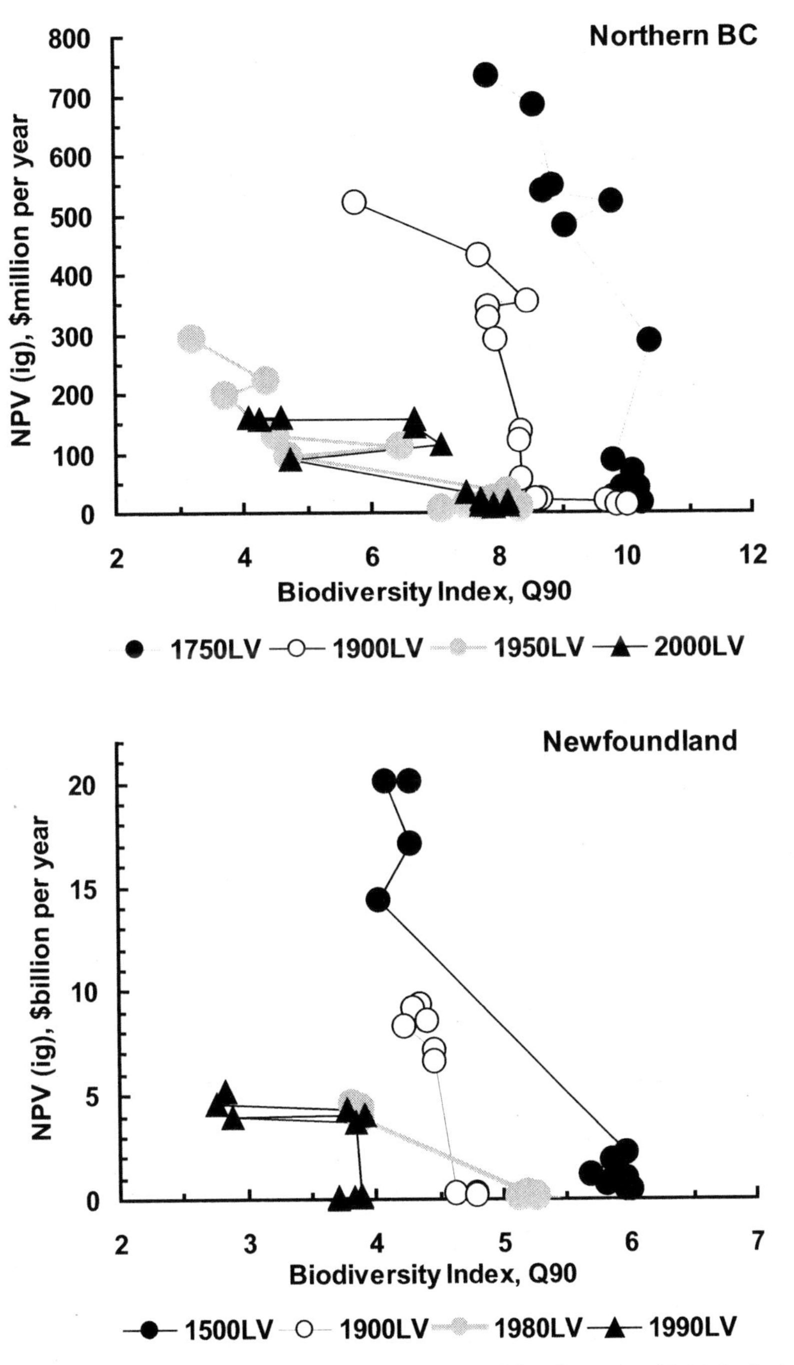

Figure 7. Trade-off plots between economics (NPV) and biodiversity (Q90 index) values for a series of sustainably fished Lost Valley ecosystems (end state of 50-year simulations) for northern British Columbia (top) and the Newfoundland shelf (lower). Lost valleys are based on precontact ecosystems (solid circles); 1900 ecosystems (open circles); mid-20th century ecosystems (gray circles); and the present day (solid triangles). Objectives for each optimal policy search are not listed here, but may be found in Ainsworth and Pitcher (2004b).

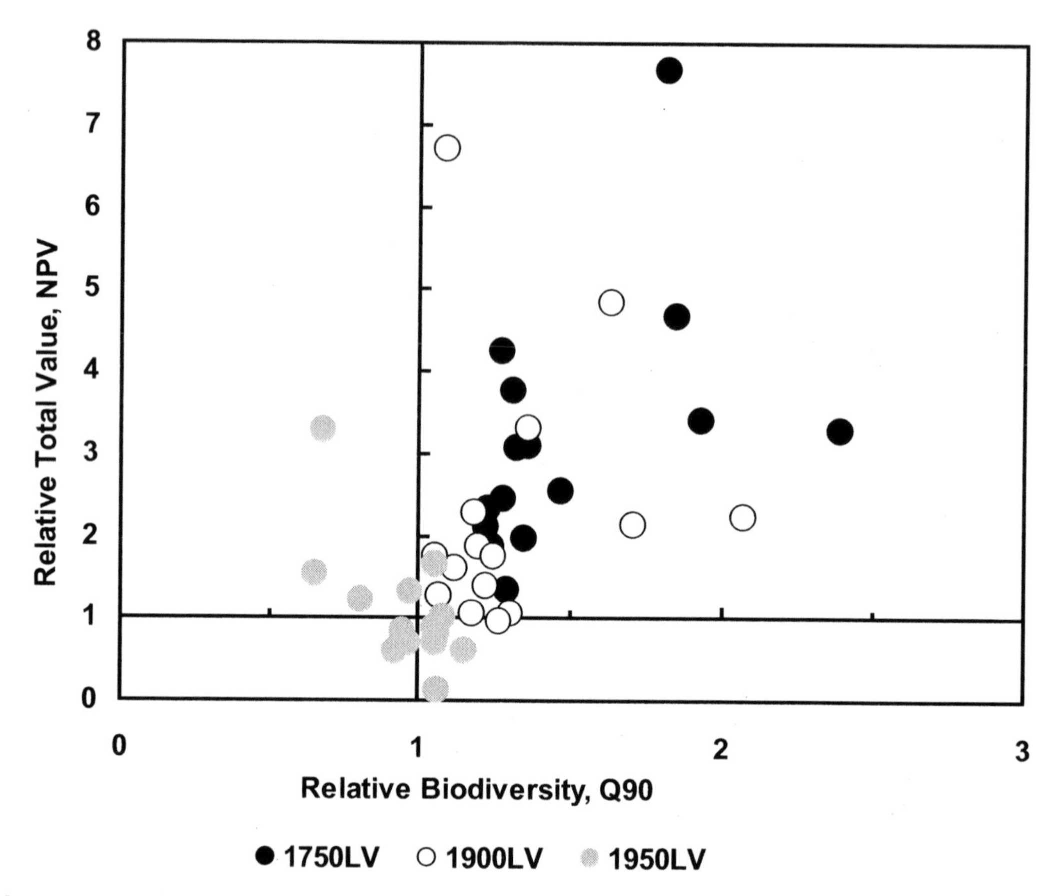

Figure 8. Simulations showing when Back-to-the-Future restoration may be expected to produce benefits. Plot of gains and losses in economic value (NPV) and biodiversity (Q90) relative to a present day northern British Columbia ecosystem that is sustainably and responsibly fished (baseline = 1). Simulations were based on 1750 Lost Valley ecosystem (solid circles); 1900 Lost Valley ecosystem (open circles; and 1950 Lost Valley ecosystem (gray circles). Values in left and lower quadrants signify losses compared to the present day; upper right quadrant represent policies that exhibit gains that might be offset against the costs of restoration.

tios higher than 1 relative to managing sustainable, responsible fisheries in the present day. Points in this quadrant therefore represent policies with tangible gains to be made from BTF restoration that may be offset against the inevitable costs of the recovery process. Table 2 shows that, for the 1750 restoration (and adding both modalities), 13/30 values are greater than twofold; for the 1990 goal, 8/30 are more than twofold; but only 1/30 values for a restoration based on 1950 are as large as that. The table also indicates that large economic gains are more likely than large biodiversity gains, but this is partly an artifact caused by the way that biodiversity is measured.

Recovery costs are considered by Ainsworth and Pitcher (2007, this volume), who consider a variety of restoration plans aiming to rebuild commercial fish biomass in northern British Columbia. All plans recover biomass, but "fast" plans neglect the needs of resource users in favor of conservation and rebuilding,

Table 2. Numbers of gains of different magnitude from BTF restoration policies from Figure 8.

Modality of gain	Economics					Biodiversity				
Magnitude	1	2	3	4	5	1	2	3	4	5
1750LV	15	12	8	3	1	15	1	0	0	0
1900LV	14	6	3	2	1	15	1	0	0	0
1950LV	5	1	1	0	0	8	0	0	0	0

while "slow" plans continue to generate significant annual income from fisheries. A preliminary cost–benefit analysis of restoration suggests that, under ideal circumstances, many scenarios of ecosystem restoration can offer a financial return comparable to bank interest, providing a direct economic incentive in support of restoration. Moreover, a convex relationship of the trade-offs between end-state profit and biodiversity (see Figure 8 in Ainsworth and Pitcher 2006, this volume) suggests that there may be an optimal rate of recovery where human needs and ecological needs can both be met (Nash equilibrium). It remains to be seen if these two findings will apply to other systems where BTF modeling has been undertaken.

The underlying question of this Fourth World Fisheries Congress has been to ask whether fisheries can be reconciled with conservation. Unfortunately, Carl Walter's keynote paper (this volume), demonstrates that in most existing fisheries analyses, concave trade-off relationships will prejudice any attempt at reconciliation, and our only option appears to be to deal directly with trade-offs that are unpleasant for all, leading to the horse-trading, bitterness and ineffective top-down control that characterizes the present-day fisheries world. In contrast, trade-off comparisons that show win–win solutions and convex trade-off boundaries offer the exciting possibility that Back to the Future can provide a genuine reconciliation, a candidate policy solution to the fisheries problem.

Acknowledgments

Back to the Future research has been supported by small grants from the Peter Wall Institute for Advanced Studies and from the Fisheries Centre at the University of British Columbia; the World Wild Fund for Nature Canada; and the Comparative Dynamics of Exploited Ecosystems in the Northwest Atlantic program of the Department of Fisheries and Oceans, Canada. A major interdisciplinary grant from the National Science and Environmental Research Council and the Social Science and Humanities Research Council of the Government of Canada to Professor Rosemary Ommer, University of Victoria, for the Coasts Under Stress' project, supported the research that made possible most of the quantitative work, interviews and workshops outlined in this paper. We are also grateful for enthusiasm and commitment from BTF team members Dr. Sheila (J.J.) Heymans, Dr. Marcelo Vasconcellos, Dr. Rashid Sumaila, Dr. Melanie Power, Eny Buchary, Aftab Erfan, William Cheung, Robyn Forrest, Nigel Haggan, Hector Lozano, Telmo Morato. Dr Villy Christensen and Professor Carl Walters provided invaluable and long-suffering support for the modeling. Cameron Ainsworth has been funded by Coasts Under Stress, a Natural Sciences and Engineering Research Council of Canada Operating Grant to TJP, and the World Wide Fund For Nature Canada.

References

Alcock, A., D. Ings, and D. C. Schneider. 2003 From local knowledge to science and back: evolving use of local ecological knowledge in fisheries science. Pages 20–39 *in* J. J. Heymans, editor. Ecosystem models of Newfoundland and southeastern Labrador: additional information and analyses for Back to the Future. Fisheries Centre Research Reports 11(5), Vancouver.

Ainsworth, C. H., and U. R. Sumaila. 2003 Intergenerational discounting and the conservation of fisheries resources. A case study in Canadian Atlantic Cod. Pages 26–33 *in* R. Sumaila, editor. Three essays on the economics of fishing. Fisheries Centre Research Reports 11(3), Vancouver.

Ainsworth, C. H., and U. R. Sumaila. 2004a. Economic valuation techniques for Back-To-The-Future optimal policy searches. Pages 104–107 *in* T. J. Pitcher, editor. Back to the Future: advances in methodology for modelling and evaluating past ecosystems as future policy goals. Fisheries Centre Research Reports 12(1), Vancouver.

Ainsworth, C. H., and U. R. Sumaila. 2004b. An employment diversity index used to evaluate ecosystem restoration strategies. Page 108 *in* T. J. Pitcher, editor. Back to the Future: advances in methodology for modelling and evaluating past ecosystems as future policy goals. Fisheries Centre Research Reports 12(1), Vancouver.

Ainsworth, C. H., and T. J. Pitcher. 2004a. Using local ecological knowledge as a data supplement for ecosystem models. In R. I. Perry, editor. Assessment and management of new and developed fisheries in data-limited situations. Alaska Sea Grant, Anchorage.

Ainsworth, C. H., and T. J. Pitcher. 2004b. Evaluating marine ecosystem restoration goals for Northern British Columbia. In R. I. Perry, editor. Assessment and management of new and developed fisheries in data-limited situations. Alaska Sea Grant, Anchorage.

Ainsworth, C. H., and T. J. Pitcher. 2005a. Using local ecological knowledge in ecosystem models. Pages 289-321 *in* G. H. Kruse, V. F. Gallucci, D. E. Hay, R. I. Perry, R. M. Peterman, T. C. Shirley, P. D. Spencer, B. Wilson, and D. Woodby, editors. Fisheries assessment and management in data-limited situations. Alaska Sea Grant, Fairbanks.

Ainsworth, C. H., and T. J. Pitcher. 2005b. Evaluating marine ecosystem restoration goals for Northern British Columbia. Pages 419-438 *in* G. H. Kruse, V. F. Gallucci, D. E. Hay, R. I. Perry, R. M. Peterman, T. C. Shirley, P. D. Spencer, B. Wilson, and D. Woodby, editors. Fisheries assessment and management in data-limited situations. Alaska Sea Grant, Fairbanks.

Ainsworth, C. H., and T. J. Pitcher. 2008. Back to the future in northern British Columbia: evaluating historic marine ecosystems and optimal restorable biomass as restoration goals for the future. Pages 317–319 *in* J. L. Nielsen, J. J. Dodson, K. Friedland, T. R. Hamon, J. Musick, and E. Verspoor, editors. Reconciling fisheries with conservation: proceedings of the Fourth World Fisheries Congress. American Fisheries Society, Symposium 49, Bethesda, Maryland.

Ainsworth, C. H., and T. J. Pitcher. 2006b. Modifying Kempton's species diversity index for use with ecosystem simulation models. Ecological Indicators (*in press*).

Ainsworth, C. H, J. J. Heymans, T. J. Pitcher, and M. Vasconcellos. 2002. Ecosystem models of Northern British Columbia for the time periods 2000, 1950, 1900 and 1750. Fisheries Centre Research Reports 10(4), Vancouver.

Angelsen, A., and U. R. Sumaila. 1996. Hard methods for soft policies: environmental and social cost benefit analysis. Forum for Development Studies 1: 87-114, Tromso, Norway.

Angelsen A., O. H. Fjeldstad, and U. R. Sumaila. 1994. Growth, development and sustainability. Development Matters 1(2): 19-22.

Baum, J. K., R. A. Myers, D. G. Kehler, B. Worm, S. J. Harley, and P. A. Doherty. 2002. Collapse and conservation of shark populations in the northwest Atlantic. Science 299:389–392.

Bundy, A. 2002. Adaptation of a Newfoundland-Labrador ecopath model for 1985–1987 in Statistical Areas 2J3KLNO to the Area 2J3KL. Pages 13–21 *in* T. J. Pitcher, M. Vasconcellos, J. J. Heymans, C. Brignall, and N. Haggan, editors. Information supporting past and present ecosystem models of Northern British Columbia and the Newfoundland shelf. Fisheries Centre Research Reports 10(1), Vancouver.

Burke, C., G. K. Davoren, W. A. Montevecchi, and I. J. Stenhouse. 2002. Winging Back to the Future: an historic reconstruction of seabird diversity, distribution, and abundance in the northwest Atlantic, 1500–2000. Pages 27–37 *in* T. J. Pitcher, M. Vasconcellos, J. J. Heymans, C. Brignall, and N. Haggan, editors. Information supporting past and present ecosystem models of Northern British Columbia and the Newfoundland shelf. Fisheries Centre Research Reports 10(1), Vancouver.

Christensen, V. 2002. Ecosystems of the past: how can we know since we weren't there? Pages 26–34 *in* S. Guénette, V. Christensen, and D. Pauly, editors. Fisheries impacts on North Atlantic ecosystems: models and analyses. Fisheries Centre Research Reports 9(4): Vancouver.

Christensen, V., and C. J. Walters. 2004. Ecopath with Ecosim: methods, capabilities and limitations. Ecological Modelling 172:109–139.

Christensen, V., S. Guénette, J. J. Heymans, C. J. Walters, R. Watson, D. Zeller, and D. Pauly. 2003. Hundred-year decline of North Atlantic predatory fishes. Fish and Fisheries 4:1–24.

Cheung, W. L., and T. J. Pitcher. 2004. An index expressing risk of local extinction for use with dynamic ecosystem simulation models. Pages 94–102 *in* T. J. Pitcher, editor. Back to the Future: advances in methodology for modelling and evaluating past ecosystems as future policy goals. Fisheries Centre Research Reports 12(1), Vancouver.

Cheung, W. W. L., T. J. Pitcher, and D. Pauly. 2005. A fuzzy logic expert system to estimate intrinsic extinction vulnerabilities of marine fishes to fishing. Biological Conservation 124: 97-111.

Clark, C. W. 1973. The economics of overexploitation. Science 181:630–634.

Coward, H., R. Ommer, and T. J. Pitcher, editors. 2000. Just fish: the ethics of Canadian fisheries. Institute of Social and Economic Research Press, St John's, Newfoundland, Canada.

FAO (Food and Agriculture Organization of the United Nations). 1995. Code of Conduct for Responsible Fisheries. FAO, Rome.

Fearnside, P. M. 2002. Time preference in global warming calculations: a proposal for a unified index. Ecological Economics 41:21–31.

Fulton, E. A., A. D. M. Smith, and C. R. Johnson. 2003. Effect of complexity on marine ecosystem models. Marine Ecology Progress Series 253:1–16.

Furness, R. W. 2002. Management implications of interactions between fisheries and sandeel-dependent seabirds and seals in the North Sea. ICES Journal of Marine Science 59:261–269.

Goulder, L. H., and R. N. Stavins. 2002. Discounting: an eye on the future. Nature (London) 419:673–674.

Haggan, N. 2000. Back to the Future and creative justice: recalling and restoring forgotten abundance in Canada's marine ecosystems. Pages 83–99 *in* H. Coward, R. Ommer, and T. J. Pitcher, editors. Just fish: ethics in the Canadian coastal fisheries. ISER Books, St. Johns, Newfoundland, Canada.

Haggan, N., and A. Beattie, editors. 1999. Back to the Future: reconstructing the Hecate Strait ecosystem. Fisheries Centre Research Reports 7(3), Vancouver.

Haggan, N., J. Archibald, and S. Salas. 1998. Knowledge gains power when shared. Pages 8–13 *in* D. Pauly, T. J. Pitcher, and D. Preikshot, editors. Back to the Future: reconstructing the Strait of Georgia Ecosystem. Fisheries Centre Research Reports 6(5), Vancouver.

Hays, G. C., A. C. Broderick, B. J. Godley, P. Luschi, and W. J. Nichols. 2003. Satellite telemetry suggests high levels of fishing-induced mortality in marine turtles. Marine Ecology Progress Series 262: 305–309.

Heymans, J. J., editors. 2003a. Ecosystem models of Newfoundland and southeastern Labrador: additional information and analyses for 'Back to the Future'. Fisheries Centre Research Reports 11(5), Vancouver.

Heymans, J. J. 2003b. First Nations impact on the Newfoundland ecosystem during pre-contact times. Pages 4–11 *in* J. J. Heymans, editor. Ecosystem models of Newfoundland and southeastern Labrador: additional information and analyses for 'Back to the Future'. Fisheries Centre Research Reports 11(5), Vancouver.

Heymans, J. J. 2003c. Fitting the Newfoundland model to time series data. Pages 72–78 *in* J. J. Heymans, editors. Ecosystem models of Newfoundland and southeastern Labrador: additional information and analyses for 'Back to the Future'. Fisheries Centre Research Reports 11(5), Vancouver.

Heymans, J. J. 2003d. Revised models for Newfoundland for the time periods 1985–87 and 1995–97. Pages 40–61 *in* J. J. Heymans, editors. Ecosystem models of Newfoundland and southeastern Labrador: additional information and analyses for 'Back to the Future'. Fisheries Centre Research Reports 11(5), Vancouver.

Heymans, J. J. 2004. Evaluating the ecological effects on exploited ecosystems using information theory. Pages 87–90 *in* T. J. Pitcher, editors. Back to the Future: advances in methodology for modelling and evaluating past ecosystems as future policy goals. Fisheries Centre Research Reports 12(1), Vancouver.

Heymans, J. J., and T. J. Pitcher. 2004. Synoptic methods for constructing models of the past. Pages 11–17 *in* T. J. Pitcher, editors. Back to the Future: advances in methodology for modelling and evaluating past ecosystems as future policy goals. Fisheries Centre Research Reports 12(1), Vancouver.

Lilly, G. R. 2002. Ecopath modelling of the Newfoundland Shelf: observations on data availability within the Canadian Department of Fisheries and Oceans. Pages 22–26 *in* T. J. Pitcher, M. Vasconcellos, J. J. Heymans, C. Brignall, and N. Haggan, editors. Information supporting past and present ecosystem models of Northern British Columbia and the Newfoundland shelf. Fisheries Centre Research Reports 10(1), Vancouver.

Lucas, S. 2004. Aboriginal values. Pages 114–116 *in* T. J. Pitcher, editor. Back to the Future: advances in methodology for modelling and evaluating past ecosystems as future policy goals. Fisheries Centre Research Reports 12(1), Vancouver.

Orchard, T. J., and Q. Mackie, Q. 2004. Environmental archaeology: principles and case studies. Pages 64–73 *in* T. J. Pitcher, editor. Back to the Future: advances in methodology for modelling and evaluating past ecosystems as future policy goals. Fisheries Centre Research Reports 12(1), Vancouver.

Morato, T., R. Watson, T. J. Pitcher, and D. Pauly. 2006. Fishing down the deep. Fish and Fisheries 7(1):23–33.

Myers, R. A., and B. Worm. 2003. Rapid worldwide depletion of predatory fish communities. Nature (London) 423:280–283.

Padilla, E. 2002. Intergenerational equity and sustainability. Ecological Economics 41:69–83.

Pauly, D., P. J. B. Hart, and T. J. Pitcher. 1998a Speaking for themselves: new acts, new actors and a new deal in a reinvented fisheries management. Pages 409–415 *in* T. J. Pitcher, P. J. B. Hart, and D. Pauly, editors. Reinventing fisheries management. Chapman and Hall, London.

Pauly, D., T. J. Pitcher, and D. Preikshot, editors (1998b) Back to the Future: reconstructing the Strait of Georgia ecosystem. Fisheries Centre Research Reports 6(5): Vancouver.

Pauly, D., V. Christensen, T. J. Pitcher, U. R. Sumaila, C. J. Walters, R. Watson, and D. Zeller. 2002. Diagnosing and overcoming fisheries' lack of sustainability. Nature (London) 418:689–695.

Pitcher, T. J. 1999. Rapfish, a rapid appraisal technique for fisheries, and its application to the Code Of Conduct For Responsible Fisheries. FAO Fisheries Circular No. 947.

Pitcher, T. J. 2001. Fisheries managed to rebuild ecosystems: reconstructing the past to salvage the future. Ecological Applications 11(2):601–617.

Pitcher, T. J. 2005. 'Back to the Future': a fresh policy initiative for fisheries and a restoration ecology for ocean ecosystems. Philosophical Transactions of the Royal Society Biological Sciences 360:107–121.

Pitcher, T. J. 2004a. Introduction to the methodological challenges in 'Back-to-the-Future' research. Pages 4–10 *in* T. J. Pitcher, editor. Back to the Future: advances in methodology for modelling and evaluating past ecosystems as future policy goals. Fisheries Centre Research Reports 12(1): Vancouver.

Pitcher, T. J., editor. 2004b. Back to the Future: advances in methodology for modelling and evaluating past ecosystems as future policy goals. Fisheries Centre Research Reports 12(1), Vancouver.

Pitcher, T. J. 2004d. Why we have to open the lost valley: criteria and simulations for sustainable fisheries. Pages 78–86 *in* T. J. Pitcher, editor. Back to the Future: advances in methodology for modelling and evaluating past ecosystems as future policy goals. Fisheries Centre Research Reports 12(1), Vancouver.

Pitcher, T. J. and P. J. B. Hart. 1982. Fisheries ecology. Chapman and Hall, London.

Pitcher, T. J. and D. Pauly. 1998. Rebuilding ecosystems, not sustainability, as the proper goal of fishery management. Pages 311–329 *in* T. J. Pitcher, P. J. B. Hart, and D. Pauly, editors. Reinventing fisheries management. Chapman and Hall, London.

Pitcher, T. J., and N. Haggan. 2003. Cognitive maps: cartography and concepts for ecosystem-based fisheries policy. Pages 456–463 *in* N. Haggan, C. Brignall, and L. Wood, editors. Putting fishers' knowledge to work. Fisheries Centre Research Reports 11(1), Vancouver.

Pitcher, T. J., and R. Forrest. 2004. Challenging ecosystem simulation models with climate change: the 'Perfect Storm.' Pages 29–38 *in* T. J. Pitcher, editor. Back to the Future: advances in methodology for modelling and evaluating past ecosystems as future policy goals. Fisheries Centre Research Reports 12(1), Vancouver.

Pitcher, T. J., N. Haggan, D. Preikshot, and D. Pauly. 1999 'Back to the Future': a method employing ecosystem modelling to maximise the sustainable benefits from fisheries. Pages 447–466 *in* Ecosystem approaches for fisheries management. Alaska Sea Grant, Anchorage.

Pitcher, T. J., M. Vasconcellos, J. J. Heymans, C. Brignall, and N. Haggan, editors. 2002a. Information supporting past and present ecosystem models of Northern British Columbia and the Newfoundland shelf. Fisheries Centre Research Reports 10(1), Vancouver.

Pitcher, T. J., J. J. Heymans, and M. Vasconcellos, editors. 2002b. Ecosystem models of Newfoundland for the time periods 1995, 1985, 1900 and 1450. Fisheries Centre Research Reports 10(5), Vancouver.

Pitcher, T. J., M. Power, and L. Wood, editors. 2002c. Restoring the past to salvage the future: report on a community participation workshop in Prince Rupert, BC. Fisheries Centre Research Reports 10(7), Vancouver.

Pitcher, T. J., J. J. Heymans, C. Ainsworth, E. A. Buchary, U. R. Sumaila, and V. Christensen. 2004. Opening the lost valley: implementing a "back to the future" restoration policy for marine ecosystems and their fisheries. Pages 173–201 *in* E. E. Knudsen, D. D. MacDonald, and Y. K. Muirhead, editors. Sustainable management of North American fisheries. American Fisheries Society, Symposium 43, Bethesda, Maryland.

Pitcher, T. J., C. H. Ainsworth, E. A. Buchary, W. L.

Cheung, R. Forrest, N. Haggan, H. Lozano, T. Morato, and L. Morissette. 2005a. Strategic management of marine ecosystems using whole-ecosystem simulation modelling: the 'Back-to-the-Future' policy approach. Pages 199–258 *in* E. Levner, I. Linkov, and J. M. Proth, editors. The strategic management of marine ecosystems. NATO Science Series: IV: Earth and Environmental Sciences 50, Springer, Berlin.

Pitcher, T. J., C. H. Ainsworth, H. Lozano, W. L. Cheung, and G. Skaret. 2005b. Evaluating the role of climate, fisheries, and parameter uncertainty using ecosystem-based viability analysis. ICES CM 2005\M: 24:1–6.

Rogers, K. 2003. Micro-level historical reconstruction of the Newfoundland fisheries between 1891–2000: findings and issues. Pages 12–19 *in* J. J. Heymans, editor. Ecosystem models of Newfoundland and southeastern Labrador: Additional information and analyses for 'Back to the Future'. Fisheries Centre Research Reports 11(5), Vancouver.

Roman, J., and S. R. Palumbi. 2003. Whales before whaling in the North Atlantic. Science 301:508–510.

Schindler, D. E., T. E. Essington, J. F. Kitchell, C. Boggs, R. Hilborn, R. 2002. Sharks and tunas: fisheries impacts on predators with contrasting life histories. Ecological Applications 12(3):735–748.

Sinclair, A. R. E., D. S. Hik, O. J. Schmitz, G. G. E. Scudder, D. H. Turpin, and N. C. Larter. 1995. Biodiversity and the need for habitat renewal. Ecological Applications 5:579–587.

Simeone, W. 2004. How traditional knowledge can contribute to environmental research and resource management. Pages 74–77 *in* T. J. Pitcher, editors. Back to the Future: advances in methodology for modelling and evaluating past ecosystems as future policy goals. Fisheries Centre Research Reports 12(1), Vancouver.

Sumaila, U. R. and C. J. Walters. 2003. Intergenerational discounting. Pages 19–25 *in* R. Sumaila, editor. Three essays on the economics of fishing. Fisheries Centre Research Reports 11(3), Vancouver.

Sumaila, U. R., and C. J. Walters. 2004. Intergenerational discounting: a new intuitive approach. Ecological Economics 52:135–142.

Sumaila, U. R., T. J. Pitcher, N. Haggan, and R. Jones. 2001. Evaluating the benefits from restored ecosystems: a Back to the Future approach. In R. S. Johnston and A. L. Shriver, editors Proceedings of the 10th International Conference of the International Institute of Fisheries Economics & Trade, Corvallis, Oregon.

Ulanowicz, R. E. 1999. Life after Newton: an ecological metaphysic. BioSystems 50:127–142.

Walters, C. J., and S. Martell. 2004. Fisheries ecology and management. Princeton University Press, Princeton, New Jersey.

Walters, C. J., V. Christensen, and D. Pauly. 2002. Searching for optimum fishing strategies for fishery development, recovery and sustainability. Pages 11–15 *in* T. J. Pitcher and K. Cochrane, editors. The use of ecosystem models to investigate multispecies management strategies for capture fisheries. Fisheries Centre Research Reports 10(2), Vancouver.

Walters, C. J., V. Christensen, S. Martell, and J. Kitchell. 2004. Single species versus ecosystem harvest management: is it true that good single species management would result in good ecosystem management? Oral presentation. Quantitative Ecosystem Indicators for Fisheries Management. International Symposium. March 31–April 3, 2004. Paris, France.

Watson, J. C., G. Ellis, and K. B. Ford. 1997. Population growth and expansion in the British Columbia sea otter population. Sixth Joint U.S.-Russia sea otter workshop. Forks, Washington.

Winship, A. 1998. Pinnipeds and cetaceans in the Strait of Georgia. Pages 53–57 *in* D. Pauly, T. J. Pitcher, and D. Preikshot, editors. Back to the Future: reconstructing the Strait of Georgia ecosystem. Fisheries Centre Research Reports 6(5), Vancouver.

American Fisheries Society Symposium 49:385–386

Session Summary

Reconciling Fisheries with Conservation and Maintaining Biodiversity

Yvonne Sadovy
University of Hong Kong, Department of Ecology and Biodiversity
Pok Fu Lam Road, Hong Kong, China

This session attracted a lot of interest. In addition to the five oral presentations, there were 30 scheduled posters from 11 countries. Together they included criteria to identify and delist species according to extinction risk, the concept and importance of biodiversity value (both economic and inter-generational), conservation status (particularly in freshwater and for salmonids), and case studies of genetic, population, species and ecosystem maintenance as policy or management goals. Understanding the specific effects on biodiversity of selective fishing, bycatch, fish culture, and so forth was identified as important for seeking conservation and management approaches likely to be effective. Monitoring and enforcement were also addressed. However, the necessary institutional support and cooperation have a long way to go to meet many conservation needs in exploited marine species.

It was notable that all of the species addressed were vertebrates, mainly fishes but there were two turtle, one dolphin, and one bird case studies. This focus partly reflects our poor understanding of conservation status of marine invertebrates but should also indicate where attention might be needed, given what we know of declines in a number of commercially important mollusks. The following is a summary of the talks and posters presented in Vancouver: specific details can be seen in the individual printed abstracts.

There persists a divide between scientific consensus that commercially exploited marine species can become extinct and a public perception that this hardly seems possible. Moreover, policies that favor economic development at the expense of biodiversity conservation are commonplace worldwide. Even if threat to species or populations is acknowledged, progress is hampered by a disjunct among instruments and institutions charged with natural resource management, on the one hand, and those responsible for conservation, on the other. The current debate about whether threatened, commercially exploited, marine species should be listed under the Convention on International Trade in Endangered Species (CITES) or, alternatively, subject to Food and Agriculture Organization of the United Nations (FAO) oversight, for example, illustrates best the problems involved, and sharks were presented as a telling case study.

To move forward, we must better define and communicate the value of maintaining biodiversity, whether for immediate economic gain, as a responsibility to future generations (intergenerational equity), or because of the long-term practical implications and negative consequences associated with eroding genetic and ecosystem biodiversity. We also need to better identify and prioritize those species that most need conservation or management attention and follow through to realize protection. Useful patterns are emerging but they are not always consistent across habitats. For example, while large size seems to be an important correlate of extinction risk in fishes in the sea, small size may be more critical in freshwaters. Tools are being developed that will greatly help in realizing greater protection, enforcement, and mitigation. One example is the use of genetic means of controlling nonnative fish where there are clear advantages for the introduction of such species. Another is the application of molecular techniques to identify traded shark fins to species level. It is also important that culturing, conducted for conservation purposes, be accompanied by habitat protection and reduced pressure on larger species.

Although the tendency is to focus on species, increasingly studies are addressing genetic and ecosystem diversity: the impacts of, say, deforestation, on freshwater faunas, of selective fishing on marine food webs, and of introductions on genetic diversity. The problem is that we must make key decisions under what are typically data-poor situations. Species databases, like FishBase and the Sea Around Us Project, are contributing to information gaps and access. Fishery studies of the impacts of temporal and spatial patterns of fishing on threatened species, taken either as target catch or as bycatch, could greatly assist in fine-tuning management decisions.

American Fisheries Society Symposium 49:387–389

Development of Methods to Evaluate Management Options for Achieving the Recovery of Endangered Salmon Stocks

LYNSEY R. PESTES* AND RANDALL M. PETERMAN

Simon Fraser University, School of Resource and Environmental Management Fisheries Science and Management Research Group 8888 University Drive, Burnaby, British Columbia, V5A 1S6, Canada

The abundance of some Pacific salmon *Oncorhynchus* spp. stocks has decreased dramatically in recent years. Canada's new Species at Risk Act (SARA) requires that a recovery strategy and an action plan be developed for species listed under the act for which recovery is "technically and biologically feasible" (Canada 2002). However, insufficient research has been done on methods to determine which management actions will allow salmon stocks to recover. Our goal is to fill this gap by developing quantitative methods that will evaluate management options to achieve recovery of salmon stocks. We use the Cultus Lake, British Columbia, sockeye salmon *O. nerka* as a case example.

The Cultus Lake sockeye stock was being considered for listing under SARA in 2004. However, the Canadian Minister of Environment chose not to list this stock under SARA because that would have triggered a ban on bycatch of the stock, which in turn would have reduced the harvest rate of all late-run Fraser River sockeye salmon (which comigrate with the Cultus Lake stock), a fishery that is worth millions of dollars annually. Regardless of the legal status of the Cultus Lake sockeye stock under SARA, the Canadian Department of Fisheries and Oceans has committed to continuing to develop a recovery strategy and a subsequent action plan for this stock. The research reported here is therefore still directly relevant to identifying management options that will meet the management objective of recovery. Furthermore, our procedures and our qualitative findings may be more widely applicable to recovery planning for other low-abundance salmon populations or even other species.

The situation for Cultus Lake sockeye salmon can be characterized by three hypothetical management objectives: (1) a recovery objective, which states that the probability that the abundance of Cultus sockeye spawners, S, will reach some recovered level, X, by some year, T, must be greater than Z_1, symbolized as Pr ($S > X$ by year T) $> Z_1$; (2) a long-term survival objective, which states that once the recovery target, X, is reached, the probability that the abundance of Cultus sockeye spawners will fall below some level Q over the next Y years must be less than Z_2, or Pr ($S < Q$ over next Y years) $< Z_2$; and (3) an economic objective, which seeks to maximize the long-term economic yield from the harvest of late-run Fraser

*Corresponding author: lpestes@alumni.sfu.ca

River sockeye under the condition that the first two objectives are met.

Our research objectives are to:

1. Determine the rank order of various harvest rules for the Cultus Lake sockeye population.

2. Explicitly take into account uncertainties in the Cultus Lake sockeye population's dynamics and its management to estimate risks associated with various levels of harvesting during different stages of recovery.

3. Identify the harvest strategies that best meet the management objectives while also being robust to changes in various assumptions.

We use Bayesian decision analysis (Punt and Hilborn 1997; Peterman et al. 1998) to find combinations of harvest parameters L (recruit abundance below which the minimum harvest rate is used), U (recruit abundance above which the maximum harvest rate is used), and H_{max} (maximum harvest rate) (Figure 1) that meet these three management objectives for Cultus sockeye. As part of our decision analysis, we stochastically model the dynamics of the Cultus Lake sockeye population and apply a harvest rule based on annual abundance of recruits (Figure 1). This model incorporates uncertainties in parameters of the spawner-to-smolt relationship, depensation in freshwater stages, marine survival rate, prespawning mortality, and implementation error (the difference between desired and realized harvest rates). We also model the population dynamics of other major late-run Fraser River sockeye stocks in order to calculate the long-term economic value of the catch of late-run sockeye.

Combinations of the harvest-rule parameters (Figure 1) that meet both of the first two management objectives will be ranked according to the third objective. We will perform extensive sensitivity analyses of the rank order of management options to different assumptions about parameter values, functional forms in the model's component processes, and management objectives (e.g., different values of Z_1, Z_2, X, and T). The abbreviated decision tree (Figure 2) shows two of the many uncertain states of nature that will be included.

The decision analysis also allows for ranking of management options using different objectives, given that these are often unclear. Sensitivity analyses will help to set priorities for future research by indicating the parameters and assumptions to which the rank order of management actions (harvest rules) is most sensitive. The model will also generate results that will allow managers to deal quantitatively with one of the most difficult challenges that they face, namely making tradeoffs between the economic value of late-run Fraser River sockeye stocks and the probability of long-term survival and recovery of the Cultus Lake sockeye salmon stock.

Acknowledgments

Fraser River sockeye salmon data were supplied by the Department of Fisheries and Oceans and the Pacific Salmon Commission. This research is funded by a Natural Sciences and Engineering Research Council grant and a Department of Fisheries and Oceans Subvention Grant, both awarded to Randall Peterman, and a Simon Fraser University graduate fellowship awarded to Lynsey Pestes. Special thanks to Chris Wood and Mike Bradford for their contributions to this work.

References

Canada. 2002. Species at Risk Act (SARA). Bill C-5, 1st Session, 37th Parliament. Government of Canada, Ottawa.

Peterman, R. M., C. N. Peters, C. A. Robb, and S. W. Frederick. 1998. Bayesian decision analysis and uncertainty in fisheries management. Pages 387–398 *in*

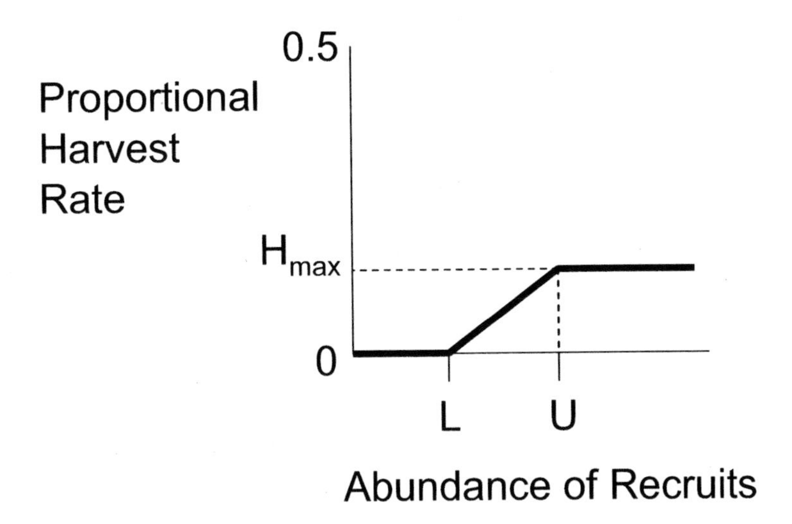

Figure 1. Harvest rule, where *L* is the lower limit of recruits below which no harvest will be taken. The maximum proportional harvest rate, H_{max}, will be applied above recruit abundance *U*.

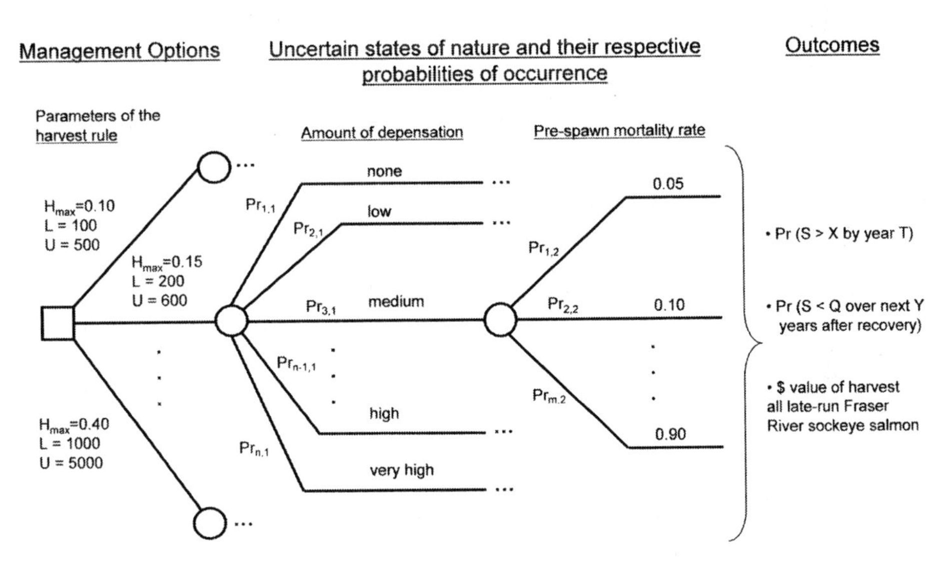

Figure 2. Decision tree illustrating the major structure of this analysis. Branches emanating from the square node represent different harvest rules, each one described by a set of parameters, as illustrated in Figure 1. Branches emanating from round nodes are uncertain states of nature. The relative weighting (or probability, $Pr_{i,j}$) on each uncertain state is generated by the stochastic model.

T. J. Pitcher, P. J. B. Hart, and D. Pauly, editors. Reinventing fisheries management. Kluwer Academic Publishers, London.

Punt, A. E., and R. Hilborn. 1997. Fisheries stock assessment and decision analysis: the Bayesian approach. Reviews in Fish Biology and Fisheries 7:35–63.

American Fisheries Society Symposium 49:391–398

Canada's Wild Pacific Salmon Policy and the Maintenance of Diversity

JAMES R. IRVINE*
Fisheries and Oceans Canada, Pacific Biological Station
3190 Hammond Bay Road, Nanaimo, British Columbia V9T 6N7, Canada

G. ALEX FRASER
Fisheries and Oceans Canada, 200–401 Burrard Street
Vancouver, British Columbia V6B 5G3, Canada

Abstract.—Canada's Wild Pacific Salmon Policy (WSP) addresses the need to protect genetic diversity within each taxonomic salmon species. Local adaptations that salmon develop in different environments are the basis for the production and survival of populations and species, their ability to adapt to change, and hence their long-term viability. Unfortunately, it is not feasible to operate major fisheries and guarantee that all diversity will be preserved. Trade-off needs to be made between the protection of diversity, and social and economic consequences. In determining how much diversity to protect, Fisheries and Oceans Canada is aligning its objectives under the WSP with those specified in Canada's Species at Risk Act (SARA). Under SARA, a species is any "species, subspecies, variety or geographically or genetically distinct population." Units that are genetically distinct from one another, or that are sufficiently disjunct to minimize the likelihood of dispersal with other units, can be listed under SARA. Conservation units identified under the WSP will consist of one or more populations that, if extirpated, will probably not be replaced through natural processes within an acceptable time and will generally conform to the minimum species unit under SARA. This will help to ensure that the goals of the WSP are consistent with this legislation and will assist Fisheries and Oceans Canada to meet its legal responsibilities. Of equal importance, by implementing the WSP, action can be taken in response to conservation concerns in advance of legal listing under the SARA. Fishery managers will adopt a precautionary approach when planning fisheries and be cognizant of the implications of fisheries on diversity. The WSP is a significant change in management to the approach used throughout the 20th century and is a reasonable way of reconciling the continuation of socially and economically important fisheries with the maintenance of genetic diversity.

Introduction

Salmon harvesting in British Columbia has undergone several major transformations. Before Europeans arrived in the Pacific Northwest, large numbers of salmon were caught annually by First Nations (Native Americans) for food and other purposes; these precontact catches may have exceeded peak catches in the 20th century (Glavin 2002). Some First Nations fisheries were near the coast and harvested mixtures of salmon species and stocks, but most First Nations fisheries occurred inland and, consequently, tended to be relatively selective.

Commencing in the 1870s, major new mixed stock and species fisheries began that supplied commercial salmon canneries established in the lower

*Corresponding author: IrvineJ@pac.dfo-mpo.gc.ca

Fraser, Skeena, and other British Columbia rivers. Since these canneries processed sockeye *Oncorhynchus nerka* and large Chinook salmon *O. tshawytscha* almost exclusively, there was enormous wastage of other species harvested (Meggs 1991). The size and the geographic extent of mixed stock fishing efforts increased dramatically over the decades. Combined with habitat degradation and events like the 1913 Hell's Gate slide that restricted the passage of salmon in the Fraser River for over 20 years, this expanding fishery resulted in the extirpation of some and declines for many salmon populations.

Rebuilding of salmon populations became a priority in the 1970s. The federal government implemented various measures to reduce and control fishing effort and a major program was initiated to increase salmon abundance through enhancement (e.g., hatcheries, managed spawning channels, habitat restoration, and lake fertilization). About this time, scientists became much more aware of the role of changing marine conditions on salmon productivity and the need to preserve biodiversity. In addition, legal decisions affirmed the constitutionally protected legal status of the traditional First Nations fishery for food, social, and ceremonial purposes. These factors resulted in dramatic changes to salmon fisheries management during the past two decades. The traditional focus throughout the 20th century on managing for maximum sustainable yield in commercial fisheries (pre 1985) evolved towards managing for a better balance among harvest, stock protection, and the maintenance of regional biodiversity (Hyatt and Riddell 2000). Management increasingly recognized that the long-term survival of salmon depended upon their store of genetic variation (NRC 1996). Also, provision of fishing opportunities for First Nations on the full range of populations became an important management priority. As a result, increasingly more attention is being placed on the maintenance of weaker and less productive populations in mixed stock fisheries.

In 1998, the release of the policy document "A New Direction for Canada's Pacific Salmon Fisheries" by the Department of Fisheries and Oceans Canada (DFO) began to codify these evolving management changes. This document identified three key elements for the future management of Canada's Pacific salmon fishery: conservation, sustainable use, and improved decision making (DFO 1998). In addition, 12 key principles were identified to guide the development of more detailed operational policies on the management of the salmon resource. To date, in keeping with this new direction, operational policies, with respect to the allocation of harvests and selective fishing, have both affirmed salmon conservation as DFO's primary management objective. In addition, extensive work was undertaken to design and develop a wild salmon policy (WSP) intended to secure the long-term viability of Pacific salmon populations in natural surroundings. The WSP clarifies DFO's long term conservation objectives and documents how it intends to achieve these in practical terms. The purpose of this paper is to discuss this new policy and some of its approaches to maintaining salmon diversity.

The Importance of Salmon Fishing

In spite of recent management changes and changing economics, salmon fishing remains an important social, cultural, and financial contributor to British Columbia and many individual British Columbians. The opportunity to harvest salmon is a focal point of First Nations life and culture in the province. Since First Nations food, social, and ceremonial fisheries are constitutionally protected, their fishing rights cannot be unjustifiably infringed.

In addition, in spite of recent harvest reductions and declining prices, commercial salmon harvesting is still a significant contributor to the provincial economy. Several thousand British Columbians depend upon this fishery for at least a portion of their livelihood. Significant First Nations participation in commercial fisheries provides important jobs and income for the residents in many remote First Nations communities where there are limited alternative employment opportunities. Present investment in salmon fishing licenses, gear, and equipment amounts to more than CAN$300 million and commercial salmon fishing expenditures are important contributors to numerous supporting businesses (Gislason 2004).

Finally, recreational salmon fishing is a key contributor to the lifestyle of many British Columbians. More than 250,000 tidal recreational fishing licenses are issued each year to resident anglers and much of the related fishing activity is focused on salmon. In addition, recreational fishing is an important element of the provincial tourist industry. As one specific example, many fishing lodges and charter businesses are directly supported by tourist demands for a salmon fishing experience (Gislason 2004). All of this implies that a policy for the conservation of salmon cannot be developed in isolation. The implications of policy choices for the continued harvesting of salmon need to be carefully considered.

Canada's Wild Salmon Policy

A draft wild salmon policy discussion paper released publicly in the year 2000 proposed six principles to protect the genetic diversity of wild salmon and protect their habitat. These included the aggregation of spawning stocks into "conservation units" (CUs) and the determination of minimum and target levels of abundance for each unit(DFO 2000). From April to July 2000, DFO consulted with 28 stakeholder organizations about the draft WSP and held 16 community forums and open houses. During October–November 2000, information sessions were carried out in nine First Nations communities. Fisheries and Oceans Canada received 43 written submissions and 110 response forms, representing input from 850 individuals. In addition, DFO held three science-based workshops examining biological reference points for wild salmon. Support for the preservation of genetic diversity of wild salmon was widespread, although there was disagreement about how much effort to spend protecting individual spawning units. In addition, the public wished for considerably more detail on how the policy would be implemented. Fisheries and Oceans Cananda responded to the comments received and released a much revised draft WSP in December 2004. During 2004 and the early part of 2005, DFO carried out further consultations and held information sessions with stakeholders and First Nations. In June 2005, the final WSP was released (DFO 2005). Important elements of this final policy are described below.

Biological Units

Salmon have a complex hierarchical population structure extending from groups of salmon at individual spawning sites all the way to taxonomic species (Figure 1). Their precise homing to natal streams and death after spawning restricts gene flow among spawning locations. Persistent spawning locations (demes) can be aggregated into populations that often exhibit local adaptations. Further along this continuum of increasing genetic diversity are conservation units (CUs), which are isolated enough from each other that if they are extirpated, they will probably not recolonize naturally within an acceptable time. Individuals within demes are more similar to each other than individuals within populations, and so on. Definitions for these terms follow.

A deme is a group of salmon at a persistent spawning site or within a stream comprised of individuals that are likely to breed with each other (i.e., well mixed). A single population may include more than one deme and demes may be partially isolated from one another. Their partial isolation may or may not be persistent over generations. There will always be at least as many demes as populations.

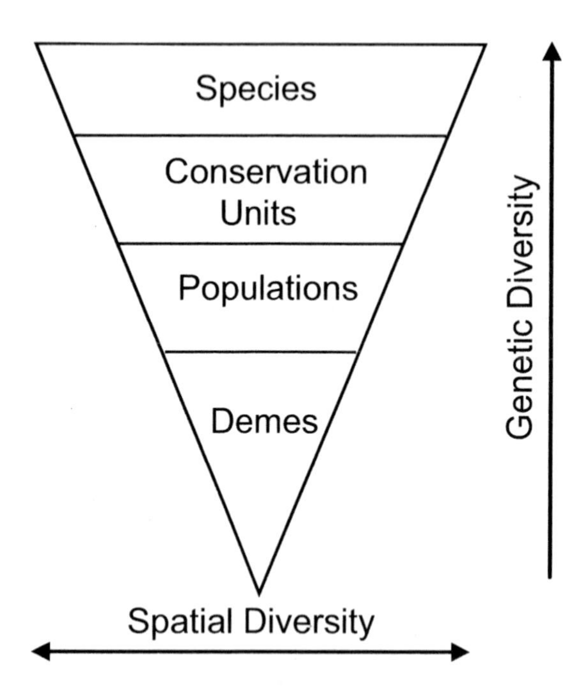

Figure 1. Schematic representation of diversity in Pacific salmon (adapted from Riddell 1993). Within this genetic hierarchy, individuals within demes are more similar to each other than are individuals within populations and so on.

A population is a group of interbreeding salmon that is sufficiently isolated from other populations so that there will likely be persistent adaptations to the local habitat[1]. Local adaptations improve survival in specific habitats and consequently increase the productivity of the population. Connectivity among populations and the maintenance of genetic diversity within populations are necessary to the long-term viability of the species. There will always be at least as many populations as CUs.

A conservation unit (CU) is a group of one or more populations that, if extirpated, will probably not be replaced through natural processes[2] within a reasonable time (i.e., a human life time). If a CU is comprised of more than one population, then the populations will share a common genetic lineage. A CU is a biological unit, but it is also a tool used by the WSP to protect diversity. Conservation units will generally correspond to minimum units eligible for protection under Canada's Species at Risk Act (SARA) as "wildlife species." A wildlife species as defined by SARA is a named species, subspecies, variety, or geographically or genetically distinct population. There will always be at least as many CUs as named taxonomic species.

Numbers and the areal extent occupied by CUs will vary among species. For instance, pink salmon *O. gorbuscha* and chum salmon *O. keta* tend to be genetically similar over greater distances than other salmon species and therefore the geographical size occupied by their CUs will be relatively large. Coho salmon *O. kisutch* also rarely exhibit marked genetic boundaries, so the size occupied by coho CUs will tend to be large and at times difficult to define. Chinook salmon, with more varied life histories, will have more CUs than pink, chum, and coho salmon, but sockeye salmon will have the most. A sockeye CU will typically be at the level of individual rearing locations (usually lakes), although sometimes it may turn out that several small nearby lakes will constitute one CU or that different timed runs within lakes may be separate CUs.

Application of the WSP

While the WSP seeks to protect the full range of populations and demes of wild salmon, their associated habitat and connectivity, by focusing on

[1] A variety of information types can be used to delineate populations under the WSP, including results from genetic surveys, measures of phenotypic variation, and ecological criteria.

[2] While it may be possible to replace extirpated salmon from a CU with other salmon, perhaps by enhancement, the new salmon will differ from the extirpated salmon. Because the new fish will not have gone through the same evolutionary processes as the original fish, the new fish may not be fully adapted to the habitat and consequently may be less productive.

the preservation of CUs, the policy strives to identify a balance between risk to the resource and benefits from harvest and other uses. In effect, the policy allows for the possibility of temporary loss of some lower level biological units within CUs as a risk whose impacts must be weighed and considered against possible benefits of harvesting. The policy recognizes that maintenance of CUs requires preservation of demes and populations, but not necessarily all of them, all the time.

Protecting CUs will require their ongoing monitoring and assessment and the development of appropriate resource management responses to changes in status. Indicator systems (ISs) are composed of fish from one or more persistent spawning locations or populations (perhaps enhanced) that are assumed to be representative of some aspect of a CU. An IS may be an index site or stream selected to detect annual changes in abundance and survival or an extensive site or stream selected to monitor species distribution and general habitat status. The status of the surrounding CU is inferred, in part, by comparing measures of abundance gathered by monitoring the IS to bench marks. There are no general rules about the number of ISs relative to the number of CUs.

Preliminary results from the Fraser River watershed illustrate the application of the WSP. There are several hundred sites where sockeye spawn on a regular basis. These persistent spawning sites or demes aggregate into more than 25 populations that consist of riverine rearing populations, populations rearing in major lakes and, in some cases, run timing groups within lakes. The number of sockeye CUs may be fewer than the number of populations because of genetic mixing among timing groups within lakes.

Individual sockeye populations and CUs need not be managed separately, nor will each unit necessarily be assessed individually. Populations and CUs may continue to be aggregated for practical management purposes. For example, if two or more CUs have similar productivity and are subject to the same risk factors (e.g., a common fishery), one common IS may be sufficient for ongoing status determination. Similarly, for fisheries management purposes, Fraser sockeye populations might remain aggregated into the four major run timing groups: early run (late June to late July), early summer run (mid-July to mid-August), summer run (mid-July to early September), and the late run (early September to mid-October). However, in developing management plans, fisheries managers will need to recognize that the harvest from each of these timing units is composed of CUs that may vary in productivity and status. For example, the late run Fraser timing unit includes CUs of varying productivity including the Harrison/Lillooet, Shuswap, Adams, and Cultus. The status of the various CUs harvested (rather than the status of the aggregate run timing group) will become the key input to fisheries management planning and decision making, as well as the postseason evaluation of fisheries success.

Nonsockeye will have fewer CUs in the Fraser River watershed than sockeye. There may only be five or six CUs for Chinook salmon (Candy et al. 2002) that are sufficiently genetically distinct, geographically disjunct, or both, that they would probably not replace themselves through natural processes in a reasonable time. Coho salmon from above the Fraser canyon are isolated from coho below the canyon and these groups constitute separate CUs (Irvine 2002). Odd and even year returning pink salmon rarely, if ever, exchange genetic material and therefore constitute separate CUs. Fraser chum salmon may all belong to one or several CUs.

Under the WSP, proposed listings of biological entities, including CUs, will be peer reviewed through the established Pacific Scientific Advice Review Committee (PSARC). This structured review process includes provision for participation by outside experts and other interested parties, including First Nations, fisheries stakeholders, and the public.

Linkages with the Species at Risk Act

The concept of CUs in the wild salmon policy fit well with the purpose of Canada's recently ratified (June 2004) SARA—"to prevent Canadian indigenous species, subspecies, and distinct populations from becoming extirpated or extinct." The Species at Risk Act gave legal status to the Committee on the Status of Endangered Wildlife in Canada (COSEWIC). The COSEWIC is responsible for recommending biological units to be afforded protection under SARA and assessing the biological status of these units. Where appropriate, status is assigned at the taxonomic species level. However, status may also be assigned to subspecies, varieties, or geographically or genetically distinct units in cases where a single designation is not sufficient to accurately portray the status of the taxonomic species. Units designated under the SARA legislation are similar to Evolutionarily Significant Units (ESUs) used by the National Marine Fisheries Service for the American Endangered Species Act, except that in Canada, units do not have to contribute to the evolutionary legacy of the species (Irvine et al. 2005).

Conservation units under the WSP should generally conform to the smallest unit that might be considered under the SARA. For instance, four Pacific salmon CUs recommended by COSEWIC for legal listing under the SARA (Sakinaw sockeye, Cultus sockeye, interior Fraser coho, and Okanagan Chinook) will also be CUs under the WSP (coho will probably be >1 CU. Consistency between the WSP and SARA will help to ensure that the goals of the WSP are consistent with this legislation and will assist DFO to meet its responsibilities under this legislation. Of equal importance, it should facilitate taking action in response to conservation concerns in advance of legal listing under the SARA. A proactive response in advance of legal listing can maintain a degree of flexibility with respect to recovery planning that may not otherwise be available. This could provide significant advantages in reducing and managing the social and economic impacts that may arise from necessary conservation measures.

Discussion

Genetic differences underlie the various adaptations that biological entities develop through evolutionary processes. This diversity of adaptations is the basis for the production and survival of populations and species, and hence their ability to adapt to change, and to withstand harvest. Protecting salmon diversity can be viewed as insurance against the impacts of future change. For example, Hilborn et al. (2003) examined the highly productive Bristol Bay sockeye fishery in Alaska and demonstrated that the runs supporting the fishery since the mid-1990s were generally minor producers in the preceding 70 years. Some historically low producers adapted to climate-induced changes that occurred in the 1990s. Failure to protect the genetic diversity of the system could have been disastrous for the Bristol Bay sockeye.

However, deciding how much genetic diversity to preserve is not strictly a scientific question. If we were to preserve the maximum genetic diversity, we would effectively eliminate human harvesting of salmon. Given the social and economic importance of salmon harvesting this is highly undesirable (and in the specific case of First Nations constitutionally protected fisheries, this may simply not be possible). At the other extreme, preserving taxonomic species while ignoring other biological entities may reduce within species biodiversity and seriously compromise their potential for long-term survival. Making the right choices and finding an acceptable balance among ecological, social, and economic benefits and costs requires a structured process where the positive and negative impacts of fisheries are considered in an open and transparent way.

To maintain salmon diversity, DFO proposes to implement a resource management strategy that consists of linked stock assessment and harvest management components. The identification and assessment of CUs will be science-based, and results will inform managers on the consequences of harvesting and other management actions to their status. Fisheries managers will adopt a precautionary risk management framework to guide harvesting decision making. This will involve identifying and prioritizing fisheries management goals and developing and evaluating the full range of available management options in consultation with First Nations, local community groups, and other fishery stakeholders including nongovernmental environmental organizations. Engaging these various interests throughout the planning process—from the establishment of planning priorities through to the evaluation and selection of the preferred management alternative—is intended to help build consensus on the most appropriate management approach and facilitate improved understanding of the final management decisions. In turn, this will ultimately lead to more successful implementation of conservation plans and better protection of wild salmon.

Salmon evolved in a diverse and complex landscape, developing many adaptations to their local environments. Connectivity among diverse populations is necessary for the production and survival of salmon, their ability to adapt to change and to withstand harvest. The WSP is intended to safeguard salmon diversity so that that future generations of Canadians will be able to use and enjoy wild salmon.

Acknowledgments

We thank all the people who have provided input into the development of the ideas expressed in this paper, especially Pat Chamut, Blair Holtby, Brian Riddell, Mark Saunders, and Chris Wood.

References

Candy, J. R., J. R. Irvine, C. K. Parken, S. L. Lemke, R. E. Bailey, M. Wetklo, and K. Jonsen. 2002. A discussion paper on possible new stock groupings (Conservation Units) for Fraser River chinook salmon. DFO Canadian Science Advisory Secretariat Research Document 2002/085. Available http://www.dfo-mpo.gc.ca/csas/. (November 2005).

DFO (Fisheries and Oceans Canada). 1998. A new direction for Canada's Pacific salmon fisheries. Statement by Minister of Fisheries and Oceans, Canada, 14 October 1998 Available: http://www-comm.pac.dfo-mpo.gc.ca/publications/allocation/st9808e.htm. (November 2005).

DFO (Fisheries and Oceans Canada). 2000. Wild salmon policy discussion paper (draft). Available: http://www-comm.pac.dfo-mpo.gc.ca/pages/consultations/wsp-sep/wsp/wsp-paper_e.pdf. (November 2005).

DFO (Fisheries and Oceans Canada). 2005. Canada's policy for conservation of wild Pacific salmon. Available: http://www-comm.pac.dfo-mpo.gc.ca/publications/wsp/default_e.htm. (November 2005).

Gislason, G. S. and Associates Ltd. 2004. British Columbia seafood sector and tidal water recreational fishing: a strengths, weaknesses, opportunities, and threats assessment. Final Report prepared for BC Ministry of Agriculture, Food, and Fisheries, Victoria British Columbia. Available: http://www.agf.gov.bc.ca/fisheries/index.htm. (November 2005).

Glavin, T. 2002. History of the salmon fisheries of the Pacific Northwest coast. Pages 261–276 *in* K. D. Lynch, M. L. Jones, and W. W. Taylor, editors. Sustaining North American salmon: perspectives across regions and disciplines. American Fisheries Society, Bethesda, Maryland.

Government of Canada. 2004. Species at Risk Act Public Registry. Available: http://www.sararegistry.gc.ca/default_e.cfm. (November 2005).

Hilborn, R., T. P. Quinn, D. E. Schindler, and D. E. Rogers. 2003. Biocomplexity and fisheries sustainability. Proceedings of the National Academy of Sciences of the United States of America 100 (11):6564–6568.

Hyatt, K. D., and B. E. Riddell. 2000. The importance of stock conservation definitions to the concept of sustainable fisheries. Pages 51–62 *in* E. E. Knudsen, C. R. Steward, D. D. MacDonald, J. E. Williams, and D. W. Reiser, editors. Sustainable fisheries management: Pacific salmon. Lewis Publishers, Boca Raton, Florida.

Irvine, J. R. 2002. COSEWIC status report on the coho salmon *Oncorhynchus kisutch* (Interior Fraser population) in Canada, in COSEWIC assessment and status report on the coho salmon *Oncorhynchus kisutch* (Interior Fraser population) in Canada. Committee on the Status of Endangered Wildlife in Canada. Available: http://www.registrelep.gc.ca/virtual_sara/files/cosewic/sr%5Fcoho%5Fsalmon%5Fe%2Epdf. (November 2005).

Irvine, J. R., M. R. Gross, C. C. Wood, L. B. Holtby, N. D. Schubert, and P. G. Amiro. 2005. Canada's Species at Risk

Act: an opportunity to protect "endangered" salmon. Fisheries 30 (12):11–19.

Meggs, G. 1991. Salmon—the decline of the British Columbia fishery. Douglas and McIntyre, Vancouver, British Columbia.

NRC (National Resource Council). 1996. Upstream: salmon and society in the Pacific Northwest. Committee on Protection and Management of Pacific Northwest Anadromous Salmonids. National Academy Press, Washington, D.C.

Riddell, B. E. 1993. Spatial organization in Pacific salmon: what to conserve? Pages 23–41 *in* J. G. Cloud and G. H. Thorgaard, editors. Genetic conservation of salmonid fishes. Plenum Press, New York.

American Fisheries Society Symposium 49:399–411

Reconciling Fisheries with Conserving Biodiversity

Yvonne Sadovy*

University of Hong Kong, Department of Ecology & Biodiversity
Pok Fu Lam Road, Hong Kong, China

Abstract.—That commercial fishing can erode biological diversity both directly and indirectly, and at the genetic, population, species and ecosystem levels, is now widely accepted in scientific circles. This perspective, however, has yet to fully permeate a wider public consciousness or translate into much needed institutional changes. For stronger public and political support, it is essential to communicate how important are the practical, economic, nutritional, and social imperatives for maintaining and protecting biological diversity at all levels. Development and improvement of means and methods to identify species or systems at risk are valuable for assigning conservation priorities. However, vigilance is needed to avoid untested approaches in the headlong quest for solutions: the issues are complex, the solutions challenging. There is a real need to harmonize the various institutions and instruments charged with conservation and management to ensure a more effective treatment of exploited species that are threatened. Such work must embrace a greater compliance with codes and conventions that address fishery and biodiversity issues, and commitment from many different sectors of society, from consumers to governments.

Introduction

Among the most urgent of many challenges facing biologists, fisheries managers, and conservationists today, given growing economic pressures and globalization of market forces, is ensuring the preservation of biological diversity in fished marine systems. The challenge is considerable because the scientific, moral, and economic importance of meeting it is often not fully understood, may not be widely appreciated, or the appropriate institutional support may be deficient. Biodiversity refers to the sum of genetic, species, and ecosystem diversity, all of which are being eroded directly and indirectly by fishing. Fishing is probably the major factor in biodiversity losses in the marine environment (Dulvy et al. 2003). This is in marked contrast to the situation in birds, mammals, and plants, for which habitat degradation is the major cause of biodiversity loss. Hunting and collecting account for most other losses. For fishes in freshwaters, the greatest threats are habitat and water degradation and invasive species (Bruton 1995).

Since fishing is a key cause of biodiversity decline, the focus of this article, therefore, is on marine systems and commercial species. I refer mainly to fishes because the attention they attract has lagged seriously behind that paid to turtles or marine mammals and birds; we know relatively little about the conservation status of most marine invertebrates. However, I occasionally use examples from a wider range of organisms to illustrate specific points. I concentrate on three key issues: loss of marine diversity and the role of fishing in such decline; means of

*Corresponding author: yjsadovy@hku.hk

identifying vulnerable species; and approaches and mechanisms for stemming the losses.

Fishing Down Marine Biodiversity

Only within the last decade has a vigorous debate emerged on the extent to which marine species, and especially fishes, might be subject to extinction from exploitation. Recent reviews question the basis for long-standing assumptions (e.g., Hutchings 2001; Sadovy 2001) that, on both economic and biological bases, exploited marine fishes and invertebrates cannot go extinct, or are less likely to do so compared with most nonmarine organisms. The reviews conclude that assumptions about high natural resilience to extinction being associated with high fecundity and wide dispersal of pelagic eggs and larvae find little empirical basis. Also, density-dependant population growth, so long an assumed foundation to fishery science, does not invariably apply at low population levels (Hutchings 2000a, 2000b, 2001; Sadovy 2001). Indeed, it may be that our perception of the apparently incredible capacity of fish stocks to persist in the face of intense fishing pressure, or recover from much reduced population levels, comes from a single, economically important group, the clupeids. It turns out that this one group may not be representative of most exploited fishes (Dulvy et al. 2003).

Just as there is no biological reason for assuming exceptional resilience to declines in many exploited fishes, nor do economic safety factors necessarily protect reduced fish populations, as is often assumed. It has been argued, for example, in economic models, that commercial exploitation to extinction is not possible because, as individuals get harder to catch with population declines, fishing them is no longer viable well before they disappear completely. This economic "safety valve" may apply to certain large industrial-scale fisheries with high fleet costs. However, the growth in luxury international seafood trade, increasing demand for fish, and high prices paid for uncommon fishes in some markets, mean that such checks and balances may no longer be enough to protect vulnerable exploited species from severe declines (Clarke 1973; Sadovy and Vincent 2002; Frank et al. 2005).

The conservation status of most fishes is considerably less well understood than for other vertebrates. Fishes make up about half of all vertebrates, yet few marine species have been assessed for conservation status (Figure 1). Of those that have we know of local extinctions (sometimes referred to as extirpations), directly or indirectly attributable to fishing. Dulvy et al. (2003) reviewed the literature for evidence of global, regional and local extinctions of marine fishes and invertebrates, noting 133 cases. Of these, exploitation was probably a major causative factor in 55%, with 37% of cases associated with habitat effects. Furthermore, the authors concluded that marine extinctions may have been underestimated, given the difficulties of identifying extinctions in the sea.

Local or global species losses are only one aspect of biodiversity decline; more subtle, but also important, is reduction in genetic diversity and range contractions from fishing. Except in extreme circumstances, these are likely to occur before species become globally extinct and are important signs of change; local extinctions are the first step to wider scale losses, and ultimately, global extinctions. As just one example, significant range contractions were concluded to have resulted from fishing on six fish species studied off Canada (Shackell and Frank 2003). Loss of genetic diversity has emerged as a concern, with mounting evidence of genetic structuring among and within marine populations. These were once widely considered to be homogeneous in many

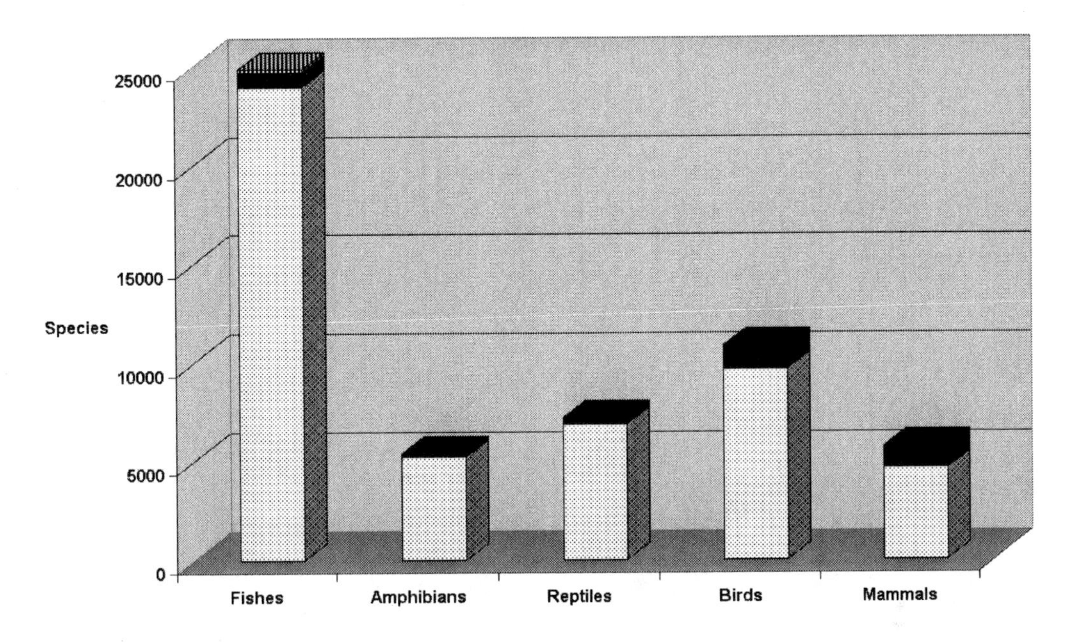

Figure 1. Of the total number of species in each of the five major vertebrate groups, indicated by the combined colors on each bar, fishes comprise about 50%. The dark shaded blocks on the top of each bar represent the number of species assessed as threatened (critically endangered, endangered or vulnerable) using IUCN (World Conservation Union) criteria. Threatened fish species are divided into marine (stripe) and freshwater (black bars). Of the five major groups, most species of mammals and birds, many reptiles, some amphibians, but very few fishes have been assessed for their conservation status (www.redlist.org).

species because of an assumed interconnectedness conferred by common water bodies and a dispersive egg or larval phase. Such characteristics could, but do not always, permit movements far from source populations or allow for recolonization of depleted populations. Losses of within-population genetic diversity is further exacerbated, as for organisms in many environments, in small or reduced breeding units by genetic drift and inbreeding. Marine examples are to be found among salmonids, sturgeons, turtles, and whales (Kenchington 2003).

Loss of genetic diversity is not only a conservation issue but can have important consequences for fisheries over the long term, because selective fishing can cause heritable differences in desirable yield-related life history traits (Smith et al. 1991; Conover and Munch 2002; Kenchington 2003). For example, removing large individuals can favor genotypes that become sexually mature at a smaller size or younger age than in unfished populations, resulting in overall lower economic value of catches because remaining fish attain smaller maximum size (e.g., Smith et al. 1991; Kenchington 2003). A growing number of examples demonstrate heritability of characteristics valued in commercially exploited species suggesting that, if mortality due to fishing greatly exceeds natural mortality, then fishing can be a strong selective force against such characters (Bohnsack and Ault 1996). Specific examples are few but some studies show clear selectivity effects. A controlled experiment on the guppy *Poecilia reticulata* over 11 years, showed that predator pressure influenced age and size of sexual maturity and that these have a genetic basis (Reznick et al. 1990). Over the long term, evolutionary changes can occur that reduce yield; work on the Atlantic silverside *Menidia menidia* clearly showed that harvested biomass evolved in the opposite direction to the intensity of size-dependent fishing mortality. Over four generations, this led to lower yields in experimental captive groups that had been selectively harvested for large-sized fish (Conover and Munch 2002).

Changes in species diversity as a result of fishing can also have profound effects at the level of ecosystem diversity and function, with possible long-term implications for fisheries. Although ecosystem level interactions are not yet well understood, evidence of fishery effects is accumulating. Differential losses of species from higher trophic levels can have knock-on effects along the food chain (Pauly et al. 1998; Myers and Worm 2003). Declines of large predators or loss of keystone species (i.e., those that can have a major influence on community structure), for example, may have profoundly changed coastal ecosystem structure and function with implications both for species diversity and long-term yields (e.g., McClanahan 1992, 1995; Jennings and Lock 1996; Jackson et al. 2001). Such changes can translate to negative impacts on the economic and nutritional value of the fishery, reduce options for future generations, threaten species diversity, and hinder recovery of overexploited stocks. There is, thus, a need to incorporate ecosystem-based concepts or perspectives to complement existing over-fishing paradigms that still tend to focus on individual species (e.g., Pauly et al. 1998; Murawski 2000).

In extreme cases, fishing "down the food web" could leave a fishery with irreversible changes or yielding little more than lower trophic level species (Pauly et al. 1998). Ecosystem shifts could hinder recovery in, or profoundly alter, some systems (e.g., Kaiser and Jennings 2002). For example, the removal of predatory benthic fishes by fishing led to unexpected replacements by cephalopods, crustaceans, sea combs, and jellyfishes, shifting groundfish to invertebrate catches in the North Atlantic (Caddy and Rodhouse 1998; Scheffer et al. 2001). In Norway, the valuable capelin *Mallotus villosus* fishery collapsed, leading to problems in coastal cod fisheries due to overexploitation of herring stock and a resulting imbalance of predator–prey relationships (Hamre 1994). An ecosystem once dominated by Atlantic cod *Gadus morhua* has undergone major changes and probable trophic cascade in the western Atlantic, which could be irreversible (Frank et al. 2005).

More recent work suggests that diversity itself may enhance ecosystem stability and related productivity. Removal of species can lead to pronounced changes in community composition and function, increasing the possibility of destabilization and collapses (Chapin et al. 2000). Indeed, experimental (nonfish) systems suggest that maintaining biodiversity may act as biological "insurance" against the loss of predictability or reliability (Naem and Li 1997). Conserving biodiversity is best considered the preferred, or default, goal, until its role in ecosystem stability, productivity, and recoverability is better understood: we rarely know a priori which species are critical to current ecosystem functioning or provide resilience and resistance to environmental change. It has even been suggested, for coral reef-associated multispecies fisheries, that fishing is best practiced by ensuring a balanced removal across species and size ranges rather than focusing heavily on a few valuable species. This approach could help to maintain biodiversity and ensure persistence of a productive fishery, while avoiding losses of functional groups of keystone species (Jennings and Polunin 1996; Halpern 2003; Bellwood et al. 2004; Kulbicki et al. 2004).

Identifying Vulnerable Species

Given that there are no grounds for assuming greater resilience to extinction for marine fishes than for other exploited vertebrates, and accepting how little we know of the biology and fishery of the great majority of exploited marine species, how can we efficiently and readily identify those that are particularly vulnerable and potentially threatened? The approach has been to establish guidelines, thresholds, and biological criteria to identify species actually or likely to be, at risk of extinction and determine readily measurable biological correlates of vulnerability.

Thresholds and Biological Criteria

Carefully derived thresholds and criteria can be used to indicate vulnerable species. The World Conservation Union (IUCN) categories and criteria are the oldest and best known globally and were originally developed to apply to all plants and animals. They are used here to illustrate the concepts of assessment thresholds and criteria because they are the most widely respected and applied and have probably been subjected to the greatest scrutiny. In the IUCN system, thresholds and criteria are used to assess the conservation status of species, subspecies, and so forth, at a local or global scale, in particular to focus attention on taxa threatened with extinction. This system is designed to determine the relative risk of extinction, based on a widely agreed set of criteria (below), which place species into various categories of conservation status. There are several "threatened" categories, Critically Endangered, Endangered, and Vulnerable, denoting a decreasing order of threat. For listing in these categories, there are five criteria. The criteria are quantitative and either refer to absolute thresholds or involve decline thresholds over specified time frames, for population size or geographic range. Also factored into assessments are prevailing conditions that could affect current and future threat; examples include whether a threat is increasing, is controlled, or is declining. As just two possible sets of conditions when population decline is considered, a listing of Critically Endangered would apply following

(1) "An observed, estimated, inferred or suspected population size reduction of ≥90% over the last 10 years or three generations, whichever is the longer, where the causes of the reduction are clearly reversible, and understood, and ceased;"

(2) "An observed, estimated, inferred or suspected population size reduction of ≥80% over the last 10 years or three generations, whichever is the longer, where the reduction or its causes may not have ceased, or may not be understood, may not be reversible" (www.redlist.org). The percentage applied is a "decline threshold." Assessed species are added to the IUCN Red List and those that are determined to be vulnerable are flagged for attention.

For more than 40 years, the IUCN criteria and categories were applied to a wide range of organisms, but it was not until 1996 that they were specifically used for marine fishes of commercial importance (Hudson and Mace 1996). Prior to the 1990s, the general view was that exploited marine fishes were unlikely to be threatened with extinction and that fishery management, rather than conservation action, could and should deal with population declines. However, following major fishery collapses and growing conservation concerns for a few exploited species, the IUCN criteria were applied to commercial fishes in a landmark workshop in 1996 (Hudson and Mace 1996; Mace and Hudson 1999).

The redlisting of several important commercial species at that workshop, such as southern southern bluefin tuna *Thunnus maccoyii* and Atlantic cod, led to vigorous debate among fishery scientists. Many still considered, and consider today, that exploited fish species differ from other vertebrates and are able to withstand and recover from much greater declines before they can be considered threatened. The result was the adoption of less conservative "decline threshold" criteria (i.e., a greater population decline in a given time period is required to trigger a given conservation category than for unexploited species) in several assessment systems, such as that of the American Fisheries Society, and a caveat for commercially exploited fishes in the IUCN criteria (Music 1999) The debate continues and will only be resolved as we learn more about the recovery potential of exploited populations in response to different levels of depletion.

Biological Correlates of Vulnerability

In addition to decline criteria (above) used to assign species to various levels of conservation concern or status, there are also biological correlates clearly associated with risk of extinction, high vulnerability to fishing and low recovery potential. For marine species, these include large maximum body size and attributes often closely associated in species that attain large size, such as longevity, delayed sexual maturation, low productivity, and ecological specialization (e.g., Jennings et al. 1998; Hawkins et al. 2000; Hutchings 2000a, 2000b; Sadovy 2001; Denney et al. 2002; Jones et al. 2002; Dulvy et al. 2003). Of these, large body size (or some correlated factor) is especially appropriate as a guideline to predict potential vulnerability because it is readily observable. Larger skates declined more rapidly than smaller species, and relative body size was a good predictor of population trend for 33 heavily fished species of parrotfishes, snappers, and groupers in Fiji, once fishing selectivity as a possible conflicting factor had been considered (Jennings et al. 1999; Dulvy and Reynolds 2002). Among the groupers, 37 species have been listed as threatened or at some level of concern, of 45 species assessed, according to three sets of criteria; most were larger species (Figure 2).

Other characteristics actually or potentially associated with extreme vulnerability to fishing in exploited species are high recruitment variability, hermaphroditism, and aggregation-spawning, Allee effects, and small range size. Species that are long-lived often have sporadic and infrequent recruitment, making them particularly susceptible to overfishing because a fishery can depend for many years on just one or two age classes (Warner and Chesson 1985; Russ et al. 1996; Roberts 1996). Hermaphroditism expressed as sex change might make species more vulnerable under size- (therefore sex) selective fishing, if fishing rate exceeds the capacity of the population to replace the removed sex by sex change or through earlier sexual maturation (Sadovy and Vincent 2002; Alonzo and Mangel 2004). Spawning in aggregations that are highly concentrated in time and space can also increase vulnerability if the aggregations are intensively exploited and service a large proportion of fish in a population (Coleman et al. 1999; Sala et al. 2003; Sadovy and Domeier 2005). Reduced reproductive capacity at very low densities due to mate availability or social facilitation issues (Allee or depensatory effects), may mean that severely reduced populations have particularly low recovery potential (e.g., Hutchings et al. 1999; Walters and Kitchell 2001; Gascoigne and Lipcius 2004). Geographic range size is sometimes negatively associated with risk of extinction, although wide-ranging species can also be vulnerable (Hawkins et al. 2000; Sadovy et al. 2003).

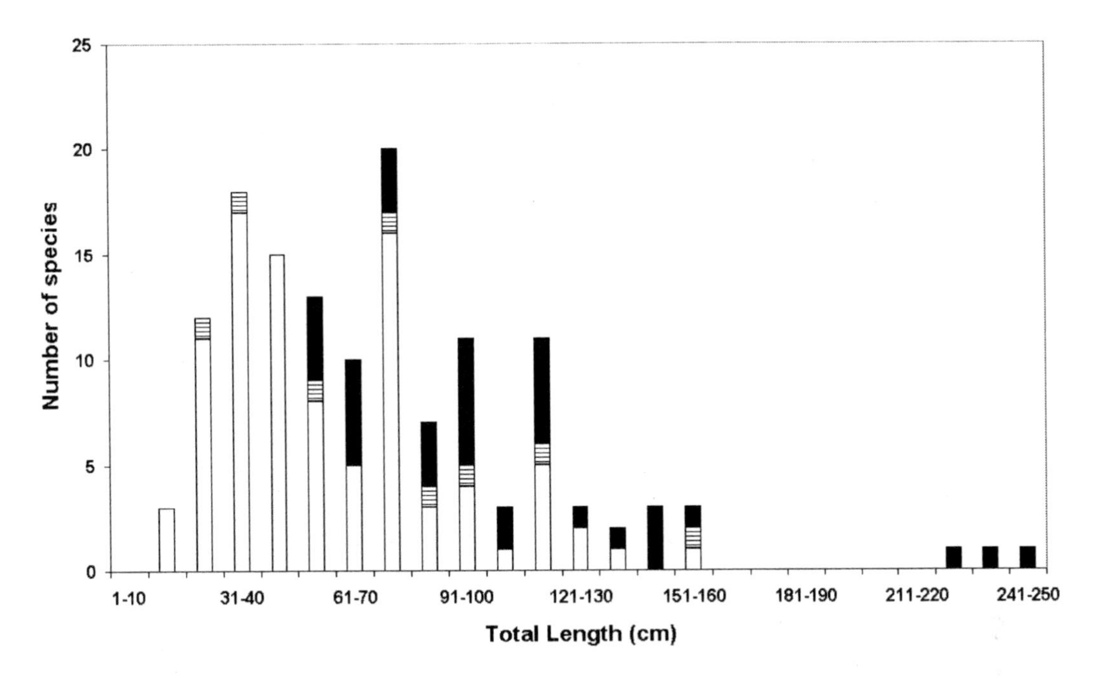

Figure 2. Size–frequency distribution of maximum known body lengths for 137 groupers (Serranidae). Sizes are the total lengths, in cm, for each species, according to Heemstra and Randall (1983); several species are not included as maximum length is not known. Forty-five species have been assessed as of 2003 according to established criteria and categories in different classsification systems. Of the 45 species, 37 were classified as either threatened (vulnerable, endangered, or critically endangered; N = 24) or near threatened/conservation dependent (N = 13) (black bars) and 8 at lower risk levels or that were assessed but found to be data deficient (striped bars) by end 2005. In general, larger species had classifications at higher levels of threat than smaller species; if different systems gave different assessments of threat, the assessment based on the most comprehensive dataset was selected for the graphic. Threat categories and criteria were based on IUCN (World Conservation Union), American Fisheries Society, and Australian Society for Fish Biology criteria (www.redlist.org; Music et al. 2000; Pogonoski et al. 2002). Nonassessed species are colored light gray; the only bias in those species selected for assessment was to focus on species for which there was already some conservation concern, not according to body size.

Reducing Negative Impacts on Biodiversity

Major threats to marine biodiversity are exploitation pressures, habitat degradation, institutional and information shortfalls, poor public awareness, and little political will for change. Given the massive scale of such problems, how can we most usefully assign and fund conservation priorities with severely limited funding and other resources? How do we reduce the negative impacts on biological diversity in exploited marine species and ecosystems when exploitation must continue? Possible solutions for species to ecosystem level protection include redirecting fishing pressure, generating alternative means of protein production, creating appropriately positioned protected areas, elucidating the causes of extreme declines and poor recovery, and developing institutional will to act and structural means of doing so.

Conservation and management at the level of individual species have long been mainstays of both conservation intervention and fishery science; the approach is still valid, although ecosystem level considerations are a relatively new development and challenge. Criteria and other guidelines (above) can be applied to identify threatened species, while fisheries managers use various thresholds to manage for sustainable use. One problem in fishery science is that there is little understanding of what happens once populations reach the low population levels associated with conservation concern and collapse (i.e. we understand little of the gray gap between unsustainable extraction and irrecoverable decline). Fishery assessments, which rely largely on mortality, growth rates, and size or age of sexual maturation, and how these interact with fishing pressure, do not factor in many of the subtleties of behavior or biology that can make certain species more or less vulnerable to overfishing (Sadovy and Vincent 2003). Clearly, there is a need to consider such biological correlates as hermaphroditism and density-independence to better understand the appropriate decline or fishery "thresholds" for exploited species that distinguish recoverable from irrevocable declines. Practicing conservation and fishery management at the ecosystem level, although not well understood, is widely discussed and increasingly a focus for funding, research, and action (Valdemarsen and Suuronen 2003).

Marine protected areas (MPAs), in the sense of "no-take" (i.e., no extractive use) areas, are widely advocated, not just for conservation but increasingly for their possible fishery benefits; they may prove to be one of the few examples of reconciling fisheries with conservation. Marine protected areas are particularly applicable to sedentary species and help to maintain refugia, ensure a broad size and age distribution, and protect critical (such as nursery or spawning) habitat (e.g., Bohnsack 1998; Polunin 2002). Proposed fishery benefits come from fish moving from protected to unprotected areas and larval export from fish in protected areas. Although evidence for significant or geographically widespread fishery benefits associated with MPAs is not strong (Russ 2002; Gell and Roberts 2003), income from tourism and increased catches from areas adjacent to MPA boundaries have fostered substantial public and community support for MPAs. Therefore, it is likely that, for certain fisheries or specific circumstances, MPAs will be an important conservation and management tool. In many situations, however, such as when fishing pressure is already high, or species are particularly mobile, MPA establishment is not enough, or is inappropriate. Other approaches, such as effort reduction, will be essential to avoid redistribution of fishing effort and to release fishing pressure on stocks (Walters and Bonfil 1999). Moreover, in most areas, MPAs are currently too small or insufficiently managed to be effective; in others they may not be socially acceptable or feasible (e.g., Hilborn et al. 2004; Sale et al. 2005). Despite such shortcomings, MPAs are one of the more important tools for conservation and management to emerge in recent years.

Greater awareness is needed of the potential economic, biological, and social risks of losing or degrading biodiversity to generate stronger public pressure, foster political support for legislation, encourage compliance with international codes and conventions, and increase funding for appropriate conservation goals (e.g., Turner et al. 2003; Toit et al. 2004). The long-term importance of keeping biodiversity options open for maintaining ecosystem function, stability, and productivity could be better disseminated and understood than it is at present. For the general public, despite a great increase in the number of articles over the last few years on fishery issues, the scale of the problem has not reached as far as it could and should. Personal experience with friends, family, students, and colleagues has shown me that many people are aware of problems but not necessarily why the problems exist nor the consequences. Importantly, many people are not aware that they, as consumers, could be a powerful force for change. Without this wider awareness, the important economic role of consumers cannot be effectively leveraged through consumer insistence for sustainable food sources. Better and broader awareness can be achieved in many ways. Examples include biologists making their work more accessible to the media and wider public by publishing and presenting it in a broad range of forums and formats. Many NGOs (nongovernmental organizations) are well-placed to conduct training and outreach activity, keep issues in the public domain, and push government departments to fulfill national and international mandates and agreements. For example, NGOs could do more to urge the implementation and practice of international conventions and codes (such as the Convention on Biological Diversity, or the FAO Code of Conduct for Responsible Fisheries). Finally, economic analyses of the short- and long-term consequences of poor fishing practices are few but invaluable for highlighting the costs to society of shortsightedness in fishery management (e.g., Cesar et al. 1997); if nothing else, governments respond to economic imperatives.

Key to advancing an agenda that jointly addresses biodiversity conservation and fisheries activities is harmonization of the various inter- and intragovernmental institutions and instruments that deal with species conservation and fisheries management. Historically, conservation and management were treated as two completely unrelated activities, nowhere more so than in marine exploited populations (with notable exceptions of turtles and whales). This is unsurprising given that only recently have certain commercially exploited marine fishes become widely acknowledged to be of conservation concern. Indeed, initial discussions on the theme of the Fourth World Fisheries Congress, "Reconciling Fisheries with Conservation," stoked a vigorous debate because, for many biologists and fishery scientists, fished species should rarely be of conservation concern. The long separation between "fish" and "wildlife" in some governments is symbolic of this perspective (C. Safina, Blue Ocean Institute, personal communication).

Perceptions are changing and, while much institutional change is needed to accommodate these shifts, some are already apparent. The Food and Agriculture Organization of the United Nations (FAO), for example, historically the global leader for fisheries development, has moved steadily towards sustainable use and precautionary practices. Increasingly, it is involved in discussions of species of conservation concern. On the other hand, the Convention on International Trade in Endangered Species of Wild Flora and Fauna (CITES), long strongly associated in the public mind with endangered species of high profile such as tigers and elephants and with trade bans, but rarely with exploited fishes, now includes, on its Appendix II, several commercially important fish species. This change is particularly poignant because it acknowledges the importance, long misunderstood or underplayed, of the Convention in fostering sustainable use in threatened or potentially threatened species that are traded internationally.

In the rush to address problems of declines and threats, there is a constant search for solutions that are sometimes applied or advocated before proven success or applicability; some may even be counterproductive. Governments in southeast Asia, for example, consider that the solution for conserving threatened marine species is through culturing and restocking despite the fact that most known threatened species cannot be sufficiently well-cultured at present, and notwithstanding a complete lack of proven success with restocking (e.g., Southeast Asian Fisheries Development Centre). In affluent conservation communities, one example of a hotly debated alternative approach for establishing marine conservation priorities is the recent focus on biodiversity "hotspots." The concept is that areas of high diversity in species number or where endemism is high merit conservation action largely because they bring most "bang for the buck" (i.e. protect biodiversity most cost-effectively). Appealing as the notion is, the extent to which species hotspots correlate with higher threat of extinction, or harbor more vulnerable species, compared to areas of lower diversity, and are thus a priority for conservation action, is not scientifically demonstrated. Moreover, whether the approach is applicable in a fishery context is far from clear; areas of lower species diversity may be more susceptible to overfishing (and possible threats associated with overfishing) than those of higher diversity because low biodiversity areas may less stable and more vulnerable to species losses (e.g., Bellwood et al. 2004). Designating areas to protect, based on species diversity alone, also ignores the complexities of the social and political issues that must be addressed for successful interventions. Hotspots are probably important for their ability to attract public and government attention and funding to conservation issues. For meaningful planning purposes, however, their value is far from clear and they may distract funds and attention from more urgent needs (Reid 1998).

We have much to improve in our ability to conserve exploited marine species and to document their status and history. Many fisheries are not monitored at all, or have no long history of data-gathering, making it difficult to gauge status changes over time for species likely to be vulnerable. Examples include significant, albeit nonconventional, fisheries that involve species such as certain sharks and reef fishes taken for food, and a range of species taken for other purposes such as the marine aquarium trade, teaching and science, bycatch species, etc. Approaches being applied to collate information more informally for use in management is the use of interviews to reconstruct fishery histories (Neis et al. 1999; Sadovy et al. 2003; Sadovy and Cheung 2003; Saenz-Arroyo et al. 2005), or the application of a precautionary, "data-lean" approach such as the use of precedent and traditional knowledge (Johannes 1998). Key biological questions that need addressing include how populations are connected or sub-structured, critically important for implementing measures such as MPAs and fishery management units. We need to better understand the spatial and temporal interactions of fishing with critical life history phases, or critical habitats, in such a way that could be useful for seeking ways to reduce the impacts of fishing. High on our set of priorities should be to fine-tune our criteria for assessing differing levels of threat to better understand the critical lower levels from below which populations might never recover. A supremely challenging range of nonbiologi-cal issues, such as reduction of fishing pressure, alternative protein sources, among others, also await fundamental advances in the way in which humans interact with their environment and use its resources wisely for food and livelihoods.

Summary

With expanding human populations and growing pressures on marine resources, biodiversity at all its levels is increasingly compromised. International trade through globalization is creating a bewildering range of new markets and casting ever wider and intensively for species to trade and novel sources of seafood. Means are needed to counter some of these pressures to ensure productive and intact marine ecosystems. Changes are needed at multiple levels in society. Consumers are part of the solution, and must be made more aware of where their fish come from and of their responsibility as consumers. Institutional changes must occur to accommodate our shifting perspectives of the vulnerabilities of exploited marine populations, and to find means of addressing the problems. Biologists and the conservation communities can play a powerful role by keeping issues topical and accessible, applying good science, and by pressuring governments to respect national mandates and abide by international codes and conventions. While some progress has occurred to address conservation issues in fisheries contexts, the challenge remains to act fast and effectively enough in a way that will avoid irreversible changes and compromise resource options for the future.

Acknowledgments

I am grateful to reviewers of the original submission for insightful and constructive comments. I also thank Bob Gillet for discussions on many fishery-related issues which have been very helpful.

References

Alonzo, S. H., and M. Mangel. 2004. The effects of size-selective fisheries on the stock dynamics of sperm limitation in sex-changing fish. Fishery Bulletin 102:1–13.

Bellwood, D. R., T. P. Hughes, C. Folke, and M. Nystrom. 2004. Confronting the coral reef crisis. Nature (London) 429:827–833.

Bohnsack, J. A. 1998. Application of marine reserves to reef fisheries management. Australian Journal of Ecology 23:298–304.

Bohnsack, J. A., and J. S. Ault. 1996. Management strategies to conserve marine biodiversity. Oceanography 9(1):73–82.

Bruton, M. N. 1995. Have fishes had their chips? The dilemma of threatened fishes. Environmental Biology of Fishes 43:1–27.

Caddy, J. F., and P. G. Rodhouse. 1998. Cephalopod and groundfish landings: evidence for ecological change in global fisheries. Reviews in Fish Biology and Fisheries 8:431–444.

Cesar, H., C. G. Lundin, S. Bettencourt, and J. Dixon. 1997. Indonesian coral reefs—an economic analysis of a precious but threatened resource. Ambio 26:345–350.

Chapin, F. S., E. S. Zavaleta, V. T. Eviner, R. L. Naylor, P. M. Vitousek, H. L. Reynolds, D. U. Hooper, S. Lavorel, O. E. Sala, S. E. Hobbie, M. C. Mack, and S. Diaz. 2000. Consequences of changing biodiversity. Nature (London) 405:234–242.

Clarke, C. W. 1973. The economics of overexploitation. Science 181:630–634.

Coleman, F. C., C. C. Koenig, A. M. Eklund, and C. B. Grimes. 1999 Management and conservation of temperate reef fishes in the grouper-snapper complex of the southeastern United States. Pages 233–248 *in* J. A. Musick, editor. Life in the slow lane: ecology and conservation of long-lived marine animals. American Fisheries Society, Symposium 23, Bethesda, Maryland.

Conover, D. O., and S. B. Munch. 2002. Sustaining fisheries yields over evolutionary time scales. Science 297:94–96.

Denney, N. H., S. Jennings, and J. D. Reynolds. 2002. Life-history correlates of maximum population growth rates in marine fishes. Proceedings of the Royal Society of London B 269:2229–2237.

Dulvy, N. K., and J. D. Reynolds. 2002. Predicting extinction vulnerability in skates. Conservation Biology 16:440–450.

Dulvy, N. K., Y. Sadovy, and J. D. Reynolds. 2003. Extinction vulnerability in marine populations. Fish and Fisheries 4:25–64.

Du Toit, J. T., B. H. Walker, and B. M. Campell. 2004. Conserving tropical nature: current challenges for ecologists. Trends in Ecology and Evolution 19(1):12–17.

Frank, K. T., B. Petrie, J. S. Choi, and W. C. Leggett. 2005. Trophic cascades in a formerly cod-dominated ecosystem. Science 308:1621–1623.

Gell, F. R., and C. M. Roberts. 2003. Benefits beyond boundaries: the fishery effects of marine reserves. Trends in Ecology and Evolution 18:448–454.

Gascoigne, J., and R. N. Lipcius. 2004. Allee effects in marine systems. Marine Ecology Progress Series 269:49–59.

Halpern, B. S. 2003. The impact of marine reserves: do reserves work and does reserve size matter? Ecological Applications 13:S117–137.

Hamre, J. 1994. Biodiversity and exploitation of the main fish stocks in the Norwegian-Barents Sea ecosystem. Biodiversity and Conservation 3(6):473–492.

Hawkins, J. P., C. M. Roberts, and V. Clark. 2000. The threatened status of restricted-range coral reef fish species. Animal Conservation 3:81–88.

Heemstra, P. C., and J. E. Randall. 1993. FAO species catalogue. volume 16. Groupers of the world. (Family Serranidae, Subfamily Epinephelinae). An annotated and illustrated catalogue of the grouper, rockcod, hind, coral grouper and lyretail species known to date. Food and Agriculture Organization of the United Nations, Rome.

Hilborn, R., K. Stokes, J. Maguire, T. Smith, L. W. Botsford, M. Mangel, J. Orensanz, A. Parma, J. Rice, J. Bell, K. L. Cochrane, S. Garcia, S. J. Hall, G. P. Kirkwood, K. Sainsbury, G. Stefansson, and C. Walters. 2004. When can marine reserves improve fisheries management? Ocean and Coastal Management 47:197–205.

Hudson, E., and G. Mace. 1996. Marine fish and the IUCN Red List of threatened fish species. Zoological Society of London, London.

Hutchings, J. A. 2000a. Collapse and recovery of marine fishes. Nature (London) 406:882–885.

Hutchings, J. A. 2000b. Numerical assessment in the front, ecology and evolution in the back seat: time to change drivers in fisheries and aquatic sciences? Marine Ecology Progress in Series 208:299–302.

Hutchings, J. A. 2001. Conservation biology of marine fishes: perceptions and caveats regarding assignment of extinction risk. Canadian Journal of Fisheries and Aquatic Sciences 58:108–121.

Hutchings, J. A., T. D. Bishop, and C. R. McGregor-Shaw. 1999. Spawning behaviour of Atlantic cod, *Gadus morhua*: evidence of mate competition and mate choice in a broadcast spawner Canadian Journal of fisheries and Aquatic Sciences 56:97–104.

Jackson et al. 2001. Historical over-fishing and the recent collapse of coastal ecosystems. Science 293:629–637.

Jennings, S., and J. M. Lock. 1996. Population and ecosystem effects of reef fishing. Pages 193–218 *in* N. V. C. Polunin and C. M. Roberts, editors. Reef fisheries. Chapman and Hall, London.

Jennings, S., and N. V. C. Polunin. 1996. Impacts of fishing on tropical reef ecosystems. Ambio 25:44–49.

Jennings, S., J. D. Reynolds, and S. C. Mills. 1998. Life history correlates or responses to fisheries exploitation. Proceedings of the Royal Society of London B 265:333–339.

Jennings, S., J. D. Reynolds, and N. V. C. Polunin. 1999. Predicting the vulnerability of tropical reef fishes to exploitation using phylogenies and life histories. Conservation Biology 13:1466–1475.

Johannes, J. E. 1998.The case for data-less marine resource management: examples from tropical nearshore finfisheries. Trends in Ecology and Evolution 13:243–246.

Jones, G. P., P. L. Munday, and M. J. Caley. 2002. Rarity in coral reef fish communities. Pages 81–101 *in* P. F. Sale, editor. Coral reef fishes: dynamics and diversity in a complex ecosystem. Academic Press, San Diego, California.

Kaiser, M. J. and S. Jennings. 2002. Ecosystem effects of fishing. Pages 342–366 *in* P. J. B. Hart and J. D. Reynolds, editors. Handbook of fish biology and fisheries. Blackwell Scientific Publications, Oxford, UK.

Kenchington, E. L. 2003. The effects of fishing on species and genetic diversity. Pages 235–253 *in* M. Sinclair and G. Valdimarrson, editors. Responsible fisheries in the marine ecosystem. Food and Agriculture Organization of the United Nations, Rome.

Kulbicki, M., P. Labrosse, and J. Ferraris. 2004. Basic principles underlying research projects on the links between the ecology and the uses of coral reef fishes in the Pacific. Pages 119–158 *in* L. E. Visser, editor. Challenging coasts. Transdisciplinary excursions into integrated coastal zone development. Amsterdam University Press, Amsterdam.

Mace, G. M., and E. J. Hudson. 1999. Attitudes towards sustainability and extinction. Conservation Biology 13:242–246.

McClanahan, T. R. 1992. Resource utilization, competition and predation: a model and example from coral reef grazers. Ecological Modelling 61:195–215.

McClanahan, T. R. 1995. A coral-reef ecosystem-fisheries model- impacts of fishing intensity and catch selection on reef structure and processes. Ecological Modelling 80:1–19.

Murawski, S. A. 2000. Definitions of over-fishing from an ecosystem perspective. ICES Journal of Marine Science 57:649–658.

Musick, J. A. 1999. Criteria to define extinction risk in marine fishes. Fisheries 24(12):6–14.

Musick, J. A., M. M. Harbin, S. A. Berkeley, G. H. Burgess, A. M. Eklund, L. Findley, R. G. Gilmore, J. T. Golden, D. S. Ha, G. R. Huntsman, J. C. McGovern, S. J. Parker, S. G. Poss, E. Sala, T. W. Schmidt, G. R. Sedberry, H. Weeks, and S. G. Wright. 2000. Marine, estuarine, and diadromous fish stocks at risk of extinction in North America (exclusive of Pacific salmonids). Fisheries: 25(11):6–30.

Myers R. A., and B. Worm. 2003. Rapid worldwide depletion of predatory fish communities. Nature (London) 423:280–283.

Naem, S., and S. B. Li. 1997. Biodiversity enhances ecosystem reliability. Nature (London) 390:507–509.

Neis, B., D. C. Scheider, F. Lawrence, R. L. Haedrich, J. Fischer, and F. A. Hutchings. 1999. Fisheries assessment: what can be learned from interviewing resource users? Canadian Journal of Fisheries and Aquatic Sciences 50:1949–1963.

Pauly D, V. Christensen, J. Dalsgaard, J. Froese, J., and F. Jr. Torres. 1998. Fishing down marine food webs. Science 279:860–863.

Pogonoski, J. J., D. A. Pollard, and J. R Paxton. 2002. Conservation overview and action plan for Australian threatened and potentially threatened marine and estuarine fishes. Environment Australia, Canberra.

Polunin, N. V. C. 2002. Marine protected areas, fish and fisheries. Pages 293–318 *in* P. J. B. Hart and J. D. Reynolds, editors. Handbook of fish biology and fisheries. Blackwell Scientific Publications, Oxford, UK.

Reid, W. V. 1998. Biodiversity hotspots. Trends in Ecology and Evolution 13(7):275–280.

Reznick, D. A., H. Bryga, and J. A. Endler. 1990. Experimentally induced life history evolution in natural population. Nature (London) 346:357–359.

Roberts, C. M. 1996. Settlement and beyond: population regulation and community structure of reef fishes. Pages 85–112 *in* N. V. C. Polunin, and C. M. Roberts, editors. Reef Fisheries. Chapman and Hall, London.

Roberts, C. M., and J. P. Hawkins. 1999. Extinction risk in the sea. Trends in Ecology and Evolution 14:241–246.

Russ, G. R., D. C. Lou, and B. P. Ferreira. 1996. Temporal tracking of a strong cohort in the population of a coral reef fish, the coral trout, *Plectropomus leopardus* (Serranidae: Epinephelinae), in the central Great Barrier Reef, Australia. Canadian Journal of Fisheris and Aquatic Science 53:2745–2751.

Russ, G. R. 2002. Yet another review of marine reserves as reef fisheries management tools. Pages 421–443 *in* P. F. Sale, editor. Coral reef fishes dynamics and diversity in a complex ecosystem. Academic Press, San Diego, California.

Sadovy, Y. 2001. The threat of fishing to highly fecund fishes. Journal of Fish Biology 59 (Supplement A):90–108.

Sadovy, Y. 2005. Trouble on the reef: the imperative for managing vulnerable and valuable fisheries. Fish and Fisheries 6:167–185.

Sadovy, Y. J., and A. C. J. Vincent. 2002. Ecological issues and the trades in live reef fishes. Pages 391–420 *in* P. F. Sale, editor. Coral reef fishes dynamics and diversity in a complex ecosystem. Academic Press, San Diego, California.

Sadovy, Y., and W. L. Cheung. 2003. Near extinction of a highly fecund fish: the one that nearly got away. Fish and Fisheries 4:86–89.

Sadovy, Y., M. Kulbicki, P.Labrosse, Y. Letourneur, P. Lokani, and T. J. Donaldson. 2003. The humphead wrasse, *Cheilinus undulatus*: synopsis of a threatened and poorly known giant coral reef fish. Reviews in Fish Biology and Fisheries 13:327–364.

Sadovy, Y., and M. Domeier. 2005. Are aggregation fisheries sustainable: reef fish fisheries as a case study. Coral Reefs: 254–262.

Saenz-Arroyo, A., C. M., Roberts, J. Torre, and M. Carino-Olvera. 2005. Using fishers' anecdotes, naturalists' observations and grey literature to reassess marine species at risk: the case of the Gulf grouper in the Gulf of California, Mexico. Fish and Fisheries 6:131–133.

Sala, E., O. Aburto-Oropeza, G. Paredes, and G. Thompson. 2003. Spawning aggregatons and reproductive behavior of reef fishes in the Gulf of California. Bulletin of Marine Science 72(1):103–121.

Sale, P. F., R. K Cowen, B. S. Danilowicz, G. P Jones, J. P Kritzer, K. C Lindeman, S. Planes, N. V. C Polunin, G. R. Russ, Y. J Sadovy, and R. S Steneck. 2005. Critical science gaps impede use of no-take fishery reserves. Trends in Ecology and Evolution 20(2):74–80.

Scheffer, M., S. Carpenter, J. A. Foley, C. Folke, and B. Walker. 2001. Catastrophic shifts in ecosystems. Nature (London) 413:591–596.

Shackell, N. L., and K. T. Frank. 2003. Marine fish diversity on the Scotian Shelf, Canada. Aquatic Conservation: Marine and Freshwater Ecosystems 13:305–321.

Smith, P. J., R. I. C. C. Francis, and M. McVeagh. 1991. Loss of genetic diversity due to fishing pressure. Fisheries Research 10:309–316.

Turner, R. K., J. Paavola, P. Cooper, S. Farber, V. Jessamy, and S. Georgiou. 2003. Valuing nature: lessons learned and future research directions. Ecological Economics 46:493–510.

Valdemarsen, J. W., and P. Suuronen. 2003. Modifying fishing gear to achieve ecosystem objectives. Pages 321–341 *in* M. Sinclair and G. Valdimarrson, editors. Responsible fisheries in the marine Ecosystem. Food and Agriculture Organization of the United Nations, Rome.

Walters, C., and J. F. Kitchell. 2001. Cultivation/depensation effects on juvenile survival and recruitment: implications for the theory of fishing. Canadian Journal of Fisheries and Aquatic Sciences 58:39–50.

Walters, C. J., and R. Bonfil. 1999. Multispecies spatial assessment models for the British Columbia groundfish trawl fishery. Canadian Journal Fisheries and Aquatic Science 56(4):601–628.

Warner, R. R., and P. L. Chesson. 1985. Coexistence mediated by recruitment fluctuations: a field guide to the storage effect. American Naturalist 125:769–787.

American Fisheries Society Symposium 49:413–424

Susceptibility in Exploited Marine Fishes and Invertebrates: Deconstructing Myths about Resilience to Extinction

ELODIE J. HUDSON*
IUCN Species Survival Commission, Division of Biology, Imperial College at Silwood Park Buckhurst Road, Ascot, Berkshire, SL5 7PY, United Kingdom

AMIE BRÄUTIGAM
IUCN Species Survival Commission, Perry Institute for Marine Science 3626 Warren Street, NW, Washington, D.C. 20036, USA

Introduction

Marine fishes and invertebrates are becoming increasingly vulnerable to depletion and extinction. Their vulnerability is indicated by widespread severe declines, local and global extinctions, and management failures, as well as by emerging science. Fisheries management has in many cases failed to account for the vulnerability of marine species due, in part, to underlying misperceptions about the resilience of marine species to severe depletion and extinction. These misperceptions arise because marine species are highly fecund, wide-ranging, and fast-growing, as well as able to withstand high levels of exploitation and recover rapidly from low numbers. In this paper, these misperceptions are explored and deconstructed in the light of the emerging science, which is improving our understanding of susceptibility in the marine environment.

Extinction, Depletion, and Susceptibility

The issue of extinction in the sea is contentious because it is so hard to measure. Nevertheless, there is evidence to suggest that many marine species are at risk from extinction (Roberts et al. 1998; Musick et al. 2000), and present rates of extinction in aquatic species are argued to be higher than background rates of extinction (Malakoff 1997; Roberts and Hawkins 1999).

All available well-documented examples of local extinctions (extirpations) and global extinctions in marine species have been recently compiled by Dulvy et al. (2003). There have been a total of 113 local and 18 global marine species extinctions, including 3 global and 62 local extinctions of fish. Exploitation is the most common cause of extinction in marine species (57%), followed by habitat loss (40%). Although the number of global extinctions is

*Corresponding author: elodie.hudson@imperial.ac.uk

small, there have been many local extinctions, particularly in fishes, the importance of which should not be underestimated.

In addition to local extinctions, there are countless examples of severe declines, especially in exploited marine fishes (FAO 2000). These severe declines are of great importance when considering the extinction risk of an exploited species. A lack of recovery from low numbers and low potential recovery rates from low numbers are both logical, defensible metrics of extinction risk. It is often assumed that fishing mortality is the primary factor inhibiting population recovery. However, this is not the case, as illustrated by the failure by Atlantic cod *Gadus morhua*, as well as many other commercially exploited species, to recover following its collapse (Hutchings 2001).

Identifying Susceptibility

A key step in preventing extinctions in marine species, and thus in protecting marine biodiversity, is identifying those populations or species that are most susceptible to extinction. Many factors affecting the susceptibility of marine species to severe depletion, extirpation or global extinction have been identified, based on theoretical understanding and empirical evidence (including Dye et al. 1994; Jones and Kaly 1995; Stone et al. 1996; McKinney 1997; Ponder and Grayson 1998; Roberts and Hawkins 1999).

Susceptibility factors can be split broadly into intrinsic factors (i.e., biological and life history characteristics inherent to a particular species or population) and extrinsic factors (i.e., exploitation or habitat changes that occur irrespective of the species' biology). A particular species' response to an extrinsic factor will depend upon the intrinsic factors affecting its susceptibility—hence, there are usually interactions between intrinsic and extrinsic factors. Despite this, the distinction is conceptually and practically useful in understanding susceptibility and informing conservation and management decisions. There is often high correlation between intrinsic factors, which can help simplify the description and identification of susceptible species.

Based on a comprehensive review of the literature, we have compiled a list of the most salient intrinsic and extrinsic factors linked to susceptibility in marine species (Table 1). As well as improving understanding of susceptibility, and thus deconstructing myths, these factors can be used as a simple tool for identifying susceptible species.

Intrinsic Biological Factors Affecting Susceptibility

The intrinsic biological factors used to describe species are often important determinants of population dynamics, and understanding them is central to understanding how different populations will respond to extrinsic factors, such as exploitation or habitat change, and how susceptible they are to depletion or extinction.

Population Growth Rate and Productivity

The population growth rate, often referred to as r (the intrinsic rate of increase) or productivity, is a key parameter in the dynamics of depletion and extinction and the assessment of susceptibility. Productivity is a complex function of fecundity, survival rates, age at maturity, and longevity and hence is a measure that embodies many life history characteristics. Both productivity and r determine a species' ability to recover from low numbers, if extrinsic factors are not limiting, as well as the level of harvest that can be taken from a population sustainably.

Table 1. A list of intrinsic (biological) and extrinsic (threat) factors that can be used to identify susceptibility to extinction or severe depletion in marine species.

Intrinsic factors
1. Low actual or potential population growth rate, indicated by any of the following: (a) Low fecundity (e.g., live bearing sharks, marine mammals, certain reef fish). N.B. the converse does not apply: high fecundity not associated with high recovery potential. (b) Late age at maturity (e.g., most sharks, some groupers) (c) Large size at maturity (d) Long generation time/longevity (e) Low natural mortality (f) Depensatory dynamics at low numbers (e.g., white abalone) (g) Low intrinsic rate of population increase, *r* 2. Ecological specialisation, for example: (a) spawning site specificity at a threatened site (b) sessile adults and/or poor dispersal (c) diadromy (where population is forced through a bottleneck of threatened habitat) (d) specialist feeding or breeding habitat 3. Range size, for example small or restricted range size
Extrinsic factors
1. Exploitation, for example: (a) overexploitation (b) severe depletion (a population can be depleted but no longer be exploited) (c) significant levels of illegal, unreported and unregulated fishing (d) significant levels of bycatch *Additional factors may exacerbate the effects of exploitation, for example*: (e) high market value (f) poor management 2. Habitat destruction, for example: (a) destructive fishing techniques (b) fragmentation (c) alteration (d) coral bleaching

The key life history characteristics that determine productivity are natural mortality, age and size at maturity, individual growth rate, fecundity, and maximum body size. These characteristics are related to one another (Beverton and Holt 1959; Pauly 1980) because in all species, there is a trade-off between the benefits of reproducing early and the costs to growth and survival of reproduction. These relationships can be used to estimate parameters that are typically hard to measure (such as natural mortality) from those that are easier to measure (such as growth rates and asymptotic sizes). These life history characteristics can also be used as surrogate measures of population growth rate (which is hard to estimate in marine species, particularly exploited ones) and can therefore be used as surrogate measures of susceptibility to extinction.

Theoretical studies have shown that stocks with higher productivity, as indexed by either high M (natural mortality), high K (individual growth rate), or high slope at the SRR (stock–recruit relationship) origin, can sustain higher harvest rates at lower relative biomass and that the harvest rate corresponding to the extinction threshold also increases with productivity (Mace 1994). The fishing mortality level needed to drive a population to extinction is greatest (and the population most resilient) for highly productive species (Punt 2000). In general, the faster the individual growth rate and the higher the level of natural mortality, the higher the potential yield is. However, Hudson (2002) has shown that populations with low natural mortality are sensitive to the shape of the stock–recruitment curve, whereas populations with high natural mortality are sensitive to variability in recruitment. These sensitivities are significant given the lack of known relationships between recruitment parameters and other life history parameters.

Empirically, there is little evidence of rapid recovery from prolonged declines in exploited fish stocks (Hutchings 2000). This contradicts the perception that marine fishes are highly resilient to large population reductions. Hutching's analysis of 90 stocks reveals that many gadids and other nonclupeids have experienced little, if any, recovery as much as 15 years after 45–99% reductions in reproductive biomass. Although the effects of exploitation may be reversible, the actual time required for recovery may be a lot longer than anticipated.

The steepness parameter, widely used in fisheries management, is linked to productivity. It is defined as the proportion of recruitment, relative to recruitment at equilibrium with no fishing, when the spawning stock biomass is reduced to 20% of the unexploited level. There is no empirical evidence of an association between steepness and natural mortality, growth rate, asymptotic length, and age at maturity (Hudson 2002), even though a theoretical link is expected (Mace 1994; Punt 2000). The findings of Hudson (2002) are consistent with those of Myers et al. (1999), who remark on the unexpected generalization that there is relatively little variation, within or between species, in maximum annual reproductive rate and, hence, in steepness. They find that almost all species have maximum annual reproductive rates of between 1 and 7, which are similar to terrestrial species (Myers et al. 1999).

Empirical work has also related life history parameters to population responses and these in turn to overexploitation (Jennings et al. 1999; Dulvy et al. 2000; Hudson 2002). Analysis of pairs of stocks in the northeast Atlantic has shown that those fishes that have decreased in abundance, compared with their nearest relatives, mature later, attain a larger maximum size, and exhibit significantly lower potential rates of population increase (Jennings et al. 1998). Body size has been implicated as a correlate of extinction susceptibility in a study of fishing in marine reserves (Russ and Alcala 1998). Dulvy et al. (2000) found that in species-specific surveys of a relatively stable skate fishery in the northeast Atlantic, the larger species had declined while two smaller species had shown increases in abundance. The increase in abundance in smaller species had masked the disappearance of the larger species in the catch information (which is usually aggregated), which poses a conservation problem for the larger, more vulnerable species.

High fecundity is an attribute often implicated to support the argument that marine species are resilient to extinction (Musick 1999; Powles et al. 2000). Highly fecund pelagic spawners (which include many commercially exploited species) often exploit rich plankton resources that fluctuate enormously in space and time. This variation in resources can also select for longevity, as species may undergo many years of low recruitment interspersed with high levels of recruitment when environmental conditions are favorable. Longevity may therefore be important to the persistence of fecund species through periods of poor recruitment, through the so-called storage effect (King et al. 2000; Hutchings 2001; Sadovy 2001; Dulvy et al. 2003). This prediction is supported by the dominance of single age cohorts for extended periods in both tropical and temperate species (Dulvy et al. 2003). It is for this reason that even highly fecund long-lived species are particularly vulnerable to overexploitation, through the erosion of older year-classes. Many long-lived fecund species have undergone severe declines due to overexploitation, including Atlantic cod, southern bluefin tuna *Thunnus macoyyi*, giant sea bass *Stereolepis gigas*, several species of rockfish *Sebastidae* spp., and the Queensland grouper *Epinephelus lanceolatus* (Dulvy et al. 2003). Fecundity per se is not an indicator of resilience, as supported by the observation that maximum rates of recruitment are fairly consistent across fish species of varying fecundities (Myers et al. 1999).

Marine species with naturally low fecundity, including most elasmobranchs, marine mammals, and seahorses, have reproductive rates that are less dependent on variable resources and, therefore show less population variability themselves, meaning that the capacity for rapid recovery from low numbers does not exist. In such cases, low fecundity is an important determinant of susceptibility to extinction.

One of the most important factors in assessing vulnerability is knowing what happens to populations at very low numbers. Depensation occurs when per-capita population growth rates are reduced at low numbers, so accelerating the path to extinction. Very little evidence for depensation has been found in exploited marine fish populations (Myers et al. 1995). Of 128 fish stocks studied, only 3 stocks showed significant depensation. The same data have also been analyzed using a hierarchic Bayesian meta-analysis approach (Liermann and Hilborn 1997), but again no strong evidence for depensation was found. However, the authors suggest that analysis of stock–recruitment data should incorporate spawner–recruit curves that include the possibility of depensation. In contrast, depensation will be particularly important for marine invertebrates, many of which are broadcast spawners whose fertilization success is highly dependent on population density Jennings (2001). Low population densities can lead to recruitment failure because excessive dilution of propagules greatly reduces the probability that the gametes will meet and be successfully fertilized. As well as the well-documented case of the white abalone (Davis et al. 1996), depensation has contributed to the extirpation of giant clams from several regions of the Indo-Pacific (Wells 1997; Roberts and Hawkins 1999).

Ecological Specialization

A specialization to a particular habitat can indicate susceptibility to extinction, if the habitat in question is threatened. This factor indicates vulnerability only if both the intrinsic (specialization) and extrinsic (threat to habitat) factors can be invoked. Habitat specificity is implicated in four recent extinctions of marine gastropods, including the eelgrass limpet *Lottia alveus* (restricted to sea grass; Carlton 1993). The diversity and abundance of reef fishes is strongly determined by habitat structure and reef organisms are thought to exhibit strong habitat specificity (Sale 1980; Friedlander and Parrish 1998).

Diadromy (migration between saltwater and freshwater), exhibited by many fishes, is considered a factor associated with increased risk because populations forced through habitat bottlenecks are often threatened. Many anadromous fishes (those that move from the sea to freshwater to spawn), such as salmon and sturgeon species, have become threatened because of disruption to migration routes and spawning sites, as well as direct exploitation (McDowall 1992). For example, the shortnose sturgeon *Acipenser brevirostrum* was listed on the U.S. Endangered Species Act as endangered following declines due to pollution and overfishing. In particular, construction of dams and pollution of many large northeastern U.S. river systems during the early 20th century have resulted in substantial habitat loss. Catadromous fishes (which migrate from freshwater to saltwater to spawn) are threatened for the same reasons (Musick et al. 2000; Billard and Lecointre 2001).

The complex reproductive strategies of territoriality, sequential hermaphroditism, and density-dependent mating systems displayed by many reef fishes make them difficult to model using fisheries assessment tools and therefore more liable to overexploitation, particularly when combined with relatively low reproductive output (Sadovy 1996; Vincent and Sadovy 1998). It has been suggested that sequential hermaphroditism (protandry in particular) is a factor indicating vulnerability to extirpation or extinction (Roberts et al. 1998), but there is currently no empirical evidence to support this assertion. The existence of spawning aggregations per se cannot be defended as a susceptibility factor unless an extrinsic threat is tightly coupled with the aggregating behavior.

Range, Distribution, and Abundance

The traditional view of marine species, both fishes and invertebrates, has been that broadcast spawners exist as large single populations in large homogenous habitats that contain no barriers to dispersal. The emerging view is that there is a finer spatial scale of population structure and habitat heterogeneity that restricts dispersal and population growth, and hence range, of many marine species (Ponder et al. 2002). Several studies have shown unexpectedly high levels of genetic structure over regional scales in some taxa with planktonic larvae, including solitary corals (Hellberg 1996), limpets (Johnson and Black 1984), and abalone (Huang et al. 2000). Many marine broadcast spawners exhibit relatively small geographical range sizes. From a sample of 1,677 species, 24% of coral reef fishes were restricted in distribution to less than 800,000 km^2 and 9% were restricted to less than 50,000 km^2 (Hawkins et al. 2000). Of marine invertebrate species, 11.3% of 1,063 hermatypic coral species, 41% of black corals, and 28% of 316 cone shell species were restricted to ranges of less than 800,000 km^2 (Hawkins et al. 2000).

Extrinsic Factors Affecting Susceptibility

Most documented marine extinctions (of populations or species) are the result of exploitation (56%) and habitat loss and degradation (36%), the two major extrinsic factors affecting marine species (Dulvy et al. 2003). In contrast, for birds, mammals, and plants, the main threat is habitat loss (87%), followed by exploitation (21%) and invasion of alien species (18%) (Hilton-Taylor 2000).

Exploitation

Exploitation is the most important threat factor for marine species. There are many examples of marine fishes and invertebrates substantially reduced by exploitation, many to the point of local or even global extinction (Vincent and Hall 1996). There are countless fish stocks that have undergone severe declines under heavy and unsustainable exploitation. The Food and Agriculture Organization of the

United Nations (FAO) estimates that 75% of the world's fish stocks are fully or overexploited (FAO 2000). The FAO analysis also shows a striking shift from undeveloped to senescent fisheries over the last 50 years. Specifically, 60% of the world's major fish resources are mature or senescent and therefore in need of urgent management and conservation attention (FAO 1997).

Stocks of Atlantic cod in the waters around Newfoundland famously collapsed to historically low levels in 1992 (Hutchings and Myers 1994), causing the fishery to close. Ten years later, there is still no significant recovery, either because of the low intrinsic rate of increase of a stock with a relatively slow life history (Myers et al. 1997) or because depensatory dynamics are operating at low stock sizes (Shelton and Healey 1999). The same is expected to happen to cod in the North Sea (Cook et al. 1997). Severe declines have also been observed in, among others, bluefin tuna *T. thynnus* and southern bluefin tuna (Polacheck et al. 1995; Safina 1993). The failure of fisheries management to prevent severe declines is a real conservation concern.

Two species of large ray have been described as near extinct over large parts of their original ranges: the common skate *Raia batis* from the Irish Sea (Brander 1981) and the barndoor skate *Dipturus laevis* from St. Pierre Bank, Sydney Bight and the southern Grand Banks (Casey and Myers 1998). There is some uncertainty as to how close to biological extinction these species are. However, there is no doubt that their low fecundity, late maturation, and susceptibility to death by trawling from an early age render them vulnerable to overexploitation, either directly or as bycatch.

Almost all fisheries are multispecies fisheries, by default or design. For unmarketable bycatch species, biological extinction may occur if reached before the point of economic extinction is reached in the target species. For bycatch species with market value, it may be economically viable to continue catching rare bycatch species so long as the fishery for the target species is still viable (Sadovy and Vincent 2002). These dynamics place some bycatch species, at least, at particular risk of extirpation and extinction. This has happened in the Irish sea where three species of skate and an angel shark have been extirpated, as a result of being taken incidentally in the trawl fishery for Atlantic cod, plaice *Pleuronectes platessa*, common sole *Solea solea* (also known as sole *Solea vulgaris*) and others (Dulvy et al. 2000). The future of two Gulf of California endemics both listed as critically endangered by the International Union for the Conservation of Nature (IUCN 1996), the vaquita *Phocoena sinus* and totoaba *Totoaba macdonaldi*, is singularly dependent on efforts to reduce incidental mortality in fishing operations.

Many fish species attract premium prices because of their high social or cultural value (Reynolds et al. 2002). For example, raw tuna or sashimi is highly coveted by rich restaurant-goers in Japan. The most valuable species is the southern bluefin tuna, which set a new price record in 2001 when US$178,000 was paid for one individual (Watts 2001). Bluefin tuna is almost as valuable, fetching US$83,500 for a single individual, which provided 2,400 servings of sushi, worth $180,000 (Reynolds et al. 2002). Another high-value fish is the Chinese bahaba *Bahaba taipingensis*, which has been exploited for its swim bladder in the South and East China seas from Shanghai to Hong Kong. In recent years, the swim bladder, or maw, has been called "soft gold" due to market prices of $20,000–$64,000 per kilogram. It is highly valued for its medicinal properties and despite the near-extinction of this species, 100–200 boats still target this species' spawning sites (Sadovy and Leung 2002). The humphead wrasse *Cheilinus undulatus* (also known as Napoleon or Maori wrasse) is the most valuable fish harvested for the Asian Live Reef Food Fish Trade, fetching around $130 per kg (Sadovy et al., unpublished data). As Southeast Asian sources have been decimated, fishing activity has spread to other areas as far away as the Seychelles, Fiji, and Kiribati, and the high market price ensures that it remains economically viable to export live fish by air to markets in China, Taiwan, and Singapore (Reynolds et al. 2002). These examples illustrate the intense pressure faced by valuable species, a situation analogous to that of rhinos, elephants, and other highly valued terrestrial species. The unremitting nature of commercial pressure means that these fishes might well be fished to extinction.

Threats to Habitat

Anthropogenic habitat loss can be caused by coastal reclamation and development, pollution, sediment loading, and nutrient input. Habitat can also be lost or damaged through exploitation. An example is the damage caused to sea beds by trawling (Dayton et al. 1995).

Habitat loss can have serious implications for marine species and has been cited as a cause of local extinctions in many cases. These include local extinctions of fifteen-spined stickleback *Spinachia spinachia*, Atlantic herring *Clupea harengus* and many species of algae all in the Wadden Sea, several species of cardinalfish, blackfin blenny *Paraclinus nigripinnis*, trunkfish *Lactophrys trigonus*, and fangtooth moray *Enchelycore anatina* in Bermuda. Global extinction of the rocky shore limpet *Collisella edmitchelli* has also been caused by habitat loss Dulvy et al. (2003).

Mangrove habitats are important because they provide juvenile habitat for a number of crustaceans and reed and lagoon fishes as well as stabilizing sediment. Mangroves habitats are being lost at a rate of at least 1% a year world-wide because they are cleared to support shrimp aquaculture and wood chip industries, as well as artisanal uses such as grazing fodder, medicine, and building material. Local losses of mangroves can be much higher (e.g. 16–32% in Thailand between 1979 and 1993 and 18% in Mexico over 16 years; Kovacs et al. 2001; Dulvy et al. 2003). It is not known whether the loss of mangrove habitat has been associated with any global extinctions of marine species, but species reliant on this threatened habitat must also be negatively affected.

Habitat degradation caused by exploitation is problematic in bottom-trawling fisheries, which are estimated to cover half the global continental shelf area (Dayton et al. 1995; Dulvy et al. 2003). The trawled area is estimated to be 150 times larger than the land area clear felled by forestry each year (Watling and Norse 1998). Benthic fishing gears such as trawls and mollusk dredges have altered benthic species composition, structural complexity, trophic structure, size structure, and the productivity of benthic communities (Jennings and Kaiser 1998; Kaiser et al. 1998; Jennings et al. 2001). The effect of fishing on habitat can be very patchy; areas can be fished up to eight times a year in the North Sea or 25–141 times a year in the Clyde estuary (UK) (Rijnsdorp et al. 1998).

Ongoing repeated coral bleaching events are causing unprecedented global scale loss of coral reef habitat (Glynn 1996; Wilkinson 2000). Coral bleaching is the whitening of diverse invertebrate taxa, which results from the loss of symbiotic zooxanthellae, a reduction in photosynthetic pigment concentrations in zooxanthellae residing within scleractinian corals, or both. Coral reef bleaching is caused by various anthropogenic and natural variations in the reef environment, including sea temperature, solar irradiance, sedimentation, xenobiotics, subaerial exposure, inorganic nutrients, freshwater dilution, and epizootics. Coral bleaching events have increased worldwide in both frequency and extent in the past 20 years. Global climate change may play a role in the increase of coral bleaching events and could cause the destruction of major reef tracts and the extinction of many coral reef species (Glynn 1996).

The scales and rates of habitat loss in the sea are comparable to those for terrestrial ecosystems. For example, between 30% and 60% of mangroves have been lost throughout large areas of Southeast Asia (Spalding 1998), around half of the world's salt marshes have disappeared (Agardy 1997), and roughly 10% of the world's coral reefs have been degraded beyond recovery (Jameson et al. 1995).

Conclusions

There is little evidence to support the assertion that marine species are resilient to severe depletion, extirpation, or extinction. On the contrary, there is plenty of evidence to suggest that many marine species are susceptible for a variety of reasons. While marine species are generally more fecund, the fecundity of broadcast spawners does not appear to correspond to higher maximum reproductive rates but reflects an adaptation to the variable environment in which these species live. While many marine species are believed to be widespread, in fact many are restricted by the heterogeneous nature of the marine environment and therefore may not have the capacity to rapidly colonize new areas. The long life, late maturity, and slow growth of many groups of marine species put them at an inherent disadvantage when subjected to exploitation. Exploitation is a truly powerful force in the decimation of marine populations and the multispecies nature of most fisheries, and the high value of others means that economic extinction will not necessarily protect species from biological extinction. Likewise, habitat loss is also an important factor, as the influence of human activity, directly or indirectly, encroaches on marine habitats.

References

Agardy, T. S. 1997. Marine protected areas and ocean conservation. Academic Press, Georgetown, Texas.

Beverton, R. J. H., and S. J. Holt. 1959 A review of the lifespans and mortality rates of fish in nature, and their relation to growth and other physiological characteristics. Pages 142–180 *in* G. E. Wolstenholme and M. O'Connor, editors. CIBA colloquia on ageing, volume 5. The lifespan of animals. J. and A. Churchill, Ltd., London.

Billard, R., and G. Lecointre. 2001. Biology and conservation of sturgeon and paddlefish. Reviews in Fish Biology and Fisheries 10:355–392.

Brander, K. 1981. Disappearance of common skate *Raia batis* from Irish Sea. Nature 290:48–49.

Carlton, J. T. 1993. Neoextinctions of marine invertebrates. American Zoologist 33:499–509.

Casey, J. M., and R. A.Myers. 1998. Near extinction of a large, widely distributed fish. Science 281:690–692.

Cook, R. M., A. Sinclair, and G. Stefansson. 1997. Potential collapse of North Sea cod stocks. Nature 385:521–522.

Davis, G. E., P. L. Haaker, and D. V. Richards. 1996. Status and trends of white abalone at the California channel islands. Transactions of the American Fisheries Society 125:42–48.

Dayton, P. K., S. F. Thrush, M. T. Agardy, and R. J. Hofman. 1995. Environmental effects of marine fishing. Aquatic Conservation—Marine and Freshwater Ecosystems 5:205–232.

Dulvy, N. K., J. D. Metcalfe, J. Glanville, M. G. Pawson, and J. D. Reynolds. 2000. Fishery stability, local extinctions, and shifts in community structure in skates. Conservation Biology 14:283–293.

Dulvy, N. K., Y. Sadovy, and J. D. Reynolds. 2003 Extinction vulnerability in marine populations. Fish and Fisheries 3.

Dye, A. H., D. M. Branch, J. C. Castilla, and B. A. Bennett. 1994 Biological options for the management of the exploitation of intertidal and subtidal resources. Pages

131–154 *in* W. R. Siegfried, editor. Rocky shores: exploitation in Chile and South Africa. Springer-Verlag, New York.

FAO (Food and Agriculture Organization of the United Nations). 1997 Review of the state of world fishery resources: marine fisheries. FAO, Rome.

FAO (Food and Agriculture Organization of the United Nations. 2000 The state of world fisheries and aquaculture. FAO, Rome.

Friedlander, A. M., and J. D. Parrish. 1998. Habitat characteristics affecting fish assemblages on a Hawaiian coral reef. Journal of Experimental Marine Biology and Ecology 224:1–30.

Glynn, P. W. 1996. Coral reef bleaching: facts, hypotheses and implications. Global Change Biology 2:495–509, Blackwell Publishing, Oxford, UK.

Hawkins, J. P., C. M. Roberts, and V. Clark. 2000 The threatened status of restricted-range coral reef fish species. Animal Conservation 3, Blackwell Publishing, Oxford. UK.

Hellberg, M. 1996. Dependence on gene flow on geographic distance in two solitary corals with different larval dispersal capabilities. Evolution 50:1167–1175.

Hilton-Taylor, C. 2000. IUCN red list of threatened species. International Union for the Conservation of Nature, Gland, Switzerland.

Huang, B. X., R. Peakall, and P. J. Hanna. 2000. Analysis of genetic structure of blacklip abalone (*Haliotis rubra*) populations using RAPD, minisatellite and microsatellite markers. Marine Biology 136:207–216.

Hudson, E. J. 2002 Conservation status assessment of exploited marine fishes. Doctoral dissertation. University of London, London.

Hutchings, J. A. 2000. Collapse and recovery of marine fishes. Nature 406:882–885.

Hutchings, J. A. 2001. Conservation biology of marine fishes: perceptions and caveats regarding assignment of extinction risk. Canadian Journal of Fisheries and Aquatic Sciences 58:108–121.

Hutchings, J. A., and R. A. Myers. 1994. What can be learned from the collapse of a renewable resource? Atlantic cod, *Gadus morhua*, of Newfoundland and Labrador. Canadian Journal of Fisheries and Aquatic Sciences 51:2126–2146.

IUCN (International Union for the Conservation of Nature). 1996. 1996 IUCN red list of threatened animals. IUCN, Gland, Switzerland.

Jameson, S. C., J. W. McManus, and M. D. Spalding. 1995 State of the reefs: regional and global perspectives. International Coral Reef Initiative Executive Secretariat, Background Paper, Washington D.C.

Jennings, S. 2001. Patterns and prediction of population recovery in marine reserves. Reviews in Fish Biology and Fisheries 10:209–231.

Jennings, S., T. A. Dinmore, D. E. Duplisea, K. J. Warr, and J. E. Lancaster. 2001. Trawling disturbance can modify benthic production processes. Journal of Animal Ecology 70:459–475.

Jennings, S., and M. J. Kaiser. 1998 The effects of fishing on marine ecosystems. Advances in Marine Biology 34.

Jennings, S., J. D. Reynolds, and S. C. Mills. 1998. Life history correlates of responses to fisheries exploitation. Proceedings of the Royal Society of London Series B 265:333–339.

Jennings, S., J. D. Reynolds, and N. V. C. Polunin. 1999. Predicting the vulnerability of tropical reef fishes to exploitation with phylogenies and life histories. Conservation Biology 13:1466–1475.

Johnson, M. S., and R. Black. 1984. Pattern beneath the chaos: the effect of recruitment on genetic patchiness in an intertidal limpet *Siphonaria jeanae.* Evolution 38:1371–1383.

Jones, G. P., and U. L. Kaly. 1995 Conservation of rare, threatened and endemic marine species in Australia. Pages 183–191 *in* L. P. Zann and P. Kailola, editors. State of the Marine Environment Report for Australia. Technical Annex I: the marine environment. Great Barrier Reef Park Authority for the Department of Enviornment, Sport and Territories. University of California Press, Berkley.

Kaiser, M. J., D. B. Edwards, P. J. Armstrong, K. Radford, N. E. L. Lough, R. P. Flatt, and H. D. Jones. 1998. Changes in megafaunal benthic communities in different habitats after trawling disturbance. ICES Journal of Marine Science 55:353–361.

King, J. R., G. A. McFarlane, and R. J. Beamish. 2000. Decadal-scale patterns in the relative year class success of sablefish (*Anoplopoma fimbria*). Fisheries Oceanography 9:62–70.

Kovacs, J. M., J. F. Wang, and M. Blanco-Correa. 2001. Mapping disturbances in a mangrove forest using multi-date landsat TM imagery. Environmental Management 27:763–776.

Liermann, M., and R. Hilborn. 1997. Depensation in fish stocks: a hierarchic Bayesian meta-analysis. Canadian Journal of Fisheries and Aquatic Sciences 54:1976–1984.

Mace, P. M. 1994. Relationships between common biological reference points used as thresholds and targets of fisheries management strategies. Canadian Journal of Fisheries and Aquatic Sciences 51:110–122.

Malakoff, D. 1997 Extinction on the high seas. Science 277:486–488.

McDowall, R. M. 1992. Particular problems for the conservation of diadromous fishes. Aquatic Conservation—Marine and Freshwater Ecosystems 2:351–355.

McKinney, M. L. 1997. Extinction vulnerability and selectivity: combining ecological and paleontological views. Annual Review of Ecology and Systematics 28:495–516.

Musick, J. A. 1999. Criteria to define extinction risk in marine fishes. The American Fisheries Society initiative. Fisheries 24(12):6–14.

Musick, J. A., M. M. Harbin, S. A. Berkeley, G. H. Burgess, A. M. Eklund, L. Findley, R. G. Gilmore, J. T. Golden, D. S. Ha, G. R. Huntsman, J. C. McGovern, S. J. Parker, S. G. Poss, E. Sala, T. W. Schmidt, G. R. Sedberry, H. Weeks, and S. G. Wright. 2000. Marine, estuarine, and diadromous fish stocks at risk of extinction in North America (exclusive of Pacific salmonids). Fisheries 25:6–30.

Myers, R. A., N. J. Barrowman, J. A. Hutchings, and A. A. Rosenberg. 1995. Population dynamics of exploited fish stocks at low population levels. Science 269:1106–1108.

Myers, R. A., K. G. Bowen, and N. J. Barrowman. 1999. The maximum reproductive rate of fish at low population sizes. Canadian Journal of Fisheries and Aquatic Sciences 56:2404–2419.

Myers, R. A., G. Mertz, and P. S. Fowlow. 1997 Maximum population growth rates and recovery times for Atlantic cod *Gadus morhua*. Fisheries Bulletin 95:762–772.

Pauly, D. 1980. On the interrelationships between natural mortality, growth parameters, and mean environmental temperature in 175 fish stocks. Journal du Conseil International pour l'Exploration de la Mer 39:175–192.

Polacheck, T., K. Sainsbury, and M. Klaer. 1995 Assessment of the status of the southern bluefin tuna stock using virtual population analysis. Convention for the Conservation of Southern Bluefin Tuna, Shimizu-shi, Japan.

Ponder, W., P. Hutchings, and R. Chapman. 2002 Overview of the conservation of Australian marine invertebrates. Report for Environment Australia, Australian Museum, Sydney.

Ponder, W. F., and J. E. Grayson. 1998 The Australian marine molluscs considered to be potentially vulnerable to the shell trade. Report for Environment Australia, Australian Museum, Sydney.

Powles, H., M. Bradford, R. Bradford, W. G. Doubleday, S. Innes, and C. Levings. 2000. Assessing and protecting endangered marine species. ICES Journal of Marine Science 57:669–676.

Punt, A. E. 2000. Extinction of marine renewable resources: a demographic analysis. Population Ecology 42:19–27.

Reynolds, J. D., N. K. Dulvy, and C. M. Roberts. 2002 Exploitation and other threats to fish conservation. Pages 319–341 *in* P. J. B. Hart and J. D. Reynolds. Handbook of fish biology and fisheries, volume 2. Blackwell Publishing, Oxford, UK.

Rijnsdorp, A. D., A. M. Buijs, F. Storbeck, and E. Visser. 1998 Micro-scale distribution of beam trawl effort in the southern North Sea between 1993–1996 *in* relation to the trawling frequency of the sea bed and the impact on benthic organisms. ICES Journal of Marine Science 55: 403–419.

Roberts, C. M., and J. P. Hawkins. 1999. Extinction risk in the sea. Trends in Ecology and Evolution 14:241–246.

Roberts, C. M., J. P. Hawkins, N. Chapman, V. Clarke, A. V. Morris, R. Miller, and A. Richards. 1998. The threatened status of marine species. World Conservation Union (IUCN) Species Survival Commission, and Center for Marine Conservation, Washington, D.C.

Russ, G. R., and A. C. Alcala. 1998. Natural fishing experiments in marine reserves 1983–1993: roles of life history and fishing intensity in family responses. Coral Reefs 17:399–416.

Sadovy, Y. 1996. Reproduction of reef fishery species. Pages 15–59 *in* N. V. C. Polunin and C. M. Roberts, editors. Reef fisheries. Chapman and Hall, London.

Sadovy, Y. 2001 The threat of fishing to highly fecund fishes. Journal of Fish Biology 59A.

Sadovy, Y., and C. W. Leung. 2002. Near extinction of a highly fecund fish: trouble among the croakers. Fish and Fisheries 3:1–14.

Sadovy, Y., and A. J. Vincent. 2002 Ecological issues and the trades in live reef fishes. Pages 391–420 *in* P. F. Sale, editor. Coral reef fishes: dynamics and diversity in a complex ecosystem. Academic Press, San Diego, California.

Safina, C. 1993. Bluefin tuna in the west Atlantic: negligent management and the making of an endangered species. Conservation Biology 7:229–234.

Sale, P. F. 1980. The ecology of fishes on coral reefs. Oceanography and Marine Biology Annual Reviews 18:367–421.

Shelton, P. A., and B. P. Healey. 1999. Should depensation be dismissed as a possible explanation for the lack of recovery of the northern cod (*Gadus morhua*) stock. Canadian Journal of Fisheries and Aquatic Sciences 56:1521–1524.

Spalding, M. D. 1998 Patterns of biodiversity in coral reefs and mangrove forests: global and local scales.

Doctoral dissertation. University of Cambridge, Cambridge, UK.

Stone, L., E. Eilam, A. Abelson, and M. Ilan. 1996. Modelling coral reed biodiversity and habitat destruction. Marine Ecology Progress Series 134:299–302.

Vincent, A. C. J., and H. J. Hall. 1996. The threatened status of marine fishes. Trends in Ecology and Evolution 11.

Vincent, A. C. J., and Y. Sadovy. 1998 Reproductive ecology in the conservation and management of fishes. Pages 209–245 *in* T. Caro, editor. Behavioural ecology and conservation biology. Oxford University Press, Oxford, UK.

Watling, L., and E. A. Norse. 1998. Disturbance of the seabed by mobile fishing gear: a comparison to forest clearcutting. Conservation Biology 12:1180–1197.

Watts, J. 2001. Loadsa tunny- £570/kg price record. The Guardian (London) 1 June:16.

Wells, S. M. 1997. Giant clams: status, trades and mariculture, and the role of CITES in management. International Union for the Conservation of Nature, Gland, Switzerland.

Wilkinson, C., editor. 2000. Status of coral reefs of the world: 2000. Australian Institute for Marine Science, Townsville.

Session

Jurisdictional Equity

American Fisheries Society Symposium 49:427–441

Addressing Illegal, Unreported, and Unregulated Fishing in the Philippines

MARY ANN PALMA*

The Australian National Centre for Ocean Resources and Security (ANCORS) (formerly the Centre for Maritime Policy), University of Wollongong New South Wales 2522, Australia

Abstract.—The problem of illegal, unreported, and unregulated (IUU) fishing is an increasing concern in the Philippines, resulting in an economic loss of about US$1 billion annually. Illegal fishing in the country includes unsustainable fishing practices, poaching by foreign vessels, and transshipment of illegally caught fish. Unregulated fishing is composed of unmanaged fisheries within its jurisdiction and unregulated fishing of unregistered Filipino vessels in areas under the jurisdiction of other states. While catches resulting from these activities are unreported, there are instances when catches are deliberately misreported. IUU fishing has negative impacts, not only on the sustainability of fisheries resources in the Philippines, but also on its marine biodiversity and environment.

Although the Philippines has yet to implement the International Plan of Action to Prevent, Deter, and Eliminate Illegal, Unreported, and Unregulated Fishing, it has adopted some of the most important flag and coastal state measures in its fisheries laws and regulations to address IUU fishing problems. These measures include the implementation of international fisheries-related instruments, registration of fishing vessels, maintenance of fishing records, issuance of authorizations to fish, application of sanctions, implementation of a monitoring, control, and surveillance (MCS) system, and cooperation with other states. However, IUU fishing in the Philippines continues to proliferate due to the inadequacy of the fisheries legislation, lack of enforcement of fisheries regulations, and lack of capacity to implement an effective MCS system.

The Philippines should therefore review and strengthen its flag state and coastal state responsibilities and adopt applicable port state measures to effectively address IUU fishing concerns. Such measures can be introduced either as amendment to its fisheries law and regulations or adopted in a national plan of action to combat IUU fishing in the Philippines.

Introduction

The problem of illegal, unreported, and unregulated (IUU) fishing has been identified by the secretary-general of the United Nations in his report to the general assembly on oceans and the law of the sea in 1999 as "one of the most severe problems affecting world fisheries" (UNGA 1999, paragraph 249). It is recognized that IUU fishing undermines the conservation and management measures implemented by coastal states and regional fisheries management organizations (RFMOs). In the Philippines, the economic losses resulting from IUU fishing is estimated at US$1 billion a year (Aliño 2002). Illegal fishing activities in the country include unsustainable fishing practices, unauthorized fishing activities

*Corresponding author: mpalma@uow.edu.au

of foreign fishing vessels, and transshipment of illegally caught fish. The Philippines is also faced with problems of unmanaged fisheries in its exclusive economic zone (EEZ) and unregulated fishing of unregistered Filipino vessels in areas under the jurisdiction of other states. Although catches resulting from illegal and unregulated fishing activities are unreported, there are also instances when fishers avoid their reporting obligations by either misreporting or underreporting their catches.

To address these problems, the International Plan of Action to Prevent, Deter and Eliminate Illegal, Unreported, and Unregulated Fishing (IPOA-IUU) was formulated in 2001. It is a voluntary instrument that was elaborated within the framework of the Code of Conduct for Responsible Fisheries and anchored on relevant rules of international law (FAO 2001, paragraphs 4–5; FAO 1995, article 1.1). The IPOA-IUU provides measures that may be used by all states, flag states, port states, coastal states, "market states," or states that engage in the international trade of fish and RFMOs to address all aspects of IUU fishing. Such measures are diverse and may be used by states, individually or in collaboration with other states, in addressing the problem. Because of this, the IPOA-IUU is referred to as a comprehensive toolbox (FAO 2002) that has a full range of measures that can be used to deal with various manifestations of IUU fishing. A state should then be able to find an appropriate tool or a combination of tools in the IPOA-IUU to reduce any incident of IUU fishing.

This article examines the adequacy of the current Philippine legal policy and regulatory framework in addressing IUU fishing against the flag state, coastal state, and port state measures provided under the IPOA-IUU. It focuses particularly on the problems of unsustainable fishing practices, poaching, and unregulated fishing of Filipino vessels outside the jurisdiction the Philippines. The paper also recommends specific measures that the Philippines may adopt to improve its legal and regulatory framework in order to effectively address IUU fishing challenges.

The IUU Fishing Terminology

Illegal, unreported, and unregulated fishing is a new terminology in fisheries management. However the concepts of illegal fishing, unreported fishing, and unregulated fishing are not new. The term IUU fishing was first used in a session of the Convention for the Conservation of Antarctic Marine Living Resources and has subsequently found its way in the meeting reports of the Food and Agriculture Organization (FAO) of the United Nations and other international and regional fisheries bodies (Doulman 2000). According to these organizations, some of the most pressing global IUU fishing concerns include

- Unauthorized fishing activities of vessels flying the flag of a state in an area of jurisdiction of another state (Bray 2000);

- Fishing of member states of an RFMO in the management area of that organization, without regard to agreed conservation and management measures (Bray 2000);

- Unauthorized fishing of vessels of states or entities that are not members of an RFMO in an area under the competence of that organization (Bray 2000); UNGA 1998, paragraphs 135 and 139; UNGA 1999, paragraph 250; UNGA 2000, paragraphs 154 and 158);

- Illegal fishing activities of vessels formerly registered in a state member of an RFMO but were subsequently reflagged in a nonmember state (UNGA 1999, paragraph 254);

• Unregulated fishing of vessels flying the flags of open register states (Bray 2000; UNGA 1999, paragraph 250; UNGA 2000, paragraph 151);

• Uncontrolled fishing of distant water fishing fleets (UNGA 1998, paragraphs 105–106); and

• Unreported catches of vessels conducting illegal and unregulated fishing (Bray 2000; UNGA 1996, paragraph 63; UNGA 1997, paragraph 78; UNGA 1998, paragraph 94).

Other factors that contribute to the proliferation of IUU fishing include the existence of overcapacity in the fishing industry, open access of high seas fisheries, lack of effective flag and port state control, lack of implementation of international fisheries management instruments, lack of cooperation among states, and lack of operational capacity of coastal states (UNGA 1999, paragraph 249).

The formulation of the IUU fishing terminology under the IPOA-IUU led to a more formal and collective classification of some of the longstanding issues confronting fisheries management. The term IUU, as defined in paragraph 3 of the IPOA-IUU, encompasses activities of all fishing vessels, with or without nationality, and conducted within a state's jurisdiction, in areas under the management of an RFMO and on the high seas. It also covers a wide range of fishing activities that can be considered illicit or in violation of, or without regard to, applicable international or national fisheries rules and standards.

Unreported fishing refers to nonreporting, misreporting, or underreporting of catches. This comes with or without the intent of providing an accurate report of catches to the proper authorities (Edeson 2000). It may also refer to reporting of catches, but not in accordance to the reporting procedures of states or RFMOs.

Illegal fishing is the most common aspect of the terminology. In areas under national jurisdiction and under the competence of RFMOs, illegal fishing can be in the form of fishing without license or authorization to fish (LOSC 1982, article 62[4][a]; FAO 1995, paragraph 8.2.2; UN 1995, articles 18[3][b] and 21[11][a]; FAO 1993, article III[2]) or conduct of fishing activities contrary to what is provided in a valid authorization to fish. The latter can include the use of explosives and poisons, small-meshed fishing nets, highly destructive fishing gears, methods and techniques, and traps and weirs, as well as the willful destruction of corals in reef fisheries and catching of juvenile or immature fish (FAO 1995, paragraph 8.4.2; United States Conference on Environment and Development 1992, paragraph 17.53). Illegal fishing may also include fishing beyond the catch limit, taking of prohibited fish species, and fishing during a closed season, among others. In the high seas, illegal fishing generally take the form of fishing contrary to high seas fisheries conservation and management measures adopted by RFMOs according to international law.

According to the IPOA-IUU, unregulated fishing in areas under national jurisdiction does not only refer to the absence of a management regime for a particular fishing area or fish stock. It also means that while a management regime exists, much of the fishing activity is not controlled or regulated (Edeson 2000). The IPOA-IUU notes that certain unregulated fishing may occur in a manner that is not in violation of applicable international law and may not require the application of measures formulated under the instrument (FAO 2001, paragraph 3.4). For example, a fisher who has conducted an unregulated activity merely because the relevant state or states have not adopted any regulatory measures for that particular activity may not be construed as engaging in an illicit act.

The IPOA-IUU provides several measures that states could adopt in order to effectively address the different forms of IUU fishing. All states have the primary responsibility to fully and effectively implement international fisheries instruments (FAO 2001, paragraphs 10–15) and formulate a national plan of action (NPOA) to combat IUU fishing (FAO 2001, paragraph 25). The IPOA-IUU also calls on states to incorporate the following measures in their national legislation: control over a state's nationals; control over vessels without nationality; application of sanctions for IUU fishing activities; reduction or elimination of economic incentives to companies or individuals conducting IUU fishing; and effective monitoring, control, and surveillance (MCS) (FAO 2001, paragraphs 16–24). States also have the responsibility under the IPOA-IUU to adopt internationally agreed market related measures to address fish trade problems associated with IUU fishing (FAO 2001, paragraphs 65–76).

The IPOA-IUU calls on flag states to exercise effective control over fishing vessels flying their flags. It enumerates the responsibilities of flag states with respect to the registration of fishing vessels, maintenance of record of fishing vessels, issuance of authorization to fish, transshipment and other fishing support activities, and information dissemination on fish catch and transport (FAO 2001, paragraphs 34–50). Port states also have the responsibility to establish measures that will prevent IUU fishing in relation to port access and landing of catches (FAO 2001, paragraphs 52–64). Last, coastal states should cooperate with other states, ensure institutional and policy strengthening, formulate measures to properly identify vessels engaged in IUU fishing, and raise public awareness of such issues (FAO 2001, paragraph 51).

Fisheries Profile of the Philippines

The Philippines has a total marine area of 2.2 million square kilometers with a total coastline length of 17,460 km (FAO 2003). Fish provide approximately 50% of the animal protein in the Philippines (CRMP et al. 1998). The country is an important producer of fish, being the 12th among the top fish producing countries in the world for 1999 (BFAR 1997–2001). Fisheries resources also contribute significantly to the economy of the Philippines. In 1998, the fishing industry generated an employment of approximately 1 million people (BFAR 1997–2001). Trade in fish for the country has likewise resulted in a surplus of US$383.1 million in 2001 (BFAR 1997–2001).

The fisheries industry of the Philippines is composed of three sectors, namely (a) municipal fisheries (i.e., fishing in inland or coastal waters within 15 km from the coastline) using vessels not greater than 3 gross registered tons (GRT), as well as fishing without the use of vessels; (b) commercial fisheries or fishing with passive or active gear utilizing fishing vessels of greater than 3 GRT; and (c) aquaculture, or fishery operations involving all forms of raising and culturing fish and fishery species in fresh, brackish, and marine water areas. The total fish production of these three sectors has leveled off at around 2.7 million tons per year since 1990 (BAS, Fisheries Statistics Division 2001–2002).

Among these sectors, municipal fisheries are the only ones that show a declining trend of production since 1992 (FAO 2000). However, the number of municipal fishing boats has significantly increased from 1980 to 2000 (BFAR 1997–2001), resulting in increased pressure on the resources. The 10 major municipal fish catch are the frigate mackerel *Auxis thazard*, yellowfin tuna *Thunnus albacares*, fringescale sardinella (also known as fimbriated herring) *Sardinella fimbriata*, anchovies, bigeye scad *Selar crumenophthalmus*, Indian sardines *Sardinella indicus*, blue swimmer crab *Portunus pelagicus*, squid *Loligo* spp., round scad *Decapterus punctatus*, and skipjack tuna *Katsuwonus pelamis* (BFAR 1997–2001).

The commercial fish catch has considerably increased over the years (BFAR 1997–2001). The commercial fishing industry was given direct and indirect subsidies, tax breaks, and rebates on fuel oil tax, which improved the capacity of vessels to travel farther offshore (Green et al. 2003). As a result, fish catch in international waters has increased to 3.18% of the total commercial fisheries production (BFAR 1997–2001). The number of commercial fishing vessels has also increased from 3,416 in 1998 to 3,601 vessels in 1999, although the gross tonnage has decreased from 299,885.82 to 270,261 in the same years (BFAR 1997–2001). Most of these fishing vessels use purse seines. The number of purse seiners varies from 30 to 80 units, which are within 100–200 gross tons, depending on the season (PCAMRD 1993). Only around 17 Philippine tuna purse seiners are operating in international waters, around the vicinity of Papua New Guinea (PCAMRD 1993). Commercial fishers generally catch the same species as those of the municipal fishers, except that their major catches are the round scad and Indian sardines (BFAR 1997–2001).

Illegal, Unreported, and Unregulated Fishing in the Philippines

Illegal, unreported, and unregulated fishing is a major challenge in the proper conservation and management of fisheries resources in the Philippines. Illegal fishing in the Philippines includes unsustainable fishing practices, poaching by foreign vessels, and transshipment of illegally caught fish. Unregulated fishing is composed of fishing activities of unregistered Filipino vessels in areas under the jurisdiction of other states. Catches derived from these illegal and unregulated fishing activities are most often unreported. There are also occasions when Filipino fishers deliberately misreport their catches.

Unsustainable Fishing Practices

Recorded illegal fishing incidents in the Philippines average 550 incidents a month, 81% of which come from blastfishing, while the rest include the use of destructive fishing methods such as trawling, use of fine meshed nets, purse seine, cyanide poisoning, and *muro ami*[1] (Aliño 2002). Areas in the Philippines that are known to have extensive illegal fishing activities include Puerto Princesa, Palawan; Batanes; Tacloban City; Cebu City; Zam-boanga City; San Fernando, La Union; and Casiguran Bay, Quezon (PCG 2001). Most of these areas are rich fishing grounds with ineffective fisheries enforcement. It is estimated that the Philippines loses 50 billion Philippine pesos, or almost US$1 billion dollars, annually due to illegal fishing activities (Aliño 2002). However, the accuracy of this information is uncertain. The estimated economic loss based on the recorded incidents of such activities only account for the value of the seized fish. It does not necessarily reflect the actual loss that may

[1] Department of Agriculture Fisheries Administrative Order No. 203 (Republic of the Philippines 2000b), Series of 2000, defines *muro-ami* as a Japanese fishing method used in reef fishing, consisting of a movable bagnet, detachable wings, and scarelines having plastic strips and iron, steel, and stone weights, affecting fish capture by spreading the net in an arc around reefs or shoals, and with the use of the scarelines, a cordon of fishermen drive the fish towards the waiting net while pounding the corals by means of heavy weights like iron, steel, stone, or rock, making it destructive to corals.

result from the impact of illegal fishing on fish habitat, marine environment, and social dislocation of marginal fishers.

Unauthorized Fishing of Foreign Vessels

The Philippines also suffers heavy economic losses due to repeated and continuous poaching of foreign vessels in areas under its national jurisdiction. Foreign fishing vessels, mostly from Taiwan, Malaysia, Indonesia, Vietnam, China, and Japan, have been poaching not only in the EEZ, but also within Philippine territorial seas in search of high value marine resources such as tuna (De Jesus 1999). Majority of the foreign intrusions occur particularly in the municipal waters of the northern or southern Palawan, Sulu Seas, and Batanes Seas. This problem causes the depletion of municipal fishing grounds and deprives municipal fisherfolks of their livelihood. Conduct of unauthorized fishing within the jurisdiction of the Philippines also poses threats to its marine biodiversity and environment.

There is an increasing trend of poaching incidents in the Philippines. Based on the records of the Palawan Council for Sustainable Development, there were 43 recorded arrests of foreign fishing vessels, involving a total of 639 foreign fishers from 1995 to 2002. Sixty-seven percent of these poachers are Chinese nationals (PCSD 2001). While most of the arrests of these foreign offenders result in judicial and administrative proceedings, not all cases against poaching are prosecuted. Of the cases filed from 1995 to 2002, only 34% involved poaching in Philippine waters, while a majority of 41% was sanctioned for violation of immigration laws (PCSD 2001). Other offenders were either convicted for a lesser offense or acquitted. This problem illustrates a weak record of prosecution against poachers in the Philippines.

Unregulated Fishing of Filipino Vessels

The Philippines is also confronted with the challenge of unregulated fishing activities by Filipino pumpboats or handline fishing vessels[2] in areas under the jurisdiction of other states, particularly in Indonesia, Palau, and Papua New Guinea. This problem indicates the country's lack of effective control over the activities of its nationals. There are about 2,500 unregistered Filipino pumpboats employing more than 30,000 fisherfolks (Congress of the Philippines, Representative Satur Ocampo 2003). The activities of these pumpboats have been a major source of strain in the bilateral relations of the Philippines and its neighboring countries. Records at the Philippine Department of Foreign Affairs (DFA) show that about 2,140 fishers have been arrested in and repatriated from Palau, Indonesia, and Micronesia since 1995 (Indonesia repatriates 15 Filipino fishermen, 2002). The cost of repatriation of these Filipino fishers captures a substantial portion of the funds of the government, which results in economic loss for the country.

Unreported Fishing

There are no comprehensive and accurate reports of illegal and unregulated fishing in the Philippines. While catches resulting from these activities are unreported, there are also instances when fish catches are deliberately misreported. Data on the transshipment of fish derived from IUU fishing activities is also unknown. The lack of accurate reports on these fishing activities results in poor fisheries management decisions and loss of revenue from fisheries in the Philippines.

---

[2] Pumpboats are traditional boats that are approximately 20 to 30 m in length, with a wooden hull, outriggers on each side, run by converted diesel truck engine, and use handlines for fishing.

The Philippine Legal, Policy, and Regulatory Framework in Addressing IUU Fishing

The utilization, conservation, and management of fisheries resources in the Philippines is governed particularly by three laws, namely the 1987 Philippine Constitution, the Philippine Fisheries Code of 1998 or Republic Act (RA) 8550, and the Local Government Code of 1991 or RA 7160. There are also fisheries administrative orders issued by the Department of Agriculture (DA). These are mostly regulatory in nature and implement RA 8550 (Republic of the Philippines 1998b).

The Philippine Fisheries Code of 1998 applies to all waters under the sovereignty and jurisdiction of the Philippines, including the country's EEZ and continental shelf (section 3a). The objective of RA 8550 is to ensure the rational use of the resources in its adjacent seas (Republic of the Philippines 1998a, section 2c). Republic Act 8550 also provides the basic fisheries management framework for all types of commercial fisheries in the Philippine EEZ. The Local Government Code of 1991, on the other hand, devolved the management of municipal fisheries to local government units. Under this code, provincial governments may issue licenses for the operation of fishing vessels, proscribe the use of explosives, noxious substances, and other deleterious methods of fishing, prescribe criminal penalty, and prosecute any violation of fisheries laws within their jurisdiction (Philippines 1991, section 149).

Other relevant legislative enactments and executive issuances addressing fisheries management include the following: RA 8435 or the Agriculture and Fisheries Modernization Act; RA 7586 or the National Integrated Protected Areas System Act; Presidential Decree 1152 or the Philippine Environment Code; Administrative Order 201 directing the implementation of the MCS system; and Implementing Rules and Regulations of the National Committee on Illegal Entrants (NCIE) on the investigation of cases involving illegal entry of foreign nationals and vessels. Policy documents such as the National Marine Policy and Philippine Environment Policy also provide governing principles applicable in fisheries management.

Fisheries administration in the Philippines reflects the complexity of managing the wealth of fisheries resources in the country. It is composed of several government agencies that execute several functions related to fisheries conservation and management. The formulation of fisheries laws, policies, and regulations is the primary responsibility of the Department of Agriculture Bureau of Fisheries and Aquatic Resources. Its major functions include the issuance of fishing licenses, monitoring and review of fisheries access by Philippine fishing vessels to international waters, establishment of a comprehensive fisheries information system, and implementation of an inspection scheme for import and export fishery products (Republic of the Philippines 1998b, section 65). The Bureau of Agriculture Statistics deals with the collection and compilation of fisheries information (Republic of the Philippines 1987a) while the Philippine Fisheries Development Authority is responsible for postharvest fishing facilities and activities (Republic of the Philippines 1987a). The registration of fishing vessels, on the other hand, is carried out by the Maritime Industry Authority (MARINA; Republic of the Philippines 1987b, sections 12 d and 12e) while the enforcement of fisheries laws is shared among the Philippine Navy, Philippine Coast Guard, and the Philippine National Police-Maritime Command (Republic of the Philippines 1998b; section 124; Republic of the Philippines 1987a, book IV title VIII chapter 8 section 53; Republic of the Philippines 1974, section 2; Republic of the Philippines 1990, section 35b).

Other relevant government agencies are the Department of Environment and Natural Resources, DFA, Department of Justice, Department of National Defense, DILG, Bureau of Immigration, Bureau of Customs, and National Intelligence Coordinating Agency. The NCIE, which is composed of most of these government agencies, serves as a forum for the effective coordination of the investigation, prosecution, and final disposition of all cases of illegal entry, including illegal fishing committed by foreign nationals and vessels (Republic of the Philippines 1995).

Implementation of International Instruments

The Philippines is a signatory and party to numerous international instruments related to the management of fisheries resources. It also participates actively in organizations under the United Nations, particularly the FAO, which look into the development and implementation of these international conventions. These instruments include the United Nations Convention on the Law of the Sea (LOSC), Agenda 21, UN Fish Stocks Agreement, Code of Conduct for Responsible Fisheries, 1995 Kyoto Declaration and Plan of Action on the Sustainable Contribution of Fisheries to Food Security, and Rome Consensus on World Fisheries. However, the Philippines has yet to accede to the 1995 FAO Compliance Agreement. It also needs to fully implement the IPOA-IUU by adopting an NPOA to address IUU fishing at the soonest possible time.

Fishing Vessel Registration and Licensing

The IPOA-IUU emphasizes the need to ensure that fishing vessels (and fishing support vessels) flying the flags of a state do not engage in IUU fishing activities (FAO 2001, paragraph 34). A genuine link must therefore exist between the flag state and the vessel (LOSC 1982, article 91[1]; FAO 1993, article III[3]) to enable the former to exercise effective control over the activities of the latter. The IPOA-IUU also provides that flag state control may be implemented through the proper registration of fishing vessels (FAO 2001, paragraphs 34–41), maintenance of their records (FAO 2001, paragraph 42), and issuance of authorization to fish (FAO 2001, paragraphs 44–50).

As part of its flag state responsibilities, the Philippines issues certificates of registry, fishing licenses, and fishing gear licenses to commercial fishing vessels. Certificates of registry are issued to vessels entitled to fly the flag of the Philippines. These vessels should be owned by Filipino nationals, who are either individuals or corporations with at least 60% capital stock of which are owned by Filipino citizens (Republic of the Philippines 1998b, section 27; Republic of the Philippines 2000e, section 3). There are also other requirements related to seaworthiness that commercial fishing vessels must comply with before they can register. Fishing companies or owners should provide vessel plans prepared by a licensed naval architect prior to construction of vessels (MARINA 1995a, section 5B). All vessels should also meet the standards of manning and certification requirements for their crew (MARINA 1995b, section B). Only after vessels have met these standards can commercial fishing vessels be registered.

Only registered fishing vessels are given licenses to fish in the Philippine waters. In order to ensure the proper conservation of fisheries resources in its EEZ, the Philippines issues fishing licenses and permits based on the limits of maximum sustainable yield (Republic of the Philippines 1998a, rule 26.3 and 1998b, section 7). The number of licenses issued to commercial boats is determined for each major fisheries, for each major fishing area, by vessel categories and by type of fishing gear, and corresponding fishing quota for each fishing boat (Republic of the Philippines 1998a, rules 7.5 and 26.1). In issuing fishing permits, priority is given to those vessels that have no record of violations of the terms and conditions of the licenses (Republic of the Philippines 1998b, rule 7.6). Any breach of the conditions of the authorization may result in license withdrawal, suspension, or cancellation.

The Philippines also issues international fishing licenses. Under the Philippine Fisheries Code of 1998, a fishing vessel should secure an international fishing permit and certificate of clearance from the Philippine government before it can undertake fishing operations in international waters or in areas under the jurisdiction of other states (Republic of the Philippines 1998b, section 32 and 2000e, section 19). The law also provides that fishing vessels under the Philippine flags can only land their fish caught in international waters in authorized landing sites (Republic of the Philippines 1998a, rule 32.1 and 1998b, section 32).

As part of the commercial fisheries development of the Philippines, support is provided to commercial fishing vessels to conduct operations in international waters. Preferential allocation of licenses is given to large commercial vessels that fish in the country's EEZ and in the high seas (Republic of the Philippines 1998a, rule 7.7). Several economic incentives are also provided to those vessels. These incentives include the grant of long term loans to acquire and improve fishing vessels and equipment, exemptions from tax and duty for a limited period, and entitlement to duty and tax rebates on fuel consumption (Republic of the Philippines 1998b, section 35). However, based on the IPOA-IUU, the Philippines should ensure that once a Philippine-flagged vessel engages in IUU fishing, such economic incentives be discontinued (FAO 2001, paragraph 23). The country needs to be more cautious in conferring such incentives to the fishing industry as it may lead to the increase in fishing capacity, which in turn encourages IUU fishing (Gréboval 2000). While fisheries subsidies may be permitted under both international and national laws, there can be no justifications to continue providing economic support to persons and entities engaging in IUU fishing activities.

Commercial fishing vessels are also required to maintain fishing records by the Philippine government. These records should include detailed information on the quantity and value of fish caught daily by fishing trip and fishing area, off-loaded for transshipment, and sold or disposed (Republic of the Philippines 1998a, rule 38.1 and 1998b, section 38). This information is then reported to the nearest designated landing points. These measures under RA 8550 are consistent with those of the IPOA-IUU.

The problem of unregulated fishing of unregistered Filipino tuna handline fishing vessels or pumpboats in areas under the jurisdiction of other states not only arises from the lack of effective implementation of the flag state responsibilities of the Philippines, but also with the inadequacy of its fisheries registration and licensing system. The rules and regulations on fishing vessel registration, however, exclude Filipino pumpboats or handline fishing vessels from being qualified to register as commercial fishing vessels. Most pumpboats are built by traditional boatbuilders and not by licensed naval architects. These fishing vessels also do not meet the manning requirements imposed by the Philippine government. Most of the employed fishing crew of pumpboats have only gained knowledge through years of fishing experience. Failure to comply with the requirements of vessel construction and manning prevents Filipino pumpboats from getting registered and securing licenses to fish in international waters. This lack of appropriate regulation to control a large number of fishing vessels that do not meet the requirements for fishing vessel registration also encourages the proliferation of unregulated fishing activities of Filipino pumpboats.

Application of Legal Sanctions

Consistent with the provisions of the 1987 Philippine Constitution, RA 8550 limits access to fishery and aquatic resources for the exclusive use and enjoyment of Filipino citizens in order to protect the rights of fisherfolks against foreign intrusions (Republic of the Philippines 1987c, article XII section 2 and article XIII section 7, and 1998b, sections 2b, 2d, 2e, and 5). Thus, the conduct of IUU fishing activities in Philippine waters directly undermines the basic state policy to protect the country's offshore fishing grounds and exclusive rights of Filipinos over the Philippine fisheries resources.

To discourage foreign intrusion, the Philippine Fisheries Code of 1998 establishes the rule that mere entry of foreign fishing vessels in the country's waters is considered as prima facie evidence that the vessel is engaged in fishing in Philippine waters, except in cases of force majeure and exercise of the right of innocent passage (Republic of the Philippines 1998b, section 87, and 2000c, section 3). This prohibition is strengthened by the imposition of a fine of US$100,000, in addition to confiscation of catch, fishing equipment, and fishing vessel (Republic of the Philippines 1998b, section 87). Section 87 of RA 8550 further states that an administrative fine for this violation ranging from US$50,000 to US$200,000 can also be imposed. These regulations are consistent with the measure provided in the IPOA-IUU, which encourages states to formulate and apply sanctions to prevent the conduct of IUU fishing activities (FAO 2001, paragraph 21; UN 1995, article 19(2); FAO 1993, article III(8), and 1996, paragraph 8.2.7). However, these stringent measures do not seem to effectively deter poaching in Philippine waters. This can be attributed to the weak monitoring, control, and surveillance system in the Philippines.

International law is not clear on how severe the penalty schedule should be for illegal fishing activities. Although the penalties for unsustainable fishing practices and poaching incidents have increased since the enactment of RA 8550, there is still the question of whether or not the punishments are stringent enough to effectively control such activities and discourage repeat offenders. Some states have higher penalties for poaching by foreign fishing vessels (e.g., Australia—US$500,000; New Zealand—US$290,000; Canada—CAD500,000–CAD750,000; and Belgium—US$112,000). The economic losses that the Philippines incur, due to illegal fishing, suggest that penalties may be inadequate or enforced ineffectively. While there had been significant additions to the penalty schedule, it can still be improved. Fines should reflect all economic and environmental costs of an illegal fishing activity. The schedule should at the very least consider the value of the resource affected by unsustainable fishing practices, the effect of the loss of that resource to the ecosystem, and the impact of the activity to the environment.

The provisions on fisheries violations found under RA 8550 address some of the characteristics of IUU fishing. These prohibitions include fishing using explosives, noxious or poisonous substance, and electricity (Republic of the Philippines 1998b, section 88, and

2001b); use of fine meshed nets (Republic of the Philippines 1986, and 1998b, section 89); use of active gear (Republic of the Philippines 1998b, section 88, and 2001a); coral exploitation and exportation (Republic of the Philippines 1998b, section 91, and 2001a); use of *muro-ami* and other fishing methods and gears known to be destructive to coral reefs and other marine habitat (Republic of the Philippines 1994, 1998b, section 92, and 2000d); fishing in overfished areas during closed seasons (Republic of the Philippines 1998b, section 95); fishing in fishery reserves, refuge, and sanctuaries (Republic of the Philippines 1998b, section 96); fishing or taking of rare, threatened or endangered species (Republic of the Philippines 1998b, section 97, and 2001a); and violation of catch ceilings (Republic of the Philippines 1998b, section 101). Under RA 8550, these offenses are distinct from poaching and are generally applied to fishing vessels flying the flags of the Philippines. However, these unsustainable fishing practices may also be committed by foreign fishers, within Philippine waters, who may be punished under the same violations.

Monitoring, Control, and Surveillance

One of the key measures to enhance the capability of the Philippines to enforce its fisheries laws and policies is to strengthen monitoring, control, and surveillance (MCS) (FAO 2001, paragraph 24). The Philippine MCS system is established at the national and regional levels to strengthen the fisheries law enforcement of the country. It has some of the necessary elements of effective fisheries compliance and data management, such as the procurement of communications and physical equipment, creation of an integrated database management system, review of the fisheries licensing system, training of personnel, and establishment of a coordinating mechanism that will implement the MCS (BFAR, no date). However, the components of the Philippine MCS system mostly include fisheries enforcement and data collection activities and do not fully conform to the FAO definition of MCS (FAO 1981). Based on the definition, MCS also encompasses the enactment of legislative instruments and implementation of the management plan through participatory techniques and strategies (FAO 2003b).

To date, the Philippines has only implemented MCS in a few municipalities. The lack of funding in the country has hindered its full implementation. The Philippines needs to find a way to ensure the development of such technology through a sustained financial support. Aside from using license fees to fund MCS operations, the country can also seek to privatize certain elements of MCS system, such as fisheries data collection. The Philippines could also look into the possibility of instituting procedures that would establish the admissibility of electronic evidence and new technologies in courts such as vessel monitoring systems, consistent with the measures provided by the IPOA-IUU (FAO 2001, paragraph 17).

Cooperation with Other States

In fulfillment of its duty to cooperate, the Philippines participates in several regional fisheries bodies. It has adopted the Convention on the Conservation and Management of Highly Migratory Fish Stocks in the Western and Central Pacific Ocean, acceded as noncontracting party to the Indian Ocean Tuna Commission, International Convention for the Conservation of Atlantic Tunas, and indicated its intention to accede as a cooperating nonmember to the Convention on the Conservation of Southern Bluefin Tuna. The Philippines is also an active member of various regional and international bodies concerned with fisheries management such as the Southeast Asian Fisheries Development Centre, Brunei Darussalam, Indonesia, Malaysia, Philippines' East Association of Southeast Asian Nations Growth Area, Asia-Pacific Fishery Commission, and Asia-Pacific Economic Cooperation.

The Philippine government is also conducting a major initiative that would assist significantly in the formulation of its NPOA. This involves the implementation of an Australian-funded project between the Philippines and Indonesia in combating IUU fishing in the shared waters of the Sulawesi Sea. This research project aims to obtain a deeper understanding on the causes of IUU fishing problems in the area; evaluate existing measures such as fishing vessel registration, fisheries data management, and MCS; and formulate a regional plan of action that would address the problem in the Sulawesi Sea. The results of this research could form the basis of drafting an NPOA that will address IUU fishing in all areas under the jurisdiction of the Philippines.

Port State Measures

Port state measures play a crucial role in the prevention of IUU fishing. The current port State control implemented in the Philippines does not effectively determine catches derived from IUU fishing activities. As a port state, the Philippines should require fishing vessels and vessels involved in fishing-related activities to provide relevant information that would help detect IUU fishing; deny port access to vessels involved in such illicit activities (FAO 2002, paragraph 62); ban the landing, transshipment, and selling of catches derived from IUU fishing UN Fish Stocks Agreement, (UN 1995, article 23[2]; FAO 2001, paragraph 56; FAO 2002, paragraph 62); and detain vessels, if necessary (Paris Memorandum on Port State Control, Paris, 26 January 1982). Implementation of these measures necessitates effective cooperation among relevant government agencies, particularly the Philippine Ports Authority and the Philippine Coast Guard.

Conclusion

The problem of illegal, unreported, and unregulated fishing in the Philippines parallels the global IUU fishing concerns. Some of the major IUU fishing issues in the country are unauthorized fishing activities of foreign fishing vessels, unsustainable fishing practices, unregulated fishing of unregistered Filipino vessels in areas under the jurisdiction of other states, and unreported fishing. These problems pose a great threat to the conservation and management of fisheries resources and further results in economic losses for the country. Although the Philippines has yet to implement the IPOA-IUU, it has adopted some IUU fishing measures in its fisheries laws and regulations. The country implements some of the most important flag state and coastal state obligations, such as the registration of fishing vessels, maintenance of fishing records, issuance of authorizations to fish, application of sanctions, and implementation of an MCS system, which help address the various IUU fishing concerns in the country. The Philippines also carries out its duties to implement international fisheries-related instruments and cooperate with other states. However, IUU fishing problems continue to proliferate due to the inadequacy of the fisheries legislation, lack of enforcement of fisheries regulations, and lack of capacity to implement an effective MCS system.

The Philippines should therefore review and strengthen its flag state and coastal state responsibilities and adopt applicable port state measures to effectively address IUU fishing in the country. Such measures can be either be introduced as amendment to the Philippine Fisheries Code or adopted in an NPOA to prevent, deter, and eliminate IUU fishing in the Philippines.

Acknowledgments

I would like to thank my supervisor, Professor Martin Tsamenyi, for providing guidance in the writing of this paper. Financial support for this paper was provided through the Indonesian/Philippine IUU Fishing Project, funded by the Australian Centre for International Agricultural Research.

References

Aliño, P. 2002. Fisheries resources of the Philippines. Presentation during the Australian consultation with the Philippines and Indonesia on the identification of researchable options for the development of policy and management frameworks to combat illegal, unreported, and unregulated fishing activities in Indonesian and Philippine waters. University of Wollongong, Centre for Maritime Policy and Oceans and Coastal Research Centre, Final Report on ACIAR Project No. FIS/2000/163, New South Wales, Australia.

BAS (Bureau of Agricultural Statistics), Fisheries Statistics Division. 2001–2002. Fisheries situation years 2001–2002, 2 reports. BAS, Quezon City, Philippines.

BFAR (Bureau of Fisheries and Aquatic Resources). 1997–2001. Philippine fisheries profile years 1997--2001, 5 years. BFAR, Quezon City, Philippines.

BFAR (Bureau of Fisheries and Aquatic Resources). No date. Monitoring, control, and surveillance (MCS) system for the Philippines.

Bray, K. 2000. A global review of illegal, unreported and unregulated (IUU) fishing. Expert consultation on illegal, unreported and unregulated fishing organized by the Government of Australia in cooperation with Food and Agriculture Organization of the United Nations (FAO), AUS:IUU/2000/6, Sydney.

Congress of the Philippines, Representative Satur Ocampo. 2003. Resolution directing the Committee on Agriculture, Food and Fisheries to conduct an inquiry, in aid of legislation, into the appropriateness and relevance of the regulations promulgated by the Maritime Industry Authority (MARINA) which fail to distinguish traditional, simple handline fishing vessels engaged in large tuna fishing from sophisticated fishing craft and thus consequently jeopardize the tuna handline fishing industry in the Philippines and the livelihood of over 30,000 tuna handline fisherfolk, especially in Mindanao. House Resolution 1132, Manila, Philippines.

CRMP (Coastal Resource Management Project Philippines), Fisheries Resource Management Project, and Department of Agriculture. 1998. Coastal resource management for food security. CRMP, Document No. 39-CRM/1998, Quezon City, Philippines.

De Jesus, N. 1999. Fishing monsters eating small fishermen. Greenfields (September):29.

Doulman, D. J. 2000. Illegal, unreported, and unregulated fishing: mandate for an international plan of action. Expert consultation on illegal, unreported and unregulated fishing organized by the Government of Australia in cooperation with Food and Agriculture Organization, AUS:IUU/2000/4, Sydney.

Edeson, W. 2000. Tools to address IUU fishing: the current legal situation. Experts consultation on illegal, unreported and unregulated fishing organized by the Government of Australia in cooperation with Food and Agriculture Organization, AUS:IUU/2000/8, Sydney.

FAO (Food and Agriculture Organization). 1981. Report on an expert consultation on monitoring, control and surveillance systems for fisheries management. FAO/Norway Cooperative Programme, FAO/GCP/INT/344/NOR, Rome.

FAO (Food and Agriculture Organization). 1993. Agreement to promote compliance with international conservation and management measures by fishing vessels on the high seas. FAO, Rome.

FAO (Food and Agriculture Organization). 1995. Code of conduct for responsible fisheries. FAO, Rome.

FAO (Food and Agriculture Organization). 2001. International plan of action to prevent, deter, and eliminate illegal, unreported, and unregulated fishing. FAO, Rome.

FAO (Food and Agriculture Organization). 2002. Implementation of the international plan of action to prevent, deter and eliminate illegal, unreported, and unregulated fishing. FAO, Technical Guidelines for Responsible Fisheries No. 9, Rome.

FAO (Food and Agriculture Organization). 2003a. Fishery country profile. FID/CP/PHI Rev 5. Available: www.fao.org/fi/fcp/en/PHL/profile.htm. (April 2003).

FAO (Food and Agriculture Organization). 2003b. Guide to monitoring, control, and surveillance systems for coastal and offshore capture fisheries. FAO Fisheries Technical Paper 415.

Gréboval, D. F. 2000. The international plan of action for the management of fishing capacity and selected issues pertaining to illegal, unreported and unregulated fishing. Expert consultation on illegal, unreported and unregulated fishing organized by the Government of Australia in cooperation with Food and Agriculture Organization, AUS:IUU/2000/13, Sydney.

Green, S. J., A. T. White, J. O. Flores, M. F. Carreon, and A. E. Sia. 2003. Philippine fisheries in crisis: a framework for management. Coastal Resource Management Project of the Department of Environment and Natural Resources, Cebu City, Philippines.

Indonesia repatriates 15 Filipino fishermen. 2002. Manila Bulletin (12 March):4.

LOSC (United Nations Convention on the Law of the Sea). 1982. LOSC, Montego Bay, Jamaica.

MARINA (Maritime Industry Authority). 1995a. Implementing guidelines for vessel registration and documentation. MARINA, Memorandum Circular No. 90, Manila.

MARINA (Maritime Industry Authority). 1995b. Schedule for penalties and/or administrative fine relative to vessel registration/licensing/documentation and vessel safety regulation. MARINA, Memorandum Circular No. 109, Marina.

PCAMRD (Philippine Council for Aquatic and Marine Research and Development). 1993. Status of the Philippine tuna fisheries. PCAMRD, Manila, Philippines.

PCG (Philippine Coast Guard). 2001. Accomplishment report first semester 2001. PCG, Quezon City, Philippines.

PCSD (Palawan Council for Sustainable Development). 2001. Internal report on poaching incidents 1997–2001. PCSD, Puerto Princesa, Palawan, Philippines.

Philippines. 1991. Local government code of 1991. Republic of the Philippines, Republic Act 7160, Manila.

Republic of the Philippines. 1974. Revised Coast Guard Law of 1974. Presidential Decree 601, Manila.

Republic of the Philippines. 1990. Department of Interior and Local Government Act of 1990. Republic of the Philippines, Republic Act 6975, Manila.

Republic of the Philippines. 1986. Regulating the use of fine meshed nets in fishing. Department of Agriculture Fisheries and Administrative Order No. 155, Manila.

Republic of the Philippines. 1987a. Administrative code of 1987. Executive Order 292, Manila.

Republic of the Philippines. 1987b. Amending EO No 125, entitled "Reorganizing the ministry of transportation and communications, defining its powers and functions and other purposes," Executive Order 125-A, Manila.

Republic of the Philippines. 1987c. The Constitution of the Philippines. Republic of the Philippines, Manila.

Republic of the Philippines. 1994. Regulations governing *Pa-aling* fishing 0peration in Philippines waters. Department of Agriculture Fisheries and Administrative Order No. 190, Manila.

Republic of the Philippines. 1995. Organizing the National Committee on Illegal Entrants. Executive Order 236, Manila.

Republic of the Philippines. 1998a. Implementing rules and regulations of RA 8550. Department of Agriculture Administrative Order No. 3, Manila.

Republic of the Philippines. 1998b. Philippine Fisheries Code of 1998. Republic Act 8550, Manila.

Republic of the Philippines. 2000a. Ban on fishing with active gear. Department of Agriculture Fisheries and Administrative Order No. 201, Manila.

Republic of the Philippines. 2000b. Banning fishing by means of "muro-ami" and the like destructive to coral reefs and other marine habitat. Department of Agriculture Fisheries and Administrative Order No. 203, Manila.

Republic of the Philippines. 2000c. Guidelines and procedures in implementing section 87 of the Philippine Fisheries Code of 1998. Department of Agriculture Fisheries and Administrative Order No. 200, Manila.

Republic of the Philippines. 2000d. Restricting the use of superlights in fishing. Department of Agriculture Fisheries and Administrative Order No. 204, Manila.

Republic of the Philippines. 2000e. Rules and regulations on commercial fishing. Department of Agriculture Fisheries and Administrative Order No. 198, Manila.

Republic of the Philippines. 2001a. Conservation of rare, threatened, and endangered fishery species. Department of Agriculture Fisheries and Administrative Order No. 208, Manila.

Republic of the Philippines. 2001b. Disposal of confiscated fish and other items in fishing through explosives and noxious or poisonous substances. Department of Agriculture Fisheries and Administrative Order No. 206, Manila.

UN (United Nations). 1995. Agreement for the implementation of the provision of the United Nations Convention on the Law of the Sea of 10 December 1982 relating to the conservation and management of straddling fish stocks and highly migratory fish stocks. UN, New York.

UNGA (United Nations General Assembly). 1996. 51st session, agenda item 24(c). Oceans and the law of the sea: large-scale pelagic drift-net fishing and its impact on the living marine resources of the world's oceans and seas; unauthorized fishing in zones of national jurisdiction and its impact on the living marine resources of the world's oceans and seas; and fisheries by-catch and discards, and their impact on the sustainable use of the world's marine living resources. Report of the Secretary-General, A/51/404, New York.

UNGA (United Nations General Assembly). 1997. 52nd

session, agenda item 39(c). Oceans and the law of the sea: large-scale pelagic drift-net fishing, unauthorized fishing in zones of national jurisdiction and fisheries by-catch and discards. Report of the Secretary-General, A/52/557, New York.

UNGA (United Nations General Assembly). 1998. 53rd session, agenda item 38(b). Oceans and the law of the sea: large-scale pelagic drift-net fishing, unauthorized fishing in zones of national jurisdiction and on the high seas, fisheries by-catch and discards, and other developments. Report of the Secretary-General, A/53/473, New York.

UNGA (United Nations General Assembly). 1999. 54th session, agenda item 40 (a) and (c). Oceans and the law of the sea: law of the sea; results of the review by the commission on sustainable development of the sectoral theme of "oceans and seas." Report of the Secretary-General, A/54/429, New York.

UNGA (United Nations General Assembly). 2000. 55th session, agenda item 34(b). Oceans and the law of the sea: large-scale pelagic drift-net fishing, unauthorized fishing in zones of national jurisdiction and on the high seas, fisheries by-catch and discards, and other developments. Report of the Secretary-General, A/55/386, New York.

United Nations Conference on Environment and Development. 1992. Protection of the oceans, all kinds of seas, including enclosed and semi-enclosed seas, and coastal areas and the protection, rational use and development of their living resources. Chapter 17, Agenda 21. United Nations Conference on Environment and Development, Rio de Janeiro, Brazil.

American Fisheries Society Symposium 49:443–444

Session Summary

The Role of Sport Fishing in Reconciling Fisheries with Conservation

IAN G. COWX
University of Hull, International Fisheries Institute
Hull, East Yorkshire, United Kingdom

An overview of the importance of recreational fisheries to society, and the benefits and conflicts of the sector with conservation of ecosystems was emphasised in this session. The value of the sector in the developed world was estimated at tens of billions of dollars and often could be more important than commercial fisheries, especially in inland waters and to rural and urban economies (Cowx, UK). Recreational fisheries were also shown to be important in protecting aquatic ecosystems because proponents demand fisheries resources are maintained and improved to the benefit of the environment, although this is sometimes offset by considerations of animal rights, wildlife disturbance, and damaging management practices.

The sport fish sector conflicts with conservation of ecosystems in several ways, not least of which is disruption of habitat (trampling of vegetation, disturbance of wildlife and modification of systems to allow access to the fisheries), but also can lead to decline in stocks through overexploitation. For example, J. Post (Canada) examined the possibility of anglers fishing down stocks to extinction in his oral presentation. He argued that the cost of fishing should increase in relation to the benefits as the resources become depleted, and eventually they must move on to other resources, giving stocks the opportunity to recover. However, this is not necessarily the case in recreational fisheries, and angling can fish down stocks to such an extent they cannot recover, possibly because ecosystem functions are disrupted. Evidence of fishing down stocks was provided by Richardson et al. (UK) where the size and weight of fish caught per angler has declined in marine sport fisheries in Wales, but they were unable to show whether external factors (e.g., climate change) contributed towards the changes observed, or the role of commercial fishing in this decline. However, recreational sea angling was recognized as very important to the economy in the region, and it was recommended that the economic gains from an improvement in the resource could more than compensate for losses in commercial landings, should management decide to regulate the latter.

Further changes in stock structure and impacts are also prevalent in recreational fisheries from stock enhancement and angler induced selection pressures. Stocking and introduction of fish for enhancement of the angling experience has caused untold damage to ecosystems through disruption of food webs, predation, competition, spread of pathogens, and loss of genetic integrity. Cooke et al. (Canada) suggested that fishing selection pressure could also alter the phenotypic (possibly genotypic) characteristics of the fish stocks towards the less accessible (smaller size, slower growing, and lower fecundity) components which have reduced capability to replace the stocks, possibly making the populations vulnerable to decline. Thus the evolutionary consequences of fishing-induced selection pressures are a concern and may represent a potential risk to the sustainability of the fisheries resources.

Some evidence of the role anglers can play in conservation was derived from their preferences towards retention of catch (Gentner; USA). Survey results suggested angler responses to whether they would retain their catch was considerably higher than reality. Reasons for the divergent behaviour was not offered but questions suggest this was because anglers were unwilling to modify their exploitation problems in fear of losing the possibility of retaining the catch should the resources become depleted. This suggests anglers do not necessarily consciously think about conservation but their behaviour contributes towards it. Anglers also are also key stakeholders in protecting the environment from degradation and should be more proactive in doing so.

To resolve the conflicts between fisheries and conservation of the ecosystem, there is a need to engage all stakeholders and make efforts to understand the motives, modes of operation and incentives of others. This should lead to optimization of the utilization of the resources and reduction of conflicts, ultimately leading to conservation of the ecosystem and sustainable exploitation of the fisheries for recreational purposes. In summary, science, management, and policy working within a larger social, economic, and political framework need to have common goals to improve the image and sustainability of recreational fisheries (Cowx, UK).

American Fisheries Society Symposium 49:445–450

Evaluation of the Importance of Recreational Fisheries on a Mediterranean Island

Beatriz Morales-Nin*, Joan Moranta, Cristina García, and Pilar Tugores

CSIC/UIB Instituto Mediterráneo Estudios Avanzados
Miguel Marqués 21, 07190 Esporles, Islas Baleares, Spain

Antoni Maria Grau

DG Pesca
Foners 10, 07006, Palma, Spain

Introduction

Coastal ecosystems are a cause for concern due to pervasive human disturbance, which is more acute in areas of rapid development and where comprehensive studies to estimate and regulate human activities are lacking. An example is the Mediterranean, where the economic transformation experienced over the last century–from a primary economy based on agriculture and cattle raising to a tertiary one based on tourism—has dramatically altered its coastal areas (Morey et al. 1992, Özhan 1996; van der Meulen 1996; García and Servera 2004). As a result of this transformation there is a greater use of the shoreline all year round with a considerable proliferation of, inter alia, aquatic leisure activities such as scuba diving, water-skiing, sailing, and fishing, which also affect the coastal and maritime ecosystems. Of these, commercial fishing plays a particular economic, social, and cultural role in the Mediterranean where it is largely made up of small-scale concerns operating in coastal areas. Over the last 20 years, catches of several key commercial stocks have been in sharp decline despite the increase in fishing effort.

Among the most frequent aquatic leisure activities, recreational fishing is an important activity in coastal zones; this implies great numbers of people and, consequently, albeit not so exactly determined recreational fishing effort could potentially even surpass the effort of commercial fishing (Dunn et al. 1989). Recreational fisheries have scarcely been studied in the Mediterranean, although the related fishing mortality must be taken into account to determine the current exploitation rates (Coll et al. 2004; Morales-Nin et al. 2005).

The island of Majorca, located in the northwestern Mediterranean Sea (Figure 1), has an area of 3,620 km^2, with a population density of 725,000 inhabitants and millions of tourists every year. Along its 623 km of coastline, there are 39 ports and 14,196 moorings. The recreational fishery is managed by traditional passive-style measures, such as seasons and bag size limits designed to limit the catch. A fish-

*Corresponding author: ieabmn@uib.es

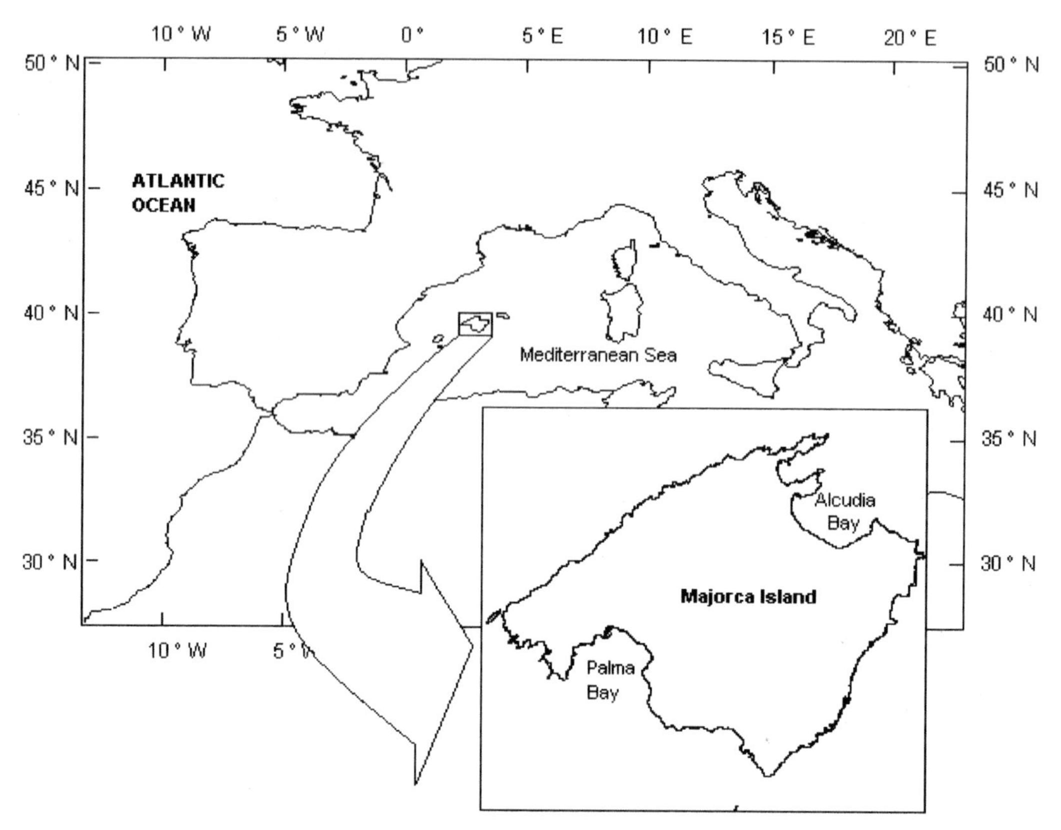

Figure 1. Situation of the island of Majorca in the Mediterranean Sea.

ing permit is required, but it is not restricted in numbers and costs few euros. Some active measures such as consensus-driven protected areas have been recently enforced and are currently in development. The number of recreational fishing licences gives an initial indication of the importance of the activity (22,000 licenses in 2002). Apart from these numbers, there was little known about recreational fishery activities around the Island of Majorca. Thus, in 2002, a study was designed to obtain the real numbers of recreational fishers and describe their habits and measure fishing effort and harvest. The study was composed of a telephone survey, on-site personal interviews, and voluntary logbooks.

From the telephone survey, 1,271 positive household interviews were obtained, corresponding to 3,632 persons. 5.14% of the people interviewed admitted to being recreational fishers. A veracity test was carried out with the members of a fishing club, albeit 4.91% of the people interviewed denied being active fishers. Thus, applying this veracity factor, it was estimated that the true percentage of the Majorcan population that were recreational fishers lies around 10.05% or more. Thus, it was estimated that 73,000 people on the island were recreational fishers. Or in other words, considering the number of registered current fishing licences, only about 30% of the recreational fishers followed the fishing rule.

The data from the surveys showed that recreational fishing was a leisure activity mainly practiced by men (91% of the total) who were middle-aged (mean = 46 years, SD = 15 years). The majority of recreational fishers were found to be between 40 and 50 years old (29.65% of the total), although the percentage of recreational fishers over 60 years was also high (20.40%). These middle-aged men usually fished alone (43.7%) or in pairs (39.2%). The percentage of recreational fishers that went fishing in a group was rather low (12.5% went fishing with 3 people, 3.1% with 4, and only 1.5% with more than 4 people).

Recreational fisheries were very diverse depending on the season, the species and the fisherman. For instance, boat fishermen might use a pelagic trolling line targeting dolphinfish and pilot fish in the fall, a bottom-trolling line for sparids and groupers, a hand-line for rocky bottom fish (small sparids), and another hand-line for razor fish on sandy bottoms, among others. In the study, three of the major fishing methods were sampled to obtain the first evaluation of the fishery: shore fishing, boat fishing and spear diving fishing. This last method was practiced without air tanks by law. The fishing method (Table 1) with the greatest number of amateurs was boat fishing while spear fishing was the least practiced (only 3.6%). The great majority of recreational fishers (95.8%) always practiced the same method and only a few practiced more than one (4.2%). Since the majority of the boat fishers (90%) used their own boat, they had a greater flexibility in their habits, with 28% practicing other fishing methods. In contrast, almost all shore fishers practiced this method only (92.76%).

The boat fishers mainly had their boats based at a marina (81.5%), while only a few (18.5%) towed their boats. Divers commonly used a towed boat (64.71%) rather than one moored in a harbor (35.29%). This implied that fishing from a boat is more site linked, while shore fishing and diving were more mobile, depending on weather and the fishermen's preference.

Recreational fishermen from all methods might use simultaneously more than one fishing rod or spear while fishing (mean = 1.27, SD = 0.59) (Table 1). The mean time recreational fishers spend on a trip is 3.86 h (SD = 1.43), although diving trips were longer (Table 1). An analysis of the daily activity showed that most trips start in the morning (83.8%), less frequently in the afternoon (13.8%), and infrequently at night (2.4%). The seasonal distribution of fishing activity (Figure 2) showed a common trend with maximum activity in summer (34.88%) and a minimum activity in winter (16.95%) (Table 2).

Fishing activity was also concentrated at weekends (68% of the interviewees); although in significant proportion fishermen fish both on weekdays and at weekends (26%). Only 6% of the interviewees only went fishing on weekdays. The percentage

Table 1. Summary of effort and harvest by fishing methods. In brackets: standard deviation for mean fishing gears per fisher and trip length; standard error for mean daily harvest.

Method	*Relative participation (%)*	*Mean gears per fisher*	*Trip length (hours/day)*	*Number of trips per year*	*Mean daily harvest (kg/day.fisherman)*
Shore	33.4	1.31(±0.70)	3.42(±1.62)	402,664	1.09(±0.07)
Boat	62.9	1.24(±0.52)	4.05(±1.25)	758,112	2.39(±0.08)
Diving	3.6	1.49(±0.57)	4.18(±1.60)	43,723	2.70(±0.78)

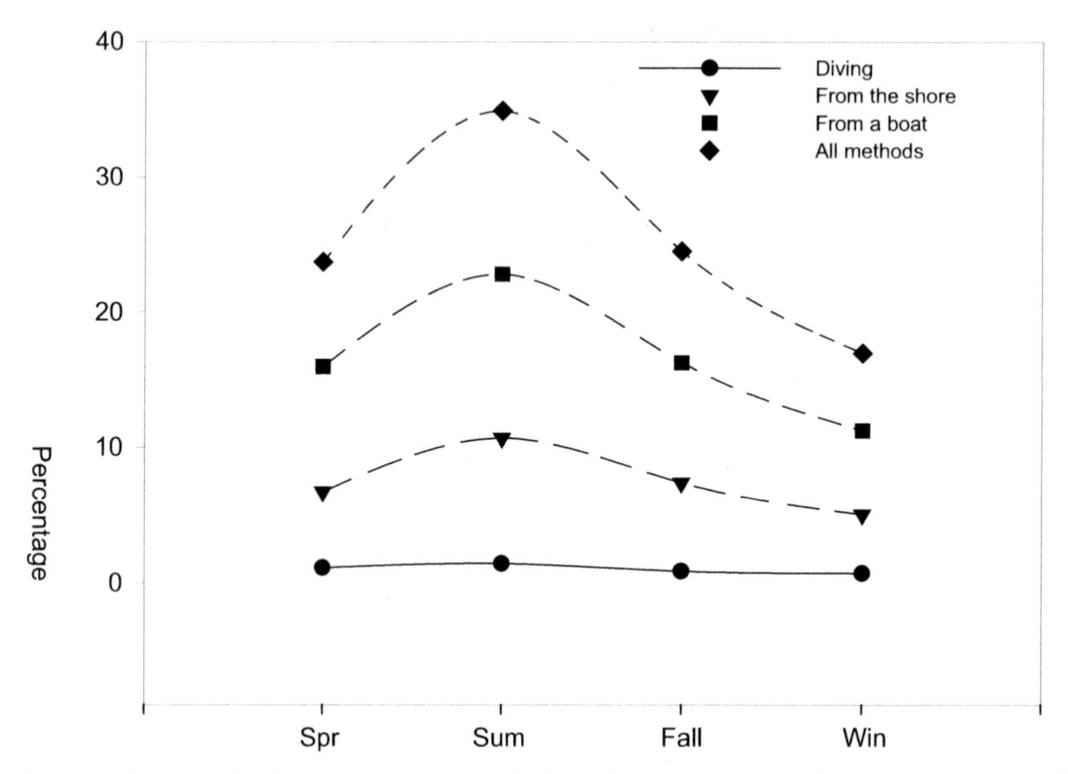

Figure 2. Temporal distribution of recreational fishing activity in Majorca during the year 2002: total percentage of fishing activity and percentage of fishing activity by fishing method.

of fishing trips on weekdays was directly proportional to the monthly activity. Moreover, recreational fishers increased the number of fishing days during their holidays and on bank holidays. Although the mean fishing frequency was 5.5 d per month (SD = 0.15), there was a clear variability in the level of activity. 46.7% recreational fishers were found to go fishing between 4 and 6 d per month, 27% fished between 6 and 10 d per month, 7.6% went between 10 and 15 d per month, and only 1.46% went more than 15 d per month. The percentage of recreational fishers that went fishing less than 4 d per month accounted for 17.2%.

From the information regarding the estimated number of recreational fishers (73,000 people), the seasonal distribution of the fishing activity by modality (Table 2), and the mean times per month that a recreational fisher went fishing, the annual mean fishing effort was estimated to be 1,204,500 fishing trips. The main part of

Table 2. Temporal distribution of the fishing activity (%) by fishing method.

Season	Boat	Shore	Diving
Spring	24.07	22.43	26.92
Summer	34.44	35.88	34.62
Fall	24.54	24.80	21.15
Winter	16.95	16.89	17.31

this effort was carried out from boat fishing while diving was the least active (Table 1).

Species composition of recreational catches included 28 families and 60 species of marine fishes and cephalopods. Despite this seemingly high diversity, only a few species comprised the bulk of the catch: painted comber *Serranus scriba*, comber *Serranus cabrilla*, rainbow wrasse *Coris julis* and annular seabream *Diplodus annularis*. The two species with closed seasons (pearly razorfish *Xyrichths novacula* and greater amberjack *Seriola dumerili*) were among the twelve most frequently captured species.

Twenty-eight percent of recreational fishers interviewed answered that they usually catch between 25 and 50 units (fishes and cephalopods) per a fishing trip. The mean abundance of the harvest by each fishing method was highest for boat fishing (29.08 units; SE = 3.64) and lowest for diving (7.24 units; SE = 0.63). For shore fishing, the mean number of captured fishes per day was 9.55 (SE = 0.82). From these figures, mean biomass extracted has been calculated for the three methods: for boat fishing = 2.39 kg (SD = 0.08 kg); for shore fishing = 1.09 (SD = 0.07 kg); and for diving = 2.70 kg (SD = 0.53 kg). From the daily capture and fishing effort, the total annual capture has been estimated to be 2,368.85 metric tons (mt), corresponding to 566.28 mt in spring, 822.37 mt in summer, 578.78 mt in fall and 401.72 mt in winter. In terms of fishing method, 1,811.89 mt is attributed to boat fishers, 438.90 mt to shore fishers, and 118.05 mt to diving. The importance of the harvest was underlined by comparison with the commercial fishery on the Island which account for almost 4,000 mt.

The great diversity of captured species, with some differences between fishing methods, showed a very varied exploitation of the littoral fauna. The greatest effort was concentrated in shallow waters (less than 30 m in depth) and from the littoral to a few nautical miles (1 nautical mile = 1.852 km) offshore (mean = 1.78 nm; SD = 1.61 nm). Besides the amount of biomass extracted, the perturbation to the environment caused by 1,204,500 fishing trips per year (nearly 5 million hours fishing) cannot be negligible due to the small dimensions of the Island. Given that this evaluation did not take into account the small-scale commercial fishery—acting in part of the same waters—the littoral fauna exploitation was high. Although the lack of previous data and the study's own limitations precluded evaluating the available biomass and the degree of overexploitation, it was clear that recreational fisheries must be taken into account in any management of the stocks. In Majorca, the recreational fishery was, in fact, an open resource with a low degree of compliance with regulations (e.g., only 30% fishermen had a legal fishing permit). In open-access sport fisheries (Cox et al. 2002) the typical regulations are not drastic enough to affect total exploitation. Thus, in order to manage the recreational fishery, stronger enforcement of the measures, and additional regulations should be considered.

Although there is a general lack of elemental data on the importance of recreational fisheries in the Mediterranean and in Europe (Morales-Nin et al. 2005), the available evidence shows that recreational fishing is one of the most popular recreational activities undertaken by Europeans. This fishery involves millions participants and presumably generates a relevant economic activity, thus it should be fully recognized as an important, social, economic and ecological factor in Europe.

Acknowledgments

This work was supported by a D. G. de Pesca (Conselleria d´Agricultura i Pesca, Illes Balears) grant cofunded by the European Union Project IFOP ES.R.BAL.5.1.3. We are grateful to the interviewers and volunteers.

References

Coll, J., M. Linde, A. García-Rubiés, F. Riera, and A. M. Grau. 2004. Spear fishing in the Balearic Islands (west central Mediterranean): species affected and catch evolution during the period 1975–2001. Fisheries Research 70: 97–111.

Cox, S. P., T. D. Beard, and C. Walters. 2002. Harvest control in open-access sport fisheries: hot rod or asleep at the reel? Bulletin of Marine Science 70:749–761.

Dunn, M. R., S. Potten, A. Radford, and D. Whitmarsh. 1989. An economic appraisal of the fishery for bass *Dicentrarchus labrax* L. in England and Wales. Volume I: an economic appraisal of the fishery for bass in England and Wales. CEMARE reports, University of Portsmouth, Hampshire, UK.

García C., and J. Servera. 2004. Impacts of tourism development on water demand and beach degradation on the island of Mallorca (Spain). Geografiska Annaler 85 A:287–300, Blackwell Publishing, London.

Morales-Nin, B., J. Moranta, C. García, P. Tugores, A. Grau, F. Riera, and M. Cerdà. 2005. The recreational fishery in Mallorca Island (Western Mediterranean): implications for coastal resources management. ICES Journal Marine Science 62:727–739.

Morey, M., M. J. Bover, and J. A. Casas. 1992. Change in environmental stability and the use of resources on small islands: the case of Formentera, Balearic Islands, Spain. Environmental Management 16:575–583.

Özhan, E. 1996. Coastal zone management in Turkey. Ocean and Coastal Management 30:153–176.

Van der Meulen, F., and A.H.P.M. Salman. 1996. Management of Mediterranean coastal dunes. Ocean and Coastal Management 30:177–195.

American Fisheries Society Symposium 49:451–459

The Practical Application of Mark-Selective Fisheries

Annette Hoffmann* and Patrick L. Pattillo
Washington Department of Fish and Wildlife
600 Capitol Way, N. MS 43200, Olympia, Washington 98501, USA

Abstract.—Mark-selective fisheries offer a promising tool to coho *Oncorhynchus kisutch* and Chinook salmon *O. tshawytscha* fisheries managers for expanding recreational harvest opportunities on healthy hatchery stocks while minimizing fishing impacts on natural stocks. The concept is simple: hatchery fish are marked with a visual identifier, e.g. an adipose fin clip, and marked fish are retained in a mark-selective fishery while unmarked fish must be released. Although simple in concept, the development of mark-selective fisheries has not been without controversy. Concerns have been raised over loss of management information (e.g. bias in exploitation rate estimates, through the conduct of mark-selective fisheries). That loss leads to more stringent requirements in managing the fishery. Therefore, practical application of this tool should consider many factors including the encountered mark rates, stocks impacted, alternative management approaches, benefits for harvest opportunity as well as stock conservation, and loss of management information. Because of these factors, not all fisheries are wise choices for selective regulation. Based on several years of having implemented selective regulations and several years of contributing to an intense evaluation process, we offer guidelines to help fishery managers design and understand the impacts of their mark-selective fisheries.

Introduction

Salmon management in the U.S. Pacific Northwest can be complex, needing to consider harvest objectives as well as incidental impacts on nontarget species or stocks. Fisheries that are mark-selective are a relatively new tool in coho *Oncorhynchus kisutch* and Chinook salmon *O. tshawytscha* management that offer the promise of protecting weak or endangered stocks while allowing for expanding recreational harvest opportunities on healthy hatchery stocks in areas where hatchery and wild stocks commingle. Conceptually, hatchery fish that are marked with an easily distinguishable mark, such as an adipose fin clip, may be retained while unclipped fish must be released. One might consider implementing a mark-selective fishery for several reasons: (1) lessen impact on weak wild stocks, (2) remove hatchery strays from rivers where straying rates are too high, and (3) increase fishing season length under impact quotas.

*Corresponding author: hoffmah@dfw.wa.gov

Controversy over Mark-Selective Fisheries: Viability of the Coded Wire Tag System

Although simple in concept, the development of mark-selective fisheries has not been without controversy and that controversy has focused on mark-selective impacts to the viability of information developed from the coded wire tag (CWT) system over the past decade. The CWT system is made up of adipose fin

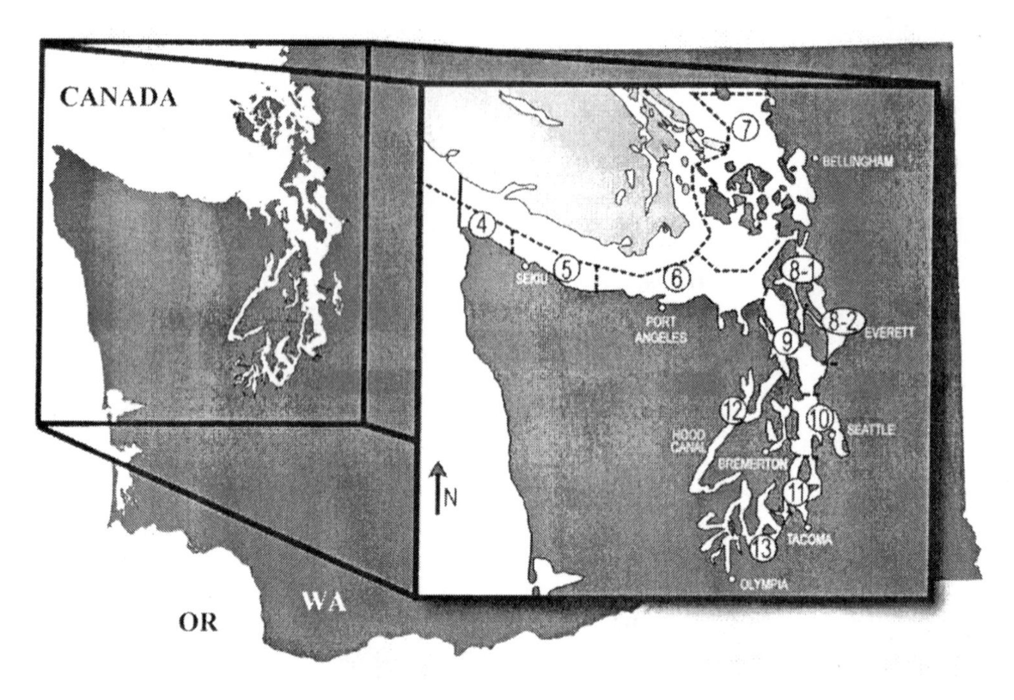

Figure 1. Map of Puget Sound marine catch areas 4–13 of Washington State.

clipped hatchery salmon that are coded wire-tagged before out-migration and subsequent CWT recoveries in fisheries, in hatcheries, and on spawning grounds. The recovered CWTs are used to calculate exploitation rates on indicator hatchery stocks and those rates are presumed to be the same as for associated wild stocks (ASFEC 1995). For the purpose of this paper, the indicator stock assumption is presumed correct. With the indicator stock assumption, the CWT system has served as a hatchery-based tool for monitoring the impact of fisheries on wild salmon stocks. With the advent of mark-selective fisheries, the exploitation rates are intentionally increased on hatchery stocks so as to purposely exploit hatchery stocks at a greater rate than wild stocks, invalidating the indicator stock assumption. Greater discrepancies are expected with larger mark-selective fisheries.

In response to this concern the Ad-hoc Selective Fishery Evaluation Committee (ASFEC) of the Pacific Salmon Commission (PSC) devised the double index tag (DIT) system where releases of hatchery salmon included both marked and unmarked coded wire-tagged fish (ASFEC 1997). Under the basic indicator stock assumption, the exploitation rates on the unmarked but tagged hatchery fish would still be applicable to associated wild stocks. In nonselective fisheries, the tags from the unmarked but tagged hatchery DIT fish could be recovered through sampling. However, in mark-selective fisheries there are expected to be incidental hook-and-release mortalities of unmarked fish that are encountered and would no longer be directly observed. Drop-off mortalities and their impacts as discussed in Lawson and Sampson (1996) are another form of incidental mortality but are not considered here. The Selective Fishery Evaluation Committee (SFEC) devised indirect estimation methods to estimate the incidental mortalities (SFEC 2002) but those methods require additional unknown input parameters such as the unmarked to marked ratios by stock and age and the incidental hook and release mortality rate. Therefore, the accuracy of the unmarked mortalities, and therefore the exploitation rates that will be used to monitor fishery impacts on wild stocks, will be impacted by the inaccuracy of those input parameters. The introduction of this risk of bias is often referred to as the information cost to the viability of the CWT system of mark-selective fisheries.

Practical Considerations

Concerns over the loss of information has lead agencies such as the Washington Department of Fish and Wildlife (WDFW) to slowly develop mark-selective fisheries and has lead to more stringent requirements in managing those fisheries than there have been in other traditional nonselective fisheries. Based on the experience gained by implementing such fisheries and contributing to the evaluation and assessment of the impacts of those fisheries, we provide a description of the key features of mark-selective fisheries to help fishery managers appreciate the importance of careful design and planning of these fisheries. We offer practical guidance to be considered when proposing a mark-selective fishery, including the importance of expected mark rate, the likelihood of angler satisfaction and participation, incidentally killed nontarget fish, and the types of sampling programs that should be implemented to help recover lost information.

Mark Rate

The mark rate is the proportion of hatchery salmon encountered by a fishery that can be visually distinguished by an external mark such as an adipose fin clip. The adipose fin clip is the mass mark used that has minimal effect on fish health (Bergstedt 1985; Thorsteinsson 2002) and can be readily identified by anglers (ASFEC 1995). Mark rates vary in the same manner that stock mixtures vary, depending on migration patterns over time and area. Hatchery production comprises half or more than half of all Chinook and coho salmon *Oncorhynchus kisutch* stocks originating from Washington, including stocks from the Columbia River, Puget Sound and the Washington coast (PFMC, 2006). However, since not all hatchery salmon are mass marked, and due to migration pattern differences, actual observed mark rates for coho salmon in Washington show significant variation between fishing areas (Table 1). They range from very low levels, generally in areas having high abundance of natural origin production but low hatchery production (e.g., the northern Washington Coast) or low marking levels of hatchery production (e.g., Puget Sound Area 8–2), to quite high levels (near 80%) in areas near the origin of large hatchery production (e.g., the Columbia River). Sampling for mark rates of coho salmon encountered in Puget Sound and Washington coastal mark-selective and nonselective recreational fisheries occurring during the summer of 2002 demonstrates this variation (Table 1).

Large variation in mark rates has also been observed within a single fishery, over the duration of the season (Table 2). This variation may be due to changes in the actual stock, noting that natural fish may have more protracted run timing than marked hatchery fish, but this observation may also be an artifact of sampling variation. The experience of our 2003 mark-selective coho sport fishery in the Strait of Juan de Fuca demonstrates this phenomenon (Table 2).

In a mark-selective fishery, a portion of unmarked fish that are released will die from stress or injury experienced during the process of handling or playing the fish (CTC 1997). If a mark-selective fishery is conducted in an area

Table 1. Observed mark rates in ocean area recreational fisheries, 2002.

Area (see Figure 1)	Observed marked percent
Neah Bay (Area 4)	39%
LaPush (Area 3)	28%
Westport (Area 2)	56%
Columbia River Ocean (Area 1)	58%
Columbia River Mouth (Buoy 10)	74%
San Juan Islands (Area 7)	46%
Port Susan Gardiner (Area 8-2)	25%
Seattle - Bremerton (Area 10)	62%
Tacoma (Area 11)	65%

Table 2. Coho mark rates from Area 5 test fishery, 2003 (data from PFMC 2004).

Week	Mark percent
27 July	25%
28 July	37%
29 July	32%
30 July	13%
31 July	31%
32 August	29%
33 August	18%
34 August	52%
35 August	29%
36 August	30%
37 September	43%
39 September	27%
Weighted average (by catch estimates)	35%

with a low mark rate, anglers may expend more effort and encounter more natural fish in pursuit of their daily bag limit than if the mark rate were greater. How high should the mark rate be?

One way to make this assessment is to compare the number of unmarked mortalities expected from the fishery assuming a nonselective status to the number expected assuming a mark-selective status. Because the comparison is made under different assumptions on the same fishery, the vulnerable population and the landed catch is kept constant. If the fishery were designated nonselective, then one would expect the number of landed unmarked fish to be in the same proportion as in the vulnerable population. On the other hand, if the fishery were designated mark-selective, then one would expect the number of unmarked mortalities to be a function of the mark rate and the incidental mortality rate (Table 3). For a particular stock, the savings of unmarked fish in a mark-selective fishery in absolute numbers from stock i (e.g. from a DIT group) is

Savings of unmarked fish =

$$U_i^{NSF} - U_i^{MSF} = CP_i \lambda_i (MR - sfm)$$

Table 3. Calculations for theoretical unmarked mortalities by type of fishery.

Basic input parameters	Common terms	
Number of fish vulnerable to fishery	V	
Mark rate	MR	
Proportion of marked fish belonging to stock *i*	P_i	
Total catch in fishery	C	
Incidental mortality rate (selective fishing hook and release mortality)	sfm	
Unmarked to marked ratio for stock *i*	λ_i	
Number of stock *i* marked vulnerable fish	$V^{M(i)} = V \cdot MR \cdot P_i$	
Number of stock *i* unmarked vulnerable fish	$V^{U(i)} = V^{M(i)}\lambda_i = V \cdot MR \cdot P_i \lambda_i$	
Derived parameters	**Nonselective fishery**	**Mark selective fishery**
Number of fish encountered to land catch *C*	$Enc^{NSF} = C$	$Enc^{MSF} = Enc^{NSF}/MR$
Encounter rate	$EncR^{NSF} = C/V$	$EncR^{MSF} = EncR^{NSF}/MR$
Total marked mortalities	$M^{NSF} = V \cdot MR \cdot EncR^{NSF} = C \cdot MR$	$M^{MSF} = V \cdot MR \cdot EncR^{MSF} = C = M^{NSF}/MR$
Total unmarked mortalities	$U^{NSF} = V(1\text{-}MR)EncR^{NSF} = C(1\text{-}MR)$	$U^{MSF} = V(1\text{-}MR)EncR^{MSF}sfm = U^{NSF}\frac{sfm}{MR}$
Stock *i* marked mortalities	$M_i^{NSF} = M^{NSF}P_i = C \cdot MR \cdot P_i$	$M_i^{MSF} = M^{NSF}P_i \quad \frac{M_i^{NSF}}{MR}$
Stock *i* unmarked mortalities (total = stock *i*)	$U_i^{NSF} = M_i^{NSF}\lambda_i = C \cdot MR\, P_i \lambda_i$	$U_i^{MSF} = M_i^{MSF}\lambda_i\, sfm = U_i^{NSF}\frac{sfm}{MR}$
Marked harvest rate (total = stock *i*)	$HR^{M,NSF} \quad \frac{C}{V}$	$HR^{M,MSF} \quad \frac{HR^{M,NSF}}{MR}$
Unmarked harvest rate	$HR^{U,NSF} \quad \frac{C}{V}$	$HR^{U,MSF} \quad HR^{U,NSF}\frac{sfm}{MR}$

The savings are essentially a function of the exploitation rate (as measured through catch numbers, CP_i) and the difference between the mark rate and the incidental hook and release mortality rate ($MR - sfm$). The break-even point, where the number of unmarked mortalities is the same under both nonselective and mark-selective statuses is when the mark rate is equal to the incidental mortality rate and the savings of unmarked fish is zero. The savings of unmarked fish increases with mark rate. However, because the incidental mortality rate is an unknown quantity and the mark rate to be encountered must be presumed before the fishery is conducted, a more conservative approach would be to only designate mark-selective fisheries where it can be argued that the mark rate encountered is expected to be significantly greater than the incidental hook-and-release mortality rate.

Angler Satisfaction

Low mark rates may result in increased encounters of unmarked fish as anglers pursue rare marked fish they are allowed to retain, offsetting one of the benefits of selective fisheries – the reduction of mortalities to unmarked fish. However, angler behavior may limit the downside of this potential effect. Angler participation is directly related to catch success and low mark rates translate into fewer fish for anglers to retain. So although anglers who participate in selective fisheries with very low mark rates will encounter greater numbers of unmarked fish, fewer anglers will participate, reducing the total number of unmarked fish encountered. Since one purpose of conducting a selective fishery is to provide meaningful recreational opportunity, fishery managers need to be aware of the potential that low mark rates may result in low angler satisfaction and low participation.

Exploitation Rates

Exploitation rates are key management parameters that describe the proportion of fish mortalities due to a particular fishery or set of fisheries. The proportion can be relative to the brood, year, life history, or the mature subpopulation, depending on the application. They are generally defined by stock and sometimes by age.

A successful mark-selective fishery will remove more marked fish from the population than unmarked fish. How many more will depend on the exploitation rate of the fishery and can be considered the mark-selective impact. Because more marked fish are removed from the population than unmarked, one expects that the mark-selective impact would be evident in the escapement. Therefore, monitoring the differences in DIT escapement recoveries is one way to monitor the degree of the mark-selective impact. Using Table 3, the mark-selective impact can be calculated as:

Mark-selective impact $= M_i^{MSF} - U_i^{MSF} = CP_i (1 - sfm\lambda_i)$

The impact on a specific stock is essentially a function of the exploitation rate (as measured through catch numbers, CP_i). Interestingly, it is not a function of the general mark rate (MR) but only of the stock specific unmarked to marked ratio λ_i.

The SFEC has indicated concern with mark-selective fisheries large enough to produce a detectible mark-selective impact (SFEC 2002) because larger mark-selective fisheries have greater potential to incur bias on CWT-based exploitation rates for unmarked fish. However, if the extra parameters necessitated by the conversion of a fishery to mark-selective status are accurate, then the DIT system of estimating exploitation rates for any size mark-selective fishery should be adequate. Therefore, either the fishery manager should be able to make the argument that the mark-selective impact is too small to be detected, or be able to defend the additional input parameter values.

In 2003, a joint workgroup involving the WDFW, the Western Washington Treaty Tribes, and the Northwest Indian Fisheries Commission analyzed the returning DIT data for coho salmon for brood years 1995–1997. They found that although the estimated simple exploitation rates for 33 coho salmon mark-selective fisheries in 1998–2000 ranged from essentially 0–15%, the mark-selective impacts were virtually undetectable in the hatchery escapements (NWFRB 2003).

Stocks Impacted

A mark-selective fishery that is likely to intercept stocks of concern to other jurisdictions will most likely require coordinating with those jurisdictions. For example, the PSC monitors exploitation rates for several chinook and coho salmon stocks as part of implementing the Pacific Salmon Treaty. WDFW conducted a midsummer mark-selective fishery on Chinook salmon in the preterminal Strait of Juan de Fuca in 2003. Historically, the fisheries in that time and area intercept many different stocks, several of which are of concern to the PSC. Therefore, this fishery underwent a lengthy and thorough review by the SFEC before implementation.

Carry-Over Impacts

Depending on the method used to estimate the unmarked incidental mortalities, there may be carry-over loss of information impacts to other fisheries that encounter the same subpopulations of fish. For example, mark-selective fisheries in preterminal areas will differentially exploit immature marked and unmarked fish. The population changes caused by that differential exploitation will persist to other fisheries and for Chinook salmon, to future years. The carry-over means that the other fisheries impacted may need to add sampling programs, that those other fisheries exploitation rates may also be biased, or that their season lengths may need to be adjusted.

Sampling Programs

A management agency can mitigate for potential loss of information by implementing additional or expanded sampling programs. For example, using a DIT program, all fish in all affected fisheries, regardless of mark status, are interrogated for CWTs. This complete sampling procedure has required agencies to invest in electronic tag detection equipment and training programs to ensure effectiveness.

To be effective and efficient, sampling programs using electronic tag detection equipment must be tailored to the logistics of the sampling situation (Blankenship and Thompson 2003). In situations with relatively low volumes of fish, hand-held wands are usually sufficient and samplers have the time to ensure quality interrogation of fish sampled. However, in most commercial sampling situations, for example those for WDFW, where volumes are high and access time for sampling is limited, stationary tube detection devices and V-type detection devices (Blankenship and Thompson 2003) are employed in addition to wands. Species, size variation of fish sampled, and mobility needs should be considered in designing the most efficient sampling program. For example, sampling recreational fisheries requires more mobility, so wands would be more desirable and mouth wanding techniques may be required to prevent tags from being undetected (Vander Haegen et al. 2002) for larger Chinook.

In addition to CWT detection and recovery, sampling programs applied to mark-selective fisheries should collect a variety of other information useful to managers as they analyze stock-specific fishing effects. Although the types of information collected will vary with the circumstances of the particular fishery, the following general objectives for sampling programs are suggested:

1. Estimates of total fishing encounters. Shore-based sampling programs cannot provide direct sampling information for fish encountered and subsequently released. However, participant surveys can be designed to provide estimates of the total number of fish encountered, caught or released. Angler surveys have been conducted with sport fisheries, both nonselective and selective, in Washington for sev-

eral years, with the intention of estimating the effect of size limits, mark-selective retention regulations, and species nonretention regulations. For mark-selective fisheries, upon completion of a fishing trip, anglers are asked to recall the number of fish released, by species and mark status.

2. Estimates of mark rates. On-water sampling of the fish population encountered by anglers provides mark rate information that is unobtainable from shore-based sampling. Given differences in run timing and inconsistent mark rates by stock and age, sampling designs should be stratified by time period. On-water sampling methods should reflect the gear types used and consider the potential for species or size selectivity.

Alternative Management Strategies

Alternative management strategies that may achieve similar objectives as mark-selective regulations, but with fewer resource demands, could also be considered. For example, lower bag limits can be used to extend fishing seasons for the same number of encounters or to lower encounter rates for the same fishing season. On the other hand, increased fishing effort may be useful in removing greater numbers of hatchery fish from a river system.

Summary and Discussion

Guidelines for fishery managers considering mark-selective fishing regulations are as follows:

• The mark rate encountered should be significantly greater than the incidental hook and release mortality rate. Otherwise, for a fixed total catch, more unmarked fish may be exploited through prosecution of a mark-selective fishery in pursuit of that catch than through prosecution of a nonselective fishery.

• Fisheries with small impacts on individual stocks have less potential for incurring bias in exploitation rate parameters. Examples of such fisheries are small terminal fisheries exploiting few stocks or large preterminal fisheries exploiting many stocks so the impact on individual stocks is small.

• Fisheries with large impacts on individual stocks have more potential for incurring bias in exploitation rate parameters. This potential is tied to the accuracy of supplied parameters such as the incidental hook and release mortality rate and the ratio of natural to hatchery fish by stock and age. Examples of such fisheries are very large preterminal fisheries or large terminal fisheries exploiting few stocks.

• Impacts caused by fisheries on mature fish (i.e., terminal fisheries) are not carried over to fisheries in other years.

• Projected bias in exploitation rates for all affected stocks in fisheries on immature fish should be calculated to determine carry over effects.

• Additional sampling efforts may be needed (in the particular fishery and elsewhere) to recover encounter information on unmarked fish. For example, the mark rate may be monitored through test fishing or through additional angler interviews.

• Alternative management strategies for achieving the same objectives for less cost should also first be considered.

Acknowledgments

We thank the WDFW Ocean Sampling and the Puget Sound Sampling Programs for collecting and summarizing the weekly mark rate data presented.

Reference

Ad-hoc Selective Fisheries Evaluation Committee (ASFEC). 1995. Joint Ad-hoc Selective Fishery Evaluation Committee Report to the Pacific Salmon Commission, Vancouver, BC, TCASFEC (95)-1, June 1995. Pacific Salmon Commission, Vancouver.

Ad-hoc Selective Fisheries Evaluation Committee (ASFEC). 1997. Reliability and feasibility of using electronic detection for recovery of coded wire tags in coho salmon. Joint Ad-hoc Selective Fishery Evaluation Committee Report to the Pacific Salmon Commission, Vancouver, BC, TCASFEC (97)-1, February 1997. Pacific Salmon Commission, Vancouver.

Bergstedt, R. A. 1985. Mortality of fish marked by finclipping; an annotated bibliography, 1934–1981. U.S. Fish and Wildlife Service, Great Lakes Fishery Laboratory, Administrative Report No. 85–3, Ann Arbor, Michigan.

Blankenship, H. L., and D. A. Thompson. 2003. The effect of 1.5-length and double-length coded wire tags on coho salmon survival, growth, homing, and electronic detection. North American Journal of Fisheries Management 23:60–65.

Chinook Technical Committee (CTC). 1997. Incidental fishing mortality of Chinook salmon: mortality rates applicable to Pacific Salmon Commission fisheries. TCCHINOOK (97)-1. January, 1997. Pacific Salmon Commission, Vancouver.

Lawson, P., and D. B. Sampson. 1996. Gear-related mortality in selective fisheries for ocean salmon. North American Journal of Fisheries Management 16:512–520.

North West Fisheries Resource Bulletin (NWFRB). 2003. Analysis of coho salmon double index tag (DIT) data for the brood years 1995–1997. Project Report Series No. 12 of the Northwest Fishery Resource Bulletin, Northwest Indian Fisheries Commission, Olympia, Washington.

Pacific Fishery Management Council (PFMC). 2006. Review of 2005 ocean salmon fisheries. Pacific Fishery Management Council, Portland, Oregon.

Selective Fisheries Evaluation Committee (SFEC). 2002. Investigation of methods to estimate mortalities of unmarked salmon in mark-selective fisheries through the use of double index tag groups. Joint Selective Fishery Evaluation Committee Report to the Pacific Salmon Commission, Vancouver, BC, TCSFEC (02)-1, February 2002. Pacific Salmon Commission, Vancouver.

Thorsteinsson, V. 2002. Tagging methods for stock assessment and research in fisheries. Report of Concerted Action FAIR CT. 96.1394 (CATAG). Reykjavik. Marine Research Institute Technical Report (79).

Vander Haegen, G. E., A. M. Swanson, and H. L. Blankenship. 2002. Detecting coded wire tags with handheld wands: effectiveness of two wanding techniques. North American Journal of Fisheries Management 22:1260–1265.

American Fisheries Society Symposium 49:461

Who Owns the Fish and What Are They Worth to Society?—Defining Ownership, Resolving Conflict, and Evaluating Costs and Benefits to Society While Reconciling Harvest Fisheries with Conservation

Steve Dunn*, Douglas J. Chapman, and Steven J. Kennelly
NSW Fisheries, Post Office Box 21, Cronulla, 2230, New South Wales, Australia

Abstract.—Everyone on our blue planet faces common fisheries management dilemmas. The world's fisheries are under increasing pressure from harvest by recreational, commercial, traditional, illegal, and unregulated fishing. Increasing coastal populations want to enjoy access to natural resources. Coastal development is destroying nursery grounds and other habitats. Urban and industrial pollution goes into our waterways and oceans, at best modifying ecological communities, at worst destroying them. Unscrupulous sectors of the industry operating both legally and illegally are targeting remote and high seas stocks—often with the blessing of their flag jurisdictions. In an environment of uncertainty, and with a paucity of meaningful data, the world is allowing fisheries resources to be placed under incredible pressure, we are fishing down food webs, and they simply cannot sustain it. Identifying who owns the fish is an important first step in the fisheries management process, but it makes little difference if the owners are ambivalent to, or unaware of, the consequences of unconstrained impacts, which in most cases they are. It is naïve to imagine that creating issue awareness among owners will somehow result in the necessary changes. In many cases, it just forces poorly resourced bureaucracies to move clumsily between agendas, in reaction to the latest "flavor of the month" campaign.

However, it may not be all bad. In some places in the world, systematic yet opportunistic approaches are showing positive signs of delivering outcomes that address at least some of these concerns. This paper describes the processes that one jurisdiction is following to strike an appropriate balance. It reflects on political opportunity, places the work being done at a localized level within a broader geographic context, and reflects on the costs and benefits.

Fisheries management needs to be administered under modern legislation that provides a structured hierarchy of objectives that embraces ecologically sustainable development and places conservation ahead of exploitation. An approach needs to be taken to develop an overarching framework that provides for issues to be managed in a proactive and timely way as they emerge. This case study comes from Australia, where, unconstrained by much of the historical baggage of Europe and North America, jurisdictions determinedly set about rights-based fishery management in the 1980s such that now all significant fisheries are limited access and strictly regulated, either through transferable quotas or broad controls on effort. By the early 1990s and concurrent with this trend towards limited access, rights-based management, debate was raging about the ecologically sustainable management of harvest sectors, the establishment of marine protected areas, controlling aquatic pests, the protection of threatened species, and resolving aquatic habitat protection issues.

*Corresponding author: steve.dunn@fisheries.nsw.gov.au

Session

Across Boundaries

American Fisheries Society Symposium 49:465–481

Barriers to World Trade in Fisheries Products: Are Markets Fair?

HENRI MOTTE

Sectoral Advisory Group in International Trade (SAGIT): Fish and Seafood Products

DANIEL E. LANE*

Telfer School of Management, University of Ottawa; and SAGIT: Fish and Seafood Products
55 Laurier Avenue, Ottawa, Ontario, K1N 6N5, Canada

Abstract.—This paper inquires about the advantages to coastal states and their ability to secure a preferred position important to fish markets. In this study, the worldwide trade of fish products over the past decades is examined. The result of this study has led to the realization that lines of trade are well established and favor particular blocs of importers and exporters in trade affinities and implied trade barriers. In particular, European markets now represent a significant and lucrative trading bloc for fisheries products. However, competition among some European suppliers has effectively divided these markets into a patchwork of sufficient market divisions for fish products at the exclusion of nonEuropean countries. The reasons for this evolution include increasing costs of air transport to markets, a lower cost of production to some states, and resource crises, in particular, for groundfish among traditional suppliers. It is suggested that fair trade can be carried out on fish products as long as the environment is sustainable and that this implies shared responsibility within trade blocs and between developing and developed trade partners. Furthermore, faced with barriers to fair trade from trading arrangements, it is concluded that commercial bilateral agreements between local trading partners have more possibility of occurring than large-scale political and economic driven changes designed to promote trade and remove these same barriers.

Introduction

Fish and seafood products have long been an essential food source for coastal marine communities. Advances in marine harvesting, food preservation, transportation, and trade linkages have meant that today, products from the world's seas can be found throughout the world's markets based on supply and demand. Fish landings and consumption grew 30% faster than the population in the second half of the 20th century, increasing from 20 million metric tons in 1950 to 80 million metric tons in 2001 (FAO 2003).

Global fisheries supply and demand have however been subject to fluctuations from a variety of sources, including changes in weather to shifts in fashionable diets and health trends. The disappearance of lucrative cod in the 1400s were attributed to storms over the English Channel and caused the push of Euro-

*Corresponding author: dlane@uottawa.ca

pean fishermen toward the coast of Iceland and west to North America in search of more stable supplies (Kurlansky 1997). These historical fluctuations led to modern international trade in fish and seafood products. Similarly, the practice of salting fish for safe transit has endured since the 1500s and helped establish ongoing lines of international trade. New principles of trade have since developed with some countries' comparative advantage in transportation, together with differences in the cost of labor.

Notions of fairness in international trade is a normative issue that evokes aspects of justice, equality, morality, and equity. Unfortunately, these normative terms are very difficult to define since one group's definition of equality may imply inequality to a disadvantaged group. Fair trade is most often associated with defining normative definitions within trading practices especially in the trade between developing and developed countries. Also known as trade justice, the active fair trade movement promotes innovative, market-based solutions to sustainable development and removal of access barriers to international markets by developing countries. Advocates of fair trade practices contend that commodity price fluctuations do not allow producers in developing countries to reduce their debt since market prices do not reflect real production costs, including environmental and social costs, and that the rich regions, including North America and Europe, effectively subsidize domestic producers (Fair Trade Federation 2004; DATA 2005; IFTA 2005). The notion of fair markets considered in this paper implies that trading lines are approximately evenly distributed (i.e., that one region's ability to export is roughly proportional to the ability of other regions' to import and that each region's imports are similarly proportional to all other regions' ability to export). In particular, the globalization debate is concerned about whether freer global trade lines are making the rich regions richer and the poor regions poorer as a consequence of expanding multinational firms (Suranovic 2007). With respect to this concern then, fairness in international trade markets corresponds in this study to a possible narrowing of the trade lines that exist between the different regions of the globe for fish and seafood products.

In this paper, recent trade patterns of international regions representing the major trading countries in fish and seafood products are analyzed. This analysis reveals emerging trade patterns and trade affinities among some countries and points out implied trade barriers that limit the potential growth and development of trade in fish and seafood products for countries in some regions. Improved opportunities for social and economic growth, from trade by countries like Canada, can allegedly be promoted through global political mechanisms such as the General Agreement on Tariffs and Trade (GATT; WTO 2003) or the World Trade Organization (WTO 2005). However, the evidence of frustration of some countries with global approaches suggests that more localized political decisions through bilateral agreements between commercial trading agents in different regions and countries are more effective in realizing liberalized trade opportunities.

Description of the World Trade in Fish and Seafood Products

International trade in fish and seafood products are grouped into eight dominant regions that approximate the Food and Agriculture Organization (FAO) GlobeFish demarcations (FAO 2000). The regional definitions used in this analysis are assumed to cover global fish trade. This paper also assumes that trade statistics derived from the FAO are in partial equilibrium (i.e., balanced global imports equal global exports in monetary value with the inclusion of an arbitrary other trade region). Regional country memberships and summary global trade statistics by region are identified in Table 1.

Global Export–Import Trade by Regions

In terms of the value (in thousands of current U.S. dollars, unless otherwise noted) of total world trade, fish and seafood products exports increased to a value of more than $55 billion in 2000. In the global context, trade in fish and seafood products represents an important source of foreign exchange for many coastal countries. Thailand (represented as part of the Asia regional group in Table 1) continues to be the leading single exporting country followed by China whose exports have increased dramatically since the early 1990s. The FAO State of the World Fisheries and Aquaculture Report for 2002 (FAO 2003) also notes that China reprocesses imported raw material for export in a value-added process. World export activity by regional group as defined in Table 1 is summarized in Figure 1 for the period 1976–1901 (FAO 2004).

Developed countries accounted for 80% of total fish imports by value in 2000. Japan has been the largest single importer nation followed by the United States. The European Community has become the third largest importer of fish products while the Europe regional group (including the principle fishing nations of Norway, Denmark, Iceland, Spain, the United Kingdom, The Netherlands, and France) is the largest of the regions in imports by value (Table 1) at 40% of all exports. World import activity by region is summarized in Figure 2 for the period 1976–1901 (FAO 2004). Figure 3 summarizes the regional trade balance for fish and seafood products for the

Table 1. Regional country memberships for international trade in fish and seafood products. (Source: FAO 2004.)

No.	Region identifier	Key member nations	Import value ($US B)	Imports % by value	Export value ($US B)	Exports % by value
1*	North America	USA, Canada	$10.978	19%	$6.129	11%
2*	Oceania	Australia, New Zealand	$0.592	1%	$1.668	3%
3*	Europe	Norway, Denmark, Iceland, Spain, UK, Netherlands, France	$22.670	40%	$16.854	29%
4*	Japan	Japan (only)	$14.846	25%	$1.396	2%
5	Africa	Morocco, South Africa, Tanzania	$0.556	1%	$3.042	5%
6	Central and South America	Chile, Peru, Argentina, Mexico	$1.023	2%	$7.910	14%
7	Asia	Thailand, Korea, India, Indonesia, Viet Nam	$3.615	6%	$12.304	21%
8	China	China (only)	$3.212	5%	$4.967	9%
9	Other	All other nations	$0.508	1%	$3.765	6%
Total trade ($US M, 5-year average)			$58.000	100%	$58.000	100%

* Denoted regions of "developed" nations, all regions (Oceania excepted) have negative balance of trade in fish and seafood products; all other regions are composed of "developing" nations.

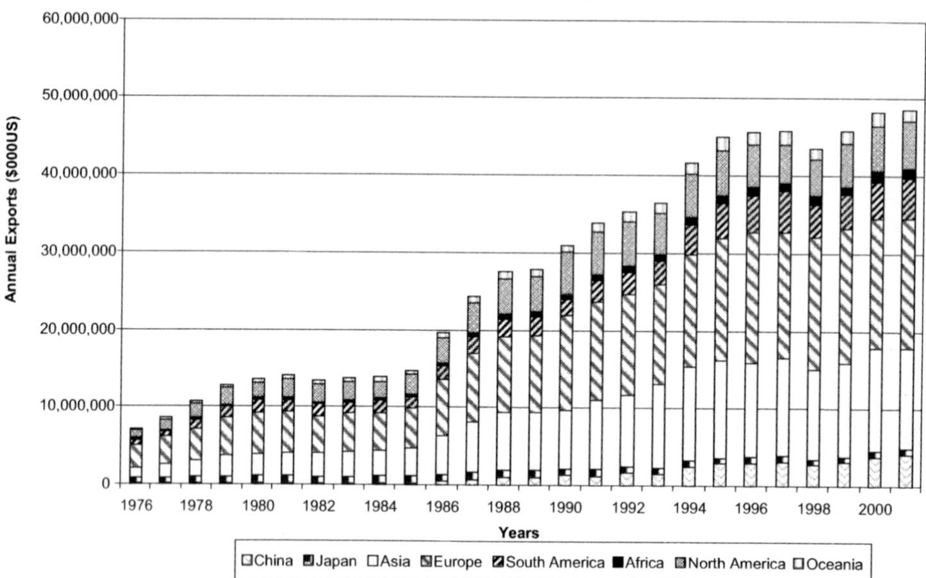

Figure 1. World export value by regional group, 1976–2001. (Source: FAO Annual Statistics, FAO 2004 and FishStat database.)

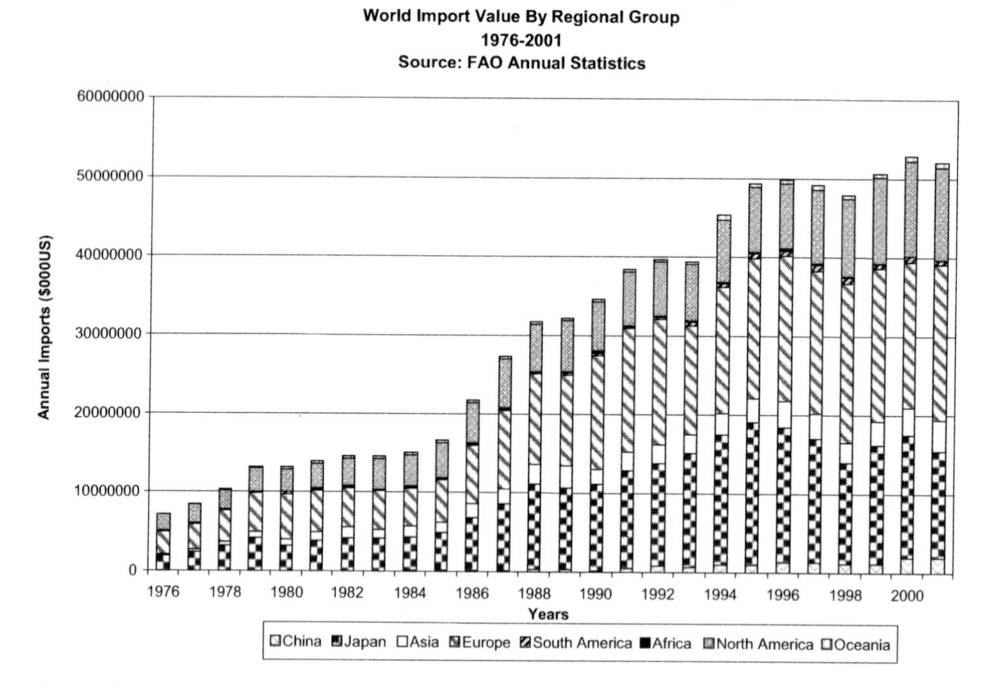

Figure 2. World import value by regional group, 1976–2001. (Source: FAO Annual Statistics FAO 2004 and FishStat database.)

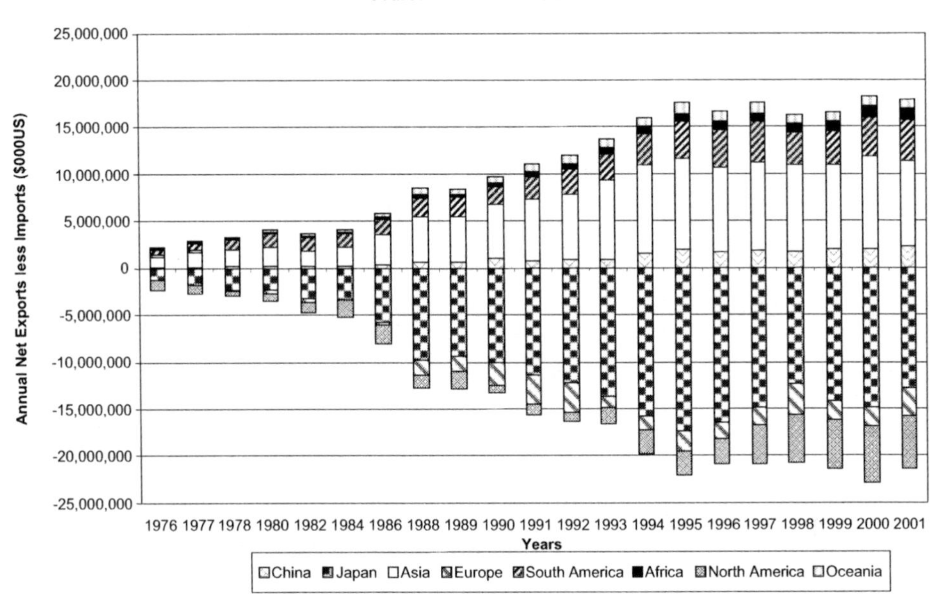

Figure 3. World net export less import value by regional group, 1976–2001. (Source: FAO Annual Statistics, FAO 2004 and FishStat database.)

period 1976 to 1901. Fisheries trade surplus is significant for the regions of South and Central America, Africa, China, and Oceania. Deficits in fisheries trade are experienced by the developed regions of Japan, North America, and Europe, all of whom are historically net importers of fish products.

The FAO State of the World Fisheries and Aquaculture Report for 2002 (FAO 2003) noted that in recent years, developing countries supply more than 50% of the total value of global exports. Trade value now flows from the less developed regions (South and Central America, Africa, Asia, China) to the more developed regions (North America, Europe, Japan). In proportional terms, nearly three-quarters of trade value of imports in fish and seafood products arises from the three developed regions: North America, Europe, and Japan. Import trade by value among the developing regions of the world amounts to only about 20% in recent years.

Table 2 presents the region by region breakdown of global fish trade exports and imports by all regional groups. Trade flows take place from the developing regions (net exporters) to the developed regions (net importers) of the world (i.e., generally from south to north). Assuming similar difficulties worldwide with assessing and estimating stock status, one potential concern of this directional trade flow is that there are conservation concerns for some exploited stocks. Stocks that are exploited for international demand are not directly associated with stock status concerns, and nations not wanting to lose economic markets may be reluctant to adjust fishing capacity until it is too late to do so from a conservation or sustainability perspective.

Commodities Trade

Fish and seafood products are defined in specific terms of the Standardized International

Table 2. Regional description of global fish trade exports and imports. (Source: FAO 2004.)

No.	Region	Trade description
1	North America	U.S. and Canada membership in this region already commercially linked; Canada is USA's largest trading partner; USA (and region) is the world's second importer by value; 66% of imports are from developing regions, notably from Central and South America, Asia, and China; actual annual regional import and exports increased appreciably after 1986; region's export at 20% of global value during late 1970s and 1980s fallen off post-1986 period to less than 15%; Imports increased to 25% in the mid-1980s falling off to less than 20% by the early 1990s, over 20% by 2001.
2	Oceania	Trade by region only 1% of total global fish trade by value; rate of growth of positive balance of trade in fish products increasing since 1976; International agreements in landings concluded between New Zealand and foreign fleets fishing in the exclusive economic zone; more than 50% of exports in high-valued fish products (e.g., shrimps and prawns).
3	Europe	Leading region in global fisheries trade attributed to coastal proximity, cultural and historical dependences, dedicated national fleets; post-1977 and United Nations Convention on theLaw of the Sea adapt to changes leading up to the Common Fisheries Policy and devolution of fishing capacity (Holden 1994); from 1976 to 1986, exports doubled, and doubled again by 1996 in nominal value (annual growth rate of 8–10% over this period); increases led by Norway, Denmark, Iceland, Spain.
4	Japan	Rapid growth in fishing capacity, 1950–1988; catches declined appreciably in recent years, landings returned to 1950s levels; global regional import leader; Japan more dependent on imports due to increasing domestic demand for seafood; post-1986, imports to Japan doubled in value then fell by mid-1990s leveling off by 2000; Japan trades with all regions, but principle trading partners include Asia, China, and North America.
5	Africa	Difficulties of low capacity and industrialized resources for capacity expansion; fisheries trade improved due to worldwide demand for higher valued shrimps and prawns; African exports increased steadily after 1986; trade flows primarily with European trading partners.
6	Central, South America	Region claims significant marine resources; active exporter to North American region trading partners; exports increased after 1986 to a maximum in 1997; negative effects on trade in 1998 due to El Niño; exports risen to near the high 1997 level.
7	Asia	Region's fisheries, led by global leading exporting country Thailand, shown appreciable growth post-1986; growth fuelled by major commodity groups, including shrimps, tunas, and fish; imports rise after 1986, but at a slower rate than export activity of the region.
8	China	Trade evolved slowly until the mid-1990s; exports in shrimps and fish increased dramatically after; from mid-1900s and 2001, exports triple in value; since growth, China a significant participant in international trade in fish and seafood products.

Trading Commission (SITC) divisions (FAO 2004). The major groupings of the SITC are noted as follows: (1) groundfish, (2) crustacean and mollusks, (3) fish, (4) flat fish, (5) pelagics, (5) salmon, (6) tuna, and (7) other. Figure 4 illustrates the export trade by major SITC commodity for the regional groups of Table 1. Crustacea and mollusks are the leading traded fish commodity by value followed by fish (not including groundfish or flatfish products), tuna, salmon, and small pelagics. Europe, primary exporter (29%), exports appreciable amounts of all commodities, including the leading export region for groundfish (accounting for 60% of global groundfish exports), fish and pelagics (40%), and crustacean and mollusks (20%) based on 1997–2001 export value averages. Central and South America export approximately 20% of crustacean and mollusks followed closely by exports from Asia and China. North America is the leader in exports value of salmon products at 42%. The leader in exports of tuna by value is Asia at 53%. As a whole, Asia is the second leading exporting region at 21%. Africa and Asia are export leaders in pelagics by value at about 20%.

Imports reflect the demand for fish and seafood by the developed regions that account for more than 80% of total imports (Table 1). Figure 5 illustrates the import trade by major SITC commodity by the regional groups. Europe dominates proportional imports of commodity groups (with the exception of crustacean and mollusks and tuna) with an average by commodity of 55% of total imports. Japan dominates the crustacean and mollusks and tuna commodities with approximately 30% of total imports by commodity group. Together, North America, Europe, and Japan account for nearly 90% of all imports of crustacean and mollusks, 96% of salmon imports, 85% of fish imports, 88% of pelagic imports, and 83% of tunas. These figures indicate the clear responsibilities of the domi-

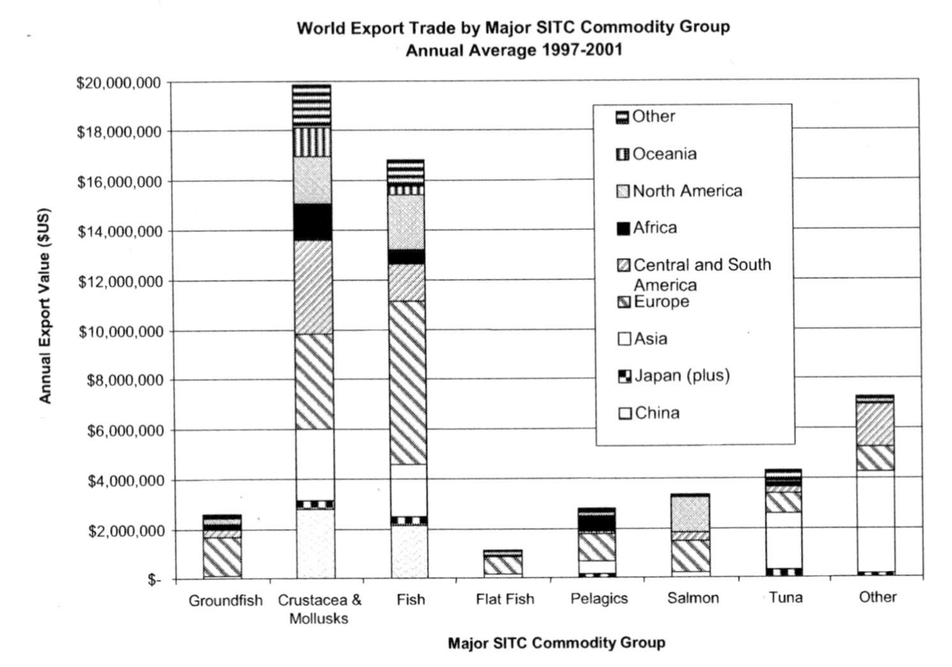

Figure 4. World export trade by major SITC commodity group, annual averages, 1997–2001. (Source: FAO Annual Statistics and FAO 2004 and FishStat database.)

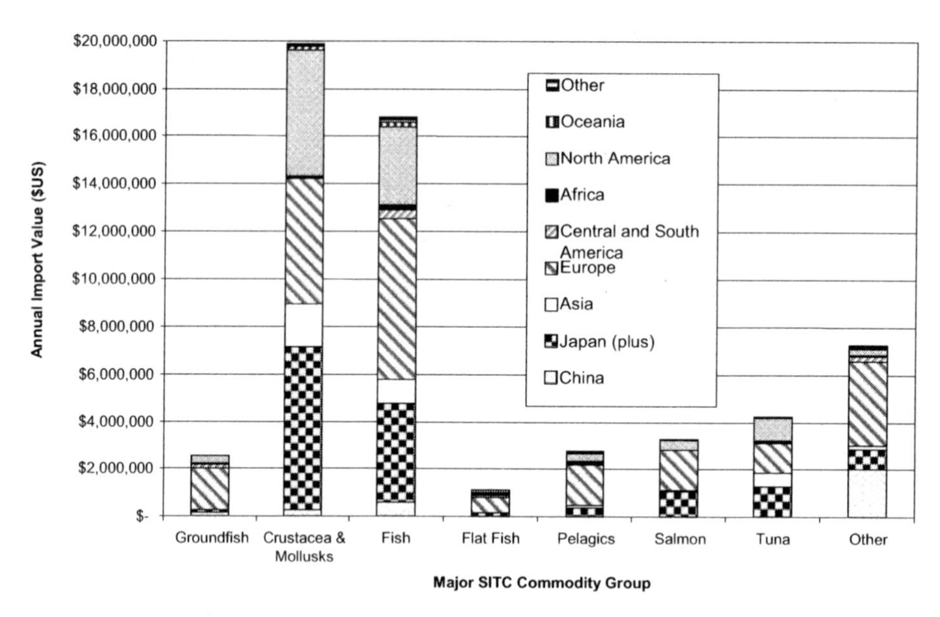

Figure 5. World import trade by major SITC commodity group, annual averages, 1997–2001. (Source: FAO Annual Statistics and FAO 2004 and FishStat database.)

nant, developed, importing regions to exact conservation impacts from their exporting developing nations trading partners while assisting them in sustaining their resources and to assure accountability and monitoring of the resources at global political fora as well as at the level of local commercial decision making.

Trade Affinity

The division of developing net exporting countries and developed net importing countries accentuates well-defined trade partners. The major trading partners for exporting regions and importing regions, respectively, as well as the identified trade affinities for international trading partners are discussed below. The particular case of Canada's position in the North American region trading block is also presented in further detail below.

International Trade Relationships

From Figure 6, in value terms, North America and Europe are developed country exporters, exporting 90% of their fish and seafood to developed regions within and across North America, Europe, and Japan. North America exports three-quarters of its fish and seafood split evenly between either Japan or the countries within its own region (i.e., between Canada and the United States) and another 15% of its exports by value to Europe. Europe's exports are dominated by trading within the European community (78% or $13 billion), with less than 8% each to North America and Japan. The remaining 10% of North American and European exports are shipped to all other regions. European internal total export trades are six times higher than North America's while they trade approximately $1 billion of total fish exports value to each other.

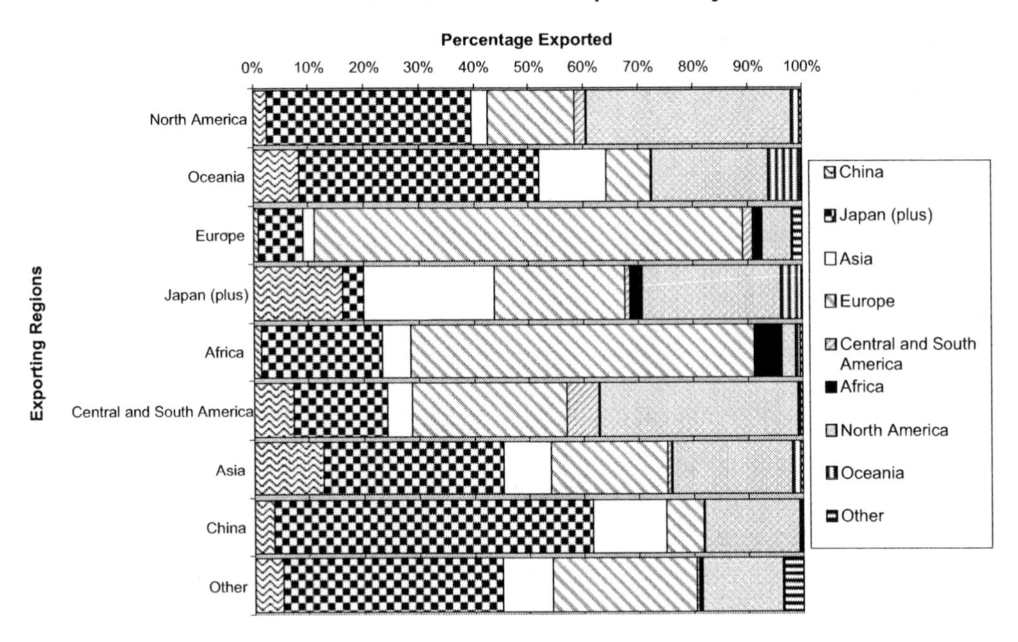

Figure 6. Total annual trade export value affinity by regional group. (Source: FAO Annual Statistics and FAO 2004 and FishStat database.)

As a developing region, the Africa region (including principle fishing nations Morocco, South Africa, and Tanzania) trades a significantly larger proportion of its exports (63%) with Europe than with any other region. This proportion is three times what Africa as a region exports to the leading global importer and second leading importer from Africa, Japan. Africa, Asia, and Central and South America export about $2 billion to Europe while China exports only $0.5 billion to Europe or 7% of its total exports by value. Central and South America export slightly more to North America than to Europe, while Asia exports about the same amount by value to there same regions. China exports nearly 60% of all its export value to Japan, with under 20% being exported to North America.

To meet its highest demand for fish and seafood products ($22 billion), Europe as a region imports a majority of $13 billion (58%) from within its region (Figure 7). The remaining 40% of its imports come from Asia, Central and South America, and Africa. Total imports originating from each of these regions exceed European imports from North America (4%). North American imports are distributed roughly evenly across Central and South America, Asia, and North American countries (average of 24% from each region). North America imports only about 8% ($0.75 billion) from Europe, or approximately twice by percentage of what Europe imports from North America (4% or $1 billion). Japanese imports of $14 billlion are distributed across Asia (27%), China (20%), North America (15%), and Europe and Central and South America (9% each).

These trade patterns demonstrate regular trade affinities between trading regions and also reveal trade barriers between other regions. For example, with respect to total products, strong

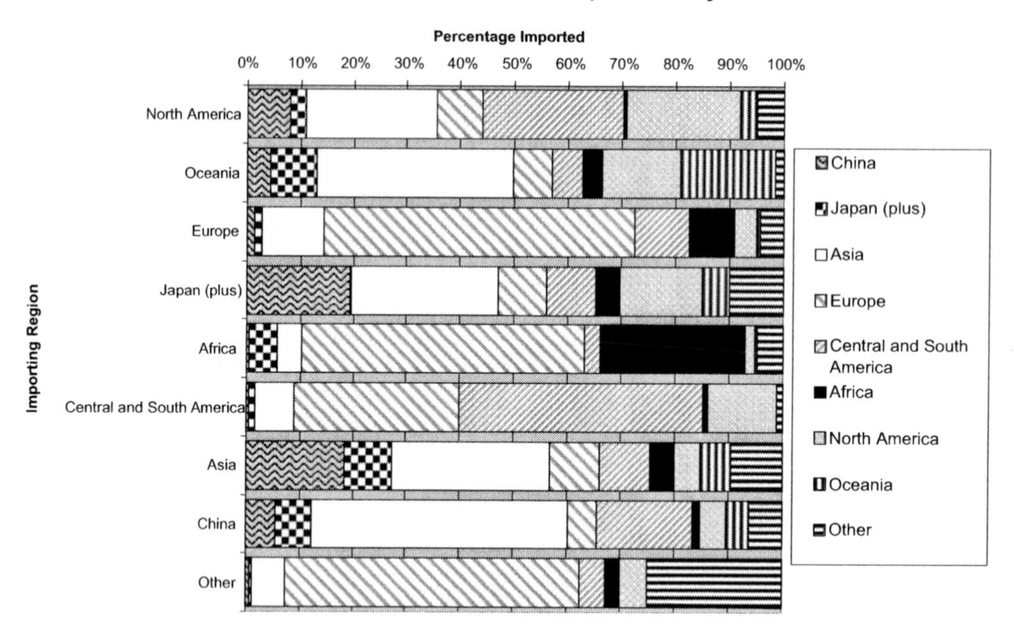

Figure 7. Total annual trade import value affinity by regional group. (Source: FAO Annual Statistics and FAO 2004 and FishStat database.)

trade affinities exist between Europe and Africa, where more than 60% of African fish and seafood are destined for European markets. However, Africa exports less than 22% to Japan, the worlds' leading import nation, and less than 3% of its $3 billion products to North America. Similarly, a significant share of exports by Oceania to Japan (44%) represents a strong trade geographical-based affinity for Oceania (even though the total trade exports of this region is relatively low at $1.7 billion or only 3% of total global export trade).

Analysis of International Trade Affinities

In order to consider fairness among actual international trade in fish and seafood products, we examine the total flow of imports and exports among the regions and then consider how these totals might be expected to be traded according to even proportional flows. Consider an example of three trading regions, A, B, and C, with total fish and seafood annual proportional exports by value (percent per millions of U.S. dollars) of (PEV_A, PEV_B, PEV_C) = (50%, 30%, 20%); and total fish and seafood annual proportional imports by value (percent per millions of U.S. dollars) of (PIV_A, PIV_B, PIV_C) = (20%, 30%, 50%). The expected flows from region i to region j are the simple products of $PEV_i \times PIV_j$, where i and j are trading regions A, B, and C, (i.e., the expected trade from region A to region C are PEV_A*PEV_C = 50%*50% = 25% of total trade flows). Expected trade flows for all eight regions (including other) by percentage of value in fish and seafood products for 2001 are presented in Table 3 as a function of total flows. Variances from these expected flows are evidence of stronger than expected trade affinities and, at the same time, may be indicative of potential trade barriers.

Examining the expected trade versus the actual total trade exchanges between regions

Table 3. Expected proportional trade flows of fish and seafood products among the regions based on total percentage imports and exports for 2001. (Source: FAO 2004. Note: Trade flows to and from the same region represent within region flows.)

	Importing regions									
Exporting regions	North America	Oceania	Europe	Japan	Africa	CSA	Asia	China	Other	Total exports (%)
North America	2.1%	0.6%	5.5%	0.4%	1.0%	2.7%	4.0%	1.7%	1.1%	19%
Oceania	0.1%	0.0%	0.3%	0.0%	0.1%	0.1%	0.2%	0.1%	0.1%	1%
Europe	4.4%	1.2%	11.6%	0.8%	2.0%	5.6%	8.4%	3.6%	2.4%	40%
Japan	2.8%	0.8%	7.3%	0.5%	1.3%	3.5%	5.3%	2.3%	1.5%	25%
Africa	0.1%	0.0%	0.3%	0.0%	0.1%	0.1%	0.2%	0.1%	0.1%	1%
CSA	0.2%	0.1%	0.6%	0.0%	0.1%	0.3%	0.4%	0.2%	0.1%	2%
Asia	0.7%	0.2%	1.7%	0.1%	0.3%	0.8%	1.3%	0.5%	0.4%	6%
China	0.6%	0.2%	1.5%	0.1%	0.3%	0.7%	1.1%	0.5%	0.3%	5%
Other	0.1%	0.0%	0.3%	0.0%	0.1%	0.1%	0.2%	0.1%	0.1%	1%
Total imports	11%	3%	29%	2%	5%	14%	21%	9%	6%	100%

with significant trade reveals substantial differences in trade patterns between trading regions in pairs. As noted above, in cases where expected trade amounts are well below expectations, then trade barriers can be said to exist. In cases where trade from a region to another is substantial and surpasses expected proportions relative to total trade (Table 4), then trade affinities are noted as being strong. Table 4 provides the comparison of trade barriers and affinities that arise from this analysis by major commodity type and for major regional groups.

In many cases, trade affinities express geographical realities and general trade arrangements (e.g., the trade affinity between North America and Central and South America is an expression of the shared geography of the Americas and the movement toward free trade in the Americas as evidenced by the North American Free Trade Agreement among Canada, the United States, and Mexico [Canada 2003]). Of note as well from the FAO fish and seafood statistics is the geographical affinity among Japan as a net importer and the developing exporting regions of Asia and China. However, the evidence of trade barriers exist between Japan and Europe, who, as leading importer and exporter, respectively, do not engage in trade to the extent expected by their totals. Europe is also noteworthy as a region for its apparent barriers to trade with the Japanese partners Asia and China. Europe as a region also presents a barrier to trade with the North America region and vice versa by virtue of the fact that trade between these leading trading regions is well below expectations (e.g., Europe imports 40% of global fish and seafood (nearly $23 billion) and North America exports 11% of global fish and seafood (more than $6 billion), however, Europe imports much less than $1 billion (3%) to North America, as opposed to an expected value of 11% of $23 billion times or greater than $2.5 billion.

Finally, the trade affinity between Europe and Africa regions is noteworthy. While African exports are valued at only greater than $3 bil-

Table 4. Trade affinities with commodities and implied trade barriers among major regional groups.

Exporting regions	Importing regions: North America	Europe	Japan
North America	–	Barrier	Salmon
Europe	Barrier	–	Barrier
Africa	Barrier Groundfish, crustacea and mollusks, fish, salmon, pelagics	Affinity	
Central and South America	Affinity Groudfish, crustacea and mollusks, fish, salmon, tuna	Salmon	
Asia	Barrier Crustacea and molluks, fish, tuna	Affinity	
China	Barrier Crustacea and mollusks, fish	Affinity	

lion or 5% of global exports, the region of Europe captures more than 60% of this trade as imports. Europe averages import values ($2 billion) from Africa that are nine times the export values ($0.26 billion) Europe trades there on an annual basis. Nevertheless, African imports from Europe represent the largest share (approximately 50%) of total African imports.

Trade affinities exist among various regions by product. These affinities reflect the total trade affinities previously noted. Similarly, the trade barriers expressed above apply to most of the major commodity groups, with few exceptions (Table 3). North America affinities with Central and South America seek out groundfish and crustacean and mollusks for import to North America. North America, along with major importing regions Japan and Europe, rely heavily on Asia for tuna. North America and Japan secure pelagics from Asia, while Europe looks to its African trading partner for its source of pelagics to import. North America is a primary exporter of salmon to Japan. China is the primary source of crustacea and mollusks and fish for import to Japan.

Africa exports more than 85% of groundfish, more than 70% of pelagics and salmon, and more than 50% of crustacean and mollusks directly to Europe, its largest trading partner. Central and South America export its salmon by more than 20% to North America and more than 45% to Europe as evidence of the strong presence of salmon aquaculture in this region. North America and Asia export over one-half of their salmon to the Japanese. Asian tuna are exported by more than 40% to Japan.

North American Trading Partnerships: the place of Canada in Global Fisheries

In this section, we consider the place of Canada in the North America region, including the United States' and Canada's particular place in world fisheries trade. This discussion also examines the notion of fair trade and trade barriers in fish and seafood products relative to Canada's position in global markets.

Canadians have a general misconception that their Atlantic and Pacific fisheries have a place of importance in world fisheries. However, Canadian fisheries have generally not followed the evolution of global fisheries and fish products due in large part to their fixation and dependence on the U.S. markets (Figure 8). Technological developments and trade affinities have contributed to a worldwide integration in trading blocks for fish and seafood products. Such developments have also led to situations that explain why fish products landed in Canada can be shipped to Asia for processing into final products that in turn are found in Canadian grocery stores. In the face of global markets, the fishing industry in Canada can no longer be considered as a source of employment creation and local processing for the resource producing country (Kirby 1982). Potential revenues have instead become a source for integrated international companies who take advantage of available supplies and relatively low-cost workers to add value to marine products for sale in the international marketplace.

As well, Canada has not been able to reestablish effectively its fishing fleets since the crash of its significant offshore groundfish fleet in the early 1990s and the subsequent moratoria that persists in the northwest Atlantic for white fish stocks. Valued species such as mackerel, skates, flatfish, and monkfish were available; however, these have had only marginal appeal in traditional markets. Canadians have had little appetite to seek out new markets (e.g., in Europe and Asia) for fish and seafood markets in what politicians had for a long time considered as a source of employment is regional despair areas of the country (Kirby 1982). As such, Canadians did not put them-

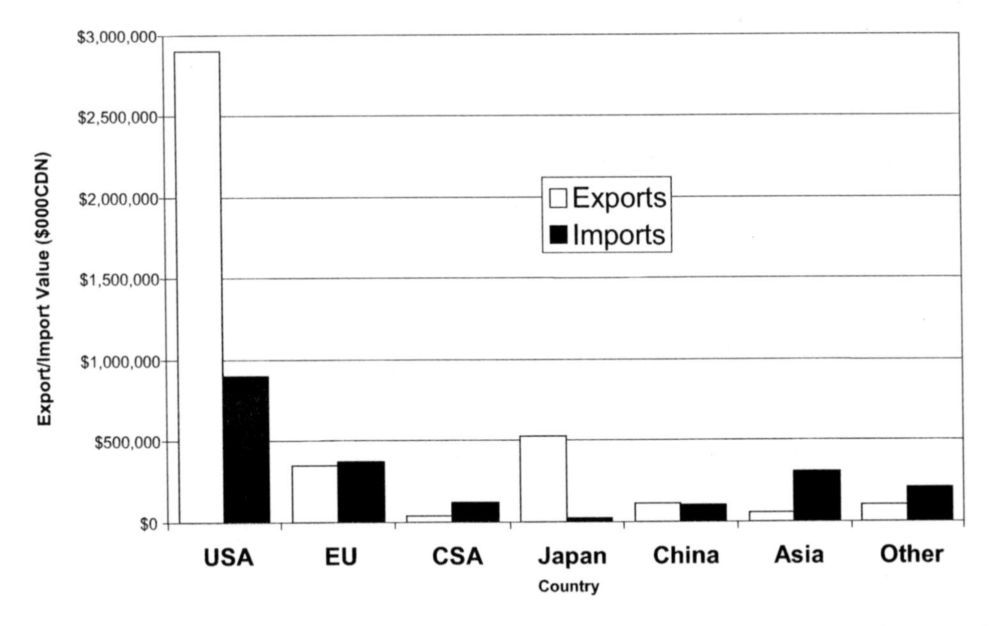

Figure 8. Canadian exports and imports to the United States and to regional groups for year 2000. (Source: Canada 2004.)

selves in a position to respond to new marine fish and seafood products at their disposal as well as new means of production and marketing for the 21st century.

Large integrated organizations such as Fisheries Products International of Newfoundland and Labrador or National Sea Products of Nova Scotia were largely subsidized by provincial and federal governments during the turbulent years of the 1980s and 1990s and allowed themselves to be attracted to easy socialist-style solutions rather than the more difficult problem of trying to compete in the evolving competitive global fish and seafood marketplace. Industries were attracted to selling marine fish and seafood products that required little or no processing (such as live lobster or crabs) or the continuation of shipping frozen round fish out of the country to be processed and exported again elsewhere as finished product. In the most recent case of shrimp, where Canadian marine inventories of this highly-valued resource have soared in recent years (Canada 2004), Canada has permitted Danish enterprises the possibility of direct marketing of northern shrimp into Europe and Asia, whereas Canada has conserved the sale of shrimp throughout North America, especially into the lucrative U.S. market. Since the start of 2004, Canada has witnessed requests for the purchase of frozen bulk products (e.g., shrimp) from Canada by Chinese businesses who are willing to import products to be processed with low labor costs for reexport and Chinese—not Canadian—value added.

Canadian exports to the United States have been established historically to meet the demand for white fish. The place of groundfish in exports has been severely reduced due to the groundfish moratoria that has effected Canadian stocks since the early 1990s. However, exports have subsequently increased appreciably in crustacean and mollusks, crabs, and pelagics (Figure 9).

Despite the increase in the production of species such as northern coldwater shrimp, North American imports of crustacean and mollusks have exceeded exports. Thus, this item is a net deficit to the region in spite of the large increase in Canadian shrimp product that cannot respond to the American demand for shrimp-like products that, through efficient marketing, come from Asia and China. Within the North American region, Canadian exports nevertheless have a strong surplus primarily due to their orientation and trade with the United States, each other's largest trading partner.

Toward Fair Markets

Canada exhibits a very large dependency on the United States to maintain its substantial trade surplus. This is true not only for fish and seafood products, but also for other primarily resource products as well. Historically, Canadians harvest their vast resources for export principally to the United States and, in turn, purchase a portion of finished products from the United States. Canada has enjoyed this trade arrangement for many years and, nationalist sentiments and resource concerns aside, acknowledges that this trade affinity has been beneficial and fair to Canada.

By acknowledging the trade affinity Canada has with the United States in fish and seafood, Canada must also acknowledge that it is less of a player in international markets outside North America. Thus, long-time trade affiliations with the United States can be said to soften Canada's competitive ability to compete in the global market. Similarly, member countries of regional blocks like Europe create an ability to trade internally that circumvents international competition and protects member countries (e.g., in the European

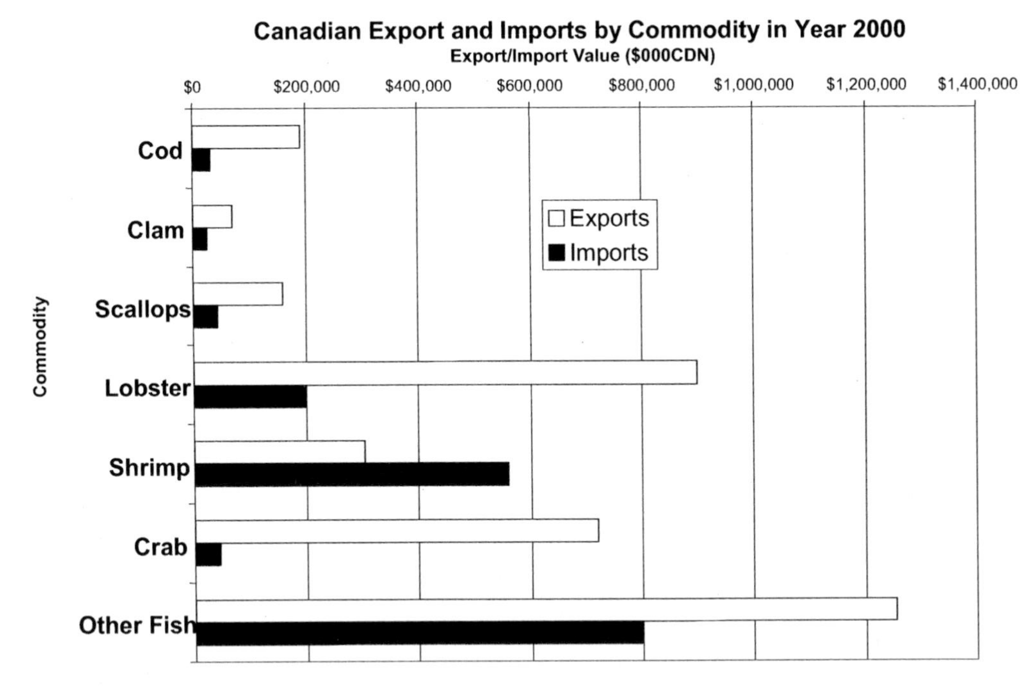

Figure 9. Canadian exports and imports by commodity for year 2000. (Source: Canada 2004.)

Union). While internal trading can be considered fair, it can also create effective barriers that limit trade from select international regions. The rise of North American free trade is at once a boon to member organizations that may currently be familiar with that trade culture, as well as a bane to outside international competitors who may seek to develop partnerships with the larger group. Similar statements can be made about European free trade operatives that are seen as a joint protectionist group maintaining trade flow more exclusively among the European community and affiliated members of that trading club (European Free Trade Association, http://www.efta.int/)

With respect to fish and seafood trade, the exploitation of marine resources requires human exploitation interventions in the environment and ecosystem. Furthermore, responsible trade in these products requires management, monitoring, and enforcement of responsible fishing activities (FAO 1995). For developed countries organized into trade blocs, these tasks can be carried out by mutual agreement. In cases of substantial trade and trade affinities between developing and developed regions (e.g., between Europe and Africa), a shared responsibility must exist between these trading partners to share mutual and comparative advantage. This shared responsibility may require a form of economic protectionism to ensure sustainability and appropriate monitoring of resource status based on accepted international practices (DATA 2005; Degnbol 2005).

In fish and seafood trade, interregional affiliations are among the most important especially for the developed regions of North America and Europe. The European countries may even be considered to be nearly self-sufficient given their elevated within trading statistics as totals and by commodity. This too may be considered fair trade with respect to the regulations of the trading agreements in place. However, in cases where obvious trade affinities engender, by necessity, trade barriers to others, it may not be possible for large scale political organizations (e.g., World Trade Organization or GATT) to rule on fairness from any purely economic perspective. In these cases, it is not effective for member countries, such as Canada, to promote large-scale political solutions to circumvent trade barriers. Empirical evidence demonstrates that it is more important to maintain consistent supply and customer relations and that these will do more to ensure stability in markets than what can be achieved hypothetically from interventions and negotiated agreements between governments.

Conclusions

The issue of determining whether trade practices are fair will remain a normative and subjectively defined evaluation. It is observed from the present analysis that fair trade can be carried out by regional groups of fish and seafood products traders across regional lines assuming that trade protectionism and legal barriers are not the reasons for the differences that exist in actual versus expected trading proportions. Trade affinities may exist legitimately for geographical or historical reasons.

Moreover, where trade affinities exist, especially between developing and developed trading regions, it is assumed that developing regions contribute to aiding in the practice of environmentally sustainable exploitation of marine resources. This implies shared responsibility within trade blocks and between developing and developed trade partners.

Finally, faced with fair trade arrangements, commercial bilateral agreements between local trading partners have more possibility of occurring than large-scale political and economic driven changes designed to promote trade and remove barriers.

The analyses presented here provide evidence of trade affinities and of potential barriers to trade among major global trading regions. Further work is required to investigate the reasons behind these affinities and potential barriers for fish and seafood products on the basis of commercial trading practices within regions, the legal implications surrounding trade among regions, and the particular cases of member countries trading practices (e.g., Canada) within the international trade for fish and seafood products.

References

Canada. 2003. Canada and the North American Free Trade Agreement (NAFTA). Available: http://www.dfait-maeci.gc.ca/nafta-alena/menu-en.asp (December 2003).

Canada. 2004. Canada's fish and seafood exports stable in 2003. Department of Fisheries and Oceans news release, NR-HQ-04–17E, Ottawa.

DATA (Debt AIDS Trade Africa). 2005. DATA report. Debt AIDS Trade Africa. Available: http://www.data.org/issues/data_report.html (July 2007).

Degnbol, P. 2005. A brighter future for cod and other fish stocks? International Council for the Exploration of the Sea (ICES). Available: http://www.ices.dk/marineworld/brighter_future_fish_stocks.asp (May 2005).

Fair Trade Federation. 2004. 2003 report on fair trade rrends in US, Canada & the Pacific Rim. Available: http://www.fairtradefederation.org/2003_trends_report.pdf (June 2004).

FAO (Food and Agriculture Organization). 1995. Code of conduct for responsible fisheries. FAO, Rome.

FAO (Food and Agriculture Organization). 2000. FISH INFONETWORK. Available: http://www.globefish.org/ (December 2003).

FAO (Food and Agriculture Organization). 2003. The state of the world fisheries and aquaculture 2002. Available: http://www.fao.org/docrep/005/y7300e/y7300e05.htm (December 2003).

FAO (Food and Agriculture Organization). 2004. Fisheries statistics. Appendix III—trade flow by region. 1997–1999, 1998–2000, 1999–2001. http://www.fao.org/ (June 2004).

Holden, M. 1994. The common fisheries policy: origin, evaluation, and future. Fishing News Books, Oxford, UK.

IFTA (International Fair Trade Association). 2005. World Fair Trade Day, May 14, 2005. International Fair Trade Association. Available: http://www.wftday.org/ (December 2005).

Kirby, M. J. L. 1982. Navigating troubled waters: a new policy for the Atlantic fisheries. Report of the Task Force on the Atlantic Fisheries. Supply and Services Canada, Ottawa.

Kurlansky, M. 1997. Cod: the biography of the fish that changed the world. Penguin Books, New York.

Suranovic, S. 2007. International trade theory and policy. The International Economics Study Center, Ó1997–2007. http://internationalecon.com/Trade/Tch125/Tch125.php (July 2007).

WTO (World Trade Organization). 2003. GATT 1994: what is it? World Trade Organization. Available: http://www.wto.org/english/thewto_e/whatis_e/eol/e/wto02/wto2_4.htm (December 2003).

WTO (World Trade Organization). 2005. What is the WTO? World Trade Organization. Available: http://www.wto.org/english/thewto_e/whatis_e/whatis_e.htm (December 2003.

American Fisheries Society Symposium 49:483–494

Three Case Studies of Cooperative Research and Management of Shared Salmon Resources in Southeast Alaska and Northern British Columbia

Kathleen Jensen*

*Alaska Department of Fish and Game, Commercial Fisheries Division
Post Office Box 240020, Douglas, Alaska 99824, USA*

Sandy R. A. C. Johnston

*Fisheries and Oceans Canada, 419 Range Road, Suite 100
Whitehorse, Yukon Territory Y1A 3V1, Canada*

Abstract.—Under the Pacific Salmon Treaty, the Transboundary Technical Committee is responsible for cooperative research programs to provide data to manage fisheries during the fishing season and to estimate escapements of transboundary river salmon stocks. We present three case studies that illustrate evolution of research programs and management structures and demonstrate responses to conservation concerns. The case studies 1) document the evolution of research programs that provide data for cooperative Stikine River sockeye salmon management, 2) document fishery managers' response to low Alsek River sockeye salmon *Oncorhynchus nerka* escapements and the development of research programs in response to the concern, and 3) outline multiagency research programs that will attempt to determine causes underlying a decline in Taku chum salmon *O. keta* abundance. We describe some of the processes that have worked and situations where problems still remain. Within the committee, the United States and Canada evaluate transboundary river salmon research programs, develop cooperative programs where personnel from both departments are actively involved in data collection and analysis, and develop a protocol where fishery managers use a common data set on which to base their management actions. The committee meets prior to each fishing season to finalize research plans, establish harvest guidelines, and refine management models. During the fishing season, researchers from each country exchange data and managers confer on planned fishery openings.

Introduction

United States and Canadian fishers harvest Chinook *Onchorynchus tshawytscha*, sockeye *O. nerka*, coho *O. kisutch*, pink *O. gorbuscha*, and chum *O. keta* salmon from the Stikine, Taku, and Alsek rivers. Concern for equitable sharing of the salmon harvest between the two nations and the need to cooperate in the conservation of transboundary salmon arose in the late 1970s and early 1980s. Negotiations between the United States and Canada resulted in the ratification of the Pacific Salmon Treaty in 1985. The Transboundary Technical Committee was established to provide the information needed to cooperatively manage the salmon stocks in the transboundary Stikine,

*Corresponding author: kathleen_jensen@fishgame.state.ak.us

Taku, and Alsek rivers to comply with the harvest sharing, conservation, and enhancement agreements of the treaty.

We present three case studies that illustrate the evolution of cooperative research and management and the responses to conservation concerns. The first case study, for Stikine sockeye salmon, describes how unilateral and bilateral research programs have provided data used during the fishing season to manage fisheries based on the abundance of the salmon runs. The second case study, for Alsek sockeye salmon, documents how fishery managers responded to low escapements and how research programs have addressed that concern. The third case study, for Taku chum salmon, outlines multiagency research programs for 2004 and 2005 that will attempt to determine causes underlying the decline of Taku chum salmon.

Study Area

The Stikine, Taku, and Alsek rivers are transboundary rivers that originate in northern British Columbia and Yukon, Canada (near the communities of Telegraph Creek, Atlin, and Haines Junction) and flow through the Coast Range mountains into the United States before entering the Pacific Ocean, near the southeastern Alaska communities of Petersburg, Juneau, and Yakutat, respectively (Figures 1, 2, and 3). All three rivers are large and glacially occluded but include a wide diversity of habitats, including main river channels, lakes, ponds, side sloughs, and clear headwater tributaries. U.S. fisheries are located in marine waters near or within lower reaches of the transboundary rivers; the marine fisheries harvest a mix of stocks that include variable portions of transboundary river fish. Canadian fisheries are located directly upstream of the U.S. and Canadian border and in upper reaches of the drainages.

Treaty Negotiations for the Transboundary River Annex

The initial Transboundary River Annex of the Treaty included harvest-sharing agreements for the 1985 and 1986 fishing seasons. Negotiations to renew the annex in 1987 were unsuccessful, and both countries set their own harvest levels. In 1988, new harvest sharing agreements were reached for the 1988–1992 period that were contingent on developing a bilateral enhancement program to increase the numbers of sockeye salmon available for harvest. Harvest sharing agreements were reached for the Stikine sockeye salmon (for 1993–1995) in 1989 and for Taku sockeye salmon (through 1999) in 1990. The remaining Transboundary River Annex provisions were extended through 1993, however negotiations to renew expiring portions of all the treaty annexes were unsuccessful from 1994 to 1997. During this period, both countries generally managed their transboundary river fisheries consistent with treaty principals and the committee continued to meet. In 1999, harvest-sharing agreements were reached for the 1999–2008 fishing seasons.

The Committee Process

The Committee meets in the spring to discuss and finalize research plans, exchange preseason forecasts of salmon run abundances, discuss plans for initial fishery openings, update models to be used for fishery management, and write a management plan for the three rivers (TTC 2003). Field camps, where personnel from both countries work together to collect data used for fishery management, are established on each river prior to the fishing season. During the fishing season, researchers from both countries are in frequent (often daily) contact with each other to exchange data, consult on problems that may arise in

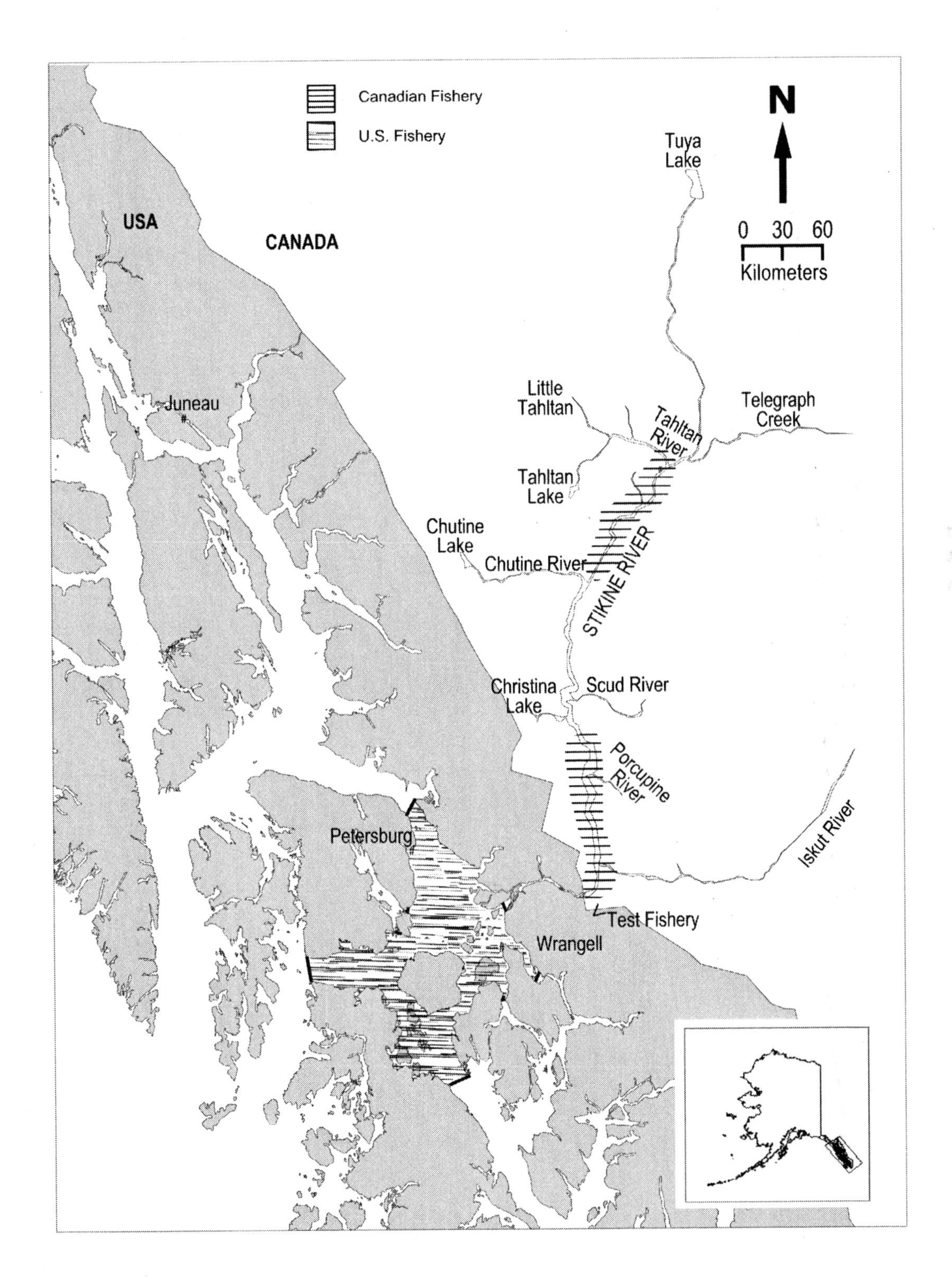

Figure 1. The Stikine River and principal U.S. and Canadian fishing areas.

the field camps, and evaluate project operations. Managers contact each other before they announce the fishery openings or extensions for the following week and notify each other of anticipated fishing time. The committee meets every fall to generate postseason estimates of harvests and escapements, to review and evaluate management actions and their effect on harvest sharing compliance, to determine if escapement goals were met, and produce an annual report (TTC, in press)

Case Study 1: Stikine Sockeye Salmon—Evolution of Cooperative Research Programs

The treaty mandated that both countries manage Stikine River sockeye salmon to comply with harvest sharing agreements and provide adequate escapement. However, there was very little data to indicate how many fish from which stocks were being harvested by each country or to set escapement goals (Table 1). To address this fundamental problem, the agencies, the Alaska Department of Fish and Game (ADF&G) and the Department of Fisheries and Oceans Canada (DFO), worked both independently and within the committee process to develop programs to collect data that could be used to estimate harvests and escapements, to set escapement goals, and to develop estimates of catch to meet harvest sharing agreements (Marshall et al. 1989; TTC 1990a; Table 1). As illustrated below, a key feature of the effort was joint evaluation of methods for identifying stocks, estimating escapements, forecasting run sizes, managing harvests, and developing fisheries enhancement programs.

Each country developed its own stock assessment tools to estimate harvest and escapement of the Tahltan Lake and mainstem Stikine sockeye salmon stock aggregates. Department of Fisheries and Oceans Canada developed an analysis of egg diameters and explored a combination of genetic stock identification, parasite analysis, fish age, and circuli counts, referred to as a mixture model, for estimating the stock composition of in-river harvests (Craig 1985; Wood et al. 1987a, 1987b), while ADF&G developed scale pattern analysis (Jensen et al. 1989). All three methods were used in the mid-1980s for the inriver stock composition estimates. Alaska Department of Fish and Game again used scale pattern analysis for estimating the numbers of Stikine River sockeye salmon in U.S. marine harvests (Oliver et al. 1985; Jensen et al. 1989) while DFO applied a mixture model to the same harvests (Wood et al. 1988, 1989). The merits of the different stock composition methods were debated extensively during committee meetings. The committee recognized that no method of stock identification was without error, and found that in situations where validation of estimates with known origin fish was possible, the methods had similar accuracies (TTC unpublished data). The results produced by the various methods were similar enough that management actions would not have differed based on any of the stock composition methods. This led to concurrence that the method preferred by DFO (egg diameter analysis) would be used for in-river catches and the method preferred by ADF&G (scale pattern analysis) would be used for marine catches. This is an example of contested stock assessment (Starr et al. 1998), where scientists from different parties work together to resolve stock assessment issues. The first version involves two parties providing competing stock assessments and in the second version, there is an agreed assessment. The stock assessments for Stikine sockeye salmon evolved from the first method to the more cooperative second approach.

As with the stock composition estimates,

Table 1. Annual catch and escapement data collected for the Stikine, Taku, and Alsek rivers. Years are underlined for programs initiated during the Pacific Salmon Treaty process.

	Stikine			Taku			Alsek		
Data collected	Chinook	Sockeye	Coho	Chinook	Sockeye	Coho	Chinook	Sockeye	Coho
Systemwide									
Escapement	1996	1983	2000	1989	1984	1987	1997	2000	
Canada catch	1972	1972	1972	1979	1979	1979	1976	1976	1976
U.S. catch	2003	1983	2001	1996	1984	1992	1961	1961	1961
Migratory timing	1996	1983	2000	1989	1984	1987			
Run reconstruction	2003	1983	2000	1996	1984	1992	1997	2000	
Smolt	2000		2001	1993		1991			
Stock-specific									
Escapement	1985	1959			1983		1976	1976	
Canada catch		1979			1986				
U.S. catch			1985			1986			
Migratory timing		1995	1979			1986			
Run reconstruction		2003	1985			1986			
Smolt			1984			1996			
Escapement Index	1979	1984	1984	1975	1980	1984	1976	1976	1976

[a] Programs to estimate Taku Chinook salmon escapement and migratory timing were not run during 1991–1994.

DFO and ADF&G preferred different methods to forecast the salmon runs (TTC 1990b; Jensen 1992). The committee compared the accuracies of the forecasts and agreed to use an average of smolt and sibling forecasts for Tahltan sockeye salmon and a sibling forecast for main-stem sockeye salmon (TTC 1997)

The committee developed a method for estimating the main-stem escapement based on in-river catch per unit effort (CPUE) and stock identification. Stock specific harvests and escapements, run reconstructions, and spawner–recruit relationships were calculated, stored in a shared database, and used to develop escapement goals and a management model (TTC 1993).

Alaska Department of Fish and Game developed a model to forecast the Stikine sockeye salmon run size based on catches and fishing effort during the current fishing season compared to historical data. The committee refined the model to produce run forecasts for both Tahltan and main-stem sockeye salmon (TTC 1988b). Changes in fishery boundaries, fishery effort patterns, and fish migration timing contributed to forecast error, and the committee made adjustments to compensate for some of these changes during annual evaluation of model performance. In-season scale pattern analysis was used in the model for 4 years in the late 1980s to provide stock composition estimates of marine harvests. Although this method worked for 3 years it failed the fourth year, leading to overprojections of run size (Jensen 1994) and discontinuance of the program. Because the model overforecasts of run size in 1997–1999 resulted in failure to meet escapement goals, DFO and ADF&G

worked together to develop a mark–recapture program to provide alternative inseason abundance estimates to fishery managers. Personnel from both countries now fish a set gill net below the border, capture, mark (with individually numbered spaghetti tags), sample, and release sockeye salmon. This program is also used to validate the escapement estimates generated from other data.

Although the model did not perfectly forecast Stikine sockeye salmon run sizes, in most years it provided information to fishery managers that allowed them to increase harvest rates when the runs were large and decrease harvest rates when runs were small. In years when the model failed, the documentation of efforts by fishery managers to meet their harvest sharing goals, as indicated by the model, alleviated some of the acrimony that might otherwise have occurred in these years. When both countries use the same data and model to decide what their fishery openings should be, there is mutual incentive to improve model accuracy.

In 1989, a bilateral cooperative enhancement program was implemented as a resolution to an impasse in harvest sharing negotiations. A fry-planting enhancement program was initiated for sockeye salmon (TTC 1988a) in the Stikine River. Department of Fisheries and Oceans Canada collects gametes from Tahltan sockeye salmon and transports them to a hatchery in Alaska. Alaska Department of Fish and Game provides hatchery incubation, thermal marking of otoliths of the incubating sockeye salmon, and transportation of resultant fry back to Tahltan and Tuya lakes (Figure 1). Tuya Lake is inaccessible to upstream migration because of falls and other velocity barriers. Each country collects otoliths from its fisheries to estimate the contribution of enhanced fish to its harvests. This, combined with samples collected from the escapements, allows estimation of the total return of fish from the enhancement program.

Currently, the enhancement program for Stikine sockeye salmon that involves planting fry into Tuya Lake is contentious. Dissension has arisen as strengths and shortcomings of the programs have been observed and perspectives held by the various participants have changed. While fishers from both nations have benefited from having more sockeye salmon available for harvest, concerns have been raised about possible genetic effects of straying and the costs of the program. In addition, the Tahltan and Iskut First Nations of Canada have cultural concerns that fish are being produced that are not harvested and cannot migrate to Tuya Lake and spawn naturally. Efforts to harvest sockeye salmon from the Tuya Lake fry plants at the barriers on the Tuya River have been only moderately successful with an estimated maximum harvest rate of 70% and an average of 33% of the terminal run harvested in the Tuya River (TTC in press). The fate of the remaining fish is unknown and concern has been raised within the committee that the unharvested sockeye salmon may be migrating downstream and intermingling with mainstem Stikine sockeye salmon on the spawning grounds. The results of a study to determine if interstock spawning occurs are currently being analyzed. The apparent simple solution of producing more fish to resolve an impasse in harvest sharing negotiations has become complicated.

The committee's work on the Stikine River illustrates two essential features of cooperative research and management programs. First, the members of the committee were committed to a cooperative approach to resolving technical issues. Second, decisions of the committee consistently reflected a focus on use of the best available science. The committee provided a forum for the development of different techniques, methods, and analyses. Although discussions of the relative merits were extensive they were generally constructive. However, the preference of each agency to use methods it developed sometimes extended discussion past the point of technical merit. Interest by each country in accurate accounting of the other country's harvest encouraged critical evaluation of results and discouraged bias.

Case Study 2: Alsek Sockeye Salmon—Response to a Conservation Concern

Low sockeye salmon counts throughout the Klukshu River (a tributary of the Alsek River) weir and low perceived escapements to the entire Alsek River (Figure 2) in the late 1990s caused fishery managers in both countries to restrict sockeye salmon harvests. Contention arose about how much impact the U.S. fishery in the lower Alsek River had on the Klukshu sockeye salmon stock. Without scientific estimates of run sizes, migration timing, and the numbers of Klukshu fish caught in the lower river, discussions were unresolved. Other than one pilot study in 1983 (McBride and Bernard 1984), the Alsek River sockeye salmon escapement for the entire drainage was unknown. It was assumed to be correlated with the CPUE in the lower river commercial fishery or with the numbers of sockeye salmon counted through Klukshu River weir.

Klukshu sockeye salmon escapement data were analyzed and escapement goals for this tributary were tentatively agreed to for 1999; goals were modified and adopted in 2000 (Clark and Etherton 2000). In 2000, ADF&G and DFO initiated a mark–recapture program to estimate the total sockeye salmon escapement in the Alsek River, to determine the migratory timing of Klukshu sockeye salmon in the lower river, and to determine the contribution of Klukshu sockeye salmon to the Alsek run. Personnel from both counties captured and marked sockeye salmon in the Alsek River downstream of the border. Tags were recovered from above-border spawning locations, fishery harvests, and the Klukshu River weir. In conjunction with the mark–recapture program, ADF&G and DFO conducted a radio telemetry program from 2001 to 2003. Alaska Department of Fish and Game supplied the radio tags that were applied in the lower river fish marking program and DFO supplied stationary radio tracking gear and flew aerial tracking flights. These programs determined spawning distribution and migratory timing of sockeye salmon spawning throughout the Alsek River and showed that the Klukshu sockeye salmon contribute a variable fraction of the entire Alsek sockeye run (14–38%) and are present in the lower river through out the Alsek sockeye run.

This case study illustrates characteristics of a successful cooperative program. First, all parties recognized the need for improved data to adequately manage the fisheries. They recognized that management of the fishery and allocation of harvest needed to be based on understanding the total escapement and the escapement to various portions of the river system. Second, all parties cooperated in design and management of the research program needed to determine spawning distribution and migratory timing in the river system.

When escapements were low, fishers in both counties endured the financial and recreational cost of resource loss as well as cultural hardship. Both countries shared the cost of developing research programs to improve knowledge of the dynamics of the Klukshu sockeye salmon in an effort to minimize impacts on the resource and resource users in years of low abundance. Sharing of the cost of addressing conservation helps promote an atmosphere of bilateral cooperation rather than animosity. While the cooperative research provided data needed and used for fisheries management no method to manage the U.S. fishery specifically for Klukshu River sockeye salmon has been found. However, there is potential to develop a management regime based on the relationship derived from this study between the performance of the U.S. fishery (i.e. CPUE, and overall inriver abundance).

Case Study 3: Taku Chum Salmon—An Example of Interagency Cooperation

A decline in the fall chum salmon run in the Taku River has raised concern within the committee, other fishery agencies, and fishers who depend on this resource. The decline is evidenced by a 78% decrease in average Taku Inlet harvests between the 1960–1989 period and the 1990–2003 period, a decrease of gillnet CPUE of 76% for the same periods and a decrease in in-river fishwheel catches of 60% from the first decade of operation (1984–1993) to the last decade (1994–2003) (TTC in press). Some of the decline in chum salmon harvest in the last decade is a result of fishery restrictions to conserve the stock. Several theories have been proposed as potential causes of the decline, including predation, competition, and environmental changes. Tobler (2002) found no evidence that climate change, ocean productivity, the abundance of coho salmon, or commercial fishing had caused the decline in chum salmon.

The Taku chum salmon decline is coincident with increased production of U.S. hatchery chum salmon in areas near the Taku River. There is a correlation between decrease in cumulative CPUE in the District 111 gillnet fishery and the increase in millions of hatchery chum salmon fry released, logarithmic regression R^2 = 0.73, (ADF&G unpublished data). Potential interactions between hatchery and wild fish are highly controversial among fishery professionals. Debates among resource users, scientists, and Pacific Salmon Commission members often did not include an attempt investigate any process that might link the two trends. With minimal experience in early marine life history studies, the committee was unable to address this issue. In 2004 and 2005, the University of Alaska Fairbanks (UAF) undertook a program to fish trawls and beach seines to capture out-migrant Taku chum salmon juveniles during their residency in the near and offshore environments near the Taku River (Figure 3). The National Marine Fisheries Service (NMFS) will collect additional samples with offshore trawls. The Douglas Island Pink and Chum Hatchery, one of the hatcheries that produce chum salmon, will analyze otoliths for thermal marks to determine if juveniles are of wild or hatchery origin; thermal marks are applied in the hatcheries during chum salmon incubation. University of Alaska Fairbanks will evaluate spatial and temporal overlap of wild and hatchery juvenile chum salmon in Taku Inlet. The National Marine Fisheries Service will analyze diets and energetics to determine the diet composition and degree of diet overlap between hatchery and wild fish and the effect of overlapping distribution on the energetic condition of the chum salmon. Alaska Department of Fish and Game will compare predation rates on hatchery and wild chum salmon juveniles to estimate the relative abundance to the two types of chum salmon in the areas sampled in Taku Inlet and will analyze spatial and temporal frequency of chum salmon in predator diets.

The Taku chum salmon decline also coincides with habitat changes. Known chum salmon spawning areas on the Taku River are subject to dramatic annual physical alteration due to natural processes. The water was clear in the principal known spawning area, locally known as "chum salmon slough" (Figure 3), during aerial surveys in the 1970s and 1980s but

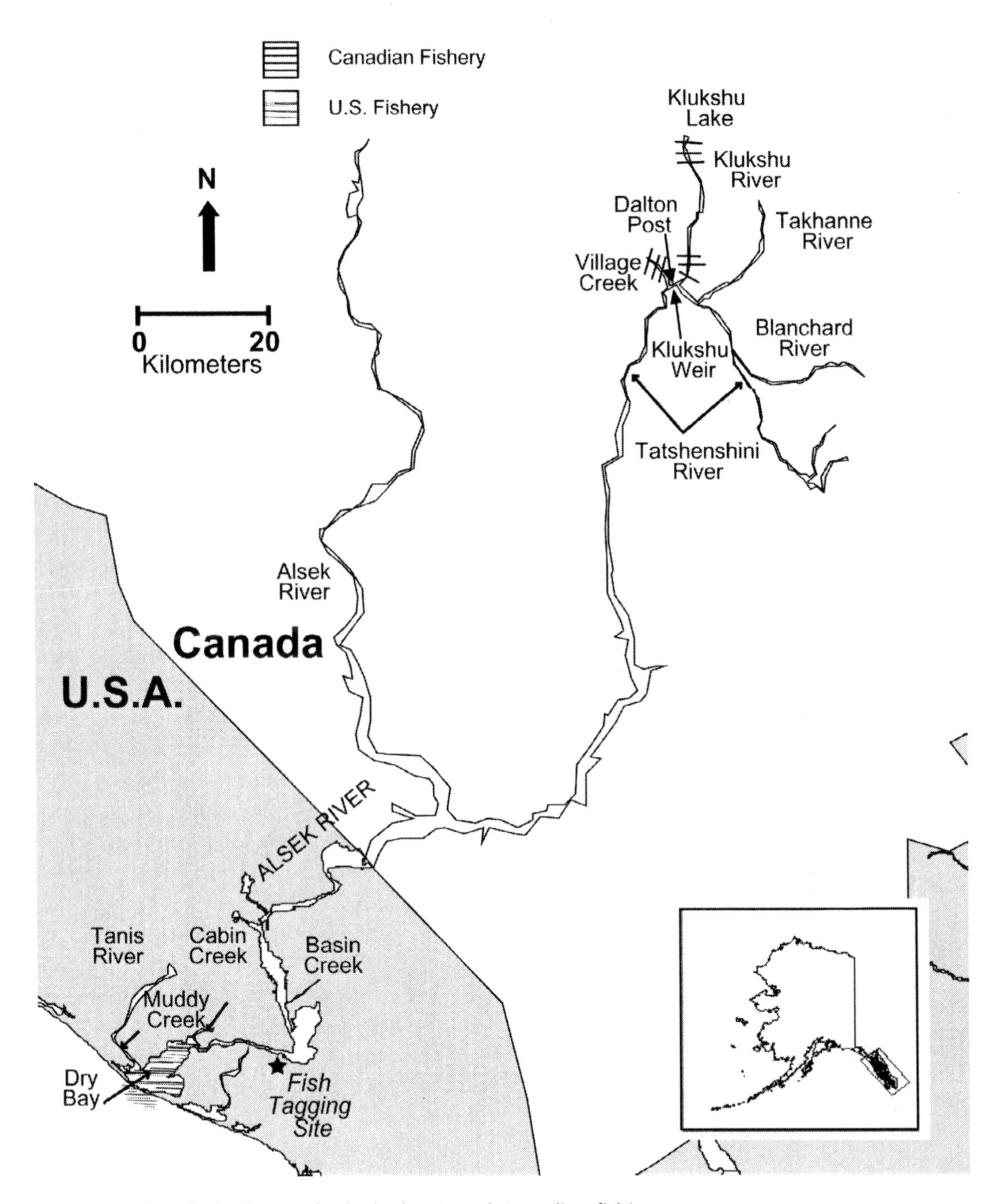

Figure 2. The Alsek River and principal U.S. and Canadian fishing areas.

has become glacially occluded due to a shift in the mainstem channel. Habitat change was proposed as one potential cause of the decline but little is known about where Taku chum salmon currently spawn. Department of Fisheries and Oceans Canada conducted a pilot study in 2002 and tracked 23 radio tagged chum salmon to spawning grounds throughout the mainstem Taku River; only two spawners were tracked to the chum salmon slough area. Improved knowledge of the spawning distribution and location of Taku chum salmon will provide information on habitat preference of spawners and may lead toward development of a population estimation program either through mark–recapture methods or index survey methods. Alaska Department of Fish and Game plans to use radio telemetry data to explore migratory timing and spawning distribution of returning adult Taku chum salmon.

This case study illustrates the Committee's willingness to go outside of its member agencies to acquire and utilize the technical capabilities of other organizations and to develop basic research programs to answer fundamental questions about a fishery resource. The inclusion of other agencies in the Taku chum salmon studies has garnered expertise and resources not available within the committee that increase the likelihood of finding the underlying cause of the chum salmon decline.

Discussion

The committee process of developing bilateral research programs, databases, and management models has worked more smoothly than negotiations within the Pacific Salmon Commission process as a whole. Even during periods when there were no successfully negotiated treaty agreements, both the United States and Canada continued the unilateral and bilateral transboundary river research programs. The fact that the countries often managed these fisheries consistent with prior agreements displayed trust that the treaty process was viable and that agreements could be reached. It also demonstrated a belief that the resource would not benefit from over harvest to provide short-term gain and that unilateral fishery actions would not be well received in the political area.

These three case studies demonstrate some of the processes that have worked well for the committee and show some of the areas where problems remain. Difficulties we have experienced include 1) differing domestic policies of collaborating agencies, for example with respect to human resources, health and safety, and data management; 2) differing priorities between counties, often based on the different needs of the various groups dependant on the resource; 3) lack of resources to develop adequate programs for some species, partially due to the high costs associated with working in remote areas; 4) lack of a guaranteed long-term commitment for bilateral support of specific projects, for example one country may have 3 years' funding for a program and the other country may suddenly loose funding for the last 2 years of their portion of that program and it must be discontinued; and 5) questions and management dilemmas remaining after successful completion of research programs. Although problems remain, the committee's implementation of the treaty has often been successful because 1) members are committed to a cooperative approach to resolve technical issues; 2) all parties recognize the need for improved data to adequately manage fisheries; 3) costs related to conservation concerns (i.e., fishery restrictions and research program development) are shared by all involved parties; and 4) benefits from sustainable fisheries are shared by all parties. In these aspects, the committee process might function as a model for the development of other cooperative resource management programs.

Acknowledgments

We are indebted to biologists, scientists, technicians, and other staff from ADF&G, DFO, First Nations, NMFS, and UAF who provided information used in this paper. We thank

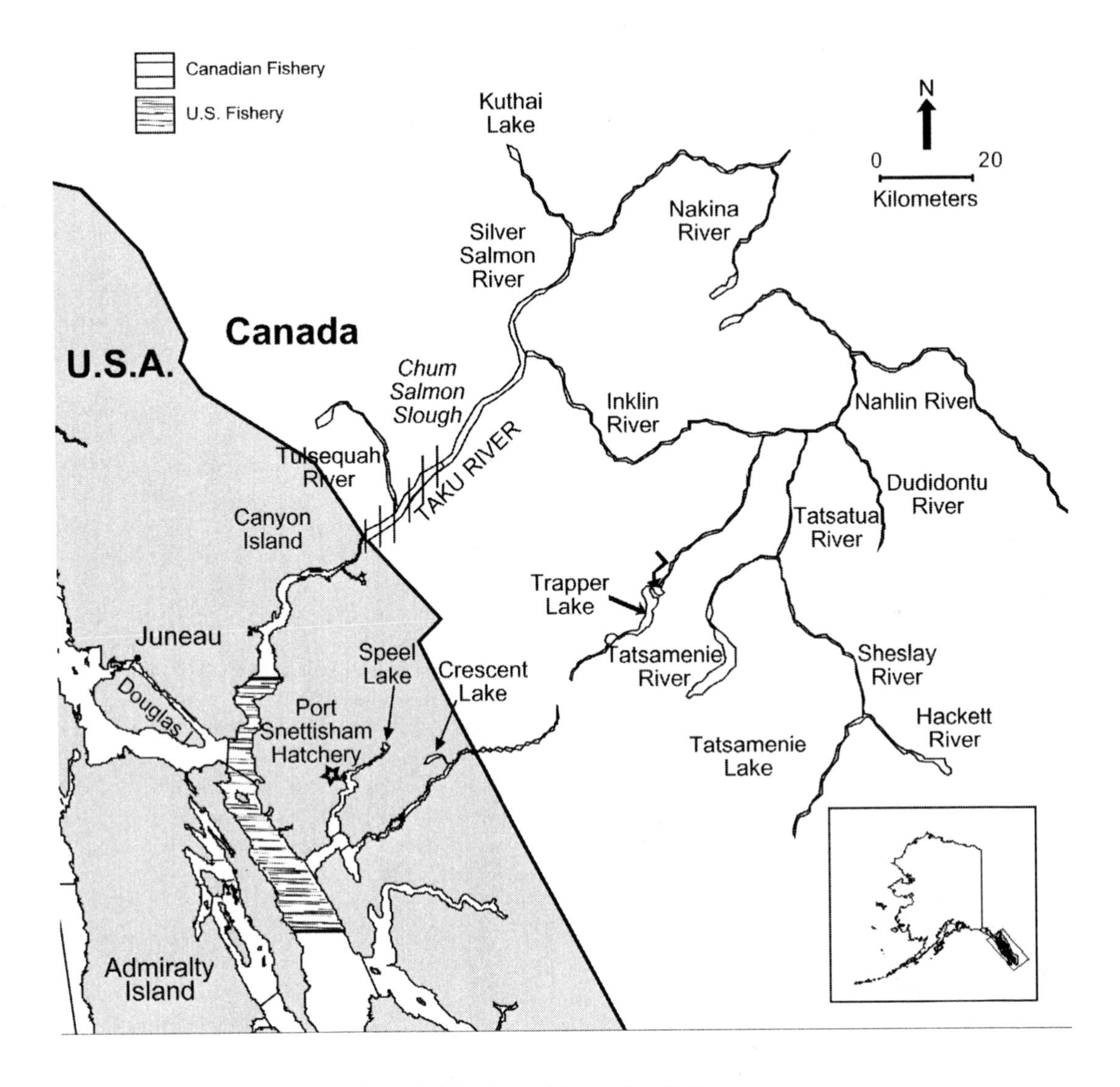

Figure 3. The Taku River and principal U.S. and Canadian fishing areas.

members of the committee, and especially the people who participated in the research programs. We are grateful for editorial and technical reviews by Leon Shaul, Andy McGregor, John Clark, Jud Monroe, and two anonymous reviewers. This paper is contribution PP-240 of the Alaska Department of Fish and Game, Division of Commercial Fisheries, Juneau. This work was supported in part by NOAA award NA17FP1081 and by the Pacific Salmon Coastal Recovery Grants NA16FP2993 and NA06FP0387.

References

Clark, J. H. and P. Etherton. 2000. Biological escapement goal for Klukshu River system sockeye salmon. Alaska Department of Fish and Game, Division of Commercial Fisheries Regional Information Report 1 No. 1J00–24, Juneau.

Craig, P. C. 1985. Identification of sockeye salmon *Oncorhynchus nerka* stocks in the Stikine River based on egg size measurements. Canadian Journal of Fisheries and Aquatic Sciences 42:1696–1701.

Jensen, K. A. 1992. Forecasts for the 1992 Stikine River sockeye salmon run. Alaska Department of Fish and

Game, Division of Commercial Fisheries, Regional Information Report 1J92–14, Douglas.

Jensen, K. A. 1994. Differences between inseason and postseason stock composition estimates for sockeye salmon in gillnet catches in 2 districts in southeast Alaska and in the Stikine River, 1986–1989. Alaska Fishery Research Bulletin 1:107–124, Juneau.

Jensen, K.A., I. S. Frank, and G. T. Oliver. 1989. Contributions of principal sockeye salmon stock groups to catches in southeast Alaska's Districts 106 and 108 and Canada's Stikine River fisheries, 1986, estimated with scale pattern analysis. Alaska Department of Fish and Game, Division of Commercial Fisheries, Technical Fisheries Report 89–01, Juneau.

Marshall, S., N. J. Sands, A. McGregor, and K. A. Jensen. 1989. Data and programs for the transboundary Stikine, Taku, and Alsek rivers needed to implement the Pacific Salmon Treaty. Alaska Department of Fish and Game, Division of Commercial Fisheries, Regional Informational Report 1J89–09, Douglas.

McBride, D. N., and D. R. Bernard. 1984. Estimation of the 1983 sockeye salmon (*Oncorhynchus nerka*) return to the Alsek River through analysis of tagging data. Alaska Department of Fish and Game, Division of Commercial Fisheries, Technical Data Report 115, Juneau.

Oliver, G. T., S. L. Marshall, D. R. Bernard, S. A. McPherson, and S. L. Walls. 1985. Estimated contribution from Alaska and Canada stocks to catches of sockeye salmon in southern southeast Alaska, 1982 and 1983, based on scale pattern analysis. Alaska Department of Fish and Game, Division of Commercial Fisheries, Technical Data Report 137, Juneau.

Starr, P. J., J. H. Annala, and R. Hilborn. 1998. Contested stock assessment: two case studies. Canadian Journal of Fisheries and Aquatic Sciences 55:529–537.

Tobler, P. 2002. Investigating the decline of the Taku River chum salmon: an evaluation of current knowledge. Report of Environmental Dynamics Incorporated to Taku and Atlin Area Community Fisheries Working Group, Atlin, British Columbia.

TTC (Transboundary Technical Committee). In press. Estimates of transboundary river salmon production, harvest, and escapement and a review of joint enhancement activities in 2002. Pacific Salmon Commission, Vancouver.

TTC (Transboundary Technical Committee). 1988a. Sockeye salmon enhancement feasibility studies in the transboundary rivers. Pacific Salmon Commission, TCTR (88)-1, Vancouver.

TTC (Transboundary Technical Committee). 1988b. Salmon management plan for the Stikine, Taku, and Alsek Rivers, 1988. Pacific Salmon Commission, TCTR (88)-2, Vancouver.

TTC (Transboundary Technical Committee). 1990a Long-term research plans for the transboundary rivers. Pacific Salmon Commission, TCTR (90)-3, Vancouver.

TTC (Transboundary Technical Committee). 1990b Salmon management plan for the Stikine, Taku, and Alsek Rivers, 1990. Pacific Salmon Commission, TCTR (90)-2, Vancouver.

TTC (Transboundary Technical Committee). 1993. Salmon management and enhancement plans for the Stikine, Taku, and Alsek Rivers 1993. Pacific Salmon Commission, TCTR (93)-2, Vancouver.

TTC (Transboundary Technical Committee). 1997. Salmon management and enhancement plans for the Stikine, Taku, and Alsek Rivers 1995. Pacific Salmon Commission, TCTR (97)-1, Vancouver.

TTC (Transboundary Technical Committee). 2003. Salmon management plan for the Stikine, Taku, and Alsek Rivers 2003. Pacific Salmon Commission, TCTR (03)-1, Vancouver.

Wood, C. C., D. T. Rutherford, and S. McKinnell. 1989. Identification of sockeye salmon *Oncorhynchus nerka* stocks in mixed-stock fisheries in British Columbia and Southeast Alaska using biological markers. Canadian Journal of Fisheries and Aquatic Sciences 46:2108–2120.

Wood, C. C., G. T. Oliver, and D. T. Rutherford. 1988. Comparison of several biological markers used for stock identification of sockeye salmon *Oncorhynchus nerka* in Northern British Columbia and Southeast Alaska. Canadian Technical Report of Fisheries and Aquatic Sciences 1624.

Wood, C. C., B. E. Riddell and D. T. Rutherford. 1987a. Alternative juvenile life histories of sockeye salmon *Oncorhynchus nerka* and their contribution to production in the Stikine River, northern British Columbia. Pages 12–24 *in* H. S. Smith, L. Margolis, and C. C. Wood, editors. Sockeye salmon *Oncorhynchus nerka* population biology and future management. Canadian Special Publication of Fisheries and Aquatic Sciences 96.

Wood, C. C., B. E. Riddell, D. T. Rutherford, and K. L. Rutherford. 1987b. Variation in biological characters among sockeye salmon populations of the Stikine River with potential use for stock identification in mixed-stock fisheries. Canadian Technical Report of Fisheries and Aquatic Sciences 1535.

American Fisheries Society Symposium 49:495–497

Session Summary

Reconciling Aboriginal Fisheries with Conservation: Themes, Concepts, and Examples

NIGEL HAGGAN
UBC Fisheries Centre
Vancouver, British Columbia, Canada

KLAH-KIST-KI_IS, CHIEF SIMON LUCAS
Hesquiaht Nation Council
British Columbia, Canada

Aboriginal people do not generally see themselves as a threat to conservation. Myths and teachings about the dire consequences of greed, waste, and disrespect are universal in the Pacific Northwest and are found all over the world. Aboriginal cultures stress the unity of life and the spiritual world (Lucas 2004). Australia's "saltwater people" make no separation between land and "Sea Country'" (Sheppard et al. 2004). Hawaiian concepts of territory extend from the mountaintops to the ocean (Smith and Pai 1992).

First Nations fishing technology was certainly powerful enough to wipe out salmon runs. Strict rules were developed to ensure that enough fish got through the fishery to spawn (Jones and Williams-Davidson 2000; Jones 2004). The prevalence of warnings about taking more than are needed suggests that hard lessons were learned over thousands of years of interdependence. This ecosystem consciousness rooted in the ancient past, enabled the chiefs and elders to avoid actions that would foreclose future options. This is what is meant by "7th generational thinking." Langdon and Kompkoff (2004) provide a grim example of how the *Exxon Vadez* oil spill cut their tribe off from resources vital to their cultural, spiritual and economic survival and the steps they have taken to raise awareness, rebuild their resource economy, and incorporate the lessons in curriculum for future generations.

The oral and poster presentations from this session addressed conservation in the context of cultures, belief systems and worldviews based on interdependence and respect for natural resources. They have, in common, the concept that traditional ownership and ecological knowledge have been superseded by "modern" scientific and management frameworks. The customary marine tenure of Indigenous people in Oaxaca, Mexico, incorporates culture and practice shaped by interaction between the community and the environment. If governments are serious about ecosystem management they need to strengthen such systems, not overlook them (Robles 2004). The presentations represented points on a continuum between the dawn of recognition of rights and responsibilities, comanagement arrangements (Charles 2004), and treaty settlement.

In Canada, Aboriginal and treaty rights are entrenched in the Constitution of 1982, but, in spite of several favorable Supreme Court decisions, little has changed in how resources are distributed. Harris (2004) described how Canadian courts are turning to Indigenous laws, traditions and customs, as well as the common law, to resolve access conflicts.

Pacific salmon are an ecological and cultural keystone species (sensu Garibaldi and Turner 2004). Jones (2004) identified local stewardship and return to more terminal harvest as essential conservation elements. Steward (2004) described litigation between U.S. tribes and the owners and operators of hydroelectric dams in Washington State that impede salmon and proposed a new, nonconfrontational approach to relicensing of these facilities. Wright and Machin (2004) describe how the Okanagan Nation took the lead role in ecosystem planning for the reintroduction of sockeye salmon to Skaha Lake as the first step in restoring populations.

On Canada's other coast, Davis et al. (2004) describe how oral history of the relationship between people, land, and water rights is beginning to define treaty entitlements. "Kat," the American eel *Anguilla rostrata*, is a key resource for the Paq'tnkek Mi'kmaq Nation, as salmon are in the Pacific Northwest. Documentation of this relationship is at the heart of an "ecosystem stewardship" approach to governance and sustainable livelihoods.

In Australia, marine tenure in the Torres Strait Islands was superseded by industrial fisheries leading to depletion and exclusion. A 1985 treaty acknowledged cultural importance, needs and rights, but participation in management is limited. Islanders are developing innovative monitoring and multisector harvest management models based on traditional stewardship (Prichard et al. 2004).

In Queensland, Sheppard et al. (2004) describe the tension between government and public recognition of Indigenous spiritual and cultural rights and responsibilities and existing fisheries management and marine protected areas (MPA) initiatives. Indigenous relationships to the sea, biological and management systems are critical elements for successful ecosystem management that includes the knowledge of commercial and sports fishers.

Australia has no treaties, but the 1992 Mabo decision recognizing nonexclusive title to lands and waters, provided a context for participation in integrated management. As in Torres Strait, entitlement to economic opportunities is limited. A 2003 conference in Perth set forth principles for participation in management, benefit sharing in fisheries and aquaculture, and capacity building (Wright et al. 2004).

New Zealand achieved full and final settlement of commercial and customary fishing claims in 1992. The Crown and Maori have developed customary fishing regulations to ensure adequate resources for traditional access. The traditional trustees appoint guardians to manage noncommercial fishing. Data collected aids local management and is included in the overall total allowable catch (TAC) processes. Regulations evolve over time (Arney et al. 2004).

Licensing and individual transferable quota (ITQ) systems lead to corporate concentration, control of science and setting access rules to the detriment of Indigenous and small scale fishers in Alaska. Alaskan First Nations have gained economic power and protected subsistence and culture through TAC rights to all fisheries. Subsistence and commercial fisheries require separate management. Science should be at arm's length from both. Economic stability can be assured through access rights to all fisheries with a 20-year nontransferability provision (Jensen 2004).

Global resource depletion and the resurgence of Aboriginal rights and title are driving a reexamination, reaffirmation and application of Indigenous knowledge, management and value systems to ecosystem management.

References

Garibaldi, A., and Turner, N. 2004. Cultural keystone species: implications for ecological conservation and restoration. Ecology and Society 9(3):1.

Jones, R. R. and T. Williams-Davidson. 2000. Applying Haida ethics in today's fishery. Pages 100–117 *in* H. Coward, R. Ommer, and T. J. Pitcher, editors. Just fish: ethics and Canadian marine fisheries. ISER Books, St. Johns, Newfoundland.

Lucas, Chief S. 2004a Aboriginal Values. Pages 114–115 *in* T. J. Pitcher, editor. Back to the future: advances in methodology for modeling and evaluating past ecosystems. Fisheries Centre Research Reports 12(1).

Smith, M. K., and M. Pai. 1992. The Ahupua'a concept: relearning coastal resource management from Ancient Hawaiians. NAGA April 1992:11–13.

American Fisheries Society Symposium 49:499–504

Northern Australia—Indigenous Fisheries Management

Rebecca Sheppard*

*Queensland Fisheries Service, Department of Primary Industries and Fisheries
Post Office Box 1085, Oonoonba, Queensland 4810, Australia*

John Beumer and Scott McKinnon

*Queensland Fisheries Service, Department of Primary Industries and Fisheries
GPO Box 46, Brisbane, Queensland 4001, Australia*

Abstract.—Indigenous people were the first custodians and managers of Australia's fisheries resources (Coleman et al. 2003). While European colonization of Australia occurred in 1788, it has been only in recent history that indigenous peoples' affiliations with the land and sea have been formally identified and recognized by governments. These affiliations result from indigenous ownership, occupation, use, and management of land and sea country. For indigenous people, fishing and hunting are not only important for food and nutrition, but are also important for ceremonial occasions, exchange, trade, and barter (Coleman et al. 2003). Many of the coastal clans of Australia's indigenous peoples identify as "saltwater people" and their traditional estates typically extend beyond the coastal zone and into the adjacent seas. Spiritual and cultural rights and responsibilities underpin the relationships of indigenous peoples with the land and the sea. Commonwealth and state legislation, particularly in relation to marine protected areas and commercial and recreational fishing, have all influenced indigenous fisheries management in Australia. In this paper we summarize fisheries legislation as it applies to indigenous fishing and hunting in Queensland, the Northern Territory and Western Australia. A case study from Injinoo, located on the Eastern Cape York Peninsula, Queensland, examines indigenous peoples' relationships to the sea, biological knowledge, and their community's management approach in relation to size limits, bag limits, spawning times, and access to areas for fishing. An overview of marine protected areas, in particular declared fish habitat areas and the Cape York Peninsula Fish Habitat Area project is also presented.

Introduction

Australia is the world's largest island and has 12,000 smaller islands within its territories. It also has one of the longest coastlines in the world, over 69,600 km (43,200 mi) with a wide variety of estuarine and marine ecosystems. Australia consists of six states, Western Australia, Queensland, South Australia, New South Wales, Victoria, and Tasmania and two territories, the Northern Territory and Australian Capital Territory. The focus of this paper is on Northern Australia, which includes Queensland (and the Torres Strait Islands), the Northern Territory, and Western Australia. Northern Australia is not only unique in

* Corresponding author: rebecca.sheppard@dpi.qld.gov.au

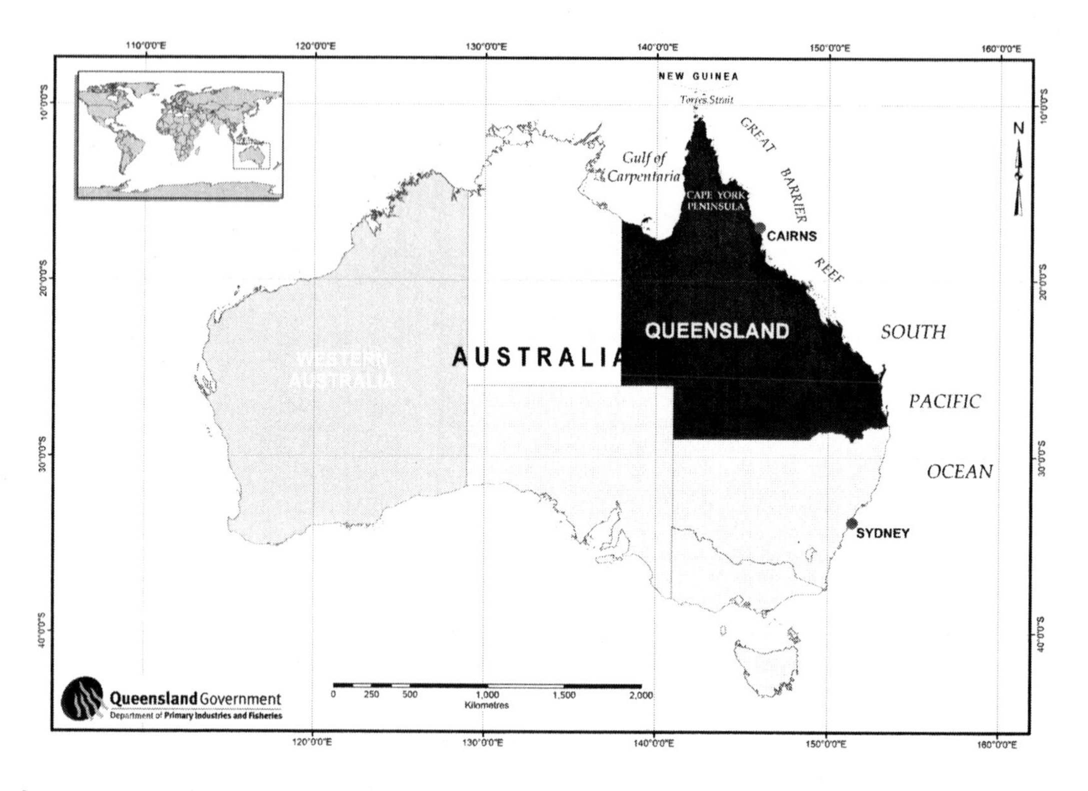

Figure 1. Map of Australia.

terms of its fisheries resources but also is home to the majority of Australia's indigenous population. The fish habitats of Northern Australia are diverse, extensive, and relatively pristine. These important habitats include freshwater systems with intense seasonal flooding, estuarine systems with extensive intertidal saltcouch, mangrove, and seagrass communities, mudflats, sand shoals, and adjacent coastal systems, including the Great Barrier Reef Marine Park (World Heritage Area). The most common species targeted by indigenous people in Northern Australia are dugong, turtle, barramundi, mullet, catfish, sea perch, snapper, bream, crayfish, prawns, crabs, mussels, and baitfish (Coleman et al. 2003).

History of Indigenous Australia

It is generally accepted that aboriginal people have occupied the Australian continent for 40,000–60,000 years. Torres Strait Islanders have occupied the northern islands of Queensland and northern part of Australia for at least 1000 years. Approximately half of Australia's indigenous population live on or near the coast (Smyth 2001). Many of these people and their communities rely heavily on coastal and marine resources for their livelihood and ongoing cultural identity.

Indigenous peoples' marine interests continue to include hunting, gathering, fishing, collecting, ceremonial, management, enforcement of customary laws, access, development and maintenance of a rich, spiritual knowledge and culture. Modern dependence on marine resources is clearly seen on remote, northern Australian coastal communities where traditional activities such as hunting for turtle and dugong continue to be widely practiced (Smyth 2001).

Until the 1990s, indigenous involvement and participation in management of fisheries resources were extremely limited. In the last decade governments have begun to recognize and understand the significance of fishing to indigenous communities not only from nutritional and health perspectives but also for its importance to the social, cultural and economic fabric of these communities.

Indigenous Legal Issues–A Summary

1967—Right to vote—Australia

Aboriginals were given the right to vote though a national referendum.

1992—Mabo case—Australia

The Mabo case overturned the premise that Australia was uninhabited (*terra nullius*) prior to 1788. The high court recognized that at common law, the Meriam People of Mer (Murray) Island always held native title over their traditional lands.

1993—Native Title Act—Australia

Legal recognition of aboriginal domain and rights.

1999—Yanner decision—Queensland

The decision in the Yanner case confirmed that aboriginal and islander people may claim a right under native title to hunt living resources according to local customary laws. The outcome was that most agencies and jurisdictions recognized indigenous rights to obtain and consume traditional foods.

2001—Croker Island—Northern Territory

The High Court found that communal native title exists in relation to the sea and the sea-bed within the claimed area and that native title rights and interests do not confer possession, occupation, use and enjoyment of the sea and sea-bed.

2001—Torres Strait Islanders—Queensland

Torres Strait Islanders approached a licensed commercial fishing vessel and demanded their catch on the basis that it belonged to local indigenous communities. A charge was then laid against the Islanders for taking the fish; however, a Queensland Court found the Islanders not guilty and said they "had an honest right of claim to the fish". The result was that an agreement was established between islanders and fishermen that commercial fishing will not occur within 15 nm of Murray Islands.

Today there are multiple sea right claims over numerous locations across Northern Australia that are currently being heard or waiting to be heard.

Fisheries Management

Indigenous people have always managed their own traditional lands and sea country. From stories passed through generations and traditional and current management practices, indigenous people understand fisheries resources, their habitats, and coastal processes. They recognize fish spawning times, turtle breeding areas, seasonal variations, fish sizes, and populations (Smyth 2001).

Case Study—Injinoo Fisheries Project

Injinoo is located in North Eastern Queensland, within Cape York Peninsula. The area is fished by Injinoo people, the Torres Strait Islander community, charter boat operators, nonindigenous residents and visitors. One of the main target species is blackspotted croaker

Protonibea diacanthus. After recording continual catches of smaller-sized fish and noting an increase in the effort targeting the blackspotted croaker, the Injinoo people became concerned about the long term sustainability of this species. Studies by researchers and the community, confirmed that the size and age of stock fell from three year old stock in 1999 to 2-year-old-stock in 2000 and that sexually mature fish comprised less than 1% of the catch. Injinoo traditional owners and the community council have self imposed a two year ban on the taking of this species. The ban aims to allow the local blackspotted croaker stocks to reach a mature size so that recruitment to the stocks improves. The Queensland Government has supported this ban through legislative changes (Phelan 2002).

This project is a good example of how indigenous communities, researchers, and the government, working together, can pool traditional knowledge with scientific findings for a sustainable and manageable outcome. Other fisheries management tools currently used by indigenous people throughout Northern Australia include dugong and turtle management plans, sea and land country management plans, and traditional use of marine resource agreements.

Overview of Australia's Indigenous Fisheries Management

Within Australia's southern states, namely New South Wales, Victoria, Tasmania, and South Australia, there are no explicit indigenous fishing rights. In general, indigenous people have to abide by existing fisheries regulations (i.e. they are required to hold a recreational fishing license and comply with recreational bag and size limits and seasonal closures).

In Australia's northern states, in particular, the Northern Territory and Queensland, there is a general exemption to fisheries laws for subsistence fishing. For example the Queensland *Fisheries Act 1994* states that "An Aborigine may take, use or keep fisheries resources, or use fish habitats, under aboriginal tradition, and a Torres Strait Islander may take, use or keep fisheries resources, or use fish habitats, under Island custom". In Western Australia there is a general exemption for aboriginal people from having to hold a recreational fishing license. However, all other recreational fishing rules such as bag limits, size limits, and seasonal closures currently apply. In the Northern states, indigenous people also have limited access to commercial fishery through community commercial fishing licenses and grants to buy existing commercial fishing licenses.

Western Australia is also the first jurisdiction in Australia to develop an aboriginal fishing strategy. Recommendations include addressing recognition of customary fishing rights, involving aboriginal people as key stakeholders in the management and protection of fish resources and developing options for aboriginal people to participate in the commercial fishing sector.

The Torres Strait Islands, north of Queensland's Cape York Peninsula also have special fisheries management measures and regulations. Specific legislative recognition of the rights and interest of Torres Strait Islanders in the management of fisheries and marine environments in Torres Straits is recognized through the Torres Strait Treaty signed by the Australia and Papua New Guinea governments. Islanders play an active involvement in fisheries advisory committees. Although there are no exclusive or commercial rights, commercial harvest in the trochus and pearl shell fisheries are reserved for Torres Strait Islanders only. The justification for these unique rights is based not only on historical and cultural significance but because seafood consumption by Torres Strait Islanders is among the highest in the world. Species harvested include dugong, turtle, fish and crayfish (Kwan 2001). Many Islanders also continue their involvement in marine industries in the harvesting of trochus shell, beche-de-mer, and crayfish. However, there is currently no indigenous participation in commercial prawn trawl fishery, which provides the main economic return in the Torres Strait.

Marine Protected Areas and Indigenous Communities

A marine protected area is an area of sea established by law for the protection and maintenance of biological diversity and of natural and cultural resources. A variety of marine protected areas are established throughout Australia: marine parks, declared fish habitat areas, nature reserves, national park, and fisheries reserves. Historically these areas have been established with only minimal consultation with indigenous communities and traditional owners. Fish habitat areas (FHAs) are marine protected areas declared in Queensland over a precisely defined area of key fish habitats for maintaining existing and future fishing and protecting the habitats on which fish and other aquatic fauna depend. The areas are declared with the specific intent of ensuring the continuation of productive recreational, commercial and traditional fisheries in a region. There are currently 68 FHAs covering 720,000 ha of fish habitats throughout coastal Queensland (McKinnon et al. 2002).

A project to assess three candidate areas for declaration as fish habitat areas in Cape York Peninsula has just been finalized in Northern Queensland. The project involved field work and data collection as well as extensive consultation and negotiation with the Indigenous communities and traditional owners. The project achieved strong community support for the declaration of two important FHAs, which will be formally declared later this year (Sheppard and Greene 2003). Fisheries knowledge was also enhanced and good working relationship and friendship were established with traditional owners throughout Cape York Peninsula. This was one of the first marine protected area projects in Northern Australia to involve and work integrally with the local indigenous communities and traditional owners.

Northern Australia—Moving Forward

Although indigenous issues have only come to the forefront of fisheries management during the last decade, there is now increasing involvement and participation in Indigenous fisheries management at all levels, particularly in Northern Australia. Governments are recognizing and defining customary and traditional fishing, and working with indigenous communities to develop land and sea management plans. There is a focus on building and maintaining relationships with people in indigenous communities to establish trust, friendship and understanding. There is also an emphasis and eagerness from indigenous people to move back to country and for the traditional owners to manage their land and sea country for traditional purposes.

References

Coleman, A. P. M., G. W. Henry, D. D. Reid, and J. J. Murphy. 2003. Indigenous fishing survey of Northern Australia. Pages 98–122 *in* G. W. Henry and J. M. Lyle, editors. The National Recreational and Indigenous Fishing Survey. FRDC Project No. 99/158. Commonwealth of Australia, Canberra.

Kwan, D. 2001. A collaborative, consultative and committed approach to effective management of dugongs in Torres Strait, Queensland, Australia. Conference Proceedings of Putting Fishers' Knowledge to Work. August 2001. University of British Columbia, Vancouver.

McKinnon, S., R. Sheppard, and D. Sully. 2002. Fish habitat area network in Queensland, Australia–an innovative aquatic protected area approach. Proceedings of the World Congress on Aquatic Protected Areas. Australian Society for Fish Biology, Brisbane.

Phelan, M. 2002. Development, outcomes and future of an area closure implemented by the indigenous communities of Northern Cape York. Proceedings of the World Congress on Aquatic Protected Areas. Australian Society for Fish Biology, Brisbane.

Sheppard, R., and K. Greene. 2003. Assessment and declaration of Cape York Peninsula Fish Habitat Areas – final report. Department of Primary Industries, Queensland, Australia.

Smyth, D. 2001. Management of sea country – indigenous people's use and management of marine environments. Pages 60–74 *in* R. Baker, J. Davies, and E. Young, editors. Working on country – contemporary indigenous management of Australia's lands and coastal regions. Oxford University Press, Melbourne, Australia.

American Fisheries Society Symposium 49:505–513

Everything is One

KLAH-KIST-KI_IS, CHIEF SIMON LUCAS*
Hesquiaht Nation Council, Box 2000, Tofino, British Columbia, V0R 2Z0, Canada

I am KLAH-KIST-KI_IS, seventh ranking Chief of the Hesquiaht Nation on the west coast of Vancouver. I am also an adjunct professor at the UBC Fisheries Centre. I was a commercial fisherman for 50 years, from age 12 to 5 years ago and was coastal cochair of the British Columbia Aboriginal Fisheries Commission for many years, with the mission of securing First Nations' place in the harvest and wise use of our traditional territories. So, in my time, I have done a lot of listening as well as talking.

I want to start off with this—it's been pretty well confirmed that the explorer Mr. James Cook landed in our territory, Hesquiat (Figure 1; See also http://www.clayoquotalliance.uvic.ca/Language/e-poster7.swf). This is probably what he heard when the canoe came: "***Hn-Cukḥ-Suu*** ? Who are you people? ***Ah-Kin-uptl-Suu***? What are you doing?***Ah-chu-aht-nee-sha. Henut-she-he-wah-aht?*** Who told you to come here?" So the word "reconciliation" I want talk about in terms of what happened at contact.

When Captain Cook was met by canoe, you can think that there must have been some awe from him and his deckhands, looking at a canoe and realizing later when they came close that it was made from one huge cedar tree (Figure 2).

And when the canoes came close by the sailboat, one cannot help but wonder, what did Mr. James Cook and his crew see in that canoe? There was cedar rope; there were huge baskets that held the cedar rope. There were other types of cedar baskets that were there to hold the nourishment. There were various spears that were made from yew wood and the spear ends were made from shell, different sort of shells. And they noticed that it had a sail. That the sail was made from some sort of fiber; at this point they didn't know that it was made from cedar. He also noticed there was a huge spear in there. It was made from yew wood; it was the whaling harpoon (Figures 3). But what he noticed was how the gentlemen, the warriors, dressed; wearing, at that time he didn't know, fur seal and sea otter pelts. And also he noticed in there that were some sort of floats—what were they? Where did they get these floats? They were from the sea otter or sea lions' stomachs—six of these floats when they harpooned a whale (Figures 4 and 5).

So it's safe to say that the explorers didn't come to our country looking for intelligence. What they were looking for? Well, we all know—the First Nations know—what they were looking for. We can now say that it's a crying shame that there was no dialogue with our people who knew about the entire universe (Nuu-chah-nulth Tribal Council 1999), who knew about the importance of the moon that lay

*Email: c/o Nigel Haggan, n.haggan@fisheries.ubc.ca

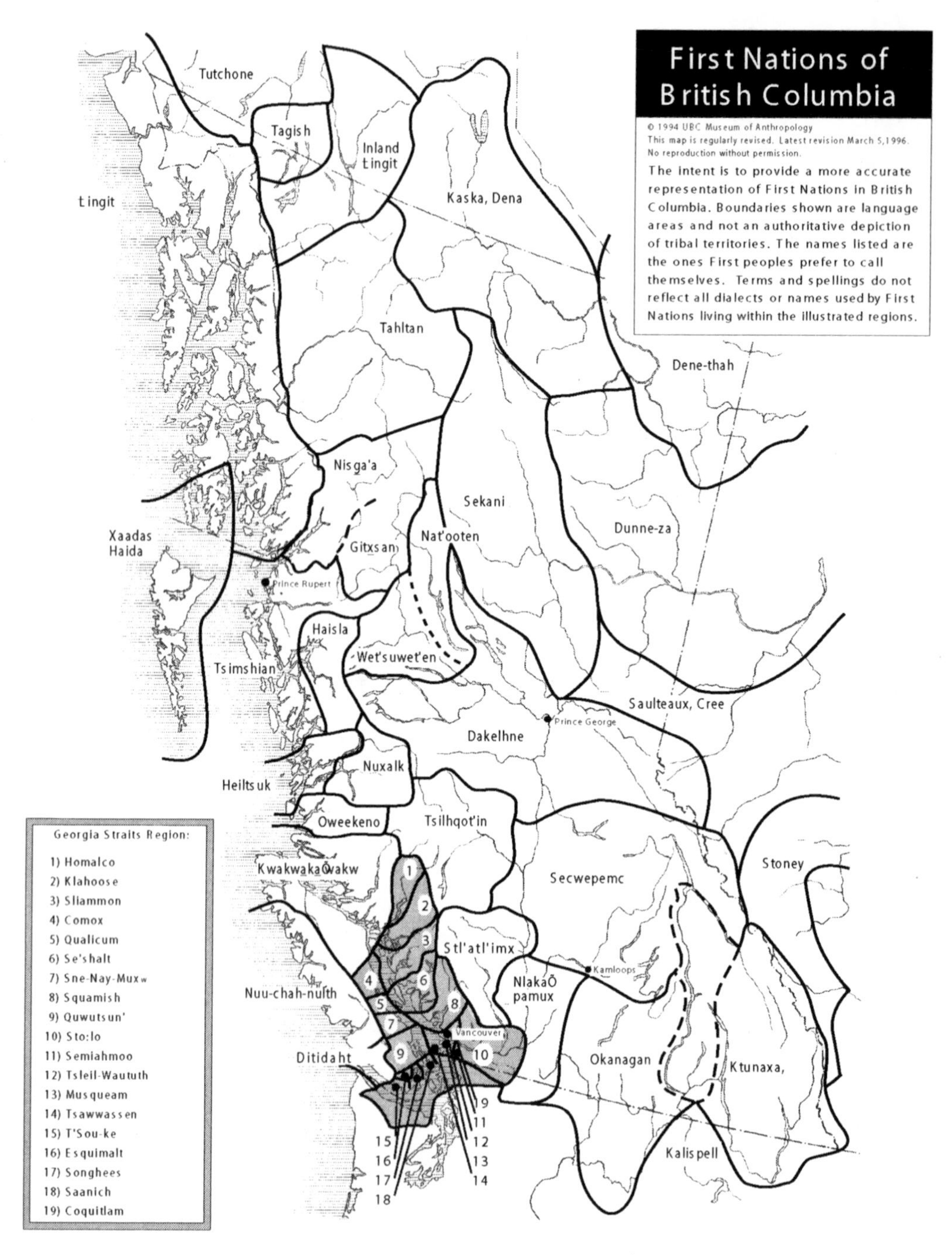

Figure 1. First Nations map of British Columbia showing Nuu-chah-nulth territories. Courtesy of UBC Museum of Anthropology.

above the skies. And what the stars meant—***Ah-čuk-Put-nit-ooh-cha-chee-eeh-waa-aht***, as we call it. And the other planets—what role do they play to this earth; and, of course, the sun. They would have found out that nothing stays the same on this great earth of

Figure 2. Nuu chah nulth canoe.

Figure 3. Harpoon. Photo: Edward S. Curtis.

Figure 4. Nuu chah nulth whaling floats. Photo: Edward S. Curtis.

Figure 5. Partially cut up whale. Photo: Edward S. Curtis.

ours. Because of what happens out in the universe, things change everyday for our people.

If they talked to our people, they would have found that they survived three great floods. Our people still know the mountains that we survived on. They would have heard that there are ten chiefs in the territories that were literally responsible for management of their own territories. That no other nation could go into their territory unless they had special grants to do so (Nuu chah nulth Tribal Council 1999). So we had a management system that was 5, 10, maybe as much as 15,000 years old. Our forefathers didn't have to worry about overexploitation (Golla 2000; Sumpter et al. 2002). And they made sure that they didn't overexploit themselves because they believed, as we do today, that we have to think about the seventh generation. So that tradition of our people bringing back the salmon bones back into the river every day: "***ƛaa-Toos***, can you please come back in live form" is part of our belief system (Nuu-chah-nulth Tribal Council 1999).

Well, what they would have also learned if there had been proper dialogue was what ritual did the *witwaak* (wolves) do? What was their belief system: belief in the Great Spirit of the bottom of the ocean, ***ƛuk-ƛuk-kʷa***; the great god from the sky, ***Hiƛ-suu-iis***; and the great god from beyond the mountain, ***Haa-ƛaa-pii-Hawiƛ***.. Then they would have found out our link to the animal world, what it meant. In the worlds of one of our elders:

> Nuu-chah-nulth see people and territory as inseparable—as interdependent. Territory and all the living creatures in that territory are also seen as inseparable and interdependent. The Nuu-chah-nulth concept of territory is fully inclusive in a way that no English word can describe. (Nuu-chah-nulth Tribal Council 1999)

Well, what he saw was what you now see (Figures 6–8). What did you see? Are you seeing the same things that Captain Cook did? Are you seeing any kind of intelligence in these lives that you see? People have survived here for thousands of years, recognizing that there had to be some sort of management, recognizing that the seventh generation was important. That's past history. What do we do now collectively around the world and at home? Well, we have some examples about what it means to work together. In my territory, the chiefs that make up Nuu-Chah-Nulth got together and said, "There are other people that live in our territory. It's our responsibility to look after whoever lives in our territory." So a Regional Aquatic Management Board was created from the communities outside of the First Nations, the regional districts, Fisheries and Oceans Cananda (DFO), environmentalists, and processors. They created a board to work together for the future, working together to make sure that the communities—all of the communities—benefited like my grandfathers benefited. We have other examples. It's a slow process but a dialogue has been started with five different levels of the federal government. It's an extremely slow process, but we made sure that we're in the plans.

Not like when Captain Cook landed here. From the day that he was there they made sure that we were out of the plan. We say, "No more." We make sure that we're in the plan. And we recognize that we can't be isolated. Our people say, "Those people who come here, who we call ***ƛsa-wats-skwi*** , also have children and grandchildren living in our territory." So when we're talking about Nuu-Chah-Nulth, we're talking about all the people who live in Nuu-Chah-Nulth territory. We're making painstakingly making small little steps towards it. It took us 5 years or so fighting at the table. So what we believe, what the communities be-

Figure 6. Hesquiaht girl in cedar bark costume. Photo: Edward S. Curtis.

lieve, is that ***Maa-muƛ-nii***—that everything is one. The nonIndian people said, "We like that. We want that." So we have tools, each First Nations has tools that make a valuable contribution to reconciling. We have an Memorandum of Understanding (MOU) working with UBC here in Vancouver. It's a slow process, so I thought I would show you a little iota of what Captain Cook saw.

He did not see intelligence.

He did not see that our province, our part of the world, was heavily exploited. They would have found out that our ocean in front of my territory is closely linked to the mountains and the lakes; that all the nourishment that the ocean needs comes from our mountains. All the fine brush—every little thing on the mountain—has a contribution to the well being of the fish that come back home (Lucas 2003; 2004).

In the archaeological dig that our tribe did about 25 years ago, one of the interesting facts was that among the buried of our people were 75 species of fish. And also we were told, "There's no way you could have been involved in tuna." Then bluefin tuna *Thunnus thynnus* remains were found at the site (Crockford 1994; 1997). So we come back from a long

Figure 7. Photo: Edward S. Curtis.

history. Our spiritual values were very much connected to the land, the air and the ocean and the stars, the sun and the moon. You'll see always the use. You will always see in all of our art a human figure in all of those pieces.

One of those we still practice, though. <sound of whistle> I have six of these whistles. Chiefs

Figure 8.

have different amounts. It's been around for thousands of years. Captain Cook said, "What is that? What do you do with it?" Well, it's not to be shown. <whistle> In the belief of our people, the wolf and the ***Hee-šuk-iš-Ƭsa-walk,*** or whom you call the killer whale, is the same animal. We still believe that to this day. Each chief in our territory has a sea serpent dance and they own all of their songs and dances. No one else is allowed to do those songs and dances unless you get permission. Each of our chiefs still has their territories in their names. Kla-Kisht-Ke-Is is who I am. I come from the Hesquiat tribe but ***Kʷa-yuts-Ƭeek Ka-Ka-win***, I'm the chief of ***Hii-maa-yees***

Why do I have the wolf? My oral history tells me that as a little boy one of my ancestors was being watched over by six wild wolves. So that's why today I have wolves, as do all Nuu-Chah-Nulth chiefs in our territory (NCN TC 1999).

So I wanted to say that the first reconciliation that needs to be done is among ourselves. We need to recognize that each of us can make a valuable contribution to the world's management system. We need to stand up and say, "Yes, we recognize the aboriginal people and their talents and intelligence that developed over thousands of years." ***Waa-waa-čuk***, what you call the Chinook salmon *Oncorhynchus tshawystcha*, ***Tyee-suu-ḥaa*** coho salmon *O. kisutch*, ***Tšuu-wit*** ; pink salmon *O. gorbuscha*, ***Ƈaa-pii*** ; halibut *Hippoglossus stenolepis*, ***Boo-eh***; the list goes on. Those names weren't made 200 years ago. So what our ancestors tell me, "***Ickḥ-moo***—don't ever forget that you've been here since time began, that your name Kla-Kisht-Ke-Is is old as the mountains and as old as the lakes and rivers and the four shores that make up our land."

Ƛuu-ƛumpƛ-nii La Perouse bank off the West coast of Vancouver Island) was at one time the largest halibut bank of our coastline, called ***Ƛuu-ƛumpƛ-nii*** and it was called that because there was so much halibut on that bank. But unfortunately our government allowed foreign trawling on La Perouse Bank and almost decimated all the halibut. We must make sure that we have total open minds and absolute trust among each other as did our ancestors. And those nations still exist today, the ones that you just saw. All of them had governance. We weren't running around like the movies showed, where we were just running around. No! We were a society with governance! We were a society with government! We had law and order! We had monitoring programs with management systems! All of them were in my own language and other languages that make up this great province of ours. We have something to offer to the scientific world, as witnessed by the following statement from the first World Indigenous Fisheries Forum. It is our intent to see a much stronger Aboriginal and indigenous presence at the Fifth World Fisheries Congress in Japan. Who knows, maybe the sixth congress can be hosted by a First Nation?

Thank you. ***Claw-cloo.***.

Statement to the Fourth World Fisheries Congress

The First World Indigenous Fisheries Forum was held at the UBC Longhouse on 30 April 2004 to facilitate information exchange and dialogue among indigenous peoples related to the state of the world's fisheries and indigenous fishing rights.

The forum emphasized the importance of recognizing indigenous title and rights, including treaty and inherent rights. This constitutes the foundation upon which local and customary fisheries and related ecosystem knowledge must be protected.

The forum focused on management and policy approaches that embrace Indigenous resource rights. Dialogue emphasized practical ways to utilize traditional knowledge and customary management systems in the conservation and sustainable use of the world's fisheries.

The forum produced the following summary statement and framework to guide the identification and sharing of related best practices:

The Convention on Biological Diversity and related international conventions acknowledge the importance of protecting local and customary fisheries and the role of traditional knowledge, innovations and practices in the sustainable use of the world's fisheries. This is also acknowledged by, and is found in many examples of national legislation.

Forums such as the World Fisheries Congress hold great promise in generating an improved understanding of the relationship between science and traditional knowledge, and the role of customary management systems in the sustainable use of the world's fisheries.

We encourage the greater involvement of indigenous fisheries practitioners, scientists, and traditional knowledge holders in future gatherings of the World Fisheries Congress. We wish to impress upon the participants of the Congress these values in understanding fish ecosystems and in sustainable use of the world's fisheries.

The participants to the forum reflected on key matters of interest and will encourage the sharing of best practices and related indicators globally to:

1. Protect resource access rights and related customary practices, traditional knowledge and management systems;

2. Encourage appropriate access to and use of traditional knowledge and management systems;

3. Involve indigenous peoples in fisheries management and setting related objectives;

4. Develop Indigenous management capacity within both Indigenous communities and government for the management of fisheries; and

5. Examine the role of best practices in developing national policy and legislation, and related resource rights-based solutions.

Rodney Dillon, Forum Co-chair
Commissioner

Aboriginal and Torres Strait Islander
Commission, Australia

A-in-chut (Shawn Atleo), Forum Co-chair
Regional Vice Chief
Assembly of First Nations, Canada

April 30, 2004

First Nations Longhouse
University of British Columbia (Sty-Wet-Tan)
Vancouver, B.C.
Canada

References

Crockford, S. 1994. New archaeological and ethnographic evidence of an extinct fishery for giant bluefin tuna on the Pacific Northwest coast of North America. Fish Exploitation in the Past. In W. Van Neer, editor. Proceedings of the 7th meeting of the ICAZ Fish Remains Working Group. Annales du Musee Royal de l'Afrique Centrale, Sciences Zoologiques No. 274., Tervuren, Belgium.

Crockford, S. 1997. Archeological evidence of large northern bluefin tuna in coastal waters of British Columbia and northern Washington. Fishery Bulletin 95:11–24.

Golla, S. 2000. Legendary history of the Tsisha?ath: a working translation. Pages 133–71 *in* A. L. Hoover, editor. Huupukwanum Tupaat: Nuu-chah-nulth voices, histories, objects and journeys. Royal British Columbia Museum, Victoria, Canada.

Lucas, Chief Simon. 2003. A native chant. Pages 13–14 *in* N. Haggan, C. Brignall, and L. Wood, editors. Putting fishers' knowledge to work. Fisheries Centre Research Reports 11(1), Vancouver.

Lucas, Chief Simon. 2004. Aboriginal values. Pages 114–115 *in* T. J. Pitcher, editor. Back to the future: advances in methodology for modelling and evaluating past ecosystems. Fisheries Centre Research Reports 12(1), Vancouver.

Nuu-chah-nulth Tribal Council (1999) Jurisdiction and governance mandate working group. Hawilthpatak Nuu-chah-nulth – Nuu-chah-nulth ways of governance. Port Alberni, British Columbia, Canada.

Sumpter, I., D. St. Claire, and S. Peters. 2002. Mid-Holocene cultural occupation of Barkley Sound, West Vancouver Island. The Midden 34(4):10–11.

American Fisheries Society Symposium 49:515–516

Session Summary

Roles of Small-Scale Fisheries in Conservation of Aquatic Ecosystems

RATANA CHUENPAGDEE

St. Francis Xavier University, Social Research for Sustainable Fisheries
Antigonish, Nova Scotia, Canada

LISA LIGUORI

University of British Columbia, Institute of Resources, Environment and Sustainability
Vancouver, British Columbia, Canada

A roundtable discussion on small-scale and artisanal fisheries, held on Monday, May 3, 2004, was organized as part of the events at the Fourth World Fisheries Congress. The aim of the roundtable was to facilitate discussion on the roles and contributions of small-scale fisheries in conservation of aquatic ecosystems. Five speakers were invited to make brief presentations to initiate the discussion by addressing the following topics: ecological impacts of small-scale fisheries in relation to conservation of marine resources, social and economic roles of small-scale fisheries in relation to conservation of marine resources, opportunities for reconciling small-scale fishers with other sectors; success and constraints in small-scale fisheries management, and setting goals and new visions for small-scale fisheries. These speakers represented perspectives from around the world (i.e., India, Derek Johnson Center for Maritime Research, the Netherlands; the Philippines, Jovelyn Cleofe, Center for Empowerment and Resource Development, Inc., the Philippines; Barbados, Patrick McConney, Centre for Resource Management and Environmental Studies, University of West Indies, Barbados; Mexico, Silvia Salas, CINVESTAV, Mexico; and global, Daniel Pauly, University of British Columbia). Their presentations posed the following points for discussion:

- The need for new visions (e.g., refocusing our perspectives and putting small-scale fisheries at the center, as opposed to current marginalization), through social reevaluation;

- Critical analysis of alternative and supplementary livelihood options for small-scale fisheries, such as tourism and mariculture (e.g., assessing skills and capital requirements);

- Assessing impacts of global fish trade on small-scale fisheries;

- Development of an inventory of the world fisheries that reflects the quantity, diversity, and importance of small-scale fisheries globally;

- Using an indicator-based approach, rather than conventional stock assessment, to evaluate fisheries (e.g., reference directions as opposed to targets), emphasizing consensual processes, supplemented by appropriate fisheries science;

· Considering a "portfolio" approach to evaluate employment alternatives and opportunity costs, accounting for the diverse characteristics of small-scale fishing communities; and

· Phasing out large-scale fisheries and support small-scale fisheries for their roles in maintaining rural income and food security.

About 60 people participated in the discussion after the presentations, raising several issues, including the various definitions of small-scale fisheries, their social and economic contributions, social changes required in managing small-scale fisheries, and the roles of women in small-scale fisheries. The need for a shift in perspectives for small-scale fisheries was brought up again, in relation to the efficiency in energy consumption and the relatively small ecosystem impacts, as in small-negative-impact fisheries (SNIFs). It was acknowledged, however, that impacts depend, not only on scale, but also gear type. Another interesting point related to the development and modernization of small-scale fisheries, with some participants noting that it could be considered a pathway forward, and might well be the preferred path of fishers, while others stressed the importance of retaining a food fishery instead of commercialization.

Acknowledging the heterogeneity of small-scale fisheries and the need for new perspectives and social changes, the participants recommended the following research topics to understand and enhance the roles of small-scale fisheries in conservation of aquatic ecosystems:

· Inventory of small-scale fisheries to document the multidimensional features, emphasizing also gender roles;

· Comparative studies to test various hypotheses, considering the different nature of resources and responses to varying degrees of fishing pressure;

· Socioeconomic studies to fill information gaps, such as costs, benefits, and values of small-scale fisheries;

· Studies to understand social organization and power structures of small-scale fisheries; and

· Exploring the connectivity of small-scale fisheries (e.g., in terms of place, industrial fishing sector, and the process of social changes).

Authors' Notes

We wish to thank the invited speakers and the roundtable discussion participants for their contributions. This summary is based on our notes taken during the discussion and may not accurately represent the view of all participants. We accept full responsibility for any mistake and/or misrepresentation.

American Fisheries Society Symposium 49:517–529

Livelihood Strategies and Survival of Small-Scale Fisheries in the Finnish Archipelago Sea

PEKKA SALMI*

Finnish Game and Fisheries Research Institute, Saimaa Fisheries Research and Aquaculture Laasalantie 9, FIN-58175 Enonkoski, Finland

JUHANI SALMI

Finnish Game and Fisheries Research Institute, Reposaari Unit Konttorikatu 1, FIN-28900 Pori, Finland

TIMO MÄKINEN

Finnish Game and Fisheries Research Institute Post Office Box 6, FIN-00721 Helsinki, Finland

Abstract.—Social and economic characteristics of a fishery are crucial for reaching sustainability and balance in the use of fish resources. This study focuses on the survival struggle of small-scale commercial fisheries in an archipelago context. Throughout the centuries, fishing, shipping, and agriculture have been combined and interlinked in the Finnish Archipelago Sea region. Fishermen have coped with the changes in their environment by leaning on the economic flexibility provided from a variety of household income sources. As a tradition the fishermen have also relied on the support from the local community. The pluriactivity nature of the livelihood is still important, but the combinations have changed. Along with the transformation to a service-oriented welfare society, primary production from sea and land has currently less importance for the households—people are increasingly dependent on wage work especially in the public sector. In spite of this change in economic weight, fishing and fishermen are still valued among the local communities and the survival of commercial fishing is locally strongly highlighted in the political arena.

Introduction

Small-scale fishing is typically heterogeneous in terms of economic preconditions, motivations as well as cultural and ecological basis for the livelihood. This forms a special challenge for fisheries management. Kuperan and Abdullah (1994) state that "Planning and setting objectives for management of small-scale coastal fisheries requires a good understanding of what is meant by small-scale coastal fisheries, the resource attributes, the traditional values of fishing communities, the institutional arrangements and the overall environment in which the small-scale fishermen operate. Without prior knowledge of these attributes, any attempts to manage the small-scale fisheries will often be met with serious resistance and problems of noncompliance."

This chapter focuses on the survival struggle of small-scale commercial fisheries in the Finnish Archipelago Sea region, where fishing, shipping,

*Corresponding author: pekka.salmi@rktl.fi

and agriculture have been combined and interlinked for centuries (Andersson and Eklund 1999). Recently, the archipelago fisheries have faced increasing profitability problems and marginalization, although the fishermen have strong identity and local support. In order to reveal the problems and attributes of the fisheries, we study the strategies fishermen use when seeking ways to continue their livelihood.

A means for studying the strategies in fisheries from a different perspective than applied in the more usual sector analyses[1] is called livelihoods approach (e.g., Allison and Ellis 2001). Two important concepts related to sustainability of livelihoods are resilience and sensitivity. According to Allison and Ellis (2001) "Resilience refers to the ability of an ecological or livelihood system to "bounce back" from stress or shocks; while sensitivity refers to the magnitude of a system's response to an external disturbance." Ideally, the livelihood system displays high resilience and low sensitivity, while the most vulnerable displays low resilience and high sensitivity. In fisheries, adaptations to uncertainty can be obtained by flexibility within fisheries operations (targeting different species according to availability), geographical mobility, and livelihood diversification. The last alternative is an application of rural pluriactivity, which Eikeland (1999) defines as "gaining an income from more than one economic activity." We address fishermen's adaptation by fisheries operations and geographical mobility, but the strategies related to pluriactivity are of special importance in the Finnish small-scale fishing.

In Finnish small-scale archipelago fishing, pluriactivity is still important, but the combinations have changed. Since the late 1970s, many fishermen switched over to fish farming in the Archipelago Sea region (Eklund 1989). However, a more important factor was the development of car, ferry, and passenger transport system, which guaranteed job opportunities for many families in this coastal area (Andersson and Eklund 1999). Along with the transformation of a coastal region of fishermen and peasants to a service-oriented welfare society, the primary production from sea and land has currently less importance for the households. People are increasingly dependent on wage work—especially in the public sector. In spite of the diminished economic weight, fishing and fishermen are still valued in the local archipelago communities and the survival of commercial fishing is locally strongly highlighted in the political arena.

The livelihood approach provides conceptual tools for understanding the multifaceted fishing livelihood and its capability of surviving. One question is what constitutes the relevant economic unit when studying small-scale fisheries. Charles (2001) states that one of the most noticeable manifestations of a failure to examine and understand the fishery system as a whole has been a preoccupation in fishery analyses with fish and fishing 'firms' as the elements of study, rather than the wider context where the fish and fishers live. In line with the livelihood approach, we use the fisherman's household as the key unit trying to cope with changes in the numerous contextual factors and forces. Our claim is that the decisions made in fishermen's households do not only depend on the needs of the fishing operations, but rather reflect the socioeconomic situation of the whole household and the surrounding community.

The material for this study was collected during an EU funded project, Aquaculture and Coastal Economic and Social Sustainability (AQCESS, QLRT-1999–31151). The Finnish study areas in the Archipelago Sea region are situated in southwestern Finland. The empirical material is composed of personal interviews conducted with fishermen and other stakeholders in the study area. We study the factors affecting the resource use, diversification of household economy and other

[1] By sector analyses we mean the typically national and international studies, which emphasize the economics of larger fishing units. In this article, the focus is on the small-scale fishing livelihoods,, which require more diversified approaches.

livelihood strategies, fishermen's attitudes towards their work, and the support from other stakeholder groups.

Development of Fisheries and Governance

The change from a production area towards that of recreation and consumption is distinctive for the tensions and conflicts in the Archipelago Sea region. Along with this development, the stakeholder groups have multiplied and especially the interests of summer cottage owners and other recreational users of the area have become more decisive. At the same time, the increased emphasis on nature protection has produced resistance from the side of local people, who stress their option for the utilization of local natural resources. Since the late 1970s, for instance, the summer cottage owners and nature conservationists have been working against the expansion of rainbow trout farming in the area (Eklund 1996; Salmi et al. 2003).

As for primary production in general, the number of commercial fishermen declined markedly in Finland during the 20th century, along with the modernization of the society. In the archipelago region, the population, especially in the outer islands, has decreased. For instance, in the Swedish speaking parts of the Archipelago Sea, the number of full-time fishermen dropped from 2,000 in 1934 to 100 in the 1970s and, correspondingly, for the part-time fishermen from 1,450 to 80 (Åbolands Fiskarförbund 1977). Parallel to the national commercial fishing, the value of landings in the Archipelago Sea region has decreased. In the small-scale archipelago fisheries, the fishermen commonly target more than one fish species. A wide variety of fishing methods permits a rapid switch from one fish species to another. Multi-species fishery is primarily an adjustment to annual changes of catches. Also, fierce autumn storms and severe winter ice conditions are especially detrimental to fishing; for example, trawl fishery is impossible in the Archipelago Sea region in wintertime because of the ice field.

The value of the Baltic herring *Clupea harengus membras* landings in the Archipelago Sea region has been reduced to one third during last two decades (Figure 1). The main reason for diminished catches was a slump in the fur industry during the beginning of 1980s. The collapse of Baltic herring landings was remarkable in the Archipelago Sea region: in the beginning of 1980s, one-half of Finnish Baltic herring catch was harvested in the area, but by the turn of the century no more than 14%. Many herring fishermen in the archipelago area changed target species and fishing methods from trawl or trap net to gill net. As a consequence the catch of Eurasian perch *Perca fluviatilis*, zander *Stizostedion lucioperca* (also known as pikeperch), and whitefish *Coregonus lavaretus* (known as powan in North America) have become increasingly important for the commercial fishermen (other species in Figure 1).

The task of fisheries management is, in principle, in the hands of the local water owners, but the current governance system has developed towards a combination of local decision making and a strengthening top-down management system by the state (Salmi and Muje 2001). Archipelago water areas, like most of the Finnish coastal and inland waters, have traditionally been under private ownership in conjunction with possession of land. The decision maker is commonly a collective of water owners, a shareholders association, which jointly controls the interests of individual owners in fishery matters. In addition to these associations, there are also a large number of water areas managed solely by an individual owner. Even if the fisherman owns private waters or a share in the association, he may need larger waters to be able to harvest more efficiently. Fishermen in the Archipelago Sea have had problems with acquiring fishing opportunities for the small and scattered privately owned water areas. Fisheries Regions are geographically larger units, often cover-

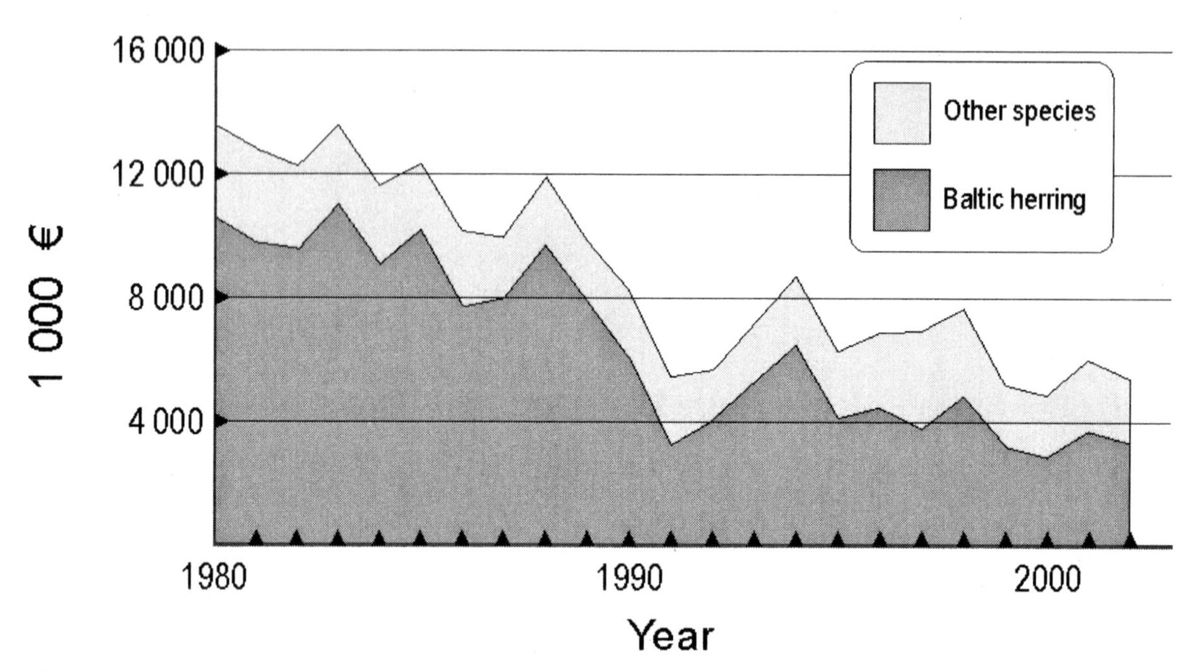

Figure1. Changes in the value (1,000 euros) of landings in the Archipelago Sea region.

ing water areas of a municipality. These organizations offer a wider forum for decision making among water owners and fishermen, but rarely arrange the supply of commercial fishing licences.

The regional level of the state fisheries administration operates under the auspices of the Ministry of Agriculture and Forestry. In addition to the national fisheries policy, the fisheries authorities have the responsibility to implement the common fisheries policy of the European Union (EU) (in practice). The total allowable catch (TAC) level and quotas of Baltic herring and Atlantic salmon *Salmo salar* for Baltic countries are decided annually by the International Baltic Sea Fishery Commission (IBSFC). However, the coastal and archipelago fisheries are chiefly regulated according to national principles because the fishermen target mostly local freshwater fish species.

The funding of the Finnish commercial fisheries is conducted according to guidelines by the EU. In line with the general tendency towards modernization and industrialization in the fishing sector (e.g., Symes 1996), the common fishery policy (CFP) by the EU has emphasized big professional fishing units, which compete in the market with the small-scale archipelago fishermen (Salmi et al. 2000). The situation is similar with Swedish fisheries, where the subsidizing policy by the EU has benefited mostly the larger units at the expense of the small-scale fisheries (Neuman and Píriz 2000). The tightened regulations together with strong competition in the market have created a general pressure to intensify fishing methods, which leads into lower employment in the field. Additionally, in order to receive fishery subsidies the proportion from fishing must exceed 30% of the total personal incomes, which has caused problems to many fishermen in the archipelago area due to the pluriactivity nature of the livelihood.

In Åland, an autonomic province and a part of the Archipelago Sea region, the official management is conducted by the provincial government and its fisheries office, mostly in line with the above-mentioned national fisheries governance principles. The provincial government of Åland has autonomy in most issues, but it also cooperates with the fisheries authorities in the Ministry of Agriculture and Forestry. The most conspicuous manifestation of the autonomic position in fisheries management is that a separate salmon quota is allocated annually to the Åland islands.

Study Area and Material

The Archipelago Sea region is situated in the Baltic Sea, which is a relatively shallow basin of brackish water. This region comprises of two main parts. The Archipelago Sea (total surface area 8,300 km^2) is closer to the mainland and situated in the southwestern province of Varsinais-Suomi. The province of Åland (6,785 km^2) is near the Swedish border. The interviews were carried out in three study sites, Northern Archipelago Sea, Southern Archipelago Sea, and Åland, each including three municipalities. Nine municipalities selected for this study cover an area of 1,110 km^2 (Kuntafakta 2003). In the inner archipelago, the islands are larger, but further from the mainland the islands are small and rocky. The islands closest to the shoreline are connected by bridges to the mainland, but traveling by car to other study sites requires using ferry connections. During the winter, the sea is covered with ice for 50–70 d in the outer archipelago and for more than 100 d in the inner parts.

There are slightly more than 7,000 inhabitants and a total of 9,000 summerhouses in the study area (Statistics Finland 2004). It has been estimated that the average number of visitors per summerhouse is four (Ministry of the Interior 2003), and consequently the total number of summerhouse visitors in the study area would be 36,000. The study area forms the center of Finnish food fish farming (rainbow trout *Oncorhynchus mykiss* cultivation) and processing, commercial fishing is productive, and measured by the number of people fishing there, the region is the leading area for recreational fishery in the country (Ahvonen 1999).

The data of commercial fishermen was collected by means of personal interviews using questionnaires with structured and open-ended questions during the beginning of the year 2002. In the sampling of informants, those fishermen who received at least 30% of their income from fishing were emphasized and thus 76 fishermen (72% of the total) belonging to this group were interviewed. In addition, 37 fishermen who received less than 30% of their income from fishery were interviewed in the selected municipalities. The official register did not include all the part-time fishermen in the area, but it can be estimated that the proportion of the interviewed part-time fishermen was 15–20% of the total.

At the same time, the fish farmers and other local people in the study areas were interviewed for the purposes of the AQCESS project. However, this chapter analyzes only the interview material collected from fishermen. Most of the questions were harmonized with the other four project partner countries to allow comparisons. The questionnaires collected background information about the interviewed person, basic quantitative data about fishing activities and economy of fishing in addition to its relative importance in the household. Attitudes towards fishing occupation were inquired mostly using a structured questionnaire form or as a combination of structured and qualitative method.

Fishermen's Livelihood Strategies

The interviewed fishermen were categorized according to their sources of income in order to reveal different adaptation strategies. This resulted in three main categories: 1) peasant combination, 2) wage work, and 3) service-oriented households. The first group represents the traditional emphasis on the use of local natural resources as basis of the livelihood. It was further divided into two subgroups with emphasis on 1) fish-oriented, including both fishing (as the main income source) and fish farming, and 2) agriculture-oriented. Households belonging to the latter group combine fishing incomes with those from farming, forestry, and horticulture, which form at least 30% of the total. Households in the first group receive, on average, 71% and the agriculture group 27% of their incomes from fishing (Figure 2). Agriculture incomes contribute 51% of the average incomes in the latter group.

The wage working combination represents fishing households, which receive at least 50% of their incomes from paid work. The incomes are generated often from work in a ferry or in some other occupation in the public sector. Older fishermen receiving pension as their main income source are included in this category. Fishing brings only 18% of the household incomes in this group (Figure 2). Nearly one-half (55) of the interviewed fishermen belonged to the wage work group, nine of whom received pensions (other income in Figure 2). The service-oriented category is composed of fisherman's households, that acquire at least 30% of their incomes straight from the tourist industry or from an own firm providing services for the leisure sector, for example, building summer cottages for the urban dwellers.

The interviewed fishermen were also categorized according to primary fishing methods in three main groups: 1) trawl, 2) trap-net, and 3) gill-net fishermen. Trawl fishermen specialized in harvesting Baltic herring. The second group includes those who combined trap-net fishery with gill-net fishery. A major part of the fishermen in the third group use only gill nets. One-half of the gill net fishermen earn less than 30% of their personal incomes from fishing. Large shares of the trawl and trap net fishermen belong to the fishing-oriented livelihood group, while gill-net fishermen dominate the other livelihood strategy groups (Table 1).

The average age of commercial fishermen in the study areas was 52 years. The age was highest among fishermen in the wage working group, while those who combine fishing and agriculture were the youngest. The occupation was dominated by men, only 7% of the interviewed persons were

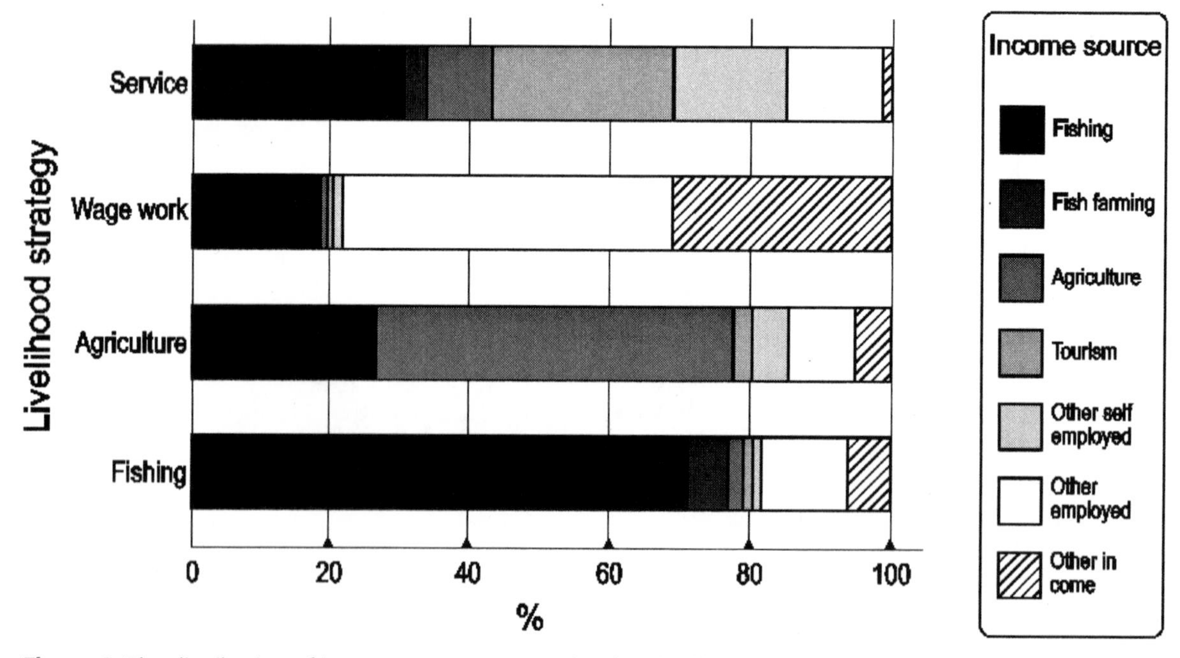

Figure 2. The distribution of income sources concerning four livelihood combinations.

Table 1. Numbers of interviewed fishermen in livelihood and fishing strategy groups.

	Livelihood strategy				
Fishing strategy	Fishing	Agriculture	Wage work	Service	Total
Trawl	7	–	5	1	13
Trap net	5	3	2	2	12
Gill net	13	15	48	12	88
Total	25	18	55	15	113

women. Most of the fishermen (73%) were married or cohabited. Average number of household members was 2.7, which is higher than the Åland average (2.3 in 2001) or that in the province of Varsinais-Suomi (2.1) (Kuntafakta 2003). The national average number of household members is 2.2 (Statistics Finland 2004). Those fishermen who received incomes from agriculture, forestry, and horticulture had more children and bigger households than the average fishermen. Majority of the interviewed fishermen had only basic obligatory education. However, there was also a significant share of persons who had acquired vocational or academic education. Only 7% had received education in the fisheries sector, although many had participated in practical courses lasting for few days.

A large majority of the fishermen were born and dwelled their whole lives in the home municipality. On average, they lived 45 years in the municipality and worked 29 years as fisherman. Fishermen usually inherited their occupation from their parents. They not only learned the professional skills, but also acquired a private fishing harbor, fishing equipment, and access rights to the waters. Access to fishing waters depends on whether the fishermen are able to fish in waters of their own or if they rent waters from outside. Three-quarters of the interviewed fishermen possessed a share or sole ownership to fishing waters of economic importance. Consequently one quarter was dependent only on rented water areas. The share of those whose main fishing waters were rented was highest among the most professional fishermen and especially among the trawlers. This is understandable because they need larger fishing areas than those who fish less actively and in case they are water owners, the areas are not sufficient. Those informants who combine fishing with agriculture and are consequently land owners use mostly their own waters for fishing.

Fishing Operations

The typical length of the fishermen's most important boat in the Archipelago Sea region vary between 6 and 9 m, with 8 m as an average. The vessel size is remarkably small when compared with commercial fishing in many other EU counties. However, the trawl vessels in the Archipelago Sea fisheries are distinctly larger than those used for other purposes, 20.5 m in length on average. Trawl and trap-net fishing was usually operated by groups of two or three persons. The gill-net fishermen often fished alone— 41% of all interviewed fishermen did not have any partners when going out to the sea.

The value of the most important fishing vessel was 72,400 euros among the trawl fishermen, 9,500 euros among the trap net fishermen, and 18,300 euros among the gill net fishermen. State subsidies to the fishery were generally of small importance, only 5% of the interviewed fishermen had received subsidies from the government or EU during the preceding year of the interview. In a previous study, one in five Finnish Baltic Sea fishermen had received subsidies for their investments during a 5-year period 1990 to 1994 (Salmi and Salmi 1998). As a response to the question "Would you apply for a decommissioning scheme?," 90% of the fishermen answered no and 8% answered yes.

The seasonal nature of the archipelago fishing is reflected in the fact that fishermen work, on average, only 147 d along the year. In study sites, where ice conditions prevent fishing for longer periods compared with other areas, the average number of fishing days is only 82. In study sites, where ice conditions are more favorable, the average number of annual fishing days sums up to 189. During the winter, fishing is mostly operated on ice and fishermen travel using snowmobiles. The highest fishing season is during the spring time. The seasonal differences are higher in trap net and trawl fishing strategies than in gill netting.

Fishing and service-oriented households worked most actively in fishing, 182 and 181 d per year, respectively. Those gill-net fishermen who received more than one-half of their income from fishing were engaged in fisheries-related work for 219 d, while those receiving less than 30% worked 107 d during the year. The trap-net and gill-net fishermen's fishing water areas are usually close to their home shore. Consequently, the trip to their fishing grounds lasts less than 50 min on average. However, there were substantial regional differences: the gill-net fishermen's trip to the fishing grounds lasted only about half an hour in the northern parts of the Archipelago Sea, but in the Åland islands, the same activity took 1 h. Trawl fishermen's fishing grounds were the most distant ones: the journey lasted, on average, 2.5 h before reaching the harvesting areas. The whole fishing trip lasted, on average, 15 h and 23 min in trawl fishing, 5 h and 3 min in trap-net fishing, and 6 h and 29 min in gill-net fishing. The fishing-oriented fishermen had the longest working days.

Trawl and trap net fishermen catch mostly Baltic herring (Figure 3). The multispecies nature of the gill-net fishermen is visible: perch, whitefish, zander, and Baltic herring have all importance. The average annual landing of a gill-net fisherman (about 4,000 kg) is less than 1% of trawl fisherman's catch. However, the producer prices per kilograms of the main

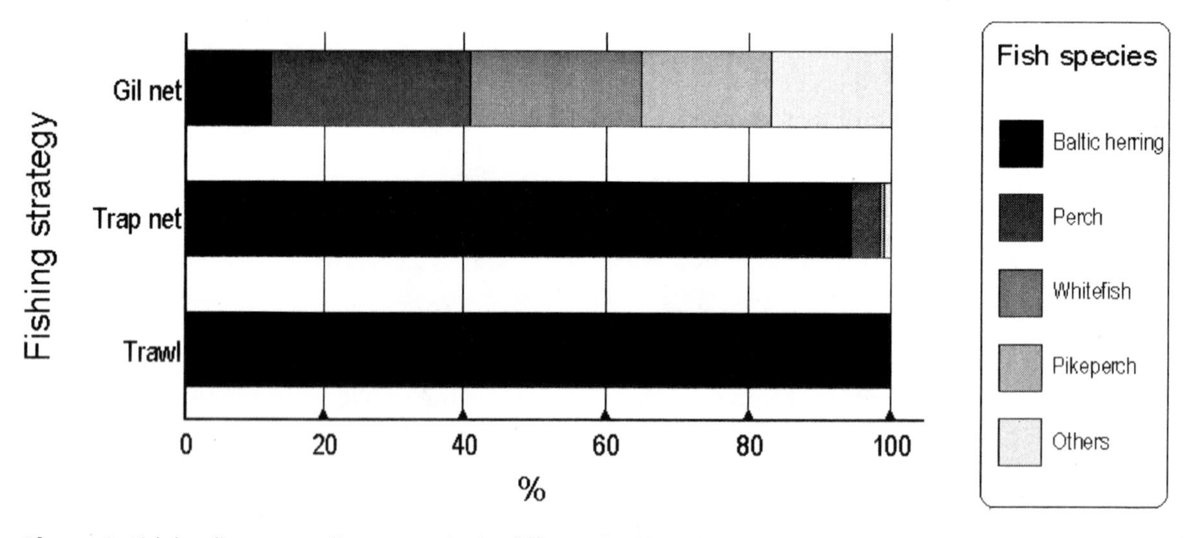

Figure 3. Fish landings according to species in different livelihood strategy groups. The average landings in the trawling group was approximately 486,000 kg, in the trap net group 123,500 kg and in gill-net group 4,000 kg.

gill netting target species are substantially higher than the price of Baltic herring.

Economy

The typical category of the annual income before taxes for the fishermen's households ranged from 16,818 to 25,228 euros. Those households where fishing had less importance earned more often higher total incomes than the others. Most (68% on average) of the household income was contributed by the interviewed persons, mostly males. This was the case especially among the most professional fishermen (79% for the interviewee), while in the group of agriculture-oriented households, the income from the spouse was of substantial importance. When studying only those households where the spouse brought incomes for the family, the average share of the spouse's contribution was 43%.

The average total income before taxes of all interviewed households was 30,770 euros (Table 2), which is lower than the national average in 2001 of 38,343 euros (Statistics Finland 2004). In the province of Varsinais-Suomi, the average total household income was 32,762 euros in the year 2000 and 43,986 euros in the province of Åland (Kuntafakta 2003). The interviewed fishermen in the fishing-oriented livelihood combination received the highest annual and hourly fishing incomes (Table 2). Although the incomes are generally low, the gained income is important as an element for the living of a household as a whole. Fishermen seldom calculate the profitability of fishing operations in terms of hourly income. The informants were asked to respond to a statement "When I think about the opportunities to continue my livelihood I compare my income per hour with other occupations" and only 22% of the interviewed fishermen agreed to that argument. They emphasized the independence and the way of life provided by the work and stated that it is enough to make a living. One informant commented, that "You can't do that, then one would stop fishing." As the fishing profession in the archipelago areas

Table2. Average annual household incomes and personal fishing incomes, working hours and earnings per hour.

	Livelihood orientation				
	Fishing	Agriculture	Wage work	Services	All
Average total income before tax	26,279	28,499	34,436	25,926	30,770
Share of interviewed person's fishing income %	61	26	17	21	29
Fishing incomes contributed by the interviewee	16,030	7,410	5,854	5,444	8,923
Annual working hours	1,656	1,120	831	1,737	1,180
Income per hour	$9.68	$6.62	$7.04	$3.13	$7.56

has a distinctive seasonal nature, the survival of the households is dependent on income combinations. The average commercial fisherman earns nearly 40% of his personal income from fishing and the corresponding share for the whole household income is 32%. Wage work forms clearly the other pillar of the household economies. The rough lines of income source distribution do not differ between those of the interviewee and those of their households, largely because the interviewed persons contribute the majority of the household incomes.

Perceptions of the Livelihood

A vast majority of fishermen have a life long commitment to their occupation and they are reluctant to leave their home islands. The commitment to the occupation is often grounded by traditions and the way of life near to nature provided by fishing. Also, lack of alternatives was put forward as a reason for continuing fishing for lifetime. Those who were not as sure about their futures as fishermen often referred to the uncertainty of the profession which, for instance, created the need for additional income sources. The rare willingness to move from the area even if fishing opportunities did not exist was motivated by general reluctance to move from their home district, additional income sources in the household, family ties, and possession of real property. Interviewed fishermen have a strong conception of a membership in the archipelago community, where the nature, sea, and fish are important elements.

Almost all of the interviewees stressed the independence and the way of life as positive aspects of the fishing occupation (Figure 4). Security of the job was regarded as positive by 41%, okay (neither good or bad) by 33%, and bad by 25% of all respondents. Those who combine fishing with agriculture rarely found time spent away from home as a positive side of the fishing activity, in contrast to the service-oriented fishermen. The attitude of people working in agriculture is understandable because the high seasons of both industries occur mainly at the same period of the year.

If the trawl fishermen should have to stop their fishing occupation, they would prefer to work in

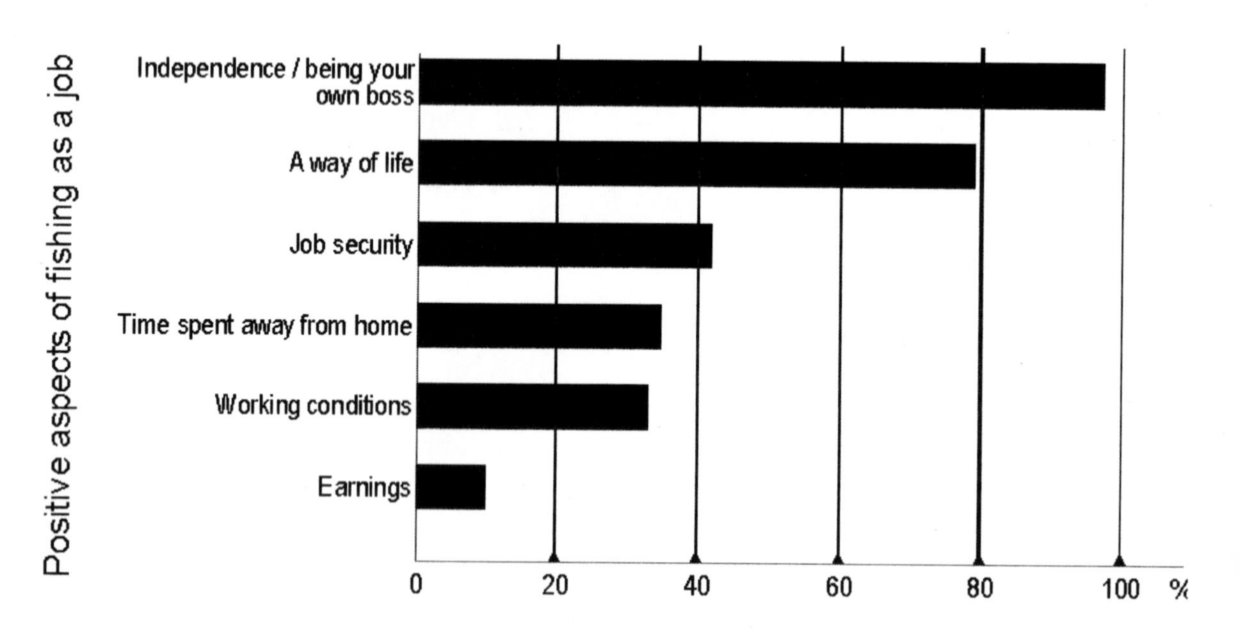

Figure 4. Different aspects regarded as positive in fishing as a job.

seafaring (46%) or tourism (31%). The trap-net fishermen, and generally those who were oriented in agriculture, were mostly interested in continuing their farming livelihood. Fishermen in the wage work-oriented group were mainly interested in seafaring, which was already their income source in many cases. The service-oriented fishermen would prefer to work in the tourism and service sector.

Most of the fishermen considered the local people to be supportive, whereas a majority felt that the authorities did not value or support their livelihood. The positive positions concerning support from local community were motivated by the traditional fit of fishing with the living in archipelago, tax revenues for the municipality, and the locals' will to rent waters for commercial fishing. Attitudes towards the recreational sector varied largely. It was no surprise that those respondents, who are in connection with, and dependent on, the leisure sector most often considered recreational users and providers of the services as supportive. The same livelihood group seldom found the locals to be nonsupportive.

The support from the EU, the state and authorities were mostly connected with financial support in investments, fish transportation, fuel costs, and compensations for damages caused by seals. According to the legislation, the financial support is coupled to the proportion of fishing incomes, which must exceed 30% of the total incomes. The interviewed part-time fishermen criticized this system and even stated that due to this limitation there would soon be no commercial fishing. However, most of the fishermen hold a view that the opportunities for financial support are not enough. They called also for support for access to fishing waters and for opportunities to stay in the area. It was also stated that authorities support only big industries and thus, due to exhaustive harvesting, fish will become extinct. The bureaucracy and paper work has increased and authorities concentrate on making new laws, quotas, and restrictions, which diminish fishermen's freedom.

Discussion

The results from the AQCESS project highlight a case where small-scale fishing livelihood shows marked resilience (see Allison and Ellis 2001) in many respects. In spite of the relatively low economic revenues, the small-scale fishermen in the studied archipelago areas have been able to sustain their livelihood by flexibility in fishing operations and livelihood strategies. In fishing operations, the fishermen have searched for flexibility by targeting a variety of highly valued fish species. Geographical mobility is a rare preference for the interviewed fishermen—most of them are neither willing nor forced to leave their home islands. The appreciation of the life mode connected to the fishing work is high. Fishermen's households have more or less successfully used various flexible and diversifying strategies in order to avoid the mobility option. However, during a longer period of time, large numbers of fishermen have stopped fishing or moved from the area and the material reveals livelihood strategies of only those fishermen who still operated at the beginning of the year 2002.

Symes (1996) has argued that "Whereas in farming, diversification of land use to nonagricultural ends has become a central theme of policies for rural restructuring, such options are not widely available to fishermen." However, when diversification is defined as a broad survival strategy, the studied case shows the importance of diversification also in fisheries. It can be concluded that the combinations of fishing and other income sources have acted as a "buffer" against fluctuations in fishing revenues in the Finnish archipelago region and thus contributed to continuity of the fishing livelihood. On the other hand, also the fishing revenues have undoubtedly acted as a "buffer" or as an important addition to other occupations. The diversification livelihood strategy is a family-based activity with the household as an economic unit. Largely due to the strengthened public service sector in the society, the wage work has become important as a part of the traditional fisherman–peasant combinations. Those who have combined fishing with wage work spend less time in fishing activities compared to other groups and operate mostly during the high fishing seasons.

Local community and culture have supported the traditional fishing life mode, although its economic basis has narrowed in competition with big national and international enterprises. Without the strong local identity and importance of the life mode and pluriactivity traditions, the small-scale archipelago fishing would have slowly disappeared. The increased recreational use of the archipelago provides options for strengthening the service-oriented livelihood strategy, which combines fishing with tourism and services (for example, for summer cottage dwellers). There are new options for marketing fish products straight to consumers and providing services, that may or may not be based on commercial fishing skills and equipment. The services fishermen provide for the recreational users of the area often include an extra value related to the long traditions and cultural values of archipelago fishing. The fishing livelihood in the Archipelago Sea region can be characterized as multifunctional: it not only produces fish, but also contributes to the upkeep of cultural heritage and rural viability in a modernizing coastal area. Fishermen may also be actors in environmental observation and management systems.

The strong ethos of independence among the interviewed fishermen is related to the long-lasting livelihood in which fishermen have considered themselves "producers" or "workers." Like their fathers, they have focused mostly on the fishing operations and stressed that they are free from the clock and outside control. This way of seeing a fisherman's role may limit choosing the service-oriented strategy, but it remains an important alternative in the future livelihood adaptations. The face-to-face interaction between the entrepreneurs and their clients improves options for the development of mutual trust and consumer orientation in the service production.

In general, fisheries are marked by a diversity of involved interests and complexity, and as van Vliet and Dubbink (1999) point out, these attributes of the sector require participatory governance rather than a hierarchical top-down one. In the studied case, the small-scale fishermen feel alienated and marginalized in the fisheries governance system and have seldom received notable benefits from the public authorities. Fishermen's arguments state that the fisheries policy preferring big harvesting units is against their ideas of social justice and, moreover, ecologically unsustainable. Consequently, it was questioned whether the sectored and bureaucratic fisheries policy in the study area enhances socio-economic sustain-ability. Although, from the administrative perspective, big and professional units provide clearer manageability (Salmi et al. 2000), the common fisheries policy of the EU and the national rules should be improved to take the variety of fishing livelihood contexts into account, as the small-scale fishing analyzed in this article. This is not a unique situation: in Sweden, for instance, fisheries policy tends to undervalue the small-scale part-time fishing (Neuman 2002). The part-time nature of the studied fisheries is not, however, equal to nonprofessionalism—it may often mean adaptation to contextual factors limiting a whole-year activity, yet operated in a skilled way learned from the fathers.

References

Åbolands Fiskarförbund 1977. Femtio år i Åboland. Åbolands Fiskarförbund 1926–1976. (Fifty years in Åboland. The Åboland Fishermen's Association 1926–1976.) Ab Sydvästkusten, Turku. (In Swedish).

Ahvonen, A. 1999. Summary by Region. *in* Regional Fisheries in Finland. SVT, Agriculture, Forestry and Fishery 1999:10, p. 57.

Allison, E., and F. Ellis. 2001. The livelihoods approach and management of small-scale fisheries. Marine Policy 25:377–388.

Andersson, K., and E. Eklund. 1999. Tradition and innovation in coastal finland: the transformation of the Archipelago Sea region. Sociologia Ruralis 39:377–393.

Charles, A. 2001. Sustainable fishery systems. Blackwell Scientific Publications, Fish and Aquatic Resources Series 5, Oxford, UK.

Eikeland, S. 1999. New rural pluriactivity? Household strategies and rural renewal in Norway. Sociologia Ruralis 39:359–376.

Eklund, E. 1989. Sommargästernas skärgård hundra år (The hundred years of the summer guests' archipelago). Nordenskiöld-samfundets tidskrift 49(Supplement): 153–160. (In Swedish).

Eklund, E. 1996. Aquaculture in the Baltic Sea: Regional development and environmental conflict. Pages 59–67 *in* C. Bailey, S. Jentoft and P. Sinclair, editors. Aquaculture development: the social dimensions of an emerging industry. Westview Press, Inc., Boulder, Colorado.

Kuntafakta (Municipal facts). 2003. Facts of Finnish municipalities. Statistics Finland.

Kuperan, K., and N. Abdullah. 1994. Small-scale coastal fisheries and co-management. Marine Policy 18:306–313.

Ministry of the Interior 2003. Saaristo-ohjelma 2003–2006. Saaret, meret, järvet, joet ja rantavyöhyke aluekehitystekijöinä (The Finnish Archipelago program 2003–2006). Saaristoasiain neuvottelukunta, sisäasianministeriö. (In Finnish).

Neuman, E. 2002. Vägar mot ett uthålligt kustfiske (Roads to sustainable coastal fisheries). Human Ecology Research Series – SUCOZOMA Report 2002:3. (In Swedish).

Neuman, E. and L. Píriz. 2000. Svenskt småskaligt kustfiske – problem och möjligheter. The Swedish small-scale coastal fisheries – problems and prospects. Fiskeriverket Rapport 2002 2:3–40.

Salmi, J., and P. Salmi. 1998. Livelihood and way of life: Finnish commercial fisheries in the Baltic Sea. Pages 175–183 *in* D. Symes, editor. Northern waters: management issues and practice. Blackwell Scientific Publications, Fishing News Books, Oxford, UK.

Salmi, J., T. Mäkinen, P. Salmi and J. Setälä. 2003. The socio-economic profile of fish farmers and their perspectives on aquaculture in the Archipelago Sea region, Finland. Pages 8–12 *in* The proceedings of the XV EAFE Conference, session 3: Coastal Zone Management, 15–16 May, 2003. Ifremer, Brest, France.

Salmi, P., J. Salmi, and A. Lappalainen. 2000. Finnish part-time fishery: a flexible local strategy of a nuisance for fisheries management? Pages 132–142 *in* D. Symes, editor. Fisheries Dependent Regions. Blackwell Scientific Publications, Fishing News Books, Oxford, UK.

Salmi, P., and K. Muje. 2001. Local owner-based management of Finnish lake fisheries: social dimensions and power relations. Fisheries Management and Ecology 8:435–442.

Statistics Finland 2004. StatFin –Online service. http://statfin.stat.fi/statweb/.

Symes, D. 1996. Fishing in Troubled Waters. Pages 3–16 *in* K. Crean and D. Symes, editors. Fisheries Management in Crisis. Blackwell Scientific Publications, Fishing News Books, Oxford, UK.

Van Vliet, M., and W. Dubbink. 1999. Evaluating governance: state, market and participation compared. Pages 11–30 *in* K. Kooiman, M. van Vliet and S. Jentoft, editors. Creative governance: opportunities for fisheries in Europe. Ashgate, London.

Åbolands Fiskarförbund 1977. Femtio år i Åboland. Åbolands Fiskarförbund 1926–1976. (Fifty years in Åboland. The Åboland Fishermen's Association 1926–1976.) Ab Sydvästkusten, Turku. (In Swedish).

American Fisheries Society Symposium 49:531–539

The Socioeconomic Structure of Fishermen of Iznik Lake

Meral Soylu*, M. Selcuk Uzmanoglu, and Ünal Erdem
Marmara University, Vocational School of Technical Sciences
Aquatic Products Department, Istanbul, Turkey

Ayse Çinar and Z. AysunAltikardes
Marmara University, Vocational School of Technical Sciences
Computer Technology and Programming Department, Istanbul, Turkey

Abstract.—Although fishing for freshwater foods has long been common in Turkey, this and related industries have not achieved targeted production levels in recent years. In this study, we explore the socioeconomic characteristics of Iznik Lake fishermen in order to identify conditions affecting their production. Overfishing of the seas and lakes of Turkey has contributed to a significant decline in the production of freshwater resources. Production was 676,004 metric tons in 1988. It declined to 364,661 metric tons in 1991. While this value rose to 582,376 metric tons in 2000, this increase is nonetheless considered insufficient relative to the rising population. Iznik Lake is located very near Istanbul, the largest city and metropolitan area in Turkey. With an area of 298 km^2, Iznik is the largest lake of the Marmara Region. These features contribute to the pertinence of our investigation. Furthermore, the needs and conditions of fishermen who live in the region are also important issues. Our research subjects are fishermen who live and work in and around Iznik Lake. Individual informants were asked questions about their family and their work in order to assess various characteristics of the workforce community, including general demographic information, average age, number of workers, education level, production capacity, production technology, export-import situation, and transportation. The socioeconomic structure of the Iznik Lake fishing community is described using the data we collected. This study will provide important baseline data and will serve as a methodological example for further research in this field.

Introduction

In the Marmara region of Turkey, there are many natural and dammed lakes of various sizes with a combined total area of 850 km^2 (Inandik 1965). One of these, Iznik Lake, was formed at the end of the third period and at the beginning of the fourth period as a result of tectonic movements (Saraçoglu 1990; Figure 1). This lake lies among mountains more than 500 m in height. In some sections around the lake, the steep, vertical sides of these mountains reach the shore, while the western and eastern sides of the lake are formed by flat valleys. It is elliptical in shape, 32 km in east–west length, and 11.5 km in north–south width (Saraçoglu 1990). The distance from Iznik Lake, at 85 m above sea level, to the Marmara Sea is 15 km (Rahe and Worthmann 1986). Its surface area is 298 km^2 (Saracoglu 1990). Its depth is 10–15 m in the northern

*Corresponding author: msoylu@marmara.edu.tr

Figure 1. Iznik Lake.

portion, somewhat less in the eastern and western sides, and 65 m at Karacakaya in the south. Fed by many rivers and streams, Iznik Lake has a water basin area of 936 km^2 (Rahe and Worthmann 1986). Water level in the lake is adjusted by a regulator, and the overflow is connected to Garsak Stream, which flows through to the Marmara Sea as a strait (Saraçoglu 1990).

Iznik Lake and its surrounding lands are primarily used for agriculture and tourism. Vegetable and fruit farming predominate in the northern regions, and olive farming in common in the southern part.

Besides these agricultural activities, the town of Iznik is famous for its ceramic tiles and for Hagia Sophia Church, which hosted meetings of the first and seventh consuls and is consequently an important place for the Christian world (www.kultur.gov.tr). Iznik Lake is located between the towns of Iznik and Orhangazi, whose populations are 44,770 and 68,902, respectively (SIS 2002). Females and males make up 50.36% and 49.64% of the population of Iznik, and 49.70% and 50.30% of the population of Orhangazi, respectively. Iznik Lake does not contain a wide variety of fish (FAO 1971; Rahe and Worthmann 1986; Geldiay and Balik 1988). The most common species are as follows: bleak *Alburnus alburnus* Linnaeus, 1758; Mediterranean sandsmelt (also known as big-scale sand smelt) *Atherina boyeri* Risso, 1810; *Barbus cyclolepis* Heckel, 1840; freshwater blenny *Blennius fluviatilis* Asso, 1801; Danube bleak *Chalcalburnus chalcoides* Güldenstaedt, 1772; common carp *Cyprinus carpio* Linnaeus, 1758; threespine stickleback *Gasterosteus aculeatus* Linnaeus,

1758; Black Sea chub *Leuciscus borysthenicus* Kessler, 1859; European chub *L. cephalus* Linnaeus, 1758; Angora loach *Nemacheilus angorae* Steindachner, 1897; tubenose goby *Proterorhinus marmoratus* (Pallas, 1811); kutum *Rutilus frisii* Nordmann, 1840; Adriatic roach *R. rubilio* Bonaparte, 1837; roach *R. rutilus* Linnaeus, 1758; wells *Silurus glanis* Linnaeus, 176; tench *Tinca tinca* (Linnaeus, 1758); vimba *Vimba vimba* Linnaeus, 1758; and Turkish crayfish *Astacus leptodactylus* (Esch.).

There are four fishing cooperatives around the lake with 200 members total. The total aquatic foods caught from the lake in 2003 totaled about 815 metric tons.

In this study, the aim is to bring out the socioeconomic structure of Iznik Lake fishermen, to show their present problems, and to determine the precautions for their more rational working style.

Methods

In this study, the socioeconomic structure of the Iznik Lake fishing community is examined. Five visits were made to the lake between January 2002 and October 2003 in order to interview the fishermen. Eighty three of the 200 fishermen were questioned. The results of the questionnaires are presented in the tables below.

Results

The ages of Iznik Lake fishermen ranges between 20 and 74. Fishermen between 40 and 49 years of age are most common (32.53%), and fishermen between 60 and 69 are least common (2.41%, Table 1). It is worthy to note that three fishermen between the age of 70–79—one aged 71 and two aged 74—are still engaged in fishing. All fishermen are male.

The marital status of the fishermen is 90.36% married and 9.64% bachelor (Table 2). For most (61.44%), graduation from primary school is the highest level of formal education received. Only two people (2.41%) are literate but without some school training. None of the fishermen have alternative incomes that are more important than fishing to their livelihood.

Primary school graduate is also the most common (82.67%) education level among the fishermen's wives, and university graduates are least common with 1.33% (Table 4). Although no fisherman graduated from university, one fisherman's wife graduated as a nurse.

Only 8.00% of the wives have jobs outside the home (Table 5).

The number of children in a fisherman's family was also investigated. Sixteen percent of

Table 1. Age distribution of Iznik Lake Fishermen.

Ages	Frequency	%
20–29	8	9.64
30–39	25	30.12
40–49	27	32.53
50–59	18	21.69
60–69	2	2.41
70–79	3	3.61
Total	83	100.00

Table 2. Marital status of Iznik Lake fishermen.

Marital status	Frequency	%
Bachelor	8	9.64
Married	75	90.36
Total	83	100.00

Table 3. Education status of Iznik Lake fishermen.

Education status	Frequency	%
Literate	2	2.41
Primary School	51	61.44
Secondary School	16	19.28
High School	14	16.87
Total	83	100.00

Table 4. Education status of Iznik Lake fishermen's wives

Education status	Frequency	%
Literate	2	2.67
Primary School	62	82.67
Secondary School	6	8.00
High School	4	5.33
University	1	1.33
Total	75	100.00

Table 5. Working Status of Iznik Lake Fishermen's wives.

Working status	Frequency	%
Housewife	69	92.00
Employed outside home	6	8.00
Total	75	100.00

families have one child. Two children are most common (58.67%, Table 6). Three children were reported to have died.

The education status of fishermen's male and female children were investigated separately, though the results are similar for both. Primary education or lower is most common, while completion of high school or higher is considerably less so (Table 7).

The number of persons in the fishermen's household is given in Table 8. Families with three persons are the most numerous with 55.42%. The most crowded home has seven persons (1.21%). Only 11 fishermen work on the lake with their

Table 6. The number of Iznik Lake fishermen's children.

Number of children	Frequency	%
1	12	16.00
2	44	58.67
3	11	14.67
4	6	8.00
5	1	1.33
6	1	1.33
Total	75	100.00

Table 7. Education status of Iznik Lake fishermen's children.

Educationstatus	Female	%	Male	%
Pre-school	15	17.24	12	14.81
Primary Education	47	54.02	41	50.62
High School	23	26.44	26	32.10
University	2	2.30	2	2.47
Total	87	100.00	81	100.00

Table 8. The number of persons in Iznik Lake fishermen's houses (in addition to himself).

Number of persons	Frequency	%
2	14	16.87
3	46	55.42
4	13	15.66
5	7	8.43
6	2	2.41
7	1	1.21
Total	83	100.00

own families. All the others employ nonrelatives. Most fishermen usually work with at least two persons in their vessel. Some fishing boats are made entirely of wood. The other wooden boats are covered with fiberglass. The lengths of the boats range between 3.00 and 8.50 m (Table 9).

The fishermen utilize engines from 5.5 to 15.0 hp in their fishing boats (Table 10).

The age of the vessels is primarily from 7 to 12 years of age (50.60%, Table 11). Vessels aged 16–18 are uncommon with 3.62%. The newest boat is 1 year old and the oldest is 18 years old.

Iznik Lake fishermen use crayfish traps, gill nets, dredging nets, hook and line, and long lines as fishing gear. While some fishermen use only one type of fishing equipment in the

Table 9. The length distribution of Iznik Lake fishermen's vessels.

Length of vessels (m)	Frequency	%
3.00–3.99	1	1.20
4.00–4.99	15	18.07
5.00–5.99	23	27.71
6.00–6.99	26	31.33
7.00–7.99	14	16.87
8.00–8.99	4	4.82
Total	83	100.00

Table 10. The power of the engine of Iznik Lake fishermen's vessels.

Power of engine (HP)	Frequency	%
Without engine	12	14.46
5.5	2	2.41
7.0	9	10.84
9.0	11	13.25
10.0	32	38.55
12.0	2	2.41
13.0	12	14.46
15.0	3	3.62
Total	83	100.00

Table 11. The age distribution of Iznik Lake fishermen's vessels.

Age of vessel	Frequency	%
1–3	13	15.66
4–6	18	21.69
7–9	21	25.30
10–12	21	25.30
13–15	7	8.43
16–18	3	3.62
Total	83	100.00

boat (such as crayfish traps), others use multiple types of gear. Eight vessels use only crayfish traps. Two use gill nets, crayfish traps, hook and line, and dredging nets (Table 12).

Gill net and crayfish trap combinations are most commonly used (47.00%). Dredging nets, gill nets, and a combination of a crayfish trap and dredging net are each uncommon at 1.20%. Fishermen are generally provided fishing materials from their own cooperatives. Mesh sizes of the nets vary between 45 and 90 mm. The lengths of these nets are 200 m, the width between 1 and 10 m. Mesh sizes of nets used for Mediterranean sandsmelt are 5–6 mm. The lengths of dredging nets vary between 150 and 300 m. The diameter of crayfish traps is about 70 cm.

Table 12. Fishing gear distribution of Iznik Lake fishermen.

Fishing gear	Frequency	%
Crayfish trap	8	9.64
Gill net	1	1.20
Gill net+crayfish trap	39	47.00
Dredging net	1	1.20
Gill net+hook and line	8	9.64
Gill net+crayfish trap+dredging net	6	7.23
Gill net+crayfish trap+hook and line	17	20.49
Gill net+crayfish trap+hook and line+ dredging net	2	2.40
Crayfish trap+ dredging net	1	1.20
Total	83	100.00

Table 13. Daily amount of aquatic products catch.

Amount of aquatic products (kg/d)	Frequency	%
1–30	49	59.04
31–60	19	22.89
61 and over	15	18.07
Total	83	100.00

According to the fishermen, their daily catch ranges from 4 to 120 kg. Daily catch of 1–30 kg is most common, and 61 kg and over the least (Table 13).

For 54.22% of fishermen, fishing is their exclusive occupation. Some also work as farmers. Others find additional employment as merchants (Table 14).

Iznik Lake fishermen work on the lake 2–3 d in a week in the winter and every day during the summer, except for restricted periods.

There are four cooperatives around the lake in the towns of Çakirca, Narlica, Iznik, and Orhangazi. Most fishermen are members of these cooperatives and they sell their own catch. Fish are generally consumed locally. As such, when catches are high, they are sent to neighboring cities for selling. Crayfish and Mediterranean sandsmelt are given to the cooperatives and are exported to France, Italy, Greece, and other countries. The expenses of fishermen are transport, food, wages for workers, and boat and gear repair. As of 2003, their annual income is between US$1,440 and $4,600.

The values of the various species in the lake ranges from $0.30 to $3.00/kg (Table 15).

Discussion

A 1986 study performed at Iznik Lake reported data from four fishing cooperatives with a total of 200 members (Rahe and Worthmann 1986). A more recent study reported on 340 fishermen (Anonymous 1994).

Table 14. Additional employment of Iznik Lake fishermen.

Jobs	Frequency	%
Fishing only	45	54.22
Farmer	24	28.91
Merchant	14	16.87
Total	83	100.00

Table 15. Selling prices of aquatic products (prices are United States dollars).

Aquatic products	Price
Silurus glanis	\$2.67/kg
Atherina boyeri	\$0.30/kg
Cyprinus carpio	\$2–2.34/kg
Astacus leptodactylus	\$2–2.67/kg
Rutilus frisii	\$1.34/piece

In the present study, the number of fisherman and the number of cooperatives more closely resembles the sample reported in 1986.

According to the data obtained in 1994, the length, age, and engine power of fishing boats were 3.0–8.0 m, 0–18 years old, and 3–18 hp, respectively. In this study, boat length, age, and power of fishing boats are quite similar: 3.0–8.0 m, 1–18 years old, and 5.5–15.0 hp, respectively, and suggest that no significant changes in boat technology have been employed.

In 1993, the total catch in Turkey's inland waters was 41,575 metric tons. The Bursa and Iznik Lake catches were 1,770 and 1,050 metric tons, respectively. The nationwide total freshwater catch increased to 54,460 metric tons in 1997. Local catches for the same year in Bursa and Iznik were 2,329 and 1,207 metric tons, respectively. Since that time (i.e., as of 2002), the total freshwater catch in Turkey has decreased to an amount of 43,938 metric tons/year, with 2,127 metric tons/year from Bursa Lake and 815 metric tons/year in Iznik Lake (SIS 1995, 1998, 2004a).

National per capita revenue was estimated to be \$2,609 in 2002 and \$3,366 in 2003 (SPO 2004). Annual income obtained from olive production was \$3,335–\$60,000, and \$1,620–\$6,300 from green houses. When compared to the income from agricultural activities, the fishermen's income of \$1,440–\$4,600 for 2003 clearly belongs in the lower income group.

Nationwide, the literacy rate in Turkey is 86.57%. In Bursa, however, it is 98.99%; in Iznik it is 90.16%, and in Orhangazi, the literacy rate is 92.55%. (SIS 2003, 2004b). All of the fishermen are literate, but none of them had education beyond high school.

According to fishermen, the ban on Mediterranean sandsmelt fishing resulted in an increase in the number of sandsmelt, causing a decrease in the recruitment of other fishes. For this reason, the ban has been removed since 2003.

Increasing agricultural activities and the discharge of domestic waste into the lake has polluted the lake waters. In addition, poor fishing practices due to insufficient education have resulted in decreased fish production.

Finally, a decrease in fishing income and the corresponding high income of agricultural activities has resulted in fishermen working in other areas.

Acknowledgments

This study was supported by Marmara University, Scientific Research Project Commission. Project number: SOC-103/131200. Project coordinator: Dr. Ayse ÇINAR.

References

Anonymous. 1994. Türkiye'deki Içsular ve Balik Çiftlikleri Incelemesi, Bazi Göllerin Su Ürünleri Istatistikleri. Cilt: 6, T.C. Tarim ve Köyisleri Bakanligi Tarimsal Üretim ve Gelistirme Genel Müdürlügü.

FAO (Food and Agriculture Organization of the United Nations). 1971. European inland water fish, a multilingual catalogue. Fishing News (Book) Ltd., England.

Geldiay, R., and S. Balik. 1988. Türkiye Tatisu Baliklari, Ege Üniv. Fen Fak. Kitap Serisi No 97.

Inandik, H. 1965. Türkiye Gölleri (Morfolojik ve Hidrolojik Özellikleri), Ist. Üniv.Yayin No: 1155, Cogr. Enst. Yayin No: 44.

Rahe, R., and H. Worthmann. 1986. Marmara Bölgesi Içsu Ürünlerini Gelistirme Projesi, Sonuç Raporu. PN 78. 2032. 7, Eschborn.

Saraçoglu, H. 1990. Bitki Örtüsü Akarsular ve Göller, ISBN 975. 11. 0366. 5, Milli Egitim Basim Evi, Istanbul.

SIS (State Institute of Statistics).1995. 1993 fisheries statistics. SIS printing division, Publication No. 1732, Ankara, Turkey.

SIS (State Institute of Statistics). 1998. 1997 fisheries statistics. SIS, Publication No. 2154, Ankara, Turkey.

SIS (State Institute of Statistics). 2002. 2000 census of population social and economic characteristics of population Province Bursa. SIS, Publication No. 2718, Ankara, Turkey.

SIS (State Institute of Statistics). 2003. 2001 household labour force statistics. SIS, Publication No. 2713, Ankara, Turkey.

SIS (State Institute of Statistics). 2004a. 2002 fisheries statistics. SIS, Publication No. 2883, Ankara, Turkey.

SIS (State Institute of Statistics). 2004b. 2002 household labour force statistics province centers. SIS, Publication No. 2875, Ankara, Turkey.

SPO (State Planning Organization). 2004. Economic and social indicators (1950–2003). SPO, Ankara, Turkey.

American Fisheries Society Symposium 49:541–546

Artisanal Small-Scale Fishery and Community-Based Fisheries Management by "Jangadeiros" in Northeastern Brazil

René Schärer*

Amigos de Prainha do Canto Verde, Caixa Postal 52722
Fortaleza, Ceará, 60151-970, Brazil

Michelle T. Schärer

University of Puerto Rico, Department of Marine Sciences
Post Office Box 908, Lajas 00667, Puerto Rico

Introduction

Brazil is the largest country in South America with a land surface area of 8.5 million km^2 and a coastline of more than 8,000 km. The coast of the state of Ceará in the northeastern part of Brazil runs west to east and spans 573 km. This area is characterized by a series of coastal communities that depend on the coastal and marine resources for their survival. Among these resources the lobster fishery has provided much needed wealth to many of these communities. After an initial boom of lobster exports from foreign investment in local fishery operations since the 1950s the stock has collapsed and generated a bust for all who depended on this resource. This study focuses on one of these communities and how an opportunity for co-management of coastal and marine resources has been taken towards sustainable alternatives in the Brazilian political, social and cultural context.

Prainha do Canto Verde (PCV) is a small village located in the state of Ceará approximately 126 km east of the capital city Fortaleza. The village is part of the Municipality of Beberibe, and comprises approximately 1,200 native residents. The history of this coastal community can be traced to 1870, and it first became known in 1928 (Newspaper "Jornal O Povo"), when three fishermen took up a challenge and sailed by "jangada" (artisanal handcrafted sailraft, Figure 1) to Belém on the Amazon River, over 1,600 km away, in 14 d. At present the community is well organized and takes over many of the obligations of the Municipal Government, in both social and economic development. The local economy depends heavily on the local fisheries for consumption and income. Today, fishing and small commerce account for 50% of local income, the rest is more diversified than in the past because of local development initiatives such as the community based Tourism Project.

The community's understanding and perception of ecological processes, fishery dynamics, responsible fishery practices and sustainable principles have led to a better awareness of the "jangadeiros" (artisanal, or small scale fishers of "jangadas") role in the future of North-

*Corresponding author: fishnet@uol.com.br

Figure 1. Artisanal sailraft locally known as "jangada" being launched for a fishing expedition.

eastern Brazilian fisheries. The process was slow among the older (more illiterate fishers) but is now accelerating among younger fishers that are soon to graduate from the local Fishery School. The community has undergone a long term capacity building initiative with international and local NGOs, which has provided significant alternatives for their involvement in the management of the coastal and marine resources upon which their livelihood depends. This case study may serve to understand some aspects of the complex social, economic and cultural considerations that hinder effective comanagement of marine resources in developing countries.

The Fishery

Two main species compose the lobster catch, Caribbean spiny *Panulirus argus*, which is the predominant species, and smoothtail spiny *Panulirus laevicauda* lobster. The main fishing gears used by artisanal fishers are relatively small wooden and monofilament nylon traps (0.5 m × 0.75 m) called "covo" or "manzuá"; and gill nets are also being used since 1995 (Fonteles-Filho 2000). The lobster fishery in northeastern Brazil began in the mid 1950s, in the state of Ceará, when fishing companies began buying catches from "jangadeiros," for industrial processing and export, mainly as frozen tails. This very quickly became the main fish export in the country and led to the development of an industrial fleet which expanded quickly and whose production reached a maximum catch of whole lobster in 1979, with a capture per unit effort (CPUE) of 0.37 kg/trap/d (Fonteles-Filho 2000). But this bounty did not last, due to decreasing yields in traditional areas (Ceará and Rio Grande do Norte) and increased operating costs, a great part of the industrial fleet (vessels over 15 m) has been retired or directed to other fishing activities. At the same time, the fleet of small and medium scale boats grew in a disorderly and illegal fashion ensuing in conflicts with traditional sail "jangadas." Additionally, the need to compensate for decreasing catches led to increased effort and a great expansion of the fishing area, which extends from the State of Espíritu Santo in the south all the way to the State of Amapá in the north (Fonteles-Filho 1997). In 2001, the fleet operating in the whole area was composed of 3,760 vessels (GTT Lagosta 2003) and is composed of industrial steel hull ships (1.4% of fleet = 6.31% of fishing effort), medium scale (53% of fleet = 92% of fishing effort) motorized boats (built of wood, whose length varies between 6 m and 14 m) and small scale boats or "jangadas" (45% of fleet = 1.50% of fishing effort). Illegal fishing methods, such as diving with the aid of compressors (hookah) as well as mesh-size of traps below the 5 cm minimum established by law, are commonly observed and continue to provide lobsters for export. This has resulted in a significant catch of undersized lobsters, and along with excessive fishing effort, has contributed to destabilize lobster stocks. The trend of diminishing CPUE (0.14 kg/trap/d) suggests the fishery is unlikely to be sustainable (Fonteles-Filho 2000). Lobster processing and exporting plants are now down to eight operations in the whole country from 25 in the 1980s. (Schärer and Aragão Negreiro 2004, unpublished).

In PCV, 74 "jangadas" comprise the fishing fleet, of which almost half (35) are smaller 1–2 crew "jangadas," 21 are medium sized and 18 are large, which are operated by 4–5 crew members. The "jangadas" of PCV are crewed by at least 60 community members of whom many are owners of their vessel and fishing gear. Lobster season begins in May after a 4-month country-wide capture prohibition. Lobsters are caught and landed live by artisanal fishers and tails reach the cooling facilities within the same day. Lobster heads are the main food for local residents during the lobster season and are often donated to poor people from neighboring communities. The lobsters caught by artisanal fishers are a better quality product than those taken by motor boats that spend an average of 14 d at sea, resulting in badly conserved lobster tails. The lobster heads of the motor boats are discarded at sea, which is absurd from a food security viewpoint.

During the closed season, PCV fishers target deep water snappers, reef fish and pelagic species for subsistence and export. These include more than 20 species of which snappers, mackerel and tarpon compose more than 50% of catch by weight. Most of the catch is sold locally or remains in the community for tourism and local consumption. Fish production fluctuates between 30,000 and 50,000 kg/year, almost equally divided into pelagic (mackerel, tuna, and shark) and demersal species (snapper, triggerfish, grunt, and stingray). This provides some diversification from the lobster fishing and much needed food and cash for local villagers.

Fishers from PCV and other communities along the coast of Ceará share part of the blame for the reduction of lobster stocks, because they succumbed to the temptation offered by lobster buyers to catch undersized lobsters. After continuing decrease of catches and under the threat of an excessive fleet (both legal and illegal) with the gentle encouragement of well-meaning development workers, fishers in PCV began to assume their responsibility and called upon fishers from other communities to join them in a campaign to combat illegal fishing and the lethargy of government agencies responsible for fisheries management.

Responsible Fisheries

At sea the "jangadeiros" have been resisting illegal fisher actions that have invaded their fishing waters since 1985 causing conflicts (and fatalities) at sea. Most fishers have a good understanding of ecological processes, commonly referred to as traditional ecological knowledge (TEK). Fishers respond to stimulation of TEK and understand the concept of overfishing and sustainability. If fishers become predatory fishers (use of illegal gears) it is usually not because of hunger or lack of awareness, but because of incentives offered by lobster buyers. Experiences in the lobster fishery along the eastern seaboard of Ceará over the last 10 years have shown that investments in fishers' awareness and technical training has increased self-esteem and enhanced cooperation with fishery regulators in enforcement actions. Some artisanal fishing communities in the state of Ceará have been actively involved in efforts to curb predatory fishing since 1993, through community meetings, partnerships, and financial contributions to pay for enforcement at sea, and actions to discuss and promote responsible fisheries.

Instituto Terramar (a local nongovernmental organization [NGO]) together with International Collective in Support of Fishworkers (ICSF) organized the First International Conference on the FAO Code of Conduct with over 300 participants and support of IBAMA (Brazilian Environmental Government Agency) in 1997 (Schärer 1997). A year later the education and outreach campaign named "Lobster Caravan" was launched. During 2 months, 20 community road-shows were performed to raise awareness of fishers and their families for the preservation of the lobster fishery. The "Lobster Caravan" was organized by a team led by Instituto Terramar and fisheries and education department of IBAMA and, with fishers and other NGOs, along with promotional support from local government and stakeholder groups. In a new community Fisheries School in PCV, younger fishers have developed self-confidence and projects constructing fishing gear, artificial reefs, or mounting structures for seaweed farming. These ideas have helped to construct knowledge for the future role of fishers on the way to responsible fishing. Rather than being the cause of the problem, as fishing industry officials are quick to point out, "jangadeiros" are part of the solution for sustainable fisheries management in Northeastern Brazil.

Since 1995, the fishers of PCV maintain local fisheries regulations which are continually updated and have been upheld by authorities like the Fishermen's Colony and the Navy's Port Authority. These regulations include an appeal to fish responsibly and it also requests a contribution of one lobster tail a week to help fund enforcement expenses of government agencies. The community imposes strict regulations regarding who can fish, what time of day fishing can occur, what can be caught, gear restrictions, closed seasons, and "jang-adeiros" are working hard to patrol their own fishing territory to ensure that overfishing and predatory fishing do not occur. There are penalties for those who violate the local fishing regulations ranging from losing permission to fish for given periods of time to having either fishing gear or boats confiscated. Additionally, since 2002, local fishers are collaborating with fisheries scientists in research to monitor their fishing area (50 square nautical miles [Nm^2]), mapping fishing grounds, collecting data on catch and production in order to present supporting data to the federal government as part of an initiative to create an artisanal fishing reserve in PCV with gear restrictions, and effort limitations in order to develop a comanagement arrangement of their local fisheries.

The 1993 protest trip of the "jangada" SOS Survival (from PCV to Rio de Janeiro in 74 d) was an important step to create awareness with regard to fisher's responsibility in the fishery, the need and the potential benefits of cooperating with fishers from other communities. As a result of the activism created by the SOS

Survival and a fisher's march to state government headquarters in 1995, it was much easier to demonstrate the role of "jangadeiros" in the sustainability of the lobster fishery. This helped break the monopoly of external lobster buyers that was financed by exporters who had total control over the price of lobsters until 1994. In 1995 lobster prices at PCV increased from the equivalent of US$10.00 to US$20.00 per kg of tail. The record catch in 1995 and 1996 brought a lot of money to fishers, but the dramatic drop in catches in 1997 and 1998 showed that there was an urgent need for responsible fishing. That miraculous catch of 1996 never returned, but fishers now realized that just as they were part of the problem, they are also part of the solution. That is how local fishing laws began and in some communities so called beach courts were implemented where fishers "convicted" colleagues who failed to respect the rules with stiff punishment that could bring up to 15 d of suspension from fishing.

Future of the Fishery

Prainha do Canto Verde was one of 10 communities worldwide chosen by the World Wildlife Fund (WWF) to test the Marine Stewardship Council (MSC) certification concept for small-scale fisheries (Schärer 1997). A preappraisal conducted in 2000 concluded that the MSC certification process is not suitable for small-scale fisheries, since the fishers have no control over the management of the lobster fishery and thus very little influence over the whole fishery (Chaffe 2000 and 2001). It would however be desirable to develop alternatives for artisanal and small-scale fisheries certification which would reward responsible behavior by fishers by granting them preferential access to marine resources and support for local community management as well as recognition and respect. Responsible behavior will also result in better handling of fish, better quality, and thus better prices on local markets. Especially in Brazil's lobster fishery, considering the extremely high prices that fishers fetch today, the guaranteed and preferential access to the resources is necessary for an improved quality of life for thousands of fisher families in coastal communities.

Although the MSC certification was not granted to the PCV lobster fishery, the assessment exercise proved beneficial in the "jangadeiros" case. Thanks to the WWF intervention the preappraisal could be considered a catalyst in gathering the different extremes of the fishery into a single discussion forum. The fishing community, management agency, commercial interests, and export firms were brought together to discuss and accept the problems that the fishery is faced with. This has helped bridge communications barriers and unites stakeholders in the development of a possible solution to the lobster fishery crisis. Now realistic solutions can be assessed for the fishery and all parties help support decisions made in regulating the fishery.

Led by the "jangadeiros" of PCV, neighboring fisher communities and NGOs have been discussing the concept of fisheries comanagement in coastal areas reserved for artisanal fishers. Since 1996 a pilot project is gathering data for habitat mapping of fishing areas as a basis for a sustainable development plan. The community has already submitted a request for the creation of a Marine Extractive Reserve under law 9.985 (18 July 18 2000) of the National System of Conservation Units of Brazil. This law is the legacy of Chico Mendes and the National Union of Rubber Tappers. Extractive reserves in general represent the first conservation units which specifically involve local communities in their design and management. This initiative has enormous potential for conserving coastal areas and securing the livelihoods of coastal populations (Pinto da Silva 2004). A marine extractive reserve is a mechanism through which a marine area is assigned to the exclusive use of a certain number of small-scale fishermen in a type of comanagement with government authorities. The community of PCV hopes that soon a marine extractive reserve will enable the community to continue their tradition as "jangadeiros" dependent on the fishery and the natural resources of their region.

Conclusion

The experience in PCV and neighboring communities in Ceará has shown that despite poor formal education, fishers TEK provides a valuable basis for sustainable development of small scale fisheries. With continuous education programs and incentives such as the marine extractive reserve along with recognition for responsible fishing behavior, improved gear, preferential access, fisher schools, access to low-impact technology, protection from predatory fishing, better markets, and so forth, other communities will join efforts for the conservation of marine resources in the coastal region. While fishers have demonstrated leadership in awareness building for responsible fishery practices (FAO Code of Conduct), the lobster fishery is depleted because of a lack of political will to enforce fishing laws and a lack of vision of fishing entrepreneurs. The development of management partnerships with "jangadeiros" and the creation of marine extractive reserves may help recover nearly depleted stocks. The great hope for fishermen is that the government agencies will understand that fishers are not welfare cases, but part of the solution to overfishing and a key player for the sustainable development of marine resources.

Acknowledgments

Financing from Amigos de Prainha do Canto Verde was essential in the development of this project since 1991. The milestones of the "jangadeiros" have been supported in full by a number of philanthropic and fisher support organizations including the ICSF and WWF. Instituto Terramar has been instrumental in organizing community and fisher activities for outreach and education. The organization of the PCV residents and fishers is acknowledged for their courage and dedication to maintain the "jangadeiros" tradition alive. The authors thank Foundation Avina for providing a travel grant to attend the Fourth World Fisheries Congress.

Reference

Chaffee, C. 2000. A partial pre-assessment of the Prainha do Canto Verde community-based lobster fishery in Brazil. Report to the WWF. Scientific Certification Systems, Oakland, California.

Chaffee, C. 2001. Lobbying for lobsters. This is a partial pre-assessment of the Prainha do Canto Verde Community-based lobster fishery. Samudra (August):30–38.

Fonteles-Filho, A. A. 1997. Spatial distribution of the lobster species *Panulirus argus* and *P. laevicauda* in northern and northeastern Brazil in relation to the distribution of fishing effort. Ciência e Cultura Journal of the Brazilian Association for the Advancement of Science 49(3):172–176.

Fonteles-Filho, A. A. 2000. The state of the lobster fishery in northeast Brazil. Pages 121–134 *in* B. Phillips and J. Kittaka, editors. Spiny lobsters: fisheries and culture. Iowa State Press, Ames.

Pinto da Silva, P. 2004. From common property to co-management: lessons from Brazil's first maritime extractive reserve. Marine Policy 28:419–428.

Schärer, R. 1997. Ceará fishermen: Sailing for a cause. Samudra (July):32–37.

American Fisheries Society Symposium 49:547–555

The Conservation–Exploitation Paradox in a Mexican Coral Reef Protected Area

LOURDES JIMÉNEZ-BADILLO, VIRGILIO ARENAS FUENTES,
AND HORACIO PÉREZ ESPAÑA

Centro de Ecología y Pesquerías, Universidad Veracruzana
Calle Hidalgo #617, Col. Río Jamapa, Boca del Río, Veracruz, 94290, México

Abstract.—The Mexico Veracruz reef system was recently declared a natural protected area. One of the most important users of this area (52,239 ha) are fishermen whose extractive activity conflicts with conservation goals. At least a thousand families depend directly on the artisanal fishery in the area, while others depend indirectly. With the objective of finding an adequate balance between exploitation and conservation of fish resources, an annual study was performed, evaluating biological, economical, and social aspects. The results showed that climatic conditions were favorable for fish only 230 d per year. A diversity of 89 finfish, 4 sharks, 2 rays, 1 lobster, and 2 octopus species were fished by 18 varieties of nets, lines, and harpoons. The catch per unit effort fluctuated between 4 and 344 kg/fishermen/d, with fishermen spending from 2 to 24 h per fishing trips, and the daily profit varied between US$2 to US$40 per fisherman. The regulations for a sustainable use of the resources are inefficient due to the scarce basic information on the dynamics of the species and the multispecies nature of the fishery making integrated management difficult. In general, artisanal fishing is not a profitable activity such that it is necessary to promote an aggregated value for fishing products. Fishermen participation in resource conservation will be viable only if the quality of their lives is increased. In this sense, octopus culture is proposed as an option to reduce the fishing effort on the reefs and to promote socioeconomic development in the fishing communities.

Introduction

Several research papers about marine reserves (Clark et al. 1989; Davis 1989; Bohnsack 1990; Roberts and Polunin 1991, 1993; J. P. Gibson, paper presented at the Fourth World Fisheries Congress, 2004) have suggested that protecting habitats in their natural state and establishing fishing closures to limiting access to spawning or nursery grounds may increase overall fishery harvest. But how long is it necessary to make that possible? While the benefits provided for the marine protected areas to fisheries is yet in debate, as Fisher et al. (2003) pointed out, human population, food demand, and pressure on artisanal fisheries are all increasing. At present, roughly 70% of fish stocks for which data are available are fully exploited or overfished (Berkes et al. 2001). Also, activities like domestic and industrial waste disposal, tourism, recreation, and maritime transportation impact the coastal areas daily, altering the environment, the biodiversity, and, consequently, the fisheries.

*Corresponding author: loujim@gmail.com

Artisanal fisheries account for nearly one-third of the food fish harvested worldwide (Holland 1995). It constitutes an important socioeconomic component of the Latin American and Caribbean fisheries. These fisheries provide an important source of employment and represent a key source of high quality food, generating essential direct incomes to artisanal communities (Defeo and Castilla 1999). Of the more than 51 million fishers in the world, over 99% are small-scale fishers (FAO 1999). In Mexico, around 250,000 artisanal fishers, who employ 97% of the nations' fishing craft, provide almost two-thirds of the total catch (Comisión Nacional de Pesca 2001). Support services such as marketing, processing, boat building, and transportation indirectly employ an increasing number of people.

From the biological point of view, protected areas are valuable in conserving habitat and biodiversity and providing amenities; from the socioeconomic point of view, the survival of fishing communities and indirect employment, such as the consequences for the local economy, are frequently forgotten. The protected areas will not solve the problem of overfishing in the short term. Alternative productive activities should be considered before a fishing closure is established.

With the purpose of identifying a scenario congruent to the conservation and use of fishery resources in the natural protected area Veracruz reef system in Mexico, a biological, economical, and social evaluation of the fishing activity was made.

This is the first integrated fishery evaluation of the zone since being declared a protected area in 1992 and recategorized in 2000 (Ministry of Government 1992, 2000). Because of that, it would be a basic point of reference for the future when the government establishes new laws to regulate their exploitation in the new context of a protected area.

The information presented here shows the first results of that evaluation, focused on the fishing effort and on some basic economic information, which could be useful in appropriately designing the management plan of the protected area.

Methods

The Veracruz reef system in México is located in the Gulf of Mexico (Figure 1) between latitude 19°15'N and latitude 19°02'N and longitude 96°12'W and longitude 95°46'W. It spans the coastal area of three provinces: Veracruz, Boca del Río, and Antón Lizardo, an area of 52,239 ha. It consists of 24 reefs, which have a high biodiversity (Ministry of Government 2000).

This study was initiated due to a government requirement to design a management plan for the recently declared protected area.

Considering the importance of whatever regulation has consensus from the majority of coastal residents, the first job was to identify all the users of the area. Different points of view represented by different users were consulted in specific workshops. As a result of these, we detected that one of the most important users of the area was fishers, whose extractive activity is conflicting with the recent conservation goals. Because of that, we focused this study on fishing activity in the area. Some information available on this activity in the area was isolated and scattered. So, we decided to evaluate the status of the fishery considering ecological, economic, and social concerns. To promote the fishers participation in the design of the proposals and to get their acceptance, a collaborative agreement with fishers was established.

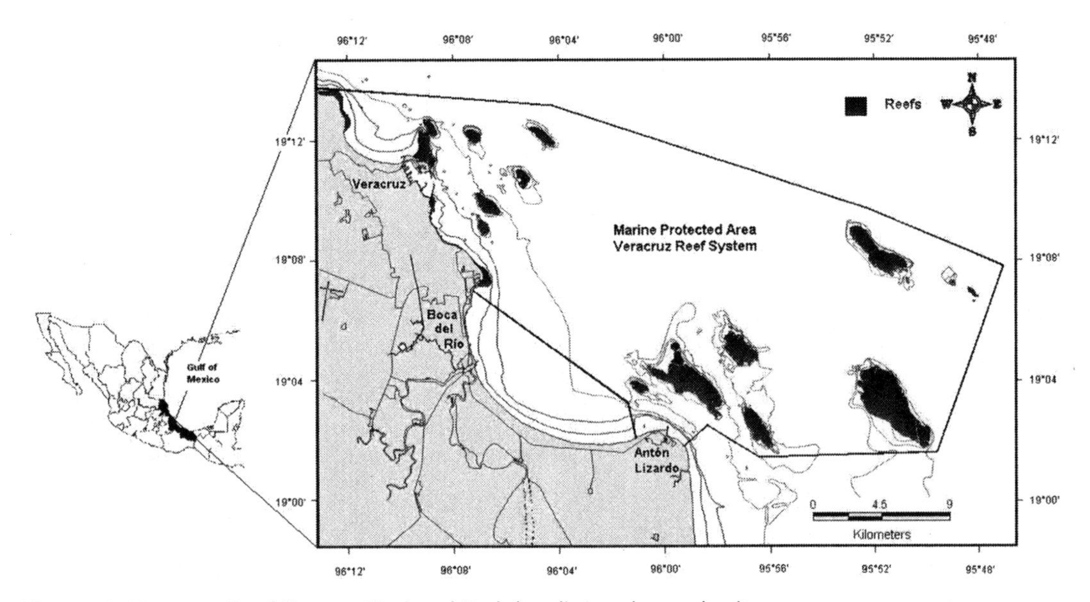

Figure 1. Veracruz Reef System National Park locality and sample sites.

During one year, we developed monthly samples of the commercial captures in several landing places in Antón Lizardo, Veracruz, and Boca del Río. Data about catch volumes of each fish species, fishing gear used in its capture, time invested by fishing trip, expenses incurred during the activity, value of the captures, and profits by fishery per fishermen were registered. Biometrics (total length and individual weight) of a subsample of each species were taken to estimate length structure of the population and to extrapolate according to Sparre and Venema (1995).

Some fishers and purchasers provided daily information from 3 years of their captures.

A register of the captures from the three provinces during the last 11 years was obtained from the Official Statistics in the Veracruz Office. It would be used to analyze the spatial and temporal trend of the fishery.

A survey was developed with two objectives: first, to know the fishers' perceptions in the new context of a protected area; and second, to have specific information about the operation of their fishing gear to identify efficiency.

A survey to quantify some indicators of quality of life of fishers' communities was developed too.

During the development of this project, several workshops with fishers were developed in order to keep continuous communication with them.

In order to get the generated information organized, a database in Access was designed.

A catalog of species and fishing gear containing ecological and fishery information was established to support academics and authorities.

Estimations of catch per unit effort (CPUE) by fishing gear were made, using the number of fishermen as unit effort.

Estimations of the profit by fisherman for each fishery were made, considering money and time invested by fishing trips, as well as the income when the product was sold.

Estimations about the impact of fish activity on the reef systems are yet in progress.

Advances in the analysis of this information allow us to identify a project, which will be proposed as an alternative activity for fishers.

Results

Fishing activities in the Veracruz reef system support at least 1,000 families, who depend directly or indirectly on this activity.

Several fisheries take place in the area during the approximately 230 d/year, in which weather conditions are adequate. At least 18 varieties of nets, lines, and harpoons were used by fishermen, without a clear spatial or temporal pattern yet established. The captures were multispecific, including 89 species of finfish, 4 sharks, 2 rays, 1 lobster, and 2 octopuses. The taxonomic position of those species was determined and some of them were registered in a catalog of species, which included a description of the varieties of fishing gear. This catalog also included some ecological and fishery information for each species. Data such as distribution, habitat, food, and reproductive habits, as well as the importance, minimum, maximum, and average length of capture, fishing gear used in the catch, monthly, and annual patterns of the magnitude of captures over 11 years, and resilience according to Froese and Pauly (2003), were included in the catalog (Jiménez-Badillo et al. 2006). Because of that, it will become a source of basic reference for academics and authorities.

The total mean catch was estimated at 1.9 metric tons a day. The fishing community of Antón Lizardo produced 86% of the captures. The monthly catch per unit effort (CPUE) for all fisheries during 2000, 2001, and 2002 fluctuated between 4 and 344 kg/fisherman/d (Figure 2). Comparing the 3 years, no pattern was established. When the CPUE was separated per fishery, two groups of fishing gear efficiency were shown (Figure 3). The first one was constituted by all varieties of gill nets, longline, octopus hook, and a variety of longline, locally called cimbra. They produced a higher CPUE than the other group constituted by fish hooks, lures, harpoons, and a local kind of line called rosario.

The most important results to date in this investigation are shown in Table 1. The time that fishermen invest in each fishery, the money that they earn after their product is sold, and the daily profits after fishing expenses are shown. Fishermen spent from 2 to 24 h per fishing trips and the daily profit varied between US$2 and US$40 per fisherman. With this money, they must survive all year, including the days in which they cannot fish. The large amount of time invested in the majority of fisheries make it almost impossible to take part in other economic activities. Artisanal fishing is far from being an economically productive activity. Nevertheless, it is necessary for the survival of fishermen and their families.

In the surveys and workshops completed, we were able to identify an ecological conscience in fishermen. Their perceptions showed a real concern about the decrease in fish. They know that resources are not inexhaustible and need protection. The majority of them are willing to respect the declaration of a protected area, but they need alternatives to survive (Jiménez-Badillo 2007).

Surprisingly, the most efficient fishery from the economic point of view was the octopus fishery. But, it was also the most damaging fishery to the reef systems because fishing gear tears the coral reefs when used to fish. Therefore, efforts are going to be made to explore octopus farming as an alternative activity before an area is closed to fishing (Méndez et al., in press).

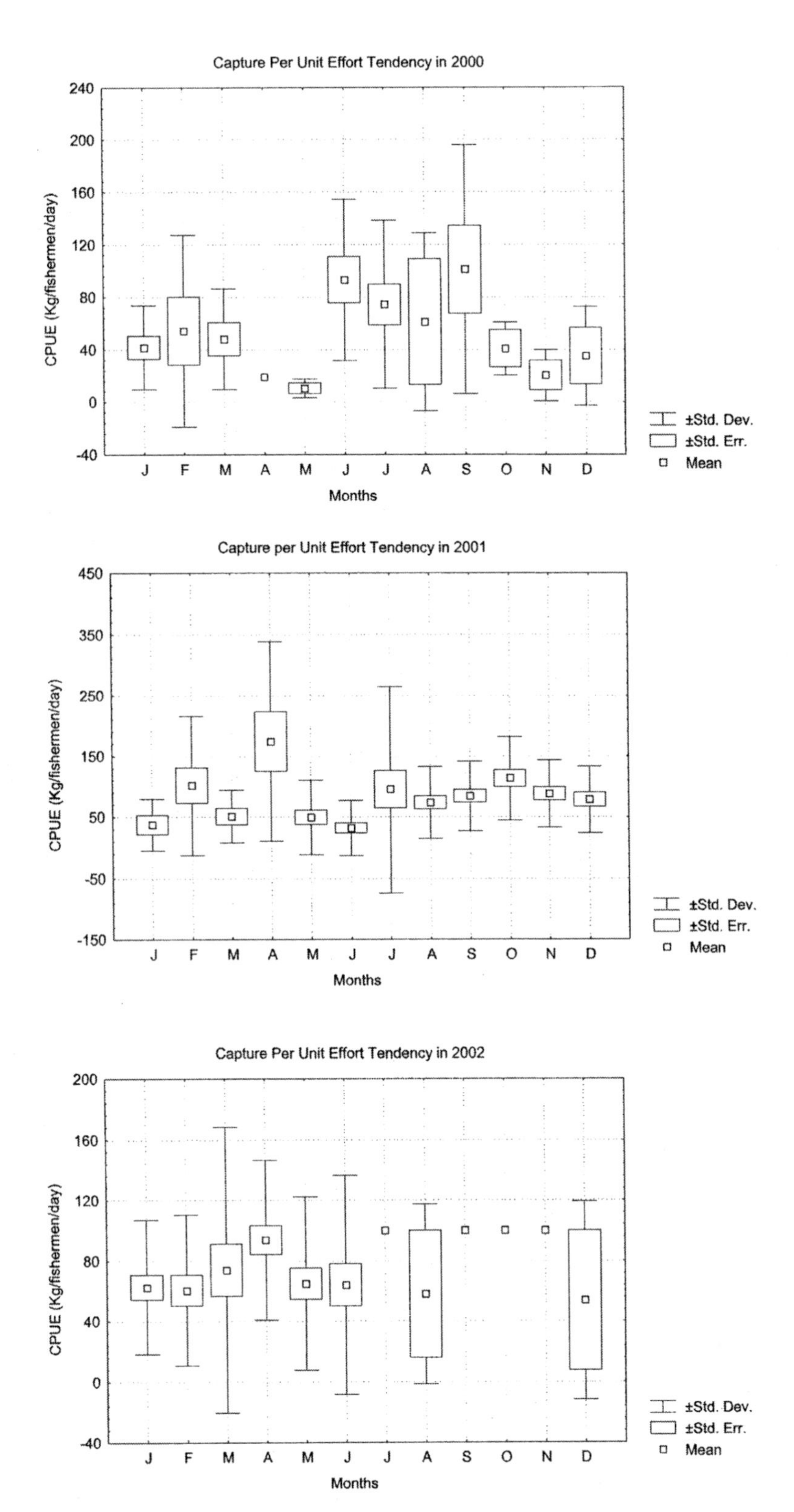

Figure 2. Catch per unit effort (kg/fishermen/d) for 2000, 2001, and 2002.

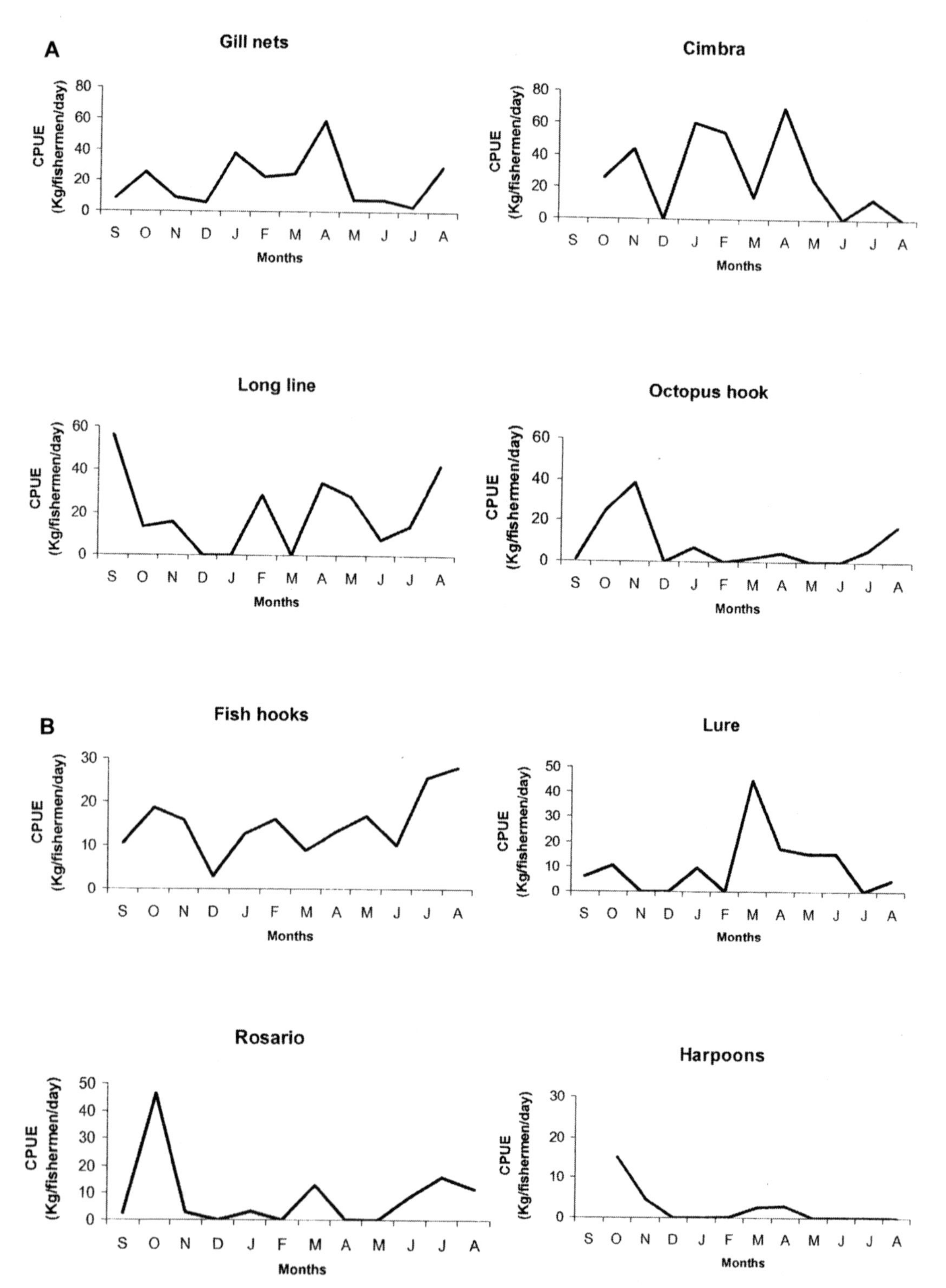

Figure 3. Efficiency of the fishing gears in terms of catch per unit effort (kg/fishermen/d): A) Fishing gears that produced the highest catch per unit effort; B) Fishing gears that produced lower catch per unit effort.

Table 1. Time and money invested by fishing outings and profit for each fishery.

Fishery	Time invested (h/d)	Capture value daily ($USD)	Spending by travel ($USD)	Profit fishermen/d ($USD)
Gill nets	6	1,867	60	8
Fish hooks	10, 24	1,573	30	2
Longline	12.5, 19	1,501	43	24
Cimbra	24	1,320	45	2
Rosario	11, 12	883	34	20
Lure	7.5	667	33	7
Harpoon	6.5	568	33	5
Octopus hook	6	563	27	40

Discussion

The Veracruz Reef System National Park is in the process of changing the rules to operate as a reserve area. We have the opportunity to make proposals for the policymakers, based on scientific evaluations, as has been the case in other protected areas.

According to Defeo and Castilla (1999), one of the main causes for the failure of the management plan is that fishers have not been included in the discussions about management. That experience has been a valuable lesson, and it has been taken into account.

Regulations for sustainable use of resources are inefficient, due to the scarcely basic information, to the dynamics of the species, and to the multispecific nature of artisanal fisheries, which are difficult to manage. As Holland (1995) indicates, reef fisheries are difficult to manage with methods that limit catch and effort because they use multiple target species and fishing methods. The management of these fisheries requires an understanding of human behavior and how people use and misuse marine resources. They also require a different kind of management regime that goes beyond command-and-control measures (Berkes et al. 2001). These fisheries depend on ecological, technological, socioeconomic, and political variables too. So, a complete understanding of the process that regulates that variation is necessary. Because of that, we grouped the information according to fishing gear as a technical criterion to reduce the variability in the analyses. Other criteria that are being explored are ecological, grouping species by food and reproductive habits, and economical, grouping them according to their market prices. From these analyses, we expect to derive patterns that can help us to understand such variability, as well as the interrelationship between variables, and help us in making predictions.

Another failure identified by Holland (1995) is that the government used the marine reserves to conserve marine habitat and biological diversity, rather than to promote fisheries production. Although as Berkes et al. (2001) indicate it is becoming increasingly clear that governments, with their finite resources, cannot solve all the problems that citizens face. The communities will need to take more responsibility for solving local problems. In order to do this, however, they must be given the power and resources to make decisions locally and to take actions to meet local opportunities and problems. In our opinion, conservation resources should be compatible with human socioeconomic development throughout the implementation of productive projects.

The fishery and the fishing communities may benefit from the closed fishing area only after an extended transition period. Thus, we are exploring the viability of some productive projects for the area. Actually, the octopus farming is in an experimental phase. If its results are viable, it will be proposed to the government as an alternative. We are conscious that fishers need to be assured of their survival if we expect them to respect regulations.

In some areas, ecotourism projects have been used as alternative activity for fishermen. In this case, fishermen's idiosyncrasies make its implementation difficult and could produce conflicts with other tourism workers. Nevertheless, initiatives in this sense could be promoted too.

In order to involve fishing communities and government in the search of those kind of alternative activities, experiences related with comanagement process, applied in other marine protected areas (Indab and Suarez-Aspilla 2004; Mahon et al. 2004; A. Satria, Y. Matsuda, and M. Suno, paper presented at the Fourth World Fisheries Congress, 2004) should be explored too.

Conclusions

In order to reduce conflicts regarding the conservation versus exploitation of the resources, ecological, economic, and social factors should be considered.

The artisanal fishery is not a profitable activity, but it is necessary to the survival of fishermen and their families. The fishermen's participation in resource conservation will be viable only if their life quality is increased. Octopus farming is being evaluated as an option to reduce fishing effort on the reefs and to promote socioeconomic development in the fishing communities.

Acknowledgment

Juan Manuel Vargas Hernández conducted the management plan for the Veracruz Reef System. We are grateful to Luis Gerardo Castro Gaspar, Surya Garza Garza, and Juan Carlos Cortés Salinas for their valuable help and encouragement during the field stage.

References

Berkes, F., R. Mahon, P. McConney, R. Pollnac, and R. Pomeroy. 2001. Managing small-scale fisheries: alternative directions and methods. International Development Research Centre, Ottawa.

Bohnsack, J.A. 1990. The potential of marine fishery reserves for reef fish management in the U.S. southern Atlantic. National Oceanic and Atmospheric Administration (NOAA) Technical Memorandum, NMFS-SEFC-261, Miami.

Clark, J. R., B. Causey, and J. A. Bohnsack. 1989. Benefits from coral reef protection: Looe Key Reef, Florida. Pages 34–39 *in* O. T. Magoon, H. Converse, D. Miner, L. T. Tobin, and D. Clark, editors. Coastal Zone 89: Proceedings of the Sixth Symposium on Coastal and Ocean Management, Charleston 11–14 July 1989. American Society of Civil Engineers, New York.

Comisión Nacional de Pesca. 2001. Anuario Estadístico de Pesca. Comisión Nacional de Pesca y Acuacultura de SAGARPA, México, D.F.

Davis, G. E. 1989. Designated harvest refugia: the next stage of marine fishery management in California. California Cooperative Oceanic Fisheries Investigations (CalCOFI) Report 30:53–57.

Defeo, O. and J. C. Castilla. 1999. A co-management approach to artisanal fisheries/Chile and Uruguay. Wise Coastal Practices for Sustainable Human Development Forum. 8 June 1999. Available: www.csiwisepractices.org/?region=4 (June 1999).

FAO (Food and Agriculture Organization of the United Nations). 1999. Indicators for sustainable development of marine capture fisheries. FAO Technical Guidelines for Responsible Fisheries No. 8, Rome.

Fisher, J., P. Murphy, and W. Craik. 2003. Marine protected areas and fishing closures as fisheries management tools. Pages 14–18 *in* J. P. Beumer, A. Grant, and D. C. Smith, editors. Aquatic protected areas: what works best and how do we know? Proceedings of the World Congress on Aquatic Protected Areas, Cairns, Australia, August 2002. Australian Society for Fish Biology, North Beach, Washington.

Froese, R., and D. Pauly, editors. 2003. FishBase. Concepts, design, and data sources. International Center for Living Aquatic Resources Management, Los Baños, Laguna, Philippines.

Holland, S. D. 1995. Management of artisanal fisheries: the role of marine fishery reserves. University of Wisconsin, EPAT/MUCIA, Policy Brief No. 11, Madison.

Indab, J. D., and P. B. Suarez-Aspilla. 2004. Community-based marine protected areas in the Bohol (Mindanao) Sea, Philippines. NAGA, WorldFish Center Quarterly. Volume 27(1–2).

Jiménez-Badillo, M. L. 2007. Management challenges of the small-scale fishing communitiess in a protected reef system of Veracruz, Gulf of Mexico. Journal of the Fisheries Management and Ecology, Blackwell Publishing Ltd. doi: 10.1111/j.1365-2400.2007.00565.x.

Jiménez-Badillo, M. L., H. Pérez, J. M. Vargas, J. C. Cortés, and P. Flores. 2006. Catálogo de especies y artes de pesca del Parque Nacional Sistema Arrecifal Veracruzano. CONABIO, Universidad Veracruzana, Conabio, México.

Mahon, R., McConney, P., Parks, J. and R. Pomeroy. 2004. Reconciling the needs of fisheries and conservation in coral reefs. Paper presented at the Fourth World Fisheries Congress. Reconciling Fisheries with Conservation: The Challenge of Managing Aquatic Ecosystems. May 2–6. Vancouver, British Columbia, Canada.

Méndez, A. F. D., M. L. Jiménez-Badillo, and V. Arenas. In press. Cultivo experimental de pulpo *Octopus vulgaris* (Cuvier 1797) en Veracruz y su aplicación al PNSAV: investigaciones actuales. In B. A. Granados, L. Abarca, and J. M. Vargas, editors. Investigaciones científicas en el Sistema Arrecifal Veracruzano. EPOMEX-Universidad Autónoma de Campeche, Campeche, México.

Ministry of Government. 1992. Diario oficial de la federación. Tomo CDLXVII no. 16. 24 de agosto de 1992. Secretaria de Gobernación, Mexico, D.F.

Ministry of Government. 2000. Diario oficial de la federatión. Primera sección. Tomo DLXIII no. 20. Lunes 28 de agosto de 2000. Secretaria de Gobernación, Mexico, D.F.

Roberts, C., and N. Polunin. 1991. Are marine reserves effective in management of reef fisheries? Reviews in Fish Biology and Fisheries 1:65–91.

Roberts, C., and N. Polunin. 1993. Marine reserves: simple solutions to managing complex fisheries. AMBIO 22(6):363–368.

Sparre, P., and S. Venema. 1995. Introduction to tropical fish stock assessment. Part. 1. Manual. FAO Fisheries Technical Paper 306.

American Fisheries Society Symposium 49:557–568

Allocation Issues and the Management of a Mexican Shrimp Fishery

Jose Ignacio Fernandez-Mendez*
National Autonomous University of Mexico, Faculty of Sciences
Circuito Exterior s/n, Ciudad Universitaria, Mexico City 04510, Mexico

Introduction

The shrimp fishery is the most important of all the Mexican fisheries in economic terms. Although it comprises only 2.66% of the total national catches, income derived from it reaches 21.22% of the 1.1 billion USD gross value of Mexican fishing production. The shrimp fleet comprises 67% of the industrial fishing fleet in Mexico (CONAPESCA 2001). The Gulf of Mexico fishery catches 22% of the shrimp caught in the country, and nearly 70% of this comes from the brown shrimp *Farfantepenaeus aztecus* fishery off Tamaulipas, just south of the U.S. border, and northern Veracruz (farther south) (Figure 1). Artisanal shrimp fisheries are very important for local economies in many Mexican coastal states.

Shrimp reach a maximum age of one to one and a half years. After a 3- to 5-month stay in nursery areas (coastal lagoons or shallow sea areas in some species), shrimp juveniles migrate to the open sea where they reach adulthood and spawn after reaching an age of 6 months. After a month as planktonic larvae, postlarvae migrate to nursery areas. Shrimps spawn all year long in tropical zones, but marked seasonal peaks of spawning activity (usually, a spring–summer and an autumn–winter peaks) result from variations in environmental factors like temperature (García and Le Reste 1981; García 1985). Usually, cohorts originated in the autumn–winter spawning event are of low abundance and spawn at an age of around 6 months during the spring–summer spawning event. This latter spawning produces cohorts that are far more abundant than the parental stock that generated them. These cohorts, in turn, spawn during the autumn–winter spawning period. This pattern results from seasonal variations in postlarvae survival related to changes in oceanographic conditions that determine their entrance to nursery areas, a phenomenon typical among Penaeid shrimps (Rotlisberg et al. 1983; Mathews et al. 1984). This pattern has been observed in the most important Mexican shrimp fisheries in the Gulf of Mexico (Fernandez-Mendez et al. 2000).

The temporal distribution of catch and effort is greatly influenced by this alternating recruitment pattern. In Tamaulipas, an average of 47% of the catch of the year in coastal lagoons is caught during the April–June period (mostly individuals of ages lower than 4 months). The main shrimp migration from lagoons to the open sea occurs in early June, coincident with the first period of full or new moon of the month (Fernandez-Mendez et al. 2000; INP 2000). Almost 51% of the off-

*Corresponding author: jfernan@servidor.unam.mx

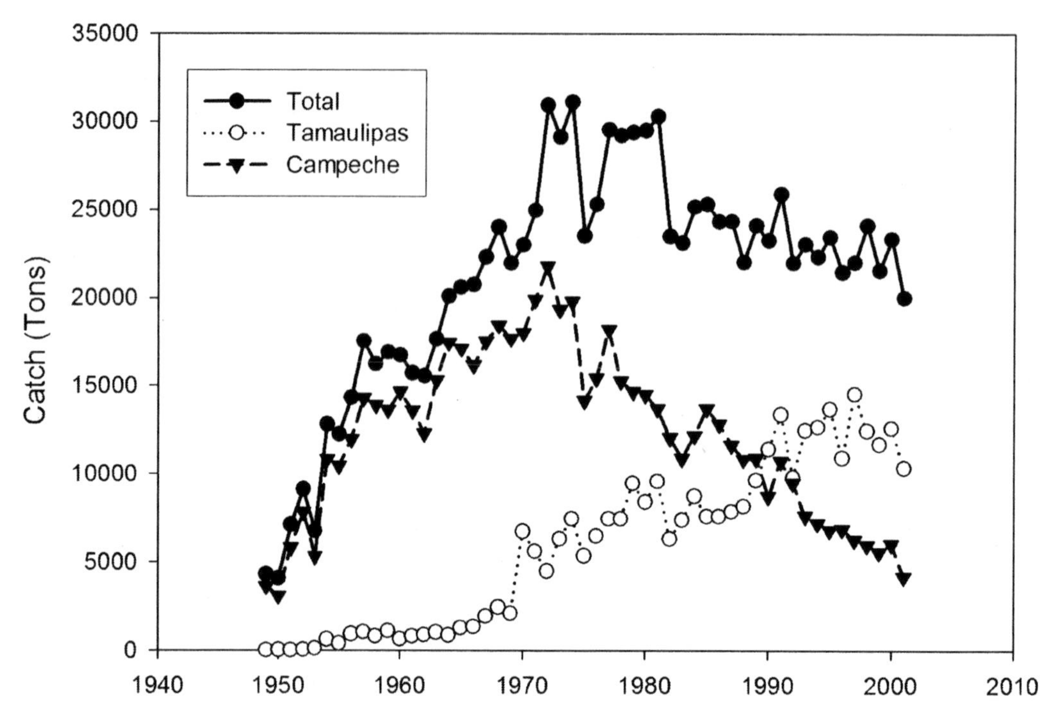

Figure 1. Trend in total shrimp catches in the Mexican shrimp fishery in the Gulf of Mexico (total in the graph) and in the Tamaulipas brown shrimp and Campeche white and pink shrimps fisheries since 1949.

shore catch is obtained during August–October (individuals older than 4 months of age). As such, most of the catch of the artisanal (lagoon) and industrial (offshore) fisheries comes from the cohorts produced during the spring–summer spawning period. Furthermore, the industrial fishery depends heavily on shrimp migrating to the open sea in early June.

As juvenile shrimp are caught in coastal lagoons by artisanal fishermen and adult shrimp are caught by offshore industrial vessels, the shrimp caught by artisanal fishermen will not be available to the industrial fleet. The shrimp fishery is thus a sequential fishery with two competing sectors catching the same stock and depending heavily on the migration of a particular cohort.

Several factors have contributed to making this competition increasingly intense in the Tamaulipas-Veracruz fishery. After the depletion of the white shrimp *Litopenaeus setiferus* and pink *F. duorarum* shrimp stocks in the Campeche Sound, more and more industrial fishing effort concentrated off Tamaulipas and northern Veracruz in the last decade as vessels previously operating off Campeche moved there from July to September of each year, searching for better catches. On the other hand, before 1993, the mostly unregulated artisanal fishery in Tamaulipas and Veracruz was obtaining more than 50% of the shrimp catches in the region.

Management of the Fishery

Before 1993, shrimp fisheries management in the Mexican Gulf of Mexico was quite lax. Although mesh size regulations were in place before that year, it was not until 1993 that a seasonal closure was established for this fishery.

Seasonal closures are the main management tool in Mexico for managing shrimp fisheries. These have been in effect since the early fifties along the Pacific shore and since 1993 in the Gulf of Mexico, although experimental closures were tried in the Gulf in the 1970s (Castro and Santiago 1976). The closure scheme applied to the Tamaulipas-Veracruz zone was apparently inspired by the Texas shrimp fishery (45–60 d of closure in May–July).

The closure in Tamaulipas-Veracruz is set usually from late May or early June to mid-July (as discussed in the next section) for the lagoon artisanal fishery and from May to late or mid July for the offshore industrial fishery.

The objective of the Texas closure was to restrict the harvest of small shrimp to obtain a larger catch of bigger shrimp offshore (Klima et al. 1982; Nichols 1982; Nance et al. 1994). The plan was aimed at reducing growth overfishing (i.e., catching individuals at a lower size than could otherwise be possible) (Leary 1985). Its implementation resulted in a higher economic value of the catch (Caillouet et al. 1979). The increase in catch was 9% in 1981 and 6% in 1982 (Leary 1985).

Despite being successful in increasing catches, two aspects of this seasonal closure scheme require critical discussion:

1) Protection of the spawning events. Apparently, in designing the closed seasons discussed so far, no consideration was made of the possibility of recruitment overfishing (i.e., the reproductive capacity of the stock being reduced by fishing) by not including in the closure the periods when spawning events occurred (February–April and September–October), a situation that has raised some objections (Condrey and Fuller 1992; Coleman et al. 2004).

However, shrimp closed seasons can be also established with such considerations in mind, as is the case in other places. For example, seasonal closures in the Mexican Pacific have been usually designed according to the alternating recruitment pattern mentioned in previous paragraphs, and they have pursued two main objectives: 1) protect the spring–summer spawning and 2) protect the growth of individuals of the main cohort (the one supporting the fishery) to have a larger number of bigger shrimp (Fernandez-Mendez et al. 2000). To comply with these objectives, the closed season in the Pacific begins in April–May, and ends in September–October (usually 6 to 7 months).

Despite the fact that the closed season in the Tamaulipas-Veracruz zone does not protect the main spawning events, but only the migration of juvenile shrimp to the open sea, its official objectives have been stated as the same as in the Pacific fishery, as can be seen in the publication of the official registrar announcing the dates of the closed season for 1999, "to protect (the shrimp species) during their spawning periods and the recruitment of their new generations to the fisheries" (DOF 1999)[1]. It seems then that the objectives of the management scheme in the Tamaulipas-Veracruz shrimp fishery have been borrowed from other closed season schemes applied elsewhere, without careful consideration of its actual implementation.

2) Allocation issues. The Texas closed season not only increased catches but also resulted

[1] DOF (Diario Oficial de la Federación, National Registrar). 1999. "AVISO por el que se da a conocer el establecimiento de épocas y zonas de veda para la captura de las especies de camarón en aguas marinas y de los sistemas lagunarios estuarinos de jurisdicción federal, del Golfo de México y Mar Caribe." Secretaría de Gobernación. México. April 30.

in a redistribution of catch and revenue from the small to the big vessel sector of the fishery, reducing the catch in bays and lagoons as it increased it offshore (Nance et al. 1994.). The same result was obtained in the Mexican fishery. As previously mentioned, closed seasons in the Gulf of Mexico were implemented only in 1993. By then, a well developed artisanal shrimp lagoon fishery existed in Tamaulipas, catching as much or even more than the industrial fishery (Figure 2).

As the closed season takes place during the period when artisanal fishermen traditionally obtained a sizable portion of their catches, shrimp not caught in the lagoon fishery migrate offshore (at a bigger size) to be caught by the industrial fishery. The proportion of the total catch obtained by the artisanal and industrial fisheries thus depends on the date when the closed season starts in the lagoons, as will be discussed in the next paragraphs.

Recent Catch History of the Tamaulipas-Veracruz Artisanal and Industrial Shrimp Fisheries Related with Closed Season Starting Dates

The 1993 closure was set to begin and end at the same time for the artisanal and offshore fisheries, from 30 May to 15 July. From roughly 50% of the total catch, the artisanal share went down to 28%. Total catch went up around 20%, and the offshore catch rose an unprecedented 54%. After protests from the artisanal fishermen (what Gracia 1997 calls "nonbiological reasons"), the closure dates in 1994 were set differently for the

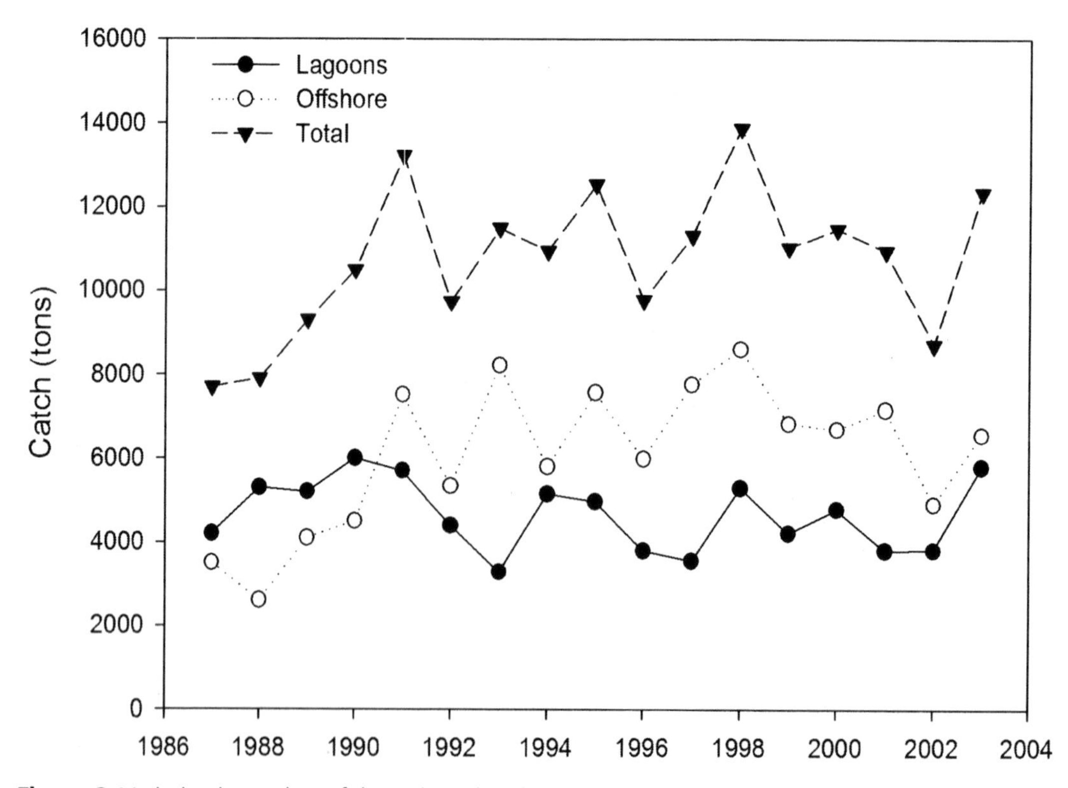

Figure 2. Variation in catches of the artisanal and industrial shrimp fisheries in Tamaulipas since 1977. The closed season was applied for the first time in 1993.

artisanal (15 June–31 July) and the industrial (25 May–31 July) fisheries. As a result, total and offshore catch fell from 1993 levels 5% and 30%, respectively, against a 56% increase in artisanal catch. Industrial and artisanal shares of the catch were 53% and 46%, respectively, just slightly different from the preclosure period. The 1994 closure allowed the artisanal fishery to operate during most of the main migration period in early June.

In its third year, the closure scheme settled on what became the pattern maintained until 2001. Closures began in early June. Roughly 65% of the total catch went to the industrial fishery and around 35% to the artisanal fishery (Figures 2 and 3). After 2001, for reasons discussed below, closures began in mid or late May.

The Decision-Making Process

Hernández and Kempton (2003) describe in detail the process of decision making on closed season dates for the Mexican shrimp fishery from 1996 to 2000. Meetings were held annually, between fishermen and state and federal government representatives (including fisheries managers). In those meetings, relevant research results were presented by the Instituto Nacional de la Pesca (INP), National Fisheries Institute and consultants to the industrial and artisanal sectors. Discussions were held to choose the dates to begin and end the closed season. Proposals on closed season's dates were usually presented by the artisanal and industrial fishermen. The latter usually proposed early or mid-May to start the closed season for the artisanal fishery, while mid-July was usually proposed by the artisanal fishermen.

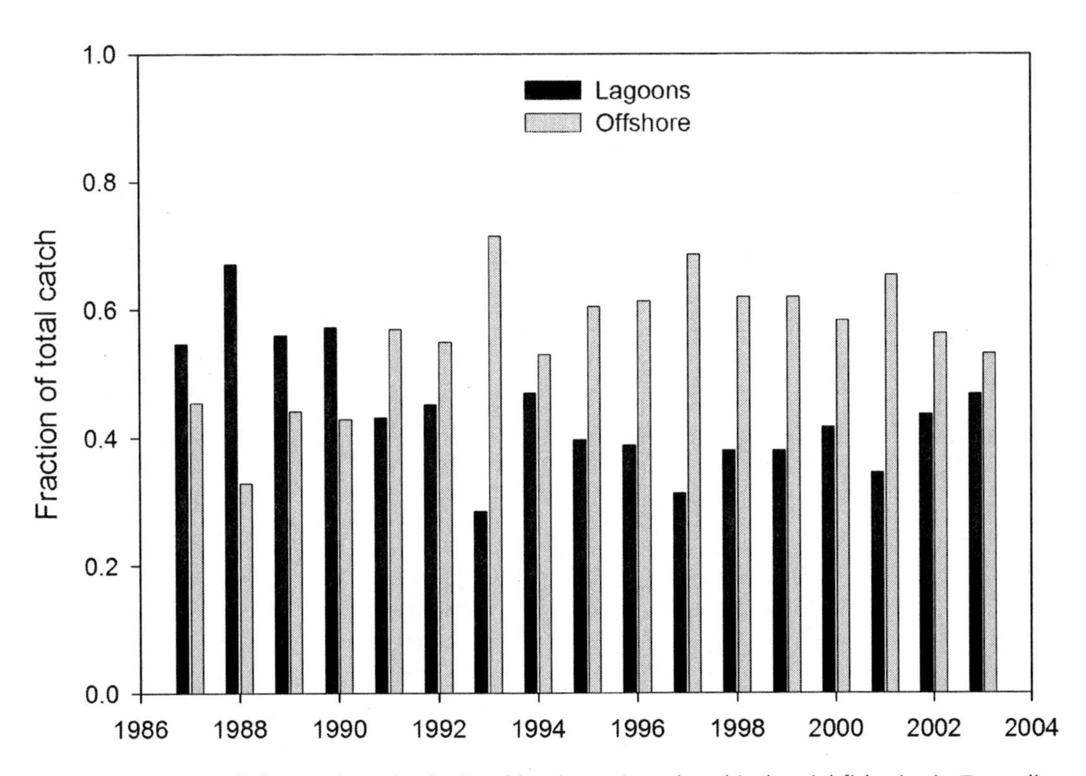

Figure 3. Fraction of the total catch obtained by the artisanal and industrial fisheries in Tamaulipas since 1997. Squares mark years with a closed season, Xs mark years previous to its establishment.

The INP developed an age-structured model (Fernandez-Mendez et al. 2000) to assess the effect on proportional catches for both fisheries of different dates of the start of the closed season. Simulations made with this model showed that the dates proposed by industrial fishermen resulted in 70–80% increases in catch over the scenario without a closed season with only 20% of the total catch going to the artisanal fishery. Simulations using the artisanal fishermen's proposed dates usually resulted in 20% increases in total catch with up to 47% of that going to artisanal catches.

The final decision was made by the Deputy Secretary of Fisheries based on the examined scenarios. The adopted scenario resulted in around 65% and 35% of the total catch for the industrial and artisanal fisheries, respectively, for a number of years.

Hernández and Kempton (2003) discussed several problems faced in the decision making process. One of them was that the artisanal fishermen were not happy with their reduction in catches as a result of the establishment of the closed season. It can be said that artisanal fishermen had little reason to be satisfied with a management scheme that only resulted, in their perception, in a lower level of catches with the industrial fishermen reaping the benefits.

Changes in Fishing Patterns in Recent Years

As a result of the establishment of the closed season, the proportion of individuals of ages 4 to 5 months (the ones that migrate in early June, during the closed season in lagoons) in the offshore catches at the start of the fishing season rose sharply in comparison to previous years (seen as a peak in age frequencies in Figure 4). The increased abundance of these individuals was the main cause of the increase in catches of the offshore fleet after 1993.

However, after 1997, that peak was reduced until its disappearance in 2000, with the frequency distributions looking remarkably similar to those recorded before the establishment of the closed season. In parallel, catches have been decreasing steadily since 1998 (Ramirez-Soberónet al. 2001).

A likely explanation for this observation is the occurrence of fishing during the period of migration supposedly protected by the closed season. A simulation of expected age frequencies under this scenario, performed with the model used in the decision-making process during the 1997 to 2001 period, shows results consistent with what has been observed in the age frequencies in 2001 (Figure 5). This leads to the conclusion that growth overfishing, the problem the closed seasons was designed to solve, is once again present.

It is important to note that the portion of the total catch obtained by the artisanal fishery in 2002 and 2003 rose to nearly 46% (CONAPESCA 2003), despite the fact that closed seasons established in those years started in mid or late May. They were thus more restrictive for the artisanal fishery by including an extra full or new moon period. This suggests that fishing is taking place in the lagoons despite the closed season.

A problem yet to be addressed is the effect of the increase in effort of the industrial fleet, from 250 vessels before 1993 to nearly 400 in 1997 (Fernandez-Mendez et al. 2000; INP, 2000). Estimates of fishing mortality caused by the industrial fleet that averaged 0.32/month for 1992 rose to 0.52/month in 1999. With a fishing season starting in mid-July and a natural mortality of 0.3/month only 5.66% of individuals would reach the October spawn-

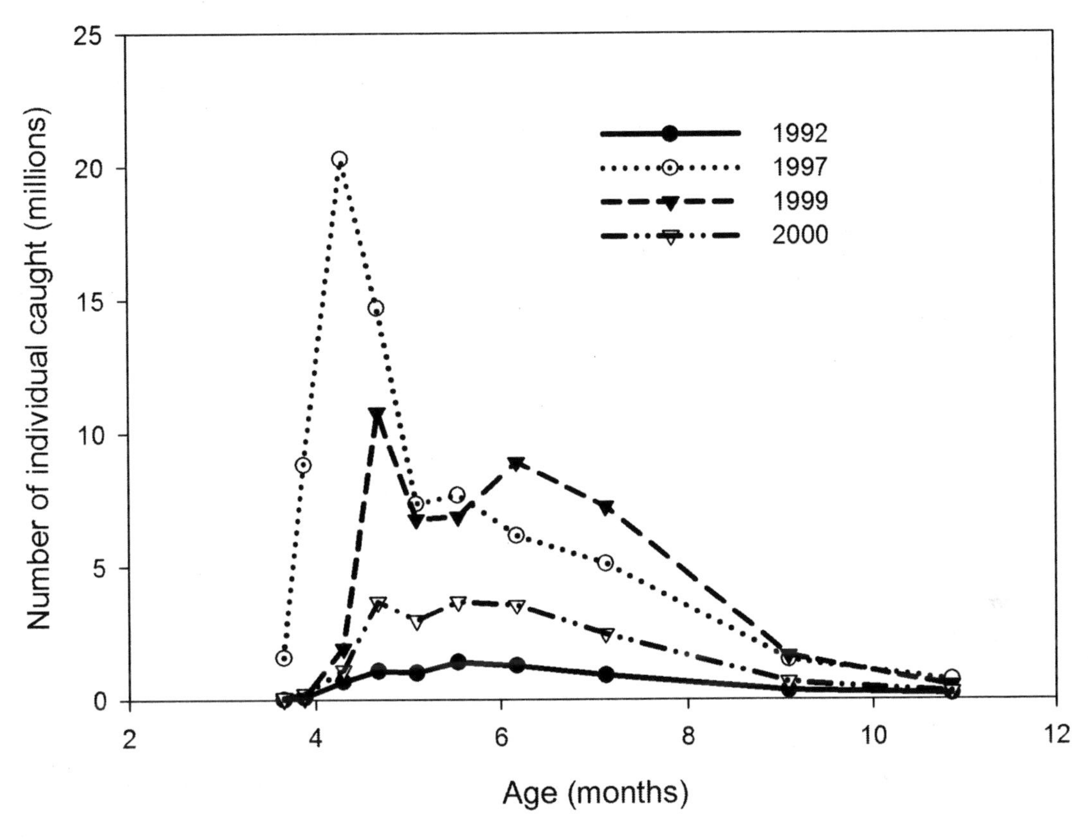

Figure 4. Age-frequency distribution in catches of the industrial fishery at the start of the fishing season (July–August) off Tamaulipas in several years. In 2001, the age distribution is similar to years previous to the establishment of the closed season.

ing season. Estimates of fishing mortality around 0.6/month have been associated with the decline of pink shrimp stocks in the Campeche Sound (Fernandez-Mendez 2001; Ramirez-Rodriguez and Arreguin-Sanchez 2003).

A recent increase in the proportions of seabob *Xiphopenaeus kroyeri*, a near-shore, shallow-water species (from less than 5% to around 15%), as well as a higher proportion of small brown shrimp in the catch of the offshore fishery at the start of the fishing season, has led some researchers to think that industrial fishermen are compensating reductions in catch by fishing in shallow waters, at present not allowed by Mexican fishing regulations (Ramirez-Soberón et al. 2001). This can result in the short term in decreasing values of yield per recruit in the offshore fishery by catching small individuals. In the long term it can result in recruitment overfishing, as these small shrimp are not allowed to reach their spawning size. As official catch statistics only show a total catch figure, without presenting catch by species, decreases in brown shrimp catches may be more serious than the trend in total catch may suggest. We must also consider that the fishing mortality in the lagoon fishery is high, a kilogram of the catch obtained in the lagoon fishery contains as much as six times the number of individuals than in the offshore fishery. So although recruitment overfishing has not been yet demonstrated, we cannot

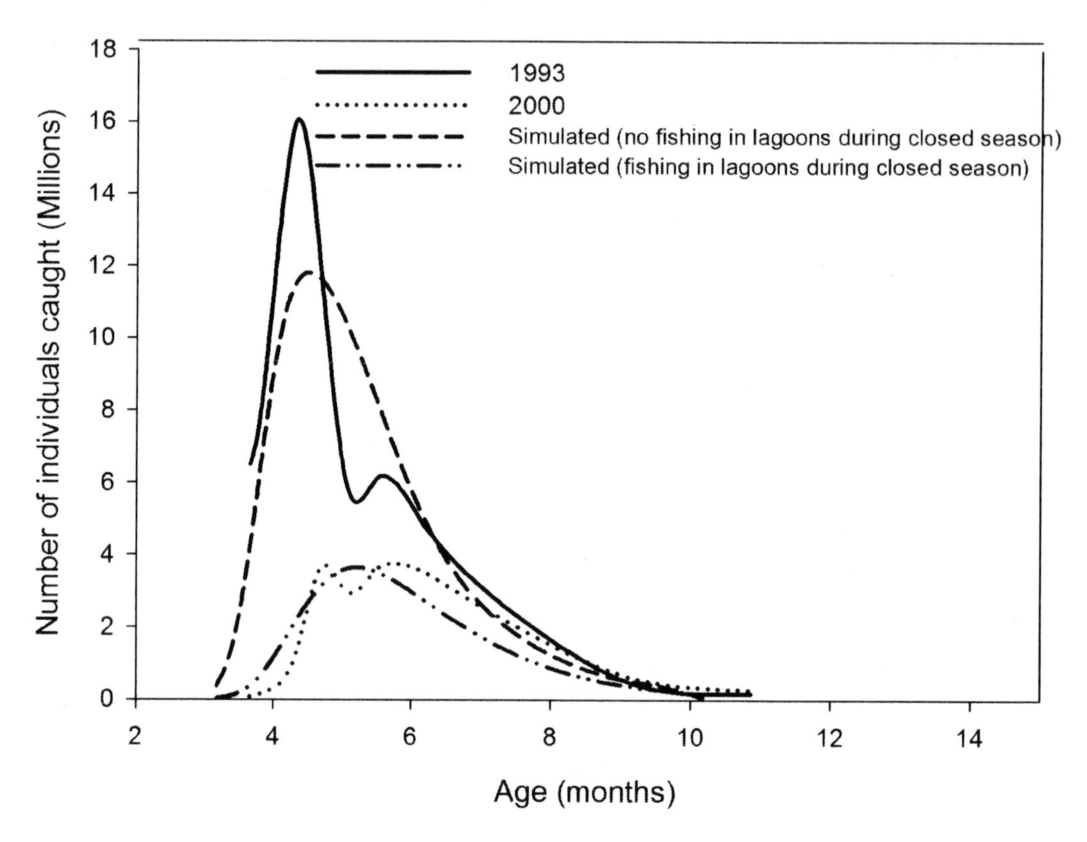

Figure 5. Simulated age frequencies, obtained using the model used in the decision making process since 1997–2001, compared to observed age frequencies in 1993 and 2000. The two simulated scenarios include fishing and no fishing during the closed season in the lagoon fishery during the migration period of early June.

disregard the fact that we have had fishing mortalities that can be considered high, together with catches that have had a descending trend in recent years.

Discussion

Only long-term solutions can lead to increased (or maintained) catches or profitability in the brown shrimp fishery in Tamaulipas-Veracruz. These long term solutions should involve two important aspects, interactions between stakeholders and conservation of the resource.

In the present situation, involving competition between two sectors in a sequential fishery, increases in total catch and value can be obtained only at the expense of the artisanal fishery. In fact, the closed season scheme presently applied affects mostly the artisanal fishery as it restricts fishing during the period (June–July) when artisanal fishermen obtained up to 35% of their annual catch. During the first 5 years after the establishment of the closed season, the average catch in the lagoon fishery was 21% lower than in the 5 years previous to 1993. In contrast, the offshore catch rose by an average of 54%.

Both groups of stakeholders, artisanal and industrial fishermen, have apparently not been satisfied with the portion of the total catch they have been obtaining. Some authors (like Olstrom 2000) discuss the role of incentives in the compliance of stakeholders with rules set by managers as an important prerequisite for success of management schemes. The disappearance from the offshore catches of the shrimp that are supposed to be protected by the closed season in the lagoons after 1998 indicates that the artisanal fishermen are not complying with the restrictions imposed upon them. The proportion of artisanal catches is again close to 50% of the total catch This may be a result of the artisanal fishermen's perception that the management scheme benefits only the industrial sector of the fishery and may also be one of the causes of reductions in industrial catches in recent years.

It also appears that the industrial fleet has adopted short-term, self-defeating strategies like fishing in shallow waters to compensate for these reductions. It may be that, as half of the fleet operates off Tamaulipas and Veracruz only for a relatively short period of time (3 months) before returning to the Campeche Sound, these short-term strategies are deemed as sensible by at least a part of the industrial fishing fleet.

Fishing authorities have responded to catch reductions by imposing closed seasons that are more restrictive for the artisanal fishery. However, the failure of the authorities (in the view of industrial fishermen) in further restricting the artisanal fishery has led to accusations of "not managing by scientific criteria," meaning that any concession to the artisanal fishery should be considered a political stance (El Sol de Tampico, 29 April 2000). Nevertheless, further restrictions without giving artisanal fishermen some economic alternatives should be considered as only a stop-gap strategy that can cause negative social results and may not prove effective in the long run. Certainly, it has not been enough to ameliorate the situation of the offshore fleet since 2001.

Although allocation has been a driving factor in the management of this fishery, it has never been explicitly stated. In our opinion, this has resulted in the possibility of discretional decisions by authorities being taken without the possibility of open negotiations by stakeholders. This, in turn, has resulted in the stakeholders reacting in a negative fashion, as has been previously discussed.

Some obvious conservation measures, like reductions of fishing effort, have not been applied in managing this fishery. In Texas, a bay shrimp license buyback program has been in place for some years (Texas Parks and Wildlife Department 2000). Probably a similar program in Mexico should be considered, although due attention should be paid to the difficult economic conditions of Mexican artisanal fishermen that drive them to violate fishing regulations (Lugo 2004).

Reductions in industrial fishing effort should also be considered, and not only because the possibility of reducing recruitment overfishing discussed previously. Several authors, when speaking about the Texas fishery, point out that without a regulation of offshore effort, any beneficial economical effect would be short-lived as the rent (the difference between the market price and the cost of harvest of the catch) would tend to dissipate (Houston et al. 1989; Griffin et al. 1993; see Iudicello et al. 1999 or Hannesson 2004 for the process of dissipation of rent). The Mexican offshore fishery has been described as "highly sensitive to changes in prices and production costs and operate at low utility margins;" in other words, it is highly overcapitalized (FIRA, 2003). Earlier analysis (Goudet and Goudet 1987; FAO/World Bank 1988) recommended a fleet reduction of 50–66%. Such a reduction would not only bring economic benefits but also reduce the fishing pressure on the resource. Maybe a regulation of numbers of industrial vessels based on fishing mortality and profitability considerations should be established. Furthermore, an individual transferable quota system should be considered for the industrial fishery in the long term (see Iudicello et al. 1999 or Hannesson 2004 for the advantages and disadvantages postulated for this management regime). Perhaps a "quota by sector" system (meaning a transferable quota explicitly allocated to the artisanal and industrial fisheries) should be examined. Mid-term solutions, like the offshore fleet "buying out" artisanal effort by paying the equivalent of what the artisanal fishery would obtain by fishing during the main migration period, have been informally proposed. However, such options have so far involved only interstakeholder negotiations, without official intervention, and have proved themselves fruitless. In our view, making the allocation of the resource an explicit management objective (as it would be under a quota system) would help in regulating effort and reduce (or at least make easier to negotiate) interstakeholder conflicts.

A measure yet to be taken is the protection of spawning events, particularly during the first quarter of the year. This lack of protection to spawning occurs in spite of such protection being an official objective of management. Spawning is protected by closed seasons in other Mexican shrimp fisheries. Although including the spawning of February–March in the offshore closed season has been recommended several times (Fernandez-Mendez et al. 2000), and it was to be applied in 2002, it has never been implemented. In 2002 an additional 1 month closed season was applied to protect the October spawning. However, the alternating nature of shrimp recruitment and the high fishing mortality exerted before that month during the fishing season argue against the effectiveness of such a measure. That closure was not been applied again after 2002.

In our view, the management of the brown shrimp fishery in Tamaulipas-Veracruz has several weaknesses

1) Originally, it was designed to reduce growth overfishing (like the Texas closed season). However, it officially aims at preventing recruitment overfishing without protecting spawning events.

2) It has had an excessive impact on allocation despite the fact that allocation is not an official and explicit management objective.

3) It has not reduced fishing effort in either the artisanal or industrial fisheries. It has relied solely on closed seasons to reduce fishing mortality.

4) It has not prevented competition between the artisanal and industrial fishery sectors. This has resulted in both sectors adopting strategies that negatively affect the sustainability of the fishery.

The sequential nature of the shrimp fishery results in allocation issues being unavoidably linked to management by closed seasons. However, conservation should be the main concern in management, as only it ensures the existence of the fishery. We suggest that a lack of careful consideration of the issues discussed here have resulted in a loss of effectiveness of management in preventing growth or recruitment overfishing. The case study presented here illustrates the need for careful statements of management objectives as a requisite for designing management strategies. The reactions of stakeholders (in particular the artisanal fishermen) to the management measures taken in this fishery, illustrates the importance of stakeholders' perceptions of their "fairness" in achieving compliance with management strategies.

References

Castro Melendez, R. G., and R. Santiago. 1976. Veda experimental de camarón en las costas de Tamaulipas en 1974. Mem. del Simposio sobre la biología y dinámica poblacional del camarón. S.I.C. Pesca. Guaymas, Sonora, México.

Caillouet C. W., F. J. Patella, and W. B. Jackson. 1979. Relationship between marketing category (count) composition and ex-vessel value of reported annual catches of shrimp in the eastern Gulf of Mexico. Marine Fisheries Review 41(5–6):1–7, Seattle.

Coleman, F. C., P. B. Baker, and C. C. Koenig. 2004. A review of Gulf of Mexico marine protected areas: successes, failures, and lessons learned. Fisheries 29(2):10–21.

CONAPESCA (Mexico's National Commission of Aquaculture Fisheries). 2001. Anuario estadístico de pesca 2001. CONAPESCA/SAGARPA (Fisheries and Food of the Mexican United States), México.

CONAPESCA (Mexico's National Commission of Aquaculture Fisheries). 2003. Anuario estadístico de pesca 2003. CONAPESCA/SAGARPA (Fisheries and Food of the Mexican United States), México.

Condrey, R., and D. Fuller. 1992. The U. S. Gulf shrimp fishery. Pages 89–119 *i*: Glantz, M. H., editor. Climate variability, climate change and fisheries. Cambridge University Press. Cambridge, UK.

DOF (Diario Oficial de la Federación, National Registrar). 1999. AVISO por el que se da a conocer el establecimiento de épocas y zonas de veda para la captura de las especies de camarón en aguas marinas y de los sistemas lagunarios estuarinos de jurisdicción federal, del Golfo de México y Mar Caribe. Secretaría de Gobernación, Tlalpan, C. P., México.

FAO (Food and Agriculture Organization of the United Nations)/World Bank. 1988 Mexico: fishery sector review. FAO/World Bank Cooperative Program, Rome.

Fernández-Mendez, J. I. 2001. Administracion de la pesqueria de camarón en el Golfo de México. Un esquema completo? Segundo Foro de Camarón del Golfo de México y Mar Caribe. Instituto Nacional de la Pesca, México.

Fernández-Mendez, J. I., L. Schultz-Ruiz, A. T. Wakida-Kusunoki, M. Medellín, M. E. Sandoval-Quintero, G. Núñez-Márquez, J. A. Uribe-Martínez, R. G. Castro- Meléndez, A. González Cruz, M. E. González, J. Santos, G. Marcet, F. Aguilar, B. Delgado, and G. Chale. 2000. Camarón del Golfo de México y Mar Caribe *in* Sustentabilidad y pesca responsable en México; evaluación y Manejo, 1999–2000. Instituto Nacional de Pesca, México.

FIRA (Fideicomisos Instituidos en Relación con la Agricultura). 2003. Perspectivas del camarón 2003. Fideicomisos Instituidos en Relación con la Agricultura. Banco de México, Morelia, México.

García, S. 1985. Reproduction, stock assessment models and population parameters in exploited penaeid shrimp populations. Pages 139–158 *in* P. C. Rothlisberg, B. J. Hill, and D.J. Staples, editors. Second Australian prawn seminar, Cleveland, Australia.

García, S., and L. Le Reste, 1981. Cycles vitaux, dynamique, exploitation et amenagement des stocks de crevettes penaeides côtieres. FAO Fisheries Technical Paper (203).

Gracia, A. 1997. Simulated and actual effects of the brown shrimp, Penaeus aztecus, closure in Mexico. Marine Fisheries Review 59(2):18–24.

Griffin, W., H. Hendrickson, C. Oliver, G. Matlock, C. E. Bryan, R. Riechers, and J. Clarck. 1993. An economic analysis of Texas shrimp season closures. Marine Fisheries Review 54(3):21–28.

Goudet and Goudet, F. 1987. La Racionalidad Económica de la Captura de Camarón en el Golfo de México; Análisis Teórico y Evidencia Empírica. Graduate thesis. Instituto Tecnológico Autónomo de México, México.

Hannesson, R. 2004. The privatization of the oceans. The MIT Press, Cambridge, Massachusetts.

Hernández, A., and W. Kempton. 2003. Changes in fisheries management in Mexico: effects of increasing scientific input and public participation. Ocean & Coastal Management 46:507–526.

Houston, J. E., A. E. Nieto, J. E. Epperson, H. S. Li, and G. W. Lewis. 1989. Factors affecting local prices of shrimp landings. Marine Research Economics, Kingston, Rhode Island.

INP (Instituto Nacional de la Pesca). 2000. Fundamento Técnico para el establecimiento de vedas en el Golfo de México en 2000. Instituto Nacional de la Pesca. Programa Camarón del Golfo de México y Caribe, Dictamen Técnico,

Iudicello, S., M. Weber and R. Wieland. 1999. Fish, markets and fishermen. The economics of overfishing. Island Press, Washington, D. C.

Klima E. F., K .N. Baxter, and F. J. Patella, Jr. 1982. A review of the offshore shrimp fishery and the 1981 Texas closure. Marine Fisheries Review 44(9–10):16–30.

Leary, T. R. 1985. Review of the Gulf of Mexico Management Plan for shrimp. Pages 267–274 *in* P. C. Rothlisberg, B. J. Hill, and D. L. Staples. 1985. Second Australian National Prawn Seminar (NPSR), Cleveland.

Lugo, R. 2004 La pesca artesanal del camarón y los derechos humanos. El Varejon No. 59, May 2004. pp. 13–20.

Mathews, G. P., M. Al-Hosseini, A. R. Abdul-Gaffar, and M. Al-Shoushani. 1984. Assessment of short lived stocks with special reference to Kuwait's shrimp fisheries: a contrast of the results obtained from traditional and recent size-based techniques. In D. Pauly and G. R. Morgan, editors. Length-based methods in fisheries research. ICLARM Conference Proceedings 13, Manila, Phillipines.

Nance, J. E., A. Martinez, and E. F. Klima. 1994. Feasibility of improving the economic return from the Gulf of Mexico brown shrimp fishery. North American Journal of Fisheries Management 14:522–536.

Nichols, S. 1982. Impacts on shrimp yields of the 1981 Fishery Conservation Zone Closure off Texas. Marine Fisheries Review 44(9–10):31–37, Seattle.

Olstrom, E. 2000. El gobierno de los bienes comunes; La evolución de las instituciones de acción colectiva. Universidad Nacional Autónoma de México/Centro Regional de Investigaciones Multidisciplinarias/Fondo de Cultura Económica, México.

Ramirez-Rodriguez M. and F. Arreguin-Sanchez. 2003 Spawning stock-recruitment relationships of pink shrimp Farfantepenaeus duorarum in the Southern Gulf of México. Bulletin of Marine Science 72(1):123–133.

Ramirez-Soberón, G., J. I. Fernandez-Mendez, and M. Medellín- Avila. 2001. Análisis de la veda aplicada a la pesquería de camarón café, *Farfantepenaeus* (Ives,1891), en Tamaulipas. Segundo Foro de Camarón del Golfo de México y Mar Caribe. Instituto Nacional de la Pesca, México.

Rotlisberg, P., J. A. Church, and A. M. G. Forbes. 1983. Modelling the advection of vertically migrating shrimp larvae. Journal of Marine Research 41:511–538.

Texas Parks and Wildlife Department. 2000. Texas Commercial Fishing Guide 2000–2001. Texas Parks and Wildlife Department, Austin, Texas.

American Fisheries Society Symposium 49:569–573

Turning the Tide: Toward Community-Based Fishery Management in Canada's Maritimes

ANTHONY CHARLES*

Saint Mary's University, Management Science/Enviornmental Studies
923 Robie Street, Halifax, Nova Scotia B3H 3C3, Canada

Abstract.—This paper describes a unique nongovernmental initiative, *Turning the Tide: Communities Managing Fisheries Together*, that is supporting community fishery approaches in Canada's Maritime provinces and helping to build linkages between native and non-native participants in the fishery. The initiative has been motivated by the desire of participants for community-based fishery management and, in particular, the recent opportunity for aboriginal First Nations to enter the commercial fishery on a community basis. The paper describes the role played in *Turning the Tide* by three types of activities: 1) workshops, networking, and capacity-building; 2) exchange visits; and 3) development of resource materials. Finally, the paper discusses the need for grassroots initiatives such as this to be complemented by governmental policy and logistic support for community-based management.

Introduction

Community-based management is rooted in the idea that fishery sustainability and the overall benefits the fishery produces can be enhanced when fishers (fishermen), together with others in coastal communities—those living closest geographically to the fish stocks—have a significant level of responsibility for and control over managing those resources. Thus, community-based management is a form of comanagement in which fishers and their communities participate in decision making on the various aspects of marine resource stewardship and fishery management (e.g., Berkes et al. 2001; Wiber et al. 2004).

A range of both practical experience and social research indicates that, while community-based management cannot be expected to work in all circumstances, where appropriate it can provide a strong vehicle for resource conservation and sustainable development. This occurs in particular through 1) better utilizing traditional ecological knowledge, 2) empowering local resource users and their communities (leading to better acceptance of conservation measures and compliance with regulations), and 3) providing a vehicle to resolve conflicting local resource uses (Pinkerton 1989, 1999; Charles 2001). As an alternative to top-down governmental management, community-based fishery management fits within the broad trend found in many societies toward devolution of governance. It is also in keeping with traditional aboriginal approaches to natural resource management (see, e.g., Barsh 2002), a key point of relevance here.

This paper explores the role of grass-roots linkages among fishery communities in supporting community-based fishery management, through examination of an initiative within Canada's Maritime provinces, *Turning the Tide: Communities Managing Fisheries Together*. Specifically, the paper discusses development of linkages among local communities, across the region, across Canada, and internationally.

Turning the Tide

Turning the Tide grew out of two realities. First, a substantial set of fishery participants in the Canadian

*Corresponding author: tony.charles@smu.ca

Maritimes has come to see the desirability of a local approach to resource management as a means to improve the sustainability of fisheries, support marine conservation, and enhance the well-being of coastal communities. Second, in many parts of the Maritimes, there has been a need to develop cooperative endeavors that link aboriginal (particularly Mi'kmaq) and non-aboriginal peoples. The latter became crucial in the aftermath of the Canadian Supreme Court's 1999 Marshall Decision, which recognized the right of aboriginal people to fish commercially, and which led to an increasing presence of Mi'kmaq people in the fishery (Coates 2000), a move with parallels elsewhere in the world, such as the United States (e.g., Singleton 1998) and New Zealand (e.g., Memon et al. 2003).

Thus, *Turning the Tide* pursues simultaneously the twin goals of building community-based fishery management and building linkages between native and non-native fishery participants. These two goals are interconnected, in that community-based management is the traditional approach of Mi'kmaq First Nations, tied to aboriginal values linking humans and nature, and this inherent philosophy is increasingly being recognized by non-native fishers as supportive of their own aspirations. Thus, *Turning the Tide*—funded through the Pew Charitable Trusts—builds on the common desire of participants to work together for better fishery management so as to reinforce well-being in their communities, and this in turn provides a rationale for cooperative initiatives among aboriginal and non-aboriginal fishing communities.

The initiative began with four major participants, all located in the province of Nova Scotia, Canada: the Bay of Fundy Marine Resource Center (a civil society organization that links a range of fishery and community groups and provides facilitation, networking, and GIS services), Acadia First Nation (one of the two Mi'kmaq aboriginal bands in southwest Nova Scotia), the Center for Community Based Management (an education and support centre at St. Francis Xavier University), and Saint Mary's University in Halifax. Over the course of the initiative, five additional participants were formally added: Bear River First Nation (the second of the Mi'kmaq bands in southwest Nova Scotia), the Mi'kmaq Confederacy of Prince Edward Island (an organization linking the two Mi'kmaq bands, Lennox Island, and Abegweit, in the province of Prince Edward Island), and three community-based fisher-men's associations—the Bay of Fundy Inshore Fishermen's Association (a multispecies community-based fishermen's organization in the Digby-Annapolis area), Local 9 of the region's Maritime Fishermen's Union, and the Fundy Fixed Gear Council (a community-based groundfish management board covering the Nova Scotia side of the Bay of Fundy).

In addition, members of several other organizations have been involved in the initiative, drawn from other First Nations, community and regional fishery organizations, and resource centers. Important connections have also been developed with 1) the Saltwater Network, an organization linking marine conservation and local management groups on the Canadian and U.S. sides of the Gulf of Maine and supporting a range of ecological and fishery oriented projects, and 2) the North Atlantic Marine Alliance, a New England based but increasingly transboundary organization focusing on the Gulf of Maine from community- and ecosystem-based perspectives.

Turning the Tide has helped participating organizations by 1) bringing together partners to discuss key issues, principles, and values, and how these relate to community-based management, as well as to identify common issues, priorities, and strategies; 2) supporting educational and capacity-building efforts, including workshops, fishery exchange visits, training courses, and development of resource materials; and 3) undertaking concrete activities related to community-based fisheries and coastal management, from those at a local level (e.g., supporting community development of fishery management plans) to those with a potentially global impact (e.g., the development of a "community fishery management handbook," described below). In particular, the major activities within *Turning the Tide* included

• Bringing together aboriginal and non-aboriginal fishermen and communities;

• Sharing information and ideas on community-based management;

• Identifying common issues, outcomes, priorities, and strategies;

• Supporting educational and capacity-building efforts;

• Highlighting the need for community-based management within government policy;

• Supporting concrete, local-level, ecologically oriented management initiatives;

• Documenting community-based management practice and insights obtained;

• Undertaking a range of educational activities promoting community-based management; and

• Establishing regional, national and international links.

These various categories of activity in *Turning the Tide* can be grouped under three headings:

1. Workshops, Networking, and Capacity-Building

Perhaps the single most important tool used by *Turning the Tide* is the apparently simple act of "convening"—bringing people together, particularly those from groups that rarely interacted in the past. This has taken place through a series of meetings and workshops (which served a dual role in facilitating informal development of cooperative community-based management ideas), as well as in several "special events"—for example, a workshop on managing multiple ocean uses was organized as part of a pilot project on "place-based" integrated management of human activities in St. Mary's Bay and the Annapolis basin, two key fishing locations on the Nova Scotia side of the Bay of Fundy. *Turning the Tide* also enabled partner organizations to attend a workshop on community-based natural resource management at St. Francis Xavier University and an associated training program. In addition, *Turning the Tide* has supported the capacity-building of its partners, such as a planning process of one First Nation partner to develop a community-based fisheries plan, in part through connections with non-native fishers.

2. Exchange Visits

A key component of *Turning the Tide* has been a set of exchange visits, in which participants travel and share experiences together, building linkages among themselves and with those they visit. Sets of First Nation and non-native participants from the Maritime provinces traveled twice across Canada to British Columbia, specifically to the west coast of Vancouver Island and to Alert Bay on the northeastern corner of the island. This provided opportunities to interact and develop stronger relationships internally, to consolidate linkages with like-minded groups on Canada's Pacific coast (including native bands, community-based commercial fisher groups, and local aquatic management boards) and to incorporate capacity-building through a range of presentations to and by the participants. A third exchange took place around the Gulf of Maine (including the Bay of Fundy) in 2003, crossing between Canada and the United States and involving participants of both countries. The trip involved stops in four provinces and states: Nova Scotia (Yarmouth, Digby), New Brunswick (St. Andrew's), Maine (Eastport, Stonington, and Saco), and Massachusetts (Cape Cod, Boston). Organized in partnership with the North Atlantic Marine Alliance and the Saltwater Network, the tour involved a combination of networking, thematic workshops to examine needs, and challenges of community-based management, field testing of resource materials under development, and occasions to promote community-based management as an appropriate conservation-minded approach for fisheries and other coastal resources. These three exchange visits required considerable effort and cost, but they turned out to be remarkably effective in achieving multiple goals—development of bonds among participants, the discovery of linkages regionally and nationally, and an innovative use of "mobile workshops" to produce valuable new directions. The concept of exchanges is nothing new, but the diversity among their participants, and among the learning and relationship-building opportunities, was exceptional, as was the good will the exchanges encouraged.

3. Resource Materials

Raising the public profile of community-based fishery management and its potential in the Canadian Maritimes and beyond is a significant goal of *Turning the Tide*. This motivation led to production of a broadly-distributed brochure and a Web site (www.turningthetide.ca) describing the initiative. In addition, *Turning the Tide* participants saw the need for resource materials on community-based fishery management to support local initiatives in "northern" countries such as Canada, analogous to those available in developing countries, where the role of and potential for this approach is better recognized. Accordingly, a practical community-based fisheries handbook is being produced to support communities in such locations pursue community-based management (see www.turningthetide.ca). This handbook draws on the knowledge base and on-the-ground experiences of native and non-native partners, to share insights obtained in initiating community-based fishery management across the Canadian Maritimes.

Discussion

Community-based comanagement as an approach to fishery management has become a major topic of study and practice in recent years. Its focus lies in bringing fishers, their organizations, and fishing communities more extensively into the management process—in other words, as more equal partners with governments. Given the many past failures of top-down government controlled fishery management, this is an important endeavor and has led to research directed at the challenge of developing suitable "vertical linkages" between community-level fishery management and governments. At the same time, there is a need for community fisheries and their local management systems to interact with and learn from one another—for "horizontal linkages" between these entities and vertical linkages providing coordination among them.

The *Turning the Tide* initiative has sought to provide a vehicle for such grassroots linkages in the Canadian Maritimes and the Gulf of Maine area. This initiative operated informally, without involvement of state governments, and with linkages built over time through a range of meetings, workshops, exchange opportunities, and the like. While it cannot be claimed that *Turning the Tide* changed in a major way the manner by which fisheries are managed in the region, it has helped to highlight an alternative vision, one that builds on important attributes of community-based management to help achieve both marine conservation and the well-being of coastal communities. In addition, personal and organizational bonds have developed that will likely support further efforts to build community-based fishery management and that will be important in helping aboriginal and non-aboriginal coastal residents cooperatively engage in sustainable resource use and environmental conservation.

The fact that *Turning the Tide* operated on a grassroots level, without governmental involvement, had implications both positive and negative. On the one hand, the initiative provided a "safe place" for a diverse set of participants to meet informally, with "no strings attached"—leading to important team-building, development of relationships with one another, and the exploration of ideas and approaches. This provided an important alternative to formal fishery stakeholder processes, in which government chairs a formal meeting at which each party states and defends their position, discussions ensue, but no real social capital is developed. Instead, *Turning the Tide* provided a more elaborate illustration of the value of "sharing a meal together"—facilitating the creation of social capital through convening informally.

On the other hand, the absence of government, particularly the Canadian Department of Fisheries and Oceans, in the development and implementation of *Turning the Tide* meant that the initiative was not within the mainstream of government-led fishery discussions. Would it have had greater impact in official circles if government participants had been involved? Or would that have merely eliminated the initiative's comparative advantage as a grassroots effort? It does not seem possible to answer these questions precisely, but one lesson does seem clear. *Turning the Tide* demonstrated not that interaction with government is unnecessary, but rather that there is value in complementing such interaction with another approach—a nongovernmental one built from the bottom up. The linkages between people and organizations developed in this manner may well be more robust and long-term in nature than could have been developed in any other way.

While it is well established that development of community-based management takes significant time, it is clear from this experience that the availability of financial resources, to facilitate the processes of positive human interaction, can speed up the process considerably. But if there is value in developing fishery linkages informally through a nongovernmental process such as *Turning the Tide*, how is this to be supported? In the Canadian context, the federal government—with its responsibility for fishery management and interest in comanagement—would seem an appropriate avenue to provide support for local level development of linkages that can enhance the effectiveness of comanagement. Indeed, the federal government supports similar activities abroad through its agencies such as the International Development Research Centre. However, in the case of *Turning the Tide*, support was provided not by the national government, but rather by a U.S. foundation, the Pew Charitable Trusts. In reality, the independence of this funding may well have helped ensure the credibility of the initiative, but such support cannot be counted on in general, so a lesson for the future is the need for governmental support for such nongovernmental linkage development.

Turning the Tide, as a people-oriented initiative, produced results that could not be measured quantitatively (e.g., in terms of the number of fish tagged), but rather were more nebulous, in terms of capacity-building, education, and, most crucially, the positive interactions developed.

It was thus a "leap of faith" on the part of the funding body to support this kind of initiative, requiring recognition that efforts such as *Turning the Tide* are crucial both for conservation and for improving the well-being of coastal communities. In particular, they bring together motivated people who must work cooperatively to achieve marine conservation and community improvement—and who, in the case of native and non-native communities, historically had little interaction with one another. The range of positive results from *Turning the Tide* has reinforced, among participants, a mutual dedication to managing fisheries in a local, community-based, and ecologically-oriented manner, and has inspired them to continue building momentum in support of community-based fishery management, through their new friendships and organizational ties.

Acknowledgments

The author is grateful to two anonymous referees for very useful and constructive suggestions that have improved the paper, to *Turning the Tide* partners in the Canadian Maritimes for their ideas, inspiration, and support, and to the Social Science and Humanities Research Council of Canada and the Pew Fellows Program in Marine Conservation for financial support.

References

Barsh, R. L. 2002. Netukulimk past and present: Míkmaw ethics and the Atlantic fishery. Journal of Canadian Studies 37:15–42.

Berkes, F., R. Mahon, P. McConney, R. Pollnac, and R. Pomeroy. 2001. Managing small-scale fisheries: alternative directions and methods. International Development Research Centre, Ottawa.

Charles, A. T. 2001. Sustainable fishery systems. Blackwell Scientific Publications, Oxford, UK.

Coates, K. S. 2000. The Marshall decision and native rghts. McGill-Queen's University Press, Montreal and Kingston, Canada.

Memon, P. A., B. Sheeran, and T. Ririnui. 2003. Strategies for rebuilding closer links between local indigenous communities and their customary fisheries in Aotearoa/New Zealand. Local Environment 8:205–219.

Pinkerton, E. W. 1989. Cooperative management of local fisheries. University of British Columbia Press, Vancouver.

Pinkerton, E. W. 1999. Directions, principles, and practices in the shared governance of Canadian marine fisheries. Pages 340–354 *in* D. Newell and R. Ommer, editors. Fishing places, fishing people: traditions and issues in small-scale fisheries. University of Toronto Press, Toronto.

Singleton, S. 1998. Constructing cooperation: the evolution of institutions of comanagement. University of Michigan, Ann Arbor.

Wiber, M., F. Berkes, A. Charles, and J. Kearney. 2004. Participatory research supporting community-based fishery management. Marine Policy 28:459–468.

American Fisheries Society Symposium 49:575–583

Small is Beautiful? A Database Approach for Global Assessment of Small-Scale Fisheries: Preliminary Results and Hypotheses[1]

Ratana Chuenpagdee*
International Ocean Institute-Canada, Dalhouise University
1226 LeMarchant Street, Halifax, Nova Scotia B3H 3P7, Canada

Daniel Pauly
University of British Columbia, Fisheries Centre
2202 Main Hall, Vancouver V6T 1Z4, Canada

Abstract.—Many aspects of small-scale fisheries hinder our efforts to understand their dynamics, as required for sustainable management. For example, there is no widely accepted definition of small-scale fisheries or global data on number of small-scale fishers and their catches. One reason for this is that most research on small-scale fisheries is done at community level and tends to focus on immediate problems, such as poverty reduction, maintaining traditional lifestyle, and conflict resolution with large-scale fisheries. Many national fisheries development plans claim, and in some cases do, favor the small-scale fisheries sector, though it is generally the larger scale fleets which benefit from such plans. Several studies suggest, moreover, that a "large" small-scale fisheries sector can be as detrimental to fisheries resources and marine ecosystems as large-scale fisheries. Undoubtedly, quantifying the impacts of the small-scale fisheries sector relative to that of the large-scale sector would help improve our ability to manage the fisheries. In this paper, we present global estimates of small-scale fisheries based on aggregation of available data on catch and number of fishers from national level statistics. The initial data were from FAO country profiles, which were used to interpolate for countries with missing data. We conclude with discussion about how to improve these estimates using field data from regional and national sources.

Introduction

"The most striking about modern industry is that it requires so much and accomplishes so little. Modern industry seems to be inefficient to a degree that surpasses one's ordinary powers of imagination. Its inefficiency therefore remains unnoticed."—E.F. Schumacher, Small is Beautiful, 1999.

For many countries worldwide, fisheries mean important sources of food, livelihood, employment, and income. However, their degree of importance in different countries varies, de-

*Corresponding author: ratana.chuenpagdee@dal.ca
[1]Based on the presentation at the Fourth World Fisheries Congress, Vancouver, May 2004. Since this was originally presented, catch and effort estimates, which update the values presented herein, have been computed, using the methods described, applied to non-FAO, local data. These updated estimates are available at www.seaaroundus.org.

pending notably on the basic features of their economy. For countries with highly developed fisheries, operating as a large-scale industry, the focus is on income, particularly through export earnings. In developing countries, the contributions of fisheries also impact on food security, as fisheries products often represent the main source of animal protein. Small and large-scale fisheries generally co-exist in many parts of the world and the extent of their interactions and conflicts depend on the relative scale and intensity of their operations (Pauly 1997). The ecosystem impact of small and large-scale fisheries also differs, depending on the gear used (Chuenpagdee et al. 2003) and overall fishing effort. For example, industrial bottom trawling, covering a large fraction of a country's continental shelf, and extracting large catches, is likely to result in greater ecosystem impacts than small-scale inshore traps. It could be argued, however, that one large-scale fishing vessel may be less destructive than many small-scale fishing boats. Further, some small-scale fishing methods can be very destructive, such as dynamite and cyanide fishing, practiced illegally in many developing countries (e.g. of Southeast Asia [Saeger 1993] or Africa [Vakily 1993]). Thus, overfishing can occur with both large-scale and small-scale fisheries (World Bank et al. 1991). Indeed, a worldwide comparative analysis of these two sectors is urgently required to assess these and related issues.

Most of the research and systematic data collection efforts have been focused on industrial fishing in developed and developing countries. As a consequence, a large body of information and knowledge about large-scale sector exists, to the extent that the common complaint about lack of data as the reason for ineffective management measures leading to overfishing is now largely unjustified. The same cannot be said about small-scale fisheries. The Food and Agriculture Organization of the United Nations (FAO), for example, coordinates and publishes fisheries statistics, such as landings from capture fisheries by species groups, from member countries on an annual basis. However, it can be largely assumed that data reported by member countries to FAO fail to include catches from subsistence and artisanal fishing (both of these are components of small-scale fisheries; recreational fisheries may also be included in the small-scale category, although, along with freshwater fisheries, they won't be considered here).

Many studies of small-scale fisheries have been conducted, but these studies have tended to emphasize the anthropological, social, and cultural aspects of small-scale fishing and generally attempted to capture their unique situations at particular locations. Information about small-scale fisheries at a country level is rare; one important exception being the fisheries country profiles published by FAO (http://www.fao.org/fi/fcp/fcp.asp), which attempt to provide a description of the large and small fishing sectors of most maritime countries. Researchers and scientists working in small-scale fisheries, however, do not always appreciate such broad overviews, claiming that natural and social systems are too complex, and that each small-scale fishing community is distinctively different from others. Another common view is that small-scale fisheries are so different between countries that global, or even regional, definitions and comparisons are impossible, again implying uniqueness for each individual fishery.

The problem with these notions, which often appear convincing at first sight, is that in effect, they add further to the marginalization of small-scale fisheries, already disadvantaged by their physical, socio-economic, political and cultural remoteness from urban centers (Pauly 1997). Small-scale fishing communities in developing countries often operate in areas away from the locations of political power, lacking landing facilities and other infrastructure, and direct access to markets. Compared with large-scale industrialized fishery sector, small-scale sector usually receives far less support (e.g., subsidies) from the governments. Also the lower economic status, on a per caput basis, of small-scale fishers, marginalizes them further and undermines the political power, that, at least in what are formally democracies, their numbers would imply.

At the onset, an attempt to counter this marginalization of small-scale fisheries would include an amount of research and a data collection effort comparable to those devoted to large-scale fisheries with aggregate catches of similar magnitude. This would help not only to provide a quantitative framework for the sociological and anthropological work performed so far, but also to allow for comparative analysis of social and economic contributions of the two sectors, as well as their relative impacts to marine and coastal ecosystems.

This contribution illustrates the reframing of research on small-scale fisheries suggested above by presenting a quantitative approach for deriving regionally stratified, global estimates of their catches and number of fishers and vessels based on data in the FAO country profiles. The following sections describe the iterative approach we have developed to achieve this, and preliminary results, followed by a discussion which emphasizes the next iterations, where the locale-specific knowledge embedded in the primary and gray literature will be used to improve our database, and the results based thereon.

Methods

Small-scale fisheries are sometimes described as subsistence and artisanal, with fishers using traditional and simple gears, some without a boat and some with boats lacking engines. These fisheries normally contribute food for household consumption, with small amount of catches used for barter or trade. In other instances, small-scale fisheries involve use of modern gears and boats with outboard or inboard motor. They are considered commercial fisheries, as catches are landed and sold either by fishers or their family members at the market or through marketing systems involving "middlemen" (often women). Concerns regarding the definition of small-scale fisheries are related to the wide range of fishing and marketing practices, framed in a great variety of cultural and political settings. Thus, the first in our effort to derive global estimates was to review the definitions of small-scale fisheries in the FAO country profiles. We used FAO data as for the initial estimates because it is the only data set that provides coverage of small-scale fisheries in a consistent format across countries.

Of 137 countries that provide report of marine fisheries to FAO, 50 offer definitions of their small-scale fisheries, in terms of boat type and size, engine size, gross boat tonnage (GRT), number of crew, gear used, and various combinations of these statistics. Table 1 illustrates the overall consistency of what is considered small-scale fisheries in developing and developing countries. It is this overall consistency which made possible the use of the database approach we present below.

The database was given the following features. First, it contains all countries with marine fisheries, grouped into two main categories, based on the UNEP human development index (HDI) (UNEP 2000) (i.e., high [H-HDI] and medium–low [ML-HDI]. Human development index measures a country's status in terms of life expectancy, educational attainment of its citizen, and adjusted real income, and is considered more appropriate than gross

Table 1. Characteristics of small-scale fisheries provided in FAO country profiles

Characteristics	General range
Boat size	between 5 and 8 m; less than 12 or 15 m; up to 21 m
Boat type	no boat canoe, sail, un-powered, outboard engine, open boat, no deck
Boat GRT	between 10 and 20 GRT; less than 20 GRT
Size of engine	15–40 hp; less than 60 hp
Number of crew	2–3; 5–6
Gear type	handline, longline, dive, traps, nets, gill nets, push nets, small trawlers
Combination	less than 10 m or 20 GRT, 5–6 m with 2–3 crew

domestic product (GDP), often used for ranking and grouping countries and their national fisheries. The grouping of the countries is done to allow for statistically improved estimates of global catches, as available data are averaged within groups of countries ("strata") and computation for missing values (i.e., their replacement by within-stata averages) is performed for countries within the same categories. Medium and low HDI countries are grouped together since their estimates are similar and they are very different from H-HDI countries. The grouping results in 40 H-HDI and 97 ML-HDI countries.

Then, for each country with data, the number of small-scale fishers, vessels, and catches were entered as reported in the FAO country profiles, along with other reported features of the small-scale fishery of each country. These data largely cover the late 1990s and 2000. As well, we entered for each country an estimate of its small-scale fishing area, defined as the area of its shelf ranging from the shoreline to 50 km offshore or 200 m depth—whichever comes first providing the limit. These limits were selected on the assumptions that small-scale fishers usually a) perform day trips (a few hours sailing, a few hours fishing, and a few hours sailing back) and, hence, the limit in terms of distance from shore and b) do not fish in very deep waters, except in areas where the shelf is very narrow (e.g., around oceanic islands) and, hence, are restricted to on-shelf (neritic) waters and resources.

Global estimates (e.g., of number of fishers) were then obtained by
1) Using available estimates by countries to compute within-strata estimates of mean number of fishers per square kilometer of small-scale fishing area;
2) Multiplying these means by the country-specific values of small-scale fishing area to obtain preliminary estimates of fishers in countries without such number;
3) Adding fisher numbers across countries, by strata. Note that this approach, which was applied in similar fashion to number of vessels and catch per fisher (and hence to absolute catches by small-scale fisheries), implies that per stratum and global estimates obtained thereby emerge from summing a reasonably high number of largely independent products.

Consequently, we can assume that underestimates in certain countries will compensate for overestimates in others (Sokal and Rohlf 1995). Technically, this approach also allows for estimating formal confidence intervals for the global estimates, although we have abstained here from dealing with issues of precision, given the dubious accuracy of some of our source estimates.

Preliminary Results and Discussion

The current database includes existing information from eight H-HDI and 27 ML-HDI countries, or 20% and 28% of the total countries, respectively. Our preliminary estimates of the number of small-scale fishers and vessels are given in Table 2 and the global catches in Table 3. The estimates show that there are five times more small-scale fishers per area in ML-HDI countries than in H-HDI countries and that they use twice the number of boats. Conversely, individual small-scale fishers in H-HDI catch five times more per year than those in ML-HDI. These ratios can be taken to reflect the level of dependency on resources and the social well-being of small-scale fishers in these countries. Different estimates can be made using reported data within strata, to differentiate between geographical regions, as shown in Table 4 for ML-HDI countries. Table 4 suggests that these estimates vary only slightly from the estimates obtained in the first procedures (see Tables 2 and 3). Yet, regional differences found within the ML-HDI statum can be informative, particularly for discussion about fisheries sustainability at the regional level. In the Asia-Pacific region, for example, there are more small-scale fishers and more boats, resulting in relatively less catch per fisher and smaller total catches, compared to other regions. Curbing of small-scale fisheries in Asia-Pacific countries may be an issue that needs further exploration.

These preliminary estimates provide information about small-scale fisheries that can be used to compare with large-scale sector, similar to the broad comparisons performed by D. Thompson (in Pauly 1997). For example, the number of crew on large-scale fishing vessels reported therein is about half a million, while our estimates suggest that there are over 10 million small-scale fishers in the world. The contribution of small scale fisheries to global fisheries catches is, at 31 million metric tons (mt, Table 3), roughly half that of the large-scale sector, which ranges from 50 million to 80 million mt, depending on the proportion of commercial small-scale fisheries reported by FAO member countries as part of their national landings (see below). Also important about small-scale fisheries is the fact that income generated from this sector is likely

Table 2. Global estimates of small-scale (SS) fishers and boats

Countries	SS fishers per km^2	SS boats per km^2	Total SS fishers ('000)	Total SS boats ('000)
H-HDI	0.19	0.15	1,380	397
ML-HDI	0.97	0.31	9,100	2,050
Total			10,480	2,447

Table 3. Estimates of global landings by small-scale fisheries

Countries	Catch per SS fisher (mt/fisher/year)	Total SSF catch (1,000 mt/year)
H-HDI	10	13,467
ML-HDI	2	17,962
Total		31,429

to stay at the local level and contribute to local well-being (Sen 1999).

When considering fishing activity in terms of food efficiency, almost all small-scale fisheries catches are used for human consumption, as opposed to only 57% in the case of large-scale fishing (Pauly 1997). The contribution of small-scale fisheries to human food security is therefore greater than that of the large-scale sector; similar analyses can be made for fuel efficiency or return on investment. Thus, the catch per metric ton of fuel consumed in small-scale fishing is 4–5 times higher than for large-scale fishing and the number of fishers employed per $1 million investment in fishing vessels is at least 100 times higher in small-scale than in large-scale fisheries (Pauly 1997).

The estimates reported in this paper are very preliminary and will require continuing update and improvement. First, efforts must be made to replace the data from FAO country profiles by data from local studies documented in the primary or report literature. Our database provides an appropriate framework for the systematic data collection required here both at local and national levels. Global estimates will be improved with the increase in data quality and quantity. Indeed, explicit consideration of uncertainty (through estimated confidence intervals, or more crudely through

Table 4. Regional breakdown of key statistics on small-scale fisheries in ML-HDI countries

Region (*n*)	Fishers/ km^2	Total fishers ('000)	Boats/ km^2	Total boats ('000)	Catch/fishers (mt/year)	Total catches (1,000 mt/year)
American/ Caribbean (25)	0.85	1,220	0.36	242	0.43	547
Africa (36)	0.93	980	0.21	187	2.21	2,040
Asia/Pacific (18)	1.35	6,060	0.57	1,490	0.72	4,360
Europe/Near East (18)	0.86	1,810	0.23	438	2.60	4,710
Total (mean)	(0.97)	10,070	(0.31)	2,357	(1.88)	11,657

ranges for the different inputs, or through a Monte-Carlo approach) will allow evaluation of the uncertainty in all estimated outputs (Figure 1).

Next, the database must be expanded to include explicitly the catches taken by women and children, which are hardly ever included in national statistics, another expression of the marginalization alluded to above. Contrary to a widespread belief, women and children in many countries do take active roles in catching fish and coastal invertebrates, rather than only in processing and marketing. Examples are reef gleaning, widespread in Southeast Asia and the Pacific (Chapman 1987), or gathering of estuarine bivalves and other invertebrates in West and East Africa (Williams 2002) and in El Savaldor (Gammage 2004). Consideration of these catches will not only add to the reported amounts, but also highlight an income source so far largely neglected in accounts of the coastal economies.

Some published social science (anthropology, sociology, economics) studies of small- scale fisheries report information on their catch composition. These data, if available for several time periods, will be useful to determine the extent of small-scale fisheries on their supporting ecosystems (e.g. by computing the mean trophic level of their catch) (Pauly et al. 1998), thus providing a basis for discussions on whether, in such cases, small is beautiful. Moreover, catch composition data can be used to infer whether the small-scale fisheries catches of a given country are included in the catch statistics it submits to FAO. Presently, many island countries with substantial small-scale fisheries exploiting near shore reefs report only tuna and or lobster catches (i.e., exported commodities), a sign that they do

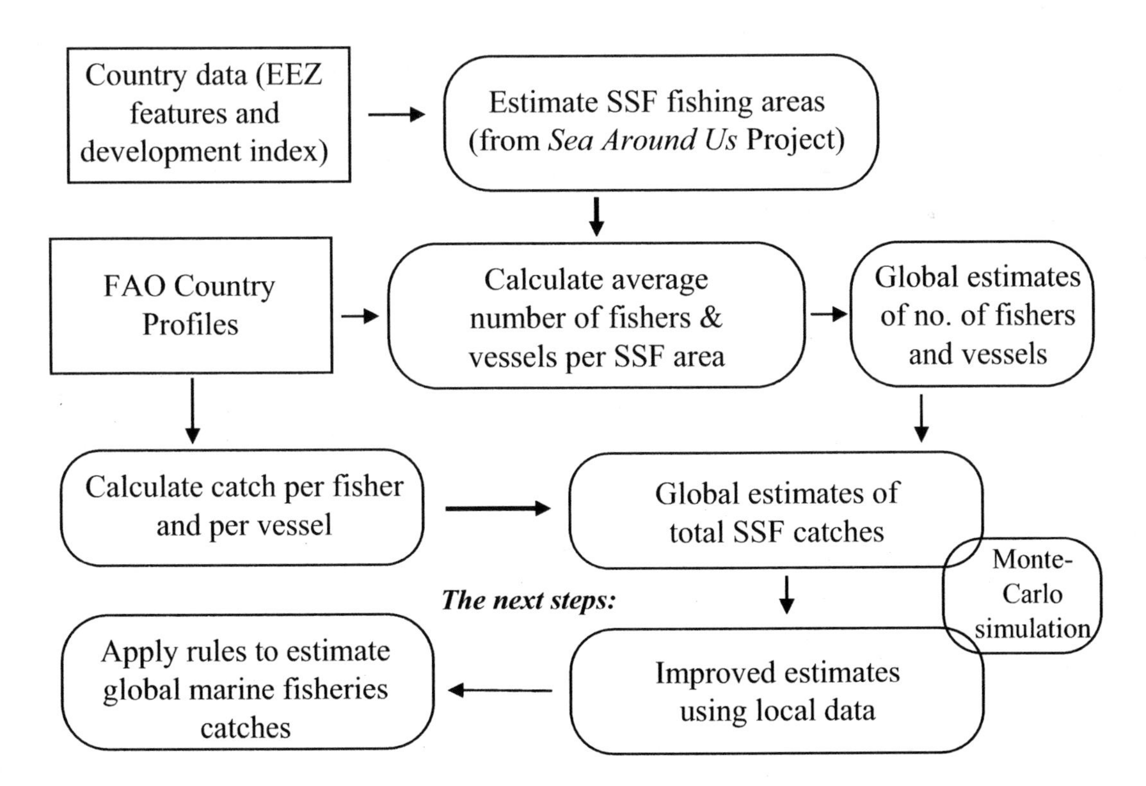

Figure 1. Schematic diagram for global estimation of small-scale fisheries.

not consider their small-scale fisheries worth attention, though it is they which feed the local populations.

Additional rules can be devised to infer whether small-scale fisheries catches are included in the FAO catches of different countries. For countries reporting "zero catches" in small-scale fisheries, our estimates should be added to FAO catches since small scale fishing occurs in all maritime countries, whether or not government officials went to the beaches to record their catches. On the other hand, in cases where small-scale fisheries catches are roughly equal to the nonidentified catches (e.g., the frequent "miscellaneous" or "other species"), then they might have been included in the FAO statistics. Further, some countries do not report catches of boats below a certain tonnage. In most cases, the entire estimate of the small-scale catch of these countries will have to be added to the FAO-based global catch estimates for marine fisheries.

Finally, estimates of value of fisheries catch should be made to suggest economic and social importance of small-scale fisheries to local communities. One possibility is to apply global average price for fish species (available at www.seaaroundus.org) to small-scale catch composition data. Alternatively, for commercial small-scale fisheries, price data can be obtained directly from field studies or inferred from prices of corresponding species in other, similar fisheries. Interviews and surveys will be required, however, to assess other values of fish, such as for household consumption and cultural and spiritual uses (e.g. in aboriginal communities).

Conclusion

The database approach presented here to obtain global estimates of sall-scale fisheries, and thus to document, rather than assert their importance, is an initiative to construct a new "mental map" for small-scale fisheries, where they are put at the front and center of research and management interests, instead of remaining at the margin. Similar to the large-scale fisheries, small-scale fisheries share features across broad regions, which render them amenable to comparative quantitative studies, and to global inferences. As the global small-scale fisheries database presented here is structurally similar to the global (large-scale) fisheries, ecosystem and biodiversity database assembled by the *Sea Around Us* Project (see www.seaaroundus.org), global comparative analyses will therefore be possible. In addition to comparing directly the contribution of small- and large-scale fisheries to the economy and food security of various countries, it will be possible to assess impacts of small-scale fishing on ecosystems and biodiversity, especially when trends can be observed. This will allow us to tell, in rigorous fashion, whether, in fisheries, small is beautiful.

Acknowledgments

We thank the Fourth World Fisheries Congress organizers for the invitation to present this paper. We acknowledge the help of Adrian Kitchingman, the *Sea Around Us* Project, University of British Columbia, for the estimates of small-scale fishing areas. For financial support, we thank the Pew Charitable Trusts of the *Sea Around Us* Project. R. Chuenpagdee also thanks Anthony Davis and Social Research for Sustainable Fisheries project and D. Pauly acknowledges funding from Canada's Natural Scientific and Engineering Research Council (NSERC).

References

Chapman, M. D. 1987. Women's fishing in Oceania. Human Ecology 15(3):267–288.

Chuenpagdee, R., L. E. Morgan, S. M. Maxwell, E. A. Norse, and D. Pauly. 2003. Shifting gears: assessing collateral impacts of fishing methods in the U.S. waters. Frontiers in Ecology and the Environment 10(1):517–524.

Gammage, S. 2004. The tattered net of statistics. Pages 36–40 *in* K. G. Kumar, editors. Gender agenda—women in fisheries: a collection of articles from *SAMUDRA* Report. ICSF (International Collective in Support of Fishworkers), Chennai, India.

Pauly, D. 1997. Small-scale fisheries in the tropics: marginality, marginalization and some implications for fisheries management. Pages 40–49 *in* E. K. Pikitch, D. D. Huppert, and M. P. Sissenwine, editors. Global trends: fisheries management. American Fisheries Society, Symposium 20, Bethesda, Maryland.

Pauly, D., V. Christensen, J. Dalsgaard, R. Froese, and F. C. Torres, Jr. 1998. Fishing down marine food webs. Science 279:860–863.

Saeger, J. 1993. The Samar Sea: a decade of devastation. NAGA, the ICLARM (International Centre for Living Aquatic Resources Management) Quarterly 16(4):4–6, Penang, Malaysia.

Sen, A. 1999. Development as freedom. Oxford University Press, Oxford, UK.

Sokal, R. R., and F. J. Rohlf. 1995. Biometry: the principles and practice of statistics in biological research. 3rd edition. Freeman, New York.

The World Bank, United Nations Development Programme, Commission of the European Communities, and Food and Agriculture Organization. 1991. Small-scale fisheries: research needs. World Bank Technical Paper Number 152. Fisheries Series.

UNEP 2000. Human Development Report 2000. United Nations Development Programme. Oxford University Press, New York.

Vakily, M. 1993. Dynamite fishing in Sierra Leone. NAGA, the ICLARM (International Centre for Living Aquatic Resources Management) Quarterly 16(4):7–9, Penang, Malaysia.

Williams, S. B. 2002. Making each and every African fisher count: women do fish. Global Symposium on Women in Fisheries. WorldFish Center, Penang, Malaysia.

American Fisheries Society Symposium 49:585–599

Handling the Legacy of the Blue Revolution in India—Social Justice and Small-Scale Fisheries in a Negative Growth Scenario

MAARTEN BAVINCK AND DEREK JOHNSON*
*University of Amsterdam, Nieuwe Prinsengracht 130
1018 VZ Amsterdam, The Netherlands*

Introduction

The blue revolution[1] represented a forceful attempt by different levels of the Indian state to modernize Indian capture fisheries from around the mid-1950s onward. It was justified with recourse to the rhetoric of development: India's abundant marine resources would be marshalled to address the protein deficiency of the masses. Moreover, the blue revolution would help to uplift the country's poor fishing population and bring them into the development trajectory.

In terms of stimulating production in Indian fisheries, blue revolution efforts were very successful. This is highlighted by production figures. In 1950–1951, marine fish production was 534,000 metric tons; by the late 1990s, it had peaked in the range of 2.8–3 million metric tons (Yadava 2001). As was the case in many other developing countries (Platteau 1989), the marine fisheries had meanwhile divided into two parts: a harbor-based trawler fishing industry and a very sizeable small-scale subsector spread out along the beaches. The conflicts between the two categories of fishers were, and are, numerous, as they tend to ply the same fishing grounds and target the same species.

Social justice and distribution issues have not been at the forefront in planning discussions with regard to capture fisheries in India, as they have been in agriculture (cf. Byres 1998). As we shall see, fisheries development planning also bore little concern for resource management until very recently when fish stocks have shown increasing signs of stress. The root cause for both phenomena is the myth of large and unexploited fish stocks. As long as development planners could adhere to the imaginary idea of ever-increasing growth in production and earnings, the disparities created by the blue revolution could be waved away as irrelevant. After all, couldn't every small-scale fisher aspire to joining the league of trawler owners?

Figure 1 demonstrates the leveling of fish catches in India since the 1990s, a trend which is considered to be related to an ecological crisis in the marine sector (Kurien and Achari 1994; Mathew 2000; Salagrama 2001). As the over-exploited condition of fish stocks has become ever clearer, with the necessary implication that absolute fishing effort has to decrease, the al-

*Corresponding author: dsjohnson@marecentre.nl

[1] We refer to what is now known as the first blue revolution in capture fisheries, the second blue revolution taking place three decades later in aquaculture (cf. Stonich and Bailey 2000).

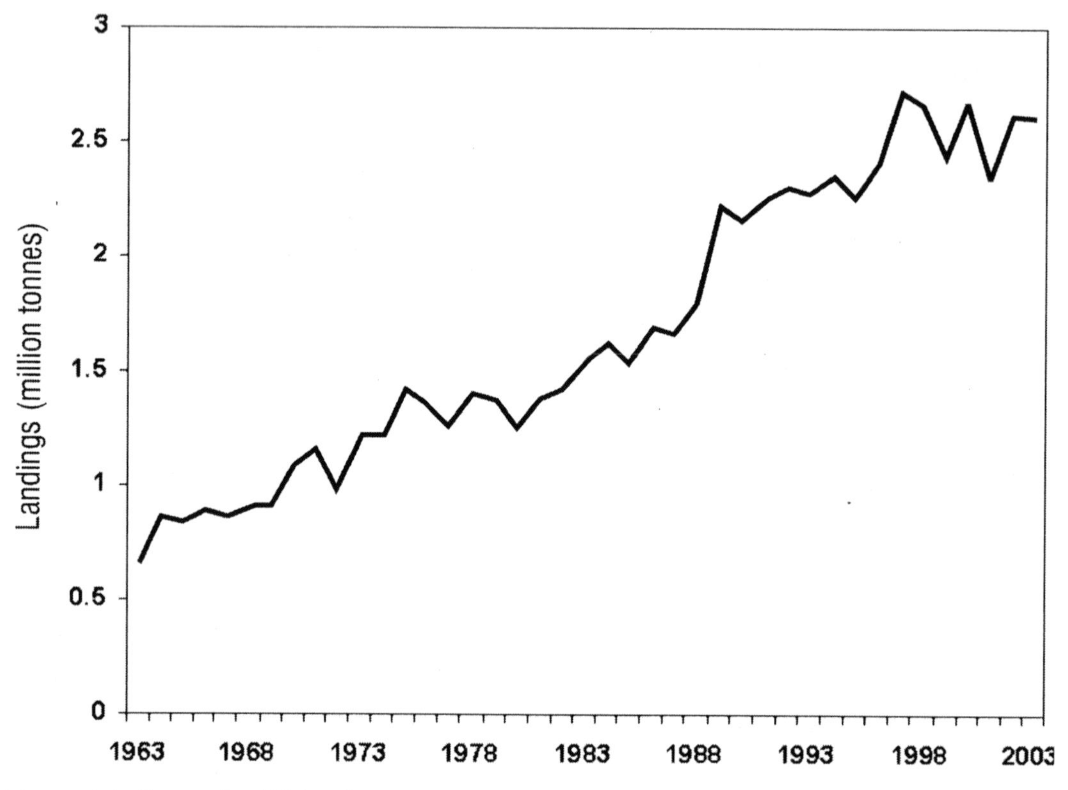

Figure 1. Indian marine fish landings, 1963–2003 (source: CMFRI 2005).

location of returns from remaining stocks becomes critical. This is an issue, nonetheless, that policymakers in India prefer to sidestep because it so pregnant with possibilities for conflict. We argue, however, that without a conscious and firm effort to incorporate well-formulated and explicit policies of distribution based on social justice, fisheries in India stand little chance of avoiding continued degradation and social upheaval.

The next section discusses the issue of social justice in fisheries from a general perspective. This is followed by an analysis of the social justice implications of the growing crisis in two major fisheries regions in India: Tamil Nadu and Gujarat. We conclude the paper with a discussion of possible policy options for fisheries governance at the state level that confront the legacy of the blue revolution head on by incorporating a strong concern for conservation and for social justice in distribution, with particular attention to the small-scale sector.

Social Justice in Fisheries

Philosophers have long debated the meaning of social justice and what principles should be applied in determining the fairness of a particular distribution of rights and responsibilities. Anthropologists have contributed to this discussion the insight that justice varies from one society, and societal position, to the next. There is thus no true or correct meaning of justice (Campbell 2001)—instead, there are many emic perspectives[2].

[2] The emicetic distinction is a classic element of anthropological theory. Emic refers to the culturally specific view, or the view from within. Etic refers to the view of the outsider, often in reference to a supposedly more objective or value neutral position (Pike 1954).

But social justice is also discussed from the viewpoint of commonly accepted—or universal—principles or criteria. International law, agreements, and declarations contain many examples hereof. For the purpose of an etic assessment of social justice, however, as Sen puts forward, one minimally requires "a working agreement on some basic matter of identifiable intense injustice or unfairness" (Sen 1999), a statement that acknowledges that there are also possible alternative etic perspectives. In fisheries, the most widely acknowledged reference point for an etic approach to social justice is the Code of Conduct for Responsible Fisheries (CCRF). The code includes several clauses that refer to social justice. The most prominent of these is Article 6.18, which, recognizing "the important contributions of artisanal and small-scale fisheries to employment, income and food security," voices a call for the protection of small-scale fishers. This appeal ties into other expressions of international law, such as those furnished by the International Labor Organization, and is of direct relevance to our topic.

Social justice is recognized to apply at various time and scale levels. One distinction is between intergenerational justice and justice among people in the present time. Concerns about ecosystem health and the sustainability of extractive practices are closely linked to notions of intergenerational justice, where development "meets the needs of the present without compromising the ability of future generations to meet their own needs" (World Commission on Environment and Development 1987; Sumaila 2003). In the present day, depending on the scale and the units of focus, analysts may discuss social justice as an issue of north and south, or, at the other end of the continuum, among individuals, households, or genders. In this paper, our concern is primarily with justice at the subsectoral level or, in other words, among categories of small-scale and semi-industrial fishers. We follow the suggestion that members of these categories may have different perspectives on what constitutes justice and injustice.

It is reasonable to suppose that perceptions of justice and injustice are linked to the character of an occupation. Capture fisheries is one of the last hunting professions and shares with other hunting societies a number of traits such as an emphasis on skill, luck, and competition (Acheson 1981; Van Ginkel 2003). The marine environment being uncertain and dangerous, and the rights to wild fish not obvious in advance, fishers display a great tolerance for difference. At the same time, ethnographies reveal an egalitarian ethos and a concern for issues of justice.

This connects in part to conceptions of fishing rights. Researchers have demonstrated that fishing societies, whether 'traditional" or newly established, design and enforce rules for fishing practice. This is logical: "resource allocation is never unstructured because continuity in the production of basic goods is never unimportant" (Benda-Beckmann 1995; following Dalton 1967). Such regulatory practices take the form of boundary and access rules (Schlager and Ostrom 1993) and also identify agencies and processes of rule making and adjudication. The conclusion is that as long as fishers operate within a common framework of rights and responsibilities, social justice issues are capable of being addressed and handled. From this perspective, problems are most likely to arise when fishers do not share the conceptions of just fishing practice (cf. Charles 2001).

In the following section, we discuss the nature of the blue revolution in two Indian states, Tamil Nadu and Gujarat. Our observations are structured by analysis of the place of intergenerational and intersector redistributive justice in the two states. What have the two states done to promote justice in these two areas, and what new policies are they proposing to address them?

Before we proceed with this task, it is important to indicate the methodological basis of our data and arguments. This paper is based on the findings of long-term ethnographic fieldwork conducted by both authors in their respective regions. Arguments, where not substantiated by secondary sources, are derived from interviews, both formal and informal, personal observations, and surveys. The number of informants consulted by each author runs into the hundreds. In Tamil Nadu, data were collected over the periods 1994–1996 and 2001–2005. In Gujarat, data were gathered in 1997–1998 (Gujarat Commissioner of Fisheries, Government of Gujarat 1998) and in 2004–2005 (Gujarat Commissioner of Fisheries, Government of Gujarat 2004). The locations of Tamil Nadu and Gujarat are indicated in Figure 2.

The Blue Revolution and the Creation of Social Injustice at the Subsector Level

Tamil Nadu

In response to a thrust by the Union Government of India (India 1951), the government of Tamil Nadu initiated in the 1950s the first phase of what was to become a blue revolution of capture fisheries in the state (Tamil Nadu 1972). The term "revolution" is actually not too grand for the effort it undertook; the supplanting of the "primitive" small-scale fisheries, by a new and modern fishing indus-

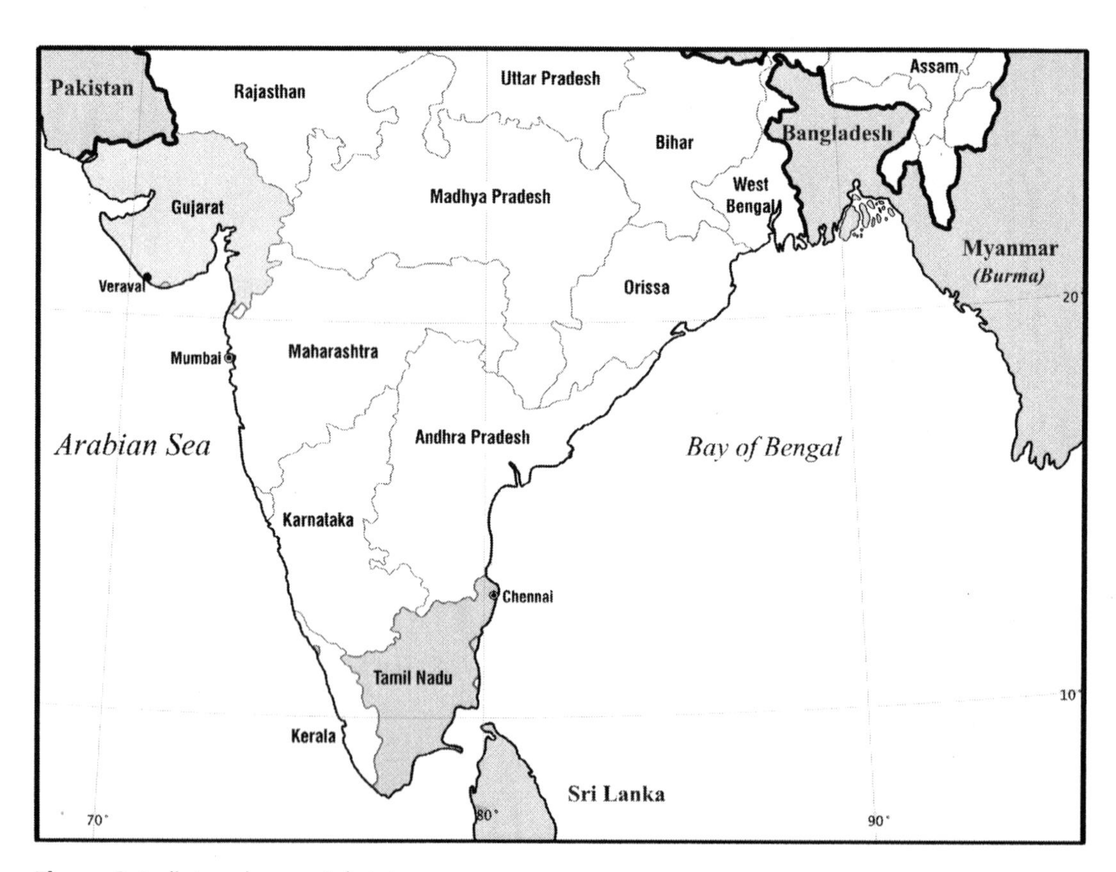

Figure 2. India's major coastal states.

try. Trawlers were the cornerstone of the revolution. In the next decades, the Fisheries Department promoted trawler fishing through massive subsidy schemes and training programmes, as well as via the construction of harbours, jetties, boat-building yards, and cooling facilities (MPEDA 1978, Bay of Bengal Programme 1983; Bavinck 2001b).

Its exertions were successful in two important respects. First of all, the program quickly caught on among fishers and nonfishers alike. The number of trawlers rapidly increased, and private entrepreneurs soon stepped into what was initially a governmental monopoly. The Fisheries Department was thus able to discontinue its programs in support of trawler fishing in the 1970s, without this affecting the further development of the industry. In 2000, the state counted upwards of 8000 small trawling vessels divided over 12 harbors and piers, spread evenly along the 1,000-km-long coast. Second, the new trawling subsector made an important contribution to the annual increase of fish harvests in the state. Table 1 provides figures on the number of small-scale and trawler craft operating in Tamil Nadu in the period of 1948–2000, as well as the growth of marine fish production.

Three outcomes surprised planners of the first hour, however (Bavinck 2001b:65, 67). The first was that trawlers did not come to supplant small-scale fishing in the state. To the contrary, as Table 1 also demonstrates, the small-scale subsector in fact has continued to grow. It predominates in the 591 fishing villages that dot the state's coastline, altogether feeding and employing approximately 600,000 people (Tamil Nadu 2000). The persistence of small-scale fishing does not mean that the technology on which it relies has stagnated. Largely ignored in the early phase of the blue revolution, small-scale fisheries underwent change through the introduction—from public and private sources—of nylon twines and new kinds of netting and, much later, of outboard engines and new vessel designs (Bavinck 1997). The natural increase of the fishing population contributed to an absolute increase of fishing effort also in this sector.

The second surprise was that the new trawler fishers did not proceed to underexploited offshore and deep sea areas, but instead concentrated on the same inshore fishing grounds as small-scale fishers. With the advantage of hindsight, it is clear that this choice was rational in view of the concentration of fish stocks, as well as the experience and expertise of the people involved. It resulted, however, in massive competition between the modern and the small-scale sectors, as well as in the large-scale destruction of gear.

Third, the supposed infinite fish stocks proved not to be as limitless as expected, with catches

Table 1. Craft and production figures from Tamil Nadu (1948 and 2000).

Year	Small-scale vessels	Trawler vessels	Trawler catches (metric tons)	Total fish catch (metric tons)
1948	13,204	0	0	27,135
2000	41,770	8,009	200,468	377,483

leveling off in the 1990s (cf. Tamil Nadu 2004), and catches per unit of effort declining (Sanpath 2003:103). Stakeholders in the fisheries now agree that overfishing is taking place in the Tamil Nadu inshore zone, and that there is urgent need for a limitation on growth, if not a real reduction of fishing effort (Sridhar 2005; fieldwork notes). This brings us to the core of this paper, the investigation of social justice.

From an emic perspective, the blue revolution in Tamil Nadu had important social justice effects. Small-scale fishers are, according to their own perception, the major victims. Along this coastline, fishing is an old activity, resulting also in the development of special fishing castes. Left alone by government until the launch of the blue revolution, fishing communities themselves have taken charge of regulation. This resulted in a chain of preferential fishing areas, covering most of the inshore waters of the state, and a practice of banning, or regulating, harmful fishing technologies. Each fishing hamlet has its own council that decides on fishing matters and resolves disputes (Bavinck 2001a, 2001b).

The introduction of trawling threw these fishing communities into turmoil. Here was a new category of fishers that fish at will, paying no attention to the prerogatives of hamlet councils or to the rights and responsibilities prevalent in small-scale fishing. According to local perception, they behave like bandits, using their superior engine power to prowl the inshore seas, damaging gears, appropriating high-value shrimp, and disappearing into the night. This is experienced, to quote Sen (1999), as acts of "intense injustice," and continuously protested as such. Such objections are voiced in court, via demonstrations and petitions, and frequently also via direct action, for example through the hijacking and ransoming of trawler vessels (Sundar 1999; Bavinck 2003).

For trawler fishers, the world of justice is constituted differently. They and their professional associations tend to remember that the Indian Constitution provides every citizen the right to work where and how he or she pleases and that it was the Indian government that in fact urged their conversion to trawler fishing. They point out their economic contribution, the many difficulties of their trade, and their continuing efforts to settle the problems that arise with their compeers, the small-scale fishers. Finally, they note the many ways in which the government has been making their lives difficult: first of all through the Tamil Nadu Marine Fishing Regulation Act (1981), which designates the inner zone of three nautical miles out of bounds for trawlers, even though it is barely enforced. More recently, they bemoan the annual monsoon ban of 6 weeks that the government has implemented for trawler fishing, presumably for conservation purposes. From a social justice standpoint, they therefore emphasize that the trawler fishing subsector has also been short-changed (Bavinck 2001b).

And what would an etic view of social justice in Tamil Nadu's marine fisheries look like? It might stress the employment and income-generating potential of small-scale fisheries in a societal context of poverty and job scarcity. It could highlight the historical rights of marine fishing castes to the occupation and the natural resources on which it depends. Finally, it could examine the extent to which small-scale fishers were and are capable of harvesting the resources presently captured by trawler fishers. While supporting the position of small-scale fisheries, an etic perspective could also, however, emphasize the rights of trawler fishers to fair settlements.

The government of Tamil Nadu has wrestled with the heritage of the blue revolution ever since its real contours emerged. In the period between 1970 and 2000, its main concern has been with defusing the conflicts that occurred between the small-scale and trawler subsectors. This was the main intention of the Marine Fishing Regulation Act. As this act failed, for various reasons, to fulfill its purpose, the authorities have instead built on agreements between the conflicting parties to establish very local, temporary settlements (Bavinck 1998, 2003).

Recently, however, conservation has become a point on the governmental agenda. It figures most clearly in what has become known throughout India as the monsoon ban, a closed season in Tamil Nadu of 6 weeks for trawlers, which is meant to enhance the recruitment of fish stocks. There is evidence (Haastrecht and Schaap 2003) that this measure is, however, disputed.

Gujarat

The history of fisheries development in Gujarat has been marked by the state's sociocultural distinctiveness. While fishing in many parts of the world is socially marginal, this has been particularly true of Gujarat where a devout Hindu vegetarian population has dominated the state. As a consequence, the hinterland of Gujarat's coasts did not historically provide a significant market for fish products, which meant that the fishery was comparatively small considering its large potential and already export oriented (Bombay State 1958; Johnson 2002). This distinctive cultural context has had two important implications for the development of the Gujarat fishery. First, as fishing became more lucrative, Gujarat's fishers faced little competition from other Gujaratis interested in fishing. Second, the small size of the fishery and the region's abundant resources meant that, in comparison to other states, indigenous institutions for resource management and allocation were few. For such a religious state, it is unsurprising that Hinduism provided the exceptions: certain religious days of the month were prohibited for fishing, as was the sacred month of Sravan at the end of the monsoon.

Gujarat's fishery has changed dramatically since the 1950s. In the early years of that decade, the most lucrative economic sector on the coast of present day Gujarat was shipping, the fishery was unmechanized, total catch was at most 5% of current levels (around 22,000 metric tons), and the fish trade was entirely domestic and focused on low-value dried fish (see Johnson 2002). Large-scale change was in the offing, however, as central government development policy emphasized the high potential for growth in food production in agriculture and fisheries to support India's modernization (India 1951). As the first Indian five year plan goes on to state, this was to be growth with equity. In addition to technical development of fisheries and processing, a vibrant cooperative sector was to be established to ensure that the benefits of development were distributed fairly.

The heyday of state development planning for fisheries in Gujarat lasted from the mid-1950s to around 1970. In the first years of intervention, local state agencies were faithful to the broad vision of growth with equity. The introduction of motorization and modern gears proceeded alongside the establishment of cooperatives (Bombay State 1958, 1962; Gujarat Department of Fisheries 1963). The balance shifted rapidly in favor of growth in the1960s, however, as the considerable export potential of prawns became apparent. Cooperatives failed to take root except as vehicles through which the state was able to channel a variety of subsidies and, early on, engage in fish marketing (Gujarat Department of Fisheries 1963; Johnson 2002).

The key parameters of growth-oriented intervention by the state in fisheries were the provision of subsidies, technological extension, and training for the adoption of motorization, new boat designs, and new materials for gear, particularly synthetic nets. The first target of these interventions from the mid-1950s to the early 1960s was the existing small-boat sector. It was only after 1962 that the push began to promote trawling and trawler fishing did not become accepted by local fishers until the late 1960s (Johnson 2002). The Gujarat Department of Fisheries continued to encourage mechanization and innovation in the small-scale sector as well as the trawler sector, likely because the largest population of fishers in the state lived in coastal villages where berthing trawlers is impossible.

Fisheries development through the late 1990s seemed a tremendous success for both small scale fishers and trawler fishers. By the mid-1970s, a strong, primarily urban, and harbor-based trawler sector had emerged, and initiative for fisheries development passed to the fishers from the state. Under fisher leadership, with strong state support, the Gujarat fishery boomed from the late 1980s through to the 1997–1998 season. Catches peaked in that year, reaching the level of 702,355 metric tons and exports of fish products grew in value to US$165 million[3] (Gujarat Commissioner of Fisheries, Government of Gujarat 2004). A sense of the magnitude of the growth in the Gujarat fishery can be seen in Table 2.

As with Tamil Nadu, however, the small scale sector has had a dynamic development history as well since the 1960s. It is spread among the 188 fishing villages and harbors along the Gujarat coast and likely accounts for the majority of the fishing population of about 300,000 in 1998[4] (Gujarat Commissioner of Fisheries, Government of Gujarat). Technological evolution has been substantial, with a complete revolution in craft and gear types, patterns of fishing, and species targeted (Johnson 2002). Unlike Tamil Nadu, Gujarat small-scale fishers were early adopters, from the early 1950s, of motorization. They continue to contribute a significant proportion of total catch, 31% in 1998–1999, as Table 2 shows.

[3] This figure is calculated from the export value from Gujarat of 6,378,500,000 rupees given in Gujarat (Gujarat: 57) using an exchange rates of 38.61, which is an average of rates taken at two monthly intervals throughout the main fishing season of September 1997 to mid-May 1998.

[4] The population figure is a rough estimate based on Gujarat Department of Fisheries data from its 1998 census (Gujarat 2000), which do not divide the marine fishing population into small scale and large scale.

Table 2. Craft and production figures, Gujarat, 1960–1961 and 1998–1999 (Gujarat Commissionerate of Fisheries).

Year	Small-scale vessels	Trawler vessels	Trawler catches (metric tons)	Total fish catch (metric tons)
1960–1961	3,531	0	0	79,412
1998–1999	15,199[a]	6,749	380,645	551,660

[a] This figure includes nonmechanised craft (9,222) and outboard motor powered canoes (6,242) but excludes the inboard motorboat sector using gill nets and bag nets in which there are 4,117 craft.

The boom ended abruptly in the 1998–1999 season with a collapse in prices and a 21.6% decline in catch. The ensuing years have been grim: total catches have stagnated after four decades of steady increase, export revenues have declined, there is ample evidence of resource degradation (Mathew 2000), and most fishers are facing a serious financial crunch. The crisis has been a particularly hard for the trawler sector, which has suffered an abrupt reversal of fortunes. Many trawler owners have been forced to reduce their fishing time to a few months of peak season and have been driven to sell off their assets to meet their costs. The economic trigger for the crisis was the Asian economic flu of the late 1990s, which drove prices down as demand evaporated. Demand and prices have since recovered, but rising input costs, particularly of diesel, have continued to press fishers financially. The ecological side of the crisis had deeper roots and warning signs that are now evident in retrospect. Among these was the overfishing of high value species, such as penaeid prawns and pomfret, by the early 1980s (Johnson 2001; Kizhakudan and Thumber 2003; Kizhakudan et al. 2003; Nair et al. 2003).

The judgment by small-scale fishers of the social justice of the current situation in Gujarat is likely to vary regionally[5]. In South Gujarat and the Gulf of Kachchh regions, small-scale fishers are facing the loss of their livelihoods due to competition for coastal space and resources with industrial and agricultural development, which, incidentally, pose a significant threat to coastal ecology (Hiraway 2000; Kizhakudan et al. 2003). In the ocean-facing area of peninsular Gujarat, small-scale fishers are more concerned about overfishing and point to trawlers as being the principal culprits. As in Tamil Nadu, small-scale fishers and trawler fisher operations overlap in coastal waters of up to about 40 fathoms in depth. Small-scale fishers are resentful of damage to their gear caused by trawling and the dangers that trawlers pose to small craft. The frustrations of the small-scale sector occasionally manifest themselves in incidents of boarding, violence, and boat capture directed towards offending trawler vessels, although this has been a less common and less organized phenomenon than in South Indian waters (cf. Kurien 1991; Bavinck 2001b).

Many boat owners in the trawler sector have been stunned by the reversal of their fortunes since the fishing crisis began in 1997–1998. The crisis has been particularly hard on those with small numbers of boats, old boats, or those with low reserves of capital. The rapidity with which the economic crisis has undermined their financial security has caused many to wish that they could leave the sector. Indeed, evidence from a number of interviews conducted in 2005 indicates that a process of consolidation has been taking place in the sector, with fish traders buying the boats of their insolvent trawler-owning creditors. Leaders of the trawler sector are now talking about measures for conservation and have accepted the 2004 government ban on the construction of new vessels. They are clear, however, that the small-scale sector has to pull its weight also, and stop fishing in the monsoon, a practice that has increased in recent years. As yet, the only sense of injustice voiced by trawler fishers is against the licenses granted by the Indian government to industrial fishing boats for operation in Gujarat's waters.

[5] The basis for this paper is research conducted along the most important fishing area of coastal Gujarat from Navabandar to Okha. The authors have only indirect reports of conditions in other areas.

[6] In the two major fishing ports of peninsular Gujarat, large numbers of migrant crew for trawler are hired from South Gujarat and other Indian states, especially Andhra Pradesh.

Major tensions over social justice between the sectors have not yet emerged in the Gujarat fishery. This may reflect the common experience of generalized development and growth and the now generalized depression that all sectors have faced. The continental shelf in Gujarat is also the widest in India, which reduces intersectoral conflicts over space. Finally, Gujarat's fishers and the fisher owners of trawler vessels[6] are linked through economic and caste relationships, which may defuse conflicts.

By supporting growth in all motorized sectors, the State Department of Fisheries implicitly pursued a strategy based on equality of opportunity. It had the good fortune to implement its policies in a context of an underexploited fishery with abundant resources, which brought economic success across the board, with low levels of conflict, even if the trawler sector got the largest share of the resource. The crisis since 1998 has shaken the Department of Fisheries into a more proactive role, a major success of which has been to finally get the Gujarat Government to pass the Gujarat Fisheries Act in 2003. The act includes explicitly conservationist and redistributive provisions. First, it prohibits the catching, processing, and sale of juvenile and "under-sized" fish and lists a number of protected species. Second, it prohibits bottom trawling within a 9-km zone from shore and in Chapter III, regulation 21, it calls upon fisheries officers to "protect the interest of traditional fishermen such as country crafts or canoes." These provisions could make the fishery more sustainable and equitable. They would necessitate a good deal of adjustment and financial hardship for the dominant class of trawler owners, however, as they stand the most to lose in such a re-equilibrating of the fishery. As this group is so powerful in the main fishing zones of Gujarat, they are unlikely to abide by the provisions of the act unless the state is able to convince them of its necessity and its utility for them.

Despite the very different contexts, an etic perspective on justice in the Gujarat fishery could come to similar conclusions as those for Tamil Nadu about the preferability of strengthening the small-scale sector at the expense of the trawler sector. The same caveat applies, however, that the trawler fishers will need fair settlements. It is important also that destructive gears in the small-scale sector be eliminated and efforts continue to be made to build their capacities to diversify their livelihoods away from dependence on fishing.

Comparison

The fisheries of Tamil Nadu and Gujarat are in similar positions at present, both are facing stagnation and possible decline due to overfishing and other threats to their marine ecology. At the heart of their difficulties is the legacy of India's blue revolution; the promotion of growth-oriented fisheries development policies with relatively little attention to issues of conservation and distribution. Primary in both cases was the promotion of export-oriented trawler fisheries, which have come to dominate fisheries production in the two states. At the same time, the pattern of development in the two states has differed in ways that could become significant for future ecological conditions and intersectoral relations in each state.

The first major difference between the two states is the degree to which fishing has been integral to social, economic, or even mental space. While in both states, fishers have low social status; in Gujarat, fishing was historically marginal to the point that the fishery had a domestic export orientation even prior to modernization. This meant that fishing in Tamil Nadu in the 1950s was relatively more economically and institutionally developed than in Gujarat. Most notably, the small-scale fishing sector had developed a system for management and resource allocation. The historical marginality of the Gujarat fishery has left a relative absence of such institutions but also a freedom from having to cope with outsiders moving into the fishery. Gujarat's fishery has also not faced the intensity of intersectoral conflict over allocation rights as in Tamil Nadu. The second major difference between the development histories of Tamil Nadu and Gujarat has been the role of the state fisheries departments. The Tamil Nadu department has had to focus its energies on intersectoral conflict management, an issue that has been much less present for the Gujarat department. Since the end of its leadership position as fisheries developer in the 1970s, the Gujarat Department of Fisheries has acted primarily as provider of subsidies, issuer of registrations, and collector of data. There is a parallel between the institutional histories of the departments in the two states, however, in that both settled into relatively passive roles in relationship to their fisheries after initially strongly interventionist periods.

Fisheries in Tamil Nadu and Gujarat are now both facing the stark reality of stagnating catches, few new possibilities for expansion, and rapidly growing coastal populations. For the two states, this is a key period of fisheries transition during which hard choices will have to be made in order to reconcile conservation with employment and livelihood needs. Each state has blue revolution legacies and particular, institutional, sociocultural, political, and economic realities that shape their room for action. In both states, fishers and their organizations have power and legitimacy that state organizations lack. At the same time, however, fisher populations are fragmented along lines of caste, class, sector, and religion, and their position has been shaken by the current ecological crisis. Should the state be able to rise to the challenge, these conditions present an opportunity for charting a new course in fisheries management in Tamil Nadu and Gujarat. In the following section, we speculate on possible futures for these fisheries, the more optimistic of which envisage the state, the fishers, or the two together taking a proactive stance.

Evaluation

Policy Alternatives

1. Status quo scenario
All parties stick their heads in the sand and wait and see what happens. Price increases might cushion the effects of increasing environmental problems or a process of silent emigration from the sector might reduce fishing effort sufficiently. Alternatively, there may be severe civil unrest that will have to be suppressed. The motto, however, is "problems frequently solve themselves. In any case, we hope they will not arise in our period of tenure."

2. Technocratic scenario
Efforts are made to address problems of resource overexploitation and livelihood erosion within existing policy parameters. Thus, government sidesteps issues of participation and fairness, and formulates a set of limited measures, such as the implementation of closed seasons. Trawler and small-scale fisher organizations may take steps, such as the establishment of artificial reefs, or agree not to use gear types that are recognized as particularly deleterious.

3. Radical scenario 1
The government decides to abolish trawler fishing and obtains subsidies from international organizations to buy out existing trawler fishermen.

4. Radical scenario 2
Trawler and small-scale fisher organizations realize the gravity of the situation and frame a joint action plan that is also aimed at "waking up" the government to its responsibilities. This plan proposes limitations on new entrants to fisheries, market regulation, resource monitoring, and government support for the growing class of jobless fishermen.

5. Radical scenario 3
Separately, or as a result of scenario three or four, a comprehensive comanagement process (cf. Wilson et al. 2003) is established to reduce fishing effort and engage the contentious issue of sectoral resource redistribution. Fisher organizations are formally recognized and involved in the policy process.

Each of these scenarios has implications for intersectoral social justice. Scenario one does nothing to redress current patterns, with the likelihood that inequality and social conflict will increase. If effectively implemented, scenario two could extend the life of the current patterns of social and economic organization of fishing. In the longer term, however, it is unlikely to counter current trends to overfishing and distributional inequity. From an etic perspective, the third scenario is most promising in terms of ecological sustainability and distribution of the resource among the largest number of fishers. Scenario four would not necessarily result in greater distributional justice within the fishery, as that outcome would depend on which group was to lead the formulation of the action plan. Scenario five also holds no guarantees in increasing social justice, although its collaborative potential at least holds out the possibility of a more inclusive process.

In both Tamil Nadu and Gujarat, evidence would lead us to argue that the scenario most likely to inform policy making in the near future is the technocratic one. State departments of fisheries do seem to be moving beyond scenario one, even if many members of the departments still would prefer to preserve the institutional status quo. Scenarios three and four are both unrealistic under current conditions. Trawler owners in both states have too much power for an outright ban on trawling while numerous and entrenched divisions between fisher groups make highly unlikely their collaborating to develop an action plan. Implementation of the fifth scenario in any well-developed way seems unlikely. The Tamil Nadu government is not at all in favor of devolution and there is little to indicate a more favorable policy in Gujarat. On an informal level, however, consultations within a framework of, for example, responsible fisheries might be held as a way of opening a dialogue on building more innovative management processes. A more inclusive process would also have to face the challenge of engaging the deep-seated social divisions mentioned above.

In Gujarat, the most compelling evidence of an acknowledgment of the need to change is the new fisheries act. It recognizes that there are limits to fishing and contains some promising elements from a social justice perspective. Nonetheless, the main regulatory thrust of the document places it within the technocratic scenario. It advocates a host of technical restrictions on fishing such as mesh size limitations, the banning of fishing with electricity and explosives, licensing, and area restrictions, and lays the foundation for a possible future quota management system. The degree to which these regulatory elements will be implemented remains to be seen and depends a great deal on their acceptance by fishers. The Fisheries Department itself is insufficiently staffed and motivated to push through the changes on its own.

In Tamil Nadu, representatives of the Fisheries Department are now orally acknowledging that there are major environmental problems to be addressed. There is some movement towards a technocratic scenario, trying to make changes by introducing a closed season for trawling; promoting the shift of trawlers to longlining; targeting new species; prohibiting specific extremely deleterious gears such as purse seines and pair trawling; declaring a marine park in Ramnad district mainly for biodiversity reasons; and, together with trawler owner associations, limiting the number of new trawlers. There are also some smaller programs for awareness building on the need of small fishers to move to other occupations. All this in the context of a department that is suffering cuts in budgets and has lost a lot of its earlier luster and motivation. It is not yet poised for a larger and more substantial policy shift.

Conclusion

The long-term legacy of the blue revolution in Tamil Nadu and Gujarat has now become evident with serious problems of overfishing due in large measure to powerful trawler sectors in both states. An ecologically sustainable future for these fisheries that also meets distributional equity criteria justice would require reversing the trend to ever-intensified production that the blue revolution has triggered. In the final section of our paper, we have spelled out a number of "radical" scenarios that might form the basis for such a policy reversal. For the reasons we enumerated, none of these radical proposals are likely to be implemented. They are useful thought exercises, nonetheless, as in pushing the bounds of the possible, they may create more room for action in practice.

Realistically, the political, economic, and social constraints on radical state and fisher action mean that, at least under current conditions, technocratic solutions are likely to dominate fisheries policy in Tamil Nadu and Gujarat. It should be remembered that these are an improvement over past regimes premised on limitless growth and they should be supported to the degree that they effectively relieve ecological pressure while supporting livelihoods. At the same time the clear limits of technocratic policies for addressing the causes of overfishing and social injustice provide a strong rationale for intensified efforts by researchers, fisher leaders, nongovernmental organization workers, and concerned government officials to press for bolder actions to reconcile Indian fisheries with conservation and social justice.

Acknowledgements

The authors would like to acknowledge the support of the Social Sciences and Humanities Research Council of Canada, the International Development Research Centre, and the Indo-Dutch Program on Alternatives in Development. Thanks also are due to two anonymous reviewers of the paper who provided helpful comments.

References

Acheson, J. M. 1981. Anthropology of fishing. Annual Review of Anthropology 10:275–316.

Bavinck, M. 1997. Changing balance of power at sea. Motorisation of artisanal fishing craft. Economic and Political Weekly 32:198–200, Mumbai, India.

Bavinck, M. 1998. "A matter of maintaining the peace", state accommodation to subordinate legal systems: the case of fisheries along the Coromandel coast of Tamil Nadu, India. Journal of Legal Pluralism 40:151–170, University of Birmingham, School of Law, Birmingham, UK.

Bavinck, M. 2001a. Caste panchayats and the regulation of common pool resource usage in India: fisheries along Tamil Nadu's Coromandel coast. Economic and Political Weekly 36:1088–1094, Mumbai, India.

Bavinck, M. 2001b. Marine resource management: conflict and regulation in the fisheries of the Coromandel coast. Sage Publications, New Delhi.

Bavinck, M. 2003. The spatially splintered state: myths

and realities in the regulation of marine fisheries in Tamil Tadu, India. Development and Change 34:633–658, Blackwell Publishing, Oxford, UK.

Bay of Bengal Programme. 1983. Marine small-scale fisheries of Tamil Nadu: a general description. Bay of Bengal Programme, Madras, India.

Bombay State. 1958. Annual report of the department of fisheries Bombay state for the year 1956–57. Department of Fisheries Bombay State, Nagpur, India.

Bombay State. 1962. Department of fisheries annual report 1959–1960 part ii Gujarat region. Bombay State Department of Fisheries, Bombay, India.

Byres, T. J., editor. 1998. The Indian economy: major debates since independence. Oxford University Press, Delhi, India..

Campbell, T. 2001. Justice. St. Martin's Press, New York.

Charles, A. 2001. Sustainable fishery systems. Blackwell Scientific Publications, Oxford, UK.

CMFRI (Central Marine Fisheries Research Institute). Production trend: all India marine fish landings. 2005. Available: http://www.cmfri.com/cmfri_trend.html. (April 2005).

Dalton, G. 1967. Traditional production in primitive African economies. Pages 61–80 *in* G. Dalton, editor. Tribal and peasant economies, readings in economic anthropology. University of Texas Press, Austin.

Gujarat Department of Fisheries. 1963. Annual administration report of the department of fisheries (1-7-1961 to 30-6-1962). Ahmedabad, India.

Gujarat Commissionerate of Fisheries. 1998. Present status of the fisheries sector in Gujarat state [western India]. Gandhinagar, India.

Gujarat Commissioner of Fisheries, Government of Gujarat. 2004. Gujarat fisheries statistics 2002–2003. Gandhinagar, India.

Hiraway, I. 2000. Dynamics of development in Gujarat. Economic and Political Weekly August 26-September 2:3106–3120, Mumbai, India.

India. 1951. "Food policy for the plan", first five year plan. Planning Commission, Government of India, Delhi.

Johnson, D. 2001. Wealth and waste: Contrasting legacies of fisheries development in Gujarat since 1950s. Economic and Political Weekly 36:1095–1102, Mumbai, India.

Johnson, D. 2002. Emptying the sea of wealth: globalisation and the Gujarat fishery, 1950 to 1999. Doctoral dissertation. University of Guelph, Ontario.

Kizhakudan, J. K., and B. P. Thumber. 2003. Fishery of marine crustaceans in Gujarat. Pages 47–56 *in* M. R. Boopendranath, R. Badonia, T. V. Sankar, P. Pravin, and S. N. Thomas, editors. Sustainable fisheries development: focus on Gujarat. Society of Fisheries Technologists, Cochin, India.

Kizhakudan, S. J., J. K. Kizhakudan, H. M. Bhint, D. T. Vaghela, and K. N. Fofandi. 2003a. Investigations on the creeks of Saurashtra. Pages 87–92 *in* M. R. Boopendranath, R. Badonia, T. V. Sankar, P. Pravin, and S. N. Thomas, editors. Sustainable fisheries development: focus on Gujarat. Society of Fisheries Technologists, Cochin, India.

Kizhakudan, S. J., K. V. S. Nair, and M. S. Zala. 2003b. Demersal finfish resources of Gujarat. Pages 57–66 *in* M. R. Boopendranath, R. Badonia, T. V. Sankar, P. Pravin, and S. N. Thomas, editors. Sustainable fisheries development: focus on Gujarat. Society of Fisheries Technologists, Cochin, India.

Kurien, J. 1991. Ruining the commons and responses of the commoners: coastal overfishing and fishermen's actions in Kerala state, India. United Nations Research Institute for Social Development, Geneva, Switzerland.

Kurien, J. and T. R. T. Achari. 1994. Overfishing the coastal commons: causes and consequences. Pages 218–244 *in* R. Guha, editor. Social ecology. Oxford University Press, Delhi, India.

Mathew, S. 2000. Gujarat fisheries: time to move from exploitative to conservation and management regime. Workshop on Current Situation in Fisheries Sector in Gujarat, Gujarat Institute of Development Research, Ahmedabad, India.

MPEDA (Marine Products Development Authority). 1978. Export potential survey of marine products in Tamil Nadu. MPEDA, Cochin, India.

Nair, K. V. S., J. K. Kizahakudan, and S. J. Kizhakudan. 2003. Marine fisheries in Gujarat-an overview. Pages 1–10 *in* M. R. Boopendranath, R. Badonia, T. V. Sankar, P. Pravin, and S. N. Thomas, editors. Sustainable fisheries development: focus on Gujarat. Society of Fisheries Technologists, Cochin, India.

Pike, K. L. 1954. Language in relation to a unified theory of the structure of human behavior. Summer Institute of Linguistics, Glendale, California.

Platteau, J. P. 1989. The dynamics of fisheries development in developing countries: a general overview. Development and Change 20:565–597, Blackwell Publishing, Oxford, UK.

Salagrama, V. 2001. Coastal area degradation on the east coast of India: impact on fishworkers. Paper presented to the conference Forging Unity: Coastal Communties and the Indian Ocean's Future, Chennai, 9–13 October 2001.

Sanpath, V. 2003. India national report on the status and development potential of the coastal and marine environment of the east coast of India and its living resources. GF/PDF Block, B Phase of FAO/BOBLME Programme. Food and Agriculture Organization of the United Nations, Rome.

Schlager, E., and E. Ostrom. 1993. Property-rights regimes and coastal fisheries: an empirical analysis. Pages 13–41 *in* T. L. Anderson and R. T. Simmons, editors. The political economy of customs and culture: informal solutions to the commons problem. Rowman & Littlefield, Lanham, Maryland.

Sen, A. 1999. Development as freedom. Oxford University Press, Oxford, UK.

Sridhar, V. 2005. 'A good occasion for change': interview with Prof. John Kurien, fisheries economy expert. Frontline 22(3):18–20, Chennai, India.

Stonich, S. C., and C. Bailey. 2000. Resisting the blue revolution: contending coalitions surrounding industrial shrimp farming. Human Organization 59:23–36.

Sundar, A. 1999. Sea changes: organizing around the fishery in a South Indian community. Pages 79–114 *in* J. Barker, editor. Street-level democracy: political settings at the margins of global power. Kumarian Press, Toronto.

Sumaila, U. R. 2003. Three essays on the economics of fishing. Fisheries Centre, Vancouver.

Tamil Nadu. 2004. Policy note 2004–2005. Animal Husbandry and Fisheries Department, Chennai, India.

Tamil Nadu. 2003. Statistical Handbook 2003. Department of Economics and Statistics, Chennai, India.

Tamil Nadu. 2000. Tamil Nadu marine fisherfolk census year 2000. Department of Fisheries, Chennai, India.

Tamil Nadu. 1972. Towards a blue revolution: Report of the task force on fisheries 1972–1984. Plan document 5. State Planning Commission, Madras, India.

Van Ginkel, R. 2003. Whatever happened to maritime anthropology: the European experience. Proceedings of the MARE Conference, People and the sea: threats, conflicts and opportunities, August 30-September 1, 2003, Amsterdam.

Van Haastrecht, E., and M. Schaap. 2003. A critical look at fisheries management practices: the 45-day ban on trawling in Tuticorin District, Tamil Nadu, India. Master's thesis. University of Amsterdam, Amsterdam.

Von Benda-Beckmann, F. 1995. Property rights and common resources. Paper read at Conference on Agrarian Questions, May 22–24, Wageningen, The Netherlands.

Wilson, D. C., J. R. Nielsen, and P. Degnbol, editors. 2003. The fisheries co-management experience. Accomplishments, challenges and prospects. Kluwer Academic Publishers, Dordrecht, The Netherlands.

World Commission on Environment and Development. 1987. Our common future. Oxford University Press, Oxford, UK.

Yadava, Y. S. 2001. Report of the national workshop on the code of conduct for responsible fisheries. Bay of Bengal Programme, Chennai, India.

American Fisheries Society Symposium 49:601–608

Reconciling Ghanaian Fisheries with Conservation by Minimizing Impacts of Continuous Overfishing in the Country's Waters through Science-Based Participatory Management

FRANCIS K. E. NUNOO* AND AYAA K. ARMAH

University of Ghana, Department of Oceanography and Fisheries
Post Office Box LG 99, Legon, Ghana

Abstract.—The increasing demand for fish and fish products to serve both the local and export markets presents a challenge to both fisheries science and management of fish stocks, and thereby make it really difficult to reconcile fisheries with the conservation of living marine resources. Fish stocks off the coast of Ghana in West Africa are continuously being overexploited with impacts evidenced by massive dwindling of the fish stocks and average fish size. The observed decline in both pelagic and demersal fish resources in Ghana's coastal waters, especially in the past decade, is akin to what has happened to cod stocks off Newfoundland. The situation is illustrated using case histories of four major contributing factors: bycatch and discards from shrimp trawl fishery; activities of local and foreign trawlers on demersal stocks; beach seine fishing in nearshore areas; and increasing export of raw and processed fish. A paradigm shift to science-based management with increased participation by fishers, scientists, and policy-makers, is advocated as an extension to Ghana's new Fishery Act 625. This will provide a positive structural reorganization of the fishery sector to ensure that fisheries in Ghana are reconciled with conservation.

Introduction

Ghana is endowed with rich natural marine and inland fish resources. Quite like many African and Asian nations, the fisheries sector in Ghana contributes significantly to national economic development goals related to food security, employment, poverty reduction, gross domestic product (GDP), and foreign exchange earnings (Directorate of Fisheries 2003). Fish and fish products are important to Ghanaians contributing 60% of dietary animal protein (Aggrey-Fynn 2001). Fish is a cheap source of high quality protein that is available in a variety of acceptable forms, and can be procured in convenient quantities throughout the year in all parts of the country. Though per capita consumption of fish in Ghana, by both rural poor and urban rich, has decreased from 27.6 kg in 1980 to 23.2 kg in 2003, it is still among the highest in West Africa. There is a high demand for fish and fish products to an annual estimated average of about 700,000 metric tons (mt). The high demand for fish is fueled by a relatively high national average population growth rate of 3% (Ghana Statistical Service 2002). However, the past decade recorded an average annual production of 335,000 mt with wide annual and seasonal fluctuations. This implies

*Corresponding author: fkenunoo@ug.edu.gh

an annual fish deficit averaging 265,000 mt, with the shortfall made up by imports. To the national economy, the fisheries sector contributes 3% of GDP and between 5% and 7% agricultural GDP; employs about 10% of population (FAO, 1998); and provides raw materials for poultry and livestock industry. Earnings from fish exports from Ghana in 2002 was US$24.5 million.

Ghana's fishery sector comprises a diverse spectrum of fishing enterprises ranging in scale from subsistence to the industrial. Within this broad range, fish stocks are exploited from rivers, lakes, coastal lagoons, and shallow seas and offshore on the high seas. By far the marine subsector is the most important source of local fish production, delivering more than 80% of the total supply. The inland sector including capture from Lake Volta and aquaculture makes up the rest. The marine fishing industry in Ghana has three sectors which are the industrial, semi-industrial and artisanal (Armah et al. 2007, this volume).

State of Resources and Impacts

Fish stocks off the coast of Ghana in West Africa are continuously being overexploited with impacts evidenced by massive dwindling of the fish stocks and average fish size. For instance, biomass, density and catch rates of demersal species experienced a decline from 1963 to 1997 and has only began to increase in 2000 (Quaatey 2004). The large pelagic fishery exploits skipjack tuna *Katsuwonus pelamis*, yellowfin tuna *Thunnus albacres*, and bigeye *Thunnus obesus* using pole and line and purse-seines gears. Mean size range of these tunas caught in Ghanaian waters between 1990 and 2002 (33–62, 40–118, and 32–120 for skipjack, yellowfin, and bigeye tunas respectively) are far below the maximum attainable lengths of 80, 200, and 180 cm respectively. Estimates of potential yield in the eastern Atlantic Ocean indicate that there is too much pressure on juvenile bigeye tuna (Bannerman 2004a). The huge effort exerted on small pelagic fish resources depicted by large numbers of canoes, increased motorization, unorthodox fishing practices, and the uncertain climatic changes have led to huge annual fluctuations in catch (Bannerman 2004b). A similar trend of decline in production is observed in the inland fish resources. For instance, the average yield from the Volta Lake declined from 46.8 kg/ha in 1976 to 32.6 kg/ha in 1998.

The major challenges confronting the sector, which have contributed to unsustainable exploitation of fish resources, are population pressure, demand pressure, high cost of inputs, falling profit margins, open access, increased competition, excessive effort, technological intensification, insanitary conditions, inadequate access to credit, tree cover depletion, conflicts, and institutional weakness towards enforcement. The observed decline in both pelagic and demersal fish resources in Ghana's coastal waters, especially in the past decade, is akin to what has happened to cod stocks off Newfoundland, Canada. This paper illustrates the situation using case histories of four major contributing factors: bycatch and discards from shrimp trawl fishery; activities of local and foreign trawlers on demersal stocks; beach seine fishing in nearshore areas, and increasing export of raw and processed fish.

Bycatch and Discards from Shrimp Trawl Fishery

The capturing of marine living resources and discarding portions of it is common occurrence in fish exploitation in Ghana. The industrial shrimp fishery which targets penaeid shrimps in shallow coastal waters is particularly guilty because of the general nonselectivity of the gear.

Characteristics of the gear such as use of heavy attachments to the net that rake the sea bed, the small mesh size of the cod end of the net (usually 20 mm instead of the legal 40 mm), and, in recent years, the high demand for quality shrimps for export results in an enormous but variable amount of bycatch and discards. The fishery exploited a highly diverse assemblage of fish comprising 78 fish taxa from 46 families. Two-thirds of the catch was discarded (mean discards to landings ratio during night fishing was 2.3 ± 0.4: 1), with most of the small sized shrimps, fish, and other invertebrates already dead (Nunoo and Evans 1997; Nunoo 1998). The large number of species found in the bycatch and discards increases interaction with other species targeted in other fisheries, thereby compounding existing pressure on fish stocks. The continuous high mortality of a range of species, removal of immature individuals of commercial species, and continuous pressure of shrimp trawling on seabed may have had harmful ecological consequences on fish stocks. By 2002, the fishery has become unprofitable from poor catches such that the fleet size of shrimpers that increased to 17 in 1995, had decreased to 2 (Table 1).

Trawlers and Demersal Stocks

The last decade has seen a decline in importance of the inshore fleet due to aging fleet and the adoption of new technologies such as echo sounders and light fishing, which coupled with increasing numbers of such locally built wooden vessels, have intensified fishing pressure on demersal stocks. The introduction of pair trawling in 2000 and numbers increasing have added to increasing pressure on stocks (Table 1). Trawling within the 30 m depth reserved for artisanal fishermen using nets of small mesh sizes to harvest juvenile fish has contributed to the sad state of demersal fish stocks and an increase in trade conflicts among operators (Koranteng 2004). It is a known fact that West Africa ocean fish are now plummeting after European industrial trawlers switched to African waters following collapse of North Atlantic fisheries. Most affected countries are Mauritania, Guinea-Bissau, and Senegal. Though there has not been any capture of foreign fleets in Ghanaian waters in the past 2 years due to inefficient monitoring, it is the view of the authors that "pirating" of fish resources is a major contributor to observed de-

Table 1. Changes in numbers of operational fleets in Ghana in last decade.

Year	Industrial trawl	Pair trawl	Shrimper	Inshore	Canoes
1990	27		10	183	8,052
1991	22		12	148	8,052
1992	35		5	152	8,688
1993	24		5	161	8,688
1994	34		11	141	8,688
1995	45		17	182	8,641
1996	35		16	165	8,641
1997	40		13	149	8,610
1998	36		8	173	8,610
1999	35		7	173	8,610
2000	30	3	8	167	8,610
2001	37	7	6	178	9,981
2002	48	7	2	200	9,981

cline in Ghana. One European Union nation vessel was a culprit in 2002 and was made to face the law.

Beach Seine Fishing

The principal craft of the artisanal fisheries has been the dug-out canoe. There were about 9,981 dug-out canoes, with 52.3% of them motorized, in the country in 2001 (Bannerman et al. 2001). Most artisanal fishermen are small scale, commercial operators using labor intensive and relatively unsophisticated variety of gears including hook and line, purse seines, set nets, drifting gill nets, and beach seines (Bannerman et al. 2001; Moses et al. 2002), and in recent times using light to attract fish as well as use of monofilament nets in coastal waters. The most common gears used in the inland waters are fish attracting devices (e.g. brush park, set net with meshes less than the approved meshes of 50 mm, bamboo and bottle fishing, gill nets, various traps, explosives, and chemicals including dynamite and carbide).

The future of Ghana's fisheries is further threatened by human impacts, especially from a nationally popular fishing practice of beach seining in nearshore marine waters that uses undersized mesh nets (10 mm at the bag) to harvest juvenile fish. It contributes 12% to the total artisanal fisheries production. There are 790 beach seine units operating from 154 landing beaches spread along the Ghanaian coastline of about 550 km (Bannerman et al. 2001). Studies have reported that the catch from beach seines in Sakumono near Tema was species rich and consisted of 62 fish species from 29 taxonomic families, 13 crustacean species from six taxonomic families, eight other invertebrate species from three taxonomic families, and 18 species of macroalgae. The catch composition of beach seines by weight usually consisted of about 90% fish, 8% of crustaceans and other invertebrates, 1.5% macroalgae, and 0.5% marine debris. Generally juveniles of larger species exploited by offshore fisheries (e.g., giant captainfish *Pseudotholithus elongatus* and European barracuda *Sphyraena sphyraena*) were recorded in the catches and their sizes fluctuated within a year. Over 90% of the fish catch were relatively small-sized (1.2 and 10 cm total length) and these sizes fluctuated markedly within a year (Nunoo 2003). These findings further stress the importance of the marine nearshore area as a critical nursery location for a diverse assemblage of fishes that are worth conserving.

Export of Raw and Processed Fish

Fish export has gained importance in the Ghanaian economy with increasing trends over the past decade (Figures 1 and 2). Fish is classified as a nontraditional export (NTE) commodity. It is the second most important NTE after horticultural products. Fish and seafood increased their share of nontraditional agricultural export products from 25% in 2000 to 33% in 2001 (ISSER 2003). Major species exported include tuna, cuttlefish, lobsters, and shrimps.

Science-Based Participatory Management

The once rich fish resources of Ghana have been heavily exploited by increasing human population, economic forces driving fish exports, irresponsible fishing through huge effort, unorthodox fishing methods and foreign fleets, to unsustainable levels. These are occurring in the midst of inadequate data on the natural environment of fish, biology of various fish populations, regular assessments of fish populations, impacts of various exploitation techniques on resource and what

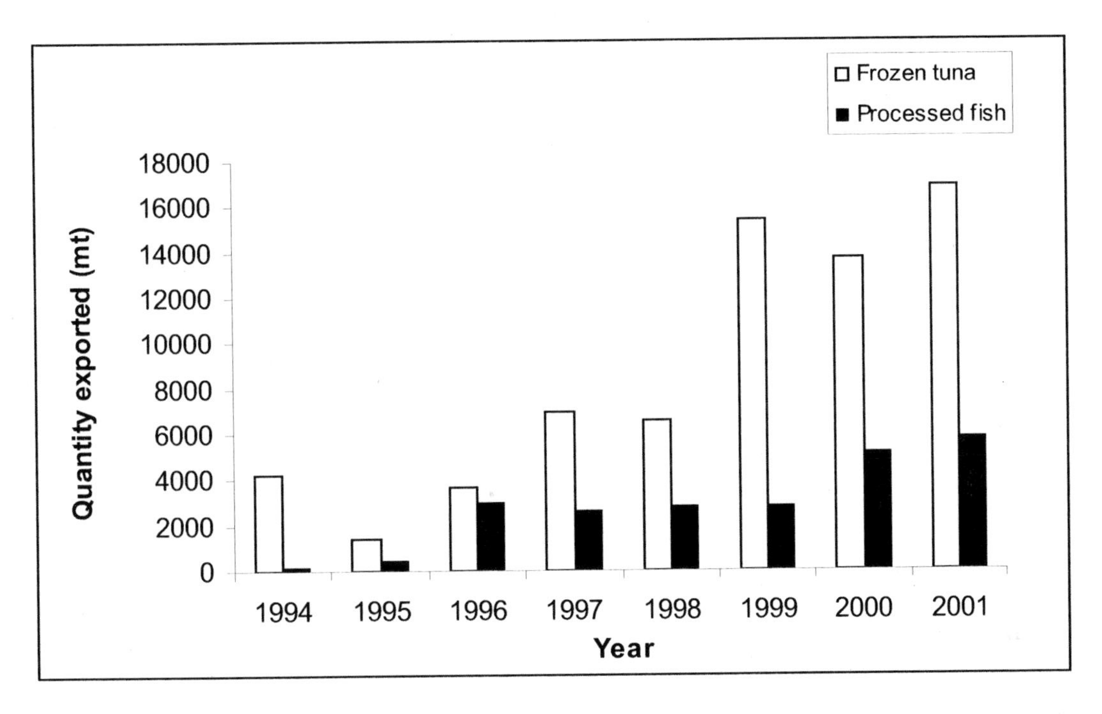

Figure 1. Quantity of frozen fish and processed fish (smoked/dried/salted) exports from Ghana for 1994 to 2001 (Source: Ghana Export Promotion Council data in ISSER 2003).

management strategies would sustain fish populations. Johannes (1998) stresses the need for conservation efforts through implementation of efficient management techniques even in the absence of adequate data in the spirit of the precautionary principle. In the midst of an almost uncontrolled over-exploitation of fish resources there is the need for new innovative approaches to fisheries management in Ghana.

Such has been the case for Ghana's fish stocks over the years. The Fisheries Directorate operates within the Ministry of Food and Agriculture as the executive arm of the Fisheries Commission by providing advice on sound management policies and efficient technical services to stakeholders in the fisheries sector for development and sustainable management of the fisheries. It is empowered by national and international fisheries policies such as the FAO Code of Conduct for Responsible Fisheries (FAO 1995) to regulate the industry. The Fisheries Act 2002, Act 625 (Government of Ghana 2002) amends and consolidates existing laws on fisheries. It provides for regulation and management of the fisheries, the development of the fishing industry, and the sustainable exploitation of the resources. It attempts to streamline legislation to respond directly to chronic and emerging issues and to conform to the national and international fishery resource development and management strategies.

Due to institutional and capacity weaknesses in Ghana, attempts at managing the national fishery resource from the top has proved difficult in the past in the face of a vast shoreline of marine coast, and numerous other communities (198 coastal and 1,232 lake communities). Capacity weaknesses include lack of adequate equipment and numbers of trained staff to conduct research and monitoring surveil-

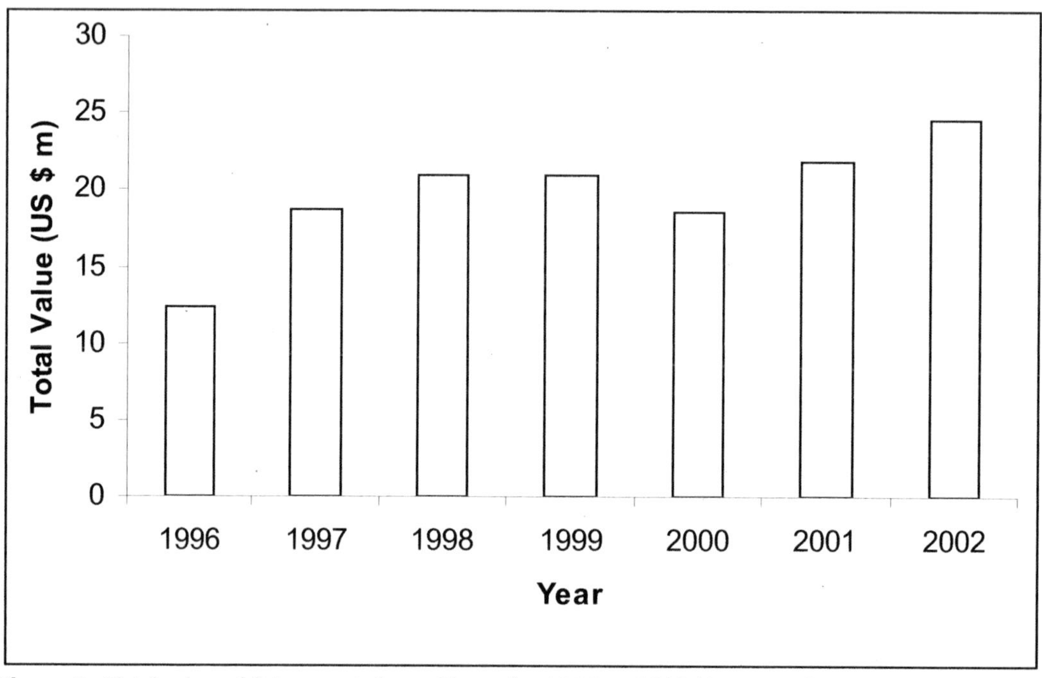

Figure 2. Total value of fish exports from Ghana for 1996 to 2002 (Source: Ghana Export Promotion Council data in ISSER 2003).

lance on fishing operations and enforcement of fishing laws.

It is suggested that Ghana implements simultaneously, and most efficiently, three major aspects of fisheries management that has seen some success in other countries such as South Africa, namely, research and development, resource management, and enforcement and compliance to achieve sustainable fish stocks. A paradigm shift to science-based management with increased active participation by fishers (e.g., communities, community-based fisheries management committees [CBFMC], and fishers organizations), scientists (universities and research institutions), policy makers (Minister of state, Fisheries Commission, Department of Fisheries, Regional and District Agriculture Development Units [RADU and DADU]), and nongovernmental organizations is advocated as an extension to Ghana's new Fishery Act 625 (Government of Ghana 2002). These institutions could collaborate to set up fisheries research boards or scientific working groups directly under the Fisheries Commission that would oversee the activities of well equipped research stations as well as mobilize funds for the conduct of proactive and pragmatic research to solve problems in the industry. Trained graduate fisheries personnel from the universities could be deployed at all fish landing sites both inland and along the coast to supervise quality biological, environmental, and socioeconomic data on the industry that would inform management policies as compared to the present use of fisheries technical officers, with little scientific background to appreciate the dynamics of fish stocks. Such capacity in addition to intensive mass education on wise exploitation of fish resource would give meaning to huge expenditures by government (such as the institution of satellite-based vessel monitoring system, subsidies for fishing inputs and landing beach infrastructure) for fishery development.

Conclusions

Increased pressure on fish stocks in Ghanaian waters from man's exploitation of the resource such as use of illegal fishing gear at unauthorized portions of water bodies, discarding captured fish, and export of raw and processed fish has led to overfishing evidenced by dwindling fish stocks, reduction in mean size of fish, and low catch rates in the midst of inadequate science-based data for management. There is the need for a shift towards a new approach that emphasizes science-based participation of relevant stakeholders from industry, policy makers, and scientists for sustainable exploitation of resources. With the necessary political will and education of the entire populace, the fish stocks of Ghana can be conserved for future generations in the midst of wise exploitation.

Acknowledgments

We are grateful for funding from the Government of Ghana through the Ministry of Food and Agriculture, Ghana and the support of the University of Ghana, Legon.

References

Aggrey-Fynn, E. 2001. The contribution of the fisheries sector to Ghana's economy. Sustainable Fisheries Livelihoods Programme (SFLP). DFID/FAO, Cotonou, Benin.

Bannerman, P. O., K. A. Koranteng and C. A. Yeboah. 2001. Report on the canoe frame survey, Ghana. Marine Fisheries Research Division, Information Report 31. Tema, Ghana.

Bannerman, P. O. 2004a. Status of the large pelagic fishery in Ghana. Report to the inaugural lectures on the state of marine environment and fishery resources in Ghana 30 March–1 April 2004. Marine Fisheries Research Division of the Directorate of Fisheries, Tema, Ghana.

Bannerman, P. O. 2004b. Small pelagic fishery in Ghana–current status. Report to the inaugural lectures on the state of marine environment and fishery resources in Ghana. 30 March–1 April 2004. Marine Fisheries Research Division of the Directorate of Fisheries, Tema, Ghana.

Directorate of Fisheries. 2003. Ghana: post-harvest fisheries overview. Directorate of Fisheries, Ministry of Food and Agriculture, Accra, Ghana.

FAO (Food and Agriculture Organization of the United Nations). 1995. Code of Conduct for Responsible Fisheries. FAO, Rome.

FAO (Food and Agriculture Organization of the United Nations). 1998. Responsible fish utilisation. FAO Technical Guidelines for Responsible Fisheries No. 7. FAO, Rome.

Ghana Statistical Service. 2002. 2000 Population and housing census. Special report on 20 largest localities. Ghana Statistical Service, Accra, Ghana.

Government of Ghana. 2002. Fisheries Act 2002. Government Act 625. Government Printer, Assembly Press, Accra, Ghana.

ISSER (Institute of Statistical, Social and Economic Research). 2003. The state of the Ghanaian economy in 2001. ISSER, University of Ghana, Legon.

Johannes, R. E. 1998. The case for data-less marine resource management: examples from tropical nearshore fin-fisheries. Trends in Ecology and Evolution 13:243–246.

Koranteng, K. A. 2004. The Ghana fishing industry and the need for research. Report to the inaugural lectures on the state of marine environment and fishery resources in Ghana. 30 March–1 April 2004. Marine Fisheries Research Division of the Directorate of Fisheries, Tema, Ghana.

Moses, B. S., O. M. Udoidiong, and A. O. Okon. 2002. A statistical survey of the artisanal fisheries of south-eastern Nigeria and the influence of hydroclimatic factors on catch and resource productivity. Fisheries Research 57:267–278.

Nunoo F. K. E., and S. M. Evans. 1997. The by-catch problem in the commercial shrimp fishery in Ghana. Pages 187–196 *in* S. M. Evans, C. J. Vanderpuye, and A. K. Armah, editors. The coastal zone of West Africa: problems and management. Proceedings of the 1st International Seminar on Coastal Zone Management in West Africa, Accra, 25–29 March, 1996. Penshaw Press, Sunderland, UK.

Nunoo F. K. E. 1998. By-catch: a problem of the Industrial Shrimp Fishery in Ghana. Journal of the Ghana Science Association 1:17–23.

Nunoo, F. K. E. 2003. Biotic, abiotic, and anthropogenic

controls of nearshore marine fish assemblage caught in beach seines at Sakumono, Ghana and their management implications. Ph.d. dissertation. University of Ghana, Legon.

Quaatey, S. N. K. 2004. State of demersal fishery resources in Ghana. Report to the inaugural lectures on the state of marine environment and fishery resources in Ghana. 30 March–1 April 2004. Marine Fisheries Research Division of the Directorate of Fisheries, Tema, Ghana.

American Fisheries Society Symposium 49:609–626

Reef and Lagoon Fish Prices: The Transition from Traditional to Cash-Based Economic Systems—Case Studies from the Pacific Islands

MECKI KRONEN* AND SAMASONI SAUNI
Reef Fisheries Observatory, PROCFish/C
BP D5 98848 Cedex, Noumea, New Caledonia

JOELI VEITAYAKI
University of the South Pacific, Marine Affairs Programme
Post Office Box 1168, Suva, Fiji

Abstract.—In this paper, case studies from the Pacific Islands are selected to demonstrate that the price paid for reef fish is an indicator of the stage of transition a community has reached in changing from a traditional to a cash economy-driven one. This transition process in the Pacific Island context also represents a merger between the nonmonetary, traditional economic system and the modern, cash-based system, rather than a purely economic valorization mechanism. Our case studies from French Polynesia, Tonga, Vanuatu, and Fiji also show that the monetary valorization of reef fish varies significantly. First, they show a connection between the degree to which a community is connected to a major market centre and reef fish prices. Second, the connection to a major market center among all communities included in this study is influenced by geographical isolation. And third, geographical isolation is determined by distance from and access to a major urban market. These findings apply regardless of living costs or living standards, which vary considerably between countries. Given these three principles, it can be concluded that reef fish remains a nonmonetary commodity in most traditional communities while market mechanisms have already replaced it in urban centers throughout the Pacific. Economic calculations reveal that the commercialization of reef fishery produce at the community level is often not viable or offers only limited revenue potential. As a result lower, socially acceptable prices are paid at the community level that do not reflect production or opportunity costs.

Introduction

Reef and lagoon resources are acknowledged as important assets to ensure protein supply to and maintain the livelihoods of people living in coastal communities of the Pacific Islands (Zann and Vuki 2000). Fishing of reef and lagoon resources in rural coastal areas differs substantially depending on the distance from and access to a major urban market center. Generally, the more isolated from urban markets, the more traditional the community, and the more fishing is performed for subsistence purposes.

Subsistence fisheries are that part of the economy where production primarily aims at the satisfaction of the fisher's own consump-

*Corresponding author: meckik@spc.int

tion and basic needs, rather than commercial principles (Fisk 1992). In many locations fish is one, if not the major, source of food security. This is highlighted by the fact that subsistence fisheries account for an estimated 80% of total coastal catches in the Pacific Islands (Gillet and Lightfoot 2001). However, the value of coastal fisheries and the resources used are little known and often not acknowledged in economic terms (Ebbers 2003). Subsistence and small-scale fishing in the Pacific Islands is restricted by a number of factors, including limited investment and inconsistent performance. Fishing is rarely done to maximize profits (Veitayaki 1995). These factors may explain why subsistence and small-scale artisanal fisheries are often neglected as a serious economic activity.

Traditionally, fish and other subsistence goods are exchanged among members of the community on a nonmonetary basis. With the introduction and increasing importance of cash economies in Pacific Islands, fishers have started to sell their catch. They are now faced with the challenges of a westernized economic production system. The implications of such socioeconomic changes for fisheries and resource management are increased fishing pressure and the replacement of traditional by more efficient fishing techniques and strategies, hence an increased risk of overexploitation. Thus, a better understanding of the value of marine resources may help to anticipate where significant changes in the demand for and exploitation of reef fish may accrue. Fisheries management strategies can then be designed to take into account these changes and also to maintain and improve the livelihoods of Pacific Islanders.

In this paper, case studies from the Pacific Islands are selected to demonstrate that the price paid for reef fish is an indicator of the stage of transition a community has reached in changing from a traditional to a cash-economy-driven one. This transition process in the Pacific Island context also represents a merger between the nonmonetary, traditional economic system and the modern cash-based system rather than a purely economic valorization mechanism. We have chosen this method of study to demonstrate some issues typical to Pacific Island situations where the traditional subsistence system and the westernized, introduced economic system coexist. The locations of the case studies selected are depicted in Figure 1. Major parameters determining urbanization processes are highlighted, and some characteristics and consequences of traditional nonmonetary systems and biases resulting in the transition to and adoption of cash-economy systems are discussed. These issues are prevalent throughout the Pacific Islands and consequently have implications for fisheries development planning in the region. The research was undertaken as part of two projects funded by the MacArthur Foundation and the European Union, respectively.[1]

Results—Pacific Island Case Studies

Urbanization—French Polynesia

Many parts of the Pacific Islands have experienced urbanization over the last 50 years. The effect on coastal, particularly small-scale artisanal, fisheries development has never been clearly understood. In some of the countries, fisheries development strategies have tried to respond to the increased demand for reef products at the national level, as well as to export fisheries that have emerged. In this first case

[1] The DemEcoFish project funded by MacArthur Foundation has been implemented by the Secretariat of the Pacific Community (SPC) in cooperation with the Institut de Recherche pour le Développement (IRD); the European Union-funded project, PROCFish-C, is being implemented by SPC .

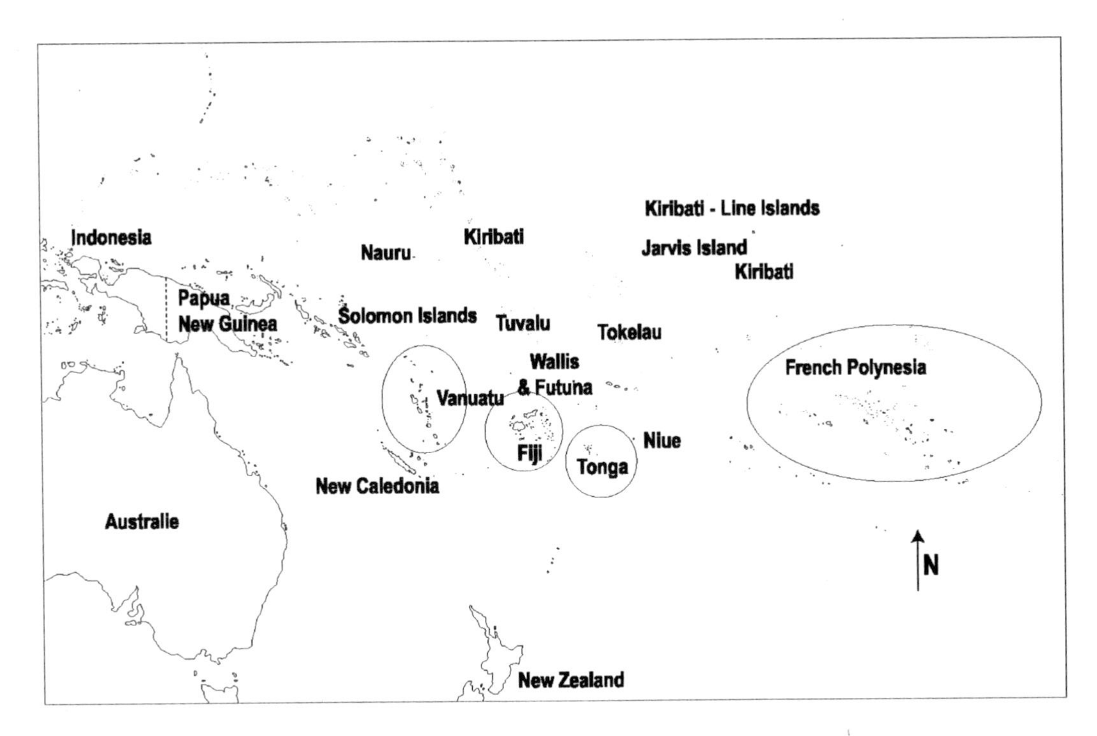

Figure 1. Location of Pacific Island countries selected for case studies.

study, from French Polynesia, we explore the influence of urbanization on fisheries development within the territory.

Generally, the standard and cost of living in French Polynesia are better and higher than those of other Pacific Island countries, particularly in nonfrancophone countries. However, local prices for reef and lagoon fish vary significantly. Our case study compares local fish prices at four different sites (Figure 2) against prices paid in Papeete, the capital city of French Polynesia.

The prices for reef fish at the different sites are listed in Table 1 and included in Figure 2. We can see that fish is either not sold in the community, or when it is sold, the prices are generally lower in those communities furthest away from the major urban market of Papeete. Minor exceptions may occur due to other reasons. For example, in the case of Fakarava, the local price paid for reef fish is relatively high because it is partly imported from neighboring atolls which involves a considerable amount of additional fuel costs.

In contrast, the price of canned fish (Table 1), a commercial product, generally increases with remoteness and distance from the country's major urban market (i.e., it reflects the increased transport and marketing costs incurred to make it available). If we assume price as a proxy for value, the value of canned fish increases the more traditional and isolated a community is. Reef fish is only of a higher value than canned fish under urban market conditions.

Data obtained from these sites (Table 2) also suggest a relationship between population size, the degree to which marine resources account for cash income, and distance from Papeete. The data show that the smaller the community and the closer it is to Papeete, the higher its dependency on fisheries as first

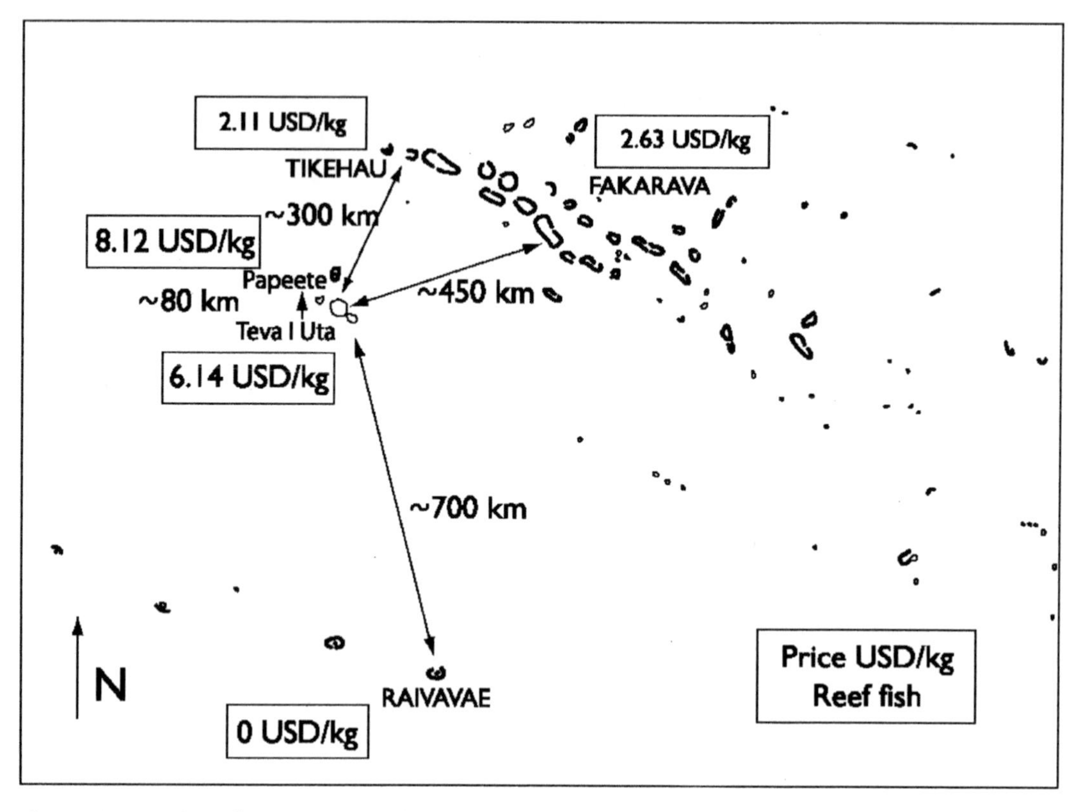

Figure 2. Location of communities, distance to Papeete and prices of reef fish in French Polynesia.

income source. In Raivavae, the most distantly located island from Papeete, reef fish is exchanged only on a nonmonetary basis. If communities earn any income from fisheries, it is from either pelagic fisheries or invertebrate export to Papeete.

The generation of income from reef fisheries varies across the communities studied. The risk of fish poisoning and the lack of marketing facilities for export to Papeete are the major reasons that limit commercial fisheries. Tikehau is the only site where a substantial proportion of reef fish is used commercially by the community members, who produce for export (Table 3). The small export volumes from Fakarava and Raivavae are due to use of reef resources for nonmonetary gifts rather than for commercial purposes. In contrast, the proportion of reef fish caught for commercial purposes (~9% of the total catch) in the community of Teva I Uta, where people lead a more urban lifestyle with ample access to alternative income sources, is negligible. This argument is supported by the fact that almost 90% of the population in Teva I Uta earns income from salaries. It is also by far the largest community of those studies, with over 7,800 inhabitants, and is well connected to Papeete. By comparison, Raivavae, Tikehau, and Fakarava represent more traditional communities (Table 2).

We also found that the more traditional the community, the higher the per capita consumption of reef fish (Table 4). The highest consumption is found in Tikehau and Fakarava. In both cases the atoll traditionally offers little in the way of agricultural production. Correspondingly, consumption in Raivavae is sub-

Table 1. Prices (US$/kg) for reef fish in five different locations in French Polynesia.

Site	Reef fish (US$/kg)	Difference from Papeete (%)	Canned fish (US$/kg)	Difference from Papeete (%)
Raivavae	0	–100	5.51	187
Tikehau	2.11	–74	5.77	196
Fakarava	2.63	–68	8.03	272
Teva I Ua	6.14	–24	3.21	109
Papeete	8.12	100	2.95	100

Table 2. Degree of urbanization at four sites in French Polynesia using selected parameters.

Site	Popula-tion	Geographi-cal location	Distance (km) from Papeete	First source of income (%)			
				Fish-eries	Agri-culture	Salary	Other
Raivavae	1,000	atoll	~700	7	10	47	36
Tikehau	350	atoll	~300	40	12	24	24
Fakarava	700	atoll	~450	12	16	48	24
Teva I Uta	7,840	Tahiti	~80	3	0	87	10

Table 3. Proportions of reef fish caught, consumed, and sold in four French Polynesian communities.

	Tikehau	Fakarava	Raivavae	Teva I Uta
Total catch (metric tons [mt]/year)	152.6	45.6	65.8	445.0
Total commercial catch[a] (mt/year)	142.3	0.0	0.0	40.0
Total subsistence catch[b] (mt/year)	10.3	45.6	65.8	374.0
Total import (mt/year)	0	7.6	0	5.0
Total supply (mt/year)	152.6	53.2	65.8	450.0
Total consumption (mt/year)	33.8	60.2	65.1	450.0
Estimated export (mt/year)	118.8	3.0	3.6	0.0

[a] Includes semi-commercial fishers.

[b] Includes "gifts" of reef fish (i.e., nonmonetary exchange and distribution).

Table 4. Reef fish consumption at four sites in French Polynesia.

Site	Reef fish consumption kg per capita/year
Tikehau	80.5
Fakarava	71.7
Raivavae	54.2
Teva I Uta	47.1

stantially lower due to the abundance of agricultural land on this high island. The lowest consumption of reef fish is found in Teva I Uta, which offers agricultural production potential but where the people enjoy an urban lifestyle. However, few differences are found regarding the percentage of households that at times consume fish they have received as a gift. This observation applies more than to 32–45% of all households surveyed in the four communities.

These results from French Polynesia suggest that distance from and access to the country's main market in Papeete influence how far a community adopts an urbanised, cash-driven economy. In parallel, the local price for reef fish increases substantially and the per capita consumption decreases. For example, Raivavae is the most distant site, with an only recently established flight service to Tahiti (2–3 weekly flights). Here, reef fishing is not commercialized at all. Tikehau and Fakarava are located on atolls but are well connected by air to Tahiti (10–12 weekly flights). Hence, their traditional systems have been partly replaced by commercial reef fisheries. The high proportion of commercialized fishing activity among Tikehau's people is a spillover effect of the site's export-oriented fisheries. However, the low local price, which is 70–75% below the prices paid in Papeete, demonstrates that traditional, subsistence, and noncommodity-driven influences are still strong. Prices for reef fish are highest in Teva I Uta, where they reach 75% of the price paid in Papeete. Teva I Uta is located just 1 h drive away from the capital. Our observations also show, however, that the traditional nonmonetary exchange system persists in all of the four communities despite major differences in local fish prices and market access.

Economic Viability of Reef Fisheries—Tonga

The economic viability of reef fisheries is an important issue for rural coastal development and sustainable fisheries management strategies in the Pacific Islands. To explore the economic viability of reef fisheries, we take a look at the situation in the Kingdom of Tonga.

In general, reef and lagoon subsistence and small-scale artisanal fisheries in Tonga are best characterized as hand-operated, multigeared, and multispecies. Fishing is mostly restricted to the nearby local coastal area, and involves small informal groups, small fishing vessels, low capital investment and correspondingly low productivity (Sabri 1977; Veitayaki 1993; Tu'avao et al. 1994; Passfield 2001). There is no clear distinction between commercially oriented fishers (also referred to as small-scale artisanal fishers) and subsistence coastal fisheries. Fishers are distinguished by their strategy rather than their status. Our survey exposed four different fisher groups across six Tongan communities. The location of the six communities surveyed is shown in Figure 3. The four fisher groups, which we will compare concerning the economic viability of their production, are characterized in Table 5.

The major distinguishing parameter for these fisher groups is ownership or use of nonmotorized or motorized boat transport. Availability of boat transport determines access to and choice of fishing grounds, choice and use of fishing gear and, to a certain degree, market choice.

For comparing the economic viability of the production of the four fishery groups, the net present value (NPV) calculation is used. NPV calculation was selected as it allows a comparison of a wide range of strategies and systems. It calculates the present net value of an investment, using a discount rate (interest rate) and a series of future payments (cost) and incomes (revenue) over a given time period (Kronen 2004). Here, we calculate hourly NPV to take into account productivity and fishing efforts as

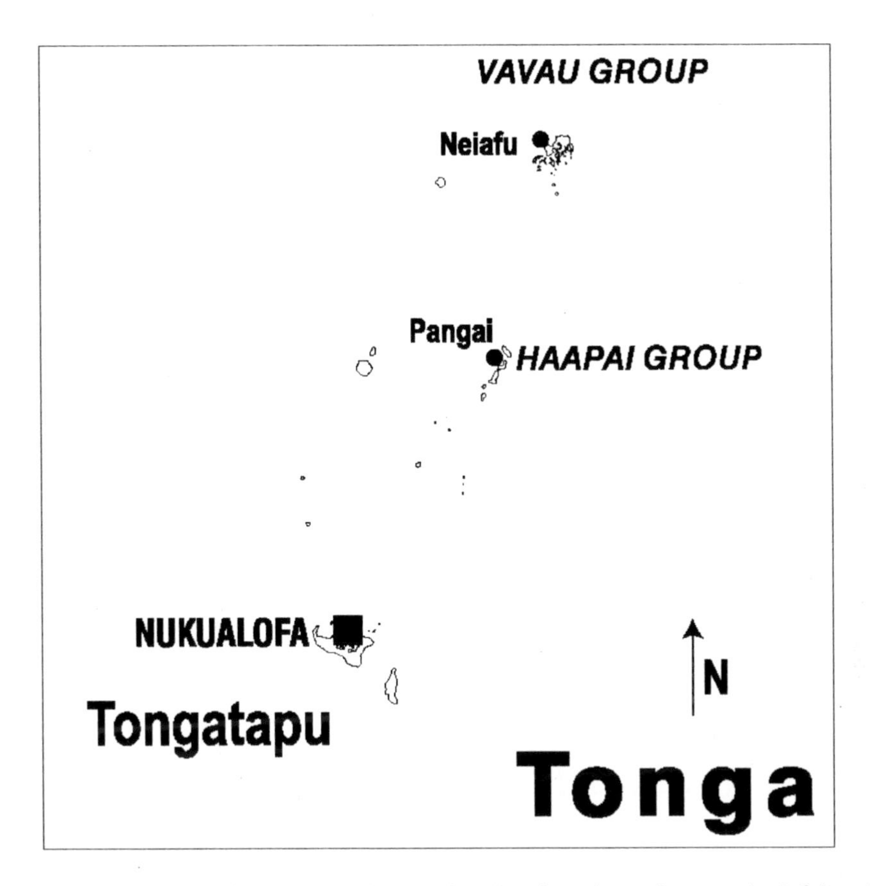

Figure 3. Location of Nuku'alofa urban market, and regional markets of Vava'u (Neiafu) and Ha'apai (Pangai), Tonga.

well as investment, maintenance and operational costs, interest rate, and possible revenues. Catch is valued according to local market price and an interest rate of 14% (Tongan Development Bank rate 2002) applies.

The results of economic assessment of the four types of Tongan subsistence and small-scale artisanal fishery show great variability. Net present values vary substantially from US$1.86 to US$6.50 per h (Table 6) if labor costs are included. Negative values accrue because local fishers do not usually account for labor costs. The figures in Table 6 also demonstrate how sensitive net revenues from artisanal fisheries are to labor costs alone. For example, the simple coastal fishery system represented in Group I with low productivity (catch per unit effort (CPUE) ≤3 kg/h, CPUE equaling total volume caught per time unit) and low catch volumes (≤50 kg/week) cannot offset labor costs. Similarly, productivity among the spear fishers in Group II is too low (CPUE 2.8 kg/h) to yield revenues that exceed labor and input costs. Spear fishers have high input costs as they need to rent all means of transport commercially.

Net returns for Groups II and IV show the highest variability. Both groups represent single or multigeared fishery systems with motorized boat transport and, for fishers in Group IV, choice of markets. Fishers who receive payment for boat transport services rendered make competitive net incomes, as do fishery systems with productivity and catch volumes exceeding 6 kg/h and 95 kg/week, respectively.

Table 5. Four Tongan subsistence and small-scale artisanal fisher groups.

Group I	Simple coastal fishery systems (no use of boat transport or canoe transport only; use of one technique only, mostly handline); CPUE ≤3 kg/h (CPUE equaling total volume caught per time unit), weekly catch: 40–50 kg, total hours fished: 16–20 h/week; no external market.
Group II	Use of single-to multi-geared fishing techniques, owner or user of motorised boat transport, boat owners may receive payments from other accompanying fishers for using boat transport; CPUE ≥ 3.3 to ≥ 6 kg/h, weekly catch: 20–60 kg, total hours fished: 8–24 h/week; no external market.
Group III	Exclusive spear fishers with rented boat transport; CPUE 2.8 kg/h, weekly catch: 75 kg, total hours fished: 8–24 h/week; no external market.
Group IV	Use of single to multi-geared fishing techniques, owner or user of motorized boat transport, boat owner may receive payments from other accompanying fishers for using boat transport; CPUE ≥ 3.3 to ≥ 6 kg/h, weekly catch: 105–160 kg, total hours fished: 18–48 h/week; market choice between community and Nuku'alofa (Tonga's capital city).

Table 6. NPV (US$/h) calculated for fishing systems of Groups I–IV, Tonga.

	Group I	Group II	Group III	Group IV
a) labor costs excluded[a]				
median NPV/h	1.70	3.41	1.30	3.79
minimum NPV/h	1.46	1.30	0.66	0.17
maximum NPV/h	2.10	7.85	1.93	7.46
b) labor costs included[a]				
median NPV/h	0.16	2.06	–0.24	2.31
minimum NPV/h	0.11	–0.05	–0.89	–1.86
maximum NPV/h	0.75	6.50	–0.41	6.45

[a] Labor costs US$2.95/h.

To put net revenues achievable from fishery and marketing systems into perspective, alternative income sources were examined. The simple comparison to an average hourly wage rate of US$2.03 in Tonga shows that fishers falling into Groups I and III (simple coastal fishery system and spear fishers) are clearly not economically viable. Some fishers in Groups II (single to multigeared, motorized boat owners and users) and IV (single to multigeared, motorized boat owners and users and market choice) operate on at least equal if not better terms than wage-based net income earners.

The results demonstrate that Tongan reef fisheries represent a small-scale economy that is vulnerable to production costs and prices. Costs for labor (time spent for fishing and transport of catch to and from regional markets) and motorized boat transport, productivity (CPUE) and market price of catch sold determine the economic viability of production and marketing sys-

Table 7. Price of reef fish at various Tongan locations.

Site	Selling price to agents and shops (US$/kg)	Consumer price (US$/kg)
Pangai, regional market Ha'apai island group	0.88–0.98	0.98–1.23
Communities on Ha'apai	0.88	0–0.98
Nei'afu, regional market Vava'u island group	0.88–1.47	1.47
Communities on Vava'u	0.88–1.47	0–1.47
Nuku'alofa, capital city Major urban market	1.96–2.21	2.45–2.70
Communities on Tongatapu	0	0–0.98

tems. Analysis shows that only fishers with access to the capital's urban market, where price levels and thus valorization of reef fish is high, attain higher revenues and adopt efficient fishing strategies (high CPUE, high catch volume).

The remoteness of a fishing community and the further it is from an urban market, the lesser the chances that its fishers will be compensated for the additional transport and labor costs required to serve regional rural markets (Ha'apai, Vava'u) where fish prices are still comparatively low (Table 7). As a result, the traditional system of regarding reef fish as a nonmonetary commodity prevails. The closer to an urban market a fishing community is, the more reef fish are accepted as a monetary commodity. However, in the case of Tonga, the introduction of reef fish as a monetary commodity within the community is subject to the perceived necessity that nowadays a fisher—like everybody else—needs to generate cash income rather than recover costs or maximize profits. Thus, prices paid among rural community members prove to be much below urban market prices.

The Tongan case study highlights two issues. All four fishery systems are part of the subsistence economy, although to varying degrees. A proportion of the catch is consumed by the fisher or shared with his extended family and does not enter the market (Hunt 1997; Tu'avao et al. 1994). Fish that enters the market is a result of cash need and not a response to market demand. If the basic tool for effective market operations does not exist, prices do not respond to the balance of supply and demand (Iwariki and Ram 1984). However, given an urbanized setting, price, and thus value, for reef fish is determined by supply and demand. The higher the degree of urbanization and the more prevalent salary-based income, and hence cash, the higher the chances to increase value of a resource that is in limited supply. In less urbanized communities, although they may be in proximity to an important urban market, price, and thus valorization, of reef fish do not follow economic rules, instead they follow social rules. Thus, choice of and access to urban markets determine possible revenues and may lead to the adoption of more efficient fisheries strategies.

Market Access—Vanuatu

The importance of market access increases as communities in the Pacific Islands develop into cash-based economies. Access to markets influences prices paid to fishers and the development of structures and services that need to be taken into account for fisheries development planning. A case study from Vanuatu illuminates some influences of market access on the value of reef fish.

The influence of market access on prices paid for reef fish within communities is assessed by comparing two villages on Vanuatu's main island, Efate (Figure 4, Map 4). Prices for reef fish are also evaluated by comparing them with those of canned fish, a much-preferred alternative in both communities.

Paunangisu village is located 1–1.5-h drive from Port Vila, the capital of Vanuatu. Moso is situated on a small island off mainland Efate. While total travel time from Moso to Port Vila may be comparable to that from Paunangisu to Port Vila, boat transfer (~20 min) is required to get from Moso to road transport (~60 min). The two villages are similar in their overall, relatively low dependence on marine resources. This is expressed by the low percentage of households with fisheries as first source of income (Table 8) and the relatively low per capita reef fish consumption (Table 9). However, consumption in Moso is about 60% higher than in Paunangisu.

The price per kg at which fish is sold among Moso's community members is about US$0.60 lower than the price of fish sold in Paunangisu (Table 9). Although overall travel time to the capital's main market is comparable, the need for boat transfer to and from Efate, although short in time and distance, renders Moso more isolated and hence less favored for the development of marketing structures. This situa-

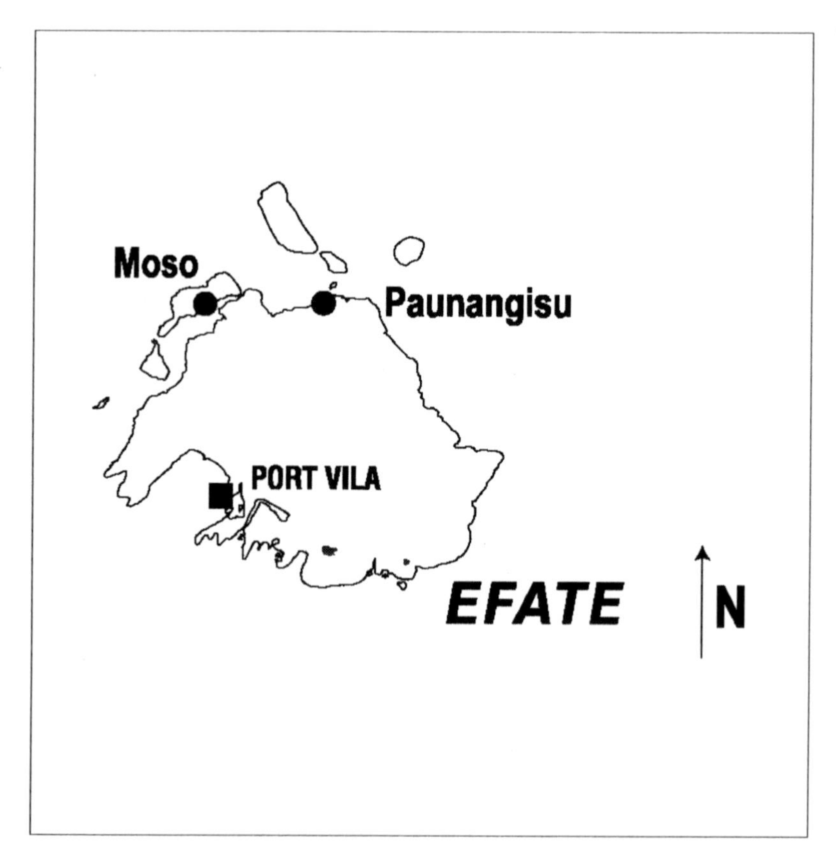

Figure 4. Location of Moso Island and Paunangisu village in relation to Port Vila, capital of Vanuatu.

Table 8. First sources of income (%) for Paunangisu and Moso households, Vanuatu.

Source of first income	Paunangisu	Moso
Fisheries	29	15
Agriculture	55	54
Salary	5	19
Handicrafts and others	11	11

Table 9. Per capita reef consumption in Paunangisu and Moso, Vanuatu.

	Paunagisu	Moso	Port Vila
Consumption of reef fish (kg per capita/year)	20.30	32.30	
Price of reef fish (US$/kg)	2.84	2.27	3.41

tion is aggravated by the fact that access to good fishing grounds requires substantial motorized boat transport or 6 h of canoeing. Paunangisu fishers reach adequate fishing grounds much easier and faster.

Paunangisu has a village shop equipped with a generator-driven refrigeration system where reef fish is bought and sold. Based on a small average catch of 10 kg, Table 10 shows that the additional costs incurred by Moso fishers who sell their catch at Port Vila are not compensated for by the urban market's higher prices. However, assuming that the additional costs did not change, a positive balance of ~US$17.00 could be achieved if Moso's fishers sold double to triple the average catch on each visit to Port Vila. It has to be taken into account, however, that marketing at Port Vila requires an additional work day in comparison to direct sales at Paungasiu.

Both Efate communities show a surprisingly high consumption of canned fish (Table 11). The price comparison raises the question of why the people do not consume more reef fish than canned fish. As shown in Table 12, average household expenditure could be reduced

Table 10. Comparison of gross revenues for fishers at Paunangisu and Moso, Vanuatu.

	Average catch		
Price and cost in US$	10 kg	20 kg	30 kg
Fish sold at Paunangisu village	20.45	40.91	61.36
Fish sold by Moso fishers at Port Vila market (minus additional cost 17.05)[a]	17.04	51.13	85.22
Balance	–3.41	10.22	23.86

[a] Additional cost for Moso fishers selling at Port Vila market in US$/trip:

- transport		5.68
- market fee	4.55	
- ice	6.82	
TOTAL	17.05	

Table 11. Consumption and cost of canned fish in Efate communities, Vanuatu.

	Paunangisu	Moso
Consumption of canned fish (kg per capita/year)	12.60	13.70
Price of canned fish (US$/kg)	6.82	9.09

Table 12. Importance of canned fish in Paunangisu and Moso, Vanuatu, measured in % of household[a] expenditure.

	Canned fish consumption kg/week	Canned fish consumption US$/week	Reef fish substitution US$/week[b]	Savings US$/week	% of average household expenditure
Paunangisu	1.31	6.52	4.47	2.05	7
Moso	1.71	13.49	4.67	8.83	26

[a] Average household size: Paunangisu = 5 people, Moso = 6 people.
[b] Reef fish substitution takes into account 20% of nonedible fresh fish parts.

by 26% in Moso if reef fish purchased at local cost was substituted for canned fish. In the case of Paunangisu, possible savings reach only about 7% of average weekly household expenditure.

The results from the Vanuatu case study highlight two major arguments. First, geographical isolation requires additional efforts (labor, transport, preservative measures) to allow individual marketing of reef fish at a major urban market. Given the small scale, these additional costs render small-scale artisanal fisheries noncompetitive. Second, geographical isolation results in lower prices paid by community members for reef fish and higher prices for canned fish. This bias may be explained by the fact that the more isolated the community, the more reef fish is still considered as traditional, nonmonetary produce. Thus, if reef fish is sold among community members, a much lower price applies as compared to an urban market, where prices are determined by costs and demand-supply balance. Canned fish was never part of the traditional, nonmonetary system but has been introduced and accepted as commercial produce.

Marketing Strategies—Fiji

The establishment of operational and effective marketing structures is one of the major objectives for coastal fisheries development in the Pacific Islands. Marketing strategies are usually closely related to the balance of demand and supply, accessibility and costs. A case study from Fiji demonstrates that reasons beyond economic rationality may determine the marketing strategies of Pacific Island fishers.

As shown in Figure 5, the communities of Lakeba and Nakawaqa are both located in Vanua Levu. Although Lakeba is situated on the main island, travel from there to the major market of Labasa is on very bad roads with limited connections and/or a long boat trip. People from Nakawaqa village, which is situated on a small island, depend on boat trans-

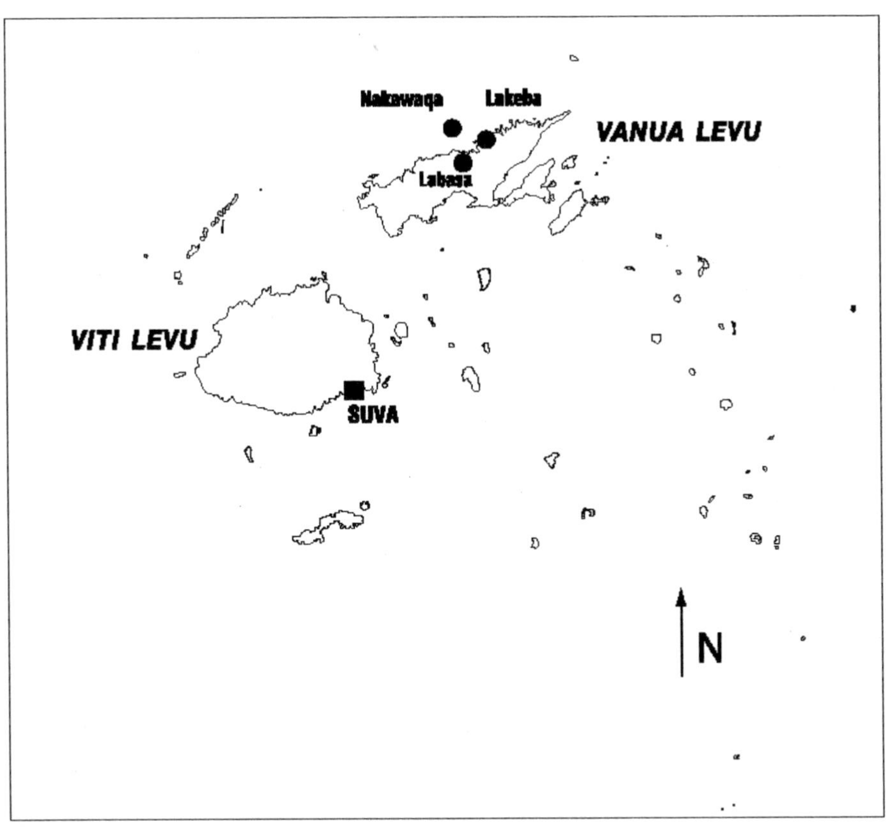

Figure 5. Location of Lakeba, Nakawaqa and Labasa, Vanua Levu, Fiji.

port to reach the Labasa market. Boat travel time and costs of transport from Nakawaqa to Labasa are less than the time and costs of getting from Lakeba to Labasa.

Dependency on marine resources for subsistence and income is high. While Lakeba people have access to agricultural land, Nakawaqa Island does not offer any cultivatable land. However, the percentage of households receiving cash income from a salary is slightly higher in Nakawaqa than Lakeba. The importance of reef fisheries may explain why reef fish consumption is relatively high in each village. The difference in access to agricultural produce may contribute to higher per capita consumption of it in Nakawaqa (2.8 kg per capita/week) than Lakeba (2.4 kg per capita/week). In neither of the two villages is fish sold among community members instead, it is exchanged and distributed on a nonmonetary basis.

Reef fisheries are the first source of revenue for 96% of all households in Lakeba, and for 81% in Nakawaqa. In both cases, reef fish is exclusively sold in Labasa, either to agents or shop owners or at the market. The same prices and thus value systems of buying reef fish apply to all fishers from both communities. However, marketing strategies and revenues obtained for reef fish catch differ substantially between the villages.

Lakeba offers the choice of marketing fish by using either the local cooperative or an individual system. The cooperative pays a lower price for the fish but accepts all fish upon landing and takes full responsibility for its marketing. The individual system involves road transport arrangements and costs, additional labor input, and the risk of losing catch and income through an insufficient cooling chain, as well as marketing risks due to fluctuations in demand and supply.

In Nakawaqa, two marketing systems are available. One is the infrequent visit of a reef fish buying agent who offers a lower price than Labasa market. The other is an individual marketing system that involves the same requirements and risks as in the case of Lakeba.

A simple comparison of opportunity costs (Table 13) illustrates the reason why two-thirds of all fishers (83%) in Lakeba use the cooperative marketing system. Under the given scenario, and not taking into account that an additional day is needed to market fish at Labasa, the average transport cost alone will consume any additional profit from the higher prices at Labasa. Of course, this simple scenario is based on relatively low-catch volumes. The small proportion of fisheries marketed individually at Lakeba contrasts with the 91% of all fishers in Nakawaqa who individually sell their catch in Labasa. Applying the same calculation for Nakawaqa, a small profit is possible for fishers who individually market their catch at Labasa. However, the calculated difference equals about US$1.50 compensation for an average of 6 h additionally required. Thus, the question arises as to why even 10% of Nakawaqa fishers do not make use of the fish buying agent or adopt a similar community strategy to Lakeba. A cooperative system in Nakawaqa would face lesser costs and therefore could pay higher prices for its clients' produce.

Field observations suggest that the marketing strategies of Nakawaqa people do not follow an economic rationale, but aim at avoiding social conflict. For example, boat transport is limited and shared among community members. Marketing is done individually to ensure control over a certain percentage of the cash income generated and thus to gain more control over the obligation to share fish and profits with family and community members. Earlier attempts to install a cooperative at Nakawaqa failed due to these social obligations, also known as the Pacific "share and care" system (i.e., to share any surplus to subsistence needs without receiving cash payments but as an investment in future favors to be returned. This share and care system contrasts with an economically efficient marketing sys-

Table 13. Simple comparison of possible revenues for fishers selling within the community or at Labasa market, Fiji.

	Nakawaqa	Lakeba
Average catch per fisher kg/trip	22.0	15.0
Catch sold at cooperative/agent: 50% grade A 1.53 US$/kg 50% grade B 1.18 US$/kg	29.9	20.4
Catch sold at Labasa: 50% grade A 1.77 US$/kg 50% grade B 2.36 US$/kg	45.4	31.0
Average transport cost return trip to Labasa (US$)	7.1	10.6
Difference in revenue (US$) including transport	8.4	0.0
Time requirements for marketing at Labasa (hours)	6.0	9.0

tem such as a cooperative, which is based on continuous and reliable cash flow. A cooperative system must therefore fail, as it will not be able to render fuel and other services without the necessary payments. On the other hand, cash income is needed. Individual marketing offers a conflict-free way to control the share of fish donated or sold at a price below market value. The use and redistribution of cash generated is also important and can be better controlled on a personal rather than communal basis. The conflicting natures of the traditional share-and-care system and rational economic principles are considered a pitfall of cooperative systems in the Pacific Islands. However, the case of Lakeba shows that there are ways to successfully combine both systems if economic principles are accepted and complied with by the community.

Discussion

Our case studies from four Pacific Island countries demonstrate that the monetary valorisation of reef fish varies significantly. Firstly, they show a connection between the degree to which a community is connected to a major market centre and reef fish prices. Secondly, the connection to a major market centre among all communities included in this study is influenced by geographical isolation. And thirdly, geographical isolation is determined by distance from and access to a major urban market. These findings apply regardless of living costs or living standards which vary considerably between countries.

Given these three principles, it can be concluded that reef fish remains a nonmonetary commodity in most traditional communities. Market mechanisms have already replaced the traditional, nonmonetary exchange system in urban centres throughout the Pacific. In urban markets, prices for reef fish represent limited supply and transport costs. Communities in transition between both extremes, where the traditional nonmonetary system and the cash-based economic system somehow intermingle, support usually significantly lower, but socially acceptable, prices for reef fish produce. These lower prices do not take into account transport or other production costs but accept the need for fishers to generate some cash income.
Economic calculations reveal that the commercialization of reef fishery produce at the community level is often not viable or renders only limited revenue potential. This observation supports the above argument that the lower, socially acceptable prices paid at the community level do not reflect production or opportunity costs. The case studies also illustrate that canned fish has become popular among people living in more isolated and less developed communities. Unlike for reef fish, a much higher monetary value is accepted for canned fish.

Observations made in our case studies also show that despite the fact that rural and isolated communities are adopting a cash-based economy more and more, traditional, nonmonetary attitudes and values (share and care) continue to influence reef fishery strategies and marketing of fishery produce (Kronen 2004). It may be argued that the continuing strong influence of the traditional nonmonetary system will limit the introduction and adoption of a capitalistic profit-maximization strategy (Halapua 1982). This explains the introduction of lower but socially acceptable prices for reef fish among community members.

Although our case studies suggest that the increased adoption of a cash-based economy is closely related to reducing reef fish consumption (Kronen et al. 2003; Adger et al. 2002), reef fish will continue to be essential to the livelihoods of rural coastal people in the Pacific Islands (Zann and Vuki 2000; Ruddle et al. 1992). Reef resources represent food security insurance in times when cash supply is low or when the supply of alternative food may fail due to natural events (cyclones) or logistical problems (islands that are entirely dependent on shipments for any alternative food).

The question remains as to how far the higher prices paid for reef fish by customers at urban Pacific markets compensate fishers and communities for their possibly weakened social network and social resilience. Social networking and accordingly social resilience will be reduced with increased commercial fishing activities and increasing adoption of profit-maximizing strategies. Direct impacts show in a reduced proportion of catch distributed noncommercially among community members. Increasing fishing efficiency and productivity may result in medium- to longer-term detrimental effects on a community's resources. Furthermore, analysis of the dynamics between reef fish price development in urban markets, the value of limited reef fish resources and increasing investment in more efficient equipment and more sophisticated fishing methods is needed.

In summary, the price level of reef fish can be used as an indicator of the stage of transition a society has reached in changing from a reciprocity exchange and direct sustenance system into a cash-based economy. Only if a Pacific Island society becomes urban in character is it no longer exclusively dependent on primary commodities and thus resource dependent (Hamilton and Otterstad 1998). With increasing substitution of subsistence production by salary-based income, social services and securities formerly provided by traditional social networking gain monetary value as well. Thus, food such as reef fish needed for subsistence reflects demand, and also modern costs not included in traditional systems for instance, transport to serve these urban centers.

Proxies such as price differences for reef fish in local communities and urban markets can be useful for the design of fisheries and rural development management strategies, as they help to anticipate where changes of value and thus demand and exploitation of reef fish resources may accrue.

However, our case studies also suggest that price is not the only indicator that can be used to express the value of reef fish in the Pacific Islands context. Under more traditional conditions, it is an acknowledgment that cash income needs to be generated rather than a recognition of production costs. In urban settings, price levels rise substantially, reflecting increased demand or shortage of supply and transport cost of reef fish. Possible detrimental effects due to weakened social networks and resilience of fishing communities are not considered in monetary terms.

Neither traditional nor western economic systems consider resource cost or replacement cost for reef fish. Both systems regard reef fish as a "free" resource, with its value mainly determined by its use, demand for it, and its transport cost. Monetary value is placed on all those production factors that are not accommodated by traditional systems, hence transport but not production. In contrast, the monetary value of canned fish an industrial and mostly imported product is uncontested, as highlighted by its widespread acceptance in the Pacific Islands. The price for canned fish usually exceeds that of reef fish. However, as demonstrated in the case of French Polynesia, under peri-urban and urban conditions, the respective prices for reef fish are comparable to or exceed those for canned fish. The widening gap between the higher prices paid for imported canned fish and the lower prices paid for locally caught fresh fish in rural, geographically isolated communities illuminates two factors: first, the possible misinterpretation of local value placed on reef fish resources, and second, the shortcomings of applying western monetarism to traditional Pacific Islands conditions.

The results also show a need to further investigate the possible consequences of adopting an economic rationale and seeking profit-maximization strategies in the Pacific Island context.

Acknowledgments

The results and experiences presented here were made possible by the kind support and cooperation of numerous people. Special thanks are due to Mr Bruno Ugolini, Chef de service, and his staff from the Service de la pêche in French Polynesia; Mr 'Ulunga Fa'anunu, Acting Secretary, and his staff from the Tonga Ministry of Fisheries; Mr Moses Amos, Director, and his staff from the Vanuatu Department of Fisheries; and Mr Malakai Tuiloa, Acting Director, and his staff from the Fiji Fisheries Division. We would also like to express our special gratitude to the authorities, chiefs, elders, and people of the communities we visited, who shared relevant information on their fisheries. We also recognize the MacArthur Foundation and the European Community for providing the necessary funding, and the Secretariat of the Pacific Community for its administrative and technical support.

References

Adger, W. N., P. M. Kelly, A. Winkels, L. Quang Huy, and C. Locke. 2002. Migration, remittances, livelihood trajectories, and social resilience. Ambio 31(4):358–366.

Ebbers, T. 2003. To whom the fish belongs: a review of rights-based fisheries and decentralization. Fish for the People 1(2):2–18, Bankok, Thailand.

Fisk, E. K. 1992. The subsistence sector in Forum Island countries. A report to the Forum Secretariat.

Gillet, R., and C. Lightfoot. 2001. The contribution of fisheries to the economies of Pacific Island countries. Asian Development Bank, Pacific Studies Series, Manila, Philippines.

Halapua, S. 1982. Fishermen of Tonga: their means of survival. Institute of the Pacific Studies in association with the Institute of Marine Resources, University of the South Pacific, Suva, Fiji.

Hamilton, L., and O. Otterstad. 1998. Demographic change and fisheries dependence in the Northern Atlantic. Human Ecology Review 5(1):16–22.

Hunt, C. 1997. Cooperative approaches to marine resource management in the South Pacific. Pages 145–164 *in* P. Larmour, editors. The governance of common property in the Pacific region. National Centre for Development Studies. Pacific Policy Paper 19 and Resource Management in Asia-Pacific. Australian National University, Canberra.

Iwariki, S., and V. Ram. 1984. An introductory study of the socio-economic aspects of household fisheries in the small islands economies of the Pacific Islands. Memoirs of the Kagoshima University Research Center for the South Pacific 5(1):53–65.

Kronen, M., E. Clua, B. McArdle, and P. Labrosse. 2003. Use and status of marine resources: a complex system of dependencies between man and nature: case studies from Tonga and Fiji, Pacific Islands. Centre for Maritime Research (MARE), People and the Sea II, Conference Proceedings, 4–6 September 2003, Amsterdam.

Kronen, M. 2004. Fishing for fortunes: a socio-economic assessment of Tonga's artisanal fisheries. Fisheries Research 70:121–134.

Passfield, K. 2001. Profile of village fisheries in Samoa. Samoa Fisheries Project. Technical report. Government of Samoa, Fisheries Division, Ministry of Agriculture, Forests, Fisheries and Meteorology.

Ruddle, K., E. Hviding, and R. E. Johannes. 1992. Marine resources management in the context of customary tenure. Marine Resource Economics 7:249–273.

Sabri, J. 1977. Small-scale fisheries and development in Peninsular Malaysia: problems and prospects. Pages 63–74 *in* B. Lockwood and K. Ruddle, editors. Small-scale fishery development: social science contributions. East-West Center, Honolulu, Hawaii.

Tu'avao, T., V. Kava, and K. Udagawa. 1994. The present

situation of fisheries in the Tongatapu Island group. Fisheries Research Bulletin Tonga 2: 27–42.

Veitayaki, J. 1993. Village-level fishing in the Pacific. Pages 73–97 *in* G. R. South, editor. Marine resources and development. PIMRIS, University of the South Pacific, Suva, Fiji.

Veitayaki, J. 1995. Fisheries development in Fiji: the quest for sustainability. Institute of Pacific Studies and Ocean Resources Management Programme, University of the South Pacific, Suva, Fiji.

Zann, L. P., and V. C. Vuki. 2000. The status and management of subsistence fisheries in the Pacific Islands. International Ocean Institute, University of Chicago Press, Ocean Yearbook (14): 163–175.

American Fisheries Society Symposium 49:627–629

Session Summary

Can Fair Allocation Impact the Reconciliation of Fisheries with Conservation?

SERGE GARCIA

FAO Fisheries Department, Fisheries Resources Division
Viale delle Terme di Caracalla, Rome 00100, Italy

Introduction

The panel assembled for this session was opened by S. M. Garcia with a general presentation of the conceptual and practical interplay between allocation and conservation and its consequences in fisheries. The presentations dealt with 1) a methodology to estimate the socially optimal allocation among competing recreational and commercial subsegments of fisheries, presented by P. McCleod (Australia); 2) the role of allocation in integrated fisheries management in an Ecologically Sustainable Development (ESD) framework in Australia, presented by P. Rogers (Perth, AU); 3) the performance of the Individual Transferable Quota (ITQ) system in relation to equity objectives in postapartheid South Africa, presented by M. Hara (Capetown, ZA); 4) a comparative analysis of performance of two systems of allocation in Alaska and Antarctic fisheries, presented by M. Lundsten (Seattle, WA); and 5) the role of a scientific advisory body (the European Inland Fishery Advisory Commission, EIFAC) in resources conservation by Mueller (Switzerland).

It is very generally agreed that fisheries resources depletion represents a good example of the "tragedy of the unmanaged commons" (Hardin 1998) because access to resources is relatively cheap if not free, and easy if not open, leading to a "race to fish" often promoted and supported by subsidies. The conventional command and control conservation measures to regulate inputs and outputs have often failed to meet their conservation objectives and a number of alternative or complementary strategies have been proposed. The establishment of taxes as a means to internalize costs and improve economic rationality, proposed by Pigou (1920) is still being considered in some areas, but are viewed by many economists as impractical for fisheries and not providing the appropriate incentive. The establishment of strong user rights, as proposed by Coase (1960) seems to be becoming the instrument of choice to increase the user's sense of long term responsibility. The performance of the approach depends, inter alia, on 1) the allocated factor such as catch, effort, time, or space; 2) the rights' attributes such as security, and transferability; 3) the right holder such as an individual or a community; 4) the level of consensus about the initial allocation; 5) the effectiveness of the rights' administration; 6) the value attached to the privilege; and 7) the destination of the rent.

The 1982 United Nations Convention of the Law of the Sea established the states' sovereign rights and the basis for their reallocation to lower jurisdictional levels. The latter process is still incomplete and the present situation rather unsatisfactory. At the regional level, fishery management organizations often have conservation as their main objective, but most of them have neither the power to allocate resources among final users nor the capacity to enforce decisions. In many cases they have failed to achieve their objective. The situation may improve if the 1995 UN Fish Stock Agreement is effectively implemented and illegal fishing is reduced. At the national level, states face the dilemma of choosing between forms of private or communal property or a combination of the two. The main dilemmas are in allocation among 1) consumptive and non-consumptive use; 2) fishers and other sectors (including conservation); 3) sub-sectors of fisheries; 4) national and foreign fishers; 5) as well as between present and future generations. These issues are further complicated in the ecosystem approach to fisheries (EAF), which implies harmonizing ecosystem and sociocultural sustainability.

Determining the socially optimal allocation among competing uses (active versus passive, commercial versus recreational, and extractive versus nonextractive) is a delicate and critical task requiring an understanding of the net benefits of the various options within an integrated management framework accounting for the values associated with the various uses of a resource, presently and in the future. McLeod proposed a methodological framework in which allocation is considered socially optimal when the marginal net benefit from each of the competing uses is the same. He illustrated the proposition in Australian crab, abalone, and wet line fisheries.

As societal demands on resources increase, single fishery and even sectorial management plans start showing their limits and need to be geographically integrated to ensure that total impacts on the broader ecosystem are taken into account, encompassing not only commercial, recreational, and other fishers but also aquaculture and nonextractive uses of these resources. Rogers proposed nine broad principles underpinning allocation decisions in the integrated fisheries management (IFM) framework being progressively introduced across Western Australia. The principles relate to: 1) State stewardship and future generations; 2) sustainability as a paramount allocation criteria; 3) information and uncertainty; 4) explicit harvest allocation for all fisheries; 5) accounting for all sources of mortality; 6) strict mandatory harvest limitations; 7) management structures and processes, including preagreed courses of action; 8) need for optimal allocation; and 9) accounting for natural variations.

Private and communal rights have different properties in relation to equity. In the context of postapartheid South Africa, Hara reviewed the outcome of the ITQ system in terms of its performance in re-allocating resources to previously disadvantaged groups and communities. He stresses that the capacity of formerly disadvantaged groups to benefit from the policy depends on their skills, availability of capital and policy stability. Expressing his own doubts on the effectiveness of the ongoing re-allocation process, he stressed that change was considered too slow, and that the information available did not allow drawing any conclusion, partly because of the opacity of the private sector.

Different types of fishing rights perform differently in relation to compliance. Describing the longline fisheries for Alaskan Sablefish and South Georgia Patagonian toothfish *Dissotichus eleginoides*, Lundsten stressed their numerous common features but significantly different allocation systems. The Alaskan fisheries (managed by the North Pacific Fishery Management Council) is regulated through a highly participative system of revocable individual fishermen quotas (IFQs) designed for and with the fishers, with strong monitoring and enforcement. The South Georgia fishery (managed by the Commission for the Conservation of Antarctic Marine Living Resources) uses a strongly regulatory system as well as nontransferable annual vessel quotas in exchange for an annual fee, with no long-term security. Both fisheries have successfully dealt with the problem of bird bycatch. The Alaskan fishery has largely resolved its discard issue, improved reproduction and ended the "race-to-fish." Lacking the long-term entitlement and the strict enforcement system, the effectiveness of the South Georgian fishery management is negatively affected by a high level of noncompliance (illegal, unreported, and unregulated fishing, [IUU]), potentially jeopardizing conservation.

The value of fishing rights may decrease with time if the quality of the derived fishery products decrease (e.g. due to contamination from pollution). Kazmierczack, (USA) measured the impact of oil industries on the value of fishing rights in the Louisiana oyster fishery, showing the link between the loss of interest for the oyster fishing leases and the increasing environmental impact. He concluded that the harvest rights structure of the oyster industry will evolve rapidly under the pressure of the coastal restoration activities in part due to this impact.

Regional mechanisms may also play a useful role in conservation, even when they have no mandate for allocation as described by Mueller described the functioning of the European Inland Fishery Advisory Commission (EIFAC) in relation to conservation of eel and sturgeon, two resources with particularly serious conservation problems.

References

Coase, R. 1960. The problem of social cost. Journal of Law and Economics 3(1):1–44.

Hardin, G. 1998. Extensions of the "Tragedy of the commons." Science 280:682–683.

Pigou A.C. 1920. Economics of welfare. MacMillan, London.

American Fisheries Society Symposium 49:631–658

Allocation and Conservation of Ocean Fishery Resources: Connecting Rights and Responsibilities

SERGE M. GARCIA*

FAO Fisheries Department, Fisheries Resources Division
Viale delle Terme di Caracalla, Rome 00100, Italy

JEAN BONCOEUR

CEDEM, University of Western Brittany
12 rue de Kergoat, CS 93837, 29238, Brest Cedex 3, France

"Resources, their scarcity, their depletion, and their conservation are concepts of the social sciences, par excellence... Property rights in resources are a primary economic institution of paramount importance in the economics of conservation." (Ciriacy-Wantrup 1968).

Abstract.—It is generally agreed that overcapacity, overfishing, habitat destruction, and wastes relate closely to the "tragedy of the commons" (i.e. to the fact that access to the resources is relatively cheap, if not free, and easy, if not open). The implication is that the establishment of a system of clear use rights will eliminate the incentive to "race for fish" and increase the user's sense of responsibility, leading to conservation. This paradigm has significant conceptual and operational implications. The concepts of conservation and allocation are well defined, and the factors of success and failure are generally well known. The performance of allocation in terms of conservation depends on the control variable (allocated factor), the rights attributes, the selection of private versus communal property, the initial allocation, the effectiveness of the rights' administration, the value of the fishing privilege, and the destination of the rent.

The interplay between allocation and conservation is iconized in the "Tragedy of Unmanaged Commons", and the solutions to the tragedy involve the use of either taxes or use rights. The interplay has a number of institutional underpinnings related to the 1982 Convention of the Law of the Sea, the Convention on Biological Diversity, the UN Fish Stock Agreement, and the regional fishery bodies and arrangements. There is a close connection between the systems of rights and the social structures using them and the two have co-evolved in time. One of the main challenges faced by governments today appears to be the choice between communal and private property.

The performance of fisheries in terms of conservation appears to be dependent on the solution of a number of allocation dilemmas between 1) consumptive and nonconsumptive use, 2) fishers and other sectors including conservation, 3) subsectors of fisheries, 4) national and foreign fishers, and 5) present and future generations.

The issues become even more complicated in an ecosystem perspective with issues related to 1) allocation between fishers (and human consumers) and other marine top predators, 2) allocation in marine protected areas , 3) the interplay between conservation and social reproduction, and 4) contrastive evolution of the spatial units selected respectively for ecosystem management and fisheries resources allocation.

*Corresponding author: serge.garcia@fao.org

Introduction

The interplay between allocation and conservation is probably as old as life on Earth. Its origin and implications have been brilliantly addressed by numerous eminent scholars and part of standard university text books. In the early 1980s, considering the issue, the late John Gulland said that it was like "telling grand-mother how to suck an egg." However, looking at the appalling state of ocean resources more than 20 years after Gulland's remark, it would seem our grandmothers still need to be told again.

The interplay between allocation and conservation has probably affected the life of our ancestors for millennia, conditioning their survival. It has excited the intellect of scores of eminent philosophers, economists, sociologists, and ecologists for centuries. It has been central to fisheries management for decades and is, today, as modern and relevant as ever as society moves towards an ecosystem approach to fisheries. The growing societal awareness of the fishery resources' distress and of the inexorable human population growth, the question of indigenous populations' rights (a sequel of colonization), and the institutional quakes of the 1982 UN Convention on the Law of the Sea (hereafter, the 1982 Convention) and the 1992 Convention on Biological Diversity (CBD) have dramatically focused the international limelight on this old but most modern issue. The old debates among economists regarding the respective efficiency and roles of individual and communal rights, of the state and of the market, are still a source of hot ideological debate, exacerbated by the ongoing process of globalization. Indeed, the problem may not reside so much in the concepts as in their practical implementation and not so much in deciding about a theoretical long-term equilibrium but in the process of transition and co-adaptation of ecosystems, human communities, and the institutions connecting them.

Conscious to be treading on an old battered track, the paper aims at providing a background to the Fourth World Fisheries Congress on allocation and conservation, with the aim of reminding this audience of some fundamentals and ongoing controversies. This review paper focuses on the interplay between allocation and conservation, rights and responsibilities, recognizing that it depends significantly on many factors, internal or external to fisheries that may not all be covered here. In order to call back to mind what the issue has always been, how it was formulated, and what was proposed to resolve it, the first section of the paper recalls the concepts and definitions related to conservation and allocation (with the support of a glossary). The interplay between the two is addressed in the second part of the paper in which conceptual aspects are described together with a series of practical examples.

Definitions and Concepts

It may be useful, for the purpose of this review paper, to clarify a multidisciplinary terminology which remains sometimes confusing (Hardin 1998). In this section, we will therefore recall the definitions and concepts of conservation seen as a central objective of management and of allocation seen as a means to achieve it. The following additional concepts and terms often used in the paper are defined in Annex 1: common pool; public resources; res communis; res nullius; discount rate; equity; efficiency; property rights; and sovereign rights.

Conservation

Conservation has ecological, economic, social, ethical, metaphysical, and emotional connotations. It tends to have two main meanings.

Firstly, it is the preservation and protection of something from change, loss, or damage, including through the prevention of exploitation. It is the maintenance of the status quo. This seems to be the meaning most generally accepted by francophones for this word, taking conservation and preservation as synonyms. Second, it is the planned management and care of (used) natural and cultural resources to prevent their destruction or neglect. It reflects wise use or sustainable use (Ciriacy-Wantrup 1968). It is "the long term maintenance of the flow of natural goods and services" (Charles 1992). In the 1980 World Conservation Strategy, it is "the management of human use of the biosphere so that it may yield the greatest sustainable benefit to present generations while maintaining its potential to meet the needs and aspirations of future generations. Conservation is positive, embracing preservation, maintenance, sustainable utilization, restoration and enhancement of the natural environment" (IUCN et al.1980). This seems to be the meaning most generally accepted by the anglophone and the economists in general. It corresponds to the concept of "strong sustainability" and is the only one making sense in this paper.

Conservation failure results in overfishing and depletion. Overfishing has been recognized and described centuries ago (Tiphaigne de la Roche 1760; Pauly and Chua 1988), and since Warming (1911), Graham, (1935), and the London Conference on Overfishing (1946), it has been clear that fishery resources could be depleted. During the last 50 years, conventional management strategies aimed formally at conservation through regulation of boats' and gears' characteristics (to limit fishing power and improve selectivity), operational constraints (e.g., closed areas and seasons, biological rest) and harvest limitations (e.g., total allowable catches, minimum landing size). The fishery management instruments, such as the 1982 Convention, have institutionalized the conservation objective, specifying that the stock size corresponding to the maximum sustainable yield (MSY)[1] is the limit below which, by convention, conservation has failed and biological overfishing occurs. This is confirmed, by necessity, in the 1995 UN Fish Stock Agreement (FSA) and the 1995 FAO Code of Conduct. More recently, the framework has been reinforced by the recent societal demand for more attention to associated and dependent species, critical habitats and, in general, the exploited ecosystem requirements (FAO 2003; Garcia et al. 2003; WWF 2003).

Constrained by the need to maintain social peace and not addressing the root causes of the phenomenon (see below), these measures attempted and failed, to mitigate the consequences of growing overcapacity. Economically inefficient, they were unable to resolve long-term use conflicts and to ensure conservation. The global consequences have been obvious for the last decades; overfishing, shortened fishing seasons, lower catch rates, decreasing harvest of lower quality, uncertain resource availability, excess harvesting and processing capacity, economic inefficiency, and social stress. The resulting high long-term costs have often been shouldered by society through direct or indirect subsidies intended to mitigate the social costs of management, compounding the problem.

It is now common knowledge that this poor sustainability is characterized by, and results from, the interaction of many interrelated factors, including:

1) The absence of guaranteed rights;

2) The supremacy of short-term socioeconomic considerations over long-term ones;

[1] Despite and accepting all its scientific and operational shortcomings.

3) A perverse incentive structure reinforcing and allowing externalization of private costs;

4) Rising demand from a growing human population and consequent rising prices;

5) Poverty and lack of alternatives in many areas;

6) Ineffective governance and weak enforcement;

7) Disturbances such as pollution, climate oscillations, and wars;

8) Scientific and administrative uncertainty; and

9) Competition between users, within and between sectors.

It is usually agreed that, in line with the general theory of sustainable development, the corrective action required includes:

1) The granting of more effective rights of use;

2) Improved transparency;

3) More participation in decision making;

4) Better understanding of the resources and the communities depending on them;

5) A more precautionary approach to management;

6) More active consideration of the ecosystem interrelationships;

7) Better monitoring and enforcement;

8) More equitable distribution of benefits;

9) Integrated development and management policies; and

10) A stronger role of consumers.

The conservation issue has taken a new and stronger development with the adoption of the Convention on Biological Diversity, signaling a change in the foundation of the conservation concept, from ethics (the need to preserve species for future generations; to consider biodiversity as a patrimony; the respect for life forms) to sustainable use. One important consequence has been the general adoption of the Ecosystem Approach to Fisheries (FAO 2003; Garcia et al. 2003), the significant consequences of which for the meaning and approach to conservation are still to be fully apprehended.

Allocation

Allocation is both an object and an act. As an object, it is a share of something set aside for a specific purpose. Synonyms with this meaning include portion, share, proportion, and quota. As an act, it is the distribution of something among selected recipients such as splitting a total allowable catch among fishing nations or assigning coastal areas to different uses. Synonyms are apportionment, allotment, appropriation, distribution, division, and repartition. The beneficiaries could be contemporary (intragenerational allocation) or belong to successive generations (intergenerational allocation). Both aspects are central to conservation.

Allocation involves the granting of rights (appropriation[2]). It may affect and be condi-

[2] According to Ostrom (1990), appropriation is the way in which the flow of resource units (e.g., catch) can be allocated in a fixed or time-independent manner so as to avoid rent dissipation and reduce uncertainty and conflict over the assignment of rights. It relates also to the allocation of spatial and temporal access to the resource, accounting for space–time heterogeneity and related uncertainties. Issues of relevance are commitment, equity, compliance, credibility, and deterrence.

tioned by preexisting ones. Allocation implies also the assignment of the responsibility of controlling the way in which these rights are exerted (provision). The solution of one set of problems must be congruent with the solution of the second. The structure of the problems, and their solution, depends on the particular configuration of variables (resources, climate, and habitat) the rules in use, and the attributes of the individual involved (Ostrom 1990).

The outcome of an allocation process (e.g., in terms of economic efficiency, conservation, and equity (see Glossary, Appendix 1) depends on a number of aspects briefly examined below.

The Objective of the Allocation

Granting access and allocating rights (and many of the characteristics of the allocation process) may be aimed at various interrelated objectives such as promoting development of an otherwise underused resource or protecting an endangered one, reaching maximum economic efficiency, maintaining self-sufficient and healthy social groups in remote rural areas, reducing use conflicts and, obviously, improving conservation. While the objective of conservation (or ecosystem reproduction) is usually clearly stated, that of social reproduction is not yet as commonly referred to and taken into account as ecosystem reproduction is, particularly in the conservation arena. There seems to be little resistance to the concept of strong sustainability in relation to the environment (implied in the definition of "conservation") but the intrinsic value of "social reproduction" may not be yet as fully understood as its environmental equivalent. It is sometimes by environmentalists, with the argument that human constructions are, by essence, evolving structures (an argument that applies indeed also to ecosystems), the possible disruptions introduced by a new system of rights on the social structure should not be a problem. The societal costs of a brutal degradation of social relationships produced by a radical shift in rights allocation (e.g., in terms of exclusion) may not yet be as well perceived as the societal costs of environmental collapse (e.g., following the cod collapse in Canada).

Allocated Factor

The first level of allocation among potential users is achieved through granting of access, usually controlled through a more or less comprehensive system of permits, ranging from simple and quasi-free administrative registration to more elaborated licensing and limited entry systems. Limited entry has been used since the early 1960s and its main problem seems to be the difficulty to reduce the initial overcapacity fast enough to mitigate the effect of technological progress despite difficult and costly buy-back schemes.

In most cases, because of their fugitivity, the resources themselves cannot be allocated between those having been granted access rights. As a consequence, the factor that is allocated (i.e., the control variable) is the best possible albeit necessarily imperfect proxy to the resources. This proxy could be:

- Space (e.g., through territorial use rights, with potential problems with mobile resources);

- Effort (e.g., through effort quotas, with potential problems with fishing capacity; time sharing, etc).

- Catches (e.g., through catch quotas, individual or not, transferable or not, with problems of declarations, highgrading, discards, etc).

Rights' Attributes

The attributes of a right determine its effectiveness in a particular situation. These attributes should relate to the characteristics of the resource or resource system such as its location, geographical structure, distribution and mobility, variability, abundance, resilience and value. They should also relate to the characteristics of the fishers community, including its history and culture, and be compatible with fishing operations.

The main attributes include exclusivity, duration, security, and transferability. Other attributes such as flexibility and divisibility are also sometimes mentioned (Ciriacy-Wantrup 1968; Scott 2000; Stokes 2000). Exclusivity gives protection from legal interference and provides the facility to exclude others. Duration encourages the sense of ownership of the user, provides conservation incentives, and facilitates the use of the right as collateral. Security facilitates the defense of the right, increasing the assurance of the holder that it will keep it, and contributing to the conservation incentive. Transferability of allocations to buyers, renters, or descendants is usually considered necessary to provide the flexibility needed for future evolution of the fishery and population of right holders, as well as to protect the family in case of death of the right holder. It provides a mechanism to deal with potential new entrants, increases economic efficiency but may lead to undesirable concentration if provisions are not made to avoid it. Mechanisms exist to permit state intervention on the market to re-establish equity as needed. However, full transferability is often opposed to ensure that the original allocation to nationals or disadvantaged groups is not modified. In France, transferability is formally illegal (Boncoeur and Troadec 2004), but in practice fishing rights are transferred with the vessel, the price of which includes the implicit right to fish. Flexibility will allow the system to adapt to changing conditions of the resource or the sector. Transferability and divisibility add to flexibility. For highly variable resources, allocating proportions of the resource flow (e.g., % of the Total Allowable Catch [TAC]) instead of fixed quantities will permit the adjustment of withdrawal to natural conditions. Divisibility allows a right to be split into components (species, seasons, subareas) that may be allocated to different right holders. The possibility to market some elements of the bundle of rights without affecting the property allows a significant reduction of transaction costs (Weber 2002). Enforceability is self-explanatory and could be considered as part of security.

The Right Holder: Private versus Communal Property

The right holder could be a country (as per the 1982 Convention), a community, a fishing company, a vessel or an individual. Selecting individual rights (private property[3]) or communal rights (common property) is probably the hottest policy issue of the allocation process, and the relative advantages or drawbacks of the two options in terms of efficiency, equity and sustainability have been debated for centuries (Ostrom 2000). While some scholars consider contemporary examples of common property as remnants of an irrelevant past bound to disappear, recent research challenges that presumption, and there is substantial evidence that, in the past, in resilient rural communities, communal property has been able to solve a wide diversity of problems at low interaction costs (Ostrom 2000: 333). Under the modern conditions of market globalization, however, it is not obvious that the conditions for this performance still exist or can be re-established[4].

[3] See definition in Annex 1.

[4] At the beginning of the 20th century, traditional rights have been voluntarily weakened by national and colonial powers, generalizing a situation of quasiopen access and common property (Ostrom 2000; Weber 2002).

The choice is rich in long-term consequences, politically delicate and conditioned by the objectives. The neoclassical economists advocating private property (e.g., individual transferable quotas) tend to stress maximum long-term economic efficiency and benefits (including conservation), with a risk on equity. The socioeconomists stress that communal property rights may be more indicated when controls are difficult, and one of the aims is to maintain human settlements in remote rural areas where livelihood alternatives are scarce. In this case the exclusion often implied in individual rights may be unacceptable (Ostrom 1990: 22), even though this solution may not be economically optimal. The proponents of communal rights argue for other sets of values including equity, ethics[5] and social reproduction[6].

The proponents of private forms of property argue also that: 1) by setting everyone's share they eliminate the "race for fish" and reduce the incentive for "stuffing in" technology, reducing costs and overcapacity; 2) they allow each individual to fix the timing of his harvest, choosing times where the weather is better (reducing life risk), the species is of higher quality, the bycatch or concentration of juveniles is lower, and so forth; 3) they turn the right to fish into a transferable asset of fairly well established value; and 4) to the extent that they allow stock rebuilding, they reduce the uncertainty of the fisher as to what fish he will find where and facilitate contractual arrangements with buyers.

For their detractors, these forms of property presents drawbacks such as: 1) unintended social effects such as nonfishermen owners and concentration of ownership, a phenomenon largely observed and apparently accepted, however, in all other industries[7]; 2) difficulty to accurately adjust the quotas to inter-annual changes in reproductive success, abundance or availability because of uncertainty, reducing therefore the accuracy of the management response (even when quotas are defined as percentages of a variable allowable annual catch)[8]; 3) political risk for elected officers in lowering the total allowable catch, reducing management response effectiveness[9]; 4) increased incentives for under-reporting and black market of unreported catches; 5) increased incentive to discard (highgrading; discards of species when its quota has already been reached); 6) difficulty to reduce capacity fast enough in highly overcapitalized fisheries; and 7) inadequacy for an ecosystem approach to fisheries.

The two sets of proponents differ also in the way they treat exclusion. On the one hand, limiting access to establishing individual rights is likely to result in the exclusion of the least efficient actors (perhaps the poorer), either initially or during the process of economic "optimization". The market is the mechanism used for exclusion and compensation. Equity is generally not considered explicitly. On the other hand, granting exclusive access to a whole "community" reduces the initial exclu-

[5] In the long-term, equity, a concept rarely specified, may not be in contradiction with economic efficiency and is implied in the concept of long-term optimal social welfare.

[6] Collet (2002) stresses that in traditional society's cooperation, solidarity, participation and democracy, morality (ethics), dignity, stewardship, and respect are instruments and processes which are operational in community property rights.

[7] Concentration of rights in a few hands is often perceived as a threat to artisanal enterprises, although it has already started, even in the present conditions of ownership, for reasons of economies of scale (Lequesne 2003).

[8] Even when relative quotas are used, quota holders may still exert pressure to ensure the highest allowable catch possible. Such pressure may be attenuated, however, by 1) transferability, because the market value of the property right (if complete) depends on the state of the resource and 2) associating the quota holders to the determination of conservation measures in a transparent manner.

[9] The argument is as weak as this risk is, of course, just as high in a conventional management system.

sion problem and allows explicit consideration of equity. However, exclusion of noncommunity members, often violently, is also a characteristic of kinship-based traditional fishery management systems. Whether market-based exclusion is less ethical than exclusion based on racial or ethnic criteria is debatable.

While individual transferable quotas (ITQs) are systematically used in countries like Australia, New Zealand, Canada, Iceland and Netherlands, they seem to be considered as unacceptable in principle by a large part of the EU fishers in Italy, Spain, and Greece. In France, a recent law provides that fishing authorizations cannot be transferred, and fisheries are considered as a "common heritage" (patrimoine collectif)[10]. This last terminology may relate to the "Common Heritage of Mankind"[11], an asset defined in the UN in the 1960s and which, by definition, goes beyond a *res communis* (see Appendix I) and cannot be appropriated. In the USA, the 1996 Sustainable Fisheries Act established a moratorium on ITQs. In Iceland, however, where a similar terminology was used in the law, the Supreme Court decided that it should not impede the free allocation of ITQs based on fishing history. In countries were ITQs have been adopted some time ago, the debate is still raging (Collet 2002).

The common property challenge needs to be faced by governments if small-scale fisheries are to survive globalization. The conditions for successful common property include (Ostrom 1990: 90): 1) clearly defined right holders (and hence exclusions) and resource boundaries; congruence between allocation/conservation rules[12] and local conditions; 2) collective choice arrangements in participative framework, allowing fine tuning of the institutions to local conditions and improved enforcement; 3) reliable monitoring, in which those in charge of monitoring are accountable to the right holders; (iv) graduated sanctions, possibly inflicted by the right holders themselves or their institutions, and social recognition of good behavior; 4) effective (rapid-access, low-cost, equitable) conflict-resolution mechanisms; 5) recognition by higher authorities of the right to organize, implying an effective transfer of the management right; and 6) institution nesting to effectively address cross-jurisdictional problems or problems at different scales.

Conversely, factors influencing success or failure in developing appropriate institutions include: 1) the number of decision makers (the less the better); 2) the number of participants (a minimum is needed to achieve collective benefits but large numbers may be a problem); 3) the diversity of participants (similar profiles, perceptions, interests and discount rates help achieving the necessary consensus); 4) presence of local leadership. Communal right holders in common pool resources systems have shown capacity to develop sustainable institutions and achieve conservation but they have not always done so. A number of inexplicable failures or successes indicate that the present understanding is not sufficient to elaborate unambiguous policy advice. It is advisable therefore to proceed with incremental action, starting at small scale, building on existing institutions and their self-transforming nature (Ostrom 1990:188–191).

The Initial Allocation

The initial allocation requires identifying the future right-holders. A lottery would random-

[10] Loi d'orientation sur la pêche maritime, Journal Officiel de la République Française 97051, du 19/11/1997.

[11] The term was created in November 1967 when Ambassador Arvis Pardo, Permanent Representative of Malta to the United Nations, urged delegates to consider the resources of the oceans beyond national jurisdiction as "the common heritage of mankind".

[12] Ostrom refers to appropriation and provision.

ize the selection and reduce corruption without ensuring equity. A sale (e.g., an auction) would probably be economically efficient, but equity would also not be addressed, as wealthier elements of the community or rich foreigners would be favored[13]. Alternatively, a regulatory or political process could be used with criteria related to historical antecedents, existence of alternative livelihoods, vulnerability, maintenance of rural communities, ethnic origin, and so forth.

Preexisting allocation patterns must be actively looked for as the new allocation of rights is likely to interfere with existing ones, potentially leading to accidental exclusion, social unrest, etc. The process should be fast enough to avoid massive entry before the law is operational. The so-called Coase Theorem (Coase 1995) holds that, if transaction costs are negligible, the economic outcome will be optimal, regardless of the initial allocation. It also stresses, however, that the outcome might not ensure equity. By symmetry, if there are significant transaction costs, the efficiency of the outcome should depend on the initial arrangement and the solution is to minimize transaction costs (Edmund-son 1995). In other words, according to Coase, it is the law that should define property in such a way as to minimize the costs of interaction between incompatible uses.

The initial allocation to new entrants is a general and recurrent problem in international and many domestic fisheries. Open-ended rights, which for existing right holders is a sword of Damocles, seriously reduce the security (and value) of the right and the incentive to conserve. The "real interest" of the fishing State applying for entry into an existing international sharing arrangement needs to be assessed but the basis for such an assessment is uncertain (Munro 2003).

The physical size of the share allocated to each right holder, if not left to the market through which each holder acquires what can be afforded, can also be negotiated using arguments such as historical antecedents (e.g., catch records), local knowledge, the position of spawning, nursery and feeding areas, concentrations of juvenile, biomass distribution and migration (defining the zonal attachment[14] in the case of straddling stocks). The agreement (e.g. on shared stocks, may include compensations and side payments as negotiation facilitators). Highly participative systems seem to be preferable and most likely to reach acceptable agreements (Munro 2003).

Administration of the Rights

Rights need to be protected in order to maintain their conservation and socio-economic efficiency through time, ensuring the existence and enforcement of rules, defining who has the right to do what and how the returns of that activity should be allocated (Ostrom 2000: 334). This requires the establishment of institutions (e.g. for monitoring the evolution of the fishery and the resource, verify conservation and economic performance as well as equity, forecast natural oscillations, ensure compliance, etc). The less effective the enforcement is, the higher the uncertainty about the future returns from the right and the lower the incentive to conserve for the future. An important element of any allocation system is the availability of dispute resolution mechanisms. With quasi property rights, ordinary tribunals may be able to resolve any dispute. In the case of international fisheries (global commons) the International Tribunal for the Law of the Sea

[13] In many countries, the allocation would probably be done with some expression of a national preference.

[14] (i.e., the degree of connection between a stock and a particular exclusive economic zone)

(ITLOS) is competent and it has recently intervened in the dispute related to the management of the Pacific Southern bluefin tuna. Local institutional capacity-building is essential for a decentralised administration.

Value of the Privilege

Obtaining an exclusive share of a resource belonging to the Nation as a whole is obviously a privilege. This privilege has a value and indeed often leads to a request for explicit or implicit "compensation" when the privilege is lost. If those fishers who may have to be excluded from the allocation need to be compensated for a loss, those obtaining the privilege might obviously be requested to pay for it. This is of course a hot issue, as fishers usually ask for compensation if forced to leave a fishery but are reluctant to pay for the privilege to continue fishing, particularly when the stocks are low and profitability is marginal. This question, too complex to be treated here, relates to the difficult one of the allocation of the rent (see below).

Destination of the Rent

The (economic) rent generated by any natural renewable resource in general and a fishery resource in particular is the difference between the total revenues obtained from the fishery resource and the total costs of production (i.e. capital and labor valued at their opportunity costs). The total costs of production include a reasonable profit and the rent is often considered as a "surplus" profit, over and above what would be considered a "normal" rate of return (FAO Fisheries Glossary, http://www.fao.org/fi/glossary/default.asp*)*. The rent varies with the intensity of fishing and can be maximized. In overfished fisheries, the rent is dissipated through over-use of capital and labor. Through resource rebuilding, the rent can be re-established progressively as profitability goes up with stock size. It could be recuperated by the State through some form of taxing or fishing-right fee under the user pays principle and used for the benefit of society. If the rent is left to the right holder, it is capitalized at the first sale of the right by its first holder (as a windfall gain) and is passed to the next right holder as an additional investment cost, possibly replacing overcapacity with over-investment.

Allocation and Conservation Interplay

Allocation and conservation are functionally interdependent. While conservation (or sustainable use) is the fundamental objective, allocation is one of the means to achieve it. If conservation fails, and resources are depleted, there will be no stable flow of resource units to allocate. Conversely, without adequate allocation of rights and responsibilities, there is no incentive for conservation and resources are depleted.

In fishery management literature, conservation is usually not a conspicuous part of the argument that tends to focus essentially on economic efficiency and marginally on equity. One of the reasons may be that rent maximization, the objective of reference of most economic studies, is supposed to maximize the long-term flow of wealth to society (Hannes-son 1993; Boncoeur and Troadec 2004), an outcome that cannot be obtained—under strong sustainability constraints—without conservation of the resource flows. However, the transitional costs related to marginalization and exclusion, in certain forms of allocation may result in social stress, civil unrest, noncompliance (piracy, illegal fishing), destructive practices (dynamite, poison), and so forth that may not be properly accounted for in performance assessments and may seriously compromise the conservation outcome of the allocation process.

The evolution of the environmental capital (towards conservation or depletion and failure of ecosystem reproduction) is convoluted with that of the sociocultural capital (towards development or poverty and failure of social reproduction) and both can be at the same time cause and consequence of the other. Despite the recognition of this strong functional link, most international agreements appear reluctant to link explicitly allocation and conservation. For instance, allocation is mentioned only once in the Code (Art. 10.2.2) and once in FSA. By comparison, access to resources is mentioned six times in the Code and eight times in the UN Agreement. The term is also used in the CBD. Rights are mentioned ten times in the Code, referring to sovereign rights of states (eight times) or people (twice) and five times in the FSA (in reference to sovereign rights of states' rights). By contrast, conservation is mentioned 61 times in the Code and 26 times in FSA and is a keyword throughout the CBD. In conclusion, while states refer easily to conservation as an agreed universal objective or type of use, they seem to be more reluctant to explicitly relate it to allocation or rights. A similar remark can be made regarding poverty eradication (easily agreed as a universal objective) and equity (a much more sensitive concept to agree with).

Allocation institutions have a consequence on social structures and vice versa (Collet 2002). This points to the need to study existing social structures and rights before introducing modifications (Ostrom 2000). We have stressed above that existing traditional rights had often been suppressed in the early 20th century in developed countries and their colonies. During the colonial era, many developing countries' borders were arbitrarily established and these were maintained at independence. The establishment of nationhood has often been obtained at the expense of traditional (tribal) rights, replacing or super-imposing traditional rights with the new State ones, weakening traditional institutions to strengthen modern ones. In some of these countries, the concept of reallocation of property and use rights to indigenous or traditional communities raises for policy-makers the spectra of renewed ethnic conflicts, and solutions will be needed to achieve the benefits of re-allocation without endangering the nations' unity. The issue is not theoretical, as demonstrated by the relation between the outbursts of civil (ethnic) wars and the discovery of rich resources (oil, diamonds). This may not always be a violent evolution, however, as shown in New Zealand, where the modern recognition of Maori rights (contained in the 1840 Treaty of Waitangi) is said to have led to the development of "tribal capitalism" institutionalizing ethnic differences in that country (Rata 2003).

Allocation and Conservation Are Connected with Equity and Poverty

The growing identification of conservation with sustainable use or sustainable development connects allocation and conservation directly to poverty and equity (Weber 2002). Lack of equity may lead disadvantaged actors (excluded from allocation) to poverty, high discount rates, noncompliance and predatory behavior. As a consequence, achieving conservation requires reduction or eradication of poverty and extreme forms of inequity (WCED1987). This, in turn, requires allocations of rights. "The way out of poverty starts from the formal recognition of secured rights of access to land, resources and public goods" (Barbault et al. 2002). The progressive destruction of agricultural soils and forests, despite apparently complete forms of property, may indicate that property is a necessary but not sufficient condition for sustainable use. The outcome of an allocation scheme will depend both on the quality of the right and the social and institutional conditions within which it is exerted, particularly the mechanisms of redistribution of the benefits.

The Tragedy of Unmanaged Commons[15]

The failure of many international instruments to explicitly connect the desired conservation and the necessary allocation may be consequential. The 1946 London Conference on Overfishing failed to lay the basis of the sustainable development of fisheries in the Northern hemisphere because states could not agree on effort and capacity allocation. The present poor state of fisheries in that region can be directly connected to that failure. It would be naïve, however, to assume that this happened out of ignorance.

According to the literature, the fact that local natural renewable resources could be depleted through wasteful competition because of lack of ownership has been known for centuries, since Aristotle, the Greek philosopher, and Justinian, the Roman emperor. This understanding has been recurrently rediscovered, reformulated, completed by Lloyd (1833); Warming (1911); Graham (1935); Whitehead (1948), Gordon (1954) and Hardin (1968). By the end of the 1960s, the "tragedy of the commons", an expression attributed by Hardin to Whitehead, was already common knowledge (Ciriacy-Wantrup 1968:142–145). This signaled that, contrary to what was conveniently assumed, the freedom of the sea was a source of problems[16] and the ocean resources were exhaustible[17].

The emerging axiom, central to sustainable development, could be expressed in many popular ways: no stewardship without ownership; no responsibilities without rights; no environment well-being without human well-being; no conservation without allocation. In more technical terms, the economic theory indicates that, as the fisheries resources most often have the characteristics of a common good, undivided and subject to subtractive use, their fishing generates a set of crossed negative externalities between fishermen. In a competitive context with no or weak regulations, this leads to divergence between individual and collective rationality, the individually optimal effort level being systematically higher than the collectively optimal one. The divergence grows with increasing pressure on stocks, leading to a "race for fish", chronic overcapacity, overfishing, and social conflicts.

As Ostrom (1990) strongly argued, the Tragedy of the Unmanaged Commons (Hardin 1998) does not imply that all resources owned in common are bound to be wasted. In small rivers, lakes, estuaries, very coastal waters, and around small islands, the evidence of resources exhaustibility, the effects of the resulting scarcity and an easier defence than in the open sea, may explain the early adoption of forms of common property regimes (*res communis*) by traditional communities. Often area-based, combined with technical regulations and taboos aiming at conservation, some of these systems persisted for centuries but many disappeared in the processes of modernization, colonization, independence[18], construction of the Nation State, and globalization of the market economy (Christy 1982; Johannes 1981; Ostrom 2000; Feeny et al. 1990; Berkes 1985; Collet 2002; Kurien 2000; Mathew 2003; Weber 2002).

---

[15] The Tragedy of the Commons (Hardin 1968) has been recast as the Tragedy of Unmanaged Commons (Hardin 1998).

[16] The freedom of the sea instituting free and open access in the oceans, conveniently elaborated by Hugo Grotius (1609) on the assumption that ocean resources were inexhaustible (there was no scarcity) and ocean property would be too difficult to defend anyway.

[17] The inexhaustibility of the sea was assumed by Grotius (1609) and supported by T. H. Huxley. Doubts about this opinion were expressed however by Alfred Marshall who, at the end of the 19th century recognized the productivity of the seas but expressed concern about the impact of the steam trawlers development forecasting that "the future population of the world will be appreciably affected ... by the available supply of fish" (Marshall 1920).

[18] On the African continent the laws expropriating traditional communities were adopted around 1929–1930 (Weber 2002).

The Tragedy's Solution: Pigou's Taxes or Coase's Property Rights

The conventional economic treatment of externalities originates in the theory of social costs of Arthur Cecil Pigou in his "*Wealth and welfare*" in 1912 and "*The economics of welfare*" in 1920 which brought social welfare into the scope of economic analysis. According to his analysis, an individual using a (scarce) common-pool resource creates negative externalities for other users (i.e., additional costs that are not borne by himself, but by society). This, in turn, generates a societal inefficiency in that the responsible activity will be developed beyond what would be optimal. In a Pigovian perspective, externalities are seen as a market failure in that competition fails to lead to efficiency. The public administration needs to correct the failure, forcing the producer to "internalize" the societal cost of its activity, taking it into account together with his more private costs of production. The Pigovian instrument of choice is a tax equal to the cost of the externality to society (or a subsidy in the case of a positive externality).

In environment management, the polluter-pays principle is a direct application of the theory. In fisheries, the theory would lead to taxing effort or catches, increasing the real cost of fishing (of already economically weak industries) to bring it back the level of withdrawal compatible with the socially optimal resource level, modulating the tax on the various fisheries to redirect efforts to less pressurized resources. Once instated, a tax would need to be increased as the resource improves to maintain its efficiency. However, taxes are not often used to regulate access to fisheries resources, and present practices to use subsidies reflect instead an opposite principle of negative taxation.

In stark contrast with Pigou's theory, Coase (1960) considers that externalities do not reflect market failures but result from an incomplete definition of rights without which a market cannot effectively operate, generating high interaction costs. According to him, the solution is not in state interference with the market through taxes but in freeing the market forces through definition of rights and establishment of institutions and mechanisms to exchange the rights (market) and reduce interaction costs (e.g., regulations and tribunals). For Coase, numerous externality issues could be solved through decentralised negotiation between actors. In environmental management, this is reflected by the establishment of transferable "pollution rights". In fisheries, it may lead to individual transferable quotas (ITQs) or other kinds of transferable individual fishing rights. According to the so-called "Coase theorem", the initial allocation of rights has no bearing on the final economic efficiency, since in any case the most efficient economic agents will appropriate the rights through the market. Of course, this has nothing to do with equity, which means that the process does not guarantee a societal acceptable (ethical or equitable) outcome. Moreover, as it was long ago demonstrated by Coase, property rights may be an inefficient framework of organization of production when transaction costs are high (Coase 1937).

Allocation Dilemmas

The explicit allocation of resources among individuals or groups of the present generation is one of the most complex and delicate of the tasks faced by a government because of the high potential financial and political costs incurred. Not surprisingly, these decisions are among those that governments often do not like to take or publicize.

Consumptive and nonconsumptive uses compete for resources. As the impact of unsustainable forms of use becomes more conspicuous, a movement in favor of nonconsumptive uses develops. This materializes through requests for allocation of resources to forms of ecotourism (e.g., on coral reefs), whale watching, and so forth at the expense of fisheries and other forms of exploitation, even though some forms of ecotourism are sometimes criticized by environmentalists for their impact on the environment. Such allocation may take place in the framework of integrated coastal areas management (ICAM) or marine protected areas (MPAs). When the objective is to exclude fisheries (e.g., in no-take zones), opposition and noncompliance is to be expected, unless decided within a participatory approach in which transitional difficulties are taken into account and mitigated. This issue is getting more emphasis in the context of the ecosystem approach to fisheries.

A second level of allocation is among consumptive uses (i.e., between the fishery and other sectors in a coastal zone or ecosystem and within the fishery sector itself, between concurrent subsectors or fisheries segments.

Allocation between fishery and other sectors needs to take place within a spatially integrated management framework such as the coastal zone in which economic activities compete for scarce space and resources within the exclusive economic zones (EEZs), and this highly competitive behavior leads to resources degradation, particularly in the coastal area. Management integration requires a framework within which choices are made and policies are implemented. The main issues relate to 1) the selection of compatible activities and exclusion of others, 2) the allocation of space to activities or groups of compatible activities within a comprehensive coastal space-use planning, and 3) the allocation of resources other than space among competing activities. The FAO Code of Conduct contains specific guidance for the integration of fisheries into coastal area management (Chua and Fallon Scura 1992; Fallon Scura 1994; FAO 1995). The allocation to conservation may be best decided in that framework. It might take the form of an allocation of quotas of preys to predators and the establishment of sanctuaries or marine protected areas. It may often involve a reallocation of resources. With some exceptions, the Integrated coastal area management frameworks have not been very successful yet, perhaps because they represent one of the most complex area of application of the Coase theorem in which 1) resources are numerous and of different types, often mobile and impossible to delimit precisely, 2) users are numerous and with different objectives competing for space and resources, 3) severe externalities are imposed by land-based activities, and 4) uncertainty and interactions costs are very high and cannot easily be reduced. Applying the Coase theorem *a contrario* would imply that 1) property rights may not be a workable solution unless transaction costs are reduced (e.g., through zoning) and (2) the initial allocation (and some agreement about it) is fundamental to the outcome.

Competition within the fishery sector is one of the most serious issues for the future of fisheries. Conflicts exists between fisheries of similar scale, for space (e.g., gear competition), resources (biological interactions, bycatch) or both (e.g., between aquaculture and capture fisheries). Space can be allocated through zoning to reduce interactions. Resources can be allocated within multispecies fisheries management, taking into account gear and species interactions, but success in that direction has been very limited. Major conflicts exist between 1) small- and large-scale fisheries, 2) subsistence and export fisheries, 3) recreational and professional fisheries, 4) capture fisheries and aquaculture (e.g., extensive aquaculture; ranching). In countries where aquaculture is rapidly developing, that industry is looking for more and more complete and secure rights

(Harte and Bess 2000). As these rights crystallize and harden, flexibility and reversibility are reduced, and the potential for conflicts increases as conditions change.

In countries with important and dominant nonindigenous population (USA, Canada, Australia, New Zealand), the issue of allocation between sub-sectors is compounded by ethnic considerations. In this highly sensitive context, the question of traditional rights and respect of colonial treaties plays a key role.

The European Union "common pond" offers a particularly interesting case. The European Commission struggles with the issue of widespread overfishing and rampant overcapacity. The management process is in two stages: 1) a conservation-based scientific analysis and negotiation leading to advice on TACs (including precautionary TACs when appropriate) at ICES level; followed by 2) a strictly political negotiation in which the maintenance of the socio-economic status quo (e.g., preservation of traditional fishing), social peace and the equilibrium allocation between countries (under the "principle of relative stability") overrides the question of conservation (Boncoeur and Mesnil 1999; Lequesne 2003). It is an example in which allocation of access and outputs does not lead to conservation despite a complex system of allocation and suballocation, officially within States and unofficially through quota hopping. It is interesting in that respect to see the development of a debate on "nationalization" of resources opposed by those with more mobile fleets (in Spain, France and Netherlands) and favored by those where local fleets dominate (in Portugal, Ireland, Scotland).

Allocation between nationals and foreigners is another thorny issue, particularly in developing and developed coastal countries, endowed with more resources they could harvest themselves. Applying the concept of "surplus", these countries gave access and quantitative harvest rights[19] to fleets belonging to Distant Water Fishing Nations (DWFNs). Following the extension of their jurisdiction, these coastal countries have rapidly developed their own harvesting and processing capacity, often without cutting down on the previous agreements, precious sources of foreign exchange. The consequence has been a huge duplication of highly subsidized investments and development of a large overcapacity that the world fishery sector is still in the process of "digesting." In the process, conservation took a plunge while the majority of the world stocks were intensively exploited or overfished.

Allocation of shared resources (resources totally or partially outside EEZs) remains a serious issue and stumbling stone. The spectacular damage of the resources in the Donut hole (Bering Sea), Peanut hole (sea of Okhosts), Nose and Tail of the Grand Banks (North America), and so forth, and more recently of sea-mounts, illustrate the problem. The use and management of transboundary stocks must be negotiated with neighboring EEZ owners. While scientific collaboration usually exists, often under the aegis of FAO, the main difficulty remains the negotiation for the allocation of shares to the various countries concerned. The straddling, highly-migratory, and purely high-sea resources remain global commons with decreasing preferential rights allocated to the coastal state. They need to be exploited and managed in collaboration through regional fishery management bodies and arrangements and management measures inside the EEZ and outside it shall be compatible (Article 7.2).

Intergenerational allocation is a key to conservation. Sustainable development and re-

[19] Expressed in vessel or catch tonnage.

sponsible fisheries management require maintenance of the resources and the ecosystem productive capacity for future generations. This is a case of intertemporal allocation of resources. While it has been proposed in the past by economists that some resources might be foregone in exchange for man-made capital offering equivalent or higher opportunities to future generations (soft sustainability), the present international consensus as illustrated by the 1982 Convention and its aliases (including the FAO Code of Conduct) is that resources should be passed unchanged to future generations (strong sustainability). Impact on these resources by present generations should be reversible within an acceptable timeframe. Considering the general disapproval of overfishing, the implication is that present generations are allowed to use the productivity of the stocks (the "interest") while the natural capital should be preserved for future generations. From a stock perspective, this has been attempted for the last 50 years at least, albeit with generally poor results. From a general ecosystem perspective, habitat destruction, persistent pollution, deforestation and global warming do not indicate a better performance.

Allocation and Conservation in Ecosystems

The allocation and conservation question becomes a lot more complex in the context of ecosystems, and a number of new issues are emerging. Ill defined and poorly understood ecosystems represent a complex, variable and changing pool of interrelated resources. Critical elements of the ecosystem (e.g., critical habitats) may be fully allocated to conservation. Animals attached to the bottom (e.g., oysters, clams, coral reefs) may be firmly allocated on a territorial basis together with a thin layer of water above it. The problem increases with the fluidity of the resources and the number of interacting sectors. The fact that ecosystem management objectives, parameters, indicators and reference values are not yet clear is reflected in the fuzziness of the allocation issue in ecosystem-based management. The "reflex" of environmentalists may be to ask for exclusion of any (consumptive) use (e.g. establishing a network of reserves as a precautionary device). This can, however, be a solution only in a small part of the biosphere such as no-take zones in the middle of managed MPAs. Obviously, more work is urgently needed in ecology and social sciences to resolve the issue. In an ecosystem perspective, allocating a resource leads to allocating a productivity chain, with its preys and predators. A network of causal links appears between rights holders fishing in the same area at different trophic levels or in different areas. The ecosystem approach to rights-based fisheries requires therefore an allocation matrix including shares for fishermen, predators, conservation and insurance purposes, as well as for other uses. The difficulty of managing shared ecosystems is one order of magnitude more complex and was already raised by Gulland (1984) 20 years ago, when he wrote that "there are no international arrangements to facilitate the necessary trade offs ...to manage ...according to an ecosystem approach".

Allocation between fisheries and other predators is progressively gaining importance as resource-rebuilding strategies are being studied to comply with the requirements of the World Summit on Sustainable Development (WSSD). Fishermen are human predators operating on behalf of fish consumers. As such, they compete with other natural predators for space and resources. In Australia, of about 1,000,000 metric tons (mt) of fishery resources being extracted from the sea, 20% are taken by fishermen, 78% are taken by marine mammals and birds, the rest being taken (allocated) to other uses, including conservation (Kearney 2003). The implication is that other predators consume more fish than humans, a fact that society might not have "registered" and formally analyzed yet. The argument has been used to suggest that reducing the abundance of predators would free resources for humans (Tamura 2003), a proposal raising serious concern and opposition in various parts of society. At the same time, the present depletion of natural predators has affected the quality and level of fisheries harvest (Garcia and Newton 1997), as well as the ecosystem structure and functioning, with important consequences for fishers themselves (Pauly et al. 1998; Jackson et al. 2001).

Competition between fishers and fish also occurs for instance when nets set for other purposes interfere with the passage of large cetaceans, entangling them. This interaction leads to damage to the fishing gear and can be fatal to the animals. The response sought by conservation has been the establishment of sanctuaries as well as the banning of drift nets. In both case, the economic activity is simply obliterated. Competition for food resources occurs when the fishermen and an animal predator look for the same prey. The competing predator may be targeted by fisheries (e.g., cetaceans, tunas, sharks and groupers) or not (e.g., seabirds). In both cases, an allocation of preys to some predators to ensure their conservation has been proposed and implemented, e.g. in the North Atlantic.

Marine protected areas may be seen as a test bed of the ecosystem approach to fisheries and of the interplay between the CBD and the 1982 Convention. They imply in most cases a nonexplicit total or partial re-allocation of existing of fishing communities, often without compensation. While their impact on local biodiversity is evident, their systemic impact on the resource systems and on fisheries and communities is improving but remains poorly documented. There is still some way to go before full consensus is developed between fishery scientists and ecologists (Hilborn et al. 2004) and a fortiori between fishers and conservationists, and more testing is needed. As MPAs grow in size and include more uses, their use becomes as problematic as in coastal area management, requiring, for instance the development of alternative source of livelihood for displacement of existing activities to other areas.

Spatial boundaries are central to both allocation and conservation. Garcia and Hayashi (2000) have stressed that, during the last 50 years, the allocation of world resources has progressed through progressive splitting of the resource pool into successively smaller bundles[20], from global commons to ITQs. During the same period, however, the requirements of conservation, particularly the Ecosystem Approach to Fisheries (EAF), demand integrated management by larger and larger areas and a change of the management scale from single species to multispecies stocks, from assemblages to whole ecosystems, from MPAs to coastal areas and Large Marine Ecosystems.

The contrast between the two evolutions is striking. The first leads to more and more specific and complete rights at the smaller possible scale (individual or communal). The second, on the contrary, leads to the definition of objects the growing complexity and uncertainty of which do not easily lead to manageable rights, except through total exclusion. The first is required for decision making to establish security and liability and generate responsibility. The second is required for scientific understanding and awareness-raising. If the two processes of allocation and conservation

[20] 1) between high seas and the 200 miles zone; 2) into EEZs; 3) into fisheries; and 4) into individual quotas resources.

are to be integrated towards sustainable development governance, it will be necessary to develop institutional bridges between the two processes at the appropriate scale.

Institutional Underpinnings

The interplay between allocation and conservation in fisheries is governed by international instruments, regional mechanisms, and national legislation. Some of the related issues are briefly addressed below.

The UN Convention on the Law of the Sea is the legal foundation of sustainable development in the ocean. In 1945, the unilateral extension of USA's jurisdiction to its entire shelf (Truman Proclamation), followed in 1952 by Chile, Peru and Ecuador claiming over a 200-mi jurisdiction, allegedly to protect their coastal resources from depletion by DWFNs, ignited a progressive but revolutionary process of re-allocation of ocean resources between Nations. The process took about half a century, culminating with the adoption, in 1982, of a legally binding Convention which entered into force in 1994. The convention is the legal foundation of sustainable ocean development and it provides that "In the exclusive economic zone, the coastal State has... sovereign rights[21] for the purpose of exploring and exploiting, conserving and managing the natural resources, whether living or nonliving, of the waters superjacent to the seabed and of the seabed and its subsoil...*(*Article 56.1)". Sovereign rights include the rights to legislate, manage, exploit, control access, and to determine property regimes applicable to the resources. The right of alienation (or right to sell, grant, lease), which together with the others constitutes complete property, is granted in Article 72 while Article 62.4.1 stipulates that the state can fix quotas.

The convention establishes therefore the conditions for states to reallocate these resources within their jurisdiction and negotiate allocations for shared resources (whether highly migratory, transboundary, or straddling). It does not provide any explicit guidance about allocation except in relation to the surplus[22]. In addition, the coastal state has an obligation of conservation, a duty of stewardship, under a concept of strong sustainability[23].

The rights resulting from the harshly negotiated convention are complex and reflect a gradient in property depending on the resources types. They appear to be exclusive and close to full ownership when the resource is either fixed to the shelf or entirely circumscribed in the EEZ. If eggs and larvae are dispersed beyond the EEZ, the ownership cannot be complete as the coastal state loses control of one or more life stages of its resource. Ownership of the coastal state is also ensured for anadromous resources that originate in the coastal state's inland waters such as salmon and sturgeon.

The convention explicitly provides for joint ownership when the resources form a common pool extending through two or more EEZs (transboundary stocks) extend into the adjacent high seas (straddling stocks) or migrate on large distances (highly migratory species).

---

[21] See Glossary, Annex 1, for a definition.

[22] The concept of surplus qualified the sovereign rights. They are not completely exclusive as the coastal state shall give to other states access to any surplus it does not have the capacity to harvest itself (i.e., the difference between the national catch and MSY). This provided a mechanism to facilitate access of DWFNs to EEZs and ensure that all stocks are exploited at maximum sustainable yield (MSY) and it has been used in the past by some fishing nations during fishing agreements negotiations. The problems raised by this provision have been discussed by Garcia et al. (1986). Following the 1995 FSA, however, MSY should be considered as a limit (not to reach) and not as a target, and the concept has lost any practical relevance.

[23] The state does not have the right to exhaust (extinguish) the natural renewable resource put under its jurisdiction by the convention.

As the high seas portion of the stock is accessible to all, the convention established in fact a regime of more or less regulated global common. For these shared resources, the convention requires instead that divergences be solved through international cooperation. The respective shares must be negotiated and the resources should be managed in a bilateral agreement or a regional fishery body or arrangement. Ideally, they should be managed as a common resource to ensure conservation and equity. Disagreements should be resolved through statutory dispute resolution mechanisms, or through the Tribunal on the Law of the Sea (ITLOS). The purely high seas resources (e.g., on sea mounts) are not explicitly addressed, leaving them in a basic regime of free and open access. Despite this, thousands of transboundary stocks are still not jointly managed.

During the 1990s, a number of international agreements and arrangements[24] have progressively specified and strengthened this regime, increasing the coastal states prerogative for high-sea resources adjacent to EEZs or strongly connected to it (i.e., the UN Fish Stock Agreement, FSA) or entirely circumscribed between EEZs (e.g., in the Barentz Sea Loophole, Bering Sea Donut hole and Okhotsk Sea Peanut hole). The process may not be stabilized yet and some countries contemplate a substantial extension of their jurisdiction beyond 200 mi such as in the Chilean concept of Presential Sea[25].

The Convention on Biological Diversity, adopted ten years later (in 1992) does not seem to modify the institutional context of the allocation/conservation interplay in any way. Focusing on intellectual property rights and patents of genetically modified organisms, it does not really deal with the natural resources, leaving them under the sovereign rights of the countries, the implications of which have already been discussed above. Considering, however, that it is extremely difficult to detect, control or impede bioprospecting, the implication is that in the CBD framework natural resources are de facto under open access, unless already appropriated by the country to specific holders (Weber 2002).

Regional fishery bodies and arrangements (RFBAs) are the institution of choice for the management of shared resources (both straddling and transboundary) but, with few exceptions, their performance in relation to conservation has been poor. These institutions have generally been established to foster international collaboration as a means to achieve conservation and it must be concluded that collaboration is necessary but cannot replace allocation[26]. Many of them do not have any mechanism of allocation of resources among their members. The most developed do and use TACs, often subdivided into national quotas, leaving the responsibility of the sub-division of national quotas (if any) to their members. They nonetheless fail to achieve their conservation objective because of 1) lack of power to adjust fishing capacity and effort to quotas; 2) persisting "race to fish" between and within national fleets when the TACs or national quotas are not suballocated; 3) inaccurate statistics; 4) lack of enforcement capacity

[24] (e.g., the 1999 Barentz Sea Loophole Agreement; the 1994 Convention on the Conservation and Management of the Pollock Resources in the central Bering Sea [Donut Hole Agreement] and the 1995 UN Fish Stock Agreements.)

[25] Mar Presencial, in Spanish, a concept proposed in 1992 by the Chilean Admiral Martinez and confirmed by President Frei in 1994.

[26] The first international fisheries congress was held in 1896 and the first international conventions were established in the late 1800s and early 1900s: North Sea Convention (1882); Spitzbergen Convention (1920); Baltic Sea Convention (1929); Whaling Convention (1931). The first international convention for fisheries was proposed in 1900 in Paris, without success (Ciriacy-Wantrup 1968).

leading to low level of compliance (underreporting, illegal, unreported, and unregulated (IUU) fishing by nonparties; and 5) difficulty to allocate quotas to new entrants, potentially leading to IUU.

In the context of an ecosystem approach to fisheries, regional institutions meet with additional (perhaps not exclusive) difficulties. For example, they may not have a comprehensive enough mandate. This is the case of the International Whaling Commission, the mandate of which does not include small cetaceans. It is also the case of the Indian Ocean Tuna Commission, the mandate of which does not cover bycatch species.

The UN Fish Stock Agreement (FSA) is the institution of reference to manage shared resources in the high seas and it intends to strengthen the arm of the coastal state and regional fishery management bodies and arrangements. It deals with the strategic confrontation between coastal states and DWFNs, both roles being often played by the same nation. Effective regional cooperative management is complex due to the potential number of participants and the differences in 1) legal status (states, geographical entities, private firms); 2) socioeconomic status (developed and developing nations); and 3) objectives and histories. Potential new entrants (and fishing nonparties) represent potentially numerous free riders and a source of illegal, unreported and unregulated fishing (IUU fishing) weakening management schemes. The FSA intends to overcome the difficulty by allocating, de facto, special use rights to members of the RFBAs, requesting nonmembers to join such institutions and in any case to apply the measures adopted by it. If such measures include quotas, the absence of an allocation to nonmembers or new entrants, including through the market, is a major impediment. To counter free-riding and abuse, the thorny question of the real interest of the potential new entrants is being examined. The implementation of the FAO International Plan of Action on IUU (IPOA-IUU) is fundamental for the performance of RFBAs.

Discussion and Conclusions

Conservation and allocation are the two main functions of fisheries management. The interplay between allocation and conservation has been known for centuries. Scientists have described the problem and advised the governments recurrently for the last 50 years. Beyond a few miles from shore, however, resources remained practically unallocated and accessible to all until the mid-1970s, when EEZs began to be unilaterally expanded[27]. During that period, governments have experimented technical regulatory measures aiming at conservation, through the regulation of gears, vessels, and operations and constraints on effort, including through limited entry and constraints on removals quantity and quality. In the process they learned, the hard way, that the approach did not really perform as hoped.

It is now quasi-generally agreed (at least by scholars) that forms of high-quality individual or communal rights, as appropriate, are needed, together with adequate institutions for their administration. The movement has tended to connect the allocation (devolution) of rights to that of responsibilities under a principle of subsidiarity[28]. The growingly accepted paradigm is that there will be no stewardship without ownership; no responsibilities without rights; no environment well-being without human well-being; and no conservation without allocation. Following the

[27] Even though they remained formally unallocated until the 1982 Convention came into force in 1994.
[28] (i.e., allocating rights and responsibilities at the lowest level possible).

general trend towards rights-based fisheries, a wide movement has started in favor of various forms of partnership management, including comanagement, even though the capacity of fishers' communities to effectively undertake the tasks concerned is limited[29].

While the lessons of the last 2–3 decades are only becoming familiar to fishery managers, the need to add ecosystem considerations seriously complicates the adjustment task. While the sector is still struggling with the implementation of the 1982 Convention, a number of new international instruments and initiatives[30] have precipitated the shift to an ecosystem approach to fisheries adding many dimensions to the allocation and conservation problem.

Various degrees of conservation are possible, depending on the balance of objectives between social and ecosystem reproduction, with significantly different cost implications for rights holders and society. Various forms of allocation are also possible, the outcome of which, in conservation terms, depends to a large extent on the nature of the rights granted (e.g., their security and transferability); the asset allocated (e.g., time, space, catch, effort, etc.) and the method used to regulate the access (e.g., through regulatory, command-and-control, measures or through economic incentives such as taxes or rights).

Experience has shown that taxes are not favored in fisheries, and negative taxes (i.e., subsidies) are generally considered as having substantially contributed to overcapacity, overfishing, and failure of conservation in general.

The rights-based fisheries concept is getting growing attention in fisheries. It is much less developed in relation to ecosystem management but the ecosystem approach to fisheries may accelerate its general application in the present context of globalization. One of the main challenges is in choosing between individual and communal rights and in finding the right balance between conservation, economic efficiency and equity. Globally, the process will need to pay attention to two global objectives: food security and poverty alleviation.

The interplay between allocation and conservation is very relevant for the interplay between ecosystem management and fisheries management. As the two types of management aim at sustainable use, there is a significant potential for synergy and the Code of Conduct articulates indeed both types of requirements. However, each type of management is underpinned by different instruments, implemented by different national, regional or global institutions with partially overlapping objectives and constituencies. The reality is that the apparent agreement on the concept of sustainable use hides significant differences in the understanding of the concept, the means to achieve its ends and above all, perhaps, in the resources allocation it implies. This situation sets the scene for misunderstanding, wasteful competition for power and significant interaction costs.

The adoption of systems of rights is not, in itself, a sufficient guarantee of long-term rent maximization, social welfare and conservation (Boncoeur 2003). The interplay between allocation and conservation, affected by climatic vagaries, is also influenced by the time horizon of the actors and their vision of the future. Their weighing of short-term versus long-

[29] The heralded decentralization, expected to reduce management costs has, itself, costs that are often "forgotten" by its champions.

[30] The 1992 UNCED Agenda 21 and CBD, the 1995 FAO Code of Conduct for Responsible Fisheries, the 1995 CBD Jakarta Mandate, the 2002 Reykjavik Declaration 33, and the 2002 World Summit on Sustainable Development (WSSD)

term benefits is conditioned by their personal discount rate, depending, inter alia, on the economic and financial context in which they operate. A discount rate higher than the intrinsic growth rate of the resource will normally induce nonsustainable extraction rates.

A practical conclusion might be that that if society attaches an existence value or an option value to resource conservation per se, it may be necessary to impose to right holders and to enforce a specific constraint of sustainability. This constraint is enshrined in the sovereign rights provided by the 1982 Convention and must be passed on during the successive process of suballocation and enforced by the coastal state under its original obligation.

In terms of policy changes, most governments face two main options: 1) to maintain social structure as a priority, through the historical fishery model, assuming that this model is still functional, at the expense of economic efficiency, injecting public resources through rehabilitation and redistributive programs; or 2) support the outright liberalization, concentration and competition for the sake of economic efficiency, injecting public resources to finance the social costs of transition, including that of exclusion. These two options might not be mutually exclusive in a transition period (Ostrom 2000) and might be used simultaneously either in a layered system of coastal and offshore rights as in Japan (Asada and Hirasawa 1983), radically transferring to coastal communities complete property rights together in a concerted effort to develop their capacity to effectively use and defend them[31]. Political shifts towards large-scale attribution or simply recognition of "hard" property rights to coastal communities might have very significant consequences for the sector, particularly in highly populated countries. There are, however, very few analyses of that option yet and the ongoing globalisation will not facilitate that outcome.

[31] Where appropriate, of course, and after having verified that the local conditions exist for the success of the institutional change.

References

Asada, Y., Y. Hirasawa, and F. Nagasaki. 1983. Fishery Management in Japan. FAO Fisheries Technical Paper No. 238.

Barbault, R., A. Cornet, J. Jouzel, F. Mégie, I. Sachs, and J. Weber, editors. 2002. Johannesburg. Sommet mondial du développement durable. 2002. Quels enjeux? Quelle contribution des scientifiques? Ministère des Affaires Êtrangerès, Paris.

Berkes, F. 1985. Fishermen and "The Tragedy of the Commons." Environmental Conservation 12(3):199–206.

Berkes, F., F. Feeny, B. J. McCay and J. M. Acheson, 1989. The benefits of the commons. Nature 340:91–93.

Boncoeur, J. 2003. Le mécanisme de la surexploitation des ressources halieutiques. Pages 57–70 *in* L. Laubier, editor. Exploitation et surexploitation des ressources marines vivantes. Rapport Science et Technologie. Académie de France No. 17.

Boncoeur, J., and J. P. Troadec. 2004. Americanénagement des pêcheries: les instruments économiques de régulation de l'accès à la ressource. Paper presented at the Colloque Pêche et Aquacülture, Nantes, France, 21–23 January 2004. Ifremer, Paris.

Boncoeur, J., and B. Mesnil. 1999. Surexploitation des stocks et conflits dans le secteur des pêches: une discussion du "triangle des paradigmes" d'Anthony Charles dans le contexte européen. Informations et commentaires 107:10–17.

Charles, A. T. 1992. Fisheries conflict: a unified framework. Marine Policy 16(5):379–393.

Christy, F. T., Jr. 1982. Territorial use rights in marine fisheries: definitions and conditions. FAO Fisheries Technical Paper No. 277.

Chua, T. E., and L. Fallon Scura. 1992. Integrative framework and methods for coastal area management. ICLARM, Manila, Philippines.

Coase, R. 1960. The problem of social cost. Journal of Law and Economics 3(1):1–44.

Coase, R. 1937. The nature of the firm. Econometrica 6.

Collet, S. 2002. Appropriation of marine resources: from management to an ethical approach to fisheries governance. Social Science Information 41(4):531–553.

Correa, C. 1995. Sovereign and property rights over plant genetic resources. Proceedings of the FAO Commission on Plant Genetic Resources. Rome, Italy, 7–11 November 1994. Background Paper No. 2. FAO, Rome.

Ciriacy-Wantrup, S. V. 1968. Resource conservation: economics and Policies, 3rd edition. University of California Press, Berkeley and Los Angeles.

Edmundson, W. A. 1995. The "Coase Theorem." Torts Handouts. Available: http://law.gsu.wedmundson/Syllabi/Coase.htm. (April 2007).

Fallon Scura, L. 1994. Typological framework and strategy elements for integrated coastal fisheries management. Report FI:DP/INT/91/007. FAO, Rome.

FAO (Food and Agriculture Organization of the United Nations). 2003. Fisheries management. 2. The ecosystem approach to fisheries. FAO Technical Guidelines for Responsible Fisheries 4(Supplement 2).

FAO (Food and Agriculture Organization of the United Nations). 1995. Code of conduct for responsible fisheries. FAO, Rome.

Feeny, D., F. Berkes, B. J. McCay, and J. M. Acheson. 1990. The tragedy of the commons: twenty-two years laters. Human Ecology 18(1):1–19.

Fichte, J. G. 1800. Der Geschlossne Handelsstaat, J.G. Cotta, Tübingen, Germany.

Furubotn, E. G., and R. Richter. 1991. The new institutional economics: an assessment. In E. G. Furubotn and R. Richter, editors. The new institutional economics: a collection of articles from the Journal of Institutional and Theoretical Economics. Mohr Publisher, Tubingen, Germany.

Garcia, S. M., A. Zerbi, C. Alliaume, T. DoChi, and G. Lasserre. 2003. The ecosystem approach to fisheries. Issues, terminology, principles, institutional foundation, implementation and outlook. FAO Fisheries Technical Paper No. 443.

Garcia, S. M., and M. Hayashi. 2000. Division of the oceans and ecosystem management: a contrastive spatial evolution of marine fisheries governance. Ocean and Coastal Management 43:445–474.

Garcia, S. M., and C. Newton. 1997. Current situation, trends and prospects in world capture fisheries. Pages 3–27 *in* E. L. Pikitch, D. D. Huppert, and M. P. Sissenwine, editors. Global trends: fisheries management. American Fisheries Society, Symposium 20, Bethesda, Maryland.

Garcia, S., Gulland, J. A., and E. Miles. 1986. The new law of the sea and the access to surplus fish resources: bioeconomic reality and scientific collaboration. Marine Biology July (192–200).

Gordon, H. S. 1954. The economic theory of a common property resource: the fishery. Journal of Political Economy 62:124–143.

Graham, G. M. 1935. Modern theory of exploitating a fishery and its application to the North Sea trawling. Journal du Conseil International pour l'Exploration de la Mer 10:264–274.

Gulland, J. A. 1984. Looking beyond the Golden Age. Marine Policy 8:137–150.

Hannesson, R. 1993. Bioeconomic analysis of fisheries. FAO and Fishing News Books, Rome and Oxford, UK.

Hardin, G. 1998. Extensions of the "Tragedy of the Commons." Science 280:682–683.

Hardin, G. 1968. The tragedy of the commons. Science 162:1243–1248.

Harte, M., and R. Bess. 2000. The role of property rights in the development of New Zealand's marine farming industry. Pages 331–337 *in* R. Shotton, editor. Use of property rights in fisheries management. Proceedings of the FishRights99 Conference, Freemantle, Western Australia, 11–19 November 1999. FAO Fisheries Technical Paper No. 404(2).

Hilborn, R., K. Stokes, J. J. Maguire, T. Smith, L. W. Botsford, M. Mangel, J. Orensanz, A. Parma, J. Rice, J. Bell, K. L. Cochrane, S. M. Garcia, S. Hall, G. P. Kirkwood, K. Sainsbury, G. Stefansson, and C. Walters. 2004. When can marine protected areas improve fisheries management? Ocean and Coastal Management 47.

IUCN/UNEP/WWF (International Union for the Conservation of Nature/United Nations Environment Programme/World Wildlife Fund). 1980. World conservation strategy: living resource conservation for sustainable Development. IUCN, Gland, Switzerland.

Jackson, J. B. C., M. X. Kirby, W. H. Berger, K. A. Bjorndal, L. W. Botsford, B. J. Bourque, R. H. Bradbury, R. Cooke, J. Erlandson, J. A. Estes, T. P. Hughes, S. Kidwell, C. B. Lange, H. S. Lenihan, J. M. Pandolfi, C. H. Peterson, R. S. Steneck, M. J. Tegner, and R. R. Warner. 2001. Historical overfishing and the recent collapse of coastal ecosystems. Science 293:629–638.

Johannes, R. E. 1981. Words of the lagoon: fishing and marine lore in the Palau district of Micronesia. University of California, Berkeley.

Kearney, R., B. Foran, F. Poldy, and D. Lowe. 2003. Modelling Australia's fisheries to 2050: policy and management implications. Fisheries Research and Development Corporation, Canberra, Australia.

Kerrest, A. 2002. Outer space: res communis, common heritage or common province of mankind? Proceedings of the Nice ECSL courses ESA-ECSL. European Centres for Space Law, Paris.

Kurien, J. 2000. Community property rights: re-establishing them for a secure future for small-scale fisheries. Pages 288–296 *in* R. Shotton, editor. Use of property rights in fisheries management. Proceedings of the FishRights99 Conference, Freemantle, Western Australia, 11–19 November 1999. FAO Fisheries Technical Paper No. 404(1).

Lequesne, C. 2003. Mer communautaire et polique commune des pêches. Pages 97–121 *in* L. Laubier, editor. Exploitation et surexploitation des ressources marines vivantes. Rapport Sciences et Technologie. Academie des Sciences de France, volume 17. Lavoisier Publishing, Paris.

Lloyd, W. F. 1833. Two lectures on the checks to population. Oxford University Press, Oxford, UK.

Marshall A. 1920. Principles of economics, 8th edition. MacMillan, London.

Mathew, S. 2003. Small-scale fisheries perspectives on an ecosystem-based approach to fisheries management. Pages 47–63 *in* M. Sinclair and G. Valdimarsson, editors. Responsible fisheries in the marine ecosystem. FAO and CABI Publishing, Romeand Wallingford, UK.

Munro, G. 2003. On the management of shared fish stocks. Paper presented at the Norway-FAO Expert Consultation on the Management of Shared Fish Stocks, Bergen, Norway, 7–10 October 2002. FAO Fisheries Report No. 695.

Ostrom, E. 2000 Private and common-property rights. II. Civil law and economics. Pages 332–379 *in* B. Bouchaert and G. De Geest. Encyclopedia of law and economics. Edward Elgar Publisher, Cheltenham, England.

Ostrom, E. 1990. Governing the commons. The evolution of institutions for collective action. Cambridge University Press, Cambridge, UK.

Pauly, D., V. Christensen, J. Dalsgaard, R. Froese, and F. Torres, Jr. 1998. Fishing down marine food webs. Science 279:860–863.

Pauly, D., and T. E. Chua. 1988. The overfishing of marine resource: socioeconomic background in Southeast Asia. Ambio 17(3):200–206.

Pigou, A. C. 1920. Economics of welfare. MacMillan, London.

Rata, E. 2003. An overview of neotribal capitalism. Ethnologies Comparées 6.

Scott, A. 2000. Introducing property in fishery management. Pages 1–13 *in* R. Shotton, editor. Use of property rights in fisheries management. Proceedings of the FishRights99 Conference, Freemantle, Western Australia, 11–19 November 1999. FAO Fisheries Technical Paper, No. 404(1).

Stokes, A. 2000. Property rights on the high seas: issues for high seas fisheries. Pages 107–117 *in* R. Shotton, editor. Use of property rights in fisheries management. Proceedings of the FishRights99 Conference, Freemantle, Western Australia, 11–19 November 1999. FAO Fisheries Technical Paper No. 404(2).

Tamura, T. 2003. Regional assessment of prey consumption and competition by marine cetaceans in the world. Pages 130–170 *in* M. Sinclair and G. Valdimarsson, editors. Responsible fisheries in the marine ecosystem. FAO and CABI Publishing, Romeand Wallingford, UK.

Tiphaigne de la Roche, G. F. 1760. Essai sur l'histoire économique sur les mers occidentales de France. C.J.B. Bauche, Paris.

Warming, J. 1911. Vor grundrenteaf fiskgrunde. Nazional Okonmiske Tidsskrift 49:499–505.

Weber, J. 2002. Enjeux économiques et sociaux du développement durable. *In* R. Barbault, A. Corner, J. Jouzel, F. Mégie, F. Sachs and J. Weber, editors. Johannesburg. Sommet mondial du développement durable. 2002. Quels enjeux? Quelle contribution des scientifiques? Ministère des Affaires Étrangères, Paris.

WCED (World Conference on Environment and Development). 1987. Our common future. Oxford University Press, Oxford, UK.

Whitehead, A. N. 1948. Science and modern world. Mentor, New York.

WWF (World Wildlife Fund for Nature). 2003. An ecosystem approach to fisheries or ecosystem-based management of marine capture fisheries? A briefing to governments for the Twenty-fifth Session of COFI, 24–28 February 2003, Gland, Switzerland.

Appendix 1: Glossary

Common Pool Resources

Ostrom (1990) defines common pool resources "as natural or man-made resources systems which are sufficiently large to make it costly - but not impossible – to exclude potential beneficiaries from obtaining benefits from its use." Berkes et al. (1989) give a perhaps clearer definition of "a class of resources for which exclusion is difficult and joint use involves substractability". In Ostrom terminology, fishing grounds or stocks may be common pool resource systems while catches taken from them are resource units. Withdrawal of resource units from a resource system is act of appropriation and who does it is an appropriator. While a common pool resource system (a fishing ground) can be jointly appropriated, the resource units withdrawn from it (e.g., the catch) are private. It may be costly to exclude any appropriator from benefits deriving from improvements of a common pool resources system, even if he does not contribute to its management (free rider) and undermines the capacity for others to do so (subtractive or competitive use). Ostrom (1990) stresses the need to avoid confusing the characteristics of a resource (as discussed above) with that of the system of use. A common property (or common) is a system of joint use underpinned by jointly agreed rules, including access rules, and regulations.

Discount Rates

When they invest or act, individuals usually attribute less value to future benefits than to present ones, and the farther away the future benefits are, the lower its present value. They tend to "discount" the future. The discount rate depends inter alia on the information available, the perceived probability to reap future benefits and on present alternatives for investment or action. Because vulnerable and insecure people usually have particularly high discount rates, poverty leads to the degradation of the only resources they have access to. Uncertainty about rights may also lead, for the same reason, to degradation of private resources. Property rights, as well as social norms of behaviour (religion, ethics), may help reducing uncertainty about the future and related discount rates (Ostrom 1990:35).

Efficiency

Economists view equity as part of the equity-efficiency axis along which the performance of policies is assessed. From that angle, one needs to distinguish: (i) Private from collective efficiency: accounting for the externalities of the first; and (ii) Short-term from long-term efficiency: accounting for conservation in the second. When facing multiple objectives or conflicting interests, a relatively frequent situation, a program is considered "efficient", or "Pareto-optimal" when no alternative exist that could improve the degree of fulfillment of an objective or the satisfaction of an interest without compromising the others. Efficiency is a condition of optimality but it should not be confused with it.

Equity

In fisheries and environmental management contexts, "equity" relates to fairness, justice, (e.g. in the allocation of rights or determination of claims). It may also relate to impartiality, freedom from bias of favoritism. It requires that similar options be available to all parties. Equity is a principle of stewardship by governments and the community. A number of sub-concepts have been referred to but may not meet with consensus. Inter-generational equity, for instance, is widely referred to and requires that future generations be given the same opportunity as the present ones to decide on how to use the resources[32]. It can be sought through avoidance of actions that are not potentially reversible on some agreed time scale (e.g., a human generation), consideration of long-term consequences in decision-making and rehabilitation of degraded physical and biological environments. Lack of intragenerational equity (i.e., equity among segments of the present generation) is recognized as one major source of conflict and source of noncompliance. Intersectoral equity would require, for instance, that the fishery sector be fairly treated when its interests conflict with those of other sectors. Cross-boundary equity may be a condition to successful shared stocks agreements between countries. Intercultural equity is relevant when allocating resources to different cultures or defining rights of minorities (e.g., between indigenous and other populations). Lack of equity leads to dissatisfaction or poverty or both and may lead to poor compliance.

Property Rights

Resource allocation is about property rights, the legal rights to control resources. Property is generally defined as a right on a thing such as an object, a good or a service[33]. Ronald Coase (1960) views property, instead, as a right to take a specific action such as the right to dump pollutants in the atmosphere or to produce agricultural products in a particular area[34]. For lawyers, property is the right to dispose of a thing in every legal way, to possess it, to use it and to exclude everyone else from interfering with it. For economists, property refers to a "bundle" of rights including: 1) access: the right to enter and enjoy non subtractive benefits; 2) withdrawal: the right to harvest, substract; 3) management: the right to regulate); 4) exclusion: the right to defend the property; and 5) alienation: the right to transfer, lease, sell all or part of this bundle of rights (Ostrom 1990; 2000). Some of these rights may be further subdivided[35]. The granting or acquisition of all four main rights characterizes full property or ownership.

In the Roman Law, property rights compound the right to use a tangible or intangible asset (*usus*), the right to harvest and appropriate the returns from the asset (*fructus*)[36], the right to give, sell and destroy the asset or its returns (*abusus*). This latter right is the foundation of complete ownership. Since these rights are valid vis-a-vis anybody, they are also called absolute property rights. From both perspectives, although only one or more of the first three rights (access, management, and exclusion or *usus, fructus, and abusus*) may be transferred for

---

[32] This, in turn, raises inter alia, the difficult question of the socially optimal discount rate.

[33] A particular problem is that marine resources are difficult to "own" because of their invisibility, mobility, and fugitiveness.

[34] Curiously, this approach that has an obvious interest for fisheries is the theory defended by Johann Gottlieb Fichte, at the beginning of the 19th century, when defending the concept of a "commercially closed state" in a violent advocacy against economic liberalism and apology of integrated planning (Fichte 1800).

[35] For instance, the right of access might be subdivided by seasons. The right of withdrawal might be subdivided by type of gear or species.

[36] The concept of usufruct (*usufructus*), defined as the right of using the returns of someone else's asset is also potentially useful in fisheries.

some time, establishing relative property rights (the right to sell remaining with the initial owner), full ownership will only be transferred if all the mentioned four absolute rights together are transferred (Kerrest 2002).

The theory indicates that the structure of property rights influences the allocation and utilization of resources in specific and predictable ways. The transfer of incompletely specified rights and the resulting attenuation of property—whether intended or not—generates uncertainty about the asset and does not provide the incentives for its optimal use. It reduces the owner's expectations about the uses of the asset (shrinkage of economic options) and therefore decreases its economic value (Furubotn and Richter 1991).

Public Resources

Resources that can be simultaneously utilized by many users without reduction of the respective availability (nonsubtractive or noncompetitive use) are considered public resources or public goods (e.g., a maritime landscape, Internet). A public good cannot, in principle, be appropriated although its use could lead to saturation. Attempts to appropriate it (e.g., by coastal developments or aquaculture) often lead to conflict. Whether a resource is public or common depends, however, on the type of use. A coastal area may be considered a public resource system for a set of "nonconsumptive" uses (e.g., conservation, diving, bathing, scientific observations) while it would be a common resource for other uses (e.g., fisheries and aquaculture). The concept of sustainable development, for instance, implies that natural renewable resources be treated simultaneously as:

- *Public resources* when considered across time and generations, inasmuch as their use by some generations does not reduce the possibility of use by others, at least under a strong sustainability concept. They need to be enhanced and preserved; and

- *Common resources,* when considered at a given time, inasmuch as the use by some members of the present generation affects the use by other members. They need to be managed.

Res Nullius/Res Communis

The concept of common property corresponds to the legal concept of *res communis* as a "thing which belongs to a group of persons, may be used by every member of the group, but cannot be appropriated by anyone" (Kerrest 2002). The high seas are a good example of modern res communis. It should not be confused with a system of nonproperty (or *res nullius*) in which resources have no owner until they have been captured. "A res nullius is a thing which does not belong to anybody and may be appropriated by anybody" (Kerrest 2002). However, for the property so generated to be legal, capture must be done in conformity with the relevant regulations. As such regulations could indeed restrict access, *res nullius* should not be automatically confused with open access. The confusion between *res nullius* and *res communis* in the *Tragedy of the Commons* (Hardin 1968) has been abundantly clarified, including by Hardin himself (1998), as *the Tragedy of unmanaged commons,* recognizing that the problem was not so much the common pool nature of the resources but the lack of regulation of its access and use.

Sovereign Rights

The 1982 Convention and the CBD recognize the sovereign rights of States over the natural resources placed under their jurisdiction but sovereign rights (a legal institution) and property rights (an economic institution) are apparently not considered synonymous. The following comes mainly from Gerald Moore (personal communication, Head of the Legal Office and Legal Counsel of FAO)[37] and Correa 1995.

Sovereign rights are the rights of independent sovereign states to legislate, manage, exploit, and control access to the natural resources under their jurisdiction. They include the right to determine property regimes applicable to those resources. Sovereign rights imply independence and exclusivity. The rights appertain only to the sovereign power concerned even though it may be subject to limitations or restrictions, in particular when States agree to exercise their sovereign rights in a particular way and subject to agreed rules, which then become binding on them. In the 1982 Convention, for instance, the statements recognizing the sovereign rights of states over their natural resources are coupled with affirmations of their responsibilities to manage those resources in such a way as to ensure their conservation. Sovereign rights are not property rights but they imply the right to establish property regimes. In the process of doing so, a state may very well determine that certain natural resources are the property of the wtate. It may also decide to alienate property, within its system of property law. As a consequence, while the 1982 Convention does not refer to property rights and has not explicitly required their establishment, it has explicitly given to the states the right to do so.

[37] Former FAO legal counsel

American Fisheries Society Symposium 49:659–670

Socially Optimal Allocation of Fish Resources Among Competing Uses

PAUL MCLEOD*
University of Western Australia, UWA Business School, 35 Stirling Highway, Crawley, Western Australia 6009, Australia

JOHN NICHOLLS
Economic Research Associates, 97 Broadway Nedlands, Western Australia 6009, Australia

ROBERT LINDNER
University of Western Australia, School of Agriculture and Resource Economics 35 Stirling Highway, Crawley, Western Australia 6009, Australia

Abstract.—Reconciling fisheries with conservation has a lot to do with allocation decisions and choices made between alternative uses, including passive and nonpassive, as well as extractive and nonextractive. These choices tend to be made implicitly rather than explicitly and in situations where the net benefits of alternative allocations and the socially optimal allocations are usually unknown. This paper argues for the application of an allocation model where the net benefits of allocating fish stocks to various competing uses are explicitly compared within a formal allocation framework to determine optimal allocations. The use of such an approach will make allocation decisions more transparent and accountable and will lead to better policy decisions that are consistent with integrated fisheries management. The paper illustrates the application of the model to the Roe's abalone fishery in Western Australia.

Introduction

Fisheries are a renewable resource. In most cases, they are typically seen as a common property resource. The responsibility for stewardship of the resource generally resides individually or collectively with governments.

Successful management of fish and other marine resources, like the management of other natural resources (water, forests, land, air, and native flora and fauna), requires the ability to make choices about the allocation of the resource among various alternatives. These alternatives can be viewed as including both passive and nonpassive uses, as well as extractive and nonextractive uses.

A decision to conserve or use a fish (or other marine) resource is inherently an allocation decision. As such it needs to be assessed within a resource allocation framework that can evaluate alternatives on a sound and consistent basis. Within such a framework, alter-

*Corresponding author: paul.mcleod@uwa.edu.au

native allocations will include whether people can access a fishery simply to enjoy the marine environment but not take marine resources, whether they can access the fishery to enjoy a fishing experience but not retain catches, or whether those that access the fishery can actually catch and keep the fish (and other marine) resources. These allocation decisions may even need to encompass whether fish stocks are utilized today or allocated for the benefit of future generations. In the case of catch and retain, the allocation decision entails making choices among various user groups, primarily between those who take fish resources for commercial purposes or those who wish to catch fish for their own consumption (or use) as part of a recreational fishing activity.[1]

Allocation decisions necessarily involve choices and trade-offs among the alternatives. For instance, a decision to prevent fishing (commercial and recreational) activity to conserve a fish (or other marine) resources is an allocation decision. Such a decision involves trade-offs between the social benefits derived from the conservation value that society associates with the protection of the marine resource and its environment and the social and economic benefits forgone as a result of the exclusion of fishing activities. Such a decision involves a judgment that the benefits to society of creating the preserved marine environment (e.g., marine park) outweigh the opportunity costs, that is, the forgone benefits from commercial and recreational use, and that this will result in a socially optimal outcome.

The same logic needs to be applied to both inter- and intrasectoral allocation decisions. In the intersectoral case, increased allocation to the recreational sector within a total sustainable catch may result in increased benefits from recreational use, but the reduced allocation to commercial fishing results in a loss of net benefits to society from commercial use. The question is whether the gain in net benefits from the increased recreational use outweighs the loss of benefits from reduced commercial use and therefore whether the overall net benefit to society from the combined use rises or falls. However, most allocation decisions are made in situations where the net benefits to society of alternatives have not been systematically assessed and where the socially optimal allocations are unknown.

Allocation theory is based on the concept that resources should be allocated among and between uses (whether passive or nonpassive) in a way that maximizes the net benefits to society, and this requires explicit consideration of the marginal net benefit from alternative uses. The management frameworks within which allocations are made should have the flexibility to allow allocated resources to move to the use or users that value the resource the most. Experience around the world shows that failure to adopt socially optimal allocation policies that explicitly and openly deal with competing demands invariably leads to overexploitation of the resource and to continued political pressures for changed allocations and increased total allocation (Rudd et al. 2002).

This paper documents a methodological framework that we developed to illustrate how resource allocation options between commercial and recreational fishers can be evaluated on a sound, consistent and like-with-like basis. The framework provides a basis for assessing the net benefits from existing catch shares and for assessing the socially optimal allocation at a point in time.

[1] Indigenous use can be viewed as part of higher order allocation issues about exclusivity of access and use or a subset of lower order intrasectoral allocation within the commercial and recreational sectors.

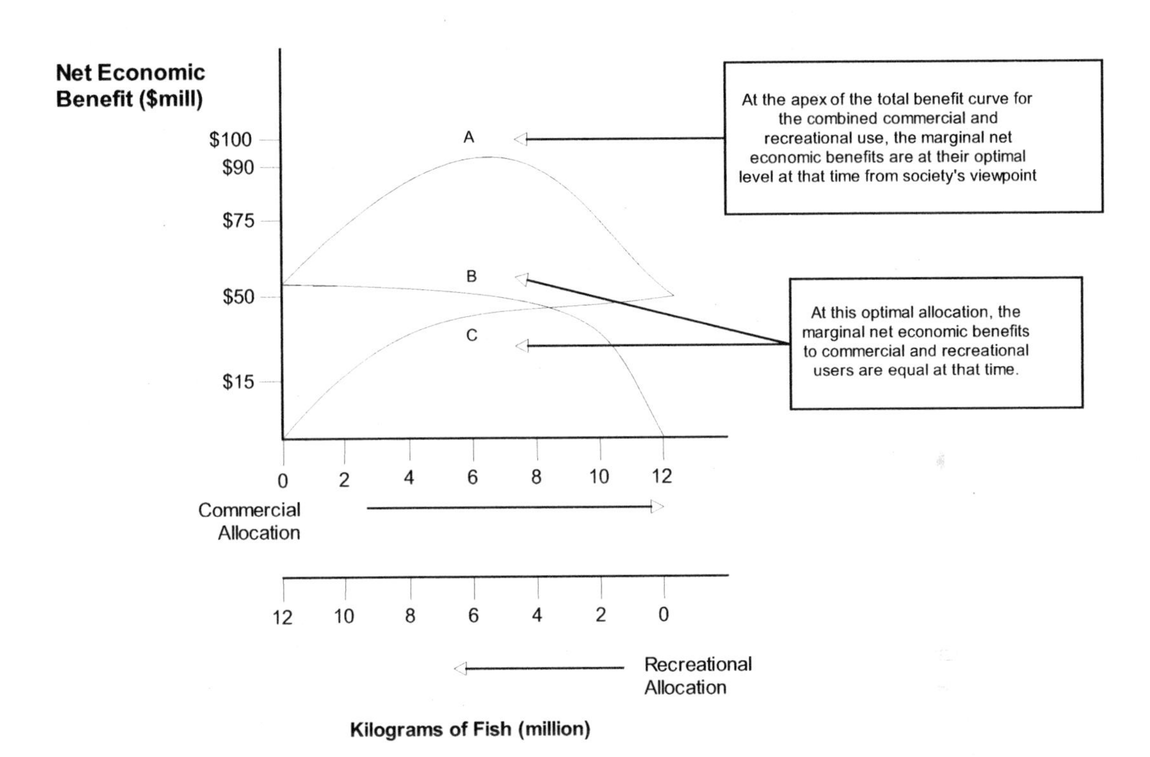

Figure 1. Total net economic benefits from allocation options between commercial and recreational use.

Assessing the Socially Optimal Allocation

The diagram above (Edwards 1990) illustrates the economic interpretation of the optimization of total net economic benefits from allocations between commercial and recreational fishing activties.

The vertical axis shows net economic value for the commercial seafood sector, the recreational fishing sector and the two sectors combined. The horizontal axis records total allowable catch to be shared among these two competing uses. The possible commercial and recreational fishing shares run in opposite directions, such that at any point along the horizontal axis, the sum of the two shares is 12 million kilograms.

From left to right, the total net benefits from the commercial seafood sector increase as the industry's catch share increases from zero to 12 million kilograms. This curve reflects the underlying assumption of diminshing marginal net benefit from each additional kilogram allocated to the commercial sector. From right to left, the net benefits from recreational sector's use of the resource increases as its share of total allowable catch increases from zero to 12 million kilograms. Again this reflects the underlying assumption of diminishing marginal net benefit from each addiitonal kilogram allocated to recreational activity.

Total net economic benefit across both sectors is maximized at point A. In the example illustrated, this is where the recreational sector's share is about 5.6 million kilograms of fish and the commercial seafood sector's share is 6.4 million kilograms.

Any deviation from these shares would reduce

total combined net economic value even though one particular sector may be better off with an increased allocation. In this sense a deviation includes any allocation based on the notion of "fair" allocations such as an equal apportionment or a system that is proportional to historical or current use. This is because the gain in net benefits by the sector that is allocated an increased share does not outweigh the loss of benefits by the sector faced with a reduced allocation.

The framework provides a basis for establishing a set of general conditions to guide allocation decisions in the direction of improved net economic benefits. The maximally efficient allocation, which achieves the greatest total net economic benefit, occurs where the marginal net economic benefits for the competing uses are equal, illustrated in the diagram at points "B" and "C."

At this point, it is of little significance, in economic terms, whether the gross value of production of the commercial seafood sector or the gross value of expenditure on recreational fishing for that fishery is greater than the other. What is important is that, from the broader community viewpoint, the overall net economic benefits of the combined uses are at their optimum and any other allocation would reduce the overall net benefit from a society's viewpoint (Hamel et al. 2000; Sharma and Leung 2001).

Achieving an increase in net benefits to society through allocation changes may mean that some fishers are better off and others worse off. In such cases, the optimality condition requires that those who are better off are able to compensate those who are worse off and still be better off as a result of an allocation change.[2] In this assessment, it may also be the case that distributional considerations need to be taken into account. These may reflect gains and losses to particular well-defined groups or particular regional locations for example. If such considerations are relevant, then it may be appropriate to weight these groups within the net benefit calculation. In this circumstance, the basic allocation principles and model as set out above do not change, but the marginal net benefits calculated and equated are weighted marginal net benefits where the groups defined, and the weights applied reflect some underlying social welfare function. In this sense the, "economic value" concept used in our analysis below is broad enough to encompass specific allowance for a variety of social considerations in assessing allocation options.

An explicit allocation analysis requires an implementation of the above model. This, in turn, requires the use of appropriate value concepts and the collection of appropriate value data.

Values Needed for Valid Comparisons

Economic value is derived ultimately from the tastes and preferences of consumers.[3] In this context, recreational fishing, regardless of whether the catch is consumed or not, is a consumptive activity.

The total economic value of fish is defined and measured in terms of what someone is willing-to-pay for fish—either as seafood or to pursue recreational catch—rather than spend the same amount of money (and time) on other goods and

[2] This compensation test is fundamental in economics. Its use within an explicit framework of allocation decisions ensures that the critical issues of compensation for those users made worse off is not ignored.

[3] The term "consumers" is used here in its broadest sense to include those who buy and those who catch fish. Hence, economic value is in fact a broad measure that encompasses any activity or change in an activity that positively or negatively affects the well being of a consumer. If, reflecting their underlying preferences, a member of society derives value from leaving a fish resource in situ then this is a legitimate economic value.

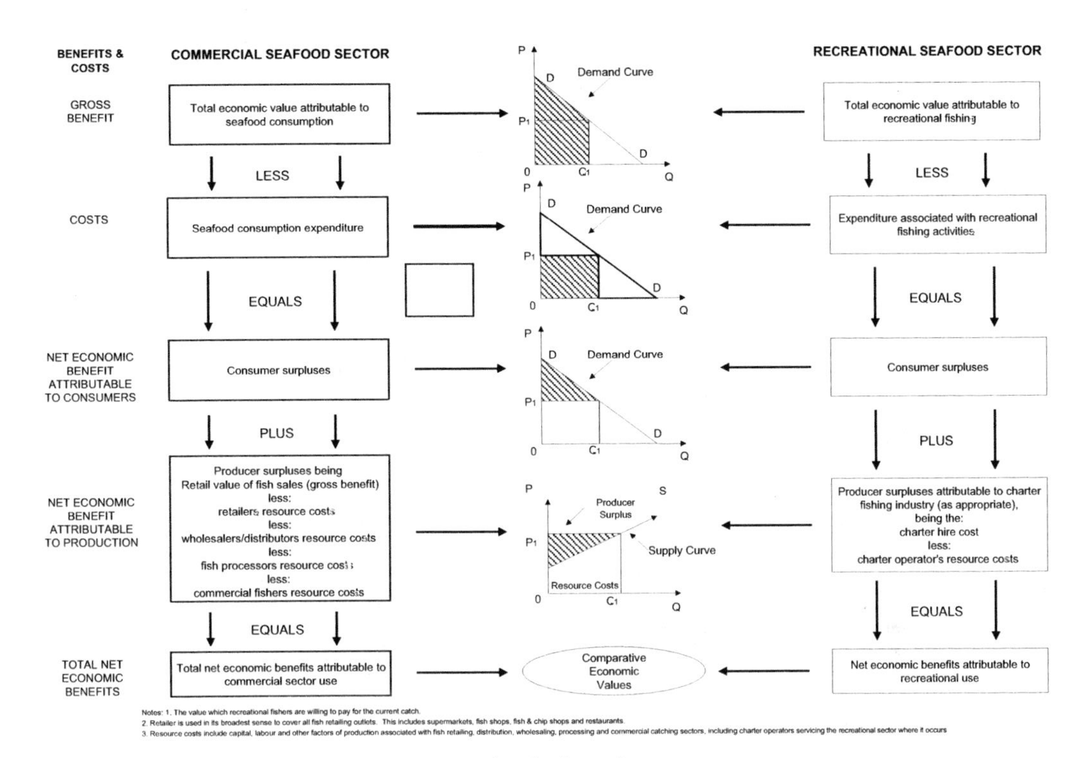

Figure 2. Structure of evaluations to assess optimal Allocation

services which are also positively valued because they satisfy underlying preferences.[4] This assumes that someone will only buy fish or go recreational fishing if the benefits of doing so exceed the costs.

In the commercial seafood sector, net economic benefit comprises the value enjoyed by consumers in excess of the sacrifice to buy fish, plus the benefits derived by the harvest and postharvest business activities in excess of the sacrifices they make in supplying fish to meet this consumer demand. These values can be derived from observed price–quantity relationships in retail seafood markets and from well-established, cost–volume relationships in the harvest and postharvest sectors.

Likewise, for the recreational fishing sector, the net economic benefit is the value enjoyed by recreational fisheries in excess of the sacrifice involved in going on recreational fishing trips plus, where relevant, benefits derived from the commercial charter fishing industry. Typically, there are no established markets for recreational fishing where these values can be observed. In such cases, there are surrogate and simulated (or experimental) market valuation methods and techniques that can be used to derive sound and reliable estimates of these recreational fishing values[5] (Gentner and Lowther 2002).

[4] In economics, money is simply a medium of exchange that allows the measurement and comparison of gains and losses on a consistent basis.

[5] Common approaches to obtaining valuations are the survey based contingent valuation method (CVM) and the travel cost method. The contingent valuation method was used in the application of this framework by the authors to their case study analysis of the crab, abalone, and wetline fisheries in Western Australia. Travel cost data was collected to allow additional analysis. The contingent valuation method was relied on for two reasons. First, the experiential and catch or consumption aspects needed to be addressed explicitly and CVM allowed this to be done. Second, in the event the use of the travel cost model was hindered because there was not a significant range in the distances traveled by the urban population being studied to access the fishing experience.

To achieve optimal resource allocation requires that these values be compared on a sound, consistent, and like-with-like basis for comparing the benefits from these alternative allocations of the resource. The following flow chart illustrates the steps in achieving such a comparison (McLeod and Nicholls 2003a).

Application of the Theoretical Framework

Three case studies have been used to test the application of the above framework. The case study applications were designed to emphasize the methodological and practical issues in applying the general theoretical framework rather than the actual results. The case study fisheries were selected to cover a range of valuation and allocation situations. They were a metropolitan crab fishery, an abalone fishery, and a demersal wetline fishery.[6]

Each of the case study fisheries operates under a different management regime with a varying mix of input and output controls. Commercial activity is a mix of domestic and export, and recreational activity involves use (catch and keep), experiential, and option values. However, all three fisheries have one aspect in common, in that they are all subject to intensifying resource allocation pressures between commercial and recreational users and competing claims about the "optimality" of the current allocations. In the subsequent discussion of the application of the model, we focus on the abalone fishery.[7]

[6] The three case study reports are available on our Web site at: http://www.daa.com.au/~era/reports/.

[7] The abalone is a metropolitan fishery. Commercial license holders dive for the abalone on the reefs. Recreational fishers hold a license that entitles then to fish on six designated Sundays for 1.5 and take a maximum of 20 abalone per day, 120 in total for the season. They are not allowed to dive and must fish by walking onto the reef and fishing in the shallows. More information on the fishery can be found in Fisheries Western Australia (2002).

Underlying Assumptions

The application of the intersectoral resource allocation model is based on certain assumptions.

While none, or some, of these assumptions may be exactly the case in every fishery, and this was certainly the case for the case studies, this was not found to detract from the insights to be gained from applying the framework to those fisheries. The assumptions turned out to be a reasonable starting point for thinking about resource allocation issues in a structured and disciplined framework. However, these assumptions are critical and need to be explicitly addressed as part of establishing an explicit allocation regime.

The model assumes that

• The current combined commercial and recreational take is all that is sustainable and available for intersectoral allocation;

• The fish resource being shared between the commercial and recreational uses come from the same stock;

• The combined commercial and recreational catch can be taken as an explicitly defined total allowable catch across both sectors;

• A zero-sum game can be played by changing share allocations between the commercial and recreational uses within that defined total allowable catch. That is, an explicit reduction in catch share in one sector is reflected in an immediate and commensurate increase in the catch share taken by the other user groups;

• All recreational participants are subjected to binding constraints (catch limits), that is, there is no spare or unused catching capacity;

• For all commercial operators, it is optimal to take their share of the defined total sustainable catch, that is, there is no unused or spare catching capability; and

• All commercial operators are internally structured to maximize producer surpluses from catches taken from a fishery.

Valuation Methods

Commercial Values

Marginal net benefit from the commercial sector is defined as the additional commercial fishing producer and consumer surplus from an increased allocation of the fish resource to the commercial fishing sector. In general, for commercial activities, data needed to estimate the net benefits from commercial use is available through market processes. On the demand side, there are well-established seafood markets in which consumers reveal their preferences. These preferences are reflected in observable price–quantity relationships. On the supply side, industry cost structures and cost–volume relationships for harvest and postharvest activities are also usually observable.

The commercial catch of abalone is virtually all exported, so no estimate of consumer surplus was needed for this fishery. For the crab and wetline fisheries, local market demand functions were estimated from survey data. The change in consumer surplus associated with the local seafood market for these species, with a change in commercial allocation, was estimated using these demand functions.

On the supply side, costs had to be obtained directly from commercial operators. Adjustments were made to apportion costs and revenue to the case study fishery, to remove transfer payments from cost data, and to ensure consistent valuation of capital and labor inputs. Commercial fishers do not always account correctly for the fisher's labor input. Average weekly earnings were used in this study to measure the opportunity cost of owner–operator labor. A particular problem arose in determining how costs would change with volume because most operators were of similar size. Commercial fishers were therefore asked directly to estimate how total costs would likely vary with volume of catch and to identify fixed and variable costs for both the harvest and postharvest activities. These data were used to estimate the marginal change in producer surplus from a reallocation of catch. Aggregate resource costs for commercial activity based on survey returns were scaled-up to derive estimates of total resource costs for harvest and postharvest activities. Scaling factors used were based on volume, individual quota unit holdings, and aggregate returns as appropriate.

Recreational Values

On the recreational side, there are generally no well-established markets where the preferences of recreational fishers can be observed. Hence, stated preferences and contingent valuation techniques were used to establish recreational values in each of the case study fisheries.[8]

Development of the contingent valuation scenario used in the three surveys was based on the following widely held perceptions:

• Recreational fishers had homogeneous preferences;

[8]Past data collections have concentrated on recreational effort and catch for recreational fishing in the Perth area, but not on valuation.

• The main motivation for recreational fishing for crab and abalone was consumptive use, while the fishing experience also was valued by recreational fishers in the wetline fishery; and

• Existing catch (daily bag) and time limits in the case study fisheries were binding and that recreational fishers had unsatisfied catch demand;

The contingent valuation scenario was designed to elicit recreational values (in the form of willingness to pay) for changes to catch limits for all three case study fisheries.[9] For the wetline fishery, an attempt was made to also elicit the experiential value from recreational fishing. Furthermore, because the wetline fishery is not a single species fishery, an attempt also was made to test whether the composition of the catch limits across the three preferred species of dhufish *Glaucosoma hebraicum*, pink snapper *Lutjanus goreensis*, and baldchin groper *Choerodon rubescens* also influences willingness to pay.

Results

As Figure 1 makes clear, the important values for like-with-like comparison purposes are the marginal net benefits from the respective commercial and recreational uses, that is, establishing whether the marginal benefit of an extra fish taken by the recreational sector was worth more or less than if that fish were caught by the commercial sector.

The recreational surveys revealed a positive aggregate and marginal willingness to pay for recreational entitlements among the survey respondents.[10] For the crab and abalone fisheries, use values (catch and keep) dominated. That is, recreational fisher participation was primarily to catch fish for direct consumption. Experiential and option values appeared to play a more important role in the values recreational fishers ascribed to fishing in the demersal wetline fishery.

The contingent valuation modeling in all three fisheries indicated that, given their current preferences and budget (money and time) constraints, many survey respondents were currently optimizing utility or well-being within current catch limits. That is, they chose to cease fishing activity with retained catches less than the proscribed catch limits and were generally satisfied with their actual (retained and released) catch. It was expected that there would be individual fishers who might place high values on extra retained catch and those who would not. This distribution of recreational values was confirmed by the survey results.

Logistic regressions were estimated using the contingent valuation survey data and the results used to estimate the mean and marginal willingness-to-pay for catch entitlements. Consistent with economic theory, the marginal values declined for each additional unit of catch.[11]

In all three case study fisheries, harvest and postharvest businesses provided data that allowed the estimation of the relevant supply

[9] The surveys used two scenarios, a willingness to pay to obtain an increase in catch limit and a willingness to pay to avoid a reduction in catch limits (McLeod and Nicholls 2003b).

[10] The survey of recreational fishers in all three case study fisheries was developed and implemented against a widely held community belief that the various existing restrictions (daily bag limits, limits on fishing days and access time, etc.) were binding constraints and that there was a universally unsatisfied demand for extra catch among recreational fishers. However, the actual survey results challenged the validity of these beliefs in all three case studies, although willingness to pay was positive.

[11] This estimate was taken to be indicative of the marginal values actual and potential recreational fishers would place on an extra catch in each of the case study fisheries.

and demand conditions. For each of the fisheries, the demand and supply functions were estimated from collected industry data and these turned out to be generally consistent with economic theory.

Using these functions, the aggregate and marginal benefits from commercial use were determined across a defined range of commercial catch volumes. These values included the net benefits attributable to harvest and postharvest production and, where appropriate, the net benefits attributable to local retail consumption.

The aggregate net benefit functions were consistent with the theoretical framework outlined in Figure 1. Consistent with economic theory, the marginal net benefits decline as the catch volume increases,

Optimizing the Net Benefits from Intersectoral Allocations: Illustrative Result for Abalone

In each case study fishery, the two net benefit functions intersected at a positive allocation to both sectors, indicating that, at the socially optimal allocation at a point in time, both sectors should receive catch allocations. The results also indicate the extent to which the actual current allocation appears to be nonoptimal and the direction and quantity of change required.

Space does not permit a full reporting of survey and allocation results from the studies cited previously. However, the final product is summarized in the diagram below for optimal allocation in the Perth Roe's abalone *Haliotis roei* fishery. The abalone fishery is a limited entry fishery on the commercial side. Commercial fishers have a total allowable catch quota and individual quota can be traded. On the recreational side, there is no restriction on the number of recreational licence holders. However, each recreational license limits the recreational fisher to fishing on six specified Sundays with a fixed 2 h on each Sunday and a bag limit of 20 abalone taken on each occasion (120 in aggregate). Commercial fishers and recreational fishers access the same reef areas, although commercial fishers are allowed to dive while recreational fishers cannot dive and must harvest from the top of the reef. Minimum size limits apply to both recreational and commercial fishers.

The relationship between the marginal benefits for commercial and recreational use is for a defined volume range. This reflects the underlying supply and demand conditions on the commercial side and the underlying preferences on the recreational side for Perth roe's abalone. The marginal producer surplus for the commercial side is a constant, based on the assumption that the export-oriented Perth Roe's abalone industry is a "price taker" (that is, supply changes will have no impact on price received) and that survey evidence that the marginal costs change little over the volume range.[12]

The right hand origin is the current allocation with existing catch shares of around 40,000 kg for recreational and 36,000 kg for commercial. The analysis shows that, working from right to left, for the next additional abalone, the marginal benefits to recreational use are higher than the marginal benefit from commercial use. If the combined existing catch levels are accepted as defining the total sustainable catch in the fishery, then a reallocation of up to another 4,500–5,000 kg of abalone to the recreational sector (based on the inverse equation) is indicated. This would increase the overall benefit to society for the com-

[12]The producer surplus estimates are sensitive to movement in U.S./Australian dollar exchange rate. Resource allocation decisions should be based on producer surplus derived by using the underlying longer-term rate and not short-term movements.

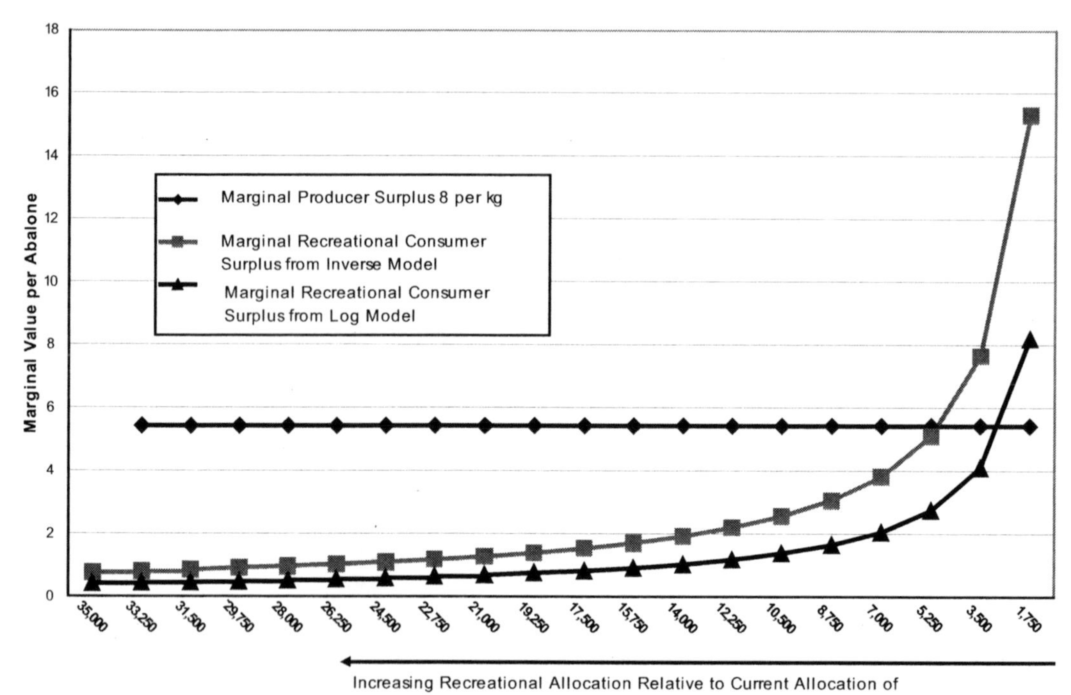

Figure 3. Optimal allocation of abalone.

bined commercial and recreational use of the resource.[13] Beyond this extra 4,500 kg, the marginal benefit to commercial use exceeds those from recreational use across the remainder of the volume range.

A critical issue in the successful application of the theoretical framework is the ability to specify the aggregate sustainable catch to be allocated between the competing uses. For the purposes of our analysis, we used the combined actual commercial (36 metric tons) and recreational (40 metric tons) catches as being indicative of the sustainable (76 metric tons) catch. However, the survey results indicated that not all recreational fishers licensed to fish for abalone used their maximum entitlements. We estimate that if the participating recreational abalone fishers had exercised their full entitlement and had fished on all the available fishing days and achieved bag limit catches within the available fishing time, then probably the actual recreational catch would have been closer to 70 metric tons (instead of the actual 40 metric tons) and the combined catch around 106 metric tons (instead of 76 metric tons). The latent effort implied by the current management regime is considerable.

The results presented above relate to the determination of the optimal allocation at a point in time, that is, a static analysis. Underlying conditions, including economic and social values, will change over time and these changes will imply changes in the optimal allocation. The model is based on the idea that instantaneous reallocation is possible and that each of the competing user groups (recreational and commercial fishers) is optimized within its current allocation.

[13]This is because the marginal benefit of an extra abalone allocated to recreational use is greater than the loss in producer surplus at the marginal up to this tonnage

Any adjustment to achieve the socially optimal outcome must ultimately work through the stock abundance so that the speed of any reallocation will depend on how quickly reduced effort for one sector shows up as improved stock abundance and catch outcomes for the other. This issue proved to be especially relevant in the wetline fishery. In this case, the underlying stock abundance issues are such that, for example, there is some uncertainty as to how any commercial effort reduction would show up as improved catch rates for recreational fishers.

Each sector is assumed to be optimizing its own allocation. In all three fisheries, this turned out clearly not be the case on the recreational side. This means that, if individual recreational fishers were given greater choice within a fisheries management regime to better match daily catches and access times to individual preferences, there would likely be reallocation of bag limits within the recreational allocation from those fishers not pushing on existing fishing limits and those that are. Such intrasectoral reallocation would increase overall benefits from recreational use without the need to address intersectoral allocation options.

The prices received by commercial fishermen were typically based on kilogram of fish caught, while recreational fisher's willingness to pay is for a fish. This means that commercial and recreational use values may be different for various fish sizes. For example, in the abalone fishery, the larger sized abalone may have a higher value in commercial than recreational use. The recreational fisher may be indifferent to abalone size so long as the catch is enough for a decent feed.

The opposite is true of the wetline fishery, where recreational fishers place a higher value on catching and keeping bigger fish than commercial fishers. The commercial fisher may be indifferent to fish size so long as the required catch volume is achieved given the market price received is on a dollar per kilogram basis.

Ideally the model needs to be extended to deal with these issues. Most importantly, any allocation framework will need to integrate a dynamic element that allows allocation to be looked at over time and will need to allow for stock abundance to be modeled, allowing for a better analysis of how actual allocations will be experienced and therefore valued by fishers.

Conclusion

The results indicate that the resource allocation model can be applied in a way that produces meaningful results about the net benefits from existing allocations and useful insights into the socially optimal allocations at a point in time.

They provide a reasonable starting point for developing resource allocation policy. They should also lead to more informed decisions about the socially optional allocations and analysis of this kind will be essential input into the development of an integrated management regime.

The analysis has identified a number of areas where issues arise and further work is needed. Most important of these is the need to assess allocation over time using a dynamic model that will allow the assessment of optimal allocation to take account of the way economic and biological variables change over time and the way stock abundance affects the way each sector experiences reallocation impacts and the timing of these for that sector.

Ultimately, the framework will have to be extended to encompass competing uses beyond commercial and recreational use. Indigenous use and environmental preservation encompassing marine parks are prime examples. The advances in recent years in applying formal valuation approaches to eliciting meaningful nonmarket valuations, especially in relation to existence and option values for natural environments, gives confidence that such an extension can be meaningfully achieved.

Acknowledgments

This paper arises from the research project on Socio Economic Evaluation of Resource Allocation Options funded by the Fisheries Research Development Corporation (FRDC), Australia. The authors would like to acknowledge the support of the FRDC.

References

Edwards, S. F. 1990. An economic guide to allocation of fish stocks between commercial and recreational use. U.S. Department of Commerce, NOAA Technical Paper NMFS 94, Washington, D. C.

Fisheries Western Australia. 2002. State of the Fisheries Report 2000–2001. Department of Fisheries, Perth, Australia.

Gentner, B., and A. Lowther. 2002. Evaluating marine sport fisheries in the USA. Pages 186–206 *in* T. J. Pitcher and C. Hollingsworth, editors. Recreational fisheries, ecological, economic and social evaluation. Blackwell Scientific Publications, Oxford, UK.

Hamel, C., M. Herrman, S. T. Lee, and K. R. Criddle. 2000. An economic discussion of the marine sport fisheries in Central and lower Cook Inlet. IIFET (Institute of Fisheries Economics and Trade) Conference, Corvallis, Oregon.

McLeod, P., and J. Nicholls. 2003a. A socio-economic valuation of resource allocation options between recreational and commercial uses: part one, a general theoretical framework. A report prepared for the Fisheries Research Development Corporation. Economic Research Associates, Perth, Australia. Available: http://www.daa.com.au/~era/reports.

McLeod, P., and J. Nicholls. 2003b. A socio-economic valuation of resource allocation options between recreational and commercial uses: part three, the Perth abalone fishery case study. A report prepared for the Fisheries Research Development Corporation. Economic Research Associates, Perth, Australia. Available: http://www.daa.com.au/~era/reports.

Rudd, M. A., H. Folmer, and G. C. van Kooten. 2002. Economic evaluation of recreational fishery policies. Pages 34–52 *in* T. J. Pitcher and C. Hollingsworth, editors. Recreational fisheries, ecological, economic and social evaluation. Blackwell Scientific Publications, Oxford, UK.

Sharma, K. R., and P. Leung. 2001. Economic impact of catch reallocation from commercial fishery to recreational fishery in Hawaii. North American Journal of Fisheries Management 21:125–134.

American Fisheries Society Symposium 49:671–676

Alaskan Sablefish and Patagonian Toothfish: A Study of Allocation and Conservation

Mark Lundsten*

Queen Anne Fisheries, 1939 Eighth Avenue West, Seattle, Washington 98119, USA

Introduction

Recently, numerous scientific reports have highlighted world's fisheries problems such as overfishing and bycatch mortality. In contrast, the Alaskan sablefish *Anaplopoma fimbria* fishery in Alaska and the Patagonian toothfish *Dissostichus eleginoides* fishery in South Georgia Island near South America and Antarctica have found solutions to these problems. Though distinct in some ways, these two fisheries are very similar. Significantly, both have functional, complex management systems based on scientific principles of conservation, and each has an individual allocation system that has eliminated the "race for fish." Fishermen in both fleets thus have the operational tools and the economic incentives not only to avoid overfishing, but also to meet other conservation goals, notably reduced seabird bycatch.

These two commercial fish stocks occur at high latitudes in opposite hemispheres and compete for the same high-value market niche in Asia (Chambers et al. 2004). Patagonian toothfish (also Chilean sea bass, or *mero* in Japan) and Alaskan sablefish (also black cod, or *gin dara* in Japan) are both oily, white-fleshed, deepwater fish harvested by offshore fleets mostly composed of longliners, with some trawlers and a smaller number of pot, or trap, vessels. The North Pacific Fishery Management Council (NPFMC) manages sablefish in the Alaska Region of the U.S. exclusive economic zone (EEZ) exclusively for U.S. fishermen. The majority of global toothfish stocks are managed by the Commission for the Conservation of Antarctic Marine Living Resources (CCAMLR). The CCAMLR manages an international fleet in both the international and territorial waters of the CCAMLR convention area. This area extends completely around Antarctica, mostly between 50 and 60 degrees latitude. The sablefish fleet fishes the same latitudes, but in the north, and across less than 60 degrees of longitude, roughly one sixth the area of CCAMLR waters. Most of the sablefish grounds are within a day or two of a delivery port and shore-based, owner-operated vessels catch most sablefish (Witherell 2000). Most of the toothfish grounds are over a week away from a port of any kind and most toothfish are caught by distant-water catcher-processing vessels (Holt et al. 2004).

The CCAMLR clearly states its two principles of management. One is the precautionary principle that assumes a great degree of uncertainty in fisheries management, particularly in the matter of setting harvest limits to avoid overfishing. The other is ecosystem-based management that considers not only target species but other species and the habitat as well (Kock 2000). Though the NPFMC has not stated such a policy of ecosystem management, it is currently developing one, and its management history is "consistent with a precautionary approach and ecosystem-based management, and ha[s] resulted in sustainable fisheries." (Witherell et al. 2000) A measure of the recognition of each fishery's conservation success is their certification by the Marine Stewardship Council (MSC). The MSC has certified the South Georgia Island toothfish fishery, in CCAMLR area 48.3, a subset of the overall fishery. Area 48.3 management is implemented by

*Corresponding author: mlundsten@comcast.net

the Government of South Georgia and the South Sandwich Islands, but directed ultimately by CCAMLR. (Holt 2004) The MSC also is currently reviewing certification of the Alaskan sablefish fishery (www.msc.org/html/content_785.htm).

The allocation of sablefish in Alaska is by an individual fisherman's quota (IFQ) system. The National Marine Fisheries Service (NMFS) Alaska region's restricted access management (RAM) Division implemented this system in 1995 simultaneously with an almost identical IFQ system for Pacific halibut *Hippoglossus stenolepis*, as the sablefish fleet is almost completely a subset of the halibut fleet. Individual fisherman's quota regulations for the halibut and the sablefish fisheries incorporate almost identical socioeconomic considerations to retain the character of the owner–operator fleet. At the inception of the IFQ program, fishermen received revocable harvest privileges—"quota shares"—based on their history in the fishery over a certain number of qualifying years. Quota shares are a fixed percentage of an annually calculated and variable total allowable catch (TAC). The fixed percentage translates into a different poundage of fish allocated to individual fishermen each year as the TAC fluctuates. If the TAC increases, then everyone's IFQ increases by the same percentage. The quota shares are transferable, with a number of qualifications, and thus have monetary value, their sale and purchase providing exit from and entrance into the fishery. This system requires a very comprehensive management system and reliable administration able to provide extensive monitoring and enforcement. Every pound of harvested fish must be documented, and every vessel must have an IFQ holder aboard while fishing and unloading the catch. The U.S. Coast Guard (USCG), NMFS Enforcement Division, and RAM all must cooperate to jointly administer the IFQ fishery. The USCG conducts periodic boarding and inspection of vessels, NMFS enforcement monitors deliveries of fish, and RAM tracks deliveries and shipments of fish (NMFS 1997).

Elimination of the so-called "race for fish" through the establishment of IFQs in the Alaska sablefish fishery has provided economic, social, and conservation benefits for the fishermen and the fishery. (Pautzke and Oliver 1997) Prior to the implementation of IFQs, fishermen raced to catch as many fish as possible in a limited time. As markets developed in the 1980s, and with no restrictions on entry into the fishery, more fishermen entered the fishery and fishing grounds became crowded. Fishermen had the incentive to set and haul as much gear as possible, often more than could be hauled back within the time limits of a season, and often in conflict with other fishermen's gear. The result was gear left or lost on the grounds and fish dead on the hooks.

After the implementation of IFQs, fishermen had a different incentive. Since IFQs provide a predetermined quota of fish for each fisherman, the resultant incentive is to maximize the value received for that fish. Setting unnecessary gear or losing or abandoning gear were no longer customary costs of doing business, but costs a fisherman could avoid and thereby increase profits. Lost gear mortality almost disappeared with the IFQ program.

Halibut bycatch mortality also has disappeared almost completely from the sablefish fishery since the beginning of the IFQ program. In the open-access race for fish, halibut and sablefish seasons had to be separate in order to keep track of total catches of each species. Even though most sablefish fishermen also fished for halibut, halibut was not retainable in the pre-IFQ sablefish fishery. In fact, the NPFMC established a limit of halibut bycatch in the sablefish fishery. As grounds became more and more crowded in the open-access fishery, fishermen had to fish in more marginal areas, many with high halibut bycatch. In 1994, the last year of the race for fish, managers closed the sablefish season before reaching the TAC due to the fleet exceeding the bycatch limit of halibut. Individual fisherman's quotas eliminated this dilemna. First, with an 8-month season, the grounds were no longer crowded, allowing the fleet to avoid high halibut bycatch areas if they chose. Second, the sablefish fleet almost entirely owned halibut quotat shares in addition to their sablefish quota. Fishermen could now target either species or could target and retain both species at the same time. The IFQ regulations state that if a fisherman catches a legal-sized halibut while fishing for sablefish and the fisherman has IFQ for that fish, he not only may retain that fish, he must retain it, by law. Like sablefish mortality due to lost gear, halibut discards also have almost completely disappeared from the sablefish fishery since the inception of IFQs (Pautzke 1997).

Eliminating the race for fish also increased the spawning potential of sablefish (Sigler and Lunsford 2001). If a fisherman's incentive is to maximize pounds of catch per season, and not to maximize the value per pound of fish landed, then, as fishermen fished marginal grounds in a crowded race for fish, they would inevitably catch smaller, immature fish along with halibut. But since the incentive under an IFQ system is to target the larger, more valuable sablefish, more fishermen have done so. They have "decreased [the level of] harvest of immature fish...[and] improved the chance that individual fish will reproduce at least once. Spawning potential of sablefish, expressed as spawning biomass per recruit, increased 9%" after the implementation of the IFQ program (Sigler and Lunsford 2001).

Iindividual fisherman's quotas have allowed managers to stay within TAC limits without having to open and close the fishery in-season, as they had to do in the open-access fishery, often with the result of exceeding the TAC. In the nine seasons of the IFQ fishery since its inception in 1995, the six areas of the sablefish fishery in the Alaska region have generated 54 annual catch totals. In one of those cases, in the year 2000 in the West Yakutat area, the fleet overharvested the TAC, by 3,762 lb more than the roughly 4 million pounds allowed, an overharvest of less than 0.1%. By statute, that amount was deducted from the next year's TAC. In 53 of those 54 cases, the fleet underharvested the TAC (NMFS 2004). The National Marine Fisheries Service allows each fisherman to over- or underharvest his annual allocation by 10% and deducts any overage from, or adds any underage to, each fisherman's allocation for the following year. Any overage above 10% results in a substantial fine and forfeiture of fish (NMFS 1997). The fleet appears to have determined how to fish most efficiently, harvest your IFQ exactly, or a take a little less. With a fine for anything over 10%, and underages compensated the following year, catching any amount greater than your annual IFQ only adds the risk of a fine and makes no sense. Though IFQs certainly do not provide any mechanism to improve the science used to determine catch limits, they have provided managers and the fleet with a mechanism to stay consistently within those limits.

The IFQ system also increased fishermen's incomes, both by decreasing operating costs in an overcrowded race for fish and by increasing fish prices. No longer did fishermen have to fish as hard as possible, with maximized crew, equipment, and supplies for an uncertain return on an unknown quantity of fish. Also, with IFQs, fishermen did not have to take a low price while crowded at the dock and looking for a buyer. Rather, with deliveries spread out, vessels arrived at uncrowded docks with buyers looking for boats, and fishermen asking for the best price. Under open-access, many fishermen sold at one time to a few processors, who thus had leverage to keep prices low. Under IFQs, fishermen come to town throughout the season and have their pick of the processors each time. Now fishermen have the leverage to keep prices high. Fishermen now make more money per pound of fish and spend less to catch it.

All of the benefits discussed so far reflect only one aspect of the IFQ fishery: the elimination of the race for fish by the establishment of a designated access privilege (DAP). Other DAPs, such as community development quotas and cooperatives, also eliminate the race for fish through different means, and accomplish similar benefits by doing so (U.S. Ocean Commission 2004). Individual Fisherman's Quotas are distinct among these allocation systems because they establish revocable ownership of harvest privileges and thereby provide other conservation effects.

The case of the endangered short-tailed albatross *Diomedea albatrus* illustrates how the value of IFQs provided the extra incentive for the sablefish fleet to achieve a conservation mandate of the law. According to the Endangered Species Act (ESA) as interpreted in a biological opinion by the U.S. Fish and Wildlife Service (USFWS), Alaska longline fishermen are allowed an incidental take of no more than four short-tailed albatrosses in any 2 years. If they do take four birds, a new consultation would be initiated between NMFS and USFWS, with the possibility of interruption or even closure of the whole $300 million longline fishery (Melvin et al. 2001). Inidividual fisherman's quotas forced fishermen to take this problem seriously by making fishermen more vulnerable to the consequences of breaking this law. The value of IFQs enhanced both the risk of not solving this problem and the reward for solving it. Before IFQs, each fisherman's stake in the fishery was much smaller and more tenuous. With IFQs, the fleet had tangible reasons, often including a large mortgage, to be committed to the future. Violating the ESA's rules could revoke that future, all because of the short-tailed albatross.

The fleet now had the time and the freedom to figure out how to avoid penalties from the ESA. This was not only a job they had to do in order to keep fishing, this was also a job that they could do while fishing because they didn't have to race for fish. They could experiment with different seabird deterrents in the course of fishing and they were able to look for solutions with no loss of the fish they were allowed to catch. A boat could run test trials on gear modifications and their income would remain the same.

Once the fleet had worked out a few promising techniques, the Washington Sea Grant Program suggested that they choose a few of the most effective techniques and test them in a cooperative research project as a basis for seabird deterrent regulations. A select number of vessels participated for 2 years, 1999 and 2000. The vessels fished their IFQ poundage, with no special permits required, while staying within the protocols of the scientific method (Melvin et al. 2001). The poundage of each vessel's catch during the research applied to their IFQ account. Not having to race for fish, the fleet easily accommodated the research requirements with no loss of income and minimal loss of time.

The result of this project was a set of recommendations for regulations, which were passed by the NPFMC and which require vessels to use streamer lines as bird deterrents that meet performance standards established by the research. Now fishermen must deploy and retrieve these deterrents with each set and ensure that they are working properly and meeting the performance standards. These streamers are simple devices, of negligible expense, so the cost to the fleet is nothing more than the time it takes to do these jobs. No fishing income is lost.

Though the regulations are fairly new, they appear to be a success. The sablefish fleet catches about 80% of its fish in the Gulf of Alaska where they and the halibut fleet comprise most of the longline effort (NMFS 2004). The bycatch of seabirds in that area has varied since 1995. But since the completion of the survey in 2000, the rates have remained below historically low levels of 0.007 birds per 1,000 hooks in 2002, a fleetwide bycatch of only a few hundred seabirds from the millions in that area. (Boldt 2003)

The IFQ program likely deflected any potential lawsuit intended to close the sablefish fishery due to the threat that an open-access race for fish would have posed to the short-tailed albatross. In contrast to the IFQ fishery, in which the fleet has a stake in the future of the fishery, and the opportunity and incentive to solve this problem, the pre-IFQ fishery provided none of those things. If each fisherman had little or no stake in the future of an uncertain fishery, had to fish as hard as possible for an unknown amount of fish, and had to compete against other fishermen and the clock, someone easily could have made a case that such a fishery should not be allowed to open without severe restrictions, if at all, due to the risk to this endangered bird.

By giving value to sablefish, managers gave value to the other species in the ecosystem. With IFQs, and the administrative ability to enforce them, managers made fishermen vulnerable to a much greater loss than before IFQs if they did not follow conservation mandates. By eliminating the race for fish, IFQs also gave fishermen the tools to meet these mandates.

The South Georgia Island toothfish fishery also has achieved a low level of seabird bycatch: 6 birds in the 2001/2002 season, a rate of 0.0015 birds per 1,000 hooks (CCAMLR 2003). They have done so less by the incentives provided by IFQs and the ESA, but rather by command and by restriction. They allow fishing only in the winter, when the vulnerable birds are not in the area, though they also have requirements similar to those in Alaska for bird deterrent devices and practices while vessels set and haul gear. This success is one of the reasons stated for the certification by MSC.

Another reason given by MSC for certification is the elimination in South Georgia of the race for fish and the consequent ability of the fishery managers to more accurately stay within the TACs. Since 2001, managers have allocated an annual, nontransferable TAC to each vessel, like a "one season IFQ," instead of setting a single TAC for all the vessels. As with sablefish TACs since IFQs, the toothfish TACs since 2001 have been met with much more accuracy than before (Holt 2004).

Each vessel that applies for and receives this nontransferable quota is required to pay a license fee. Managers collect these fees from an international fleet of catcher and processing vessels, nothing like the Alaskan fleet composed almost entirely of owners and operators. Since the aim of the local government is to maximize its earnings from these fees, each year managers reevaluate the fees to accurately reflect market value. The managers consider the history of the vessel applying for the license, mainly whether or not the vessel has a history of illegal fishing and how well the vessel has complied previously with CCAMLR measures, based on the records of inspectors and observers (Holt 2004). A vessel needs a clean record to receive a license.

The South Georgia fleet thus fishes with a variation of the same incentives as the sablefish fleet. They have a revocable stake in a highly regulated and well-administered fishery. Any vessel not fishing "clean," and according to the regulations, risks losing the right to fish in the future. Through a licensing and fee system, with different purposes and in a completely different socioeconomic context than Alaska, South Georgia has established another kind of dedicated access privilege that allows fishermen and managers to avoid the race for fish and its accompanying problems. As in Alaska, the vessels in this area thus not only have the incentives to fish legally, they also have the mechanisms to meet conservation mandates.

Unfortunately, illegal, unregulated, and unreported (IUU) fishing for toothfish in and around the CCAMLR region not only threatens both toothfish stocks and seabird populations but also undermines the incentives for conservation and for fishing within the law (Holt 2004). By causing overfished stocks and by killing seabirds and other bycatch species, IUU fishing diminishes rewards for "legal" fishermen, who are left with smaller TACs and more onerous bycatch restrictions and requirements. Illegal, unregulates, and unreported fishing for toothfish has also caused market sanctions and boycotts, diminishing the value of the whole fishery, legal and illegal.

As these effects mount, the economic reasons for not fishing illegally diminish and the incentives for any given vessel to participate in IUU fishing increase. The CCAMLR waters are vast and the weather forbidding, making enforcement difficult and keeping the risk of IUU fishing relatively low. Conversely, the potential reward is high with a fish this valuable. Anywhere that stocks face an uncertain future because of illegal fishing, any fisherman has an incentive to fish illegally in order to get his share of a return.

If a fishery resource is like a bank account, then the TAC corresponds to the interest earned every year. For good conservation—as in Alaska and South Georgia Island—responsible managers only allow the interest, not the principal, to be withdrawn. A sound system of allocation is a method of distributing that interest systematically to the account holders: the fishermen. But IUU fishermen are like bank robbers who degrade the integrity of the bank; they deplete the principal. Thus, legitimate account holders stand to lose return unless they rob the bank too. Just as allocation can provide incentives to follow conservation mandates, IUU fishing provides the opposite. It gives an incentive to fish outside the regulations and thereby obviates conservation mandates.

Fishermen fish illegally for the money. One obvious way to deter illegal fishing is to offer a soundly administered, enforceable, and sustainable fishery that gives a higher long-term return than poaching. In order to achieve these two goals, a solid conservation management framework obviously is required. The South Georgia Island toothfish fishery, like the Alaskan sablefish fishery, meets these conditions, and succeeds, while other CCAMLR areas suffer from IUU fishing.

As the basis of law that allows governments to manage fish, political jurisdiction over the ocean seems essential to conservation. On the other hand, the legacy of open oceans—fishing without any rules—seems only to disable conservation. Most CCAMLR's waters are international and its governing convention is recognized by a limited number of nations. Many do not recognize it or they flout it. Fisheries managers in South Georgia and Alaska could not allocate the privilege to fish, or succeed at conservation, if they could not determine, and decide, what was legal, or not legal, on the ocean. Without that jurisdiction, those managers could not manage fish at all.

In South Georgia Island and in Alaska, fisheries managers do practice conservation and use similar tools. In their respective toothfish and sablefish fisheries, both governments have allocated dedicated access privileges so that neither fleet races for the fish. Consequently, both fleets have been able to meet conservation requirements, such as bird deterrence, with flexibility and without losing fishing income. These allocation systems also enable conservation because they allow managers to require conservation as a prerequisite to fishing. In each fishery, managers have the authority to close a fishery or deny access or a license if a fleet or a vessel does not meet conservation mandates. In short, the allocation systems in each area enhance conservation because they make conservation measures the economical thing to do.

References

Boldt, J., editor. 2003. North Pacific fishery management council Bering Sea/Aleutian Islands and Gulf of Alaska stock assessment and fishery evaluation report. North Pacific Fishery Management Council, Anchorage, Alaska.

Chambers, S., J. Wyman, and B. Warren. 2004. Market report. Pacific Fishing, March, 2004, Seattle.

CCAMLR (Commission for the Conservation of Antarctic Marine Living Resources). 2004. CCAMLR's Management of the Antarctic. CCAMLR, Hobart, Australia.

CCAMLR (Commission for the Conservation of Antarctic Marine Living Resources). 2003. CCAMLR's work on the elimination of seabird mortality associated with fishing. Hobart, Australia.

CCAMLR (Commission for the Conservation of Antarctic Marine Living Resources). 2003. "Schedule of conservation measures in force 2003/04 Season." CCAMLR, Hobart, Australia.

Holt, T, P. Medley, J. Rice, J. Cooper, and A. Hough. 2004. Certification report for South Georgia toothfish longline fishery, Moody Marine, Ltd. Derby, UK.

Kock, K., editor. 2000. Understanding CCAMLR's Approach to Management. CCAMLR, Hobart, Australia.

Marine Stewardship Council. 2004. Marine stewardship council fisheries undergoing assessment. Available: www.msc.org/html/content_785.htm.

Melvin, E., J. Parrish, K. Dietrich,. and O. Hamel. 2001. Solutions to seabird bycatch in Alaska's demersal longline fisheries. Washington Sea Grant Program, Seattle.

National Marine Fisheries Service, Restricted Access Management Division. 1997. Report to the IFQ review committee of the National Academy of Sciences National Research Council Ocean Studies Board. NMFS, Juneau, Alaska.

National Marine Fisheries Service, Restricted Access Management Division. 1995–2003. Individual fishing quota allocations and landings. Available: http://www.fakr.noaa.gov/ram/ifqreports.htm.

Pautzke, C. G., and C. W. Oliver. 1997. Development of the individual fishing quota program for sablefish and halibut longline fisheries off Alaska. North Pacific Fishery Management Council, Anchorage, Alaska.

Sigler, M. F., and C. R. Lunsford. 2001. Effects of individual quotas on catching efficiency and spawning potential in the Alaska sablefish fishery. Canadian Journal of Fisheries and Aquatic Sciences 58:1300–1312.

U.S. Commission on Ocean Policy. 2004. Preliminary Report of the U.S. Commission on Ocean Policy Governors' Draft, U.S. Commission on Ocean Policy, Washington, D.C.

Witherell, D., C. Pautzke, and D. Fluharty. 2000. An ecosystem-based approach for Alaska groundfish fisheries. ICES Journal of Marine Science 57:771–777.

Witherell, D. 2000. Groundfish of the Bering Sea and Aleutian Islands area: species profiles 2001. North Pacific Fishery Management Council, Anchorage, Alaska.

American Fisheries Society Symposium 49:677–685

Estimating Bycatch Survival in a Mark-Selective Fishery

CHARMANE E. ASHBROOK*, JAMES F. DIXON, K. W. HASSEL, AND ERIK A. SCHWARTZ

Washington Department of Fish and Wildlife
600 Capitol Way North, Olympia, Washington 98501, USA

JOHN R. SKALSKI

University of Washington 1325 Fourth Avenue Suite 1820
Seattle, Washington 98101, USA

Abstract.—Pacific Northwest fishery managers have recently adopted mark-selective commercial fisheries to concentrate harvest on healthy spring Chinook salmon *Oncorhynchus tshawytscha* stocks while conserving the weaker stocks. Mark-selective fisheries are possible where a mark, such as a clipped adipose fin, enables fishers to visually separate fish that may be retained from those that must be returned to the water. Selective fishing allows a fishery to avoid nontarget species or stocks, or when encountered, to capture and release them in a manner that minimizes mortality. Previous research comparing tangle nets, which capture fish by the snout rather than by the gills, to the conventional gill nets, revealed high postrelease survival of tangle netted fish. Consequently, beginning in 2003, Columbia River fishery managers adopted mark-selective tangle net fisheries during part of the commercial fishery. Because fishery managers must now account for the survival of the released nontargeted bycatch stocks as well as the total harvest of targeted stocks, we have developed methods to estimate the immediate, postrelease, and total survival of fish captured and released from commercial fishing nets. Treatment spring Chinook salmon were captured and released from 4.25-in to 4.5-in tangle nets and control fish were captured and released from an adult fish trap just upstream. Captured fish were tagged and released for recovery in succeeding fisheries, at hatcheries, and during spawning ground surveys. The results showed that immediate survival was 98%, postrelease survival was 70.1%, and total survival was 68.6% following capture in tangle nets. Total survival combines the results of the immediate and postrelease survival. These estimation methods can be adapted to other fisheries, such as hook and line fisheries, provided an appropriate control is available.

Introduction

Severe declines in Pacific salmon *Oncorhynchus* species and stocks have often resulted in fishery closures. Typically these fishery closures occur in areas that contain mixed stocks of both abundant and protected animals. At times, fisheries can continue by instituting gear restrictions as time and area closures. However, in areas where many stocks require protection, the fishing industry can suffer great economic losses even though abundant stocks are also present. To provide fishing opportunity on abundant stocks while meeting con-

*Corresponding author: ashbrcea@dfw.wa.gov

servation goals for nontarget species and stocks, Pacific Northwest fishery managers have begun using selective fishing methods and regulations (more accurately described as "live capture selective harvest"). In mark-selective fisheries, fishers distinguish stocks that may be retained from stocks that must be released by harvesting only fish with clipped adipose fins, which represent fish from healthy hatchery stocks. For a selective fishery to be successful, both a harvest and a conservation goal must be met. Although Chopin and Arimoto (1995) noted that immediate and delayed mortality caused by encounters with commercial fishing gears is often high, Vander Haegen et al. (2004) have recently shown that the use of a tangle net combined with careful handling techniques can significantly reduce spring Chinook salmon *O. tshawytscha* mortality compared to the conventional gill net. Tangle nets are similar to a gill net but have a smaller mesh size (typically 3.5–4.5 in) and multifilament web. Tangle net gear is fished in the same manner and locations as gill nets but tangle nets capture fish by the maxillary or teeth. This allows fish to continue respiring in the net, so they can be released live. The successful outcome of tangle net research prompted the next question for responsibly adopting a selective fishery, "what survival rates should managers apply to the nontargeted, released salmon?"

Although Vander Haegen et al. (2004) estimate immediate and postrelease survival, they did not estimate total mortality or provide final standard errors. Previously, survival of fish captured in gill nets has been measured by confining the fish in net pens following their capture (coho salmon *O. kisutch* [Farrell et al. 2001a]; lake trout *Salvelinus namaycush* [Gallinat et al. 1997]; sockeye salmon *O. nerka* [Thompson et al. 1971]; spotted seatrout *Cynoscion nebulosus* [Murphy et al. 1995]). Because fish held in net pens do not experience stresses that released fish may encounter, the survival estimates provided in these studies are unlikely to reflect the true survival. Survival estimates for sport (Bendock and Alexandersdottir 1993; Gjernes et al. 1993; Muoneke and Childress 1994) and seine captured fish (Candy et al. 1996) have been estimated, but because the capture method is very different, they are unlikely to apply to tangle net fisheries.

Ricker (1954, 1958) was among the first to develop a tag-recovery model to estimate survival using control and treatment releases. The method is referred to as the "relative recovery method," because survival is estimated as the ratio of the two tag recovery rates. Seber (1965) and Youngs and Robson (1975) extended the method to include multiple mark releases and multiple recovery times. Burnham et al. (1987) developed an extensive theory for release–recapture models that use paired upstream-downstream releases. They classified the Ricker relative recovery method as a "first capture history" protocol and labeled the method as model $H_{1\phi}$. Mathur et al. (1996) developed a variation of the paired-release method to estimate acute mortality through hydroturbines that uses information from both alive and dead recoveries. More recently, single release-recovery (Brownie et al. 1985) and single release–recapture (Skalski et al. 1998) models have been used to estimate survival inriver. Statistically based survival studies in the Columbia River began with passive tag techniques (Skalski et al. 1998) such as the PIT tag and have been extended to include active tag technologies such as radio and acoustic tags (Skalski et al. 2001; Townsend et al. 2006) in recent years. While single-release models are useful in estimating survival over longer expanses of time or distance, paired-release models remain the preferred method when attempting to isolate sources of mortality. For this reason, a paired-release model was used in this study to estimate postrelease delayed mortality between control and test fishery caught fish.

The primary objective in this study was to provide incidental mortality estimates for nontargeted salmonid stocks captured and released from a tangle net fishery. The spring Chinook salmon fishery in the Columbia River basin was chosen for our study. Severe declines have resulted in several stocks of Columbia River spring Chinook salmon being listed as threatened or endangered under the federal U.S. Endangered Species Act. Of the five salmon species in the Columbia River (chum salmon *O. keta,* coho salmon, rainbow trout *O. mykiss,* sockeye salmon, and Chinook salmon) the first returning adult salmon, spring Chinook, are arguably the most valuable to commercial gill-net fishers because of their large size, flesh quality, and high fat content. Small run sizes of spring Chinook salmon resulted in closure of the commercial fishery from 1977 to 2000. This closure had a large economic impact on the commercial fishing fleet, as well as on the small communities along the Columbia River, where most of the commercial fishers live.

The potential for commercial fishing returned in 2001, with the largest return of spring Chinook salmon to the Columbia River since 1938. The abundant return of both hatchery and wild spring Chinook salmon provided an opportunity to evaluate a commercial spring Chinook salmon mark-selective fishery. Removal of the adipose fins of most hatchery stocks has provided the capability for a mark-selective fishery. However, management of a selective fishery differs from the historic fishery because the bycatch can include nontargeted stocks and species of concern that must be incorporated into the harvest allocation. In addition to achieving a harvest goal, to be considered successful, selective fisheries must also achieve a conservation goal. Given that the tangle nets could allow harvest and consequently an economically viable fishery, managers needed a survival rate to apply to selective fisheries to ensure the conservation goal for the stocks of concern is met. Consequently, an estimate was needed for the total adult spring Chinook salmon survival following their release from tangle nets.

Methods

Study Site

The Columbia River is the second largest river in the United States, draining an area of 668,200 km^2 and flowing 2050 km from its source in British Columbia, Canada, to its mouth between the states of Washington and Oregon in the United States.

Adult salmon returning to the Columbia River encounter the first mainstem hydroelectric dam, Bonneville Dam, at river kilometer 234. As they migrate up the Columbia River, they can encounter up to eight additional mainstem dams before reaching the impassable Chief Joseph Dam at river kilometer 955. Fish venturing up the largest tributary to the Columbia River, the Snake River, encounter up to seven additional dams. Spawning grounds for spring Chinook salmon are dispersed throughout the Columbia River basin, as are a number of hatcheries that produce spring Chinook salmon for supplementation and harvest. Spring Chinook salmon returning to the Columbia River belong to a number of stocks. These stocks are mixed as they enter the river's mouth, and disperse as they migrate upstream. Harvest on these stocks occurs throughout the river and consists of commercial, recreational, and tribal ceremonial and subsistence fisheries.

Fishing Methods

The fishing methods we used are similar to that described in Vander Haegen et al. (2004.) We contracted local fishers to fish approximately 11.3 km downstream of the Bonneville Dam during April and May of 2003. Fishing vessels were equipped with a hydraulic reel mounted in the bow that was used to deploy and retrieve the nets. The nets were set by reeling them across the river (typically in a curved pattern) and allowing both ends to drift freely. The fishers used tangle nets that were 274.4 m long and constructed of 10.8 cm or 11.4 cm mesh tangle net (1.5 mm × 5 strands, hung at a ratio of 2:1). The hang ratio describes the number of fathoms of mesh per fathom of cork line. The gear was hung to a suitable depth (10.7 m) and was light green.

All vessels were equipped with a revival box similar to that described by Farrell et al. (2001a, 2001b). We asked the fishermen to mimic the fishery pertaining to the location and as to how nets were deployed. We also asked the fishers to cover both sides of the river to ensure a representative sample of the various spring Chinook salmon stocks present in the river. During each fishing session, we alternated the end of the net that was closest to shore so that the fishing effort of each net type was as similar as possible for each area fished. Fishers removed fish from the net carefully, particularly to avoid touching the gill area. As possible, fishers also looked over the bow as the net was being retrieved so they could lift fish over the roller. Fish were placed immediately into a tank of freshwater located near the bow.

Two observers were on board each vessel. One observer primarily recorded data, while the other observer handled fish. For each net set, observers recorded the following: the time when the net was placed in and removed from the water, the longitude and the latitude for the set (using a Garmin handheld global positioning system unit), which net type was put in the water first, and which net type was removed from the water first. Observers also recorded the date, skipper's name, boat name, observer names, set number, weather conditions, water and surface temperatures, presence of marine mammals, and any other observations pertaining to each particular set. Observers informed fishers when to start picking up nets to ensure short soak times (the time from when the first cork goes in the water until the last cork is removed from the water). For each spring Chinook salmon caught, the observer noted whether the adipose fin was missing, measured the fork length, and tagged the fish with a jaw tag covered with a plastic sheath and printed with a number. Concurrently, fish were collected and tagged with a differently colored jaw tag at the adult fish trap located in the fish ladder at Bonneville Dam on the Washington shore of the Columbia River. The fish caught at the dam served as control fish for our study.

Standard procedures at the fish trap required that fish in the holding tank be anesthetized. Consequently, although we did not anesthetize the treatment fish and although anesthetic was not needed to successfully tag our control fish; all our control fish were anesthetized. Each fish in the control group was measured (fork length) and tagged. The observer recorded whether the fish was missing its adipose fin and its visible injuries. Fish were then transferred to another holding tank with freshwater until they revived back to a lively condition. Following this, they were released into a chute and diverted back to the fish ladder to continue their migration. Both treatment and control fish were recovered in succeeding fisheries, at hatcheries, and during spawning ground surveys by a variety of state, federal and tribal groups and recreational fishing public.

Estimating Survival

The study was designed to estimate immediate, postrelease, and total survival probabilities following capture in a tangle net. Immediate survival $\left(\hat{S}_I\right)$ was estimated as the binomial proportion

$$\hat{S}_I = \frac{a}{n},$$

where n = total number of fish captured with the tangle net, and

a = number of fish retrieved from the tangle net that lived.

The sample error for immediate survival was estimated by

$$\sqrt{\hat{\mathrm{Var}}\left(\hat{S}_I\right)} = \sqrt{\frac{\hat{S}_I\left(1-\hat{S}_I\right)}{n}},$$

with confidence intervals estimated by the normal approximation (Zar 1984:379.) Postrelease survival $\left(\hat{S}_P\right)$ was estimated using the Ricker relative recovery model where

$$\hat{S}_P = \frac{\left(\frac{t}{T}\right)}{\left(\frac{c}{C}\right)}$$

and where T = number of tangle net fish released,

t = number of tags recovered from tangle net fish,

C = number of control fish released, and

c = number of tags recovered for control fish.

The variance of $\hat{S}_P$ was estimated by the delta method where

$$\hat{\mathrm{Var}}\left(\hat{S}_P\right) = S_P^2\left[\frac{1}{c} - \frac{1}{C} + \frac{1}{t} - \frac{1}{T}\right].$$

Total survival $\left(\hat{S}_T\right)$ was calculated as the product of the immediate and postrelease survival estimates where

$$\hat{S}_T = \hat{S}_I \cdot \hat{S}_P$$

with approximate variance estimator

$$\hat{\mathrm{Var}}\left(\hat{S}_T\right) = \hat{S}_I^2 \cdot \hat{\mathrm{Var}}\left(\hat{S}_P\right) + \hat{\mathrm{Var}}\left(\hat{S}_I\right) - \hat{\mathrm{Var}}\left(\hat{S}_I\right) \cdot \hat{\mathrm{Var}}\left(\hat{S}_P\right).$$

Results

Immediate Survival

We fished from 24 March until 8 May 2003, and captured 1,173 adult spring Chinook salmon. Of those, 25 could not be revived following capture for an estimate of immediate survival of 98% ($\widehat{SE}$ = 0.42%) (Table 1). Of the fish captured by tangle net that survived capture, all but 34 were tagged, so that 1,114 salmon were released following capture in the tangle nets. A total of 1,124 control fish were captured and released from the Bonneville Dam fish trap between 27 March and 19 May 2003. The later start and finish in capturing control fish was thought to coincide with migration timing from the fishing area to the upstream control site. No control fish died during handling.

Postrelease Survival

We jaw-tagged and released a total of 1,114 adult spring Chinook salmon from the tangle-net gear and 1,124 spring Chinook salmon from the control group. Jaw tags were recovered throughout the Columbia River in succeeding fisheries, at hatcheries and during spawning ground surveys between 31 March and 2 October 2003 (Figure 1). Recoveries

Table 1. Immediate survival of adult spring Chinook salmon captured during the 2003 test fishery in each treatment on the Columbia River. The number of adult spring Chinook salmon captured (*n*) and the number recovered alive (*a*) reported for both treatment and controls.

Treatment	*n*	*a*	Immediate survival	SE
Tangle	1,173	1,148	98.87%	0.42%
Control	1,124	1,124	100.0%	0%

were clumped in areas with the most intensive sampling, popular sport fisheries and at hatcheries, but do not indicate that tagged fish did not return to other areas. We assumed that treatment fish captured in the nets and control fish captured at the Bonneville Dam were from the same populations, and therefore their tags were equally likely to be recovered, so that observed differences in tag recovery rates were due to survival differences. Further, we assumed that reporting rates were the same among the survivors of the two groups. To test this assumption, we looked at the proportion of recoveries above Bonneville Dam between the treatment and control groups. A chi squared test failed to show a significant difference for treatment and control recoveries for these regions (Tables 2 and 3; $\chi_2^2 \geq 0.8506$; P = 0.6536). Using the relative recovery rates for fish that were linked to a geographic area, we estimated that 70.74% of the fish survived their tangle net postrelease experience (Ta).

Total Survival

Incorporating the immediate and postrelease survival, we estimate a 68.57% ($\widehat{SE}$ = 6.69%) survival for adult spring Chinook salmon captured in tangle nets during 2003. Considering the 95% confidence around this point estimate, the actual survival is between 55.46% and 81.69%.

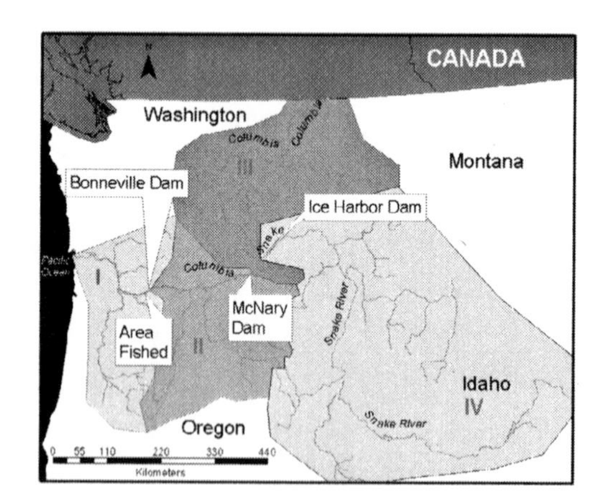

Figure 1. Map showing geographic areas where fish were recovered: I—Below Bonneville Dam; II—Between Bonneville and McNary dams; III—Upper Columbia above McNary Dam; and IV—Snake River above Ice Harbor Dam.

Discussion

With large returning runs of salmon to the Pacific Northwest since 2001, the political incentive for resumed harvest heightens. However, harvesting fish in mixed stock fisheries requires a balancing act between providing harvest opportunity and protection. Otherwise, future harvest opportunity may not be possible. In addition to allowing harvest to occur, another economic benefit to selective harvest is that fish harvested using less stressful and damaging gears should realize higher market prices because they will have fewer net marks and bruising on their bodies.

Research comparing the tangle net to a gill net indicates the tangle net is a commercially

Table 2. Spring Chinook salmon recovery numbers by geographic area in 2003.

Region	Group	Tangle net	Control
I	Recoveries below Bonneville Dam	0	1
II	Recoveries Bonneville Dam to McNary Dam	84	118
III	Recoveries Columbia R. above McNary Dam	24	42
IV	Recoveries Snake R. above Ice Harbor Dam	42	55
Totals		150	216

Table 3. Recovery of tag groups from hatcheries, fisheries, and spawning grounds for recovered fish linked to a geographic area.

Group	Number tagged	Number recovered	Percent recovered	Survival rate	SE
Control	1,124	216	19.22%	70.07%	6.83%
Tangle net	1,114	150	13.46%		

viable substitute for gill nets and meets the criteria for selective fishing when combined with modifications in fishing practices, including the use of fish recovery boxes, short soak times, and careful fish handling. Furthermore, external and associated internal injuries are reduced using the tangle net (Vander Haegen et al. 2004.) We have taken this work a step further toward application in management by conducting a study using the tangle net gear chosen by fishery managers and estimated total mortality as a result of capture in commercial gear. This estimation method should be useful for selective harvest of other species in other areas. Although immediate mortality was less than 2.0%, the postrelease and total mortality estimates were much larger. To achieve a more precise confidence interval in the overall survival estimate, tagging more fish, evaluating the study over many years, or using tags with a greater recovery rate will be necessary. Our research shows that although it has been common to estimate total mortality by releasing fish into holding pens (Gallinat et al. 1997; Farrell et al. 2001a, 2001b; Geoff Whisler, Oregon Department of Fish and Wildlife, personal communication), incidental survival for released fish in a fishery is very different and should not be ignored.

Another method of evaluating postrelease survival (Buchanan et al. 2002) relied on swim tests. Here the researchers found fish attained a velocity comparable to speeds of physiologically recovered fish (Farrell et al. 2001a). Because stress differences among species exist (Schreck et al. 2001), it would be useful to evaluate the physiology of spring Chinook salmon following their capture in tangle nets to learn if a difference exists among species.

Key to estimating incidental survival is a suitable control. The adult fish trap at Bonneville Dam is the best control we know of because fish are not held in place or released such that overt exposure to pinniped predation can occur. There were no tributaries between the fishing site and the trap where fish may have turned away. Although control fish were anesthetized with clove oil, the available literature does not indicate that this difference would influence survival (Anderson et al. 1997; Schreck et al. 2001; Wagner et al. 2002; Pirhonen and Schreck 2003).

Some of us (Ashbrook and Hassel) recently evaluated spring Chinook and steelhead salmon following their capture in purse seines in conjunction with careful handling techniques. All fish captured in this manner were observed to be in excellent condition with respect to vigorous swimming and lack of net marks, descaling, and bruising. However, the use of purse seines as a control for estimating survival during selective fisheries needs further evaluation. For instance, capture with a purse seine would hold fish in place, increasing their vulnerability to pinniped predation. Ideally, the purse seine should be compared to a control such as the adult trap at Bonneville Dam.

During the commercial spring Chinook fishery that occurred lower in the Columbia River during 2002, many steelhead salmon were captured in 5.5-in mesh monofilament nets and evidence indicates this mesh size may act as a gill net for steelhead salmon (Dan Rawding, personal communication, Washington Department of Fish and Wildlife 2002). Whether or not the 4.5-in mesh net would similarly act as a gill net for steelhead salmon is unclear. The survival estimate for this bycatch species requires evaluation to ensure responsible harvest in the presence of weak and endangered stocks. Further, because the stress response can be maladaptive to reproductive fitness (Schreck 2001), although released spring Chinook and steelhead salmon may survive capture and release, their ability to reproduce may be impaired, countering the potential conservation benefits of increased survival. Consequently, we recommend experiments examining the physiological responses of spring Chinook salmon to capture and the resulting effects on reproduction.

Acknowledgments

Funding for this work was provided by the Bonneville Power Administration (Contract #00004684) and a Saltonstall Kennedy Grant (Contract #NA03NMF4270133). Steve Schroder, Geraldine Vander Haegen, and Annette Hoffmann of the Washington Department of Fish and Wildlife provided intellectual discussion and policy and planning assistance. Personnel from the following groups returned tags: University of Idaho; Idaho Department of Fish and Game; Nez Perce Tribe; Oregon Department of Fish and Wildlife; U.S. Fish and Wildlife Service; Washington Department of Fish; and Wildlife Yakama Indian Nation. Heuker Brothers, Inc. provided test fishing.

Reference

Anderson, W. G., R. S. McKinley, and M. Colavecchia. 1997. The use of clove oil as an anesthetic for rainbow trout and its effects on swimming performance. North American Journal of Fisheries Management 17:301–307.

Bendock, T., and M. Alexandersdottir. 1993. Hooking mortality of Chinook salmon released in the Kenai River, Alaska. North American Journal of Fisheries Management 13:540–549.

Brownie, C., D. R. Anderson, K. P. Burnham, and D. S. Robson. 1985. Statistical inference from band recovery data–a handbook. U.S. Fish and Wildlife Service Resource Publication 156.

Burnham, K. P., D. R. Anderson, G. C. White, C. Brownie, and K. H. Pollock. 1987. Design and analysis methods for fish survival experiments based on release-recapture. American Fisheries Society, Monograph 5, Bethesda, Maryland.

Buchanan, S., A. P. Farrell, J. Fraser, P. Gallaugher, R. Joy, and R. Routledge. 2002. Reducing gill-net mortality of incidentally caught coho salmon. North American Journal of Fisheries Management 22:1270–1275.

Candy, J. R., E. W. Carter, T. P. Quinn, and B. E. Riddell. 1996. Adult Chinook salmon behavior and survival after catch and release from purse-seine vessels in Johnstone Strait, British Columbia. North American Journal of Fisheries Management 16:521–529.

Chopin, F. S., and T. Arimoto. 1995. The condition of fish escaping from fishing gears—a review. Fisheries Research 21:315–327.

Farrell, A. P., P. E. Gallaugher, and R. Routledge. 2001b. Rapid recovery of exhausted adult coho salmon after commercial capture by troll fishing. Canadian Journal of Fisheries and Aquatic Sciences 58:2319–2324.

Gallinat, M. P., H. H. Ngu, and J. D. Shively. 1997. Short-term survival of lake trout released from commercial gill nets in Lake Superior. North American Journal of Fisheries Management 17:136–140.

Gjernes, T. A., A. R. Kronlund, and T. J. Mulligan. 1993. Mortality of Chinook and coho salmon in their first year of ocean life following catch and release by anglers. North American Journal of Fisheries Management 13:524–539.

Mathur, D., P. G. Heisey, E. T. Euston, and J. R. Skalski. 1996. Turbine passage survival estimates for chinook salmon smolt (*Oncorhynchus tshawytscha*) at a large dam on the Columbia River. Canadian Journal of Fisheries and Aquatic Sciences 53:542–549.

Muoneke, M. I., and W. M. Childress. 1994. Hooking mortality: a review for recreational fisheries. Reviews in Fisheries Science 2:123–156.

Murphy, M. D., R. F. Heagey, V. H. Neugebauer, M. D. Gordon, and J. L. Hintz. 1995. Mortality of spotted seatrout released from gill-net or hook-and-line gear in Florida. North American Journal of Fisheries Management 15:748–753.

Pirhonen, J., and C. B. Schreck. 2003. Effects of anesthesia with MS-222, clove oil and CO_2 on feed intake and plasma cortisol in steelhead trout (*Oncorhynchus mykiss*). Aquaculture 220:507–514.

Ricker, W. E. 1954. Stock and recruitment. Journal of the Fisheries Research Board of Canada 11:559–623.

Ricker, W. E. 1958. Handbook of computations of biological statistics of fish populations. Fisheries Research Board of Canada, Vancouver.

Schreck, C. B., W. Contreras-Sanchez, and M. P. Fitzpatrick. 2001. Effects of stress on fish reproduction, gamete quality and progeny. Aquaculture 197:3–24.

Seber, G. A. F. 1965. A note on the multiple recapture census. Biometrika 52:249–259.

Skalski, J. R., J. Lady, R. Townsend, A. E. Giorgi, J. R. Stevenson, C. M. Peven, and R. D. McDonald. 2001. Estimating inriver survival of migrating salmonid smolts using radiotelemetry. Canadian Journal of Fisheries and Aquatic Sciences 58:1987–1997.

Skalski, J. R., S. G. Smith, R. N. Iwamoto, J. G. Williams, and A. Hoffmann. 1998. Use of PIT-tags to estimate survival of migrating juvenile salmonids in the Snake and Columbia Rivers. Canadian Journal of Fisheries and Aquatic Sciences 55:1484–1493.

Thompson, R. B., C. J. Hunter, and B. G. Patten. 1971. Studies of live and dead salmon that unmesh from gillnets. International North Pacific Fisheries Commission Annual Report 1969:108–112.

Townsend, R. L., J. R. Skalski, P. Dillingham, and T. W. Steig. 2006. Correcting bias in survival estimation resulting from tag failure in acoustic and radiotelemetry studies. Journal of Agricultural Biology and Environmental Statistics 11:183–196.

Vander Haegen, G. E., C. E. Ashbrook, K. W. Yi, and J. F. Dixon. 2004. Survival of spring Chinook salmon captured and released in a selective commercial fishery using gill nets and tangle nets. Fisheries Bulletin 68:123–133.

Wagner, E., R. Arndt, and B. Hilton. 2002. Physiological stress responses, egg survival and sperm motility for rainbow trout broodstock anesthetized with clove oil, tricaine methanesulfonate or carbon dioxide. Aquaculture 211:353–366.

Youngs, W. D., and D. S. Robson. 1975. Estimating survival rate from tag returns: model tests and sample size determination. Journal of the Fisheries Research Board of Canada 32:2365–2371.

Zar, J. H. 1984. Biostatistical analysis. Prentice Hall, Englewood Cliffs, New Jersey.

American Fisheries Society Symposium 49:687–703

Transformation in the South African Fishing Industry and Its Ability to Redistribute Fishing Rights

Moenieba Isaacs* and Mafaniso Hara

University of Western Cape, School of Government, Program for Land and Agrarian Studies Private Bag X17, Bellville, Western Cape Province 7515, South Africa

Abstract.—The new democracy, new constitution, bill of rights and a new fisheries policy created an action space for all previously disadvantaged individuals or blacks[1] to benefit from the redistribution and transformation processes in South Africa. The redistribution of fishing rights to coastal communities forms one of the corner stones of South Africa's new fisheries policy. In the quest to achieve this, the government has tried a number of formulas to increase the number of previously disadvantaged individuals into the fishing sector. Initially from 1994 to 1998, entitlement formed a key strategy. After 1998, with the enactment of new Marine Living Resources Act (MLRA—Republic of South Africa 1998), race and gender were instrumental in how established industry maintained their fishing rights. In 2001 the government tried a social policy through the introduction of subsistence permits to address poverty along the coast. In 2002 a medium term was introduced to create stability with fishing rights. The success of the new entrants depended on the opportunities and constraints and their ability to take advantage of the new space created by the MLRA. This depended on their entrepreneurship skills, ability to raise capital, link to an organization, and access to information. This paper reviews the various formulas the fisheries department (Marine and Coastal Management) embarked on in restructuring and transforming the industry, how communities have attempted to take advantage of the action space and whether the new dispensation has the potential to contribute towards the socioeconomic upliftment of formerly marginalized coastal communities.

Introduction

The new democracy and the new constitution paved the way to formulate and implement the new fisheries law, the Marine Living Resources Act (MLRA). This law firmly entrenched the goals of transformation and redistribution of fishing rights in South Africa, thereby creating an action space for all South Africans to apply for fishing rights. The law, however, had a dialectical nature—while it enabled the target actors to apply for fishing rights, it also constrained them. The rules and regulations that were set to form organizations or entities were beyond the level of skills and capacities of many of the target actors. At the same time, the shift in the government's macroeconomic strategy from the reconstruction and development (RDP) to growth employ-

*Corresponding author: misaacs@uwc.ac.za

[1] South African realities unfortunately force a resort to racial categories. Thus "African" refers to indigenous inhabitants whose ancestors' presence in the region pre-dated the arrival of European and other settlers; 'coloured' refers to people of mixed race origins and slaves from the East (Indonesia, Malaysia and Philippines); "Indian" refers to descendants from South Asia; "white" refers to descendants of European settlers; "black" refers inclusively, in the manner of Black consciousness (BC), to all South Africans who are not "white." The researcher falls into the category of descendants from South Asia but classified as "coloured"; however, she classifies herself as black in the same manner of the BC.

ment and redistribution (GEAR), which focused on job creation and employment, restricted any mass redistribution of fishing rights and opened the door for litigations by the established industry, and also resulted in streamlining community organizations into closed corporations and private companies. Not only did this have the effect the action space for the poor, it also made the arena contested or a site of struggle where the dominant discourse of GEAR was set on creating new entrepreneurs with the intention that it would have trickle-down effects in the community with regard to job creation and employment. The entrepreneurs and the established companies, both competing for the same fishing rights, dominated the contested arena or action space. The entrepreneurs legitimized their role in the action space arena through the political transformation in the fishing industry. Since the goals relating to transformation in the fishing industry were so vague, these entrepreneurs technically formed part of the target group. The established industry legitimized their role by transforming their companies through shareholding schemes, change in ownership by bringing on board Black economic empowerment groups and through joint venture agreements. In the end, the legitimate target actors (marginalized poor) found themselves outside this action space.

The action space became a site of struggle between the established industry and the organizational entrepreneurs, with the dominant discourse favoring business. On the one side were the ambitious entrepreneurs who usually lacked credit, business skills and infrastructure while on the other side were the mighty established industry with all the skills and infrastructure competing for the same limited resources. The state entered the action space by reinforcing the dominant discourse of favoring business above charity (job creation, creating stability in the industry, and creating employment) on the one hand, while on the other hand it wanted to show that it was doing the right thing by reinforcing its political goals of transforming the industry. The race and gender complement of established industry became the mainstream transformation agenda while the marginalized poor were catered for in the subsistence category to address poverty alleviation in fishing communities.

This paper analyses the attempts involved in transforming the fishing industry. It is divided into three main parts. The first part looks at why transformation of South African fishing industry. The second part focuses on policy processes and actions meant to redistribute and transform the fisheries. The last part discusses the rationale for Individual Transferable Quotas what the key features of redistribution in two sectors that were meant to benefit those marginalized coastal settlements.

Transformation in South African Fisheries

South Africa produces an average of 580,000 metric tons (mt) of fish in live-weight (valued at an estimated 400 million U.S. dollars) annually. Fisheries contribute only about half a percentage point to the country's gross domestic product (GDP). The sector employs about 30,000 fish workers (around 20,000 fishers and about 10,000 in the processing). Management of the fishery has mainly been through the distribution of an annual global total allowable catch (TAC) for all the main commercial species (hake, abalone, rock lobster, anchovies and pilchard) and permits (for tuna and squid) from its exclusive economic zone (EEZ). Due to the discriminatory political, economic and structural policies of the apartheid and colonial governments towards blacks South Africa has the rare distinction of having developed an industrial fishery at the expense of an artisanal or small-scale fishery, unlike the rest of Africa, Asia or Latin America. The forced removals and the revoke or denial of fishing rights had a profound effect on black coastal communities. In 1994, whites or white owned companies owned 93% of all commercial fishing vessels, 96% of the licenses for commercial fishing and disposed of 99.25% of the total allowable catch (TAC) (Hersoug 1996). Like in most other parts of the country, the black coastal communities are characterized by very high levels of poverty, unemployment and general economic deprivation. Having gained political power, African National Congress (ANC) government was left with the arduous task of reversing these effects of unfairly disproportionate allocation of resources and wealth.

Policy Processes in Redistributing Fishing Rights

The Reconstruction and Development Programme (RDP) formed the main policy document of the African National Congress (ANC) government as the party ascended to power in 1994 (African National Congress 1994). The RDP document stated that "Marine resources must be managed and controlled for the benefit of all South Africans, especially those whose livelihoods depend on the resources from the sea." It further stated that the democratic government had to assist people to gain access to the resource through the introduction of new and appropriate legislative measures (FAWU 1997).

The government initiated the task of transforming the industry in November 1994 by forming a multi-stakeholder Fisheries Policy Development Committee (FPDC) that was mandated to work out a Green Paper. Notably, the disenfranchised communities were poorly represented in this process since they were not an organized sector like the established fishing industry (Hersoug 1996). For obvious historical reasons, the issue of access to fishing quotas and rights dominated the debate. Three issues were instrumental in the policy deliberations;

- *Nature of property right*—Here, contestation pertained to the rights of the existing individual rights holders against the state's prerogatives for redistribution of the resource.

- *Position of user group and organized interests*—The organized interests within the industry wanted greater influence on the determination and award of quotas and rights other than leaving this solely to ministerial discretion.

- *Principles and mechanisms of redistribution*—Here, the question was how and the extent to which the previously disadvantaged groups/individuals[2] (PDG/Is) would be allowed into the industry.

Thus in reviewing and revising the fisheries policy, the key question had been 'How could redistribution be achieved without adversely affecting the industry?' As the struggle over what form access rights should take became protracted, the minister decided to appoint an Access Rights Panel, composed of people supposedly without any connection to the industry, whose task was to review the access rights options for the industry. The panel recommended the introduction of real, long-term transferable rights. Following the panel's report, a Portfolio Committee was instituted in order to move the political process forward. Apparently, the committee was unhappy with idea of real, long-term transferable rights. It was thought that such rights would have the

[2] Previously disadvantaged individuals, a technical term used in the allocation criteria to describe the position of blacks as compared to whites in the attempt to transform the fishing industry.

effect of maintaining the status quo and might actually further entrenches the dominant position of the established commercial fishing industry. Such a development would have been quite the opposite of the political and economic vision of the government for the industry as stipulated in the RDP. With slight editing, the recommendations of the Access Rights Panel were written into the White Paper, which was presented to Parliament in May 1997 (MFWP 1997). The White Paper was distributed widely so as to contribute further to the policy debate over the proposed changes. Contrary to the normal procedure, however, the writing of the bill for the new fisheries policy did not await the responses to the White Paper. Due to pressure to produce speedy results, the "Marine Living Resources Bill" was prepared in a parallel process with the White Paper (Hersoug and Holm 1998). After this lengthy process, a new fisheries act, the Marine Living Resources Act (MLRA), became law in September 1998. What can be said is that while there were compromises on all the contentious issues, the minister retained most of the powers of final decision. Exercise of such powers had to be based on the underlying principles of the MLRA which sort to achieve equity in the fishing industry, stability in the market place, political stability in the communities, and biological sustainability of the resource. The following were the main outcomes of the process as enacted in the act: the discredited Quota Board that had been inherited from the previous administration was scrapped; a Consultative Advisory Forum (CAF) was formed to act as restraint on the minister's powers in case he wanted to exceed the safe biological limits; and a Fisheries Transformation Council (FTC) was mandated to replace the Commercial Public Company that had been proposed in the Green Paper. The FTC was supposed to be a holding and processing vehicle for rights to the PDG/Is. It was further agreed that the minister would lease out fishing rights, not sell them; rights would be leased for a maximum period of 15 years; and finally that rights would be leased as a proportion of the of the quota rather than a portion.

From a perspective of fisheries management, three broad institutional choices had been available to government; community, market and state (Hersoug and Holm 1998). The *community* approach was never really considered as a serious option mainly due to lack of traditional communities and also the experience with the failed community quotas that the government had attempted to implement in 1994. In any case, this route was not favored by any group of the primary stakeholders. In addition, those from the PDG/Is viewed this as being a ploy to keep them out of quotas for commercial fishing. The use of market mechanisms was favored by the established fishing industry and in this context they sort to influence the introduction of long-term individual transferable quotas (ITQs). They tried to influence policy towards this direction by arguing that there was need for keeping stability in the industry and that cutting their quotas would result in job losses. During the deliberations and experience from other countries implementing ITQs such as New Zealand and Iceland, it became clear that the market solution would continue and even further entrench the existing (white) industrial dominance and monopoly. For ANC, to allow this would have meant abandoning its RDP goals and objectives. From a political point of view, the only way to increase the potential for redistribution was through structural reform of the industry other than relaying on the market place. In the end, the government opted for state control (Hersoug and Holm 1998).

One cannot say though that the market solution was completely abandoned especially with the government's switch of macroeconomic policy from RDP to the growth employment and reconstruction (GEAR 1996). Most commentators and critics have pointed out that the change represented a near reversal of the RDP principles, as GEAR is a more neoliberal approach than socialist. As Marais (2001) points out, "three key tests have to be applied to GEAR. Would it trigger growth, would it stimulate job creation, and would it increase socioeconomic equality?" The author concludes (Marais 2001) that "Sadly, the verdict is negative on all three counts". In the fishing industry, the arguments for economic stability and need for job creation were in line with the principles of GEAR and favored the position of the established industry rather than mass reallocation of fishing rights to coastal communities and marginalized fishers.

Quota and Rights Reallocations

As Hersoug and Isaacs (2001) have demonstrated, quite a sizeable reallocation of quotas and rights had occurred between 1992 and 2000 (Table 1). Apparently, the Quota Board appointed in 1994 that included members from the PDG/Is made heavy impact on the restructuring process partly on its own initiative and partly due to pressure from the new political forces then coming to power. The Quota Board started to bring in a large number of new entrants even before the new act was passed. In addition, the established industry knew that with the fall of the apartheid regime transformation was inevitable even though they might not have been very sure the extent to which this would happen and what form it would take. Several strategies have been used by the established companies in order to try and retain their dominant positions in the industry: sell of part of their share holding to black interest groups; offering limited shareholding ownership for their black employees; bringing in black business leaders or prominent politicians into company management positions; improving labor conditions (salaries, fringe benefits, pension schemes, training and skills up-grading schemes) for their largely black shop floor workers; and finally, entering into joint venture agreements with new entrants. While in some cases such changes have been genuine, there are also cases whereby it has largely been what can be said to be "window dressing." All the same, these tactics have been used to strengthen claims for maintaining their existing quotas holdings or even for increased amounts (Isaacs 2004).

The reallocations shown in Table 1 can be explained as follows: in the *Abalone* sector, the new entrants had been allocated 47%, with 8% set-aside for subsistence fishers by 2000. If shareholding changes in the established companies are taken into consideration the share for the PDG/Is had moved up to 77% of the TAC. In the West Coast rock lobster *Janus lalandii*, which is one of the most disputed species since it is easily accessible to small-scale fishers, new quota holders (including the subsistence fishers) controlled 44% of the TAC while the established companies controlled 43%, with the 13% set aside for recreational fishers a sector that also included new entrants. In 2000 the 4% increase for new rights was mainly due to a special allocation to subsistence fishers. If restructuring of the established companies is taken into consideration, the PDG/Is controlled 70% of the TAC. In South Coast rock lobster *Palinurus gilchristi*, new entrants controlled 38% of the TAC, or 62% if the restructuring of established companies is taken into consideration by 2000. In the pilchard (the pelagic sector), new entrants controlled 48% of the TAC by 2000 and if shareholding restructuring of established companies is taken into account this had increased to 72%. In anchovy, new entrants controlled 45% of the TAC by 2000 or 71% if

Table 1. Proportion of quotas and sharing holdings that had been reallocated from established companies to previously disadvantaged groups/individuals by the year 2000.

Sector	TAC allocated to PDG/Is by 2000 (as % age of total) (1)	Company shareholdings sold to PDG/Is by 2000 (as % age of total) (2)	Total interests held by PDG/Is by year 2000 (as % age of total) (1) + (2)
Abalone	55	22	77
West Coast rock lobster	44	26	70
South Coast rock lobster	38	24	62
Pilchard	48	24	72
Anchovy	45	26	71
Hake	28	18	44

shareholding restructuring of the established companies is factored in. In the most valuable hake sector, 28% of the TAC had been reallocated to new entrants, a figure that goes up to 44% if the transformation of the established companies is taken into consideration (Isaacs 2004).

It seems from the preceding table as if a major shift had taken place in terms of quota allocations. There are problems that can be attributed to the way transformation is being interpreted, especially by the established fishing industry who have most to gain from such number crunching. Firstly, it is not clear to what extent the new entrants' only function as front companies for the old established ones. With considerable experience from the apartheid era in setting up front companies, daughter companies, holding companies etc. in order to evade sanctions, Manning (2000) has shown, in the case of Namibia, that it is very difficult to really ascertain who is owning or controlling what. Secondly, until it is possible to evaluate how the new entrants operate, it is not possible to know whether they are investing in the industry so as to be able to dispose of their quotas let alone process and market the product on their own. In other words, to what extent are they receiving the quotas just to lease them out? Thus although the reallocation figures might look impressive at first glance they do not tell the full story as the extent to which the benefits of such reallocations are resulting in economic and social benefits of the formerly disadvantaged coastal communities. Since the old Quota Board gradually started the reallocations in 1994 and the process gathered pace after the MLRA was passed in 1998, there has been an obsessive interest in applying for entry into the fishery. Potential entrants created cooperatives, closed corporations, private companies and holding companies in the thousands. Unfortunately only a fraction (less than 10%) seemed to have succeeded, resulting in recriminations against the few who had been successful and loss of faith in the MCM and the transformation process itself. With such a huge number of unsuccessful applicants, the widely held belief within the fishing communities has been that those who needed to gain most from the transformation process did not in the end do so (Isaacs and Mohammed 2000; Isaacs and Hersoug 2000; Hersoug and Isaacs 2001).

Besides, those who had succeeded in getting quotas or permits soon discovered that gaining a quota or permit was only the first hurdle. They had to compete successfully with the established companies at sea and on the markets, in order to get a firm foot hold in the industry. As a result, most had ended up as 'paper quotas' or had to enter into joint ventures with the established companies on less than favorable terms due to their weak negotiating positions, having entered into a very unfamiliar business environment without investment capital or the required skills. How will these problems affect groups or individuals characterized by such passive ownership in their ability to qualify for consideration for the more permanent quota allocations the next round remains unanswered. What had become evident by the year 2000 was that redistribution is only one step in the process of transformation, and that it is in fact possible to have a massive reallocation (of quotas), without much transformation in terms of empowering the new entrants to be active participants in the new industry (Hersoug and Isaacs 2001).

By mid 2000, the MCM administration admitted that the reallocation process was not going very well. According to an internal document from the (MCM) Rights Allocation Unit (quoted in Hersoug and Isaacs (2001);

"In the historic absence of a defined fishing rights allocation procedure, the department has been unable to call upon any dedicated capacity to manage the allocation process, and this lack of capacity has meant that the department has become so immersed in the mechanics of fishing rights allocation that the development of an effective allocation policy and plan has been severely compromised. This has lead to a virtually permanent state of crisis management with regard to rights allocation and a lack of proper planning" (DEAT 2000).

Some of the problems that had dodged the process included; challenges on the minister's powers to allocate a portion of the commercial species TAC to the Fisheries Transformation Council (FTC); litigations against the minister by established industry against cuts to their quotas. For example, the 1999 allocations were prepared according to the new Act and its accompanying regulations. In the case of West Coast rock lobster the establish industry challenged the reduction to their quotas that the minister had done in order to accommodate some new entrants. As a result, the fishery was halted for some months as the due legal process went through its paces. Due to legal technicalities (applications for the 1998–1999 season had to be made according to the previous Sea Fishery Act of 1988), the rights of the new entrants were suspended and the shares of the old participants were restored. In other fisheries, such as Cape hake and abalone, a negotiated settlement was reached, based on the fact that the Minister could force a stronger cut the next year if the former rights holders did not accept the "new deal" (Hersoug 2000); bureaucracy and costs involved in quota applications and their processing; legal and constitutional problems of defining previously disadvantaged individuals and small and medium enterprises (SMME); allegations of corruption and graft within MCM; allegations of corruption and graft among community leaders who had claimed to be acting on behalf of communities; contentious and cumbersome allocation procedures; lack of reallocation targets and lack of political commitment. In terms of the last point, the government had been reluctant to build commitments and targets for redistribution into the institutional system. Lastly, the PDG/Is had not been able to organize and establish the institutions that could strengthen their claims and interests in order to challenge the existing order. One of the main problems around self-organization among the PDG/Is had been the heterogeneity within communities, economic differentiation among community members and sociopolitical factionalism, individualism and graft. Government realized that there was urgent need to come up with a new formula for the allocation of access rights.

The work on a new formula for rights allocation started mid 2000, led by a professional trouble-shooter from the Ministry of Finance (Hersoug and Isaacs 2001). His mission was simply to get the allocation process in order by mid 2001 in time for the 2001–2002 seasons by which time the allocation of long-term rights would start. The new proposed application structure involved three stages. During the first phase, the intention of the allocation procedures would be to reduce the number of applicants dramatically by discouraging less serious applicants from applying. This would be done by increasing the application fee for commercial rights from R100 (100 South African rands) to R6000; making it a requirement to submit a tax form showing that all outstanding taxes had been paid; submission of all the relevant documentation such as company particulars, founding statements, loan agreements, shareholders agreements, business plans, transformation goals, job creation plans and any other relevant documentation; and an oral presentation of the business plan upon request. Applicants would then be graded into specific categories before the applications went to the second stage of evaluation. During this second stage, all applicants would be evaluated based on the Rule Book criteria. The final phase would implement incentive-based transformation, driven by the "follow the buck" principle, specifically directed to established industries. At this stage, the applicants could be ranked according to a transformation factor. This transformation factor could include all funds which are paid to HDI based staff, contractors, and so forth through salaries, share schemes, etc. The highest transformation factor-ranked applicant will have the first opportunity to select additional quota of varying sizes from the transformation pool. Then the second ranked applicant will select quota from the pool, then the third, etc until all quota blocks within the pool were exhausted. This would be implemented annually during Phase 1. Transformation would take place during a 3–5 years period and quotas will be issued accordingly, as from the 2001–2002 fishing season. Thereafter, long-term rights (up to 15 years) would be issued for the second phase. New entrants would be phased in through annual attrition (ibid.). For example, if it is decided that a subsector should be restructured so that 25% should be allocated to new entrants, 5% will be given each year based on the quantity reverting back to the state in the 5-year period (Isaacs and Hersoug 2001).

Towards Individual Transferable Quota System—Medium Term Rights Allocations

The new formula for issuing of rights was ready by mid-2001. Thus on 27 July 2001, MCM put out an invitation (Government Gazette No. 22517—Republic of South Africa 2001) for applications for commercial fishing rights. The notice indicated that the rights would generally be awarded for a longer term of up to 4 years (up to 2 years for abalone) for all species. The department also announced that no subsistence fishing rights would be issued for the West Coast rock lobster and abalone. Thus any fishers who were previously involved in subsistence fishing for West Coast rock lobster or abalone and wished to continue with their activities had to apply for a commercial fishing right. The issuing of limited commercial rights for 850 kg or less abalone and 1.5 mt or less West Coast rock lobster (termed small-scale commercial) was specifically introduced to accommodate small-scale operators. An application fee of R500 (nonrefundable) per application would be charged for such Limited Commercial applications. A fee of R6000 (nonrefundable) per application would be charged for all commercial applications. An independent verification unit (in 2001 this task was awarded to Deloitte and Touché Private Limited and Sithole AB&T Chartered Accountants) was appointed and tasked to receive, process, verify application forms, and investigate information submitted by all applicants. The directorate hoped that such an independent verification process would result in a more transparent and credible process that would afford the applicants an insight into the adjudication process. It was also meant to ensure the factual correctness of the submitted applications in order to eliminate front companies, corruption and nepotism. The rights verification process would provide the minister with a comprehensive picture of the companies and individuals that had applied for rights. The verification process would also enable the department to gain access to vital information to assist it in the adjudication of applications and to enable the department to carry out a more credible allocation process. Finally, the verification process was meant to do away with the widespread selling of fishing rights or holding of so called "paper quotas" which had been a common practice in the industry over the last few years.

The deadline for applications was 13 September 2001, just a month and half after the notice went out. Of particular note was that people could only apply as individuals, commercial companies, closed corporations or trusts. Forums and Associations, which had been one of the major requirements for applying for subsistence fishing permits the previous season, had fallen away as required organizational form for communities. This resulted in frantic and panicky activities as the former forums and associations for subsistence fishers formed closed corporations so as to be able to apply for the limited commercial quotas. Although MCM attempted to make the application forms available as widely as possible by establishing various distribution points along the coast where people could collect the forms, the personnel from MCM were instructed that it was not part of their duties to assist applicants in filling in the forms (a help-line was made available all the same). Thus applicants had to sort help elsewhere for completion of the forms if they could not do it on their own, which was the case in most cases, as experience from previous seasons would have shown. According to MCM (DEAT 2002) 7,100 forms had been issued within the first three days of the invitation, showing that there was still a huge amount of interest in acquiring rights.

All applications for fishing rights were scrutinized to determine whether they were properly lodged, whether they were materially defective and whether they fulfilled the essential requirements of the sector. An application was not properly lodged if it had been received late, the fee was not paid in full, the applicant had applied for more than one sector using the same application form, the application was not received on the correct application form, there was no authentic number on the application form, or the applicant had provided insufficient copies of the application form. An application was materially defective if the application was not signed and commissioned, more than one application was submitted for the sector, the applicant did not provide sufficient information for the purposes of evaluating the application, or the applicant did not adequately demonstrate that he or she was a South African citizen. The applications that were not properly lodged or materially defective or did not fulfill the essential requirements were rejected. The remaining applications were divided into two streams, namely previous right-holders ("2001" right-holders) and the potential new entrants. The 2001 right-holders were assessed against one another in accordance with assessment criteria based on degree of transformation; degree of involvement and investment in the sector and vessel; capacity to harvest and market the resource; past performance; legislative compliance; and degree of paper quota risk. The potential new entrants were assessed against one another in accordance with assessment criteria based on degree of transformation or the degree to which the application was constituted by persons from historically disadvantaged community; degree of knowledge, involvement and community investment in the industry; degree of business acumen and financial capability; capacity to harvest and market the resource; legislative compliance; and degree of paper quota risk.

Rights were granted for a longer period than in the past with the intention of encouraging investment in the industry, ensuring sustainable development of small- and medium- sized commercial enterprises and alleviating poverty. The department anticipated that the allocation of 4-year rights would create a climate of stability in the South African fishing industry and enhance the capacity of the industry to build stable markets for South African products abroad. It was also hoped that following the conclusion of this rights allocation process, the department would be well placed to devote more resources and effort to control, monitoring and surveillance activities so as to eradicate over fishing and illegal harvesting of fisheries resources.

The policy guidelines stated that as much as possible, applicants were to demonstrate their initiative, experience, and capability to participate in the fishery they were applying for. Applicants were required to come up with a business plan, operational plan and investment strategy. Another important aspect was the extent to which an applicant was going to meet the objectives of equity, transformation, and empowerment. The degree of transformation would be evaluated by looking at the ownership structure or equity of proposed entities, the distribution of wealth that would be achieved, creation of employment for members of the historically disadvantaged sectors of the community and the manner and extent to which gender inequity would be addressed. Another policy consideration would be the impact of fishing on resources and the environment. A point emphasized was that any conviction or transgression committed by an applicant would negatively impact on their future applications. Finally, the policy guidelines emphasized that regard would be given to accommodate new entrants, particularly those from historically disadvantaged communities in order to redress the historical imbalances and achieve equity. It was noted that such new entrants would not be unfairly disadvantaged or disqualified because of non-ownership of fishing equipment due to lack of access to capital. All the same, new entrants had to be able to demonstrate their involvement in the fishing industry, show a clear commitment to making it in the industry, business acumen, knowledge of the sector or fishery they were applying for and the capacity or ability to catch, process and market the right applied for. The brief indicated that the transfer of fishing rights, or part thereof, would not generally be favored during the first 2 years from the date of allocation in order to discourage "paper quotas". It was reiterated that the allocation of rights would be the responsibility of the minister or his delegated official.

Many potential entrants were unhappy about these requirements as they saw them as mechanisms by government to keep them out of the industry. They felt that the system favored the established companies and those who are literate other than entrepreneurs from formerly marginalized communities. Thus among the majority of the PDG/Is, there existed a great amount of pessimism and anger about government's perceived inability to deliver on the new policy and legislation. In the short-term therefore, hopes and dreams seem to have been subdued, with very few happy new entrants and still many unhappy unsuccessful applicants.

Results of the 2001 Medium-Term Allocations

The quota allocations for the four seasons 2001–2002 to 2004–2005 are give in Table 2 and 3. The data and information was sourced from the official Department of Environmental Affairs and Tourism (DEAT 2002a–2002h as stated in Isaacs 2004) documents on the quota allocations for the various species and sectors.

Discussion

There are, however, important differences among the sectors. In the low capital-intensive sectors, such as WCRL, transformation had been achieved mainly through increasing access to new entrants and by reallocating quotas formerly reserved for the established industry. In the more capital-intensive sectors, such as deep-sea hake trawling, transformation had taken place through the active buy-in of Black empowerment companies. What is also interesting, as demonstrated in Isaacs (2004) is that transformation in the first period (1994–1998) before the MLRA was not insignificant. However, the old Quota Board implemented transformation on an ad hoc basis, as quotas were largely allocated without guidelines. With the MLRA, guidelines and criteria became more specific, but verification was largely nonexistent, which allowed room for those with the most resources and skills, and not necessarily those who were most in need of fishing rights, benefiting. Following the passing of the MLRA in 1998, MCM embarked upon a policy encouraging increased Black ownership in the fishing industry in the form of mergers, joint ventures and takeovers of the established industries. With the allocation of medium-term fishing rights in 2001–2002, this trend continued. With the exception of limited-commercial rights in abalone and WCRL, the new allocation system produced no dramatic changes among the established companies. By and large the new entrants who entered between 1994 and 2000 had their quotas renewed, together with most of the old established companies (the historical companies).

Another lesson from those seasons was the issue of forums and associations. The forums or associations as business entities seem to have had a lot of problems. The executive committee of a given forum or association was supposed to act as the business management en-

Table 2. Redistribution of fishing rights for historically disadvantaged individuals for the seasons 2001/2002–2004/2005 in the different sectors.

Sector	HDI-managed	HDI-owned
Abalone	16%	84%
West coast rock lobster	34%	66%
South coast rock lobster	71%	29%
Pilchards & Anchovy	53%	47%
Hake	74%	16%

Table 3. Fishing Rights allocations for the seasons 2001/2002 to 2004/2005[a] in the different sectors.

Sector		Amount allocated	As a percentage
Abalone	Full commercial	313.96	72.8
	Limited commercial	74.39	17.2
	Reserved for appeals	43.15	10.0
	TAC (metric tons)	**431.5**	**100**
Hake (deep sea trawling)	Established operators[b]	136205	98.4
	New entrants	-	-
	Reserved for appeals	2290	1.6
	TAC (metric tons)	**138495**	**100**
Hake (longline)	Established operators	3700	34.1
	New entrants	4875	45.0
	Reserved for appeals	1015	20.9
	TAC (metric tons)	**10840**	**100**
Hake (inshore)	Established operators	9665	95.1
	New entrants	-	-
	Reserved for appeals	500	4.9
	TAC (metric tons)	**10165**	**100**
Hake (sole)	Established operators	784.8	90.0
	New entrants	-	-
	Reserved for appeals	87.2	10.0
	TAC (metric tons)	**872**	**100**
Pelagi	Established operators	351878	89.4
—pilchards136500 metric tons	New entrants	1181	0.3
—anchovy 222600 metric tons	Bait catchers	1181	0.3
—Bycatch 34500 metric tons	Reserved for appeals	39360	10.0
	TAC (metric tons)	**393600**	**100**
West Coast rock lobster	Full commercial	1513.4	71.2
	Limited commercial	218.2	10.3
	Reserved for:		
	·Appeals	162.6	7.62
	·Hangclip area	50.0	2.4
	·Witsands area	181.0	8.5
	TAC	**2125.2**	**100**

[a] For abalone, the rights are for 2 years, while these are for 4 years for all the other species.

[b] Established operators are those who had quotas in 2000/2001, including the new entrants from PDG/Is.

tity. In an environment where someone has to take tough decisions on a daily basis, this type of structure proved to be very difficult in most instances, resulting in accusations of theft, corruption and incompetence by their members. As a result, most groups broke up within a given season. The other problem with the forums and associations was the sheer number of members. In most instances, members of a single forum or association exceeded fifty and in a few cases run into hundreds. The large number of people brought about a lot of problems when it came to participation in the actual fishing, which was one of the conditions of the permits, and in the sharing of profbeen the optimal number of people per forum or association?

The short fishing seasons also meant that those who got fishing rights still had to find other sources of income out of season. If they could not, then it became tempting to continue fishing illegally out of season. The decline in the resource, whether due to over fishing or due to environmental problems means that seasons might get even shorter or might result in the actual closure of some fisheries in some specific locations. Having invested in the expensive required equipment, such occurrences would be disastrous for the small-scale fisher solely dependent on one seasonal fishery.

While the system of individual, Closed Corporation and Trust applications as applied for the medium term rights for the period 2001–2005 might help to overcome the problems related to forums and associations, this is likely to immediately raise the issue of exclusion. Most of those that had been formerly disadvantaged under the previous dispensation are the ones that are likely to have been disadvantaged by the new conditions of application. While most had expressed the desire to get commercially viable individual quotas, they were also the most ill-equipped to take advantage of any new opportunities unless government could have ensured financial and institutional support in the early years of the operations of new entrants. In any case, there will still a large majority of unhappy people who will have failed to make the cut.

The broader reform agenda of the national government's redistribution of access rights in the fishing industry encompasses the promotion of black economic empowerment, shareholding schemes for employees as well as creating local entrepreneurs. These strategies limited the action for the marginalized poor fishers and shifted the fishing rights away from the target audience (Isaacs 2004).

Individual Transferable Quotas in South Africa

From the 2001–2002 season, the government introduced medium term rights (2 years for abalone and 4 years for all the other species) with the view of moving to long term (15 years) rights in the next round of rights allocation. The idea of long-term rights is based on the theory of individual transferable quotas (ITQs). The aim of introducing ITQs is to maximize economic efficiency. It is testing a concept that has been widely advocated by economists but has not yet been generally accepted by the world fisheries management community. One of the assumptions of the ITQ system is that since the fisheries will be more profitable, the bargaining position of fishing crews will improve, resulting in higher wages (Arnason 1991). For fisheries economists such as Scott (1988) and Arnason (1991), the privatization of common fisheries resources is not a side effect of ITQs, but rather the most important objective of the system. The forerunners in the use of ITQs are New Zealand and Iceland.

Even if the New Zealand and Icelandic experiences were successful (Gissurarson 2000), the conditions in many other countries do not favor such ambitious ITQ systems. New Zealand and Iceland had essentially ignored equity and distributional issues. In South Africa, these remain key concerns of fisheries managers and their political masters. It is therefore imperative that South Africa moves with caution towards this direction of long term (tradable) rights.

Marginalized Groups in ITQ System

According to the works of Scott (1985, 1990) and Webster and Engberg-Pedersen (2002) marginalized groups do not possess many assets to really influence politics, while current dominant discourses favor the more affluent groups with the necessary skills. The facilitatory role of local organizations in promoting the political agency of the poor is also not very clear. The action space created by the MLRA for marginalized poor can, according to Scott, be reflected on two levels, the onstage arena and the offstage arena. On the onstage arena there would be those that would publicly accept the new allocation of fishing quotas, as they were included in the process. There would also be those that would publicly defy the new allocation process through public protest. In the offstage arena there would be some that would show their acts of defiance through poaching. There would also be those who did not receive any fishing rights who would be discontent with the new allocation regime, but they would disguise their acts of defiance through gossip, euphemisms, etc. Through the open and disguised forms of resistance, the marginalized poor declare their agency from below. Scott's hidden transcripts (onstage and offstage discourse) help us to understand why the discrepancy did not result in massive demands for reallocation. There were certainly protest groups and a few of them succeeded. The established industry, together with the successful new entrants, had gained ideological domination by claiming that fishing was a business and not a charity! The unsuccessful remained negative about the whole process, but their protest made little impact on the reform process, with one important exception, namely poaching (Isaacs 2004).

Poaching may be seen as one form of weapon or tool by the fishers who had tried to go the prescribed legal way, but who had lost out.[3] Added to this perspective of offstage discourse could be the fact that even among the poor bona fide fishers there would be different interests. Not all were similarly dependent on fishing and not all were equally poor. As indicated in Isaacs (2004), the unsuccessful group was also very diverse. By adding these elements to the more technical explanations of implementation, we start to get a picture of why the realities on the ground turned out to be different from the first optimistic visions of improving the lot of the poor fishers in the coastal communities. In short, policy goals were unclear, the administrative apparatus weak, the conditions unstable, the opposing forces strong and the potential beneficiaries weak. Due to the agency of all groups involved, the power relations and hence the action space would change over time, so that a policy script with the best intentions could end up with a number of unintended consequences.

Conclusion

Much as the RDP had been formulated with the best intentions of a liberation movement intent on achieving the political, social, and economic objectives of the struggle for its constituency once it had gained power, the ANC government soon discovered the painful realities of the world economic system (Marais 2001). As Wallerstein (1996) quoted in Marais (2001) caustically summarizes:

"......Numerous liberation movements have taken the theory of sovereignty at its face value, and have assumed that sovereign states are autonomous. In fact they are not and have never been....All modern states, without exception, exist within the framework of the interstate system and are constrained by its rules and its politics. The productive activities within all modern states, without exception, occur within the framework of capitalist world-economy and are constrained by its priorities and its economics...."

But as Joseph Stiglitz, the World Bank vice-president argues, (quoted in Marais 2001) the growth paths of China and the South East

[3] This is not to say that all these fishers poach, as there is a large group of professional poachers who have never intended to take the legal route.

Asia's "tiger economies" remind us that no single formula exists for economic growth. Here was a cluster of countries that had not closely followed the Washington Consensus prescriptions but had somehow managed the most successful development in history (Stiglitz 1998)

Thus within the constraints of macroeconomic policymaking and implementation, it is expected that the role of the South African state is to ensure social and economic justice through transformation. Emphasis as far as possible should be given to generating new productive capacity in small black enterprises of all kinds, with the related training and technical support in order to create an expanded industrial base in a nonracial and democratic economic structure. Hence, encouragement should be given to small enterprises that create jobs within communities. These small firms need to be supported through low-cost finance, business training, technical and marketing advice, provision of infrastructure and administrative services. Thus subsistence, limited commercial and small-scale commercial fishers should be assisted through government support. It is clear that strong government intervention is needed to achieve a more equitable economic structure. As Turok (1996) warns us socio-economic transformation requires South Africa to transform the policy environment and production sector, the economic empowerment of blacks and the poor through access rights, access to credit for enterprises; investment in human development, and equality of opportunity in gaining access to means of production (Turok 1996).

The central issue in transforming South Africa is not a reduction in state involvement but the redirection of effort to ensure equity and greater or broader participation in production that will benefit the previously disadvantaged. The economic strategy should not simply be based on growth or creation of short-term employment. An economic strategy is needed which should focus on greater empowerment of fishing communities and individuals to provide them with new capacities to create sustainable livelihoods. Transformation should empower the people and build institutional and organizational capacity to enable people to participate fully in the political and economic life of the society. An economic strategy for transition must be driven by political considerations to ensure legitimacy. Politics must be seen to deliver material advances (Turok 1996). The transformation process in the fishing sector is in many ways a microcosm of transformation process in the broader South African society. The definition and meaning of transformation should be premised on strong state intervention, with parallel social and economic reforms and not in the context whereby social delivery only depends on economic growth.

In this paper, we have sort to look at the issues pertaining to the transformation of the South African fishing industry meant to achieve redistribution of the resource. The government would like to see increased political, economic and structural transformation of the fishing industry in order to address the past imbalances so that the industry can contribute towards the social uplift of the coastal communities. For the politicians, the only way to increase the potential for redistributive capacity is through structural reform of the industry other than relaying on the market place through the interests of the established industrial sector. With the adoption of the GEAR policy, real transformation might decline to snail's pace. For most PDG/Is, it remains incomprehensible as to why, having 'captured' political power, the ANC cannot deliver on the promised redress of the past injustices immediately.

Sen (1999) captures the transformation in the South African fishing industry and its ability to deliver restitution and redistribution neatly through providing us with an indirect route of alleviating poverty. Sen argues that by concentrating on a "capability approach", which he relates to five types of freedoms governments would be able to politically, economically and socially transform their countries. These freedoms include political freedom, economic facilities, social opportunities, transparency guarantees, and security of society. Political freedom encompasses civil rights, an uncensored press, and free democratic elections. The poor fishers in South African fishing industry have achieved political freedom via the democratic elections, a new constitution, and new fisheries policy, which were necessary but not sufficient condition. The second type of freedom is that of economic facilities include both free markets and access to the resources necessary to participate in a market economy. The coastal fishing communities are experiencing a lack of economic infrastructure and credit facilities to start-up viable businesses. The third type of freedom refers to the social opportunities that include education, health care, and so forth which greatly affect a person's ability to improve his situation. Most of the poor in the coastal fishing communities lack the basic social opportunities of education and health care, which negatively affects their ability to improve their situation. The fourth type of freedom is that of transparency guarantees, which require that people deal openly with each other under guarantees of disclosure and lucidity, help to prevent corruption and financial irresponsibility. Many of the poor fishers viewed the reform process (allocation process) as not being transparent, which resulted in charges of corruption and mismanagement against the Marine and Coastal Management department, and thus loss of trust and legitimacy in the reform process at local community level. The fifth type of freedom refers to the protective security of society, the social safety net (e.g., unemployment benefits, income supplements for the poor) that prevents abject poverty. The freedom of protective security of society of unemployment benefits and income supplements are not present for the poor.

References

African National Congress (ANC) 1994. Reconstruction and Development Programme (RDP):A Policy Framework. ANC, Johannesburg.

Arnason, R. 1991. Efficient management of ocean fisheries. European Economic Review 35:408–417.

DEAT (Department of Environmental Affairs and Tourism). 2000. Draft discussion document for fisheries management plan to improve the process of allocating fishing rights. August 2000. DEAT Chief Directorate: Marine and Coastal Management. Cape Town, South Africa.

DEAT (Department of Environmental Affairs and Tourism). 2002a. Summary of recommendations, considerations and decisions in respect of Hake Long- line rights for the 2002–2005 seasons. DEAT Chief Directorate: Marine and Coastal Management, Cape Town, South Africa.

DEAT (Department of Environmental Affairs and Tourism). 2002b. Recommendations, considerations and decisions in respect of Hake Deep-sea trawl rights for the 2002–2005 seasons. DEAT Chief Directorate: Marine and Coastal Management, Cape Town, South Africa.

DEAT (Department of Environmental Affairs and Tourism). 2002c. Recommendations, Considerations and Decisions in Respect of Squid rights for the 2002–2005 seasons. DEAT Chief Directorate: Marine and Coastal Management, Cape Town, South Africa.

DEAT (Department of Environmental Affairs and Tourism). 2002d. Recommendations, Considerations and Decisions in Respect of Abalone rights for the 2002–2003 seasons. DEAT Chief Directorate: Marine and Coastal Management, Cape Town, South Africa.

DEAT (Department of Environmental Affairs and Tourism), 2002f. Recommendations, Considerations and Decisions in Respect of Pelagic rights for the 2002–2005 seasons. DEAT Chief Directorate: Marine and Coastal Management, Cape Town, South Africa.

DEAT (Department of Environmental Affairs and Tourism). 2002g. Recommendations, Considerations and

Decisions in Respect of Mari culture for the 2002–2005 seasons. Chief Directorate: Marine and Coastal Management, Cape Town, South Africa.

DEAT (Department of Environmental Affairs and Tourism), 2002h. Where have all the fish gone? Measuring transformation in the South African fishing industry. October 2002. DEAT Chief Directorate: Marine and Coastal Management, Cape Town, South Africa.

FAWU (Food and Allied Workers Union). 1997. A Response to the Fisheries White Paper. Labour Research Service. FAWU's Sea Fisheries Development Project, Cape Town, South Africa.

GEAR (Growth, Employment and Redistribution). 1996. A Macro-economic Strategy. Available: http://www.policy.org.za/govdocs/policy/growth.html.

Gissurarson, H. H. 2000. The politics of enclosures with reference to the Icelandic ITQ system. In R. Shotton, editor. Use of property rights in fisheries management. FAO Fisheries Technical paper 404(2)1–16.

Hersoug, B., and M. Isaacs. 2001. 'It's all about money!': Implementation of South Africa's new fisheries policy. Land reform and agrarian change in Southern Africa occasional paper no. 18. Programme for Land and Agrarian Studies, University of the Western Cape, Capetown, South Africa.

Hersoug, B. 1996. Fishing in a sea of sharks–reconstruction and development in the South African fishing industry. The Norwegian College of Fishery Science. University of Tromsoe, Tromsoe, Norway.

Hersoug, B., and P. Holm. 1998. Change without redistribution: an institutional perspective on South Africa's new fisheries policy. Norwegian College of Fishery Science, University of Tromsoe, Tromsoe, Norway.

Isaacs, M., and N. Mohammed. 2000. Co-managing the commons in the 'new' South Africa: room for manoeuvre? Paper presented at the 8th Biennial Conference of the International Association for the Study of Common Property entitled, Constituting the Commons, May 31–June 4, 2000. Available: http://iodeweb1.vliz.be/odin/bitstream/1834/761/1/Isaacs23.pdf. (June 2007).

Manning, P. 2000. Review of the Distributive Aspects of Namibia's Fisheries Policy Draft. Namibian Economic Policy Research Unit, Nepru Research Report, Windhoek, Namibia.

Marais, H. 2001. Limits to change: the political economy of transition (revised and expanded edition). Zed Books, London.

MFWP (Marine Fisheries White Paper). 1997. A summary of the White Paper by the Chief Marine Fisheries White Paper for South Africa–Directorate of Sea Fisheries of the Department of Environmental Affairs and Tourism. Available: http://www.polity.org.za/govdocs/white_papers/marine.html. (September 2002).

Republic of South Africa. 1998. Marine Living Resources Act of 1998, volume 395, No. 747.

Republic of South Africa. 2001. Government Gazette No. 22517 of 27 July 2001.

Scott, A. D. 1988. Conceptual origins of rights based fishing. In Neher et al., editors. Rights based fishing. Kluwer Academic Publishers, Dordrecht, The Netherlands.

Scott, J. C., 1985. *Weapons of the Weak. Everyday Forms of Peasant Resistance.* Yale University.

Scott, J.C., 1990. Domination and the arts of resistance–hidden transcripts. Yale University, New Haven, Connecticut.

Sen, A . 1999. Development as freedom. Oxford University Press, Oxford, UK.

Stiglitz, J. 1998. More instruments and broader goals: moving toward the post-Washington consensus, WINDER annual lecture (January 7). United Nations University/World Institute for Development Economics Research, Helsinki, Finland.

Stuttaford, M . 1992 – 2000. Fisheries Industry Handbooks. *Exbury Publications,* Cape Town, South Africa.

Turok, B. 1996. The debate about reconstruction and development in South Africa as stated in Adedeji, A. 1996. South Africa and Africa: within or apart? Zed Books Ltd, London.

Wallerstein. I. 1996. 'The ANC and South Africa: the past and future of liberation movements in the world system', address to South African Sociological Association, Durban (7 July). Available: http://www.binghamton.edu/fbc/iwsoafri.htm. (June 2007).

Webster, N., and L. Engberg-Pedersen, editors. 2002. In the name of the poor: contesting political space for poverty reduction. Zed Books Ltd, London.

American Fisheries Society Symposium 49:705–706

Session Summary

Reconciling Fisheries with Conservation and the Valuation of Fisheries

Ussif R. Sumaila, Gakuski Ishimura, Yajie Liu, and Dale Marsden
University of British Columbia Fisheries Centre, Fisheries Economic Research Unit
Vancouver, British Columbia, Canada

This session was made up of five oral presentations and six posters. Three of the oral presentations—Sumaila, Alcock, and Ainsworth and Sumaila—focused on economic valuation of fishery resources, emphasizing the role of discounting (that is, the process by which future dollars are reduced to today's dollars) in valuation. One oral presentation was on integrated, spatial assessments of fisheries and marine conservation (Scholz et al.), while the fifth was a presentation on fisheries subsidies and how they impact fisheries sustainability (Clark, Munro, and Sumaila).

In his opening presentation, Sumaila (Canada) made the point that the reconciliation of fisheries to conservation is all about values and valuation. It is about what individuals and society ("we") value, how we value the present and the future, how we value here and there, and so forth. A key conclusion of this contribution was that the theory of economic valuation is broad enough to help ensure that fisheries are reconciled with conservation. The problem lies with the practice of valuation. A quick survey of the literature revealed that only about 1% of papers published in nine leading environmental and resource economics journals in the last decade considered values other than those traded in the market. For economics to help society reconcile fisheries with conservation, the discipline needs to get marine ecosystem values and valuation right!

Alcock (USA) compared the characteristics and values of the large-scale sector (responsive to global markets, weak sensitivity to externalities, and high discount rates), the small-scale sector (responsive to local markets, stronger sensitivity to externalities, and lower discount rates), the recreational sector (ecosystem service value) and environmental nongovernmental organizations (NGOs) (passive and option values). He concluded that the objectives of national fisheries policy are more in tune with the large-scale sector's values, with the consequence that fisheries valuation fails to help reconcile fisheries with conservation.

Ainsworth and Sumaila (Canada) showed that where the conventional methods of discounting advocates aggressive harvest policies, the intergenerational discounting method of Sumaila and Walters (2005) might have made the historic gross overfishing of Atlantic cod economically unappealing compared to a more conservative long-term strategy. To derive this result, the authors compared the historic harvest trend from 1985 with theoretical optimal harvest profiles determined by an ecosystem model using the two discounting approaches (Ainsworth and Sumaila 2005).

Scholz, Mertens, Sohm, Steinbeck, and Bellman (USA) presented work on a United States west coast-wide analytical framework for integrated, spatially explicit assessments that link areas of the ocean to coastal communities. These authors demonstrated that with this framework it is possible to derive the value of different areas of the ocean to different fishing fleets, and that it is possible to jointly consider socioeconomic and conservation concerns.

Clark, Munro and Sumaila (Canada) demonstrated that, contrary to the conventional wisdom, decommissioning subsidies or buyback schemes can be decidedly negative to conservation if the vessel owners anticipate that this subsidy will be paid to them in the future. The basis for this result is "rational expectations," which captures the fact that fishers will take into account their knowledge and understanding of future policy into their current decision-making on how much to invest in fishing capacity.

Posters presented in this session included

1. Msiska (Namibia), Sumaila (Canada), and Ithindi (Namibia)—developed economic indicators for Namibian fisheries, and applied these indicators to explore the economic performance of the fisheries in Namibia over the period 1995–1999;
2. Wagey (Canada), Buchary (Canada), Supangat (Indonesia), and Kepel (Indonesia)—presented a poster on the analysis carried out by the Indonesian government before the introduction of new management practices, such as creating MPAs, to help the government make better decisions;
3. Liu (Canada) and Sumaila—sought to identify and quantify waste discharges from salmon aquaculture operations, the ecological impacts and costs of these discharges, and to incorporate these costs into profitability calculations for the operations;

4. Luque (Columbia) and Bonilla (Columbia)—proposed an ecotourism project to provide income, reduce fishing pressure, and subsidize research on the possible effects of a reduction in water volume that would occur with a proposed hydroelectric project;

5. Buchary (Canada) and Sumaila—presented work in progress on a global database of exvessel fish prices. The aim of the work is to compile price data that will inform a wide range of fisheries studies (Sumaila et al. 2007);

6. Wakeford (USA)—used four North American historical case studies to examine if fisheries buyout programs are an appropriate solution to fisheries management and conservation problems. Each of these poster presentations were set in the context of reconciling fisheries with conservation.

References

Ainsworth, C. H., and U. R. Sumaila. 2005. Intergenerational valuation of fisheries resources can justify long-term conservation: a case study in Atlantic cod (*Gadus morhua*). Canadian Journal of Fisheries and Aquatic Sciences 62: 1104-1110.

Sumaila, U. R., D. Marsden, R. Watson, and D. Pauly. 2007. Global ex-vessel fish price database: construction and applications. Journal of Bioeconomics 9:39–51.

Sumaila, U. R., and C. Walters. 2005. Intergenerational discounting: a new intuitive approach. Ecological Economics 52:135–142.

American Fisheries Society Symposium 49:707–712

Getting Values and Valuation Right: A Must for Reconciling Fisheries with Conservation

Ussif Rashid Sumaila*

Fisheries Economics Research Unit/Sea around Us Project, Fisheries Centre Vancouver, British Columbia, Canada

Abstract.—This paper addresses three key issues: is the economic theory of valuation adequate enough to support the reconciliation of fisheries to conservation; does the practice of valuation enhance our ability to reconcile fisheries to conservation; and what new approaches are being currently developed to deal with any inadequacies?

Introduction

My point of departure in this contribution is that the reconciliation of fisheries to conservation is all about values and valuation. It is about what individuals and society (we) value, how we value the present and the future, how we value the here and there, and how we value market and nonmarket values. A key challenge to economics is how to value benefits from marine ecosystems in a comprehensive manner and in a way that captures their long-term conservation and sustain the benefits they provide. Three key questions asked in this paper are

(1) Does the economic theory of valuation capture the general need to reconcile fisheries with conservation?

(2) Does the practice of fisheries valuation enhance our ability to reconcile fisheries to conservation?

(3) What new valuation approaches are being developed to support efforts to reconcile fisheries with conservation?

Theory of Valuation

The economic theory of valuation is based on what people want—their preferences (Brown 1984; Arrow et al. 1993). The theory is therefore based on individual preferences and choices. People's preferences are expressed through the choices and tradeoffs they make given the resource and time constraints that they face. It is therefore extremely important that we capture a given population's preferences fully into the decision-making process regarding the use and nonuse of marine ecosystem resources, in particular, and natural and environmental resources, in general.

The economic theory of valuation of natural and environmental resources calls for a comprehensive compilation of all values into a total economic value (Goulder and Kennedy 1997). The theory stipulates that the total

*Corresponding author: r.sumaila@fisheries.ubc.ca

economic value (TEV) should include market[1] and nonmarket values[2], which consists of direct use value[3], indirect use value[4], option value[5], existence value (Krutilla 1967)[6], and bequest value (Young 1992)[7]. One can assert that if all these values are fully incorporated into policy decision making, then there is a good chance that fisheries will be successfully reconciled with conservation. Market values bring into the equation the need for some level of fishing to continue, while nonmarket values will incorporate the need for conservation through time. Hence, one may conclude that the economic theory of valuation is broad enough to reconcile fisheries with conservation. The problem seems to lie in the practice of valuation.

The Practice of Valuation

To get an idea of whether the practice of valuation is as broad as the theory, I carried out a survey of nine leading natural and environmental resource economics journals, namely, (1) Journal of Economics and Environmental Management, (2) Ecological Economics, (3) American Journal of Agricultural Economics, (4) Canadian Journal of Agricultural Economics, (5) Land Economics, (6) Journal of Agricultural Economics, (7) Food Policy (8) Australian Journal of Agricultural economics, and (9) Journal of Agricultural and Resource Economics. These journals were selected based on their position in the *Rakings of Academic Journals and Institutions in Economics* (Kalatzidakis et al. 2001).

The number of articles published in the above journals from 1994 to 2003 was determined. Then, the articles that contain the words "nonmarket" or "existence value" or "option value" or "bequest value" were noted. This second step is meant to determine the extent to which values other that market are covered by the articles published in the leading natural and environmental economics journals in the past 10 years. The result from the second step is then split into articles published in the periods 1994–1998 and 1999–2003. This last step is meant to reveal trends in the application of the theory of economic valuation of natural and environmental resources.

The Results of the Survey

Results from the survey turned out to be much stronger than expected. It was discovered that less than 1% of the 4,705 articles (or 43) published in the nine journals between 1994 and 2003 mentioned nonmarket, direct use, indirect use, option, existence, or bequest values (Figure 1). This is such a small percentage that one can conclude that these values are virtually not part of economic analysis. So, even though the theory of valuation talks about total economic value, in practice, virtually all natural and environmental economic studies and analysis incorporate only market values. A more encouraging result of the survey is that 85% of the articles that mentioned nonmarket values appeared in the past 5 years of the survey time period (Figure 2), indicating that economists are beginning to pay some attention to these values.

[1] Values traded in the market (e.g., the value of fish caught and sold in the market).

[2] Values not traded in the market (e.g. existence value [see below]).

[3] Value of ecosystem goods and services that are directly used for consumptive purposes (e.g., the value of commercial output such as fish harvest).

[4] Value of ecosystem goods and services that are used as intermediate inputs to production (e.g., services such as water cycling and waste assimilation).

[5] The potential that the ecosystem will provide currently unknown valuable goods and services in the future.

[6] This is the value conferred by humans on the ecosystem regardless of its use value. This essentially is what is described as nonuse value in the literature. An environmental good may be valuable merely because one is happy that it exists, quite apart from any future option to consume it, visit it, or otherwise use it. This value may arise from esthetic, ethical, moral, or religious considerations.

[7] This value captures the willingness to pay to preserve a resource for the benefit of one's descendants (future generations).

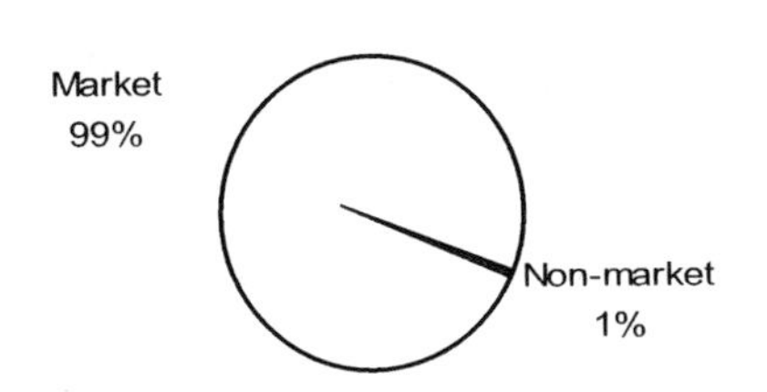

Figure 1. Percentage of studies published in 9 leading environmental and natural resource economics journals from 1994 to 2003, which mentioned market and nonmarket values, respectively.

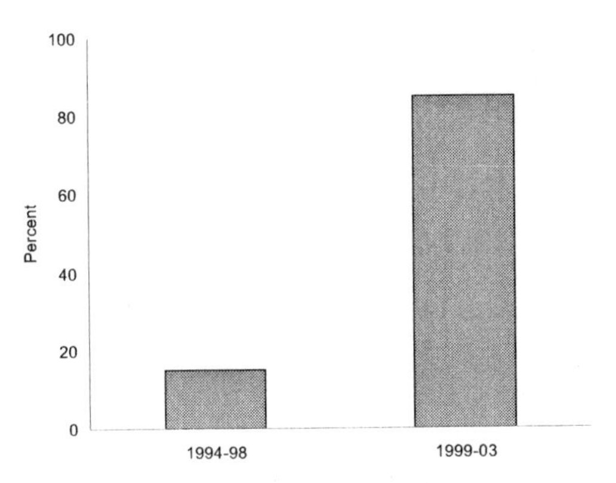

Figure 2. Percentage of studies published in nine leading environmental and natural resource journals in the first and second 5-year periods within the last decade that mentioned nonmarket values.

Measuring Market Values

Market values are the values of marine ecosystems that are included in most economic studies and analyses. But even here, there is still much to be desired. Data on fish prices and the costs of fishing are very difficult to come by. Global fisheries statistics are available from only very limited sources. One key source is the Food and Agriculture Organization of the United Nations (FAO). But the FAO does not compile exvessel prices, which are arguably the relevant prices for management purposes, since this is the price that motivates fishers to go fishing. The *Sea Around Us* project (www.seaaroundusproject.org) and the Fisheries Economics Research Unit (www.feru.org) are currently developing a global exvessel price database that will help fill some of the gaps here.

Measuring Option and Existence Values

A possible approach to providing reasonable estimates of option and existence values is to employ a modified version of the contingent valuation method (CVM) to seek the opinions of all major stakeholder groups (the fishing industry, government, the scientific community, and the general public) on their own estimate of the potential option and existence values of a given marine ecosystem.

The modified CVM approach is based on the simple idea that fish in the ocean is worth more than the same fish on the plate because of the option and existence values that the former may still carry. The idea then is to use survey and/or opinion poll techniques to ask different stakeholders in society to assign values to fish in the ocean relative to the same fish on their plate, taking into account the option and existence value of the fish still in the ocean. The collected information is then aggregated in a statistically acceptable manner to estimate the option and existence values of the given marine ecosystem.

Figure 3 illustrates how this method was used to explore the option and existence values of commercial fish species in the North Sea. The market value is calculated using the potential catches of all the commercial species from the North Sea and multiplying this by the prices of these species over a given time horizon. The total cost of fishing is then deducted and the result discounted to obtain the net present value. The final result is then normalized to one and plotted as the first dot on the figure.

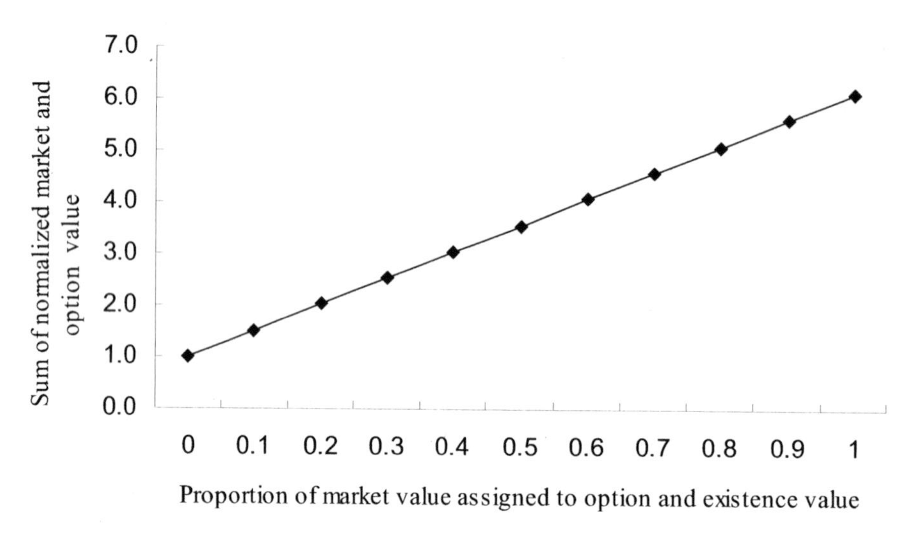

Figure 3. Sum of normalized market, option and existence values (of commercial species in the sea as a function of market value).

The option and existence values are then determined by finding out how much the society in question places value on the option and existence values of fish in the ocean in relation to the computed market value of the same fish. The line plotted on the figure expresses the sum of market, option, and existence values. Assigning an option and existence value of zero (as it is usually done in practice) implies that the sum of market, option, and existence values of the fishery is equal to the market value. On the other hand, if a survey or opinion poll were to show that stakeholders assign equal value to fish on the plate and fish in the ocean because the latter preserves the option and existence values of fish, then the sum of market, option, and existence values would be about six times the market value in this example.

Incorporating Bequest Value

So far, most of the literature on this issue has treated bequest value as if it was just another element/component of the total economic value (see Loomis and Richardson 2001). I will argue that bequest value is a very special component of the TEV. It is about the flow of *all* values to future generations, as distinct to those to the current generation. As a result, the valuation of bequest values requires much more than a simple addition to the TEV. It needs a complete restructuring of the way flows of benefits are counted. It requires that the time perspectives of future generations with respect to benefit flows from marine ecosystems need to be explicitly incorporated. Such an approach will explicitly provide a measure of intergenerational equity, unlike the current approach, where only the current generation's time preference is taken into account through the practice of discounting (Sumaila 2004). In other words, we need to adjust the way we conventionally discount flows of values from marine ecosystems, in particular, and natural and environmental resources, in general. The key point here is that just like the current generation, future generations will like to enjoy direct use value (V^d), indirect use value (V^i), option value (V^o), existence value (V^e) and bequest value (V^b). So, instead of summing all these values and discounting them to obtain the net present value (NPV) using the discounting clock of only the current generation, as expressed below (Angelsen and Sumaila 1997; Loomis and Richardson 2001):

$$NPV = \sum_{t=\infty}^{T} d^t (V_t^d + V_t^i + V_t^o + V_t^e + V^b)$$

where d denotes the discount factor given the prevailing standard discount rate, r, that is,

$$d = \frac{1}{(1+r)}$$

we rather take the sum of the direct and indirect use, option, and existence values and discount them using the discounting clocks of all generations to ensure that bequest values are explicitly incorporated. Mathematically, this is expressed as

$$NPV = \sum_{t=\infty}^{T} w^t (V_t^d + V_t^i + V_t^o + V_t^e)$$

where, from Sumaila and Walters (2005),

$$w = d^t + \frac{d_{fg} d^{t-1}}{G} \left[\frac{1-\Delta^t}{1-\Delta} \right]$$

$$\text{if } d_{fg} \neq d \text{ otherwise w} = d^t + \frac{d_{fg} d^{t-1} t}{G}$$

In the above equations, G denotes the generation time, d_{fg} stands for the future generation discount factor, and

$$\Delta = \frac{d_{fg}}{d},$$

the ratio between the intergenerational and the standard discount factor. The second term in w captures the bequest value, in a much more comprehensive way than V^b.

Concluding Remarks

In conclusion, one can say that the economic theory of valuation is broad enough to help ensure that fisheries are reconciled with conservation. The problem lies with the practice. In this regard, more valuation approaches are needed to help implement what the economic theory of valuation stipulates, that is, to include the TEV of environmental and natural resources in the economic study and analysis of the use and nonuse of these resources.

Acknowledgments

I thank Yajie Liu and Gakushi Ishimura for their assistance in conducting the literature survey. My appreciation goes to OCEANA, the *Sea Around Us* project, and the Pew Charitable Trusts, Philadelphia for their generous support.

References

Angelsen, A., and U. R. Sumaila. 1997. Hard methods for soft policies: environmental and social cost benefit analysis. Pages 20–42 *in* F. A. Wilson, editor. Towards sustainable project development. Edward Edgar, Cheltenham, UK.

Arrow, K. R. Solow, P. R. Portney, E. E. Leamer, R. Radner, and H. Schuman. 1993. Report of NOAA panel on contingent valuation, Federal Register 58:10(15 January 1993):4601–4614.

Brown, T. C. 1984. The concept of value in resource allocation. Land Economics 60:231–246.

Goulder, H., and D. Kennedy. 1997. Valuing ecosystem services: Philosophical bases and empirical methods. Pages 23–48 *in* G. C. Daily, editor. Nature's services: societal dependence on natural ecosystems. Island Press, Washington, D.C.

Kalatzidakis, P., T. P. Mamuneas, and T. Stengos. 2001. Rankings of academic journals and institutions in economics. University of Cyprus, Department of Economics Discussion Paper 2001–10, Cyprus, Greece.

Krutilla, J. 1967. Conservation reconsidered. American Economic Review 57:787–796.

Loomis, J. B., and R. Richardson. 2001. Economic values of U.S. wilderness systems: research evidence to date and questions for the future. International Journal of Wilderness 7:31–34.

Sumaila, U. R. 2004. Intergenerational cost benefit analysis and marine ecosystem restoration. Fish and Fisheries 5:329–343.

Sumaila, U. R., and C. Walters .2005. Intergenerational

discounting: a new intuitive approach. Ecological Economics 52:135–142.
Young, M. D. 1992. Sustainable investment and resource use: equity, environmental integrity and economic efficiency, volume 9. Man and the Biosphere Series. UNESCO, Paris.

American Fisheries Society Symposium 49:713–726

Subsidies, Decommissioning Schemes, and Their Implications for Effective Fisheries Management

Ussif Rashid Sumaila*

University of British Columbia, Fisheries Economics Research Unit, Fisheries Centre, 2259 Lower level Mall Vancouver, British Columbia V6T1Z4, Canada

Gordon Munro

University of British Columbia, Department of Economics and Fisheries Centre 2259 Lower level Mall Vancouver, British Columbia V6T1Z4, Canada

Colin W. Clark

University of British Columbia, Department of Mathematics, University of British Columbia 2259 Lower level Mall Vancouver, British Columbia V6T1Z4, Canada

Abstract.—There is general agreement that many, if not most, fisheries subsidies are detrimental to effective resource management. There is not, however, general agreement about subsidies used for decommissioning or buyback purposes. One school of thought argues that such subsidies can have a positive impact upon resource management, by removing excess fleet capacity from the fisheries. An opposing school of thought criticizes the use of these subsidies on the grounds that the subsidies are, by and in the large, ineffective. This paper argues that decommissioning or buyback subsidies can have a positive impact upon resource management, but only if they are wholly unanticipated by the fishing industry. If, on the other hand, the subsidies are anticipated by the industry, the subsidies can have a decidedly negative impact, intensifying both economic waste in the fisheries and overexploitation of the fishery resources.

Introduction

The impact of subsidies upon the conservation and management of capture fishery resources worldwide has been the focus of steadily increasing concern and debate for well over a decade and a half, within and among national governments and international bodies. For example, the Food and Agriculture Organization of the United Nations (FAO), which had a detailed discussion of the issue over a decade ago (FAO 1992), has in the more recent past held two Expert Consultations on the issue, in 2000 and 2002 (FAO 2000, 2003). A Technical Consultation on subsides took place in late June and early July 2004 (FAO 2004). To take a second example, fisheries subsidies have come up for detailed discussion in the current Doha Round of the World Trade Organization[1].

* Corresponding author: r.sumaila@fisheries.ubc.ca

[1] One of the authors of this paper, Ussif Rashid Sumaila, was asked to present a talk on the issue of fisheries subsidies at the Fifth WTO Ministerial Conference in Cancun, in September 2003.

One reason for the growing intensity of the debate is the magnitude of these subsidies. Estimates vary widely. One of the more conservative estimates, arising from the carefully researched study of Matteo Milazzo for the World Bank (Milazzo 1998), places the level of fisheries subsidies during the mid-1990s at US$15–20 billion per year. Earlier estimates, pertaining to the late 1980s, placed the number as high as US$50 billion a year, perhaps more (FAO 1992). Even if one rejects the high estimates as unreasonable, no one can contest the fact that fisheries subsidies are very large indeed.

Subsidies Defined and Their Implications for Fishery Resource Management and Conservation

A guide on fisheries subsidies recently prepared for the FAO (Westlund 2003) defines fisheries subsidies as follows:

> Fisheries subsidies are government actions or inactions that are specific to the fisheries industry and that modifies—by increasing or decreasing—the potential profit by the industry in the short, medium, or long term (Westlund 2003).

The fact that there is in the definition reference to government actions or inactions, which would decrease fishing industry profits, simply means that the author is incorporating negative, as well as positive, subsidies in her definition. An example of a negative subsidy would be a tax. Be that as it may, in the discussion to follow we shall focus solely upon positive subsidies, examples of which would be fuel tax rebates and vessel investment grants (see Westlund 2003).

Not all fisheries subsidies are harmful to resource management and conservation. The OECD, another international body to have studied fisheries subsidies extensively, deems government expenditures on fisheries research, such as stock assessment, to be a form of indirect financial transfer benefiting the fishing industry (OECD 2000). Unquestionably, these expenditures would fall within the Westlund definition. The OECD maintains that such subsidies can be seen as having a beneficial impact upon resource management and conservation (Flaaten and Wallis 2000; OECD 2000). Few, if any, would disagree.

There is also little disagreement, among economists at least, that many subsidies are, in fact, seriously detrimental to resource management and conservation. Capture fisheries worldwide have historically been noted for their "common pool" nature, in the sense that property rights to the resources, public or private, have been ineffectively implemented, or have been simply nonexistent. It has been recognized for half a century that the common pool nature of fisheries leads to the emergence of a system of perverse incentives confronting the fishers, which in turn leads to resource overexploitation and general economic waste (Gordon 1954; Bjørndal and Munro 1998; Clark 1990). When common pool conditions prevail in a fishery, it will prove to be the case that any subsidy, which 1) increases price for harvested fish received by fishers, or 2) reduces either their vessel operating costs, or costs of acquiring vessel capital, will be detrimental to resource management and conservation. Such subsidies have the effect of making a bad situation worse (Munro and Sumaila 2002)[2].

Where there is not agreement, however, is over the use of subsidies for vessel decommission-

[2] What if the common pool aspect of the fisheries is removed? It might appear that, if a full-fledged rights-based fishery were in place (i.e. a privatized fishery), subsides would cease to be a problem. Munro and Sumaila (2002) have demonstrated that, while subsidies would be less of a menace than they would be in a common pool fishery, they can still do extensive damage.

ing schemes (buybacks). In order to understand the disagreement, we must first digress and review the concepts of "malleable" and "nonmalleable" vessel capital, and then review the reasons why excessively large fleets in fisheries may, in fact, really matter for management and conservation purposes.

The concepts of malleable and nonmalleable vessel capital have now been adopted by the FAO (Gréboval 1999). Malleability refers to the ease with which vessel capital can be removed from a fishery. If such capital is perfectly malleable, it can quickly and costlessly be redeployed out of the fishery, or sold off, with no risk of capital loss to its owners. We will come to define perfect malleability of vessel capital for analytical purposes as the situation in which the resale price (unit "scrap" value) of such capital is equal to its purchase price. By way of contrast, perfectly nonmalleable vessel capital is capital that has no alternative uses whatsoever, and which never depreciates (see Gréboval 1999). So called quasimalleable capital lies in between these extremes. While cases can be found of fisheries in which the vessel capital employed therein can be deemed close to being perfectly malleable, these are the exceptions which prove the rule.

It is when vessel capital is less than perfectly malleable that excess fleet size becomes problematic for management purposes. Consider fisheries, which are subject to some degree of regulation over the global harvests. It has long been recognized that, if there are no effective controls on the fleet size, redundant vessel capital will emerge—given that the vessel capital is less than perfectly malleable (Gréboval and Munro 1999).[3] Redundancy of capital implies pure economic waste. The presence of this excessive amount of nonmalleable vessel capital in the fishery will, in addition, stand as a constant threat to effective resource management. Consider such a fishery, in which the harvests have been effectively controlled in the past. Now suppose that the resource is subject to a negative environmental shock. The appropriate resource management policy response is to reduce the total allowable catch (TAC), or equivalent thereof. The vessel owners, having few if any alternative means of employment for their vessels, can be expected to resist bitterly the called for TAC reductions. Indeed, the proposed TAC reduction could well confront many of the vessel owners with the threat of bankruptcy (National Research Council 1999).[4] As a consequence of the industry pressure brought to bear, the TAC may be reduced by less than that required by the demands of effective resource conservation, or may not be reduced at all.

Subsidies and Decommissioning Schemes

If an excessively large fleet constitutes economic waste and stands as a potential threat to future resource conservation, then it would seem clear that effective resource management demands a reduction of the fleet size. If the vessels have little or no alternative use, then it would seem to be equally obvious given the quote below that subsidized decommissioning schemes may be called for. As Matteo Milazzo, in his World Bank study on fisheries subsidies, states:

[3] As we shall demonstrate in the discussion to follow, if vessel capital is perfectly malleable, there is no reason for the fleet to be excessively large.

[4] An example is provided by the lucrative snow crab fishery in the Gulf of St. Lawrence, in early May of 2003. An announcement by the Canadian Department of Fisheries and Oceans on May 2 that the TAC for the coming season would have to be set approximately 23 % below that of the previous season led to more than just protests by affected New Brunswick crab fishers. It led, as well, to widespread rioting and arson (The Globe and Mail, May 5, 2003).

> ... many commentators have noted how difficult it is to induce the exit of capital from fishing because these assets ... have little other practical use. For that reason ... disinvestment in fisheries has to be actively promoted with economic incentives, i.e. subsidies (Milazzo 1998).[5]

Thus, the argument can be made, and has been made, that subsidies for decommissioning purposes should be seen as having a beneficial impact upon resource management, which can serve to further the reconciliation fisheries with conservation (Milazzo 1998).

Having said this, however, many economists in the past have been exceedingly skeptical of the benefits to be derived from decommissioning subsidies. The essence of their arguments is that decommissioning subsidies often prove to be ineffective. Vessel capital, upon being removed from the fishery, tends to seep back in (see, for example, Holland et al. 1999). There is another round of economic waste as yet more investment in vessel capital returns the fleet to an excessively large size. The threat to resource conservation reemerges (Weninger and McConnell 2000). Indeed, Cunningham and Gréboval warn in a study prepared for the FAO that, while decommissioning schemes appear to offer an ideal way to reduce fleet capacity, resource managers must hold off introducing such schemes until the conditions necessary for ensuring the scheme's long term effectiveness are in place (Cunningham and Gréboval 2001).

These arguments imply the following. If the postdecommissioning scheme seepage of vessel capital back into the fishery can in some way be blocked, then decommissioning subsidy schemes could indeed have a positive impact upon resource management and conservation.

It is the contention of this paper that subsidies for decommissioning schemes can have a decidedly negative impact, even if the problem of seepage of vessel capital back into the fishery is completely eliminated. This negative aspect of decommissioning subsidies is linked to the expectations of vessel owners investing in nonmalleable vessel capital, an aspect of the problem that has largely been ignored by economists heretofore.[6]

Investors in nonmalleable vessel capital cannot, upon acquisition of the capital, look forward to being able to dispose of it quickly and costlessly. The investors are thus unable to afford the luxury of myopia. They have no choice but to formulate expectations about the future profitability of their investments.

Economists specializing in macro-economics have long made a distinction between expectations that are adaptive and those that are "rational." Adaptive expectations essentially mean that agents base their expectations upon extrapolations from the immediate past. Rational expectations, on the other hand, implies that agents, in forming their expectations, do not restrict themselves to recent history, but rather take into account all available information about future events, including future government policy actions. Needless to say, no one now takes adaptive expectations models very seriously.

[5] It is worth noting that the European Commission submitted a proposal to the World Trade Organization, which called for the elimination of subsidies used to increase fishing capacity, but which also called for the retention of subsidies used for decommissioning/buyback purposes (Megapesea 2003).

[6] Adopting the assumption that vessel capital is non-malleable does introduce formidable complexities to the economic models of the fishery. This has encouraged fisheries economists, when formulating their models, to avoid the explicit assumption that vessel capital is non-malleable. The avoidance has come at a cost, however.

The acceptance of rational expectations has, in turn, led to the recognition of the problem of time inconsistency in policy actions. Time inconsistency refers to the fact that policy actions, which appear to be optimal at the time that they are put into effect, may prove to be inconsistent with the long-run best interest of society. Economists F. E. Kydland and E. C. Prescott (1977) authored a seminal article on the time inconsistency problem that state the nature of the problem as follows: "...there is no mechanism to induce future policy makers to take into consideration the effect of their policy, via the expectations mechanism, upon current decisions of the agents" (Kydland and Prescott 1977:481).

The Kydland and Prescott dictum does, as we shall attempt to illustrate, have direct and immediate relevance to decommissioning subsidy schemes.

Decommissioning Subsidies: A Model

The authors have, among them, written three papers on the issue at hand. Two of these have been published, while the third is at the discussion paper stage (Munro and Sumaila 2002; Clark and Munro 2003; Clark et al. 2003). The discussion to follow draws heavily upon these three papers.

In these papers, the authors examine decommissioning schemes under two separate conditions: 1) introduction of vigorous resource management to a fishery, after a history of pure open access; and 2) regulated open access. The term "regulated open access" refers to a situation in which the resource managers effectively control the global harvest in the fishery, but have proven in the past to be ineffective in controlling the fleet size (see Homans and Wilen 1997). The discussion in this paper will be restricted to the simpler case of regulated open access. In so doing, we shall focus on the economic consequences of an excessively large fleet. It must always be kept in mind, however, that an excessively large fleet poses a potential conservation threat (e.g. in the face of unanticipated environmental shocks).

We first require, as background, an economic model pertaining to the fishery, explicitly incorporating nonmalleable vessel capital. This is provided for us by the article by Clark et al. (1979), which was the first to deal explicitly with nonmalleable vessel capital.

In following the Clark et al. model, let us denote fishing effort by $E(t)$ and the stock of vessel capital by $K(t)$, where $K(t)$ can be thought of in terms of the number of standardized fishing vessels. We then have

$$0 \leq E(t) \leq E_{max} = K(t), \tag{1}$$

which asserts that maximum fishing effort is determined by $K(t)$ and that actual fishing effort cannot exceed E_{max}. The sense of equation (1) is that the existing fleet may, or may not, be fully utilized.

Given an initial stock of vessel capital $K(0) = K^0$, adjustments in the stock of capital are given by:

$$\frac{dK}{dt} = I(t) - \gamma K, \tag{2}$$

where $I(t)$ is the gross rate of investment (in physical terms), and γ (a constant) is the rate of depreciation.

Let c_1, a constant, denote the unit purchase price of vessel capital, and let c_s denote the unit "scrap value" (unit resale value) of capital. To repeat our earlier definitions, we deem vessel capital to be perfectly malleable if:

$$c_s = c_1, \quad (3)$$

which implies that the risk to the vessel owner of capital loss is nonexistent. Continuing with our earlier definitions, we deem vessel capital to be perfectly nonmalleable if:

$$c_s = \gamma = 0. \quad (4)$$

The capital has no resale value and never depreciates. Investment in vessel capital is forever.

The more realistic intermediate cases, which we refer to as quasi-malleable capital, are given by the following:

$$c_s = 0; \gamma > 0 \quad (5)$$

$$0 < c_s < c_1; \gamma ^3 0 \quad (6)$$

In the case of equation (5), one can disinvest (slowly) through depreciation, while in the case of equation (6) one can sell off the vessels acquired in the past, but only at a capital loss.

With these preliminaries completed, we turn to our case of regulated open access. Let us assume that the resource managers specify an annual total allowable catch (TAC), or the equivalent thereof, which remains fixed for all future time. Let Q denote this fixed annual TAC in metric tons (mt). Entry into the fishery is initially unrestricted; the variable K denotes actual entry of vessels into the fishery. The catch rate of fishing is q metric tons (mt)/d/vessel. Thus, if K vessels fish for D days during the year, the fleet's total annual catch, or harvest, is equal to qKD mt.

Let D_{max} denote the maximum possible length of the annual season. If the fleet size is such that $qKD_{max} \leq Q$, then the fishing season will be at its maximum length. *If* $qKD_{max} > Q$, then the season must be reduced below its maximum length in order to ensure that the TAC is not exceeded. Thus:

$$\text{Total annual catch} = \begin{cases} qKD_{\max}, & \text{if } K \leq Q / qD_{\max} \\ Q, & \text{otherwise} \end{cases}.$$

Now, let the price of harvested fish be denoted as p, a constant. Let the daily operating costs for a given vessel be denoted as c. The fleet's annual operating profits are thus given by:

Fleet annual operating profits =

$$\begin{cases} (pq - c)KD_{\max}, & \text{if } K \leq Q / qD_{\max} \\ (pq - c / q)Q, & \text{otherwise} \end{cases}.$$

Next, recall that the unit, or purchase price of vessel capital is denoted by c_1. Let the annual rate of interest be denoted as r. Now let us suppose, to begin with, that the rate of depreciation (γ) is zero, but that vessel capital is perfectly malleable. This implies that $c_s = c_1$. At any moment in time, the vessel owners have the option of selling the vessels for exactly what they paid for them. This is the equivalent of leasing the vessels on a minute by minute basis. The cost of vessel capital, under these circumstances, is a *rental* cost. The rental cost is a type of opportunity cost, namely what the vessel owners could have earned on the funds invested in the vessels, had they invested the funds elsewhere, at the annual rate of interest r. Given our assumption that $\gamma = 0$, the annual rental cost of the vessel capital for a fleet of size K is simply

$$Kc_1 r \cdot \frac{D}{365}.$$

Now let K_0 denote the number of vessels that would be required to take Q, if $D = D_{max}$. Thus, $K_0 = Q/q\ D_{max}$. Fleet annual operating profits would then equal: $(pq - c)K_0\ D_{max}$, and fleet annual rental costs would equal

$$K_0 c_1 r \frac{D_{\max}}{365}.$$

We shall assume that annual operating profits

exceed annual rental capital costs, otherwise, the fishery is not viable. Given this assumption, fleet annual operating profits and fleet annual rental costs can be depicted as functions of K (Figure 1).

Total fleet annual net profits obviously achieve a maximum at $K = K_0$. Suppose that actual $K > K_0$. Fleet annual operating profits and rental costs, and thus net profits, would be identical to what they would have been, had actual $K = K_0$. The basic reason is that, in this situation, the rental cost of capital is really just another form of operating cost. Hence suppose for the moment that $K_0 = 200$, and $D_{max} = 365$. Annual operating profits are: $(pq - c)200 \cdot 365 = (p - c/q) \cdot Q$, while vessel capital rental costs are

$$200c_1 r \frac{365}{365} = 200c_1 r .$$

Now suppose that K is quintupled to 1,000 and that, as a consequence, D is reduced to 73. Annual operating profits obviously do not change, while simple arithmetic shows that vessel capital rental costs continue to equal $200c_1 r$. Thus, annual net profits (resource rent) remain unchanged for the fishery as a whole. From this, we can conclude the following. Given that the resource managers are able to exercise iron control over the total harvest, there is, under regulated open access, with perfectly malleable fleet capital, no unique optimal fleet size, and obviously no need for decommissioning schemes. Thus, nonmalleability of vessel capital does indeed lie at the heart of the problem of excessive fleet size.[7]

With all of this in mind, let us now alter our assumptions and suppose that vessel capital is nonmalleable. For ease of exposition, we shall adopt the assumption (admittedly extreme) that vessel capital is *perfectly* nonmalleable (i.e. $c_s = \gamma = 0$). Now, fleet size most assuredly mat-

[7] The one important qualification is that, if the fleet is very large, the resource managers' monitoring problem increases. Restricting the fleet size should, however, be relatively easy, and costless since, in these special circumstances, equally good opportunities for the vessel owners exist elsewhere.

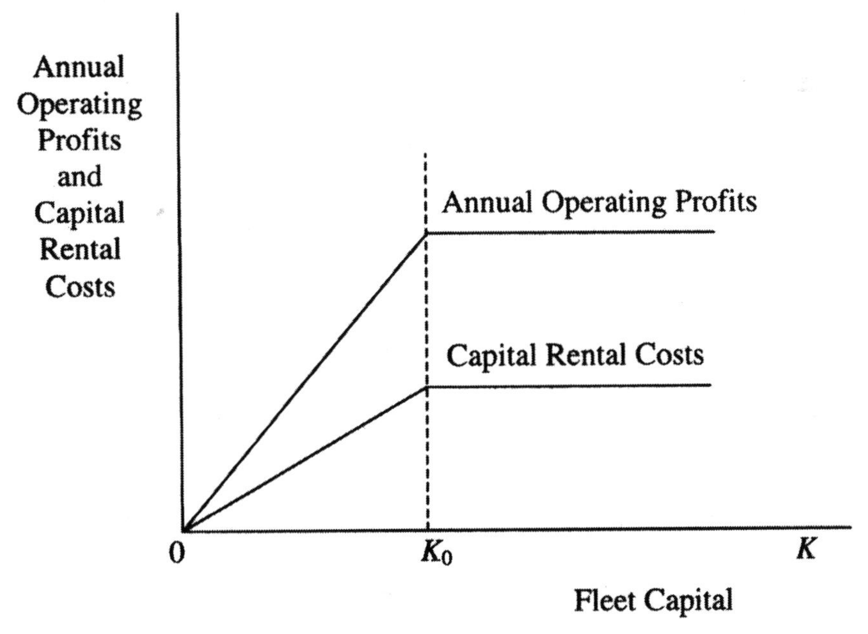

Figure 1. Annual operating profits and capital costs for the fishery.

ters, and there is without question an optimal fleet size, namely $K = K_0$. From here on, let us refer to K_0 as K_{opt}.

Given that vessel capital is perfectly nonmalleable, the would-be investor must think in terms of the expected present value of operating profits from the fishery and his or her possible share of these profits. We shall assume that the vessels (and crews) are identical. Consequently, the owner of a single vessel can be assumed to enjoy an average share of the present value in question (i.e. total present value of operating profits divided by K).

If the total annual harvest Q is taken, then the present value of fleet operating profits will be equal to: $[(p - c/q)Q](1 + r)/r$. The present value of fleet operating profits, on a per vessel basis, is simply

$$\left[(p-c)Q\right]\cdot\left(\frac{1+r}{r}\right)\frac{1}{k} \tag{7}$$

If we denote

$$\left[(p-c/q)Q\right]\cdot\left(\frac{1+r}{r}\right) \tag{8}$$

as Γ, we can observe that the *maximum* present value of operating profits per vessel is simply

$$\omega = \Gamma / K_{opt}. \tag{9}$$

From this, it follows that no investment in fleet capacity will take place unless $c_1 \leq \omega$ If it is in fact the case that $c_1 \leq \omega$, then investment in fleet capacity will continue up to the point that

$$c_1 K_{ROA} = \Gamma\ , \tag{10}$$

where K_{ROA} denotes the regulated open-access equilibrium level of fleet capital.

If $c_1 = \omega$ (i.e., we have a breakeven fishery), then $K_{ROA} = K_{opt}$. If, on the other hand, $c_1 < \omega$, we shall be assured that $K_{ROA} > K_{opt}$. Excess fleet size, when $c_1 < \omega$, is simply: $K_{ROA} - K_{opt}$, and the economic loss imposed upon society by this excess fleet size can be expressed equally simply as:

$$c_1(K_{ROA} - K_{opt}) = \Gamma - c_1 K_{opt} \tag{11}$$

Thus, the economic measure of excess fleet size under regulated open access is equal to present value of dissipated resource rent. We can refer to the L.H.S. of equation (11) as the redundancy deadweight loss of regulated open access. Let it be noted that the redundancy deadweight loss is incurred the *instant* that the excess capital is acquired. Once incurred, the Loss cannot be reversed, by a decommissioning scheme, or anything else.

Nonetheless, let us concede that there is at least the appearance of a prima facie case for a decommissioning scheme. Let us consider, therefore, the effect of a decommissioning scheme, introduced after the regulated open access equilibrium is achieved. Existing vessel owners are licensed, and entry is strictly limited. The resource managers then persuade vessel owners to sell their vessels (and licenses) to the managers, and go on doing so, until the fleet is reduced to the optimal level K_{opt}. The accompanying limited entry program is carefully and effectively designed to prevent the fleet from once again exceeding its optimal size.

The impact of the decommissioning and limited-entry program will depend critically upon whether the program is, or is not, anticipated by the vessel owners. The vessel owners' anticipation of the program, or lack thereof, will determine whether a time inconsistency problem does, or does not, arise.

Let us illustrate with the aid of a simple numerical example (Clark and Munro 2003). Let

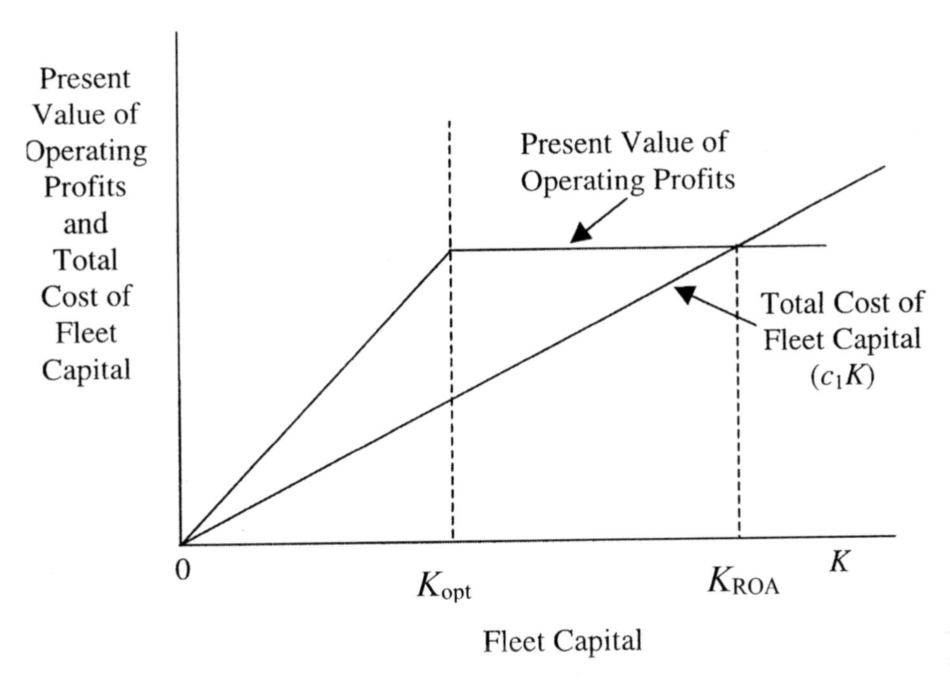

Figure 2. Present value of fleet operating profits and capital costs.

it be supposed that D_{max} = 200 d. We assume, in addition that:

Q = 10,000 mt;
q = 1 mt per vessel per day;
p = \$1,000 per metric ton;
c = \$500 per vessel per day;
c_1 = \$500,000 per vessel;
r = 0.10 – (i.e., 10% per annum).

The present value of total annual fleet operating profits will be:

$$\Gamma = (p - c/q)Q\frac{(1+r)}{r} = \$55 \text{ million}. \quad (14)$$

Also,

$$K_{opt} = \frac{Q}{qD_{max}} = \frac{10{,}000}{200} = 50 \text{ vessels}, \quad (15)$$

which implies that ω = \$1.1 million. Since c_1 = \$0.5 million, maximum resource rent is \$(1.1 – 0.5) million × 50 = \$30 million.

Now, let it be supposed that, if an excessively large fleet emerges, the resource managers will, by t = 10, introduce a buyback program with the objective of reducing K to 50 vessels. Let us commence by also assuming that at t = 0 resource managers' future plans are wholly unanticipated by vessel owners. They assume, incorrectly, that regulated open access will continue forever. We can thus anticipate that at t = 0, investment in vessel capital will be given by:

$$K_{ROA} = (p - c/q)\frac{Q(1+r)}{c_1 r} = 110 \text{ vessels}. (14)$$

Thus there is an excess of 60 vessels in the fleet, representing a Redundancy Deadweight Loss of \$30 million, and resource rent (in present value terms) of \$0 million. At t = 10, the resource managers introduce a "sudden death" buyback program to the complete surprise of the vessel owners. The vessel owners are, however, convinced (correctly) that the authorities will do whatever is necessary to reduce the fleet to 50 vessels and are further convinced (also correctly) that the accompanying limited entry program will be effective forever.

The present value of the operating profits of each of the remaining 50 vessels, discounted back to t = 10 will be \$1.1 million (i.e., ω). Thus, we can be assured that the resource managers cannot offer less than \$1.1 million per vessel. We shall assume, somewhat unrealistically, that the managers are able to achieve their goal by offering a purchase price of \$1.1 million and the accompanying limited entry program is indeed fully effective. The fleet remains at $K = K_{opt}$ from henceforth.

Let us suppose that the decommissioning scheme is financed by the government drawing upon its general revenues. If one can assume that the resultant increase in taxes, increased government borrowing, or reduced government expenditures on other activities causes no perceptible loss to the economy, we can say that each vessel owner will enjoy a windfall gain of \$600,000 and that the redundancy deadweight loss to the economy remains at \$30 million. The initial loss to the economy cannot be undone by the decommissioning program, but at least no further damage is done.

Now let us change the example by supposing that, at t = 0, the vessel owners have perfect foresight, with respect to the resource managers' policy. They are certain that the resource managers will react to the emergence of any excess capacity by introducing a sudden death buyback program at t = 10 and that the managers will offer a price of \$1.1 million per vessel. The vessel owners also know that the fleet will be stabilized at 50 vessels and that the accompanying limited entry program will be entirely successful.

We shall also assume that vessel owners are aware that at t = 10, the resource managers will declare that only vessels operating in the fishery since t = 0, or before, will be deemed to be bona fide participants in the fishery. Any vessels entering the fishery after that time will be denied licences and forced out of the fishery without compensation. The reason for this seemingly artificial assumption will become apparent in due course.

We can now calculate the level of investment in vessels at t = 0, which we shall denote by K'_{ROA}. Equilibrium will then be achieved when:

$$c_1 K'_{ROA} = \sum_{i=0}^{10} (p - c/q) \cdot \frac{Q}{(1+r)^i} + \frac{c_2}{(1+r)^{10}} \cdot K'_{ROA}, \quad (15)$$

where c_2 denotes the resource managers' offer price at t = 10. Observe that it is a matter of indifference whether an individual vessel owner sells his or her vessel at t = 10 or continues to operate it as one of the remaining 50. Also observe that equation (15) can be rewritten as:

$$c_3 = \left[\sum_{i=0}^{10} (p - c/q) \frac{Q}{(1+r)^i} \right] \Big/ K'_{ROA}, \quad (15a)$$

where

$$c_3 = c_1 - c_2/(1+r)^{10}, \quad (16)$$

where $c_2 = \omega$ is the buyback price. In our example, we have:

$$K'_{ROA} = \frac{\$35{,}722{,}836}{\$75{,}093} = 476 \text{ vessels}. \quad (17)$$

The implication is that the eminently successful buyback program will lead to a redundancy deadweight loss of \$500,000 (476 – 50) = \$213 million. Recall that, if the resource managers had done nothing (i.e., had forgone a buyback program), net resource rent from the fishery (in present value terms) would have been \$0 million (as the standard fisheries economics theory would predict). In our example of the anticipated buyback program, the resource rents (net economic benefits) from the fishery are \$55 million–\$213 million = –\$158 million. In terms of the goal of increased economic efficiency, the decommissioning scheme, when fully anticipated, is an unmitigated disaster.

The reason that the anticipated buyback program induces a large investment in fleet capacity is made transparent by equation (16). The effective supply price of fleet capital is no longer c_1, but is rather $c_3 = c_1 - c_2/(1 + r)^{10}$. Thus, would-be vessel owners are effectively being subsidized in their purchase of vessels. Indeed, as the reader can verify in our example, exactly the same outcome could have been produced under a do-nothing policy (i.e., $K_{ROA} = 476$) by having the government offer the vessel owners at $t = 0$ a subsidy per vessel equal to 77% of the purchase price c_1.

Now note the following. Our choice of $t = 10$, as the initiation time of the buyback program, is entirely arbitrary. If the buyback program is anticipated α years in the future, we now have:

$$c_3 = c_1 - c_2/(1+r)^a \qquad (18)$$

Recall that $c_2 = \omega$, and recall further that, in order for any investment in fleet capacity to take place, we must find that $c_1 \leq \omega$.

In our numerical example where $c_1 < \omega$, c_3 is negative for $\alpha \leq 8$ years. In this case, the prediction is that $K^*_{ROA} = +\infty$; there is no limit to the redundancy deadweight loss as the initiation time approaches zero.

If it is indeed the case that $c_1 < \omega$, and α is small enough, investors in fleet capital can achieve a positive return on their investment, even if they never actually catch any fish. This is perhaps the most perverse aspect of decommissioning subsidies. If anticipated, such subsidies may encourage a large scale increase in fleet capacity, unrelated to any prospect of actually participating in the fishery.

The reason for our seemingly artificial assumption should now become clear. If there was no such restriction, vessel owners, with perfect foresight, would wait until the last possible moment before $t = 10$ to invest in capacity. Then we would indeed have $K^*_{ROA} = +\infty$.

Next note that we have $c_2 = \omega$ because of a perfectly enforced limited-entry and decommissioning program from $t = 10$ onwards. If, on the other hand, it is anticipated that the program will be less than perfect, then we would have $c_2 < \omega$. We conclude that the more successful is the limited-entry and decommissioning scheme, the greater is the potential for vast economic waste, in the event that the program can be anticipated in advance.[8]

Be that as it may, we are confronted with what might be termed a classic time inconsistency problem in that what may appear to be an optimal policy at $t = 10$ is inconsistent with

[8] If the buyback program is not successful, in that there is seepage of capacity back into the fishery, there will, of course, be further economic waste. Consider an extreme case, in which capacity, once removed, is promptly replaced by new investment. Allow this exercise in futility to be repeated over and over, and let the buybacks be anticipated. We first note that c_2 will be very low indeed because operating profits per vessel will be kept to a minimum by the ever-restored capacity. Second, we note that the PV of fleet investment costs, while high, can never approach infinity, given that the discount rate is positive.

the long-run best interests of society. The problem arises directly from the "rational expectations" of the vessel owners. To rephrase the Kydland and Prescott (1977) dictum, we can say that there is no mechanism to induce resource managers at $t = 10$ to take into consideration the effect of their policy, via the expectation mechanism, upon decisions of the vessel owners at $t = 0$.

The example, which we have used, is, of course, extreme in that we have assumed perfect foresight of vessel owners with respect to resource managers' policies and perfect nonmalleability of vessel capital. Yet the point remains. Even in a world in which foresight is considerably less than perfect, and in which vessel capital is quasimalleable, rather than perfectly nonmalleable, decommissioning programs, and their accompanying subsidies, can be expected to have a major impact upon vessel owner's investment decisions.

Owners of nonmalleable vessel capital, as we have emphasized, cannot afford the luxury of myopia. They must make educated guesses of the future profitability of the fishery. It makes little sense to suppose that they will not also make educated guesses with respect to future policies of resource managers.

Finally, it is appropriate to ask whether there is any empirical evidence in support of the claim that decommissioning subsidies will stimulate the expansion of fleet capacity. The answer is that there is. An empirical study, undertake a few years ago by two Danish economists on the impact of EU decommissioning subsidies (Jörgensen and Jensen 1999), concludes that such subsidies do indeed act as a stimulus to investment in fleet capacity. These subsidies not only influence EU investors in fleet capacity directly but also influence the investors' bankers. The evidence shows that these subsidies lead to the bankers offering more generous credit terms to would-be investors in fleet capacity, than would otherwise be the case. The authors of the study point out that their results confirm those arising from an earlier Danish-Dutch study (Frost et al. 1996).[9]

Conclusions

If decommissioning schemes can overcome the seepage problem, they can have a beneficial, or at least nondamaging, impact upon fisheries management—provided that vessel owners' expectations pertaining to resource managers' policy are myopic. If the schemes are anticipated, however, then, even with no seepage, we run into a time-consistency problem in that the subsidies can have a strong negative impact on the management of the fishery. The conclusion is not particularly radical, and is really no more than an application of "rational expectations" to fisheries management, and recognition of the fact that it is folly to assume that vessel owners are myopic in their investment decision making.[10]

We have no easy solutions to offer to the problems of capacity and conservation, other than to say that there appears to be no way out other than to adopt what the FAO refers to as

[9] Up to this point, we have assumed implicitly that the resource, and hence the TAC, are stable through time. As we have acknowledged before, however, this calm could be broken by a negative environmental shock, which takes both resource managers and vessel owners by surprise. It would seem reasonable to argue that, in spite of what we have said about decommissioning schemes, they are still desirable as a protection against unpleasant environmental surprises. Yet, the anticipation of decommissioning schemes can be relied upon to intensify the excess fleet overhang, and the conservationist threat, which the overhang imposes.

[10] Clark et al. (2003) discuss in detail the case in which decommissioning schemes are a component of a vigorous management program introduced into a fishery subject, heretofore, to weak, or non-existent, management. The authors demonstrate that, in this case, anticipated decommissioning schemes can have a direct and unambiguous negative impact upon resource conservation.

an Incentive Adjusting (as opposed to an Incentive Blocking) approach to management (FAO, 1998). This would involve using taxes, and or, some form of rights based system, such as ITQs, cooperatives, or community based management (See, for example, Anderson 2002; Clark and Munro 2002; Felthaven 2002; Jensen 2002). How these are crafted together will certainly depend on the type of fishery being managed.

Acknowledgments

The authors wish to express their gratitude for the generous support provided by the Sea Around Us Project, of the Fisheries Centre, University of British Columbia, which is, in turn, sponsored by the Pew Charitable Trust of Philadelphia, USA.

References

Anderson, L. G. 2002. A micro-economic analysis of the formation and potential reorganization of AFA coops. Marine Resource Economics 17:269–290.

Bjørndal, T., and G. Munro 1998. The economics of fisheries management: a survey. In T. Tietenberg and H. Folmer, editors. The international yearbook of environmental and resource economics. Edward Elgar Publishing Limited, Cheltenham, UK.

Clark, C. W. 1990. Mathematical bioeconomics: the optimal management of renewable resources, 2nd edition. Wiley-Interscience, New York.

Clark, C. W., F. H. Clarke, and G. R. Munro. 1979. The optimal management of renewable resource stocks: problems of irreversible investment. Econometrica 47:25–47.

Clark, C. W., and G. R. Munro. 2002. The problem of overcapacity. Bulletin of Marine Science 70:473–483.

Clark, C. W., and G. R. Munro 2003. Fishing capacity and resource management objectives. Pages 13–34 *in* S. Pascoe and G. Gréboval, editors. Measuring capacity in fisheries. FAO Fisheries Technical Paper 445.

Clark, C. W., G. R. Munro, and U. R. Sumaila. 2003. Subsidies, buybacks and sustainable fisheries. Journal of Environmental Economics and Management 50:47–58.

Cunningham, S., and D. Gréboval. 2001. Managing fishing capacity: a review of policy and technical issues. FAO Fisheries Technical Paper 409.

Felthaven, R. G. 2002. Effects of the American Fisheries Act on capacity, utilization and technical efficiency. Marine Resource Economics 17:181–206.

Flaaten, O., and P. Wallis. 2000. Government financial transfers to fishing industries in OECD Countries. in Microbehavior and Macroresults, Proceedings of the Tenth Biennial Conference of the International Institute of Fisheries Economics and Trade, Corvallis, (CD Rom).

FAO (Food and Agriculture Organization of the United Nations). 1992. Marine Fisheries and the Law of the Sea: a decade of change. FAO Fisheries Circular No. 853.

FAO (Food and Agriculture Organization of the United Nations). 1998. Report of the Technical Working Group on the Management of Fishing Capacity, La Jolla, USA, 15–18 April 1998. FAO Fisheries Report No. 585.

FAO (Food and Agriculture Organization of the United Nations. 2000. Report on the Expert Consultation on Economic Incentives and Responsible Fisheries, Rome, Italy, 28 November – 1 December 2000. FAO Fisheries Report No. 638.

FAO (Food and Agriculture Organization of the United Nations). 2003. Report on the Expert Consultation on Identifying, Assessing and Reporting on Subsidies in the Fishing Industry, Rome, 3–6 December 2002. FAO Fisheries Report No. 698.

FAO (Food and Agriculture Organization of the United Nations). 2004. Technical Consultation on the Use of Subsidies in the Fisheries Sector, Rome 30 June–2 July 2004. Available: http://www.fao.org/fi/NEMS/events/Home_search_events.asp.

Frost, H., R. Lanters, T. Smit, and P. Sparre. 1996. An appraisal of the effects of the decommissioning scheme in the case of Denmark and the Netherlands,. University Press, Esberg, South Jutland.

Gordon, H. S. 1954.The economic theory of a common property resource: the fishery. Journal of Political Economy 62:124–142.

Gréboval D., editor. 1999. Managing fishing capacity: selected papers on underlying concepts and issues. FAO Fisheries Technical Paper 386.

Gréboval, D., and G. Munro. 1999. Overcapitalization and excess capacity in world fisheries: underlying economics and methods of control. Pages 21–48 *in* D. Gréboval, editor. Managing fishing capacity: selected papers on underlying concepts and issues. FAO Fisheries Technical Paper 386:21–48.Holland,

D., E. Gudmundsson, and J. Gates. 1999. Do fishing vessel buyback programs work: a survey of the evidence. Marine Policy 23:47–69.

Homans, F. R., and J. E. Wilen. 1997. A model of regulated open access resource use. Journal of Environmental Economics and Management 32:1–21.

Jensen, C. L. 2002. Reduction of fishing capacity in 'common pool' fisheries. Marine Policy 26:155–158.

Jörgensen, H., and C. Jensen. 1999. Overcapacity, subsidies and local stability. Pages 239–252 *in* A. Hatcher and K. Robinson, editors. Overcapacity, over-capitalisation in European fisheries. Centre for the Economics and Management of Aquatic Resources, University of Portsmouth, Portsmouth, UK.

Kydland, F. E., and E. C. Prescott. 1977. Rules rather than discretion: the inconsistency of optimal plans. Journal of Political Economy 85:473–491.

Milazzo, M. J. 1998. Subsidies in world fisheries: a reexamination. World Bank Technical Paper No. 406, Fisheries Series, Washington D.C.

Munro, G. R., and U. R. Sumaila. 2002. Subsidies and their potential impact on the management of the ecosystems of the North Atlantic. Fish and Fisheries 3:233–250.

NRC (National Research Council). 1999. Sustaining marine fisheries. National Academy Press, Washington D.C.

OECD (Organization of Economic Cooperation and Development). 2000. Transition to responsible fisheries: economic and policy implications. OECD, Paris.

Weninger, Q., and K. E. McConnell. 2000. Buyback programs in commercial fisheries: efficiency versus transfers. Canadian Journal of Economics 33:394–412.

Westlund, L. 2003. Draft guide for identifying, assessing and reporting on subsidies in the fisheries sector. Appendix E *in* Report of the Expert Consultation on Identifying, Assessing and Reporting on Subsidies in the Fishing Industry, FAO Fisheries Report 698.

American Fisheries Society, Symposium 49:727–734

Can We Get More Fish or Benefits from Fish and Still Reconcile Fisheries with Conservation?

Yingqi Zhou*

Shanghai Fisheries University

334 Jungong Road, Shanghai, 200090, China

Keynote Speaker
4 May 2004

The world's fishery is undergoing a historic change which can be called the second great fishery revolution, transferring the fishery from a capture methodology into aquaculture in our effort to "get more fish." It has extended its range into other areas as well, such as recreation fisheries, including angling, viewing (snorkel and scuba), fish civilization (aquarium ecosystems), and so forth. The benefits from fish can be generally classified into four groups: 1) food supply for demand of protein; 2) economic benefit for revenue; 3) employment and society benefit; and 4) meet the requirements of recreation, civilization and culture, and so forth.

In this paper, I review the current characteristics of fisheries, fishery resources, and fishery science in relationship to conservation. Fish, as a resource, represents a biologically renewable and mobile nature resource. Under scientific preservation and rational exploitation, the sustainable utilization of the world's fishery resources can be achieved. Fish, like humans, birds, and other migratory mammals, are a mobile resource with harvest and exploitation attributes similar to water, oil and air. Their ownerships cannot be clearly or easily defined. In most cases, "occupation" means "ownership," resulting in the larger size of fishing fleets with fishing capacity greatly exceeding the limitations of sustainable fisheries. Overinvestment and overcapitalization of the fishing industry exist. The charge paid by fishing companies represents only the costs of exploitation of the fishery resources, not the costs of resources generation or sustainability. Low costs and higher returns attract the capital and labor force gathering in today's fishery, which results in continuously expanding fishing capacity. Therefore, most of economically viable fishery resources have been destroyed around the world. It is necessary to set up and implement regulations requiring those who engage in fishing or aquaculture to pay the true resource costs for fish or water resources, and to promote the development of landbased or "desert aquaculture," where no water is discharged into the natural system.

The current goals of fishery management are to strengthen the monitoring, control, and surveillance (MCS) system on the exploitation of fisheries resources and the elimination of illegal, unregulated, and unreported (IUU) harvest through the application of quota and licenses systems on capture fisheries. Specify the property rights and jurisdiction over certain fisheries resources should be implemented by means of exclusive economic zones and regional international fishery agreements. There are still many problems to be studied before effective implementation of these measures is possible.

We should actively transfer new technologies into fisheries since most are postindustry complex systems and

*Corresponding author: ygzhou@shfu.edu.cn

represent a drive force in the development of related industries. Modern society has to actively support the development of aquaculture while raising serial, strict requirements and regulations on this activity. Aquaculture should not only supply aquatic products, but also be required to take effective measures for the protection and conservation of the environment and other resources, food safety, provide for a healthy and energy saving fishery, and actively pursue ecologically sound production. Integrative and intensive technologies provide the means to meet these requests. The species and stocks of fish used for aquaculture should be standardized. Attention should be paid to "aquaculture up the food web," where "fish feed fish" and there is strong concern for energy saving. We might get more from fish by means of improved fish processing, utilization on marine organisms in medicines and biologics, reducing the waste or discard of fish during harvest, expansion in the production of pleasure fish for aquaria, promotion and the development of the game fisheries, and fish art and related cultural activities.

There are some basic misunderstandings in the exploration of fishery resources. The exhaustion of natural resources has be neglected, concealed, and covered up by the development of new scientific technology where exploitation could be carried out for low costs and with higher efficiency, providing higher value products at low prices. This approach ignores the true capital characteristics of nature resources. Fisheries assessments based on harvest yield is flawed, because while catch is increasing fishermen or managers of fishing fleets will consider the fish resources abundant and unlimited, continually increasing fish power and capacity until the resource collapses. Fishers and managers typically do not care about disappearing of species or changes in species variation; they do not consider the whole food chain and continue to catch large amounts of unique species which hold important links on the food chain.

My suggestions to address this problem include state, market, and civil societies working together and imposing new controls on fisheries and aquaculture which support their sustainable development. Studies of fishing gear selectivity are some of the best ways to give technical support to the administration of fishery regulation in a practical and feasible manner. In marine fisheries there is no more room to increase total catch. The waste and discard of catch should be reduced. Fishing capacity should be reduced immediately. Attention should be put on "fishing down the food web." Pelagic species, such as herring, mackerel, squids, sardine, and krill, which are food for many larger fishes and marine mammals, should be given special care. Increased value will come from sports and recreation fishing, and observational activity of "treasure" fish. Fishing technology should transferred or developed in cooperated with aquaculture, such as cage farming, and with similar stow or set net fisheries. Trawling with damage-limiting codends will provided higher catch value for aquarium or scientific requirements. Keeping catch alive to grow and survive and in cage culture for market or aquaria, will increase the value of certain fish. Fishing with damage-limiting techniques will contribute to the environment and other species, such as birds, dolphins, and marine mammals. This approach is called ecological friendly fishing.

There are three questions to answer in this session; first, what do we as a society want from fish or fishing? The current benefits from world fisheries can be generally classified into four groups: 1) food supply for demand of protein; 2) economic benefit for revenue; 3) job and social benefit; and d) meet the requirements of recreation, civilization, and culture. Secondly, is the fishing industry on the way of sustainable development? Finally, what is our future in fisheries? Each of these issues will be addressed in this paper.

Situation of Fishery Resources in General

In spite of the fact that humans obtained great achievements during the 20th century, the fact remains that many natural resources, which accumulated through 3.8 billion (1×10^9) years, have been rapidly spent out and depleted over the last 50 years. That is a tremendous loss. The situation in fisheries is more serious. Food and Agricultural Organization of the United Nations (FAO) statistic data in 1999 shows that 30% of the fish resources had been overfished and only 5% of them might have a chance at recovery, that 5% were in danger and could not be possibly be recovered, that 30% of the fish resources had been fully exploited, and only 25% of fish species might have the possibility to sustain continued exploitation. More recent data show that only 25% of all fisheries aquatic resources are underexploited, 45% fully exploited, 18% overexploited, and 10% of all fish stocks are depleted. This situation is getting worse.

It has been estimated that global nature resources related to fishery will in a few short decades lose up to 70% of available reef and cay habitats, which will have serious negative impacts on the marine ecosystem because species living in the cay represent about 25% of total marine living resources in the oceans. These data also suggest that the globe is losing 6% of its total freshwater ecosystem annually and losing 4% of the seawater ecosystem each year. Scientific surveys show that biological variation is decreasing in most marine areas, leading to what has been called "the sea becomes the desert." All these studies have indicated that human society might be on the path of un-sustainable development and human modes of production and harvest produce problems in constitutive property. In other words, the ecosystems of our earth are changing.

Characteristics of Fishery Resources

Fishery resources have common economic characteristics with other nature resources. I discuss three aspects of this continuity below.

1) While fishing, the costs paid by fishermen or fishing companies are for exploitation or fishing operation only; nothing is paid for factors leading to fish growth and nothing is paid for the services provided by the ecosystem. This is similar to the costs incurred by other extractive activities such as the oil industry. The oil industry pays only the costs for drilling and exploration, nothing is paid for the generation of oil, which was done by the earth when trees and vegetation were buried under high pressure for thousands of years. Only on rare occasions do resource explorers pay for or consider replacement time or value of the extracted resource. These are considered a part of the services provided by the earth's "free" ecosystems. It is well know that total capital consists of several elements: labor capital, financial capital, and the conversion costs. Labor capital includes labor force (energy), intellect, cultural and organizational structure, and so forth. Financial capital is presented as currency, investment, monetary measures (i.e., circulating capital), etc. The conversion costs are invested on facilities, machines, tools and manufactory (i.e., fixed capital), and so forth. However, the allocation of total costs in nature capitalism should include an accounting of all resources, living systems, and the ecological system upon which these resources depends. Nature capitalism should be calculated using a certain and clear currency as the common capital and provide a return to nature. For instance, what are the ecological values of different fish and what is their true price on the larger scale of nature capitalism?

All of economic activities, the nature capital and the ecological service system have a certain "value of money." However, in most situations, it is free to use existing natural resources without feedback costs. In comparison, we assume that the costs for land use, seeds, and fertilizer should be paid for by the farmer, yet the fisherman makes no payment for the factors leading to growth in the fish he or she harvests. Activities in hunting and miner are also undertaken with relatively lower costs as it the industry directly conveys nature resources into profit. Resource extraction industries often cite the "risks" associated with exploration of unpredictable commodities as a "cost," however, successful extraction seldom replaces or returns equally unpredictable ecosystems. Therefore, during the last century, it can be said that human society was developed on the back of despoiled natural resources, and, through the destruction of the balance among ecological systems, it has gone a long way down the road of unsustainable development.

How much should be paid for the value of services derived from ecological systems of the earth? A rough estimation has made as. It is well known that the experimental biosphere of the biological circle II was carried out during 1991 and 1993 in Alular, Arizona. This study cost US$200 million over 9 years. A 130,000-m^2 building was set up with a life support system for eight people to live in the sealed space. The system was artificial, imitative, and was supposed to simulate the earth's ecological environment. However, the system only run 17 months and had to stop. The biosphere experiment failed because it could not provide sufficient air, drinkable water, or food for the people to survive. If we hypothetically transfer this experiment to the total human population, 6 billion people, similarly equipped facilities with a value US$200 million each, it would cost US$39 trillion to produce an equivalent ecological system (which still would not work). Another calculation published in the journal *Nature* conservatively estimated the value of all the earth's ecosystem services to be at least US$33 trillion a year. That's close to the gross world product.

Yet the free-of-charge approach is still used in fish resource management—the lower costs have attracted more capital and a greater labor force dedicated to the capture fishery, which results in an over expanded fishing capacity and a harvest that greatly exceeds any limitation the fishery resources could bear and remain sustainable. The free-of-charge approach also results in the destruction of fish resources, overexploitation, and the wasting of resources.

2) Fish resources represent a kind of renewable biological resource. Because it is renewable, we have suffered for a long time under the basic misunderstanding that fish in the vast sea are unlimited. It was common in 1950s that if one species disappeared due to overfishing, we would expect another species to appear, providing another lucrative fishery. This theory resulted in fishermen hunting one species after another, destroying one species and another, and ignoring variation in fish abundance, size, or species composition (i.e., ignoring their fishing down the food web).

3) Fish, like humans, birds, and other migratory animals, are a mobile resource. Boundaries of ownership are not clearly defined. In most cases, "occupation" means "ownership." This results in larger sized fishing fleets and overcapacity of fishing effort which greatly exceeds any limitation the fishery resources could bear and remain sustainable. Overinvestment and overcapitalization in the fishing industries exist for the rapidly possession of more resources in a short period of time. Continued low costs and higher returns attract the capital and labor force gathered in the fishery, resulting in an over expanded fishing capacity. Therefore, the most of economic fishery resources have been destroyed around the world. It is necessary to set up new regulations determining who may engage in fishing or aquaculture and who should pay the subsequent resource costs for fish or water resources. One approach is to promote the development of land-based aquaculture and "desert aquaculture," where there is no waste-water discharged.

Some Misleading Information and Misunderstandings

1) Misleading concepts resulting from the development of science technology

The development of science technology might conceal and cover up the exhaustion of natural resources because exploitation can be carried out at low costs and with higher efficiency by powerful tools, even providing a "feeling" or impression that the stock is more abundant than it actually is. Similarly, the oil industry can extract the last drop of oil by many modern technical devices, yet the price of patrol may not go up equivalently. Similar extraction technologies in fishing, using echo sounders, fishing with lights, or harvest of fish gathering at rafts can appear to increase the catch without appreciably adding any information on standing stock.

2) Misleading efforts in fisheries management

In some cases, fishery resource assessments are based only on data from fishing yield. While the catch is increasing, the fishermen or managers of fishing fleets will consider the fish resources to be abundant and continue to increase fishing power and capacity, even when the resources is nearly broken down, but yield continues to rise due to modern fishing technology. For fishing fleets to survive, some of the fishing efforts have turned to smaller species or younger and smaller fish, which reduces the size of recruitment in fish stocks. Fish populations are sharply reduced which impacts the life histories and case the age at sexual maturation, as the age of first capture gets earlier and earlier.

3) Overfishing and fishing down the aquatic food web

Due to overfishing, the smaller species of fish and shrimp are increasing in the catch of trawl and set nets, which indicates the ecosystem has lost balance. Fishing down the aquatic food web disregards the phenomenon of disappearing species and the destruction of species diversity. These situations now occur around the world.

Can We Get More fish?

Because fish are renewable biological resources under scientific preservation and rational exploitation, if there is sufficient recruitment into different stocks, sustainable utilization of fishery resources could be maintained. However, action should be taken, as FAO recently noted, to reduce at least one-third of the current fishing capacity, to cut down all subsidies on fisheries and related activities, and to set up policies for "resource-rents" to pay for the costs of using different resources. Countries have to change the priority of their fisheries (i.e., turning away from the main goal of supporting established capture fisheries as a national strategic target and turning away from the unitary goal of economic benefit to goals for food supply and food safety through fisheries).

Some Thoughts and Questions on Fishery Management

1) As a common practice, are maximum sustainable yield (MSY), total allowable catch (TAC), quotas, and individual transferable quotas (ITQs) processes the right way to go? There are some questions addressing this issue:

a) Maximum sustainable yield is based on an estimate developed from a single species perspective, but fish live in a multispecies ecological world impacted by environment and climate change. Does MSY theory really work or should it be changed?

b) In practice, scientists are always complaining about data shortage limitations and that they require more data for estimating TAC or doing fish stock assessments. Why do some TAC estimations fall within wide ranges, for instance 36 million metric tons (mt) –300 mt?

c) Estimations of stock or population abundance are not sufficiently accurate for fishery assessments. Why are accurate estimates too complicated and not good enough for fishery management?

d) Difficulties exist in fisheries monitoring. Monitoring requires a lot of manpower and facilities, such as observers on board ships at sea or placing inspectors at fish ports. Why are fisheries data always delayed in their application to fishery statistics?

2) Because quotas are based on species under strict regulation, fishermen make every effort to fully utilize their quota and cannot avoid bycatch, therefore discarding up to 25% of the total catch at sea. This is a waste since most of the bycatch will not survive. Bycatch survive rates are important and should be observed and meaningfully quantified. Fisheries quotas should not only be set for the target species in the catch but also for discard.

3) The world's fishing industry will exist forever

When people complain about fisheries having negative effects on the ecosystem, we need to realize that the fishing industry will exist forever because it is a basic production activity of human beings. What can we do? We should realize that fishermen need jobs to survive. So the important thing is to give effective technical support to fishermen to maintain an appropriate fishery through effective regulations.

Supporting studies of fishing gear selectivity is one way to give technical support to fishermen and fisheries officers to effectively carry out fishery regulations and to make regulations more feasible and practical. Gear studies could help us meet the requirements of "catch larger and release smaller" fishes and reduce bycatch by implementation of devises designed to enhance target species or to eliminate species in danger. Selectivity studies should involve analyses of fish body length with mesh size and shape (such as square mesh) and fish swimming speed and endurance with trawling speed and cable attach angle. Studies are needed on fish reaction to multipanel trawls or turtle excluder devices (TEDs), which could separate fish species during fishing. We need additional studies on the selectivity of hooks in varied shapes, visibilities of different types of netting, selectivity of trap entrance geometry and position of crab tape, and so forth. Trawling with harm-free, rigged codends will provided a higher valued catch for aquarium or scientific requirements. Keeping catch alive to growth in cages raises the value of fish. Fishing without damage to the environment and to other species, such as birds, dolphins, marine mammals, is called "ecological friendly fishing." In addition, ecologically friendly fishing includes light contact trawling and squid jigging, but not bottom trawling.

4) No more is available from capture fishery

Fishermen hope to catch more fish and fishing industries hope to get more fish and increased yield from the catch. However, we should notice that there are more than enough fishing gear and fishing vessels, perhaps even too much fishing gear and fishing vessels. There is too much fishing capacity but not enough fish stocks. The contribution from capture fisheries will mainly depend on the effectiveness of fisheries management and stock recovery. From a technical point of view, the yield of capture fishery has no more room to increase since the sharp decline in fishery resources with two-thirds of the species overfished. Fishing capacity should be reduced immediately and the waste and discard of bycatch should also be reduced.

So the policy of zero increase in marine fishing, as practiced in China, and the measurement of "closed season or closed area" can lower costs and provide effective methods for MCS. China has implemented a three months closed season in all Chinese coastal waters and has extended this measure into the largest freshwater lake, Boyang Hu, since 2002.

5) How to "cut-the-cake" in the marine food web?

The answer to this question depends on the structure of food web and the role of any individual species plays in the food web. Management regulations should maintain sufficient recruitment into the fish stock to sustain adequate population size. Special attention should be paid to certain pelagic species, such as sardine, herring, squids, krill, etc. These fishes, representing only a few species, have great fluctuations in stock abundance that are greatly influenced by environment and atmospheric change. However, if abundance declines, it will have serious impacts on the ecosystem, perhaps even breaking down the food web, because they are important food for many other fishes, marine mammals, and other predators. The amount of catch of these species should be carefully controlled even when their population is abundant.

How Can We Get More from fish?

At first, there will be no more fish, but increased value and economic benefits from fish. For instance, it is important to fully utilize the fish as an industrial raw material by processing for value added and fully utilizing every part of the fish without waste. We suggest learning from agriculture, to process wheat into flour, and then producing thousands of products. Full utilization of fish meal, fish filet, and their derived products is highly recommended.

Secondly, we recommend avoiding the practice of frequently changing the species in the market—if species change too fast, there will be less research and higher costs.

We might get more from fish if we standardize brood stocks, creating a few highly productive species, and get more from fish by fine processing to produce marine medicines and biologics, which are called "no weight with higher value" products.

Thirdly, we suggest getting more from fish through fisheries management. Through sustainable development or sustainable fisheries we could get more fish forever. A key fact should be focused on definitive ownership or definitive property rights of the fishery resources. The owner can run and utilize the fish resources in a proper way over the long term. The quota system, licenses system, exclusive user rights, and so forth should make clear and definite parameters of ownership of fishery resources or the user rights of fishery resources. Because fish are mobile and movable, they belong to nobody, not only in the high seas, but also in the areas where fish migrate or pass through. Therefore we should strengthen the power and function of regional international fisheries management organizations or RIFMOs. This organization structure would apply the reporting system and observer program. One of the key measures needed to eliminate IUU fishing includes action taken by market and end-user countries. There is a very simple reason for this, if there is no market to sell and no buyer or salesman, then no fishing activities will occur. Fighting IUU is like fighting drugs, it is a long war.

Get More Social Benefits from Fish

Today's fishing industry is a kind of postindustrial system which closely relies on the development of other industries and scientific technologies, such as the technology and products from the ship building industry, electronics, synthetic fiber, hydraulic machines and cold storage. Products from other industries are the fundamental base for the modern fishing industry. Therefore, we can say that fisheries is an important market for innovation from other industries and the development of fisheries works as a driving force in the development of those industries and the society related to fisheries. Another important social benefit is providing jobs. Land-based aquaculture as a modern fishery industry is a complex system that requires a lot of technical support. It will stimulate many new developments.

Get More Benefits from Fish by Facilitating the Development of Cultural and Nonmaterial Products

Fisheries have extended into other economic realms; fish are not only important as human food, but also as spiritual nonmaterial products. Products from fish and fishery culture appear as major themes throughout nonmaterial requirements of humans. They are found in fish and fishery arts, recreational sports fishing (including catch-and-release fishing), fishery museums and aquariums, fish photography, and other cultural related activities. One of the biggest businesses in the world is the reproduction of pleasure fish for aquaria and ponds. We can expect that pleasure fish in aquaria will be as basic a "furnishing and ornament," as flowers are in a house. Organizing bio-ecological tours, taking the fishery and ocean as major topics, will be a part of the tourist industry. This has been called the 6S activities (i.e. seafood, sports fishing, see fish, scuba diving, sea viewing, and sunshine. More benefits can be derived from launching education programs on environment science and ecology to enhance, heighten, and improve the oceanic consciousness of the public. Cultural products associated with fish and fisheries will have active and positive affects promoting the protecting our blue earth and oceans, and advance conservation understanding and knowledge of the natural ecology and environment in the general public.

Get More Fish from Aquaculture

There is no doubt that more fish will come from the aquaculture in the future. Right now, the yield of fisheries consists of two parts—capture fisheries and aquaculture. The production value and yield from capture fishery currently occupies first place in the world fisheries, representing about 90% of total yield at the end of the 20th century. Most of the capture fisheries come from our oceans, so that the marine fishery is the most important component of the world fishery. However, fishery statistics obviously show that continued overfishing has led to production declines. In fact, global marine catches have declined since reaching a peak of about 80 million metric tons in the late 1980s. Most abundant resources, such as pollock in the Pacific and cod and herring in the Atlantic, have experienced precipitous declines. On the other hand, since the 1980s, aquaculture has played an important role in world supplies of aquatic food resources. Aquaculture production increased from 7 million mt in 1984 up to 33 million mt in 1999 and provides over 30% of the world's total food fish. At a growth rate of 11% per year on average, aquaculture could surpass beef production by 2010 and will provide over half of the fish consumed by the world's people by 2030. Projections of world fishery production in 2010 were indicated by FAO; they range between 107 and 144 million mt, including about 30 million mt intended for fish meal and oil for nonfood use. So, estimated quantities available for human consumption might range between 74 and 114 million mt. However, the total catch had already reached 92.36 million mt in 2001.

Fishery is an industry subject to historic change. If we recall that human activity began with fishing as hunting in ancient times representing the first fishery revolution, then the fast development of aquaculture could be called as the second fishery revolution. The recent innovation of fishery industrial construction is similar to activity several thousands years ago (i.e:, human activities changed and transformed from hunting into planting and agriculture). Aquaculture will be an important turn in the course of events in the history of fishery science and fishery industry.

At the same time, we should also note that fisheries have extended their range into other area, such as recreation fishery, including angling, viewing and fishery or fish civilization. This means the fish is not only important to humans as food, but also as spiritual and nonmaterial products characterized by recreation sports and relaxing activities. As the world's standards of living increase and there is further recognition of the need for harmony and coexistence between humans with nature, recreational fisheries will experience greater development and increasing economy. Therefore, aquaculture will provide more fish products, consumptive and nonconsumptive, in the future. Fish farming in cages and land based aquaculture—modern marine industry—represent an integration of modern technology and engineering with great economic potential.

As opposed to fishing down the food web, aquaculture has the tendency to "aquaculture or farm up the food web," based mainly on the culture of predators, such as perch, weever, and so forth, because of their good taste. However, if we consider the energy flow through the food web, it would be better to recommend the culture of lower energy-consuming species such as herbivore fishes. They will consume less energy. If so, it is a big job to study how to improve their taste using different feeds or biotechnologies or by selecting suitable species from salt water or freshwater and changing the water salinity in their aquaria. Aquaculture consumes a lot of fish meal feeding cultured fish, so aquaculture, as well as the fishing industry, needs to release the pressure imposed on fisheries resources. Using fish meal in aquaculture can stimulate fishing fleets to catch and retain everything including small fish. So using fish to feed fish in aquaculture is not recommended.

Efforts are needed to restore or recover the ecosystem of habitats and fishing grounds by means of artificial reefs, planting aquatics, and so forth. Artificial reefs will prevent or stop improper trawling and improve, recover, or create new sustainable fish grounds. In freshwater lakes, we are successfully improving or restoring the ecosystem by eliminating blue alga and planting aquatics and beneficial grasses on the lake bottom.

There is a new word "Aqua-cap-culture" which means development of multifunctional fishing cages, such as a set net or a seine net, that can be operated as a fishing gear to catch fish and then as a fish cage for farming, similar to stow set nets. Development is underway of autosubmerged fish cages, which can survive storms with fish guides and collect systems.

If we set the target for reconciling fisheries and conservation as sustainable fishery development with the subsequent adjustments and regularization of fisheries and aquaculture, progress will require the support from governments, the whole society, other industries, and from new developments in fisheries science and technology.

American Fisheries Society Symposium 49:735–736

Session Summary

Food Web Constraints on Getting More Fish

HIROYUKI MATSUDA
University of Tokyo, Ocean Research Institute
Tokyo, Japan

The main subject of the "Food Web Constraints" section was the relationship between harvesting and trophic structure. Some authors investigated impacts of harvesting on trophic structure and ecosystem status (Matsuda [Japan], Zetina-Rejon [Mexico], España [Mexico], Abrams [Canada], Logerwell [USA], Gaichas [USA], Grabowski [USA], Salomon [USA], and Bolotova [Russia]), while others investigated effects of trophic interactions on harvesting (Matsuda, Zetina-Rejon, and Gaichas).

Matsuda argued possibilities of increased sustainable yield from the world fisheries. Matsuda also demonstrated concerns that arise from food web constraint in maximal sustainable yield theory and adaptive management.

España calculated temporal and regional variations in the average trophic levels in landings. He found a negative relationship between trophic level and catch amount among regions in Mexico. Especially in small and local fisheries, the average trophic level in landings did not decrease, which suggests that fishing down the food web does not appear to result from all coastal fisheries. However, the mean trophic level found in landings in specific fishing areas does not always reflect the potential magnitude of overexploitation. Abrams summarized cases using food web models that demonstrated different predictions than those derived from single species models. He portrayed a model that showed the possibility of increasing stock and catch with increasing fishing effort. However, conclusions by España and Abrams do not support the possibility of increasing total landings from the world fisheries. Abrams demonstrated that equilibrium population sizes ultimately decline rapidly toward extinction as conditions continue to worsen in some cases.

Human impacts on ecosystems are often complex and include positive and negative effects. Logerwell investigated interactions between commercial fisheries and endangered Stellar sea lions. She suggested that commercial fishing operations could impact the foraging success of sea lions either through disturbance of prey schools or through direct harvest of highly connected species. In contrast, Grabowski et al. investigated a positive contribution of herring bait on farm lobsters in Maine.

All speakers in this session referred to the importance of trophic interactions of a target species with other species. In this sense, fisheries management theory based on single species models is not sufficient. Several presentations presented the theoretical possibility that trophic interactions may have an important role in fisheries management and ecosystem conservation that are not obtained from single species models. However, it is often difficult to show full evidence for the importance of trophic interactions.

How powerful are ecological models to investigate full evidence? Zetina-Rejon suggested that predation probably had an important role in the shrimp stock depletion because of the "mixed control scenario" in which vulnerability was 0.6, closer to historical trends in shrimp fishery. Ecological modeling is powerful and can obtain heuristic implication and quantitative projections. However, we need to investigate the robustness of conclusion drawn from ecological modeling using sensitivity analysis or other methods.

The impact of fisheries on ecosystems is often evaluated by change in population abundance of target species. Fishing down (decrease of the average trophic level in landings) is used as an indicator of overfishing. España estimated trophic levels of landings in Mexican fisheries. The average trophic level value in both the Gulf of Mexico (3.14) and Mexico's Pacific coastal areas (2.94) has not decreased during the study period, rather it has increased since captures where stabilized in the 1980s. Matsuda also argued that the average trophic level in marine ecosystem is a better indicator for overfishing. Matsuda stated that we can get more fish if we catch bioresources at lower trophic levels.

We might expect that premodern fisheries were sustainable. However, Salomon et al. challenged this assumption by describing the roles of subsistence harvest and natural factors on coastal food web structures and ecosystem productivity in south central Alaska. Alternation in traits of target species is caused by both fisheries and predators. Thurman et al. (USA) investigated spatial variations in growth rate of roughtongue bass *Pronotogrammus martinicensis*. These variations maybe caused by predation risks. Growth rate differs among sites, probably caused by variation in predation risks. Bass at predator-free sites grow faster than bass with predators. Bolotova considered change in species composition associated with harvest.

Gaichas proposed a new idea of food web structure with fisheries management and applied it to the North Pacific ecosystem. The implication that food webs exhibit a specific type of structure implies that certain species are essential to maintaining the community in its current configuration. She classified food webs according to network properties (e.g., a randomly connected network versus a scale-free network containing very few highly connected nodes and many nodes with few connections). She would use this approach to open a new page of community ecology and ecosystem management. In a scale-free network, selective removal of highly connected species results in the fragmentation of the network, suggesting the need to focus on protecting the highly connected ("key stone") species to avoid structural impacts on the food web. She suggested that current fisheries management should focus on protecting highly connected species to avoid structural impacts of fishing on marine food web.

American Fisheries Society Symposium 49:737–744

Can We Say Goodbye to the Maximum Sustainable Yield Theory? Reflections on Trophic Level Fishing in Reconciling Fisheries with Conservation

HIROYUKI MATSUDA*

Faculty of Environment and Information Sciences, Yokohama National University 79-7 Tokiwadai, Hodogaya-ku, Yokohama, Kanagawa 240-8501, Japan

PETER A. ABRAMS

Department of Zoology, University of Toronto, 25 Harbord Street, Toronto, Ontario M5S 3G5, Canada

Abstract.—Ecosystems, including those that contain fisheries resources, are characterized by uncertainty, dynamic properties, complexity, and evolutionary responses of the component species. However, the classical maximum sustainable yield (MSY) theory does not include any of these. Thus, it is perhaps not surprising that the MSY theory and its derivatives have not worked for fisheries management. Food and Agriculture Organization of United Nations (FAO 2000) noted that about three-fourths of stocks are either fully exploited or overexploited. It should be noted, however, that although the total landings of demersal fishes had reached a plateau by the 1970s, those of pelagic fishes increased until the late 1980s. Some of these pelagic fish species naturally fluctuate greatly in stock abundance, even without fisheries. The collapse of Japanese sardine in the 1990s was almost certainly caused by natural variation in the environment. When the stock is at a low level, the impact of fisheries on pelagic fishes prevents the stock from recovering. Therefore, in this century, we need to consider how to use nonstationary bioresources. We make 11 recommendations that could both increase the food resources derived from fish and reduce the chances of overexploitation or extinction: 1) catch fish at lower trophic levels; 2) do not use fish as fish meal; 3) reduce discards before and after landings; 4) establish food markets for temporally fluctuating fishes at lower trophic levels; 5) improve the food-processing technology used on small pelagic fishes; 6) switch the target fish to correspond to the temporally dominant species; 7) conserve immature fish especially when the species is at a low stock level; 8) develop technologies for selective fishing; 9) conserve both fish and fisheries; 10) say goodbye to traditional MSY theory; and 11) monitor not only the target stock level but also any other indicator of the "entire" ecosystem.

Introduction

It has long been felt that maximum sustainable yield (MSY) theory is one of the fundamental bases of fisheries management (e.g., Clark 1990). Using single-species stock dynamic models, MSY theory argues that under full knowledge of stock dynamics, stock persistence is guaranteed. Despite this, MSY theory recently has been criticized for a variety of failings (e.g., Hilborn 2002). It is increasingly well known that ecosystems are uncertain, nonstationary, and likely to have complex dynamics (Matsuda and Katsukawa 2002). The classical MSY theory ignores all of these three char-

*Corresponding author: matsuda@ynu.ac.jp

acteristics. Over the past decade, ecologists and managers have increasingly recognized the importance of species interactions, nonstationary properties, and adaptive response in traits with climate change and human impacts (Collie and Spencer 1994; Merrick 1997; Köster et al. 2001; Heino and Godø 2002; Matsuda and Abrams 2004). In this paper we will show that, when applied to multispecies systems, MSY theory is likely to produce extinctions because of food web interactions. Adaptive management has increasingly been recognized as one of the best ways to reconcile fisheries and conservation, given uncertain ecosystem dynamics (Walters 1986). In addition, we will show the mathematical possibility of undesired outcomes from at least "passive" feedback control and will discuss another caution for application of adaptive management.

Total Landings of Small Pelagic Fish Are Still Increasing

The Food and Agriculture Organization of the United Nations (2000) reported that about three-fourths of stocks are either fully exploited, overexploited, have been depleted, or are recovering from depletion. This is one of the strongest reasons for the pessimistic view of the future total world landings. However, we note that these statistics differ for different groups of species. The world landings of marine demersal fishes has not grown since the 1970s (Figure 19 of FAO 1996). In contrast, the world landings of marine pelagic fishes looks to be still increasing at least until approximately 1990 (Figure 20 of FAO 1996). The species composition of the pelagic fish landings in the world has changed from decade to decade.

During the mid-1960s to mid-1970s, the Peruvian anchoveta (also known as Peruvian anchovy) *Engraulis ringens* was abundant. The catch of Peruvian anchoveta in 1970 was 12 million metric tons, while the total world landings from all marine bioresources was approximately 100 million metric tons. More than 10% of the world landings in 1970 was accounted for by one species in one country. Also in the 1970s, more than 10% of the total world landings consisted of Japanese pilchard (also known as Japanese sardine) *Sardinops sagax melanostictus* and Chilean sardine (also known as South American pilchard) *S. sagax*. Stock collapses of Peruvian anchoveta in the mid-1970s and sardines in the early 1990s were probably not because of overfishing. It is well known that small pelagic fishes such as anchovy and sardine have fluctuated since prehistorical time (Baumgartner et al. 1992). Therefore, we were able to get more Peruvian anchoveta in approximately 1970 and more Pacific sardines *S. sagax caeruleus* in the 1980s. Now we are able to get more Japanese pilchard and Pacific saury *Cololabis saira*.

Drops in Landings Are Not Always Caused by Overexploitation

Another major reason for the pessimistic view of the future world landings is called "fishing down the food web" (Pauly et al. 1998). Pauly et al. (1998) evaluated the mean trophic level of all fishes in landings, and they found a gradual decrease of the mean trophic level. The average trophic level decreased from approximately 3.4 in 1950 to 3.1 in 1995 (Pauly et al. 1998, 2002). This is recognized to be a consequence of significant alteration of the marine trophic structure caused by harvesting. However, the mean trophic level of landings depends on both the marine trophic structure and trophic selectivity of fisheries. The mean trophic level of all individuals in marine ecosystems is definitely much smaller than three. Bioresources at higher trophic levels are selectively harvested by the world fisheries. If the change in average trophic level in landings reflects alteration of the marine trophic structure and if the trophic structure shift is only caused by overharvesting, this is again a reason for pessimistic view of the world landings. There is no doubt that many predatory species have been overfished (Jackson et al. 2001; Myers and Worm 2004), but it is also true that reductions in the mean trophic level of landings have been influenced by shifts in the trophic level targeted by the largest fisheries. For example, the average trophic level in landings decreased during the 1960s, mainly because the catch of Peruvian anchoveta increased (Pauly et al. 1998).

Harvesting lower trophic levels likely produces a smaller impact on marine ecosystems than does harvesting higher trophic levels, given the same total catch. Trophic pyramids imply that harvesting of lower levels produces a smaller proportional change in population densities. We could get more fish while reconciling fisheries with conservation, if harvesting were directed more heavily at lower trophic levels. If this recommendation were adopted, the apparent fishing down in landings would not signal a change in the trophic structure in marine ecosystems.

We Can Get More Fish that Are Consumed by Top Predators

Tamura and Ohsumi (1999) estimated the total consumption by whales and concluded that consumption by whales is at least three to five times larger than the total fisheries landings. According to data obtained from the Japanese whale research program under special permit in the North Pacific, northern Pacific minke whales *Balaenoptera bonaerensis* consumed anchovy and Pacific saury in 1990s. Yodzis (2001) also estimated consumption by marine mammals and birds and top predatory fishes and concluded that consumption by top predators is much larger than the fisheries landings in all of the marine ecosystems he analyzed.

The fact that the total consumption by whales is much larger than the total fisheries landings leads to the idea of fisheries–whale competition. The Japan Fisheries Agency and Institute for Cetacean Research in Tokyo argue on this basis that a complete harvesting ban on whales destroys the ocean. However, Yodzis (2001) used a general ecological argument to show that culling top predators (e.g., whales) may have either positive or negative indirect effects on the abundance of target fish. Since the catch of anchovy in northern Pacific is not very large, northern Pacific minke whales consume nontarget fish species and therefore there is no apparent direct competition between fisheries and whales. In addition, the estimated biomass of whales in the beginning of the 20th century is much larger than the current biomass even if the minke whales increased after moratorium of commercial whaling began in 1986. Whales have always consumed more fishes and other bioresources than fisheries.

More recently, another controversy has arisen regarding the impact of fisheries on marine trophic structures. Fisheries directly exploit top predators such as tunas and sharks. Myers and Worm (2004) evaluated compiled data of the catch per unit effort (CPUE) of longline fisheries and found that the CPUE decreased by 90% during the second half of the 20th century. They also concluded that the decline of CPUE reflects the equal magnitude of stock decline. Combining substantial decline of top predators by Myers and Worm (2004) with substantial decline of the mean trophic level by Pauly et al. (1998), the argument that fisheries alters marine trophic structures has increasingly been recognized. Therefore, it is uncertain whether harvesting more fish that are consumed by top predators reconciles fisheries with conservation.

Lessons from Nature in Fishing Low Stock Target Pelagic Species Instead of Low Market Dominant High Stock Species

As shown in Figure 1, the landings of each species of small pelagic fish fluctuate greatly. Based on the catch statistics of Japan, Japanese sardine was dominant in the 1930s and 1990s, Japanese anchovy *Engraulis japonica* was dominant from the mid-1950s until the 1960s and again in the 1990s, and Pacific chub mackerel *Scomber japonicus* was dominant in the 1970s (Figure 1). Alternation of dominant pelagic fishes is known as spe-

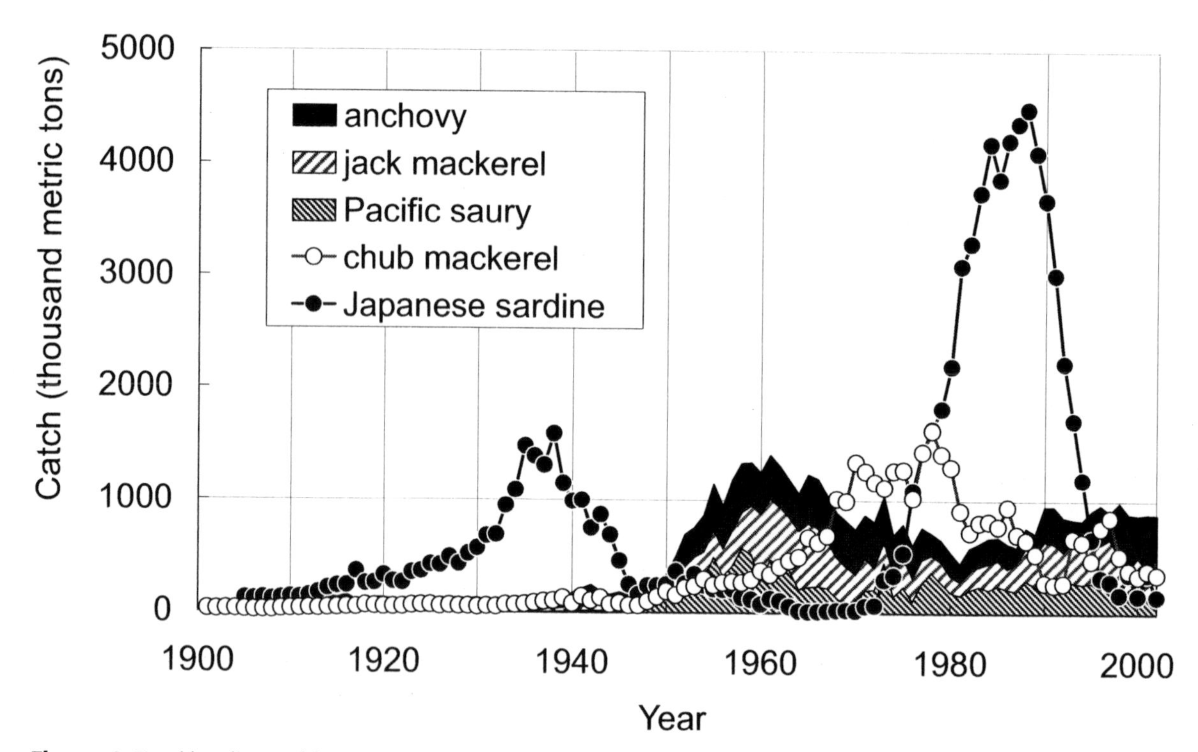

Figure 1. Total landings of five major pelagic fish species in Japan. The catch of chub mackerel includes Pacific chub mackerel and spotted mackerel *S. australasicus*. The catch of jack mackerel includes Japanese scad *Trachurus japonica* and amberstripe scad *Decapterus muroadsi* (Source: Matsuda et al. 2002; Japan Ministry of Agriculture, Forestry and Fisheries 2003).

cies replacement (Matsuda et al. 1992a; Matsuda and Katsukawa 2002).

The collapse of sardine in 1990s was almost certainly caused by natural variation in the environment. Overfishing usually decreases the mean age and size of both catch and stock. The mean age and size of sardine increased in the 1990s (Watanabe et al. 1995). The sardine stock collapse was accompanied by successive failure of recruitment. In the 1980s, the estimated stock of the northwest Pacific sardine was at least 10 million metric tons (Yatsu, Nagasawa and Wada, unpublished report). We were able to harvest more sardine in the 1980s and more anchovy and Pacific saury in the 1990s.

However, the rate of exploitation of some of the small pelagic fish was remarkably large when they were at low stock levels. In the 1990s, Pacific chub mackerel was heavily exploited, especially as immature individuals in the Pacific Ocean off Japan (Kawai et al. 2002). The Japanese Government has regulated the total allowable catch of Pacific chub mackerel in the Pacific Ocean off Japan since 2003 for the sake of stock recovery. Overexploitation of Pacific chub mackerel decreased the medium-term catch and stock (Matsuda et al. 1992b) and prevented the stock from recovering (Kawai et al. 2002). Therefore, in this case, switching the target species based on what is known of species replacement should have increased the total average yield (Matsuda and Katsukawa 2002; Katsukawa and Matsuda 2003). The north Pacific minke whales change their prey based on prey abundance. Their main prey item was Pacific chub mackerel in the 1970s, sardine in the 1980s, and anchovy and Pacific saury in the 1990s (see Matsuda and Katsukawa 2002). Unlike northern Pacific minke whales, Japanese purse seine fisheries did not switch their target from sardine to anchovy in 1990s.

The MSY Theory Does Not Guarantee Species Coexistence

There are many dimensions in the criticism of the MSY. Ecosystems are characterized by 1) temporal fluctuations, 2) uncertainty in assessment and ecosystem processes, 3) an open system concerned with straddling stocks and international fisheries management (Matsuda and Katsukawa 2002), and 4) evolutionary responses (Matsuda and Abrams 2004). However, the classical MSY theory does not include any of these. In addition to these dimension of criticisms, here we illustrate a criticism based on the constraints posed by food web interactions. In single stock dynamics, it is clear that neither zero catch nor complete exploitation of a bioresource can be an optimal exploitation policy, because neither one produces any catch. The MSY is achieved at an intermediate harvesting effort. Therefore, it has been thought that MSY theory always guarantees species persistence. However, food web interactions imply a very different story. The optimal harvesting rate of a particular species that maximizes the sustainable yield depends on abundances of its prey and predator. If fishers exploit both a prey and its predator, the optimal harvesting rate of the prey must depend on the harvesting rate of the predator and vice versa.

Clark (1990) has investigated the total sustainable yield from a predator–prey system using a simple mathematical model; he concluded that either the prey is exploited after the predator is removed or only the predator is exploited. Therefore, the MSY theory does not always guarantee species coexistence even for simple predator–prey systems.

In order to investigate when and where it is appropriate to harvest multiple trophic levels, Matsuda and Abrams (2006 and unpublished report) investigated simple three and six species food web models and defined unconstrained MSY and constrained MSY as the maximal yields from the entire biological community when extinction of some species is either allowed or is not allowed. At the unconstrained MSY solution, it is rare for all species to coexist. By definition, the constrained MSY for a particular food web is equal to or smaller than the unconstrained MSY for the same food web. The ratio of constrained MSY to unconstrained MSY ranged from 100% to less than 10%, although its average and range depends on how one constructs food webs and trophic interactions. Loss of biodiversity probably reduces the total yield and sustainability in the long run, although the simple unconstrained food web models do not exhibit this long-term effect of species extinction. However, we have not yet seen any ecosystem modeling used in fisheries management that explicitly accounts for the long-term advantages of biodiversity conservation on the sustainable yield. Therefore, even with complete knowledge and complete compliance with regulations, MSY fishing policies in ecosystems may require explicit provisions for conservation.

We Must Pay Attention to Adaptive Management Based on a Single Stock Model

There is a cognitive gap between different groups of fisheries scientists regarding the MSY. Some feel that the MSY exists but is unknown; the second group feels that the MSY exists but is unknowable; and the third group argues that the MSY does not exist because ecosystems are not stationary. Although the term MSY is still alive, at least in the United Nations Law of the Sea, the concept of MSY has evolved from the single stock dynamic pool model to biological reference points in fishing mortality and biomass, and even MSY-based reference points are now criticized (Hilborn 2002).

As the sun apparently sets on MSY theory, adaptive management has increasingly been applied to a variety of situations that involve uncertainty. If the harvesting rate changes with the most recent estimate of stock abundance, we can maintain the stock level close to any target using a single stock dynamic model. We do not need to link the target stock size to the MSY level. Such feedback control is effective against uncertainty in the MSY level if it exists, process errors (instability and stochasticity in stock dynamics), and measurement errors to some extent. Adaptive learning by successive monitoring also evaluates the stock–recruitment relationship and the magnitude of various sorts of uncertainty.

Although it has been expected that the feedback control stabilizes a broad range of target stock levels, this is again a result of relying on single stock dynamic models. Suppose there is a predator–prey system with harvesting of the prey. Mathematical models for prey–predator systems suggest that harvesting of prey decreases the predator abundance rather than the prey abundance. If a fisheries control rule reflects a target stock abundance that is lower than the predator's requirement, harvesting rate of the prey does not decrease until the predator goes extinct. For larger food web models, feedback control in harvesting rate may result in extinction of another species if the harvesting rate reflects only the target stock size (H. Matsuda and P. A. Abrams, unpublished report). If the harvesting rate also reflects the status of a species that interacts with the target species, we may avoid extinction of these species. Therefore, we must monitor not only stock level of target species, but also the "entire" ecosystem (i.e., all species potentially at risk due to food web connections), if we wish to reconcile fisheries with conservation.

Recommendations and Observations Based on Food Web Constraints on Getting More Fish

As mentioned above, (1) we can harvest more fish at lower trophic levels, especially small pelagic fish such as sardine and anchovy. (2) We can eat more fish if we do not use fish as fish meal. Aquaculture often produce fish and marine invertebrates at higher trophic levels from fish meal and agricultural products. This is an inefficient way to utilize primary production to foods. This is economically reasonable if the ratio in price of fish at higher trophic levels to lower trophic level is larger than the energetic conversion rate in trophic interactions. However, it is never reasonable if the goal is to maximize human food supply, and increasing food-supply will become increasingly important as the human population increases. (3) We can eat more fish if we reduce discards before and after landings. Landings have an impact on removal of organic materials from marine ecosystems. Therefore, discard after landings may be worse than discard in the sea. Irrespective of economic incentives, reduction of discards definitely increases sustainable food supply. We should reduce discards after landings for ethical reasons.

As we also mentioned above, small pelagic fish are often characterized by natural fluctuations in stock abundance. We can get more fish by concentrating on those species at a high stock level, while present day fisheries often overexploit the same species when it is at a low stock level. We must protect small pelagic fish species when they are rare. Therefore, the species composition of landings must change from decade to decade or even from year to year. Fish markets definitely dislike such fluctuation. Throughout human history, almost every sort of industry has sought stability. In spite of this, foreign currency markets have not yet been able to reduce fluctuation. By reducing discards and spares in fast foods markets and convenience stores, their markets have suceeded in growing. In addition, even under a constant harvesting rate policy, fish markets are vulnerable to fluctuation in species composition and catch amount. Therefore, (4) we can get more fish if we establish food markets for temporally fluctuating fishes at lower trophic levels. Although it may be difficult, (5) we must establish such dynamic markets for reconciling fisheries with preservation of marine ecosystems. One of the possible solutions is a long-term direct contract between fishers and consumers. If consumers are willing to buy any species of seafood, fishers can easily switch their target into temporally abundant species.

Despite health benefits of fish species at lower trophic levels (low levels of toxins and high levels of necessary fatty acids), there is little demand for these fish. Those who eat beef every day are unlikely to eat fish every day. Because of the tendency of human consumers to focus on a few species, the Shannon-Wiener index of diversity in seafood items is much smaller than that of biological communities. (6) We can get more fish if food-processing technology of underexploited fish species is developed, although this may be less economically valuable than food-processing of fishes at higher trophic levels.

As we again mentioned above, we can get more fish if we switch to target fish that are temporally dominant (Matsuda and Katsukawa 2002). Purse seine fisheries can exploit chub mackerel, sardine, and anchovy if fishers change their mesh size. Although it is costly for a fishing boat to carry several purse seine nets with different mesh size, they can change nets from year to year. In the early 1980s, the fishers changed mesh sizes and switched their main target from chub mackerel to sardine, but they did not later change mesh size to exploit anchovy when that species became abundant.

We need to give up the viewpoint that catching bigger fish is always better. Fish at lower trophic levels are usually smaller than fish at higher trophic levels. We still need to protect immature individuals. Impact per unit weight landing of immature fish on the stock sustainability is usually bigger than the impact of mature fish. (7) We can get more fish if we conserve immature fish especially when the species is at a low stock level.

In order to switch a target fish, to save immature individuals and to reduce bycatch, (8) we must improve technology for selective fishing. Technology has usually been improved to increase the amount of catch, to find schools of fish more effectively, and to reduce the costs of fishing. As results of such a mass consumption technology, CPUE has decreased, discards and bycatch have increased, and there have been increased impacts on the target stock and the entire ecosystem. For reconciling fisheries with conservation, we need to improve technology that contributes increasing CPUE, decreasing bycatch, and reducing discards.

Before a target fish is threatened, its fishery and fishers are often threatened. It is qualitatively reasonable that the value of natural capital and ecosystem services is much bigger than the yield from fisheries (Costanza et al. 1997). This means that fisheries are possible if we maintain the ecosystem integrity of the fishing grounds. Therefore (9), a sustainable fishery is a good indicator of ecosystem integrity. To conserve a sustainable fishery is often as effective as, and more cost-effective than, conserving an umbrella species.

It is still unclear whether we can get more fish whether or not fisheries are reconciled with conservation. In spite of the fact that neither the sardine nor anchovy were overfished when they were at high stock levels, it is possible that getting more small pelagics may have a big impact on top predators or other components of marine ecosystems. We should be aware of the fact that (10) the MSY theory does not guarantee the coexistence of species and may drive some species to extinction. Instead of a constant harvesting rate policy based on MSY reference points, we usually recommend adaptive management for a fishery of nonstationary pelagic fish. However, passive adaptive management that depends on single stock assessment sometimes results in undesired outcomes such as extinction of other species. Therefore, 11) we must monitor not only the target stock level but also any other indicator of the "entire" ecosystem and test hypotheses after agreement of the stakeholders.

Acknowledgments

We thank the Fourth World Fisheries Congress Organizing Committee, especially Yvonne Sadovy and Tony Pitcher, for arrangement of this chapter. We also thank T. Katsukawa, K. Sainsbury, and K. Sakuramoto for valuable discussion and anonymous reviewers for valuable comments. The work was supported by a grant from the Japan Ministry of Education, Science, Sports, Culture, and Technology to Hiroyuki Matsuda and a Strategic Project grant from the Natural Sciences and Engineering Research Council of Canada to Peter Abrams.

References

Baumgartner, T. R., A. Soutar, and V. Ferreira-Bartrina. 1992. Reconstruction of the history of the Pacific sardine and northern anchovy populations over the past two millenia from sediments of the Santa Barbara basin, California. CalCOFI (California Cooperative Oceanic Fisheries Investigations) Report 33:24–40.

Clark, C. W. 1990. Bioeconomic modelling and fisheries management, 2nd edition. Wiley-Interscience, New York.

Collie, J. S., and P. D. Spencer. 1994. Modeling predator-prey dynamics in a fluctuating environment. Canadian Journal of Fisheries and Aquatic Sciences 51:2665–2672.

Costanza, R., R. d'Arge, R. de Groot, S. Farber, M. Grasso, B. Hannon, K. Limburg, S. Naeem, R.V. O'Neill, J. Paruelo, R. G. Raskin, P. Sutton, and M. van den Belt. 1997. The value of the world's ecosystem services and natural capital. Nature (London) 387:253–260.

FAO (Food and Agriculture Organization of the United Nations). 1996. The state of world fisheries and aquaculture. Available: http://www.fao.org/sof/sofia/. (May 2004).

FAO (Food and Agriculture Organization of the United Nations). 2000. The state of world fisheries and aquaculture. Available: http://www.fao.org/sof/sofia/. (May 2004).

Heino, M., and O. R. Godø. 2002. Fisheries-induced selection pressures in the context of sustainable fisheries. Bulletin of Marine Science 70:639–656.

Hilborn, R. 2002. The dark side of reference points. Bulletin of Marine Science 70:403–408.

Japan Ministry of Agriculture, Forestry and Fisheries. 2003. White paper of Japanese Fisheries. The Government of Japan, Tokyo (in Japanese).

Katsukawa, T., and H. Matsuda. 2003. Simulated effects of target switching on yield and sustainability of fish stocks. Fisheries Research 60:515–525.

Köster, F. W., C. Möllmann, S. Neuenfeldt, M. Plikshs, and R. Voss. 2001. Developing Baltic cod recruitment models I: resolving spatial and temporal dynamics of spawning stock and recruitment for cod, herring and sprat. Canadian Journal of Fisheries and Aquatic Sciences 58:1516–1533.

Kawai, H., A. Yatsu, C. Watanabe, T. Mitani, T. Katsukawa, and H. Matsuda. 2002. Recovery policy for chub mackerel stock using recruitment-per-spawning. Fisheries Sciences 68:961–969.

Matsuda, H., and P. A. Abrams. 2006. Maximal yields from multi-species fisheries systems: rules for harvesting top predators and systems with multiple trophic levels. Ecological Applications 16:225–237.

Matsuda, H., and T. Katsukawa. 2002. Fisheries management based on ecosystem dynamics and feedback control. Fisheries Oceanography 11:366–370.

Matsuda, H., T. Wada, Y. Takeuchi, and Y. Matsumiya. 1992a. Model analysis of the effect of environmental fluctuation on the species replacement pattern of pelagic fishes under interspecific competition. Researches on Population Ecology 34:309–319.

Matsuda, H., T. Kishida, and T. Kidachi. 1992b. Optimal harvesting policy for chub mackerel in Japan under a fluctuating environment. Canadian Journal of Fisheries and Aquatic Sciences 49:1796–1800.

Merrick, R. L. 1997. Current and historical roles of apex predators in the Bering Sea ecosystem. Journal of Northwestern Atlantic Fisheries Science 22:343–355.

Myers, R. A., and B. Worm. 2004. Rapid worldwide depletion of predatory fish communities. Nature (London) 423:280–283.

Pauly, D., V. Christensen, D. Johannes, F. Rainer, and T. Francisco Jr. 1998. Fishing down marine food webs. Science 279:860–863.

Pauly, D., V. Christensen, S. Guénette, T. J. Pitcher, U. R. Sumaila, C. J. Walters, R. Watson, and D. Zeller. 2002. Towards sustainability in world fisheries. Nature (London) 418:689–695.

Tamura, T., and S. Ohsumi. 1999. Estimation of total food consumption by cetaceans in the world's oceans. Institute for Cetacean Research, Tokyo.

Watanabe, Y., H. Zenitani, and R. Kimura. 1995. Population decline of the Japanese sardine *Sardinops melanostictus* owing to recruitment failures. Canadian Journal of Fisheries and Aquatic Sciences 52:1609–1616.

Walters, C. J. 1986. Adaptive management of renewable resources. McMilllan, New York.

Yodzis, P. 2001. Must top predators be culled for the sake of fisheries? Trends in Ecology and Evolution 16:78–84.

American Fisheries Society Symposium 49:745–757

The Role of Predation in the Shrimp Stock Collapse in the Southern Gulf of Mexico

MANUEL J. ZETINA-REJÓN*, FRANCISCO ARREGUÍN-SÁNCHEZ, AND VÍCTOR H. CRUZ-ESCALONA

Centro Interdisciplinario de Ciencias Marinas, Instituto Politécnico Nacional AP 592, 23000, La Paz, Baja California Sur, Mexico

Abstract.—The Campeche Sound, in the southern Gulf of Mexico, has historically been an important area for the exploitation of shrimp. Highest yields were obtained in the early 1970s with nearly 25,000 metric tons (mt) per year. However, catches declined 50% by 1980 and 90% by the late 1990s. Currently yields amount to around 1,500 mt. Several hypotheses have been suggested to explain the stock decline, including high fishing intensity, environmental changes, impact of the oil industry, and degradation of coastal areas. In this contribution, we explore the influence of predation in the shrimp stock collapse. The approach is based on a trophic model using Ecopath for the period of 1978–1981, previous to the drastic decline in shrimp yields. The model considers explicitly the interaction of two interdependent ecosystems, the Terminos Lagoon and the adjacent continental shelf. Shrimp life history was partitioned to juveniles living in the lagoon and adults on the continental shelf. They are linked through a stock–recruitment relationship. The model provides the starting point for dynamic simulations using Ecosim. Simulations were calibrated using the historical exploitation pattern. Several scenarios show how shrimp is vulnerable to predators assuming bottom up, top down, and mixed control. When bottom up control is assumed, the model does not simulate the historical stock depletion. Under top down control, depletion was higher than recorded by the fishery. Mixed control scenario was closest to historical data trend. Results suggest that predation has an important role in the shrimp stock depletion.

Introduction

The continental shelf off Campeche Sound, in the southwestern Gulf of Mexico (Figure 1), constitutes an ecosystem in which several human activities occur, among the most important of which are the fishing and oil industries. In this area, the oil industry extracts around 70% Mexico's total oil and natural gas production. Before the oil industry was established in the early 1980s, fishing, particularly for shrimp, was the base of the regional economy. However, this fishery has collapsed (Figure 2); annual harvests have fallen from 27,000 metric tons (mt) during the early 1970s, to 3,000 mt or less in recent years (Arreguín-Sánchez et al. 1997). The main problem in the area is the decline of the pink shrimp *Farfantepenaeus duorarum* stock, which constituted in the past almost 90% of the total catch. Several hypotheses have been proposed to explain the decline in the shrimp stock (Arreguín-Sánchez et al. 1997a, 1999); one frequently argued by the

*Corresponding author: mzetina@ipn.mx

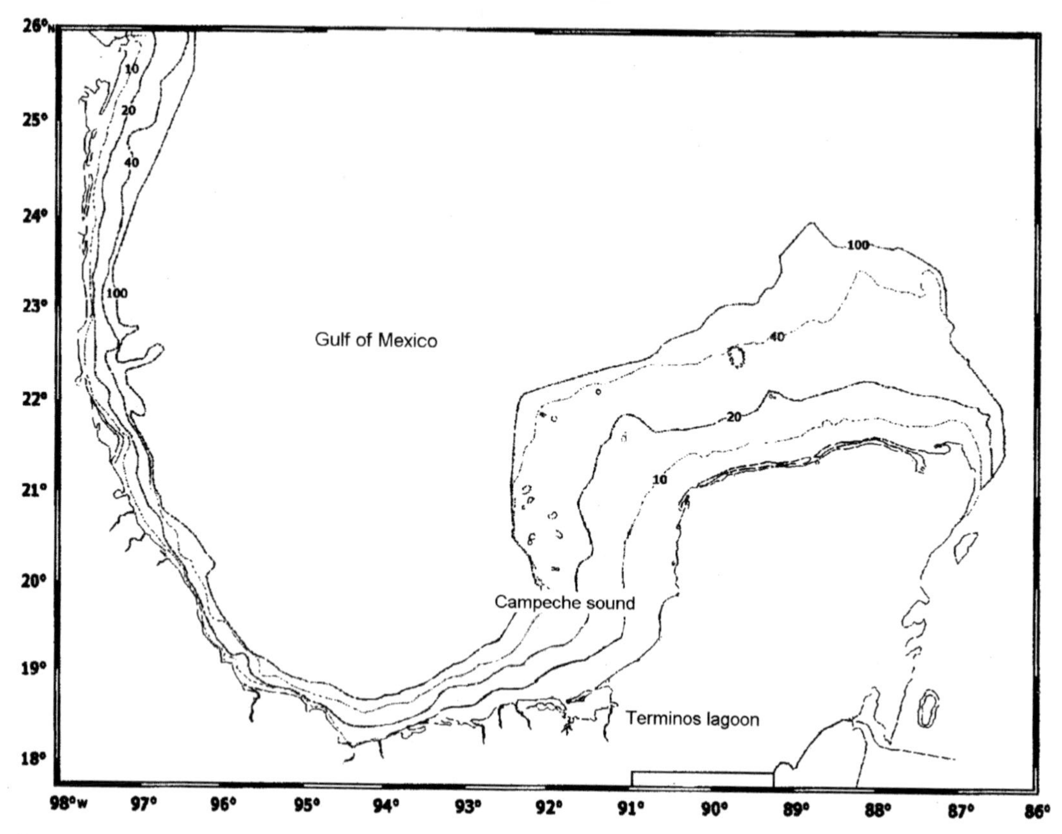

Figure 1. Study area, southern Gulf of Mexico.

public is the impact of the oil industry on recruitment, nursery areas and habitat degradation due to pollution or physical damage. Another hypothesis underlines a strong impact by the catches of juveniles in lagoons and estuaries (Gracia 1997). Recently several authors (for example, Ramírez-Rodríguez et al. 2000, 2003; Ramírez-Rodríguez and Arreguín-Sánchez 2003) concluded that changes in environmental conditions over the last decades have affected recruitment in the long term and may be an important factor for stock depletion. Currently, fishing effort in the area focuses on other, less valuable species, particularly seabob *Xiphopenaeus kroyeri*, finfish, and octopus, at the artisanal level.

Development of fishery management strategies requires an evaluation of the response of exploitation to the target species and of the ecosystem. Cochrane (2000) has pointed out the necessity of evaluating alternative strategies under specific objectives in order to construct an appropriate framework for decision making. In this contribution, we attempt to evaluate the impact predation in the collapse of the pink shrimp fishery in Campeche Sound.

Materials and Methodology

Model Description

The ecosystem of the continental shelf off Campeche, on the southwestern Gulf of Mexico, has an area of 65,000 km^2 (Figure

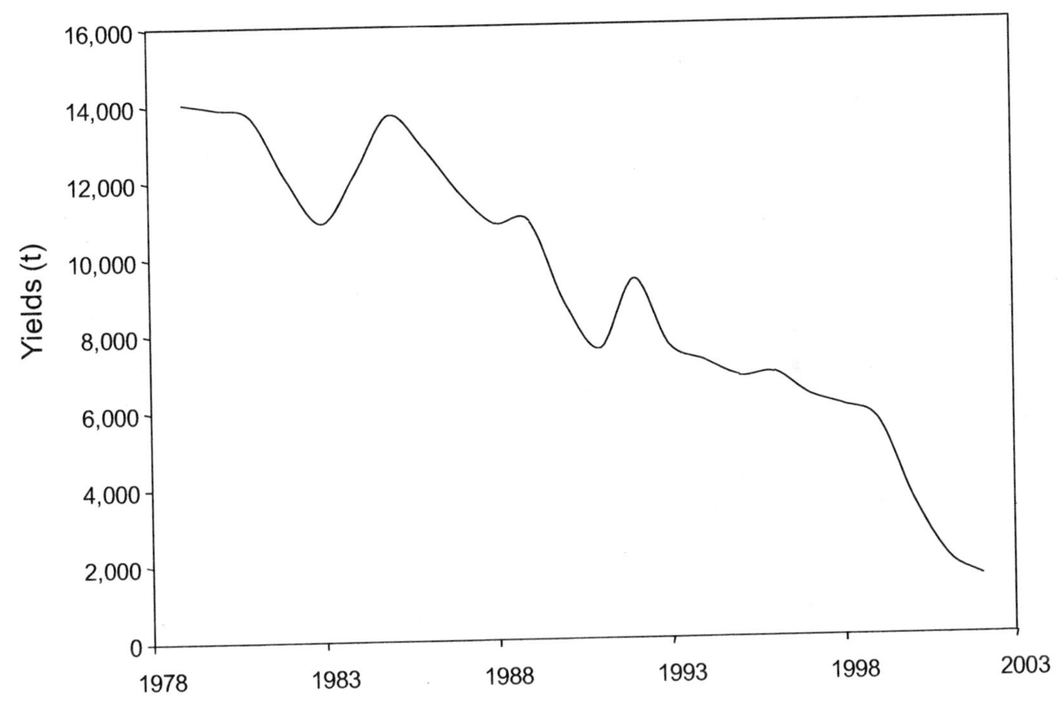

Figure 2. Trends in shrimp yields in Campeche Sound.

1). In this area are some 320 shrimp trawlers and more than 4,400 boats belonging to the artisanal fleet.

The analysis is based on a trophic model using the "Ecopath with Ecosim" approach (Pauly et al. 2000; Christensen and Walters 2004) representing 1978–1981 period, previous to the drastic decline in shrimp yields. The model considers explicitly the interaction of two interdependent ecosystems, the Terminos Lagoon and the adjacent continental shelf (Figure 1) through diet matrix and juvenile–adult dynamic relationship explained thereafter. The trophic functioning of the ecosystem was represented through Ecopath mass balance equation (one per each functional group):

$$B_i \cdot \left(\frac{P}{B}\right)_i \cdot \mathrm{EE} = \sum_{j=1}^{n} B_j \cdot \left(\frac{Q}{B}\right)_j \cdot \mathrm{DC}_{ji} + B_i \cdot \left(\frac{P}{B}\right)_i \cdot (1-\mathrm{EE}_i) + Y_i$$

where B_i = biomass of group i; $(P/B)i$ = producction/biomass ratio of i, which is equal to the total mortality coefficient (Z) under a steady-state condition (Allen 1971; Merz and Myers 1998); EE_i = ecotrophic efficiency; B_j = biomass of predator j; $(Q/B)_j$ = consumption/biomass ratio of predator j; DC_{ji} = proportion of prey i in the diet of predator j; EX_i = the export of group i, represented here bycatch data. The input parameters for each group in the model are B, P/B, Q/B, and EE. Because the model is balanced, if one parameter is unknown, it can be estimated from the model. In addition, the diet composition of all consumers is required. Data used in this work come from published information and were supplemented with that of unpublished reports. All rates and biomasses were expressed as wet weight (ww) and standardized for an average year and by the area of both ecosystems (Terminos lagoon: 3,670 km^2 and Campeche Sound: 33,000 km^2). We define several functional groups marine mammals

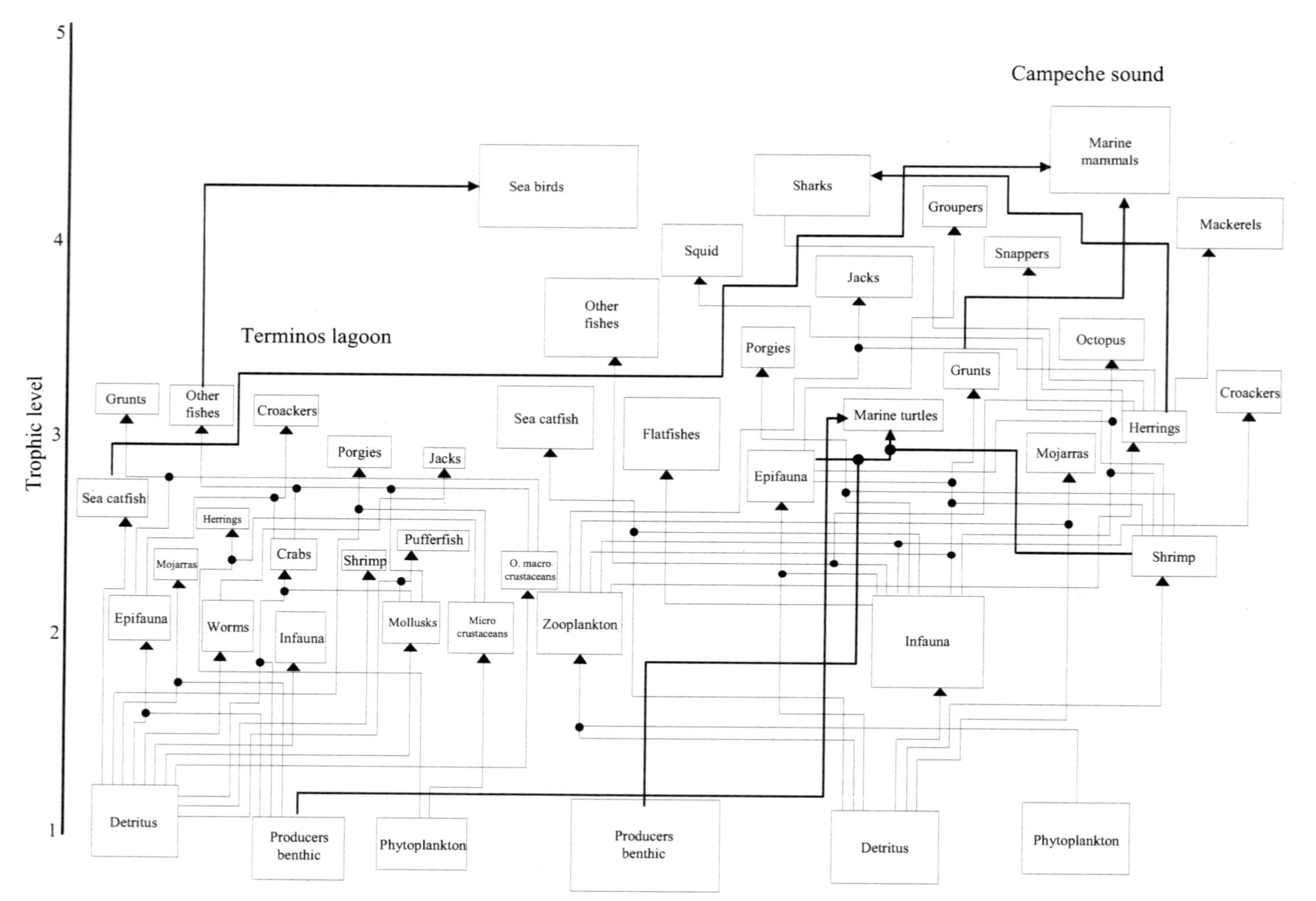

Figure 3. Energetic flow diagram showing the interactions between main functional groups in Términos lagoon and the adjacent continental shelf, Campeche sound.

(dolphins), sea birds, and marine turtles; 12 groups of fish and sharks; and several invertebrate groups, including shrimp, octopus, squid, epifauna, infauna, zooplankton, and as primary producers, phytoplankton, benthic producers, and detritus. The shrimp life history is split into juvenile and adult stages. Juveniles are assumed to live entirely within the lagoon and adults life on the continental shelf. These stages are linked through a stock–recruitment relationship. The fisheries were divided into shrimp trawling and artisanal fisheries (Arreguín-Sánchez et al. 1997b; Gracia 1997). The latter was divided in four groups: the demersal fleet that mostly uses short longlines and manual hook-and lines, targeting groupers and snappers; the pelagic fleet which uses mainly gillnets to catch mackerels and jacks, and hook-and-lines for sharks; the beach seine fishery; and the octopus fleet, which uses live crabs as bait (Solís-Ramírez et al. 1997; Arreguín-Sánchez et al. 2000). The ecosystem model also incorporated anomalies in primary productivity as forcing factors that affect the shrimp stock as shown by Arreguín-Sánchez (2001). These anomalies in primary productivity are related to environmental factor as sea surface temperature and salinity (Ramírez-Rodríguez et al. 2000, 2003; Ramírez-Rodríguez and Arreguín-Sánchez 2003).

Dynamic Model Simulations

The model provides the starting point for dynamic simulations using Ecosim; this model describes the dynamics of groups in the ecosystem through the equation:

$$\frac{dB_i}{dt} = g_i \sum Q_{ji} - \sum Q_{ij} + I_i - (M0_i + F_i + e_i) \times B_i$$

where dB_i/dt represents the production during the time interval (1 year) in biomass units (B_i), g_i is the net growth efficiency (production/consumption ratio), I_i is the immigration rate, $M0_i$ is the other mortality rate not due to fishing and predation, F_i is the fishing mortality and e_i is the emigration rate. The two summations estimate consumption rates (Q), the first expresses the total consumption by group i, and the second the predation by all predators j feeding on prey i.

Based on Ecosim, we tested the consistency of the input data the model in order to represent observed trends in the fisheries for which we had data. Time series of fishing mortality and relative abundance of pink shrimp and red snapper, both expressed as catch per unit of effort, were used to calibrate the dynamic model simulations. The fitting procedure consisted of minimizing residuals of estimated versus observed trends. Once historical tendencies were reasonably well represented by the model, we considered that model was also reasonably well calibrated to continue with simulations to analyze how the shrimp could be vulnerable to predators and to measure how much predation influences the stock depletion.

Ecosim model allow explore the implications on system dynamics of trophic flows control. Two extreme scenarios are top-down control (where predators regulate abundance of preys) and bottom-up control (resources or preys regulates the population size of predators). To model this aspect of predator–prey interactions, the group biomasses (B) on the underlying Ecopath model were splitted in Ecosim as consisting of two components, one vulnerable (B) and the other invulnerable to predation (V). Further, it is assumed that there is an exchange rate (v_{ij}) of organisms from the invulnerable to the vulnerable stage (vulnerability), and conversely (v_{ji}), with the assumption $v = v_{ij} = v_{ji}$ (Walters et al. 1997, 2000).

Each component consumption rate Q_{ij} of prey type i by pool j is predicted with the "foraging arena" consumption rate as:

$$Q_{ij} = v_{ij} a_{ij} B_i B_j / \left(2v_{ij} + a_{ij} B_j\right)$$

where v_{ij} represents "flow" of prey from behaviorially (or locationally) invulnerable to vulnerable states (which can limit predation rates to be far lower than would be predicted from simpler mass-action models, and create strong "apparent competition" among predators through the denominator term of the equation), a_{ij} represents rate of effective search for predator i on prey type j. The a_{ij} parameters can be calculated from Ecopath diet input data, while the v_{ij} parameter was defined. Low vulnerability means that an increase in predator biomass will not cause any noticeable increase in the predation mortality the predator may cause on the given prey. A high vulnerability indicates that if the predator biomass is for instance doubled, it will cause close to a doubling in the predation mortality it causes for a given prey. We tested for three scenarios representing bottom up (v = 0.1), top down (v = 0.6) and mixed controls (v = 0.3).

Results and Discussion

We use time series analysis to check model consistency and to calibrate subsequent simulations. We use historic fishing mortalities and a recruitment index to model shrimp dynamics, and also a series of observed relative biomass and relative fishing effort of snapper. The model fit to the time series of relative abundance for pink shrimp is shown in Figure 4 and the model for red snapper in Figure 5.

Using an automatic routine included in "Ecopath with Ecosim" software we create an estimated primary production time series in order to get a better fit of observed and predicted values during the fitting process. Then to test whether this series was reasonable we analyzed the output with records of sea surface temperature (SST) and salinity. Sea surface temperature and salinity are physical variables that affect primary production. The correlation between primary production with SST and salinity are shown in Figure 6. The results indicate a negative relationship between temperature and primary productivity, and a positive relationship between salinity and primary production.

Bottom-Up, Top-Down, and Mixed Scenarios

To evaluate the effect of predation, we test the effect of changing shrimp vulnerability to predators. Three scenarios represent possible predation effect on shrimp, bottom up, top down, and a mixed scenario. Additionally, we run simulations including past years with observed data until we complete 100 years in order to explore long term consequences of predation and a potential recovering of the collapsed shrimp stock.

In the bottom up scenario, vulnerability of shrimp was set up to a low value (v = 0.1), implying that the low availability of pink shrimp vulnerable biomass reduce the predation rate on this population. Resulting biomass tendencies of adult and juveniles stocks of shrimp are shown in Figure 7. In this scenario there is no stock collapse, apparently due to the assumption that shrimp are not subject to high predation mortality. In fact, juveniles and adult biomasses could be higher than those observed, and in the long term they stay stable in intermediate levels even with the high fishing mortality used.

In the top down scenario, vulnerability of shrimp was set up to a high value (v = 0.6),

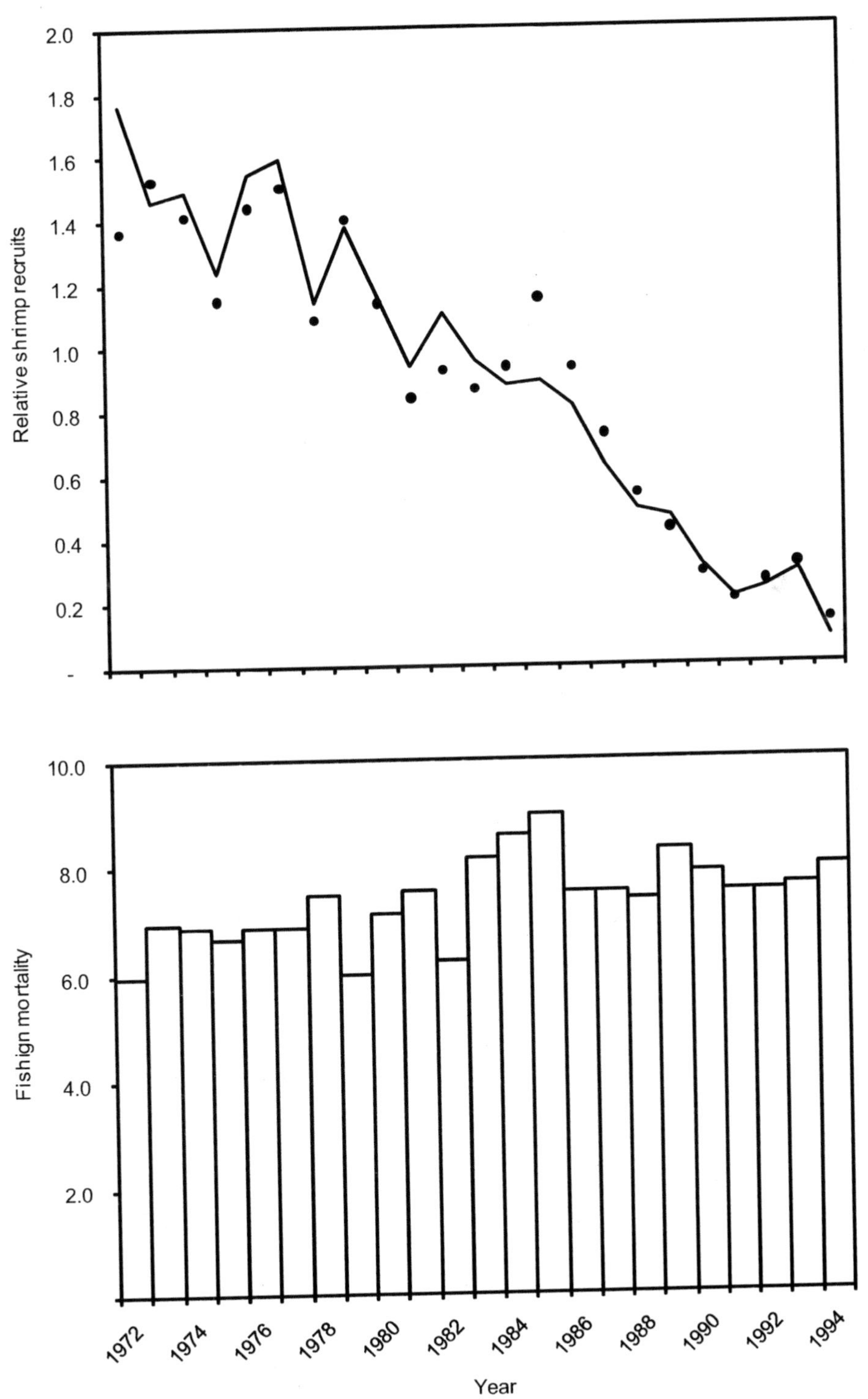

Figure 4. Historic trends in shrimp recruits biomass (top panel) and fishing mortality pattern (bottom panel) used to calibrate the model. Dots represent observed biomass and lines biomass simulated by the model.

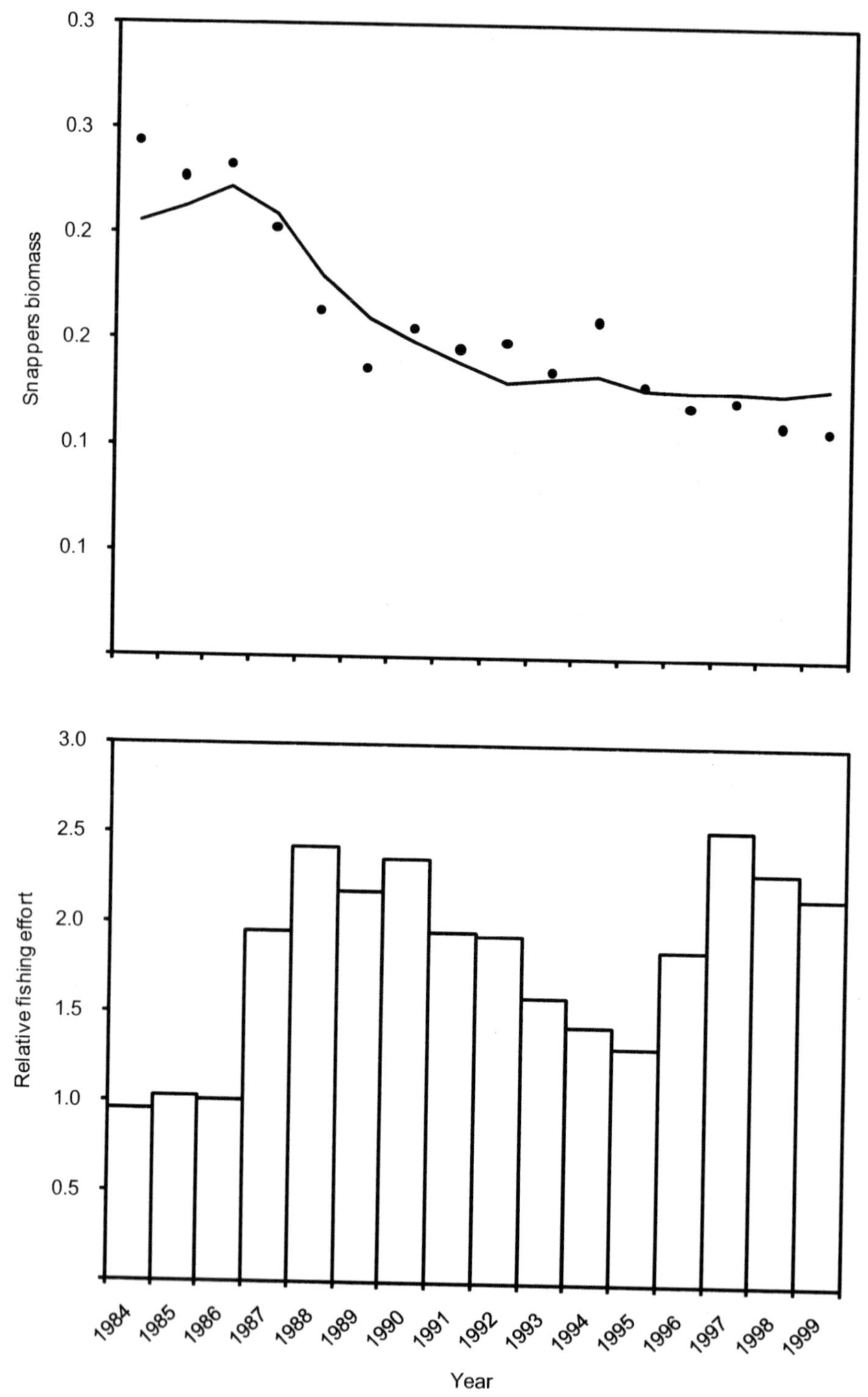

Figure 5. Historic trends of red snapper biomass (top panel) and fishing mortality pattern (bottom panel) used to calibrate the model. Dots represent observed biomass and lines biomass simulated by the model.

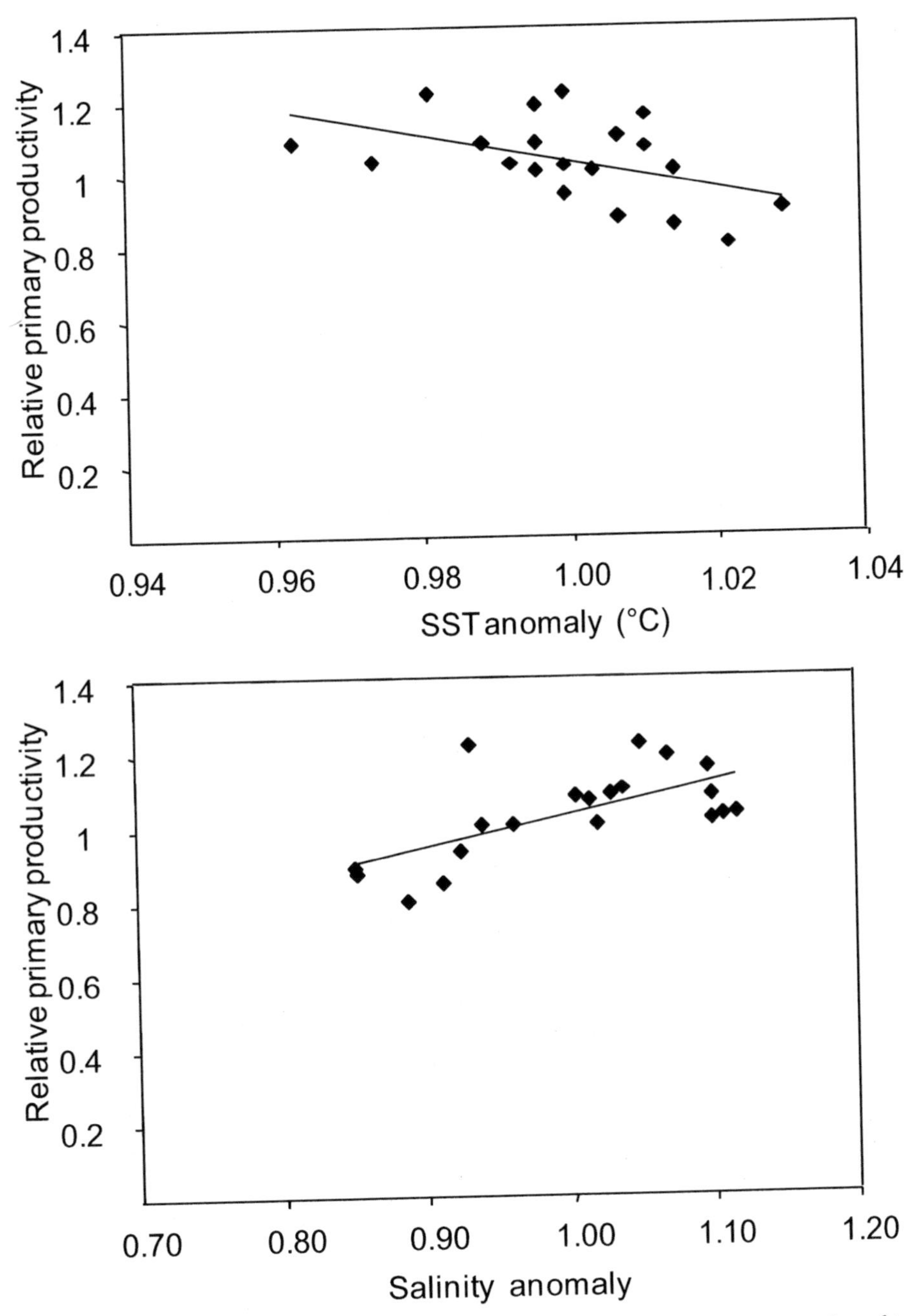

Figure 6. Relationship between primary productivity time series with observed anomalies of temperature and salinity.

implying that the high availability of pink shrimp biomass increase the predation rate upon this population. Resulting shrimp biomass tendencies of adult and juveniles stocks are shown in Figure 8. The collapse in this scenario is abrupt and even faster than the historic observed tendency, maybe due to the assumption that shrimp are highly vulnerable

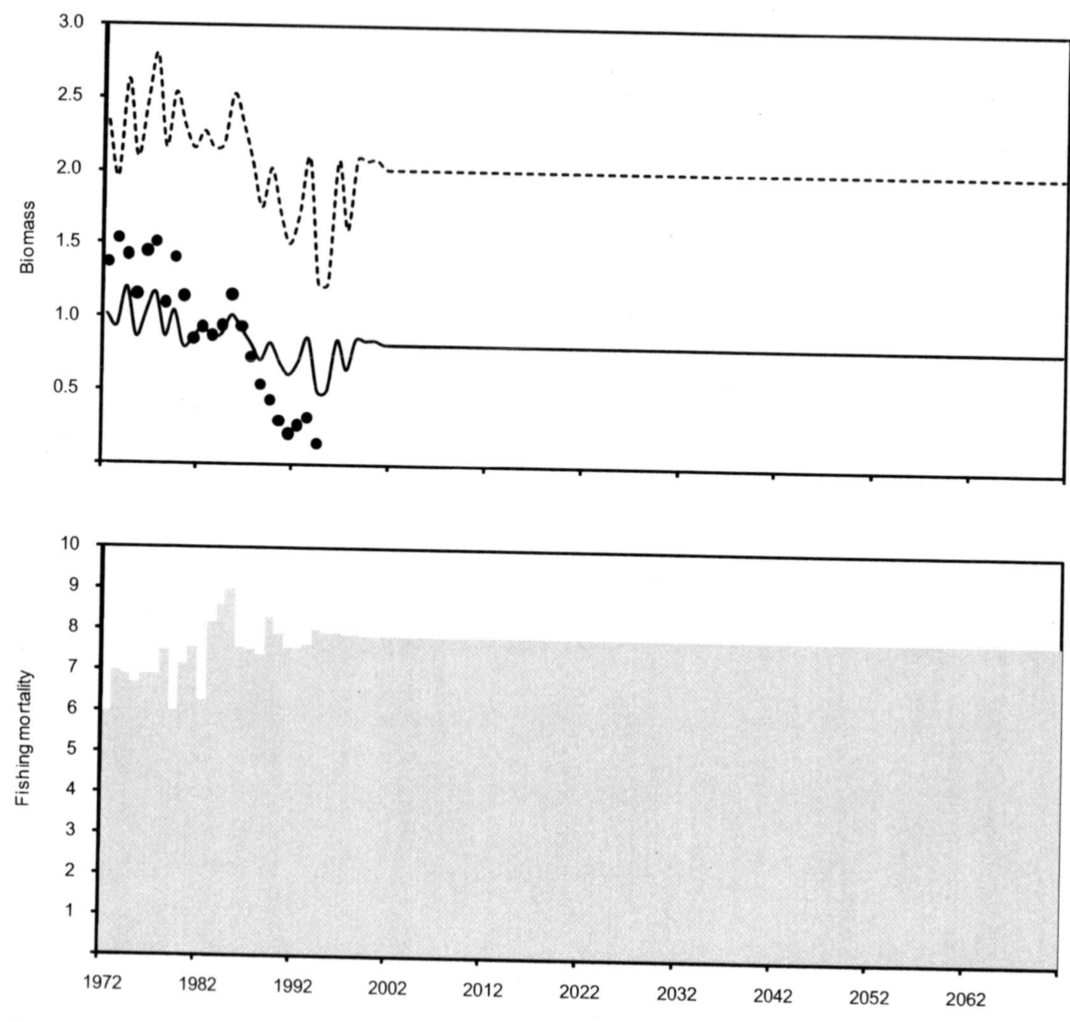

Figure 7. Bottom up scenario using historic fishing mortality from 1972 to 1994 and last 3 years average fishing mortality for the rest of simulation. Solid line is the simulated adult biomass and dash line is the juvenile biomass (dots = observed value and line = simulated values).

to predation mortality. In the long term biomasses never recover, which can also be due to the added fishing mortality effect.

The mixed control scenario, with vulnerability of shrimp set up to an intermediate value (v = 0.3) results in model projections that closely mirror the observed data (Figure 9). The collapse is similar in nature and timing to the stock collapse that has been recorded. In the long term, biomass recovers after about 50 years assuming high fishing mortality.

We can conclude that both natural predation and fishing could have an important role in shrimp stock collapse. An alternative to regulate predation on shrimp is to establish management policies to exploit shrimp's predators, like groupers, snappers and croaker to diminish abundances and potentially diminish predation in shrimp. This management scheme was also suggested by Arreguín-Sánchez et al. (2004) based on the exploration of shrimp stock recovery management strategy by controlling fishing fleets. Our results comes from

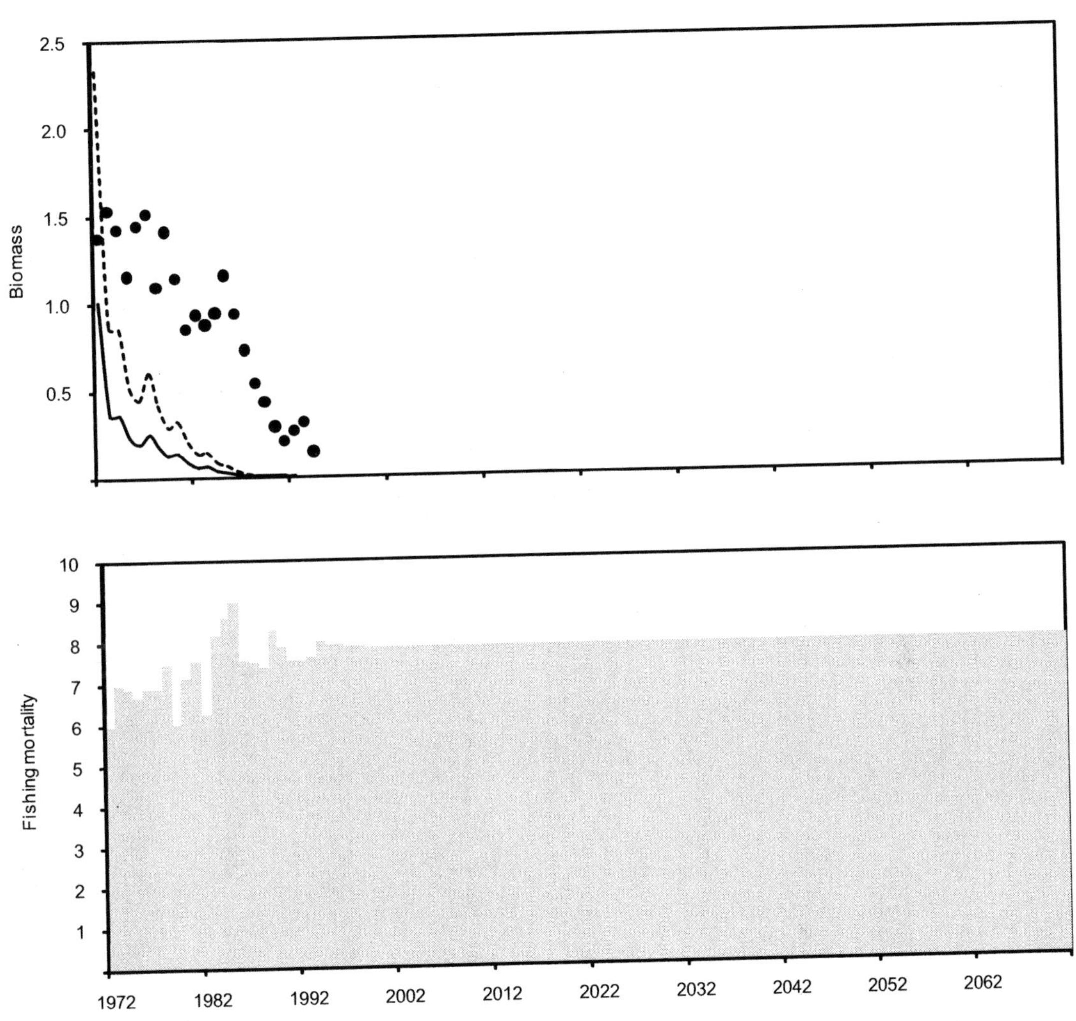

Figure 8. Top down scenario using historic fishing mortality from 1972 to 1994 and last 3 years average fishing mortality for the rest of simulation. Solid line is the simulated adult biomass and dash line is the juvenile biomass (dots = observed value and line = simulated values).

simulated data, however fishermen argue that increase of abundance of some predators could impact shrimp stock, particularly in coastal waters. The next step is to meet and incorporate evidence of predators' roles in the ecosystem and on the processes behind the pink shrimp stock collapse.

Acknowledgments

Author thanks support through projects SIP-20071286, CONACYT-SAGARPA 2003–O–157, and SEMARNAT-CONACyT C01–1231; and grants through COFAA and EDI, both programs of the National Polytechnic Institute.

Reference

Arreguín-Sánchez, F. 2001. Impact of harvesting strategies on fisheries and community structure on the Continental Shelf of the Campeche Sound, southern Gulf of Mexico. Fisheries Centre Reports 10(2):127–135.

Arreguín-Sánchez, F., L. E. Schultz-Ruíz, A. Gracia, J. A.

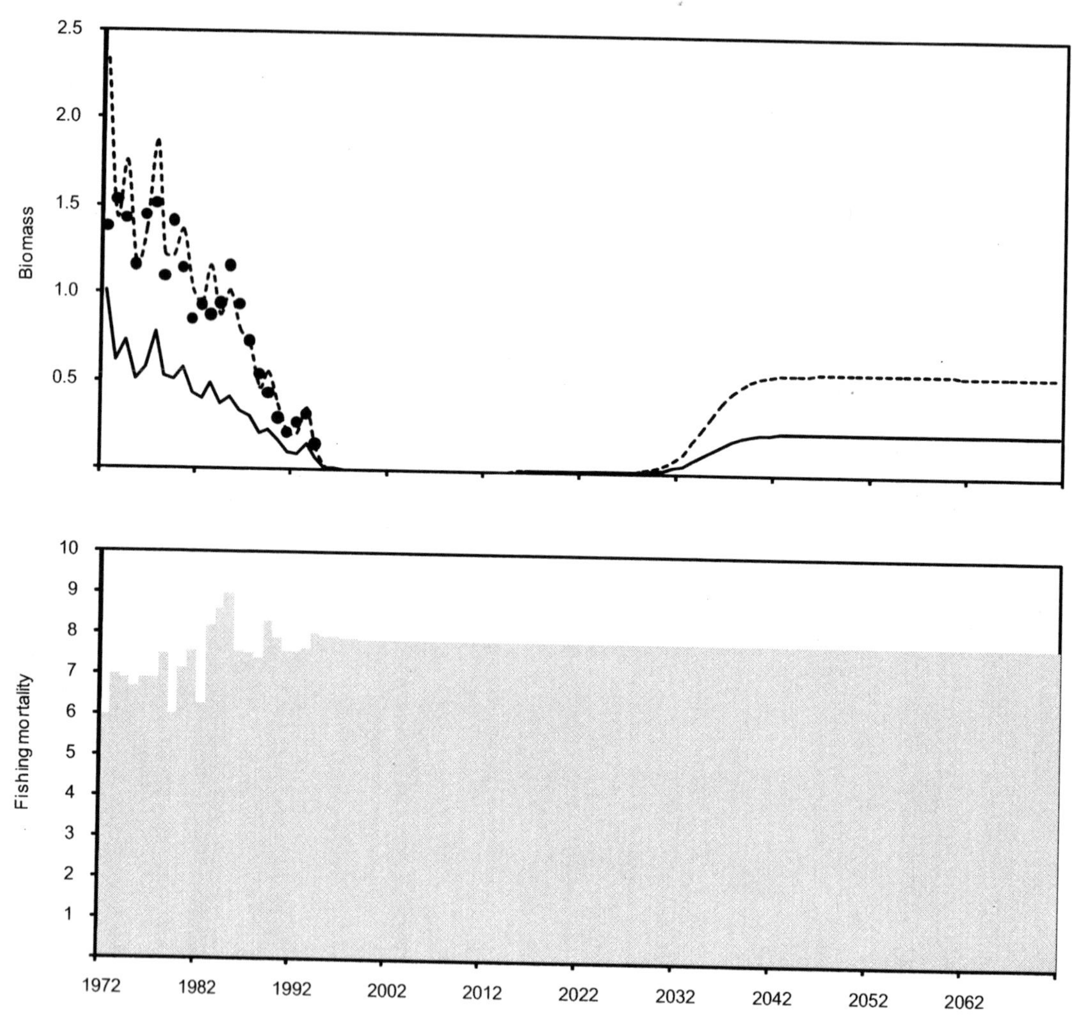

Figure 9. Mixed control scenario using historic fishing mortality from 1972 to 1994 and last 3 years average fishing mortality for the rest of simulation. Solid line is the simulated adult biomass and dash line is the juvenile biomass (dots = observed value and line = simulated values).

Sánchez and T. Alarcón. 1997a. Estado actual y perspectiva de las pesquerías de camarón. Pages 185–203 *in* D. Flores-Hernández, P. Sánchez-Gil, J.C. Seijo, and F. Arreguín-Sánchez, editors. Análisis y diagnóstico de los recursos pesqueros críticos del Golfo de Méxco. EPOMEX Serie Científica 7.

Arreguín-Sánchez, F., L. E. Schultz-Ruíz, A. Gracia, J. A. Sánchez and T. Alarcón. 1997b. Las pesquerías de camarón de altamar, explotación, dinámica y evaluación. Pages 145–172 *in* D. Flores-Hernández, P. Sánchez-Gil, J. C. Seijo and F. Arreguín-Sánchez, editors. Análisis y diagnóstico de los recursos pesqueros críticos del Golfo de Méxco. EPOMEX Serie Científica.

Arreguín-Sánchez, F., J. A. Sánchez, D. Flores-Hernández, J. Ramos-Miranda, P. Sánchez-Gil and A. Yáñez-Arancibia. 1999. Stock-recruitment relationships (SRRs): a scientific challenge to support fisheries management in the Campeche Bank, Mexico. Pages 225–235 *in* H. Kumpf and K. Sherman, editors. The Gulf of Mexico large marine ecosystem. Blackwell Scientific Publications, Oxford, UK.

Arreguín-Sánchez, F., M. J. Solís-Ramírez, and M. E. de la Rosa-González. 2000. Population dynamics and stock assessment for *Octopus maya* (Cephalopopda: Octopodidae) fishery of the Campeche Bank, Gulf of Mexico. Revista de Biología Tropical 48(2/3):323–331.

Arreguín-Sánchez, F., M. J. Zetina-Rejón, S. Manickchand-Heileman, M. Ramírez-Rodríguez, and L. Vidal. 2004. Simulated response to harvesting strategies of an exploited ecosystem on the southwestern Gulf of Mexico. Ecological Modelling 172(2–4):421–432.

Christensen, V., and C. Walters. 2004. Ecopath with Ecosim: methods, capabilities and limitations. Ecological Modelling 172:109–139.

Cochrane, K. 2000. Reconciling sustainability, economic efficiency and equity in fisheries: the one that got away? Fish and Fisheries 1(1):3–21.

Gracia, A., 1997. Pesquería artesanal de camarón. Pages 173–184 *in* D. Flores-Hernández, P. Sánchez-Gil, J. C. Seijo and F. Arreguín-Sánchez, editors. Análisis y diagnóstico de los recursos pesqueros críticos del Golfo de Méxco. EPOMEX Serie Científica 7.

Pauly, D., V. Christensen, and C. Walters. 2000. Ecopath, ecosim and ecospace as tools for evaluating ecosystem impact of fisheries. ICES Journal of Marine Science 57:697–706.

Ramírez-Rodríguez, M., E. A. Chávez, and F. Arreguín-Sánchez. 2000. Perspective of the pink shrimp (*Farfantepenaeus duorarum* Burkenroad) fishery of Campeche Bank, Mexico. Ciencias Marinas 26(1):97–112.

Ramírez-Rodríguez, M., F. Arreguín-Sánchez, and D. Lluch-Belda. 2003. Recruitment patterns of the pink shrimp *Farfantepenaeus duorarum* in the southern Gulf of Mexico. Fisheries Research 65:81–88.

Ramírez-Rodríguez, M., and F. Arreguín-Sánchez. 2003. Spawning stock recruitment relationship of pink shrimp *Farfantepenaeus duorarum* in the southern Gulf of México. Bulletin of Marine Science 72(1):123–133.

Solís-Ramírez, M., F. Arreguín-Sánchez, and J. C. Seijo. 1997. Pesquería de pulpo en la plataforma continental de Yucatán. Pages 61–80 *in* D. Flores-Hernández, P. Sánchez-Gil, J. C. Seijo, and F. Arreguín-Sánchez, editors. Análisis y diagnóstico de los recursos pesqueros críticos del Golfo de Méxco. EPOMEX Serie Científica 7.

Walters, C., V. Christensen, and D. Pauly. 1997. Structuring dynamic models of exploited ecosystems from trophic mass-balance assessments. Reviews in Fish Biology and Fisheries 7(2):139–172.

American Fisheries Society Symposium 49:759–761

Session Summary

Reconciling Fisheries with Conservation and the Ecological Footprint of Aquaculture

JOHN P. VOLPE
University of Victoria, School of Environmental Studies
Victoria, British Columbia, Canada

Almost one-third of the global supply of fish, crustaceans, and mollusks are currently produced by aquaculture. This trend is expected to increase significantly as pressure on wild stocks continues to build. Aquaculture production is currently increasing at an average compounded rate of 9.2%, compared to 1.4% and 2.8% for capture fisheries and terrestrial meat production respectively. Given the substantial and growing profile of aquaculture it is intriguing that this was the only one of the 39 sessions at the Fourth World Fisheries Congress to explicitly consider aquaculture. Perhaps as a result, the six oral presentations making up this session adopted a broad perspective, with presenters emphasizing widely applicable lessons from their particular research. Specific regional, species, and technological issues were featured in the 24 posters associated with this session.

Oral presentations were initiated by session leader John Volpe who suggested that recent lessons from ecosystem ecology and systems theory are applicable to decisions of how aquaculture development should proceed. He noted that a goal of "sustainable development" is to safeguard biological processes and their products, energy transfer, nutrient flow, and so forth. Stability and resilience of such systems can be predicted by the number and strength of interactions among the constituent members of the system. Using hypothetical biological communities, Volpe argued that those communities characterized only by strong interactions are inherently less stable than those characterized by coupled strong and weak interactions; the weak interactions providing functional resilience to the system in the event of the loss of the coupled dominant member. Volpe illustrated the principle on micro and macro scales to underline its usefulness. At a fine scale, the issue of ectoparasitic sea lice in coastal British Columbia and the contentious role of salmon farms in driving the apparent surge in lice abundance were used to illustrate this point (this issue figured prominently in the session's poster presentations). Salmon farms, with upwards of one million salmon per farm, each a potential louse host, swamp the natural parasite–host balance. Formerly weak interactions of lice with an array of spatially diffuse wild hosts are replaced with a singularly strong interaction with farm salmon. Wild host populations are unable to deal with localized lice abundance spikes and can be severely affected. On a macro scale, Volpe commented on the proliferation of offshore aquaculture in U.S. waters that may have similar effects; the replacement of an array of relatively small capture production systems with a handful of major aquaculture species produced in offshore sea cages. By example, Volpe illustrated the collapse of wild salmon fisheries from Oregon to Alaska in the wake of global commoditization of salmon. Stability and resilience of coastal communities lay in maintaining a diversity of economic pathways. In particular, sustainable development of aquaculture will only be possible if development is coupled with, rather than pitted against, extant capture fisheries. Failure, Volpe concluded, will jeopardize long-term viability of fisheries-dependant coastal communities globally.

Shinya Otake (Japan) followed this up with his presentation on the application of an ecosystem modeling approach to evaluate the effects of intensive Pacific oyster culture in southern Korea at Jinhae Bay. This bay represents a natural experiment in that the western half supports intensive farming while farms are absent from the eastern half. Intensive spatial measurement of standard environmental indicators such as dissolved oxygen (DO), nutrients, chemical oxygen demand (COD), sediment depth and composition, phytoplankton abundance, and species diversity were conducted throughout the Jinhae Bay. Also, a particle tracking experiment was conducted to quantify transport and turnover of matter and if rates differed appreciably as a result of bivalve culture. These data led Otake and coauthors to conclude seawater exchange had a minor effect on hydraulic characteristics of the whole Jinhae Bay, which maintained a trophic balance owing to plentiful phytoplankton production within the bay, but the authors noted that increasing eutrophication in the farming area would eventually necessitate intervention to supply oxygen to the seabed in the future if the apparent balance was to be maintained.

Daniel Lane (Canada) then presented a decision-making model for evaluating potential development activities in coastal zones. Lane and his coauthors illustrated the use of their model for informing on aquaculture tenure applications. Such tenure applications can be particularly complex when one must consider the array of biological and social dynamics involved, which are often idiosyncratic to the site in question. Thus one size fits all types of models are likely to fail. Lane et al.'s approach was to use an iterative process of stakeholder involvement to identify key ecosystem components and interactions involving coastal resources and activities. These data are translated into a geographical information systems (GIS) format, which enables spatial overlays of natural processes such as tide and current profiles and commercial activities such as marine transport, fishing, and aquaculture. Stakeholder input, which parameterizes the model, can be weighted by each stakeholder participant, yielding a unique valuation hierarchy. The model then compares the alternative evaluations of the diverse perspectives and provides a framework for a group decision evaluation process to aid, in this case, decisions regarding aquaculture tenure location. Lane concluded by relating the successful application of the model and process for Grand Manan Island in New Brunswick's Bay of Fundy.

Following this, Peter Tyedmers (Canada) impressed the audience with his capacity to ad lib, as the technical folks struggled to overcome a seemingly complete meltdown of the audio/visual equipment. Once dealt with, Tyedmers presented an incisive series of data exploring the energy consumption and global expansion of aquaculture production. The main thrust of the talk was to challenge the blanket perception that aquaculture augments capture fisheries by increasing net protein availability. In fact, the data presented by Tyedmers demonstrated that aquaculture comes in many forms, some far more efficient than others. By combining time series data from 1970 to 2001 of global aquaculture's reliance on net primary production (NPP) together with corresponding data of functional trophic level (FTL) of species cultured, a number of interesting trends emerged. Foremost was that while aquaculture production outside of Asian countries grew at a mean annual rate of 5.6%, the appropriation of NPP grew annually by 11%. Appropriation of natural resources per unit of product is growing, not declining. Tyedmers noted that this is in a sense, the opposite trend to that occurring in capture fisheries where the mean FTL of target species has declined with time; "fishing down/farming up the food web." Additionally, striking regional differences were uncovered in both the FTL of fish destined for reduction (ending up in formulated feeds) and the FTL of aquaculture products. For instance, the northeast Atlantic reduction fisheries target species nearly half a trophic level higher than those targeted in the southeast Pacific. The mean FTL of Asian aquaculture products has remained steady over the past three decades while non-Asian products have climbed steadily, largely driven by explosive growth of intensive salmon culture destined for urban markets in developed countries. These factors combined to paint a rather unflattering picture of the recent expansion of aquaculture in terms of energy efficiency and sustainability. Tyedmers concluded by emphasizing that "aquaculture" encompasses a diverse array of production systems and support should be emphasized for those systems with demonstrable positive efficiencies.

A case history of ecofriendly rice-prawn polyculture in southwest India was presented by Joseph Paimpillil (India). Following harvest, rice paddies are flooded and seeded with young prawns, which feed on the decaying stubble. Prawn excreta fertilize the paddy in anticipation of the next year's rice crop. Particularly striking is that this system uses absolutely no pesticides or fertilizers, making it vastly different from the prevalent prawn farming practice. The prawns only receive external feed during the final phase of growth when protein requirements are particularly high. Therefore, because the prawns require almost no feeding, profit is very high and the polyculture cycles nutrients through the landscape, and maintains marsh lands, promoting both ecological and social stability.

The final presentation was by Theresa Bert (USA), representing her nine other coauthors who collectively explored ecological impacts of aquaculture on a global scale with emphasis on introduced species. Not surprisingly, Bert et al.'s work demonstrated that what is construed as an "impact" is context dependant and types of impacts vary widely across the globe. China, the world's leading aquaculture producer, is dealing with the consequences of many introduced species. The proliferation of dams associated with the growth of hydroelectric power in Korea has provided many grow out options but has resulted in widespread eutrophication of waterbodies. The spread of exotic Pacific oysters in Australian marine waters has severely impacted the once lucrative rock oyster fishery. The spread of feral salmonids through Patagonia is balanced by economic opportunities of the resultant sport fishery. Bert noted that magnitude of impact did not necessarily correlate with the area affected, a common theme in terrestrial invasion literature. Bert concluded by demonstrating mitigation procedures are as varied as the problems they are meant to address, varying from the application of high-tech tools to folk wisdom.

The take home message here, as was reiterated in nearly all oral and poster presentations of this session, was that the many challenges to aquaculture development are not static in time or space; therefore, off-the-rack solutions are unlikely be broadly applicable.

American Fisheries Society Symposium 49:763–777

The Influence of a Bivalve Farming System on the Water Quality in Coastal Waters of Korea

Shinya Otake*, Shingo Hiroshi, Ryuji Kondo, and Takashi Yoshida

Fukui Prefectural University, Department of Marine Bioscience Gakuen-cho, Obama, Fukui 917-0003, Japan

Moon Ock Lee, YangHo Yoon, Hyeon-Seo Cho, and Il Huem Park

Yosu National University, Department of Ocean System Engineering San 96-1, Doondeok-dong, Yeosu, Jeonnam 550-749, Korea

Abstract.—Using an ecosystem model, the influence of a bivalve farming system on water quality has been examined because bivalve shellfish are known to have a function of water purification in terms of filtrating suspended matter in the water when it takes in food. The material budget was calculated, and then, water quality parameters and plankton populations were compared to find out the difference between farming and nonfarming areas. The numerical model used here is the Prinston Ocean model, but it is modified somewhat in order to express moving boundaries due to tides in the coastal zone. Some current data, water quality factors, and biological properties have been observed in Jinhae Bay, one of the places where marine aquaculture has been densely conducted for about 30 years, from 1998 to 1999. The results showed no essential difference in the water quality between farming and nonfarming areas. However, the species and distribution of zooplankton and phytoplankton were remarkably different from each other. For instance, the population and flux of phytoplankton in the farming area appeared to be approximately twice that of those in the nonfarming area. These results have proved that red tides (i.e., Akashio) have not yet occurred in bivalve-farming areas because they are under a good condition of material circulation, and as a result, the farming area creates a sound ecosystem.

Introduction

The southern sea of Korea has a Rias coastline and a lot of aquaculture facilities as, recently, people have greatly demanded marine products. Also, according to abrupt economic development, part of the coast has been reclaimed in order to construct an industrial complex so that urbanization is drastically being undertaken. As a result, the marine environment is becoming polluted, and it has not been unusual to see algal blooms in coastal waters. Therefore, the marine environment, including water quality, urgently needs to be preserved.

Farms refer to an addition of new system to the aquatic biosphere in terms of ecology. Thus, they can be built within a carrying capacity, even though they lead to an increase in the organism loads of the aquatic environment. However, when they exceed a carrying

*Corresponding author: otakes@fpu.ac.jp

capacity, the sea gets eutrophicated. Consequently, the aquatic environment turns anaerobic since oxygen is required for the decomposition of extra organic matter.

This study elucidates the mechanism of material circulation in a semienclosed bay, where the industrialization has progressed but the algal blooms have not occurred in the past few decades, even though there have been many farms. Thus, this study aims to investigate how material circulation takes place and how aquaculture contributes to material circulation in the bay.

Also, this study attempted to replace the farming areas with systems of material circulation in the sea, reproducing precisely a physical environment. Then, applying an ecosystem model to the study area, environmental characteristics in a farming area were examined.

Materials and Methods

Study Site

The study area is located in the southwest portion of Jinhae Bay in Korea, where not only industrial development but also shellfish farming are vigorously carried out (see Figure 1). Jinhae Bay is a semienclosed bay with a few small inlets and has a dimension of approximately 40 km east westward and 40 km south northward. The mouth of the bay is located at the southeast toward the Tsushima Strait. The bay also has a very narrow channel, Gyeonnaeryang, having a seawater exchange

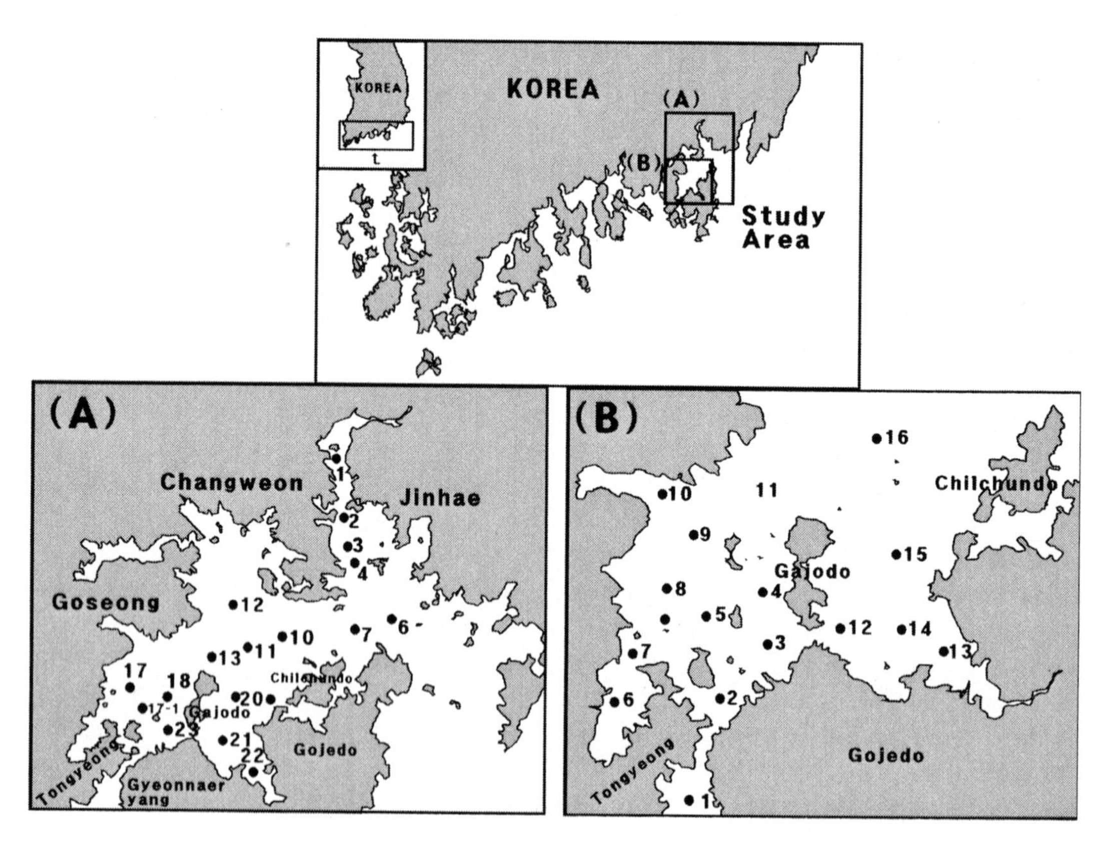

Figure 1. Study area and sampling stations of Jinahe Bay conducted in 1998 (A) and 1999 (B).

with the outside at the southwest. The north region of the bay from this channel is an area where oyster farming is densely carried out.

This study examined a function of farms through the comparison of fishing environment between a farming area (western area) and a nonfarming area (eastern area), divided by the island of Gajodo in the southwest region of Jinhae Bay, as indicated in Figure 1B. Then, a material circulation has been evaluated in both farming and nonfarming areas, based on such factors as the mechanism of seawater exchange, water quality, seabed conditions, and population of phytoplankton.

Model of Seawater Exchange

A three-dimensional numerical model, named Prinston Ocean model (POM), has been used in order to elucidate seawater behavior in the bay. The grid nets for computational domain consist of 150 × 164 with a rectangular shape of $x = y = 200$ m in a horizontal direction and five levels in a vertical direction. Tidal records, which were observed at Nampo (35°03.897'N'128°36.840'E) from 1 December 2000 to 10 January 2001 and at Gyeonnaeryang(34°52.952'N,128°28.612'E) from 30 November 2000 to 10 January 2001 were employed for open boundary conditions.

Stations of Field Survey and Analysis of Water Quality and Sediments

A field survey has been conducted three times (23 June, 12 August, and 23 November 1999) in order to grasp the environment of water quality and seabed in 16 stations of the west of Jinhae Bay, as already shown in Figure 1. The seawater was taken from the surface, intermediate (5 m below the surface), and bottom layers (1 m from seabed), and sediments were taken from the same place as the seawater sampling. Then, all sampling data were chemically analyzed in the laboratory. In particular, at stations 2, 11, and 15, the seawater of a full day of 25 h was sampled every hour at three layers in order to see diurnal variations of water quality. A gum bucket and Van Dorn water sampler for water sampling and also a gravity type of corer for sediments were adopted, respectively.

All the data were analyzed in accordance with standard instructions suggested in general test processes. The analytical items were pH, suspended solids, dissolved oxygen (DO), chemical oxygen demand (COD), total organic carbon (TOC), NH_4^+-N, NO_2^--N, NO_3^--N, PO_4^{3-}-P and $Si(OH)_4$-Si in case of water quality and IL, COD, H_2S and TOC in case of sediments.

Phytoplankton

The population of phytoplankton has been investigated four times (August 1998; June 1999, August 1999 and November 1999) using a training ship of Yosu National University at stations, as shown in Figure 1. The sampling was carried out by a Van Dorn water sampler to take water from surface to seabed with an interval of 5 m, but 1 m near the bottom. The species of plankton was identified, and number of appearance cells was calculated using an optical microscope. Thus, a standing crop of phytoplankton was expressed in terms of the number of appearance cells per unit volume. The appearance species of phytoplankton was classified by methods proposed by Parke and Dixon (1976), Hartley (1986), and Thomas(1997). Biomass of phytoplankton was measured in terms of chlorophyll *a* using a submersible fluorometer (Alec Electronic Co., Ltd, ACL 1151-D). These data were then calibrated using a spectrophotometry, which extracts a chlorophyll for the exact measured value (SCOR-UNESCO 1966).

Finally, based on this observed data, a principal component analysis (PCA) was conducted in order to consider the characteristics of the marine environment (Yoon and Park 2000). The analysis was carried out with 70% principal component, as the limit of the analysis has cumulative contribution rates.

Ecosystem Model

As the constituent elements of ecosystem, this model contains four organic elements, phytoplankton (*P*), zooplankton (*Z*), particulate organic matter (detritus, POM), and dissolved organic matter (DOM); two inorganic elements, dissolved inorganic phosphorus (DIP) and total dissolved inorganic nitrogen (DIN); and two water quality elements, DO and COD. The seabed system, such as benthic animals or sediments, is also involved in the model as a variable of environmental elements. These processes can be expressed by equations to describe the rate that a standing crop of constituent elements, that is, B, varies with time at an arbitrary point. This model is connected with a multilayered model that calculates the material transport due to the flows in the bay. That is,

$$\frac{\partial B}{\partial t} = \text{advection + diffusion +}$$
$$\text{biological or chemical variation}$$
$$= -u\frac{\partial B}{\partial x} - v\frac{\partial B}{\partial y} - w\frac{\partial B}{\partial z} + \frac{\partial}{\partial x}\left[K_x\frac{\partial B}{\partial x}\right]$$
$$+\frac{\partial}{\partial y}\left[K_y\frac{\partial B}{\partial y}\right] + \frac{\partial}{\partial z}\left[K_z\frac{\partial}{\partial z}\right] + \frac{\partial B}{\partial t}$$

where B = a standing crop of constituent elements; t = time; u, v, w =velocity components in x, y, z directions, K_x, K_y, K_z = eddy diffusivities in x, y, z directions, $\partial B/\partial t$= variations of constituent elements per unit time due to a total biomass and chemical process.

The variations of concentration of phytoplankton (*P*) and nutrients (DIN, DIP) with time are can be expressed as follows:

1) Phytoplankton (*P*)

As the constituent elements of phytoplankton, $P(\text{mgC}/m^3)$, we consider a single species of population as the average for dominant species in study area.

dP/dt = propagation due to photosynthesis-secretion outside the cells–respiration–uptake (or decay and sinking by zooplankton)

$$\left(\frac{dP}{dt}\right) = \left[1-\mu_3(P)\right]V_1(T)\mu_1(\text{DIP, DIN})$$
$$\bullet\mu_2(I)P - V_2(T)P$$
$$-V_3(T)Z - V_4(T)P - W_p\frac{\partial P}{\partial z}$$

2) Nutrients (DIP, DIN)

$$\frac{d}{dt}(DIP,DIN) = -\left[P:C_p\right]\bullet V_1(T)\bullet\mu_1(DIP,DIN)\bullet\mu_2(I)\bullet P$$
$$-\left[P:C_{POM}\right]\frac{1}{1+k}V_6(T)\bullet(POC)+\left[P:C_p\right]\bullet V_2(T)\bullet P$$
$$+\left[P:C_z\right](\mu-v)\bullet V_3(T)\bullet Z+\left[P:C_{POM}\right]\bullet(DOC)+q_p$$

$$\frac{d}{dt}(DIP,DIN) = -\text{uptake by phytoplankton+}$$
$$\text{respiration by plankton+excretion of zooplankton+}$$
$$\text{decomposition of suspended organic matter+}$$
$$\text{inorganization of dissolved organic matter+}$$
$$\text{supply from outside the system}$$

3) COD

$$\left(\frac{dCOD}{dt}\right) = \left[COD:C_p\right]\left(\frac{dP}{dt}\right) + \left[COD:C_z\right]\left(\frac{dZ}{dt}\right)^*$$
$$+\left[COD:C_{POM}\right]\left(\frac{dPOC}{dt}\right)^* + \left[COD:C_{DOM}\right]\left(\frac{dPOC}{dt}\right)^*$$

These equations were formed as a simulation model by Kim (1994) and modified for applying in the southwest water of Jinhae Bay. All the coefficients used in these equations are indicated in Table 1.

Table 1. Coefficients used in ecosystem model.

Symbol	Definition	Unit	Value	References
α_1	maximum growth rate of phytoplankton at 0°C	d^{-1}	1.2	Kim (1994), Baca and Arnett (1976), Choi (1993), Eppley (1972), Thomann et al. (1975)
β_1	Temperature coefficient of α_1	$°C^{-1}$	0.0633	
α_2	Respiration rate of phytoplankton at 0°C	d^{-1}	0.01	Kim (1994), Choi (1993)
β_2	Temperature coefficient	$°C^{-1}$	0.0524	
α_3	Maximum grazing rate of zooplankton at 0°C	d^{-1}	0.2	Kim (1994), Choi (1993)
β_3	Temperature coefficient of α_3	$°C^{-1}$	0.0693	
α_4	Death rate of phytoplankton at 0°C	d^{-1}	0.03	Kim (1994), Choi (1993), Richey (1977), Scavia et al. (1976), Thomann et al. (1975)
β_4	Temperature coefficient of α_4	$°C^{-1}$	0.0693	
α_5	Natural death rate of zooplankton at 0°C	d^{-1}	0.05	Kim (1994), Choi (1993), O'Connor et al. (1973), Richey (1977), Scavia et al.(1976)
β_5	Temperature coefficient of α_5	$°C^{-1}$	0.0693	
α_6	Mineralization rate of particulate organic carbon at 0°C	d^{-1}	0.02	Kim (1994), Choi (1993)
β_6	Temperature coefficient of α_6	$°C^{-1}$	0.07	
α_7	Mineralization rate of dissolved organic carbon at 0°C	d^{-1}	0.008	Kim (1994), Choi (1993)
β_7	Temperature coefficient of α_7	$°C^{-1}$	0.0693	
K_{SP}	Half saturation constant for uptake of PO_4^{3-}-P at 0°C	µg-at/L	0.06	Kim (1994), Choi (1993)
K_{SN}	Half saturation constant for uptake of DIN at 0°C	µg-at/L	0.6	Kim (1994), Choi (1993)
k_o	Dissipation coefficient of light independent of Chlorophyll a	m^{-1}	0.1	Kim (1994), Choi (1993)
v	Constant of dissipation coefficient depending on Chlorophyll a	m^{-1}(mg Chl.a/ $m^3)^{-1}$	0.0179	Kim (1994), Choi (1993)
W_P	Settling velocity of phytoplankton	m/d	0.2	Kim(1994), Choi(1993)
W_{POC}	Settling velocity of detritus (POC)	m/d	1.1	Kim (1994), Choi (1993)

Results and Discussions

Exchange of Seawaer

In the past, it has been not clear whether there is a southward residual flow in Gyeonnaeryang, but a southward flow exists in the southwest channel, Gyeonnaeryang, by POM. Here, in order to transform into a box model, the calculated results of a layer model were recalculated using a level model of five layers. The results using a level model compared favorably with the results using a layer model.

In addition, using the Marker-and-Cell method, a hundred particles were released at a cross section of Budosudo (i.e., east end of the bay) every hour for each grid of 25 and then the paths of the particles were pursued for 7 h, as indicated in Figure 2. The particles went forward 6 km at most inside the bay during 7 h. This suggests that incoming waters through the east channel, Gaduksudo, do not influence the whole flow as much as expected in a seawater exchange of the bay. Similarly, 100 particles were also released in the southwest channel, Gyeonnaeryang, every hour for each grid, and

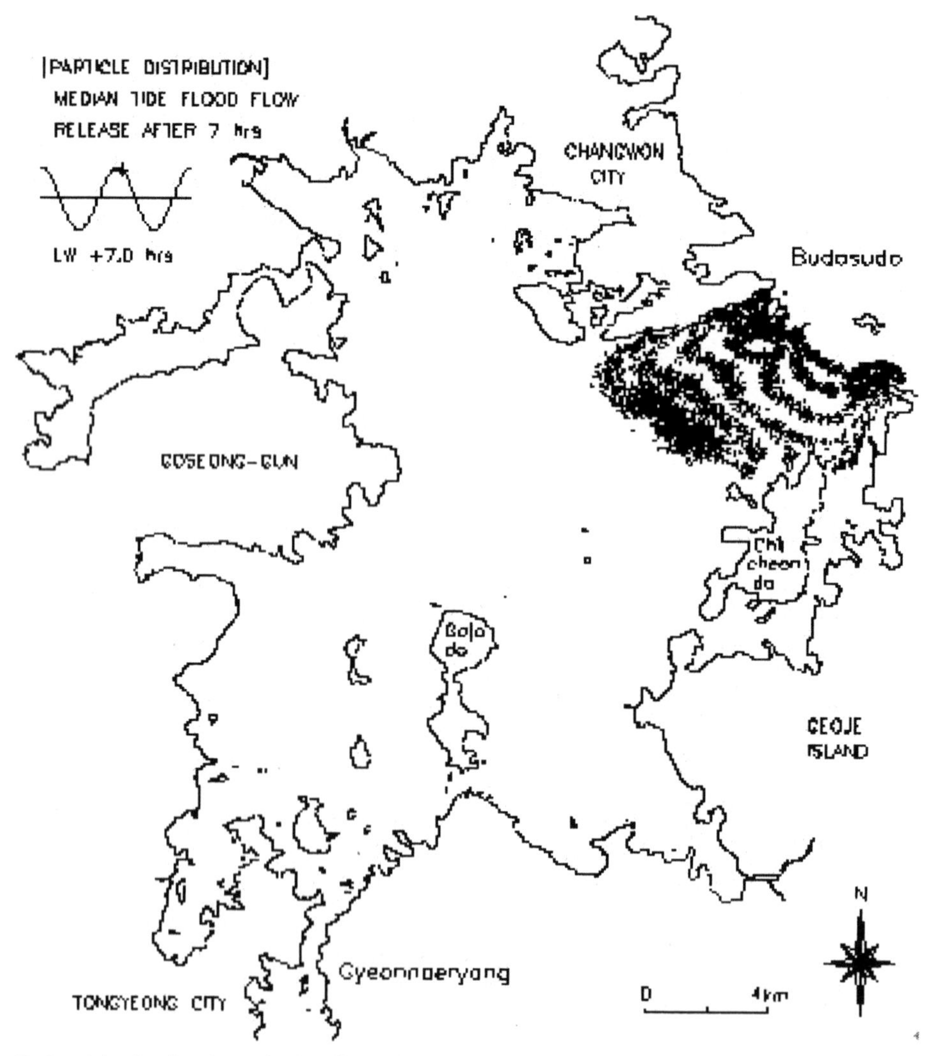

Figure 2. Particle distributions during flood flow, 7 h after being released.

then the paths of the particles were pursued for 7 h The particles went forward 2 km inside the bay during 7 h because the flow is weak compared to Gaduk-sudo. This proved that a seawater communication of Gyeonnaeryang, flowing in and out of the bay does not greatly affect the environment of oyster farms.

On the other hand, Figure 3 shows the results of distorted grids of particles 3 h after 100 particles were initially put on each side of the grids with an interval of 2 km in x and y directions and then released. The seawater started coming in or going out through the east channel, Gaduksudo, approximately 3 h after high water or low water, and reached the island of Gajodo. However, the grids of particles were not disturbed very much except for Gaduksudo, so that the seawater behavior in Jinhae Bay looked inactive as a whole.

Environment of Water Quality

Dissolved oxygen is one of the most important factors for estimating the amount of cultured fish or an environmental indicator of farm wa-

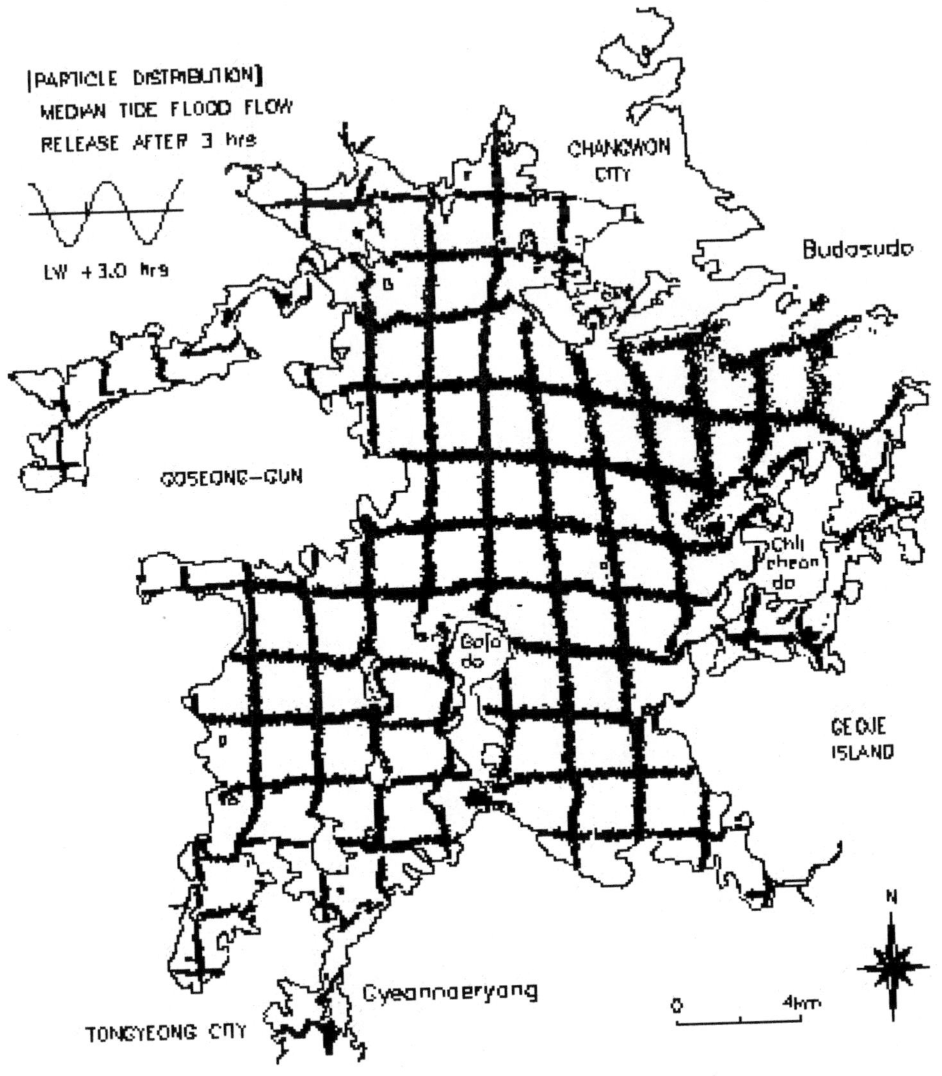

Figure 3. Particle distributions during flood flow 3 h after being released.

ter (Nakamura 1991). Figure 4 represents horizontal distributions of DO concentration at the bottom layers in August. Concentrations of DO during the period of observation were 7.32–13.14 mg/L at the surface and 0.09–8.30 mg/L at the bottom layers, respectively. However, the bottom layer of the north of the bay already started becoming anoxic at less than 1.0 mg/L by June so that finally anoxia was found at nearly all of the bottom layers in August. Nonetheless, in November, concentrations of DO were discovered to be 4–6 mg/L in many regions. In particular, concentrations of DO for the farming area tended to be a little higher at the surface layer while a little lower at the bottom layer, compared with the nonfarming area. This might occur because of a weak seawater exchange, stratification from spring to autumn, or oxygen consumption by the decomposition of organic matter.

On the contrary, Figure 5 shows horizontal distributions of Redfield ratio (=N/P), obtained from field surveys at the bottom layers in August. The Redfield ratio appeared to be less than 16 in most regions, even though it was a little higher in nonfarm-ing areas. However, it turned out to be approximately 1.0 in June in all the layers, indicating that nitrogen is a limiting factor for the growth of phytoplankton in the study area.

In June and August, the values of COD appeared to be larger in farming areas than in nonfarming areas. These values largely exceeded 20 mg/g-dry, which is a standard of sediments for fishery in Japan. On the other hand, concentrations of hydrogen sulfide were 2.07 mg/g-dry in June, 1.62 mg/g-dry in August, and 1.41 mg/g-dry in November so that they tended to gradually decrease with

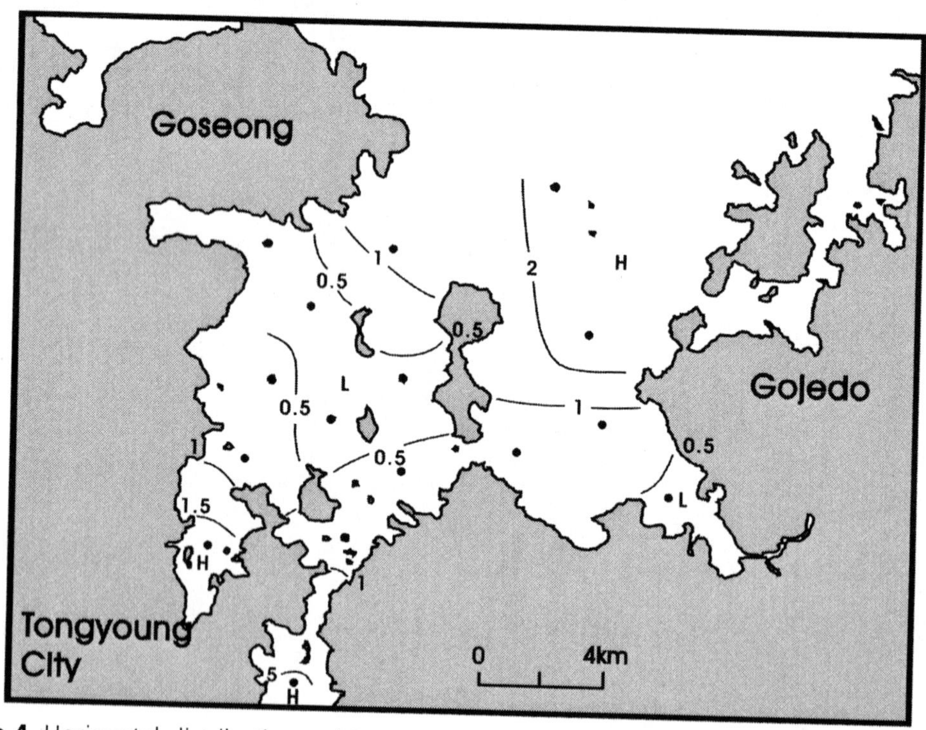

Figure 4. Horizontal distributions of DO concentration at the bottom layers in August. The farming area is the west of Gajodo island, which is center of the figure; the nonfarming area is the east.

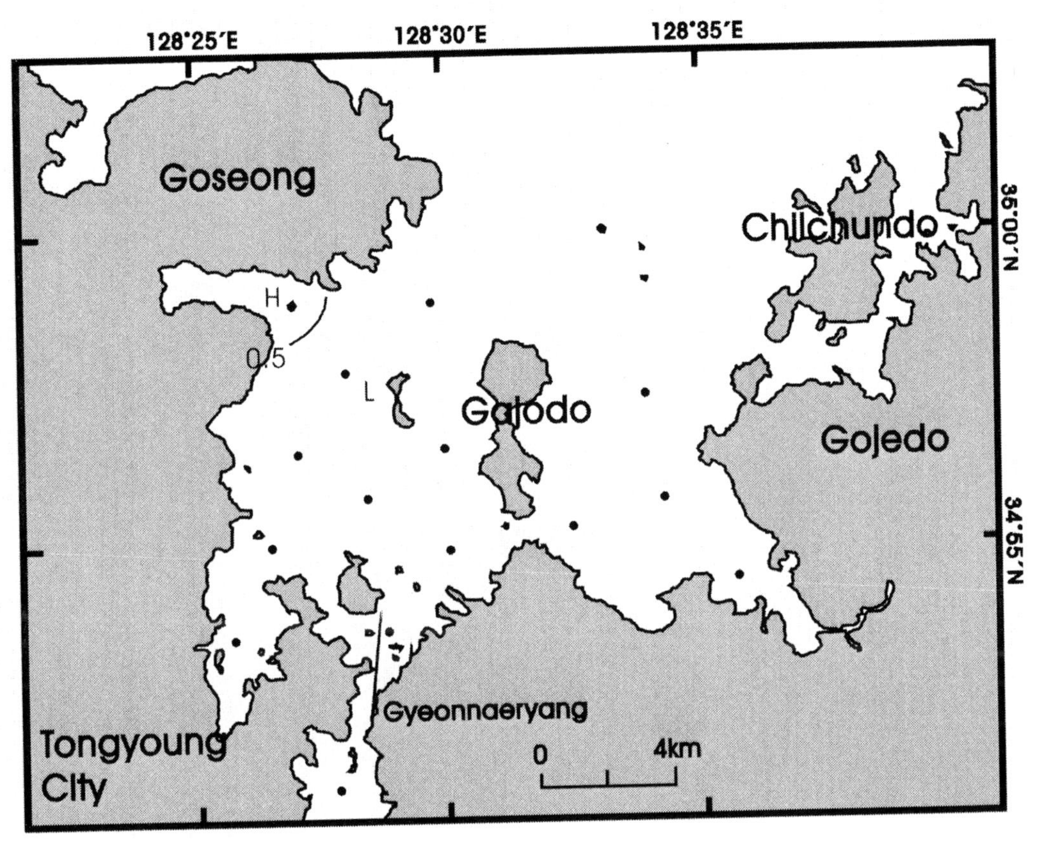

Figure 5. Horizontal distributions of Redfield ratio (=N/P) at the bottom layers in August.

time. These values also surpassed 0.2 mg/g-dry, which is a standard of sediments for fishery in Japan.

Therefore, both farming and nonfarming areas seemed inappropriate environments for benthic animals to live in because of organic contamination (Kang 1995).

Distribution of Plankton

Phytoplankton appeared to consist of 47 total genera and 94 species during the period of observation. Diatoms were the most dominant species with an appearance rate of 56%, and then, dinoflagellates held an appearance rate of 37%. In particular, dinoflagellates seemed to appear dominantly in this area when water temperature is high, compared with other coastal waters of Korea (Yoon 1995, 2000). However, *Pseudonitzschia multiseris*, which is a species known to cause Amnesic Shellfish Poisoning, was found when water temperature is low, so that the possibility of shellfish poisoning cannot be excluded (Lee and Baek 1997).

A PCA has been conducted in order to elucidate the environmental characteristics of sea, based on distributions of phytoplankton.

Accordingly, the west sea of Jinhae Bay can be largely divided into three areas, using the scores that the third principal component acquired at each station. The results are illustrated in Figure 6 and can be also summarized as follows:

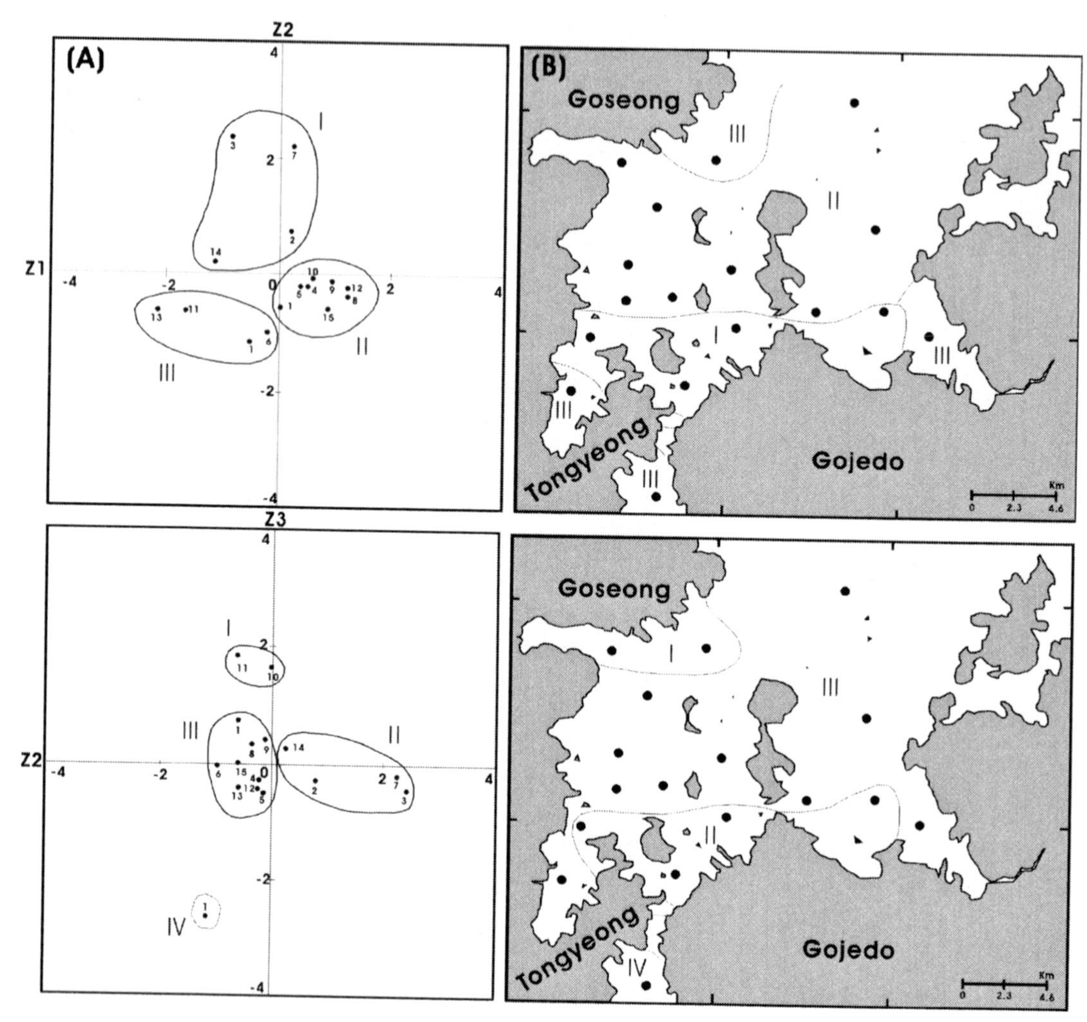

Figure 6. Distributions of scores by the PCA (left) and separated zones (right) in June 1999.

(1) Zone I has a high activity of primary organisms and inflow of freshwater from land.

(2) Zone II has a high activity of primary organisms by self-supply of nutrients due to material circulation.

(3) Zone III has a low activity of primary organisms and a high salinity.

In a word, the western area of Jinhae Bay keeps a high production of marine organisms, so that it is well used as a production area for various fishery resources such as shellfish. However, due to not only a long-term utilization of fishing grounds, but also a partially treated sewage from land, the sea could not avoid being euptrohicated. A high production of primary organisms here seems to be supported mainly by nitrogen, but the sources of supply are different from place to place. Most of them may result from internal material circulation in the area. However, phosphorus is different from nitrogen, so that it may come from land, rather than internal material circulation. These three distributed zones are actually different from the initial zoning that is divided into two areas (i.e., faming and nonfarming areas).

Thus, these results proved that both farming and nonfarming areas have the same environment as Zone II. However, this statistical treatment ignores the exchange of water and flux among environmental factors.

Accordingly, the ecosystem model should calculate the flux of environmental factors.

Environmental Assessment by Ecosystem Model

Figure 7 indicates physical processes when biomass of phytoplankton is calculated within the ecosystem model. A dark-colored area denotes an area where input is larger than output so that a pollutant is accumulating, while a light-colored area denotes an area where output is larger than input so that a dispersion occurs actively toward other regions. Consequently, this region shows that the accumulation is a little larger than the dispersion. On the contrary, Figure 8 indicates biological processes when biomass of phytoplankton is calculated within the ecosystem model. This biomass means a quantity that subtracts an uptake by zooplankton from a net primary production of phy-

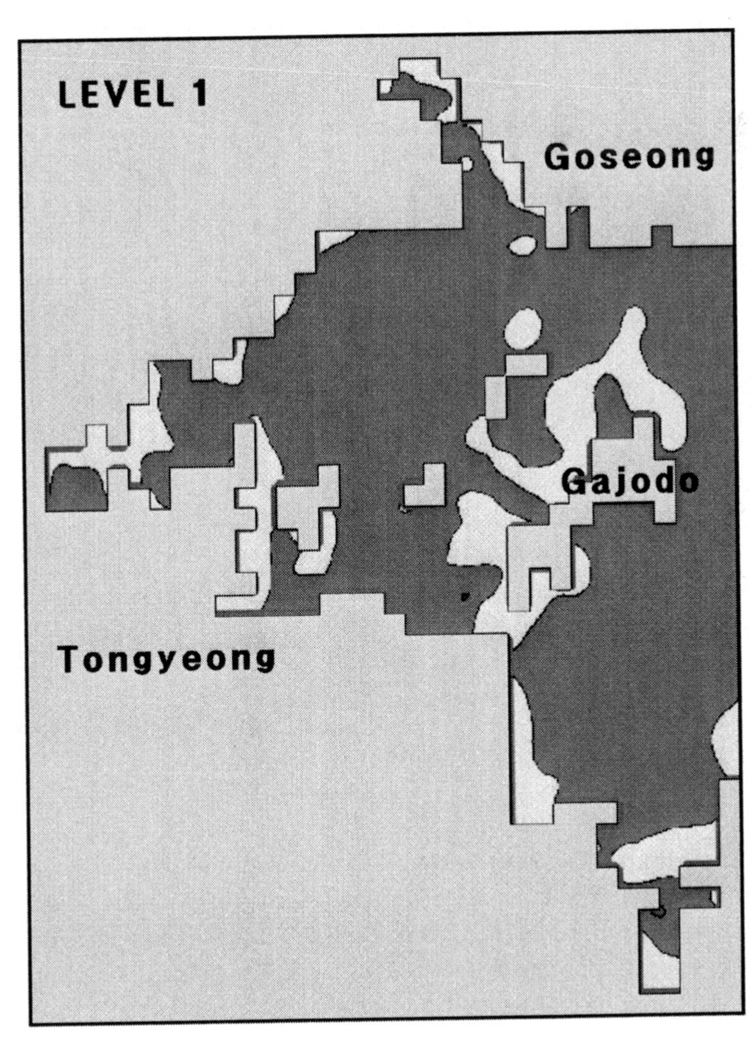

Figure 7. Physical processes for calculations of phytoplankton in the ecosystem model.

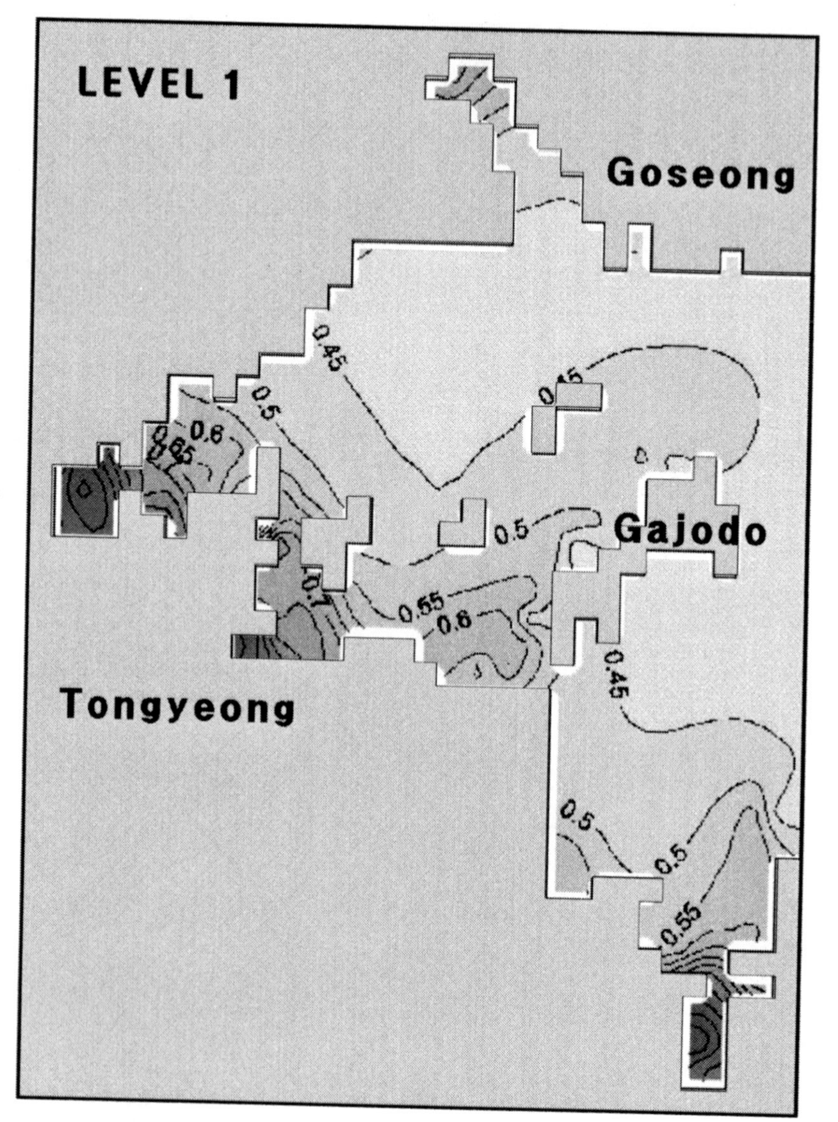

Figure 8. Biological processes for calculations of phytoplankton in the ecosystem model.

toplankton. Therefore, it can be a criterion of primary production or food supply of farm animals for each area.

On the other hand, Figure 9 shows relationships of material budget for farming and nonfarming areas at the surface layer. All the items for farming area A had approximately twice the items of the nonfarming area. Estimating the supply of food based on these results, it appeared to be 0.55 $g/m^2/d$ for area A and 0.52 $g/m^2/d$ for area B, respectively. Thus, applying a mean value of 0.8 mg C/g/d for food demand per wet tissue, which was obtained from some other areas, to this area, estimated carrying capacities for these two areas are summarized, as indicated in Table 2. Carrying capacities turned out to be 6.88 metric tons C/ha for farming areas and 6.50 metric tons C/ha for nonfarming areas, respectively. Accordingly, a higher production of oyster is expected in the farming

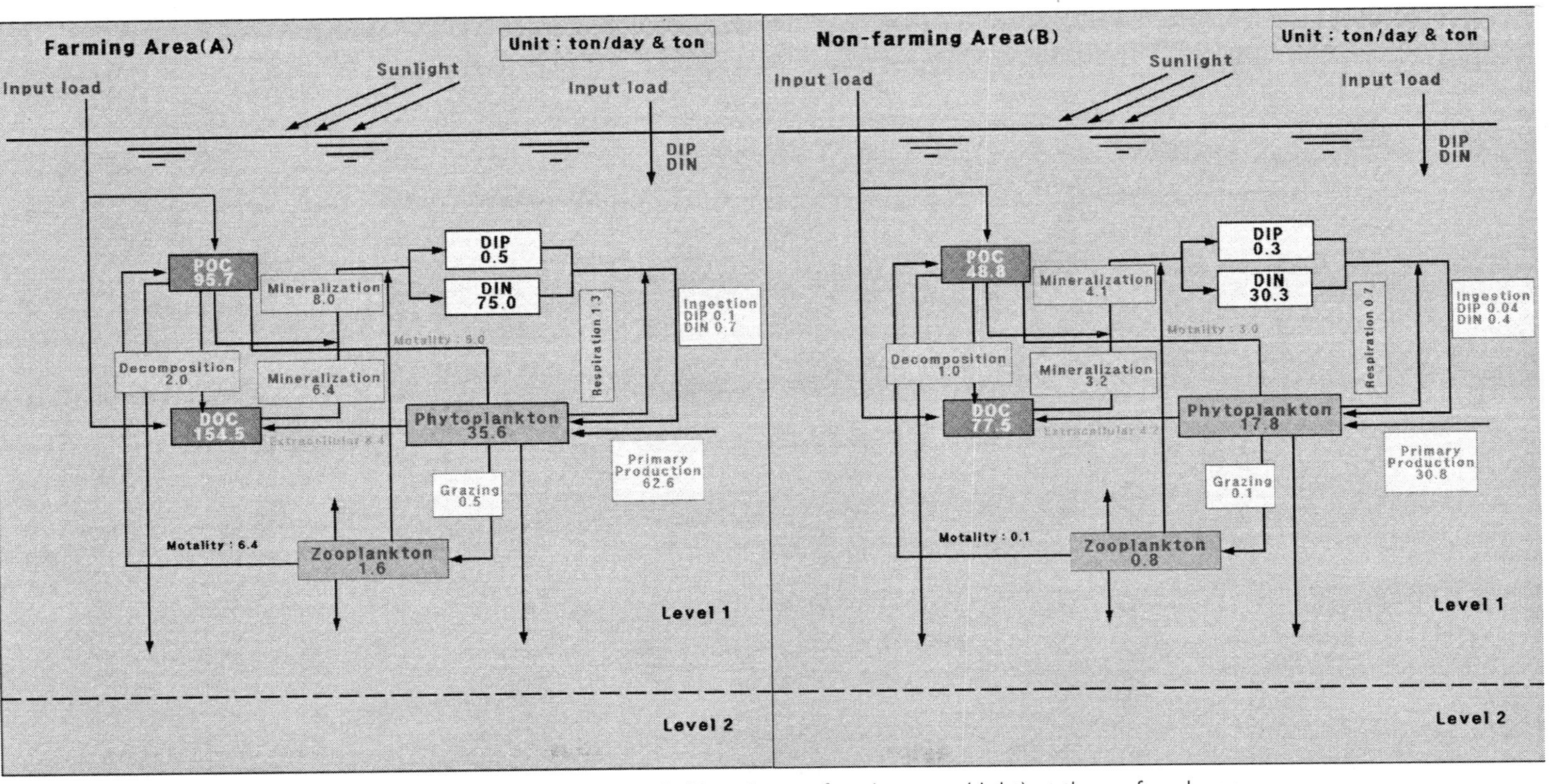

Figure 9. Relationships of material budget for a farming area (left) and a nonfarming area (right) at the surface layer.

Table 2. Carrying capacities calculated from material budget between food supply and its demand.

Area	Food supply	Food demand per wet tissue	Carrying capacity
Faming area	0.55	0.8	6.88
Nonfarming area	0.52	0.8	6.50

area than in the nonfarming area, even though the carrying capacity is not too much different from each other.

Conclusions

In the southwestern waters of Jinhae Bay of Korea, a comprehensive field survey and numerical simulation have been conducted in order to gather information on the fisheries environment around oyster farms. Then, based on these physical, chemical, and biological data, an ecosystem model was employed to analyze environmental differences in farming and nonfarming areas. The major results are summarized as follows:

(1) The flow field of the west of Jinhae Bay produced by the POM coincided well with the observed results. As a result, it proved that a seawater exchange through the west channel had a minor effect on hydraulic characteristics of the whole Jinhae Bay, but this has been not clear before.

(2) A farming area turned out to be more eutrophicated than a nonfarming area, according to field surveys of water quality conducted during from 1998 to 1999. Also, most of the seabed appeared to be anoxic, particularly in summer while the surface layers were saturated with oxygen.

(3) Using a PCA, the relationship of water quality and phytoplankton was analyzed so that the difference between farming and nonfarming areas turned out to be insignificant.

(4) According to the predicted results by ecosystem model for a standing crop of phytoplankton, the supply of the feed to the farming area was 5% larger than to the nonfarming area.

(5) Environmental carrying capacity in terms of the supply of the feed was 5% larger in the farming area than in the nonfarming area, assuming that a feed efficiency of oyster is constant. However, this area maintains a trophic balance owing to plentiful phytoplankton, so that some devices are required to supply oxygen fully to seabed in the future.

Acknowledgments

We would like to greatly express our gratitude to Professor Makoto Nakamura and Professor Yukio Uekita at Fukui Prefectural University for their help and good advice. We also thank the crews of the student training ship *Donbaek Ho* of Yosu National University for their support during our filed survey. This study has been partially supported by the Fukui Prefecture University Academic Funds.

References

Hartley, B. 1986. Check-list of the freshwater, blackish and marine diatoms of the British Isles and adjoining coastal waters. Journal of the Marine Biological Association of the United Kingdom 66:531–610.

Kang, S. H. 1995. Impact of enriched toxic organic pollutant and pollution stresses on shellfish and gasropoda inhabited in Jinhae Bay. Doctoral dissertation. Seoul National University, Seoul, Korea.

Kim, J. G. 1994. The eutrophication modeling for Jinhae

Bay in summer. Doctoral dissertation. Pukyong National University, Busan, Korea.

Lee, J. H., and J. H . Baek. 1997. Neurotoxin-producing *Pseudonitzschia multiseris* (Hasle) Hasle in the coastal waters of southern Korea II. Production of domoic acid. Algae (The Korean Journal of Phycology) 12:31–38, Seoul, Korea.

Nakamura, M. 1991. Fisheries civil engineering. Kogyozizitsushinsha Publishing Co., Tokyo.

Parke, M., and P. S. Dixon. 1976. Check-list of British marine algae-third revision. Journal of Marine Biological Association of the United Kingdom 56:527–594.

Redfield, A. C., B. H. Ketchum, and F. A. Richards. 1963. The influence of organisms on the composition of sea water. The Sea 2:26–27.

Thomas, C. R. 1997. Identifying marine phytoplankton. Academic Press, London.

Yoon, Y. H. 1995. Seasonal dynamics of phytoplankton community and red tide organisms in the northern Kamak Bay, southern Korea. Yosu National Fisheries University, Bulletin of Marine Science Institute 4, South Jeolla, Korea (In Korean).

Yoon Y. H. 2000. Distributional characteristics and seasonal fluctuations of phytoplankton community in Haechang Bay, southern Korea. Journal of the Korean Fisheries Society 33(1):43–50.

Yoon, Y. H., and J. S. Park. 2000. Analysis of variation causes for marine environment and phytoplankton in Geogeumsudo using a principal component analysis. Journal of the Korean Fisheries Society 9(1):1–11, Seoul.

American Fisheries Society Symposium 49:779–795

Rural-Based Aquaculture in Kenya: Wise Use of Indigenous Practices and Knowledge for Fish Feeding Purposes

Daniel O. Okeyo* and John F. Omollo
University of Fort Hare, Faculty of Science and Agriculture
Private Bag X1314, Alice 5700, Republic of South Africa

Abstract.—This paper looks at the current perspective on the problem faced by rural-based fish farmers in western regions of Kenya, due to expensive fish feeds. Studies concentrate on indigenous ways and means employed by fish farmers to solve such problem. Eight different types of fish species are cultured in ponds, with Nile tilapia *Oreochromis niloticus* (Linnaeus, 1758) and redbelly tilapia *Tilapia zillii* (Gervais, 1848) as predominant species. At least 8,133 fish ponds and 5,187 fish farmers exist in 15 districts of western regions of Kenya. The majority of fish ponds are taken care of by women and children. Fish farmers employ indigenous practices and knowledge, thousands of years old, of collecting and gathering, in feeding their fish. The farmers collect vegetables for human as well as for fish consumption at the same time. Eight naturally occurring, terrestrial wild plants (some poisonous to humans) and seven crop plants are utilized as food supplements for fish. Results from a 28-d laboratory feeding experiment of plants (e.g., gallant soldier *Galinsoga parviflora* Cav., black jack *Bidens pilosa* L.) and pellets, at 20°C ± 2°C, however, were proof that fish lost weight when fed on leaves of wild plants alone. Weight declines in fish fed different plants were significant ($F_{3,16} = 7.0$, $P < 0.05$).

Introduction

The history of small-scale freshwater fish farming in Africa is reviewed in depth by Bardach et al. (1972) and Pullin (1985). According to Bardach et al. (1972), aquaculture in Africa is riddled by a host of problems that are socioeconomic and sometimes political in nature. This has resulted in low-level fish farming in Africa as compared to the well-developed extensive fish farming in Southeast Asia (Li 1990, 2001).

In Kenya, fish farming was first introduced in 1905 when two species of the delicious rainbow trout—rainbow trout *Oncorhynchus mykiss* (Walbaum, 1792) and brown trout *Salmo trutta* Linnaeus, 1758—were introduced in some local rivers around Mount Kenya region (Dadzie and Oduol 1987). Then, around 1924, the first fish ponds were started, cultivating Athi River tilapia *Oreochromis niger* Günther, 1894. Information on fish farming in Kenya is scanty. However, records from the Department of Fisheries—1981–2003—indicate that aquaculture production has risen steadily, 10-fold, in the past three decades.

*Corresponding author: dokeyo@ufh.ac.za

Many fish farmers have integrated fish farming with subsistence agriculture to maximize land use (Dadzie and Oduol 1987).

Kenya has one of the fastest growth rates in human population, which is not accompanied by a commensurate increase in food production! According to 2001 census records, Kenya had a human population of 29.8 million (PRB 2001); a year later, it had jumped to 31.1 million (PRB 2002a). According to PRB (2001, 2002a) there is a prediction of 26% increase in Kenyan human population (37.4 million) by 2050. In this respect, one of the government policies, in the past decade, was to increase food production to match the spiraling population rise.

Fish has been regarded as a source of cheap animal protein (Jauncey and Ross 1982) easily obtainable. Yet, global fish catches have declined since reaching a peak of about 80 million tons in the late 1980s (CEU 2001). The situation in global decline in naturally occurring fishes therefore reflects the importance of increased food (especially protein) production through pond fish farming, besides crop production. Fish farming is considered the fastest growing food production sector in the world (Beveridge et al. 2001). Thus, fish farming is being part of the solutions to increased protein production. Fish farming is regarded as a means of recycling biodegradable natural resources and waste, and also one way of conservation of the environment and of biodiversity in Africa.

The aim of this study is to survey the status of rural-based fish farms in western regions of Kenya, with regards to indigenous practice and knowledge for fish feeding purpose, in order to analyze problems and suggest solutions. The objectives of the study are to establish the fish species commonly cultured in ponds, to estimate the number of active fish ponds and the number of fish farmers existing in districts of western region of Kenya (Figure 1), to establish and describe fish food supplements, and to determine the ingestion and growth rate by fish on plant food supplements.

Methods

Fish Species, Numbers of Active Ponds, and Number of Farmers

Records come from intensive literature search and examination of government references. Records also come from interviews with district and provincial fisheries officers. In a few cases, listed fish species, numbers of active ponds, and farmers are based on reliable interviews with local fishers. Information is also gathered from field visits to crop and harvest fish during the duration of this study. Fish are identified according to standard systematic and taxonomic methods for field as well as laboratory analysis (Eschmeyer 1990; FishBase 2000; De Vos 2001). *Oreochromis* spp. is the most commonly cultured fish in districts of western regions of Kenya

Field Food Supplements

Farmers use indigenous practices and knowledge, well established for thousands of years, of collecting and gathering, in feeding their fish. Vegetables (naturally occurring terrestrial wild plants) are collected for human as well as for fish consumption, at the same time. Specific numbers of the wild plants (some poisonous to humans) and of crop plants, utilized as food supplements for fish, are established. Wild and crop plants used as fish food supplements are identified and described according to standard methods and the literature (Watt and Breyer-Brandwijk 1962; Purseglove 1969; Gibbs-Russell et al. 1987).

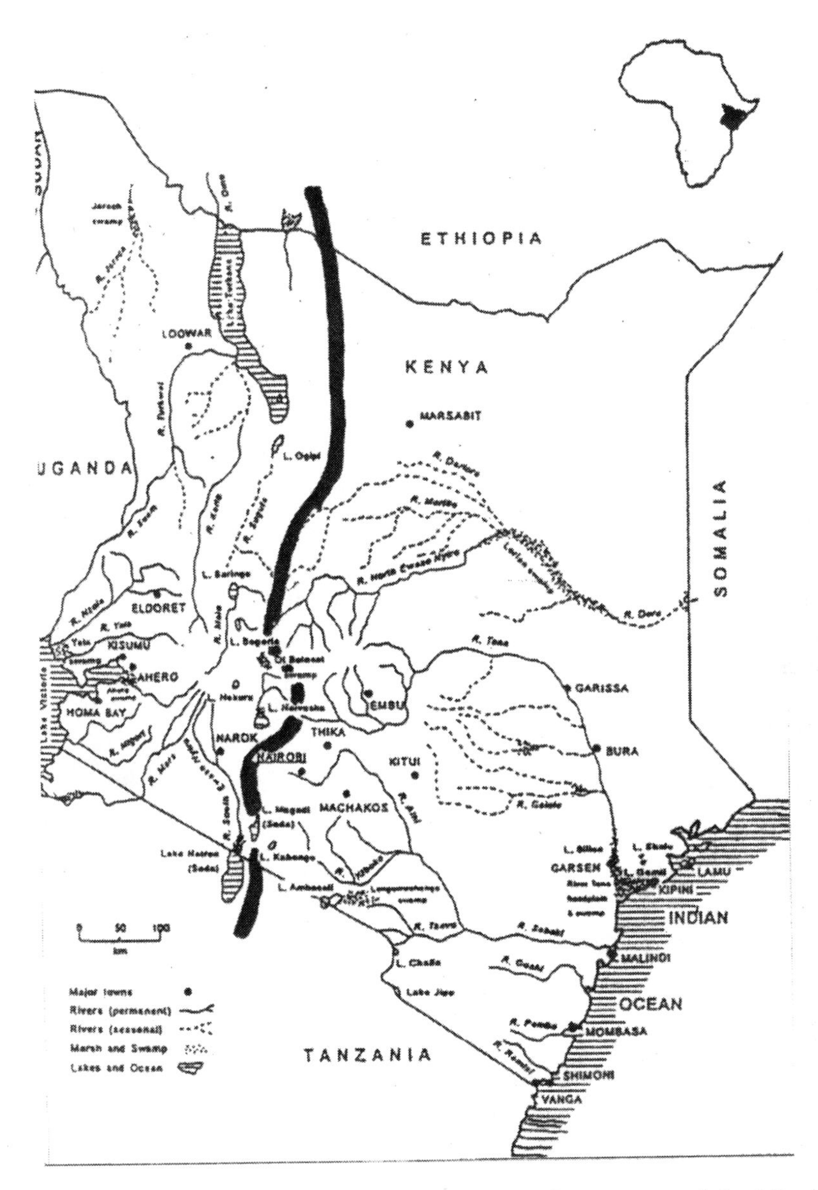

Figure 1. Map of Kenya showing major water bodies. Study area occurs left of the bold line.

Laboratory Food Supplements

Two of the naturally occurring wild plants, gallant soldier *Galinsoga parviflora* Cav. and black jack *Bidens pilosa* L., are widely and predominantly used as fish food supplements by fish farmers in districts in western regions of Kenya. The two plants form treatments that were used as food supplements in the laboratory feeding experiments of this study.

Four fish food supplements of known composition (Omollo 1990; Table 1), gallant soldier, black jack, mixed (50% gallant soldier and 50% black jack), and commercial pellet, were fed to juvenile (1.90–2.23 g mean weight) Nile tilapia *Oreochromis niloticus* (Teleostei: Cichlidae) (Linnaeus, 1758) in a 28-d experiment. Plants were washed as best as possible by moving them through the water to remove adhering sediments and encrust-

Table 1. The approximate nutrient composition of plant leaves (Omollo, 1990) and pellet (Unga Feeds Limited, Nairobi) fed to fish.

Diet composition	*G. parviflora*	*B. pilosa*	Pellet
Crude protein (%)	2.87	4.2	25
Carbohydrates (%)	52	59	5
Crude fiber (%)	2.3	5	6
Energy (KJ/kg)	1,749	1,720	2,550

ing material. The literature reports of protein, carbohydrates, crude fiber (percent dry weight) and energy, respectively, include gallant soldier, 2.87%, 52%, 2.3%, 1,749 KJ/Kg; black jack, 4.2%, 59%, 5%, 1,720 KJ/Kg; and pellets, 25%, 5%, 6%, 2,550 KJ/Kg (Omollo 1990; Table 1).

Initiation of Feedings Experiments

Fish were removed from holding tank and body weights determined to the nearest 0.1 g with a sensitive top-loading balance after 24-h starvation and prior to beginning experiments, so as to determine weight gain or loss. A feeding time schedule of starvation on day 1, weighing on day 2, and feeding on days 2–6 was established; day 7 began the cycle again with starvation. Three fish were then placed in each of four aerated, 40-L aquaria in the laboratory at 20° ± 2°C; the laboratory maintained a 12 light/12 dark photoperiod. Preliminary trials indicated cannibalism when tanks were stocked with more than three fish at a time. Water in each aquarium was continuously bubbled through cotton and charcoal filters, and feces and uneaten food were siphoned from tanks daily to reduce any opportunity for coprophagy. Fish were acclimated for at least 1 week in the laboratory while being fed ad libitum on experimental foods (Table 2).

Feeding Experiments

During the 28-d experiments, more food (at least 4 g of plant material) than fish could finish eating within 24 h (ad libitum), was supplied; wet weights of plants were measured to the nearest 0.1 g on the top-loading balance after plants had been blotted with paper towels. Experimental fish food supplement were added each day to experimental tanks and attached to rocks with rubber bands, a procedure similar to that employed by Gerking (1984) and Okeyo and Montgomery (1992). Attempts were made as much as possible to recover all uneaten food particles. Large fragments of uneaten food particles were removed from tanks by scoop nets, while fine fragments siphoned out and weighed to the nearest 0.1 g to determine daily ingestion rate. No attempts were made to assay caloric or nutrient content or food or feces.

Ingestion rate

Percent body weight of pellet fed to fish was determined. Fish were paired by weight, and then members of pairs were assigned separately to the two food treatments (gallant soldier or black jack). For example, fish receiving gallant soldier were again fed ad libitum; daily ingestion of wet plant was determined. The dry weight equivalent of a day's food was estimated from wet weight/dry weight ratios determined previously. Pellets equivalent to the calculated dry weight of ingested gallant soldier were then fed to the other member of the pair. Thus, both fish of the pair received equivalent amounts of pellet and gallant soldier on a dry weight basis during the duration of the experiment.

Table 2. Aquaria, experimental diet, and fish set-up for the 28-d experiment.

Aquarium	Experimental diet	Fish no.
1	*Galinsoga parviflora*	3
2	*Bidens pilosa*	3
3	"Mixed" (50% *Galinsoga*:50% *Bidens*)	3
4	Pellets	3

Ration is expressed as percent of body weight (BW) ingested per day:

$$\text{Ration } (\%\text{BW} \times \text{d}^{-1}) = \frac{\text{g food ingested per day}}{\text{BW(g)}} \times 100\%$$

Growth rate

All experiments followed the same repetitive timetable of starvation on day 1, weighing on day 2, and feeding on days 2–6; day 7 began the cycle again with starvation. Fish were starved for at least 24 h prior to weighing, so as to determine weight gain or loss. This is a standard procedure for fish feeding studies and limits errors due to weight of ingested food. Although it may be expected that this recurring handling may suppress growth to some degree, the literature supports the fact that juvenile tilapia fish held under identical conditions and fed dry pelleted foods either maintain weight or grow (Okeyo and Montgomery 1992).

Daily growth (loss or gain in weight) is calculated as percent of initial weight (IW):

$$\text{Growth}(\%\text{IW} \times \text{d}^{-1} = \frac{\text{change in weight}}{\text{IW}} \times 100\%$$

Two-way analysis of variance (ANOVA) (*F*-test, $p < 0.05$ Zar 1984) as well as two-sample *t*-test ($p < 0.05$ Zar 1984) are used to determine different food usages by fish. These analyses show any significant differences in ingestion and growth rates among plants by fish at 20° ± 2°C. Statistica Computer Programmes (statistical packages) are used for ANOVA and Systat for the *t*-test.

Results and Discussion

Field Surveys: Fish Species in Culture

Farmers in western Kenya use eight fish species (Table 3) in pond culture. Nile tilapia *Oreochromis niloticus*, redbelly tilapia *Tilapia zillii*, and Air-breathing catfishes *Clarias* spp. are the predominantly cultured fish. Farmers prefer these fishes due to their ability to feed on locally available plant matter (Ibrahim 1976; Trewavas 1983; Njiru 1999; Skelton 2001). The original source of pond fish is the nearby water bodies; it is also evident that the majority of cultured fish come from introductions from far-reaching places.

According to Welcomme (1967), the Nile tilapia was introduced in Kenyan waters of Lake Victoria (from River Nile) in 1957. Records in Welcomme (1967, 1988) also indicate a series of introductions between 1953 and 1955 of the redbelly tilapia, from Lake Albert into Kenyan waters of Lake Victoria, and into Lake Naivasha in 1955, to fill a vacant niche. Graham's tilapia originally belongs to Lake Victoria drainage but has been introduced to several dams and surrounding waters. Lever (1996) mentions that the blue-spotted tilapia *Oreochromis leucostictus* was introduced in 1953 or 1954 from Lake Albert into Kenyan waters of Lake Victoria; the fish species has also established a stock in Lake Naivasha (Hickley et al. 2002) and dams in the country. Rainbow trout is well established and self-sustaining in streams all over the Aberdare and

Table 3. Fish species used in pond culture in districts of western regions of Kenya.

No.	Fish name	Authority/date	Common name
1	*Oreochromis niloticus*	(Linnaeus, 1758)	Nile tilapia
2	*Tilapia zillii*	(Gervais, 1848)	Redbelly tilapia
3	*O. esculentus*	(Graham, 1982)	Graham's tilapia
4	*O. leucostictus*	(Trewavas, 1933)	Blue spotted tilapia
5	*Clarias* spp.		Air-breathing catfishes
6	*Protopterus aethiopicus*	Owen,1839	African lungfish
7	*Oncorhynchus mykiss*	(Walbaum, 1792)	Rainbow trout
8	*Labeo* spp.		Labeo

Elgon mountains. The trout is cultured in a few ponds (e.g., at Kericho) that receive cool highland fresh waters. First introduction of trout was made into Kenya around 1910 from South Africa and United Kingdom, for angling and aquaculture (Dadzie and Oduol 1987; Welcomme 1988). The identification to species level is still underway of the other cultured fish types (air-breathing catfishes, and labeo).

Four types of cultivable air-breathing catfishes (e.g., Alluaud's catfish *Clarias alluaudi*, sharptooth catfish *C. gariepinus*, smoothhead catfish *C. liocephalus* and Werner's catfish *C. werneri*) occur naturally in waters within the study area (Ochumba and Manyala 1992; Goudswaard and Witte 1997; Mugo and Tweddle 1999; Seegers et al. 2003). Two labeos (Redeye labeo *Labeo cylindricus* and Victoria labeo *L. victorianus*) occur in the study area (Ochumba and Manyala 1992; Mugo and Tweddle 1999). Complete taxonomic identification is required on these fishes so as to track any introductions.

Field Surveys: Number of Ponds and Farmers

8,133 fish ponds, of which approximately 551 are fallow, and ~5 187 fish farmers exist in 15 districts of western regions of Kenya (Table 4). Obviously, some individual farmers have multiple numbers of fish ponds; thus, the reason for higher number of existing ponds than farmers. Even though specific data are not presented herein, the fish farm workers throughout the districts tend to be women and the youth. The main reason for having fish farms is to supplement food and cash for families. The distribution of fish ponds and fish farmers in the districts follows certain patterns of fish mongering behavior by humans.

All districts of western regions of Kenya (except Kapsabet, Kerio Valley, Nandi Hills, and Nakuru) are inhabited by groups of people who believe in fish as staple food for human consumption. Very few fish ponds (50; ~6 are fallow) and fish farmers (~4) exist in the districts (Kapsabet, Kerio Valley, Nandi Hills, and Nakuru), where fish meat is not necessarily a staple food for human consumption (Table 4). Peoples of the aforementioned four districts consist mainly of cattle herders, rather than fishers. High numbers of fish ponds and fish farmers exist in districts (Kisii— 2,080 fish ponds:1,714 fish farmers; Kakamega—1,411:857; Nyamira—942:806) far away from Lake Victoria than in districts (Homa Bay—39:257; Migori—449:250; Siaya—426:267) next to the lake. Fish consumers who live near Lake Victoria get food fish supplements from the lake itself. A large contingent of fallow ponds exists in the Siaya (106), Busia (112), and Kakamega (201) districts. The

Table 4. Established numbers of fish ponds and fish farmers in 15 districts of western regions of Kenya. Dash (-) indicates no data collection. Total numbers of inactive ponds and farmers are approximated (~), due to unrecorded data.

District	No. of ponds	Inactive ponds	No. of farmers
Nakuru	40	4	–
Kericho	300	3	–
Kisii	2,080	–	1,714
Nyamira	942	–	806
Homa Bay	339	3	257
Migori	449	5	250
Kisumu	560	6	386
Siaya	426	106	267
Bungoma	882	82	536
Busia	477	112	311
Kakamega	1,411	201	857
Uasin Gishu (Eldoret)	217	27	147
Kapsabet	0	0	0
Nandi Hills	10	2	4
Kerio Valley	0	0	0
TOTAL	8,188	~551	~5,187

main reason given for this pattern is based on the introduction of sugar cane plantations in the districts; some families opt to grow sugar cane instead, for high cash income. Meeting the needs of a growing population frequently requires some form of land-use change, such as clearing forest land in order to expand food production (PRB 2002b).

Field Surveys: Fish Food Supplements

Fish farmers in districts of western regions of Kenya use the indigenous practices and knowledge, thousands of years old, of collecting and gathering, in feeding their fish. The farmers collect vegetables for human as well as for fish consumption at the same time. Eight naturally occurring, terrestrial wild plants (some poisonous when directly consumed by humans) and seven crop plants (Table 5 and Figure 2A–O) are utilized as food supplements for fish. Fish farmers uproot herbaceous plants (e.g., gallant soldier, black jack) in whole or pluck leaves off shrub plants (e.g., elephant ears *Colocasia antiquorum* Schott, Napier grass *Pennisetum purpureum* Schumach.) and tree plants (e.g., paw paw *Carica papaya* L., cassava *Manihot esculenta* Crantz) and throw them into the fish ponds. Fish can be seen rushing for the fresh feed, which are nibbled down to the veins. The utilization of wild plants for fish food is significant to the fish farmers in two ways: substitute for crop plants, which are also consumed as food by humans; and economically viable due to natural availability in nearby surrounding.

Reports about fish feeding of wild terrestrial plants are scanty in the literature; the current report constitutes one of the first for districts in western regions of Kenya. The originality of such a practice may seem to have been accidental. Wild plants species may seem to be neglected and underutilized, but more recently, awareness on the potential of these species in food and nutritional security, health, and income generation has been increasing in the face of growing environmen-

Table 5. Wild and crop plants utilized as food supplements for fish by farmers in districts of western regions of Kenya. Local bracketed names in Luo language.

No.	name	Authority	Common (local) name
	Wild plants		
1	*Galinsoga parviflora*	Cav.	Gallant soldier (osieko)
2	*Bidens pilosa*	L.	Black jack (nyanyiek-mon)
3	*Euphorbia hirta*	L.	Asthma [milk] weed (achakmalak)
4	*E. geniculata*	L.	Painted spurge (achak)
5	*Ipomea tonuirostris*	Steud. ex Choisy	Hedgeknot morning glory
6	*Amaranthus lividus*	L.	Prostate pigweed (odotwasungu)
7	*A. caudatus*	L.	Love-less bleeding (ododo)
8	*Pennisetum purpureum*	Schumach.	Napier grass (lumb-ogada)
	Crop plants		
1	*Carica papaya*	L.	Paw-paw (a[o]poyo)
2	*Manihot esculenta*	Crantz	Cassava (marieba)
3	*Ipomea batatas*	(L.) Lan.	Sweet potato (rabuon-nyaluo)
4	*Colocasia antiquorum*	Schott	Elephant ears (rabuon-nduma)
5	*Vigna unguiculata*	(L.)	Cow pea (ngor-bo)
6	*Brassica oleracea* var. *acephala*	L.	Kale (kandhira)
7	*B.oleracea* var. *capitata*	L.	Common cabbage (kabich)

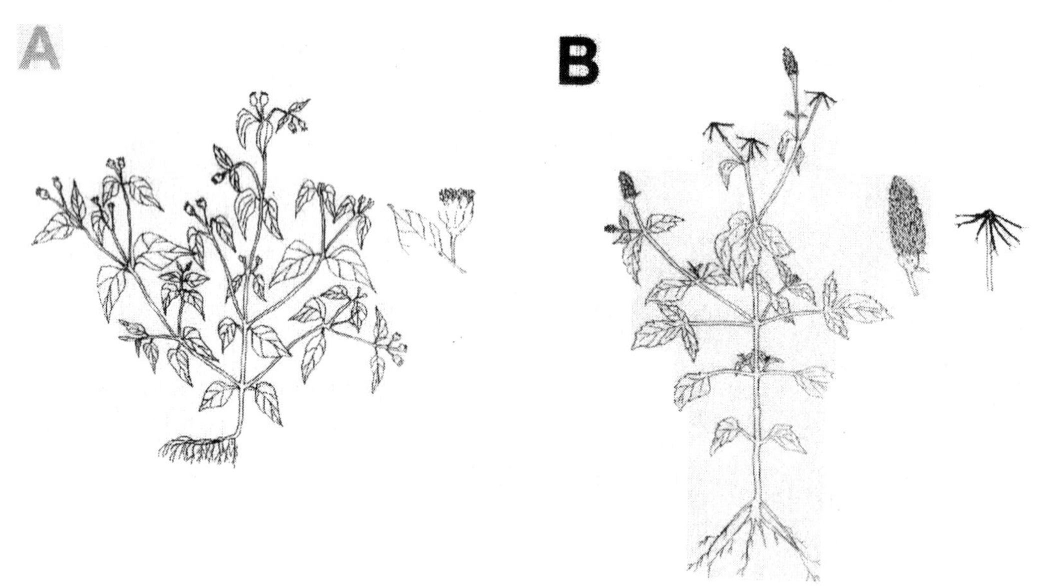

Figure 2. Wild and crop plants utilized as food supplements for fish by farmers in districts of western regions of Kenya. WILD: A = *Galinsoga parviflora* Cav., B = *Bidens pilosa* L., C = *Euphorbia hirta* L., D = *Ipomea tonuirostris* L., F = *Manihot esculenta* Crantz, G = *Ipomea batatas* (L.), H = *Carica papaya* L., I = *Vigna unguiculata* (L.) Lan., J = *Colocasia antiquorum* Schott, K = *Amaranthus caudatus* L., L = *A. lividus* L., M = *Pennisetum purpureum* Schumach.; CROP: , N = *Brassica oleracea* var. *capitata* L., and O = *Brassica oleracea* var. *acephala* L.

Figure 2. Continued.

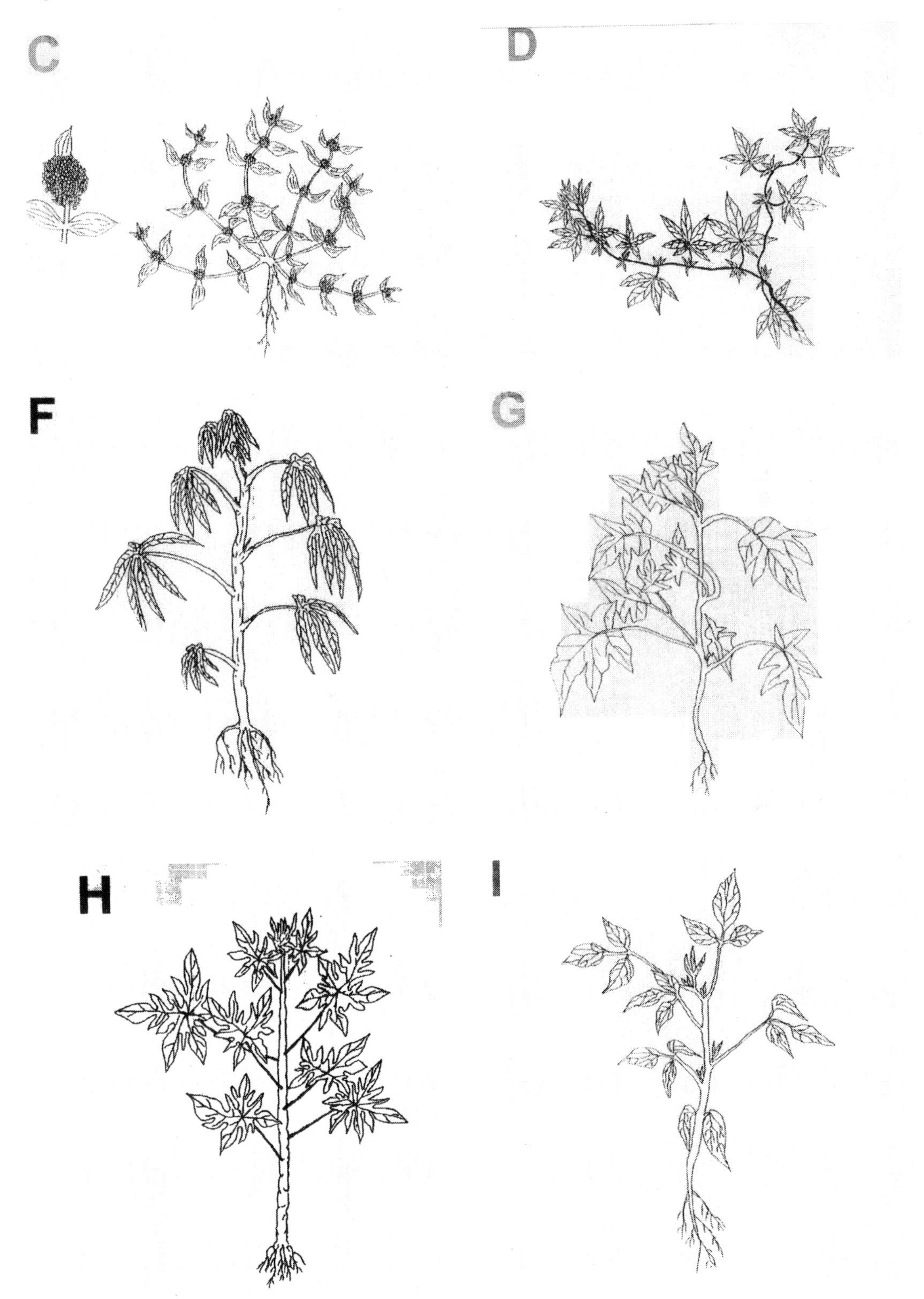

Figure 2. Continued.

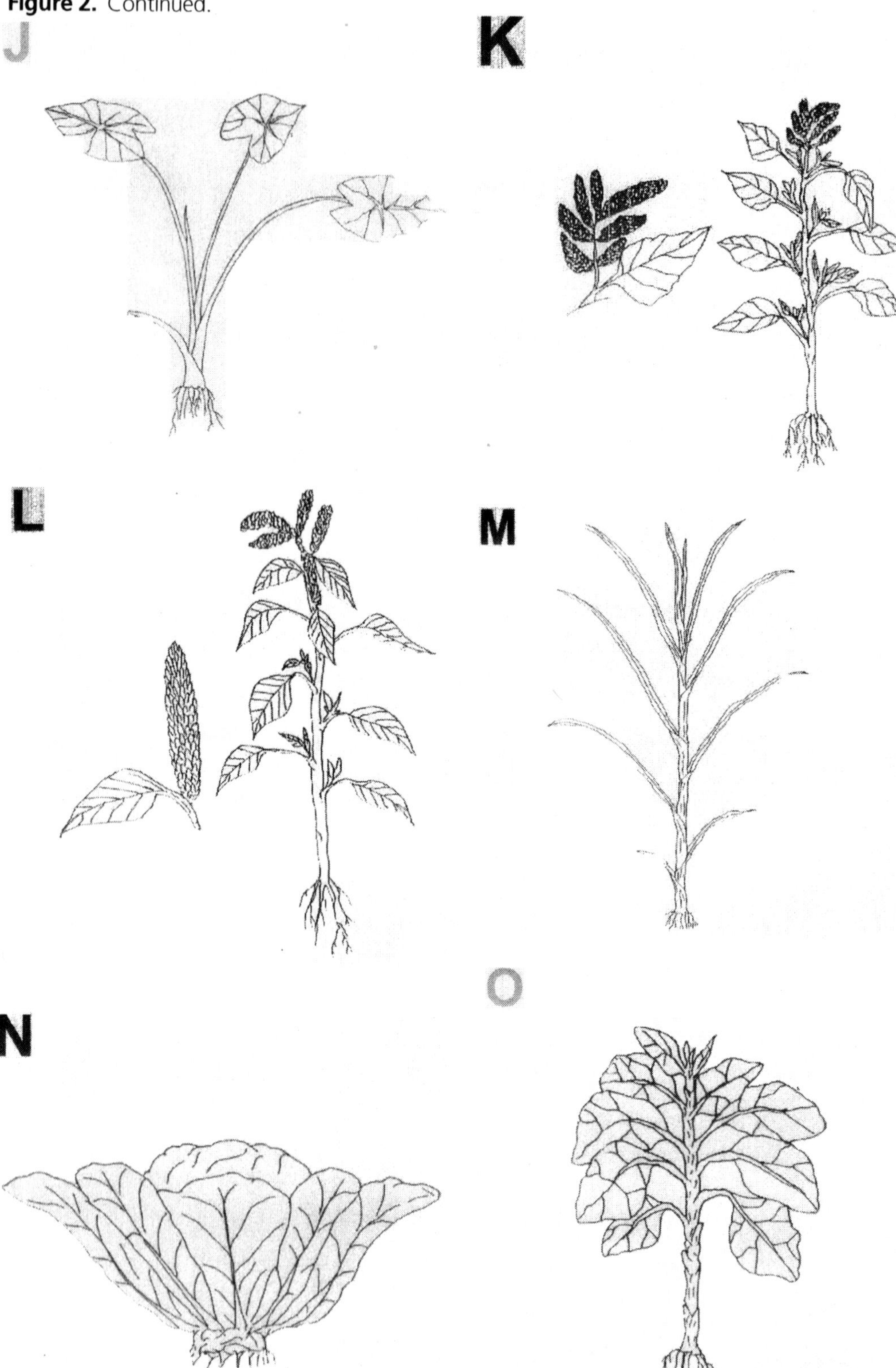

tal and socioeconomic changes (Padulosi et al. 2003).

Crop plants on the other hand are readily utilized; according to Eisawy and El Bolock (1976), banana leaves were occasionally fed to fish in ponds in Malawi. Records in Wee (1988) and Pantastico (1988) also indicate that crop materials (e.g., groundnut vines, rice straw, sugar cane tops, sweet potato vines, cassava leaves, pea forage, pumpkin leaves) were used as feeds for pond cultured fish species in regions of Asia.

Feeding Experiments: Ingestion of Plants and Pellet

Many studies involve predetermined rations rather than *ad libitum* feeding. In this study, fish had food available to them throughout their daily activity period, so that ingestion was a function of fish behavior rather than experimental protocol. Juvenile Nile tilapia (1.90–2.23 g mean weight) were chosen to run feeding experiments due to predominance in culture in most districts of western regions of Kenya. Combined weight of all four experimental foods ingested by fish throughout the 28-d experiment averaged 0.03 g/d at 20° ± 2°C (Table 6). In terms of mean percent body weight per day, juveniles appeared to consume more pellets (15.2% BW/d) than gallant soldier (13.8% BW/d) or black jack (5.8% BW/d) or mixed (50% gallant soldier and 50% black jack; 7.4% BW/d) at 20° ± 2°C (Figure 3). Nonetheless, two-way ANOVA indicates that rates of ingestion of the four experimental foods are significantly different ($F_{3, 12} = 21.7$, $p < 0.05$).

Such selectivity suggests that plants differ in some aspects of palatability or quality. Some plants have high quantities of refractory carbohydrates (Boyd 1968; Fischer 1973). Also, tilapia does not readily digest cellulose (Buddington 1979) and must ingest large quantities of plants in order to obtain sufficient nutrients for body maintenance. This study shows that tilapia generally ingested more of foods with high protein value (e.g., pellet) rather than those with low protein value (e.g., plants; Omollo 1990; Unga Feeds Ltd, Nairobi). The study also shows a slight increase in ingestion rate by juveniles of mixed plant foods than of black jack. At this point, it is difficult to identify the factors influencing selection of certain plants by fishes. These may include texture, nutrient composition, and palatability (Horn et al. 1982).

Feeding Experiments: Growth of Fish

All fish lost weight when juvenile (1.90–2.23 g mean weight) Nile tilapia were fed four experimental foods (gallant soldier, black jack, mixed (50% gallant soldier and 50% black jack), except on pellet. If one accepts that the food quality of a given diet is directly proportional to its ability to support growth (Bowen 1982), then the three plant foods offered to fish in the present experiments are low-quality foods.

Juvenile fish fed experimental plants over the 28-d experiments exclusively lost an average of 0.18 g (SD = 0.23 g), while fish fed pellets exclusively gained an average of 0.11 g (SD = 0.12) (Table 6). The large standard deviations in weight gains and losses obliterate any significant variations and put into question the significance of tested diets. Perhaps this is due to sample and data collection or size. Once-off tests were run in order to indicate whether or not fish can maintain weight when fed terrestrial plants alone. No plant used in the experiments supported any positive growth (weight gain) by juvenile fish (Table 7; Figure 4). Less weight loss occurred on mixed (mean = –0.19 g) than on black jack (mean = –0.25 g) or on gallant soldier (mean = –0.26 g). The declines in juvenile fish weight throughout

Table 6. Weights (g; mean, standard deviation, range) of wild plant material and pellet ingested by juvinile Nile tilapia *Oreachromis niloticus* during feeding experiments at 20°±2^{0}C. Mean are total amounts of food ingested for each 3 replicates during the 6 day experiments, giving $N = 3$ for every entry. Mean weights (g) of experimental fish at the end of each week are also given. Food used in this experiment included *Galinsoga parviflora*, *Bidens pilosa*, "mixed" (50% *Galinsoga* and 50% *Bidens*), and pellet.

Days	*Galinsoga parviflora*			(Fish)	*Bidens pilosa*			(Fish)	*Galinsoga & Bidens*			(Fish)	pellet			(Fish)
	Mean	SD	Range		Mean	SD	Range		Mean	SD	Range		Mean	SD	Range	
0–7	0.078	0.058	0.02–0.16	1.67	0.118	0.104	0.03–0.25	1.77	0.142	0.128	0.04–0.38	2.51	0.340	0	–	2.29
8–14	0.133	0.064	0.02–0.18	0.95	0.116	0.087	0.00–0.23	1.67	0.207	0.101	0.08–0.33	2.60	0.330	0	–	2.32
15–21	0.103	0.050	0.06–0.17	0.93	0.031	0.028	0.00–0.07	1.64	0.152	0.110	0.01–0.33	2.66	0.350	0	–	2.37
22–28	0.136	0.100	0.03–0.32	0.88	0.132	0.050	0.07–0.20	1.51	0.258	0.092	0.16–0.40	2.25	0.350	0	–	2.65

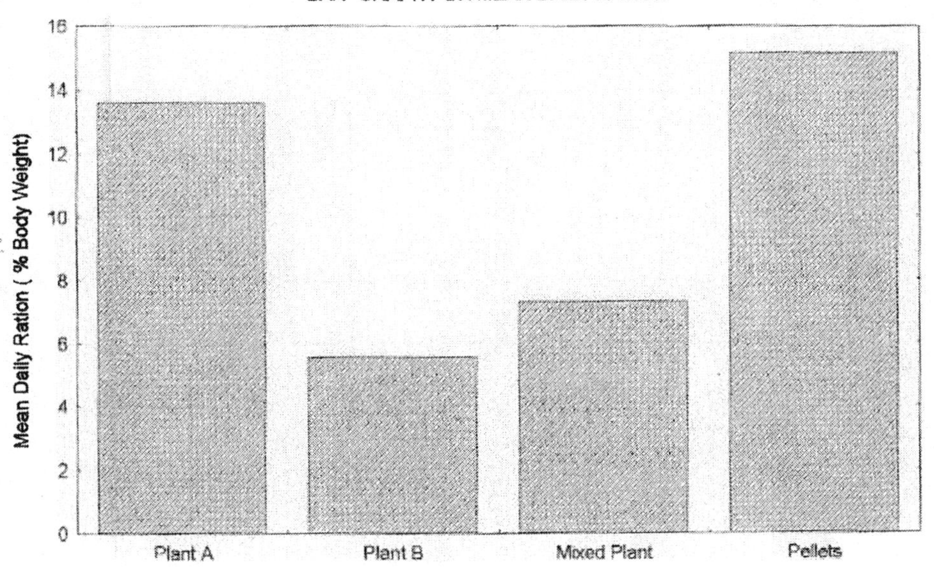

Figure 3. Rates of ingestion (mean % body weight/d) of wild terrestrial plants and pellet by juvenile *Oreochrimis niloticus* over the 28 d of feeding experiments at 20° ± 2°C. Plant A = *Galinsoga parviflora*; Plant B = *Bidens pilosa*; "mixed"=50% *Galinsoga* and 50% *Bidens*), and pellet. Means are for daily determinations for each of four replicates.

the 28-d experiments are significant ($F_{3, 16}$ = 7.00, $p < 0.05$; Figure 4). Marked differences in weight loss by juvenile fish are detected at 20° ± 2°C, using two-sample *t*-tests, regardless of the identity of the food plant (or pellet) ($t = -25.9$, $p < 0.05$; Figure 4).

Juvenile tilapia fish lost weight (relative to total body weight), perhaps due to reduced digestion of plants caused by incomplete development, and thereby function, of the gut and the inability of small fish to utilize foods that are not usually encountered in the natural environment (Mironova 1976). A transitional size at which dietary shifts occur has been shown in other species and correlated with development of a fully functional digestive system (Kawai and Ikeda 1973). It may be possible that small tilapia fish feed selectively on more nutritional food resources to compensate for digestive inadequacies, but this is yet to be fully demonstrated in nature.

Conclusions

Two fish species, Nile tilapia and redbelly tilapia, are preferred for culture by fish farmers in districts in western regions of Kenya, perhaps due to the ability of the fish to eat cost-effective plant-food supplements. More fish ponds exist in districts where fish mongering traditionally takes place; local inhabitants indigenously belief that women who eat fish give birth to smart/clever/brainy children. Due to human population increases in the Lake Victoria basin, some fish farmers migrate to places far away in such for space for new homes and farms. The farmers utilize locally available terrestrial wild and crop plants as fish food supplements.

Juvenile Nile tilapia ate more gallant soldier than black jack, thus exhibiting some measure of selectivity or willingness to ingest certain plants over others. Juvenile fish lost weight when fed all experimental plants, despite the

Table 7. Absolute mean weight (g) gain or loss by juvinile Nile tilapia *Oreachromis niloticus* during feeding experiments included *Galinsoga parviflora*, *Bidens pilosa*, "mixed" (50% *Galinsoga* and 50% *Bidens*), and pellet foods. wt = weight.

Days	(Fish on *G. parviflora*; 1.90 g initial fish wt)	(Fish on *B. pilosa*;) 1.90 g initial fish wt)	(Fish on mixed; 2.67 g initial fish wt)	(Fish on pellet; 2.23 g initial fish wt)
0–7	0.23	0.12	0.16	0.06
8–14	0.72	0.10	0.09	0.03
15–21	0.02	0.03	0.06	0.05
22–28	0.05	0.13	0.41	0.28

relatively high ingestion rate. This may be due to an inability of fish to digest plant materials or relatively low nutrient levels in plant food. Nonetheless, greater attention should be given to digestive and growth capabilities of such fishes when fed on naturally occurring plants, for such fishes may lend themselves to simple and inexpensive culture techniques applicable where additional protein in human diets would be beneficial. Locally available, rapidly growing, and partially digestible plants (e.g., gallant soldier and black jack in this experiment) could be used in integrated fish culture in permanent or temporary waters.

Recommendations

- Further inventories should be incorporated into government budgets in conjunction with

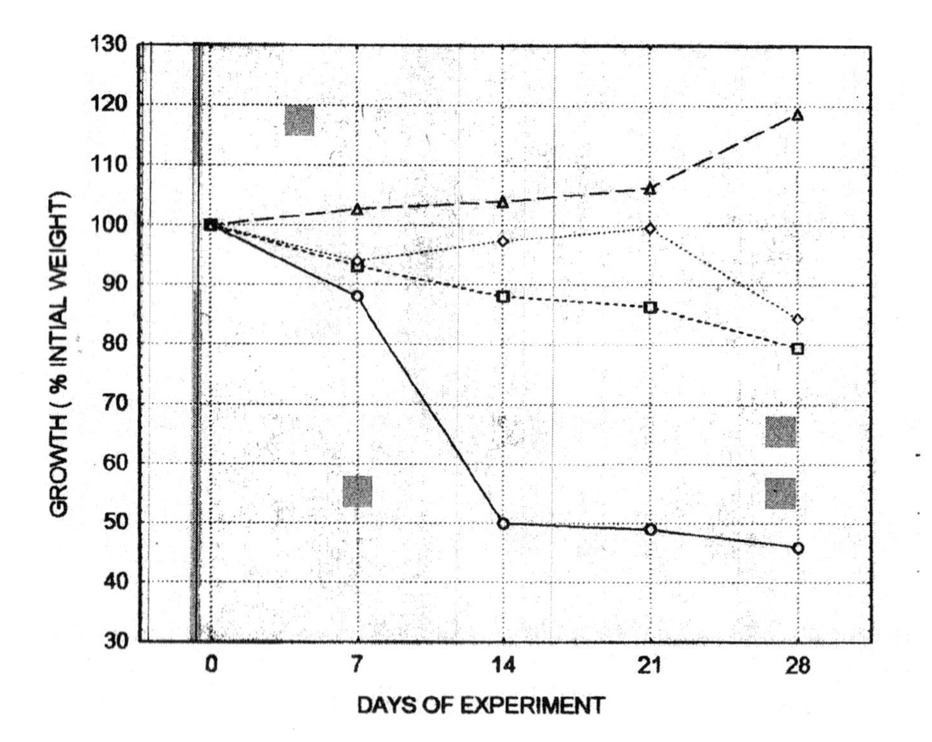

Figure 4. Growth (mean % of initial body weight of juvenile *Oreochromis niloticus* during feeding experiments with *Galinsoga parviflora*, *Bidens pilosa*, "mixed" (50% *Galinsoga* and 50% Bidens), and pellet. $N = 3$ in each case. △ = pellets; ◇ = mixed; □ = *Bidens*; ○ = *Galinsoga*.

experts from research institutions (e.g., the Lake Basin Development Authority, the Kenya Marine and Fisheries Research Institute, and local and overseas institutions of higher education).

• Curricula and facilities set up at universities and polytechnics or technicons for training of field officers in backgrounds relevant to fish culture.

• Slow-down in the rapid rate of transfers of competent and productive field officers. Research (with locally available resources) in feeding proportions and food compositions to establish economically suitable fish pellets.

• Encouragement and training in integrated farming.

• Direct and strong links with universities.

• Need of feeder roads net work to farm lands to improve lack of communication/commutation.

• Research on what an ideal pond should be for each district.

Acknowledgments

We express our gratitude to all pond fish farmers in districts of western regions of Kenya, who warmly received us in the field, at times impromptu, and helpfully responded to conversations, questions, and dialog; to project students, staff, and the Department of Zoology, Kenyatta University, Nairobi, for providing facilities and hand in running of the feeding experiments; to J. Ndege and R. Akyea-Agbadzi of the Faculty of Science and Agriculture, University of Fort Hare, for provision of statistical packages and analysis. This work was partially supported by grants from the Dean of Faculty of Science Committee Research Funds, Kenyatta University, to Okeyo and Omollo (the late). Okeyo hereby dedicates this paper in commemoration of the late John Fanuel (JF as he was well known) Omollo, with whom field work was made as easy as a breeze—may his soul rest in eternity.

References

Bardach, J. E., J. H. Ryther, and W. O. McLarney. 1972. Aquaculture: the farming husbandry of freshwater and marine organisms. Wiley, New York.

Beveridge, M., P. Bueno, H. Kongkeo, and R. Subasinghe. 2001. The Bangkok conference on aquaculture in the third millennium: outputs and the challenges ahead. EC Fisheries Cooperation Bulletin Supplement 14(1–4):41–42.

Bowen, S. H. 1982. Feeding, digestion and growth – quantitative considerations. Pages 141–156 *in* R. S. V. Pullin and R. H. Lowe-McConnell, editors. The biology and culture of tilapia. International Center for Living Aquatic Resources Management, ICLARM Conference Proceedings 7, Manila, Philippines.

Boyd, C. E. 1968. Fresh water plants: a potential source of protein. Economic Botany 22:359–368.

Buddington, R. K. 1979. Digestion of an aquatic macrophyte by *Tilapia zillii*. Journal of Fish Biology 15:449–456.

CEU (Commission of the European Union). 2001. Knowledge to lead fisheries from global crisis towards sustainability: highlights from an international workshop at the commission, 3–4 December 2001. EC Fisheries Cooperation Bulletin Supplement 14(1–4):15.

Dadzie, S., and C. S. G. Oduol. 1987. Status of aquaculture in Kenya. Pages 95–107 *in* H. Powles, editor. Research priorities for African aquaculture. International Development Research Centre, IDRC-MR-149, Ottawa.

De Vos, L. 2001. *Synodontis manni* (Teleostei: Siluroidea: Mochokidae), a new catfish from the lower Tana River, Kenya. Ichthyological Exploration of Freshwaters 12:41–50.

Eisawy, A., and A. El Bolock. 1976. Status of aquaculture in Arab republic of Egypt. CIFA Technical Paper 4(Supplement 1):5–15.

Eschmeyer, W. N. 1990. Catalogue of the genera of recent fishes. California Academy of Sciences, San Francisco.

Fischer, Z. 1973. Physiology and bioenergetics of grass carp. Polish Archives of Hydrobiology 20:529–557.

FishBase. 2000. FishBase 2000. CD-Rom. International Center for Living Aquatic Resources Management, Manila, Philippines.

Gerking, S. D. 1984. Assimilation and maintenance ration of an herbivorous fish, *Sarpa salpa,* feeding on a green alga. Transactions of the American Fisheries Society 113:378–387.

Gibbs-Russell, G. E., W. G. Welman, E. Retief, K. L. Immelman, G. Germishuizen, B. J. Pienaar, M. van Wyk, and A. Nicholas. 1987. Memoirs of the botanical survey of South Africa – list of species of southern African plants, 2nd edition, part 2, Botanical Research Institute, Department of Agriculture and Water Supply, Pretoria, Republic of South Africa.

Goudswaard, K., and F. Witte. 1997. The catfish fauna of Lake Victoria after the Nile perch upsurge. Environmental Biology of Fishes 49:21–43.

Horn, M. H., S. N. Murray, and T. W. Edwards. 1982. Dietary selectivity in the field and food preferences in the laboratory for two herbivorous fishes, *Cebidichthys violaceus* and *Xiphister mucosus* from a temperate intertidal zone. Marine Biology 67(3):237–246.

Ibrahim, K. H. 1976. Progress and present status of aquaculture in Tanzania. FAO, CIFA Technical Paper 4(Supplement 1):132–146.

Jauncey, K., and B. Ross. 1982. A guide to tilapia feeds and feeding. University of Stirling, Stirling, Scotland.

Kawai, S., and S. Ikeda. 1973. Studies of digestive enzymes of fishes IV: developments of carp and black sea bream after hatching. Bulletin of the Japanese Society of Scientific Fisheries 39:877–881.

Lever, C. 1996. Naturalized fishes of the world. Academic Press, San Diego, California.

Li, S. 1990. Recent advances in aquaculture in freshwater China. Pages 144–161 *in* L. I. Chiu, C-Z. Shyu, and N-H. Chao, editors. Aquaculture in Asia. Asian Productivity Organization, Tokyo.

Li, S. 2001. About fish breeding engineering construction. Scientific Fish Faming 5–7:3–4.

Mironova, N. V. 1976. Changes in the energy balance of *Tilapa mossambica* in relation to temperature and ration size. Journal of Ichthyology 10:120–129.

Njiru, M. 1999. Changes in feeding biology of Nile tilapia, *Oreochromis niloticus* (L.), after invasion of water hyacinth, *Eichhornia crassipes* (Mart.) Solms, in Lake Victoria, Kenya. Pages 175–183 *in* I. G. Cowx and D. Tweddle, editors. Report on the fourth FIDA WOG workshop held 16–20 August 1999, Kisumu, Kenya. Lake Victoria Fisheries Research Project Technical Document, LVFRP/TECH/99/07, Jinja, Uganda.

Hickley, P., R. Bailey, D.M. Harper, R. Kundu, M. Muchiri, R. North, and A. Taylor. 2002. The status and future of Lake Naivasha fishery, Kenya. Hydrobiologia 488(1–3):181–190.

Mugo, J., and D. Tweddle. 1999. Preliminary survey of the fish and fisheries of the Nzoia, Nyando and Sondu-Miriu rivers, Kenya. Pages 106–125 *in* I. G. Cowx and D. Tweddle, editors. Report on the fourth FIDA WOG workshop held 16–20 August 1999, Kisumu, Kenya. Lake Victoria Fisheries Research Project Technical Document, LVFRP/TECH/99/06, Jinja, Uganda.

Ochumba, P. B. O., and J. O. Manyala. 1992. Distribution of fishes along the Sondu-Miriu rivers of Lake Victoria, Kenya with special reference to upstream migration, biology and yield. Aquaculture and Fisheries management 23:701–719.

Okeyo, D. O., and W. L. Montgomery. 1992. Ingestion, growth and conversion efficiency in the blue tilapia, *Oreochromis aureus*, when fed on three aquatic macrophytes. Journal of the Arizona-Nevada Academy of Science 24–25:1–10.

Omollo, B. W. 1990. Nutritional analysis of some plants fed to tilapia in western Kenya. MSc. Thesis. Kenyatta University, Nairobi, Kenya.

Padulosi, S., J. Noun, A. Giulian, F. Shuman, W. Rojas, and B. Ravi. 2003. Realizing the benefits in neglected and underutilized plant species through technology transfer and human resources development. In O. T. Sandlund and P. J. Schei, editors. Proceedings of the Norway/UN Conference on Technology Transfer and Capacity Building, 23–27 June, Trondheim, Norway.

Pantastico, J. B. 1988. Non-conventional feed resources in aquaculture: an overview of work done in the Philippines. In S. S. G. De Silva, editor. Finfish nutrition research in Asia. Heinemann Asia, Singapore.

PRB (Population Reference Bureau). 2001. World population data sheet: demographic data and estimates for the countries and regions of the world. Population Reference Bureau, Washington, D.C.

PRB (Population Reference Bureau). 2002a. World population data sheet: demographic data and estimates for the countries and regions of the world. Population Reference Bureau, Washington, D.C.

PRB (Population Reference Bureau). 2002b. Land-use change often creates threats to human and ecological health. Making the link (Population, health and environment). PRB, Measure Communication, Washington, D.C.

Pullin, R. 1985. Tilapias: every man's fish. Biologists 32(2):84–88.

Purseglove, J. W. 1969. Tropical crops – Dicotyledons. Longman group LTD., London.

Seegers, L., L. De Vos, and D. O. Okeyo. 2003. Annotated checklist of the freshwater fishes of Kenya (excluding the haplochromines from Lake Victoria). Journal of East African Natural History 92:11–47.

Skelton, P. H. 2001. A complete guide to the freshwater fishes of southern Africa. 2nd edition. Southern Book Publishers, Halfway House, South Africa.

Trewavas, E. 1983. Tilapia Fishes of the Genera *Sarrotherodon, Oreochromis and Danakilia.* British Museum (National History), London.

Watt, J. M., and M. G. Breyer-Brandwijk. 1962. The medicinal and poisonous plants of southern and eastern Africa. 2nd edition. E&S Linvingtone LTD, Edinburgh and London.

Wee, K. L. 1988. Alternate feed sources for finfish in Asia. In S. S. G. De Silva, editor. Finfish nutrition research in Asia. Heinemann Asia, Singapore.

Welcomme, R. L. 1967. Observations on the biology of the introduced species of *Tilapia* in Lake Victoria. Revue de Zoologie et de Botanique Africaines 76 (3–4):249–279.

Welcomme, R. L. 1988. International introduction of aquatic species. FAO Fisheries Technical Paper 294.

Zar, J. H. 1984. Biostatistical analysis, 2nd edition. Prentice Hall, Inc., Englewood Cliffs, New Jersey.

American Fisheries Society Symposium 49:797–813

Integrated Systems Analysis for Coastal Aquaculture

YANLAI ZHAO, DANIEL E. LANE*, AND WOJTEK MICHALOWSKI
Telfer School of Management, University of Ottawa
55 Laurier Avenue, Ottawa, Ontario K1N 6N5, Canada

ROBERT L. STEPHENSON AND FRED H. PAGE
Department of Fisheries and Oceans, St. Andrews Biological Station
St. Andrews, New Brunswick, E0G 2X0, Canada

Abstract.—This paper presents a model for the evaluation of coastal zone sites in conjunction with supporting decision making on the use of potential sites for aquaculture as well as other site activities, including commercial fisheries, and as reserves for natural resources. The decision support model captures site specific data in a geographical information system that overlays information about selected marine geographical regions with natural resources, habitat, commercial activities including aquaculture, and influence plumes from effluents. Descriptive data for selected regions, including cumulative effects attributed to system overlays and ecosystem component interactions, are then evaluated as selected site yields. These yields provide input to a multicriteria analysis that positions different decision makers in the coastal zone with respect to their relative importance of resources, habitat, commercial activities, and influence plumes. The model compares alternative evaluations of selected regions among the diverse users, as well as providing a group decision evaluation procedure to assist in coastal zonal governance decision makers such as the awarding of fish farm site applications. The model is applied to the coastal zone of Grand Manan Island, New Brunswick situated in the Bay of Fundy.

Introduction

Aquaculture is now an important food product producer and employment provider. New Brunswick and British Columbia accounted for 83.2% of all Canadian aquaculture revenues in 2000. Total 2000 aquaculture gross output, including sales, subsidies, and growth in inventories, was $722.47 million, up $25.1 million from 1999 (CAIA 2003). The benefits of coastal aquaculture have been widely recognized and include increased income, employment, foreign exchange earnings and improved nutrition (Pullin 1989). However, as Barg (1992) noted "…aquaculture may increasingly be subject to a range of environmental, resource and market constraints. Aquaculture is competing for land and water resources, which in some cases resulted in conflicts with other resource users. Also, there is growing concern about the environmental implications of aquaculture development, comprising the adverse effects of aquaculture operations on the environment as well as the consequences of increasing aquatic pollution affecting feasibility and

*Corresponding author: dlane@uottawa.ca

sustainable development of aquaculture." Similarly, Brindley (1991) pointed out that more is not always better. He noted the necessity to evaluate coastal aquaculture from different points of view and that a multidisciplinary team, including specialists involved in decision making, should consider the economic, social, cultural, and environmental aspects to assess all possible impacts of coastal aquaculture development (Lane and Stephenson 1998).

Traditional fisheries are based on wild and spatially uncontrolled stocks living in coastal and marine environments. The difficulties associated with stabilizing wild fish stock exploitation has meant that aquaculture is becoming more important in the supply of products from the sea that can be better controlled and managed. Consequently, aquaculture is growing more rapidly than all other animal food producing sectors (Allen et al. 1992; Fletcher and Neyrey 2003). At the same time, aquaculture is challenging and being challenged by the complex interactions between resources, ecosystems and traditional resource users to share access to the marine environment. However, there are no applicable frameworks to assist in defining aquaculture's appropriate place and to reconcile the ecosystem impacts and challenges of historical and developing shared marine use. What is required is an integrated, spatially defined approach to evaluate and promote sustainable development in the coastal zone and to find new group consensus-based decision making aids toward sharing the marine environment among alternative use strategies (Nath et al. 2000; Hambrey and Southall 2002). Spatial information needs for decision makers who evaluate biophysical and socioeconomic characteristics as part of aquaculture planning efforts are illustrated by Kapetsky and Travaglia (1995). Osleeb and Kahn (1998) noted that these decision support needs cannot be effectively addressed without the use of geographic information systems (GIS). Geographic information systems is already being used effectively for scientific-based aquaculture site selection and evaluation (Carswell 1998; Ross 1998; Cross 1993).

This paper is applied to the coastal aquaculture developments around Grand Manan Island, New Brunswick, in the Bay of Fundy region of Atlantic Canada. The objective of the paper is to develop a prototype multicriteria decision support system for the evaluation of marine sites involving multiple participants in governance of the coastal zone and specifically for supporting decision-making processes for determining coastal aquaculture sites. The decision support system integrates the multiple components of the marine ecosystem together with human intervention activities through a spatial representation and valuation analysis of the marine system. The spatial description is coupled with a structured formulation of the multicriteria marine use decision problem confronting the various participants in the marine governance setting (Lane 1988; Saaty 1980). A computer interface integrates the GIS and the decision problem formulation into a single supporting software package.

Methodology

There are three methodological structures used in this modeling exercise. These are depicted in Figure 1:

1. Marine site components—describes the geophysical state of the marine ecosystem using spatial GIS indicators;

2. Selected site specific valuation—assigns an areal value to the ecosystem component inventories present in user defined selected marine sites, including the marine ecosystem

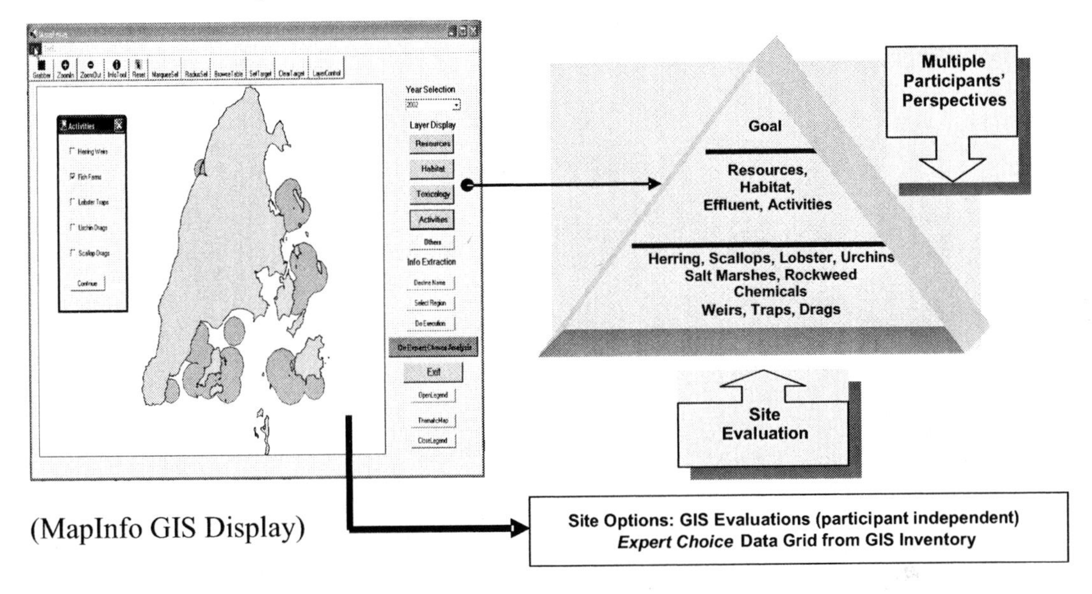

Figure 1. Methodology for the coastal aquaculture problem identifying the three main structures.

components and the human marine activities; selected marine sites are determined by the user using the GIS mapping tools and may correspond to designated areas targets as potential sites of interest by government, commercial, or community interests; and

3. Site comparison and multicriteria group decision ranking—uses tradeoff information about the marine ecosystem components and human activities (including fish farms) attributed to the various participants in the marine governance system to determine a ranking of alternative use strategies available at selected marine sites.

These structures are discussed in further detail below.

Marine Site Components

Site ecosystem inventory components are categorized into four major components: resources (R), habitat (H), socioeconomic activities (A), and effluents (E). For Grand Manan Island (Figure 2) the key ecosystem components are described as follows:

(1) Resources—thematic layers representing the natural resource stocks of importance to Grand Manan as spatial distributions of natural resources of several select species, including herring schools staging areas, inshore feeding areas, lobster molting and feeding areas, and scallops and urchins population area distributions;

(2) Habitat—static thematic layers to identify rockweed, benthic structures (in particular gravel bottom), salt marshes, and the presence (or absence) of a significant current flow;

(3) Effluents—thematic layers for different naturally occurring and activity-induced effluents denoted as specific chemicals;

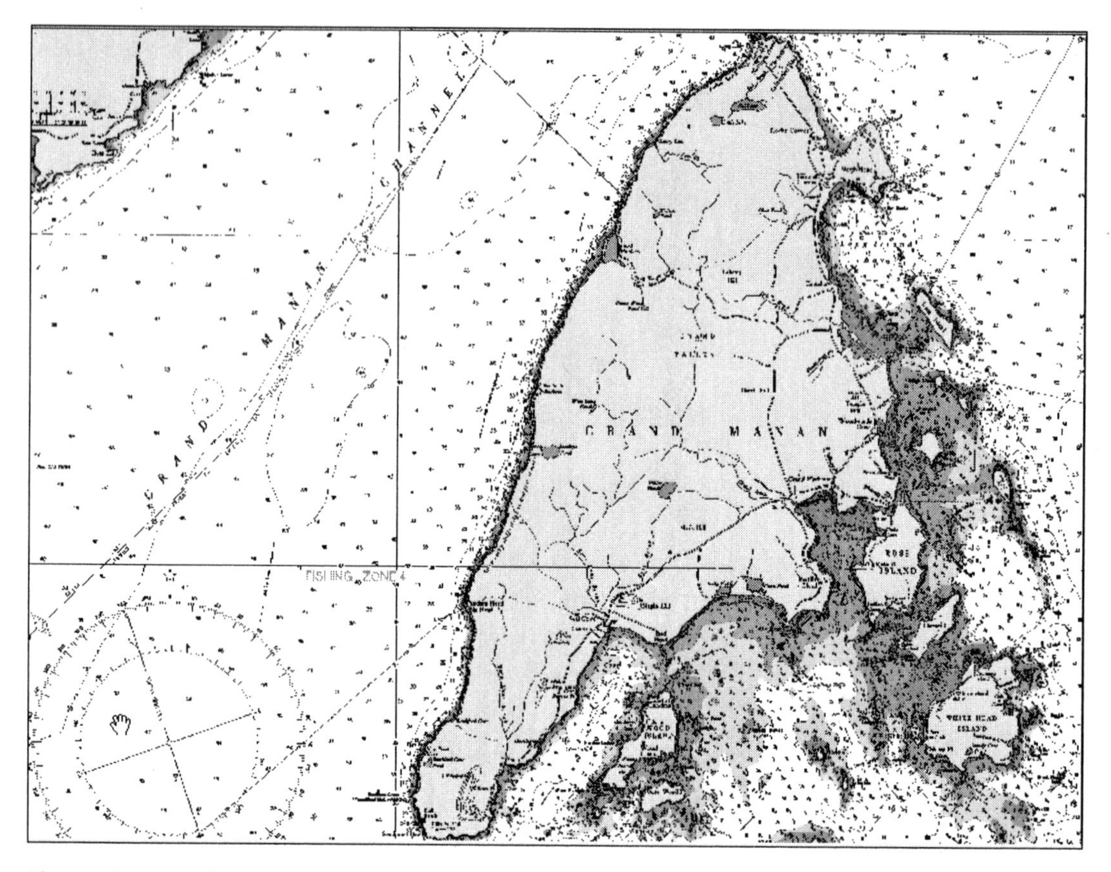

Figure 2. Map of Grand Manan Island, 35 km off the south coast of New Brunswick in the Bay of Fundy off Canada's Atlantic coast.

(4)Activities—thematic spatial layers representing socioeconomic interventions on the marine ecosystem, such as lobster traps, herring weirs, fish farm sites, scallop and urchin drags, and recreational areas.

The subcomponents of each ecosystem component set are represented by individual spatial data sets and their associated thematic data to determine the marine ecosystem inventory. The subcomponents for the case of Grand Manan Island are identified in Table 1. All information on the Grand Manan ecosystem components are organized into tables constituting a spatial mapping. Data are geocoded graphic objects (polygons) with longitude and latitude coordinates (MapInfo 2004; ESRI 2004). Local data for analysis of the case of Grand Manan Island include:

(1) Fish farm activity sites—Multiple years of fish farm site productivity data around Grand Manan from 1998 to 1902. Stationary fish farms sites are assigned a buffer size based on (disguised) harvest productivity data. (Actual productivities of the fish farms in the demonstration database have been altered for reasons of data propriety.)

(2) Urchin resource sites—Urchin resource locations and total estimated biomass data are recorded. Urchin distributions are concentrated along the east, south, and north shores of Grand Manan. Sedentary urchin resources are assigned buffers denoting loca-

Table 1. Ecosystem components and subcomponents with nonoverlapping unit yield values.

Ecosystem component	Ecosystem subcomponent	ID	Total area: (km²)	Total yield units:[b] Biomass (mt[c])	Yield per unit area: Biomass (mt[c])
Resources	Herring day	R1	200.	50,000	250.00
	Herring night	R2	99.3	50,000	400.00
	Lobster	R3	65.4	1,000	16.00
	Scallops	R4	38.4	300	7.50
	Urchins	R5	258.	30,000	116.28
		Area (km²)	*Area (km²)*	*Area (km²)*	
Habitat	Rockweed	H1	22.6	113	5.00
	Salt marshes	H2	0.5	3.5	7.00
	Current flow	H3	71.4	142	2.00
	Bottom structure	H4	395	395	1.00
		Area (km²)	*PPM[d]*	*PPM[d]*	
Effluents	Chemical A	E1	38.5	3.04	0.080
	Chemical B	E2	8.5	0.11	0.010
	Chemical C	E3	3.6	0.02	0.005
		Area (km²)	*Dollar values (000s $)*	*Dollar values (000s $)*	
Activities	Herring weir	A1	9.6	$1,500	$156.25
	Fish farms	A2	54.2	$10,000	$184.50
	Lobster traps	A3	24.1	$6,500	$266.67
	Scallop drags	A4	18.3	$550	$30.00
	Urchin drags	A5	116.4	$2,000	$17.18
	Recreational	A6	160.8	$6,700	$41.67
Total marine study area (km²)		1,602.1			

[a] Total area is relative to the Grand Manan area of interest.
[b] Total yield is estimated assuming subcomponents are nonoverlapping over entire Grand Manan area of interest.
[c] Metric tons.
[d] Parts per million.

tion; smaller buffers denote urchin harvesting areas.

(3) Herring weir activity sites—Stationary herring weirs sites are located along the east and northeast shores of Grand Manan Island. Buffers of influence (0.5 km in diameter) are added to point weir sites. The harvest of herring by weirs in 2002 around Grand Manan Island is estimated as 5,379 kg and valued at approximately $1.5 million (Waters and Clark 2004).

(4) Herring resources distribution—Schools of herring move daily aggregating in offshore waters from Grand Manan Island during the day. In the night they return in aggregations to feed closer to shore. The herring resource is partitioned into these offshore staging areas and aggregations in the inshore feeding areas. The inshore and offshore herring aggregations denoted here are based on estimated data from ongoing scientific research observations (Power et al. 2004).

(5) Habitat distribution—Habitat data are provided for Rockweed and Salt Marshes around Grand Manan Island. Rockweed grows along and almost the entire coastline of the

Island. Salt Marshes are located uniquely along the midwestern shore of Grand Manan and occupy only 0.46 km^2.

(6) Effluent distribution—Available effluent data are associated with naturally occurring sources and with the location and productivity of commercial activities including existing fish farms. Spatial information on zones of influence was taken from scientific models of surface drift analysis for a subset of Grand Manan fish farms.

GIS data compiled for the Grand Manan sources above inventoried the status of marine and coastal sites around the Island. Figure 3 illustrates the map of overlapping ecosystem subcomponents for the total study area.

Selected Site Specific Valuation

The model utilizes site specific spatial data sets described above representing resources abundance, habitat inventory, values from commercial and recreational activities, and influence plumes from sources of effluents. These data sets are processed as thematic layers in a GIS describing the marine site. Quantitative yields are assigned by layer area of the selected site. Yields are assigned to each subcomponent according to the estimated pairwise overlapping cumulative effects layers of the ecosystem subcomponent datasets based on linear percentage impact yield adjustment parameters defined for each of the two overlapping component pairs (e.g., when herring night distribution resources overlap with rockweed habitat the unit yield of herring resources increases by an estimated 9%, whereas the rockweed habitat unit yield is expected to be unaffected by this same overlap of area). A database of pairwise yield adjustment values accounts for all 153 possible overlapping pairs in the 18 subcomponent set.

Ecosystem subcomponent phenomena are represented by polygons that assume a static uniform distribution over the two-dimensional area of the polygon. For each of the subcomponents, a unit area value per square kilometer denoting "yield" for the subcomponent present is assigned (Table 1). Resource subcomponents, R, are assigned unit area values in metric tons (mt) consistent with biomass measurements of abundance of those phenomena in specific spaces at a particular instance in time (e.g., average daily biomass configuration over a year). Habitat subcomponents, H are assigned area values (i.e., km^2) representing the existence of habitat within a specific area at a particular instance in time (e.g., average configuration over a year). Effluent subcomponents, E, are assigned unit values in ppm, and represent specific areas assumed to be uniformly contaminated by effluent substances at a particular instance in time (e.g., average daily effluent configuration over a year). Finally, activity subcomponents, A are assigned currency unit values ($) representing the estimated economic value of included activities over a particular instance of time (e.g., annually).

Full yield valuation of a selected marine site takes into account the presence of singular and overlapping layers of resources, habitat, effluent and activities. If there is no overlap, the Yield values are a simple function of the area (in km^2) of the nonoverlapping layer and the Yield per unit area values input for each Grand Manan subcomponent (Table 1). When an overlap of subcomponents occurs in a selected area, area yields may increase, decrease, or be unaffected compared with the original subcomponent yields. For example, if a polygon of a Resources subcomponent overlaps another resource or a habitat subcomponent polygon, the valuation effect will be a positive increase on the overall resource yields and the habitat yield of the selected area that includes the overlap. If, on the other hand, the overlapping subcomponent belongs to an activity layer, the effect will be negative on the resource yield in the overlap portion of the selected area, and positive on the activity overlap. If an effluents polygon overlaps with any other component polygon, the yield of that overlapped area will be negatively affected, while the effluents component remains unaffected. A summary of the higher level pairwise subcomponent yields is provided in Table 2. These results are developed from discussion among fisheries scientists and represent intuitive interpretations for valuation purposes. Net yields from total subcomponent items including overlapped and non-overlapped areas for selected sites areas 1 and 2 are provided in Table 3 (see also Figure 3). Estimated ecosystem cumulative effects interactions are designed simply to integrate known and anticipated impacts in the complex marine environment among the participating decision makers. While providing a basis for describing these interactions, the procedure does not imply a comprehensive or predictive model of ecosystem interactions.

Site Comparison and Multicriteria Group Decision Ranking

A comprehensive evaluation of ecosystem components is provided from interpretations by different participants in the decision-making problem. The analytical hierarchy process (AHP) is adopted to evaluate different participants' perspectives on the importance of the ecosystem components and subcomponents (Saaty 1980). A common hierarchy model is built for all participants that specifies the goal, components, and subcomponents of the Grand Manan ecosystem. The following items are explicitly identified as the common AHP 3 level hierarchy model (Figure 1, item 3). The hierarchy illustrated in Figure 4 (from the AHP software *Expert Choice*) provides AHP generated weighted attributes for one participant group (federal scientists) as a result of feedback on pairwise comparisons of the hierarchy elements by level as follows:

Level 1. Goal: The goal of the multicriteria problem—to rank selected marine site use strategies given their respective yield valuations and each participants' importance weights assigned from pairwise trade-offs to the components and subcomponents of the ecosystem.

Level 2. Components: Representation of the ecosystem components of the marine system around Grand Manan, namely, resources, habitat, effluent, and activities.

Level 3. Subcomponents: Representation of the Grand Manan ecosystem subcomponents,

Table 2. Summary ofrules for single and multiple overlapping layers in pairs.

	Yield-affecting components			
Yield-affected components	Resources	Habitat	Effluent	Activities
Resources	+	+	–	–
Habitat	+	+	–	–
Effluent	0	0	+	0
Activities	+	+	–	–

Note: Plus signs (+) denote that the overlapping effect on yields for the layer indicated at the left hand side is positive; minus signs (–) denote negative yields; zeroes (0) denote no overlapping effect. Note also: this is not a symmetric matrix. For example, the overlapping effect of resources and activities has a negative effect on resources but a positive effect on activities.

Table 3. Tables of net yield valuations by subcomponent for the total study area and selected areas.

Ecosystem components	Ecosystem subcomponents	Total study area	Area 1 with fish farm	Area 1 without fish farm	Area 2
Resources (metric tons)	Herring day	50,362	0.0	0.0	0.0
	Herring night	40,732	678	717	3,529
	Lobster	988	75	75	211
	Scallops	203	30	30	52
	Urchins	30,023	1,283	1,313	3,047
Habitat (km^2)	Rock weed	108.44	2.80	2.87	8.70
	Salt marsh	3.23	0.00	0.00	0.00
	Current flow	142.85	4.86	4.86	11.86
	Bottomstructure	395.01	11.58	11.58	28.40
Effluents (parts per million)	ChemicalA	3.04	0.17	0.00	0.07
	ChemicalB	0.11	0.03	0.03	0.01
	ChemicalC	0.02	0.02	0.01	0.00
Activities ($000s)	Herring weirs	$1,442.10	$315.36	$368.66	$177.06
	Fish farms	$9,630.73	$835.70	$0.00	$1,312.54
	Lobster traps	$7,571.77	$615.78	$615.32	$1,965.85
	Scallop drags	$668.73	$98.94	$98.97	$221.00
	Urchin drags	$2,100.26	$112.41	$112.55	$286.62
	Recreation	$6,598.11	$106.05	$108.35	$398.44
Area (km^2)		1,602.1 km^2	12.4 km^2	12.4 km^2	30.4 km^2

namely, the resources—herring staging area, herring inshore feeding area, lobster, scallops, and urchins; habitats—rockweed, salt marshes, current flow, and bottom structure; effluents—chemicals A, B, and C; and activities—herring weirs, fish farms, lobster traps, scallop drags, scallop drags, urchin drags, and recreational use.

To complete the common AHP hierarchy model, weights are calculated for each level 2 and 3 of the hierarchy based on pairwise comparisons feedback from each of the participating decision makers. Participant organizations are represented by five groups: (1) local communities (including aboriginal community members who inhabit the coastal zone typically adjacent to selected marine areas); (2) scientists (with federally defined resource sustainability mandate); (3) industrial organizations (interested in commercial marine exploitation and use opportunities); (4) nongovernmental organizations (NGOs; focused principally on environmental protection); and (5) provincial governments (with jurisdictions to license and manage marine use sites).

To illustrate the application of the model, AHP weights (Table 4) for the relative importance of the ecosystem subcomponents are determined based on attributed pairwise comparisons for each of the participating groups. Table 4 results indicate that local communities attribute importance to effluents (0.435) and habitat (0.302) components, and apply less weight to activities and resources. Scientists weight resources highest (0.546) and activities lowest (0.075) consistent with the scientific mandate for conservation of resources. Industrial organizations attach most importance to activities (0.538) and lower relative weighting to habi-

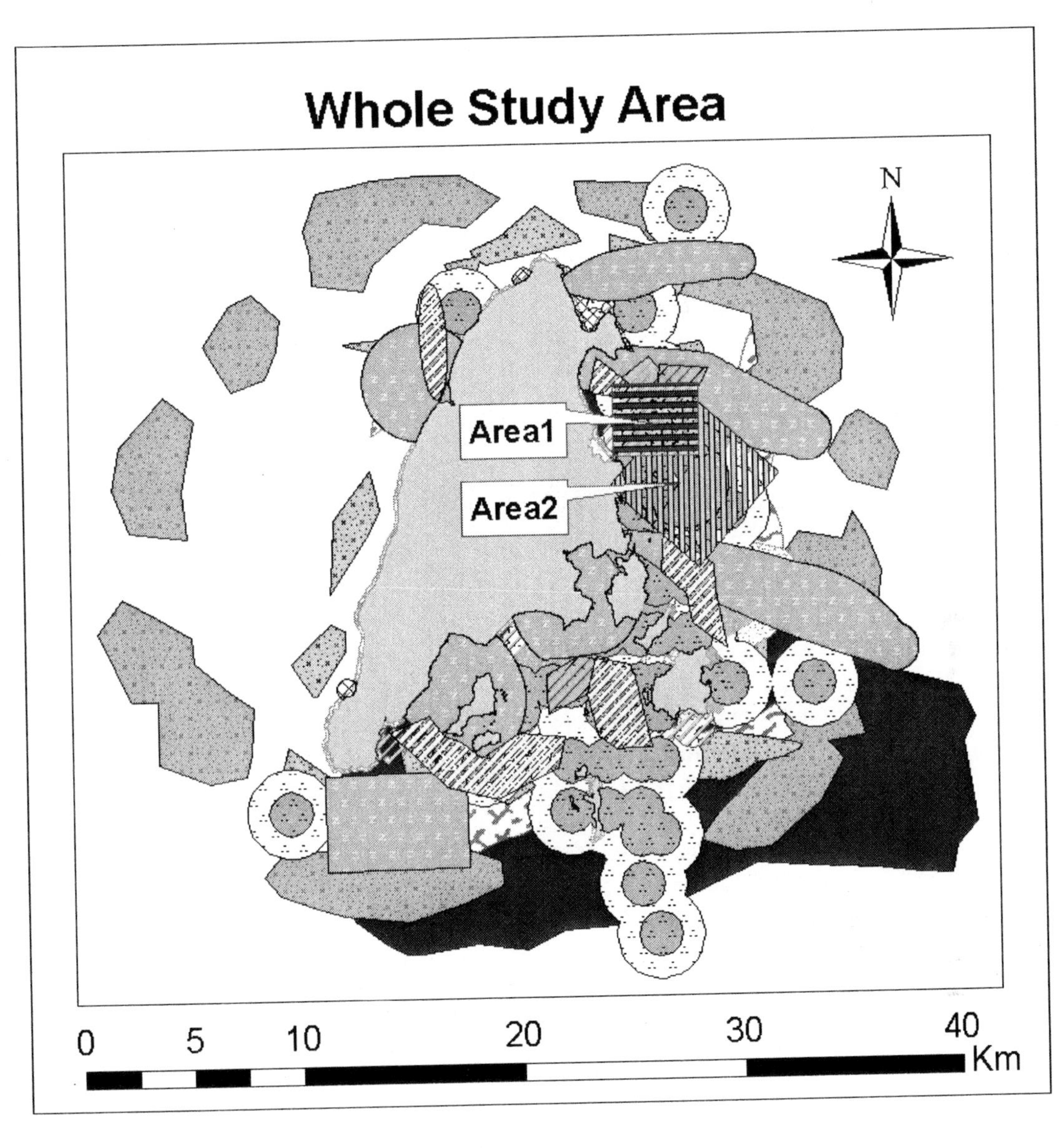

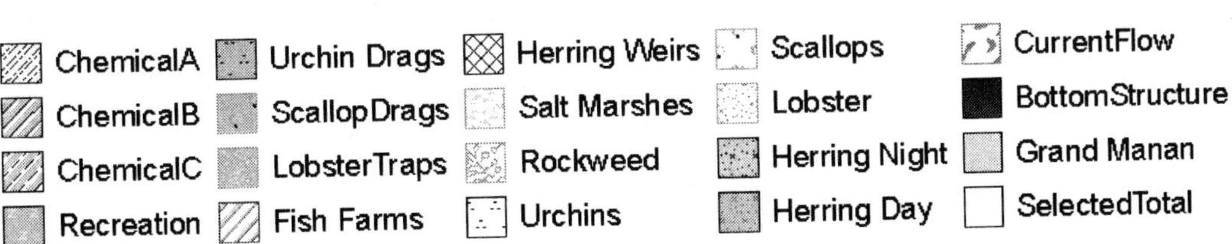

Figure 3. MapInfo overlay information for the whole study region of Grand Manan Island.

tat (0.124) and effluent (0.082) components. For NGOs, effluents are regarded as the most important (negative) component in the system (0.439). The provincial government evaluates components in close comparison to the industrial organizations with the exception of differences with respect to the weighting of the effluent and the habitat components.

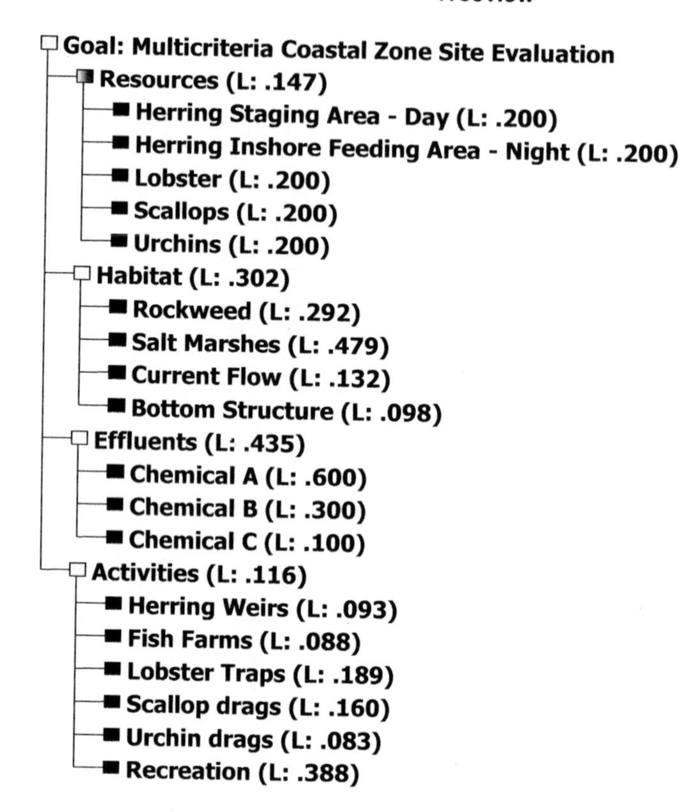

Figure 4. The hierarchical model for the aquaculture problem; weight values are attributed to the Scientist group. (Source: Output of problem formulation from *Expert Choice 2000*)

In terms of the ecosystem subcomponents, industrial organizations and the provincial government attribute more relative importance to lobster and scallops due to their relatively higher commercial value. All groups attribute relative importance (>0.250) to rockweed as a habitat subcomponent, and local communities assign salt marshes as the

Table 4. Attributed AHP importance weights to the ecosystem subcomponents for the five participants in the common hierarchy.

Components and subcomponents	Local communities	Federal scientists	Industrial organizations	Nongovernmental organizations	Provincial governments
Resources	0.147	0.546	0.215	0.235	0.226
Herring day	0.200	0.200	0.079	0.200	0.096
Herring night	0.200	0.200	0.079	0.200	0.096
Lobster	0.200	0.200	0.276	0.200	0.330
Scallops	0.200	0.200	0.523	0.200	0.330
Urchins	0.200	0.200	0.043	0.200	0.148
Habitat	0.302	0.217	0.115	0.235	0.075
Rockweed	0.292	0.250	0.375	0.375	0.301
Salt marshes	0.479	0.250	0.125	0.375	0.410
Current flow	0.132	0.250	0.125	0.125	0.171
Bottom structure	0.098	0.250	0.375	0.125	0.118
Effluent	0.302	0.217	0.115	0.235	0.075
Chemical A	0.691	0.691	0.691	0.691	0.691
Chemical B	0.218	0.218	0.218	0.218	0.218
Chemical C	0.091	0.091	0.091	0.091	0.091
Activities	0.116	0.075	0.593	0.082	0.514
Weirs	0.093	0.326	0.095	0.174	0.073
Fish farms	0.058	0.154	0.163	0.105	0.310
Lobster traps	0.189	0.277	0.248	0.262	0.155
Scallops drags	0.160	0.102	0.408	0.063	0.155
Urchin drags	0.083	0.059	0.053	0.063	0.247
Recreation	0.388	0.082	0.034	0.332	0.059

most important habitat subcomponent (0.479). NGOs do not differentiate among Activities subcomponents (all 0.200). Scientists prefer the noninvasive, passive exploitation of herring weirs (0.380) as they are judged to be passive with respect to the environment, whereas draggers' operations (for scallops and urchins) are ranked low by scientists (<0.100). For industrial organizations, subcomponent evaluations imply market values such that scallop drags weight (0.462) exceeds that of lobster traps, fish farms, herring weirs and urchin drags (0.264–0.039). Fish farms are given high consideration by the provincial government (0.358) related to the greater provincial jurisdiction in this activity as an opportunity for employment. Nongovernmental organizations and local communities weight recreation highest among all activities (over 0.3).

Application and Results

Specific experiments of selected marine areas around Grand Manan are presented to illustrate the application of the model for decision support. These applications include: (1) yield evaluation of the total study area, (2) comparison of one selected area to a designated ideal area, (3) comparison of two or more areas to each other including comparative evaluation of areas including and excluding fish farms.

(1) Yield Evaluation of the Total Study Area

The whole area around Grand Manan Island comprises a total area of 1602.1 km^2 (Table 1). There are 110 total polygons identified for the subcomponents data in the Grand Manan Island study area. The net yield values by subcomponent for the total study area and the 2 areas (area 1, with and without fish farm, and area 2) as identified in Figure 3 are derived for each of the subcomponents for Grand Manan in Table 3. The yield valuation for the entire Grand Manan Island marine study area is determined from all nonoverlapping and all overlapping layer data components. The net yields for each of the 18 subcomponents are provided in Table 3, column 3. These net yield valuations itemize the resources, habitat, effluent, and activity components inventory including the cumulative overlapping impacts for the entire area. For example, total herring (day) net valuation for the entire Grand Manan marine area in the study is over 50,000 mt, while herring (night) resources are valued at 41,000 mt. Similarly, urchins have overall calculated net yield value of over 30,000 mt. Lobster and scallop resources are both estimated at net yields less than 1,000 mt.

In the Grand Manan Island study area, rockweed is evaluated at just over 100 km^2 of inventoried area all along the coastline of the main island, whereas salt marshes are only present in 3 km^2 in a small area along the east coast of the island on Long Island Bay. Influential current flow occurs in 143 km^2 and gravel bottom structures are present in the study area in areas of nearly 400 km^2. Throughout the entire marine area, chemical effluents are evaluated in average densities of 3, 0.1, and 0.2 ppm of the Chemicals A, B, and C, respectively. Finally, fish farms, lobster fishing, and recreational activities have highest net yield values for all activity yields at $10 million, $7.6 million, and $6.6 million respectively over the whole of the study area. Urchin dragging (at $2.1 million), herring weir operations (at $1.5 million) and scallop dragging (at $0.7 million) round out the Grand Manan marine activities' net yields.

In comparison with the Table 1 nonoverlapping yield values for each of the 18 subcomponents, the results of Table 3 are adjusted for net overlapping (positive and negative) yield cumulative effects of the spatially defined subcomponents. This explains why the nonover-lapping yields of Table 1 (column 5) for the different subcomponents differ from the net yields of the whole study area that account for overlapping subcomponents and cumulative yields by subcomponent (Table 3, column 3). It is noted, for example from Table 3, that the estimated Grand Manan net yield attributed to herring weirs ($1.44 million) is slightly less than the $1.5 million nonover-lapping expected yield for this activity (Table 1) due to the negative yield impacts from overlapping fish farms and effluents. Similarly, the resource yield for urchins of 30,000t (Table 1) is evaluated at a slightly higher yield (30,023 mt) because of the positive overlap from habitat and other resources in the current configuration of the spatial data. The total study area net yield valuations provide a benchmark for the ecosystem subcomponent valuations of selected areas. Selected areas with subcomponent yields, approaching yields for the entire study area, identify characteristic areas that capture a disproportionate share of the subcomponent.

(2) Comparison of One Selected Area to a Designated Ideal Area

Another use of the yield valuations is in the comparison of selected area's yields against area targets for the different subcomponents. There may be interest, for example, in selecting areas that achieve target yields for an activity, such as lobster traps dollar yields, in order to consider designating the area as a lobster fishing area. From Table 3, subcomponent net yields for the total study area (column 3) are used to establish benchmarks on the yields from the accumulated coverage of the total area. These benchmark yields are used to establish upper bounds on subcomponent yield targets and represent ideal values for area subsets of the whole study space. Subcomponent net yield values are scored as 1 and the relative net yield scores for the alternatives are scored as the fraction of the benchmark ideal value by subcomponent. Figure 5 shows the results of the rankings of selected area 1 (square space included with fish farm in Figure 3) to an ideal area. Using the net yields for area 1 in Table 3 and calculated ideal subcomponent values as alternatives' data, the AHP ranking results show that all participants, as well as a combined input (for all equally weighted participants' inputs) rank area 1 to be between 27% to 31% compared to the ideal. Similarly, the comparison of other selected areas to the same ideal provides a direct means of interpreting area preferences relative to specified ideals as targets.

(3) Comparison of Two or More Selected Areas

Consider the yield valuation of two selected sites in the Grand Manan marine area. These sites are designated as area 1 (square) and area 2 (diamond) along the eastern coast of Grand Manan Island as shown in Figure 3. Area 1 comprises an area of 12.4 km^2 in Long Island Bay that includes a fish farm site. Area 2 is an area in Duck Island Sound of 30.4 km^2. Table 3 (columns 4 and 6) presents the yield valuations for these two areas. From the yield valuations of these two sites, it is noted that area 2 yields dominate all area 1 net yield values with the sole exception of the herring weir activity (where area 1 yields exceed area 2 yields by almost double [i.e., $315,000 versus $177,000]). In all other subcomponents, the larger area 2 dominates area 1 yield estimates (i.e., resources, habitat, and activities

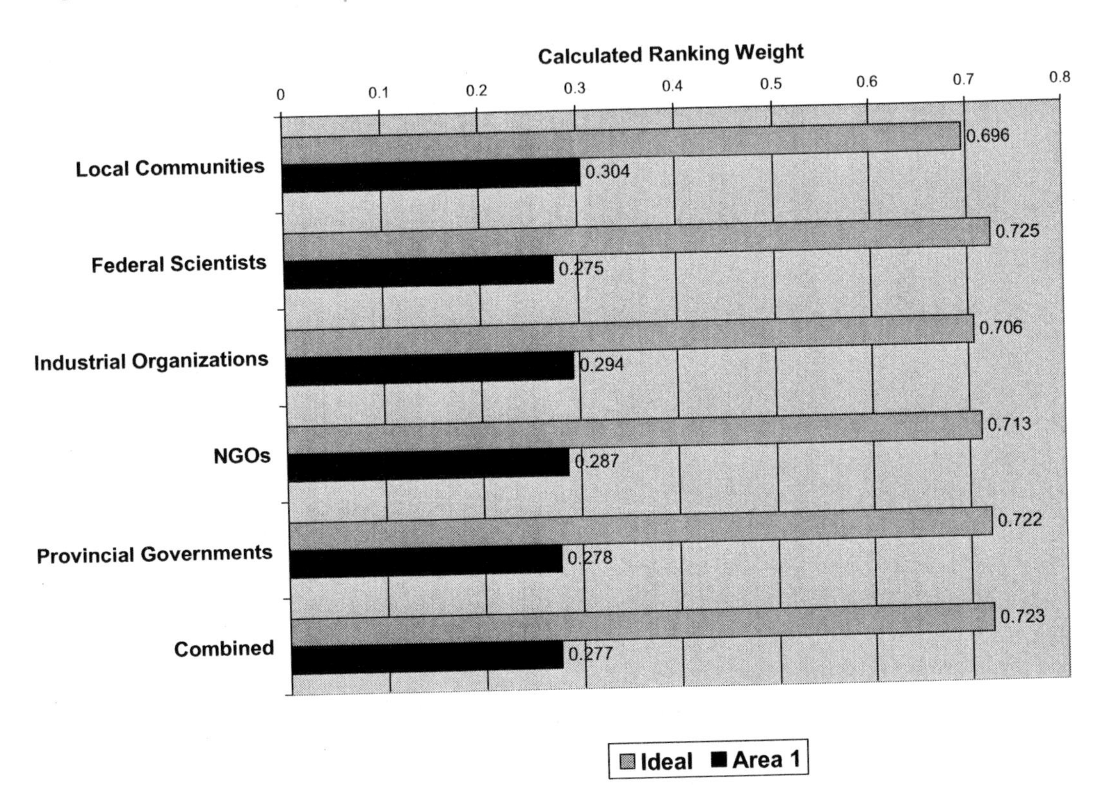

Figure 5. Evaluation summary of the participating groups and combined results in comparing the site area 1 and the ideal.

net yields for area 2 exceed those for area 1, and evaluated net yields for effluents for area 2 are less than those for area 1). On a prorated area basis (i.e., adjusting area 2 totals by 12.4/30.4 = 0.408), area 2 remains dominant in 10 of 16 subcomponents (two subcomponents for salt marshes habitat and herring day resources are not present in either area 1 or 2), and close to area 1 yields in 3 of 6 subcomponents. Area 1 dominated area prorated yields for the scallop resource, fish farm activity, and (as noted above) herring weirs. Preference for area 1 over area 2, or vice versa, will depend on the purpose of the comparison (i.e., if the areas are being evaluated for the potential designation of a new herring weir, then area 2 domination of the herring resource [herring night], 3529t for area 2 versus 678 mt for area 1, would suggest that area 2 might be a preferable site for a herring weir). The effect of a new herring weir on each of the areas could be investigated by locating space for the weir in areas 1 and 2 and re-evaluating and comparing the yields for the areas accordingly.

Alternatively, it may be considered that area 1's overall yield deficit compared to that of area 2 may be due to the presence of a fish farming activity. Removing the fish farm from area 1 and recalculating the yields would allow the areas to be compared under this assumption. Table 3, column 5 provides the net yield evaluation for area 1 with the existing fish farm removed. The effect of removing the fish farm from area 1 leads to changes in the net yield values for the subcomponents. In particular, the herring night resource increases by 5% from 678 mt with the fish farm to 717 mt without the fish farm. Marginal improvements are also observed for the urchins resource, for herring weirs activity, recreational activity, and chemical A (an effluent associated with the fish farm) improving the overall yield value of area 1 with the fish farm removed. However, these positive yield effects are offset by the large loss in the fish farm activity yield of $835 thousand (Table 3).

In comparison with the unchanged yield valuations of area 2, the area 1 (without the fish farm) net yields do not change the relative dominance of area 2 as discussed above. Ultimately, the direct comparison of the two areas depends on the relative importance of the different ecosystem subcomponents. While it can be said that area 2 does not strictly dominate area 1 in terms of the yields (since the subcomponent-wise values for area 2 do not all improve on the corresponding values for area 1), it can be argued that area 2 would appear to be preferable to area 1 given the number and extent of improvements in a majority of subcomponent yield values.

Figure 6 provides the rankings of the AHP model results for the participant groups. As expected, Area 2 having the highest calculated ranking weight is preferred by all participants. Ranking weights for area 1 with and without its fish farm vary by participant. Nongovernmental organizations, local communities, and scientists who all assign lower relative value to the fish farm activity compared to other activities (Table 4) all rank area 1 without its fish farm higher than area 1 with its fish farm. Conversely, industrial organizations and provincial governments assign a higher ranking to area 1 when its fish farm is included in its area definition. Overall the combined inputs from all participants demonstrate the relative preference of area 2 and relative indifference in area 1 with and without the fish farm.

Similarly, the model may be used to evaluate strategic options for marine use. This includes a review of alternative distributions of exist-

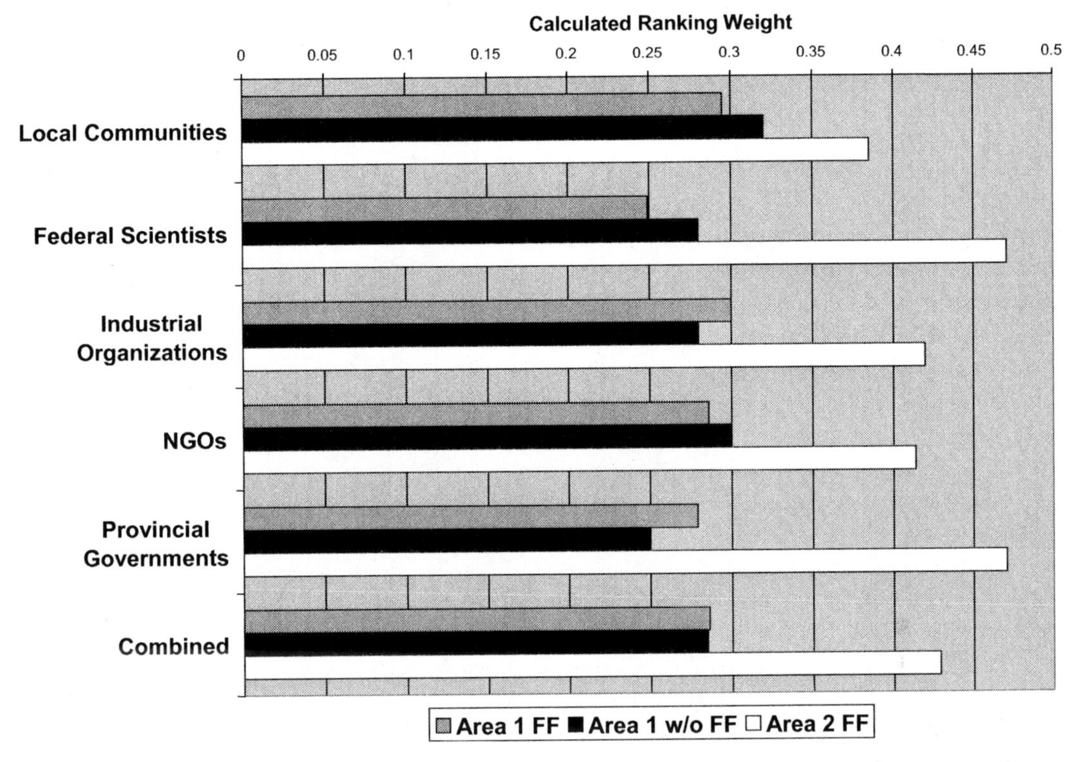

Figure 6. Evaluation summary of the participating groups and combined results in comparing the area sites area 1 (with and without fish farm) and area 2 in the vicinity of Great Duck Island Sound.

ing fish farms to maintain productivity which consider different clustering arrangements that would be more amenable to the local communities and other participants.

Conclusions

The system establishes a logical interpretation of Grand Manan spatial marine site valuation through the calculation of cumulative effects yields as a function of the ecosystem subcomponents present at the site. Attributed pairwise comparison trade-offs of the importance of the ecosystem subcomponents as interpreted by the various participants in the marine governance decision-making process, are used to develop a priority ranking system based on the yield valuation of marine sites around Grand Manan. Selected sites experiments demonstrate how the model can be used as a decision aid in actual decision problems of site use strategy evaluation. The model provides a useful methodology for developing and evaluating alternative spatial marine use strategies including the spatial assignment of fish farms in the context of making smart decision choices (Hammond et al. 1999). Similarly, the comparative ranking of alternative uses can be used to exclude certain strategies from further consideration or to exclude selected areas for particular uses (e.g., as potential aquaculture sites). Finally, the method provides shared participatory group learning concerning the complex ecosystem-based decision problems that lead to group consensus decisions on marine use.

Problem exploration and group learning aside, the actual effectiveness of this decision support system depends on (1) the accuracy and precision of the spatial input data, (2) the mutual acceptance and agreement with the cumulative effects net yield valuations by the participants, and (3) the extent to which the participants contribute their trade-off information and their commitment to a consensus resolution of a particular marine use problem. Further developments of the model framework provide clear direction on future integrated data requirements and the importance of the consensus building exercise for group decision making in the marine environment.

Acknowledgments

Data are obtained with thanks from the St. Andrews Biological Station in St. Andrews, New Brunswick. The biological station is part of the federal Department of Fisheries and Oceans, Canada and is home to federal fisheries scientists who participate in the monitoring and analyses of aquaculture in the local area, including Grand Manan Island in the Bay of Fundy.

Reference

Allen, P. G., L. W. Botsford, A. M. Schuur, and W. E. Johnston. 1992. Bioeconomics of aquaculture. Developments in aquaculture and fisheries science, volume 13. Fourth impression. Elsevier Publishers B.V, Amsterdam.

Barg, U. C. 1992. Guidelines for the promotion of environmental management of coastal aquaculture development. FAO Fisheries Techical Paper 328, Rome.

Brindley, B. 1991. What is "sustainable"? Some rules for the development road. Ceres 128:35–8.

CAIA (Canadian Aquaculture Industry Alliance). 2003. Available: http://www.aquaculture.ca. (June 2005).

Carswell, B. 1998. BCAS: an information system for aquaculture and marine resource planning. Ministry of Agriculture, Fisheries and Food, British Columbia. B.C. Government Publications Index. Available: http://www.publications.gov.bc.ca/default.aspx. (July 2007).

Cross, S. F. 1993. Assessing shellfish culture capability in coastal British Columbia: sampling design considerations for extensive data acquisition surveys. Aquametrix Research. Prepared for British Columbia Ministry if Agriculture, Fish, and Food, Victoria.

ESRI. 2004. ArcInfo. Available: http://www.esri.com/software/arcgis/arcinfo/. (June 2005).

Expert Choice. 2004. What we do? Available: http://www.expertchoice.com/aboutus/whatwedo.htm. (July 2005).

Fletcher, K. M., and E. Neyrey. 2003. Marine aquaculture zoning: a sustainable approach in the growth of offshore aquaculture. Available: http://www.olemiss.edu/orgs/SGLC/zoning.htm. (June 2005).

Hambrey, J., and T. Southall, 2002. Environmental risk assessment and communication in coastal aquaculture: a background and discussion paper for GESAMP Working Group 31 on Environmental Impacts of Coastal Aquaculture. Available: ftp://ftp.fao.org/FI/DOCUMENT/gesamp/ERAbackg_paperGESAMPWG31.pdf. (July 2007).

Hammond, J., R. Keeney, and H. Raiffa. 1999. Smart choices: a practical guide to making better life decisions. Broadway Books, New York.

Kapetsky, J. M., and C. Travaglia. (1995) Geographical information systems and remote sensing: an overview of their present and potential applications in aquaculture. Pages 187–208 *in* K. P. P. Nambiar, and T. Singh, editors. Aquaculture towards the 21st century. INFOFISH, Kuala Lumpur, Malaysia.

Lane, D. E. 1988. A partially observable model of decision making by fishermen. Operations Research 37(2):240–254.

Lane, D. E., and R. L. Stephenson. 1998. Fisheries co-management: organization, process, and decision support. Journal of Northwest Atlantic Fishery Science 23:251–265.

MapInfo. 2004. MapInfo. Available: http://www.mapinfo.com. (July 2005).

Nath, S. S., J. P. Bolte, L. G. Ross, and J. Aguilar-Manjarrez. 2000. Applications of geographical information systems (GIS) for spatial decision support in aquaculture. Aquacultural Engineering 23:233–278. Available: http://www.aquaculture.stir.ac.uk/GISAP/Pdfs/AqEngReview.pdf. (June 2005).

Osleeb, J., and S. Kahn. 1998. Integration of geographic information. Pages 161–189 *in* V. Dale and M. English, editors. Tools to aid environmental decision making. Springer-Verlag, New York.

Power, M. J., R. L. Stephenson, K. J. Clark, F. J. Fife, G. D. Melvin, L. M. Annis. 2004. 2004 Evaluation of 4VWX Herring. Canadian Science Advisory Secretariat Research Document 2004/030. Available: http://www.dfo-mpo.gc.ca/csas/Csas/publications/ResDocs-DocRech/2004/2004_030_e.htm. (July 2007).

Pullin, R. S. V. 1989. Third World aquaculture and the environment. NAGA ICLARM Quarterly 12(1):10–13.

Ross, L. G. 1998. The use of Geographical Information Systems in Aquaculture: a review. Paper presented at I Congreso Nacional de Limnologia, Michoacan, Mexico. November.

Saaty, T. L. 1980. The analytic hierarchy process. McGraw-Hill, New York.

Waters, C. L., and K. J. Clark. 2004. 2004 Summary of the Weir Herring Tagging Project, with an Update of the HSC/PRC/DFO Herring Tagging Program. Canadian Science Advisory Secretariat Research Document 2004/032. Available: http://www.dfo-mpo.gc.ca/csas/Csas/publications/ResDocs-DocRech/2004/2004_032_e.htm. (July 2007).

Session

Freshwater Habitat Improvement

American Fisheries Society Symposium 49:817–834

Restoration of Atlantic Salmon and Other Diadromous Fishes in the Rhine River System

Frank Molls* and Armin Nemitz

Fisheries Association North Rhine-Westphalia, Diadromous Fish Program
Frankfurter Strasse, D-53721 Siegburg, Germany

Abstract.—The Rhine River was once one of Europe's important large rivers for diadromous fishes such as Atlantic salmon *Salmo salar*, sturgeon, and others. Nearly all diadromous fishes disappeared from the river system during industrialization of the 19th and 20th centuries. Severe water pollution and the construction of weirs and dams are considered the major causes of the extinctions. Additional settlement, intense agriculture, and the need for flood prevention degraded river morphology from the upper catchment regions to the delta. Sewage treatment since the 1980s permitted ecological recovery to some extent. Nevertheless, most of the diadromous fishes are still missing or endangered today. However, there are several programs that assist re-establishment of the populations. The Atlantic salmon is the official flagship species of the International Rhine Action Program. This species serves as an important indicator organism because the different phases of its life cycle depend on different aspects of the river ecosystem and structure: (a) the morphology of streambeds and catchment area (reproductive phase), (b) the integrity of structural habitat in flowing waters (juvenile phase), (c) impacts on the continuity of downstream migration routes (smolt phase), and (d) barriers that hinder upstream migration (returning adults). Performance of the indicator species and restoration of major habitat factors in several Rhine tributaries are monitored annually. The ambitious Rhine River project demonstrates that fisheries and conservation agencies can act in concert to restore habitat from the upper catchment area to the river mouth.

The Rhine River

Hydrology and Geography

With a length of 1,320 km and a mean flow of 2,230 m^3/s (peak flow 11,000 m^3/s 1995), the Rhine River is one of Europe's largest rivers (Figure 1). The total catchment area is about 185,000 km^2. The Rhine's source is in the Swiss Alps. The river then passes along the French–German border and flows through Germany and the Netherlands. The catchment also includes parts of Austria and Luxembourg. Main tributaries are the rivers Aare, Neckar, Main, Moselle, Lahn, and Ruhr. In the Netherlands, the Rhine splits into the three delta branches: Waal, Lek/Nederrijn, and Ijssel. The river flows into the North Sea mainly via the New Waterway (Niewe Waterweg) and the Haringvliet and Lake Ijsselmeer sluices. The dams and sluices have resulted in freshwater domination of former estuarine zones (Haringvliet and Lake Ijsselmeer).

*Corresponding author: frank.molls@wasserlauf-nrw.de

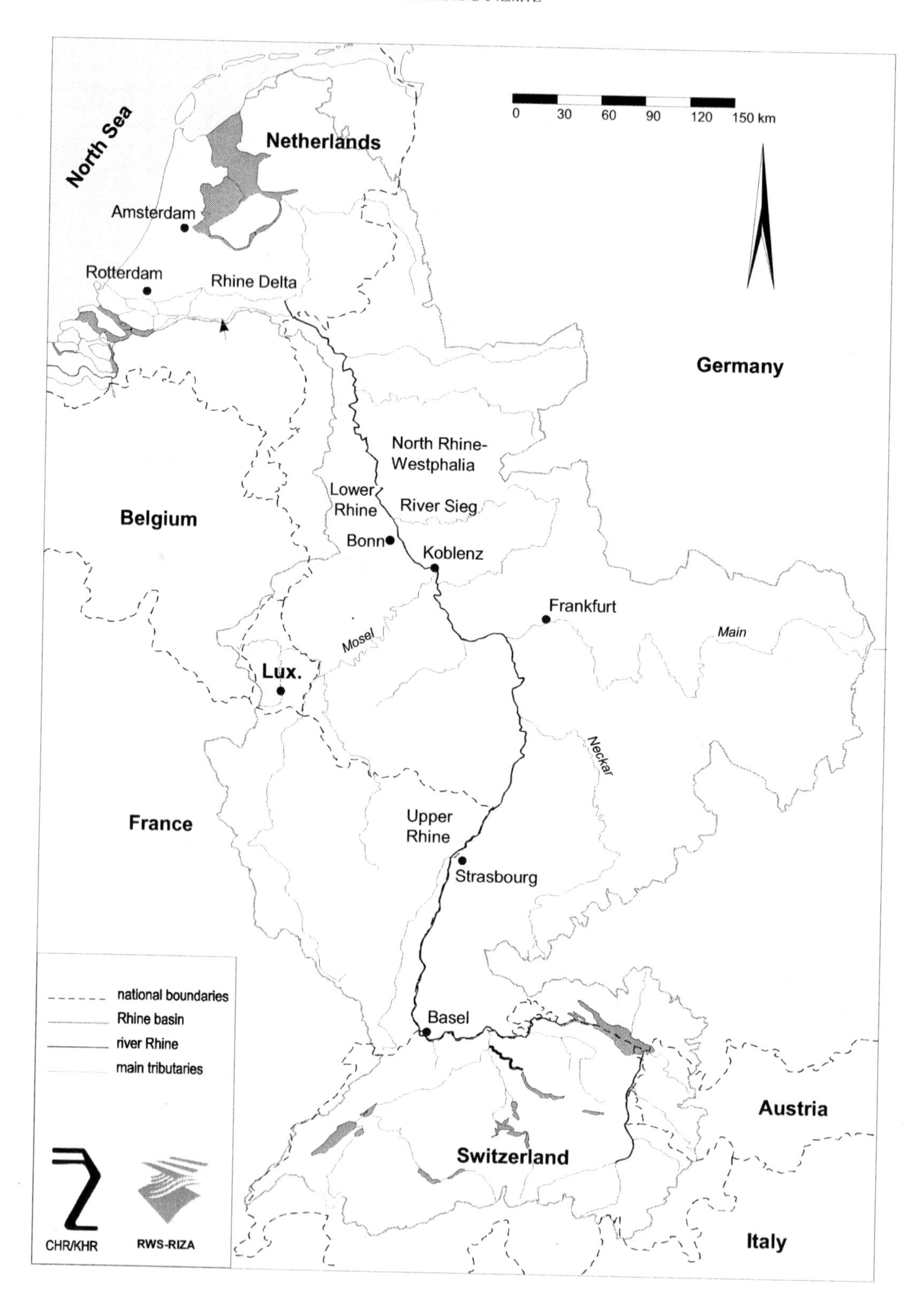

Figure 1. Map of the Rhine River catchment (changed after RIZA). North Rhine Westphalia = NRW.

Fish Species and Fisheries

The river system was originally populated with about 35 potamodromous and at least 12 diadromous fish and cyclostome species (Lelek and Koehler 1990; Lelek & Buhse 1992; de Groot 2002). The latter were sea lamprey *Petromyzon marinus*, European river lamprey *Lampetra fluviatilis*, European sturgeon *Acipenser sturio*, allice shad *Alosa alosa*, twaite shad *Alosa fallax*, Atlantic salmon *Salmo salar*, sea trout *Salmo trutta trutta*, coregonid species *Coregonus* spp., Euorpean smelt *Osmerus eperlanus*, European eel *Anguilla anguilla*, and flounder *Platichthys flesus*. Until the beginning of the 20th century, the Rhine River's diadromous fish runs supported huge fisheries. Annual catches reached 250,000 salmon in the Dutch and German parts of the Rhine at the end of the 19th century (de Groot 1989). These catches composed an unknown fraction of the total salmon run. Considering the total catchment area of the river and its supposed carrying capacity, 19th century salmon harvest levels were probably a small fraction of the former natural potential. Salmon catches declined from the beginning of the 20th century. The Rhine River salmon became extinct in the 1950s. Populations of the other diadromous fishes (other than eel) also declined, and at least sturgeon, houting *Coregonus oxyrinchus*, and allice shad were most probably extinct. Today only river lamprey, sea lamprey, eel, sea trout, twaite shad, smelt, and flounder seem to have remained in the Rhine to some extent or have reestablished. For potamodromous fish species, see Borcherding and Staas (2007, this volume).

Industrial Development and Habitat Degradation

The main reasons for decline of diadromous fish populations from the 19th century onward are considered to be river channelization in the upper Rhine River and large tributaries, isolation of most tributaries by weirs or dams, and water pollution. For example, the Wupper River was heavily polluted by the 1830s, whereupon it lost its salmon population (Ullmann 1971). From the second half of the 19th century, upstream-migrating fish in many Rhine tributaries had such strong odor from water pollutants that they were unsuitable for consumption (von dem Borne 1883). Channelization of the Upper Rhine began in 1817 (the Tulla Correction). The first hydropower stations on the river and its tributaries were built in the second half of the 19th century. In addition to habitat destruction and pollution, fishery harvest of Rhine River fish stocks (mainly Atlantic salmon and allice shad) grew rapidly along the entire river. Overexploitation was hastened by the development of very effective fishing methods, such as long, steam engine-hauled seines (Bartl and Troschel 1997).

Habitat degradation continued with the industrial development of Europe. The sad climax to this occurred in the 1960s and 1970s when disastrous water pollution in the main river removed all fishes other than a few generalist species. Even before then, pollution with toxic heavy metals and organic substances had much diminished the aquatic fauna, low oxygen concentrations in the main river caused several spectacular fish kills, and the Rhine River fisheries had collapsed.

Public longing for better protection of the environment grew. The riparian states, coordinated by the International Commission for the Protection of the Rhine (Internationale Kommission zum Schutze des Rheins, IKSR) began to take measures for better treatment of industrial and domestic wastewater. Enormous investments were made in sewer networks and sewage treatment plants all along the Rhine River.

An initial improvement in the Rhine's water quality and recolonization by many aquatic species was followed by a giant setback. In 1986, a fire broke out in a chemical factory in Switzerland (High Rhine), and water used to extinguish it flushed toxic pesticides into the Rhine. Fishes and other aquatic fauna were killed as far downstream as the lower Rhine. This tremendous accident stimulated a major ecological restoration project. In 1987, the riparian states along the Rhine decided to set up the Rhine Action Program to restore the river ecosystem (IKSR 1987, 1999; Schulte-Wülwer-Leidig 1992). The official flagship species for this program is the Atlantic salmon. The goal was to bring salmon back to the Rhine by the year 2000, so the program was also called Salmon 2000. Internationale Kommission zum Schutze des Rheins (IKSR) coordinated the work, and the European Union (Life-Project) funded it. Schulte-Wülwer-Leidig et al. (1995) and Neumann (2002) described the program's first results. It continues as the Diadromous Fish Program in most of the states along the river, and an international program, Rhein 2020 (IKSR 2001), has been agreed upon.

Reintroduction Program for the Atlantic Salmon

Goals of the Reintroduction

The Atlantic salmon reintroduction program has two main goals: to return the species to its former habitats, and in so doing, support the European salmon stocks as far as possible; and to use the species as an indicator of the ongoing process of river restoration on a systematic, whole catchment scale. The salmon project proved to be an extremely valuable accompaniment to implementation of the new water framework directive of the European Union. Salmon monitoring involved in the project has revealed some significant ecological bottlenecks. In addition, salmon is a well-known symbol for healthy water resources, and as such, it enhances public acceptance and understanding of the conservation and restoration of rivers. General awareness of the seriousness of reasons for initiating the salmon project has ensured its continuation, even though there are ecological bottlenecks that hinder rapid recolonization by this species. The goal is to manage adaptively to reestablish viable populations of Atlantic salmon (and other diadromous fishes) in the Rhine River system via iterative experiments, in which habitat and population changes are monitored and the lessons learned are applied in improved restorative procedures.

We present data on the Rhine River reintroduction program. Our detailed results from the intensively investigated Sieg River, located in the state of North Rhine–Westphalia (NRW, Germany) in the lower Rhine region (Figure 1), may give insights about restoration in the entire Rhine River system.

Assessment of Habitats

The habitat requirements of a species must be assessed before starting its reintroduction. About 323 ha of habitat suitable for juvenile salmon and 34 ha of potential spawning grounds have been detected within the project areas of the Rhine system. The natural habitat potential (before agricultural and industrial development) must have been much higher. Today most of the habitats are isolated by weirs and dams on the large tributaries and the Upper Rhine River, or they have been degraded by other river engineering works. A standardized method for the assessment of habitat types has been developed in North Rhine–Westphalia (Molls and Nemitz, State

Agency for Ecology 1999, unpublished data). Spatial distributions of juvenile rearing habitat and potential spawning grounds were recorded in a geographic information system (ArcView). The essential spatial combination of potential spawning grounds with juvenile habitats is assessed using this system.

Release of Juveniles

Reintroduction of Atlantic salmon into the Rhine River system began in 1988 as a release of juveniles into the Sieg River. Releases are now also taking place in Switzerland, France, other German states (Baden-Württemberg, Hesse, Rhineland-Palatinate) and Luxembourg. In total, about 19 million juveniles were released between 1988 and 2003 (Figure 2). In the beginning, most of the fish were released as unfed or fed fry. It turned out to be very important that the fry be well dispersed over areas of suitable habitat and that stocking be limited to densities of 0.5–1.5 fry/m^2. In the later years, a small proportion was stocked as older parr, especially in larger streams. Since 2003, in North Rhine–Westphalia, there has been additional smolt release (80,000–100,000 selected pure smolts per year). This experiment will provide more information about smolt-to-adult return rates. It will also increase the number of adults that return to those rivers that have been selected for development and monitoring as spawning rivers.

In the beginning, Atlantic salmon originating from a number of regions were released in the Rhine River system; they came from rivers in Norway, Iceland, Scotland, Ireland, Sweden, France, Spain,and Denmark (Schmidt 1996 and unpublished IKSR-reports). From 2004 onward, salmon from only three rivers are to be introduced into the Rhine River system. This reduction of sources is designed to allow investigation of the ecological performance of salmon of differing origin. It also allows for salmon of three origins to be treated (and marked) separately in artificial breeding. This could help maintain the populations' genetic integrity and avoid outbreeding depression. Only salmon originating from a single river will be released within a single monitoring unit (i.e., a tributary river system that has an adult trap), (compare Schneider et al. 2004).

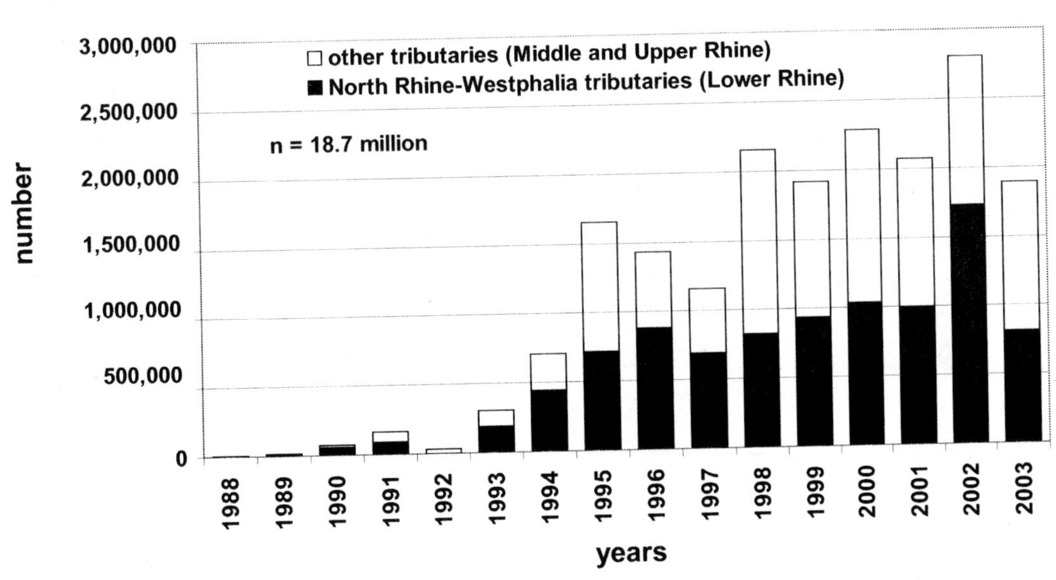

Figure 2. Number of released juvenile salmon in the Rhine system.

Survival of Juvenile Salmon

For fry released in NRW streams from April to May, we measured survival to the September parr stage by electrofishing (point abundance sampling) (Molls and Nemitz 1999). Within recorded habitats, low stocking densities (0.5–1.5 fry/m^2) resulted in fry-to-parr survival that was mostly 5–20% (Figure 3). This compares favorably to results from other re-introduction stocking (Kennedy 1988, 1.3–31.0% for unfed fry). At the end of September, most parr had reached total length of 60–110 mm, the faster growth occurring in larger streams.

Good parr densities occurred in small streams (Figure 4, data from 1997 to 2002) and ranged from 15 to 35 parr per 100 m^2. The results show good potential in some of the NRW rivers when compared to parr densities in rivers that maintain natural salmon populations (Symons 1979; Winstone 1993). On the other hand, parr densities were significantly lower in some streams and rivers, especially the larger ones (>25 m width). This probably resulted from degraded structural habitat, as well as changed hydrologic regimes, temperatures, and chemical water quality.

Smolt Production

The smolt output of the Sieg River system was estimated in the years 2000–2002 (see Figure 5 for 2002) by fyke-netting in a bypass close to the Unkelmühle hydropower station. The total smolt run was calculated by means of several mark–recapture experiments. Annual smolt output of the Sieg River system ranged from about 30,000 to 60,000.

During these years, smolt migration started in the second half of March and ended at the end of May. Migration peaks occurred at the

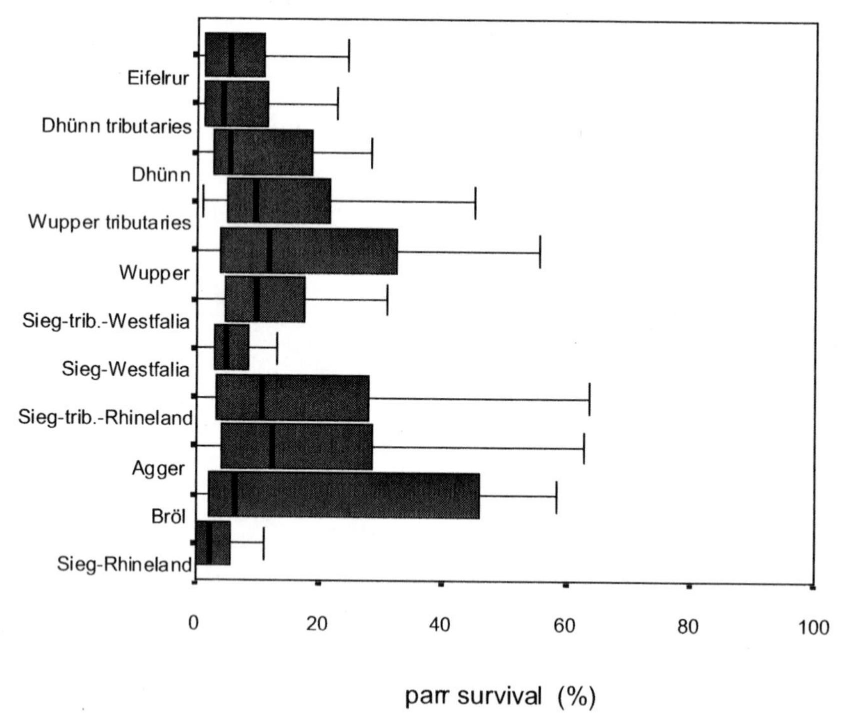

Figure 3. Survival rates of salmon fry to 0+ year parr in September (several North Rhine–Westphalia rivers, fry released in April–May).

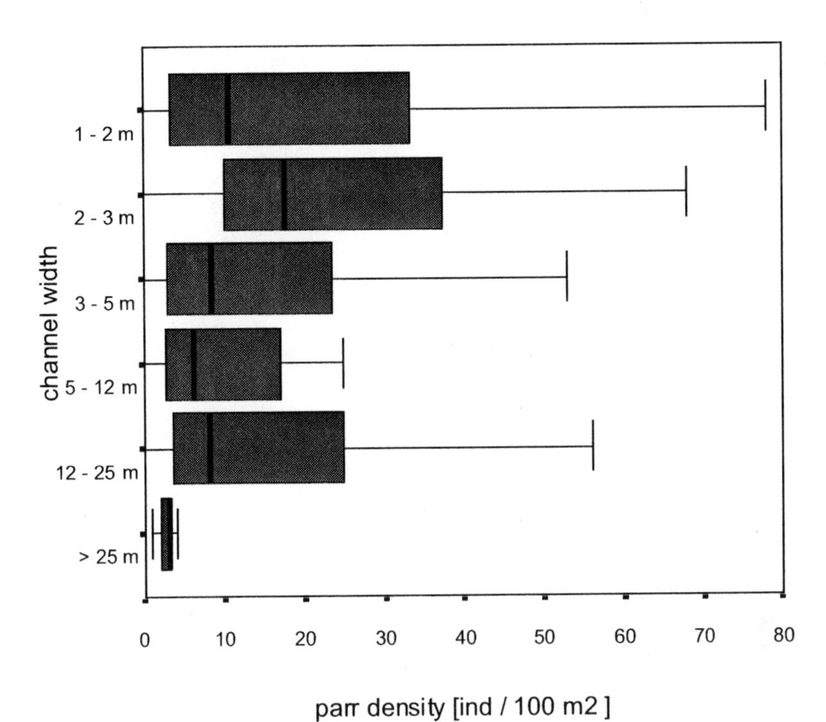

Figure 4. 0+ year parr densities in September in different-sized streams (about 80 North Rhine–Westphalia streams).

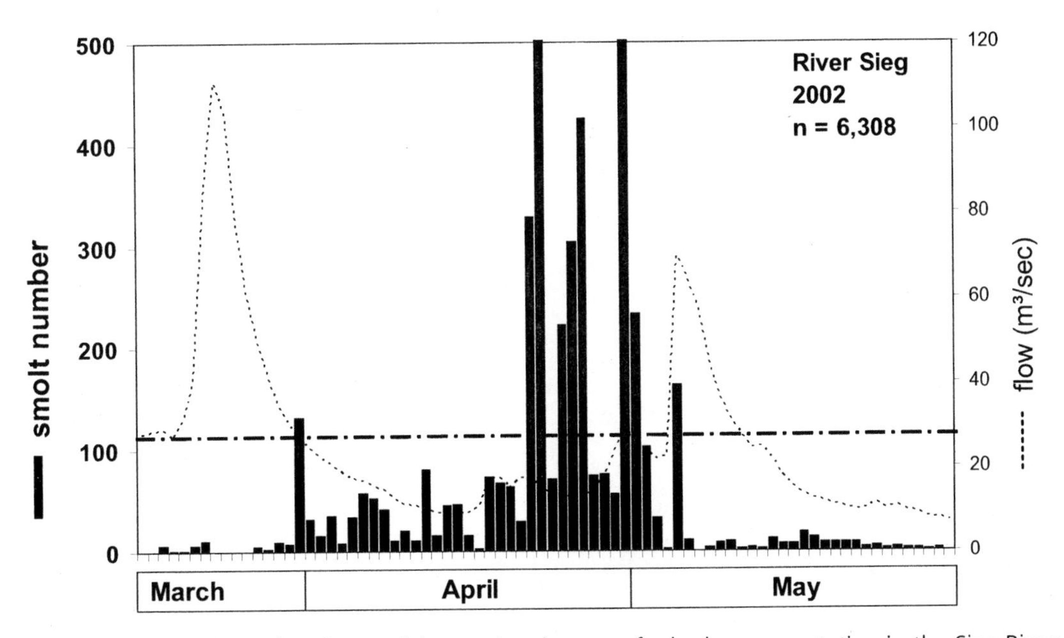

Figure 5. Daily smolt catches from a fyke-net in a bypass of a hydropower station in the Sieg River (North Rhine-Westphalia, Unkelmühle), mean flow of the river = 27 m^3*/s.

end of April, when about 50% of the total smolt number passed through within 12 d (Figure 6). The migration peaks did not directly correspond to flow conditions. During the migration periods, the river's discharge was often below mean flow. This finding is important for management of hydropower stations because nearly all of the water passes the turbines during mean flow or below. Most migration occurred at water temperatures above 10°C. There was no evidence that temperature alone triggered migration. The observations suggest that migration is controlled mainly by photoperiod. This is in agreement with the findings of Hoar (1976), Saunders and Harmon (1990) and McCormick et al. (1998).

Survival from released fry to smolt in the Sieg River system ranged from 2.8% to 6.8% (total of 1+ and 2+ year smolt). Smolt usually migrate at ages 1+ or 2+ in NRW streams.

Adult Salmon Return

The first postreintroduction record of an adult salmon in the Sieg River system occurred in 1990, 2 years after the first fry release. From 1990 to 2003, 2,436 salmon were detected in the whole Rhine area (Figure 7). The Dutch Rhine delta accounted for 30% of the salmon, the Lower Rhine River (NRW, mainly the Sieg River) for 44%, and the other regions for about 26%. No trap for recording adult return existed

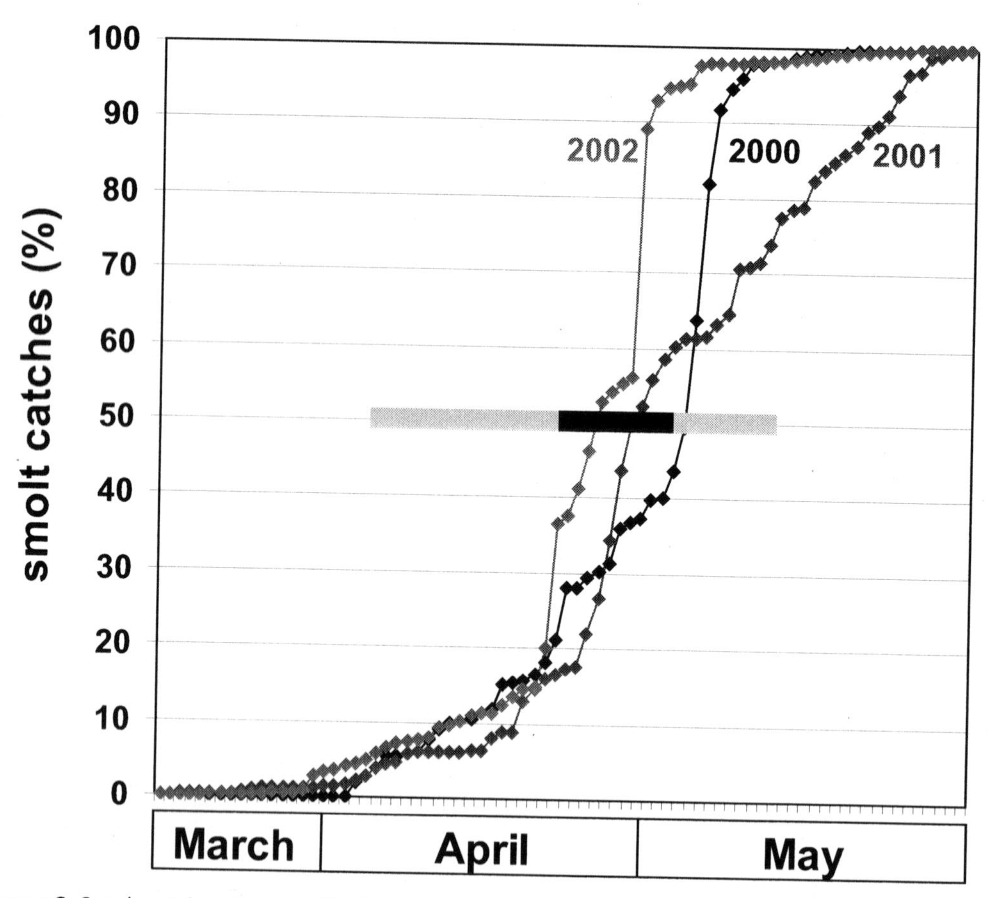

Figure 6. Smolt catches (cumulative) in the Sieg River from three successive years, gray bar indicates time span for 90 % of the smolt, black bar 50 %.

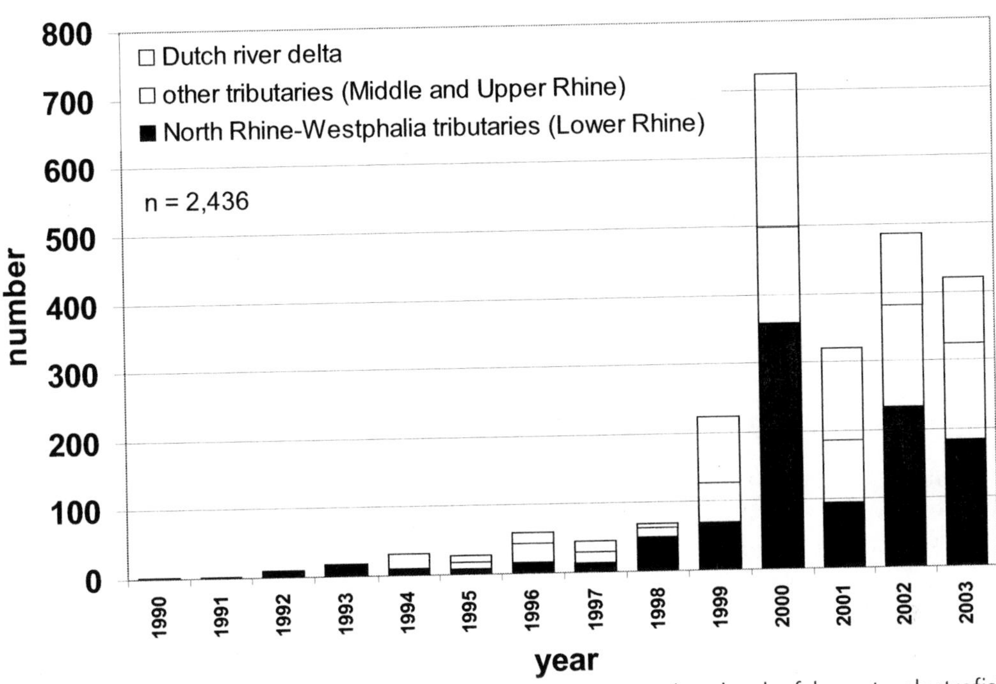

Figure 7. Number of adult salmon recorded in the Rhine system (catches by fyke-nets, electrofishing and trap stations).

until the year 2000, when permanently operated traps began monitoring at the Sieg River's Buisdorf Weir and on the Upper Rhine River at Iffezheim Dam on the French–German border). Smolt-to-adult return can be calculated for the Sieg River, even though some returning adults jump the Buisdorf Weir and pass without being recorded. Using mark–recapture methods, we estimated that the trap there caught about 50% of adults that passed the weir. On this basis, smolt-to-adult return ranged from about 0.1& to 0.7% for grilse and from about 0.06% to 0.15% for multi-seawinter (mainly two-seawinter) salmon (Table 1). These return rates are quite low compared to those of natural salmon populations (range: 2–20%, depending on exploitation; Hansen and Bielby 1988; Reddin 1988; Mills 1989). This might reflect effects of migration obstacles and fisheries, especially in the Rhine delta. Perhaps also, the introduced Atlantic salmon were poorly adapted to conditions of the Rhine River system.

Reproduction

The first evidence of natural reproduction by salmon reintroduced into the Rhine area occurred in the Siege River system in 1994 (Schmidt et al. 1994). Several records of salmon fry followed in subsequent years. Fry were also found in other Rhine tributaries (e.g., the Saynbach stream in the state of Rhineland-Palatinate; Schneider unpublished data). Systematic sampling has been done in the NRW project streams from 2000 onward. Redd surveys are made on foot or by helicopter every year. Areas where redds have been recorded or where other potential for spawning is suspected are sampled for fry in June by electrofishing combined with driftnetting (point abundance estimates). Densities of 3–12 fry per 100 m^2 occurred at some locations

Table 1. Calculation of smolt-to-adult return rate for Atlantic salmon in the Sieg River system (the total number of returning adults is calculated on the basis of an estimated trap catch rate of 50 %). MSW = multi-sea-winter.

	2000	2001	2002	2003
Smolt production	85, 364	57, 556	47, 207	no registration
returner number (Grilse)	473	103	369	228
returner number (MSW)	73	51	49	84
return rate (Grilse)		0.12	0.64	0.48
return rate (MSW)			0.06	0.15

in the Sieg River system, although mean density was only one fry per 100 m^2. The proportion of naturally reproduced smolts is very small (~1–5%) compared to that of smolts that resulted from the fry releases. Results from other Rhine River areas are similar. This indicates that salmon populations in the whole Rhine are far from self-sustaining.

Ecological Bottlenecks and Required Restoration Measures

Reproductive Phase

The status of spawning grounds is undoubtedly a key to self-sustaining salmon populations; likewise for other lithophilous species such as brown trout *Salmo trutta fario* and European grayling *Thymallus thymallus*. In 1995, gravel beds of NRW streams lacked oxygen during incubation of salmon ova and sac fry (Ingendahl and Neumann 1996; Ingendahl 2001). In 2000–2002, Sieg River potential effects of land use, sewage effluent, fish ponds, and stormwater were assessed and calculated for the whole Bröl Creek catchment (length 45 km, 217 km^2 catchment area) within the Sieg River system (Ministry of Environment and Protection, Agriculture and Consumer Protection, unpublished data). Accompanying this were detailed observations of the stream's water parameters (N, P, pH, O_2, organic and inorganic particles, conductivity) using permanently operated stations and regular sampling of the tributary network. In addition, measurements of dissolved oxygen concentrations at depths of 10, 20, and 30 cm in the gravel beds of various stream reaches from November to May (Stefan Staas, personal communication, Limnoplan Consulting) confirmed that insufficient oxygen is a serious problem. The critical dissolved oxygen value for development of salmon ova and fry is about 6 mg O_2/L (Ingendahl 2001). Another comparative study in NRW streams showed that most have the same problem (Niepagenkemper and Meyer 2002; Figure 8). Some rivers show critical values below 6 mg O_2/L even at the relatively shallow depth of 10 cm. In most streams, the critical values occurred at depths of 20 cm and greater. The Bröl Creek Study indicated severe harm from wastewater, especially untreated overflow during storm runoff, as well as from spreading of liquid manure and other fertilizers and from land uses that cause soil erosion.

Reduction of dissolved oxygen in the spawning grounds progresses as follows (Figure 9): after the spawner cleans the gravel (the investigator imitates this when burying the O_2 sensor), fine sediment rapidly clogs gravel. This

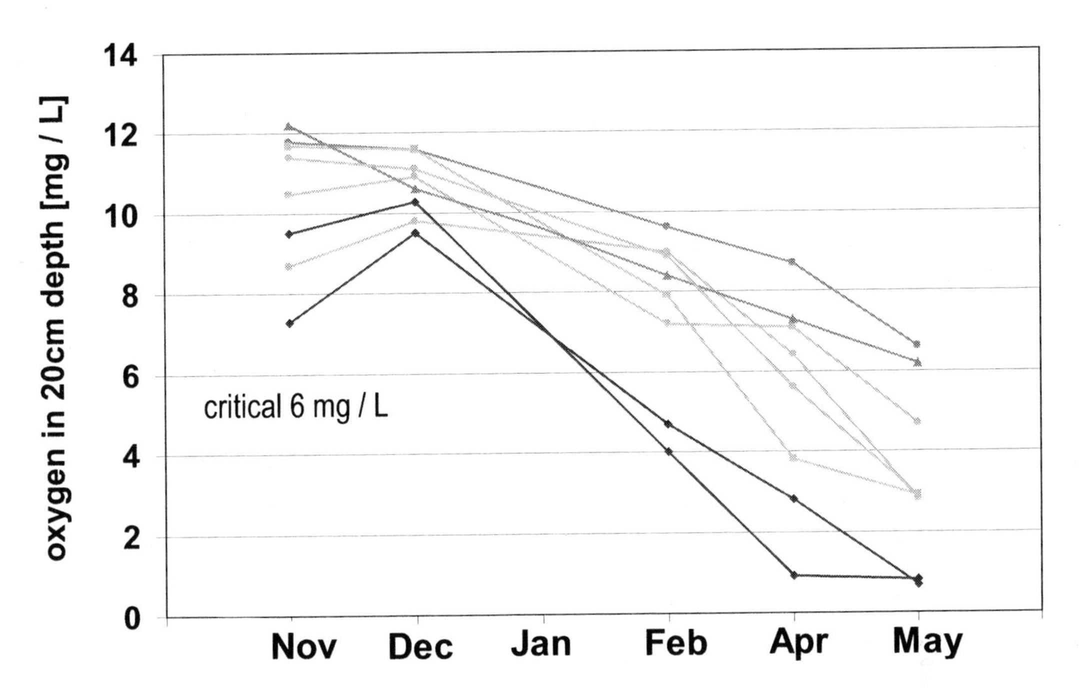

Figure 8. Concentration of dissolved oxygen in gravel beds (20 cm depth) of potential salmon spawning grounds in North Rhine–Westphalia rivers (measured in permanently exposed sensor tubes which were sampled with O2 optodes; for method, see Niepagenkemper and Meyer 2002).

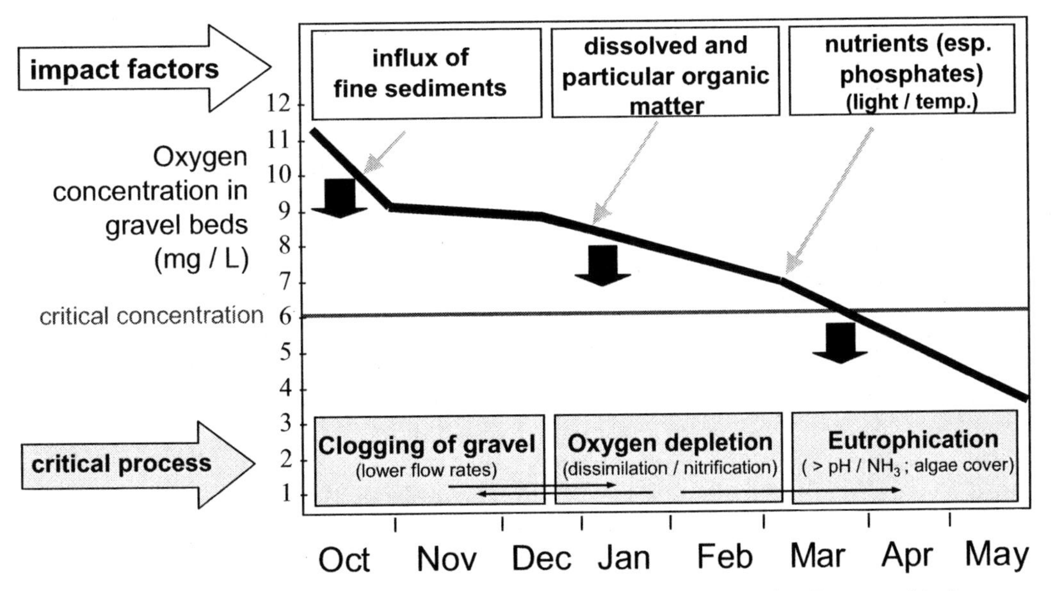

Figure 9. Simplified scheme showing the main impacts and processes leading to critical oxygen conditions in the gravel beds of salmon spawning areas.

decreases interstitial flow. Microbial dissimilation and nitrification decrease the dissolved oxygen within the hyporheic interstitial. With rising light intensities and temperatures in springtime, nutrient pollution (phosphates) leads to an intense algal bloom on the streambed surface, forming a dense algal blanket that hinders interstitial flow and, in many cases, results in critical pH-conditions (biogenic decalcification; peak over 9.0 pH; Figure 10). These effects occur before fry emergence from the gravel, which takes place at the end of April or beginning of May.

The excessive influx of fine sediments and organic matter, including nutrients for algae, comes from intense land use, as well as from stormwater runoff and inadequate sewage systems. We consider the results of the Bröl Creek Study representative of many of the streams in the Rhine system. Some of the most densely populated areas of Europe exist along the Rhine River. In the second half of the 20th century, dramatic land use changes took place, particularly intensified agriculture and urbanization, the later involving increased wastewater effluent.

Restoration measures needed for salmon spawning areas are as follows:

1) Improvement of sewer networks and of treatments for sewage stormwater. The problem of untreated wastewater from combined sewer systems must be solved (e.g., by retarding basins and soil filters).

2) Reduction of anthropogenic harm to the catchment area; land uses should be altered, well-vegetated buffer zones created along streams, and livestock excluded from riparian areas.

3) Restoration of river dynamism, including moderate lateral migration of channels and movement of bed load (see also the Juvenile Phase section). The NRW Ministry of Environment is developing new guidelines for restoring and protecting salmon rivers.

Juvenile Phase

Few ideal habitats for juvenile salmon exist in the Rhine River system today. Some nearly

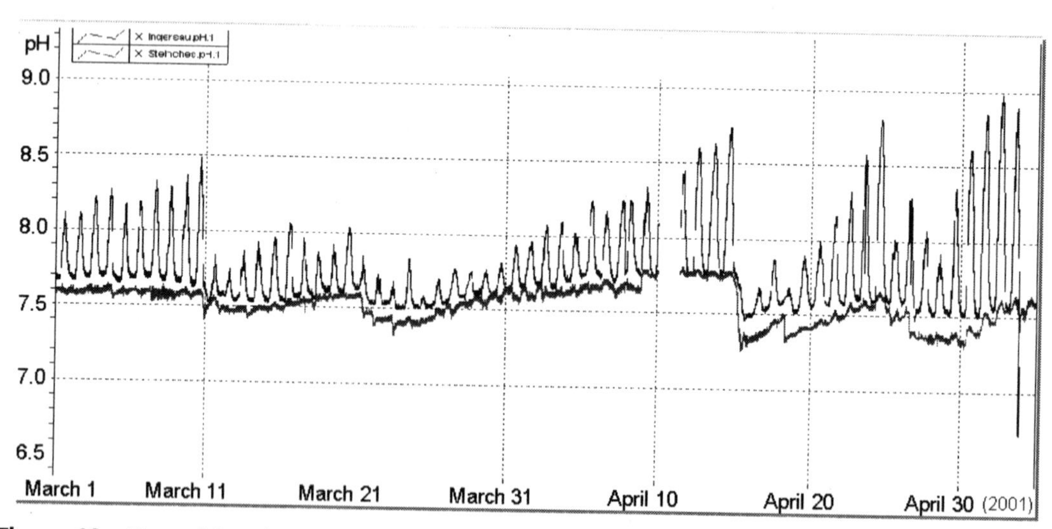

Figure 10. pH conditions in springtime in the Bröl River (upper fluctuating line) during biogenic decalcification and an affluent small creek (lower line); measured by a permanent station in the year 2001.

natural stretches can be found that offer parts of the original stream channel form. However the combination of shallow fry habitats with deeper parr habitats and overwintering structure (e.g., large woody debris and deep pools), is very rare. The natural variability of bed cross section, water depth, flow conditions, and bed material is mostly restricted by stream engineering work and changed flow conditions in the catchment. Engineerings in most streams have narrowed and incised the channel and have unnaturally stabilized the banks. This channelization limits dynamic processes like lateral migration of the channel, scouring of pools, building of riffles, and recruitment of large woody debris to the stream by the undermining and toppling of riparian trees and bushes. The hydrologic and hydrothermal regimes have become more extreme because urbanization and land drainage increased the frequency and severity of high water and decreased base flow. This has changed stream channel form and streambed structure. One consequence is "streambed armoring," the development of a layer of relatively large, tightly packed stones at the bed surface because the more frequent and severe flows stay in the incised channel. The altered hydrologic, thermal, and morphologic conditions adversely affect habitats for salmon spawning (availability of suitable sized gravel and proper riffle form) and juvenile rearing (Kondolf 1995; Waters 1995; Beechie and Sibley 1997; White 2002). This stream's relatively low and varying survival of released fry to the smolt stage (lowest recorded rate = 2.8%, see Smolt Production) probably results from this deficiency of rearing-habitat. Winter habitat for parr may be particularly inadequate.

Required restoration measures to improve juvenile habitats: 1) Create space for river dynamism in selected parts of the river system. 2) Remove artificial streambank armoring, streambed stabilizations, and weirs. 3) Add large woody debris to stream channels, and stop removal of fallen trees. 4) Remeander and reopen former river courses and lateral channels. Special NRW restoration plans incorporating such measures have been worked out for selected spawning streams (e.g., Bröl and Dhünn creeks).

Smolt Phase

Smolt migration conditions may also be a key to restoring salmon in the Rhine River system. The lengths of migration routes from potential spawning streams to the sea range from about 370 km for tributaries of the Sieg River to 720 km for the River Ill system in the French Upper Rhine region. Significant anthropogenic changes in the Rhine River system include series of hydropower stations that block the Upper Rhine River mainstem, as well as the main tributaries (e.g., the rivers Neckar, Main, Moselle, Lahn, and Ruhr). Besides the fish mortality that their turbines cause, the low stream velocities and altered fish communities of impounded reaches increase predation risk (compare Jepsen et al. 1998). In addition, the main river has artificial channel structure, an unnatural flow regime, and intense use as a navigation route. The Rhine is one of Europe's most frequented waterways, and its species composition has changed; for example, it now holds dense populations of zander *Sander lucioperca*. In addition, dams, hydropower stations (two in the Lek/Nederrijn branch), and restrictive flow management have completely changed the Rhine delta, destroying the river's formerly extensive estuaries. The zone of transition from fresh- to saltwater is now very small.

It is difficult to assess the exact influence of the single factors mentioned, which may, among others, lead to the low return rates of salmon adults to the Rhine River system (see the above Sieg River data).

However, a good estimate for the effects of hydropower stations is possible. Turbine-caused mortality of salmon smolts ranges from about 5% to 10% per station to a maximum, which can be much higher (Raben 1957; Monten 1985, Larinier and Dartiguelongue 1989). Therefore, after three to six hydropower stations (without protection measures), the total mortality will probably be more than 25%. Required restoration measuresto improve downstream migration routes: 1) Prepare for each river system, a conceptual restoration plan that accounts for all known obstacles from the spawning area down to the sea. For each obstacle, possible remedies, such as removing a dam or constructing deflection screens and bypasses, should be explored. The achievable survival rate should be calculated for the total migration route. Such plans are in preparation for the upper Rhine River and for NRW streams. 2) Install protection systems, remove inefficient hydropower stations, or undertake other management to avoid significant fish mortality (compare Dumont 2000). Screens for protecting Atlantic salmon smolts probably need 10-mm mesh width and less than 0.5 m/s approach-velocity of water at the screen's vertical component. 3) Reestablish some estuary-like regions, for example at Haringvliet and Lake Ijsselmeer, to rebuild at least a small transition zone between fresh- and saltwater. This will support other species, such as allis shad, twaite shad, and houting.

The Adult-Return Phase

As with downstream migration of smolt, adult return of salmon is adversely influenced by artificial conditions in the Rhine delta, main river, and tributaries. For salmon, the only remaining entrance to the Rhine system is the New Waterway, which lies close to the city of Rotterdam and passes through its large harbor (Figure 11). Dams block the other branches of the Rhine delta at Haringvliet, and Lake Ijsselmeer. Three other dams on the Lek/Nederrijn branch strictly control distribution of water discharge between the New Waterway, Haringvliet and Lake Ijsselmeer. Water management supports navigation, flood prevention, and water use. When discharge is low (£1,000 m/s), more than 90% of the Rhine water flows to the New Waterway. At higher discharges (e.g., 6,000 m/s), about 60% passes the Haringvliet, and Lake Ijsselmeer gets about 14%.

A field study indicated the extent to which artificial conditions in the delta hamper adult migration of sea trout and salmon (Bij der Vaate et al. 2003). Adult sea trout and salmon were tagged with highly efficient transponders in the sea in front of the Rhine River. Detection took place at about 18 fixed detection stations widely distributed within the delta and its river mouths. Another detection station existed in the German part of the Rhine close to the Dutch border, and another in the Sieg River. Of 580 sea trout tagged in front of the river mouths, only 34% were detected in the freshwater. Of the sea trout that arrived in freshwater, about 47% reached the German part of the Rhine and 12% reached the lower River Meuse. This may indicate that about 41% of sea trout (in freshwater) did not migrate inland successfully. Of the sea trout detected in the German Rhine, about 14% entered the Sieg River (Ingendahl, unpublished data). Of 75 adult salmon tagged with transponders in front of the river mouths (mainly Haringvliet), only two reached the German part of the Rhine. One was detected in the Sieg River.

Required restoration measures to improve adult migration: 1) Reopen the Haringvliet sluices. In the Netherlands, plans exist for partial reopening in 2007. 2) Build fish passages in the Lek/Nederrijn branch. The last passages should be complete by the end of

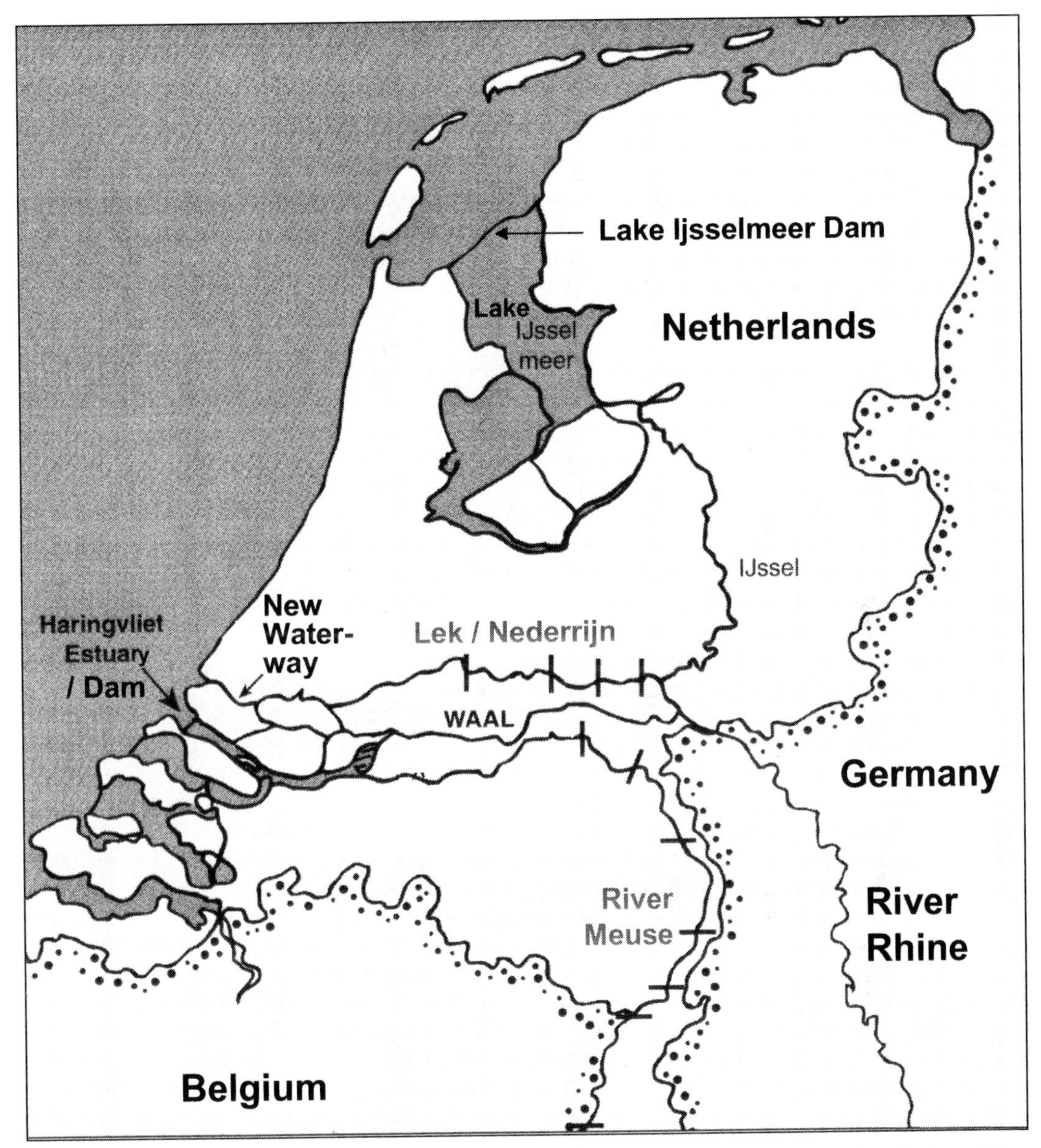

Figure 11. Map of the Dutch Rhine delta with three river mouths (Haringvliet Sluices, New Waterway, Lake Ijsselmeer Sluices); bars indicate weirs in branch Lek/Nederrijn and in River Meuse.

2004. 3. Build further fish passages in Rhine tributaries and the upper Rhine and remove weirs on the smaller streams.

Conclusions

•The International Rhine River Program has achieved enormous ecological improvements, notably in water quality and in number of invertebrate and fish species.

•The Atlantic salmon has served as an essential means for promoting restoration throughout the Rhine River catchment.

•Future reestablishment of self-sustaining

salmon populations requires much additional restorative action on spawning areas, on up- and downstream migration routes, and in the Rhine delta.

•Today, in addition to the weirs, artificial channelization and streambank armoring that have damaged structural habitat for salmon throughout most of the Rhine River system, intense agriculture and urban wastewater harm the spawning streams, and hydropower stations, flow management, and delta dams hamper migration. These anthropogenic problems differ significantly from those that extirpated Atlantic salmon from the Rhine in the beginning of the 20th century, and they must be overcome if this species and other diadromous fishes are to be reestablished.

•Improved monitoring from the upper catchment to the delta is needed to determine quantitative effects of present problems.

Acknowledgments

The NRW Diadromous Fish Program is supported by the Ministry of the Environment and Conservation, Agriculture and Consumer Protection of the Federal State of North Rhine–Westphalia (Düsseldorf) and by a contribution from anglers in that state. We thank our colleagues from the State Agency for Ecology, several fisheries organizations, and the International Commission for the Protection of the Rhine (IKSR). Several freelancers are thanked for their excellent cooperation in this program.

References

Bartl, G., and H. J. Troschel. 1997. Historische Verbreitung, Bestandsentwicklung und aktuelle Situation von *Alosa alosa* und *A. fallax* im Rheingebiet Zeitschrift für Fischkunde 4 1:119–162.

Beechie, T. J., and T. H. Sibley. 1997. Relationships between channel characteristics, woody debris, and fish habitat in northwestern Washington streams. Transactions of the American Fisheries Society 126:217–229.

Bij de Vaate, A., A. W. Breukelaar, T. Vriese, G. De Laak and C. Dijkers. 2003. Sea trout migration in the Rhine delta. Journal of Fish Biology 63:892–908.

Borcherding, J., and S. Staas. 2007. Rhine River fisheries (part I): potamodromous fishes as promoters for habitat restoration in the floodplain area. Proceedings of the 4th World Fisheries Congress (2004), Vancouver, Canada.

De Groot, S. J. 1989. Literature survey into the possibility of re-stocking the Rhine River and its tributaries with Atlantic salmon (*Salmo salar*). Published Reports on the Ecological Rehabilitation of the Rivers Rhine and Meuse 11:1–56.

De Groot, S. J. 2002. A review of the past and present status of anadromous fish species in the Netherlands: is restocking the Rhine feasible? Hydrobiologia 478:205–218.

Dumont, U. 2000. Fishways for the migration of fish in rivers–current technical solutions and international experience. Wasser & Boden 52–4:10–15 (in German, English abstract).

Hansen, L. P., and G. H. Bielby. 1988. Salmon in Norway. Atlantic Salmon Trust, Pitlochry, Scotland.

Hoar, W. S. 1976. Smolt transformation: evolution, behaviour and physiology. Journal of the Fisheries Board of Canada 33:1233–1252.

IKSR (Internationale Kommission zum Schutze des Rheins). 1987. Rhine Action Program. Report International Commission for the Protection of the Rhine against Pollution (Internationale Kommission zum Schutze des Rheins, IKSR), Koblenz, Germany.

IKSR (Internationale Kommission zum Schutze des Rheins).1999. Second International Rhine Symposium "Salmon 2000." Proceedings of the International Commission for the Protection of the Rhine (Internationale Kommission zum Schutze des Rheins, IKSR), Koblenz, Germany.

IKSR (Internationale Kommission zum Schutze des Rheins). 2001. Conference of the Rhine Ministers 2001. Program Rhine 2020. International Commission for the Protection of the Rhine (Internationale Kommission zum Schutze des Rheins, IKSR), Koblenz, Germany.

Ingendahl, D., and D. Neumann, 1996. Possibilities for successful reproduction of reintroduced salmon in tributaries of the Rhine River. Archiv feur Hydrobiologie, Supplement 113 (Large Rivers)10:333–337.

Ingendahl, D. 2001. Dissolved oxygen concentration and

emergence of sea trout fry from natural redds in tributaries of the Rhine River. Journal of Fish Biology 58:325–341.

Jepsen, N., K. Aarestrup, F. Økland, and G. Rasmussen. 1998. Survival of radio-tagged Atlantic salmon (*Salmo salar* L.) and trout (*Salmo trutta* L.) smolts passing a reservoir during seaward migration. Hydrobiologia 371/ 372:347–353.

Kennedy, G. J. A. 1988. Stock enhancement of Atlantic salmon (*Salmo salar* L.)Pages 345–372 *in* D. Mills and D. Piggins, editors. Atlantic salmon: planning for the future. Croom Helm, London.

Kondolf, G. M. 1995. Five elements for effective evaluation of stream restoration. Restoration Ecology 19:1–15.

Larinier, M., and J. Dartiguelongue. 1989. The circulation des poissons migrateurs: le transit a travers les turbines des installations hydroelectriques (The movements of migratory fish: passage through hydroelectric turbines). Bulletin Français de la Peche et de la Pisciculture 312–313:1–90.

Lelek, A., and C. Koehler. 1990. Restoration of fish communities of the Rhine River two years after a heavy pollution wave. Regulated Rivers: Research & Management 5:57–66.

Lelek, A., and G. Buhse. 1992. Fische des Rheins. Springer Verlag, Berlin.

McCormick, S. D., L. P. Hansen, T. P. Quinn, and R. L. Saunders. 1998. Movement, migration and smolting of Atlantic salmon (*Salmo salar*). Canadian Journal of Fisheries and Aquatic Sciences 55 (Supplement 1):77–92.

Mills, D. 1989. Ecology and management of Atlantic salmon. Chapman and Hall, London.

Molls, F. and A. Nemitz. 1999. Habitat quality for 0+ salmon (*Salmo salar* L.) in rivers and small streams of Northrhine-Westphalia, Germany. Pages 103–112 *in* Second International Rhine Symposium "Salmon 2000." Proceedings of the International Commission for the Protection of the Rhine (Internationale Kommission zum Schutze des Rheins, IKSR), Koblenz, Germany.

Monten, E. 1985. Fish and turbines. Norstedts Tryckeri, Stockholm, Sweden.

Neumann, D. 2002. Ecological rehabilitation of a degradated large river system–considerations based on case studies of macrozoobenthos and fish in the lower Rhine and its catchment area. Internat. Rev. Hydrobiol 87 (2–3):139–150.

Niepagenkemper, O., and I. Meyer. 2002. Messungen der Sauerstoffkonzentrationen in Flusssedimenten zur Beurteilung von potentiellen Laichplätzen von Lachs und Meerforelle. editors. Fisheries Association of Northrhine-Westphalia (ISBN 3–9809545-2- 8):1–87.

Reddin, D.G. 1988. Ocean life of Atlantic salmon (*Salmo salar* L.) in the Northwest Atlantic. *in* D. H. Mills and D. Piggins, editors. Atlantic salmon–planning for the future. Croom Helm, London & Sidney:483–511.

Saunders, R. L., and P. R. Harmon. 1990. Influence of photoperiod on growth of juvenile Atlantic salmon and development of salinity tolerance during winter-spring. Transactions of the American Fisheries Society 119:689–697.

Schmidt, G. 1996. Wiedereinbürgerung des Lachses *Salmo salar* L. in State Agency of Ecology North Rhine Westphalia, editors. Landesanstalt für Ökologie, Bodenordnung & Forsten NRW, LÖBF-Schriftenreihe, Band 11 (ISBN: 3–89174-023–9).

Schmidt, G., W. Lehmann, and G. Marmulla. 1994. Natürliche Fortpflanzung des Lachses (*Salmo salar*) wieder in Deutschland. Natur & Landschaft 69:213.

Schneider, J., L. Jörgensen, F. Molls, A. Nemitz, and K. Blasel. 2004. Notwendigkeit und konzeptionelle Ausrichtung eines effektiven Monitorings bei der Lachswiederansiedlung im Rhein – das Monitoring-Einheiten-Konzept. Fischer & Teichwirt 2/ 2004:528–531.

Schulte-Wülwer-Leidig, A. 1992. International Commission for the Protection of the Rhine against Pollution – the integrated ecosystem approach for the Rhine. European Water Pollution Control 2(3):37–41.

Schulte-Wülwer-Leidig, A., G. M. Van Dijk, and E. C. L. Marteijn. 1995. Ecological rehabilitation of the Rhine River: plans, progress and perspectives. Regulated Rivers: Research & Management 11:377–388.

Symons, E. K. 1979. Estimated escapement of Atlantic salmon (*Salmo salar* L.) for maximum smolt production in rivers of different productivity. J. Fish. Research Board Can 36:132–140.

Ullmann, F.-P. 1971. Veränderungen der Fischfauna in der Wupper unter Berücksichtigung industrieller Abwässer. Annuell Report of the „Naturwissenschaftlicher Verein Wuppertal" 24:76–88.

Von dem Borne, M. 1883. Die Fischerei-Verhältnisse des Deutschen Reiches, Oesterreich-Ungarns, der Schweiz und Luxemburgs. Hofdruckerei W. Moeser, Berlin: 1–304.

Von Raben, K. 1957. Zur Beurteilung der Schädlichkeit der Turbinen für Fische. Die Wasserwirtschaft 48: 60ff.

Waters, T. F. 1995. Sediment in streams: sources, biological effects and control. American Fisheries Society, Monograph 1, Bethesda, Maryland.

White, R. J. 2002. Restoring streams for salmonids: Where have we been? Where are we going? Proceedings of the 13th International Salmonid Habitat Enhancement Workshop, Westport / Ireland, Central Fisheries Board:1–31.

Winstone, A. J. 1993. Juvenile salmon stock assessment and monitoring by the National River Authority—a review. *in* R. J. Gibson and R. E. Cutting, editors Production of juvenile Atlantic salmon in natural waters. Can Spec. Publ. Fish. Aquat. Sci. 118:123–126.

American Fisheries Society Symposium 49:835–843

Local Riverine Fish Communities as Promoters for Habitat Restoration in the Floodplain Area of the Lower Rhine

JOST BORCHERDING*
Zoological Institute of the University of Cologne, Department of General Ecology and Liminolgy Ecological Field Station Grietherbusch, D-50932 Köln, Germany

STEFAN STAAS
LimnoPlan-Fisch-und Gewässerökologie, Römerhofweg 12 D-50923 Erftstadt, Germany

Introduction

Intensive human impact upon European rivers, including the Rhine, began more than 500 years ago. The rivers were canalized for the purpose of navigation and regulated using weirs and sluices for water resource control and flood defense. Habitats were fragmented and flood plain land was reclaimed for urban and industrial purposes (Nienhuis et al. 2002; for a section of the lower Rhine cf. Figure 1). Starting in the 19th century, major dam building activities for both hydroelectric power and drinking water supply began. In addition, massive dyke constructions substantially narrowed the floodplain area and completely changed the hydrological regime; increased current velocities, stronger altitudes in discharge, and more extreme flood peaks were the results. About 85% of the formally regularly inundated area along the River Rhine is lost today (IKSR 1998), which equals an area of about 2,450 km^2 (Engel, German Federal Institute of Hydrology, personal communication). Accordingly only less than 5% of the river forelands at the lower Rhine have been valuated as "in good ecological state" (IKSR 2003). The effects of this overall habitat loss and degradation on the fish community are well documented only for the economically valuable anadromous fish species, in particular for Atlantic salmon *Salmo salar* (e.g., de Groot 2002; cf. Molls and Nemitz 2006, this volume). Together with intensive fishing, the result was a population decline from the late 19th century onwards, which ended in the extinction of several species (de Groot 2002). How far the riverine fish communities were disturbed due to constructive alterations of the River Rhine and its floodplain remains unclear because there are no reliable reports from the early 19th century on the abundances of those fish species not of economical interest. It can be assumed, however, that not only the anadromous fish species but also the local riverine fish communities were altered due to the canalization of the river, with massive losses, particularly of limnophilous and phytophilous fishes (Lelek and Buhse 1992).

*Corresponding author: jost.borcherding@uni-koeln.de

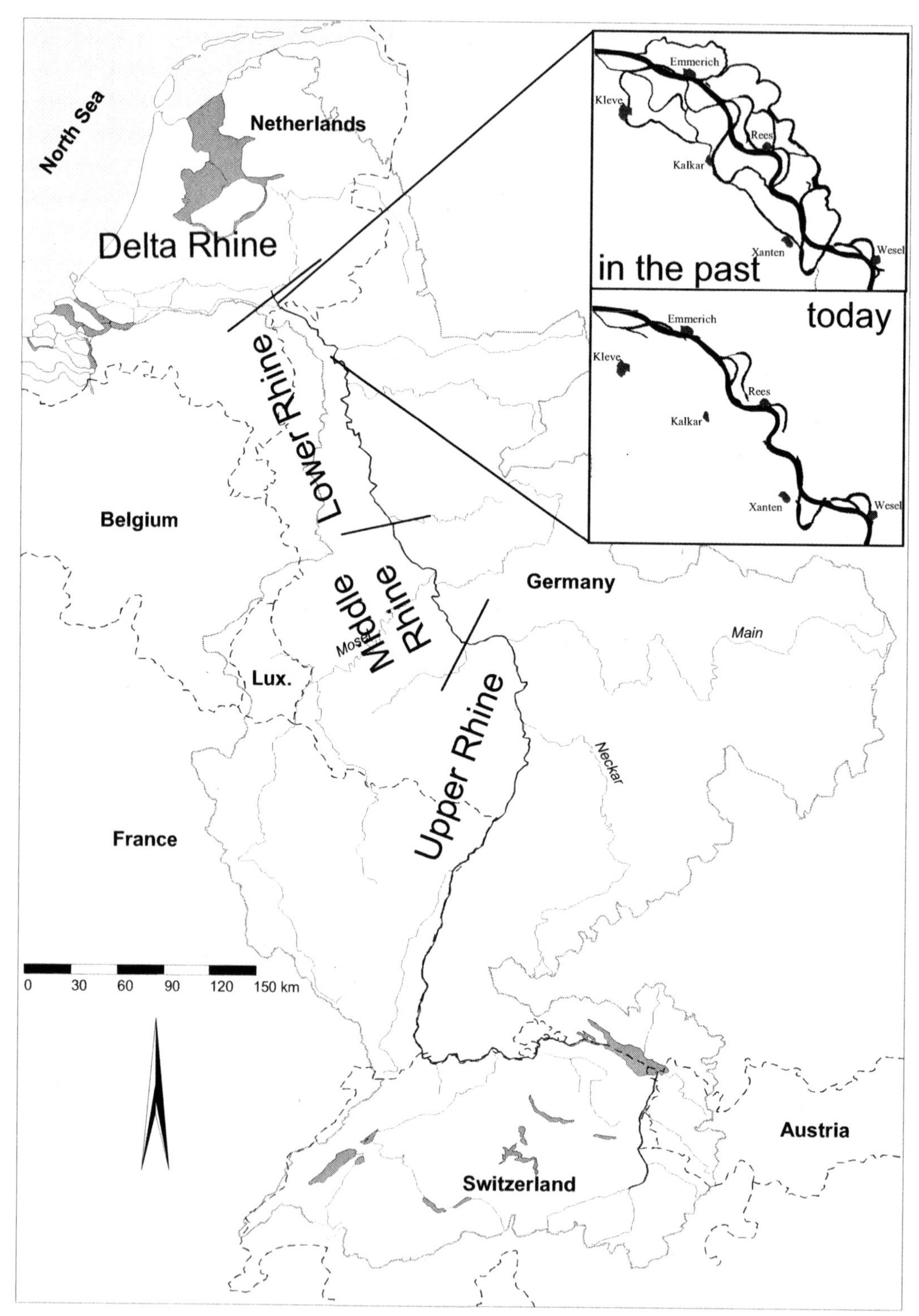

Figure 1. Map of the different sections of the River Rhine (changed after RIZA). The small inserted figures exhibited the number and size of oxbows in former times and today for a 50-km-long section of the Lower Rhine.

Table 1. Fishing methods used, the number of samples, and the number of caught fish at the different parts of the study.

Subject	Method	Number of samples	Number of fish	Source
Adult fish, River Rhine (Figure 2a)	Electro-fishing	1997: 205 stretches 1998: 118 stretches	1997: 12,581 1998: 12,650	Staas 1999
0+ fish, River Rhine (Figure 2b)	Beach seine	1992: 88 samples 1993: 80 samples 1994: 16 samples	1992: 7,375 1993: 32,020 1994: 15,849	Staas 1996
0+ fish, oxbow lakes (Figure 2c)	Beach seine and net	1993: 14 samples 1994: 91 samples 1995: 62 samples	1992: 1,212 1993: 24,743 1994: 31,568	Molls 1997
0+ fish, connected gravel pit lakes (Figure 2d)	Beach seine	1992: 29 samples 1993: 85 samples	1992: 42,589 1993: 31,269	Staas 1996
0+ fish, connected gravel pit lake, reconstructed (Figure 2e)	Beach seine	2000: 34 samples	2000: 10,737	Scharbert and Greven 2002

The Fish Community Today

In addition to the alteration of the fish community due to reconstruction and intense economical development after the Second World War, extreme water pollution disrupted the fish community of the River Rhine substantially. The fish community recovered to some extent when water quality began to improve in the 1980s (Lelek and Köhler 1989). In the following, we present our results and postulations for the lower Rhine, which represents the meander zone of the Rhine with only low drop. Today, the adult fish community of the lower Rhine is dominated by eurytopic species (e.g., bleak *Alburnus alburnus* and roach *Rutilus rutilus*) and rheophilic species (e.g., barbel *Barbus barbus* and European chub *Leuciscus cephalus*), while limnophilous and phytophilous fishes (e.g., common carp *Cyprinus carpio*, tench *Tinca tinca*, and rudd *Scardinius erythrophthalmus*) are nearly absent (Figure 2a, for details on the methods cf. Table 1). Even bream *Abramis brama*, which 1) has been quoted as being a eurytopic species (Schiemer and Waidbacher 1992), 2) occurs frequently as adult (Figure 2a), and is 3) the region-naming species for the lower Rhine ("bream zone"), is rare in the 0+ fish community of the stream's inshore habitats (Figure 2b). As described by Molls (1999), the successful recruitment of

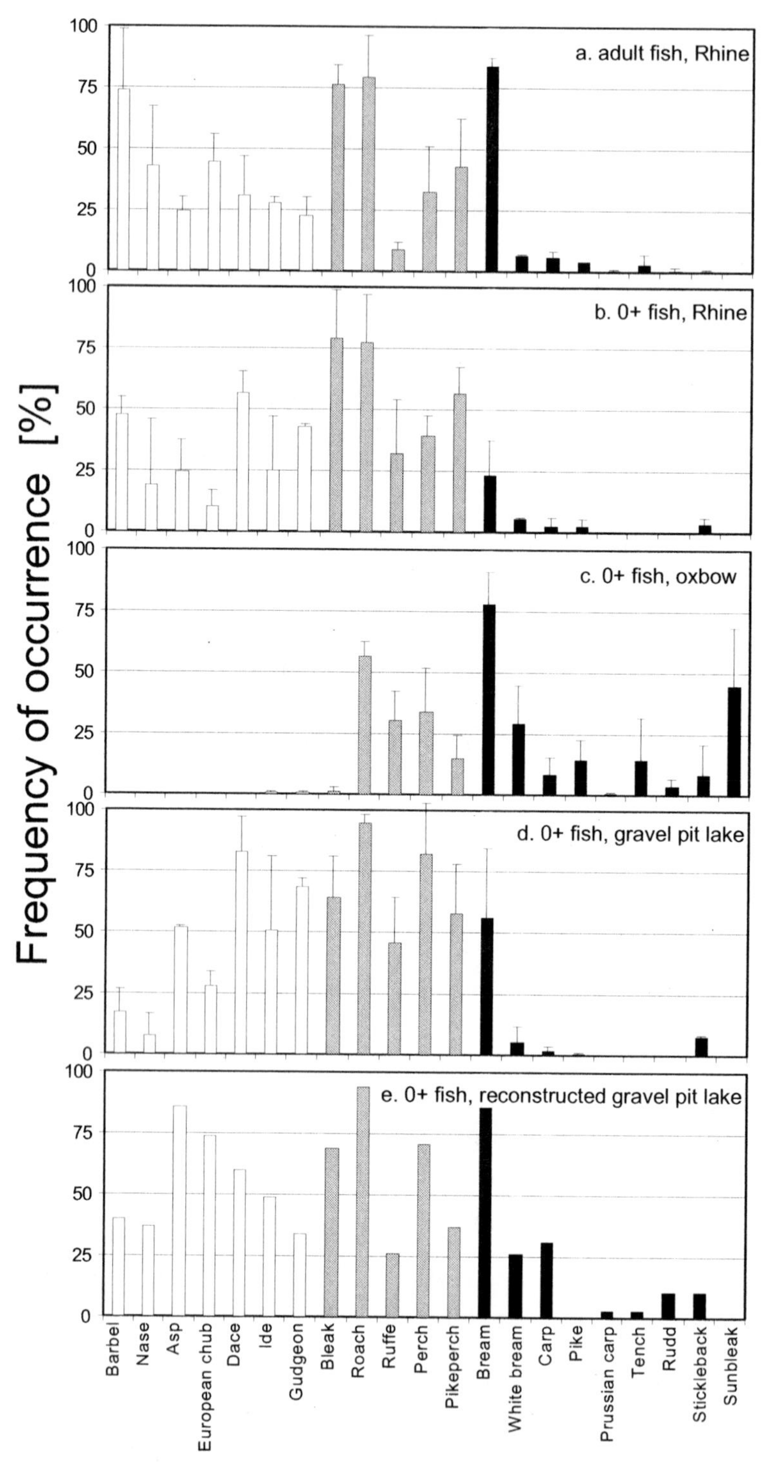

Figure 2. Frequency of occurrence of (%) of the most abundant fish at the Lower Rhine expressed as the mean (±standard deviation) of the different years. The fishing method, as well as the number of samples and caught fish are given in Table 1. White columns: rheophilic species; gray columns: eurytopic species, black columns: limnophilous/phytophilous species.

bream at the lower Rhine depends on the lateral connectivity to the stagnant waters of the floodplain. Adult bream migrate every spring from the stream to adjacent oxbows, where spawning takes place. After at least 6 months, the 0+ or juvenile bream migrate to lotic habitats in the river, where they grow to maturity at which point they return to the stagnant backwaters for reproduction (Molls 1999; Grift 2001). Bream, however, represents only the tip of the iceberg of the typical phytophilous riverine species (cf. Figure 2c). As described above, the spawning sites of these species became rare due to loss of floodplain areas. Furthermore, the connectivity to the remaining backwaters is increasingly lost due to the channel erosion in the range of 5–50 mm per year (Dröge et al. 1992). In consequence, the deepening of the riverbed of about 2 m during the last century at the lower Rhine leaves the natural backwaters terraced and elevated in the remaining floodplain area, which reduces the connectivity to the stream tremendously (Neumann and Borcherding 1998). Based on calculations made for flood protection scenarios, the flooded area of a mean annual flood in former times should equal the flooded area of a flood that today occur on average only every 20 years, assuming that all dykes are absent. In addition, the altered hydrological regime has reduced the duration of connectivity of floodplain waters with the stream (Engel, German Federal Institute of Hydrology, personal communication). Thus, the lack of limnophilous and phytophilous species in the lower Rhine reflects the massive loss of floodplain areas and the separation of the remaining oxbows from the main stream.

Postulation I: Improve Connectivity

Based on this information, it seems clear that the remaining natural backwaters of the floodplain need more constant connections to the main stream. Therefore, fish passes have to be constructed, which, together with the water management in these agricultural areas, would increase the functionality of the backwaters as important reproductive areas for typical riverine fishes. One of these new fish passes was constructed at the connection of the oxbow Bienen and the Rhine harbor Dornick (for a site map refer to Molls 1999). Beneath the protective function for the agricultural area against floods, this new type of construction should allow a steady connection at different water levels, in order to extend the connecting period under the premise that only limited water amounts are available for the operation of the fish ladder (Figure 3). The first results on the migration activity monitored in 2001 and 2002 (cf. Nemitz et al. 2002) revealed that adult bream use this fish pass to immigrate to the oxbow Bienen and that juvenile bream emigrated back to the Rhine (Figure 4). The results give clear evidence that lost or deteriorated connections between the stream and its backwaters can be improved substantially, and thus can help to maintain the recruitment of typical phytophilous riverine fishes.

Postulation II: Improve Bank Structure

However, the massive loss of natural backwaters in the floodplain area cannot be totally compensated by a few improved connections to the last oxbows. Because rheophilic species (sensu Schiemer and Waidbacher 1992) depend on suitable riverine habitats through their whole lifecycle, important parts of the fish fauna are affected mainly by deterioration of habitat structure in the main channel. Therefore, the bank structure of the "shipping channel" Rhine has to be improved. Only about 20% consisted of undisturbed gravel banks (Staas 1999), which represent essential spawning and nursery areas for some

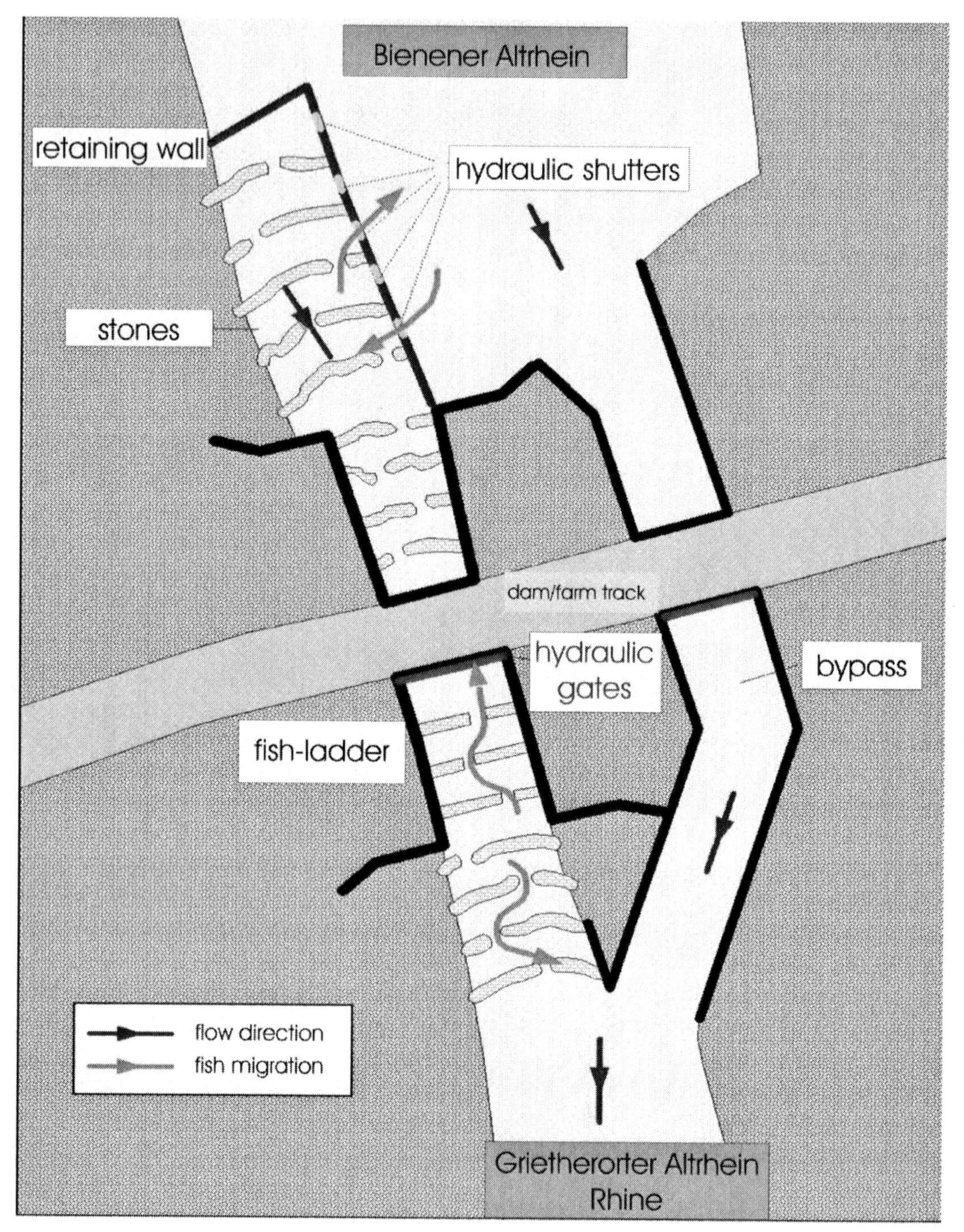

Figure 3. Schematic diagram of the fish pass constructed between the oxbow Grietherort/River Rhine and the oxbow Bienen (for a site map refer to Molls 1999). The position of the hydraulic shutters in the retaining wall differ in depth, with the deepest close to the Rhine. At high water levels in the oxbow Bienen, only the upper one is open. In relation to falling water levels each time the following hydraulic shutter is opened (with courtesy of the "Naturschutzzentrum im Kreis Kleve e.V.", Rees-Bienen).

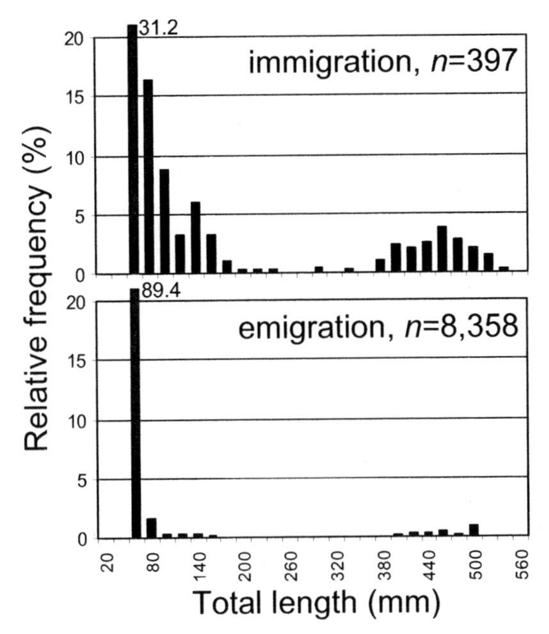

Figure 4. Total length of bream that immigrate into (above) and emigrate out of (below) the oxbow Bienen, monitored with metal fykes in the fish ladder (cf. Figure 3) in 2001/2002, data from Nemitz et al. (2002).

rheophilic species, such as barbel and nase *Chondrostoma nasus* (cf. Figure 3 in Staas 1998). Another 40% of the riverbanks belong to different kinds of groyne fields. Under the premise that these areas offer a wide variety of bank structures, they can serve as recruitment areas for several riverine fishes (e.g., asp *Aspius aspius*, bleak, pikeperch *Sander lucioperca*, Staas 1998). However, nearly 40% of the banks of the River Rhine in North-Rhine-Westphalia (226 Rhine kilometer) consist of bank revetment (Staas 1999). These areas are unsuitable for nearly all fishes and may even act as a sink habitat for the juveniles (sensu Delibes et al. 2001; Bonesi et al. 2002) due to the tremendous loss of individuals caused by the wash of the waves produced by ships (Wolter and Arlinghaus 2003). Therefore, we propose the 1) dismantling of bank revetment, 2) the construction of guide dykes to protect the banks against the wash of the waves produced by ships, and 3) the improvement of groyne fields by enhancement of the variety of bank structures.

Postulation III: Improve Gravel Pit Lakes

During the last few decades, a new kind of aquatic biotope has originated in the floodplain area by dredging for gravel and coarse sand. Gravel pit lakes connected to the stream pass for substitute habitats, which can partially take over the functionality of the former backwaters (Neumann et al. 1996). Even if these gravel pits were not reconstructed with respect to the ecological requirements of fish, these lakes are inhabited by 0+ fishes of eurytopic cyprinids and especially percids in high abundances, which form the major part of the adult fish community of the river (Figure 2 d). Thus, although not in optimal shape, these connected gravel pit lakes were a major source for the recruitment of riverine fishes. If gravel pit lakes connected to the stream are reconstructed with respect to the ecological requirements of limnophilous and phyto-philous fishes, these lakes can support the recruitment of these species in addition to those species found in nonreconstructed gravel pit lakes, which are much more common than reconstructed gravel pit lakes. One example is the Lake Spey, a former gravel pit lake inundated by the River Rhine immediately above the river's average level (Scharbert and Greven 2002). In this lake with vast shallow areas covered by a mix of flooded terrestrial and also aquatic (submerged) vegetation, the 0+ fish community was quite similar to that of other gravel pit lakes, however with the substantial addition of those species regularly found only in natural oxbows (e.g., bream, carp, and rudd, cf. Figure 2c, 2e). In consequence, we propose not only preserving gravel pit lakes connected to the stream (against the tendency towards backfilling these lakes), but also reconstructing them with respect to the ecological requirements of the rare limnophilous and phytophilous of the fish community at the Lower Rhine.

Feasibility

The application of our postulations is partly feasible in the frame of current flood protection programs or general ecological rehabilitation measures. For instance, the planed and yet partly realized flood protection measures (creation of flood retention volume by replacement of dykes at 11 localities along the lower Rhine) cover an area of 47 km^2 (IKSR 1998), which offers excellent opportunities for a vitalization of floodplain habitats. The possibilities for improvement of bank structures are of course limited due to restrictions of the waterway authorities. Nevertheless, an actual programme of a nature conservation agency includes measures like dismantling of bank revetment, structuring of groin fields and creation of side channels at selected sites (www.lebendiger-rhein.de), which demonstrates the feasibility of our postulations.

Acknowledgments

Our thanks are due to P. Beeck, A. Nemitz, F. Molls and A. Scharbert for providing unpublished data. We thank F. Bartlett for improving the English text.

References

Bonesi, L., S. Rushton, and D. Macdonald. 2002. The combined effect of environmental factors and neighbouring populations on the distribution and abundance of *Arvicola terrestris*. An approach using rule-based models. Oikos 99:220–230.

de Groot, S. J. 2002. A review of the past and present status of anadromous fish species in the Netherlands: is restocking the Rhine feasible? Hydrobiologia 478:205–218.

Delibes, M., P. Gaona, and P. Ferreras. 2001. Effects of an attractive sink leading into maladaptive habitat selection. American Naturalist 158:277–285.

Dröge, B., H. Engel, and E. Gölz. 1992. Channel erosion and erosion monitoring along the Rhine River. IAHS Publications 210:493–503.

Grift, R. E. 2001. How fish benefit from floodplain restoration along the lower River Rhine. Doctoral dissertation., Wageningen University, Wageningen, The Netherlands.

IKSR (Internationale Kommission zum Scutz des Rheins).1998. Assessment of the ecologically valuable areas of the Rhine and the first steps on the way to a biotrop network. International Commission of the River Rhine, Koblenz, Germany.

IKSR 2003. Hydrographic map of the Rhine: accompanying report. International Commission of the River Rhine, Koblenz, Germany.

Lelek, A., and G. Buhse. 1992. Fish of the Rhine in earlier times and today. Springer-Verlag, Berlin.

Lelek, A., and C. Köhler. 1989. Analysis of fish species assemblages in the Rhine, 1987–1988. Fischökologie 1:47–64.

Molls, F. 1997. Population biology of the fish species in a Lower Rhine floodplain area: reproductive success, life cycles, short distance migrations. Doctoral dissertation., University of Cologne, Cologne, Germany.

Molls, F. 1999. New insights into the migration and habitat use by bream and white bream in the floodplain of the River Rhine. Journal of Fish Biology 55:1187–1200.

Molls, F., and A. Nemitz. 2007. Restoration of Atlantic salmon and other diadromous fishes in the Rhine River system. Proceedings of the Fourth World Fisheries Congress, Vancouver, Canada.

Nemitz, A., P. Beeck and A. Scharbert. 2002. Functional check of the fish passageway "Dornicker Schleuse" at the outflow of the oxbow Bienan-Praest (Kleve District): final report–investigation on the behalf of the LOEBF. Unpublished report, Bonn, Germany.

Neumann, D., and J. Borcherding. 1998. The fish fauna of the Lower Rhine and their former floodplain area: current condition, ecological adaptations, and recommendations for future actions. LÖBF-Mitteilungen 2/98:12–15.

Neumann, D., S. Staas, F. Molls, C. Seidenberg-Busse, A. Petermeier, and J. Rutschke. 1996. The significance of man-made lentic waters for the ecology of the Lower River Rhine, especially for the recruitment of potamal fish. Archiv für Hydrobiologie Supplement 113:267–278.

Nienhuis, P. H., A. D. Buijse, R. S. E. W. Leuven, A. J. M. Smits, R. J. W. de Nooij, and E. M. Samborska. 2002. Ecological rehabilitation of the lowland basin of the river Rhine (NW Europe). Hydrobiologia 478:53–72.

Scharbert, A., and H. Greven. 2002. Remodeled gravel pit lakes: floodplain waters of the future? Verhandlungen der Gesellschaft für Ichthyologie 3:131–187.

Schiemer, F., and H. Waidbacher. 1992. Strategies for conservation of a Danubian fish fauna. Pages 363–382 *in* P. J. Boon, P. Calow and G. E. Petts, editors. River Conservation and Management. John Wiley & sons, Chichester, UK.

Staas, S. 1996. The abundance of juvenile fish in the lower Rhine and adjacent waters as affected by bank structure. Doctoral dissertation. University of Cologne, Cologne, Germany.

Staas, S. 1998. The abundance of juvenile fishes in the River Rhine and some gravel pit lakes connected to the River Rhine. LÖBF-Mitteilungen 2/ 98:15–19.

Staas, S. 1999. The ecological significance of large rivers as illustrated by the significance of structural aspects for the fish fauna of the (lower) Rhine. Laufener Seminarbeiträge 4/99:83–98.

Wolter, C., and R. Arlinghaus. 2003. Navigation impacts on freshwater fish assemblages: the ecological relevance of swimming performance. Reviews in Fish Biology and Fisheries 13:63–89.

American Fisheries Society Symposium 49:845–850

Water Quality Improvements Following Political Changes, Enhanced Fish Communities, and Fisheries in the Czech Republic

Pavel Jurajda*, Milan Penáz, and Martin Reichard
Institute of Vertebrate Biology, Academy of Sciences of the Czech Republic
Kvetná 8, 603 65 Brno, Czech Republic

Ilja Bernardová
TGM Institute of Water Research
Drevarská 12, 657 57 Brno, Czech Republic

Abstract.—Political and economic changes in 1989 in the Czech Republic (a part of the former Czechoslovakia) led to the collapse of several large industries (e.g., sugar refineries, paper mills), which utilized old technology that negatively affected water quality. Simultaneously, the financial resources available to the agricultural sector were considerably reduced, which was manifested by a decrease in the use of fertilizers and pesticides. In addition, old waste-water treatment plants were modernised and new ones constructed. Together these activities substantially increased the water quality of Czech rivers. In the River Morava (Danube basin), one of the largest rivers in the Czech Republic, we observed substantial changes in the fish community over this period. Fish species richness has increased continuously over the last 10 years and has almost reached the situation that existed 100 years ago. Anglers' statistics also document an increase in fish catches. Improvements in water quality and the absence of formerly regular seasonal acute fish poisoning have come about despite the physical structure of the river remaining unchanged. Further enhancement measures will necessitate river system revitalization measures, such as longitudinal reopening of the river channel and reconnection of the main channel with its floodplain and associated water bodies.

Introduction

The River Morava is one of the most important tributaries of the River Danube. However, it has been the subject of substantial anthropogenous impacts over the last 100 years; it was one of the most seriously affected rivers in the Czech Republic in terms of the amount of water pollution resulting from industrial, household and agricultural wastes (Penáz et al. 1986). As early as the 19th century serious pollution of its waters was recorded (Kaspar 1886), with the number of incidents increasing thereafter. Increasing levels of pollution resulted in the first fish kill in the thirties, and the frequency of sporadic fish kills peaked in the 1950s and the 1960s. The result was that its formerly rich fish fauna was impoverished, with the almost complete absence of many of the original fish species (Penáz

*Corresponding author: jurajda@brno.cas.cz

et al. 1986). The source of much pollution was waste water from the cellulose industry and periodic releases from sugar refineries. This organic pollution decreased dissolved oxygen below critical levels and thereby directly affected all fish assemblages in the river.

Originally the morphology of the River Morava, in common with other temperate rivers, was a continuous lentic-to-lotic sequence of alternating riffles, raceways, and pools, with a highly heterogeneous shoreline and an extensive floodplain in its middle and lower reaches. This rich patchwork of different habitat types offered a variety of niches for many reproductive guilds of fishes and was responsible for the original species-rich fish assemblage. However, a major environmental change to the river resulted from drastic channelization and regulation of the river for flood control, water extraction, and a planned scheme for navigation in the lower sections (Penáz et al. 1986). River channelization, which began during the last century and was completed in the 1980s, resulted in the isolation of meanders and other floodplain water bodies from the main channel (Matejícek 1990). River regulation created a uniform trapezoidal channel and few natural sections of the river remain.

Study Area, Material and Methods

The River Morava rises at 1380 m a.s.l. and flows 352 km into the River Danube. The section under study included river kms 69–347 in the territory of the Czech Republic. The total area drained by the river catchment is 26,579.7 km^2. The river's flow regime is extremely variable, averaging about 110 m^3/s, but ranging between 15 and 700 m^3/s at the confluence with the Danube.

Data for the present study were obtained from a fish monitoring program undertaken between 1985 and 2003. Samples were obtained by electrofishing 120 localities along the longitudinal profile of the River Morava. About 25% of localities were sampled annually in 1991–2003.

According to Czech angling law, all fishes taken from a water must be recorded in a catch statistic list by the licence holder. The fishing practice of catch and return is not widely practiced and most fish caught are retained and included in statistics. The Anglers association summarized all these statistics in all river sections for each fish species. Data on water quality were obtained from Project Morava II.–IV., analyzed by the TGM Water Research Institute in Brno (Bernardová 2003).

Results and Discussion

Water Quality Improvement

Political change in the Czech Republic (a part of the former Czechoslovakia) in 1989 (the so-called "Velvet Revolution") led to the collapse of several large industries that utilized old technology, which had hitherto negatively affected water quality in the country. Simultaneously, the financial resources available to the agricultural sector were considerably reduced, which was manifested by a decrease in the use of fertilizers and pesticides. Use of industrial fertilizers (NPK) in the River Morava basin decreased from 250 kg/ha/year in the 1980s to 75 kg/ha/year in 1997 (Figure 1). In addition, old waste water treatment plants servicing large conurbations were modernised, while new water treatment plants were constructed (Figure 1). The result was a significant decrease (>75%) in the organic load (BOD5) of rivers over the last 10 years. In addition, the load of toxic pollutants has been substantially limited (Bernardová 2003). All these direct and indirect activities have in-

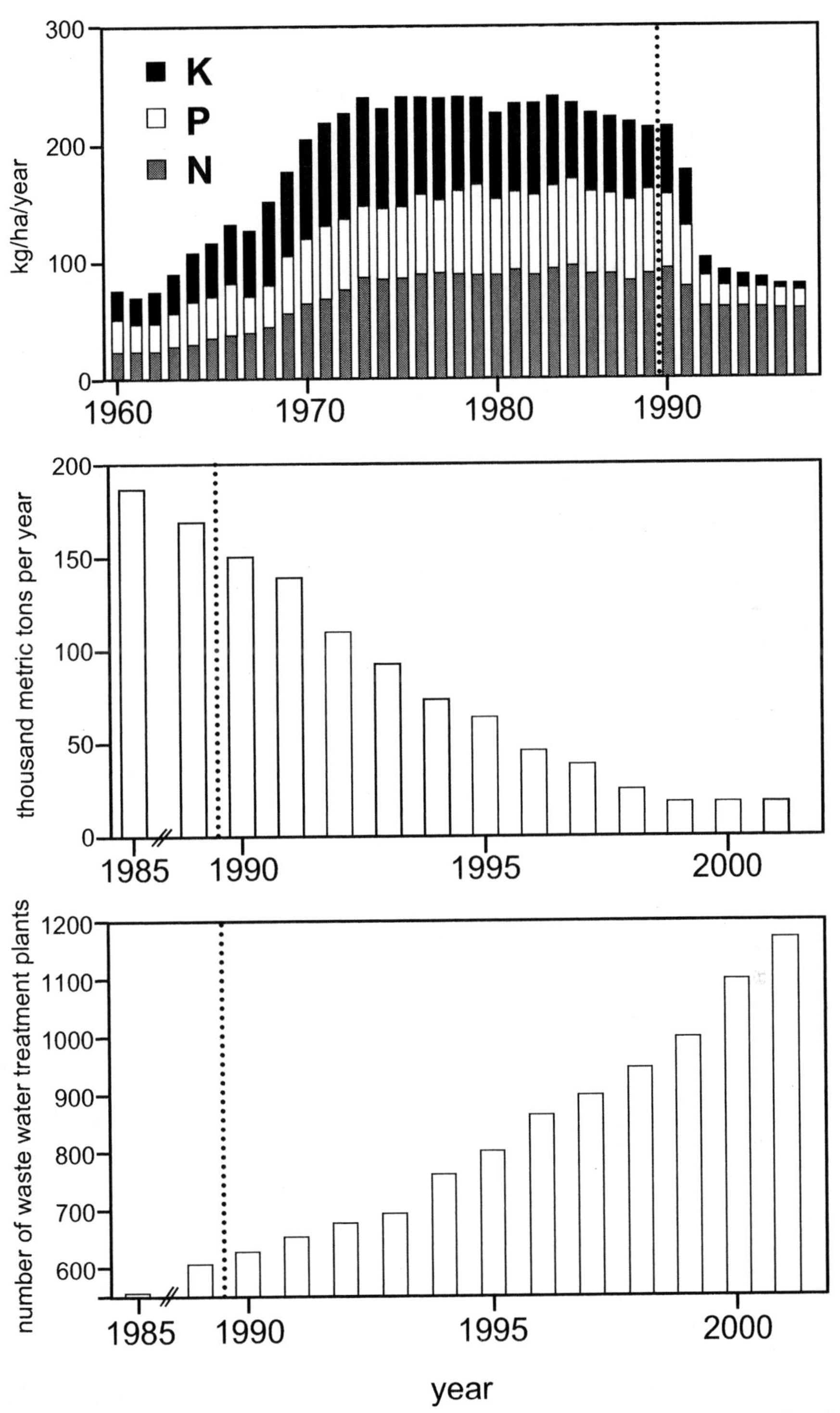

Figure 1. Application of industrial fertilisers in the River Morava basin between 1960 and 1998 (upper). Pollution discharged from point sources in the Czech Republic (BOD_5 index) between 1985 and 2001 (middle), and number of waste water treatment plants in the Czech Republic between 1985 and 2001 (lower). 1989, the year in which political changes began, is indicated by a dotted line.

creased water quality in Czech rivers substantially (Anonymous 2003).

Physical Habitat of River

No dams were built on the main channel of the River Morava, though 37 weirs were installed. These weirs, with heights up to 7.5 m, were built to regulate water levels and have had the effect of interrupting the river continuum. The general character of the modified river habitat, comprising a regulated river channel without a connection to the floodplain or adjacent water bodies, has not been substantially improved during the last 14 years and almost no rehabilitation activities have been undertaken.

Species Richness

The historical ichthyofauna of the River Morava is probably one of the best documented in central Europe. The first scientific lists of fishes were already published in the mid-1800's by Heinrich (1856) and Jeitteles (1863, 1864), followed by faunal records by Kitt (1905). Based on these historical records, 2 lampreys and 51 fish species were recorded as native to the River Morava system in the territory of the Czech Republic (upstream of river km 69). During the fifties and sixties of the 20th century, when anthropogenous impacts were most severe, many sensitive species disappeared from the main channel (Kux 1956). At the beginning of the nineties 34 fish species was recorded in the river. In subsequent years species richness has continually increased, reaching a tally of 47 native species by 2003 (Figure 2). In total, including exotic species, the ichthyofauna of the River Morava now amounts to 53 species.

Angling Statistics of Catches

In this study we present angler's catches from the main channel of the River Morava between 1985 and 2002. Considerable increases in catches by anglers were apparent approximately five years following political and economic change, and reflected water quality improvements. The total catch by anglers reached almost 25 metric tons of all species in 2002, that is about 500% higher than at the beginning of nineties (Figure 2).

Conclusions

Political and subsequent economic changes in 1989 in the Czech Republic led to sudden changes to the industrial and agricultural sectors; old industries collapsed while the financial resources available to farmers were considerably limited, manifested by a decrease in the use of fertilisers and pesticides. Simultaneously, the Czech government invested in upgrading waste water treatment facilities. Together these activities substantially improved the water quality of Czech rivers. In the River Morava, fish species richness has increased, almost attaining the same situation that existed 100 years earlier. Anglers' statistics corroborate our findings, documenting a significant increase in fish catches over the same period. These improvements to the fish community have come despite few improvements to the physical condition of river habitats.

Our results demonstrate the potentially dramatic improvements to fish community richness that can arise from substantially improved water quality. Future improvements to the fish community and fisheries of the River Morava will depend on further conservation measures and rehabilitation of the river channel. The implementation of plans to create a link between the Danube, Oder and Elbe rivers, through the River Morava and a network of canals, could considerably affected the fish fauna in the Morava and Danube river systems.

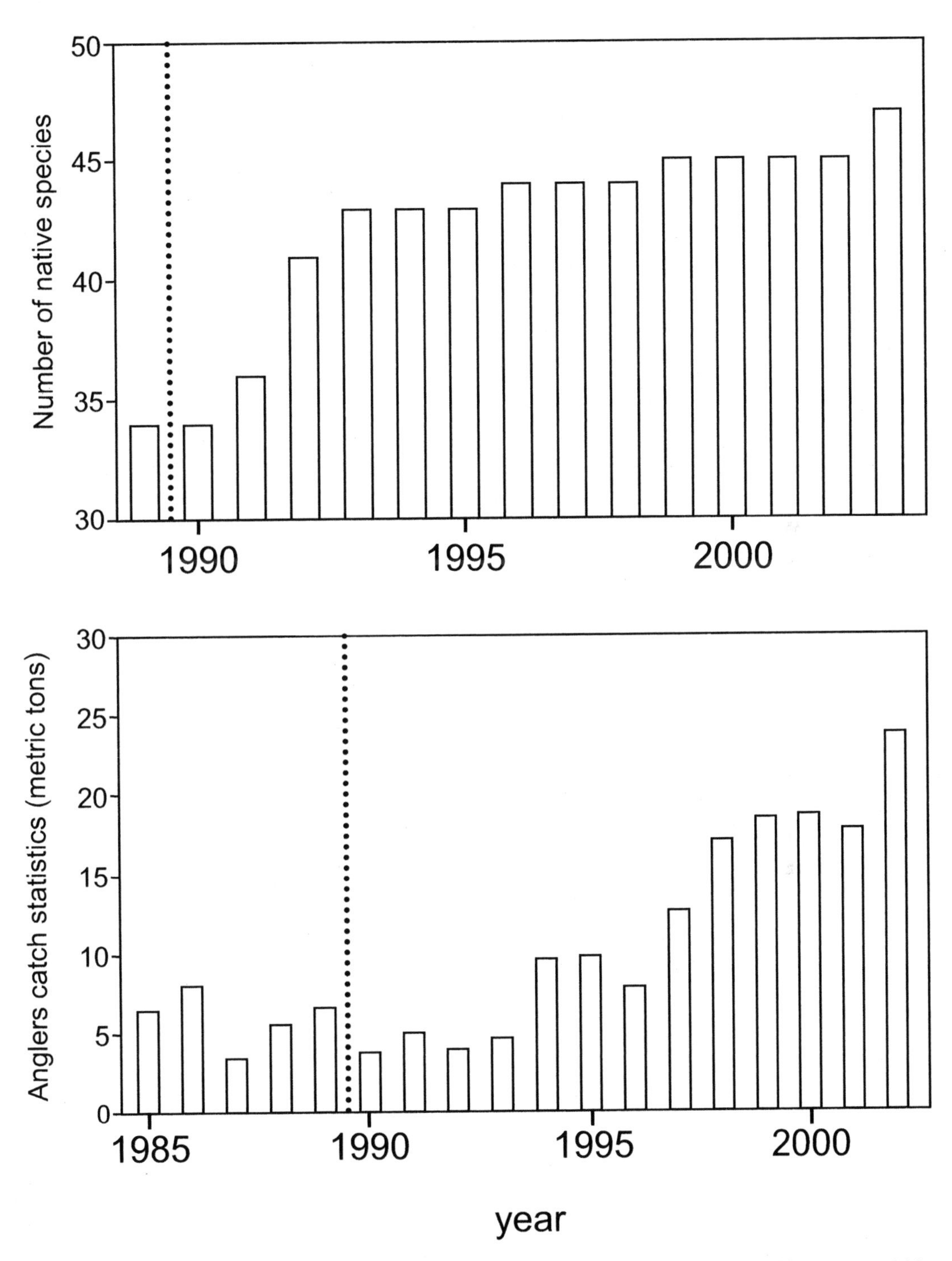

Figure 2. Cumulative species richness of fish in the River Morava (Czech Republic) between 1989 and 2003 (upper) and the dynamics of angler's catches of all fishes (in metric tons) in the River Morava (Czech Republic) between 1985 and 2002 (lower). 1989, the year in which political changes began, is indicated by a dotted line.

Acknowledgments

This study was supported by the Grant Agency of the Academy of Sciences of the CR (No. 66902 and No. IAB6093106) and Project Morava. We are grateful to the officials and managers of the Moravian Anglers Association for their cooperation, as well as all our field assistants.

References

Anonymous, 2003. Statistical yearbook of the Czech Republic. Czech Statistical office, Praha 2003.

Bernardová, I. 2003. Project Morava. Surface water quality assessment, yearly report. TGM Water Research Institute Praha, Brno branch, Brno, Czech Republic.

Heinrich, A. 1856. Mährens u. Schlesiens Fische, Reptilien u. Vögel. Brno, Czech Republic.

Jeitteles, L. H. 1863. Die Fische der March bei Olmütz. I. Abth. Jahres-Bericht über das kaiserl.-königl. Gymnasium in Olmütz während des Schuljahres 1863:3–33, Frank Slawik, Olomouc, Czech Republic.

Jeitteles, L. H. 1864. Die Fische der March bei Olmütz. II. Abth. Jahres-Bericht über das kaiserl.-königl. Gymnasium in Olmütz während des Schuljahres 1864:3–26, Frank Slawik, Olomouc, Czech Republic.

Kašpar, R. 1886. Ryby moravské a slezské. Cas. vl. sp. Mus. 3:132–134, Olomouc, Czech Republic.

Kitt, M. 1905. Die Fische der March bei Olmütz, Bericht der Naturwiss. Section des Ver Botanischer Garten in Olmütz 1905:1–15, Olomouc, Czech Republic.

Kux, Z. 1956. Príspevek k ichtyofaune dolní Moravy a Dunaje. Cas. Moravského Musea 41:93–112, Brno. Czech Republic.

Matejícek, J. 1990. Vodohospodárské úpravy na jizní Morave ukonceny. Vodní Hospodarství 3:95–101.

Penáz, M. and P. Jurajda. 1993. Fish assemblages of the Morava River: Longitudinal Zonation and Protection. Folia Zoologica 42:317–328.

Penáz, M., Šterba, O., and M. Prokeš. 1986. The fish stock of the middle part of the Morava River, Czechoslovakia. Folia Zoologica 35:37–384.

American Fisheries Society Symposium 49:851–866

Watershed Restoration to Reconcile Fisheries and Habitat Impacts at the Keogh River in Coastal British Columbia

Bruce R. Ward*

University of British Columbia, British Columbia Ministry of Enviornment, Aquatic Ecosystem Science Section 2204 Main Mall, Vanouver, Brirish Columbia V6T 1Z4, Canada

Patrick A. Slaney

PSlaney Aquatic Science Ltd., 214 Nelson Street, Coquitlam, British Columbia, V3K 4M4, Canada

Donald J. F. McCubbing

InStream Fisheries Research Inc., 223–2906 West Broadway, Vancouver, British Columbia V6K 2G8, Canada

Abstract.—Reconciling fisheries, climate, and habitat impacts with conservation may be possible with ecosystem restoration. Riparian logging decreased large woody debris in most British Columbia streams, reducing fish cover, overwinter refuges, and bank stability. Salmon survival at sea declined dramatically in the 1990s in southern British Columbia, at some places accompanied by declines in yield from freshwater life stages. Watershed restoration may offset poor survival. At the oligotrophic Keogh River, approximately 500 habitat structures were installed, off-channel ponds developed, nutrients added annually, and logging roads storm proofed and stabilized. Treatment effectiveness was evaluated stepwise by annual monitoring of juvenile fish in stream, and steelhead *Oncorhynchus mykiss* and coho *O. kisutch* salmon smolts, in comparison to the adjacent Waukwaas River. Mean juvenile sizes and densities were highest where treated with nutrients and habitat structures. By 2003, 90% (previous mean, 32%) of steelhead smolts were age-2 or younger with average previous size of age-3 smolts, as found in the nutrient-enriched river in the mid-1980s. Yields exceeded (coho) or approached (steelhead) the average smolt yield of an earlier more productive decade (1980s), despite exceptionally low returns. Steelhead smolt yield per spawner improved substantially in the Keogh River, compared to a pretreatment extirpation trajectory. Regardless, covariance analysis indicated that steelhead smolt recruits per spawner were lower than that of the earlier regime (1980s) and not significantly different from those in the recent (1990s) regime, prior to treatment. The latter was confounded by differing levels of spawners. Nonetheless, based on the density and growth studies in river, earlier age of steelhead smolts (with no change in mean size), and observations on other species, we conclude that populations are in better condition after watershed restoration to respond to low survivals of smolts in the ocean, potential harvest impacts, and climate shifts.

*Corresponding author: bruce.ward@gov.bc.ca

Introduction

Anadromous salmonids in their southern distribution of North America have undergone substantial declines over the last few decades that have been attributed to lower survival during the marine life stage (Anderson 1999; Smith and Ward 2000; Ward 2000). Reasons for these declines appear coincident with a series of El Niño events, the Pacific decadal oscillation, and global climate warming. Further, climate changes may have affected freshwater habitats for steelhead negatively by increasing summer droughts and winter storms (Ward et al. 2003b). Past habitat impacts of forest land use, particularly historical riparian logging practices, are also considered a significant factor (Slaney and Martin 1997). Moreover, lower productivity of salmonid stocks may, in part, be related to low salmon escapements in the Pacific Northwest and fewer salmon carcasses (i.e., a depletion of marine-derived nutrients transported into oligotrophic freshwater environments; Larkin and Slaney 1997).

Applied strategies of habitat rehabilitation may assist in counteracting these trends of depressed salmonid stocks by increasing stock productivity and capacity in freshwater. Programs of watershed restoration have emerged in northern California, Oregon, Washington, and British Columbia (Slaney and Martin 1997); these recovery initiatives assume improvements in salmonid productivity and capacity in freshwater. Only a few watershed and habitat restoration programs have provided sufficient monitoring and evaluation to confirm their effectiveness. A long-term evaluation of steelhead *Oncorhynchus mykiss* and coho salmon *O. kisutch* smolt yield from 14 km of large wood restoration at Fish Creek, Oregon, was inconclusive (Reeves et al. 1997). However, a well-replicated postrestoration evaluation of salmonid streams in coastal Washington and Oregon demonstrated greater juvenile abundances in rehabilitated reaches, particularly in winter (Roni and Quinn 2001). Similarly, controlled before and after evaluations of large wood restoration in two small coastal streams in Oregon confirmed improvements in coho salmon, steelhead, and cutthroat trout *O. clarkii* smolt yields in response to instream large wood and off-channel alcove restoration (Solazzi et al. 2000). In south coastal British Columbia, experimental whole river addition of nutrients in the mid-1980s increased juvenile salmonid growth and steelhead smolt yield (Johnston et al. 1990; Slaney et al. 2003), and steelhead trout and coho salmon juvenile abundance responded positively to physical habitat manipulation (Ward and Slaney 1981). Thus, past evidence supports habitat recovery measures; yet whole river ecosystem strategies to date have not been evaluated, and their effectiveness is uncertain under the current conditions of low survival of smolts in the ocean (Ward 2000).

More recently, the Keogh River received watershed-level rehabilitative treatments, including large-scale habitat restoration, nutrient additions, and logging road rehabilitation. An increasing trend in densities of juvenile steelhead in the Keogh River was documented from 1998 to 2001, as the stream rehabilitation treatments progressed from 1997 to 2001 (Ward et al. 2003a, 2003b). As a response to marine-derived nutrient replacement, steelhead and coho salmon fry size (mean weight by autumn) exceeded the reference stream by 50% and 38%, respectively (Ward et al. 2003b). By 2000, coho salmon smolt yield increased in the Keogh River to 74,500, or 20% above of the historical average of 62,000 smolts, and steelhead smolt yield increased to 2,338 fish, the highest since 1993, but still well below the historical average of 6,000 smolts (Ward 2000). Moreover, steelhead smolts per spawner, albeit at low spawner density, showed a sharp increase over four brood years up to 1998. Thus, comparisons of juvenile abundance and growth, as well as yields of steelhead in treatment and reference streams, have provided preliminary support for the contention that an extinction trajectory of steelhead may be halted or reversed (Ward et al. 2003a, 2003b).

Our preliminary results suggested that reductions in survival rate at one salmonid life stage (smolt-to-adult) can to be compensated by improvements in another stage (fry-to smolt; Ward et al. 2003a, 2003b). However, several years of additional monitoring of a highly variable juvenile response was required to confirm sustained positive trends in smolt yield, smolt yield per spawner, and returning adults as the ultimate response variables. Herein, we report on two additional years of the response by steelhead and coho salmon smolts, to 2003.

Methods

The Keogh and Waukwaas rivers, two oligotrophic fourth-order streams, are situated at the northern end of Vancouver Island, British Columbia, Canada (Figure 1). The logging history, river size, and fish species present have been described previously (Ward et al. 2003a, 2003b).

The experimental design for instream response involved two treatment types: habitat structures and slow-release nutrient addition. The plan of restoration within the Keogh watershed was adapted to a staircase-like experimental design, with structure placement commencing in the upper reaches, nutrient addition beginning in lower reaches, and both treatments expanding annually over 5 years until the whole watershed was treated (Ward et al. 2003b).

The entire watershed underwent habitat restoration treatments to address historical logging impacts, as part of British Columbia's Watershed Restoration Program (WRP) (Slaney and Martin 1997). Treatments included instream habitat and nutrient addition, rehabilitation of several off-channel ponds as overwinter flood refuges for juvenile fish, and storm-proofing of logging roads. Overall rehabilitation costs, including road deactivation of about Can$200,000, approached Can$1,000,000.

Riparian trees had been removed within about 50% of the basin during 1960–1985, resulting in simple featureless channel segments as remaining in-stream wood decayed over time. From 1996 to 2001, approximately 500 habitat structures were added to 10 km of the main stem where overwinter and summer rearing habitats were most lacking. This included 6.5 km of reaches Z and Y, a 0.5 km segment located midway between reach Y and X, 0.6 km of reach X near the Island Highway, 1 km of reach X near the mainline logging road crossing, and a 1.1 km segment of reach W in the lower river (Figure 1). In stream structures were largely based on natural templates, designed as either boulder clusters (Ward 1997) or boulder-ballasted large woody debris (LWD) structures (Slaney et al. 1997), and reconstructed riffle-pool habitats (Slaney and Zaldokas 1997).

Low-level inorganic nutrients were added as slow-release briquettes of magnesium ammonium phosphate from 1997 to 2003 to compensate for reduced marine-derived influx of nutrients (Larkin and Slaney 1997). Pink salmon *O. gorbuscha* and coho salmon escapements were at historical lows of about 15–25% of average estimated historical abundances (data on file). Initially, the lower half of the river was treated with nutrients, and by 1999, the total length (31 km) of the Keogh River main stem to the outlet of Keogh Lake was treated every 3–5 km, including ~10 km of small tributar-

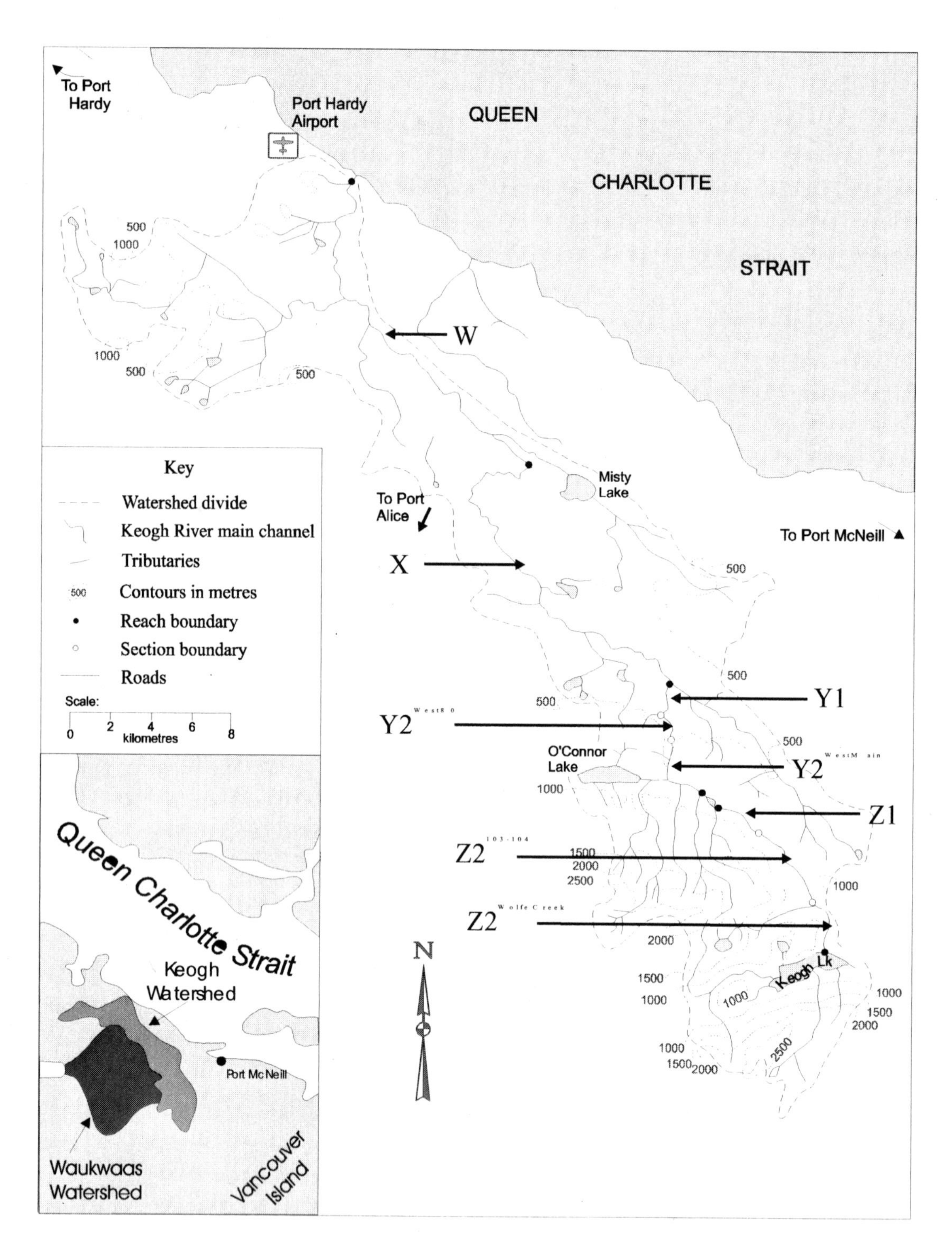

Figure 1. The Keogh River watershed, indicating reach boundaries and section boundaries used in this study, elevations, the watershed divide, and the location on the British Columbia coast (inset), as well as the relative position of the neighboring Waukwaas River, British Columbia.

ies. The objective was to moderately increase the very low concentration of dissolved inorganic phosphorus from less than 1 or 1 μg/L upwards to 3–5 μg/L (as soluble reactive phosphorus) during mid to late May to September, as the river was P-limited rather than N-limited (Johnston et al. 1990; Slaney et al. 2003). Details of methods were provided in Ashley and Slaney (1997) and briefly in Ward et al. (2003b), the latter at the Keogh River.

Several off-channel ponds were also rehabilitated to provide additional overwintering refuges for juvenile coho salmon. Between 1997 and 2000, seven ponds were expanded from remnant alcoves to elongated ponds averaging about 200 m in length by 10 m in width or about 0.2 ha in area (range 0.1–0.5 ha). Six of seven sites were connected to small water courses or groundwater inflows to elevate dissolved oxygen and provide fish access (one site had low oxygen concentrations). Ponds were located 0.5, 1.5, 4.0, and 20 km from the mouth of the river, and smolt out-migration was monitored.

Finally, an extensive network of logging roads throughout the watershed was stabilized from 1994 to 2000. About 75% of roads were associated with rolling low-slope forest lands, and about 25% were associated with steeply sloping forest lands in the headwaters (Figure 1). Overall, road segments were storm-proofed to comply with 1995 forest practice regulations to ensure adequate drainage and prevent washouts during large coastal storm events. At hill slope access roads, improved drainage was achieved using cross ditches and water bars. In addition, a small portion of roads (about 20 km), particularly at the base of the steep hill slope, was semipermanently deactivated by removing water course crossings and leaving swales. As well, several kilometers of roads were permanently deactivated and revegetated. Some stormproofing of logging roads was also undertaken in the reference Waukwaas watershed.

To evaluate the effectiveness of watershed rehabilitation techniques, four methods were utilized. First, the density of salmonid juveniles in freshwater was assessed by mark–recapture, using electroshocking and seine-netting techniques (McCubbing and Ward 1997). Fish were sampled within each reach (100 m except reach W on the Keogh where 200 m were sampled) in proportion to the frequency of occurrence of habitat classes. These results were reported by Ward et al. (2003a, 2003b). Second, steelhead and coho smolt yields were assessed through operation of a full-river counting fence near the mouth of the Keogh River, and by mark–recapture estimates using rotary screw traps on the Waukwaas River as described below and in Ward et al. (2003a, 2003b), here adding two more years to the results. Third, steelhead smolt-age data from scales were used to detect the main response to nutrient addition (age and size), similar to nutrient experiments at the same river in the mid-1980s (Slaney et al. 2003). Further, as a key variable, we analyzed the steelhead smolt yield per spawner as a function of the number of spawners in the Keogh River, in comparison to the historic record.

A permanent counting fence is located at the mouth of the Keogh River, where a total count of coho salmon and steelhead smolts has been conducted each spring from April to June since 1976, including systematic subsampling for size and age. A description of the methodology of fish sampling, smolt enumeration, and estimation of overall capture efficiency by smolt marking is provided in Ward and Slaney (1988). Fence failures during spring floods were rare, lasted only 1–2 d, and numbers were adjusted either by mark–recapture (partial trapping) or by interpolation between pre-

and postflood counts of each species (e.g., McCubbing 2001).

Numbers of adult winter-run steelhead have been estimated at the counting fence by mark–recapture and, since 1997, independently assessed by an electronic counter verified by video records (McCubbing et al. 1999). Methods of population estimation were described in detail elsewhere (Ward and Slaney 1988). Briefly, total numbers of adult male and female steelhead were estimated separately by the adjusted Petersen estimate by marking upstream adults with an opercular punch and examining kelts during their downstream migration. During December to March, commencing in 1998, a Logie 2100C resistivity fish counter was also utilized to enumerate adults. Accuracy was tested on coho salmon adults in the fall and was greater than 90% (McCubbing et al. 1999).

Smolt yields from the nearby Waukwaas River, as an external reference watershed, were estimated from 1996 to 2003 based on mark–recapture estimates derived from operation of two rotary screw traps from mid-April to mid-June. The two 1.50-m traps were spaced approximately 500 m apart, above tidewater, with the upper trap utilized for marking and the lower trap used for recapture and enumeration, and sampling for size and age. Migrant coho salmon greater than 70 mm and migrant steelhead greater than 130 mm were defined as smolts when silver colouring was most evident. Details on marking (mainly caudal fin clipping) and estimation procedure were provided in Ward et al. (2003b).

Steelhead smolt recruitment per spawner was based on the total number of smolts manually counted, their age based on scale analyses, and tabulation by brood year of origin. These data were utilized to estimate the number of smolts produced per spawner in comparison to the number of spawners contributing to their recruitment, as in Ward et al. (2003a, 2003b). Density, growth, and smolt yield of salmonid juveniles were analyzed by anova and *t*-tests. Smolt yields and smolts-per-spawner were utilized as the key response variable and tested by covariance analysis as described below because adults were affected by highly variable ocean conditions and, for coho, ocean fishing for salmon.

We tested for differences in smolt production under different watershed production regimes and among years of watershed rehabilitation. Therefore, we performed a statistical analysis to challenge the null hypotheses that the number of smolts produced per spawner did not differ among the four treatment classes (i.e., 1980s regime, fertilizer addition, 1990s regime, and WRP; Ward et al. 2003b). Our analysis structure was based on an initial scrutiny of our data that indicated that the smolt–spawner relationship for the Keogh River followed a Beverton–Holt formulation (Ricker 1975; Ward 2000). However, we modified the fundamental Beverton–Holt relationship to allow the intercepts and slopes of that relationship to potentially differ according to treatment, such that our formulation is

$$\frac{R_y}{S_{y-d}} = \frac{1}{\sum_{t=1}^{4} \beta_t Y_t + \sum_{t=1}^{4} \beta_{4+t} S_{y-d} Y_t} + \varepsilon_y$$

where R_y is the number of smolts leaving the river in year y and S_{y-d} is the number of spawners that produced these smolts d years in the past, where d can be 2 or 3. The Y_y represent the proportional contributions of each treatment effect in each year y. The coefficients to be estimated are the $\beta_t s$ (units = $\left(\frac{S_{y-d}}{R_y}\right)$), which measure the intercepts for the four treatments, and the $\beta_{4+t} s$ (units = $\frac{1}{R_y}$) which measure the slope of the influence of the number of spawners on the number of smolts produced. Scrutiny of our data also indicated that prediction error variance appeared to be proportional to the predicted mean number of smolts per spawner. Given this apparent proportionality, and also that observed values for the number of smolts per spawner cannot be less than zero, we chose to represent the distribution of model error deviates, ε_y, with a gamma distribution. This required an additional model parameter, γ, a variance scalar, such that the variance of prediction error is represented by $\left(\frac{R_y}{S_{y-t}}\right)^2 \gamma$. We used an information–theoretic approach to judge the quality of model fit and selected the best model on the basis of lowest Akaike's Information Criteria corrected (AICc) for sample size (Burnham and Anderson 2002), then used covariance analysis to judge the statistical significance of our estimates of the β_t coefficients.

Results

The age of steelhead smolts was reduced by restoration treatments at the Keogh River while abundance increased. The proportion of younger steelhead smolts was altered in the Keogh River in 2001, with age-1 and age-2 smolts representing over 67% of the smolt yield, compared to an average age-2 composition of 32% in years of no nutrient addition (Figure 2). By 2002 and 2003, age composition showed a marked shift to smolts 1 year younger; 90% were age-1 and age-2. The calculated proportion of age-2 smolts in the Waukwaas River remained similar through the 1997–2001 period (Figure 3). The difference in the proportions of age-2 smolts from the Keogh River versus Waukwaas prior to the impact of nutrient addition (1997–1999) versus after addition (2000–2003) was significant ($p < 0.05$, anova). The proportion of age-3 smolts was significantly lower in Keogh than Waukwaas after 2000 ($p < 0.05$, anova). In years where partial or full-river fertilizer addition was undertaken on the Keogh River, smolt length-at-age was clearly increased (Figure 4). Age-2 smolts averaged 173 mm and age-3 smolts, 204 mm from 1998 to 2001 (Figure 4). Length-at-age of Waukwaas smolts remained stable through the 4 years; age-2

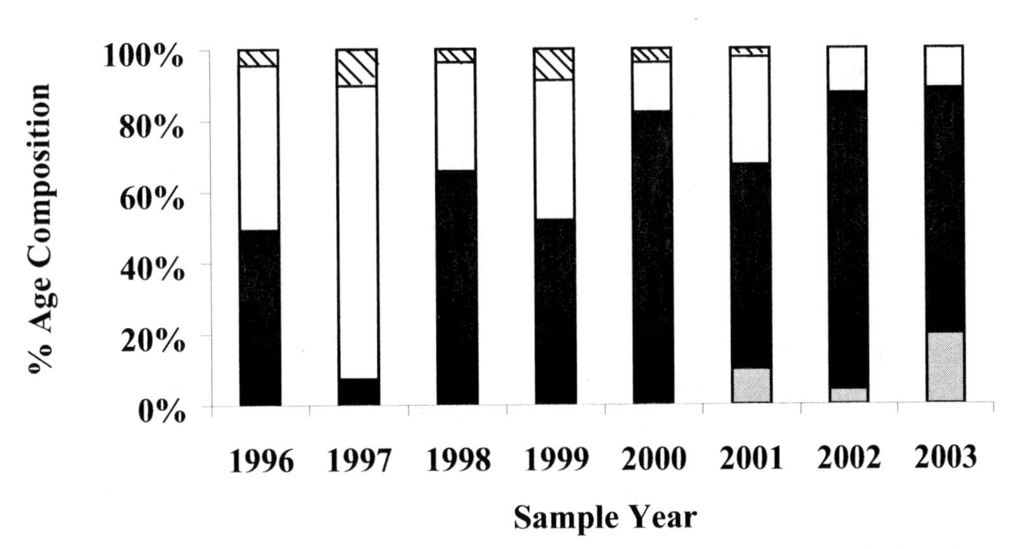

Figure 2. Percent composition of age-1 (gray bar), age-2 (black bar), age-3 (white bar), and age-4 (hatch bar) steelhead smolts in the Keogh River in 1996 and 1997 (no nutrient addition) and from 1998 to 2003 (nutrients added).

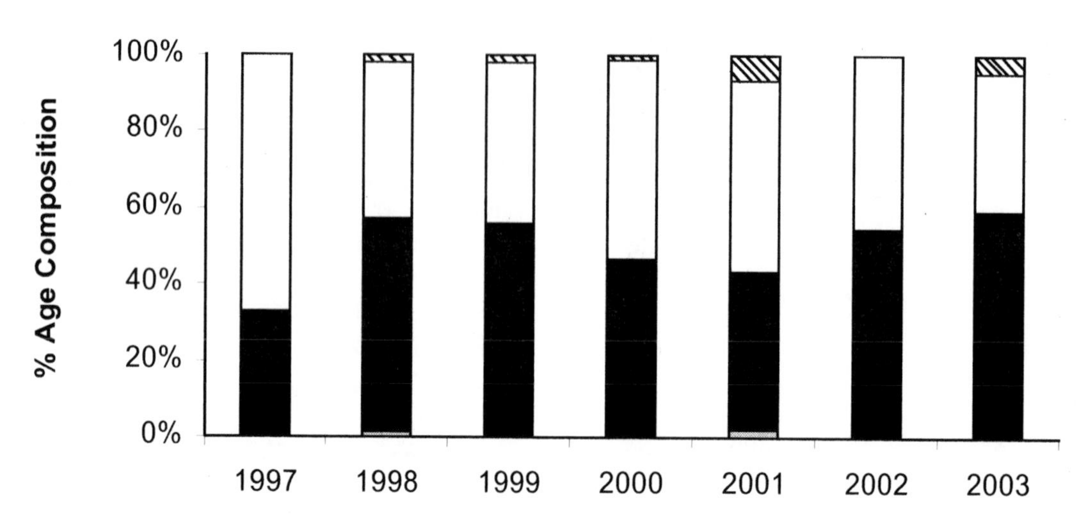

Figure 3. Proportional age composition estimates for Waukwaas River steelhead smolts, 1997–2003. Age-1 (gray bar), age-2 (black bar), age-3 (white bar), and age-4 (hatch bar).

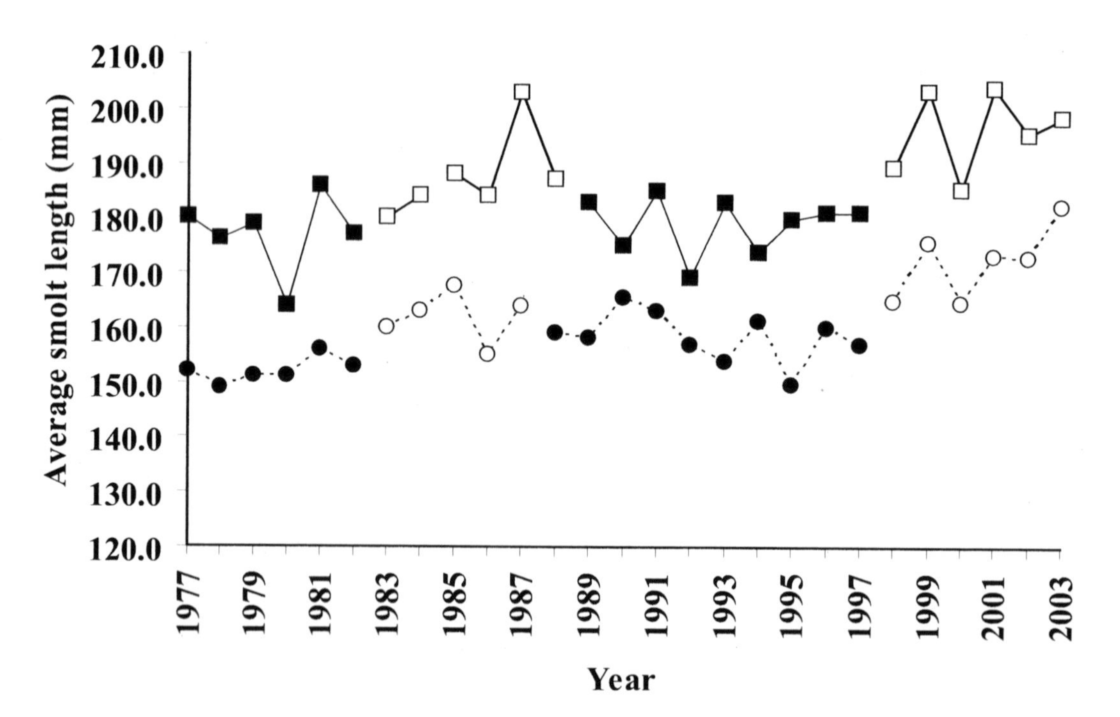

Figure 4. Mean length of 2-year-old (broken line, circles) and 3-year-old (solid line, squares) steelhead smolts. Open markers indicate years affected by nutrient addition at the Keogh River, British Columbia.

smolts averaged 165 mm and age-3 smolts, 187 mm, similar to Keogh smolts prior to nutrient addition.

Smolt yield reflected the watershed response of juvenile salmonids (Figures 5 and 6), but also the response to extreme low adult abundance. At the Keogh River, steelhead smolt yield increased in 1999–2001 to about 2,000 smolts after a pretreatment record low yield of less than 1,000 in 1998. Then, smolt yield elevated sharply to almost 5,000 in 2003, close to the historical average. In comparison, estimated yield on the Waukwaas River remained low at 1,859 fish, the second lowest in the six sample years, but developed an increasing trend to 2,800 and 4,000 in 2002 and 2003, respectively (Figure 5). The difference between steelhead smolt yield on the Keogh and Waukwaas rivers was significant ($P < 0.05$, single factor anova) for years when Keogh fish were influenced by restoration treatments (2000–2003). Mean difference in yield was higher by 355 smolts (SD = 907) at Keogh versus years prior to treatment effect (1995–1998) when the mean difference in yield was higher by 2,142 fish at Waukwaas (SD = 1,754). Yields from 1999 were treated as transitional and thus omitted from the analysis (i.e., nutrient additions were stepped up annually starting in 1997, thus smolts [majority were 2+] from treated areas were not expected until 1999 and later).

Coho smolt numbers rose steadily for both watersheds through the first 3 years of inves-

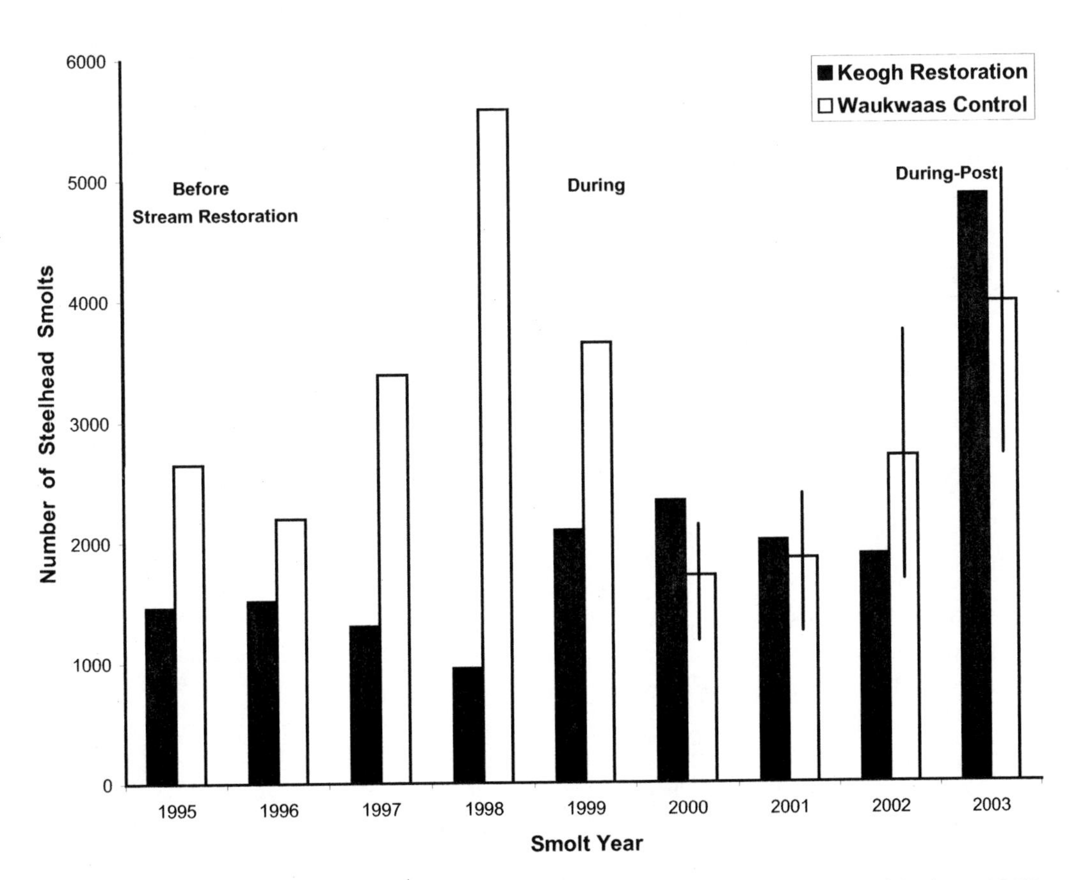

Figure 5. Steelhead smolt yield from the Keogh and Waukwaas rivers, British Columbia, from 1995 to 2003.

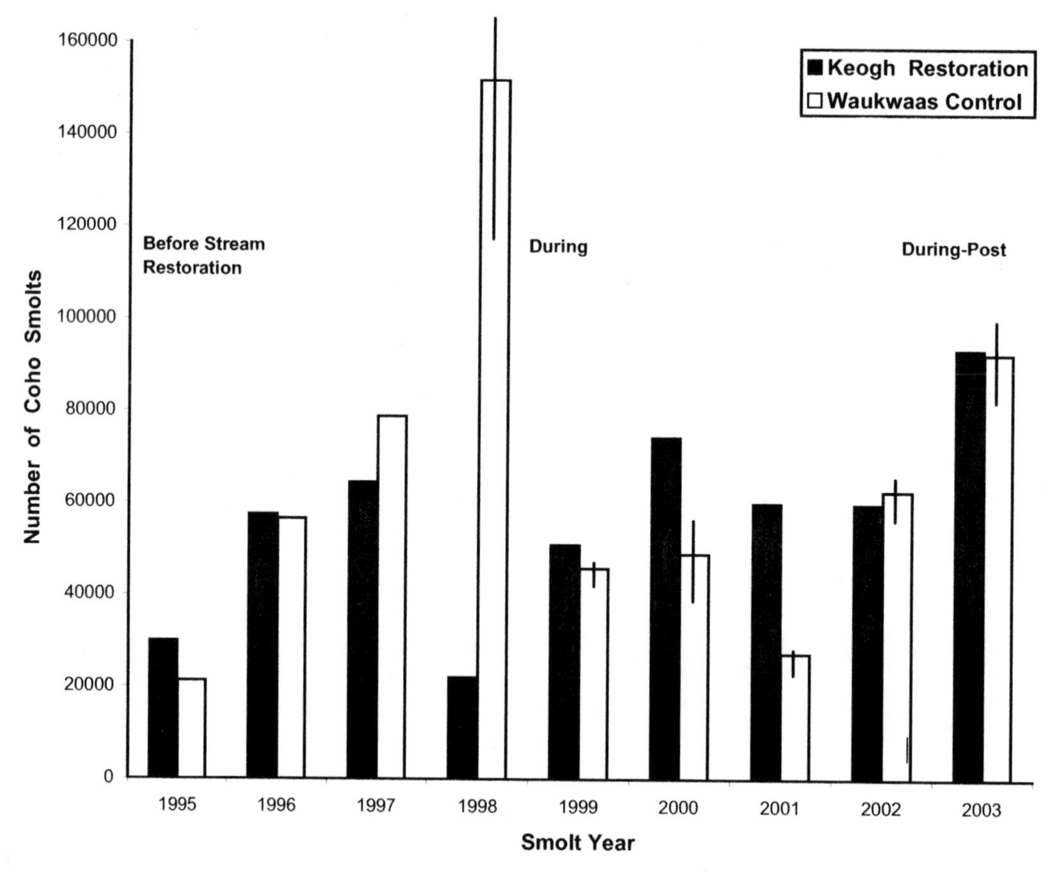

Figure 6. Coho smolt yield from the Keogh and Waukwaas rivers from 1995 to 2003.

tigation, but fell sharply in the Keogh River in 1998. Thereafter, an increasing trend developed at Keogh, reaching 94,000 in 2003 or greater than 50% above the historical average (Figure 6). In the Waukwaas River, reductions in coho smolt yield in 1999 and 2000 were followed in 2001 by a further decline, but increased thereafter to a level similar to that of the Keogh River in 2003. On average, pretreatment differences (1995–1997) were slightly higher at Waukwaas, by 1,514 smolts (SD =11,667). Posttreatment differences (1999–2003) were higher in favor of the Keogh River, by an average 13,585 fish (SD = 20,874), but differences were insignificant (P = 0.3). Since coho were mainly 1-year-old smolts, the smolt yield in 1998 was considered transitional and omitted from anova.

A substantial shift in steelhead smolt recruitment per spawner in the Keogh River was a key response in the effectiveness evaluation (Figure 7). Pretreatment data from the 1990s indicated very low yield, with less than 2 smolts per spawner. This was rapidly exceeded under even partial treatment, when greater than 14 smolts per spawner were produced from the 1996 brood year. Further increases to 42 smolts per spawner occurred from the 1997 brood year, and the highest yield in 26 years was recorded from the 1998 brood year (53 smolts per spawner). Coho yield was estimated at 9 and 21.5 smolts per spawner in 1998 and 1999, respectively, assuming 90% of smolts were 1-year-old. By 2002 and 2003, and associated with low adult escapements, coho smolts per spawner were much greater, or 52 and 33, respectively. Comparative data

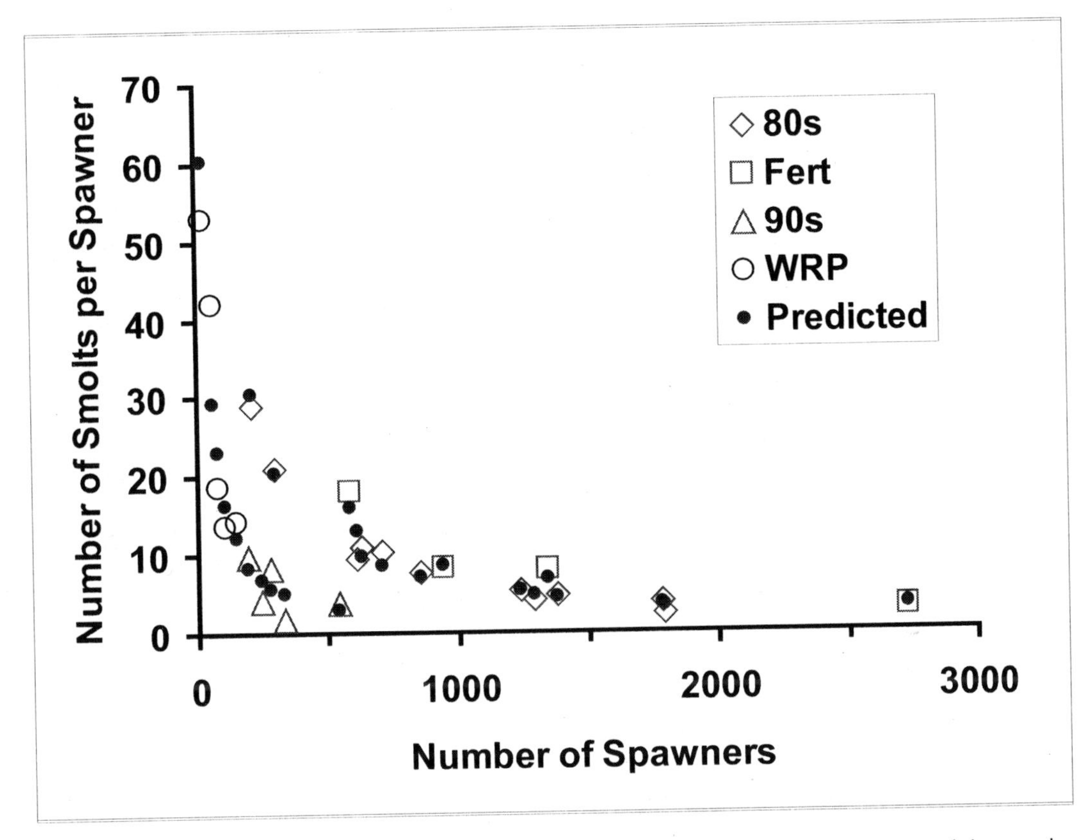

Figure 7. The relationship between the number of the steelhead smolts per spawner and the number of spawners in the Keogh River during the production regimes of the 1980s (diamonds) and the 1990s (triangles), during nutrient experiments in the mid-1980s (squares), and initial results during watershed restoration treatments to 2003 (circles). Our modified Beverton–Holt fit to these data (solid points) statistically distinguishes the 1980s and nutrient experiments from all other treatments, while the 1990s and initial watershed restoration treatments cannot be statistically distinguished from each other.

are not available from the Waukwaas River because spawners were not enumerated.

Based on lowest AICc, the best fit of our Beverton–Holt spawner–smolt relationship to our observed steelhead data retained the value of all four intercept terms (β_t), t = 1–4, at zero. This indicates there was no significant difference in the mean level of steelhead smolt yield per spawner, independent of the number of spawners, among treatments. However, there were differences in the relationship between spawner numbers and smolt production among some treatments. During the 1980s, and under the fertilization regime, there was a significant relationship between steelhead spawner numbers and smolt production, with $\beta_5 = 1.65 \times 10^{-4}$ (*SE*: 1.47×10^{-5}) and $\beta_6 = 1.08 \times 10^{-4}$ (*SE*: 1.58×10^{-5}), respectively. In other words, significantly more smolts were produced per spawner during years when nutrients were added to the stream in the 1980s. During the 1990s and the WRP regime, this relationship changed and was significantly different from the relationship identified for the 1980s and under the nutrient-addition regime (Figure 7). However, no statistical distinction could be made between the 1990s and the WRP regime, thus, $\beta_7 = \beta_8 =$

5.89×10^{-4} (*SE*: 5.17×10^{-5}). Note that due to the reciprocal structure of the Beverton–Holt spawner–smolt relationship, larger values of β_t indicate less smolt production per spawner. The model prediction error was approximately 27% (i.e., $100\sqrt{\gamma}$) of the predicted values.

Discussion

Increased smolt yield was attributed to ecosystem rehabilitation treatments. However, statistical comparisons of trends in smolt yields between the two rivers were perhaps weakened by the impact of ongoing climate change and likely differences in adult escapement between the Keogh River, draining to the east and into Queen Charlotte Strait, and the Waukwaas River, which drains into Quatsino Inlet and then to the west coast of northern Vancouver Island. Welch et al. (2000) and Smith and Ward (2000) noted major differences in survival of steelhead stocks from the east versus the west coast of Vancouver Island. As a consequence, adult recruitment to the Keogh River was below capacity as was subsequent steelhead fry recruitment in the Keogh River (Ward et al. 2003a, 2003b). Adult steelhead decreased to less than 100 spawners in the Keogh River by 2001, while two- to threefold this number were required to effectively recruit the entire river length with enough fry to produce the full benefits of the habitat rehabilitation (i.e., at or near capacity smolt yield). Experiments of this nature are best conducted when yields are at or near capacity production, thus avoiding the statistical difficulties of this study where definition of the recruitment relationship at low spawner levels confounded the analysis.

Trends of instream juvenile abundance and smolt yield in the west coast Waukwaas River likely reflected higher abundances of adult steelhead, based on estimated summer fry abundances and local observational records of adults from catch-and-release angling. The abundance of coho fry and smolts at Keogh appeared to exceed or equal that of the Waukwaas River, despite a reduction in coho adult escapement after 1998 in the Keogh River. Despite low recruitment of steelhead fry at the Keogh River, the abundance of steelhead parr in reaches treated with both restoration techniques were either higher, or at least similar, to those observed in the untreated Waukwaas River (Ward et al. 2003a, 2003b). Thus, Keogh restoration appears to have dramatically and persistently improved fry-to-parr survival rates (perhaps by as much as threefold) of steelhead. By 2003, at the Keogh River, coho and steelhead smolts increased to near and well above their historical averages, respectively.

Trends in fish age in the Keogh River indicated positive effects of watershed restoration that were incrementally increased over 5–6 years, although tempered with changes in fry density related to low and varied adult escapements. Accordingly, there has been a marked shift in dominant smolt age from age 3 to age 2, while increasing average smolt size at age and maintaining the overall average smolt size. The smolt age and size-at-age response to the restoration treatment is unequivocal and was similar to that detected earlier in 1980s after 3 years of experimental whole-stream enrichment, when steelhead smolt yield increased on average by 62% (Slaney et al. 2003).

Length at age trends in steelhead smolts from the Keogh River increased during nutrient addition to 2003, with evidence of an additional marine-derived nutrient signature due to improved pink salmon escapement (highly abundant in even years). Ward and Slaney (1988) suggested this may have been the case in the past, based on annual variation in the shape of the length frequency distribution of steelhead smolts. Steelhead smolt length-at-age increased again in 2001 for both age-2 and age-3 smolts to among the highest recorded in 27 years of study on the Keogh (Ward 2000; Slaney et al. 2003). In 2003, mean length increased even more than in 2001 among age-2 smolts, even though they comprised nearly 90% of the steelhead smolts and neared historic average abundance. An increase in length and weight of juvenile steelhead apparently improved overwinter survival, culminating in sustained increases in smolt yield. Historically, larger steelhead smolts have survived at higher rates in the ocean, producing a greater escapement of adult steelhead upon return than smaller smolts (Ward and Slaney 1988; Ward et al. 1989).

As the key response indicator, smolt yields per spawner have increased substantially up to 2003. Steelhead smolts per spawner in the Keogh River have risen from the dangerously low of less than 3 smolts per spawner from the 1996 brood (i.e., below adult replacement at 4% smolt-to-adult survival) to less than 53 smolts per spawner from the 1998 brood year, the highest production per spawner on record (Ward 2000). Thus, there were continued high levels of production from low adult escapements, as expected in a Beverton–Holt recruitment relationship. While these values were high by Keogh standards, Cramer et al. (2003) reported a range from 8 to 223 smolts-per-spawner for summer-run steelhead of the more productive Yakima River, Washington, a tributary of the Columbia River, for the period of 1985–1997.

Smolt recruits during the 1990s indicated a new recruitment relationship, or lower productivity and capacity within a different climate regime, as Welch et al. (2000) had suggested for adult steelhead returns after 1989. We could not detect a significant statistical difference in smolt recruitment from years of watershed restoration and that obtained in the 1990s regime. However, definition of the shape of the recruitment curve at both very low and high spawner abundance was poor from data of the 1990s, although clearly near the replacement line (Ward 2000). The statistical testing of WRP results (Figure 7) with this data was confounded by comparison of values at different spawner densities. Nonetheless, the most recent WRP values were above recruitment replacement (25 smolts per spawner at 4% smolt-to-adult survival) whereas previous points (1990s) were not. Thus, the restoration work may have provided sufficient recruitment of steelhead smolts to offset poor survival at sea and assist in population viability.

Historically, approximately 90% of coho smolts have migrated seaward at age 1, thus, the 1998 brood year produced about nine smolts per spawner. This increased significantly for the 1999 brood year with an estimate of 19 smolts per spawner. By 2002 and 2003, coho smolts per spawner increased substantially to 52 and 33, respectively. In comparison, coho smolt recruitment in 1999 averaged 83 smolts per female spawner at Black Creek, near Courtney on the east coast of Vancouver Island. The latter is a highly productive stream in an agricultural setting, with abundant coho habitat (K. Simpson, DFO, Nanaimo, personal communication).

Regardless of high variability to date in smolt yields, the magnitude of a recovery trend of all Keogh salmonid fish stocks, combined with coinciding size and age changes of smolts, suggested that ecosystem recovery was advancing. Although lacking statistical controls, increases were also apparent in Dolly Varden *Salvelinus malma* smolts (sevenfold) and adults (threefold) of the Keogh River (McCubbing and Ward 2003). The apparent recovery trend in smolt and adult yield of Dolly Varden char from 1998 to 2003 provided additive support that recovery was linked to improved freshwater conditions, since this species migrates for only a few months at sea before returning (Smith and Slaney 1980). Thus, negative environmental factors should be buffered, in part, by stream habitat rehabilitation and nutrient replacement. Thereby, restorative watershed treatments, combined with occasional years of improved ocean survival, may sustain salmonid populations in coastal streams against poor ocean conditions. Broader implications are that protection of the spawning stock, habitat protection and rehabilitation, and a better understanding of the limits to production in both the freshwater and the marine environment remain of profound importance. To date, whole-river restoration appears to be a useful tool towards mitigation of logging impact and adaptation to climate change for wild salmonids on the west coast of Canada.

Reconciling fisheries with conservation through ecosystem restoration, now and in the future, will need to incorporate climate considerations, even when fisheries have been temporarily closed. Can we get more fish from habitat rehabilitation in freshwater? Our results suggest it may be possible, but we could not yet confirm recruitment benefits from whole watershed treatments, largely because climate conditions worsened through the course of the work in both freshwater and the ocean, such that the numbers of spawners declined to levels that were far below previous conditions. There remains uncertainty in the steelhead smolt recruitment we currently may expect at higher levels of spawners. Regardless, the weight of evidence, from work on steelhead and coho abundance and growth in stream, as well as the positive changes in smolt age and results of previous experiments with nutrient addition and habitat alteration studies in the 1980s, suggested that smolt yields would have been much lower otherwise, to the point where the steelhead population would likely now be far below recruitment replacement, and near local extirpation. The hope is that habitat improvements may assist with population viability until climate conditions improve for these salmonids, or these fish adapt effectively.

Acknowledgments

Research on salmonids at the Keogh River has been supported through British Columbia's Habitat Conservation Trust Fund. British Columbia's Watershed Restoration Program (1995–2002) funded the watershed rehabilitation work and its in-stream evaluations. We are most thankful to the field staff and, in particular, to Mark Potyrala for early stream habitat assessments and restoration, and Lloyd Burroughs for undertaking supervision of habitat restoration, fish enumeration, and fish sampling activities in the field. Barry Smith (Canadian Wildlife Service) provided assistance with statistical analysis of the recruitment response.

References

Anderson, J. J. 1999. Decadal climate cycles and declining Columbia River salmon, Pages 467–484 *in* E. E. Knudsen, C. R. Steward, D. D. MacDonald, J. E. Williams, and D. W. Reiser, editors. Sustainable fisheries management: Pacific salmon. Lewis Publishers, Boca Raton, Florida.

Ashley, K. I., and P. A. Slaney. 1997. Accelerating recovery of stream, river and pond productivity by low-level nutrient replacement. Chapter 13 *in* P.A. Slaney and D. Zaldokas, editors. Fish habitat rehabilitation procedures. Watershed Restoration Technical Circular Number 9. British Columbia Ministry of Environment, Lands and Parks, Vancouver.

Burnham, K. P., and D. R. Anderson. 2002. Model selection and multimodel inference: a practical information-theoretic approach. Springer-Verlag, New York.

Cramer, S. P., D. B. Lister, P. A. Monk, and K. L. Witty. 2003. A review of abundance trends, hatchery and wild fish interactions, and habitat features for the Middle Columbia steelhead ESU. Contract report prepared for Mid Columbia Stakeholders. S. P. Cramer and Associates, Sandy, Oregon. Available: http://www.spcramer.com/reports/pdf/mid-col-sthd-final-June03.pdf. (November 2005).

Johnston, N. T., C. J. Perrin, P. A. Slaney, and B. R. Ward. 1990. Increased juvenile salmonid growth by whole river fertilization. Canadian Journal of Fisheries and Aquatic Sciences 47:862–872.

Larkin, G. A., and P. A. Slaney. 1997. Implications of trends in marine-derived nutrient flow to south coastal British Columbia salmonid production. Fisheries 22(11):16–24.

McCubbing, D. J. F. 2001. Adult steelhead trout and salmonid smolt migration at the Keogh River during spring 2001. Province of British Columbia Ministry of Environment, Lands and Parks, Fisheries Research and Development Section Contract Report by the Northern Vancouver Island Salmonid Enhancement Association, University of British Columbia, Vancouver.

McCubbing, D. J. F., and B. R. Ward. 1997. The Keogh and Waukwaas Rivers paired watershed study for B.C.'s Watershed Restoration Program: juvenile salmonid enumeration and growth 1997. Watershed Restoration Project Report No. 6. British Columbia Ministry of Environment, Lands and Parks, Vancouver.

McCubbing, D. J. F., and B. R. Ward. 2003. Adult steelhead and salmonid smolt migration at the Keogh River during spring 2003. Province of British Columbia, Ministry of Water, Land, and Air Protection, Aquatic Ecosystem Section Contract Report by Northern Vancouver Island Salmonid Enhancement Association, University of British Columbia, Vancouver.

McCubbing, D. J. F, B. R. Ward, and L. Burroughs. 1999. Salmonid escapement enumeration on the Keogh River: a demonstration of a resistivity counter in British Columbia. Province of British Columbia Fisheries Technical Circular Number 104.

Reeves, G. H., D. B. Hohler, B. E. Hansen, F. H. Everest, J. R. Sedell, T. L. Hickman, and D. Shively. 1997. Fish habitat restoration in the Pacific Northwest: Fish Creek of Oregon. Pages 335–359 *in* J. E. Williams, C. A. Wood, and M. P. Dombeck, editors. Watershed restoration: principles and practices. American Fisheries Society, Bethesda, Maryland.

Ricker, W. E. 1975. Computation and interpretation of biological statistics of fish populations. Fisheries Research Board of Canada Bulletin 191.

Roni, P., and T. P. Quinn. 2001. Density and size of juvenile salmonids in response to placements of large woody debris in western Oregon and Washington streams. Canadian Journal of Fisheries and Aquatic Sciences 58:686–693.

Slaney, P. A., and A. D. Martin. 1997. Watershed restoration program of British Columbia: accelerating natural recovery processes. Water Quality Research Journal of Canada 32:325–346.

Slaney, P. A. and D. Zaldokas, editors. 1997. Fish habitat rehabilitation procedures. Watershed Restoration Technical Circular No. 9. British Columbia Ministry of Environment, Lands and Parks, Vancouver.

Slaney, P. A., R. J. Finnigan, and R. M. Millar. 1997. Accelerating the recovery of log-jam habitats: large woody debris-boulder complexes. Pages 9–1 to 9–24 *in* P.A. Slaney and D. Zaldokas, editors. Fish Habitat Rehabilitation Procedures. Watershed Restoration Technical Circular No. 9. British Columbia Ministry of Environment, Lands and Parks, Vancouver.

Slaney, P. A., B. R. Ward and J. G. Wightman. 2003. Experimental nutrient addition to the Keogh River and application to the Salmon River in coastal British Columbia. Pages 111–126 *in* J. G. Stockner, editor. Nutrients in salmonid ecosystems: sustaining production and biodiversity. American Fisheries Society, Symposium 34, Bethesda, Maryland.

Smith, H. A., and P. A. Slaney. 1980. Age, growth, survival and habitat of anadromous Dolly Varden (*Salvelinus malma*) in the Keogh River, British Columbia. Province of B.C. Fisheries Management Report No. 76.

Smith, B. D., and Ward, B. R. 2000. Trends in adult wild steelhead (*Oncorhynchus mykiss*) abundance for coastal regions of British Columbia support the variable marine survival hypothesis. Canadian Journal of Fisheries and Aquatic Sciences 57:271–284.

Solazzi, M. F., T. E. Nickelson, S. L. Johnson, and J. D. Rodgers. 2000. Effects of increasing winter rearing habitat on abundance of salmonids in two coastal Oregon streams. Canadian Journal of Fisheries and Aquatic Sciences 57:906–914.

Ward, B. R. 1997. Using boulder clusters to rehabilitate juvenile salmonid habitat. Pages 10–1 to 10–10 *in* P. A. Slaney and D. Zaldokas, editors. Fish habitat rehabilitation procedures. Watershed restoration Technical Circular 9. British Columbia Ministry of Environment, Lands and Parks, Vancouver

Ward, B. R. 2000. Declivity in steelhead trout recruitment at the Keogh River over the past decade. Canadian Journal of Fisheries and Aquatic. Science 44:298–306.

Ward, B. R., and P. A. Slaney. 1981. Further evaluations of structures for the improvement of salmonid rearing habitat in coastal streams of British Columbia. Pages 99–108. *in* T. J. Hassler, editor. Proceedings: propagation, enhancement and rehabilitation of anadromous salmonid populations and habitat symposium, Humboldt State University, Arcata, California.

Ward, B. R., and P. A. Slaney. 1988. Life history and smolt-to-adult survival of Keogh River steelhead trout (*Salmo gaidneri*) and the relationship to smolt size. Canadian Journal of Fisheries and Aquatic Sciences 44:1110–1122.

Ward, B. R., D. J. F. McCubbing, and P. A. Slaney. 2003a. Stream restoration for anadromous salmonids by the addition of habitat and nutrients. Pages 235–254 *in* D. Mills, editor. Proceedings of the 6th International Atlantic Salmon Symposium, Salmon at the Edge, July 2002, Edinburgh, Scotland.

Ward, B. R., D. J. F. McCubbing, and P. A. Slaney. 2003b. Evaluation of the addition of inorganic nutrients and stream habitat structures in the Keogh River watershed for steelhead trout and coho salmon. Pages 127–147 *in* J. Stockner, editor. Proceedings of the International Conference on Restoring Nutrients to Salmonid Ecosystems, April 24 to 26, 2001, Eugene, Oregon.

Ward, B. R., P. A. Slaney, A. R. Facchin, and R. W. Land. 1989. Size-biased survival in steelhead trout: back-calculated lengths from adults' scales compared to migrating smolts at the Keogh River, BC. Canadian Journal of Fisheries and Aquatic Sciences 46:1853–1858.

Welch, D. W., B. R. Ward, B. D. Smith, and J. P. Eveson. 2000. Temporal and spatial responses of British Columbia steelhead (*Oncorhynchus mykiss*) populations to ocean climate shifts. Fisheries Oceanography 9:17–32.

American Fisheries Society Symposium 49:867–878

Riparian Habitat Restoration and Native Southwestern USA Fish Assemblages: A Tale of Two Rivers

JOHN N. RINNE*
Rocky Mountain Research Station
2500 South Pine Knoll Drive, Flagstaff, Arizona 86001, USA

DENNIS MILLER
Western New Mexico University, Department of Biology
Silver City, New Mexico 88062, USA

Abstract.—In the southwestern United States, livestock grazing has occurred on the landscapes for over a century. Because of aridity of climate and frequent drought, domestic livestock often resort to wetlands and riparian stream corridors for forage and water. One argument against this type of land use is the indirect impact on native fish species, many threatened and endangered within these aquatic habitats. Removal and reduction of livestock to improve and restore riparian vegetative habitat has been a management alternative that has been implemented across a large proportion of streams in the southwestern United States. We present data on two river systems, the Gila and the Verde, from which livestock have been removed for similar periods of time. Fish assemblages in these "improved" riparian areas have responded in two markedly different manners. In one, introduced species of fishes are now dominant. In the other, native fishes dominate fish assemblages. We compare physical habitat data and fish assemblages in these two rivers. We submit that overriding, natural hydrologic influences are perhaps more important in terms of habitat essential for sustaining fishes assemblages than is simple increase in riparian vegetation that may result from riparian restoration activities.

Introduction

Riparian areas throughout the United States have been altered with European settlement of the United States. Historically, it has been estimated that more than 60 million hectares of this habitat existed prior to immigration of European settlers. Depending on the respective region, riparian areas currently comprise from one to five percent of the landscape in the continental United States (Johnson et al. 1985; Medina et al. 2005). However, human utilization of these areas is vastly disproportionate to their relative area and value for conservation of natural resources. By the early 20th century, the demand for surface water commenced a period of development of water resources for irrigation, hydropower, and flood control especially in the arid western regions. These early impacts were followed 50 years later by mining of subsurface waters to provide for rapidly developing metropolitan areas. Rivers and streams have been dammed or diverted and wetlands drained resulting in drastic reduction in this habitat (Rinne 2002, 2003a, 2003b).

*Corresponding author: jrinne@fs.fed.us

Aquatic habitat and its quantity and quality are keystone features for conservation and sustaining native southwest fishes. In the region, livestock grazing has occurred on the landscapes for over a century (Rinne 2000). Because of aridity of climate and frequent drought, domestic livestock resort to wetlands and riparian stream corridors for forage and water (Johnson et al. 1985). One argument against this land use is the indirect impact on native threatened and endangered fish species within these aquatic habitats (Rinne 1999b).

In the late 1800s, livestock numbers in the western United States numbered in the millions and, coupled with drought, depleted vegetative resources across the landscape, including water rich wetlands (Hendrickson and Minckley 1985) and riparian areas (Young 1998). The U. S. Forest Service recognized the need to control livestock numbers on public lands and in the 1934 Taylor Grazing Act concerning livestock grazing management was enacted to provide guidelines on how to manage livestock primarily, on western public lands.

The grazing of public lands by livestock has been a very contentious issue ever since the enactment of the National Environmental Policy Act of 1969. In the 1970s, a series of acts including the Endangered Species Act and Public Rangelands Improvement Act further affected livestock management and reduced herd numbers across the western U.S. (Rinne and Medina 1996). Livestock numbers on western rangelands have been reduced by about half since the early part of the 20th century (Medina and Rinne 1999).

Impacts to riparian areas resulting from livestock grazing are viewed differentially (Rinne 1999). Livestock potentially impair various riparian functions when habitats are over utilized. Impacts of grazing of riparian areas and uplands have been well documented (Kaufman and Krueger 1984; Platts 1991; Belsky et al. 1999). Livestock grazing may alter natural riparian and channel processes such as upland and streambank erosion, channel sedimentation, and channel reach widening. Further, increased water temperature and decreased water quality may ultimately affect aquatic biota (Platts 1991; Belsky et al. 1999). Impacts on fishes and other biota are apparent, but more difficult to measure and less well documented than physical changes (Platts 1991; Rinne 1999). Indeed, it is very difficult to establish direct linkages between fish and other aquatic biota health factors and the complex interactions of livestock grazing and many environmental variables (e.g., vegetation, hydrology, geomorphology, geology, and climate). Many studies have reported negative interactions between fish and ungulates (Kauffmann and Krueger 1984; Platts 1991; Fleischner 1994; Belsky et al. 1999; Rinne 1999a), but none have established direct linkages of cause and effect. Others (Rinne 1999; 2000) suggest we need more objectivity and control in our studies of these interactions. Assessment of direct effects is often difficult, but should be performed with objectivity and validated methods.

Numerous strategies to "restore" or improve riparian areas and aquatic habitats impacted from grazing have been developed and are being implemented throughout North America and other parts of the world. Platts (1991) listed several innovative management strategies that can be used to address and corroborate livestock grazing impacts in riparian areas. While numerous and sometimes complex, all grazing strategies include control of livestock numbers, distribution, duration, timing of grazing, or control of forage use, or some combination of the these factors (Platts 1991). Fencing or complete grazing removal are the most common and potentially successfully "grazing strategies" employed to provide short- and long-term exclusion of ungulates and allow for riparian recovery and are emphasized herein.

In this paper, we present trends of change in fish numbers and assemblages recorded in research and monitoring case studies on two southwestern desert rivers, the Verde and Gila. Both have the same Gila River native fish fauna and the same suite of introduced nonnative fish species (Stefferud and Rinne 1995; Rinne 2005; Rinne et al. 2005). The two case studies are provided to address a key question relative to this Fourth World Fisheries Conference theme: "Does riparian restoration always translate into native fish restoration and conservation?" or "Does one size fit all?" We will compare selected physical habitat data and fish assemblages in these two rivers. We suggest that overriding, natural geomorphic and hydrologic influences are more important in terms of providing habitat important for sustaining fish assemblages than is simple increase in riparian vegetation.

Study Areas and Rationale for Study

The upper Verde River study area is described physically and biologically in Stefferud and Rinne (1995) and Rinne and Stefferud (1996). The upper Gila River study area is defined in Rinne et al. (2005). Both are warm water, desert rivers that contain the Gila River native fish fauna and the same suite of introduced fish species (Stefferud and Rinne 1995; Rinne et al. 2005). Flows on the Verde are both less in size and variability (Rinne 2002 in press; Neary and Rinne 1997; Figure 1A).

To improve riparian habitat for the threatened spikedace *Meda fulgida* in the upper Verde and for this species and the threatened loach minnow *Rhinichthys cobitis* in the upper Gila, livestock were removed from both river corridors in 1997–1998. Research and monitoring of fish numbers and assemblages on the upper

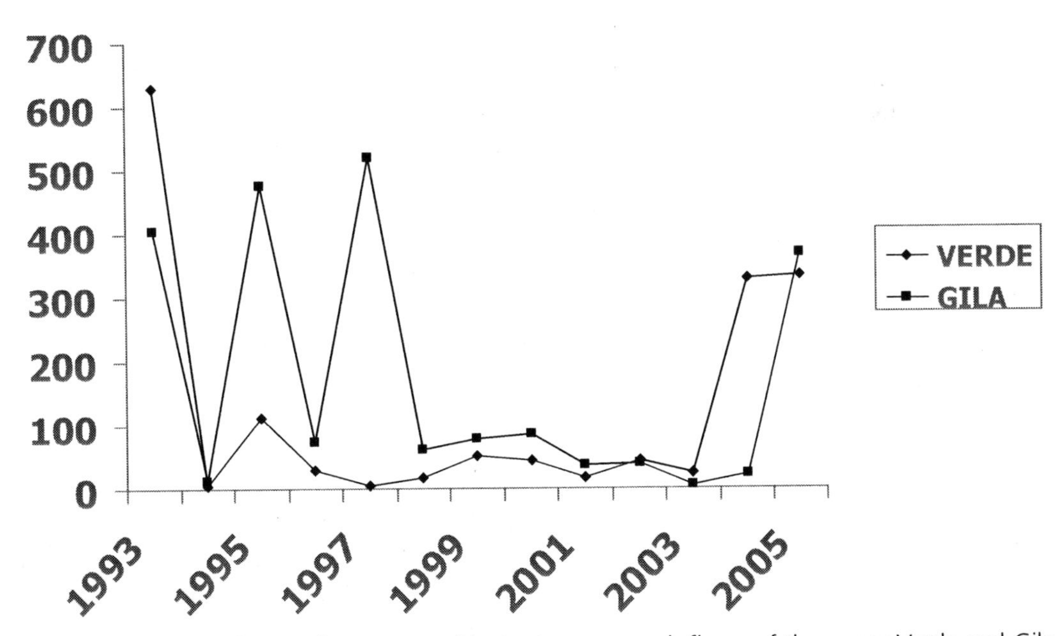

Figure 1A. Comparative maximum annual instantaneous peak flows of the upper Verde and Gila Rivers, 1993–2005.

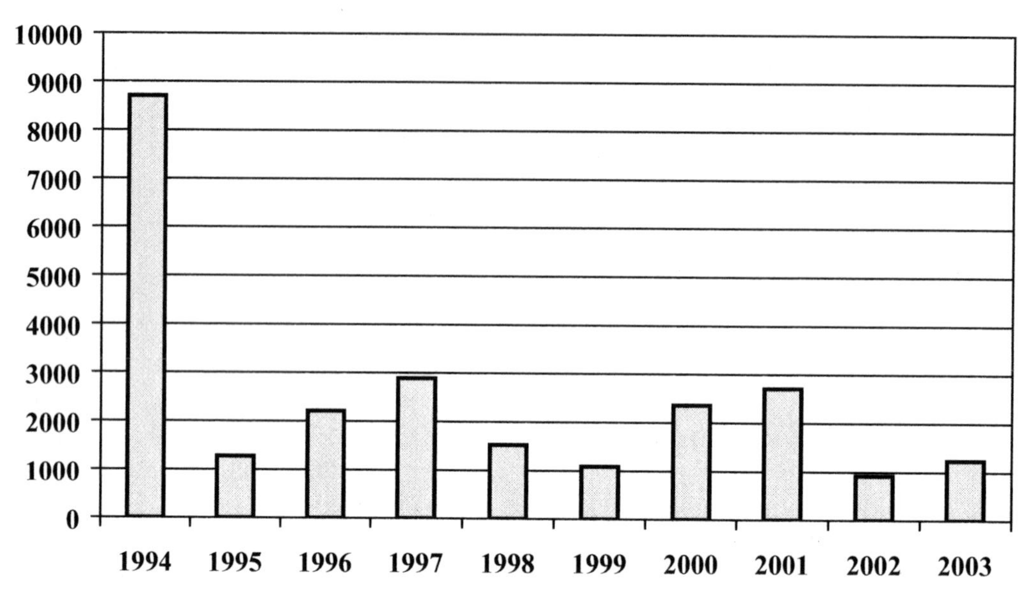

Figure 1B. Total numbers of fish collected at the seven monitoring sites in the upper Verde River, 1994–2003.

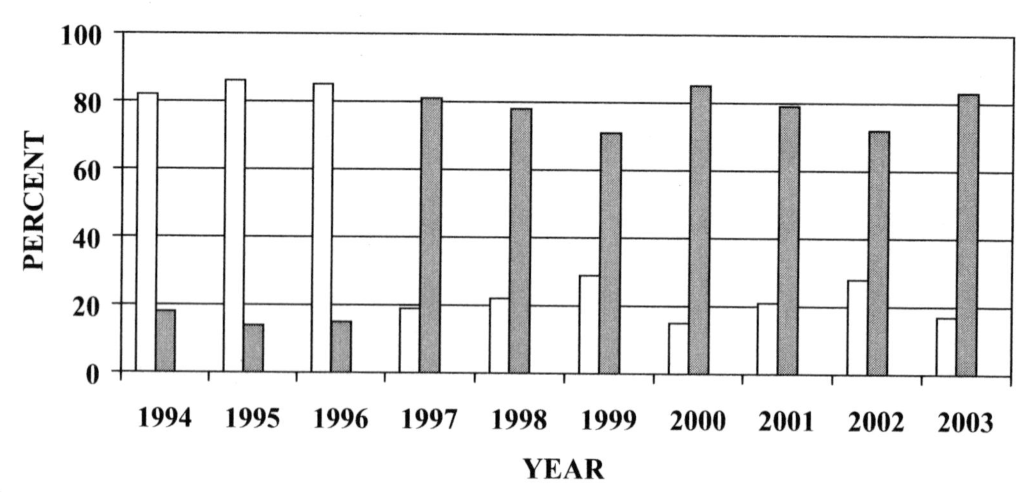

Figure 1C. Proportional distribution of native and nonnative fish components of the total fish assemblage in the upper Verde River, 1994–2003.

Verde was initiated prelivestock removal in 1994, following major flooding in 1992–1993 (Figure 1A). Monitoring in the Bird Area of the upper Gila was initiated in 1996. Livestock were removed the next year coincidental with a major flood event associated with Hurricane Linda (Figure 1A). Initially, the major objectives of study on the upper Verde River were to examine the relative roles of hydrologic regime and nonnative fishes in the sustainability of native fish assemblages. By comparison, study of the Gila Bird area on the Gila River, near Cliff, New Mexico was initiated independently by the second author in 1996 at Western New Mexico University with the primary objective to monitor the effects of the riparian restoration through grazing removal on spikedace, loach minnow and the total fish assemblage. In 1999, both authors initiated a more comprehensive, collaborative study of the two rivers for comparison of response to hydrographs and nonnative, invasive fish species. Six years later both study activities provided a novel opportunity to analyze and compare information on habitat and fishes and evaluate the success of livestock removal as a riparian enhancement activity in terms of native fish enhancement and sustainability in both rivers.

Methods

Methods of fish collection are described in Stefferud and Rinne (1995). Those for habitat assessment are in Stefferud and Rinne (1995) and Rinne and Stefferud (1996). Briefly, seven long-term monitoring sites were established in 1994 on an approximate 60 km reach of river. The Rocky Mountain Research Station initiated research and monitoring at a dozen sites on the upper Gila River fish assemblages in spring 1999 following flooding in autumn 1997. In 1996, monitoring of the effects grazing removal was initiated at six sites on a 10 km reach of the Gila River by the second author at Western New Mexico University. These two efforts provided a comparative framework of fish and habitat changes and also of riparian enhancement through grazing removal on the two low desert rivers with the same native cypriniforms assemblages of fishes.

The same sampling procedures were employed at all study sites on both rivers. Study reaches ranged from 200 to 500 m depending on attainment of 8–12 habitat units (Rinne and Stefferud 1996; Sponholtz and Rinne 1997). Sampling was performed with back electro fishing gear and pull seines, depending on characteristic substrate, depth and velocity of respective habitat units (Stefferud and Rinne 1995). All fishes were enumerated and measured except if numbers at a site exceeded 100 individuals. In this event, a subsample of 100 fish was measured and the remainder only numerated. Widths, depths, velocity, and substrates were recorded. Widths and depths with metered tapes and depths with meter rule. Normally, 3–4 transects were taken dependent upon length of respective habitat units. Flow data for the respective rivers were obtained from U.S. Geological Survey records.

Because multiple pass or depletion sampling protocol was not used, we present trend data only. No statistical analyses were applied to data. Instead, total fish numbers and the change in fish assemblages before and after grazing removal were used to determine if there was any observable response in the estimates relative to time of grazing removal on the two rivers. Of equal or more importance, the role that river hydrographs may play in confounding analyses of grazing removal and its effect on fish populations and assemblages was evaluated.

Results

Fish Assemblages

Verde

Total numbers of fishes in the upper Verde River were markedly greater in initial sampling in 1994 (Figure 1B) following winter flooding in 1992–1993 (Figure 1A). Again, a reduced level of flooding occurred in 1995 and again augmented fish abundance in 1995–1997, a year prior to livestock removal. However, fish numbers have never exceeded a third of their abundance in 1994. Fish assemblages were dominated (>80%) by native fishes from 1994 to 1996 (Figure 1C) just prior to livestock removal. By 1998, the year of livestock removal, the Verde had already inversely changed to a dominantly nonnative fish assemblage (>70%) and remained so to 2003.

Gila

In contrast to the upper Verde, total fish numbers in the Upper Gila River were low (ca. 1,000) at the six study sites at initiation of livestock removal in 1996–98 (Figure 2A). In 1999, numbers doubled, were similar to those in the upper Verde in 1994, and have remained high through 2003. In marked contrast to the upper Verde, native fishes predominated (>90–95%) fish assemblages within the study area in all years of study.

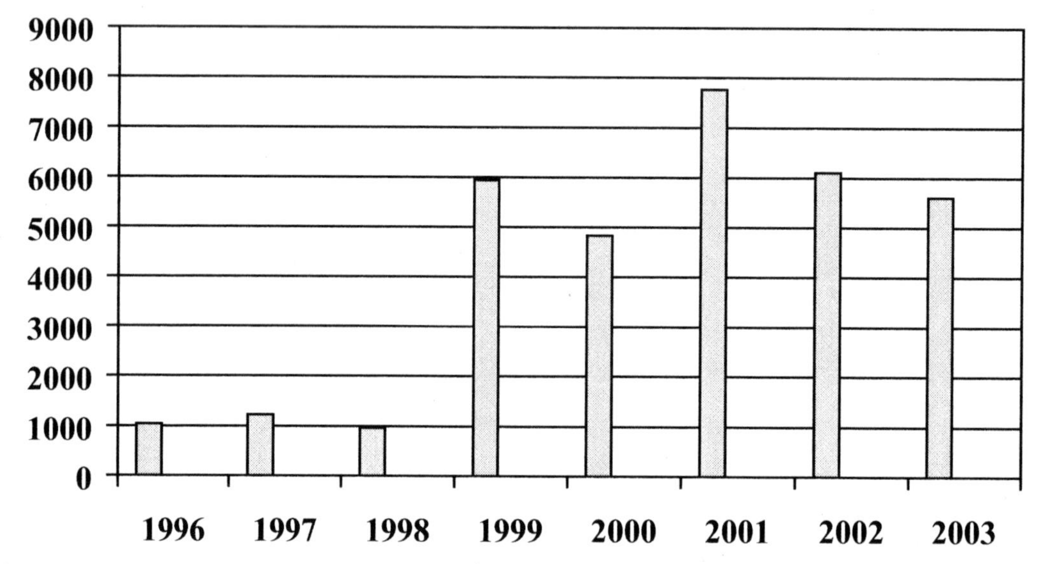

Figure 2A. Total numbers of fish collected at 6 monitoring sites in the upper Gila River, 1996–2003.

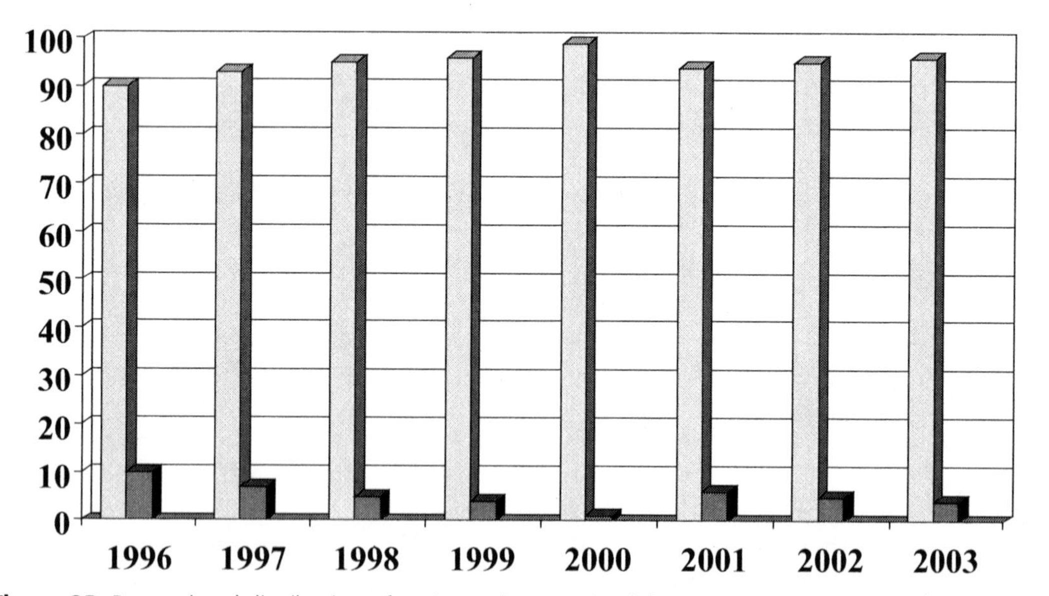

Figure 2B. Proportional distribution of native and nonnative fish components of the total fish assemblage in the upper Gila River, 1996–2003.

Habitat Characteristics

Verde

Stream channel widths and depths in the upper Verde were broad and shallow in 1996 following flooding in 1993 and 1995 (Table 1; Figure 3A). A shallow, pebble and cobble riffle (Rinne and Stefferud 1996) above the Perkins-ville bridge (Figure 3A) supported more than 60 speckled dace *Rhinichthys osculus* in spring 1996. By comparison, none were taken within this dramatically changed (Figure 3B) habitat unit in 2,000, 2 years following removal. Average stream width was about half and stream depths twice those estimated in 1996 (Table 1) and banks were lined with dense aquatic vegetation (Figure 4A). Accompanied by sustained low base flows (Figure 1A) this morphology and vegetation response was sustained until flooding in winter 2004–2005 (Rinne et al. in press).

Gila

Stream widths in the Bird Area in the upper Gila River have remained constant during all years of study and averaged a little over 6.0 m in width and about 0.2 m in depth (Table 1). In contrast to the upper Verde (Figure 4A), bank vegetation has only sparsely and intermittently developed (Figure 4B).

Flow Regimes

Stream hydrographs in the two rivers differ markedly (Figure 1A). Base flow in the upper Gila is twice that (40 cfs versus 20) of the upper Verde. Annual maximum instantaneous peak flows are both more variable (Table 3) and there is a higher incidence of large flows (Figure 1A; Table 4) in the upper Gila compared to the upper Verde.

Discussion and Conclusions

Recorded response of fish abundance and assemblages relative to grazing removal and riparian restoration were opposite in the two rivers. Following 6 years of livestock removal, native fishes dominated in the upper Gila River whereas nonnative fishes have dominated in the upper Verde. Channel narrowing, deepening, and lining with aquatic vegetation is more optimum for the three nonnative species, smallmouth bass *Micropterus dolomieu*,

Table 1. Channel morphology changes at Burn Ranch and Perkinsville monitoring sites in the upper Verde River between 1994 and 2000 and within the Gila Bird area, upper Gila River, 1996–2002

	Upper Verde River			
	1994		2000	
	Burnt Ranch	Perkinsville	Burnt Ranch	Perkinsville
W (m)	6.3	6.0	3.6	2.9
D (m)	0.19	0.19	0.35	0.38
	Upper Gila River, mean of 6 sites			
	1997		2002	
W (m)	6.1		6.4	
D (m)	0.21		0.24	

Table 2. Comparison of instantaneous peak flows (cfs) in the upper Verde and Gila rivers from 1974 to 1999.

	Upper Verde	Upper Gila
> 5,000	7	21
> 10,000	3	11

green sunfish *Lepomis cyanellus*, and yellow bullhead *Ameiurus natalis* (Rinne 2003). Deeper waters and increased vegetation are more optimum and beneficial to these "cover species" (Pflieger 1975). In contrast, shallow channel depth and the lack of vegetation that characterized the upper Verde prior to livestock removal in 1998 (Figure 3A, Table 1) was more optimum for native fishes. Similarly, these same habitat conditions have persisted in the upper Gila River every since the large flood event in September 1997, coincidental with grazing removal (Table 3; Figure 4B). It should be pointed out that although no large (>100 m^3/s) flow events have occurred since 1997 in the upper Gila (Figure 1A; Table 3) fish numbers and the dominance of native species have been sustained. Such sustained dominance and abundance might suggest that grazing removal may have benefited the native component of the fish assemblage. However, instantaneous peak flows that are four times the base flows of the upper Verde may be sufficient to repress streambank vegetation and sustain a broader channel width, both which are positive for native fishes (Rinne 2003b).

The native fish assemblage in the upper Gila River appeared to respond positively to livestock removal. However, instantaneous peak flows in combination with greater variability of flows more likely have sustained a more broad and shallower stream channel with much reduced aquatic vegetation compared to that present in the upper Verde (Figures 3 and 4B, Table 1). By contrast, lower (<50 m^3/s) peak flows have been sustained on the upper Verde since 1996 encompassing the time of livestock exclusion. Lack of flooding during

A

B

Figure 3. Comparison of change in aquatic habitat at the Perkinsville long term monitoring site from 1996 (A) to 2000 (B). Note the shallow, pebble/cobble substrate in 1996 deposited by previous (1993 and 1995) floods.

Table 3. Maximum Instantaneous peak flow (m^3/s) comparisons in the upper Verde and Gila rivers, 1993–2000.

Year	River	
	Verde	Gila
1993	630	405
1994	5	12
1995	113	476
1996	30	72
1997	6	519
1998	17	60
1999	51	79
2000	43	86
2001	17	37
2002	43	38
2003	25	6

A

Figure 4A. A characteristic reach of stream in the upper Verde (1997) with a narrow channel width (2–3 m), deeper waters (0.5 m) and abundant stream bank vegetation.

B

Figure 4B. A characteristic reach of stream in the Gila Bird Area, upper Gila River, 1999 with a more broad (6+ m), shallow stream channel with sparse aquatic vegetation.

the period of study in the upper Verde combined with livestock removal resulted in a deeper, narrower channel lined with abundant aquatic vegetation—conditions favorable to several nonnative fishes. These same species are present in the upper Gila, but have not increased because of flooding and its effect on stream channel dynamics.

Summary and Conclusions

Conventional historic management has been to reduce stocking (Medina and Rinne 1999) or more recently to totally exclude this land use from these habitats to conserve and restore them to historic conditions. This management approach has been applied on a broad scale in the Southwest. Most of the study of removal of livestock and riparian areas addresses vegetation and channel morphology response Rinne 1999a). The direct or indirect and negative impacts of livestock grazing on cypriniformes fishes have been little and inconsistently demonstrated (Rinne 1999a) and largely extrapolated from the literature on salmonid species (Rinne 2000).

Examination of these two low desert streams from which livestock have been excluded do not support conventional wisdom that restoration and sustainability of riparian areas through grazing removal directly translates into conservation, restoration and sustainability of native fishes. Vegetation increase and stream channel narrowing that generally occur in upper elevation montane riparian-stream systems occurred in the Verde, but not in the Gila. Differing hydrographs and geomorphologies in the two rivers (Rinne and Miller 2006) and the presence of introduced species of fishes during the period of study appear equally or more important for the restoration and enhancement of the native fish assemblages than does simple livestock removal.

In reality, increased vegetation as occurred with grazing removal in the upper Verde River accompanied with sustained base flow actually benefited "cover seeking" species such as smallmouth bass and green sunfish. Grazing removal in 1998 in the upper Verde at a time when native fishes were already markedly reduced (Rinne 2005) failed to enhance fish abundance or effect a change to dominance of native species in the total fish assemblage in ensuing years (Figures 1B, 1C). By comparison, following grazing removal in the upper Gila, fish abundance increased sixfold in 1999. However, timing of flood event (September 1997), sampling in 1998 (June), recruitment of natives (July–August 1998; and parallel changes in fish abundance at two contiguous sites (Rinne and Miller 2006) suggest these changes are not singularly the result of grazing removal.

Our study of fish abundance and assemblage response to grazing removal in the two rivers points again to some of the pitfalls and confounding factors (Rinne 1998) that may potentially strongly affect valid, defendable research and monitoring of riparian-stream areas that are "restored" through alteration of grazing strategy. Our efforts encompass some of these same pitfalls and further emphasize the need for vigilance in seeking to achieve valid, defendable study design in the future (Rinne 1999a).

References

Belsky, A. J., A. Matzke, and S. Uselman. 1999. Survey of livestock influences on stream and riparian ecosystems in the western United States. Journal of Soil and Water Conservation 54(1):419–431.

Fleischner, T. L. 1994. Ecological costs of livestock grazing in western North America. Conservation Biology 8:629–644.

Hendrickson, D. A., and W. L. Minckley. 1985. Cienegas, vanishing aquatic climax communities of the American Southwest. Desert Plants 6:131–175, Globe, Arizona.

Johnson, R. R., C. D. Ziebell, D. R. Patton, P. F. Ffolliott, and R. H. Hamre, editors. 1985. Riparian ecosystems and their management: reconciling conflicting uses. First North American Riparian Conference. April 16–18, Tucson, Arizona. USDA Forest Service General Technical Report RM-120, Fort Collins, Colorado.

Kauffmann, J. B., and W. C. Krueger. 1984. Livestock impacts on riparian ecosystems and streamside management implications—a review. Journal of Range Management 37(5):430–438.

Medina, A. L., and J. N. Rinne. 1999. Ungulate/fishery interactions in southwestern riparian ecosystems: pretensions and realities. Proceedings of the North American Wildlife and Natural Resources Conference 62:307–322, Washington D.C.

Medina, A. L., J. N. Rinne, and P. Roni. 2005. Riparian stream restoration through grazing management: considerations for monitoring project effectiveness. Pages 97–126 *in* P. Roni, editor. Monitoring stream and watershed restoration. American Fisheries Society, Bethesda, Maryland.

Neary, D. G., and J. N. Rinne. 1997. Base flow trends in the upper Verde River relative to Fish Habitat Requirements. Hydrology and Water Resources in Arizona and the Southwest 27:57–63, Flagstaff, Arizona.

Pflieger, W. L. 1975. The fishes of Missouri. Missouri Department of Conservation, Columbia.

Platts, W. S. 1991. Livestock grazing. Pages 389–423 *in* W. R. Meehan, editor. Influences of forest and rangeland management on salmonid fishes and their habitats. American Fisheries Society, Special Publication 19, Bethesda, Maryland.

Rinne, J. N. 1998. Grazing and fishes in the southwest: confounding factors for research. Pages 75–84 *in* D. F. Potts, editors. Proceedings American Water Resources Agency Specialty Conference: rangeland management and water resources. American Water Resources Agency/Society for Range Management Specialty Conference on Rangeland and Water Resources, Herndon, Virginia.

Rinne J. N. 1999a. Fish and grazing relationships: the facts and some pleas. Fisheries 24(8):12–21.

Rinne J. N. 1999b. The status of spikedace, *Meda fulgida*, in the Verde River, 1999. Implications for research and management. Hydrology and Water Resources in Arizona and the Southwest 29:57–64, Flagstaff, Arizona.

Rinne J. N. 2000. Fish and grazing relationships in southwestern United States. Pages 329–371 *in* R. Jamison and C. Raish, editors. Ecological and socio-economic aspects of livestock management in the Southwest. Elsevier, Amsterdam.

Rinne, J. N. 2002. Hydrology, geomorphology and management: implications for sustainability of native southwestern fishes. Hydrology and Water Resources of Arizona and the Southwest 32:45–50, Flagstaff, Arizona.

Rinne, J. N. 2003a. Native and introduced fishes: their status, threat, and conservation. Pages 194–213 *in* P. F. Ffolliott, M. B. Baker, L. F. Debano, and D. G. Neary, editors. Ecology, hydrology and management of riparian areas in the southwestern United States. Lewis Publishers, Boca Raton, Florida.

Rinne, J. N. 2003b. Fish habitats: conservation and management implications. Pages 277-297 *in* P. F. Ffolliot, M. B. Baker, L. F. Debano, and D. G. Neary, editors. Ecology and management of riparian areas in the southwestern United States. Lewis Publishers, Boca Raton, Fl.

Rinne, J. N. 2005. Changes in Fish assemblages, Verde River, Arizona, 1974–2003. Pages 115–126 in J. N. Rinne, R. M. Hughes, and R. Calamusso, editors. Historical changes in large river fish assemblages of the Americas. American Fisheries Society, Symposium 45, Bethesda, Maryland.

Rinne, J. N., and J. A. Stefferud. 1996. Relationships of native fishes and aquatic macro habitats in the Verde River, Arizona. Hydrology and Water Resources in Arizona and the Southwest 26:13–22, Flagstaff, Arizona.

Rinne J. N., and A. L. Medina. 1996. Implications of multiple use management strategies on native southwestern (USA) fishes. Pages 110–123 *in* R. M. Meyer, editor. Fisheries Resource Utilization and Policy, New Delhi, India.

Rinne, J .N., J. Simms, and H. Blasius. 2005. Changes in hydrology and fish fauna in the Gila River, Arizona–New Mexico: epitaph of a native fish fauna? Pages 127–156 *in* J. Rinne, R. M. Hughes, and B. Calamusso, editors. Historical changes in large river fish assemblages of the Americas. American Fisheries Society, Symposium 45, Bethesda, Maryland.

Rinne, J. N., and D. Miller. 2006. Hydrology, geomorphology and management: implications for sustainability of southwestern native fishes. Reviews in Fishery Science 14:91–110.

Rinne J. N., P. Boucher, D. Miller, A. Telles J. Montzingo, R. Pope, B. Deason, C. Gatton and B. Merhage. 2004. Fish community structure in two southwestern desert rivers. Pages 64–70 *in* M. J. Brouder, C. L. Springer, and S. C. Leon, editors. 2005. Proceedings of two symposia. Restoring native fish to the lower Colorado River: interactions of native and non-native fishes. July 13–14, 1999, Las Vegas, Nevada, and restoring natural function within a modified riverine environment: the lower Colorado River. July 8–0 1998. U. S. Fish and Wildlife Service, Southwest Region, Albuquerque, New Mexico.

Comparative fish community structure in two desert rivers. *in*, Leon, S.; Stine, P.; Springer, C. editors. Restoring Native Fish to the Lower Colorado River: Interactions of Native and Non-Native Fishes: A Symposium and Workshop.

Rinne, J. N., C. D. Carter, and Albert Sillas. In press. Fish assemblages in the upper Verde River: Species abundance and interactions with River hydrology, 1994–2005. Journal of Freshwater Ecology.

Sponholtz, P., and J. N. Rinne. 1997. Refinement of aquatic macro habitat definition in the upper Verde River, Arizona. Hydrology and Water Resources in Arizona and the Southwest 27:17–24, Flagstaff, Arizona.

Stefferud, J. A., and J. N. Rinne. 1995. Preliminary observations on the sustainability of fishes in a desert river: the roles of stream flow and introduced fishes. Hydrology and Water Resources in Arizona and the Southwest 22- 25:26–32, Flagstaff, Arizona.

Young, D. W. 1998. The history of cattle grazing in Arizona. Hydrology and water resources in Arizona and the Southwest 28:13–17.

American Fisheries Society Symposium 49:879–880

Session Summary

Reconciling Fisheries with Conservation Through Marine Habitat Improvement

William Seaman
University of Florida, Florida Sea Grant Program
Gainesville, Florida

The aspect of habitat improvement addressed in this session was the use of artificial reefs as modifications of benthic marine systems to alter ecological structure and function and in some cases behavior of people. Both the scientific investigation and a more diverse stakeholder application of artificial reef technologies are increasing globally. The breadth of interest is reflected taxonomically and geographically in the reports delivered in this session of the Fourth World Fisheries Congress, which included six oral and 14 poster presentations, addressing seagrasses, coral, mollusks, decapod crustaceans, and fin fishes variously from areas of Asia, Europe, and North America. Data and interpretation presented from biological, economic, social, engineering, and policy perspectives indicate the broadening integration of fishery, ecosystem, and natural resource management considerations characterizing this subject.

The historical use of artificial habitats constructed from natural and man-made materials to enhance fishing success has been supplemented widely by efforts that aim to protect ecosystems and restore biodiversity, as well as produce new biomass in aquaculture and wild-harvest situations. A case can be made for the use of artificial reefs as an instrument of mainstream fishery science research, policy, and ecosystem management. In the north and western Mediterranean Sea, artificial reefs that include structures designed to impede illegal trawling are used to protect seagrass meadows and the nursery functions they support. In Italy, mussel production is documented on concrete blocks in the Adriatic Sea, while in Portugal a pilot study established the scientific basis for a larger deployment of 20,500 modules, occupying an area of 43.5 km^2, to increase fish biomass. In these areas and elsewhere, socioeconomic evaluation at last has become a common part of habitat deployment schemes, in order to document monetary costs and benefits of reef deployment, fishing yield and other attributes.

Artificial reefs (as both a subject and a set of tools) are advancing the methods of fishery science and aquatic ecology. The contingent valuation method of economists is used to assess nonuse values of such structures, for example in a study by English and Portuguese scientists. In the eastern Gulf of Mexico, American scientists have preliminary findings that indicate that a modification of the "electron transport system assay" is a valid, nonlethal means of estimating metabolic rates in the gag *Mycteroperca microlepis*. In Italy and elsewhere, reefs may be used in certain protected areas. As a means of manipulating habitat attributes, artificial reefs offer advantages for use of experimental techniques that cannot be used in natural systems such as coral reefs.

A significant advance since the first World Fisheries Congress (1992) convened a session on artificial reefs has been the advancement of knowledge and understanding of the ecology of organisms that might utilize artificial marine habitat and, in turn, application of this information to the design of reef structures to accommodate the life history requirements of particular species. A review of the effectiveness of 145 habitat improvement structures in Japan and Korea determined that principal considerations for reef design and deployment include water depth, void space of the structure, and target species. The emphasis in these two nations is to maintain and increase sustainable seafood fishery production for national consumption. Elsewhere, independent studies of habitat preference for spiny lobsters in Australia and the Caribbean basin have converged through a series of longer term experimental programs. Thus, settlement of juvenile lobster onto structures such as limestone blocks is monitored to assess survivorship, according to varied reef designs. In both the northeastern Gulf of Mexico and the northern Mediterranean Sea, manipulation of the size of patches of concrete modules and the spacing of these (various-sized) clusters reveals differences in density and behavior of fishes (e.g., greater prey consumption of gag grouper at smaller reefs). The expected augmentation of fish biomass production due to restoration of oyster habitat in the eastern United States has been calculated based on synthesis of ecological information.

While much of the research on artificial reefs is on smaller structures deployed over relatively small areas of the seafloor, some pertains to reefs and related habitats at a larger areal extent. At an extreme scale, over 4,000 oil and gas production platforms in the northern Gulf of Mexico are viewed as artificial habitats that, in aggregate, have increased the regional carrying capacity of reef fish species of commercial importance, such as red snapper. To determine the role that artificial reefs might serve in fishery restoration in the Taiwan Strait, a comprehensive assessment of regional primary productivity, ecological efficiency, and fishery stock sizes was made. In Florida, a 260-km^2 fisheries management area is being developed on the basis of artificial reefs that will serve conservation aims. Meanwhile, ships represent large items used individually as reefs, with one notable application in creation of eco-tourism and recreational dive sites that may reduce user pressure on adjacent natural reefs.

The science and technology of artificial reef design and deployment has evolved to a point where the objectives for reefs are more specific and therefore performance is more measurable. Development of a global network for technical exchange among researchers and practitioners has promoted the scientific basis for this field. A remaining challenge is to define the influences of reefs upon sustainability in fisheries, especially at increasingly larger scales.

American Fisheries Society Symposium 49:881–889

Fishery Conservation and Habitat Improvement in Marine Ecosystems

William Seaman*

University of Florida, Florida Sea Grant Program, Department of Fisheries and Aquatic Sciences Post Office Box 110400, Gainesville, Florida 32611, USA

Margaret Miller

NOAA Fisheries, Southeast Fisheries Science Center 75 Virginia Beach Drive, Miami, Florida 33149, USA

Introduction

Perhaps the most significant challenge to aquatic sciences and resources management is to formulate responses to the twin worldwide forces of fisheries depletion and habitat degradation. The 2004 World Fisheries Congress identified two problems: 1) overexploitation yielding depletion of serial areas, species and trophic levels and 2) resource decline via major degradation of fish habitat. Cochrane (2000), among several authors who have characterized fisheries loss, cited regional statistics of overfishing for 30% of species in the United States and heavy exploitation of 57% of fish stocks in European waters in defining its magnitude. Habitat impairment, meanwhile, is the more diverse, diffused, and less easily measured of the two situations. Among its multiple causes is nutrient enrichment from a doubling of human-fixed nitrogen additions to the biosphere between 1960 and 1980 (Boesch 2002) and resultant ecosystem changes in coastal systems such as widespread anoxia events and shifts in food webs. It should also be noted that fishing activity, itself, is recognized increasingly as an agent of habitat degradation (e.g., Wheeler et al. 2005).

Given the ubiquity of these threats to fisheries resources and productivity, the Congress aim of "harmonizing fisheries and conservation" is clearly a tall order. Combating habitat degradation via active improvement measures may be one element in advancing toward this goal. This paper addresses certain principles, practices, and measures of marine habitat improvement as one aspect of responding to loss of fisheries productivity. Human-made reefs are featured in this discussion as one tool of many in the management of both fisheries and habitat. Principal subjects introduced in this brief paper include a definition of marine habitat improvement and determination of its attainment, the present applications of reef construction technology, and three examples that offer desirable attributes for incorporation into future use of this technology. Our purpose is to offer a context for related, more extensive, and specific Congress papers addressing habitat improvement and stimulate readers to consider the validity, applications and limits to artificial reef technology, a field rooted in antiquity but only re-

*Corresponding author: seaman@ufl.edu

cently becoming under girded by rigorous scientific investigation.

Context for Marine Habitat Improvement

This section establishes a context for habitat improvement by addressing the purposes for which it may be applied, considerations in using technologies available for its achievement and practices for assessing performance. The question that must be stated first and foremost, prior to any actual habitat alteration is: "Improvement to what?" We maintain that it is essential that a baseline condition be defined by fishery and habitat scientists and managers and other informed stakeholders, as a guide for design of a habitat improvement project and for its evaluation, and that this definition be established prior to implementation of any improvement practices. However, it should be acknowledged that whatever current natural conditions may prevail in a given coastal system, they likely represent a shifted baseline, given the ubiquity of historical overexploitation (Jackson 2001).

Purposes of Habitat Improvement

Habitat improvement may be undertaken from either a relatively pristine or a relatively degraded baseline condition. The former approach is generally termed "enhancement" and seeks to augment the natural level of productivity of a given system. Artificial reefs often are applied in this context. For example, the placement of artificial reef habitat in flat sandy areas may provide for the creation of new food webs within the ecosystem, thereby enhancing overall trophic throughput. The demonstration that this approach is effective has been a severe challenge as evidenced by the attraction/production controversy (Burns and Beal 1997), although documentation of a spectrum of ecological responses is readily available in recent literature.

In fact, even relatively pristine baselines in marine systems are rare and thus, most enhancement projects therefore involve some element of the latter approach, termed "restoration." Though artificial reefs are applied in this context as well, it is important to consider a wide range of actions as habitat improvement. For example, the ambitious Greater Everglades Ecosystem Restoration is seeking to return estuaries in southern Florida, USA to historical baseline conditions by restoring natural quantities, qualities, timing, and distribution of freshwater inputs that had been severely diminished and distorted by the drainage and flood control system implemented in the 1950s. Naturalizing the flow of freshwater into southern Florida estuaries will reduce current salinity extremes and foster a steadier mesohaline environment conducive to important fisheries such as pink shrimp *Farfantepenaeus duorarum*, for which growth rates are maximized at salinity of 30‰ (Browder et al. 2002). This program is of interest if for no other reason than it is possibly the largest habitat restoration ever attempted in the world, with costs estimated at US $8 billion (USACOE 1999) over 20 years. Project components include "replumbing" canals built originally for drainage, creation of large wetland systems for freshwater storage and nutrient removal, and novel approaches such as freshwater storage via re-injection into subterranean aquifers.

Clearly, different criteria for evaluation will apply in these two contexts for habitat improvement. For enhancement, the goal to be evaluated could be viewed as "more fishery yield than was present before the improvement." For restoration, the goal is to restore some prior condition of ecosystem structure, function, or yield. Here, the challenge is in defining what that prior condition looked like (as it may have existed so long ago that no one remembers) and in determining if that historical baseline is a reasonable goal given the possible loss of resilience in a given system. In the Everglades example, elevation changes over the landscape due to soil oxidation and subsidence in drained wetlands make the historical condition of freshwater flow impossible to restore, even if these areas were converted back to wetlands. Also, a substantial portion of the system is now in urban development. Thus, in the southern Florida context, pink shrimp fisheries improvement will occur in a broad ecosystem context including a host of other project goals and constraints.

Habitat Improvement Technologies

In sum, the practices addressed in this paper involve purposeful placement of either human-made or natural materials in a benthic marine ecosystem, generally on the coastal shelf or in an estuary, with a goal of modifying ecological structure and function. Methods and research for artificial reef development have been reviewed in reports from a recent international scientific conference on the subject (ICES 2002) and compilations of regional research and development (e.g., Europe–Jensen et al. 2000). Its aims can be broad and multi-faceted or quite limited, and may include individual or combined physical, biological, and socioeconomic objectives. Our assessment is that reefs can be employed as a management tool to address both of the concerns noted in the introduction, namely (1) to resolve certain fishery stock issues (e.g., establishment of nursery and reserve areas, protection of essential habitat from physical disturbance), as well as (2) in the more visible activity of creating physical habitat where biological production is expected to occur (in a way that mimics natural processes). However, extensive knowledge of population and ecosystem function, as well as careful planning and implementation, are required to accomplish either of these goals.

The role of artificial reef technology in mainstream fishery science to achieve these goals still may be under debate, even as its applications in this sector increase and its scientific underpinnings are strengthened. Beyond the scientific community, this technology is being embraced by a growing number and variety of practitioners, according to both geography and purpose. Thus, we indicate the breadth of the field as a reminder to the fishery science profession that numerous interests, perhaps allied only loosely to it, nonetheless have a stake in its practices in scores of countries bordering tropical and temperate seas. Additional goals, therefore, beyond the immediate fishery habitat purview of this paper, include: aquaculture, either on a limited site (e.g., Italy) or in the broader context of marine ranching (e.g., Korea); promotion of biodiversity (e.g., Monaco); enhancement of scuba recreational diving (e.g., Canada); ecotourism development, using submersible vehicles to visit designed reefs (e.g., Mexico, Bahamas); expansion of recreational fishing (e.g., Australia); artisanal and commercial fisheries for seafood (e.g., India, Japan respectively); and research (e.g., United Kingdom).

Indicators of Progress

In a discussion of the attributes of ecological restoration, a working group of the Society for Ecological Restoration and the IUCN Commission on Ecosystem Management addressed "indicators of restoration progress" in stating that "an ecosystem is considered to be fully restored when it contains sufficient biotic and abiotic resources to sustain its structure, ecological processes and functions with minimal assistance or subsidy" (Summary Record, Ecosystem Restoration Working Group, 2–5

March 2003, unpublished). Against this framework, the evaluation of aquatic restoration frequently has lagged. For example, while a large number of estuarine wetlands restoration and creation projects have occurred since 1980, monitoring data and evaluation in terms of performance criteria remain problematic (Desmond et al. 2002). Only recently has the field of artificial reef technology begun to emphasize definition of clear measurable objectives and utilization of consistent practices to measure progress toward objectives (Seaman 2000).

Restoration of Marine Habitats

The following three case studies are presented as a guide to current and emerging considerations for habitat improvement. In each situation, the goal addresses actual restoration of a degraded system with fisheries enhancement as one but not the only goal. These examples are provided from Atlantic, Indian, and Pacific Ocean biogeographic regions and reflect a diversity of goals including restoration of plant habitats, bivalve mollusk systems, and fisheries stocks. Desirable attributes of these situations include their recognition and utilization of baselines, measurable objectives, reef design consistent with species life history requirements, ecosystem contexts, and quantitative approaches that include pilot studies and modeling. These attributes pertain equally to habitat enhancement.

Restoring Oysters in Ecosystem Context

The Chesapeake Bay on the eastern coast of the United States is an example of a large system with a drastically shifted baseline over a fairly long historical time period. As early as 1881, a wide survey of Chesapeake Bay eastern oyster *Crassostrea virginica* beds determined that "overworked" beds had reduced structure, increased amounts of sand and mud, and were composed of 97% broken shell and debris as compared with 30% for unfished beds (Wilson 1881 cited in Kennedy and Breisch 1981). Oyster catch in the Chesapeake Bay has been on a general decline from its peak in 1874 (14 million bushels) to less than half a million bushels in recent years. The current stock of oysters in the Chesapeake Bay is estimated to be approximately 2% of the historic baseline.

The continuing declines in oyster abundance and oyster reef habitat occurring throughout the 20th century have been co-incident with other (possibly related) modes of habitat degradation such as declining water quality (e.g., increased turbidity, eutrophication, and anoxia). These declines led to cascading disturbances such as loss of sea grass habitat due to turbidity and additional oyster losses to disease. Because oysters are filter feeders, there has come to be wider acceptance that the loss of oysters (along with land-based pollution) may have been a factor in this water quality decline (Coen and Luckenbach 2000). Hence, restoration of habitat (including conducive water quality, sea grass, and oyster reefs) in this system may depend on maintaining a certain biomass of filter feeders in the system. Traditional oyster habitat restoration approaches focused narrowly on providing low artificial shell reefs to attain fisheries goals (i.e., increasing harvestable oysters). This approach has not met with success, as continued overharvest and anoxia events have not yielded increased habitat quality or productivity (Lenihan 1999; Coen and Luckenbach 2000).

A more recent scientific consensus (e.g., Coen and Luckenbach 2000 or see website http://noaa.chesapeakebay.net/habitat/hab_oyster.htm) is emerging; that for artificial oyster reefs to constitute effective habitat improvement, they need to be taller (i.e., provide more structurally complex habitat and a potential refuge from bottom anoxia events) and they need to be protected from harvesting in order to provide for persistence of the reef structure and the maintenance of sufficient oyster biomass both for filtering and for reproductive capacity (Coen and Luckenbach 2000). Recent research modeling the effects of bivalve filtering (oysters and clams) on turbidity in Chesapeake Bay has provided the prediction that maintaining an average oyster biomass of 25 g/m^2 would reduce turbidity by an order of magnitude, greatly increasing the amount of light reaching the bottom and thereby expanding the suitable area for sea grasses habitat (Newell et al. 2003). Such modeling research provides specific quantitative goals for planning and evaluating habitat improvement.

Kelp Mitigation

A more focused example involves the creation of large kelp *Macrocystis pyrifera* beds as mitigation for habitats destroyed in the coastal Pacific Ocean by the operation of the San Onofre Nuclear Generating Station in southern California, USA. In this case, the operators of the electrical power plant were mandated to create 61ha (150 acres) of new kelp bed habitat off site from the one that was destroyed. Thus, off-site mitigation has aspects of both restoration (replacing the function lost at the old site) and habitat enhancement (creating new productivity in the new site).

The placement of artificial reefs is one component of this project. However, the project managers chose to begin with a moderate sized 8.9 ha (22 acres) pilot project to gain assurance that an appropriate design for the full-scale reef would yield the habitat characteristics and functions legally required by the mitigation permit. Thus, a set of experimental reefs with different substrate characteristics (quarry rock versus concrete, different coverage of hard substrate versus sand) or other actions (e.g., transplantation of kelp) are undergoing extensive evaluation to determine the degree of habitat improvement for fishes and benthic communities provided. Specifically, the successful recruitment and growth of giant kelp (or survival and growth of transplants) onto the artificial reef structure is the primary point of concern. Very specific quantitative goals have been laid out including those related to physical structure (e.g., % cover of rock versus sand in the artificial reef patches), biotic habitat (kelp and other benthic species), and fish communities (Reed et al. 2002). Some of these performance standards are in terms of an absolute historic baseline (i.e., the total amount of kelp that was lost—ultimately 61 ha [150 acres] at a density of four adult plants/100 m^2). Others are stated relative to current status of other similar habitats in the area. For example, fish assemblage, recruitment, and production should be "similar to natural reefs in the region" (Reed et al. 2002).

Over the first 2 years of monitoring the experimental reef, kelp recruitment has been successful and both kelp density and fish recruitment (i.e., young of year juvenile fish density) compare favorably with natural reference reefs. This suggests that habitat improvement has indeed been accomplished, though monitoring will continue over a 5-year period. (See http://www.sce.com/sc3/006_about_sce/006b_generation/006b1_songs/default.htm.)

Fisheries Restoration

The third of our case studies of (structural and physical) responses to habitat and fishery degradation includes the most emphasis on simulation modeling of ecosystems, fishing and policy, and is the newest, from Hong

Kong, China. There, high trawling effort during the last quarter of the 20th century produced declining catch, high fishing mortality, greater relative capture of low-value, short-lived species, and virtual elimination of longer lived demersal species of higher value (Pitcher et al. 2002). After a peak fishery harvest of over 240,000 metric tons (mt) in 1989, catch in 1998 was under 145,000 mt, due both to high exploitation and to habitat loss and disturbance (Wilson et al. 2002). In response a multifaceted approach including fishing licenses, protected areas, and restoration and enhancement of habitats was proposed; a 5-year Artificial Reef Program started in 1996, funded at US $13,000,000, as discussed by Wilson et al. (2002). (See also http://www.artificial-reef.net/English/main.htm.) These latter authors described the planning and implementation process, consultations with stakeholders, and preliminary results for reef ecology studies. They documented juvenile fish recruitment for species of Sparidae and Lutjanidae, residence of adult Serranidae, and increased catch of small-scale fisheries for bream (Sparidae).

In brief, deployed vessels (including along park boundaries to prevent trawling), rock, tire units, and concrete units (28,000 m^3 total) were located in two marine parks, according to a voluntary no-fishing arrangement made possible by placement of additional artificial reefs for fishing in open mud areas. An area of 10% of Hong Kong waters has been set aside as a fisheries protection area. According to predictions by Pitcher et al. (2001), the value of the fishery would increase by over 50% if 10–20% of waters were managed on a no-take basis.

Forecasting the responses to this artificial reef-based fishery restoration project was done by Pitcher et al. (2002), using three ecosystem and resource models. Information from a variety of local databases and consultations allowed these authors to incorporate (1) diet, growth and mortality data for 255 reef-associated and nonreef fish species, sorted by size, and collected into 27 functional groups; and (2) descriptions of seven sectors (e.g., trawling) of the Hong Kong fishery into "Ecosim" and "Ecopath" models. In turn, these provided the basis for dynamic "Ecospace" simulations to predict fishery performance according to fishery sector and habitat. In contrast to a nonreefs scenario that depicted continuing depletion of the fishery (e.g., five of 27 functional groups reduced to almost zero after 10 years) and increase of lower-trophic level organisms, an actual increase of harvest of large reef fish is forecast when artificial reefs are deployed. In one situation, the authors forecast a total catch of reef fish of 100 mt per year, including 60 mt of large demersal reef fish.

This situation represents an early and significant application of ecosystem simulation to artificial reef performance and coastal fishery and habitat restoration. Pitcher et al. (2002) note both the advantages of such an approach, including the capabilities for analysis of trade-offs among marine protected area and reef deployment design practices and for comparison of policy options, and also potential concerns including levels of confidence and uncertainty.

Conclusion

Here we describe certain trends in marine habitat improvement as manifested in the preceding three case studies (Table 1). Each situation includes the measurable objectives necessary to successful evaluation of aquatic ecosystem restoration. Both the Chesapeake Bay and San Onofre efforts specify units of oyster biomass (25 g/m^2) and plant density (4/100 m^2), respectively, while the Hong Kong program is more general in seeking increased fish-

Table 1. Components for enhancement of fishery habitat restoration performance.

Component of habitat development	System		
	Oyster reefs	Kelp forests	Reef fisheries
1. Goal/performance measure	Oyster biomass = 25g/m^2; monitoring in progress	4 plants/100m^2; monitoring in progress	Increased fishery yield; monitoring in progress
2. Ecosystem context	Oysters as critical component of ecosystem to enhance water quality; opportunity for recovery of other habitats (e.g., sea grass)	Adjacent natural reefs as reference target and source of recruits	Considers adjacent natural reefs and open mud and sand
3. Ecological basis for design	Physical structure; anoxia	Height, spacing of reefs; predators	Species diet, growth
4. One tool of many used	Coupled to watershed management	Kelp transplantation being evaluated	Coupled to management of fishing effort
5. Advanced techniques	Modeling to predict ecosystem benefits; water quality - sea grass linkages	Experimental pilot study to ensure design most likely to attain targets	Modeling forecasts of fishery response

ery catch. In all cases, monitoring to acquire data for measurement of performance is in place. Further, ecology of organisms has been used to direct design of reef structures.

In all cases, also, reefs are being used in broader contexts. As a fishery management tool, for example, they are coupled with new fishing license measures in Hong Kong. In contrast with many typical artificial reef deployments that have relatively small areal "footprints," such as individual ships or "patch reefs" of concrete modules, each is being implemented on a relatively larger scale, from a 61-ha (150 acres) site in California to marine parks in Hong Kong to potentially large areas of Chesapeake Bay (165,760 km^2 [64,000 mi^2] watershed). Thus, in a broader ecosystem context, management of nutrients from the Chesapeake Bay watershed, along with oyster reef restoration and protection to enhance filtering, are expected to improve water quality and increase opportunity for sea grass bed recovery. The southern California kelp bed project is explicitly quantifying recruitment of kelp, other benthic species, and fishes in a spatially explicit way, cognizant of the importance of the mosaic of surrounding habitats for reference and as a source of recruits.

Finally, the use of pilot studies to test reef designs (kelp) and ecological modeling to predict reef function (oysters, Hong Kong) represents an effective step in maximizing success of the projects through rigorous scientific study design. We suggest that marine habitat and fishery restoration and enhancement now under consideration or planning could benefit from the approaches of these three systems.

Acknowledgments

We are grateful to J. Whitehouse of the University of Florida for typing drafts of the manuscript, which was prepared in part under the auspices of U.S. Department of Commerce NOAA Grant NA16RG2195 to the Florida Sea Grant College Program.

References

Boesch, D. F. 2002. Challenges and opportunities for science in reducing nutrient over-enrichment of coastal ecosystems. Estuaries 25(4B):886–900.

Browder, J. A., Z. Zein-Eldin, M. M. Criales, M. B. Robblee, S. Wong, T. L. Jackson, and D. Johnson. 2002. Dynamics of pink shrimp (*Farfantepenaeus duorarum*) recruitment potential in relation to salinity and temperature in Florida Bay. Estuaries 25:1355–1371.

Burns, D. C., and K. L. Beal, editors. 1997. Special issue on artificial reef management. Fisheries 22(4):3–36.

Cochrane, K. L. 2000. Reconciling sustainability, economic efficiency and equity in fisheries: the one that got away? Fish and Fisheries 1:3–21.

Coen, L. D., and M. W. Luckenbach. 2000. Developing success criteria and goals for evaluating oyster reef restoration: ecological function or resource exploitation? Ecological Engineering 15:323–343.

Desmond, J. S., D. H. Deutschman, and J. B. Zedler. 2002. Spatial and temporal variation in estuarine fish and invertebrate assemblages: analysis of an 11-year dataset. Estuaries 25(4A):552–569.

ICES (International Council for the Exploration of the Sea). 2002. Seventh International Conference on Artificial Reefs and Related Aquatic Habitats. ICES Journal of Marine Science 59 (Supplement) 362 pp.

Jackson, J. B. C. 2001. What was natural in the coastal ocean? Proceedings of the National Academy of Sciences 98:5411–5418.

Jensen, A. C., K. J. Collins, and A. P. M. Lockwood, editors. 2000. Artificial reefs in European seas. Kluwer Academic Publishers, Dordrecht, The Netherlands.

Kennedy, V. S., and L. L. Breisch. 1981. Maryland's oysters: research and management. University of Maryland Sea Grant Publication # UM-SG-TS-81–04. Available: http://www.mdsg.umd.edu/oysters/research/mdoysters.html. (April 2005).

Lenihan, H. S. 1999. Physical-biological coupling on oyster reefs: how habitat structure influences individual performance. Ecological Monographs 69:251–275.

Newell, R. I. E., R. R. Hood, E. W. Koch, and R. E. Grizzle. 2003. Modeling the effects of changes in turbidity on light available for submerged aquatic vegetation. A Final Report Submitted to The NOAA/UNH Cooperative Institute for Coastal and Estuarine Environmental Technology. Available: http://www.ciceet.unh.edu/files/full_project/Newell%20Hood%20and%20Koch%20final%20.pdf. (April 2005).

Pitcher, T. J., R. Watson, N. Haggan, S. Guenette, R. Kennish, R. Sumaila, D. Cook, K. Wilson, and A. Leung. 2001. Marine reserves and the restoration of fisheries and marine ecosystems in the South China Sea. Bulletin of Marine Science 66(3):543–566.

Pitcher, T. J., E. A. Buchary, and T. Hutton. 2002. Forecasting the benefits of no-take human-made reefs using spatial ecosystem simulation. ICES Journal of Marine Science 59(Supplement):S17–S26.

Reed, D., S. Schroeterand, and M. Page, editors. 2002. Proceedings from the Second Annual Public Workshop for the SONGS Mitigation Project. Report to the California Coastal Commission. Marine Science Institute, University of California, Santa Barbara.

Seaman, W., Jr., editor. 2000. Artificial reef evaluation–with application to natural marine habitats. CRC Press, Boca Raton, Florida.

USACOE (U.S. Army Corps of Engineers). 1999. Central and South Florida Comprehensive Review Study Final Feasibility Report and Programmatic Environmental Impact Statement. Available: http:/

/www.evergladesplan.org/pub/restudy_eis.cfm#fact%20sheet. (April 2005).

Wheeler, A. J., B. J. Bett, D. S. M. Billett, D. G. Masson, and D. Mayor. 2005. The impact of demersal trawling on Northeast Atlantic deepwater coral habitats: the case of the Darwin Mounds, United Kingdom. Pages 807–817 in P. W. Barnes and J. P. Thomas, editors. Benthic habitats and the effects of fishing. American Fisheries Society, Symposium 41, Bethesda, Maryland.

Wilson, K. D. P., A. W. Y. Leung, and R. Kennish. 2002. Restoration of Hong Kong fisheries through deployment of artificial reefs in marine protected areas. ICES Journal of Marine Science 59(Supplement):S157–S163.

American Fisheries Society Symposium 49:891–898

Fisheries and Their Management Using Artificial Reefs in the Northwestern Mediterranean Sea and Southern Portugal

GIULIO RELINI*

Universita degli Studi di Genova, Laboratori di Biologia Marina ed Ecologia Animale DIP. TE. RIS., C.so Europa, 26–16132 Genova, Italy

GIANNA FABI

Consiglio Nazionale dell Ricerche-Istituto Ricerca Pesca Marittima Largo Fiera della Pesca-60125 Ancona, Italy

MIGUEL NEVES DOS SANTOS

IPIMAR/CRIPSul, Av. 5 Outubro s/n-8700-305 Olhão-Portugal

ISABEL MORENO

Universidad de las Islas Baleares Campus Universitari 07071, Palma de Mallorca

ERIC CHARBONNEL

Parc Marin de la Cote Bleue Maison de la Mer. Le port. BP 37-13960, Sausset-les-Pins, France

Abstract.—Though in the Mediterranean Sea, artificial reefs (ARs) are mainly multipurpose structures, the main goal has been the protection and enhancement of fishing resources exploited by sport and commercial fishermen. All around the inshore waters, the main problem is prevention of illegal activity of otter trawlers that destroy the habitats, juveniles, and spawners of important commercial species, though trawling is forbidden up to 50 m depth or 3 mi from the shore. The ARs are built with a special anti-trawling module, armoured or very heavy and large in size. Sometimes, ARs are used to protect natural habitats and in particular valuable, *Posidonia oceanica* sea grass beds that play a fundamental role as nursery areas, biodiversity, refuge, and a source of food, and for traditional small-scale fisheries. In the western Mediterranean basin, ARs have been developed since 1960 in France, Italy, Spain, Portugal, and Monaco. A large AR complex, composed of seven artificial reef systems, has been deployed in the Algarve (southern Portugal). It occupies an area of 43.5 km^2 of seabed with more than 20,950 modules weighing 73,000 metric tons (mt) and with a volume of 103,000 m^3. These ARs have successfully increased the fishing yield and diversified the catches. On the other hand, they increase biodiversity and epibenthic biomass, and have became an interesting recreational diving spot, as well as a site for research. Spain shows the more extensive reef-building activity in terms of number of reefs (103 in 2002) and seabed protected (550 km^2). The armed modules (concrete cylinders or square blocks with four protruding iron spikes; 3–5 mt) used for

*Corresponding author: biolmar@unige.it

anti-trawling are effective to prevent illegal fishing but poor in improving hard substrata and microhabitat availability. In France, 44,000 m^3 of concrete modules were deployed in 20 sites, equally distributed in the west and east coasts. Eight sites are off the western coast (Languedoc–Roussillon) to protect static fishing gears, longlines, and natural habitats from illegal trawling. Extensive colonization by oysters, fish, and lobsters on some reefs was reported. In some other ARs, more fish than in comparable natural reefs was found. Where small modules (1–2 m^3) were placed in chaotic heaps of 50–100 m^3 or so, the greatest amount of fish was recorded. In Italy, ARs, made mainly of concrete perforated cubic blocks, were deployed in all the seas with different aims. But the best results in terms of harvestable yield were obtained in the middle Adriatic Sea, due to heavy settlement and fast growth of mollusks, in particular Mediterranean mussel *Mytilus galloprovincialis*. Within 4 years from deployment, the incomes of small-scale fisheries exceeded by three times the initial investment.

Fisheries and Their Management Using Artificial Reefs (ARs) in the Northwestern Mediterranean Sea and Southern Portugal

Artificial reefs consist of bioecological mechanisms that are able to increase fishable biomass and biodiversity inside and around the structures deployed on the sea bottom by man for this purpose. In the present paper, only this kind of AR is considered, and results obtained in Portugal, Spain, France, Monaco, and Italy are described. Rigs to reefs (oil and gas production structures), reefs and breakwaters for shore protection, surf reefs, ship wrecks, and so on, all may behave in part as true ARs, but ARs immersed for other aims are not taken into account. Also, floating ARs as fish aggregating devices (FADs) are not examined here. Fish aggregating devices are, and have been, used throughout history by fishermen to improve pelagic finfish catches, especially in the central and western Mediterranean Sea (for an overview, see Massuti and Morales-Nin 1999). In Sicily, the use of traditional FADs (Canizzi), constructed with vegetal materials (cork and palm leaf), is linked to commercial fishing for greater amberjack *Seriola dumerili* (Risso 1810) and dolphinfish *Coryphaena hippurus* Linneus 1758. During the 1996 fishing season, about 337 metric tons (mt) of dolphinfish were landed (Potoschi et al. 1999).

The main aims for ARs construction in the Mediterranean sea are

• To provide protection against illegal trawling and to protect juvenile fish, especially those of commercial interest;

• To protect sensitive habitats as *Posidonia oceanica* meadows and coralligenous rocks;

• To promote biodiversity and diversification of microhabitats;

• To contribute to the recovery of quality of environment and of coastal fishing resources;

• To promote alternative and innovative forms of management, for the enhancement of the coastal strip, a rational exploitation of the fishing resources, and the development of mariculture;

• To develop the potential of ecotourism and scuba diving; and

• To develop integrated studies on the coastal ecosystems functioning.

Following Bombace (1997), three main types of ARs can be defined:

1. Protection ARs, which protect nurseries and conserve preexisting habitats (mobile bottom

nurseries, *Posidonia* beds, coralligenous biocoenoses located on continental shelves, etc.). The first type of reef protects mainly young fish from the impact of illegal trawling and so enables more young fish to develop and, thereby, increase the biomass of the stocks in the open sea.

2. Production ARs, which form new habitats similar to rocky areas (i.e., complex ecological systems, targeted habitats, etc.), by using principles of ecobiological engineering. In this case, modules have cavities and can be assembled vertically to harness the energy of water column. Spatial discontinuity, colonization surfaces, attraction of spores and larvae, and the elevation from the substratum create the physical conditions for the diversification of the habitats and food webs. This reef structure can also be integrated with mariculture in those cases where the environmental condition (eutrophic water) favors the settling and growth of economically important bivalves.

3. Polyvalent, mixed types of ARs, which combine the characteristics of the aforementioned types and which may be integrated with mariculture.

Artificial Reefs in Different Countries

Portugal

The deployment of artificial reefs in the Portuguese coastal waters is relatively recent. In 1990, benefiting from the regional financial support the Instituto de Investigacoa das Pescas e do Mar (IPIMAR), through its regional Centre of Fisheries Research of the South, developed a large ARs complex in the Algarve coast water, consisting of seven ARs. Each one is consists of at least 2,940 concrete modules having unitary weight of 3 mt and 36 modules of great dimensions weighing about 40 mt each. The reef complex, that contains more than 20,950 modules and a volume of 103,000 m^3, occupies, in a discontinuous way, a total area of 43.5 km^2, with an estimated area of influence of 67 km^2. The Faro Ancão ARs seem to be the largest structure of this type in Europe. Finally, the structures are implanted at depths below 15 m to allow the bivalve dredge fishery, which plays an important role within the Algarve fisheries.

The aim is to have positive impacts at the following aspects

- Bioecological—new "habitats," valuing the coastal zone in a relatively enlarged extension (>65 km^2), protection of juveniles, and recovery of the coast fishing resources.

- Fisheries management—increase of the fishing yields and the readiness and diversity of the resources for the fishing activity and creation of a new conditioned fishing zones.

- Regulation of the fisheries and of other activities, namely offshore aquaculture and ecotourism.

- Socioeconomic—as a consequence of the effects mentioned above.

Further information is available in Santos et al. (1997), Santos and Monteiro (1998), and Monteiro and Santos (2000).

Spain

Spain has already more than 103 artificial reefs deployed in the Mediterranean coastal area, with the support of both the national and local fishery authorities, and has plans to continue deploying, especially near the town of Barcelona.

The aim of all these ARs is twofold: prevent illegal trawling on *Posidonia oceanica* meadows and increase substrate for epibenthos to increase or concentrate fish biomass that can be captured by the local artisanal fishing fleet. Recently, some ARs have also been planned for recreational uses, such as scuba diving.

Due to different aims, the units (modules) vary very much in shape, size, and composition and in the design of the reef itself. In spite of this diversity, it has been proved that the antitrawl-ing reefs have been highly successful as such, in particular the armed modules (of different shape with four protruding iron spikes and 3–5 mt in weight), but they are poor in improving hard substrata and microhabitat availability. In any case, trawling on *P. oceanica* meadows and in other not allowed grounds has decreased drastically, and, in many cases, the seabed has shown a significant recovery.

Epibenthos colonization of the units is rather quick; in 20 months, it seems to have reached the climax. The speed in this colonization is directly proportional to the good state of the seabed, and it takes place more quickly on the *Posidonia* bed than on bare sand; although, with time, the coverage and the species present is the same in both cases.

The epibenthos presents a high diversity and biomass, representing an important food source for foraging animals. Fish aggregation has also been proved, not only in total biomass, but also in the amount of individuals of target species. This fish abundance is likely due more to fish aggregation than to significant fish biomass increase, which has not been proved. The local artisanal fishing fleet yield has improved, and, as long as there are strict fishing regulations in place, it can be sustainable (Moreno 2000; Ramos-Espla et al. 2000; Revenga et al. 2000).

France

French experiments on ARs constituted the first trial in Europe and started earlier in 1968, with some experiments of pilot reefs made from waste materials placed (car bodies, old tires). It was only at the mid-1980s that a concerted program was developed, with 30,000 m^3 of specially designed reef shape made of concrete materials. Today, 44,000 m^3 of reefs are deployed on 20 sites along the French Mediterranean coast with the main aim of promotion of fisheries and resources, protection of natural habitats, increasing of biological production, and biodiversity.

Most of the ARs are specially designed and made of concrete, which is stable, durable, easy to shape, and has high colonization rates. However, the cost is really expensive. Several kinds of modules were tested, both for production and protection effects. Most of production reefs are composed of small cubic modules of 1–2 m^3, gathered into chaotic piles of 50–150 m^3, and also bigger modules (158 m^3), inspired by Japanese technology, but having a very low habitat complexity.

Concerning antitrawling reefs, five kinds of modules were used in the Côte Bleue Marine Park, while only one model (pipe unit) is currently deployed in Languedoc-Roussillon. To be efficient, these protection reefs have to weigh 8 mt minimum. These reefs are deployed one by one, separated from each other by 50–200 m. They can be regularly deployed, occupying all the space or being deployed on lines, creating barriers perpendicular to the coast.

Fishermen are often at the origin of AR projects. Management varies from one area to another: in Languedoc, access is permitted and all reefs are open to fishermen, while in Provence Alpes Côte d'Azur, most of the reefs are in protected areas where fishing is prohibited. In the Cote Bleue Marine Park (4,680 m^3 of reefs), the situation is intermediate, with three reefs open for fisheries purposes (both commercial and recreational) and two reefs with protective management.

The most characteristic example of success of protection reefs is the decreased incidence of illegal trawling after reef deployment and creation of the marine reserve in Cap Couronne (Côte Bleue marine park).

Concerning production reefs, previous studies showed that the fish assemblages associated with ARs are similar in species composition, density, and biomass to those occurring on natural reefs made of rocky bottoms, and most often, with superior performance, due to multimodular aspect and higher heterogeneity (Charbonnel et al. 2001, 2002). Availability of shelters may be more important, because food resources are generally not limiting. It is important to notice that ARs worked as marine reserves and have similar "reserve effects" with 1) increasing biodiversity and global species richness, due to higher frequency of rare target species; 2) increasing abundance of individuals, particularly those of species targeted by fishing; 3) appearance of protected species which have a high heritage value, such as dusky grouper *Epinephelus marginatus* and brown meagre *Sciaena umbra*; and 4) recovery of balanced demographic structure, with a higher frequency of large individuals which are potential spawners. All these phenomena are similar to those observed in reserves (Charbonnel et al. 2000, 2001) and the same supposed beneficial effect to fish numbers seen in the peripheral area can be attributed to AR.

The reef design is a crucial factor in their effectiveness, and, the more the reef is complex and large, the more the colonization can proceed over a longer period of time before reaching the maximum of maturation and of carrying capacity.

Monaco

In Monaco, small experimental ARs have been built from 1974 to 1992 to protect *Posidonia* and to allow experiments on artificial reproduction of red coral *Corallium rubrum*.

Italy

Since the 1960s in Italy, at least 41 ARs of different materials, shape and volumes were deployed along all the coasts. Three main groups of ARs can be recognized in the Adriatic Sea, in Sicily, and in Liguria.

In the Adriatic Sea, the first AR was deployed in 1974–1975 at Porto Recanati, 5.5 km offshore, on a muddy seabed, 13–14 m deep. The reef was constructed by Consiglio Nazionale delle Ricerche-Istituto Ricerca Pesca Marittima (CNR-IRPEM) with the aims of preventing trawling, restocking the area, and developing new sessile biomass, especially mussels and oysters. It is composed of a central "oasis" (3 ha) formed by 12 pyramids, each one 6 m high and built from 14 IRPEM blocks. The original basic reef module was a 8-m^3 concrete block (2 × 2 × 2 m), weighing 13 mt, which incorporated holes of different shapes and sizes. It was called the IRPEM block and was employed for many other ARs along the Italian coast. The pyramids were deployed in a rectangular arrangement about 50 m apart. The oasis is surrounded by a protected area (2,000 ha) where cubic blocks and other smaller antitrawling blocks were dispersed. A total of 453 blocks were used: 168 for the pyramid construction and 285 as isolated antitrawling blocks, giving a total volume of 3,624 m^3. The stones placed under the pyramids provided a volume of 396 m^3.

In all, more than 4,000 m^3 of material were deployed (Bombace et al. 2000).

In 1983, the Portonovo AR was built, and, between 1987 and 1989, seven other ARs were deployed along the northern and central coast of the Adriatic Sea.

Initially, reef pyramids were made of 14 stacked blocks (nine at the base, four in the middle, and one on the top) placed at 15–20 m depth. The large scale ARs (7,500–13,000 m^3), deployed in 1980s, were made by two-layer pyramids placed at lower depth (10–15 m) and associated with concrete cages for shellfish culture (6 × 4 × 5 m, Bombace et al. 2000). Other different types of modules have been tested: in the last period, Tecnoreef modules were successfully tested. Very satisfactory results were obtained in the Adriatic Sea, due also to eutrophic waters.

The biomass of mussels settled on block surfaces of ARs was estimated at about 80–100 kg/m^2 in the summer of 1977, 2.5 years after reef deployment. Biomass stabilized at about 50 kg/m^2 in the years 1977–1980. The commercial production of the whole reef was about 200–250 mt/year in the first 3 years and 160 mt/year in the following 4 years. Between 1977 and 1980, the average fishing yields obtained with trammel nets and traps were 20 mt/year of high priced fish and cephalopods and 160 mt/year of *Sphaeronassa mutabilis*, a small edible gastropod presumably attracted to the ARs. At sites far from natural rocky habitats, species richness and diversity, as well as fish abundance, increased after reef deployment. This increase was particularly notable in reef-dwelling nekto-benthic species such as brown meagre, Shi drum *U. cirrosa*, and a few sparids. The mean catch increments recorded for these species, 3 years after artificial reef deployments were 10–42 times the initial values in weight (Bombace et al. 1994; Fabi and Fiorentini 1994).

The ARs of Sicily and results obtained are reviewed by Riggio et al. (2000), Badalamenti et al. (2000), and D'Anna et al. (2000). The development of ARs in Sicily can be considered as a multipurpose technology for resource management within the coastal zone, taken in mind that fishery yield and biodiversity are enhanced wherever rocky seabed is interposed with *P. oceanica*. Sicilian reef deployment has tried to promote the re-colonization of damaged seabed habitat, beyond the immediate scope of merely increasing the fishing yield.

The ARs were assembled on a sandy seabed at depths between 10 and 30 m. Most of the reefs were made from concrete IRPEM blocks (13 mt, 8 m^3), deployed to form three-layer pyramid-shape reef units.

The ARs were successful in preventing nearshore trawling and greatly favored the protection of the *P. oceanica* beds and their recolonization wherever sea grass had been cleared or damaged. Artificial reefs have, subjectively, improved the quality of the benthic environment and increased the fishing yields, although the rise in catches was partly due to fish attraction and is, by no means, comparable to the fishery returns obtained in highly productive waters, as in the Adriatic Sea.
In the Ligurian Sea, after a negative experience with car bodies in 1970, suitable materials were tested for ARs (Relini and Relini Orsi 1989). In the Marconi Gulf AR, the annual summer sportfishing competition showed an increase in catches, in terms of weight of fish per fisherman and per trip, from 1982 to 1986, despite the increased number of participants in each competition. In the summer of 1986, in the waters off the coast at Loano, 350 concrete blocks were immersed at depths between 5 and 45 m to protect around 350

ha of seabed covered with a thin regressing *P. oceanica* meadow. In the central part, 30 pyramids of five cubic perforated blocks (2 m sides) were placed. Evaluation both of epibenthos and fishes was described (Relini 2000). Fish, cephalopods, and crustaceans have been surveyed using fishing gears and direct underwater observations. The list is composed of 76 species, 67 of which are fish (45 found in immersion), four crustaceans, and five cephalopods. There are some species of commercial interest and others are important because they have become rare along the Ligurian coast. The qualitative and quantitative composition of the fish population is the same as can be found in protected rocky areas, such as the marine parks. The most highly represented families are Sparids, Labrids, and Serranids, as is characteristic of the rocky areas of the northwest Mediterranean.

The yield of net fishing (average 2.32 kg/100 m) is higher than that obtained in ecologically similar areas in the Tyrrhenian sea and is comparable to that obtained on artificial reefs after a year of installation in a highly productive environment such as the Adriatic Sea (1.71–2.8 kg/100 m).

Conclusions

In the northwest Mediterranean Sea and Portugal, the ARs are of three types (protective, productive, and polyvalent), but most of them have, as a first aim, the protection against illegal trawling. Most of ARs have been monitored and studied for a long time so that a lot of data are available on benthic colonization, biodiversity, fishery catch and harvest, and ecological aspects.

The harvestable biomass production in particular of mollusks is much higher in the eutrophic waters of the middle and northern Adriatic Sea. In the oligotrophic waters, as in the Sicily and Ligurian seas, the biomass of benthos is lower, but biodiversity is higher and also, fish biodiversity is higher. In the Adriatic ARs, the biggest increment of fish catch yields was 30–80 times the initial value in weight, in particular for nekto-benthic fishes (Bombace et al. 1994).

The IRPEM-CNR of Ancona showed that within a 4 year period from deployment, the income of small-scale fisheries (high price bonyfish, mussels, oysters, and gastropods) would exceed by three times the initial investment.

Finally, ARs constitute a response to many problems concerning coastal resources, overexploitation of fishery resources, and ecosystem degradation. Artifical Reefs represent a good tool for coastal marine resources management and can maintain a sustainable development of fisheries and fishermen as marine protected areas (particularly no take areas). However, they are only one facet of the overall management, which must take into account all phases of the life history of the overexploited species, and especially their spawning and nursery areas (Charbonnel et al. 2000). Artificial reefs are well-adapted tools in the actual context of rules changes in Europe (new policy of fisheries "PCP") and several projects of reef deployment are ongoing.

References

Badalamenti F., G. D' Anna, and S. Riggio. 2000. Artificial reefs in the Gulf of Castellammare (North-West Sicily): a case study. Pages 75–96 *in* A. C. Jensen, K. J. Collins, and A. P. M. Lockwood, editors. Artificial reefs in European seas. Kluwer Academic Publishers, London.

Bombace, G. 1997. Protection of biological habitats by artificial reefs. Pages 1–15 *in* A. C. Jensen, editor. European artificial reef research. Proceedings of the 1st EARRN Conference, Ancona, Italy. Southampton Oceanography Centre, Southhampton, UK.

Bombace, G., G. Fabi, L. Fiorentini, and S. Speranza. 1994. Analysis of the efficacy of artificial reefs located in five different areas of the Adriatic Sea. Bulletin of Marine Science 55(2–3):559–580.

Bombace, G., G. Fabi, and L. Fiorentini. 2000. Artificial reefs in the Adriatic Sea. Pages 31–63 *in* A. Jensen, K. Collins, and A. Lockwood, editors. Artificial reefs in European seas. Kluwer Academic Publishers, London.

Charbonnel, E., P. Francour, J. G. Harmelin, D. Ody, and F. Bachet. 2000. Effects of Artificial Reef Design on associated fish assemblages in the Côte Bleue Marine Park (Mediterranean sea, France). Pages 365–377 *in* Jensen, A. C., Collins, K. J., and A. P. M. Lockwood, editors. Artificial Reefs in European seas. Kluwer Academic Publishers, London.

Charbonnel, E., D. Ody, L. Le Direach, and S. Ruitton. 2001. Effet de la complexification de l'architecture des récifs artificiels du Parc National de Port-Cros sur les peuplements ichtyologiques. Scientific Report Port-Cros Natural Park, France 18:163–217, Parc National de Port Cross, Hyeres, France.

Charbonnel, E., C. Serre, S. Ruitton, J. G. Harmelin, and A. Jensen. 2002. Effects of increased habitat complexity on fish assemblages associated with large artificial reef units (French Mediterranean coast). ICES Journal of Marine Sciences 59:208–213.

D'Anna G., F. Badalamenti, and S. Riggio. 2000. Artificial reefs in north-west Sicily: comparisons and conclusions. Pages 97–112 *in* A. C. Jensen, K. J. Collins, and A. P. M. Lockwood, editors. Artificial reefs in European seas. Kluwer Academic Publishers, London.

Fabi, G., and L. Fiorentini. 1994. Comparison of an artificial reef and a control site in the Adriatic Sea. Bulletin of Marine Science 55(2):538–558.

Massuti, E., and B. Morales-Nin, editors. 1999. Biology and fishery of dolphinfish and related species. Scientia Marina 63(3–4):181–475.

Monteiro, C. C., and M. N. Santos. 2000. Portuguese artificial reefs. Pages 249–261 *in* A. C. Jensen, K. J. Collins, and A. P. M. Lockwood, editors. Artificial reefs in European seas. Kluwer Academic Publishers, London.

Moreno, I. 2000. Artificial reef programme in the Balearic Islands: western Mediterranean sea. Pages 219–234 *in* A. C. Jensen, K. J. Collins, and A. P. M. Lockwood, editors. Artificial reefs in European seas. Kluwer Academic Publishers, London.

Potoschi A., L. Cannizzaro, A. Milazzo, M. Scalisi, and G. Bono. 1999. Sicilian dolphinfish (*Coryphaena hippurus*) fishery. Scientia Marina 63 (3-4): 439-445.

Ramos-Espla, A. A., J. E. Guillén, J. T. Bayle, P. Sanchez Jerez, and J. L. Sanchez Lisazo. 1993. Protección de la pradera de *Posidonia oceanica* (L.) Delile mediante arrecifes artificiales disuasorios frente a la pesca de arrastra ilegal; el caso de El Campello (SE ibérico). Publicaciones Especiales. Instituto Español de Oceanografía 11:431–439.

Relini, G., and L. Orsi Relini. 1989. The artificial reefs in the Ligurian Sea (N-W Mediterranean): aims and results. Bulletin of Marine Science 44(2):743–751.

Relini, G. 2000. The Loano artificial reef. Pages 129–149 *in* A. C. Jensen, K. J. Collins, and A. P. M. Lockwood, editors. Artificial reefs in European seas. Kluwer Academic Publishers, London.

Revenga, S., F. Ferdandez, J. L. Gonzalez, and E. Santaella. 2000. Artificial Reefs in Spain: the Regulatory Framework. Pages 185–194 *in* Jensen, A. C., Collins, K. J., and A. P. M. Lockwood, editors. Artificial Reefs in European Seas. Kluwer Academic Publishers, London.

Riggio S., F. Badalamenti, and G. D'Anna. 2000. Artificial reefs in Sicily: an overview. Pages 65–74 *in* A. C. Jensen, K. J. Collins, and A. P. M. Lockwood, editors. Artificial reefs in European seas. Kluwer Academic Publishers, London.

Santos, M. N., C. C. Monteiro, and G. Lasserre. 1997. A four years overview of the fish assemblages in two artificial reef systems off Algarve (south Portugal). Pages 345–352 *in* L. E. Hawkins and S. Hutchinson, editors with A. C. Jensen, J. A. Williams, and M. Sheader. The responses of marine organisms to their environment. University of Southampton, Southampton, UK.

Santos, M. N., and C. C. Monteiro. 1998. Comparison of the catch and fishing yield from an artificial reef system and neighbouring areas off Faro (south Portugal). Fisheries Research 39:55–65.

American Fisheries Society Symposium 49:899–915

Artificial Reefs as Fishery Conservation Tools: Contrasting the Roles of Offshore Structures Between the Gulf of Mexico and the Southern California Bight

ANN SCARBOROUGH-BULL*

U.S. Department of the Interior, Minerals Management Service, Pacific-Region
770 Paseo Camarillo, Camarillo, California 93010, USA

MILTON S. LOVE AND DONNA M. SCHROEDER

University of California at Santa Barbara, Marine Science Institute
Santa Barbara, California 93101, USA

Abstract.—Rigs-to-Reefs is a term used when converting obsolete offshore oil and gas platforms into designated artificial reefs. In Federal waters, there are over 4,000 platforms in the Gulf of Mexico and 23 structures in the Southern California Bight. Platforms in the Gulf of Mexico are concentrated in the north-central and northwestern regions where few natural reefs exist; they harbor unique communities bearing little resemblance to those in the natural surrounding habitat. There is evidence that the artificial habitat supplied by platforms in the Gulf of Mexico has increased the regional carrying capacity for economically important reef fish species such as red snapper. Platforms in the Gulf of Mexico are customary destinations for both commercial and recreational fishing and Rigs-to-Reefs programs of Gulf States are actively expanding each year. Contrary to the Gulf of Mexico experience, California platforms are concentrated in the Santa Barbara Channel area among natural reefs and offshore islands and they harbor fish assemblages that resemble those found in nearby habitats. Observations at natural reefs and platforms off California found that platforms have become refugia for increasingly rare and overfished species, which is thought to be, in part, a result of light to non-existent fishing pressure at those platforms. Many of the oil and gas structures off California are nearing the end of productive life as sources of oil and gas. Within the next decade important decisions will be made concerning the ultimate fate of the California platforms and the fish populations they support.

Introduction

Along the coasts of North America, the most substantial man-made structures are the offshore facilities created for the extraction of hydrocarbons and other minerals. Worldwide, there are over 6,000 large offshore platforms that extract oil and natural gas from beneath the outer continental shelf (OCS) (Schroeder and Love 2004). About 4,000 of those structures are located in U.S. federal waters off the Gulf of Mexico, with at least another 1,000 in Gulf State waters. The first offshore oil well in the United States was brought in by Kerr-McGee in the Gulf of Mexico, about 72 km south of Morgan City in the Ship Shoal Block

*Corresponding author: ann.bull@mms.gov

32 field, marking the birth of the offshore oil and gas industry. Since that time, thousands of platforms have been both installed and removed from the Gulf, and, currently, annual removals exceed installations. Off California, 23 offshore platforms have been installed within federal waters beginning in 1967. Seven platform installations occurred in state waters prior to this time. Development of the OCS on the Pacific coast was delayed by a disastrous blowout and oil spill in 1969; the last platform was sited in 1989 (Nevarez et al. 1998). There is no support for further petroleum development off California at this time, and it is highly unlikely that will change in the foreseeable future.

Petroleum producing platforms are set in place by driving steel support legs (hollow pilings) deep into the seafloor. Working machinery and personnel sit above water supported by a steel network (jacket) that is intentionally overbuilt and remarkably secure. The vertical pipes that carry the oil and gas from below the seafloor are the conductors. Near the surface, the submerged jacket is braced by crossbeams that are placed every 15–20 m, and deeper, they occur every 30–40 m. Horizontal, diagonal, and oblique crossbeams extend both around the perimeter of the jacket and reach inside and across the platform (see Figure 1).

The hard substrate of the submerged platform jacket provides habitat for many sessile invertebrates, including mussels, barnacles, scallops, sponges, tunicates, corals, and oysters. Indeed, in the northwestern and north-central Gulf of Mexico OCS region, where the amount of natural hard substrate is limited, it is estimated that offshore platforms contribute from 10% to 28% of the regional "reef" habitat (Gallaway and Lewbel 1982; Scar-

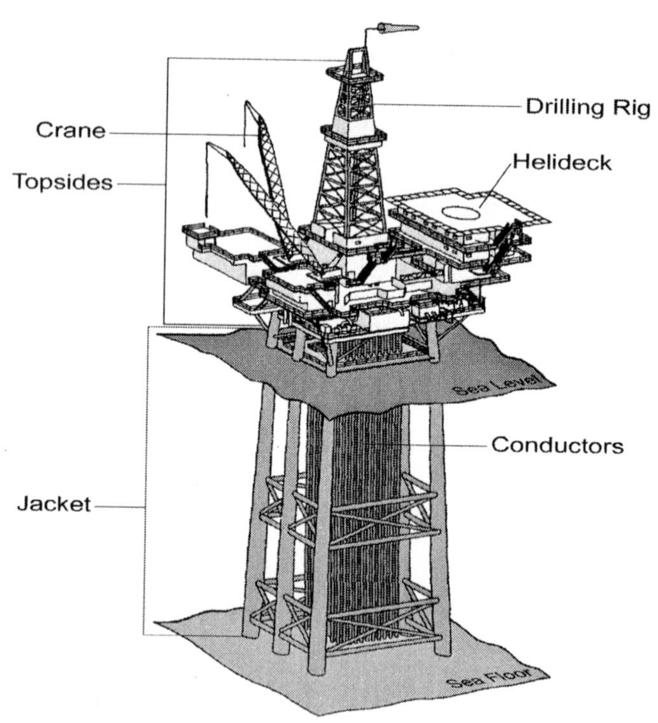

Figure 1. Schematic of a generalized offshore oil and gas platform. Adapted from Manago and Williams 1998.

borough-Bull and Kendall 1994; Stanley and Wilson 2003). There are, however, no quantitative estimates of the extent to which platforms contribute to the total amount of "reef" habitat in the Pacific OCS region (Carr et al. 2003). Estimates based on the general amount of hard substratum in shallower regions of the Santa Barbara Channel, including the Santa Barbara Channel Islands, lead to the conclusion that this contribution may be very small (Holbrook et al. 2000; Helvey 2002). However, our observations imply that rocky outcrops are relatively scarce below about 50 m in the areas where platforms occur. Thus, deeper-water platforms may locally provide considerable hard structure. For instance, it is estimated, from blueprints of the platform and a detailed multibeam survey of the surrounding area, that Platform Hidalgo supplies about 46% of the hard surface in its local area off Point Conception (within about 3 km of the platform) (Love et al. 2003). In addition, there are few natural pinnacles that rise so abruptly and no reefs in any region with the physical relief comparable to the structures. Platforms start below the seafloor and extend through the sea surface. As such, offshore platforms as artificial habitats are unique (Carr et al. 2003). According to Gallaway (see LGL and SAIC 1998), it may be more ecologically accurate to consider platform reefs as a new and distinct habitat rather than to assume that they are merely additions to existing reef systems.

Communities associated with petroleum platforms off California and within the Gulf of Mexico are obviously characterized by distinct regional faunal assemblages and species associations (Scarborough-Bull 1989b; Stanley and Wilson 2000; Love et al. 2003). There are ecological features unique to each region, and platform assemblages differ from nearby natural environments to varying degrees. In the northwestern and north-central Gulf, where most of the natural habitat is sedimentary, few representatives of local and coastal fauna occur on offshore platforms (Sonnier et al. 1976; Gallaway and Lewbell 1982; Scarborough-Bull 1989a). It is assumed that originally many of the epibiotic organisms and fish species that now inhabit Gulf offshore platforms traveled to these structures as pelagic larvae for extended periods over relatively long distances; however, they may have also arrived via shipping or large semisubmersible drilling rigs moved from South America. Historically, as a species successfully settled on a Gulf platform and reached maturity, its larvae would not necessarily need to travel far to reach a similar habitat (another platform).

Contrary to observations in the Gulf of Mexico, California platform fish assemblages tend to resemble those found on nearby natural habitats (Scarborough-Bull 1989b; Love et al. 2003). Much of the platform fish assemblages on Pacific platforms probably reflect recruitment of larval and pelagic juvenile fishes from both near and distant maternal sources as well as some attraction from natural reefs (Love et al. 2003). The fish assemblages that develop at the platforms in both the Gulf and Pacific regions do so over time. The influence that the platform fish communities may exert, and the role that the platforms as long-lived artificial reefs may play in the conservation of economically important species rests, in their vastly different history and use in regional fisheries.

Man-Made Reefs in the Gulf of Mexico

The largest unplanned artificial reef complex in the world exists in the northwestern and north-central Gulf of Mexico, where over 4,000 petroleum platforms are scattered across the region (Figure 2). Scientists have hypothesized that artificial reefs and platforms improve and/or diversify habitat, increase fishery resources, modify the assemblages of or-

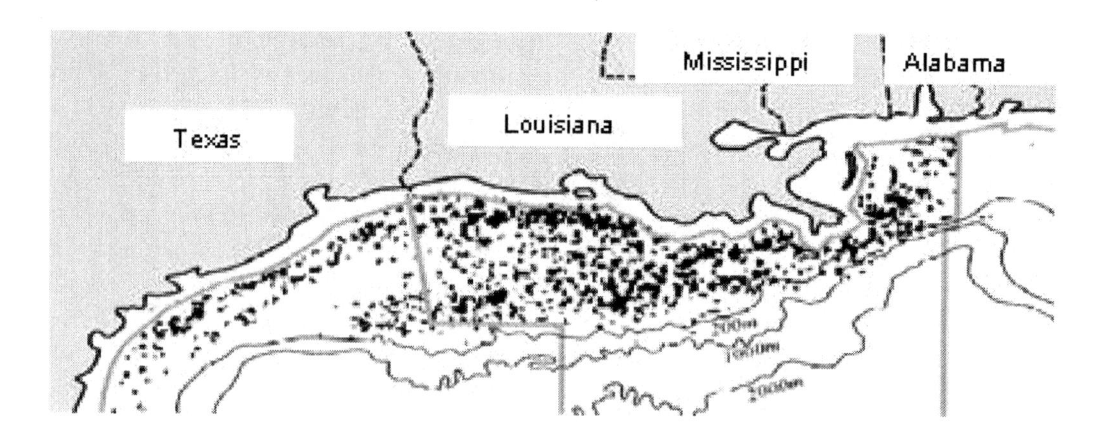

Figure 2. Distribution of petroleum platforms across the Gulf of Mexico. Gray lines denote federal waters and separate OCS planning areas for hydrocarbon production. Gross bathymetry lines at 200, 1,000, and 2,000 m are included.

ganisms in the region, or concentrate existing resources (Bohnsack et al. 1991; Seaman and Sprague 1991). One cannot doubt that placement of these artificial structures in the Gulf of Mexico over many decades has had some influence on fishery resources and has had an effect on the greater regional marine community.

In the area where platforms are concentrated, investigators have found that fish abundance at these structures varies from a few hundred to thousands, depending on platform size, location, season, and time of survey (Putt 1982; Stanley and Wilson 1991). Stanley and Wilson (1997) estimated fish abundance monthly for 1 year at a petroleum platform off Louisiana and reported significant variability between winter–summer months in some reef fish numbers. Further research by Stanley and Wilson (2000, 2003) confirms that there is variability of fish assemblages associated with petroleum platforms but also demonstrates the importance of petroleum platforms to the marine environment of the northern Gulf due to the high abundance of fishes found at the sites. The area of influence of a Gulf platform on the OCS is reported to be 16–20 m from the platform. Fish densities within 20 m of a platform are 3–60 times higher than that of soft bottom areas at greater distances, and densities observed at distances greater than 50 m are comparable to background levels of the open waters of the northern Gulf of Mexico. Although some variance was observed, 10,000–30,000 fishes were found at platforms at any one time, and since more than 1,000 platforms are found in similar water depths, it is clear that these structures influence the fish abundance of the region, especially the abundance of reef fish species that require structure (Stanley and Wilson 2003).

Of concern to fishery conservation is the red snapper *Lutjanus campechanus* a reef fish that occurs in large numbers at platforms. The red snapper stock in the Gulf of Mexico is extremely valuable, both commercially and recreationally, and it is likely the most targeted reef species in the region (Nieland and Wilson 2003). Red snapper are long-lived apex predators, predominately demersal and sedentary, and, as adults, are geographically restricted to reef and reef-like hard bottom habitats (Nieland and Wilson 2003). The Gulf of Mexico red snapper resource has been depleted by growth overfishing of adults and by trawling as bycatch of juveniles (Schirripa and

Legault 1999; GOMFMC 2000). Putt (1982) found red snapper to constitute 2–4% of the total fishes inhabiting platforms off Texas. As reported in NRC (1996), and shown by Stanley and Wilson (1996, 1997), red snapper numbers vary from 521 to 8,202 individuals at platforms off western Louisiana. Similarly, among estimated total fish populations of ~26,000 and ~13,000 individuals at two platforms off Central Louisiana, 4.4% and 19.2%, respectively, were red snapper (Stanley and Wilson 2000). Red snapper has been found to be among the dominant 10 species at offshore platforms, and, when taken together, the dominant 10 species consistently comprise more than 95% of the total fishes at platforms (Stanley and Wilson 2003).

Information about the importance of offshore platforms to Gulf-wide fisheries is found in Witzig (1986), where it was estimated that 70% of all saltwater fishing trips (recreational and commercial) in the exclusive economic zone (EEZ) (more than 4.8 km from shore) off Louisiana were destined for one or more of the offshore platforms. Reggio (1987a) reported that recreational anglers fishing at platforms caught bigger, more desirable fish, especially red snapper, than those anglers who fished in other areas. In fact, fishing at platforms had the highest catch rates of all recreational fisheries in the United States (Stanley and Wilson 1990). More recently, Hiett and Milon (2002) performed an exhaustive survey of recreational fishing from Alabama west through Texas for 1999 and indicated that out of a total of about 4.5 million marine recreational fishing trips, 52% of commercial passenger fishing vessel trips, 32% of charter fishing vessel trips, and 21% of small private boat trips fished at offshore platforms.

Research from Texas A&M University at Corpus Christi strongly indicates that platforms as artificial reefs have increased biomass in the Gulf of Mexico when one considers all species of nonharvested invertebrates and finfish and the algae growing on these structures, all key elements in the ecosystem dynamics of the Gulf (Dokken 1997). Linton (1994) reported that the amount of hard surface habitat added in the northern Gulf of Mexico, through the proliferation of offshore platforms, is a contributing factor to the increased finfish abundance Gulf-wide. Although fish associated with platforms are not all reef fish and platforms do not constitute all reefs, given the large number of red snapper at platforms and the large number of platforms, one would assume that the presence of platforms has increased the standing stock of this species in the Gulf. Indeed, it is believed that over the past 30–50 years, the installation of offshore platforms in the northern Gulf of Mexico has increased the carrying capacity of the region for reef fish species, specifically red snapper (Linton 1994; Robert Shipp, University of Southern Alabama, personal communication).

Historical fisheries data reviewed for the Gulf of Mexico Fishery Management Council (GOMFMC) supports such a conclusion (GOMFMC 1996, 2000; Schirripa and Legault 1999). Since at least 1972 (Schirripa and Legault 1999), a significant proportion of the domestic red snapper landed in Florida has been derived from areas west of the Mississippi River. The same observation is true for Alabama and Mississippi. Fish caught at platforms off Louisiana dominate Louisiana landings of red snapper. Texas commercial landings are primarily composed of fish from platforms off Texas, but some are also taken off Louisiana. The general trend for red snapper sold in a Gulf State is to have been captured in immediately adjacent waters or in waters to the west (Dimitroff 1982; Schirripa and Legault 1999). Dimitroff (1982) conservatively estimated that a group of 112 commercial reef fish fishers from the Florida panhandle regularly fished for red snapper at platforms off Louisiana and Texas and landed nearly one-half million pounds of reef fish annually. It seems clear from these data that the commercial harvest of red snapper has moved steadily westward in the Gulf since the mid-20th century and that the commercial fishery has died out in the eastern half of their historical range in the U.S. Gulf of Mexico. Since 1990, more than 90% of the Gulf-wide commercial red snapper harvest has occurred off the Louisiana and Texas coasts (Schirripa and Legault 1999), and while statistics do not exist on the percentage of red snapper caught at petroleum platforms, it is assumed to be a significant portion. Stanley and Wilson (1990) reported the composition of catches made at petroleum platforms off Louisiana shifted in late 1970s from one predominated by croaker and drum (Family Sciaenidae) to one predominated by reef fish and red snapper (Family Lutjanidae).

One should not underestimate the role that platforms may play as man-made reefs for the conservation of reef fishes at a time when many demersal reef fish populations, including economically important red snapper that are habitat-limited, are in serious decline in the Gulf of Mexico. The red snapper continues to be the subject of intense conservation and regulation as the species is overexploited and there are significant problems with the long-term viability of the stock for both commercial and recreational harvest (Schirripa and Legault 1999; GOMFMC 2000). Since 1991, both commercial and recreational fisheries for red snapper have been constrained by size limits, creel or trip limits, and quotas. Even with the best efforts of the GOMFMC, and in spite of the commercial and recreational sectors, overfishing of red snapper in the Gulf may persist (Schirripa and Legault 1999). Recognizing that red snapper resources must be conserved, GOMFMC has considered reefing platforms to form artificial reef reserves, possibly in combination with natural reef reserves or as a fishing reef alternative when natural reefs become no-take marine protected areas (MPAs) (GOMFMC 1996), but not as a substitute for the inclusion of natural reefs in protected areas. Although GOMFMC has not recognized offshore platforms as essential fish habitat for reef fish, they have recognized existing platforms as important habitat for the conservation of red snapper and have stipulated that the platforms may serve as a critical (but not obligate) link in the life history of both juvenile and adult red snapper for the Gulf-wide stock (GOMFMC 1996, 2000).

The influence of Gulf platforms on red snapper resources is accidental and secondary to their primary objective of producing hydrocarbons. Offshore platforms are not intended to be permanent. Within 1 year of an OCS lease termination, MMS requires that the lessee totally remove the platform structure to a depth of 5 m below the seafloor and the leased area must be cleared of obstructions (Dauterive 2000). Over 90% of decommissioned platforms are totally removed and taken to shore for recycling or scrapping while the surrounding seafloor is returned to the condition that existed prior to structure installation. However, MMS may waive the requirements for total removal, and, under specific circumstances, that must include transfer of ownership and liability to a recognized state-sponsored rigs-to-reefs program, MMS will accommodate conversion of a decommissioned platform to an artificial reef. Table 1 shows the total number of platform installations, removals, and rigs-to-reefs in the Gulf of Mexico since 1942.

In the 1980s, offshore platforms were recognized as having significant direct benefits on offshore recreational, commercial trolling, and hook-and-line fishing and scuba diving in the

Table 1. Installation, removal, and reefing history for Gulf of Mexico offshore oil and gas platforms.

Gulf of Mexico platform history 1942–April 2004	
Platforms installed	6,394
Platforms removed	2,403
Platforms remaining	3,991
Platforms reefed	211
Percent reefed	8.7%

northern Gulf of Mexico (Reggio 1987b). This recognition, plus the increase in platform removals relative to platform installations over time, and the understanding that Gulf reef fish resources are greatly dependent on the presence of offshore platforms prompted Louisiana and Texas to create artificial reef programs in which retired petroleum platforms are the material of choice (Wilson et al. 1987; Stephan et al. 1990). One aspect of the positive attitude that residents in the Gulf of Mexico region have toward reefing offshore platforms is that the modern industry of petroleum extraction and commercial and recreational reef fishing developed together. Most extended families in Texas, Louisiana, Mississippi, and Alabama have family members that have worked in one or the other endeavors, many for generations. The two livelihoods expanded together, each is familiar on land as well as sea, neither is generally perceived as a threat to the other, and platforms are the usual destination for most recreational and commercial fishers who are targeting reef fish (Reggio and Kasprzak 1991). The rigs-to-reef policy idea represents a recombination of an old solution (a reliance on artificial reefs to increase carrying capacity and fish stocks in habitat-limited fisheries) to a perceived new problem (a lack of natural habitat and potential consequences associated with complete removal of many OCS oil and gas structures) (Carr et al. 2003). The conversion of offshore platforms to reefs involves removing most of the superstructure and laying what remains and the entire submerged jacket on its side at its present location (topple in place), relocating the jacket (with or without part of the superstructure) elsewhere (tow-and-place), or simply removing the superstructure and uppermost 20–30 m of the jacket and leaving the rest standing (partial removal) (Figure 3).

The general practice for incorporating platforms into Gulf State sponsored reef programs involves two options. The Louisiana (LARP) and Mississippi artificial reef programs have configured many reef sites by toppling in place decommissioned platforms within predetermined artificial reef areas or by taking obsolete platforms to such an area by tow and place. Both of these options involve the use of explosives to sever the conductors and all or some of the platform legs below the seafloor. The use of explosives to either totally remove platforms or to aid toppling-in-place fatally concusses about 85% of all fish associated with the platform, kills virtually 100% of swim bladder fishes, and dislodges about 60% of the attached sessile invertebrates (Scarborough-Bull and Kendall 1994; Gitschlag et al. 2003). More recently, LARP has followed the less biologically destructive practice preferred by the Texas artificial reef program of encouraging partial removal when a platform stands within a predetermined artificial reef area. Partial removal does not require the use of explosives to severe the conductors and jacket

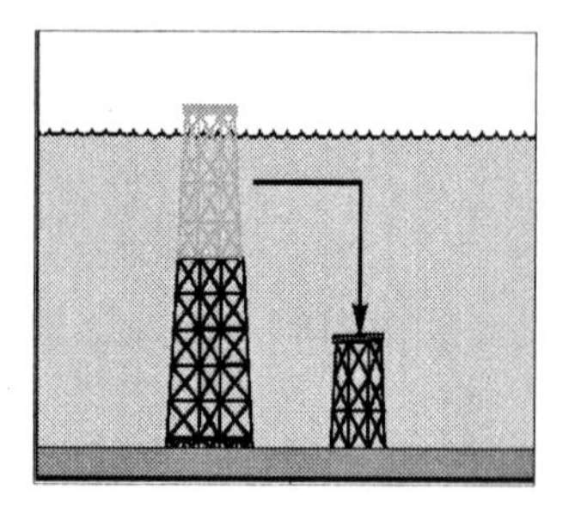

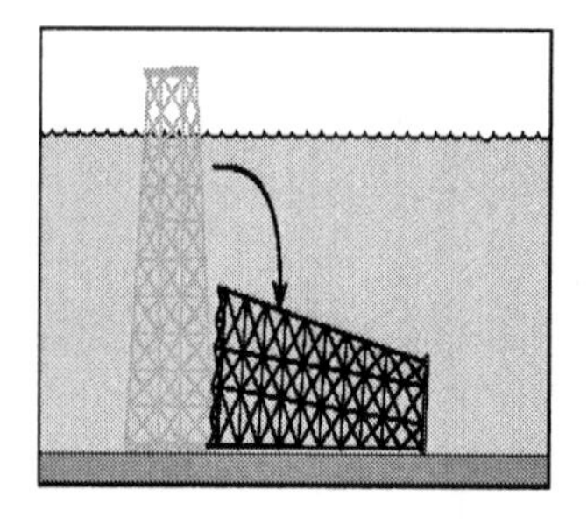

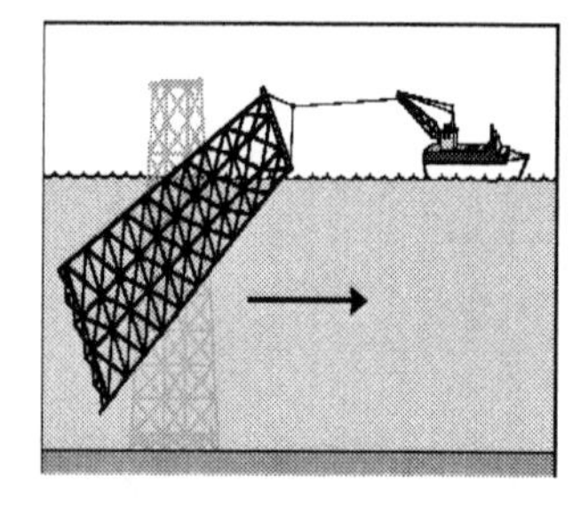

Figure 3. Three methods for standard deployment of offshore platforms as artificial reefs in the Gulf of Mexico.

legs (Schroeder and Love 2004). Partial removal to a depth of 20–30 m below sea level does require that buoys mark the site, a somewhat less expensive Coast Guard regulation when compared to requirements for structures that rise closer to or above the sea surface. There are several dozen predetermined artificial reef areas off Texas, Louisiana, and Mississippi within the region where petroleum platforms are most concentrated (see Figure 4), and partial removal (if possible) is now the preferred method for rigs-to-reefs. Gulf state rigs-to-reefs programs continue to serve the needs of red snapper conservation and the interests of both commercial and sports fishing industries by maintaining habitat that inadvertently increased the carrying capacity of the Gulf for habitat-limited species. For a complete characterization of rigs-to-reefs in the Gulf of Mexico, please refer to Reggio and Kasprzak (1991) and Dauterive (2000).

Man-Made Refugia off California

At present there are 27 offshore platforms in the Southern California Bight, 23 on the OCS and 4 in state waters (Figure 5). Platforms on the OCS are at least 4.8 km from shore and only one stands in water depths less than 50 m. Pacific platforms have individual names that are somewhat capricious but start with a designated letter by zone per U.S. Coast Guard specifications. Off California, seven obsolete platforms have been decommissioned in state waters. Of these, three platforms, Harry (in 1974), Helen (in 1988), and

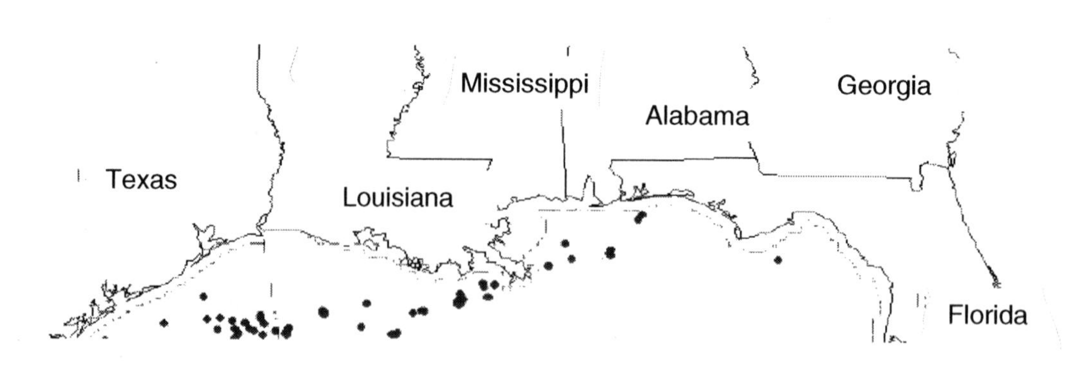

Figure 4. Distribution of artificial reefs created from decommissioned platforms, rigs-to-reefs, in the Gulf of Mexico.

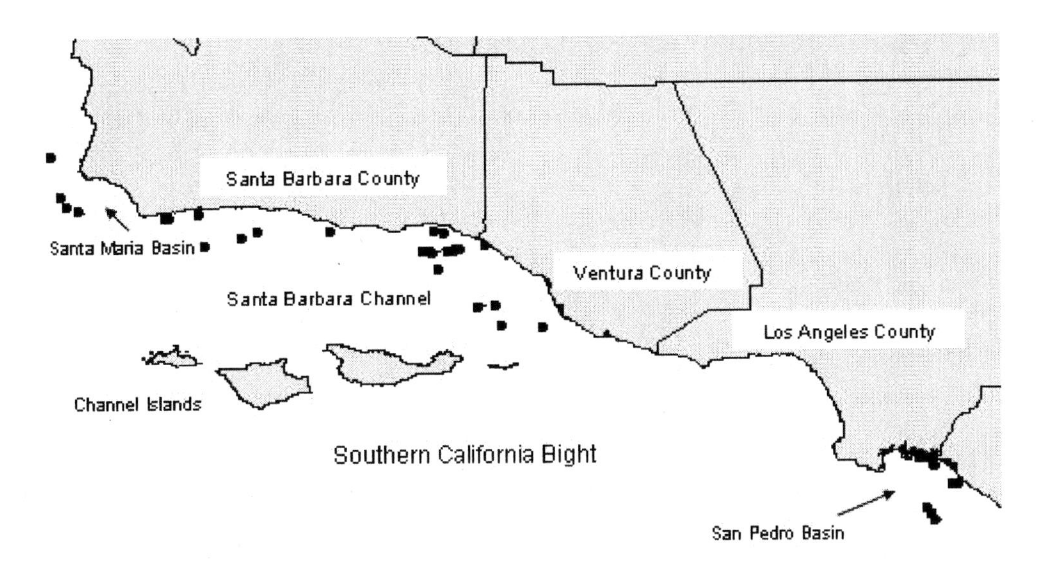

Figure 5. Distribution of petroleum platforms off the Southern California coast.

Herman (in 1988) were totally removed without a great deal of controversy. However, in 1994, late in the planning for the removal of platforms Hilda, Hazel, Hope, and Heidi, a recreational angler's group began to lobby for their retention as reefs. Ultimately, the platforms were totally removed in 1996 (Schroeder and Love 2004).

On the submerged jacket of platforms in the Southern California Bight, the sessile attached biota forms a biotic layer that may become over 0.5 m thick, extending from the intertidal height on platform legs to at least 30 m and deeper. When these encrusting animals are dislodged or broken by storms or die from predation or other natural causes, their shells and remains fall to the seafloor forming a shell mound that covers the substrate surrounding and beneath the platform jacket. Shell mounds may rise 7 m or more above the seafloor and cover over 6 sq km in area, depending upon platform age, seafloor depth, and biogeographic area (which affects species composition of attached invertebrates) (Love et al. 2003; Sea Surveyor Inc. 2003). Together, standing platforms and their shell mounds support large and viable fish communities (Love et al. 1999a, 2003).

Over a 6 year period, Love et al. (2003) studied fish assemblages at Pacific platforms over a wide range of bottom depths, ranging between 29 and 224 m, and sited from north of Point Arguello off central California to Long Beach off Southern California. The majority of the research occurred in the Santa Barbara Channel and Santa Maria basin (see Figure 5). The Santa Barbara Channel and Santa Maria basin are situated in a dynamic marine transition zone between the regional flow patterns of central and Southern California. Except for one instance, three distinct fish assemblages characterized all of the platforms: midwater, base, and shell mound. Overall, rockfishes (Family Scorpaenidae), totaling 42 species, dominated all platform habitats in density and species richness. Fish densities at most platforms were highest in the midwater reflecting the depth preferences of young-of-the-year rockfishes and a few other species. The number of young-of-the-year, in some years, at some platforms was in the hundreds of thousands. Subadult and adult rockfishes and several other species were the most abundant group at the base of platforms. The number of fishes at the platform base ranged from nearly 1,000 to over 12,000. In general, more than 90% of all the fishes around the base of the platforms were rockfishes. Shell mound assemblages were also dominated by rockfishes comprising between 53% and 98% of all fishes. The number of fishes at the shell mounds range up to several thousand (Love et al. 1999a). Depending on depth and size of platform the number of adult fishes (all species) at the studied platforms (all assemblages) ranged up to several tens of thousands (Love et al. 2003).

Important species to the region, such as bocaccio *Sebastes paucispinis*, cowcod *S. levis*, and vermilion rockfish *S. miniatus*, lingcod *Ophiodon elongatus*, and cabezon *Scorpaenichthys marmoratus*, are found in high densities at platforms. Both bocaccio and cowcod have been declared overfished, are economically important, and have been among the most targeted rockfishes in Southern California. They are long-lived apex predators, predominately demersal as adults, with low reproductive rates, and they experience only sporadically successful recruitment that is adversely affected by El Niño and Pacific Decadal Oscillation events. Many of the rockfish resources of the Pacific coast, but especially bocaccio and cowcod, have been depleted by overfishing and poor juvenile recruitment (MacCall 2003; PFMC 2004).

There has been limited recreational fishing at the platforms off California with very little, if any, commercial fishing occurring. The relative amount of fishing pressure among platforms is dependent on ease of access and local ocean conditions. Off California, it is difficult to commercially fish for rockfish and other species around the base of platforms because of strong currents and constant gear loss due to the complex submerged platform structure. Many commercial anglers also believe that platform operators do not welcome fishing vessels near their platforms and that is certainly true for all vessels during periods of heightened security alert. Some of the shallower platforms in the Santa Barbara Channel and further south near harbors and marinas have been important fishing stops for commercial passenger fishing vessels and small private boats (Love and Westphal 1990). In all of these instances, fishing effort was seasonal and directed at surface or midwaters, rather than at the platform base.

Fishing was continuously strong for most of the 20th century on natural reefs in the Santa Barbara Channel and at the Channel Islands, and it is reported that this harvest mortality has had a profound effect on the size and age structure and density of targeted populations of rockfishes (as well as indirect effects on nontargeted species) (Love et al. 2003). High fishing pressure has led to many habitats being almost devoid of large fishes and has drastically altered the species composition of many natural reefs off central and southern California (Yoklavich et al. 2000; Love et al. 2003; Milton Love, unpublished data). Over most moderate-depth and deep reefs off central and Southern California, many, or sometimes all, of the larger predatory fishes, including rockfishes, are gone.

Analysis of rockfish populations at offshore platforms compared to rockfishes observed at a majority of natural reefs off Southern California during the same years and with the same methods and effort reveals that some Pacific platforms harbor higher densities of adult bocaccio and cowcod than all natural reefs surveyed (Love et al. 2003). Some of the differences detected in population size structure, density, and rockfish assemblage structure at platforms may simply reflect the effects of both the recreational and commercial fisheries, rather than differences between habitat types or adequacy (Yoklavich et al. 2000; Love et al. 2003). Thus, the platforms may harbor much of the local and perhaps regional, adult fishes of some depleted species and are likely acting as harvest refugia for demersal fishes, including bocaccio, cowcod, vermilion rockfish, lingcod, and cabezon, and this is a factor in their relatively high densities compared to most natural reefs.

One should not underestimate the role that platforms may play as man-made refugia for the conservation of rockfish at a time when many demersal rockfish populations, including economically important bocaccio and cowcod that are recruitment-limited, are in serious decline on natural reefs. Schroeder and Love (2002) compared the rockfish assemblages at three deeper-water sites subjected to various fishing pressures. Two were natural reefs: one reef was open to both commercial and recreational fishing and one was open only to recreational fishing. The third site was Platform Gail. The reef that allowed both fishing types had the highest densities of rockfishes (7,212 fish/ha); however, the assemblage was completely dominated by dwarf species. A dwarf species is comparatively small when adult, much below the ordinary size of most rockfish species, does not enter into the commercial fishery, and is of lesser interest to recreational anglers. The reef where only recreational fishing was allowed had the lowest rockfish density (423 fish/ha), and this assemblage was also dominated by small fishes. Platform Gail possessed a relatively high density (5,635 fish/ha), and the fishes tended to be larger than individuals at either of the fished sites. Cowcod and bocaccio had 32- and 408-fold higher densities, respectively, at Platform Gail than the recreational site, and 8- and 18-fold higher densities, respectively, at the platform than the recreational and commercial fishing site. Another set of surveys made at a nonfished natural reef further north off central California showed very high densities of large predatory fishes including rockfishes (Yoklavich et al. 2000). Fish assemblages at most of the platforms contain relatively high densities of large predatory and economically important species and low numbers of dwarf species and, thus, more closely resemble the community at a nonfished site than those at many natural reefs.

Platforms on the OCS off California will soon end their service as hydrocarbon producers, and it is very likely that many, if not most, will be removed at the same time (McGinnis et al. 2001). Although their influence on the conservation of rockfish is unintentional, without the opportunity to reef at least some Pacific platforms using partial removal (no explosives), there may be important ecological results from the total removal of platforms off California. Perhaps the most important consequence may be a change in local and regional fish production (the biomass of fish accrued per year) and the composition (i.e., relative abundance) of the regional rockfish assemblages, both of which may in turn influence yields to fisheries (Carr et al. 2003). Polunin and Roberts (1993) note that the total reproductive output of relatively small habitats harboring larger females could possibly be as or more productive than much larger areas harboring smaller or fewer females. That possibility becomes even more important when considering that rockfishes are livebearers that increase fecundity with age, and over 12 rockfish species, including bocaccio and cowcod, produce multiple broods per season (Love et al. 2002). In addition, there is evidence that older, larger female rockfishes produce larvae that withstand starvation longer and grow faster than the offspring of younger fish (Berkeley et al. 2005). The higher number of large, reproductively mature bocaccio and cowcod at the base of Pacific platforms compared to the number at natural reefs indicates that those rockfish adults at platforms likely function as a significant source of biomass and larval production for the local and perhaps regional populations (Love et al. 1999b, 2001, 2003).

We calculated the potential egg output from adult bocaccio and cowcod at the base of five platforms (which includes only the bottom and less than 5 m of the water column above) and the density and potential egg output of adults of the same species at 25 reefs in the Santa Barbara Channel region. The natural reefs were equivalent to a platform base in general location, depth, and vertical relief but were larger in area. At each site, adult densities and size frequencies were calculated. Using known length–fecundity relationship, the potential egg output per individual was estimated. Comparing densities and egg output per individual, we computed the total egg production and potential larval production per square meter at each site. For both bocaccio and cowcod, the mean potential larval production was significantly greater at the platforms (Figure 6).

Rockfish continue to be the subject of intense conservation and regulation as many of the species are severely depleted and reports for the Pacific Fishery Management Council (PFMC) describe bocaccio and cowcod as historically overexploited (MacCall 2003). In January 2001, the National Oceanic and Atmospheric Administration (NOAA) received a petition to consider listing bocaccio under the Endangered Species Act and, although not listed under the act, bocaccio, cowcod, and other rockfish species have been listed as species of special concern (NOAA 2004). In 2002, the PFMC and the state of California began to restrict targeted fishing for these and other rockfish species. In addition, the state of California banned the trawl fishery for spot prawn in April 2003, in order to eliminate bycatch of bocaccio. The PFMC stock assessment for bocaccio and cowcod indicate that even with extremely low harvest rates, the stocks will take several decades to nearly a century, respectively, to rebuild (Butler et al. 2003; MacCall 2003). Although PFMC has not recognized Pacific offshore platforms as essential fish habitat for any fish, they have recognized existing platforms as important habitat for some rockfish and began to consider the creation of artificial reefs from platforms at their March 2004 meeting (Helvey 2002; PFMC 2004). The PFMC considered the designation of offshore platforms as a habitat area of special concern in a proposed alternative within the Draft Environmental Impact Statement, Essential Fish Habitat for Groundfish (PFMC 2005) and voted to recommend this designation for 13 offshore platforms during their June 2005 council meeting.

The debate over reefing decommissioned platforms is mainly political in California but has involved the intermingling of ecological information, economic factors, and polarized preferences, interests, and attitudes associated with OCS oil and gas activity. The Southern California context is much different from that of the Gulf of Mexico experience. Gulf States are willing to accept the rigs-to-reefs alternative and have developed state policy and programs. In California, state legislation that could lead to the creation of a rigs-to-reefs program under the California Department of Fish and Game has been introduced repeatedly and passed once to the governor's desk where it was vetoed in 2001. Although recreational anglers agreed that if platforms were reefed, they could remain nonfished harvest refugia, the legislation was rejected for political reasons with an expression of scientific uncertainty and mixed conclusions among marine ecologists (Carr et al. 2003).

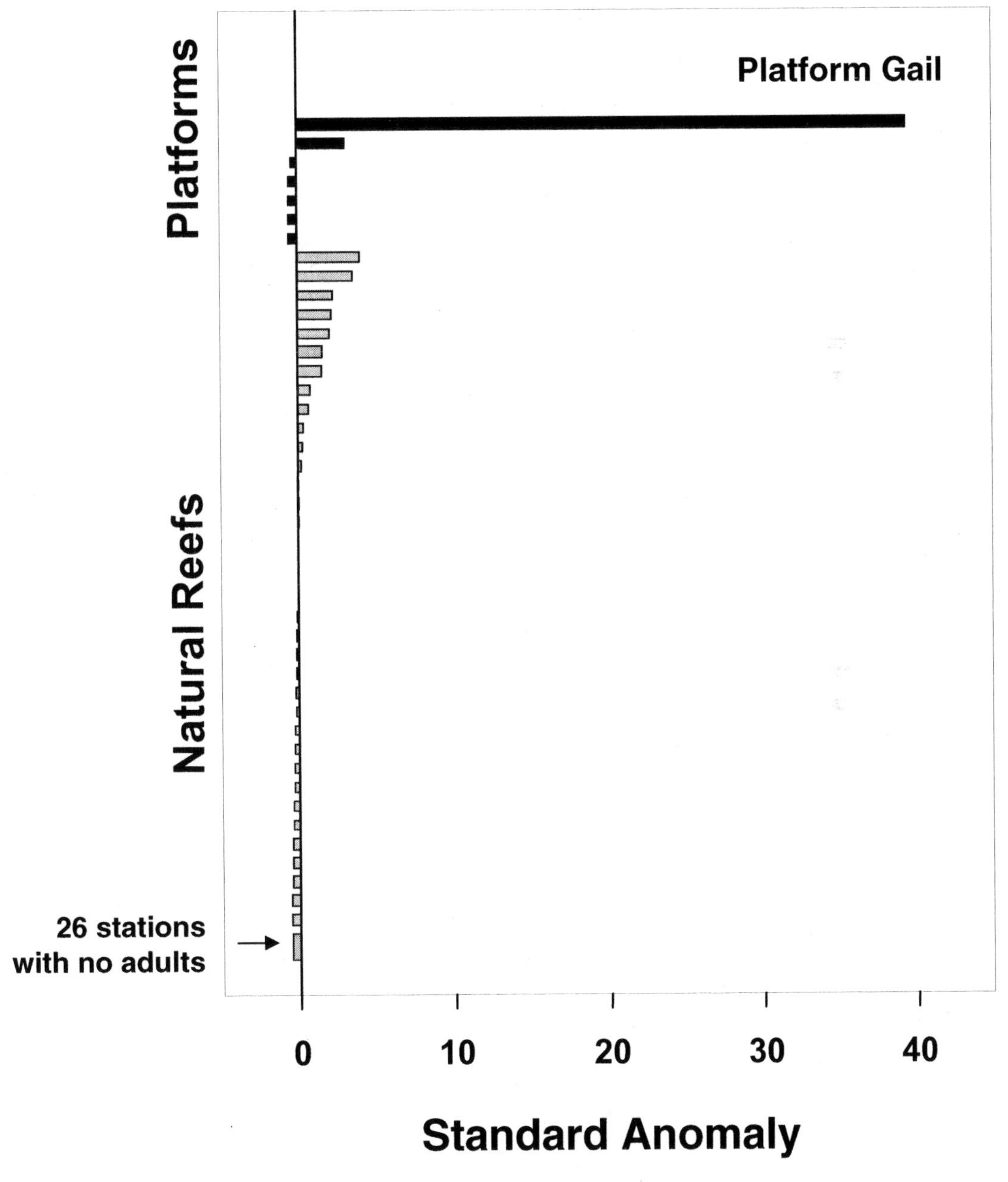

Figure 6. Comparison of mean larval production between platforms and natural reefs for bocaccio and cowcod. Differences between platforms and natural reefs are significant at the < 0.05 level.

Compared to the Gulf States, California has limited experience and limited infrastructure in decommissioning obsolete platforms, especially in deeper water. However, it is likely to be the first region where large deepwater platforms will be decommissioned, and, unlike the Gulf of Mexico, California stakeholder views are highly contentious and divided (see Manago and Williamson 1998). It is expected that several of the Pacific platforms in federal waters will be decommissioned within 10–15 years, and there is no process or politically acceptable arrangement to reef platforms off California. It appears certain that even the consideration of reefing during eventual decommissioning of Pacific platforms will be polarized and controversial. For an historical treatment of the California artificial reef program please refer to McGinnis et al. (2001), and for an indepth consideration of the rigs-to-reefs debate in California, please see Carr et al. (2003) and Schroeder and Love (2004).

Conclusion

The artificial reef complexes created from offshore structures in the Gulf of Mexico and the Southern California Bight are useful as fisheries conservation tools but in different ways. The extensive array of over 4,000 Gulf of Mexico structures raises the carrying capacity of fish species that associate with hard substrate. The offshore structures located in the Southern California Bight act as de facto marine protected areas and provide ecological benefits for severely overfished species, including a higher potential larval production when compared to nearby natural areas that are fished. Thus, the removal of offshore structures should be considered and incorporated into any fisheries management plan, as the removal of these structures will almost certainly have ecological and economic impacts.

Acknowledgments

I thank Robert Shipp for information pertaining to red snapper in the Gulf of Mexico. I also acknowledge Villere Reggio and Lester Dauterive, now retired from the Minerals Management Service. The contents do not necessarily reflect the views and policies of the U.S. Department of the Interior, Minerals Management Service.

References

Berkeley, S. A., M. Hixon, R. J. Larson, and M. S. Love. 2005. Fisheries sustainability via protection of age structure and spatial distribution of fish populations. Fisheries 29(8):23–32.

Bohnsack, J. A., D. L. Johnson, and R. F. Ambrose. 1991. Ecology of artificial reef habitats. Pages 61–108 *in* W. Seaman, Jr. and L. M. Sprague, editors. Artificial habitats for marine and freshwater fisheries. Academic Press, New York.

Butler, J. L., T. Barnes, P. Crone, and R. Conser. 2003. Cowcod rebuilding review. In Status of the Pacific coast groundfish fishery through 2003 and stock assessment and fishery evaluation, volume 1. Pacific Fishery Management Council. National Oceanic and Atmospheric Administration. Available: http://www.pcouncil.org/groundfish/gfsafe0803/gfsafe0803.html. (June 2005).

Carr, M. H., M. V. McGinnis, G. E. Forrester, J. Harding, and P. T. Raimondi. 2003. Consequences of alternative decommissioning options to reef fish assemblages and implications for decommissioning policy. University of California, Marine Science Institute, Coastal Research Center, OCS Study/MMS 2003–053, Santa Barbara.

Dauterive, L. 2000. Rigs-to-Reefs policy, progress, and perspective. U.S. Department of the Interior, Minerals Management Service, OCS Report/MMS 2000–073, New Orleans, Louisiana.

Dimitroff, F. 1982. Survey of snapper and grouper fishermen of northwest Florida coast. Pages 56–60 *in* Proceedings third annual Gulf of Mexico information transfer meeting. Minerals Management Service, New Orleans, Louisiana.

Dokken, Q. 1997. Platform reef ecological and biological productivity: fact or fiction? Pages 13–21 *in* Proceedings sixteenth annual Gulf of Mexico information transfer meeting. Minerals Management Service, OCS Study/MMS 97–0038, New Orleans, Louisiana.

Gallaway, B. J., and G. S. Lewbell. 1982. The ecology of petroleum platforms in the northwestern Gulf of Mexico: a community profile. United States Fish and Wildlife Service, Office of Biological Service, Open-file Report 82–03, Washington, D.C.

Gitschlag, G. R., M. J. Schirripa, and J. E. Powers. 2003. Impacts of red snapper mortality associated with the explosive removal of oil and gas structures on stock assessments of red snapper in the Gulf of Mexico. Pages 83–94 *in* D. R. Stanley and A. Scarborough-Bull, editors. Fisheries, reefs, and offshore development. American Fisheries Society, Symposium 36, Bethesda, Maryland.

GOMFMC (Gulf of Mexico Fishery Management Council). 1996. Reef fish stock assessment panel review of 1996 analysis by Gallaway and Gaze. Prepared by the Reef Fish Stock Assessment Panel. Gulf of Mexico Fishery Management Council, Tampa, Florida.

GOMFMC (Gulf of Mexico Fishery Management Council). 2000. Final report of the Reef Fish Stock Assessment Panel. Prepared by the Reef Fish Stock Assessment Panel. Available: http://www.gulfcouncil.org/downloads/RFSAP.e.1200-final.pdf. (June 2005).

Helvey, M. 2002. Are southern California oil and gas platforms essential fish habitat? ICES Journal of Marine Science 59:S266–S271.

Hiett, R. L., and J. W. Milon. 2002. Economic impact of recreational fishing and diving associated with offshore oil and gas structures in the Gulf of Mexico: final reportU.S. Department of the Interior, Minerals Management Service, Gulf of Mexico OCS Region, OCS Study MMS 2002–010, New Orleans, Louisiana.

Holbrook S. J., R. F. Ambrose, L. Botsford, M. H. Carr, P. T. Raimondi, and M. J. Tegner. 2000. Ecological issues related to decommissioning of California's offshore production platforms. A report to the University of California Marine Council by the Select Scientific Advisory Committee on Decommissioning, University of California, Los Angeles. Available: http://www.ucop.edu/research/ucmc_decommissioning. (June 2005).

LGL Ecological Research Associates, Inc., and Science Applications International Corporation (SAIC). 1998. Cumulative ecological significance of oil and gas structures in the Gulf of Mexico: information search, synthesis, and ecological modeling: phase 1, final report. U.S. Department of the Interior, U.S. Geological Survey, Biological Resources Division, USGS/BRD/CR—1997–0006 ND, Minerals Management Service, Gulf of Mexico Region, OCS Study MMS 97–0036, New Orleans, Louisiana.

Linton, T. L. 1994. A comparison between the fish species harvested in the U.S. Gulf of Mexico and the number of production platforms present. Page 1344 *in* Proceedings of the fifth international conference for aquatic habitat enhancement. Long Beach, California.

Love, M. S., and W. Westphal. 1990. Comparison of fishes taken by a sportfishing party vessel around oil platforms and adjacent natural reefs near Santa Barbara, California. Fishery Bulletin 88:599–605.

Love, M. S., J. Caselle, and L. Snook. 1999a. Fish assemblages on mussel mounds surrounding seven oil platforms in the Santa Barbara Channel and Santa Maria basin. Bulletin Marine Science 65:497–513.

Love, M. S., M. Nishimoto, D. Schroeder, and J. Caselle. 1999b. The ecological role of natural reefs and oil and gas production platforms on rocky reef fishes in southern California. U.S. Department of the Interior, Minerals Management Service and U.S. Geological Survey, OCS Study MMS 99–0015, Camarillo, California and Seattle.

Love, M. S., M. Nishimoto, and D. Schroeder. 2001. The ecological role of natural reefs and oil and gas production platforms on rocky reef fishes in Southern California. . U.S. Department of the Interior, Minerals Management Service and U.S. Geological Survey, Interim Report OCS Study MMS 2001–028, Camarillo, California and Seattle.

Love, M. S., M. M. Yoklavich, and L. Thorsteinson. 2002. The rockfishes of the northeast Pacific. University of California Press, Berkeley.

Love, M. S., D .M. Schroeder, and M. M. Nishimoto. 2003. The ecological role of oil and gas production platforms and natural reefs on fishes in southern and California: a synthesis of information. Final report U.S. Department of the Interior, U.S. Geological Survey, Biological Resources Division, OCS MMS 2003–032, Seattle.

MacCall, A. 2003. Status of bocaccio off California in 2003. In Status of the Pacific coast groundfish fishery through 2003 and stock assessment and fishery evaluation, volume 1. Pacific Fishery Management Council. National Oceanic and Atmospheric Administration. Available: http://www.pcouncil.org/groundfish/gfsafe0803/gfsafe0803.html. (June 2005).

Manago, F., and B. Williamson, editors. 1998. Proceedings: public workshop, decommissioning and removal of oil and gas facilities offshore California: recent experiences and future deepwater challenges. September 1997. U.S. Department of the Interior, Minerals Management Service, Pacific OCS Region, OCS Study/MMS 98–0023, Camarillo.

McGinnis, M. V., L. Fernandez, and C. Pomeroy. 2001. The politics, economics, and ecology of decommissioning offshore oil and gas structures. University of California, Marine Science Institute, Coastal Research Center, MMS OCS Study 2001–006, Santa Barbara.

NRC (National Research Council). 1996. An assessment of techniques for removing offshore structures. National Academy Press, Washington, D.C. Available: http://www.nap.edu/books/NI000141/html. (June 2005).

Nevarez, L., H. Molotch, P. Shapiro, and R. Bergstrom. 1998. Petroleum extraction in Santa Barbara County, California: an industrial history. U.S. Department of the Interior, Minerals Management Service, OCS Study MMS 98–0048, Camarillo, CA.

NOAA (National Oceanic and Atmospheric Administration). 2004. Species of concern (Formerly candidate list). Available: http://www.nmfs.noaa.gov/pr/species/concern (June 2005).

Nieland, D. L., and C. A. Wilson. 2003. Red snapper recruitment to and disappearance from oil and gas platforms in the northern Gulf of Mexico. Pages 73–82 *in* D. R. Stanley and A. Scarborough Bull, editors. Fisheries, reefs, and offshore development. American Fisheries Society, Symposium 36, Bethesda, Maryland.

Polunin, N. V. C., and C. M. Roberts. 1993. Greater biomass and value of target coral-reef fishes in two small Caribbean marine reserves. Marine Ecology Progress Series 100:167–176.

PFMC (Pacific Fishery Management Council). 2004. Artificial reefs in southern California. Briefing book materials. Available: http://www.pcouncil.org/bb/2004/0304/exg3.pdf. (June 2005).

PFMC (Pacific Fishery Management Council). 2005. Draft Environmental Impact Statement, Essential Fish Habitat for Pacific Groundfish. Available: http://www.pcouncil.org/groundfish/gfefheis/gfefheis_doc.html. (June 2005).

Putt, Jr., R. E. 1982. A quantitative study of fish populations associated with a platform within Buccaneer Oil Field, northwestern Gulf of Mexico. Master's thesis. Texas A&M University, College Station.

Reggio, Jr., V. C. 1987a. Rigs-to-Reefs. Fisheries 12(4):2–7.

Reggio, Jr., V. C. 1987b. Rigs-to-reefs: the use of obsolete petroleum structures as artificial reefs. Minerals Management Service, Gulf of Mexico OCS Regional Office, OCS Report/MMS 87–0015, Metairie, Louisiana.

Reggio, Jr., V. C., and R. Kasprzak. 1991. Rigs-to-Reefs: fuel for fisheries enhancement through cooperation. American Fisheries Society, Symposium 11, Bethesda, Maryland.

Scarborough-Bull, A. 1989a. Fish assemblages at oil and gas platforms compared to natural hard/live bottom areas in the Gulf of Mexico. Pages 979–987 *in* Proceedings of the sixth symposium on coastal and ocean management. Charleston, South Carolina.

Scarborough-Bull, A. 1989b. Some comparisons between communities beneath petroleum platforms off California and in the Gulf of Mexico. Pages 47–50 *in* V. C. Reggio, Jr. Petroleum structures as artificial reefs: a compendium. Fourth international conference on artificial habitats for fisheries, Rigs-to-Reefs special session, November 4, 1987, Miami, Florida.

Scarborough-Bull, A., and J. J. Kendall. 1994. An indication of the process: offshore platforms as artificial reefs in the Gulf of Mexico. Bulletin Marine Science 55(2–3):1086–1098.

Schirripa, M. J., and C. M. Legault. 1999. Status of the red snapper in U.S. waters of the Gulf of Mexico updated through 1998. National Marine Fisheries Service, Southeast Fisheries Science Center, Contribution MIA-97/98–5, Miami.

Schroeder, D. M., and M. S. Love. 2002. Recreational fishing and marine fish population in California. CalCOFI Report 43:182–190.

Schroeder, D. M., and M. S. Love. 2004. Ecological and political issues surrounding decommissioning of offshore oil facilities in the Southern California Bight. Ocean and Coastal Management 47:21–48.

Seaman, W., and L. M. Sprague. 1991. Artificial habitats for marine and freshwater fisheries. Academic Press, San Diego, California.

Sea Surveyor, Inc. 2003. An assessment and physical characterization of shell mounds associated with outer continental shelf shell mounds located in the Santa Barbara Channel and Santa Maria basin, California. U.S. Department of the Interior, Minerals Management Service, final report, MMS Contract No. 1435–01-02-CT-85136, Camarillo, California.

Sonnier, F., J. Teerling, and H. D Hoese. 1976. Observations on the offshore reef and platform fish fauna of Louisiana. Copeia 1:105–111.

Stanley, D. R., and C. A. Wilson. 1990. A fishery dependent based study of fish species composition and associated catch rates around petroleum platforms off Louisiana. Fishery Bulletin 88:719–730.

Stanley, D. R., and C. A. Wilson. 1991. Factors affecting the abundance of selected fishes near oil and gas platforms in the northern Gulf of Mexico. Fishery Bulletin 89:149–159.

Stanley, D. R., and C. A. Wilson. 1996. The use of hydroacoustics to determine abundance and size distribution of fishes associated with a petroleum platform. ICES Journal of Marine Science 202:473–475.

Stanley, D. R., and C. A. Wilson. 1997. Seasonal and spatial variations in the abundance and size distribution of fish associated with a petroleum platform in the northern Gulf of Mexico. Canadian Journal of Fisheries and Aquatic Sciences 54:1166–1176.

Stanley, D. R., and C. A. Wilson. 2000. Seasonal and spatial variation in the biomass and size frequency distribution of the fish associated with oil and gas platforms in the northern Gulf of Mexico. Prepared by the Coastal Fisheries Institute, Center for Coastal, Energy, and Environmental Resources, Louisiana State University. U.S. Department of the Interior, Minerals Management Service, Gulf of Mexico Region, OCS Study/MMS 2000–005, New Orleans, Louisiana.

Stanley, D. R., and C. A. Wilson. 2003. Seasonal and spatial variation in the biomass and size frequency distribution of fish associated with oil and gas platforms in the Northern Gulf of Mexico. Pages 123–154 *in* D. R. Stanley and A. Scarborough Bull, editors. Fisheries, reefs, and offshore development. American Fisheries Society, Symposium 36, Bethesda, Maryland.

Stephan C. D., B. G. Dansby, H. R. Osburn, G. C. Matlock, R. K. Riechers, and R. Rayburn. 1990. Texas artificial reef fishery management plan. Fisheries management plan series number 3. Texas Parks and Wildlife Department, Coastal Fisheries Branch, Austin.

Wilson, C. A., V. R. Van Sickle, and D. L. Pope. 1987. Louisiana artificial reef plan. Louisiana Department of Wildlife and Fisheries, Bulletin 41, Baton Rouge.

Witzig, J. 1986. Rigs fishing in the Gulf of Mexico – 1984 marine recreation fishing survey results. Pages 103–105 *in* Proceedings sixth annual Gulf of Mexico information transfer meeting. Minerals Management Service, OCS Study/MMS 86–0073, New Orleans, Louisiana.

Yoklavich, M. M., H. G. Greene, G. M. Calliet, D. E. Sullivan, R. N. Lea, and M. S. Love. 2000. Habitat associations of deep-water rockfishes in a submarine canyon: an example of a natural refuge. Fishery Bulletin 98(3):625–641.

American Fisheries Society Symposium 49:917–924

Coupling Fisheries with Ecology through Marine Artificial Reef Deployments

Stephen A. Bortone*,[1]

*Marine Laboratory, Sanibel-Captiva Conservation Foundation
900A Tarpon Bay Road, Sanibel, Florida 33957, USA*

Over the past several decades, there have been a host of conferences and compilations on the applications of artificial reefs (e.g., Colunga and Stone 1974; Buckley et al. 1985; D'Itri 1985; Seaman et al. 1989; Nakamura et al. 1991; Seaman and Sprague 1991; Secretaria de Pesca 1992; Grove and Wilson 1994; Sako and Nakamura 1995; Cenere and Relini 1995; Jensen et al. 2000; Seaman 2000; Jensen 2002; Cowan 2005, unpublished). The often cited applications are benefits to the fishing community (i.e., recreational, commercial, artisanal, etc.; aquaculture enhancement, intentional and unintentional habitat protection through trawl inhibition, and controls on fishing mortality as well as scientific research). Occasionally cited are the application of artificial reefs to conservation biology and habitat restoration and mitigation, but these appear to be overstatements as they are ancillary applications rather than the chief objectives.

Artificial reefs can have applications to the nonfishing community and these include recreational scuba diving, submarine tourism, environmental education, and the internment of human remains as memorials. While the benefits of artificial reefs have been heralded, conspicuously absent is their lack of inclusion in the management programs directed toward depleted or overfished fisheries (Bortone 2006). This is the situation in the United States, especially along its southeast Atlantic coast, including the Gulf of Mexico. The lack of inclusion of artificial reefs in fisheries management plans is significant given the large number of reefs deployed along these coastal areas and the assumption by the general public that the reefs are being used to enhance fishery resources.

It is the thesis of this paper that, while artificial reefs are often touted as having application to fisheries management, their specific incorporation into management programs has yet to be realized. Future success of this application depends on understanding the ecological role that artificial reefs can play in enhancing life history attributes and eventual survivorship of carefully-targeted, artificial reef-associated species.

Basic Artificial Reef Ecology

Foremost, artificial reefs serve to attract fishes from one area to another. Arguments have also been put forth suggesting that artificial reefs can enhance overall production (i.e., increasing the standing biomass in a given area) of

*Corresponding author: sbortone@umn.edu

[1] Current address: University of Minnesota, Minnesota Sea Grant College Program, 211 Washburn Hall, 2305 East Fifth Street, Duluth, Minnesota 55812, USA

fishes (e.g., Bohnsack et al. 1997), but these hypotheses have been difficult to test and more investigation into this feature of the functional role of artificial reefs remains (Bortone 1998). Nevertheless, the adage "if you build it, they will come" rings true for most artificial reef efforts to date. Numerous studies have documented species colonization and succession on newly deployed artificial reefs (e.g., Bortone et al. 1994). Thus, a structure, once deployed, inevitably becomes inhabited by fish.

The driving mechanism behind this adage of artificial reef ecology remains unknown but may have to do with the natural, yet generally unpredictable, dispersal of reef fish larvae, juveniles, and adults. While predicting that fish species will inhabit an artificial reef is straightforward (and almost trivial), predicting the size, abundance, and species composition of the associated assemblage is less certain. Due to the vagaries of reef colonization, the best that scientists may do is predict the structure and abundance of colonizing fishes by their respective ecological groups or feeding guilds.

Classifying fishes according to the ecological role that they play in the trophodynamics of the artificial reef biotope may prove useful for management purposes. In a study along the northern Gulf of Mexico, Nelson and Bortone (1996) were able to group 25 fish species into seven feeding guilds based on their food habits. These groups included: lower structure pickers (balistids, labrids), ambush predators (batrachoidids), lower structure crustacean predators (sciaenids), upper structure pickers (sparids), upper structure predators (haemulids), water column pickers (pomacentrids), and reef-associated, open-water predators (carangids, lutjanids).

Grouping fishes according to their trophic role on artificial reefs can assist managers and researchers in establishing generalized patterns that become more certain than when dealing with complex and highly diverse species assemblages encountered in many areas. The number of feeding guilds can vary from reef to reef, but distilling the plethora of species into meaningful ecological subunits gives direction to their application toward fisheries management issues.

Nakamura (1985) classified fishes relative to their spatial position on a reef. For example, A-type fishes are found proximate to, or inside holes and crevices on the reef (blenniids, gobiids); B-type species are found closely associated with the reef but not in direct contact with it (lutjanids, serranids); C-type species are loosely associated with the reef, often schooling above the reef (carangids, clupeids). Figure 1 depicts an expansion of this idea by adding one additional position type. D-type species have similar needs to C-type species, but they are found on or above the substrate surrounding the reef (bothids, synodontids). Thus, the sphere of attraction includes species associated with the substrate upon which the reef sits.

Regardless of how fish are classified (functional feeding group or behaviorally mediated position—neither of which is mutually exclusive of the other) the ecological impacts of artificial reef deployments are important when including them as part of a larger-scale management effort. Food studies conducted by several investigators have indicated that the preponderance of food energy consumed by artificial reef-associated fishes comes, not directly from the structure itself, but from their foraging activity that takes place away from the reef (Bortone et al. 1998). Most artificial reef-associated fishes apparently make feeding forays out from the reef, presumably on a daily basis. The affect that foraging fishes have on the local trophic structure has been confirmed

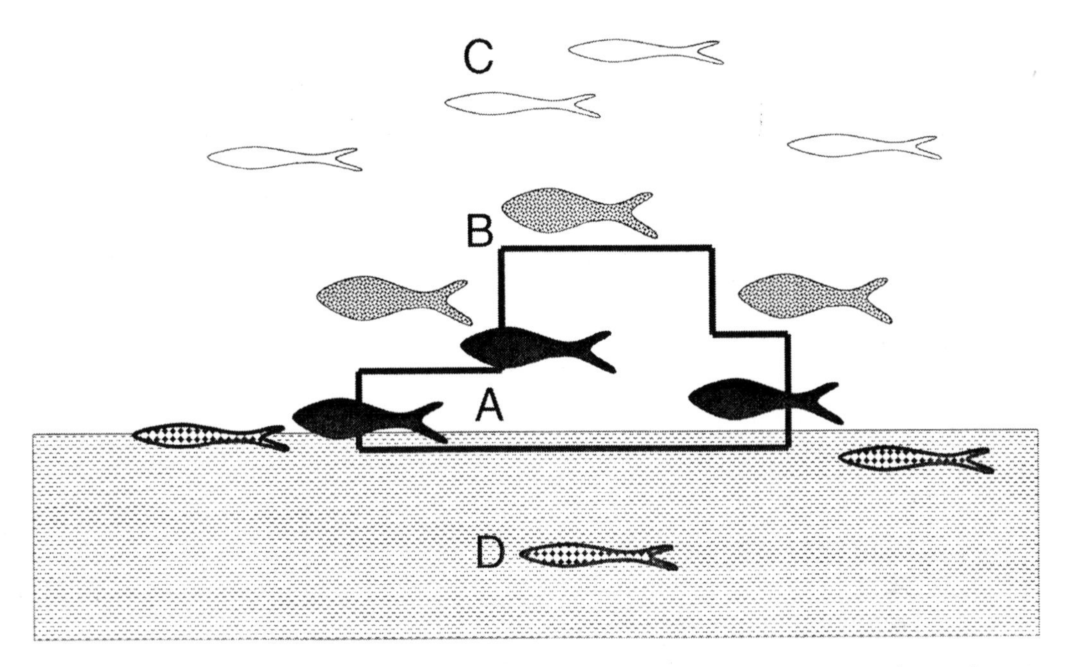

Figure 1. Classification of artificial-reef fishes based on their position relative to the reef: A-type species are directly on the reef or inside reef holes and crevices; B-type species are found directly around the reef but do not make contact with it; C-type species are found above the reef; D-type species are found on or over the substrate next to the reef. Modified from Nakamura (1985).

by studies that indicate organisms comprising benthic communities are depleted near the reef but gradually increase to the in situ biomass of background (i.e., nonreef associated bottom) areas (Figure 2; see also Lindquist et al. 1994; Bortone et al. 1998). The implications of the affects that artificial reefs have on local benthic communities are important yet not often considered in the deployment strategy.

A Model for Targeting Artificial Reef Fish Species

Artificial reef projects, when part of a fishery management program, should target species that are more or less preadapted to this biotope. Concomitantly, fisheries managers should avoid targeting species that have a low probability of benefiting from the structure. Figure 3 depicts a generalized model indicating high production and low attraction favors A-Type species, as these species are limited by reef substrate. Conversely, C- and D-Type species will only be attracted to the reef and will add little to overall reef productivity. Indeed, their attraction may lead to increased fishing efficiency and increased fishing mortality (Polovina 1991). For eastern North America, species targeted by most artificial reef fisheries are B-type species (lutjanids and serranids). These species are attracted to reefs to a limited degree but also gain some production benefit from the reef platform. B-type species may range from high attraction and production to low attraction and production. Species that more closely emulate the attraction and production features of B''-type species (i.e., higher attraction and higher production) would clearly be preferred over B'-type species (i.e., lower attraction and lower production), as seen in the model (Figure 3).

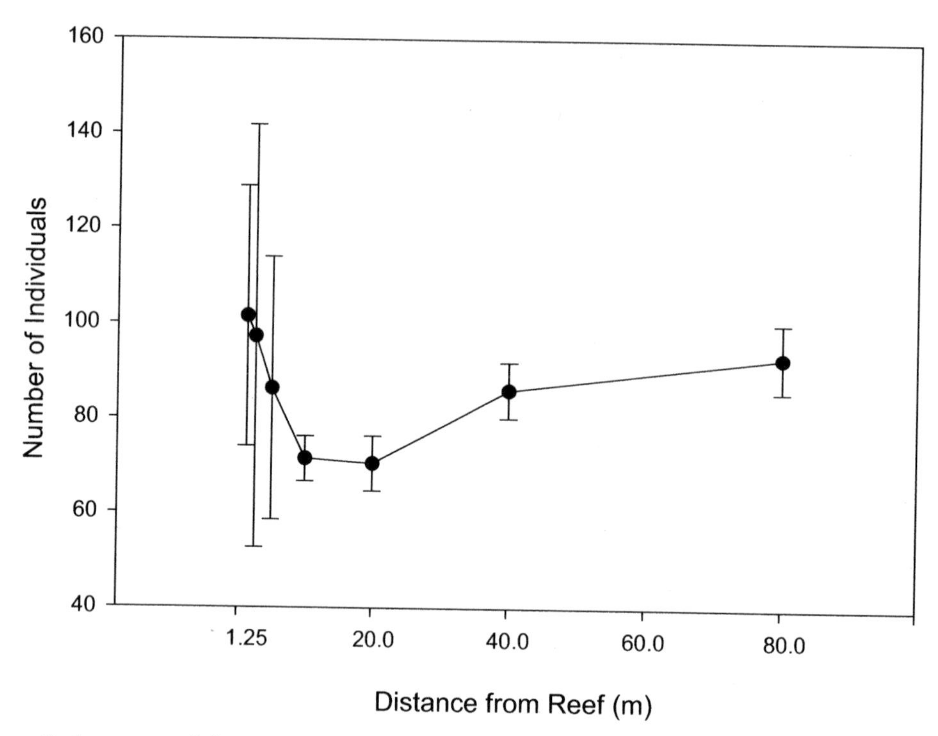

Figure 2. Summary of data presented in Bortone et al. (1998) indicating the decline and recovery of potential forage organisms as function of their distance from artificial reefs.

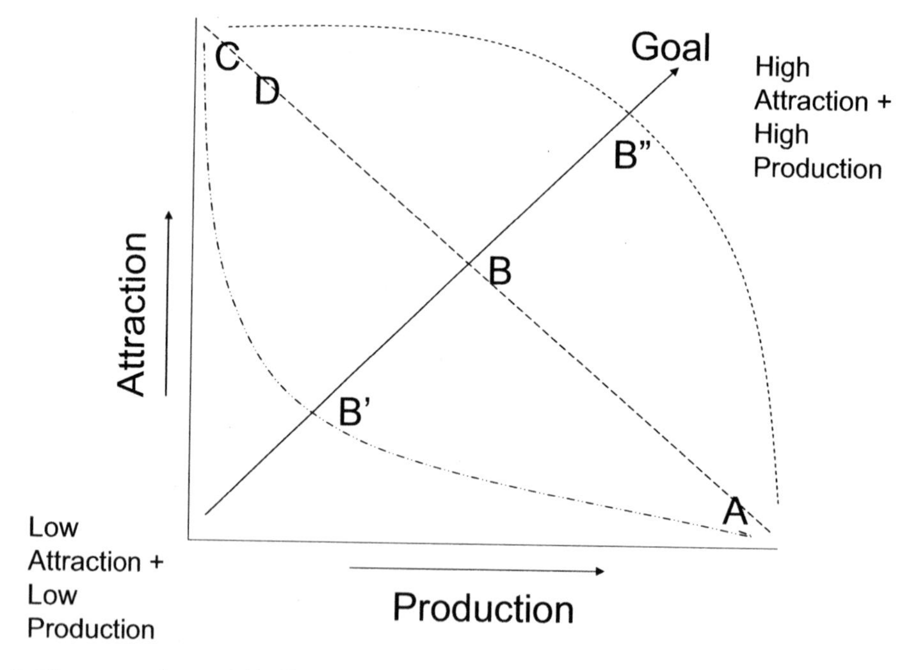

Figure 3. Diagrammatic model indicating the relative relationship of A-, B-, C- and D-type species with regard to their association with attraction versus production. The diagonal line from the origin indicates a position that species would occupy if they are balanced for attraction and production. B′ and B″ indicate the position of B-type species with varying life history strategies. The ideal target species for artificial reefs are B″-type species that respond positively to both attraction and production features of reefs.

Thus, eastern North American fisheries managers should carefully design, construct and orient artificial reefs to enhance the life history requirements of B"-type species, rather than B'-type species.

Current Role of Artificial Reefs in Fisheries Management

To date, the Caribbean spiny lobster *Panulirus argus* fishery is probably the best example of a situation where artificial reefs are incorporated into management, even if only informally by the community of commercial fishers. While the use of the *casita* in the Caribbean spiny lobster fishery is discussed in another section (Herrnkind and Cobb 2007, this volume), it is important to note that the successful application of artificial reefs in this fishery is because of the direct role that the structure plays in reducing natural mortality, especially during the juvenile stage when Caribbean spiny lobsters are particularly vulnerable to predation.

In the Gulf of Mexico, the red grouper *Epinephelus morio* is a species of concern that, along with several other hard-bottom associated fishes (including other lutjanid and serranid fish species), is regulated at both state and federal levels. Habitat requirements of red grouper are presumably more structurally specific than other grouper species. This assumption prompted local natural resource managers to deploy artificial reefs to provide a low-relief habitat that resembles a natural offshore rock-ledge reef, sparsely found in the area, but apparently preferred by adult red grouper (Chris Koepfer, Lee County [Florida] Division of Natural Resources, personal communication). This project has only recently been initiated and so far no data are available indicating the success of this project.

Another planned project in the Gulf of Mexico is the deployment of artificial reefs to enhance specific life history features of another grouper species, the gag *Mycteroperca microlepis*. The reef configuration and deployment plan is specifically designed to facilitate growth rates among gag. Growth rates are apparently greater among gag found on smaller, widely scattered reefs as opposed to larger, densely placed reefs. Also, a demographic bottleneck occurs during the transition from juvenile to adult life stages when the juveniles move from inshore grassbeds to offshore, hard-bottom biotopes (Lindberg and Relini 2000). This project, while still in the early stages of implementation, is directed toward enhancing two specific life history requirements of this B-type, artificial reef-associated species that has been influenced by overfishing.

A study directed toward facilitating colonization is being investigated in several areas of the Gulf of Mexico by the Mote Marine Laboratory in cooperation with the Gulf Coast Research Laboratory and the Oceanic Institute. The research plan is to raise red snapper *Lutjanus campechanus* under hatchery conditions and then directly deploy the fish onto artificial reefs. One objective is to evaluate the potential role that stock enhancement can play in artificial reef management of another B-type species of concern to regulatory agencies in the Gulf of Mexico (Kenneth M. Leber, Mote Marine Laboratory, personal communication). The deployments are being conducted at different densities as it is not known if density dependent conditions in snapper are related to fitness and eventual management success. Increasing colonization success through deploying hatchery-reared fish on artificial reefs is an interesting and logical combination of these two historically independent fishery management options.

The Future of Artificial Reefs as Part of Fishery Management Programs

As a benefit to a fishery, artificial reefs can help reduce both natural and fishing mortality, increase survivorship, increase fitness, and provide essential fish habitat. More specifically, artificial reefs can be used to overcome habitat-limiting, species-specific life history requirements. Artificial reefs should be designed, oriented, and deployed to increase a species' overall fitness at the larger scale of species management. At a lower scale, enhancing species-specific growth, reducing natural mortality, and facilitating social interactions are testable objectives for the roles that artificial reefs can play in fisheries management. These roles can include enhancing colonization from nursery to adult stock areas and facilitating directional movement from one area to another. For example, fish may experience reduced mortality when reefs are available as stepping stones or way stations in the midst of an area void or depauperate of optimal habitat when moving into (or out of) a marine protected area. Within the context of the marine protected area (MPA), regulated time and space-limited fishing could be allowed within an MPA enhanced with artificial reefs. These conceptually-modified MPAs then could serve to enhance fishing on a time-and-space, rotational basis (Figure 4). Fishing on a rotational basis could allow natural or artificial habitats and the associated fish community relief from fishing pressure and habitat interference. Currently, most MPAs exclude fishing, however, the allowance of limited fishing (with controls on time and space) might be an approach that would make the addition of more MPAs more politically palatable.

Facilitating colonization through artificial reef design and deployment may yet play a vital role in facilitating fisheries management. Reefs should be designed to facilitate colonization. Additionally, artificial reefs should be designed to enhance the retention of artificial reef-associated fish species–particularly those with B-type position requirements. Placement of reefs in proximity to each other or to other natural features may affect retention of fishes once recruited

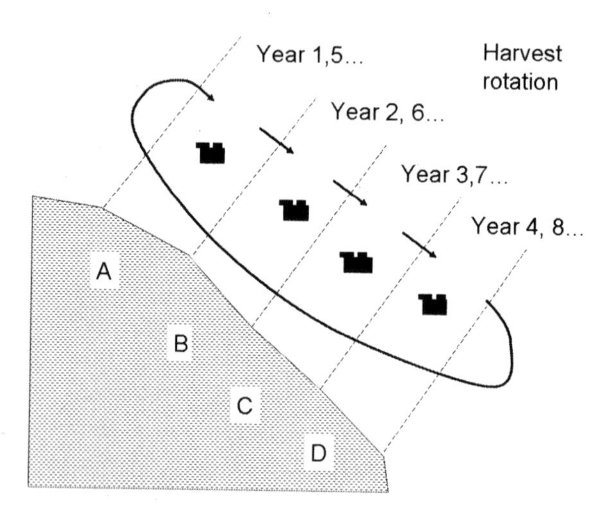

Figure 4. Diagrammatic plan for harvest rotation on time/space-limited artificial reefs within marine protected areas. Reefs in zones A to D are sequentially fished every 5 years in rotation. The reefs serve as no-take marine protected areas during the interim periods.

to the structure. Consequently, the application of ecological, founder theory may be of special advantage to fisheries managers when considering artificial reef design.

Conclusion

Artificial reefs should target species (or species guilds) that would benefit most from deployments. Investigations are necessary to determine the behavioral and life history features that would be enhanced by the addition of artificial reefs. Concomitantly, artificial reef design and deployments should accommodate species' habitat and life history requirements. Thus, reefs should be designed to the advantage of some target species while other target species should be chosen that have limiting features that artificial reefs can supplement.

Acknowledgments

I thank Chris Koepfer, Kenneth Leber, and William Lindberg for discussions regarding their ongoing research projects. I also thank William Seaman, Jr. for his continued support in developing the science behind artificial reef applications. Michael Hannan and the anonymous reviewers offered constructive criticism that improved the manuscript.

Reference

Bohnsack, J. A., A. M. Eklund, and A. M. Szmant. 1997. Artificial reef research: is there more than the attraction-production issue? Fisheries 22(4):14–16.

Bortone, S. A. 1998. Resolving the attraction-production dilemma in artificial reef research: some yeas and nays. Fisheries 23(3):6–10.

Bortone, S. A. 2006. A perspective of artificial reef research: the past, present, and future. Bulletin of Marine Science 78:1–8.

Bortone, S. A., R. P. Cody, R. K. Turpin, and C. M. Bundrick. 1998. The impact of artificial reef fish assemblages on their potential forage area. Italian Journal of Zoology 65(Supplement):265–267.

Bortone, S. A., T. R. Martin, and C. M. Bundrick. 1994. Factors affecting fish assemblage development on a modular artificial reef in a northern Gulf of Mexico estuary. Bulletin of Marine Science 55:602–608.

Buckley, R. J., J. Grant, and J. Stephens, Jr. 1985. Third international artificial reef conference 3–5 November 1983, Newport Beach, California – Forward. Bulletin of Marine Science 37(1):1–2.

Cenere, F., and G. Relini. 1995. Conegno di Loano per la difesa del mare (8–9 Iuglio 1994). Biologia Marina Mediterranea 2(1):iii.

Colunga, L., and R. Stone, editors. 1974. Proceedings of an international conference on artificial reefs. March 20–22, 1974. Houston, Texas. Texas A & M Sea Grant College, TAMU-SG-74–103, College Station.

D'Itri, F. M., editor. 1985. Artificial reefs. Marine and Freshwater Applications. Lewis Publishers, Inc., Chelsea, Michigan.

Grove, R. S., and C. A. Wilson. 1994. Introduction. Bulletin of Marine Science 55(2):265–267.

Jensen, A. C., editor. 2002. Seventh international conference on artificial reefs and related aquatic habitats. ICES Journal of Marine Science – Journal du Conseil 59 (Supplement).

Jensen, A. C, K. J. Collins, and A. P. M. Lockwood, editors. 2000. Artificial reefs in European seas. Kluwer Academic Publishers, Dordrecht, The Netherlands.

Lindberg, W. J., and G. Relini. 2000. Integrating evaluation into reef project planning. Pages 195–235 *in* W. Seaman, editor. Artificial reef evaluation: with application to natural marine habitats. CRC Press, Boca Raton, Florida.

Lindquist, D. G., L. B. Cahoon, I. E. Clavijo, M. H. Posey, S. K. Bolden, L. A. Pike, S. W. Burk, and P. A. Cardullo. 1994. Reef fish stomach contents and prey abundance on reef and sand substrata associated with adjacent artificial and natural reefs in Onslow Bay, North Carolina. Bulletin of Marine Science 55:308–318.

Nakamura, M. 1985. Evaluation of artificial reef concepts in Japan. Bulletin of Marine Science 37:271–278.

Nakamura, M., R. S. Grove, and C. J. Sonu, editors. 1991. Recent advances in aquatic habitat technology. Japan–U.S. Symposium on Artificial Habitats for Fisheries Proceedings, June 11–13, 1991. Tokyo, Japan. Southern California Edison Co., Rosemead, California.

Nelson, B. D., and S. A. Bortone. 1996. Feeding guilds among artificial-reef fishes in the northern Gulf of Mexico. Gulf of Mexico Science 14(2): 66–80.

Polovina, J. J. 1991. Fisheries applications and biological

impacts of artificial reefs. Pages 153–176 *in* W. Seaman, Jr. and L. M. Sprague, editors. Artificial habitats for marine and freshwater fisheries. Academic Press. San Diego, California.

Sako, H., and M. Nakamura, editors. 1995. ECOSET'95: International Conference on Ecological System Enhancement Technology for Aquatic Environments. Volumes I-III. International Marine Science and Technology Federation, Tokyo.

Seaman, W., Jr., and L. M. Sprague (editors). 1991. Artificial habitats for marine and freshwater fisheries. Academic Press, Inc., San Diego, California.

Seaman, W., Jr., editor. 2000. Artificial reef evaluation: with applications to natural marine habitats. CRC Press. Boca Raton, Florida.

Seaman, W. Jr., R. M. Buckley, and J. J. Polovina. 1989. Advances in knowledge and priorities for research, technology and management related to artificial aquatic habitats. Bulletin of Marine Science 44(2):527–532.

Secretaria de Pesca. 1992. I reunion international sobre mejorameinto de habitats acuaticos para pesquerias (arrecifes artificiales). Instituto Nacional de la Pesca, Centro Regional de Investigacion Pesquera en Manzanillo, Colima. 29–30 de Octubre de 1992. Manzanillo, Colima, Mexico.

American Fisheries Society Symposium 49:925–932

Artificial Shelters for Clawed and Spiny Lobsters: A Critical Review of Enhancement Efforts

William F. Herrnkind*
Department of Biological Science, Florida State University, Tallahassee, Florida 32306, USA

J. Stanley Cobb
Department of Biological Sciences, University of Rhode Island, Kingston, Rhode Island 02881, USA

Abstract.—Efforts by fishers to increase spiny and clawed lobster numbers for easier, expanded harvest by supplementing sheltering structures probably predate formal fishery studies. Harvest of natural lobster populations has increased to an apparent worldwide maximum, even as some populations have oscillated or declined from causes ranging from overfishing to disease. Yet most past enhancement attempts, whether based on addition of artificial habitat or on stocking, have failed or produced uncertain effects. Applicable knowledge about recruitment ecology has improved in the last two decades as a result of long-term, experimentally competent analyses of both enhancement attempts and underlying ecological processes. We review and highlight key habitat-enhancement efforts in light of basic research findings that explain probable reasons for success or failure.

Introduction

The term lobster applies to members of the Palinuridae (spiny and slipper lobsters), Nephropidae (clawed lobsters), and Thalassinidae (mud lobsters). Lobsters collectively constitute an annual fishery catch exceeding 200,000 metric tons (mt) (Lipcius and Eggleston 2000), primarily of spiny and clawed lobsters. Most lobster fisheries operate at or above maximum sustainable yield, and a few species have suffered severe declines from damaging natural environmental events (South African rock lobster *Jasus lalandii* in Namibia; Grobler and Noli-Peard 1997) or recruitment overfishing (European lobster *Homarus gammarus* in Norway, van der Meeren 2003). High demand for lobster drives high fishing effort, often resulting in declining harvest, and has long impelled efforts to increase lobster numbers artificially. Numerous attempts have been made to "enhance" lobster populations through the addition of hatchery-reared young or the addition of sheltering habitat in the form of crevice-bearing artificial and natural materials.

We focus here on artificial sheltering structures for spiny and clawed lobsters, but culturing and stocking with juveniles also bears mention because the two can potentially be combined to achieve greater effect. Furthermore, the underlying recruitment processes that establish and maintain populations require understanding of the interaction between the numbers of settling larvae and postsettlement sheltering effects on survival (Butler and Herrnkind 2000; Wahle 2003).

*Corresponding author: herrnkind@bio.fsu.edu

Life History and Ecology

The palinurid lobsters are mainly distributed in tropical and subtropical shallow waters worldwide, whereas commercially important nephropids are found in temperate waters of the North Atlantic. Most species occupy coastal habitats of less than 100 m depth, although Norway lobster *Nephrops norvegicus* is found on mud bottoms at considerably greater depths The American lobster *Homarus americanus* occurs from coastal waters to the shelf break from New Jersey, USA, to Newfoundland, Canada, whereas European lobster, the European clawed lobster, ranges coastally from Spain to Great Britain to northern Norway. Both have a brief planktonic larval phase in coastal waters before settling. The settlement and early benthic stages are primarily found in cobble habitat and thought to be shelter-bound. Juveniles and adults of both species typically occupy rock crevices by day and emerge at night to forage or migrate.

Numerous palinurid species across three genera are heavily fished; targets of the largest fisheries are Caribbean spiny lobster *Panulirus argus* of the Caribbean and western Atlantic and western rock lobster *P. cygnus* of western Australia, on the southeastern Indian Ocean. Western Atlantic Caribbean spiny lobster is fished from Bermuda to southern Brazil (there as a subspecies); its range is one of the largest of any benthic crustacean species and is apparently facilitated by its 6–9-month-long planktonic larval phase and the circulation patterns of western Atlantic currents. Postlarvae typically settle into vegetated or crevice-rich substrates and thereafter aggregate in dens through most of benthic life.

The similarities and differences in life histories and behavior provide insight into the rationale for the different directions of enhancement attempts in the two taxa. Hatchery rearing and stocking have dominated in *Homarus* enhancement efforts, whereas supplementing shelter is the primary method for enhancing *Panulirus*. In both taxa, the early benthic phases have myriad predators, mainly fishes. As they grow larger, both types gain some refuge from predators (Wahle and Steneck 1991, 1992; Smith and Herrnkind 1992). For recruitment at the population level, shelter is most critical to the vulnerable early benthic phases, of which predation mortality is extremely high, estimated at 96–99% in the first year after settlement in Caribbean spiny lobster in Florida (Herrnkind et al. 1997). Estimates of mortality in European lobster are considerably lower (Bannister and Addison 1998) providing a possible biological basis for different approaches to population enhancement.

The two most important differences between the two taxa are in duration of the larval period (weeks in nephropids, months to a year in palinurids) and sociality (palinurids are congregative, *Homarus* is solitary and aggressive). Because *Homarus* is easily reared, it can be cultured and stocked in large numbers. Lower early-benthic-phase mortality suggests stocking is more likely to be successful in *Homarus* than in *Panulirus*. Among constraints is the antisocial, cannibalistic behavior of clawed lobsters, which necessitates raising them solitarily. Although methods are well known, culturing of clawed lobsters to market size remains expensive and has not yet relieved pressure on the natural stocks. In contrast, spiny-lobster culture at the commercial level remains experimental, but shelter sharing by crowded palinurids facilitates application of artificial structures to attract them for both harvesting and, theoretically, increasing recruitment.

Clawed-Lobster Shelter Enhancement

Habitat supplementation for clawed lobsters has been attempted several times. In 1965, a reef of sandstone rocks up to 1 m in size was constructed in an area of sand and small cobble in Northumberland Strait, Canada (Scarratt 1968). The premise behind construction of the reef was that shelter was limiting in the area and that vagrant lobsters would populate the new artificial structure. Colonization was slow in the first year, but in the subsequent 6 years, the biomass of lobsters there exceeded that in nearby natural areas (Scarratt 1968, 1973). An artificial reef of concrete shelters specially designed to allow aggressive clawed lobsters to space themselves was installed in Rhode Island, USA (Sheehy 1976). A year later, the reef housed lobsters ranging from postlarvae to egg-bearing females at a biomass higher than that on nearby good lobster ground. In this case, presence of newly settled benthic juveniles suggested that recruitment to the local population had been increased. The utility of combining the disposal of solid waste and construction of artificial reefs was explored in the UK. Pulverized fuel ash was compacted into small bricks and placed in a 10-m by 30-m array of 8 conical 4-m by 1-m units on sandy substrate at 10 m depth in Poole Bay, England. Initial colonization of this reef complex was by large lobsters, but later small lobsters were seen, suggesting postlarval settlement on the reef (Jensen et al. 2000).

After an oil spill in 1989 in Narragansett Bay, Rhode Island, USA, an artificial reef was designed for mitigation of the loss of lobster population. A research program was undertaken that was intended to measure the impact of the new habitat on the local lobster population. Castro and associates (Castro et al. 2001) placed rock rubble on featureless mud substrate to form six replicate reefs, each 10 m by 20 m. The development and dynamics of the lobster population of the reef was monitored by trapping, tagging, and diver surveys for 5 years. Before the reefs were set out, few lobsters occupied the site, but within 3 months vagrant large juvenile and adult lobsters had moved into the new habitat. Lobster density on the reef, measured by diver surveys and by trapping, was equal to or greater than that of nearby natural areas every year after the first. Natural settlement of postlarvae was monitored, and density of young-of-the-year lobsters was not different from that in natural areas. The reefs increased natural settlement and recruitment locally, but few of the greater than 6,000 hatchery-reared, microwire-tagged juveniles released on the reefs were recovered, so further research on improved methods of rearing and release is needed (van der Meeren 2003; K. Castro, University of Rhode Island, personal communication).

Spiny-Lobster Shelter Enhancement

Artificial shelters for spiny lobsters have been used to attract and concentrate them for easy harvest (Cruz et al. 1986; Briones-Fourzan et al. 2000), to mitigate loss of natural shelter (Davis 1985; Butler et al. 1995; Herrnkind et al. 1999), and in attempts to increase lobster populations (Nonaka et al. 2000; Briones-Fourzan et al. 2000).

In Japan, large-scale efforts began in 1933 to create new lobster-fishing sites for Japanese rock lobster *Panulirus japonicus* in the form of extensive concrete-block structures (Nonaka et al. 2000). Near shore, stone beds were also built, some as large as 200 m by 7 m by 2 m high. The lobster catch paid for the construction after 2–3 years; between 111 and 331 kg were harvested from each 100 m^3 of piled rock. The catch consisted mainly of otherwise dispersed lobsters drawn to the artificial reefs, however, and did not represent an apparent increase in population recruitment (Nonaka et al. 2000). Nonetheless, the observation that small, recently settled juvenile Japanese rock lobster often appeared in crevices of artificial structures impelled a long-term, large-scale effort in the 1970s to establish artificial reefs that both induced postlarval settlement and subsequently served as nursery habitat to potentially increase recruitment at the population level. Recruitment results remain inconclusive, but Nonaka et al. (2000) urged continued efforts to determine effectiveness.

Beginning in 1985, scientists in Florida collaborated to assess the relative contribution to juvenile recruitment of puerulus number and postsettlement conditions, especially shelter, in natural Caribbean spiny lobster populations (Butler and Herrnkind 1992, 2000; Herrnkind and Butler 1994; Herrnkind et al. 1999). They took a field-experimental approach, manipulating both key variables. Captured pueruli were injected with microwire tags (Sharp et al. 2000), then released within settlement habitat. For crevice shelter, the investigators used double-stacked, 3-hole partition concrete blocks (10 by 20 by 40 cm; Butler and Herrnkind 1997), which have functional attributes similar to those of large sponges and other crevice-bearing natural structures. Data supported both the premise that shelter was key to survival of the small juveniles (<35 mm carapace length, CL) and the conclusion that large numbers of settlers were necessary to influence the eventual numbers of surviving juveniles strongly (Butler et al. 1997).

On the basis of earlier findings, Herrnkind et al. (1999) experimentally deployed 1-ha arrays of 240 double-stacked 3-hole concrete partition blocks (10 cm by 20 cm by 40 cm) to test the feasibility of replacements for sponge crevices lost in a widespread 1991–1993 sponge die-off encompassing hundreds of square kilometers in the Florida Keys. Because postsettlement juveniles have a limited movement range, small juveniles (<25 mm CL) found in the blocks had to have settled as postlarvae in the 1-ha array. After 3 months, numbers of newly recruited postalgal juveniles on the block sites exceeded those on the sponge-less control sites and approximated the numbers on sponge-rich sites.

Briones-Fourzan and Lozano-Alvarez (2001) used small shelters (0.9-m by 1.2-m rectangular frames of 3.8-cm PVC with ferrocement roofs, like scaled down versions of the *casitas* described below) designed for juveniles up to ~50 mm CL, placed in a crevice-poor, vegetated lagoon at Puerto Morelos, Yucatan, Mexico. Collectively, the experimental results in both Florida and Mexico demonstrated that adding shelter for the early benthic phases, in areas with few natural crevices, increases local juvenile recruitment by alleviating postsettlement predation mortality.

Differences in the amount of natural shelter probably explain variation in recruitment among nursery areas that receive large numbers of pueruli; emerging postalgal juveniles are highly vulnerable to predators in crevice-poor habitat, and the result is low population numbers (Herrnkind and Butler 1994; Herrnkind et al. 1999; Briones-Fourzan and Lozano-Alvarez 2001). Concomitantly, large differences in puerulus supply will be reflected mainly in shelter-rich habitat, because cohort survival is relatively high there, but these recruitment interpretations should not be applied to the highly motile and less vulnerable nomadic juvenile, subadult, and adult lobsters (Butler and Herrnkind 1997; Herrnkind et al. 1999; Briones-Fourzan and Lozano-Alvarez 2001).

After WWII, Cuban lobster fishers attempted to increase their catch of Caribbean spiny lobster by using *pesqueros*. Initially, they used raft-like structures of palm logs approximately 4 m^2 placed on the bottom in shallow water where natural crevices were scarce (Cruz and Phillips 2000). A single shelter concentrated as many as 200 lobsters, which were readily captured by divers using encircling nets. The same technique was adopted on the Mexican Yucatan coast with analogous structures, called *casitas* (Miller 1982; Moe 1991; Briones-Fourzan et al. 2000). Recently, similar structures have been tested or deployed in the Bahamas, the U.S. Virgin Islands (Quinn and Kojis 1995), Florida (W. Sharp, Florida Fish and Wildlife Conservation Commission, personal communication), and Africa (Okechi and Polovina 1995).

These inexpensive, durable, and easily harvested devices have revolutionized lobster fishing and management in the Caribbean (Seijo et al. 1991, as cited in Briones-Fourzan et al. 2000; Cruz and Phillips 2000). Some 250,000 *pesqueros* contribute 50% of the Cuban spiny lobster fishery, the largest in the Caribbean. Lobster fishermen believe that the shelters also substantially increase the lobster population (Briones-Fourzan et al. 2000). Entrepreneurs have promoted *casitas* because of assumed conservation benefits and enhancement of local lobster populations (Moe 1991; R. Monroe, Darden Industries, personal communication).

Potentially offsetting detriments occur, however. For example, the shelters may make otherwise scattered lobsters vulnerable to overfishing (Briones-Fourzan et al. 2000). High densities of lobsters may reduce local prey (food) abundance. Crowding may facilitate transmission of diseases (Evans et al. 2000), and repeated handling of undersized individuals might exacerbate injury and reduce growth or delay maturity (Davis 1981). Long-term residency and concentration inshore could alter migratory patterns. In addition, these devices are large enough to attract predators of small juvenile lobsters, especially gray snappers *Lutjanus griseus*, red groupers *Epinephelus morio*, nurse sharks *Ginglymostoma cirratum*, and octopus (Mintz et al. 1994; Arce et al. 1997). They might therefore exacerbate predation mortality on the small, highly vulnerable new settlers in the area before they move into *pesqueros/casitas* (Herrnkind and Butler 1986; Smith and Herrnkind 1992; Herrnkind et al. 1999).

Extensive field experimentation reveals that shelter selection by large juveniles and larger individuals is a function of lobster size, shelter dimension, and lobster density (Eggleston and Lipcius 1992; Ratchford and Eggleston 1998). Aggregation in shelters potentially provides antipredation benefits according to selfish-herd theory and through cooperative defense (Herrnkind et al. 2001). The limited opening prevents entry by predators of large lobsters (nurse sharks) or effective attack (by triggerfish) within the shelter (Lozano-Alvarez and Spanier 1997). Large juveniles and subadult lobsters experimentally tethered in place survive significantly better in a *casita* than in the open outside a *casita.*

Despite the theoretical benefit of being in social aggregations in *casitas*, however, tethered lobsters 30–55 mm CL survived equally in smaller, "artificial sponge" dens that could hold relatively few individuals (Mintz et al. 1994). Tethering in the open may not be an ecologically appropriate comparison to conditions outside of *casita* areas. To our knowledge, no published studies have undertaken to quantify the occurrence or fate of lobsters living in areas of scattered shelter except during mass migrations (Herrnkind 1980). Likewise few direct measures are available of predation on freely moving nomadic or migratory lobsters (but see Herrnkind et al. 2001). Such information would allow a more valid comparison of potential differences in predation risk between natural and *casita* areas.

The most convincing argument for the protective role and increased survivorship provided by *casitas* would be a direct comparison showing higher long-term survival by *casita*-resident lobsters than by same-aged lobsters roaming about large areas of sparse natural shelter. The required experiment is logistically daunting. Rapid movement through areas of sparse shelter is easily confounded with mortality, causing its overestimation in tag-recapture studies using statistical analyses such as Jolly–Seber (Wahle 2003). Assumptions of random or homogeneous movement are often not tested or not supported (R. Wahle, Bigelow Laboratory for Ocean Science, personal communication), so statistical mortality estimates of *casita* occupants and nomads in shelter-poor natural areas cannot be appropriately compared.

Evidence is presently insufficient that *casitas* used for commercial fishing increase lobster survival at the population level or provide an estimate of the magnitude of any contribution or deficit to population recruitment. Continued efforts to determine their effects by strong field methods, perhaps incorporating modeling (Butler et al. 2001), are encouraged, but any recruitment gain theoretically would be relatively small compared to that produced by reducing natural mortality on the smaller, far more vulnerable postsettlement juveniles.

Shelter Enhancement Outlook

The results of carefully crafted experimental studies on the role of added shelter for newly settled spiny and clawed lobster suggest that placing structures designed for early benthic life stages can at least mitigate loss of crevice shelter and sometimes increase local on-site recruitment through improved postsettlement survival (Herrnkind et al. 1999; Briones-Fourzan and Lozano-Alvarez 2001; Castro et al. 2001). In theory, large-scale deployment of such structures could augment the number of predator-resistant larger juveniles that would survive, potentially to enter the fishery, but several constraining issues remain: choice of device, spatial extent, expense, maintenance, deleterious ecological impacts, and sufficiency of the recruitment increase. Achieving a measurable increase in the fishable stock will require both high expense and considerable risk. We recommend that, to the extent practicable, future enhancement efforts incorporate research on the issues raised above and feasibility experiments at an adequate ecological scale.

References

Arce, A. M., W. Aguilar-Davila, E. Sosa-Cordero, and J. F. Caddy. 1997. Artificial shelters (*casitas*) as habitats for juvenile spiny lobsters *Panulirus argus* in the Mexican Caribbean. Marine Ecology Progress Series 158:217–224.

Bannister, R. C. A., and J. T. Addison. 1998. Enhancing lobster stocks: a review of recent European methods, results, and future prospects. Bulletin of Marine Science 62:369–387.

Briones-Fourzan, P., and E. Lozano-Alvarez. 2001. Effects of artificial shelters (*casitas*) on the abundance and biomass of juvenile spiny lobsters *Panulirus argus* in a habitat-limited tropical reef lagoon. Marine Ecology Progress Series 221:221–232.

Briones-Fourzan, P., E. Lozano-Alvarez, and D. Eggleston. 2000. The use of artificial shelters (*casitas*) for research and harvesting of Caribbean spiny lobsters in Mexico. Pages 420–446 *in* B. F. Phillips and J. Kittaka, editors. Spiny lobsters: fisheries and culture, 2nd edition. Fishing News Books (Blackwell), Oxford, UK.

Butler, M. J., IV, and W. F. Herrnkind. 1992. Spiny lobster recruitment in south Florida: quantitative experiments and management implications. Proceedings of the Gulf and Caribbean Fisheries Institute 41:508–515.

Butler, M .J., IV, and W. F. Herrnkind. 1997. A test of recruitment limitation and the potential for artificial enhancement of spiny lobster populations in Florida. Canadian Journal of Fisheries and Aquatic Sciences 54:452–463.

Butler, M. J., IV, and W. F. Herrnkind. 2000. Puerulus and juvenile ecology. Pages 276–301 *in* B. F. Phillips and J. Kittaka, editors. Spiny lobsters: fisheries and culture, 2nd edition. Fishing News Books (Blackwell), Oxford, UK.

Butler, M. J., IV, J. H. Hunt, W. F. Herrnkind, M. J. Childress, R. Bertelsen, W. Sharp, T. Matthews, J. M. Field, and H. G. Marshall. 1995. Cascading disturbances in Florida Bay, USA: cyanobacteria blooms, sponge mortality, and implications for juvenile spiny lobsters, *Panulirus argus*. Marine Ecology Progress Series 129:119–125.

Butler, M. J., IV, W. F. Herrnkind, and J. H. Hunt. 1997. Factors affecting the recruitment of juvenile Caribbean spiny lobsters dwelling in macroalgae. Bulletin of Marine Science 61:3–19.

Butler, M., T. Dolan, W. Herrnkind, and J. Hunt. 2001. Modeling the effect of spatial variation in postlarval supply and habitat structure on recruitment of Caribbean spiny lobster. Marine and Freshwater Research 52:1243–1252.

Castro, K. M., J. S. Cobb, R. A. Wahle, and J. Catena. 2001. Habitat addition and stock enhancement for American lobsters, *Homarus americanus*. Marine and Freshwater Research 52:1253–1261.

Cruz, R., and B. Phillips. 2000. The artificial shelters (*pesqueros*) used for the spiny lobster (*Panulirus argus*) fisheries in Cuba. Pages 200–419 *in* B. F. Phillips and J. Kittaka, editors. Spiny lobsters: fisheries and culture, 2nd edition. Fishing News Books (Blackwell), Oxford.

Cruz, R., R. Brito, R. Diaz, and R. Lalana. 1986. Ecologia de la langosta (*Panulirus argus*) al SE de la Isla de la Juventud. I. Colonizacion de arricefes artificiales. Revista de Investigaciones Marinas 7:3–17.

Davis, G. 1981. Effects of injuries on spiny lobster, *Panulirus argus*, and implications for fishery management. Fishery Bulletin, U.S. 78:979–984.

Davis, G. 1985. Artificial structures to mitigate marina construction impacts on spiny lobster, *Panulirus argus*. Bulletin of Marine Science 37:151–156.

Eggleston, D. B., and R. N. Lipcius. 1992. Shelter selection by spiny lobster under variable predation risk, social conditions, and shelter size. Ecology 73:992–1011.

Evans, L., J. Jones, and J. Brock. 2000. Diseases of spiny lobsters. Pages 586–600 *in* B. F. Phillips and J. Kittaka, editors. Spiny lobsters: fisheries and culture, 2nd edition. Fishing News Books (Blackwell), Oxford, UK.

Grobler, C., and K. Noli-Peard. 1997. *Jasus lalandii* fishery in postindependence Namibia: monitoring population trends and stock recovery in relation to a variable environment. Marine and Freshwater Research 48:1015–1022.

Herrnkind, W. F. 1980. Spiny lobsters: patterns of movement. Pages 349–407 *in* J. S. Cobb and B. F. Phillips, editors. The biology and management of lobsters, volume 1. Academic Press, New York.

Herrnkind, W. F., and M. J. Butler, IV. 1986. Factors regulating postlarval settlement and juvenile microhabitat use by spiny lobster *Panulirus argus*. Marine Ecology Progress Series 34:23–30.

Herrnkind, W. F., and M. J. Butler, IV. 1994. Settlement of spiny lobsters, *Panulirus argus* in Florida: pattern without predictability. Crustaceana 67:46–64.

Herrnkind, W. F., M. J. Butler, IV, J. H. Hunt, and M. J. Childress. 1997. Role of physical refugia: implication from a mass sponge die-off in a lobster nursery in Florida. Marine and Freshwater Research 48:759–769.

Herrnkind, W., M. Butler, and J. Hunt. 1999. A case for shelter replacement in a disturbed spiny lobster nursery in Florida: why basic research had to come first. Pages 421–437 in E. R. Irwin, W. A. Hubert, C. F. Rabeni, H. L. Schramm, Jr., and T. Coon, editors. Catfish 2000: proceedings of the International Ictalurid Symposium. American Fisheries Society, Symposium 22, Bethesda, Maryland.

Herrnkind, W., M. Childress, and K. Lavalli. 2001. Cooperative defence and other benefits among exposed spiny lobsters: inferences from group size and behaviour. Marine and Freshwater Research 52:1113–1124.

Jensen, A. C., J. A. Wickins, and R. C. A. Bannister. 2000. The potential use of artificial reefs to enhance lobster habitat. Pages 379–401 *in* A. C. Jensen, K. J. Collins, and A. P. M. Lockwood, editors. Artificial reefs in European seas. Kluwer, London.

Lipcius, R., and D. Eggleston. 2000. Ecology and fishery biology of spiny lobsters. Pages 1–41 *in* B. F. Phillips and J. Kittaka, editors. Spiny lobsters: fisheries and culture, 2nd edition. Fishing News Books (Blackwell), Oxford, UK.

Lozano-Alvarez, E., and E. Spanier. 1997. Behaviour and growth of spiny lobsters *Panulirus argus* under the risk of predation. Marine and Freshwater Research 48:707–713.

Miller, D. 1982. Construction of shallow water habitat to increase lobster production in Mexico. Proceedings of the Gulf and Caribbean Fisheries Institute 34:168–179.

Mintz, J. D., R. N. Lipcius, D. B. Eggleston, and M. S. Seebo. 1994. Survival of juvenile Caribbean spiny lobster: effects of shelter size, geographic location, and conspecific abundance. Marine Ecology Progress Series 112:255–266.

Moe, M. 1991. Lobsters: Florida, the Bahamas, the

Caribbean. Green Turtle Publications, Plantation, Florida.

Nonaka, M., H. Fushimi, and T. Yamakawa. 2000. The spiny lobster fishery on Japan and stocking. Pages 221–242 *in* B. F. Phillips and J. Kittaka, editors. Spiny lobsters: fisheries and culture, 2nd edition. Fishing News Books (Blackwell), Oxford, UK.

Okechi, J. K., and J. J. Polovina. 1995. An evaluation of artificial shelters in the artisanal spiny lobster fishery in Gazi Bay, Kenya. South African Journal of Marine Science 16:373–376.

Quinn, N., and B. Kojis. 1995. Use of artificial shelters to increase lobster production in the U.S. Virgin Islands, with notes on the use of shelters in Mexican waters. Caribbean Journal of Science 31:311–316.

Ratchford, S. G., and D. B. Eggleston. 1998. Size- and scale-dependent chemical attraction contribute to an ontogenetic shift in sociality. Animal Behaviour 56:1027–1034.

Scarratt, D. J. 1968. An artificial reef for lobsters (*Homarus americanus*). Journal of the Fisheries Research Board of Canada 25:2683–2690.

Scarratt, D. J. 1973. Lobster populations on a man-made rocky reef. Page 47 *in* International Council for the Exploration of the Sea, ICES Shellfish and Benthos Committee CM 1973/K, Copenhagen.

Sharp, W. C., W. A. Lellis, M. J. Butler, W. F. Herrnkind, J. H. Hunt, M. Pardee-Woodring, and T. R. Matthews. 2000. The use of coded microwire tags in mark–recapture studies of juvenile Caribbean spiny lobster, *Panulirus argus.* Journal of Crustacean Biology 20:510–521.

Sheehy, D. J. 1976. Utilization of artificial shelters by the American lobster (*Homarus americanus*). Journal of the Fisheries Research Board of Canada 33:1615–1622.

Smith, K. N., and W. F. Herrnkind. 1992. Predation on juvenile spiny lobsters, *Panulirus argus*: influence of size, shelter, and activity period. Journal of Experimental Marine Biology and Ecology 157:3–18.

van der Meeren, G. I. 2003. The potential of ecological studies to improve on the survival of cultivated and released aquatic organisms. Doctorate thesis. University of Bergen, Norway.

Wahle, R. 2003. Revealing stock–recruitment relationships in lobsters and crabs: is experimental ecology the key? Fisheries Research 65:3–32.

Wahle, R., and R. S. Steneck. 1991. Recruitment habitats and nursery grounds of the American lobster *Homarus americanus*: a demographic bottleneck? Marine Ecology Progress Series 69:231–243.

Wahle, R. A., and R. S. Steneck. 1992. Habitat restriction in early benthic life: experiments of habitat selection and in situ predation with the American lobster. Journal of Experimental Marine Biology and Ecology 157:91–114.

American Fisheries Society Symposium 49:933–942

Design of Artificial Reefs and Their Effectiveness in the Fisheries of Eastern Asia

CHANG GIL KIM*, HO SANG KIM, CHUL IN BAIK

Korean National Fisheries Research and Development Institute
Sirang-ri, Gijang-eup, Gijang-gun, Busan, 619–902, Korea

HIROSHI KAKIMOTO

Underwater Structures Laboratory, Japan National Coastal Fisheries Development Association
Kamakurakashi Building, 4F, 2–2-1, Uchikanda, Chiyoda-ku, Tokyo, 101–0047, Japan

WILLIAM SEAMAN

University of Florida
Building 803, Post Office Box110400, Gainesville, Florida 32611, USA

Abstract.—This study reviews structural and fishery characteristics of artificial reefs used in Korea and Japan and describes a new type of reef structure. It reviews 92 concrete and 53 steel reef designs, of which 137 are from Japan and 8 from Korea. Reef structures installed in Korea and Japan have been chiefly designed with respect to water depth, target species, and ratio of void space. Most reefs were developed on the basis of how fish species would respond to them behaviorally, and their placements were oriented accordingly. Also, ratio of void space has been enlarged to minimize the amount of materials required in construction. Japan has developed taller reefs that were designed to attract all fish that respond positively to the presence of a reef. An enlargement in bulk volume, and height, with a limited void–space ratio, characterizes these taller reefs. Korea has developed a concrete reef style, the box reef, which was designed to meet the biological requirements of both Korean rockfish *Sebastes schlegeli* and madai (also known as red porgy) *Pagrus major*, which behave differently on reefs. The box reef attracted more Korean rockfish and madai than a natural reef. Such a high-storied reef (14 × 14 × 30 m), which was installed at a depth of 75 m, attracted fish species such as false kelpfish *Sebastiscus marmoratus*, rockbream *Oplegnathus fasciatus*, greater amberjack *Seriola dumerili*, and bambooleaf wrasse *Pseudolabrus japonicus*, which behave differently on reefs.

Introduction

Countries in eastern Asia, Japan and Korea in particular, have a long history of working to increase their fish production through artificial reefs in freshwater and marine systems. In rivers and lakes, structures made of pine logs and rocks have been used to attract fish. The logs were latticed into a sort of hut without a roof 4 to 5 m high; it was floored, floated to the desired site, and sunk in depths up to 10 m by casting stones randomly within. In marine waters, rafts or floats made of bamboo about 4 to 5 m long were floated in the sea

*Corresponding author: cgkim@nfrdi.re.kr

and placed to attract pelagic fishes such as dolphinfish *Coryphaena hippurus*, buri (also known as yellow trail) *Seriola quinqueradiata*, and Pacific saury *C. saira*. The rafts were anchored via sandbags. Although these fishing devices are very primitive, they have been handed down until the present day as a valuable technique of creating artificial reefs (Ogawa 1968).

In Japan, the first recorded reef construction effort occurred in 1640. Knights (samurai) in the Kochi prefecture placed quarry rocks in random heaps to attract fish in coastal waters (Hirao et al. 1976). The next record of reef building is found in 1795. Fishermen ordinarily caught large numbers of three-banded grunts *Plectorhynchus cinctus* around the sunken boats. After a 7-year decline in the grunt harvest, fishermen sunk sandbags with overhead bamboo and log lattice-work adjacent the sunken boats. Fish catches at these new reef sites were twice those taken earlier at the sunken boats (Ogawa 1972). Thus, Japanese fishermen learned that the best fishing occurred around sunken boats and underwater objects composed of rocks, sandbags, and logs.

The first record of an artificial reef in Korea occurred in 1801. Rocks were used in western Korea to attract thigmotaxis fishes like eel (whitespotted conger *Conger myriaster*). Surface fish attractors made from rice-straw were used in eastern Korea. The most popular type of reef was old fish boats, which were sunk intact. Historically, in Korea government support for artificial reefs began in 1971. Artificial reefs have been installed to enhance fish production and deter overfishing of the resource by small otter trawlers. In the 1970s, cubes (1.0 × 1.0 × 1.0 m) were installed. By the 1980s, the number of reefs increased substantially, and the structural designs were more diverse. Four reef modules used since the 1980s are jumbo (6.5 × 5.0 × 5.7 m), dice (2.0 × 2.0 × 2.0 m), turtle (Ø1.3 × 2.0 m), and tube (Ø1.8 × 1.8 m). In the 1990s, steel reefs with a height of more than 10 m were installed. Jumbo and dice are frame structure, and turtle and tube are face structure. Frame structural reefs contain over 80% void space, while face structural reefs have less than 80% void space.

This study reviews the structural and fishery characteristics of artificial reefs used in Japan and Korea and describes a new type of reef structure. The review of the reef structures was made based upon the biological, geometric, and engineering aspects. Biological aspects examine fish response to a reef and a relation between the reef structure and fish (Nakamura 1985). Geometrical aspects consider height, bulk volume, and void space of reefs, while the engineering aspects include the weight and strength of reefs. New types of reef structure are described as the high-storied steel reef of Japan and box reef of Korea.

Review Items and Methods

Analysis of Artificial Reef Structure for Target Fish Species

The review of artificial reef structures was accomplished according to fish response to a reef and the induced stimuli for fish activities at reefs. The former are classified into three types of reef behavior (Japan Coastal Fisheries Development Association 1986). Type I species prefer physical contact with their bodies against hard objects (eel, flounder, and rockfish). Type II species like to remain in proximity to a hard object without actually touching it (porgy, black porgy *Acanthopagrus schlegeli*, and red sea-bream *Pagellus bogaraveo*), and Type III includes species that are indifferent to the presence of a hard object (wahoo *Acanthocybium solandri* and yellow tail).

In a Type III response, fish gathering around and in an artificial reef tend to perceive the presence of the reef through stimuli received by their sensory organs. These stimuli are divided into four excitations: visual, contact, stream (pressure fluctuation), and sound. Type I species respond to contact or visual excitation, and Type II species respond to visual or stream excitation. Type III species perceive stream or sound excitation (Japan National Coastal Fisheries Development Association 1986).

Comparative Analysis of Structural Characteristics of Artificial Reefs

The most prominent structural characteristics of artificial reefs include height, bulk volume, the ratio of void space, weight, and strength. These factors exert an important effect upon biological conditions, such as the attraction of fish; engineering conditions, such as construction and durability; and socioeconomic conditions, such as manufacturing price. We analyzed 145 artificial reefs, of which 92 are made from concrete and the others from steel. A total of 137 reefs are from Japan, with the remainder from Korea. We examined 145 reefs with explicit design specifications (Japan National Coastal Fisheries Development Association 1999). In practice, the number of reefs used in both countries is composed of more than 200 reefs (Japan National Coastal Fisheries Development Association 1999; Kim et al. 2002a).

Results and Discussion

Comparative Analysis of Structural Characteristics of Artificial Reefs

Figures 1 and 2 show the height of reefs used in Japan and Korea. Concrete reef designs in Japan (n = 86) have heights ranging from 1.0 to 12.0 m (Figure 1). When the range of their

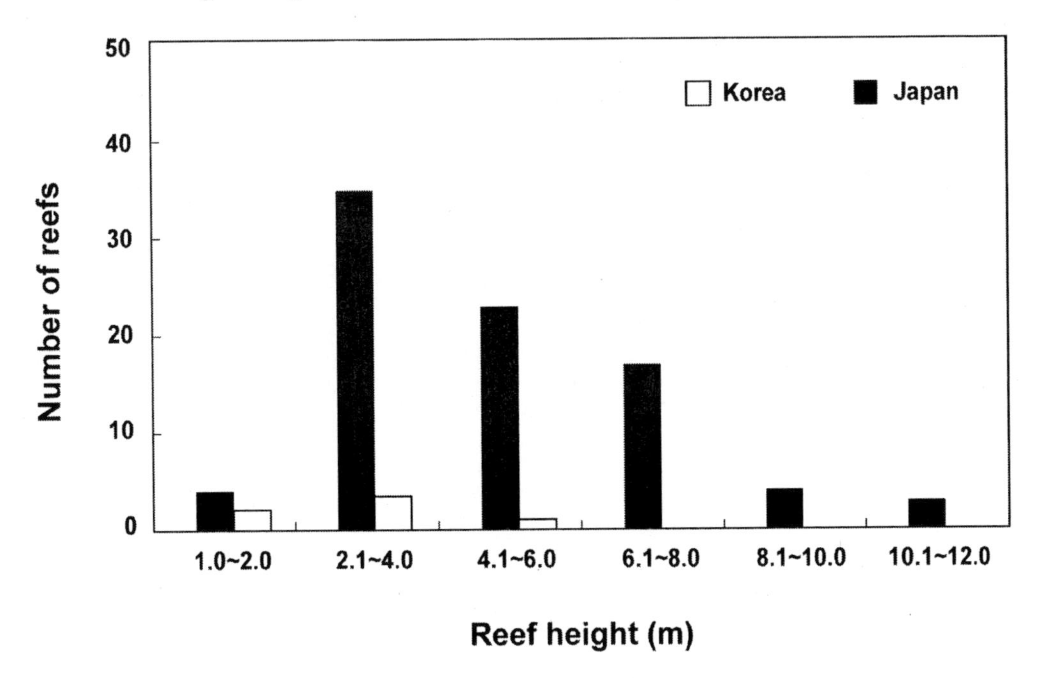

Figure 1. Comparison of the height of concrete reefs used in Japan (From Japan National Coastal Fisheries Development Association 1999) and Korea (From Kim et al. 2002b).

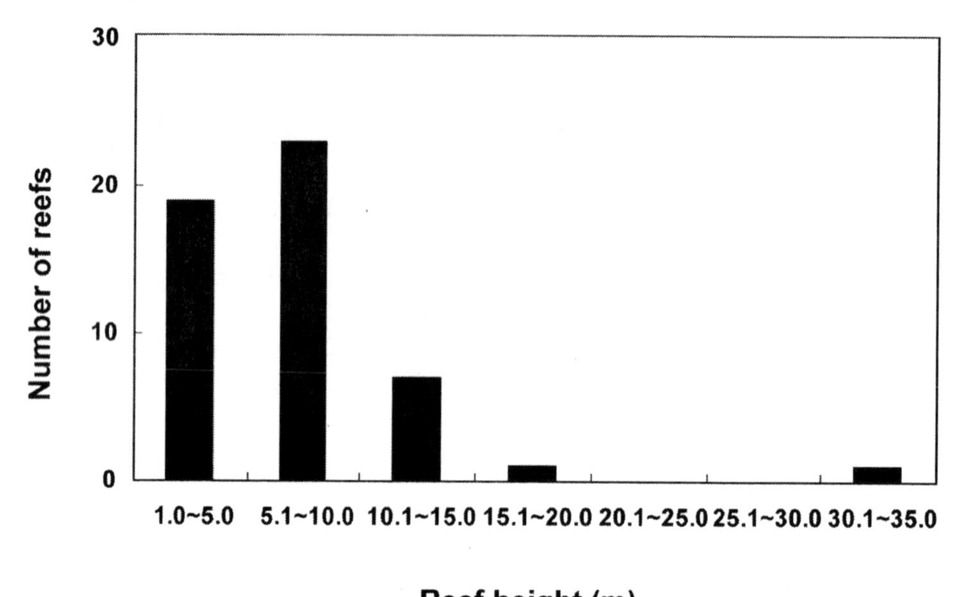

Figure 2. Comparison of the heights of steel reefs used in Japan (From Japan National Coastal Fisheries Development Association, 1999).

height is expressed by total of 40.7% range from 2.1 to 4.0 m in height, while 26.7% range from 4.1 to 6.0 m, and 3.5% range from 10.1 to 12.0 m. Meanwhile, a total of 51 steel reefs range from 1.0 to 35.0 m in height (Figure 2). Of these, steel reefs with a height of 5.1–10.0 m account for 45.1%, while those of 1.0–5.0 m comprise 37.3%. The height of concrete reefs used in Korea ranges from 1.0 to 6.0 m (Figure 1), and steel reefs range from 9.0 to 10.2 m. Concrete reefs in Japan are chiefly installed in waters of less than 40 m, while steel reefs are submerged in deeper waters. Distribution of reefs in Korea is similar.

When discussing the height of a reef, there are basically two meanings that may apply. One refers to the height of a reef unit itself, while the other indicates the topmost point of a heaped pile of reefs. As small-sized reefs have a small volume, they are placed in a chaotic heap. However, too many heaps in a single place tend to decrease overall effectiveness by the intercepted flow between the inside and outside (Uno 1966). Therefore, the installation height of small-sized reefs is suited to heaping up at two to three times the reef height (Japan Fisheries Resources Preservation Association 1976). Kakimoto et al. (1982) surveyed fish caught by hook over a 1-year period at reefs with heights of 1–12 m. Fish catches were much higher at 1–5 m reef than at other reefs. Also, scuba divers observed that black rockfish *Sebastes melanops*, Japanese sea perch (also known as Japanese sea bass) *Lateolabrax japonicus*, rock bream *Oplegnathus fasciatus*, and Korean rockfish *Sebastes schlegeli* were abundant at reefs with heights less than 5 m. Apparently, Types I and II tend to gather relatively close to the bottom irrespective of reef height (Kakimoto and Okubo 1985).

In water that is weakly stratified, a moderate current impinging upon an obstacle such as a reef generates a lee wave, which stream stimuli for fish. Type III fish tend to stay upstream of an obstacle and beyond the range of influence of the lee wave. Reef height determines suitability for type III fish. When reef height is about one-tenth of water depth, the lee wave is maximized (Nakamura 1979). This phenomenon may not apply to areas where the thermocline is prominent or to pelagic fishes.

Takaki et al. (2000) surveyed fish catches at reefs with heights of 17 and 35 m in the coastal waters of Yamakata prefecture in Japan. The 35 m reef was made equal to the height of a natural reef, while a 17 m reef stood at half the height of the natural reef. The results showed that catches of goldeye rockfish *Sebastes thompsoni* were much higher at the 35-m reef than at the 17-m reef. Also, they confirmed that the fish catch of goldeye rockfish is related to their swimming layer, water temperature, and the amount of plankton.

These results mean that the height of reefs for type III species should be designed based upon the configuration of the seabed, habitable water depth of target fish and its food organisms, and the water temperature suitable to their habitat.

New Reef Designs in Korea and Japan and Their Effectiveness

Box Reefs in Korea

There have been many reefs developed in Japan and Korea, but this study describes only two designs: Box reef of Korea (Figure 3) and high storied-reef of Japan (Figure 4). Box reef is 3 × 3 × 3 m in volume with a void space of 75%. It consists of two parts: the outer box with the face structure and the inner one with a frame structure. It was designed to meet the biological requirements of both

Figure 3. Box reef (3 × 3 × 3 m) developed in Korea.

Figure 4. Moderately high storied reef (14 × 14 × 30 m) developed in Japan.

Korean rockfish (type I) and madai (type II), which behave differently in reefs (Kim 2001). The lower portion of the reef, which extends upwards 2 m from the bottom, is designed for rockfish and their preference for face structures. The top one meter section was designed for porgy, which prefer a frame structure. Material for the box reef was selected based upon ease of construction, and thus the outer box with a simple face structure is made of concrete, while the inner structure is composed of steel plates. All of the inner spaces of the reef are less than one square meter. It was based on the fact that rockfish tend to gather in dark and small spaces, and that porgy prefer a complex reef structure with a large volume. The effectiveness was measured by resultant fish catch and analyzed to compare box reefs, natural reef, and dice reefs. The survey was made over 2 years from 2001 to 2002 at a different reef sets composed of box reefs and dice reefs.

According to a comparative survey of the box reef and a natural reef, using trammel gill nets, there was no significant difference in the number of fish species found in a box reef and a natural reef. However, the catch at the box reef was 4.4 times greater than that obtained in at a natural reef. Rockfish occurred at the box reef only, and twice as many porgy were taken there versus the natural reef (Table 1). Meanwhile, when compared with the 2 m dice reef, the catch at the box reef was 3.0 times greater. There were 2.6 as many rockfish and 1.7 as many

Table 1. Comparison of fish catch from trammel gill net at box reef, natural reef, and dice reef during 2001 to 2002 (From Kim et al. 2002c).

Fish species	Type of fish response to reef	Weight of fish caught per net (g)		
		Box reef	Natural reef	Dice reef
Paralichthys olivaceus		80.0		
Hexagrammos otakii	I	131.3		
Epinephelus septemfasciatus	I	187.5		
Sebastes schlegeli	I	1,870.0		736.7
Parapercis sexfasciata	I	30.0	63.8	15.3
Stephanolepis cirrhifer	II	35.6		23.2
Thamnaconus fasciatus	II	46.3	31.3	
Oplagnathus fasciatus	II	313.3		
Pagrus major	II	135.0	68.9	80.0
Zeus fabe	II		11.3	
Ditrema temmincki	II		66.7	50.4
Sphyraena japonica	Others	3.8		
Sillago sihama	"	6.3	6.3	4.3
Cebrias fasciatus	"	7.5	3.8	80.7
Raja porosa	"	165.0	116.3	102.6
Eopsetta griogrjewi	"	178.8	103.8	
Liparis tessellatus	"	193.8		
Pleuronichthys cornutus	"	208.9	216.3	91.7
Cynoglossus robustus	"		37.5	
Platycephalus indicus	"		97.5	
Total		3593.1	823.5	1,184.9

porgy caught at the box reef than at the dice reef. Scuba divers observed that rockfishes swam slowly about the inside and outside of the lower portion, from the bottom to a height of 2 m. Sevenband grouper *Epinephelus septemfasciatus* gathered in dark corners of the inner bottom of the reef; surfperch *Ditrema temmincki*, which is a type II species, swam at a height of 2 to 3 meters around the reef. Porgy, however, were not observed at the box reef.

We interpret these observations to mean that a box reef with a dual structure (frame and face structure) is more effective in attracting type I and II species than a dice reef with a simple structure.

High-Storied Reef in Japan

This study describes high-storied reefs installed at a depth of 75 m in the Yamakuchi prefecture of Japan. It is 14 × 14 × 30 m in volume with a weight of about 70 metric tons and bulk volume of 2,500 m^3 (Figure 4). As shown in Figure 4, it has a frame structure. The lower portion, which extends 10 m upward from the bottom, is trapezoidal, and the 10–20 m portion is rectangular, with a steel plate attached to each face. The section at 20–30 m is composed of four pillars and to one face of each pillar is attached a steel plate, which forms the middle of the reef. Steel plates on each side of the reef are attached in order to connect each frame and to disrupt the current. The survey was made in and around a reef installed at a depth of 75 m, and effectiveness was gauged according to fish catch. Observation was carried out via scuba diving and underwater video camera for 3 years from 1999 to 2001 (Japan National Coastal Fisheries Development Association 2002a).

The surveys conducted during 1999–2001 indicate that the number of fish species gathered around an ordinary, high-storied reef ranged from 18 to 24. The number increased with time (Japan National Coastal Fisheries Development Association 2002b). Type II and III species were dominant (Table 2), and Japanese scad were caught in great abundance. Angular and underwater video camera confirmed that large pelagic fishes such as buri and greater amberjack, which feed on jacks *Carangidae* spp., gathering about the reef.

Acknowledgments

We thank for the assistance of J. W. Tae. Also we thank Dr. Carter R. Gilbert and Professor Emeritus, University of Florida in the United States and Professor Antony Jensen, University of Southapmton in the UK who kindly provided constructive criticisms of the manuscript.

References

Hirao, M., D. Yamamoto, S. Yokogawa, K. Hirota and N. Shimeno. 1976. Kaizanshuu of Nonaka Kenzan, No. 4, History (3), 48–694. (In Japanese)

Japan Fisheries Resource Preservation Association. 1976. A review of artificial reef research-I. Japan Fisheries Resource Preservation Association, Tokyo (In Japanese).

Japan National Coastal Fisheries Development Association. 1986. Guide to creation of artificial reef for coastal fisheries consolidation and development projects. Japan National Coastal Fisheries Development Association, Tokyo (In Japanese).

Japan National Coastal Fisheries Development Association. 1999. Manual of artificial reefs used in Japan's coastal fisheries consolidations and development program. Japan National Coastal Fisheries Development Association, Tokyo (In Japanese).

Japan National Coastal Fisheries Development Association. 2002a. On development of offshore. Japan National Coastal Fisheries Development Association, Tokyo (In Japanese).

Japan National Coastal Fisheries Development Association. 2002b. Report of biological environment survey for fisheries foundation consolidation (On effect of height of artificial reef on fish attraction), 49–126. (In Japanese)

Table 2. Fishes confirmed by underwater observations at moderately high storied reef (14 x 14 x 30 m), installed at depth of 75 m in the Yamakuchi prefecture of Japan during 1999–2001 (From Japan National Coastal Fisheries Development Association 2002a).

Fish species	Type of fish response to a reef	Size of fish (mm)	Nos. of fish observed around a reef
Goniistius zebra	I	250	0.1
Pterois lunulata	I	150~250	0.3
Petroscirtes breviceps	I	30~50	1.0
Sebastiscus marmoratus	I	150~250	2.4
Cirrhitichthys aureus	I	50~120	2.9
Epinephelus awoara	II	200	0.2
Sebastes schlegeli	II	250~300	0.3
Parapristipoma trilineatum	II	200	0.3
Chaetodontoplus septentrionalis	II	150~200	0.4
Pterogobius zacalles	II	80~120	0.8
Pagrus major	II	100	1.0
Epinephelus septemfasciatus	II	300~400	2.2
Oplegnathus fasciatus	II	150~300	2.6
Apogonidae sp.	II	50~80	4.3
Sebastes thompsoni	II	50~200	24.8
Stephanolepis cirrhifer	II	80~250	30.4
Pempheridae sp.	II	50~70	33.3
Apogon semilineatus	II	10~120	3,291.7
Coryphaena hippurus	III	300~400	0.9
Upeneus bensasi	III	50~80	1.2
Seriola quinqueradiata	III	250~400	1.3
Seriola dumerili	III	400	4.0
Clupeiformes spp.	III	80~100	200.0
Carangidae spp.	III	80~300	17,579.2
Trachurus japonicus	III	30~200	23,002.2
Gobiidae sp.	III	50	0.1
Choerodon azurio	III	200	0.2
Paralichthys olivaceus	III	300	0.2
Parapericis pulchella	III	200	0.6
Pseudolabrus japonicus	III	80~200	6.3

Kakimoto, H., and H. Okubo. 1982. Study on size and development of reef. Fisheries Civil Engineering, volume 18(2)43–52. (In Japanese)

Kakimoto, H., and H. Okubo. 1985. Comprehensive study on the projects of artificial reef in the coastal waters of the Nikata prefecture in Japan. Nikata prefecture, Nikata, Japan (In Japanese).

Kim, C. G. 2001. Artificial reefs in Korea. Fisheries 26(12) 15–18.

Kim, C. G., B. J. Min, and J. H. Jung. 2002a. The actual state of maintainance of artificial reefs installed in Korean coastal waters and their improvement scheme. Korean Ministry of Maritimes affairs and Fisheries, Seoul (In Korean).

Kim, C. G., H. S. Kim, T. H, Kim, and B. C. Baik, 2002b. Design of box reef marine ranching in Korean Tongyeong coastal waters. Japanese Society of Fisheries Science 68(2):1903–1904.

Kim, C. G., H. S. Kim, T. H, Kim, and B. C. Baik. 2002c. Studies on the development of marine ranching program in Tong-yong, Korea. Korean Ministry of Maritimes affairs and Fisheries, Seoul (In Korean).

Nakamura, M. 1979. The science of fisheries engineering, the industrial-current affairs news agency, 401–419. (In Japanese).

Nakamura, M. 1985. Evolution of artificial reef concepts in Japan. Bulletin of Marine Sciences 37(1):271–278.

Ogawa, Y. 1968. Artificial reef and fish. Propagation of Fisheries, Special No. 7, Tokyo (In Japanese).

Ogawa, Y. 1972. Propagation of fish in shallow waters and artificial reef. Pages 83–94 *in* 1st Japan and USSR Joint Symposium on Propagation of Marine Resources of the Pacific Ocean, Tokyo. (In Japanese).

Takaki, H., A. Moriguchi, K. Kimoto, K. Arai, T. Hasuo, H. Nakamura and K. Kimura. 2000. Development of large-scale high-rise artificial reef and its effectiveness. Fisheries Engineering 22:1–14. (In Japanese).

Uno, K. 1966. On size of artificial reef. Study of artificial reef 5:3, Tokyo (In Japanese).